24th IUPAP Conference on Computational Physics

(IUPAP-CCP 2012)

Journal of Physics: Conference Series Volume 454

Kobe, Japan
14-18 October 2012

ISBN: 978-1-62993-092-3
ISSN: 1742-6588

Printed from e-media with permission by:

Curran Associates, Inc.
57 Morehouse Lane
Red Hook, NY 12571

Printed by Curran Associates, Inc. (2013)

Institute of Physics
Dirac House, Temple Back
Bristol BS1 6BE UK

Phone: 44 1 17 929 7481
Fax: 44 1 17 920 0979

techtracking@iop.org

Additional copies of this publication are available from:

Curran Associates, Inc.
57 Morehouse Lane
Red Hook, NY 12571 USA
Phone: 845-758-0400
Fax: 845-758-2634
Email: curran@proceedings.com
Web: www.proceedings.com

TABLE OF CONTENTS

FINITE-DIFFERENCE, FINITE-VOLUME, FINITE-ELEMENT METHODS

PARTICLE METHODS

MOLECULAR DYNAMICS

MONTE CARLO METHODS

QUANTUM MONTE CARLO METHODS

DENSITY FUNCTIONAL THEORY

DENSITY MATRIX RENORMALIZATION GROUP (INCLUDING DIRECT MATRIX DIAGONALIZATION, MATRIX PRODUCT STATES, PEPS, MERA)

BIO-COMPUTING

CLIMATE AND DISASTER PREVENTION

EDUCATION IN COMPUTATIONAL PHYSICS

VISUALIZATION

OTHER METHODS AND TOPICS

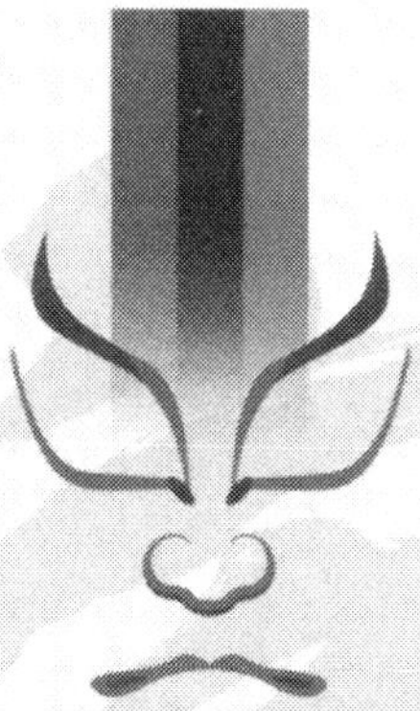

Conference on Computational Physics (CCP2012)

for Physics, Chemistry, Biology, Engineering and related academic fields and industrial applications

October 14 (Sunday)-18 (Thursday), 2012
The Nichii Gakkan Conference Center, Kobe, Japan
(next to the K-computer site)
Organized by C20, IUPAP

Journal of Physics: Conference Series **454** (2013) 011001 doi:10.1088/1742-6596/454/1/011001

Contents

Journal of Physics: Conference Series **454** (2013) 011001 doi:10.1088/1742-6596/454/1/011001

Welcome to CCP2012

You are welcome to CCP2012, held next to the K computer site in Kobe and in the best season of Japan. The Conference on Computational Physics (CCP) is organized annually under the auspices of the Commission 20 of the IUPAP (International Union of Pure and Applied Physics).

This is the first time it is held in Japan. I was asked to be the chairman about two and half years ago and when I accepted the request I decided to make the conference very unique, different from the traditional style of CCP. I was not satisfied when I attended big conferences where the parallel sessions are classified with the name of the research field. We have many chances to attend domestic and international conferences these days. There it is possible to listen to many talks on the same topics and if the topics are very new, such conference is very useful for my research. I wanted, however, to have some conference where I can listen to a variety of topics carried out with the same method.

Computational science is very unique and it is easy to organize a new type of conference with the classification in the horizontal direction of the matrix made of the names of research fields and the name of numerical methods. You may be able to enumerate easily the name of methods, finite difference, Monte Carlo, particle, molecular dynamics and so on. My dissatisfaction is found to have come from the fact that most of conferences focus on research fields and the method that brings to the scientific research is not highlighted so much. I wanted to listen to topics from fundamental physics to industrial science in a systematic way.

In order to realize such conference, a small number of experts is not enough and I asked the help of more than 100 Japanese computer scientists in a variety of fields. We called this group the Japan Advisory Board (JAB). I asked them to recommend the member of the International Advisory Board (IAB). Then, we could start making the list of plenary speakers and invited speakers. It was almost the end of March this year.

CCP2012 is organized also to celebrate the shared use of the K computer and we selected a venue next to the K computer. The shared use is of course open to the public and started on September 28th, one month earlier than previously scheduled. I hope you also enjoy the guided tour to the K computer.

Throughout CCP2012, I hope new collaborations start among scientists in different fields. It would be also my great pleasure if such an inter-disciplinary conference will encourage young scientists (with their fresh energy and skills) to challenge new topics in different fields, especially emerging ones like bio-computing, industrial applications, social sciences and so on.

Finally, allow me to express my sincere thanks to all members of the local organizing committee (LOC). Twenty scientists from three universities and one institute voluntarily worked very hard to prepare CCP2012 as LOC.

The Chairman, CCP2012

Hideaki Takabe (Aki)

Journal of Physics: Conference Series **454** (2013) 011001 doi:10.1088/1742-6596/454/1/011001

Greetings on behalf of the co-organizing academic societies

Shoji Nagamiya

President of AAPPS and ex-President of JPS

It is my great pleasure to be here at the CCP2012 conference to give an opening address.

First I would like to comment that Moore's law exists in the computer society. Namely, the computing power is increased twice per 1.5 years. If one waits ten years, the computer power is increased by 100 times, and for the 30 years by 10^6. This is amazing and this trend, which is similar to Livingston's law in my field of accelerators, is still continuing. The KEI is on this line, I guess, and it provides many useful applications including basic science. Some physics can develop only by such a high-speed computer.

In the field of physics where I am involved, a Nambu theory was published over 50 years ago. The proton mass was generated by a spontaneous symmetry breaking. However, the calculation of the Nambu theory was not possible until very recently, since it requires parallel processors at very high speeds. This is only one example. I would like to say that physics is being developed with computer power enormously, and some filed can be developed only when high-speed computer became available. Namely, Physics and Computer are benefiting each other. I feel, therefore, it very important to have this type of conference here at this time in Japan.

Secondly, in this greeting, since I am from Association of Asia Pacific Physical Societies called, the AAPPS, I would like to say a few words on this organization. First, the AAPPS is the organization for Asian Physical Society, which is similar to EPS. It has held regular conferences during the past 30 years, first in Singapore. Immediately after the 3rd meeting in Hong Kong in 1988 the organization called the AAPPS, the association, was invented by the effort of Professor C. N. Yang and many others. The 12th meeting will be held in Japan in July of the next year.

The Association consists of 17 countries and regions written shown below.

ASEAN Institute of Physics: svirulh@chula.ac.th
Australian Institute of Physics: http://www.aip.org.au/
The Chinese Physical Society: http://www.cps-net.org.cn/
The Physical Society of Hong Kong: http://www.pshk.org.hk/
Indian Physics Association: www.tifr.res.in/~ipa
Indonesian Physical Society: http://hfi.fisika.net/
The Physical Society of Japan: http://wwwsoc.nii.ac.jp/jps/
The Japan Society of Applied Physics: http://www.jsap.or.jp/english/
The Korean Physical Society: http://www.kps.or.kr/home/kor/
Malaysian Institute of Physics: kuru@um.edu.my
Mongolian Physical Society: gantsog@num.edu.mn
Nepal Physical Society: http://www.nps.org.np/
New Zealand Institute of Physics: http://nzip.rsnz.govt.nz/
Physical Society of Philippines: http://www.nip.upd.edu.ph/spp/
Institute of Physics, Singapore: http://www.physics.nus.edu.sg/~phyips
South East Asia Theoretical Physics Association: kkphua@wspc.com.sg
The Physical Society locate din Taipei: http://psroc.phys.ntu.edu.tw/
Thai Institute of Physics: http://www.geocities.com/thai_physics/
Vietnam National Institute of Physics: http://www.iop.vast.ac.vn

Next year at APPC, C. N. Yang, Japanese Nobel laureate, for example, Makoto Kobayashi, and many others join this conference. We receive support not only from AAPPS but also from Japanese Physical Society, Japanese Society of Applied Physics. We also plan to have joint session with European Physical Society, EPS. This is the third meeting between EPS and AAPPS. The conference will be held at Makuhari, close to the Tokyo Airports, both Narita and Haneda. If you are interested in, we always welcome you.

Finally, I would like to comment that the AAPPS donates for this conference the prize for the best young (but not student) poster, similar to the EPS.

Congratulation of this important conference and wish you a great success of the conference. Please also enjoy Japan.

Journal of Physics: Conference Series **454** (2013) 011001 doi:10.1088/1742-6596/454/1/011001

Greetings on behalf of the co-organizing universities

Saburo Aimoto

Vice-President of Osaka University

Good morning, everybody. I am Saburo Aimoto, the trustee and vice–president of Osaka University in charge of basic science.

I would like to extend greetings to all of you here on behalf of co-organizing four Universities of Kobe University, University of Hyogo, Kyoto University, and Osaka University.

It is a great honor for us to welcome you to the Conference on Computational Physics 2012. We would like to express our thanks to all of participants from the heart, especially ones from oversea countries for attending this conference. We also wish to express our sincere thanks to the Commission 20 of the International Union of Pure and Applied Physics, for they decided to hold this conference in Kobe, Japan on the occasion of start of the open use of the K computer. As co-organizers, we are very glad to hear that more than 400 researchers participated from 44 countries.

I am an organic chemist, not a physicist. However, I am often astonished at the rapid and remarkable progress of computational science. Scientists in this field show us deep insight into the truth hidden behind experimental data, and predict the features that we cannot elucidate otherwise. This conference covers a wide variety of topics from different disciplines. It is really fantastic that topics on computational theory, atomic nucleus, prediction of typhoon behavior, laser fusion, chemical reaction, and biological and artificial nanostructures will be discussed in one conference. This suggests that the approach based on computation should be an essential and core driving force for natural science and technology.

I suppose that this must be true for social science, too. Therefore the role of this conference and peoples expectation for the progress of this field must be enormous. In this sense, we have to recall Dr. Noyori's great effort. I believe that his foresight and leadership have realized the setting of the K computer. Without his strong faith, the construction of the K computer would have ended in an illusion. I hope this conference will send the strong message of the importance of powerful computer infrastructure to the public.

Finally, we wish to earnestly thank to the member of the International Advisory Board and all members of the local organizing committee for the preparations of this conference.

We hope that this conference will be successful and that computational physics will make brilliant progress from now on.

I am a citizen of Kobe. As one of citizens of Kobe, I am very happy if you enjoy your stay in port city Kobe.

Thank you very much for your attention.

Journal of Physics: Conference Series **454** (2013) 011001 doi:10.1088/1742-6596/454/1/011001

Greetings on behalf of the Japanese ministry, MEXT

Takahiro Hayashi

Director, Office for the Promotion of Computing Science

I'd like to say a few words to congratulate the success of CCP2012 conference. I am very pleased to learn that a lot of researchers came together from more than 40 countries and regions, which, I believe, makes this conference truly international. Then, I'd like to express my appreciation to the chair-person, Prof. Takabe of Osaka University and other supporting committee members for their substantial efforts to hold this conference here in Kobe in such a successful way.

The reason why this conference is held here is K computer, that is located at the nearby RIKEN Institute and I know you are invited to the laboratory tours to see K computer during this conference.

As many of you know, K computer has started its full service since last month for academia and industrial uses. Although the first prize of TOP500 was taken over by Sequoia of LLNL, K computer is still the most powerful computer opened to the public and is now ready for production runs to produce innovative new findings which are obtained by only K computer. This is our primary goal and I believe you will find some of those preliminary results at this meeting.

Today, the computational science is becoming a powerful tool for various research areas, and I believe national competitiveness in both academia and industries of the nations depends on the performances of supercomputers they have. In US, Europe, China, Russia, India and other countries as well as in Japan, the computational science is being recognized to be one of the most important strategic technologies for keeping the competitiveness up. And thus, a lot of countries including Japan have been promoting the computational science intensively world wide.

In order to develop such computational sciences, not only hardware developments but also application promotions are important. Such promotion programs, for example, INCITE of US and PRACE of Europe, have been conducted under the governmental supports. Of course, we have also similar one that 50 % of the K computer resources are delivered to in the fields of Life Science, Material, Prediction and Protection of Disaster, Industrial Applications and Astrophysics.

As I mentioned before, the computational science is a powerful tool for various research areas. This means researchers in various areas could interact with each other through the computational science. Therefore, I believe, the computational science has a possibility to be an engine which promotes a fusion among different research areas and then creates a new science. This is another important role of it. In that sense, the role of this conference is very significant.

Finally, I want to emphasize the following. By keeping close collaborations with US, Europe and other countries, we are anxious to continue our activities for making computational science more useful and more productive than ever. As the result, the computational science will become an indispensable tool in daily R&D activities of academia and manufacturing. And, I anticipate IUPAP C20 and CCP will play an important role at the center of it.

K is opened to international researchers through the peer review system. I hope K will play a significant role to promote international collaborations.

Journal of Physics: Conference Series **454** (2013) 011001 doi:10.1088/1742-6596/454/1/011001

I hope all of you'll have a fruitful time by sharing the latest research knowledge at this conference. And please enjoy your stay here in Kobe and Kobe Beef. Thank you for your attention.

Young Scientist Prize in Computational Physics

During CCP2012, the **Young Scientist Prize in Computational Physics** for year 2012 was awarded by IUPAP to **Professor Roger Melko** (Department of Physics & Astronomy, University of Waterloo, Canada) for *his innovative and deep achievements in developing quantum Monte Carlo methods for quantum information theory and condensed matter physics.*

Message upon receiving the award

I am honoured to receive the 2012 Young Scientist Prize in Computational Physics, and, given the list of previous recipients who came before me, am particularly humbled and grateful to the IUPAP for this recognition. I would like to take the opportunity to thank all of my many collaborators, without whom I could not have performed this research, especially Matthew Hastings who worked patiently with me to develop the first Monte Carlo measurement techniques for Renyi entropies in 2009. With the wide visibility that accompanies such a prestigious award, I hope that younger generations of scientists will be inspired to examine the connections between condensed matter and information theory through computer simulations in the future.

Summary of the talk presented at CCP2012

Title: *The Information Age in Simulations of Quantum Matter*

Abstract:
Monte Carlo simulations have been ubiquitous in efforts to simulate and characterize properties of materials, matter, and systems, since the advent of computers themselves. In the last several decades, condensed matter physicists have turned simulation technology to the study of a new set of phenomena, loosely called "emergent", present in striking examples such as quasiparticle excitations with fractional charge. Despite this interest, emergence is notoriously difficult to characterize, since it is often not manifested in traditional correlation functions. Motivated by this, a new set of tools was recently developed that allows one to probe emergent phenomena in Monte Carlo simulations through their entanglement entropy - a concept borrowed from quantum information theory. Remarkably, since certain scaling terms in the entanglement entropy appear to be universal, its utility in characterizing phases and phase transitions may be ubiquitous. Thus, Monte Carlo simulations are poised to play a central role in an upcoming paradigm shift where physicists increasingly rely on concepts of information theory to characterize correlations in condensed matter, materials, and systems.

Poster prizes

At the end of the conference three poster prizes offered by EPS (for students), AAPPS, and CCP2012 were assigned to the following participants.

1) **Mr. Francesco Calcavecchia** (Johannes Gutenberg Universität, Mainz, Germany)(student)

Message upon receiving the award

I want to thank all the organizer of the event, that was very interesting and useful, and gave me the opportunity to visit a wonderful country such as Japan.

Summary of the poster presented at CCP2012

Title: *Variational approach to hydrogen's electronic structure*

Abstract:
Hydrogen has a complex and still not well-understood phase diagram, in particular at very high pressures and temperatures. Computational studies of its phase diagram are convenient, since it is extremely difficult and often impossible to experimentally achieve such a high pressure and temperature. Being able to accurately describe the electronic structure is a key ingredient in this investigation.
The variational approach has the big advantage that it permits to control on the quality of the trial wavefunction used to describe the electronic structure, and it is therefore easy to compare different results (the function that gives a lower variational energy is to be favored). The product of the Slater Determinant given by the orbitals found from a mean field approach and a so-called Jastrow correlation function, that takes two-body correlations into account, has shown to give accurate results and to work well for different phases. Nevertheless, we have investigated also a different trial function, called Shadow Wave Function, that potentially gives an even larger flexibility and allows for even more accurate calculations.

2) **Dr. Kamal Kumar Choudharys** (Shri Vaishnav Institute of Technology and Science, Indore, India)

Message upon receiving the award

CCP2012 has provided a great opportunity and platform to share and discuss the views and ideas in the interdisciplinary fields of computational physics. Awards presented at CCP2012 will definitely motivate the researchers and young scholars to look forward in the field of their interest. I am very much grateful to organizers of CCP2012 to award me the APPS Best Poster Prize.

Summary of the poster presented at CCP2012

Title: *Quantitative analysis of thermoelectric properties of crystalline semiconductors embedded with ErAs nanoparticles.*

Abstract:

We quantitatively analyzed the thermo electric figure of merit ZT $(=S^2\sigma T/\kappa)$ which can be enhanced by nanostructuring thermoelectric materials. The key reason for increase in ZT is the reduction of thermal conductivity (κ) and increase in thermoelectric power (S) by embedding ErAs nanoparticles in $In_{0.53}Ga_{0.47}As$ crystalline semiconductors. The lattice thermal conductivity and thermoelectric power were studied by incorporating the scattering of phonons with defects, grain boundaries, electrons and phonons in the model Hamiltonian to evaluate the thermoelectric properties. We found that ErAs nanoparticles provide an additional scatterer to phonons, on inserting the nanoparticles in the crystal the phonon scattering with point defects and grain boundaries become more efficient which cause in decrease the thermal conductivity up to half and increase in thermoelectric power up to double of its value of pure crystal. The temperature dependent of thermal conductivity and thermoelectric power are determined by competition among the several operating scattering mechanisms for the heat carriers which depend on concentration of nanoparticles in the crystal. Numerical analysis of thermoelectric properties from the present analysis will help in designing better thermoelectric materials for thermoelectric applications.

3) **Dr. Muhammad Shabbir** (Department of Materials Engineering Science, Graduate School of Engineering Science Osaka University, Japan)

Message upon receiving the award

It was a great experience for me to participate in conference of computational chemistry (CCP2012) held in Kobe, Japan. The real exciting parts were the fascinating poster and plenary lecture sessions though which we have shared our knowledge as well as exchanged our ideas with world leading physicists on many cutting edge issues of computational physics. At the end, the best poster award for our poster came as pleasant surprise to me. It is not only a matter of honor for me but also for our Nakano's group in Osaka University.

Summary of the poster presented at CCP2012

Title: *Interplay between Diradical Characters and Third-Order Nonlinear Optical Properties in Fullerene Systems*

Abstract:

In the modern era, nonlinear optical (NLO) and spintronic materials are two types of hi-tech and smart materials that have versatile properties. As a pioneering attempt towards understanding of the interplay between these two properties, we proposed a new structure-property relationship between the diradical character (y_i), which is a chemical index of the bond nature, and the third-order NLO polarizability (second hyperpolarizability, γ) of open-shell singlet systems. We studied the topological dependence of diradical character and second hyperpolarizability (γ) in fullerenes. We found that the large differences between the geometry and topology of fullerenes have a significant effect on the diradical character of each fullerene as elucidated by their odd electron densities distributions. On the basis of their different diradical character, these fullerenes were categorized into three groups, that is, closed-shell (y_i=0), intermediate open-shell ($0<y_i<1$), and almost pure open-shell compounds ($y_i \cong 1$). This categorization has been found in accordance with Clar's sextet rule that has been applied on Schlegel projections of these fullerenes. For example, we found that closed-shell fullerenes include $\mathbf{C_{20}}$, $\mathbf{C_{60}}$, and $\mathbf{C_{70}}$, whereas fullerenes $\mathbf{C_{26}}$ and $\mathbf{C_{36}}$ and $\mathbf{C_{30}}$, $\mathbf{C_{40}}$, $\mathbf{C_{42}}$, and $\mathbf{C_{48}}$ are pure and intermediate open-shell compounds, respectively. Interestingly, the γ_{zzzz} enhancement ratios between $\mathbf{C_{30}}/\mathbf{C_{36}}$ and $\mathbf{C_{40}}/\mathbf{C_{60}}$ are 4.42 and 11.75, respectively, regardless of the smaller π-conjugation size in $\mathbf{C_{30}}$ and $\mathbf{C_{40}}$ than in $\mathbf{C_{36}}$ and $\mathbf{C_{60}}$. Larger γ_{zzzz} values were obtained for other fullerenes that had intermediate diradical character that is in line to our previous valence configuration interaction (VCI) results for the two-site diradical model. The γ_{zzzz} density analysis shows that the large positive contributions originate from the large γ_{zzzz} density distributions on the right- and left-extended edges of the fullerenes, between which significant spin polarizations (related to their intermediate diradical character) appear within the spin-unrestricted DFT level of theory. On the bases of this structure-property relationship, we have further constructed bucky ferrocenes with robust second hyperpolarizabilities that can be switched *on* and *off* in their *singlet* and *triplet* ground states, respectively.

Journal of Physics: Conference Series **454** (2013) 011001 doi:10.1088/1742-6596/454/1/011001

Conference organization

The **Conference on Computational Physics (CCP)** is organized annually under the auspices of **the Commission 20 of the IUPAP** (International Union of Pure and Applied Physics).

Main purpose of CCP2012

This 24th Conference on Computational Physics aims at stimulating interdisciplinary discussion and collaboration by putting together researchers interested in various fields of computational science, with focus on pure and applied Physics, Chemistry, Biology, Engineering, Climate, Weather, Earth Science and so on.

Distinctive features

1. Broad range of topics.
2. Thirteen plenary talks with comprehensive reviews from theoretical physics to industrial application.
3. Half of the parallel session organized on the basis of numerical methods and the other half about special topics deemed of particular importance.
4. Speakers are invited to present: i) a comprehensive overview of their research field, ii) the reasons why the selected numerical methods are useful or necessary for their computation, iii) numerical scheme and results, and iv) future prospects.

CCP2012 is co-organized by:

Osaka University
Kyoto University
Kobe University
University of Hyogo
The Japan Physical Society (JPS)
The Japan Society of Applied Physics (JSAP)

It is also endorsed by:

The Ministry of Education, Culture, Sports, Science & Technology in Japan (MEXT)
The Advanced Institute for Computational Science (AICS)
Association of Asia-Pacific Physical Societies (AAPPS)
European Physical Society (EPS)
American Physical Society (APS)
The Asahi Shimbun (The Newspaper Co LTD)
The Kobe Shimbun (The Newspaper Co LTD)

It is supported by:

The Institute of Laser Engineering (ILE), Osaka University
The Research Center for Nuclear Physics (RCNP), Osaka University
The Earth Simulator Center (JAMSTEC)
Chinese Physical Society (CPS)

It is financially supported by:

Japan Society for the Promotion of Science (JSPS)
Japan World Exposition 1970 Commemorative Fund (JEC Fund)
Kobe Convention & Visitor Association
Nakauchi Tsutomu Convention Promotion Foundation
Fujitsu
NEC

History of the Conference on Computational Physics

The Conference on Computational Physics was organized for the first time in 1989 in Boston, and since 2003 it takes place every year, rotating between Europe-Africa, North-South America, and Asia-Oceania. The keynote plenary talks are presented by prominent researchers in each of the several sub-fields of computational physics and its applications.

Past and future editions of the Conference on Computational Physics:

2013 – Moscow, Russia
2012 - Kobe, Japan
2011 - Gatlinburg, USA
2010 - Trondheim, Norway
2009 - Kaohsiung, Taiwan
2008 - Ouro Preto, Brazil
2007 - Brussels, Belgium
2006 - Gyeongju, Republic of Korea
2005 - Los Angeles, California, USA
2004 - Genoa, Italy
2003 - Beijing, China
2002 - San Diego, California, USA
2001 - Aachen, Germany
2000 - Brisbane, Australia
1999 - Atlanta, Georgia, USA
1998 - Granada, Spain
1997 - Santa Cruz, USA
1996 - Cracow, Poland
1995 - Pittsburgh, USA
1994 - Lugano, Switzerland
1993 - Albuquerque, USA
1992 - Prague, Czech Republic
1991 - San Jose, USA

1990 - Amsterdam, the Netherlands
1989 - Boston, USA

Journal of Physics: Conference Series 454 (2013) 011001 doi:10.1088/1742-6596/454/1/011001

CCP2012 Committees

Chair:

Hideaki Takabe (Osaka University, Japan)

Vice-Chairs: (representatives from three regions of the world)

Kimihiko Hirao (RIKEN, (AICS), Kobe, Japan)
Michele Parrinello (ETH, Zürich, Switzerland)
Robert Rosner (University of Chicago, USA)

Scientific secretary:

Luca Baiotti (Osaka University, Japan)

International Advisory Board:

Joan Adler (Technion, Israel)
Gabrielle Allen (Louisiana State University & CCT, USA)
Sinya Aoki (University of Tsukuba, Japan)
Amanda S. Barnard (CSIRO, Australia)
Manuela Campanelli (CCRG & Rochester Institute of Technology, USA)
Stefano Curtarolo (Duke University, USA)
Aysen Ergin (METU, Turkey)
Timothy Germann (Los Alamos National Laboratory, USA)
Helmut Grubmuller (MPI for Biophysical Chemistry, Germany)
Karen A. Hallberg (Centro Atomico Bariloche, Argentine)
Alex Hansen (Norwegian University of Science and Technology , Norway)
Lin-Ni Hau (NCU, China-Taipei)
Xian-Tu He (IAPCM, China)
Jisoon Ihm (Seoul National University, Korea)
Masatoshi Imada (University of Tokyo, Japan)
Chisachi Kato (University of Tokyo, Japan)
A. D. Kennedy (University of Edinburgh, UK)
Werner Krauth (Ecole Normale Superieure, France)
David P. Landau (University of Georgia, USA)
Rubin H. Landau (Oregon State University, USA)
Ewald Müller (MPI for Astrophysics, Germany)
Haruki Nakamura (Osaka University, Japan)
Witold Nazarewicz (University of Tennessee & ORNL, USA)
Jens K. Nørskov (Stanford University, USA)
David Randall (Colorado State University, USA)
Sitangshu Bikas Santra (Indian Institute of Technology Guwahati, India)
Stefano Sanvito (Trinity College Dublin, Ireland)
Katsuhiko Sato (NINS, Japan)
Lev Shchur (Landau ITP, Russia)

Ed Seidel (NSF, USA)
Junko Shigemitsu (Ohio State University, USA)
Sandro Sorella (SISSA, Italy)
Toshikazu Takada (MEXT, Japan)
William M. Tang (Princeton University, USA)
Shinji Tsuneyuki (University of Tokyo, Japan)
DingSheng Wang (Institute of Physics, China)
Kunihiko Watanabe (JAMSTEC, Japan)
Wolfgang Wenzel (KIT, Germany)
Philipp Werner (University of Fribourg, Switzerland)
Naoki Yoshida (IPMU, Japan)
Yasunari Zempo (Hosei University, Japan)

Local Organizing Committee:

Chair: Takahito Nakajima (RIKEN, AICS)
Vice-chair: Ryusuke Numata (University of Hyogo)
Vice-chair: Hideyuki Usui (Kobe University)

Luca Baiotti (Osaka University)
Atsushi Hosaka (Osaka University)
Shiaki Hyodo (University of Hyogo)
Satoshi Itoh (RIKEN, AICS)
Akira Kageyama (Kobe University)
Hiroyuki Kamano (Osaka University)
Yoshinobu Kuramashi (RIKEN, AICS)
Yoshitada Morikawa (Osaka University)
Hideo Nagatomo (Osaka University)
Shin-ichiro Shima (University of Hyogo)
Yuji Sugita (RIKEN, AICS)
Hideaki Takabe (Osaka University)
Shigenori Tanaka (Kobe University)
Seiichiro Ten-no (Kobe University)
Kei Tokita (Osaka University)
Hirofumi Tomita (RIKEN, AICS)
Shugo Yasuda (University of Hyogo)
Seiji Yunoki (RIKEN, AICS)

More than 100 computational scientists experts in fields ranging from pure physics to industrial applications of physics contributed to the completion of the scientific program.

Journal of Physics: Conference Series **454** (2013) 011001 doi:10.1088/1742-6596/454/1/011001

Guidelines for presenters and session chairpersons

We organized parallel sessions on the basis of numerical methods while including very different topics from fundamental physics to industrial applications. In order for the audience to be able to understand the presentations and join the discussion, each presenter (including plenary speakers) is warmly invited to pay attention to the following points:

1. Explain the background and motivation of your research with easy words at first. Avoid using jargon. Assume the audience has knowledge at the level of an undergraduate.
2. Show the basic equations you are going to solve numerically and show them in a form as simple as possible.
3. Explain why you use your specific numerical method and what the unique and challenging points of that numerical scheme are.
4. Describe what the scientific product and result of your simulation is.
5. Explain how to improve the present numerical methods. Then show what original results you expect with such improved or new numerical methods.

The chairperson of each session is asked to stimulate discussion, also by giving ideas on how to help improve the speaker's computations.

We hope that in this conference you will enjoy interdisciplinary discussion on forefront research and that new collaborations will start through such discussions between scientists in different fields.

Panel discussion

On Thursday 18th between 11:00 and 12:30 we organized a panel discussion on the subject:

"Will computational science be able to provide answers to important problems of human society?"

The panel started with a plenary talk (30 minutes) by a journalist, Ms Atsuko Tsuji of the Asahi Shimbun. The remaining time was left for discussion. Simultaneous English-Japanese translation will be available.

Journal of Physics: Conference Series **454** (2013) 011001 doi:10.1088/1742-6596/454/1/011001

Booths

Several booths will complement the conference, advertising scientific projects and other activities.

- **Booth 1.** Predictable life science, healthcare, and drug discovery foundation (by RIKEN)
- **Booth 2.** New materials and energy creation (by Institute for Solid State Physics of the University of Tokyo; Institute for Molecular Science; Institute for Materials Research, Tohoku University)
- **Booth 3.** Projection of global change toward the mitigation of natural disasters (by Japan Agency for Marine-Earth Science and Technology)
- **Booth 4.** Next-generation manufacturing technology (by Institute of Industrial Science, University of Tokyo; Japan Aerospace Exploration Agency; Japan Atomic Energy Agency)
- **Booth 5.** The origin of matter and the universe (by University of Tsukuba, High Energy Accelerator Research Organization, National Astronomical Observatory of Japan)
- **Booth 6.** Fujitsu Company
- **Booth 7.** Center for Planetary Science, Kobe University
- **Booth 8**. Graduate school of Simulation Studies, University of Hyogo
- **Booth 9.** International Office, Osaka University

Journal of Physics: Conference Series **454** (2013) 011001 doi:10.1088/1742-6596/454/1/011001

Program at a glance

Time	Sunday 14 October	Monday 15 October	Tuesday 16 October	Wednesday 17 October	Thursday 18 October
8:00		Registration	Registration	Registration	Registration
9:00		Opening ceremonies	Plenary: Zhu	Plenary: Okuda	Plenary: Arsuaga
9:45		Plenary: Imada	Plenary: Pietrucci	Plenary: Pourtois	Plenary: Rezzolla
10:30		Break	Break	Break	Break
11:00		Plenary: Papenbrock	Plenary: Blügel	Plenary: Jansen	Panel discussion
11:45		Plenary: Takahashi	Plenary: Melko	Plenary: Krauth	
12:30		Lunch	Lunch	Lunch	Lunch
13:50	Registration (starting at 15:00) and reception (starting at 17:00)	Parallel sessions	Parallel sessions	Parallel sessions	Parallel sessions
16:00		Poster session	Poster session	Poster session	Poster session
19:00			Banquet		

Overview of parallel sessions

Oct. 15 (Mon)	K-computer special	Quantum Monte Carlo methods	Multi-hierarchy methods	Education in computational physics	Finite-difference, finite-volume, finite-element methods	Density Matrix Renormaliz. Group	Climate and disaster prevention
Oct. 16 (Tue)	K-computer special	Quantum Monte Carlo methods	Large-scale computing	Molecular dynamics	Finite-difference, finite-volume, finite-element methods	Density Functional Theory	Particle methods
Oct. 17 (Wed)	Monte Carlo methods	Quantum Monte Carlo methods	Large-scale computing	Molecular dynamics	Community-driven codes	Visualization	Industrial applications
Oct. 18 (Thu)	Monte Carlo methods	Bio-computing	Multi-hierarchy methods	Molecular dynamics	Finite-difference, finite-volume, finite-element methods	Density Functional Theory	Particle methods

Journal of Physics: Conference Series **454** (2013) 011001 doi:10.1088/1742-6596/454/1/011001

Daily schedules

Oct. 14 (Sun)	
15:00-19:00	Registration
17:00-19:00	Reception

Oct. 15 (Mon)							
8:00-	Registration						
9:00-9:45	Opening						
9:45-10:30	**Masatoshi Imada,** *Quantum Monte Carlo for strongly correlated systems,* University of Tokyo (Japan)						
10:30-11:00	Break						
11:00-11:45	**Thomas Papenbrock**, *Computing the atomic nucleus,* University of Tennessee and Oak Ridge National Laboratory (USA)						
11:45-12:30	**Keiko Takahashi**, *Challenge toward the prediction of typhoon behavior and downpour,* Japan Agency for Marine-Earth Science and Technology (Japan)						
12:30-13:50	Lunch						
13:50-16:00 Parallel sessions	**K-computer special**	**Quantum Monte Carlo methods**	**Multi-hierarchy methods**	**Education in computational physics**	**Finite-difference, finite-volume, finite-element methods**	**Density Matrix Renormaliz. Group**	**Climate and disaster prevention**
Rooms	K-computer building (AICS), seminar room 1st floor	Kobe University, Convention Hall, 2nd floor	Nichii Gakkan Conference Center, 3rd floor hall A	Nichii Gakkan Conference Center, 2nd floor room 1	Nichii Gakkan Conference Center, 3rd floor hall B	Nichii Gakkan Conference Center, 2nd floor room 2	Nichii Gakkan Conference Center, 2nd floor room 3
16:00-17:30	Poster session (with refreshments available)						
16:00-16:45	Visit to K computer (group A)						
16:45-17:30	Visit to K computer (group B)						

Schedule of the opening ceremony
October 15 (9:00-9:45)

MC: Luca Baiotti

Welcome address:

Hideaki Takabe, Chair of CCP2012

Greeting on behalf of co-organizing academic societies:

Shoji Nagamiya, President of AAPPS and ex-President of JPS

Greeting on behalf of the co-organizing four universities:

Saburo Aimoto, vice-President of Osaka University

Greeting on behalf of our ministry, MEXT:

Takahiro Hayashi, Director, Office for the Promotion of Computing Science

Greeting on behalf of IUPAP,

Alex Hansen, Chair of Commission 20 (Computational Physics), IUPAP

Oct. 16 (Tue)							
9:00-9:45	**Shao-Ping Zhu**, *Computer simulations on laser fusion,* Institute of Applied Physics and Computational Mathematics (China)						
9:45-10:30	**Fabio Pietrucci**, *Molecular dynamics challenges: from chemical reactions to biological and artificial nanostructures,* EPF Lausanne (Switzerland)						
10:30-10:50	Break						
10:50-11:35	**Stefan Blügel**, *Computing inhomogeneous solids by density functional theory,* Forschungszentrum Jülich (Germany)						
11:35-12:25	IUPAP Young Scientist Award 2012: **Roger Melko**, *The information age in simulations of quantum matter*, University of Waterloo (Canada)						
12:25-12:45	Conference photo, in front of the building of K computer						
12:45-13:50	Lunch						
13:50-16:00 Parallel sessions	**K-computer special**	**Quantum Monte Carlo methods**	**Large-scale computing**	**Molecular dynamics**	**Finite-difference, finite-volume, finite-element methods**	**Density Functional Theory**	**Particle methods**
Rooms	K-computer building (AICS), seminar room 1st floor	Kobe University, Convention Hall, 2nd floor	Nichii Gakkan Conference Center, 3rd floor hall A	Nichii Gakkan Conference Center, 2nd floor room 1	Nichii Gakkan Conference Center, 3rd floor hall B	Nichii Gakkan Conference Center, 2nd floor room 2	Nichii Gakkan Conference Center, 2nd floor room 3
16:00-17:30	Poster session (with refreshments available)						
16:00-16:45	Visit to K computer (group C)						
16:45-17:30	Visit to K computer (group D)						
19:00-21:00	Banquet						

Oct. 17 (Wed)							
9:00-9:45	**Motoi Okuda**, *Development of K-computer and toward exascale computing,* Fujitsu (Japan)						
9:45-10:30	**Geoffrey Pourtois**, *Modeling challenges in nanoelectronics: an atomistic point of view,* Imec (Belgium)						
10:30-11:00	Break						
11:00-11:45	**Karl Jansen**, *Lattice computations for high energy and nuclear physics,* DESY (Germany)						
11:45-12:30	**Werner Krauth**, *Hard-disk melting: New algorithms, new insights,* École Normale Supérieure (France)						
12:30-13:50	Lunch						
13:50-16:00 Parallel sessions	**Industrial applications**	**Visualization**	**Large-scale computing**	**Molecular dynamics**	**Quantum Monte Carlo methods**	**Community-driven codes**	**Monte Carlo methods**
Rooms	K-computer building (AICS), seminar room 1st floor	Kobe University, Convention Hall, 2nd floor	Nichii Gakkan Conference Center, 3rd floor hall A	Nichii Gakkan Conference Center, 2nd floor room 1	Nichii Gakkan Conference Center, 3rd floor hall B	Nichii Gakkan Conference Center, 2nd floor room 2	Nichii Gakkan Conference Center, 2nd floor room 3
16:00-17:30	Poster session (with refreshments available)						

Oct. 18 (Thu)							
9:00-9:45	**F. Javier Arsuaga**, *Modeling topological changes of highly confined DNA: Applications to the genomic organization of bacteriophages and trypanosomes*, San Francisco State University (USA)						
9:45-10:30	**Luciano Rezzolla**, *Using numerical relativity to explore fundamental physics and astrophysics*, Albert Einstein Institute (Germany)						
10:30-11:00	Break						
11:00-12:30	Panel discussion: "***Will computational science be able to provide answers to important problems of human society?***"						
12:30-13:50	Lunch						
13:50-16:00 Parallel sessions	**Bio-computing**	**Monte Carlo methods**	**Multi-hierarchy methods**	**Molecular dynamics**	**Finite-difference, finite-volume, finite-element methods**	**Density Functional Theory**	**Particle methods**
Rooms	K-computer building (AICS), seminar room 1st floor	Kobe University, Convention Hall, 2nd floor	Nichii Gakkan Conference Center, 3rd floor hall A	Nichii Gakkan Conference Center, 2nd floor room 1	Nichii Gakkan Conference Center, 3rd floor hall B	Nichii Gakkan Conference Center, 2nd floor room 2	Nichii Gakkan Conference Center, 2nd floor room 3
16:00-16:20	Break						
16:20-16:50	Poster awards and closing						
17:00-17:45	Visit to K computer (group E)						

Parallel Sessions

October 15th (Monday)

Finite-difference, finite-volume, finite-element methods (October 15th)	
Location: Nichii Gakkan Conference Center, 3rd floor, hall B	Chairperson: Brian Van Straalen
13:50-14:20	**Hideo Aochi**, *Finite difference simulations of seismic wave propagation for understanding earthquake physics and predicting ground motions: Advances and challenges*, Bureau de Recherches Géologiques et Minières (France)
14:20-14:50	**Petar Mimica,** *Numerical simulations of dynamics and emission from relativistic astrophysical jets*, Universidad de Valencia (Spain)
14:50-15:10	CANCELLED
15:10-15:30	**Takayuki Umeda**, *Global Vlasov simulation on magnetospheres of astronomical objects,* Nagoya University (Japan)
15:30-15:50	**Jerome Breil**, *Multi-material reconnection-based arbitrary Lagrangian Eulerian (ReALE) method,* CELIA (France)

Journal of Physics: Conference Series **454** (2013) 011001 doi:10.1088/1742-6596/454/1/011001

Quantum Monte Carlo methods (October 15th)	
Location: Kobe University, Convention Hall, 2nd floor	Chairperson: Masatoshi Imada
13:50-14:20	CANCELLED
14:20-14:50	**Michele Casula**, *Variational Monte Carlo approaches as a route to describe strongly correlated materials from a fully ab-initio perspective*, Pierre and Marie Curie University (France)
14:50-15:20	**Ting-Wai Chiu**, *Simulation of lattice QCD with domain-wall fermions*, National Taiwan University (Taiwan)
15:20-15:40	**Nils Blümer,** *Momentum-dependent pseudogaps in the half-filled two-dimensional Hubbard model*, Johannes Gutenberg University (Germany)
15:40-16:00	**Satoshi Morita,** *Many-variable variational Monte Carlo calculations of the J_1-J_2 Heisenberg model*, University of Tokyo (Japan)

Density Matrix Renormalization Group (October 15th) (including Direct Matrix Diagonalization, Matrix product states, PEPS, MERA ...)	
Location: Nichii Gakkan Conference Center, 2nd floor, room 2	Chairperson: Naoki Kawashima
13:50-14:20	**Pieter Maris**, *No Core CI calculations for light nuclei with chiral 2- and 3-body forces*, Iowa State University (USA)
14:20-14:50	**Valentin Zauner**, *Calculating excited states of 1D lattice systems with Matrix Product States*, University of Vienna (Austria)
14:50-15:10	**Kenji Harada**, *Numerical study of incommensurability of the spiral state on spin-1/2 spatially anisotropic triangular antiferromagnets using entanglement renormalization*, Kyoto University (Japan)
15:10-15:30	**Nicolas Lucien Jean**, *Computational issues of configuration interaction frameworks describing open quantum systems*, University of Tennessee (USA)
15:30-15:50	**Takashi Abe**, *Recent development of Monte Carlo shell model and its application to no-core calculations*, University of Tokyo (Japan)

Multi-hierarchy methods (October 15th)	
Location: Nichii Gakkan Conference Center, 3rd floor, hall A	Chairperson: Ritoku Horiuchi
13:50-14:20	**Alphonse Finel**, *Inertia dominated criticality in martensites*, Laboratoire d'Etudes des Microstructures, Onera-CNRS (France)
14:20-14:50	**Ryoichi Yamamoto,** *Multiscale simulations of polymeric flow*, Kyoto University (Japan)
14:50-15:20	**Valentina Vetere,** *From ab-initio to multiscale modeling of electrochemical systems*, CEA/LITEN (France)
15:20-15:40	**Nina Elkina,** *Adaptive mesh refinement method for computational electromagnetics and plasma physics*, Ludwig-Maximilians University of Munich (Germany)
15:40-16:00	CANCELLED

Journal of Physics: Conference Series **454** (2013) 011001 doi:10.1088/1742-6596/454/1/011001

Climate and disaster prevention (October 15th)	
Location: Nichii Gakkan Conference Center, 2nd floor, room 3	Chairperson: Keiko Takahashi
13:50-14:20	**Yoshiyuki Kaneda**, *Advanced simulation research on earthquake and tsunami for disaster mitigation,* Japan Agency for Marine-Earth Science and Technology (Japan)
14:20-14:50	**Aysen Ergin**, *Computational challenges of coasts: Disaster prevention and adaptation,* Middle East Technical University (Turkey)
14:50-15:20	**Muneo Hori**, *Earthquake response simulation of structures and urban areas using HPC,* University of Tokyo (Japan)
15:20-15:50	**Masaki Satoh,** *The global cloud-resolving simulation by the Nonhydrostatic Icosahedral Atmospheric Model, NICAM,* University of Tokyo (Japan)
15:50-16:20	**Phil Cummins**, *New Bayesian approaches to geophysical data inference on parallel computers,* Australian National University (Australia)
16:20-16:40	**Takane Hori,** *Numerical experiment of sequential data assimilation for crustal deformation between Tonankai and Nankai earthquakes,* Japan Agency for Marine-Earth Science and Technology (Japan)

Education in computational physics (October 15th)	
Location: Nichii Gakkan Conference Center, 2nd floor, room 1	Chairperson: Joan Adler
13:50-14:20	**Nithaya Chetty**, *Probing the extensive nature of entropy,* University of Pretoria (South Africa)
14:20-14:50	**Steven Gottlieb**, *From many students per VAX to many cores per student: Some thoughts on teaching computational physics,* Indiana University (USA)
14:50-15:20	**Knut Mørken,** *Integrating computational methods throughout the bachelor education,* Oslo University (Norway)
15:20-15:40	**Kihyeon Cho**, *The fusion research of theory-experiment-simulation for particle physics,* Korean Institute of Science and Technology Information (Republic of Korea)

K-computer special (October 15th)	
Location: K-computer building (AICS), seminar room, 1st floor	Chairperson: Atsushi Oshiyama
13:50-14:20	**Shinobu Yoshimura**, *Petascale simulations of nuclear power plants subjected to strong earthquakes on K-computer,* University Tokyo (Japan)
14:20-14:50	**Kazuo Kitaura**, *Large scale quantum chemical calculations on biomolecules,* Kobe University (Japan)
14:50-15:20	**Hideaki Fujitani**, *High performance computing for drug development on K computer,* University of Tokyo (Japan)
15:20-15:50	**Takaharu Otsuka**, *New horizon of computational nuclear structure physics in the K-computer era,* University of Tokyo (Japan)

Journal of Physics: Conference Series **454** (2013) 011001 doi:10.1088/1742-6596/454/1/011001

October 16th (Tuesday)

Finite-difference, finite-volume, finite-element methods (October 16th)	
Location: Nichii Gakkan Conference Center, 3rd floor, hall B	Chairperson: Ian Hawke
13:50-14:20	**Kozo Fujii**, *Spectral-like schemes and their application to CFD study toward innovation,* Japan Aerospace Exploration Agency (JAXA) (Japan)
14:20-14:50	**Maxime Viallet**, *Time-accurate implicit methods for the modeling of low to moderate Mach number flows in stellar interiors,* University of Exeter (UK)
14:50-15:10	**Pedro Montero**, *BSSN equations in spherical coordinates without regularization: vacuum and non-vacuum spherically symmetric spacetimes,* Max Planck Institute for Astrophysics (Germany)
15:10-15:30	CANCELLED
15:30-15:50	**Ming-Yi Lee**, *Three-Dimensional Finite Element Simulation of Reflectance of Sub-Wavelength Structures on Silicon Nitride for Solar Cells,* National Chiao Tung University (Taiwan)

Particle methods (October 16th)	
Location: Nichii Gakkan Conference Center, 2nd floor, room 3	Chairperson: Anatoly Spitkovsky
13:50-14:20	**Luís O. Silva,** *Modeling of multiscale extreme plasma physics scenarios with the Osiris particle-in-cell framework,* Instituto Superior Técnico, Lisbon (Portugal)
14:20-14:50	**Kohji Yoshikawa,** *An alternative to N-body methods in astrophysical self-gravitating systems: Vlasov-Poisson simulations,* University of Tsukuba (Japan)
14:50-15:10	**Nils Moschüring**, *Divergence-free particle merging using energy conserving particle pushing,* Ludwig-Maximilians University of Munich (Germany)
15:10-15:30	**Sebastiano Fabio Schifano**, *Exploiting parallelism in many-core architectures: a test case based on Lattice Boltzmann Models,* University of Ferrara and INFN (Italy)
15:30-15:50	**Mingyu Zhang**, *An improved surface tension model for numerical simulation of interfacial flow by Smoothed Particle Hydrodynamics method,* Institute of Applied Physics and Computational Mathematics (China)

Molecular Dynamics (October 16th)	
Location: Nichii Gakkan Conference Center, 2nd floor, room 1	Chairperson: Ivana Savic
13:50-14:20	**Ryoji Asahi**, *Extension of applicability of molecular dynamics in Li ion battery,* Toyota Central R&D Lab. (Japan)
14:20-14:50	**Roland Faller,** *Molecular modeling as a tool for nano-biotechnology,* University of California at Davis (USA)
14:50-15:20	**Timothy C. Germann,** *Molecular dynamics studies of material dynamics: from petascale to exascale,* Los Alamos National Lab (USA)
15:20-15:40	**Titus Adrian Beu,** *Nanofluidic Transport and field-effect conductance in voltage-controlled carbon nanotubes,* University Babes-Bolyai, Cluj-Napoca (Romania)
15:40-16:00	**Tomás Miguel Sintes,** *Optimal ring size in magnetic filaments,* Institute for Cross-disciplinary Physics and Complex Systems (Spain)

Quantum Monte Carlo methods (October 16th)	
Location: Kobe University, Convention Hall, 2nd floor	Chairperson: Michele Casula
13:50-14:20	**Shinji Ejiri,** *Numerical study of QCD phase structure at finite temperature and density,* Niigata University (Japan)
14:20-14:50	**Stefano Gandolfi**, *Neutron matter equation of state, symmetry energy and neutron stars*, Los Alamos National Lab (USA)
14:50-15:10	**Americo Tristao Bernardes,** *Unveiling global innovation networks,* Universidade Federal de Ouro Preto (Brazil)
15:10-15:30	**Tooru Yoshida,** *Cluster structure obtained from Monte Carlo shell model calculation,* University of Tokyo (Japan)
15:30-15:50	**Shixun Zhang,** *A study of parallelizing O(N) Green-Function-Based Monte Carlo method for many fermions coupled with classical degrees of freedoms,* University of Tsukuba (Japan)

Density Functional Theory (October 16th)	
Location: Nichii Gakkan Conference Center, 2nd floor, room 2	Chairperson: Takashi Nakatsukasa
13:50-14:20	**Silke Biermann**, *First principles calculations for correlated electron materials -- where do we stand?,* École Polytechnique, Palaiseau (France)
14:20-14:50	CANCELLED
14:50-15:20	**Minoru Otani**, *Computer simulations on electrode-electrolyte interface in batteries,* National Institute of Advanced Industrial Science and Technology (Japan)
15:20-15:40	**Van An Dinh**, *Can a small polaron form in Olivine LiNiPO4? A Hybrid functional study on the polaron-vacancy complex diffusion,* National Institute for Materials Science (Japan)
15:40-16:00	**Yoong-Kee Choe**, *Nature of proton transport in polymer electrolyte membranes for fuel cell applications: A first-principles molecular dynamics study,* National Institute of Advanced Industrial Science & Technology (Japan)

Large-scale computing (present and future prospects) **(October 16th)**	
Location: Nichii Gakkan Conference Center, 3rd floor, hall A	Chairperson: Michael Marty Marinak
13:50-14:20	**Michael Norman**, *Large scale simulations of cosmic reionization,* San Diego Supercomputer Center (USA)
14:20-14:50	**Emanuel Gull**, *Large cluster dynamical mean field simulations for Hubbard models,* University of Michigan (USA)
14:50-15:10	CANCELLED
15:10-15:30	**Truong Vinh Truong Duy**, *A three-dimensional domain decomposition method for large-scale ab initio electronic structure calculations,* Japan Advanced Institute of Science and Technology (Japan)

K-computer special (October 16th)	
Location: K-computer building (AICS), seminar room, 1st floor	Chairperson: Hideaki Fujitani
13:50-14:20	**Kazuo Saito**, *Super high-resolution mesoscale weather prediction* Meteorological Research Institute (Japan)
14:20-14:50	**Atsushi Oshiyama**, *Real-Space-Density-Functional approach to electronic properties of nanostructures,* University of Tokyo (Japan)
14:50-15:20	**Norbert Attig,** *JUQUEEN: A multi-petaflop IBM Blue Gene/Q system at Jülich for science and engineering in Europe,* Forschungszentrum Jülich (Germany)
15:20-15:40	**Tomoaki Ishiyama**, *Petascale cosmological N-body simulations on K Computer,* Tsukuba University (Japan)

October 17th (Wednesday)

Molecular Dynamics (October 17th)	
Location: Nichii Gakkan Conference Center, 2nd floor, room 1	Chairperson: Thomas Papenbrock
13:50-14:20	**Stéphane Mazevet**, *Simulating matter under extreme conditions,* Laboratoire Univers et Théories (LUTH) (France)
14:20-14:50	**Ivana Savic**, *Molecular dynamics and Monte Carlo approaches to thermal transport in nanostructured materials,* University of California at Davis (USA)
14:50-15:20	**Enge Wang,** *Surface studies of ice,* Peking University (China)
15:20-15:40	**Vladimir Stegailov,** *Atomistic simulation of ultrafast laser ablation of gold: Effect of electronic pressure relaxation,* Joint Institute for High Temperatures, Russian Academy of Sciences (Russian Federation)
15:40-16:00	**Hongsuk Yi,** *Parallel programming in Intel MIC architecture,*

	Korean Institute of Science and Technology Information (Republic of Korea)

Monte Carlo methods (October 17th)	
Location: Nichii Gakkan Conference Center, 2nd floor, room 3	Chairperson: Lev N. Shchur
13:50-14:10	**Tor Nordam**, *The validity of the reduced Rayleigh equation*, Norwegian University of Science and Technology (Norway)
14:10-14:30	**Sakineh Hosseinabadi**, *Stochastic and fractal properties of silicon and porous silicon rough surfaces*, Islamic Azad University, East Tehran Branch (Iran)
14:30-14:50	**Ingve Simonsen**, *Photonics on the computer*, Norwegian University of Science and Technology (Norway)
14:50-15:10	**Andreas Tröster**, *Optimized Fourier Monte Carlo simulation of crystalline membranes*, Vienna University of Technology (Austria)
15:10-15:30	**Sally J. Bridgwater**, *Adapting phase-switch Monte Carlo for use with flexible organic molecules*, University of Warwick (UK)
15:30-15:50	CANCELLED

Quantum Monte Carlo methods (October 17th)	
Location: Nichii Gakkan Conference Center, 3rd floor, hall B	Chairperson: Karl Jansen
13:50-14:20	**Manolo Per**, *Calculating physical properties with electronic-structure quantum Monte Carlo*, Commonwealth Scientific and Industrial Research Organisation (Australia)
14:20-14:50	**Naoki Kawashima**, *Quantum Monte Carlo simulations of deconfined critical point*, Institute for Solid State Physics (Japan)
14:50-15:10	**Denis Perret-Gallix,** *Computational particle physics for event generators and data analysis*, IN2P3/CNRS (France)

Industrial applications (October 17th)	
Location: K-computer building (AICS), seminar room, 1st floor	Chairperson: Yasunari Zempo
13:50-14:20	**Chisachi Kato,** *Industrial applications of large-scale fluid-dynamics simulations*, University of Tokyo (Japan)
14:20-14:50	**Erich Wimmer,** *Computational materials science and engineering: achievements, challenges, and perspectives*, Materials Design, Inc. (USA and France)
14:50-15:20	**Masaya Ishida**, *Computational materials science in industry: Practical applications*, Sumitomo Chemical (Japan)
15:20-15:50	**Akira Yamaguchi**, *Simulation based approach in nuclear safety assessment*, Osaka University (Japan)
15:50-16:10	**Umar Fauzi**, *Pore space characterization and fluid flow properties estimation of digital porous materials*, Institut Teknologi Bandung (Indonesia)

Visualization (October 17th)	
Location: Kobe University, Convention Hall, 2nd floor	Chairperson: Hiroaki Ohtani
13:50-14:20	**Chandrajit Bajaj**, *Enhancing visualization of multiscale biophysical simulations,* University of Texas at Austin (USA)
14:20-14:50	**Sam Yang**, *Integrate model and data to visualize microstructures of materials non-destructively,* Commonwealth Scientific and Industrial Research Organisation (Australia)
14:50-15:20	**Akira Kageyama**, *Scientific visualization by immersive virtual reality,* Kobe University (Japan)
15:20-15:40	**Joan Adler**, *3d visualization of atomistic simulations on every desktop,* Technion (Israel)
15:40-16:00	**Xiao Li**, *Parallel visual analysis for multi-physics petascale simulations,* Institute of Applied Physics and Computational Mathematics (China)

Community-driven codes (October 17th)	
Location: Nichii Gakkan Conference Center, 2nd floor, room 2	Chairperson: Ian Hawke
13:50-14:20	**Brian Van Straalen**, *Chombo: Still mostly a Cathedral,* Berkeley University (USA)
14:20-14:50	**Synge Todo**, *The ALPS project: Open source software for strongly correlated systems,* University of Tokyo (Japan)
14:50-15:20	**Frank Löffler**, *The Einstein Toolkit: A community code for computational relativistic astrophysics,* Lousiana State University (USA)

Large-scale computing (present and future prospects) **(October 17th)**	
Location: Nichii Gakkan Conference Center, 3rd floor, hall A	Chairperson: Michael Norman
13:50-14:20	**Michael Marty Marinak**, *New frontiers in the simulation of inertial confinement fusion targets,* Lawrence Livermore National Laboratory (USA)
14:20-14:50	**Pablo Ordejón**, *Beating the size limits of first-principles calculations in nanoscale systems,* Centre d'Investigació en Nanociència i Nanotecnologia (Spain)
14:50-15:10	**Adam Padee**, *Double-layer evolutionary algorithm for distributed optimization of particle detection on the Grid,* National Centre for Nuclear Research (Poland)
15:10-15:30	CANCELLED

Journal of Physics: Conference Series **454** (2013) 011001 doi:10.1088/1742-6596/454/1/011001

October 18th (Thursday)

Finite-difference, finite-volume, finite-element methods (October 18th)	
Location: Nichii Gakkan Conference Center, 3rd floor, hall B	Chairperson: Luciano Rezzolla
13:50-14:20	**Yasuhiro Idomura**, *Computational challenges in petascale fusion plasma simulations,* Japan Atomic Energy Agency (JAEA) (Japan)
14:20-14:50	**Scott Noble,** *Frontiers in computational relativistic magnetohydrodynamics applied to astrophysical systems*, Rochester Institute of Technology (USA)
14:50-15:20	**Bart van der Holst**, *Radiation-hydrodynamic simulations of high-energy-density experiments*, University of Michigan (USA)
15:20-15:40	**Yuichiro Sekiguchi**, *General relativistic neutrino-radiation (magneto-) hydrodynamics simulations: Formulations and applications,* Yukawa Institute for Theoretical Physics (Japan)
15:40-16:00	**Ian Hawke**, *Numerical simulations of neutron star crusts,* University of Southampton (UK)

Particle methods (October 18th)	
Location: Nichii Gakkan Conference Center, 2nd floor, room 3	Chairperson: Hideyuki Usui
13:50-14:20	**Anatoly Spitkovsky**, *Kinetic simulations of astrophysical shock waves,* Princeton University (USA)
14:20-14:50	**Rainer Spurzem**, *Astrophysical supercomputing with programmable hardware in China and Germany,* Heidelberg University (Germany) & National Astronomical Observatories of China, Chinese Academy of Sciences (China)
14:50-15:10	**Yosuke Matsumoto**, *Electron accelerations at high Mach number shocks: Two-dimensional Particle-in-Cell simulations on massively parallel supercomputer systems,* Chiba University (Japan)
15:10-15:30	CANCELLED
15:30-15:50	**Go Ogiya**, *Study of the core-cusp problem in cold dark matter halos using N-body simulations on GPU clusters,* University of Tsukuba (Japan)

Journal of Physics: Conference Series **454** (2013) 011001 doi:10.1088/1742-6596/454/1/011001

Molecular Dynamics (October 18th)	
Location: Nichii Gakkan Conference Center, 2nd floor, room 1	Chairperson: Masaharu Isobe
13:50-14:20	**Richard More**, *Molecular dynamics with atomic transitions and nuclear reactions,* Lawrence Berkeley National Laboratory (USA)
14:20-14:40	**Shinichi Miura,** *Development of variational path integral molecular dynamics method with applications to molecular systems,* Kanazawa University (Japan)
14:40-15:00	**Kim Hyeon-Deuk,** *Photoexcited electron and hole dynamics in semiconductor quantum dots: phonon-induced relaxation, multiple exciton generation and recombination,* Kyoto University (Japan)
15:00-15:20	**Hideo Kaburaki,** *A molecular dynamics simulation of fracture process of metals,* Japan Atomic Energy Agency (Japan)
15:20-15:40	**Shoji Ishibashi,** *Computational study of magnetic structure of electron-doped CaMnO3,* National Institute of Advanced Industrial Science and Technology (Japan)
15:40-16:00	**Satoshi Ohmura,** *Dissociation mechanism of bromo aromatic molecules: an ab initio molecular-dynamics study,* Kyoto University (Japan)

Monte Carlo methods (October 18th)	
Location: Kobe University, Convention Hall, 2nd floor	Chairperson: Werner Krauth
13:50-14:20	**Markus Eisenbach**, *Thermodynamics of magnetic systems from first principles: Combining Monte-Carlo and Density Functional calculations,* Oak Ridge National Lab (USA)
14:20-14:50	CANCELLED
14:50-15:10	**Lev N. Shchur,** *Parallel uncorrelated streams of pseudorandom numbers: problems and solutions,* Landau Institute for Theoretical Physics (Russian Federation)
15:10-15:30	**Yoshiaki Kato,** *Modeling of hot accretion flows around the galactic center black hole,* National Astronomical Observatory of Japan (Japan)
15:30-15:50	**Paulo Martins,** *Probability distribution of the order parameter in the directed percolation universality class,* Universidade Federal de Mato Grosso, Brazil (Brazil)

Journal of Physics: Conference Series **454** (2013) 011001 doi:10.1088/1742-6596/454/1/011001

Density Functional Theory (October 18th)	
Location: Nichii Gakkan Conference Center, 2nd floor, room 2	Chairperson: Stefan Blügel
13:50-14:20	**Thomas Pruschke**, *Reduced Density Matrix Functional Theory - A novel path to treat correlations from first principles?*, Georg-August-Universität Göttingen (Germany)
14:20-14:50	**Jaejun Yu**, *First-principles investigations of strain-dependent magnetism and topological characteristics of quantum materials*, Seoul National University (Republic of Korea)
14:50-15:10	**NguyenTien Cuong**, *Numerical study on electronic and phononic properties of patterned nano pores structured graphene*, Japan Advanced Institute of Science and Technology (Japan)
15:10-15:30	**Shuichiro Ebata**, *Simulation of heavy ion collision using a time-dependent density functional theory including nuclear superfluidity*, University of Tokyo (Japan)
15:30-15:50	**Manoharan Muruganathan**, *Impact of point defects in the graphene nanoribbon on its transport characteristics*, Japan Advanced Institute of Science and Technology (Japan)

Multi-hierarchy methods (October 18th)	
Location: Nichii Gakkan Conference Center, 3rd floor, hall A	Chairperson: Ryoichi Yamamoto
13:50-14:20	**Ritoku Horiuchi,** *Multiscale simulations of magnetic reconnection*, National Institute for Fusion Science (NIFS) (Japan)
14:20-14:50	CANCELLED
14:50-15:10	**Cao Xiao Lin,** *Multi-physics petascale simulations using JASMIN infrastructure*, Institute of Applied Physics and Computational Mathematics (China)
15:10-15:30	**Keizo Fujimoto,** *AMR-PIC simulation of collisionless magnetic reconnection*, National Astronomical Observatory of Japan (Japan)
15:30-15:50	**Shu Takagi,** *Numerical simulation of the platelets adhesions on an injured vessel wall in the presence of red blood cells*, University of Tokyo (Japan)

Bio-computing (October 18th)	
Location: K-computer building (AICS), seminar room, 1st floor	Chairperson: Fabio Pietrucci
13:50-14:20	**Leonardo Guidoni**, *Tackling the electron correlation in biomolecules by Quantum Monte Carlo / Molecular Mechanics*, University of L'Aquila (Italy)
14:20-14:50	**Kaori Fukuzawa**, *Development and application of ab-initio fragment molecular orbital method for bio-macromolecules*, Mizuho Information & Research Institute Inc. (Japan)
14:50-15:10	**Chi-Tin Shih**, *Structural and functional analysis of the drosophila brain network*, Tunghai University (Taiwan)
15:10-15:30	**Busara Pattanasiri**, *Thermodynamics and structural behavior of a confined HP protein determined by Wang-Landau sampling*, Mahidol University (Thailand)

Poster sessions

MONDAY 15					
N.	SURNAME	NAME	AFFILIATION	COUNTRY	TITLE
1	Baumeister	Paul F.	PGI & IAS, FZ Julich, JARA	Germany	juRS - Massively Parallel DFT calculations in Real-Space
2	Bernardes	Americo Tristao	Universidade Federal de Ouro Preto	Brazil	Computer simulation of direct iron reduction
3	Bobrowski	Maciej	Gdansk University of Technology	Poland	Reduction of metal salts and chemical vapour deposition of parylenes
4	Chen	Kuan Peng	National Center For High-performance Computing, National Applied Research Laboratories	Taiwan	Quotient-Algebra Partition and Quantum Error-Correction Codes
5	Chiu	Pao-Hsiung	National Center for High-Performance Computing	Taiwan	GPU implementation for the flux-splitting Maxwell's equations solver
6	Choudhary	Kamal Kumar	Shri Vaishnav Institute of Technology and Science, Indore	India	Quantitative analysis of thermoelectric properties of crystalline semiconductors embedded with ErAs nanoparticles
7	de Doncker	Elise H.J.	Western Michigan University	USA	Shared memory iterated numerical integration for Feynman loop integrals
8	Endoh	Akira	Fujitsu Laboratories Ltd./NICT	Japan	Monte Carlo Simulation of InAs HEMTs Considering Strain and Quantum Confinement Effects
9	Fauzi	Umar	Faculty of Mathematics and Natural Sciences, Institut Teknologi Bandung	Indonesia	Computer rock model to study influence of clays and lamina on fluid permeability
10	Freza	Sylwia	Gdansk University of Technology	Poland	Parylene copolymerization
11	Fujii	Risa	Kobe University	Japan	Fluid simulation of plume head-on collision dynamics during pulsed laser

					ablation
12	Hieida	Yasuhiro	Computer and Network Center, Saga University	Japan	A population Monte Carlo estimation of the current distribution of an asymmetric exclusion process
13	Ishida	Toshimasa	Fukui Institute for Fundamental Chemistry, Kyoto University	Japan	A Non-adiabatic Trajectory Approach to Photoisomerization for Rhodopsin and Isorhodopsin. The Origin of Faster and More Efficient Isomerization in Rhodopsin
14	Jiang	Fu-Jiun	Physics Department, National Taiwan Normal Universiy, Taipei	Taiwan	Very High Precision Determination of Low-Energy Parameters: The 2-d Heisenberg Quantum Antiferromagnet as a Test Case
15	Jones	Thomas W	University of Minnesota	USA	MHD simulations of compressible, subsonic turbulence in galaxy cluster-like environments
16	Kaneko	Ryui	Department of Applied Physics, University of Tokyo	Japan	High-Precision Multi-Variable Variational Calculations for One-Dimensional Hubbard Model
17	Kargarian	Ameneh	Kharazmi University	Iran	Numerical study of electron acceleration by a magnetized Plasma wave
18	Kaurav	Netram	Department of Physics, Govt. Holkar Science college	India	Analysis of thermal conductivity of LaFeAsO at low temperature
19	Kim	Kyungsik	Pukyong National University	Republic of Korea	Characteristics of seismic networks in spatial scales
20	CANCELLED				
21	Lee	Jae Hwan	Soongsil Unviersity	Republic of Korea	Exact partition function zeros of a polymer on a square lattice
22	Lee	Seongsik	kyung hee university	Republic of Korea	A Mechanistic Study of Fluorination of Graphene
23	Li	FuLi	Department of Mechanical Engineering, National Chiao Tung University	Taiwan	One-Dimensional Fluid Modeling of a Planar Atmospheric-Pressure N2/O2/NH3 Dielectric Barrier Discharge with Reduced Chemical Kinetics

Journal of Physics: Conference Series **454** (2013) 011001 doi:10.1088/1742-6596/454/1/011001

24	Li	Zhi	Yukawa Institute for Theoretical Physics, Kyoto University	Japan	Spin Density Wave in Chromium under High Pressure
25	Lin	Paoan	National Tsing Hua University	Taiwan	Band structure of zigzag graphene nanoribbon with DFT calculation
26	Liu	Yun-Ping	Department of Physics, National Taiwan University	Taiwan	New Half-metallic Materials Study on double perovskite Sr_2ZnBO_6 (B=Tc and Re)
27	Madkour	Tarek M.	The American University in Cairo	Egypt	Conformational analysis investigation into the influence of nano-porosity of ultra-permeable ultra-selective polyimides on its diffusivity as potential membranes for use in the "green" separation of natural gases
28	Matsushita	Katsuyoshi	Cybermedia Center, Osaka University	Japan	Multicanonical Simulation of Coupled Folding and Binding of Intrinsically Disordered Protein on an Ising-like Protein Model
29	Mayes	Maricris L	Argonne National Laboratory	USA	Towards Large Scale Fully Ab Initio Calculations Using Fragment Molecular Orbital Method on Mira (Blue Gene/Q)
30	Medina	Stefan	University of Mainz	Germany	A mesoscale simulation method for polyelectrolytes at high salt concentrations
31	Minoshima	Takashi	JAMSTEC	Japan	Multi-moment advection scheme for Vlasov simulations of magnetized plasma
32	Mitsutake	Kunihiro	Frontier Research Center, Canon Inc.	Japan	Theory of electric transport in helical polyacetylene - a multipath pi-electron system
33	Nguyen	Huy Duy	Graduate School of Engineering, Osaka University	Japan	Transport properties of boron carbide nitride hetero-nanotubes
34	Ochi	Masayuki	Department of Physics, The University of Tokyo	Japan	Optimization of the Jastrow factor in the correlated wave function of electrons using the first-principles transcorrelated method for solid-state calculations
35	Ono	Youky	The University of Tokyo	Japan	First-Principles Molecular Dynamics Simulations for Graphene Growth Process on Stepped SiC(0001) Surface
36	Purohit	Ghanshyam	Sir Padampat Singhania University, Udaipur	India	Calculation of triple differential cross section for the electron impact ionization of noble gas targets
37	Rafiee Dastjerdi	Somayeh	Shahid Beheshti University	Iran	Finite difference time domain method for calculating the band structure of 2D photonic crystal lens and near-infrared imaging

38	Sako	Tokuei	Nihon University	Japan	Fermi- and conjugate-Fermi hole analysis in two-electron atomic systems
39	Salam	Tejeshwori	NEHU	India	Computational calculation on alkyltransferase of alkylated mutagenic DNA bases
40	Sano	Takayoshi	Osaka University	Japan	Magnetic Field Amplification Associated with the Nonlinear Growth of Richtmyer-Meshkov Instability
41	Sharma	Uttam	Shri Vaishnav Institute of Technology and Science, Indore	India	Development of tungsten coatings and characterization under plasma operations in ADITYA limiter tokamak and in future SST-1 diverter tokamak
42	Shinjo	Kazuya	Yukawa Institute for Theoretical Physics, Kyoto University	Japan	Dynamical DMRG Study of Spin Excitations in Disordered Spin-Peierls Systems
43	Skurski	Piotr	Gdansk University of Technology	Poland	Reactivity of parylene with substituted vinyl molecules
44	Su	Wei Chih	National Center for High-performance Computing	Taiwan	One decomposition method for Hordecki's 2X4 system
45	Takeuchi	Yasushi	Hinode Science Center, National Astronomical Observatory of Japan	Japan	A graph theoretical approach to fluctuating networks in glass-forming liquids: A molecular dynamics study with applications of the pebble game algorithm
46	Tang	Ping-Han	National Chiao Tung University	Taiwan	Order parameter by instantaneous normal mode analysis for melting behaviour of cluster Ag17Cu2
47	Tosa	Valer	National Institute for R&D Isotopic and Molecular Technologies, Cluj-Napoca	Romania	Modelling migration in multilayer polymer systems by a finite difference method
48	Tröster	Andreas	Vienna University of Technology Wiedner Hauptstrasse 8-10/136 A-1040 Wien	Austria	Monte Carlo Simulation of Curved Interface Free Energies
49	Wang	Kaier	School of Engineering, The University of Waikato, Hamilton	New Zealand	Underlying mechanism for the slow oscillation observed in nonREM sleep, general anesthesia and generalized seizure

50	Wardak	Khaleda Abdul Aleem	King Abdul Aziz University	Saudi Arabia	Investigation of solid state emission of a Cu(I) Br Complex using ab initio method
51	Watanabe	Takeshi	Department of Scientific and Engineering Simulations, Nagoya Institute of Technology	Japan	Kinetic energy spectrum of the low Reynolds number turbulence with polymer additives
52	Wilms	Johannes	University of Vienna	Austria	Mutual information as a tool to study correlations and phase transitions
53	Wu	Z-B	Chinese Academy of Sciences	China	Advances in Thermocapillary Droplet Migration
54	Yamada	Yuta	Osaka Institute of Technology	Japan	Numerical Study of 5-dimensional Gravitational Collapses
55	Yamanaka	Masanori	College of Science and Technology, Nihon University	Japan	Numerical study of topological crossover of protein genus
56	Yasuda	Hiroaki	National Institute of Information and Communications Technology	Japan	Non-equilibrium Green's function calculations of terahertz-quantum cascade lasers for room-temperature operations

TUESDAY 16					
N.	SURNAME	NAME	AFFILIATION	COUNTRY	TITLE
57	Bui	Kieu My Thi	NIMS	Japan	Hybrid Functional Study on Diffusion in Silicate Cathode Material Li2NiSiO4
58	Chau	Shiu-Wu	National Taiwan University of Science and Technology	Taiwan	Thermal Plasma Flow Modeling of Non-Transferred Steam Torch Using a Non-Equilibrium Approach
59	Cho	Kihyeon	KISTI	Republic of Korea	Belle II Data Handling System
60	Fujimura	Takayoshi	The Institute of Scientific and Industrial Research, Osaka University	Japan	Stable arrangement of impurities of copper in silicon
61	Fukushima	Akinori	Tohoku University	Japan	Molecular dynamics simulation of water droplet in micro pores
62	Hanaoka	Kyohei	Graduate School of Pure and Applied Sciences, University of Tsukuba	Japan	QM/MM simulation revealed a substrate mediated proton relay mechanism in DNA religation reaction catalyzed by Type II DNA topoisomerase
63	Iwata	Ryosuke	Graduate school of engineering, Gifu University	Japan	Effect of atomic adsorption of catalytic metals on mechanical properties of graphene
64	Kawaguchi	Kazutomo	Kanazawa University	Japan	Free energy profile of Hsp90-ADP binding by molecular dynamics simulations
65	Kawatsu	Tsutomu	Institute for Molecular Science	Japan	A Tunnel Pathway Analysis using the Semi-classical Instanton Approach
66	Kim	Kyungsik	Pukyong National University	Republic of Korea	Dynamical mechanism of the scaling behavior in multifractal structures
67	Kinjo	Tomoyuki	Toyota Central R&D Labs., Inc.	Japan	Coarse-grained Particle Model for Polar Solvent
68	Kirihara	Takanobu	University of Tsukuba	Japan	Resolving the outer density profile of dark matter halo in Andromeda galaxy
69	Kodera	Mitsuru	National Institute for Materials Science	Japan	First-principles study for initial stage of graphene nucleation at the step on SiC
70	Lee	Kyungeun	RIKEN	Japan	Computational simulation of flow in double curved vascular model: A test on the accuracy

Journal of Physics: Conference Series **454** (2013) 011001 doi:10.1088/1742-6596/454/1/011001

71	Lu	Hantao	YITP, Kyoto University	China	Photoinduced spin-order destructions in one-dimensional extended Hubbard model
72	Maeng	SeongEun	Inha University, Dept of Physics	Republic of Korea	Compatibility-based evolution of bipartite networks and its connectivity and topological overlap patterns
73	Masaki	Akiko	Institute for Solid State Physics, University of Tokyo	Japan	Quantum Monte Carlo Simulation of Bose-Fermi Mixtures in One-Dimensional Incommensurate Optical Lattices
74	Matsumoto	Kosuke	Department of Physics, Kyushu University	Japan	GPU-Accelerated MD Simulation for Short-Range Particle Interaction
75	Matsumoto	Masaharu	Kobe University	Japan	Two-Dimensional AMR-PIC Simulation on Solar Wind Interaction with Mini-Magnetosphere
76	Miki	Yohei	University of Tsukuba	Japan	Hunting a Wandering Black Hole in M31 Halo Using GPU Cluster
77	Mima	Toshiki	Department of Mechanical Engineering, University of Tokyo	Japan	Molecular dynamics simulation for vapor-liquid phase transition of water in nanocylinder with controlled wettability
78	Nakamura	Etsuko	Graduate School of Engineering, Gifu University	Japan	Wave-packet dynamics in gated bilayer graphene within Tight-Binding model
79	Nakatsukasa	Takashi	RIKEN Nishina Center	Japan	Mean-field calculation including proton-neutron mixing in atomic nuclei
80	Nishida	Keisuke	Kyoto University	Japan	The role of a plasmoid ejection in 3-dimensional magnetohydrodynamic simulation of a solar flare
81	Noda	Masashi	Institute for Molecular Science	Japan	Massively-Parallel TDDFT calculations based on Finite Difference Method in Real-time and Real-Space
82	Ohtani	Hiroaki	National Institute for Fusion Science	Japan	Data compression concepts on scientific visualization for time-series data of huge particle system
83	Otsuka	Yuichi	RIKEN	Japan	Quantum Monte Carlo study of the half-filled Hubbard model on the honeycomb lattice
84	Otsuka	Takao	RIKEN Quantitative Biology Center (QBiC)	Japan	Structure relaxation and binding energy calculations of FK506 binding protein complexes using order-N DFT code CONQUEST

85	Saito	Hiroaki	Kanazawa University	Japan	Binding Free Energy of Azurin-Cytochrome c551 Complex by all-atom Molecular Dynamics Simulation
86	Santra	Sitangshu Bikas	Department of Physics, Indian Institute of Technology Guwahati, Assam	India	Directed sandpile models on random and scale-free networks
87	Shabbir	Muhammad	Osaka University	Japan	Interplay between Diradical Characters and Third-Order Nonlinear Optical Properties in Fullerene Systems
88	Shahjahan	Mohammad	ISIR, Osaka University	Japan	Ferromagnetic half metals based on chalcopyrite semiconductors CuAlSe2 and CuInS2
89	Shiibashi	Tomohiro	Seikei university	Japan	Influence of the substitutional impurities in the electronic transportation properties of carbon nanotubes
90	Shimamura	Kohei	Department of physics Kumamoto University	Japan	Ab initio study of dissociation reaction of ethylene molecules on nickel cluster
91	Shinaoka	Hiroshi	AIST	Japan	First-principles study on noncollinear magnetism and effects of spin-orbit coupling in 5d pyrochlore oxide Cd2Os2O7
92	Shinaoka	Hiroshi	AIST	Japan	Unconventional spin-glass behaviors in pyrochlore Heisenberg antiferromagnets coupled with lattice distortions
93	Shiroto	Takashi	Department of Aerospace Engineering, Tohoku Univ.	Japan	Radiation Hydrodynamics Simulation of High-Z Doped ICF Targets
94	Shoji	Masafumi	Institute of Space and Astronautical Science	Japan	Precipitation of Highly Energetic Protons by Helium Branch Electromagnetic Ion Cyclotron Triggered Emissions
95	Sota	Shigetoshi	RIKEN AICS	Japan	Dynamical density matrix renormalization group study of high-energy optical conductivity in high-Tc copper-oxides
96	Suzuki	Takafumi	University of Hyogo	Japan	Edge state of two dimensional quantum spin systems
97	Tadano	Terumasa	Department of Physics, The University of Tokyo	Japan	First-principles studies on the reduced thermal conductivity of a clathrate Ba8Ga16Sn30
98	Tajima	Nobuo	National Institute for Materials Science	Japan	Carbon atom clustering and cluster growth in CVD graphene formation on nickel surface: A theoretical study

99	Takano	Yu	Osaka University	Japan	Density functional study of conformational preferences of intermediates and transition states in the alkaline hydrolysis of dimethyl phosphate
100	Tsuyuki	Hiroyoshi	Seikei University	Japan	Electronic transport properties in substitutionally doped graphene nanoribbons
101	Tung	JenChuan	Graduate Institude of Applied Physics, National Chengchi University	Taiwan	Spin and Anomalous Hall Conductivities in Co-based Heusler Alloys: A First-principle Study
102	Uemura	Naoki	The Institute of Scientific and Industrial Research, Osaka university	Japan	First-principles study of α-tetragonal boron
103	Usui	Hideyuki	Kobe University	Japan	Full Particle-In-Cell Simulation of Ion Beam Neutralization
104	Velechovsky	Jan	Czech Technical University	Czech Republic	Symmetry-preserving Remap of Vectors for Staggered ALE Hydrodynamics
105	Vu	Truong V.	Ritsumeikan University	Japan	Numerical simulations of solidification with volume change by a front-tracking method
106	Watanabe	Hiroshi	RIKEN	Japan	Theoretical study of novel insulating and superconducting states in Ir oxides with large spin-orbit coupling
107	Yagi	Yohei	Graduate school of System Informatics, Kobe University	Japan	Development of Parallelized Adaptive Mesh Refinement Particle in Cell with Dynamic Domain Decomposition
108	Yamamoto	Atsusi	AICS RIKEN	Japan	Superfluidity of One-dimensional Trapped Fermionic Optical Lattices with Spatially Alternating Interactions
109	Yoshimoto	Yuta	Department of Mechanical Engineering, The University of Tokyo	Japan	Construction of interaction models of dissipative particle dynamics by coarse-graining Lennard-Jones fluids: Investigation on the system with vapor-liquid interfaces
110	Zivny	Oldrich	Institute of Plasma Physics AS CR, Prague	Czech Republic	Calculation of Composition and Thermodynamic Properties of Non-Equilibrium Thermal Plasma at High Temperatures

Journal of Physics: Conference Series **454** (2013) 011001 doi:10.1088/1742-6596/454/1/011001

WEDNESDAY 17					
N.	SURNAME	NAME	AFFILIATION	COUNTRY	TITLE
111	Anh	Le The	Japan advanced institute of science and technology	Viet Nam	Ab initio study of Phosphorus donor states in single dopant transistor with a stub-shaped channel
112	Belosludov	Rodion V	Institute for Materials Research, Tohoku University, Sendai	Japan	Realization of Gas Storage Materials based on Clathrate Hydrate: Computational Modelling
113	Blümer	Nils	Johannes Gutenberg University Mainz	Germany	Momentum-dependent pseudogaps in the half-filled two-dimensional Hubbard model
114	Bouamama	Khellil	Univesity Ferhat Abbas of Setif, Algeria	Algeria	Ab-initio calculation of the photoelastic constants of SiC
115	Calcavecchia	Francesco	Johannes Gutenberg Universitat, MAINZ Graduate School	Germany	Variational Approach to Hydrogen's Electronic Structure
116	Chakraborty	Himanshu	Department of Physics, Indian Institute of Technology Bombay	India	Large Scale Configuration Interaction Calculations of Linear Optical Absorption of Octacene, Nonacene and Decacene
117	Chen	Chih-Huan	Department of Physics and Center for Theoretical Sciences, National Taiwan University	Taiwan	Spin-spiral waves in zigzag 3d transition metal atomic chains from first principle calculations
118	Chen	Chun-Nan	National Chiao Tung University	Taiwan	Multi-Objective Solar Cell Design Optimization Using Semiconductor Device Simulation-Based Evolutionary Algorithm
119	Chen	Tingting	Hiroshima University	Japan	Empirical Study of the GARCH model with Rational Errors
120	Chetty	Nithaya	University of Pretoria, South Africa	South Africa	The African School of Electronic Structure Methods and Applications

121	Chouakri	Sid Ahmed	Laboratoire de Telecommunications et Traitement Numerique du Signal, University of Sidi Bel Abbes	Algeria	Wavelet Transform and Huffman Coding Based ECG compression Algorithm: Application to Telecardiology
122	Doi	Shotaro	Department of Physics, Graduate School of Science, Osaka University	Japan	Development of first-principles electronic structure calculation code for large super cell systems by using screened KKR method
123	Fujii	Tetsuya	Dept. of Micro Eng. Kyoto University	Japan	Energy analysis of a cofilin-decorated actin filament using molecular dynamics simulations
124	Fujiwara	Susumu	Kyoto Institute of Technology	Japan	Shape transition of micelles in amphiphilic solution: A molecular dynamics study
125	Fukuda	Ikuo	RIKEN	Japan	A simple non-Ewald scheme: the zero-dipole summation method and its application to molecular systems
126	Heide	Marcus	Graduate School of Engineering, Osaka University	Japan	Effcient relaxation of magnetic moments in noncollinear DFT calculations
127	Higuchi	Katsuhiko	Hiroshima University	Japan	Correction method of the pair density functional theory and its application to atomic structure calculations
128	Hirokawa	Kazuya	Niigata University	Japan	Parallel-computing solutions of symmetry-unrestricted Hartree-Fock-Bogoliubov equations in the three-dimensional Cartesian-mesh representation
129	Huang	Shu-Ping	Kyushu University	Japan	DFT study of the mechanism for methane hydroxylation by soluble methane monooxygenase (sMMO): effects of oxidation state, spin state, and coordination number
130	Ikeda	Yuji	Department of Micro Engineering, Kyoto University	Japan	Perturbative Approach for Calculating Electronic Structures with Electric Currents
131	Inglis	Stephen C	University of Waterloo	Canada	Phase transition detection using Renyi entropy in classical and quantum Monte Carlo

Journal of Physics: Conference Series **454** (2013) 011001 doi:10.1088/1742-6596/454/1/011001

132	Inoue	Shinri	Japan Advanced Institute of Science and Technology	Japan	Ab initio Study on Adsorption of Gas Molecules on Graphene Nanoribbons
133	Ishimoto	Yukitaka	RIKEN CDB	Japan	Two-dimensional lattice liquid models
134	Jansen	Karl	NIC, DESY Zeuthen	Germany	A first look at Quasi-Monte Carlo methods for lattice field theory problems
135	Kaneko	Tomoaki	Computational Materials Science Unit, NIMS	Japan	Effect of device geometry on current injection from metal electrodes to graphene: DFT-NEGF calculations
136	Kaneko	Yuta	Graduate School of Frontier Science, Tokyo University, Japan	Japan	Hamiltonian dynamical structure of the Reduced MHD applied to the numerical simulation
137	Kawazura	Yohei	Graduate School of Frontier Science, The University of Tokyo	Japan	Numerical simulation of primordial vorticity generation by relativistic effect
138	Koyama	Hiroshi	NIMS	Japan	Structures and electronic states of 3C- and 4H-SiC surfaces
139	Lee	Jae Woo	Department of Physics, Inha University, Incheon	Republic of Korea	Agent-based wealth exchange dynamics and power-law distribution of the wealth
140	Linh	Nguyen Van	Ritsumeikan University	Japan	Tracking large scale structure of open channel flow using POD-based Unscented Kalman Filter
141	Miyake	Yohei	Graduate School of System Informatics, Kobe University	Japan	Development of a scalable PIC simulator and its application to spacecraft interaction with streaming plasma
142	Miyazaki	Tsuyoshi	National Institute for Materials Science	Japan	Parallel efficiency and recent progress of a linear-scaling DFT code CONQUEST
143	Moritaka	Toseo	Institute for Laser Engineering, Osaka University	Japan	Plasma particle-in-cell simulation with Monte-Carlo QED reactions on pair production experiment using high-Z target
144	Nagara	Akihide	Graduate School of System Informatics, Kobe University	Japan	Development of a portable AMR module for various numerical simulations

145	Nakamura	Kazuma	The University of Tokyo, Department of Applied Physics	Japan	Massive parallelization of ab initio RPA and GW codes
146	Nakamura	Tatsufumi	Japan Atomic Energy Agency	Japan	High power gamma-ray source from laser irradiated solid target
147	Naylor	Wade	Department of Physics, Osaka University	Japan	Photon pair creation in microwave cavities with losses and infinitely coupled systems of differential equations
148	Nikbakht	Shahla	University of Queensland	Australia	Computation of Unequal Time Correlation Function for the Asymmetric Simple Exclusion Process Using Matrix Product States
149	Oda	Akifumi	Kanazawa University	Japan	Evaluations of conformational search accuracy of CAMDAS using experimental three-dimensional structures of protein-ligand complexes
150	Ogino	Yousuke	Tohoku University	Japan	Computational code of high-enthalpy flow equations with collisional-radiative processes
151	SaiToh	Akira	National Institute of Informatics	Japan	A multiprecision C++ library for matrix-product-state simulation of quantum computing
152	Sakakibara	Hirofumi	The University of Electro Communication	Japan	First principles band structure +FLEX approach to the pressure effect on Tc of the cuprate superconductors
153	Santra	Sitangshu Bikas	Department of Physics, Indian Institute of Technology Guwahati, Assam	India	Does explosive percolation occur away from the critical point?
154	Schaerf	Daniel	University Mainz	Germany	Predicting New Crystal Structures from First-Principles
155	Senami	Masato	Kyoto University	Japan	Effects of a surrounding medium of quantum systems on Rigged QED Simulation
156	Shirai	Nobu C	Graduate School of Science, Osaka University	Japan	Multicanonical simulation of the Domb-Joyce model and the Go model: new enumeration methods of self-avoiding walks
157	Shirakawa	Tomonori	RIKEN	Japan	Theoretical studies of a three-band Hubbard model with a strong spin-orbit coupling for 5d transition metal oxides Sr2IrO4

Journal of Physics: Conference Series **454** (2013) 011001 doi:10.1088/1742-6596/454/1/011001

158	Takahashi	Masayuki	Department of Aerospace Engineering, Tohoku University	Japan	Hydrodynamics-Orbit Coupling Calculation for Flight Analysis of Actively Controlled Laser Vehicle
159	Takaishi	Tetsuya	Hiroshima University of Economics	Japan	Analysis of spin financial market by GARCH model
160	Tokita	Kei	Cybermedia Center, Osaka University	Japan	GPGPU simulations of 2D lattice neutral models in ecology
161	Torralba	Antonio Sanchez	Spanish National Cancer Research Centre (CNIO)	Spain	The allosteric couplings map of the meganuclease I-DmoI reveals long-range anticorrelations between the LAGLIDADG and DNA-binding moieties
162	Utsunomiya	Toshio	National Defense Academy	Japan	A Numerical Calculation Method for Frequency-Shift of a Plasma Wave Using Wave Digital Filters
163	Yamanaka	Shusuke	Osaka University	Japan	How to determine boundaries for QM/MM calculations? A guideline based on linear response function
164	Yamasaki	Takahiro	Fujitsu laboratories limited	Japan	Planewave based first-principles calculations on the 80,000 nodes K-computers: Application to SiC screw dislocations
165	Yoshimoto	Yoshihide	Tottori University	Japan	First-principles calculation combined with multicanonical simulation: application to the melting of ice
166	Yuasa	Fukuko	KEK	Japan	Acceleration of Feynman loop integrals in high-energy physics on many core GPUs

IUPAP

OSAKA UNIVERSITY

KYOTO UNIVERSITY FOUNDED 1897

KOBE UNIVERSITY

JPS The Physical Society of Japan

AAPPS

EPS

APS physics

朝日新聞

神戸新聞

中國物理學會 Chinese Physical Society

FUJITSU

JAMSTEC

ILE

RCNP

NEC

Conference on Computational Physics (CCP2012)

special event

(October 18, 11:00 - 12:30)

Will computational science be able to provide answers to important problems of human society?

Panel discussion on computational science, with the media and 400 participants

Overview

Shared use of the K supercomputer began on September 28, 2012. To mark this occasion, the Conference on Computational Physics (CCP2012) organized under the auspices of the International Union of Pure and Applied Physics (IUPAP) was held in Japan from October 14 to October 18. More than 400 participants from 44 countries participated. On the last day a panel discussion on the theme, "Will computational science be able to provide answers to important problems of human society?" was held as a special event. The general public was represented by members of the media and the discussion involved the researchers in the field of computational science participating to the conference. Simultaneous interpretation was available. With respect to the K supercomputer, which became widely known as a result of the government's budget screening scheme, debate will also focus on how computational science programs using the supercomputer should fulfill its accountability to the taxpayers. The keynote speech was given by Atsuko Tsuji, an editorialist with the Asahi Shimbun. The title was "Requests and expectations for computational science". This was followed by a free discussion with Ms. Tsuji.

Journal of Physics: Conference Series **454** (2013) 011002 doi:10.1088/1742-6596/454/1/011002

Requests and expectations for computational science

by
Atsuko Tsuji
(Asahi Shimbun)

Atsuko Tsuji, science journalist, is editorial writer for the Asahi Shimbun.

She earned her B.A. from Tokyo University in History of Science and Philosophy of Science in 1976. After joining the Asahi Shimbun in 1979, she has been working as a staff writer in various sections as the science news section and the American General Bureau in Washington DC. She was a Knight Science Journalism Fellow at MIT in 1989 and a Reuters Fellow at Green College of Oxford University in 2004.

Introduction

I have been told that this CCP2012 conference is attended by over 400 computational science experts and budding researchers from around the world. As a scientific journalist, normally, I would be the one visiting each of you for interviews, so please excuse me for being very nervous about standing here and addressing you, the experts. My role here is to act as a spokesperson for the general public outside and to convey their requests and expectations for computational science. Communication between expert groups and society has become increasingly important. That's why I mustered up the courage to come today, to help in this cause in some small way. I, therefore, appreciate your kind understanding.

First, the poster for this conference caught my attention. It lists everything from physics, chemistry and engineering through related academic fields and industrial applications. This broad scope of applications is one of the key features of computational science. I hear that Dr. Takabe, the chairperson of this event, made it a point to invite researchers from many different fields. In an age when interdisciplinary collaboration between various fields has become a big issue, I think that it is significant that researchers of various fields have come together in this way. In particular, while it has always been said that engineering and science should work hand in hand, we have not seen much progress in this area, perhaps hampered by cultural differences and other complex hurdles. Breaking through these barriers is a key issue, not only for computational science, but also for Japan's science and technology communities as a whole.

There are, in fact, other issues that have clearly emerged in computational science that are also commonly found in Japanese science and technology today. Some of these may be unique to Japan; others may be found in other countries too. I hope we can learn about, and get your thoughts on, these wide-ranging issues in the discussion to follow. Please look upon my speech as a prelude to that discussion.

Computational science and the press

While computational science plays a major role in society, this may not be widely known among the general public. When I checked the Asahi Shimbun article database on how many times the word "computational science" has appeared in our newspaper, I found that it has only appeared in 62 articles. In most of these instances, it was merely used in organization names; there were very few articles on the subject of computational science itself. It first appeared in 1994. This was in reference to a budget request for 2.9 billion yen lodged to the Science and Technology Agency at the time. This would equate to around 30 million dollars. This resulted in the launch of the Earth Simulator project, which became number one in the world in 2002. I also found an article about the establishment of the Frontier Research Center for Computational Sciences at the Tokyo University of Science in 1997 as Japan's first research center for computational science.

What about the word "simulation"? In the mid-1980s, there were only around 40 articles a year containing this word, but more than 100 articles made mention of it for the first time in 1989. From 2000 onwards, there have been around 400 articles a year. This means that, on average, we see the word "simulation" appearing somewhere in the papers every day. Of course, not all of these relate to computational science. Some are used in reference to professional baseball managers "simulating" the game in their mind and various other situations. Nonetheless, at present it would appear to be rare for the field of computational science to come up in the readers' thoughts. This may partly be because it is a relatively new field. Getting the field recognized by more people is a big issue for the future.

The K computer

I heard that the purpose of this conference is also to celebrate the start of the shared use of the K computer, a state-of-the-art supercomputer earmarked by the Japanese government as a key national technology. Shared use was originally scheduled to start in November, after this conference, but was moved forward and has been in operation since September 28. As a Japanese citizen, I hope that K will produce results that only it could provide.

Figure 1. The K computer, located in Kobe, Japan.

Many of you have probably seen it already. I also had the chance to see the K computer at the ceremony held prior to the start of shared use. It was quite a sight seeing 864 units lined side by side, each unit just a little taller than an ordinary person measuring 2 meters in height. It reminded me of the Terracotta Warriors and Horses in China's ancient capital, Xian. Needless to say, the terra cotta figures were an army of soldiers made of clay that were placed there to protect the body of Shi Huangdi, the first Qin Emperor, in the netherworld. Shi Huangdi commanded a powerful army and overwhelmed surrounding states in numbers during China's Warring States period, unified the country and became its first emperor in 221 BC. Even today, more than 2000 years later, the wordless statues speak volumes of Shi Huangdi's immense power. In contrast, the K computer will not endure for so long. The game is about how much it can achieve in a much shorter period of time. The person giving the commands holds the key to this. For the terra cotta army, this was Shi Huangdi, for K, it's you. It is up to you

Figure 2. The Terracotta Warriors: Warriors and horses in China's ancient capital, Xian.

Figure 3. Lawmaker Renho at a budget screening on the K computer project. Her comment on K "Why No. 1? What's wrong with being No. 2 in the world?" became very famous.

whether you can take advantage of K's capabilities. Of course, the strength of an army is not determined merely by its size. We find many examples of this worldwide, a typical example being the Battle of Red Cliffs in the *Chronicle of the Three Kingdoms*, which gained public attention through the movie *Red Cliff*. Your strategies for controlling your soldiers and your ability to read ahead like Zhuge Kongming (the chancellor of the state and military strategist of Shu Han) will determine whether you can use K wisely or let it go to waste. As researchers, you bear a grave responsibility. Whether we will see revolutionary results will depend on the ingenuity and foresight of the researchers using the computer. In that sense, I think it is particularly important to foster personnel with a wide vision.

The K computer faced many challenges before it was completed. Many of you living outside Japan may not be aware of this, but after the Democratic Party of Japan took over the government, in November 2009, it embarked on "budget screening" to filter out wasteful expenses in government projects and cut down on their budget. This was carried out in the open and attracted much public attention as a new initiative. The K computer, which was under construction at the time, was also subjected to examination. In regards to its development aim of becoming the world's fastest computer, lawmaker Renho asked the representative from the Ministry of Education, Culture, Sports, Science and Technology, "Why No. 1? What's wrong with being No. 2 in the world?" The developers were unable to provide her with a satisfactory answer and a harsh verdict was reached that the budget should be slashed. This footage was aired over and over again, and became a famous symbol of the budget screening initiative. The science community reacted strongly and Nobel laureates joined in issuing a statement in protest. They expressed their fierce objection saying, "If you don't aim to become No. 1, you cannot even become No. 2 or No. 3." In some ways, however, the lawmaker's question was very legitimate; it was something that had to be asked. In the end, the project received virtually the same budget as originally allocated and actually received an extra budget to speed up development. Last year, the K computer was ranked number one in the world's top 500 rankings, twice. It was ousted from this top position in June this year by the supercomputer, Sequoia, developed by the Livermore Laboratory in the United States, so it was no longer the fastest when shared use began.

Questions arise, first of all, regarding the method for determining these projects. It is not hard to imagine that those involved in K's development wanted to show off to the rest of the world the advanced level of Japan's supercomputer technology, which came to be known thanks to Fujitsu's Numerical Wind Tunnel and NEC's Earth Simulator. Becoming No. 1 in the world was not one of the Earth Simulator's objectives, but it took the world by surprise when it rose to that position in 2002 and remained there until 2004. As a result, the K computer project got underway with the clear-cut goal of retaking the top position, which was consistent with political and administrative intentions. It is only natural that Ms. Renho asked, "Would you spend 100 billion yen to be No. 1? Is there any point to that?".

Figure 4. Nobel laureates joined in issuing a statement of protest. They expressed their fierce objection saying, "If you don't aim to becoming No. 1, you cannot even become No. 2 or No. 3."

As an aside, the term "No. 1 in the world" reminds me of a story I heard from Mr. Sakurai, a former manager of the Honda team which was a constant winner in Formula 1 racing. He was asked by a company executive if it was possible to cut down on the expenses a little, even if the team were to drop to second place, because it was costing the company the not inconsiderable sum of several billion yen a year. To this he answered, "The moment we say it's okay to be second, you'll find we won't even be able to be No. 2". He meant that it is only when you strive desperately to become No. 1 that you have any chance of achieving this, if all goes well. Even then, you could still end up becoming No. 2 by the slightest margin. If you're going to do something, you have to aim for the top.

However, you must be able to explain why. A supercomputer is a tool. Developers should aim to become No. 1 with the results of using this tool and should begin by asking what this would require. This was not the case with K and that's where the problem lay.

Moreover, there was much criticism that K became more costly and inferior in performance to Sequoia because of the insistence on the use of Japanese parts. The aim was to provide support to the Japanese semiconductor industry but the manufacturers ended up stopping their chip production nonetheless.

Deliberation on development plans for the next supercomputer to succeed K is currently underway. I think it is vital to formulate a plan with reasonable objectives, following a series of scientific studies into what type of computer should be built and for what purpose. It is precisely because this was lacking that the developers were unable to answer the lawmaker's question properly.

Behind this issue is the lack of mechanism in Japan for delivering scientific assessments to the policy making process. In dealing with the Fukushima nuclear accidents too, scientific findings could not be combined and utilized, resulting in much confusion. Learning from this experience, discussions are presently in progress on creating a framework for incorporating scientific advice into policy making.

Raison d'être of supercomputers

What is the purpose of building expensive supercomputers in the first place? As Dr. Takabe mentioned earlier, I think that it is true to say that national defense is the primary objective of building the world's fastest supercomputer in the U.S. The main purpose of the world's fastest Sequoia is said to be to maintain and manage the performance of nuclear weapons. What then is the position of computational science and the reason for having expensive supercomputers in a country like Japan, which does not have nuclear weapons? I guess this would be for fundamental science and industrial applications. But how does that justify the enormous costs? I think computational scientists have a responsibility to answer to the taxpayers.

Needless to say, expensive supercomputers, such as K, are not necessarily essential to computational science. Today, many people have more than one personal computer, and their capacities far exceed that of Cray-1, which appeared in 1976 and is said to have marked the start of the age of supercomputers. The infrastructure for computational science is now available to anyone. The only difference between a personal computer and K is the speed and memory size. It is a known fact that an increase in numbers leads, at some stage, to a leap in quality. But will this vast growth in the number of computers result in scientific discoveries of a different nature from the past? This kind of question also emerges.

Is computational science necessary?

Computational science simulation is currently called the third mode of science after theory and experimentation. Would their results, then, merit awards such as the Nobel Prize? In his will, Alfred Nobel skillfully captured the essence of each field of science by setting out that his awards were for discovery or invention within the field of physics, chemical discovery or improvement, and discovery within the domain of physiology or medicine. How can the expensive computers in existence around the world today contribute to such inventions, discoveries and improvements? There is no denying that technical advances have supported the progress of science, and the Nobel Prize rates development of innovative devices highly. What about the role of the computer?

Looking back on the history of the computer, it goes without saying that it has its roots in the ENIAC developed for military purposes in the United States during World War II. It was developed to calculate the trajectory of ammunition fired from warships and decrypt coded messages, and also used to compute the fluid flow of the type of atomic bomb used at Nagasaki. The computer made it possible to solve non-linear fluid equations that even math geniuses of the likes of Von Neumann could not solve. In the early days, Enrico Fermi and his fellow researchers made a remarkable discovery using a computer in which interacting non-linear systems, which were expected to end up in a disorderly state, went through what is called a recurrence phenomenon and returned to their initial state. Consequently, some expected at the time that the computer would enable an ongoing succession of Nobel Prize-level discoveries. More than half a century has passed since then and computers have undergone astronomical progress in their capabilities. But it seems that computational science discoveries have not thus far produced any Nobel Prize winners. That said, we should not forget that computational science simulations on global warming contributed greatly to the awarding of a Nobel Peace Prize to the Intergovernmental Panel on Climate Change (IPCC) in 2007.

According to Dr. Takabe, computational science is the optimal research method for fundamental science on the evolution of the universe, particularly because there is no way of carrying out experiments in astrophysics. However, we must verify whether computational simulations are reproducing actual physical phenomena by checking the correctness of the mathematical models used, through simulation experiments or by other means. In light of the rapid advancement in observation technologies, to truly impress astrophysics observation and theory experts with simulations, their programming must be verified. In the materials field too, many researchers are having a go at numerical material design. How close are computer simulations to the prediction of new materials? Computational science has always had to deal with the prejudice that it is just a follow-up study to experiments and theories. To live up to its name as the third mode of science, I feel that computational science is now required to acquire predictive abilities.

Comparison with earthquake science

The day before yesterday, there was a special symposium on earthquake prediction at the Fall Meeting of the Seismological Society of Japan and it made references to computational science. Let me briefly introduce some of the interesting discussions that went on. As you know, on March 11 last year, a huge earthquake of magnitude 9.0 occurred off the coast of the Tohoku region in the northern part of Japan, and the giant tsunami that followed took many lives. It caught people completely unawares. Presently in Japan, only the area off the coast of Shizuoka Prefecture has an extensive system of observation in place for the anticipated Tokai Earthquake and offers any chance of prediction. But the Tokai Earthquake is yet to occur. Instead, we have had frequent earthquakes in other areas such as the Great Hanshin Earthquake and the massive earthquake of last year. Seismologists have lost their confidence and earthquake prediction itself has come under scrutiny. The symposium discussed the history of seismology over the past 50 years, during which observation systems have been reinforced, as well as its outlook for the future. With respect to technical progress, advancement in computational science has enabled simulations using realistic models and there are moves to predict earthquakes through simulations. But the symposium also pointed out certain limitations.

The comment by Dr. Hiroo Kanamori of the California Institute of Technology was especially noteworthy. He said that since we do not have a full understanding of physical phenomena of earthquakes, it is difficult to utilize study results as they are in real-life disaster prevention. He then went on to list three areas in which seismology can meet the demands of society. The first was to reinforce education and research activities and train human resources who have the ability to think creatively and flexibly. This is a common issue in any field, but the fact he mentioned this first reflects his shrewd discernment. This also has to do with the fact that, because new tools such as K have been developed, creativity is required in making the best use of them. Next, he explained that outstanding research results should be utilized effectively in disaster prevention, bearing in mind the uncertainty factor they still possessed, and lastly, that real-time seismology methods should be developed to prevent damage. He emphasized the importance of collaboration with engineers in this regard. Applying this to simulations, this means that models should be verified using observed data, and be utilized in society while remaining aware of their limitations. I think this is an important point.

Future applications of computer science

How then, should we make use of computational science in the years ahead? There is no doubt that it will be used increasingly in various aspects of natural science. Besides climate change research, it can be used for disaster prediction as well as studies on energy, genetic information, functions of the brain and heart, and design of new materials. The possibilities would appear to be unlimited. And it is not just in natural science. In the future, there will likely be demand for wide application in the fields of social and cultural sciences too. For example, researchers engaged in social science could input large amounts of data (big data) relating to economic crises or yen appreciation into a supercomputer and perform simulations using mathematical models. Of course, there will still be problems that would be difficult to address in real life. However, I believe that deeper research is required to facilitate evidence-based policy making. It is also important to train computational science experts who will undertake research in various fields.

Take policy making, for instance. In today's world with its vastly improved means of globalization, communication and transport, policy making is an extremely complex task. But with computational scientists working together to create mathematical models of national activities, large amounts of data (big data) held by government offices could be processed so that we can prepare efficient budgets and growth strategies without resorting to budget screening, hurriedly implemented over a short period of time. As a member of the public, I have great hopes in these developments. What do you think?

After last year's nuclear accident in Fukushima, the government asked the public to what extent they thought Japan should rely on nuclear energy for its energy requirements: 0% 15% or 30%? With this too, I thought the government could have prepared simulations in advance on various scenarios for their energy policies until 2030, and then logically explained under what conditions each of those assumptions would be realized, in a similar way to global warming simulations. It is unfortunate that they did not do this because, had they done so, their ideas would have been more widely accepted and the debate would have become much more scientific.

In terms of methodology, it is somewhat similar to statistics. Statisticians are called upon to show interest in various fields, and to try and find solutions to problems in cooperation with experts in the respective fields. Unfortunately, statisticians in Japan tend not to be so enthusiastic in getting involved in outside fields, though this seems to be changing gradually more recently. I hope that computational science experts will be different, and that they will boldly take on challenges in a variety of areas to provide clues for steering society in a more positive direction.

Here are some themes that we can possibly examine in the following discussion:

1. Computational science to address humankind's problems
2. Ideal style of supercomputer development
3. How to inform society
4. How to foster human resources

Journal of Physics: Conference Series **454** (2013) 011002 doi:10.1088/1742-6596/454/1/011002

Will computational science be able to provide answers to important problems of human society?

Part (2)

Record of discussions at the IUPAP Conference on Computational Physics 2012

(This is the record of the comments made by the audience following the keynote address by journalist Atsuko Tsuji on the above theme at the IUPAP Conference on Computational Physics 2012 that was held in Kobe from October 14 to October 18, 2012. The moderator was Professor Hideaki Takabe of Osaka University, who also chaired the conference.)

Moderator: Thank you Ms. Tsuji. Ms. Tsuji has brought our attention to a number of issues from the media perspective. I am sure that there are many computational scientists here at this conference who would like to comment. However, I suggest that the discussion focuses on the four points pointed out by Ms. Tsuji in the final viewgraph. I welcome comments from the floor.

Floor: I would like to comment on the second item regarding development strategy. Ms. Tsuji suggested that, a supercomputer is expensive because it is uniquely developed and, therefore, it should be purchased because it is merely a tool. That is one opinion. Ms. Tsuji, at the same time, also mentioned the Fukushima Daiichi nuclear power station. When nuclear energy first began in Japan, physicists in Japan such as Dr. Hideki Yukawa were not essentially opposed to it. However, their view was that nuclear energy technology should be developed independently. But what actually happened was that we bought it because it was cheap, fast and part of the U.S. global strategy. Thus, the Fukushima Daiichi nuclear power station was made possible by introducing technology from abroad. One of the causes of the recent accident was, I believe, that they bought equipment that could be activated at the flick of a switch. Installing emergency generators underground was the U.S. design and the fact that the equipment was installed that way is a lesson. Therefore, I have my doubts about how cheap and fast purchased technology is. Prior to this, I was relegated to the Old Guard by your correspondent Komiyama but the fact remains that we must develop much of the key technology ourselves because of its very nature. For this reason, I am inclined to think that we will still need to use budgets to some extent.

Figure 5. Ms. A. Tsuji responding a question from Floor.

Tsuji: I don't believe that we should buy everything either. Development is needed in key aspects. For example, we should buy the chips if we cannot make them in Japan but we need to develop key software. Given the age that we live in, I think that we should use those items that are used around the world and develop key items ourselves. The crux of my argument is that we have persisted with pure domestic products and that is why it has been expensive. Now, with the Fukushima nuclear plant as well, when we have the nuclear power from a country with a focus on natural disasters such as hurricanes and tornadoes in a country like Japan which is affected by tsunamis, even if we use it as it is initially, there is no doubt that we failed to make modifications using Japanese technology.

Floor: When nuclear reactors were supplied by the U.S., Japan should have revised the design. In my view, something in the nature of a committee should have been set up to examine this. If it were my country, India, there would be special committees to review the designs. Therefore, the problem was not buying from the U.S.; it was probably the fact that there should have been a study by a committee to ensure that they were appropriate for the location.

Floor: First of all, I would like to thank Ms. Tsuji for a wonderful keynote address. My name is Eric Rimenda. I used to head up one of the laboratories at Cray Research. When we are thinking about supercomputers, we must not forget software. The software for designing the new materials we use today was created 30 years ago for the supercomputers of that time. When it comes to that software being used by the younger generations, it must be upgraded for the supercomputers of today or even 10 or 20 years into the future. The key is to consider the hardware and software together and focus on the point that the K supercomputer is presenting new concepts for software.

Floor: On the question concerning why you need to be the best, I think the importance lies in not having the fastest but creating something that is faster than what we have today. For example, it may not be possible to make predictions about earthquakes. I don't believe it is possible to predict other natural disasters either. But, I do think it is possible to make limited predictions. For example, when strong wind and rain is initially not too violent and then unexpectedly worsens causing more deaths, it may become possible to predict developments like that with a faster supercomputer than K.

Tsuji: Typhoons can be predicted with considerable accuracy in Japan today. But with the recent changes in meteorological phenomena, there have been some fairly strange events. We have even seen tornadoes in Japan. Speaking of tornadoes, I understand that they can be predicted with fairly high accuracy in the US. However, the accuracy of tornado prediction in Japan is totally inadequate and I think we need to have better performing computers. This is my perception and I have no objections.

Floor: First of all, thank you for your presentation. I have listened with great interest. At the same time, it made me think from a different perspective. I disagree on one point in your address. That is where you said that computational science is the third mode of science. I think this is incorrect. Computational science is a modern science. Theoretical physics cannot be debated with pen and paper. It is not even possible to subject it to a simple analysis. A hundred years ago it was possible to analyze the atom but in this age, we need computers to conduct experimental analysis on subatomic particles. Computational science is a theoretical science, and supercomputers should be utilized to use this in making advances in science. The following is a comment. It was mentioned that computational science challenges the problems of humankind. But this is a fairly difficult problem. The computational science that I deal with has many theoretical aspects and areas in which we are not sure how they relate to humanity. But, let's consider this mobile phone. It functions beautifully through two theoretical developments. The first is Fourier analysis. Fourier was a mathematician who lived 200 years ago. I imagine he could not even conceive things like mobile phones. The other is GPS. GPS works on the basis of Einstein's theory of general relativity. We carry out our work without really knowing how our research will contribute to society but I think it is examples like those that motivate us.

Journal of Physics: Conference Series **454** (2013) 011002 doi:10.1088/1742-6596/454/1/011002

Moderator: I thought your point that computational science is not the third mode of science was very persuasive. As Ms. Tsuji mentioned, computational science practiced by people with flexible thinking will lead to new discoveries. That is a very important point. If we allocate computational science to a place that is independent of theory, we will end up looking at technological aspects only. I think that what we are essentially looking for in computational science are new theoretical directions and such results. If the younger generation is taught that computational science is the third mode without being properly educated about this point, the key components of theoretical insights related to the results of this science are going to be incomplete.

Tsuji: Yet, in Japan, it is frequently stated that computational science is the third mode. Maybe this is an expression of the computational scientists' conviction that they are also part of the scientific world. Perhaps there are some here at this conference who disagree and believe that computational science is the third mode of science. If so, I would like to hear their views.

Floor: I would also like to comment. First, I think we should call it the "most powerful" computer rather than the "fastest". As the participant from Cray Research pointed out earlier, it becomes the fastest only when the hardware and software work well together. Second, on the subject of whether computational science is the third mode, I think that it emerged as a new form of science. It is led by hypotheses, but is also based on experiments and observations, as well as data. It is still in its early stage. In the past, it helped advance theoretical and experimental science. But I expect that it will become more independent in the future.

Floor: I would like to touch on an extremely significant point, if I may. But first, let me talk about three smaller points. First, the Fukushima nuclear plant. Japan did not buy the plant from the US. It was constructed in Fukushima in collaboration with U.S. parties. In addition, the blunder in Fukushima was the result of being unable to respond promptly in an intelligent manner. Supercomputers may have been able to help but they were not utilized. I think Japan could have modified the location of the emergency power supply themselves but this was neglected, which is the problem. So please stop accusing the U.S. unilaterally. Second, about the Sequoia. You say it is cheaper than the K supercomputer but it was produced in series and this is the 5th model. Therefore, the development expenses should be assessed as a whole, and such assessment would show that it is very expensive. Sequoia was cheap only because improvements were made over five successive models. My third and final point: insistence on purely domestic products. I read a very interesting Japanese book, a book by Yukichi Fukuzawa. He was the first Japanese to visit San Francisco. He walked down the streets of San Francisco carrying a couple of swords at his side. Americans watched him with great curiosity because they had guns instead. However, he didn't come there to fight. He was there to learn. Besides, he was extremely observant, and perhaps much more intellectual than those he observed. He also visited Paris and other countries. This was 150 years ago but he learned that the technical and cultural revolutions were transforming the world. He wrote a book to give the Japanese public his advice. Sound advice for anyone, not just Japanese. In his book he says that the Meiji Restoration required two major changes in awareness. He used key words. One of which was "independence". He said Japan had to become independent of other countries and make its own decisions on national concerns. This also applies today and other countries should imitate this. Another word he mentioned was "numbers". When considering something, you must look at the numbers. This point is brought out in *The Autobiography of Fukuzawa Yukichi*. A large computer produced and programmed by Japanese people, functioning independently and handling numbers efficiently, embodies the spirit of Yukichi Fukuzawa.

Figure 6. A participant said that Japan should follow the way indicated by Yukichi Fukuzawa 150 years ago, when Japan opened its gate to the US and European countries.

Moderator: Let me add to the comment from our attendee from the United States. Please take a look at a 10,000 yen note, if you have one. There you'll find a portrait of Yukichi Fukuzawa. He was the first person to introduce a psychological revolution, the concept of democracy, to Japan during the Meiji Restoration some 150 years ago. We just heard comment that the problem during the Fukushima nuclear accidents was not whether the nuclear reactors, the hardware, were made in the U.S.A. or Japan, but rather if the Japanese government had sufficient software to enable it to take prompt action in response to the situation. He called to our attention the fact that Japanese politics had not yet been able to maintain a firm sense of independence in disclosing information, and to respond promptly as a democratic nation, trusting in the intellect of its people. Quoting Fukuzawa's words from 150 years ago, the commenter pointed out that Japan should reflect more on its lack of software in the form of swiftness, rather than focus on whether the reactors were made in the U.S.A.

Moderator: Our discussion began with the question of whether computational science is the third mode of science. Is there anyone who is of the opinion that, as the third way, computational science can exist independently of experiments and theories?

Floor: Yes. I think they are clearly different. Documents sometimes contain theories, results of experiments proving those theories and things that were verified using mathematical models. In other words, numerical verification. With respect to the Nobel Prize, Kenneth G. Wilson won the Nobel Prize in Physics in 1982 for his "study of critical phenomena in connection with phase transitions". This achievement is known today as "Wilson's renormalization group." He is a professor at Cornell University and has contributed to the fields of theories, simulations, science and engineering. I think this is a good example. It is significant because he won the Nobel Prize and because it shows that theories and simulations are completely different. What is mentioned there cannot be proved using mathematical models alone, but can be visualized using computer in simulations.

Floor: That's mathematics. Theorems have nothing to do with physics. In physics, there are concepts as well as experiments to confirm them. Theories change constantly.

Floor: Is computational science the third mode of science or just a scientific tool? To answer this question, I think we need to look at science from various perspectives, especially in terms of what it will enable us to do. Firstly, it allows us to study complex systems. In other words, it can help us to encapsulate nature into laws. Through computational science we can study laws inherent in nature, and examine larger complex systems. It can also satisfy our needs for faster computer resources. It enables us to handle various complicated systems in real life using the computer. Moreover, we can perform experiments on the computer. Computational science makes it possible to study subjects which would be too costly to perform experiments on, or which do not allow for experiments at all.

Journal of Physics: Conference Series **454** (2013) 011002 doi:10.1088/1742-6596/454/1/011002

Moderator: A very important point was just raised. Computational science exerts great potential when examining complex systems. With complex systems, we look not only at the results but the biggest elements affecting those results. Computational science really excels when used as a tool for performing calculations to remove various elements from the equation and identifying what the essence is. Whether scientists using the computer can achieve this is where the abilities and scientific instincts of the researchers, who we compared to Shi Huangdi earlier, come into play. It is important for computational science to continue elucidating governing mechanisms in complex systems in this way.

Figure 7. A snapshot of the discussion.

Moderator: There was another noteworthy comment. That was that computational science provided a means for simulating effects on the computer where experiments are not possible. I think that it is extremely important for computational science to keep venturing into these fields and seek to find new forms of science. Today's theme is "Computational science to address humankind's problems". As Ms. Tsuji explained, shouldn't young people be encouraged to tackle social science issues where experiments are not possible without adhering to the field of natural science, and senior researchers in the computational science field assist them in their pursuits? For instance, an easy-to-understand example would be to study how changes in the government's economic policies would affect economic activities. Simulated results of various changes in policies may be used by economic experts to implement policies that are considered to be the most logical. Wouldn't you agree that we need studies like that? I hope that young researchers in particular will be motivated to try their hands in new areas like these.

Floor: Computers are a tool, a means to an end. They lead to swift decision making and everyone reaps the benefits. Therefore, everyone in Japan should be proud to have the K computer. It was mentioned that high-performance computers play an important part in national defense, but isn't it also important to protect a country in the area of science, without using weapons and ammunition? As an international research consortium, we are grateful that the Japanese government has directed efforts into developing an infrastructure for computational science. I wish my country, Russia, would do more.

Floor: I want to stress one point. That is that it doesn't necessarily have to be supercomputers that solve humankind's problems. To give you a short example, the project I am involved in is funded by the EU. We analyze how much customers are being informed of the contents of food and other products, and also look at to what extent package labeling has filtered through among customers. A PC is sufficiently adequate for handling numerical databases for this analysis. At times, solving small problems can benefit the entire human race. Supercomputers will be used to solve larger issues. This requires hardware, software and humanware. I do not see the costs of the K supercomputer as being such a big problem. People in Japan pay three to four times as much as people in Europe for a bottle of beer.

Moderator: That is a valid point. We are not saying that supercomputers are essential if we are to tackle new issues facing humankind. As pointed out earlier, it is the software that is important. This software becomes increasingly complicated with time. But we must pass on these assets to the next generation. For that, education is required. Problems that started off manageable on a personal computer may become complicated and more widespread due to historical developments. It is then that we need supercomputers and the challenge of passing on technologies to the next generation arises. That's why the question of how to foster human resources mentioned by Ms. Tsuji becomes a major issue.

Floor: I would like to make a brief comment. I think large computers are necessary for fields such as astrophysics. We cannot control data on the evolution of the universe. Therefore to follow the evolution of the universe over billions of years, sometimes we have no choice but to rely on computers. On the other hand, CERN's LHC (Large Hadron Collider) experiment on the Higgs boson is a large-scale experiment in which everything appears to be controlled, and it requires a large-scale computer. It has 150 million sensors and countless instruments binding these together, and data needs to be accumulated every millisecond. It is no longer possible to perform theoretical analysis of this data manually. Performing computer simulations in conjunction will lead to discoveries of new particles. We are already in an age when we cannot engage in leading-edge research in this field without using a large computer.

Moderator: You touched on an important point and I think it leads on nicely to the subject of "big data" which has become a hot topic recently. Traditionally, in statistics, data collected by individuals is used for statistical analysis. Personally, I think it is inevitable that the accumulated data will eventually exceed that which can be handled on individual levels and will have to be analyzed by computer. Ms. Tsuji, what are your thoughts on this?

Tsuji: As things stand, in the U.S., President Obama announced a national project on "big data." In Japan, although the issue is being discussed by various parties, we have not seen the kind of progress that the US has had regarding national initiatives due to the vertically-divided structure of the government ministries and agencies. Even when we talk of big data, usually, those held by individuals are very small compared to those handled by organizations such as CERN. Also, there are many types of "big" data. Therefore, we must look at multiple types to consider the hardware requirements, but such discussion is yet to be held. Things are still in a state of disarray. It has been pointed out that the matter should be examined scientifically and that a consensus reached within the scientific community should be incorporated into government policies. This has not yet materialized. I think that we need to formulate a grand design, a map as it were, for this process.

Figure 8. Many attendants joined the discussion on the important subject of the special session.

Moderator: Again, perhaps it is necessary for us to take the initiative in making inroads into the areas of statistics and big data from the computational science side. We must promote joint research while passing on computational science's progress in hardware and techniques in the development of software for handling complex problems. Take huge volumes of data that would not fit inside a person's head, store it inside a computer and use exceptional software to ensure consistency. Actions like this are what we really need, right now.

Floor: I am involved in the analysis of biological data. Perhaps some of you have heard of the term "personalized medicine". If you attend conferences on cancer research, you'll find that the major breakthrough at the moment is in genetic sequencing. This was originally very costly, but it is getting much cheaper. It is now becoming possible for cancer patients to have their genetic code sequenced. However, this requires a computer with terabytes of memory that can analyze cancer tumors at genomic levels. This data can then be used to determine the appropriate treatment for that particular patient. The costs will fall to around a half or a third of what they are. By the way, let me also comment on the question, "What's wrong with being No. 2?", raised earlier on in this discussion. A few years ago, I worked at SmithKline Beecham. I talked with various researchers in the company and was told that, as a pharmaceutical company, for example, you end up a loser if you come in second place and that's why you must always stay at the top. Unless you are the first one to have access to various resources, you are going to lose out. I think this is the kind of reasoning that goes through people's mind when they are investing on something.

Moderator: It is almost time to bring this session to a close. Mr. Alex Hansen, you are the Chair of the C20 Commission of IUPAP, the host of this event. Do you have any comments?

Floor: Allow me to share my views on whether computational science is the third mode of science. In regards to Kenneth G. Wilson who was awarded the Nobel Prize in 1982 (The Nobel Prize in Physics 1982 was awarded to Kenneth G. Wilson "for his theory for critical phenomena in connection with phase transitions"), his work began by examining the model described in Onsager's thesis of 1942. At the time, nobody understood the physics contained in this thesis because it was so complicated. So, in the 1960s, Wilson kept mulling over how to interpret it. In the 1970s, he created an algorithm to simulate the contents of the thesis, first by analyzing it on paper and then with the help of a computer. As a result, he was able to understand the physics through the algorithm. The important point here is that he was able to discover physics that had been concealed in the math, by using a computer. That was over 30 years ago.

Floor: As a second example, let's take fractal and other complex systems, which

Journal of Physics: Conference Series **454** (2013) 011002 doi:10.1088/1742-6596/454/1/011002

is my area of expertise. The concept of fractal has existed for over 2,000 years but large volumes of data and calculation are required for it to come close to nature. I feel that looking at computer results gives rise to new concepts and new mathematical models. Structures you have never even thought of before sometimes emerge when you actually do the calculations. This, in turn, helps science on complex systems to evolve further. As illustrated in these examples, I think that computational science is the third mode of science. At the same time, I believe that it is a necessary element of science that is compatible with other scientific means.

Moderator: Listening to IUPAP-C20 Chairman Alex's comment, I felt that he gave us some hints on how to inform society of our study results, a point we have not dealt with yet in this discussion. We computational scientists tend to keep our wonderful results to ourselves and are not good at letting the public know in a way that they can understand. I think he used the word "fractal" to describe how we should always be mindful of this fact and strive to express ourselves. In recent years, large supercomputer development is fully funded by taxes. That is not to say that the funds come from beer tax alone. So we need to explain clearly, even to housewives who do not drink at all.

Moderator: Our time is up. Let me briefly summarize our discussion. Today we discussed the theme, "Computational science to address humankind's problems." On the second item, "Ideal style of supercomputer development", Dr. Koyanagi raised a number of issues. We could not discuss the third topic, "How to inform society," at great length but I think that this all begins with us being aware of our society and striving to convey the results of our individual and group research in a manner that is easy to understand. There is still much more to be implemented in this regard, particularly here in Japan. We must try harder. The fourth point pertaining to fostering human resources is linked to software development. No matter how fast the supercomputer is, its accomplishments depend on the software. The software will become increasingly sophisticated along with the hardware. This advancement will continue through successive generations and that is why it is so important to train up the next generation of scientists. In doing so, we must not only pass on the discipline, but also provide an environment in which knowledge can be passed down accompanied by an improvement in quality, by taking a broader view of, and collaborating with, other fields. Computational science should be able to contribute greatly to society if we continue to drive ahead with it, bearing in mind the issues that were raised in Ms. Tsuji's keynote address and the views expressed in this discussion.

Journal of Physics: Conference Series **454** (2013) 011003 doi:10.1088/1742-6596/454/1/011003

Summary of the organization of the conference

Hideaki Takabe (Aki)

Conference Chairman

Institute of Laser Engineering and Department of Physics and Space and Earth Science, Osaka University

Yamada-oka 2-6, Suita, Osaka 565-0871 Japan

Abstract:

I briefly report the story of the preparation of the conference, focusing on how we worked for organizing the Japanese committee, the international committee, and the local organizing committee, and on how we composed a very inter-disciplinary technical program with the help of about 200 computational scientists. I will briefly describe how we tried a new idea for inter-disciplinary parallel sessions in numerical methods and for a panel discussion on the relation between computational science and citizens. We had 404 attendants from 37 countries/regions. This report has been written as a reference for those who are organizing an international conference and will hopefully help them in establishing a local organizing committee that really works and in obtaining the money necessary to run the conference.

The 24th IUPAP-CCP was held on October 14-18, 2012, in Kobe, Japan. I would like to report its overview as organizer of the conference. The IUPAP is the International Union of Pure and Applied Physics (http://www.iupap.org/). It consists of physics-related academic societies from all over the world. The Physical Society of Japan (JPS) and the Japan Society of Applied Physics (JSAP) are the member societies from Japan. "CCP" stands for "Conference of Computational Physics" and is organized annually by the Commission 20 (Computational Physics) of IUPAP. This one was the 24th conference.

The number of total attendants was 404, from 37 countries. We were very happy to welcome people from far distant countries such as Brazil, South Africa, and Algeria. This diversity indicated by the number of countries of the attendants means that computational physics has grown up as a basic infrastructure for higher education in not only developed countries but also developing countries. This fact enforces IUPAP Commission 20's most intense commitment to systematic education of computational physics, even if what discipline we regard as the computational physics is still under a lot of discussion in the world community.

Journal of Physics: Conference Series **454** (2013) 011003 doi:10.1088/1742-6596/454/1/011003

Figure 1. Memorial picture taken just before lunch on the second day. Participants gathered in the space in front of the "K" supercomputer. Roughly half of the participants are in the picture.

The start of my commitment on this CCP was triggered by phone calls from a friend in our ministry MEXT (http://www.mext.go.jp/english/). He mentioned to me that it was decided that in 2012 the IUPAP-CCP would take place in Asia-Oceania and he said that Japan had never hosted the conference so far. He suggested that, since we had almost completed the then world-fastest "supercomputer K", it would have been a good chance to organize the CCP in Kobe in concomitance with the opening of the K computer for shared use, scheduled for November 2012 (http://www.aics.riken.jp/en/). After several phone calls I could not avoid accepting to become the chairman of the organization of the IUPAP-CCP in Kobe. Then, I told him that if I could organize and make the program as I wished, I would accept the charge of the chairmanship with the condition that there be no interference by our ministry except for support. Soon after, we decided the venue: beside the K computer site located in artificial island "Portopia" in Kobe.

In order to prepare the conference, I attended CCP2009 at Kaoshung in Taiwan and CCP2010 at Trondheim, Norway. Honestly speaking I was not so excited by parallel sessions, although the plenary sessions, of course, were very interesting because we could listen to many topics from all fields of physics. In fact, the parallel sessions programmed with the names of research fields are a kind of duplication of the topics we listen to in other conferences related to our field. While I had been a council member of the Physical Society of Japan (2003 - 2006), I had insisted on the reform of the structure of the annual JPS assembly meeting, namely moving from the classification based on physics fields used for a long time to one that included a mixture of fields in order to make new interdisciplinary sessions. When I formally submitted this proposal as the chair of the reform committee after a lot of discussion to the council meeting, strong negative voices had made all the others quiet and finally it was not approved. I have, therefore, decided to carry on this idea in this IUPAP-CCP conference in Kobe.

The idea is based on two new methods:

(1) Extending CCP from computational physics to computational science in general and attracting scientists from the theory and applications.

(2) Organizing parallel sessions according to numerical methods and techniques instead of research fields.

Computational science is the academic discipline spun by the warp called area of research and the woof called the calculation technique. This time, I attempted to organize the sessions following of the woof, from the foundations to applications. I had been wanting to organize a conference in such a style for a long time. However, soon after it became clear that this requires a lot of serious work. I cannot make even the program unless I borrow the knowledge of many researchers of computational science, from elementary particles to industrial applications.

I asked the help of many friends to organize the "Japanese Advisory Committee (JAC)", made of about 100 persons, centering on the Kansai area of Japan. And I asked the JAC to nominate about 40 specialists for the "International Advisory Committee (IAC)" comprised of a wide range of fields. Then, the Program Committee became consisting of a total of about 200 persons, including 22 from the “Local Organizing Committee (LOC)”. Around January 2012, we obtained from the Program Committee recommendations for the keynote speakers and the invited speakers. We composed the program and came to the opening of the home page on April 1.

Figure 2. A snapshot of the audience in a plenary session. There were two screens.

Figure 3. One poster session.

I accepted the chairmanship, but in my Institute most professors are experimentalists and the faculty working in computational science are only four. I could not organize the conference without a bigger group. In parallel I have tried to obtain the joint sponsorship of Osaka University, Kyoto University, Kobe University, University of Hyogo, and AICS (Advanced Institute for Computer Science, Riken). Then, I asked each of my acquaintances from Osaka University, Kobe University, University of Hyogo, and AICS, the last three near to the venue, to name five computational scientists each as members of the Local Organizing Committee (LOC). This functioned very well and the conference greatly rose as the result. I am glad to have had 22 friends of computational science in Kansai throughout the organization and execution of this IUPAP-CCP2012. Through the success of the conference management, we also talked about the establishment of a "Kansai Computer Science Consortium" with the members of the present LOC.

The strategy determined by the LOC was that we would not offer any financial support for traveling expenses or hotel expenses. We had to apply for funds and limit expenses. The LOC tried to increase the number of participants by organizing an attractive conference. AICS took charge of the reception, guided tour of the K computer, and banquet. University of Hyogo took charge of the conference rooms, Internet connection, and signboards. Kobe University took charge of poster session site and booths for display. Osaka University took charge of general affairs and, in part, of banquet. Everything was decided at LOC meetings held routinely at Kobe University, University of Hyogo, and AICS, and everybody prepared devotedly. We started the bimonthly LOC meetings in December 2011.

Figure 1 is the memorial picture taken on the second day of CCP2012. Figure 2 is a snapshot of a plenary session. Figure 3 shows the opening of the banquet, the so-called "Kagami-wari", a typical Japanese ceremony involving the breaking of the wooden plate of a big "sake" barrel to show the mirror surface of the 18 liters of sake. Figure 4 is a snapshot of the guided tour to the K computer near the venue.

Journal of Physics: Conference Series **454** (2013) 011003 doi:10.1088/1742-6596/454/1/011003

Figure 4. Banquet was held Tuesday night at "Kacho-en" near the venue. It is Japanese custom to break the wooden plug of the 18 liter Japanese sake bottle with wooden hammers. This is the start of the banquet.

The conference consisted of 12 plenary talks, about 50 invited talks, seven parallel sessions, and three sessions of poster viewing. Although five parallel sessions were considered at the beginning, since there were many requests for oral presentations, we decided to increase the number of sessions to seven. The sessions that had more contributions were those on quantum Monte Carlo Methods, finite difference volume element methods, molecular dynamics, Monte Carlo methods, large-scale computing, density functional methods, particle methods, visualization, and community-driven Codes.

Figure 5. A snapshot of the guided tour of the "K" supercomputer.

In the case of the Monte Carlo session, for example, we composed the program so that in the same session topics from lattice quantum color dynamics of elementary particle physics to polymer structure analysis of the worn-out tire (which is an industrial application) were presented and discussed inter-disciplinarily. If presenters gave their talks as in front of their usual specialized audience, it could have been regarded as "It's all Greek to me" by participants; namely, nonsense. Therefore, I asked five golden rules to all the presenters including the plenary speakers. They are as follows.

1. Presenters are invited to give a talk, assuming that the background of participants in a session is of the level of a university undergraduate student and without using technical terminology.
2. Presenters are invited to show all basic equations which one uses in solving problems on computers.
3. Presenters are required to explain why a certain numerical method is necessary, how one applied it, and where is the unique and challenging point.
4. Presenters are invited to show intelligibly the scientific results obtained in the simulation.
5. Presenters are invited to explain what kind of original results will be expected, if the present numerical method is improved and at what points.

I expected that joint research between different fields would start inter-disciplinarily and internationally through this conference. Therefore, I have requested all presenters to follow such golden rules in advance. During the conference, I was glad, seeing the spectacle of researchers of different fields arguing in front of a white board till late at night. Since one of those persons is a friend of mine, I heard his comments about the discussion. He said that they are still having email exchanges. He said that the interlocutor, from Germany, is an excellent person and that he discovered through her that a development he was researching had already been accomplished. He was surprised and admired her capability very much.

At this conference, we performed another new attempt. It was a debate on computational science and citizens. The aim of the discussion was how to let the taxpayers understand the value and importance of the computational science. We must be accountable to the citizens. Indeed the public political evaluation (called "shiwake") of the funding for the K supercomputer became very popular in the Japanese news. The time scheduled for the debate was only one hour and a half. In the debate, a science journalist, representing citizens, pointed out four subjects for discussion at the conclusion of her plenary talk of 30 minutes. Subsequently, I coordinated free discussion among all participants for one hour.

I asked Ms. Atsuko Tsuji, a leading writer of the Asahi Shimbun, to give the plenary talk on behalf of the Japanese mass media and finally persuaded her. She gave a really interesting talk, and you can read her speech in these proceedings. At her conclusion she showed the four subjects for the subsequent free discussion:

(1) Computational science to address humankind's problems.
(2) Ideal style of supercomputer development.
(3) How to inform society.
(4) How to foster human resources for computational science.

I took over as chairman of the discussion for one hour. I was anxious about keeping the discussion lively. However, worries ended up being groundless apprehensions. Remarks from the audience continued uninterruptedly. The utterances came in rapid succession with English accents from Britain, India, the U.S., Germany, France, and various other countries. It turned out that the scheduled time was too short. I could not coordinate the discussion for a longer time. The most important discussion was about "whether computational science is the third way to research science after theoretical and experimental ones". The conclusion was that computer science is not the third science. I was very much surprised at this conclusion. The record of this discussion is also in the proceedings and I hope the readers think again seriously about such conclusions. I felt lonesome, since there was only one comment by a Japanese participant (to discussion point 2).

The number of papers presented at the conference exceeded 300. There were five participants from South Africa. UNESCO once founded the theoretical physics center (ICTP) in Trieste, Italy, as a science base for the third world. When young excellent people educated at the ICTP return to their countries, such as Africa, they find lack of infrastructures for research. But today's personal computers are 1000 times faster than the supercomputers of my doctorate days. Therefore, since high-level science can be studied with a personal computer even in Africa, computer science is spreading. The five persons from South Africa were scientists who are promoting excellent teaching for young men of all Africa.

The subsequent conference in 2013 will be held in Moscow at the end of August (http://ccp2013.ac.ru/). Our two new attempts were welcomed very much by many participants. Many participants evaluated them positively. Furthermore, many opinions like "I should make this system the standard form of IUPAP-CCP from now on" were offered by the IUPAP-C20 commission members. If we have some new idea, we should implement them with resolution.

Figure 6. The final LOC meeting at a Japanese style restaurant near San-no-miya station, the biggest station near the venue. We thanked each other for the success of the conference organization obtained through the sharing of all powers and efforts.

Finally, I am thankful to Takahiro Hayashi, Director of the Office for the Promotion of Computing Science of the Ministry of Education, Culture, Sports, Science and Technology (MEXT). I also thank Dr. T. Takada and ex-Director Mr. Inoue for providing me such a good chance to organize the conference. I am thankful to Dr. K. Hirao, the director of AICS, who kindly helped us to arrange the "K" supercomputer guided tour for many participants over the capacity of the normal tour. Moreover, I appreciate the support of the Japan Society of Promotion of Science, the Expo 70 Memorial Foundation, the Kobe International Tourist Association, and several other associations. And I am thankful to all individuals in the committees described above. Especially Prof. Luca Baiotti who played an active role in the general affairs, programing, proceedings etc., Ms. Yuko Sakamoto for her assistance as secretary, Chairman of the LOC Dr. T. Nakajima (Riken, AICS), and Vice-Chairs Prof. H. Usui (Kobe University) and Prof. R. Numata (University of Hyogo) for their excellent partnership with us to really run the conference successfully.

Peer review statement

All papers published in this volume of *Journal of Physics: Conference Series* have been peer reviewed through processes administered by the proceedings Editors. Reviews were conducted by expert referees to the professional and scientific standards expected of a proceedings journal published by IOP Publishing.

Numerical simulations of dynamics and emission from relativistic astrophysical jets

Petar Mimica[1], Miguel Angel Aloy[1], Jesus M Rueda-Becerril[1], Siham Tabik[2] and Carmen Aloy[1]

[1] Departamento de Astronomía y Astrofísica, Universidad de Valencia, 46100, Burjassot, Spain
[2] Dept. Computer Architecture, University of Malaga, 29080 Málaga, Spain

E-mail: petar.mimica@uv.es

Abstract. Broadband emission from relativistic outflows (jets) of active galactic nuclei (AGN) and gamma-ray bursts (GRBs) contains valuable information about the nature of the jet itself and about the central engine which launches it. Using special relativistic hydrodynamics and magnetohydrodynamics simulations we study the dynamics of the jet and its interaction with the surrounding medium. The observational signature of the simulated jets is computed using a radiative transfer code developed specifically for the purpose of computing multi-wavelength, time-dependent, non-thermal emission from astrophysical plasmas. We present results of a series of long-term projects devoted to understanding the dynamics and emission of jets in parsec-scale AGN jets, blazars and the afterglow phase of the GRBs.

1. Introduction

Relativistic jets are found in a variety of astrophysical scenarios such as active galactic nuclei (AGN), microquasars, gamma-ray bursts (GRBs) and those resulting from tidal-disruption events (TDEs). While jet temporal and spatial timescales can vary from relatively small and short-lived (GRB jets) to very large and long-lived (AGN jets), what they have in common is that they are relativistic, seem to be very collimated and appear to be launched from a region very close to the accreting black hole (see e.g., [1] for a recent review and introduction to relativistic jets).

In this paper we review a series of projects whose aim is to understand the dynamics and emission from relativistic jets by focusing on specific scenarios appearing in relativistic jets of AGNs, blazars and GRBs. We begin by describing the numerical method in section 2, where we describe the equations of special relativistic magnetohydrodynamics and the code *MRGENESIS* (section 2.1), and then give some details about the treatment of non-thermal particles (section 2.2) and the radiative transfer (section 2.3), both of which are handled by the *SPEV*. In section 3 we present three of the applications of *MRGENESIS+SPEV* method, and we conclude in the section 4.

2. Numerical method

The numerical method we use consists of three parts, each dealing with a different aspect of the jet dynamics and emission. In the section 2.1 we discuss the equations of the special relativistic magnetohydrodynamics (RMHD). In sections 2.2 and 2.3 the treatment of non-thermal particles and emission, as well as that of the radiative transport is explained. Finally, the structure of the *SPEV* code is explained in 2.3.1.

2.1. Special relativistic magnetohydrodynamics

We solve the system of equations of ideal special RMHD[2], which consists of the mass and the total-energy momentum conservation[1]

$$\nabla_\alpha (\rho u^\alpha) = 0 \quad ; \quad \nabla_\alpha T^{\alpha\beta} = 0\,, \tag{1}$$

where the covariant derivative is denoted by ∇_α. ρ is the fluid rest-mass, u^α is the fluid 4-velocity, and the energy-momentum tensor is defined as

$$T^{\alpha\beta} := \rho h^* u^\alpha u^\beta + p^* \eta^{\alpha\beta} - b^\alpha b^\beta\,. \tag{2}$$

The Minkowski metric is $\eta^{\alpha\beta} = diag(-1, +1, +1, +1)$, and the total pressure is $p^* := p + |b|^2/2$, where p is the fluid pressure. b^α is the magnetic field four-vector,

$$b^0 := \Gamma\,(\mathbf{v}\cdot\mathbf{B})\;, \quad b^i := \frac{B^i}{\Gamma} + v^i b^0\,. \tag{3}$$

Here $\mathbf{B}$ is the magnetic field 3-vector and $\mathbf{v}$ the fluid velocity 3-vector, whose Lorentz factor is defined as $\Gamma := (1 - v^2)^{-1/2}$. The total specific enthalpy is defined as

$$h^* := h + \frac{|b|^2}{\rho} = 1 + \epsilon + \frac{p}{\rho} + \frac{|b|^2}{\rho}\,, \tag{4}$$

where ϵ is the specific internal energy, which depends on the equation of state $p = p(\epsilon, \rho)$. We use the TM approximation to the Synge equation of state, where the specific enthalpy h is defined as [3] $h = (5/2)(p/\rho) + \sqrt{(9/4)(p/\rho)^2 + 1}$. Finally, the magnetic field has to satisfy the induction equation

$$\frac{\partial \mathbf{B}}{\partial t} - \nabla \times (\mathbf{v} \times \mathbf{B}) = 0\,, \tag{5}$$

and the divergence constraint $\nabla \cdot \mathbf{B} = 0$.

We use the code *MRGENESIS* [4, 5, 6, 7] to numerically solve the system of equations of RMHD. *MRGENESIS* uses operator-splitting, finite-volumes, explicit method and employs approximate Riemann solvers for the computation of intercell fluxes, as well as total variation diminishing second and third-order Runge-Kutta methods for the time integration (for a recent overview of jet simulation codes see e.g. [8]). *MRGENESIS* has been massively parallelized using OpenMP and MPI libraries and has achieved reasonable scaling up to 10^4 cores on the *MareNostrum* machine at the Barcelona Supercomputing Centre.

2.2. Non-thermal particles and emission

As discussed in the section 1, a small fraction of very energetic particles ("non-thermal electrons" in the following) is responsible for the observed jet emission. In [9] we discuss in detail the method which deals with the spatial and temporal evolution of the non-thermal electrons in a magnetized relativistic fluid. Here we summarize the most important properties of the method.

[1] In this section we assume a system of units where the speed of light $c = 1$ and the factor $1/\sqrt{4\pi}$ is included in the definition of the magnetic field.

2.2.1. Non-thermal electron evolution We assume that electrons are injected at relativistic shocks and use the phenomenological injection term which denotes the number of particles injected per unit volume and unit time,

$$Q(\gamma) = Q_0 \left(\frac{\gamma}{\gamma_{\rm min}}\right)^{-q} S\left(\gamma;\ \gamma_{\rm min}, \gamma_{\rm max}\right), \tag{6}$$

where γ is the electron Lorentz factor $\gamma = E/(m_e c^2)$, E being its energy and m_e being its mass. Q_0 is the normalization of the injection energy spectrum, q is the power-law index and the function $S(x;\ a, b)$ has the value 1 if $a \le x \le b$ and 0 otherwise[2]. The temporal evolution of the electron energy distribution in the fluid comoving frame[3] is governed by the kinetic equation [10]:

$$\frac{\partial n(t;\gamma)}{\partial t} + \frac{\partial}{\partial \gamma}\left[\dot{\gamma} n(t;\gamma)\right] = Q(\gamma), \tag{7}$$

where $n(t;\gamma)$ is the number density of electrons in an interval $\mathrm{d}\gamma$ around the Lorentz factor γ at the time t. Energy gains and losses are described by the term $\dot{\gamma} := \mathrm{d}\gamma/\mathrm{d}t$. For sufficiently small intervals of time we can write the energy losses as

$$\frac{\mathrm{d}\gamma}{\mathrm{d}t} = k_a \gamma - k_s \gamma^2, \tag{8}$$

where the adiabatic gains or losses are described by the quantity

$$k_a := \frac{\mathrm{d}\ln\rho}{\mathrm{d}t},$$

which denotes the compression or expansion of the fluid element, while the synchrotron losses are denoted by

$$k_s := -\frac{4\sigma_T B'^2}{3 m_e^2 c^2}$$

where B' is the magnetic field strength in the fluid comoving frame and σ_T is the Thomson cross section.

As described in section 3.2 of [9], we discretize the electron energy space in a number of logarithmically spaced bins and use a semi-analytic solver to accurately solve the equation (7) with a minimum computational effort.

2.2.2. Non-thermal radiation processes We consider the following radiation processes in our simulations: synchrotron emission, synchrotron-self Compton (SSC) scattering, and the external inverse-Compton (EIC) scattering.

Synchrotron emission is produced when high-energy electrons gyrate around the magnetic field lines, producing a non-thermal, broadband emission spectrum. It is computed for each energy bin of the electron distribution using the interpolation method described in the sections 2.1.3 and 4.3.1 of [11].

Inverse-Compton scattering is a scattering of a low-frequency photon off a high-energy electron, whereby the frequency of the outgoing photon can be many orders of magnitude larger than that of the incoming one. In our numerical method it is computed analytically (see section 2.2.2 of [11]) assuming that the incoming emission spectrum is a power-law, as is the energy spectrum of the electrons off which the photons scatter. This method has been used in [12] to compute the EIC scattering of the ultraviolet stellar, photons, as well as in [13] to compute both the SSC and EIC emission in the blazar jets.

[2] We fix $q = 2.3$ and determine the rest of the parameters by assuming a proportionality between the number and energy density of the non-thermal electrons and those of the thermal fluid (see equations 36 and 37 of [9]).

[3] We assume that non-thermal electrons are advected with the thermal fluid. The spatial distribution of non-thermal electrons is represented by Lagrangian tracer particles.

2.3. Radiative transfer

To compute the time- and frequency-dependent synthetic image we use the algorithm described in the Appendix A of [9]. Both this algorithm and the one described in the previous section are building blocks of the code *SPEV*. *SPEV* solves the radiative transfer equation

$$\frac{\mathrm{d}I_\nu}{\mathrm{d}s} = j_\nu - \alpha_\nu I_\nu \,, \tag{9}$$

where I_ν, j_ν and α_ν are the specific intensity, emissivity and absorption coefficient at frequency ν. s is the distance along the line of sight which is seen by the observer at the time T. The relation between the time of observation and the time t in the laboratory frame (the frame in which the jet evolution is simulated) is given by

$$t = T + (D - s)/c \,, \tag{10}$$

where D is the distance of the jet injection point from the Earth and we choose that $s = 0$ at that point. As can be seen, in order to solve the equation (9), it is not enough to process each saved simulation snapshot at time t_{snapshot}. Instead, it is necessary to process *all* snapshots *simultaneously* because of the fact that distance from the observer and the time at which the snapshot has been taken are not independent, but must satisfy the equation (10). In practice, this means that the radiative transfer equation (9) cannot be solved before *all* the information along the line of sight has been gathered. This makes the problem tightly coupled and highly non-local, complicating the code parallelization and increasing the CPU and memory costs.

2.3.1. Structure and parallelization of the SPEV *code* The evolution and emission from the non-thermal particles (section 2.2) and the computation of the resulting emission (section 2.3) are done by the *SPEV* code whose structure can be sketched as follows: after creating the initial Lagrangian particles (representing the pre-existing non-thermal electrons), *SPEV* iteratively runs through the three phases: evolution of existing particles, injection of new particles, and calculation of the emission (i.e., contribution to the pixels of a virtual detector on which the emission is computed).

SPEV has two important characteristics very useful for its parallelization: the evolution of each Lagrangian particle during the simulation does not affect the evolution of the rest of the particles, and the contribution to each pixel of the detector is independent from the contribution to any other pixel. Therefore, *SPEV* is parallelized over particles *and* detector pixels. The data is partitioned into different sets and distributed among different computing units (e.g., multiple multicore-nodes connected by means of the network or by means of buses). Each set of particles can evolve separately without the need of any kind of communications during the iterative process. When the iterative process ends, data is transferred to the master node and then the radiative transfer equation (9) is solved for each pixel. More details about the parallel implementation are provided in [14, 15]. This parallel approach has shown very good scalability on different HPC-systems, especially on medium scale multi-sockets multicores of up to 50 cores.

3. Applications

In this section we outline a number of application of *MRGENESIS* + *SPEV* approach to relativistic jets. In section 3.1 we discuss the radio emission from parsec-scale jets. The section 3.2 is devoted to address the highly variable and highly energetic emission from the blazars, while section 3.3 presents results of a long-term study of gamma-ray burst afterglows.

3.1. Parsec-scale jets

Parsec-scale jets are radio features on a scale of several light years. They are seen emerging from AGNs over periods of months to years. It is widely accepted that the radio maps are not direct observations of the physical quantities (density, pressure, velocity, composition, etc.) in the jet, but are influenced by a number of relativistic effects (Doppler shift, beaming) as well as by the degradation of the image due to a finite resolution of the radio telescopes. By using numerical simulations of the jet dynamics and, subsequently, by computing the synthetic radio maps we can test theoretical jet models and directly compare them to the observations, as well as study some of the events which are expected to occur in them[4]. In a previous work [9] we studied one such event, which is an injection of a perturbation into a steady jet shown on Figure 1. This jet is over-pressured and under-dense with respect to the surrounding medium and has a number of recollimation shocks, which can be seen as knots in the density profile shown on Figure 1.

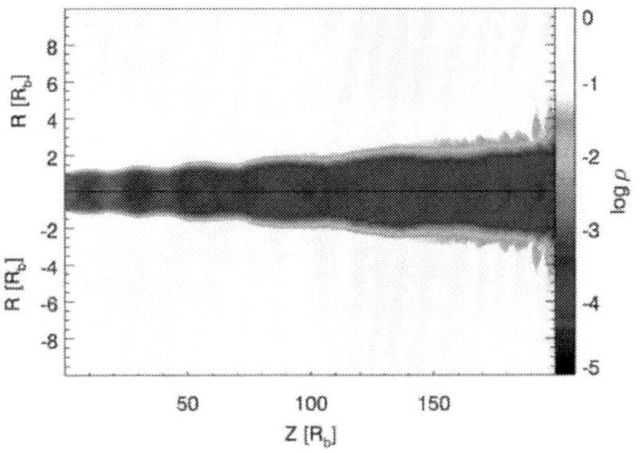

Figure 1. Logarithm of the density of a steady-state jet propagating through a stratified atmosphere. The jet is initially under-dense and over-pressured with respect to the surrounding medium. The density is shown in arbitrary units, while the length scale is in the units of the jet radius at the left grid boundary. See section 2 of [9] for the details of the hydrodynamic models. Note that the vertical scale is enlarged for the better visualization of the jet.

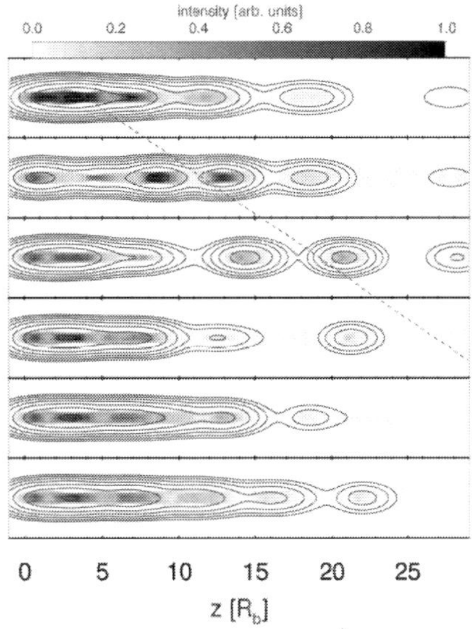

Figure 2. Radio emission at 43 GHz resulting from a temporary velocity perturbation propagating through the steady-state jet shown on Figure 1. The gray shades in the panels show the radiation intensity (in arbitrary units) at (from top to bottom) 0.02, 0.39, 0.75, 1.12, 1.94 and 4.58 years since the injection of the perturbation. The contours shows the image degraded to the resolution available with today's VLBI technique. The dashed blue line shows the approximate trajectory of the main component. See section 7 of [9] for more details.

The perturbation has a form of a temporary increase in the fluid velocity at the jet injection point located at the left grid boundary. This results in the development of shock and rarefaction

[4] Our work on blazar jets (section 3.2) as well as studies of the stability of initially very magnetized ($\sigma \geq 100$) jets (e.g. [16]) indicate that at parsec-scales distances AGN jets are expected to be at most very weakly magnetized (e.g. $\sigma \leq 10^{-4}$)

waves which propagate down the jet and interact with the recollimation shocks, resulting in a complex, non-linear evolution of the perturbation and several trailing components (see section 7 of [9] or more details). Figure 2 shows the radio maps taken at six different epochs of the jet evolution. In the first three panels we can see the main component (perturbation) as a dark region propagating through the jet. In the last three panels we see a gradual reestablishment of the steady jet.

The contours show the image degraded to the resolution available with today's radio telescopes. As can be appreciated, many of the small-scale details of the jet evolution and emission are lost in the relatively strong blurring of the image due to the low resolution of the actually observable radio maps. Therefore it is important to use the hydrodynamic simulations and possess the ability to compute time-dependent radio maps from those simulations because this enables us to compute both the full-resolution and the degraded images and thus evaluate which of the observed features of the jet dynamics and emission are intrinsic to the jet and which are the artifact of the *finite* observational resolution.

3.2. Blazar jets

Blazars are jets pointing almost directly towards us. They are characterised by high variability of their emission, often in the form of flares observed by X-ray satellites[5]. It is thought that the cause of the flares are the collisions of parts of the jet with different velocities ("shells" in the following), whereby a fraction of the shell kinetic energy is dissipated by the shocks which propagate through the shells as a result of their collision, and a fraction of the dissipated energy is radiated and observed as a flare. This is called the internal shock scenario [18, 19, 20, 21, 22, 23, 6, 24, 3, 13]. Figures 3 and 4 illustrate the internal shocks model and the typical distance scales at which shells collide in blazar jets.

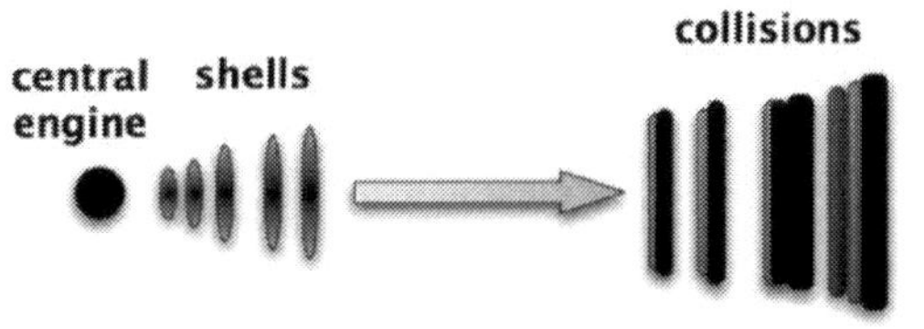

Figure 3. An illustration of the internal shocks model: an intermittently active central engine (supermassive black hole at the center of the galaxy) ejects inhomogeneous shells which collide and produce observed flares.

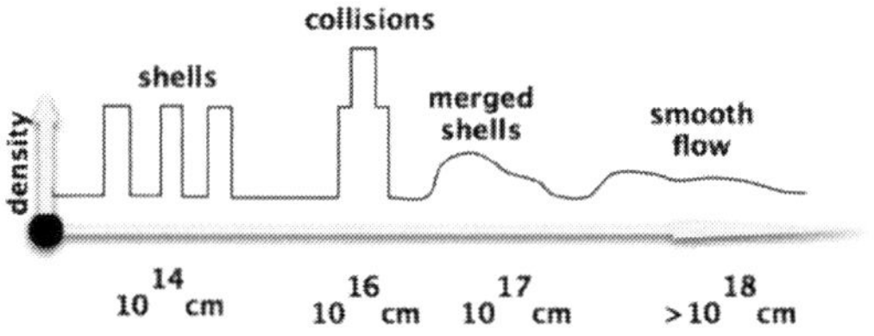

Figure 4. Distance scales in the internal shocks model in blazars: the shells, whose characteristic size is $\approx 10^{14}$ cm, collide at distances $\approx 10^{16} - 10^{17}$ cm from the central engine, merge and continue flowing as a smooth jet at parsec-scale distances ($\approx 10^{18}$ cm).

We study the dynamics and emission from single shell collisions, using one-dimensional [23] and two-dimensional [21] hydrodynamic simulations, as well as one-dimensional magnetohydrodynamic simulations [6]. We found that the lateral shell spreading is negligible on the length scales considered and that the one-dimensional approximation is accurate [21]. Furthermore, we find that the magnetization of the flow plays a crucial role in the efficiency of the conversion of kinetic into thermal energy (and subsequently into radiation). From the computational point of view we found the hydrodynamic simulations to be rather expensive and unsuitable for parameter space studies.

[5] For an example of X-ray flares from a well known blazar Mrk 421 see e.g., [17]

In the follow-up works [3, 13] we simplify the hydrodynamic approach in order to be able to devote more resources to the computation of the emission and to be able to cover larger parameter space. We study a large number of shell collisions where both shells have the same energy and size, but their velocity and the degree of magnetization can vary. We denote the shell Lorentz factor by $\Gamma := (1 - v^2/c^2)^{-1/2}$, where v is the shell velocity. The magnetization for cold and relativistic shells (e.g. $p/\rho c^2 \ll 1$ and $\Gamma \gg 1$, where ρ and p are the shell rest-mass density and the thermal pressure) is defined as $\sigma := B^2/(4\pi\Gamma^2\rho c^2)$. We compute the exact solution of the Riemann problem[6] for each shell collision and determine the strength and velocity of the shocks which are formed by the shell interaction. By covering a large parameter space of possible shell magnetization (we studied 10^6 shell collisions with σ ranging from 10^{-6} to 10^3 for each shell independently) we find that the "sweet spot" for the efficiency of the kinetic energy conversion is not for $\sigma \ll 1$, as might be expected (the smaller the σ, the easier it is to shock the fluid), but rather for ($\sigma_L = 1$, $\sigma_R = 0.1$), where subscripts L and R denote the faster (left) and the slower (right) shell, respectively [3]. This result makes the magnetized internal shock model viable. In the follow-up work [13] we compute the multi-wavelength, time-dependent emission from $\approx 10^3$ shell collisions.[7]

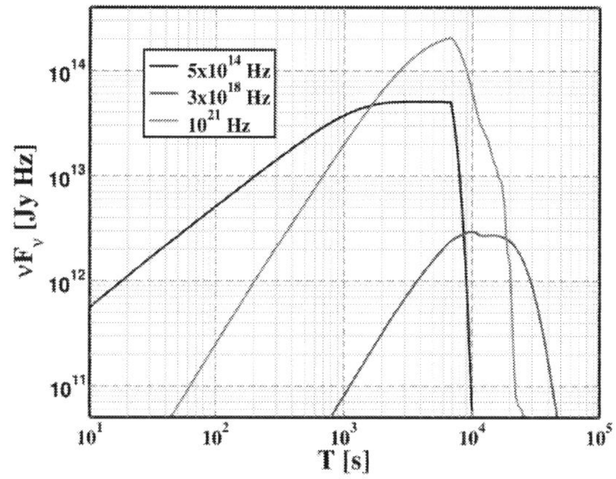

Figure 5. Optical (5×10^{14} Hz), X-ray (3×10^{18} Hz) and γ-ray (10^{21} Hz) flare (black, blue and red lines) produced by a collision of a faster shell with the initial Lorentz factor $\Gamma_L = 20$ with a slower shell $\Gamma_R = 10$. The shell magnetization is $\sigma_L = 10^{-3}$ and $\sigma_R = 10^{-1}$. The shell geometry is cylindrical (radius 3×10^{14} m, length 6×10^{11} m) and their initial energy 10^{52} erg. The flux is shown in units Jy Hz (1 Jy Hz $= 10^{-23}$ W / m^2).

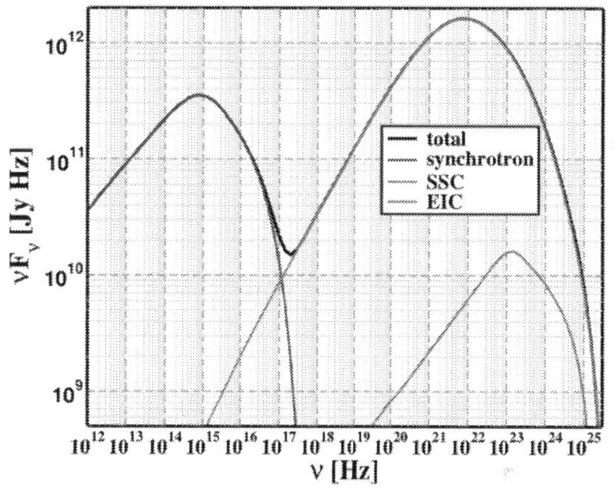

Figure 6. Average spectrum of the flare shown in figure 5. The total spectrum is shown in black, while the synchrotron, SSC and EIC contributions are shown in blue, red and green, respectively. The seed photons for the EIC are assumed to be monochromatic (frequency 2×10^{14} Hz). The viewing angle (angle between the jet axis and the line of sight) is $5°$, and the jet is assumed to be located at redshift $z = 0.5$.

Figures 5 and 6 show an example of the light curve and spectra produced by a collision of two moderately magnetized shells in a typical blazar jet (Rueda, Mimica & Aloy, 2012, in preparation). In figure 5 can see that the optical and γ-ray light curves peak somewhat earlier

[6] We use the exact Riemann solver [25] for the one-dimensional ideal relativistic magnetohydrodynamic Riemann problem with magnetic fields perpendicular to the flow velocity.

[7] We could not compute the emission from all of the 10^6 shell collisions studied in [3] because it was not feasible with present supercomputers (the computation time for the emission from 10^3 shell collisions exceeds 200 thousand hours).

than the X-ray light curve. The reason is that the SSC emission (section 2.2.2) is produced by the scattering of the synchrotron emission which needs a finite time to travel from the place where it is produced to the site of its scattering. The sharp drop in the emission is related to the moment when shocks cross the shells and cease dissipating energy. Average spectra are shown in figure 6. The low-frequency emission is dominated by the synchrotron emission, while the high-energy spectrum is dominated by the SSC component. As can be seen, the correct computation of the SSC component is essential, and since this component is computationally most expensive, it puts an upper limit on a number of models which can be computed on today's machines (see footnote 7).

3.3. Gamma-ray burst afterglows

Gamma-ray bursts (GRBs) are produced by an internal dissipation of energy in the relativistic outflow. The GRB afterglow emission is produced by the interaction of the relativistic jet with the external (circumburst) medium. It is not known to what extent the jet is magnetized. If the outflow is initially dominated by the thermal energy (fireball model [26, 27]) then it is expected that its $\sigma \ll 1$. If, on the other hand, the outflow is initially dominated by the Poynting-flux [28, 29], then it is possible to find a significantly magnetized jet at the onset of the afterglow phase.

We study the interaction of an arbitrarily magnetized GRB jet with the external medium. We model the interaction by considering the deceleration of a dense, magnetized shell as it moves through and compresses a constant density medium in front of it. One of the crucial differences between the evolution of a non-magnetized and a magnetized shell is in the existence of a reverse shock (a shock propagating through the shell): if σ is high enough, then the interaction between the shell and the external medium is not strong enough to shock the shell material itself. By analyzing timescales of the shell deceleration and the propagation of magnetohydrodynamic waves through the shell we find the conditions for the existence of a reverse shock depending on the GRB magnetization, energy, Lorentz factor, duration and the density of the external medium [30]. Figure 7 shows the parameter space spanned by the ejecta magnetization and a dimensionless parameter ξ (see figure caption for definition). As we can see, there is a substantial portion of the parameter space (shaded region), where the reverse shock does not form. Since it is absent, the particles are not accelerated and we expect to fail to observe an emission from this shock during the early afterglow, which would be consistent with the absence of such observations in most GRB optical afterglows [31].

In the follow-up works [7, 32] we confirmed these results by means of numerical simulations. Using *MRGENESIS* we performed high-resolution, long-term one-dimensional relativistic magnetohydrodynamic simulation of a shell-medium interaction, and then use *SPEV* to compute the resulting light curves. Figure 8 shows an example of the optical light curve produced by non-magnetized, weakly-magnetized and strongly-magnetized shells with the same initial energy, Lorentz factor and width. As can be seen, the strongly magnetized shell light curve does not possess the optical flash feature that the other two possess. Therefore, strong magnetization might be required for the paucity of the observed optical flashes to be explained. Since most of the observed GRBs have ξ in the range $[0.3, 3]$, this would require $\sigma > 0.1$ for almost all GRBs [30, 32]. This would imply that GRB jets could be strongly magnetized even at such late stages of their evolution as the afterglow phase, which puts constraints on the amount of magnetic dissipation that can happen at previous stage (prompt emission).

The simulations needed to correctly compute afterglow light curves are very computationally demanding due to the high resolution required in the direction of the jet motion (e.g. more than 10^4 zones in the initial shell) and the long timescales needed to compute the light curve of sufficient duration (approximately 10^7 numerical iterations). Therefore, performing this type of calculations to cover a full parameter space of feasible GRB afterglows is not possible at

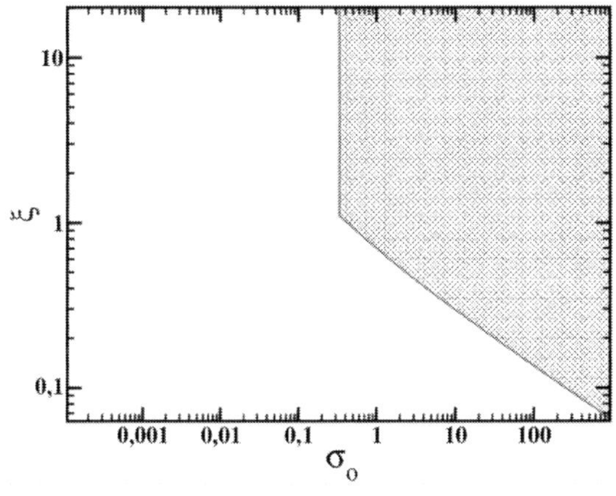

Figure 7. Parameter space of the ejecta magnetization σ_0 and $\xi := (3E/4\pi\rho_{\rm ext}c^2)^{1/6}\,\Delta_0^{-1/2}\Gamma_0^{-4/3}$, where E, Δ_0 and Γ_0 are the ejecta initial energy, width and the Lorentz factor, and $\rho_{\rm ext}$ is the external medium density. The shaded region shows the ejecta which do not form a reverse shock, while the vertical boundary at $\sigma_0 = 0.3$ indicates those ejecta whose radial spreading becomes significant and makes the formation of a reverse shock inevitable [30].

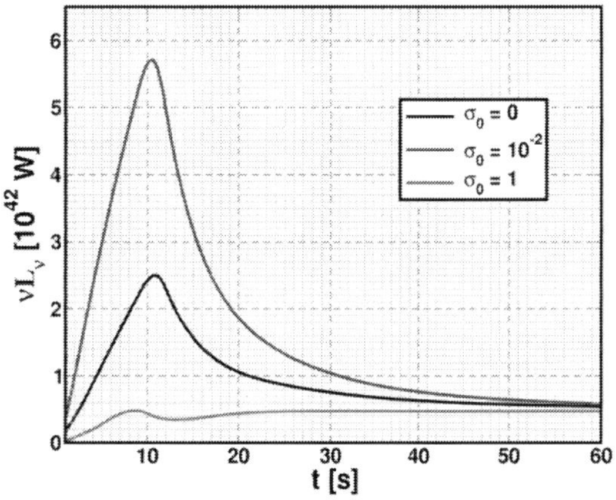

Figure 8. Optical (5×10^{14} Hz) light curve produced by the non-magnetized ($\sigma_0 = 0$), weakly magnetized ($\sigma_0 = 0.01$ and strongly-magnetized ($\sigma_0 = 1$) shells (black, blue and green lines, respectively) whose initial conditions are $E = 3 \times 10^{47}$ J, $\Gamma_0 = 300$ and $\Delta_0 = 3 \times 10^9$ m. The non/weakly magnetized ejecta have a strong reverse shock which produces an optical flash, while the strongly magnetized ejecta light curve does not have this feature [32].

the moment. However, the existence of re-scaling laws (see section 4.4 of [7]) can help us to perform a small number of simulations and then extend their results to cover a somewhat larger parameter space.

4. Conclusions

We have presented a framework consisting of a numerical relativistic magnetohydrodynamic code *MRGENESIS* and the non-thermal radiative transfer code *SPEV*. We show how its modular nature has enabled it to be successfully been applied to a number of current problems and issues in the relativistic jets physics, and how it enables us, via the computation of synthetic observations, to directly compare the results of our simulations with the observations from ground- and space-based telescopes. In the future we plan to improve the scaling of *MRGENESIS* to more than 10^4 cores, and also to add new radiative processes to *SPEV* (e.g., polarization, improved inverse-Compton, radiative transfer in curved space-times, etc.). The code scaling improvements will allow us to perform full 3D magnetohydrodynamic simulations and radiative transfer calculations, relevant for astrophysical scenarios where axial symmetry cannot be assumed.

Acknowledgments

PM, MAA and CA acknowledge the support from the ERC grant CAMAP-259276 and the grants AYA2010-21097-C03-01 and PROMETEO-2009-103. JMRB acknowledges the support from the Grisolia fellowship GRISOLIA/2011/041. PM, MAA, ST and CA acknowledge the support from the Consolider grant CSD2007-00050. The authors thankfully acknowledge the computer resources, technical expertise and assistance provided by the Barcelona Supercomputing Center

- Centro Nacional de Supercomputación.

References

[1] Böttcher M, Harris D E and Krawczynski H 2012 *Introduction and Historical Perspective;* in *Relativistic Jets from Active Galactic Nuclei (eds M. Bttcher, D. E. Harris and H. Krawczynski)* (Wiley-VCH Verlag GmbH & Co. KGaA) chap 1, pp 1–16 ISBN 9783527641741

[2] Anile A M 1989 *Relativistic fluids and magneto-fluids: With applications in astrophysics and plasma physics* (Cambridge University Press) ISBN 9780521304061

[3] Mimica P and Aloy M A 2010 *Mon. Not. R. Astron. Soc.* **401** 525–532 (*Preprint* 0909.1328)

[4] Aloy M A, Ibáñez J M, Martí J M and Müller E 1999 *Astrophysical J. Suppl. Ser.* **122** 151–166 (*Preprint* arXiv:astro-ph/9903352)

[5] Leismann T, Antón L, Aloy M A, Müller E, Martí J M, Miralles J A and Ibáñez J M 2005 *Astron. Astrophys.* **436** 503–526

[6] Mimica P, Aloy M A and Müller E 2007 *Astron. Astrophys.* **466** 93–106 (*Preprint* arXiv:astro-ph/0611765)

[7] Mimica P, Giannios D and Aloy M A 2009 *Astron. Astrophys.* **494** 879–890 (*Preprint* 0810.2961)

[8] Aloy M A and Mimica P 2012 *Simulations of Jets from Active Galactic Nuclei and Gamma-Ray Bursts;* in *Relativistic Jets from Active Galactic Nuclei (eds M. Bttcher, D. E. Harris and H. Krawczynski)* (Wiley-VCH Verlag GmbH & Co. KGaA) chap 10, pp 297–339 ISBN 9783527641741

[9] Mimica P, Aloy M A, Agudo I, Martí J M, Gómez J L and Miralles J A 2009 *Astrophysical J.* **696** 1142–1163 (*Preprint* 0811.1143)

[10] Kardashev N S 1962 *Soviet Astron.* **6** 317

[11] Mimica P 2004 *Numerical Simulations of Blazar Jets and their Non-thermal Radiation* Ph.D. thesis Max-Planck-Institut für Astrophysik

[12] Mimica P and Giannios D 2011 *Mon. Not. R. Astron. Soc.* **418** 583–590 (*Preprint* 1106.1903)

[13] Mimica P and Aloy M A 2012 *Mon. Not. R. Astron. Soc.* **421** 2635–2647 (*Preprint* 1111.6108)

[14] Tabik S, Mimica P, Plata O, Zapata E L and Romero L F 2011 *HiPC 2011* **1** 1–10

[15] Tabik S, Romero L F, Mimica P, Plata O and Zapata E L 2012 *Computer Physics Communications* **183** 1937–1946

[16] Giannios D and Spruit H C 2006 *Astron. Astrophys.* **450** 887–898 (*Preprint* arXiv:astro-ph/0601172)

[17] Brinkmann W, Papadakis I E, Raeth C, Mimica P and Haberl F 2005 *Astron. Astrophys.* **443** 397–411 (*Preprint* arXiv:astro-ph/0508433)

[18] Rees M J and Meszaros P 1994 *Astrophysical J. Lett.* **430** L93–L96 (*Preprint* arXiv:astro-ph/9404038)

[19] Kobayashi S, Piran T and Sari R 1997 *Astrophysical J.* **490** 92 (*Preprint* arXiv:astro-ph/9705013)

[20] Daigne F and Mochkovitch R 1998 *Mon. Not. R. Astron. Soc.* **296** 275–286 (*Preprint* arXiv:astro-ph/9801245)

[21] Mimica P, Aloy M A, Müller E and Brinkmann W 2004 *Astron. Astrophys.* **418** 947–958 (*Preprint* arXiv:astro-ph/0401266)

[22] Kino M, Mizuta A and Yamada S 2004 *Astrophysical J.* **611** 1021–1032 (*Preprint* arXiv:astro-ph/0404555)

[23] Mimica P, Aloy M A, Müller E and Brinkmann W 2005 *Astron. Astrophys.* **441** 103–115 (*Preprint* arXiv:astro-ph/0506636)

[24] Bošnjak Ž, Daigne F and Dubus G 2009 *Astron. Astrophys.* **498** 677–703 (*Preprint* 0811.2956)

[25] Romero R, Martí J M, Pons J A, Ibáñez J M and Miralles J A 2005 *Journal of Fluid Mechanics* **544** 323–338 (*Preprint* arXiv:astro-ph/0506527)

[26] Goodman J 1986 *Astrophysical J. Lett.* **308** L47–L50

[27] Paczynski B 1986 *Astrophysical J. Lett.* **308** L43–L46

[28] Usov V V 1992 *Nature* **357** 472–474

[29] Meszaros P and Rees M J 1997 *Astrophysical J. Lett.* **482** L29 (*Preprint* arXiv:astro-ph/9609065)

[30] Giannios D, Mimica P and Aloy M A 2008 *Astron. Astrophys.* **478** 747–753 (*Preprint* 0711.1980)

[31] Gomboc A, Kobayashi S, Mundell C G, Guidorzi C, Melandri A, Steele I A, Smith R J, Bersier D, Carter D and Bode M F 2009 Optical flashes, reverse shocks and magnetization *American Institute of Physics Conference Series* (*American Institute of Physics Conference Series* vol 1133) ed Meegan C, Kouveliotou C and Gehrels N pp 145–150 (*Preprint* 0902.1830)

[32] Mimica P, Giannios D and Aloy M A 2010 *Mon. Not. R. Astron. Soc.* **407** 2501–2510 (*Preprint* 1004.2720)

Journal of Physics: Conference Series **454** (2013) 012002 doi:10.1088/1742-6596/454/1/012002

BSSN equations in spherical coordinates without regularization: spherically symmetric spacetimes

Pedro J Montero and Isabel Cordero-Carrión

Max-Planck-Institute für Astrophysik, Karl-Schwarzschild-Str. 1, D-85748, Garching bei München, Germany

E-mail: montero@mpa-garching.mpg.de

Abstract. Brown recently introduced a covariant formulation of the BSSN equations which is well suited for curvilinear coordinate systems. This is particularly desirable as many astrophysical phenomena are symmetric with respect to the rotation axis or are such that curvilinear coordinates adapt better to their geometry. We show results from a newly developed numerical code solving the BSSN equations in spherical symmetry and the general relativistic hydrodynamic equations written in flux-conservative form. A key feature of the code is that uses a second-order partially implicit Runge-Kutta method to integrate the evolution equations, and does not need a regularization algorithm at the origin. We discuss a number of tests to assess the accuracy, numerical stability and expected convergence of the code.

1. Introduction

The 3+1 formulation of Einstein equations originally proposed by Nakamura [1] and subsequently modified by Shibata-Nakamura [2] and Baumgarte-Shapiro [3], which is usually known as the BSSN formulation, has become the most widespread used formulation in the numerical relativity community. The main drawback of the BSSN formulation in its original form resides in the fact that it is particularly tuned for Cartesian coordinates, since this involves dynamical fields which are not true tensors and assumes that the determinant of the conformal metric is equal to one. Brown [4] addressed this issue and introduced a covariant formulation of the BSSN equations which is well suited for curvilinear coordinate systems.

However, the singularities associated with the curvilinear coordinate systems are a known source of numerical problems. For instance, one problem arises because of the presence of terms in the evolution equations that behave like $1/r$ near the origin $r = 0$. Although on the analytical level the regularity of the metric ensures that these terms cancel exactly, on the numerical level this is not necessarily the case, and special care should be taken in order to avoid numerical instabilities. A similar problem appears also near the axis of symmetry in axisymmetric systems if curvilinear coordinate are used.

Several methods have been proposed to handle the issue of regularity in curvilinear coordinates. One possible approach is to apply a regularization procedure. One such regularization method, presented by [5], enforces both the appropriate parity regularity conditions and local flatness in order to achieve the desired regularity of the evolution equations. Such method has the advantage that it allows a more generic gauge choice, and has been explored by [6, 7, 8] who have performed several numerical simulations of regular spacetimes in spherical and axial symmetry. A disadvantage of such a regularization algorithm is that it is not easy to implement numerically both conditions simultaneously, and it requires the introduction of auxiliary variables as well as finding their evolution equations. This is an obstacle if one wants to perform 3D simulations of regular spacetimes with spherical coordinates.

Implicit or partially implicit methods are used to deal with systems of equations that require a special numerical treatment in order to achieve stable evolutions. The origin of the numerical instabilities may be diverse. Stiff source terms in the equations can lead to the development of numerical instabilities, and with some choices of the coordinate system, source terms may introduce factors which can be numerically interpreted as stiff terms (e.g., $1/r$ factors due to spherical coordinates close to $r = 0$ even when regular data is evolved).

Recently, partially implicit Runge-Kutta (PIRK) methods for wave-like equations in spherical coordinates [9] have been successfully applied [10] to the hyperbolic part of Einstein equations in the Fully Constrained Formulation [11]. Following this success, Montero & Cordero-Carrión [12], assuming spherical symmetry, applied a second-order PIRK method to the full set of the BSSN Einstein equations in curvilinear coordinates, and produced the first successful numerical simulations of vacuum and non-vacuum spacetimes using the covariant BSSN formulation in spherical coordinates without the need for a regularization algorithm at the origin. We discuss this approach and show a number of tests to assess the accuracy, numerical stability and expected convergence of the code; while we refer to Montero & Cordero-Carrión [12] for further details. We use units in which $c = G = M_\odot = 1$.

2. Basic equations

2.1. BSSN equations in spherical symmetry

We solve the BSSN equations in the special case of spherical symmetry [8, 12]. In particular, we follow [8] for the description of the BSSN equations below, and refer to this work for more details. Under this symmetry condition the spatial line element is written as

$$dl^2 = e^{4\chi}[a(r,t)dr^2 + r^2 b(r,t)d\Omega^2], \tag{1}$$

where $d\Omega^2$ is the solid angle element $d\Omega^2 = d\theta^2 + \sin^2\theta d\varphi^2$, $a(r,t)$ and $b(r,t)$ are the metric functions, and χ is the conformal factor defined as

$$\chi = \frac{1}{12}\ln(\gamma/\hat{\gamma}), \tag{2}$$

where $\hat{\gamma}$ is the determinant of the conformal metric. The conformal metric relates to the physical one by

$$\hat{\gamma}_{ij} = e^{-4\chi}\gamma_{ij}. \tag{3}$$

Initially, the determinant of the conformal metric fulfills the condition that it equals the determinant of the flat metric in spherical coordinates $\mathring{\gamma}_{ij}$ (i.e. $\hat{\gamma}(t=0) = \mathring{\gamma} = r^4 sin^2(\theta)$). Moreover, we follow the so called "Lagrangian" condition $\partial_t\hat{\gamma} = 0$ (i.e. choosing $\sigma = 1$). The evolution equation for the conformal factor $X \equiv e^{-2\chi}$ takes the form

$$\partial_t X = \beta^r \partial_r X - \frac{1}{3}X(\alpha K - \sigma\hat{\nabla}_m\beta^m), \tag{4}$$

K being the trace of the extrinsic curvature, α the lapse function, and $\hat{\nabla}_m\beta^m$ the conformal divergence of the shift vector β^i. The evolution equations for the conformal metric components are:

$$\partial_t a = \beta^r \partial_r a + 2a\partial_r\beta^r - \frac{2}{3}\sigma a\hat{\nabla}_m\beta^m - 2\alpha a A_a, \tag{5}$$

$$\partial_t b = \beta^r \partial_r b + 2b\frac{\beta^r}{r} - \frac{2}{3}\sigma b \hat{\nabla}_m \beta^m - 2\alpha b A_b, \tag{6}$$

where $\hat{A}_{ij}$ is the traceless part of the conformal extrinsic curvature, and

$$A_a \equiv \hat{A}^r_r, \qquad A_b \equiv \hat{A}^\theta_\theta. \tag{7}$$

Next, the evolution equation for the independent component of the traceless part of the conformal extrinsic curvature, A_a, is given by

$$\begin{aligned}\partial_t A_a = {} & \beta^r \partial_r A_a - \left(\nabla^r \nabla_r \alpha - \frac{1}{3}\nabla^2 \alpha\right) + \alpha\left(R^r_r - \frac{1}{3}R\right) \\ & + \alpha K A_a - 16\pi\alpha(S_a - S_b),\end{aligned} \tag{8}$$

where R^r_r is the mixed radial component of the Ricci tensor, R its trace. The evolution equation for K is:

$$\partial_t K = \beta^r \partial_r K - \nabla^2 \alpha + \alpha(A_a^2 + 2A_b^2 + \frac{1}{3}K^2) + 4\pi\alpha(E + S_a + 2S_b), \tag{9}$$

with the matter source terms measured by the Eulerian observers given by

$$\begin{aligned} E &= n_\mu n_\nu T^{\mu\nu}, \\ j_i &= -\gamma_{i\mu} n_\nu T^{\mu\nu}, \\ S_{ij} &= \gamma_{i\mu}\gamma_{j\nu} T^{\mu\nu}, \end{aligned} \tag{10}$$

$T^{\mu\nu}$ being the stress-energy tensor for a perfect fluid, which is written as a function of the rest-mass density ρ, the specific enthalpy h, the pressure P and the fluid 4-velocity u^μ

Finally, the evolution equation for $\hat{\Delta}^r$, the radial component of the additional BSSN variables $\hat{\Delta}^i = \hat{\gamma}^{mn}\hat{\Delta}^i_{mn}$ with $\hat{\Delta}^a_{bc} = \hat{\Gamma}^a_{bc} - \mathring{\Gamma}^a_{bc}$, is given by

$$\begin{aligned}\partial_t \hat{\Delta}^r = {} & \beta^r \partial_r \hat{\Delta}^r - \hat{\Delta}^r \partial_r \beta^r + \frac{1}{a}\partial_r^2 \beta^r + \frac{2}{b}\partial_r\left(\frac{\beta^r}{r}\right) \\ & + \frac{\sigma}{3}\left(\frac{1}{a}\partial_r(\hat{\nabla}_m \beta^m) + 2\hat{\Delta}^r \hat{\nabla}_m \beta^m\right) - \frac{2}{a}(A_a \partial_r \alpha + \alpha \partial_r A_a) \\ & + 2\alpha\left(A_a \hat{\Delta}^r - \frac{2}{rb}(A_a - A_b)\right) + \frac{2\alpha}{a}\left[\partial_r A_a - \frac{2}{3}\partial_r K + 6A_a \partial_r \chi\right. \\ & \left. + (A_a - A_b)\left(\frac{2}{r} + \frac{\partial_r b}{b}\right) - 8\pi j_r\right].\end{aligned} \tag{11}$$

We use the so called "non-advective 1+log" condition [13] for the lapse, and a variation of the "Gamma-driver" condition for the shift vector which has a proper vector on the right hand side[14, 8]. We do not find necessary to add a damping term on the right hand side of the shift condition in order to avoid strong oscillations of the shift vector(*i.e.* see left panel of Fig.2). These are written as follows

$$\partial_t \alpha = -2\alpha K, \tag{12}$$

$$\partial_t B^r = \frac{3}{4}\partial_t \hat{\Delta}^r, \tag{13}$$

$$\partial_t \beta^r = B^r. \tag{14}$$

The Hamiltonian constraint is used as diagnostics of the accuracy of the numerical evolutions:

$$\mathcal{H} \equiv R - (A_a^2 + 2A_b^2) + \frac{2}{3}K^2 - 16\pi E = 0. \tag{15}$$

Journal of Physics: Conference Series **454** (2013) 012002 doi:10.1088/1742-6596/454/1/012002

2.2. *Formulation of the hydrodynamic equations*

The general relativistic hydrodynamic equations, expressed through the conservation equations for the stress-energy tensor $T^{\mu\nu}$ and the continuity equation are:

$$\nabla_\mu T^{\mu\nu} = 0 \,, \qquad \nabla_\mu \left(\rho u^\mu\right) = 0. \tag{16}$$

Following [15], the general relativistic hydrodynamic equations are written in a conservative form in spherical coordinates. The following definitions for the hydrodynamic variables are used:

$$v^r \equiv \frac{u^r}{\alpha u^t} + \frac{\beta^r}{\alpha}, \tag{17}$$

$$W \equiv \alpha u^t, \tag{18}$$

where W is the Lorentz factor. By defining the vector of unknowns, $\mathbf{U}$, as

$$\mathbf{U} = \sqrt{\gamma}(D, S_r, \tau), \tag{19}$$

where the conserved quantities are D, S_r, τ. Then, the set of hydrodynamic equations (16) can be written in conservative form as

$$\partial_t \mathbf{U} + \partial_r \mathbf{F}^r = \mathbf{S}, \tag{20}$$

where $\mathbf{S}$ is the vector of sources and $\mathbf{F}^r$ the fluxes.

To close the system of equations, we choose the Γ-law equation of state, $P = (\Gamma - 1)\,\rho\epsilon$, where ϵ is the specific internal energy.

3. PIRK methods

Let us consider the following system of PDEs,

$$\left\{ \begin{array}{l} u_t = \mathcal{L}_1(u,v), \\ v_t = \mathcal{L}_2(u) + \mathcal{L}_3(u,v), \end{array} \right. \tag{21}$$

$\mathcal{L}_1$, $\mathcal{L}_2$ and $\mathcal{L}_3$ being general non-linear differential operators. Let us denote by L_1, L_2 and L_3 their discrete operators, respectively. L_1 and L_3 will be treated in an explicit way, whereas the L_2 operator will be considered to contain the unstable terms and, therefore, treated partially implicitly.

We use a Runge-Kutta (RK) method to update in time the previous system (21). Each stage of the method consists of two steps: i) the variable u is evolved explicitly; ii) the variable v is evolved taking into account the updated value of u. This strategy implies that the computational costs of the methods are comparable to those of the explicit ones. The resulting numerical schemes do not need any analytical or numerical inversion, but they are able to provide stable evolutions due to their partially implicit component.

PIRK methods are described in [9] and have been derived up to third-order in Δt in such a way that the number of stages is minimized. These methods are based on stability properties for both the explicit and implicit parts, recovering the optimal SSP explicit RK methods [16] when the L_2 operator is neglected, i.e., partially implicitly treated parts are not taken into account. The second-order PIRK scheme which we use is:

$$\left\{ \begin{array}{l} u^{(1)} = u^n + \Delta t \, L_1(u^n, v^n), \\ v^{(1)} = v^n + \Delta t \left[\dfrac{1}{2} L_2(u^n) + \dfrac{1}{2} L_2(u^{(1)}) + L_3(u^n, v^n) \right], \end{array} \right. \tag{22}$$

$$\begin{cases} u^{n+1} = \frac{1}{2}\left[u^n + u^{(1)} + \Delta t\, L_1(u^{(1)}, v^{(1)})\right], \\ v^{n+1} = v^n + \frac{\Delta t}{2}\left[L_2(u^n) + L_2(u^{n+1}) \right. \\ \qquad\qquad \left. + L_3(u^n, v^n) + L_3(u^{(1)}, v^{(1)})\right], \end{cases} \tag{23}$$

where Δt denotes the time-step.

This scheme is applied to the hydrodynamic and BSSN evolution equations. We include all the problematic terms appearing in the sources of the equations in the L_2 operator. Firstly, the hydrodynamic conserved quantities, the conformal metric components, a and b, the conformal factor, χ, or the quantity X (function of the conformal factor), the lapse function, α, and the radial component of the shift, β^r, are evolved explicitly (as u is evolved in the previous PIRK scheme); secondly, the traceless part of the extrinsic curvature, A_a, and the trace of the extrinsic curvature, K, are evolved partially implicitly, using explicit and implicit values of α, a and b; then, the quantity $\hat{\Delta}^r$ is evolved partially implicitly, using the updated values of α, a, b, β^r, conformal factor, A_a and K; finally, B^r is evolved partially implicitly, using the updated values of $\hat{\Delta}^r$. Matter source terms are always included in the explicitly treated parts.

4. Numerics

Derivatives in the spacetime evolution equations are calculated using a fourth-order centered finite difference approximation in a uniform grid except for the advection terms (terms formally like $\beta^r \partial_r u$), for which an upwind scheme is used. We also use fourth-order Kreiss-Oliger dissipation [17] to avoid high frequency noise appearing near the outer boundary.

We use a second-order slope limiter reconstruction scheme (MC limiter) to obtain the left and right states of the primitive variables at each cell interface, and a HLLE approximate Riemann solver [18, 19].

The computational domain is defined as $0 \leq r \leq L$, where L refers to the location of the outer boundary. We used a cell-centered grid to avoid that the location of the puncture at the origin coincides with a grid point in simulations involving a BH. At the origin we impose the conditions derived from the assumption of spherical symmetry. At the outer boundary we impose radiative boundary conditions [14] for the spacetime variables.

5. Numerical Results

5.1. Pure gauge dynamics

We first consider the propagation of a pure gauge pulse using the same initial parameters as in [8]. The main difference with respect to [8] is that we do not regularize the origin and rely only on the PIRK scheme to achieve a stable numerical simulation. The initial data are given by

$$\chi = 0, \tag{24}$$

$$a = b = 1, \tag{25}$$

$$A_a = A_b = K = 0, \tag{26}$$

$$\hat{\Delta}^r = 0, \tag{27}$$

$$\alpha = 1 + \frac{\alpha_0 r^2}{1 + r^2}\left[e^{-(r-r_0)^2} + e^{-(r+r_0)^2}\right], \tag{28}$$

with $\alpha_0 = 0.01$ and $r_0 = 5$. We evolve these initial data with a grid resolution of $\Delta r = 0.1$ and $\Delta t = 0.5\Delta r$. We use zero shift and harmonic slicing,

$$\partial_t \alpha = -\alpha^2 K. \tag{29}$$

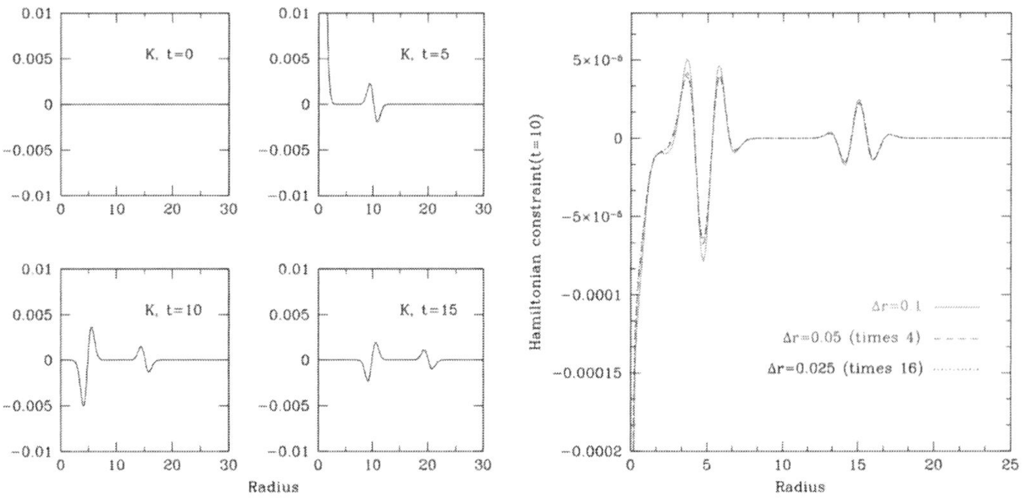

Figure 1. Left panel shows the trace of the extrinsic curvature, K, for a pure gauge pulse as a function of the radius at four different times; while in the right panel we show the Hamiltonian constraint at $t = 10$ for simulations of a pure gauge wave with three different resolutions $\Delta r = 0.1$, $\Delta r = 0.05$, and $\Delta r = 0.025$ rescaled by the factors corresponding to second-order convergence.

In the left panel of Fig. 1, we show the trace of the extrinsic curvature, K, as a function of the radius at four different times ($t = 0, 5, 10, 15$). The initial pulse separates in two pulses propagating in opposite directions. The snapshots of the evolution of the trace of the extrinsic curvature show that the evolution remains well behaved everywhere in the computational grid. We note that at $t = 5$ the value of K reaches a value of 0.1 at the origin, but later returns to zero when the pulse moves outwards as shown by [8].

In order to asses the convergence of the code, we have performed three simulations with resolutions $\Delta r = 0.1$, $\Delta r = 0.05$ and $\Delta r = 0.025$. The Hamiltonian constraint violations rescaled by the factors corresponding to second order convergence at $t = 10$ are plotted in the right panel of Fig. 1. All three lines overlap indicating that the code achieves the second-order convergence expected for the PIRK scheme used.

5.2. Schwarzschild black hole

The Schwarzschild metric in isotropic coordinates is used as initial data to test the ability of the code to evolve BH spacetimes within the moving puncture approach. The initial data are such that the 3-metric is written as

$$dl^2 = \psi^4(dr^2 + r^2 d\Omega^2), \qquad (30)$$

where the conformal factor is $\psi = (1 + M/2r)$, M being the mass of the BH, which we set as $M = 1$. Here r is the isotropic radius. Initially the extrinsic curvature is $K_{ij} = 0$.

We evolve the single stationary puncture initial data with a precollapsed lapse and initially vanishing shift vector. We use the gauge conditions given by (12)–(14), with a resolution $\Delta r = 0.05$, $\Delta t = 0.5\Delta r$ and $N_r = 30000$ grid points to place the outer boundary sufficiently far way from the puncture so that errors from the boundary do not affect the evolution.

As pointed out by [20, 21], the numerical slices of a Schwarzschild BH spacetime with these gauge conditions reach a stationary state after $t \sim 20$. This is shown in Fig. 2, where the time

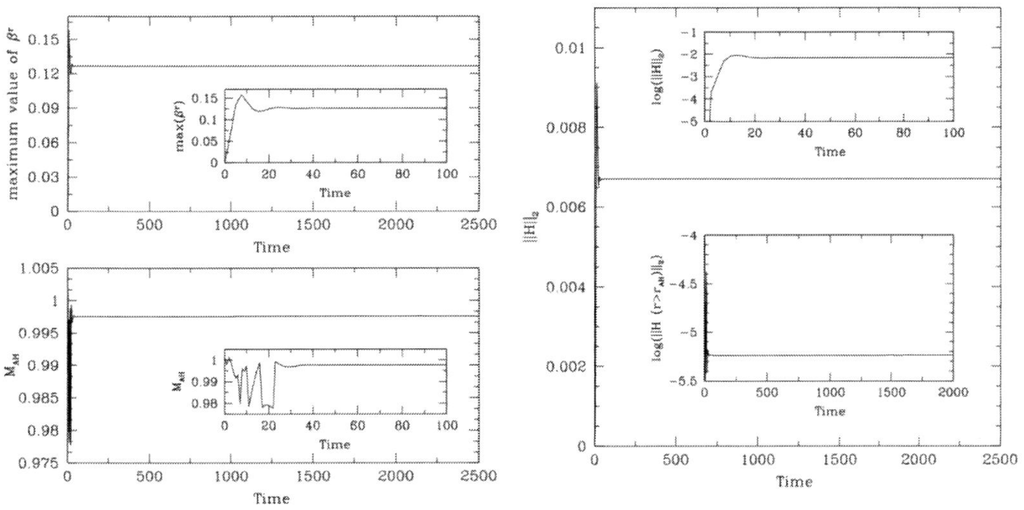

Figure 2. Left panel displays the time evolution of the maximum value of the radial shift β^r (top panel), and of the mass of the AH (bottom panel) in the single puncture BH simulation. In the right panel, we show the L2-norm of the Hamiltonian constraint in the single puncture BH simulation. The insets show the L2-norm during the initial phase, and the L2-norm computed outside the AH respectively.

evolution of the maximum value of the radial shift β^r is displayed in the upper left panel. After an initial phase in which the maximum value of the shift vector grows rapidly, it settles to a value of ~ 0.15 and we find almost no drift until the end of the simulation at $t = 2500$. In the lower left panel of Fig. 2, we show the time evolution of the mass of the apparent horizon (AH), defined as $M_{\rm AH} = \sqrt{\mathcal{A}/16\pi}$, where $\mathcal{A}$ is the area of the AH. We notice that $M_{\rm AH}$ is conserved well during the evolution and the error at $t = 2500$ is less than 0.2%

The Hamiltonian constraint violation results are displayed in the right panel of Fig. 2, which shows the L2-norm of the Hamiltonian constraint as a function of time. Both the initial phase driven by the gauge dynamics (see upper inset in the right panel of Fig. 2) and the stationary phase are clearly visible. The lower inset shows that the L2-norm computed outside the AH is about three orders of magnitude smaller than the L2-norm computed in the whole grid, which is to be expected as the largest spatial violation of the constraint occurs near the puncture (due to the finite differencing of the irregular solution).

In Fig. 3, we show the Hamiltonian constraint at $t = 100$ for simulations of single initial data with three different resolutions $\Delta r = 0.1$, $\Delta r = 0.05$, and $\Delta r = 0.025$ rescaled by the factors corresponding to second-order convergence. We find that the second-order errors associated with our PIRK method are smaller than the fourth-order error of our spatial derivatives.

5.3. Spherical relativistic stars

For our first numerical simulation of the coupling of Einstein equations and the general relativistic hydrodynamic equations, we use the Tolman-Oppenheimer-Volkoff (TOV) solution. We focus on an initial TOV model that has been extensively investigated numerically by [22, 23]. This model is a relativistic star with polytropic index $N = 1$, polytropic constant $\kappa = 100$ and central rest-mass density $\rho_c = 1.28 \times 10^{-3}$, so that its gravitational mass is $M = 1.4$, its baryon rest-mass $M_* = 1.5$ and its radius $R = 9.59$.

We evolve these initial data with our non-linear code until $t = 3000$ (~17 ms). In the upper

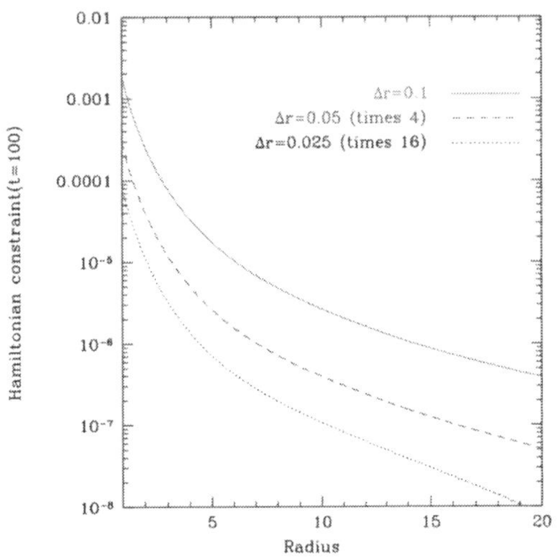

Figure 3. Hamiltonian constraint at $t = 100$ for simulations of single BH initial data with three different resolutions $\Delta r = 0.1$, $\Delta r = 0.05$, and $\Delta r = 0.025$ rescaled by the factors corresponding to second-order convergence.

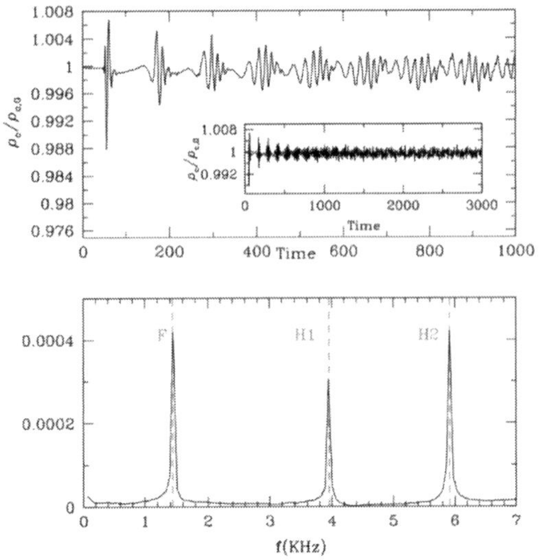

Figure 4. Time evolution of the normalized central density for an $M = 1.4$, $\kappa = 100$, $N = 1$ polytrope. Power spectrum of the evolution of the central rest-mass density is shown in the bottom panel. F, H1 and H2 represent the frequency of the fundamental mode and the first two overtones computed by [23].

panel of Fig. 4 we plot the time evolution of the central rest-mass density for a simulation with $\Delta r = 0.025$ and $N_r = 4000$ until $t = 1000$. In the inset we show the same quantity for the whole evolution. We observe that the truncation errors at this resolution are enough to excite small periodic radial oscillations, visible in this plot as periodic variations of the central density. We see that the damping of the periodic oscillations of the central rest-mass density is very small during the whole evolution, which highlights the low numerical viscosity of the implemented scheme.

By computing the Fourier transform of the time evolution of the central rest-mass density we obtain the power spectrum, which is shown with a solid line in the bottom panel of Fig. 4, while the dashed vertical lines indicate the fundamental frequency and the first two overtones computed by [23]. Note that the locations of the frequency peaks for the fundamental mode and the two overtones are in very good agreement, the relative error in the fundamental frequencies being less than 0.1%. The result of this simulation shows the ability of the scheme to maintain the numerical stability in long-term non-vacuum regular spacetime simulations in spherical coordinates without the need of an additional regularization at the origin.

5.4. Gravitational collapse of a marginally stable spherical relativistic star

We next test the capability of the code to follow BH formation with the gravitational collapse to a BH of a marginally stable spherical relativistic star. For this test, we consider a $\kappa = 100$, $N = 1$ polytropic star with central rest-mass density $\rho_c = 3.15 \times 10^{-3}$, so that its gravitational mass is $M = 1.64$ and its baryon rest-mass $M_* = 1.79$. In order to induce the collapse of the star, we initially increase the rest-mass density by 0.5%.

We present numerical results for a simulation of the gravitational collapse of a marginally

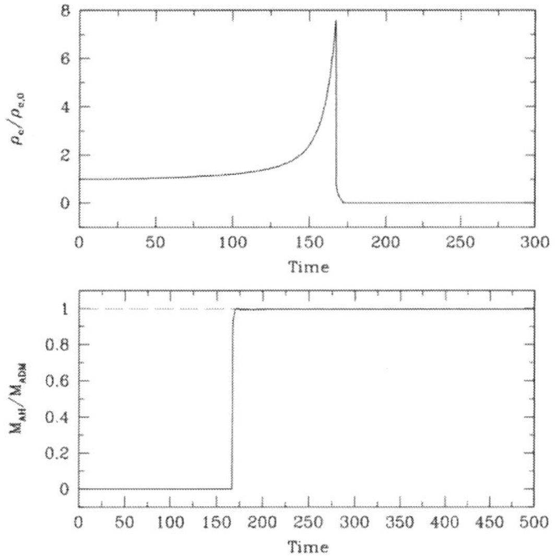

Figure 5. Time evolution of the normalized central density (top) and mass of the AH in units of the ADM mass of the system (bottom) for the collapse of a marginally stable spherical star to a BH.

stable spherical relativistic star performed with resolution of $\Delta r = 0.125$. We use the gauge conditions given in (12)–(14). We plot, in Fig. 5, the time evolution of the normalized central density until $t = 300$ (top), and of the mass of the AH in units of the ADM mass of the system until $t = 500$ when we stopped the simulation (bottom). Overall, as the collapse proceeds the star increases its compactness, reflected in the increase of the central density as shown in the top panel. The most unambiguous signature of the formation of a BH during the simulation is the formation of an AH. Once an AH is found by the AH finder, we monitor the evolution of the AH area, and also of its mass which is plotted, in the bottom panel of Fig. 5. This panel shows that approximately at $t \sim 167$, an AH is first found and that the mass of the AH relaxes to the ADM mass of the system. The difference in the ADM mass and the mass of the AH at $t = 500$, is about 0.2%. Results of this simulation indicate that the numerical scheme to integrate the evolution equations in time can handle accurately the transition between a regular spacetime (that of the star) and a irregular spacetime containing a puncture singularity at $r = 0$.

6. Summary

We have reviewed the work of Montero & Cordero-Carrión [12] who presented results from a numerical code solving the BSSN equations in spherical symmetry and the general relativistic hydrodynamic equations written in flux-conservative form. Montero & Cordero-Carrión [12] applied a second-order PIRK method to integrate in time the evolution equations in curvilinear coordinates, and produced the first successful numerical simulations of vacuum and non-vacuum spacetimes using the covariant BSSN formulation in spherical coordinates without the need for a regularization algorithm at the origin. We have discussed this approach and showed a number of tests to assess the accuracy and expected convergence of the code, namely a pure gauge wave, the evolution of a single BH, the evolution of a spherical relativistic star in equilibrium, and the gravitational collapse of a spherical relativistic star leading to the formation of a BH. Finally, we refer to [24] for details on the implementation of this approach to the BSSN equations in spherical polar coordinates without any symmetry assumptions in three spatial dimensions.

Acknowledgments

P.M. acknowledges support by the Deutsche Forschungsgesellschaft (DFG) through its Transregional Centers SFB/TR 7 "Gravitational Wave Astronomy". I. C.-C. acknowledges support from Alexander von Humboldt Foundation and AYA2010-21097-C03-01 of the Spanish MICINN.

References

[1] Nakamura T, Oohara K and Kojima Y 1987 *Prog. Theor. Phys. Suppl.* **90** 1
[2] Shibata M and Nakamura T 1995 *Phys. Rev. D* **52** 5428
[3] Baumgarte T W and Shapiro S L 1999 *Phys. Rev. D* **59** 024007
[4] Brown J D 2009 *Phys. Rev. D* **79** 1004029
[5] Rinne O and Stewart J M 2005 *Class. Quant. Grav.* **22** 1143
[6] Alcubierre M and González J A 2005 *Comp. Phys. Comm.* **167** 76
[7] Ruiz M, Alcubierre M and Nuñez D 2008 *Gen. Rel. Grav.* **40** 159
[8] Alcubierre M and Mendez M D 2011 *Gen. Rel. Grav.* **43** 2769
[9] Cordero-Carrión I and Cerdá-Durán P 2012 *Preprint arXiv:1211.5930 [math-ph].*
[10] Cordero-Carrión I, Cerdá-Durán and Ibáñez J M 2012 *Phys. Rev. D* **85** 044023
[11] Bonazzola S, Gourgoulhon E, Grandclément P and Novak J 2004 *Phys. Rev. D* **70** 104007
[12] Montero P J, and Cordero-Carrión I 2012 *Phys. Rev. D* **85** 124037
[13] Bona C, Massó J, Seidel E and Stela J 1997 *Phys. Rev. D* **56** 3405
[14] Alcubierre M, Brügmann B, Diener P, Koppitz M, Pollney D, Seidel E and Takahashi R 2003 *Phys. Rev. D* **67** 084023
[15] Banyuls F, Font J A, Ibánez J M, Martí J M and Miralles J A 1997 Astrophys. J. **476** 221
[16] Gottlieb S and Shu C W 1998 *Math. Compt.* **67** 73
[17] Kreiss H O and Oliger J 1973 *Methods for the Approximate Solution of the Time Dependent Problems*, edited by GARP Publ. Ser.
[18] Harten A, Lax P D and van Leer B 1983 *SIAM Rev.* **25** 35
[19] Einfeldt B 1988 *SIAM J. Numer. Anal.* **25** 294
[20] Hannam M, Husa S, Pollney D, Brgmann B and OMurchadha N 2007 *Phys. Rev. Lett.* **99** 241102
[21] Hannam M, Husa S, Brügmann B, Gonzlez J A, Sperhake U and OMurchadha N 2007 *J. Phys. Conf. Ser.* **66** 012047
[22] Font J A, Stergioulas N and Kokkotas K D 2000 *MNRAS* **313** 678
[23] Font J A, Goodale T, Iyer S, Miller M, Rezzolla L, Seidel E, Stergioulas N, Suen W M and Tobias M 2002 *Phys. Rev. D* **65** 084024
[24] Baumgarte T W, Montero P J, Cordero-Carrión I and Müller E 2013 *Phys. Rev. D* **87** 044026

Symmetry-preserving momentum remap for ALE hydrodynamics

J Velechovský[1], M Kuchařík[1], R Liska[1] and M Shashkov[2]
[1]Czech Technical University, Faculty of Nuclear Sciences and Physical Engineering, Břehová 7, Praha 1, 115 19, Czech Republic
[2]Los Alamos National Laboratory, Group XCP-4, Los Alamos NM 87545, USA
E-mail: velecjan@fjfi.cvut.cz

Abstract. In this paper, a symmetry-preserving remapping algorithm for vectors respecting their local bounds by components and in magnitude is presented. First, description of the Vector Image Polygon (VIP) limiter for a piece-wise linear velocity reconstruction is presented. Numerical fluxes obtained from this reconstruction lead to symmetry- and bounds- preserving remap of momentum for staggered Arbitrary Lagrangian-Eulerian (ALE) hydrodynamical methods. A novel bounds definition for vectors and corresponding modification of the VIP limiter is introduced, which fixes undershoots in a radial velocity component for polar meshes. Comparison with standard scalar-based limiters is given. Cyclic remapping is performed to numerically verify properties of the methods.

1. Introduction

The Arbitrary Lagrangian-Eulerian (ALE) [1] methods became a standard and successful approach in the hydrodynamic codes for laser generated plasmas. In our 2D code Prague ALE (PALE) [2], we are able to perform simulations of high-velocity plasma flows generated by an intense laser beam. PALE code incorporates a staggered discretization of the hydrodynamic equations, in which the thermodynamic quantities are located in the computational cells while the kinematic quantities are located on the nodes of the mesh. A typical ALE method can be divided in three steps: 1) the Lagrangian solver; 2) mesh relaxation; and 3) the remapping method interpolating conservatively all fluid quantities between the meshes. Many production codes employ the ALE machinery, next to our PALE code, let us name CHIC [3], CORVUS [4], FLAG [5], ALEGRA [6], or ALE-AMR [7].

Standard limiters for a scalar are usually applied for Cartesian components of vectors during the remapping stage. Symmetry-preserving vector remapping methods use the VIP limiting procedure [8] or the projection of the velocity in the principal flow direction [9].

2. Remapping overview

Suppose that we have two computational meshes – an old (Lagrangian) mesh $\{c\}$ and a new (rezoned) mesh $\{\tilde{c}\}$. Let us assume that both meshes have the same connectivity and are close to each other. The remap of mass (for example) can be written in the following flux form

$$m_{\tilde{c}} = m_c + \sum_{c' \in C'(c)} F^m_{c,c'}\,, \qquad (1)$$

where $C'(c)$ represents a set of all cells neighboring with cell c, and $F^m_{c,c'}$ stands for the (outwards oriented) mass flux from c to c'. Depending on the particular remapping method, the $F^m_{c,c'}$ fluxes can be constructed in different manners. In our approach, integration of the density function over all intersections of the original cell c with all its neighbors in the new mesh is used [10].

For nodal mass, we employ remap in a flux form analogous to (1),

$$m_{\tilde{n}} = m_n + \sum_{n' \in N'(n)} F^m_{n,n'} \,, \tag{2}$$

where $N'(n)$ stands for the set of nodes neighboring with node n, and $F^m_{n,n'}$ represents the inter-nodal mass flux from n to n'. In the formula, we only consider one flux per neighbor, which can in practice be split in two parts corresponding to the different attached cells.

Exact integration by intersecting the dual cells of old and new mesh is computationally expensive due to generally non-convex dual cells. One possibility is the cheap interpolation of the inter-nodal mass fluxes $F^m_{n,n'}$ from the inter-cell ones in a way similar to the one described in [11].

For remap of nodal momentum vector, formula analogous to (2) is used. This remap can be rewritten for nodal velocity $\vec{w}_n$ as

$$\vec{w}_{\tilde{n}} = \frac{1}{m_{\tilde{n}}} \left(m_n \, \vec{w}_n + \sum_{n' \in N'(n)} \vec{F}^{\mu}_{n,n'} \right) . \tag{3}$$

The new nodal mass $m_{\tilde{n}}$ is already known (2) and the momentum fluxes $\vec{F}^{\mu}_{n,n'}$ can be obtained in several ways. We prefer reusing the computed mass fluxes $F^m_{n,n'}$, and constructing the momentum fluxes by their multiplication by the reconstructed velocity vector, $\vec{F}^{\mu}_{n,n'} = F^m_{n,n'} \, \vec{w}_{n,n'}$, which is consistent with the DeBar condition [12].

Our goal is to develop momentum remapping method which is symmetric and preserves bounds on a polar grids for radial velocity field and radial grid motion. In this pseudo-1D case, we want to keep velocity vector in limits given by standard 1D bounds for its radial component (note that the angular component is 0 for the symmetric case). This is equivalent to bounds in velocity magnitude for a special case where radial velocity does not change direction along a ray from an origin, e.g. purely convergent or divergent velocity field. This property is not satisfied by standard approaches for velocity reconstructions.

3. Velocity reconstructions

Let us remind notation for variables on our staggered logically orthogonal quadrilateral mesh. We denote cell quantities (such as cell center x_c) by integer subscript c and nodal, e.g. node position $\vec{x}_n$ or velocity $\vec{w}_n$, by subscript n. A cell c is defined by a set of its nodes $N(c)$. Similarly, a node n is shared by a set of cells $C(n)$.

3.1. Piecewise constant velocity reconstruction

This reconstruction imitates the cell-based donor remapping method. A particular node is chosen with respect to the sign of an inter-nodal mass flux, which is given by the grid movement. Imagine that we have mass flux $F_{n,n'}$ from node n to n' given in some point $\vec{x}$. Then, the reconstructed velocity $\vec{w}_{n,n'}$ has the form

$$\vec{w}_{n,n'} = \begin{cases} \vec{w}_n & \text{for} \quad F_{n,n'} > 0 \\ \vec{w}_{n'} & \text{for} \quad F_{n,n'} < 0 \,. \end{cases} \tag{4}$$

Resulting remapping satisfies both conditions for symmetry and bounds preservation. However, it is only first order accurate and therefore too diffusive.

3.2. Piecewise linear velocity reconstruction

There are many ways how to construct high-order velocity reconstruction, which are almost equivalent in regions with sufficiently smooth velocity field. Our choice of piecewise linear reconstruction on a dual mesh is motivated by the fact, that we want to approach the low-order piecewise constant reconstruction by setting a slope limiter value to zero in regions near shocks. The piecewise linear reconstruction in node n is given by

$$\vec{w}_{n,n'} = \vec{w}_n + \phi_n(\nabla\vec{w})_n(\vec{x} - \vec{x}_n) \tag{5}$$

where $\vec{x}_n$ and $\vec{w}_n$ are nodal position and velocity, $\vec{x}$ stands for approximate flux position and $\phi_n \in (0, 1)$ for a slope limiter. Unlimited slopes in velocity gradient $(\nabla\vec{w})_n$ are computed from velocities in eight neighboring nodes using Green formula. This unlimited reconstruction (for $\phi_n = 1$), as well as resulting remapping, is also symmetric. However, it produces a new extrema near shocks. Limitation is therefore necessary.

3.3. Barth-Jespersen limiter for piecewise linear reconstruction

This limiter for scalars was introduced in [13]. In most of recent approaches, x- and y-components of velocity $\vec{w} = \binom{w^x}{w^y}$ are limited separately with constrains

$$\begin{aligned} w_n^{x,\min} &\le w_n^x(\vec{x_m}) \le w_n^{x,\max} \\ w_n^{y,\min} &\le w_n^y(\vec{x_m}) \le w_n^{y,\max} \qquad \forall m \in N(C(n)), \end{aligned} \tag{6}$$

where $w_n^x(\vec{x_m})$ and $w_n^y(\vec{x_m})$ are components of reconstructed velocity given by (5) and e.g. $w_n^{x,\min}$ is minimal velocity component from set given by 8 neighboring nodes of node n and the node n (this set is denoted by $N(C(n))$).

Let us make a short note on the scalar limiter ϕ_n in (5). To limit the velocity components separately, it is common to replace the scalar by a diagonal matrix. This diagonal matrix with different limiters for velocity components was also used in our numerical tests. However, regardless of the choice of scalar or matrix form, these methods are neither symmetric nor bounds preserving for radial problems as defined at the end of section 2, because its limiter value is angularly dependent (even for the usual technique of a common limiter value taken as the minimum of both Cartesian components).

4. Bounds for velocity vector

Remaining question is how to compute the nodal limiter values $\phi_n \in (0, 1)$ in (5). Similarly as in [8], we divide this question in two parts. First, how to define bounds for vector, and second, how to compute the optimal value of ϕ_n to keep reconstructed vector in the bounds. This bounds definition has to be general for any field. However, we want to satisfy conditions for radial velocity on a polar mesh as mentioned above. To be illustrative, we start with comment on the Barth-Jespersen approach applied in a standard way, component by component.

Any given set of velocity bounds defined by BJ-limiter represents a rectangle (which can degenerate to a line or a point) in the velocity space (where velocity vectors are expressed by their x- and y- components), as shown in figure 1 (a, b). However, this rectangle is not invariant with respect to rotation of the coordinate system. Just compare figure 1 (a) with (b) for the same set of velocities in rotated coordinate system. It leads to violation of symmetry in our first test case for polar grids with radial velocity, see figure 2. This situation can by clearly understood by the fact that limiter moves reconstructed velocities inside the bounds represented by the angularly dependent rectangle, so the limiter value and the therefore velocity reconstruction as well as resulting remap show angular dependency even for symmetric problems. One way how to remove this dependency on Cartesian components is to switch to local coordinate system given e.g. by the principal flow direction [9, 14].

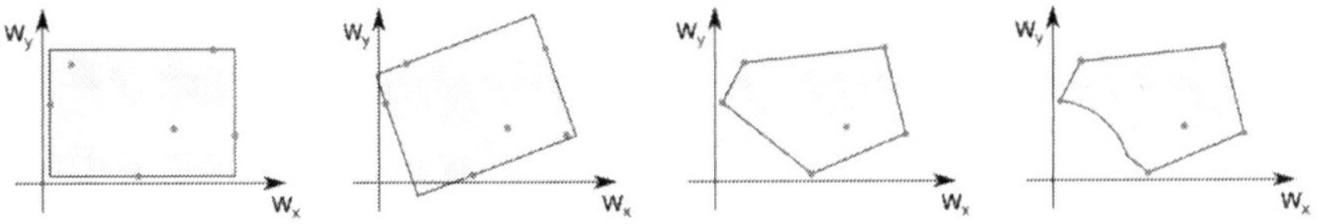

Figure 1. Yellow objects represent different bound definitions for 6 velocities given by red dots. (a, b) corresponds to BJ applied component by component, (c) VIP approach and (d) our modification (MVIP) of these bounds.

5. Vector Image Polygon approach

In [8], another definition of vector bounds is presented, which is independent of rotation of the coordinate system. These bounds are defined by the Vector Image Polygon (VIP), defined as the convex hull of a given set of velocities, see figure 1 (c). Compared to the BJ limiter applied component by component, VIP is more restrictive, because this convex hull of given velocities is equal to common intersection of all rectangles from BJ bounds for all arbitrary rotations of the coordinate system.

Our particular implementation for logically orthogonal quadrilateral mesh is slightly different from the usual approach (6). For BJ approach, we require to keep reconstruction in all neighboring nodes in the common bounds given by velocities from the same set of nodes. On the contrary for the VIP approach, we evaluate unlimited reconstruction in all cell centers around the node. For every particular cell center, we want to keep this reconstruction in limits given by nodes of the cell.

The next section describes in detail computation of the nodal limiter value ϕ_n. To determine its value, we evaluate the unlimited reconstruction in four cell centers $\vec{x}_c$ around the node

$$\vec{w}_{nc}^{\text{unlim}} = \vec{w}_n + (\nabla\vec{w})_n(\vec{x}_c - \vec{x}_n)\,. \tag{7}$$

For every cell center $\vec{x}_c$, we want to find ϕ_{nc} to keep reconstructed velocity

$$\vec{w}_{nc}^{\text{lim}} = \vec{w}_n + \phi_{nc}(\nabla\vec{w})_n(\vec{x}_c - \vec{x}_n) \tag{8}$$

in VIP bounds of its four nodes $n \in N(c)$, so we require

$$\forall c \in C(n) \quad \vec{w}_{nc}^{\text{lim}} \in \text{VIP}_c \tag{9}$$

where VIP_c represents the Convex Hull of nodal velocities $\vec{w}_n$, $n \in N(c)$. At first, for a particular cell c attached to node n find the maximal $\phi_{nc} \in (0, 1)$ which keeps $\vec{w}_{nc}^{\text{lim}}$ inside VIP_c. Let us call these ϕ_{nc}^{VIP} and $\vec{w}_{nc}^{\text{VIP}}$. Having computed all four ϕ_{nc}^{VIP} corresponding to $\vec{w}_{nc}^{\text{VIP}}$, we set the final ϕ_n to the minimum of nodal values ϕ_{nc} to satisfy condition (9) in all neighboring cells. Computation of a particular ϕ_{nc}^{VIP} can be done in the following way. We denote reconstructed linear part of velocity $\vec{u}_{nc} = \phi_n(\nabla\vec{w})_n(\vec{x}_c - \vec{x}_n)$. This one is in limits if it lies in the convex hull (CH) of set

$$\text{CH}\{\vec{w}_m - \vec{w}_n, m \in N(c)\} = P_{nc}\,.$$

Polygon P_{nc} has one to four different vertices depending on the velocity field and contains the $\vec{0}$ point. The zero velocity comes from bound in reconstruction center node and it represents piece-wise constant low-order reconstruction. Because of used linear reconstruction and because of scalar limiter ϕ_{nc}, achievable reconstruction is located on a line segment from origin to $\vec{u}_{nc}$. Possible limiter values are represented by an intersection of this line segment with the convex hull of P_{nc}. The best (least diffusive) limiter value ϕ_n^c given by constraints in the cell center $\vec{x}_c$ is the one closest to the $\vec{u}_{nc}$ and corresponds to intersection Φ_n^c.

To avoid construction of P_{nc} convex hull (CH), we discuss following two cases separately. We denote $T_{nc} = \{P_{nc} \backslash \vec{0}\}$ which represents triangle of constrains. If $\vec{0} \in T_{nc}$, then the triangle is also boundary of P_{nc} convex hull and we can find Φ_n^c as an intersection of the line segment with T_{nc}. If $\vec{0} \notin T_{nc}$, then computation of intersections Ψ_{nc}^i of T_{nc} with a ray from $\vec{0}$ to $\vec{u}_{nc}$ is also sufficient. If there is no intersection, then $\vec{u}_{nc}$ points out of the CH and $\Phi_n^c = \vec{0}$. Otherwise $\Phi_n^c = \min(1, \max_i(\Psi_{nc}^i))$, where $\max_i(\Psi_{nc}^i)$ returns intersection which is further from the origin (closer to $\vec{u}_{nc}$). If it exists, second intersection Ψ_{nc}^i lies inside of the P_{nc} convex hull and therefore it is not interesting for us. To conclude, computation of intersection of a single ray with three line segments of T_{nc} is sufficient in both cases.

Let us project elements of $T_{nc} = \{\vec{u}_{nc_i}, i \in \{1,2,3\}\}$ to the coordinate system given by $\vec{u}_{nc} = \binom{u_{nc}^x}{u_{nc}^y}$, so

$$\vec{v}_{nc_i} = \frac{1}{|\vec{u}_{nc}|^2} \begin{pmatrix} u_{nc}^x & u_{nc}^y \\ -u_{nc}^y & u_{nc}^x \end{pmatrix} \vec{u}_{nc_i}\,. \tag{10}$$

After this projection, intersections have a simple form $\Psi_{nc}^i = \binom{\psi_{nc}^i}{0}$ and they are easy to find (11). Moreover, value ψ_{nc}^i now represents ratio of bound- and reconstructed- velocity linear parts, so this value is a candidate for the limiter in (5). It is only necessary to choose the correct intersection (12) as discussed above. Finally, minimal value from bounds in the four cells of node n is taken (13)

$$\psi_{nc}^i = v_{nc_i}^x + \frac{v_{nc_i}^y}{v_{nc_i}^y - v_{nc_{i+1}}^y}\left(v_{nc_{i+1}}^x - v_{nc_i}^x\right) \text{ if } v_{nc_i}^y v_{nc_{i+1}}^y < 0, \text{ else } \psi_{nc}^i = 0\,. \tag{11}$$

$$\phi_n^c = \min(1, \max_i(0, \psi_{nc}^i))\,. \tag{12}$$

$$\phi_n = \min_{c \in C(n)} (\phi_n^c)\,. \tag{13}$$

Special care must be taken of when $|\vec{u}_{nc}|^2$ or $v_{nc_i}^y - v_{nc_{i+1}}^y$ are close to zero. Second case means that the line given by $\vec{v}_{nc_i}, \vec{v}_{nc_{i+1}}$ is parallel to the ray or that our CH is only one point. It is sufficient to check if they are on or out of the axis. If they are out, we take zero limiter value, if they are on the axis, we simply take higher coordinate from the line endpoints. In a simple 1D case (constant velocity field along one direction), this limiter approaches standard 1D Barth-Jespersen method regardless of our special choice of bounds.

Application of the limited reconstruction for remapping is straightforward extension of the low-order method. Instead of using nodal values to compute momentum fluxes according to sign of corresponding mass flux, we simply use reconstructed velocity evaluated in the flux position. Note that in presence of discontinuity, the limiter value is zero and only bounds preserving donor fluxes are used.

In our particular implementation, remapping using piecewise linear velocity reconstruction with the VIP limiter is perfectly symmetric on a polar mesh, but there is small violation of bounds for a minimum of the radial velocity component, see fig. 3 (c). This undershoot was the main motivation for our novel corrected definition of vector bounds called Modified VIP (MVIP).

6. Modified VIP

The goal of this modification of the VIP method is a general correction which removes undershoot for the radial velocity component. Note that we can not directly limit this radial component, because is it not clear how to define radial direction for general vector field. Therefore, we add a new constraint for the minimal velocity magnitude (fig. 1 (d)) which is applied only in situations where it is suitable. Imagine at first the situation with only two velocities in the opposite directions with the same magnitude. Then the VIP set is just a line segment and after the restriction on a minimal magnitude, we would not allow any contribution of any high-order term. This is an analog of 1D situation, where any magnitude smaller than the original one is acceptable. Analogous situation occurs in 2D when e.g. the origin of the coordinate system lies inside the CH of nodal velocities. Then it does not make a good sense to cut out a whole circle from the CH.

Therefore, we have decided to choose the following condition for application of the minimal magnitude correction of VIP. If we can find a rotated coordinate system in which all limiting velocities are located in one quadrant, then these velocities have a small angular discrepancy and the condition is applied. In more general case, we require limiting the angular discrepancy by any parameter φ_0. Our choice corresponds to $\varphi_0 = \pi/2$. In other words, for any pair of bounds-defining velocities, their dot product is positive. For the maximal velocity magnitude, we do not need to check this limit because velocities from VIP are always in bounds for the maximal magnitude.

Computation of the MVIP slope limiter extends the pure VIP to the following procedure. We evaluate the unlimited reconstruction in four cell centers that surround the node n (7, 8) and check limits for MVIP

$$\forall c \in C(n) \quad \vec{w}_{nc}^{\rm lim} \in {\rm MVIP}_c \tag{14}$$

where

$$\mathrm{MVIP}_c = \begin{cases} \mathrm{VIP}_c \cap \mathrm{R}_c(0, \vec{w}^{\min}) & \text{for} \quad \forall n \in N(c), \forall m \in N(c) \quad \vec{w}_n \cdot \vec{w}_m > 0 \\ \mathrm{VIP}_c & \text{otherwise}, \end{cases} \tag{15}$$

and where VIP_c represents the Convex Hull of nodal velocities $\vec{w}_n$, $n \in N(c)$, and $\mathrm{R}_c(0, \vec{w}^{\min}) = \left\{\vec{w}, |w| \geq \min\limits_{n \in N(c)} (|\vec{w}_n|)\right\}$.

For a particular cell c attached to node n, we find at first maximal $\phi_{nc} \in (0, 1)$ which keeps $\vec{w}_{nc}^{\rm lim}$ inside VIP_c. Let us call these $\phi_{nc}^{\rm VIP}$ and $\vec{w}_{nc}^{\rm VIP}$. For a velocity field with a large angular discrepancy $\mathrm{MVIP}_c = \mathrm{VIP}_c$ and therefore $\phi_{nc} = \phi_{nc}^{\rm VIP}$. For a small one, check if limited velocity $\vec{w}_{nc}^{\rm VIP}$ lies also in MVIP_c and set

$$\phi_{nc} = \begin{cases} \phi_{nc}^{\rm VIP} & \text{for} \quad \vec{w}_{nc}^{\rm VIP} \in \mathrm{R}_c(0, \vec{w}^{\min}) \\ \phi_{nc}^{i} & \text{otherwise}, \end{cases} \tag{16}$$

where ϕ_{nc}^i is the limiter value corresponding to $\vec{w}_{nc}^{\rm lim}$ computed as the intersection of $\mathrm{R}_c(0, \vec{w}^{\min})$ boundary with a line segment given by $\vec{w}_n$ and $\vec{w}_{nc}^{\rm VIP}$. Note that now $\vec{w}_{nc}^{\rm VIP}$ is not in $\mathrm{R}_c(0, \vec{w}^{\min})$ (it is inside the circle given by its boundary) and clearly $\vec{w}_n \in \mathrm{R}_c(0, \vec{w}^{\min})$, so the intersection exists and it is unique, $\phi_{nc}^i \in (0, \phi_{nc}^{\rm VIP})$ and therefore $\vec{w}_{nc}^{\rm lim} \in \mathrm{MVIP}_c$.

Having computed all four ϕ_{nc} corresponding to $\vec{w}_{nc}^{\rm lim}$, we would like to set the final ϕ_n to a minimum of nodal values ϕ_{nc} to satisfy condition (14) in all neighboring cells. However, MVIP_c's are not convex in general and therefore we need to check again $\forall c \in C(n)$, that the reconstructed velocities lie in MVIP_c. Let us note that

$$\phi_n^{\min} = \min_{c \in C(n)} (\phi_{nc})$$

and $\vec{w}_n^{\min}(\vec{x}_c)$ is the reconstructed velocity corresponding to this limiter, finally

$$\phi_n = \begin{cases} \min\limits_{c \in C(n)} (\phi_{nc}) & \text{for} \quad \forall c \in C(n) \Big(\vec{w}_n^{\min}(\vec{x}_c) \in \mathrm{R}_c(0, \vec{w}^{\min}) \vee \forall (n, m) \in N(c) \quad \vec{w}_n \cdot \vec{w}_m \leq 0 \Big) \\ 0 & \text{otherwise}. \end{cases} \tag{17}$$

The last choice of 0 for the limiter value could be replaced by a better choice. However, it would require multiple repetition of the whole procedure and the improvement of the method would not be significant.

An extension of our MVIP limiting method for polygonal mesh with any number of cell vertices is straightforward. The only difference is, that three line segments of T_{nc} components of which are used in (10) and represent 3 nodal velocities have to be replaced by line segments given by all pairs of all nodal velocities.

7. Numerical examples

A standard cyclic remapping method [10] was performed to verify properties of the described remapping techniques. Two kinds of prescribed grid motions are presented. First one is radially symmetric, so no motion in an angular and smooth one in a radial direction was used. Radial motion of a particular grid is given by

$$r(\xi, t) = (1 - \alpha(t))\xi + \alpha(t)\xi^3 \ , \ \alpha(t) = \frac{\sin(4\pi t)}{2} \ , \ 0 \le \xi \le 1 \ , \ -1/8 \le t \le 1/8 \qquad (18)$$

where ξ represents discrete value of relative nodal radius and t is the pseudotime. We use grids having $20/40/80$ cells in radial and $40/80/160$ cells in angular direction corresponding to $100/200/400$ remaps. These grids were scaled to the computational domain $(0.01, 0.51) \times (0, 2\pi)$. Second grid motion is described at the end of this section.

7.1. Constant density and radial velocity on a polar grid

A simple initial conditions are given by constant density $\rho = 1$ and radial velocity with step in magnitude $|\vec{w}| = 1$ for $r \ge 0.25$ and $|\vec{w}| = 3$ for $r < 0.25$. Contour plots in Figure 2 correspond

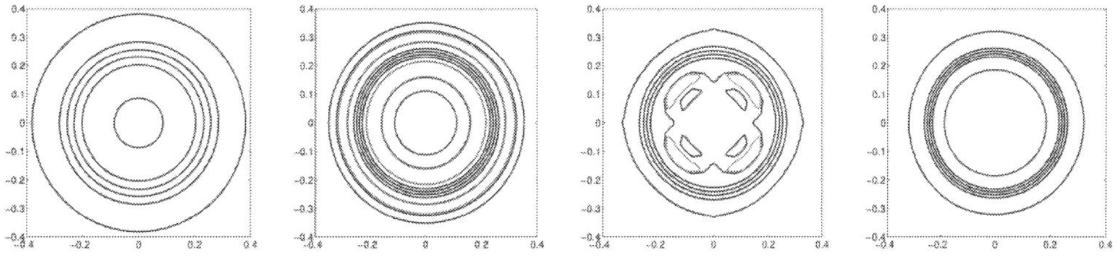

(a) (b) (c) (d)

Figure 2. Contour plots of velocity magnitude for cyclic remapping of step in radial velocity field on a polar grid. (a) low-order, (b) unlimited, (c) BJ- and (d) MVIP-limited method.

to velocity magnitude of the remapped initial function. Red contours represent values out of bounds, namely 0.99 and 3.01. The figure clearly shows violation of symmetry by the Barth-Jespersen limiter. Figure 3 (a) shows the exact solution compared to the selected methods. BJ limited reconstruction leads to non-symmetric result. These results are in figure 3 (b, c) compared to VIP, resp. MVIP methods. The only difference between last two methods is visible in sub-figure (c), showing detail for minimal magnitude, where MVIP fixes small undershoot produced by the original VIP method.

To verify the order of accuracy of the MVIP method, we tried also a smooth function for radial velocity $w_r(r) = \exp(-r)$ and again constant density. Table 1 shows second order of convergence for the new method.

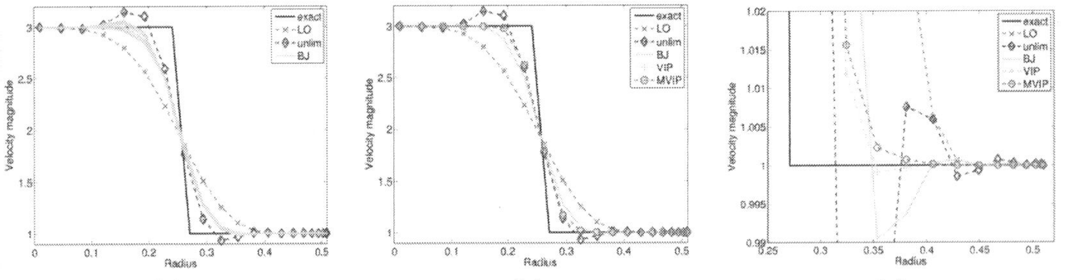

(a) (b) (c)

Figure 3. Scatter plots of velocity magnitude over angular direction. Subfigures (a, b) show the whole domain, where subfigures (b, c) show only one selected line for BJ-limiter and subfigure (c) discovers detail of the region with the minimal magnitude.

Table 1. Velocity L_1 error and corresponding convergence rate for radial velocity $w_r(r) = \exp(-r)$. Methods with piece-wise constant (LO), unlimited-, resp. MVIP-limited piece-wise linear velocity reconstructions.

remap	$E_{20\times40}$	$E_{40\times80}$	$E_{80\times160}$	$\frac{20\times40}{40\times80}$	$\frac{40\times80}{80\times160}$
LO	$6.19{\cdot}10^{-3}$	$3.27{\cdot}10^{-3}$	$1.67{\cdot}10^{-3}$	1.9	2.0
unlim	$1.07{\cdot}10^{-3}$	$2.68{\cdot}10^{-3}$	$6.62{\cdot}10^{-5}$	4.0	4.0
MVIP	$1.59{\cdot}10^{-3}$	$4.16{\cdot}10^{-4}$	$1.07{\cdot}10^{-4}$	3.8	3.9

7.2. Noh-like problem on a polar grid

Initial conditions for density ρ and velocity $\vec{w} = w_r(r)\binom{\cos\varphi}{\sin\varphi}$ corresponding to an exact solution of the 2D Noh problem in selected time are

$$\rho(r) = \begin{cases} 16 & \text{for } r < 0.375 \\ 1 + 1.125/r & \text{for } r \geq 0.375 \end{cases}, \qquad w_r(r) = \begin{cases} 0 & \text{for } r < 0.375 \\ -1 & \text{for } r \geq 0.375 \end{cases}.$$

Figure 4 (a) shows result for radial component of velocity (because of symmetry preservation of all methods, the absolute value of this component is equal to the velocity magnitude). Only the low-order and the MVIP methods keep the bounds. Convergence for this setup is shown in table 2.

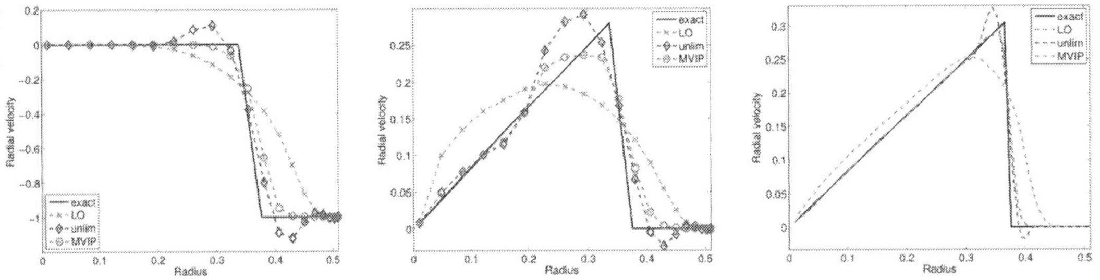

(a) (b) (c)

Figure 4. Radial velocity along radial direction for Noh-like (a), resp. Sedov-like (b,c) cyclic remapping on 20×40 (a,b), resp. 80×160 (c) cells.

Table 2. Velocity L_1 error and corresponding convergence rate for Noh-like cyclic remapping.

remap	$E_{20\times40}$	$E_{40\times80}$	$E_{80\times160}$	$\frac{20\times40}{40\times80}$	$\frac{40\times80}{80\times160}$
LO	$1.32{\cdot}10^{-1}$	$9.73{\cdot}10^{-2}$	$6.82{\cdot}10^{-2}$	1.4	1.4
unlim	$6.47{\cdot}10^{-2}$	$4.13{\cdot}10^{-2}$	$2.65{\cdot}10^{-2}$	1.6	1.6
MVIP	$4.45{\cdot}10^{-2}$	$2.70{\cdot}10^{-2}$	$1.62{\cdot}10^{-2}$	1.6	1.7

7.3. Sedov-like problem on a polar grid

The last two tests use the initial conditions which are close to the solution of the Sedov problem

$$\rho(r) = \begin{cases} 6\left(\frac{r}{0.375}\right)^8 & \text{for} \quad r < 0.375 \\ 1 & \text{for} \quad r \geq 0.375 \end{cases}, \qquad w_r(r) = \begin{cases} 0.83r & \text{for} \quad r < 0.375 \\ 0 & \text{for} \quad r \geq 0.375 \end{cases}, \tag{19}$$

where $r = \sqrt{(x-x_0)^2 + (y-y_0)^2}$ and $x_0 = y_0 = 0$. Results on the polar grid are shown in figure 4 (b, c) for two different grids. MVIP again preserve bounds and symmetry.

7.4. Sedov-like problem on a Cartesian grid

This cyclic remapping test is similar to [15]. The initial condition is the same as previous, but $x_0 = y_0 = 0.5$ in (19). We present results in half-time compared to the original paper. This pseudo-time corresponds to the maximal deformation of the grid. Our selection was motivated by the fact that some truncation errors may cancel during backward mesh movement [16]. Cartesian grid with 64×64/128×128/256×256 initially equidistantly spaced cells was used and 160/320/640 remaps were performed. Grid movement is given by

$$x(\xi,\eta,t) = \xi + t\sin(2\pi\xi)\sin(2\pi\eta)\ , \tag{20}$$

$$y(\xi,\eta,t) = \eta + t\sin(2\pi\xi)\sin(2\pi\eta)\ ,\ 0 \leq \xi \leq 1\ ,\ 0 \leq \eta \leq 1\ ,\ 0 \leq t \leq 0.25\,, \tag{21}$$

where ξ, η represent discrete position in x,y plane and t is the discrete pseudotime. Table 3 shows convergence of the selected methods and symmetry of the remapped radial velocity component is compared in figure 5. Red contours (starting on value -0.01) represent undershoots in radial velocity. Note that the MVIP method keeps symmetry better than pure low-order or unlimited method even for non-symmetric mesh movement.

Table 3. Velocity L_1 error and corresponding convergence rate for Sedov-like cyclic remapping.

remap	E_{64^2}	E_{128^2}	E_{256^2}	$\frac{64^2}{128^2}$	$\frac{128^2}{256^2}$
LO	$2.32{\cdot}10^{-2}$	$1.63{\cdot}10^{-2}$	$1.12{\cdot}10^{-2}$	1.4	1.5
unlim	$1.30{\cdot}10^{-2}$	$7.78{\cdot}10^{-3}$	$4.70{\cdot}10^{-3}$	1.7	1.7
MVIP	$1.37{\cdot}10^{-2}$	$8.21{\cdot}10^{-3}$	$4.84{\cdot}10^{-3}$	1.7	1.7

8. Conclusions

In this paper, we have presented a significant modification of the Vector Image Polygon (VIP) limiter for a piece-wise linear velocity reconstruction. Resulting remap of momentum and

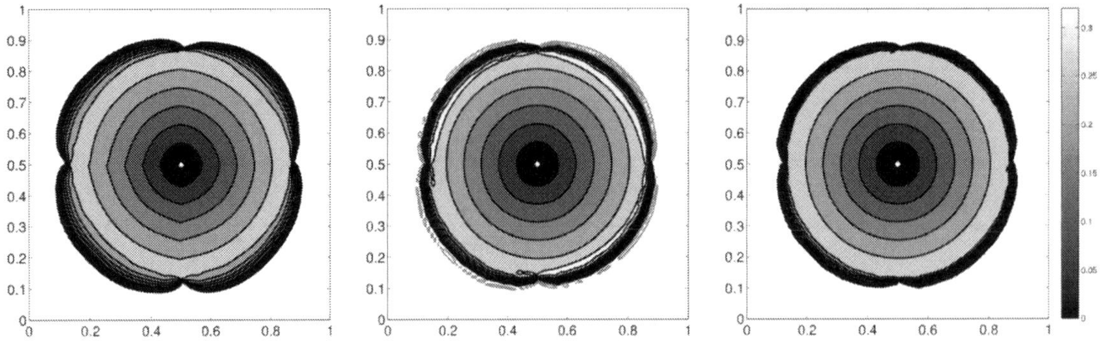

Figure 5. Contour plots of radial velocity component for cyclic remapping on a Cartesian grid. (a) low-order, (b) unlimited and (c) MVIP-limited method.

corresponding velocity are symmetric. Compared to the standard VIP method, the Modified VIP (MVIP) method shows local-bound preservation even for vector magnitude. However, there are still unresolved issues related to poor convergence for a case of small velocity gradients corresponding to almost constant fields on angularly under-resolved meshes. In this case, the VIP as well as MVIP limitations seem to be over-restrictive, which results in only first-order of convergence. We are also working on an alternative symmetry-preserving remap of momentum based on Flux Corrected Transport (FCT) ideas which overcomes this issue. Still, the Modified VIP method seems to be promising for application in ALE hydrodynamical codes.

Acknowledgments
This work was performed under the auspices of the National Nuclear Security Administration US DOE at LANL under Contract No. DE-AC52-06NA25396, and partially supported by ASCR and ASC Programs, the CTU grant SGS10/299/OHK4/3T/14, the Czech Science Foundation projects P205/10/0814 and by the Czech Ministry of Education project RVO 68407700.

References
[1] Hirt C W, Amsden A A and Cook J L 1974 *Journal of Computational Physics* **14** 227–253
[2] Liska R *et al.* 2011 *Finite Volumes for Complex Applications VI Problems & Perspectives* **4** 857–873
[3] Galera S, Maire P H and Breil J 2010 *Computers & Fluids* **46** 161–167
[4] Barlow A J 2008 *International Journal for Numerical Methods in Fluids* **56** 953–964
[5] Burton D E 1994 Consistent finite-volume discretization of hydrodynamics conservation laws for unstructured grids Tech. rep. LLNL UCRL-JC-118788
[6] Robinson A C *et al.* 2008 *Proceedings of the 46th AIAA Aerospace Sciences Meeting* AIAA-2008-1235
[7] Anderson R W, Elliott N S and Pember R B 2004 *Journal of Computational Physics* **199** 598–617
[8] Luttwak G and Falcovitz J 2011 *International Journal for Numerical Methods in Fluids* **65** 1365–1375
[9] Maire P H 2011 Contribution to the numerical modeling of inertial confinement fusion UB I, Habilitation.
[10] Margolin L G and Shashkov M 2003 *Journal of Computational Physics* **184** 266–298
[11] Pember R B and Anderson R W 2000 A comparison of staggered-mesh Lagrange plus remap and cell-centered direct Eulerian Godunov schemes for Eulerian shock hydrodynamics Tech. rep. LLNL UCRL-JC-139820
[12] DeBar R B 1974 Fundamentals of the KRAKEN code Tech. Rep. UCIR-760 Lawrence Livermore Laboratory
[13] Barth T and Jespersen D 1989 The design and application of upwind schemes on unstructured meshes Tech. Rep. AIAA-89-0366 AIAA, NASA Ames Research Center
[14] Maire P H, Loubère R and Váchal P 2011 *Commun. Comput. Phys.* **10** 940–978
[15] Liska R, Shashkov M, Váchal P and Wendroff B 2010 *Journal of Computational Physics* **229** 1467–1497
[16] Velechovsky J, Liska R and Shashkov M 2012 *Computers & Fluids* In press, http://dx.doi.org/10.1016/j.compfluid.2012.06.006

Journal of Physics: Conference Series **454** (2013) 012004 doi:10.1088/1742-6596/454/1/012004

Fluid-orbit coupling calculation for flight analysis of impulsively driven laser vehicle

Masayuki Takahashi, Naofumi Ohnishi

Department of Aerospace Engineering, Tohoku University, Sendai 980-8579, Japan

E-mail: mtakahashi@rhd.mech.tohoku.ac.jp

Abstract. Using a fluid-orbit coupling simulator, we numerically solve the three-dimensional Navier-Stokes equations with exchanging information of six-degree-of-freedom reactions for predicting impulsive flight motions of a laser propulsion vehicle driven by blast waves. By feedback of angular and translational velocities into the flowfield, pressure and viscous drags induced by the unsteady vehicle motion are introduced to provide precise motion analysis. In the impulsive-motion estimation of the laser-boosted vehicle, restoring forces and moments are underestimated if the vehicle motion effect is modeled using aerodynamic coefficients of steady flow. Also, a simple model using impulse data examined by experiments for predicting the impulsive motion is compared with our coupling approach which can reproduce instantaneous acceleration resulting from the interaction between the vehicle and the blast wave. Velocity overshoot is generated by evaluating sharp thrust through the coupling calculation, and the flight height becomes 6% larger than conventional prediction using the impulse data.

1. Introduction

For various purpose in science, business, and national security, many small artificial satellites are launched by a shared-ride piggyback transportation together with a main satellite [1]. However, the shared-ride system restricts the launch timing and injection orbit of the small satellite, and this is the main factor preventing easy planning of small- or middle-scale projects. Therefore, it is necessary to establish an exclusive launch system for small satellites at lower cost. Laser propulsion is a novel system for small-satellite launch. It saves fuel on the vehicle and increases the payload ratio [2–7]. In aero-driving type propulsion, the vehicle obtains thrust through interactions with a blast wave induced by repetitive laser pulse irradiated from a ground base. According to feasibility and cost studies of laser propulsion [8–10], launch cost can be reduced to a quarter of the present LOH/LH2 engine rocket by repaying the initial cost of the ground base through repetition launch of small-sized vehicles.

It is well known that the flight performance of the "lightcraft" proposed by Myrabo [3] is excellent among present laser propulsion vehicles. The interaction between the blast wave and a shroud of the lightcraft pushes the vehicle back toward the center of the laser beam if the incident laser has lateral misalignment against the vehicle axis, and so the lateral offset is reduced [11]. A type-200 spinning lightcraft of 50-g weight was launched and achieved 71-m altitude in 2001 [3].

In past studies of the lightcraft, some experiments were conducted to examine recentering and angular impulses for lateral offset with a single pulse [12, 13]. Also, flight simulations were performed with multiple pulses based on the impulse data for lateral offsets [11, 14]. In the modeling of the multiple-pulse flight, the vehicle velocity is incremented by adding the impulse data at the time of incidence of the pulse for simplicity of temporal integration of the time-dependent thrust, and the unsteady interactions with the strong blast wave are ignored in the motion analysis [11, 14]. However, the vehicle is rapidly accelerated during the unsteady interaction with shock waves; thus, a state-of-the-art flight analysis is required for more precise predictions of the vehicle motion using a fluid-orbit coupling approach.

Objective of the present study is analysing the unsteady motion of the lightcraft by a numerical code coupling with compressible hydrodynamics and flight orbit [15–17]. For the impulsive motion estimation, we compare the simple approach using impulse data with our coupling calculation which can reproduce the severe interaction between the vehicle and the blast wave.

2. Numerical Methods

Computational models for fluid-orbit coupling simulation are based on our previous papers [15–17], and we briefly introduce the numerical methods here.

2.1. *Flowfield calculation*

Flowfield around a spinning lightcraft is solved to estimate forces and moments acting on the vehicle. To predict an unsteady flowfield involving a laser-induced blast wave, the three-dimensional Navier-Stokes equations are numerically solved:

$$\frac{\partial \mathbf{Q}}{\partial t} + \frac{\partial (\mathbf{E} - \tilde{\mathbf{E}})}{\partial x} + \frac{\partial (\mathbf{F} - \tilde{\mathbf{F}})}{\partial y} + \frac{\partial (\mathbf{G} - \tilde{\mathbf{G}})}{\partial z} = \mathbf{0}, \tag{1}$$

where $\mathbf{Q}$ is the vector of conservative variables, $\mathbf{E}$, $\mathbf{F}$, and $\mathbf{G}$ are the inviscid flux vectors, and $\tilde{\mathbf{E}}$, $\tilde{\mathbf{F}}$, and $\tilde{\mathbf{G}}$ are the viscous flux vectors in the directions of x, y, and z, respectively.

Discretization is done through a cell-centered finite volume method. The AUSM-DV method [18] is employed for the numerical flux with the second-order MUSCL method [19]. The numerical flux is calculated with a local rotation matrix to introduce a general curvilinear coordinate [20,21]. In the MUSCL method, we choose the characteristic variables as interpolation quantities and use the minmod limiter to fulfill the TVD condition [19]. The viscous flux is estimated through a second-order central difference method, and the viscous coefficient is determined by Sutherland's formula [22]. The Spalart-Allmaras model [23] with an assumption of fully developed turbulence due to the axial spin of the lightcraft is employed to introduce the effect of the turbulent flow. Time integration for unsteady flows is performed through first-order explicit Euler method.

2.2. *Orbital calculation*

A vehicle's flight trajectory can be predicted by solving six-degree-of-freedom (6-DOF) equations of motion. The conservation of linear and angular momenta for a rigid body results in 6-DOF equations of motion expressed in the body-fixed coordinate system. The 6-DOF equations of motion for tracing the vehicle trajectory are given by

$$m(\dot{U} + QW - RV) = X_g + X_a, \tag{2}$$

$$m(\dot{V} + RU - PW) = Y_g + Y_a, \tag{3}$$

$$m(\dot{W} + PV - QU) = Z_g + Z_a, \tag{4}$$

$$I_{xx}\dot{P} + (I_{zz} - I_{yy})QR = L_g + L_a, \tag{5}$$

$$I_{yy}\dot{Q} + (I_{xx} - I_{zz})RP = M_g + M_a, \tag{6}$$

$$I_{zz}\dot{R} + (I_{yy} - I_{xx})PQ = N_g + N_a, \tag{7}$$

where m is the mass of the vehicle, U, V, and W are the velocities of the vehicle, P, Q, and R are the angular velocities of the vehicle (Fig. 1), X_a, Y_a, and Z_a are the aerodynamic forces, X_g, Y_g, and Z_g are the gravity forces, L_a, M_a, and N_a are the aerodynamic moments, L_g, M_g, and N_g are the gravity gradient torques [24] in the body-fixed coordinate system XYZ, and I_{xx}, I_{yy}, and I_{zz} are the moments of inertia for X-, Y-, and Z-directions, respectively.

When the blast wave interacts with the vehicle surface, the aerodynamic forces and the aerodynamic moments are estimated by integrating the surface pressure distribution and the viscous stress from the flowfield solution in the body-fixed coordinate system. We also introduce gravity forces and gravity gradient torques acting on the vehicle in addition to aerodynamic forces and moments. The gravity gradient torques generate the restoring moment because the distance between the center of the earth and each micro-element of the vehicle is different [24].

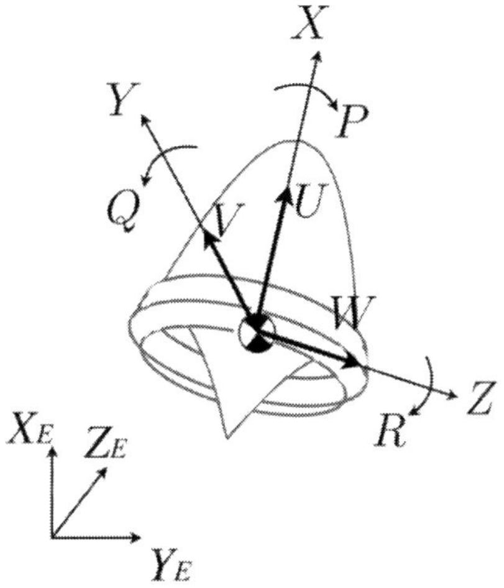

Figure 1. Definitions of Velocities and angular velocities in the body-fixed coordinate system XYZ [15, 16].

2.3. Coupling method

Because the vehicle is instantaneously driven by the strong blast wave, we must introduce an effect of vehicle motion on the flowfield. We combine the flowfield and the orbital calculation to introduce the vehicle motion effect in the body-fixed coordinate system.

In the orbital calculation, the aerodynamic forces and moments are estimated by integrating the surface pressure distribution and the viscous stress from the flowfield solution in the body-fixed coordinate system. For the flowfield computation, the time variations of vehicle velocities $\Delta\mathbf{U} = [\Delta U, \Delta V, \Delta W]^T$ and angular velocities $\Delta\boldsymbol{\Omega} = [\Delta P, \Delta Q, \Delta R]^T$ are fed back to the flowfield calculation at each step except for an axial high speed spin:

$$\mathbf{u}_{new} = \mathbf{u} - \Delta\mathbf{U} - \Delta\boldsymbol{\Omega} \times (\mathbf{r}_c - \mathbf{r}_g), \tag{8}$$

where $\mathbf{u} = [u, v, w]^T$ is the flowfield velocity, $\mathbf{u}_{new} = [u_{new}, v_{new}, w_{new}]^T$ is new velocity of the flowfield, and $\mathbf{r}_c$ and $\mathbf{r}_g$ are the position vectors for the center of each computational cell and the center of gravity of the vehicle, respectively.

For modeling effects of a rotating fluid due to the axial spin of the lightcraft, a rotational boundary condition is employed. Fluid velocity in the normal direction is set to 0 on the wall, and tangential velocity on the vehicle surface u_t is determined by an arm's length from the vehicle axis to the body surface R_w and the velocity of the gyro spin P [25]:

$$u_t = R_w P. \tag{9}$$

The pressure gradient for the normal direction on the wall boundary is estimated by

$$\frac{\partial p}{\partial n} = \rho \frac{u_t^2}{R_w}. \tag{10}$$

When the rotating fluid settles to a steady state, the laser beam is irradiated to the vehicle.

For comparison with the fluid-orbit coupling calculation, aerodynamic coefficients of a non-spinning lightcraft for steady flow are employed by the following forms [11]:

$$C_D = 0.3928\zeta^3 - 0.8463\zeta^2 + 0.0398\zeta + 0.731, \tag{11}$$

$$C_L = 0.1993\zeta^5 - 0.3317\zeta^4 - 0.5531\zeta^2 - 0.0316\zeta, \tag{12}$$

$$C_M = 0.1598\zeta^6 - 0.6619\zeta^5 + 0.8794\zeta^4 - 0.3585\zeta^3 - 0.0583\zeta^2 + 0.080\zeta, \tag{13}$$

where C_D, C_L, and C_M are the aerodynamic coefficients of drag, lift, and pitching moment, respectively. These coefficients are the functions of the angle of attack ζ.

3. Simulation Conditions

A type 200-5/6 lightcraft [3] is assessed in the present study. The vehicle diameter is 122 mm, and the vehicle mass is 32.5 g. Initial spin rate is set to 10,000 rpm based on past study [11]. The laser beam energy is deposited into the steady-state solution given by the rotational boundary using a ray-tracing method [26]. A 420-J laser beam is numerically divided into $\sim$3,000 rays for the ray-tracing, and the laser rays reflected by a parabolic mirror equipped on the lightcraft are focused inside the shroud. The laser intensity is set to be spatially uniform within a cross-section of 7.2-cm radius [11]. The efficiency of laser energy absorption by plasma is assumed to be 60% so as to adjust the generated impulse to the experimental data [11].

4. Results and Discussions

4.1. Coupling effects for angular and translational velocities

When the laser beam is irradiated to the vehicle with lateral y-offset of −50 mm and no angular offset, an asymmetric blast wave is generated from the focal point (ring) in the negative side of the y-axis (figure 2(a)). This strong blast wave rotates the vehicle in a clockwise direction. The counter-clockwise flow is consequently induced by the feedback of the angular velocity because the flowfield computation is conducted in the body-fixed coordinate system, and this counter-clockwise flow provides the restoring moment against the vehicle rotation (figure 2(b)).

Figure 3(a) shows that the angular velocity of the coupling calculation becomes about 0.4% smaller than the angular velocity in the case without the motion feedback. Non-coupling calculation is also conducted using the aerodynamic coefficients of steady flow (11)–(13). The angular velocity estimated by the coupling calculation becomes smaller in comparison with the case using the aerodynamic coefficients, that is, the restoring moments obtained by the aerodynamic coefficients are underestimated because the moment amplitude is not determined based on the rotational velocity of the vehicle. If the vehicle has no translational velocities, we can not obtain the restoring moments using the aerodynamic coefficients, and therefore, the coupling calculation is effective for predicting the vehicle rotation precisely.

The translational velocities in the body-fixed coordinate system are also fed back to the flowfield for modeling the pressure and viscous drags induced by the vehicle moving. The resulting vehicle velocity becomes small as compared with the non-coupling cases with and without aerodynamic coefficients (figure 3(b)). In case the aerodynamic coefficients are employed, the drags caused by the vehicle motion are assessed as smaller size; thus, the fluid-orbit coupling calculation is needed for an accurate estimation of rapidly accelerated reaction.

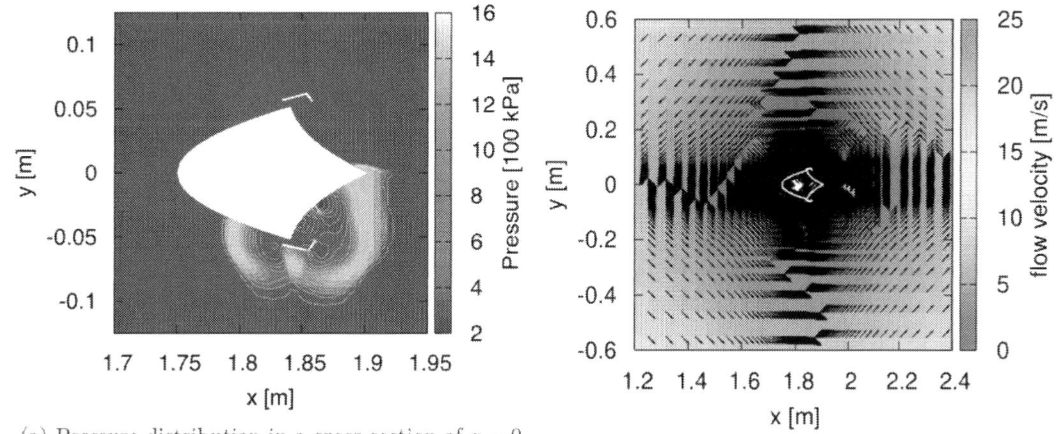

(a) Pressure distribution in a cross-section of $z = 0$ at 60 μs

(b) Velocity vector of flowfield at 1000 μs

Figure 2. Pressure distribution and velocity vector for y-offset of -50 mm and no angular offset.

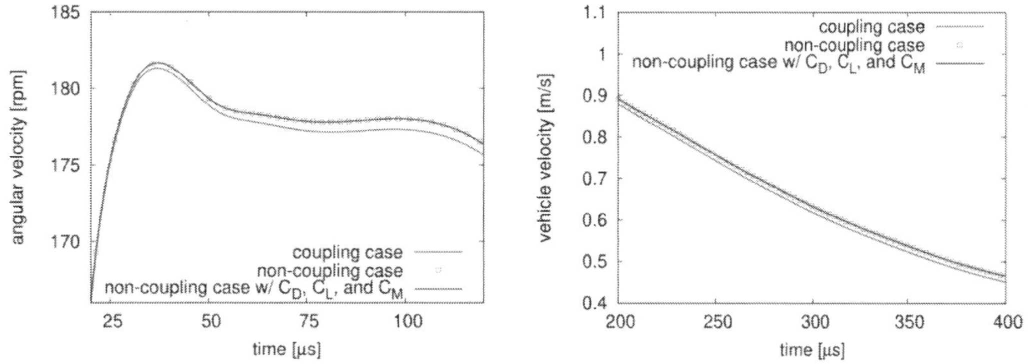

(a) Time evolutions of angular velocity $\sqrt{Q^2 + R^2}$

(b) Time evolutions of translational velocity U

Figure 3. Time evolutions of angular and translational velocities in the coupling case and non-coupling cases with and without aerodynamic coefficients.

4.2. Comparison between impulse data model and coupling calculation

In conventional schemes for modeling the impulsive vehicle motion, the impulse data examined by the experiments is simply added to the velocity at each time step to avoid the temporal integration of time-dependent thrust based on an assumption that the blast wave interaction finishes promptly. At the laser incident timing, the vehicle velocity is updated by the linear equations of motion as follows [11, 14]:

$$m\mathbf{U}_1 = m\mathbf{U}_0 + \mathbf{I}, \tag{14}$$

where $\mathbf{U}_1$ is the updated velocity vector, $\mathbf{U}_0$ is the present velocity vector, $\mathbf{I}$ is the impulse vector examined by experiments, and m is the vehicle mass. By the simple approach using the impulse data, sharp thrust during the blast wave interaction is ignored. However, because the vehicle is rapidly accelerated during the interaction with the shock wave, it is necessary to integrate the unsteady force for an accurate prediction of the impulsive reaction. We compare the simple model shown by (14) with the motion prediction based on the fluid-orbit coupling approach.

A single pulse is irradiated with no lateral and angular offsets. Using our hydrodynamic calculation code, impulse data is estimated by the integration until the blast wave interaction vanishes (1000 μs), and this impulse is added to initial condition in solving the equations of motion using the aerodynamic coefficients. On the other hand, the fluid-orbit coupling calculation is conducted until 1000 μs by evaluating the instantaneous force of the shock wave to compare with the simple model.

A positive thrust is obtained when the shock wave touches the rear surface of the vehicle at 10 μs, and a positive force is continuously generated for a longer time (figure 4(a)). As a result, the vehicle is instantaneously accelerated, and the velocity overshoot occurs as compared with the simple model using the impulse data (figure 4(b)). After 800 μs, the vehicle velocity has an almost constant speed due to the vanishing of the blast wave interaction, while the velocity amplitude decreases gradually because of the gravity and aerodynamic drag. The flight height of the simple model is underestimated in comparison with the direct integration using the coupling simulator because the rapid acceleration is not reproduced in the simple case.At 1000 μs, the height of the coupling calculation becomes 6% larger than the simple model, and this prediction error can be serious if multiple pulses are irradiated into the vehicle to fly into 10–100 km-order altitude. It is better that the time-dependent force is directly integrated during the shock interaction regime by the coupling scheme for the accurate analysis of the vehicle motion.

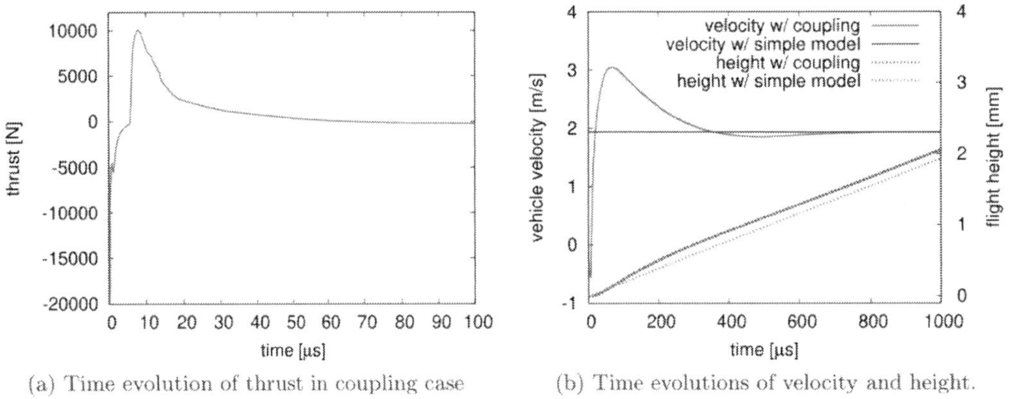

(a) Time evolution of thrust in coupling case

(b) Time evolutions of velocity and height.

Figure 4. Time evolutions of vehicle thrust, velocity, and height with coupling calculation and simple model based on impulse data.

5. Conclusions

Using a three-dimensional hydrodynamic code coupled with 6-DOF equations of motion, the blast wave propagation and the flight motion of a laser-boosted vehicle were simulated by introducing the vehicle motion effect. By the motion feedback to the flowfield, the angular and translational velocities become lower than in the non-coupling case using the aerodynamic coefficients of steady flow. The restoring forces induced by the moving vehicle are underestimated if the force is estimated through the aerodynamic coefficients. Additionally, the simple model using the impulse data was compared with the trajectory prediction by the coupling calculation. The coupling calculation yields rapid acceleration of the vehicle during the interaction with the blast wave, and the flight height becomes larger than the simply modeled case. It is essential to conduct the coupling calculation for the precise estimation of unsteady flying dynamics of the laser propulsion vehicle.

References

[1] Söderlund B 2006 *Master's Thesis, Luleå University of Technology*
[2] Kantrowitz A 1972 *Astronautics and Aeronautics* **9** 40–42
[3] Myrabo L N 2001 AIAA Paper 2001–3798
[4] Mead F B Jr, Messitt D G and Myrabo L N 1998 AIAA Paper 98–3735
[5] Scharring S, Eckel H A and Röser H P 2001 *Beamed Energy Propulsion: Seventh International Symposium on Beamed Energy Propulsion, AIP Conference Proceedings* **1402** 115–131
[6] Sasoh A 2000 AIAA Paper 2000–2344
[7] Mori K, Komurasaki K and Arakawa Y 2004 *Journal of Spacecraft and Rockets* **41** 887–889
[8] Katsurayama H, Komurasaki K and Arakawa Y 2002 AIAA Paper 2002-3778
[9] Katsurayama H, Komurasaki K and Arakawa Y 2009 *Acta Astronautica* **65** 1032–2041
[10] Phipps C R, Reilly P R and Campbell J W 2000 *Laser and Particle Beams* **18** 661–695
[11] Ballard C G, Anderson K S and Myrabo L N 2009 *Journal of Computational and Nonlinear Dynamics* **4** 041005-1–041005-8
[12] Libeau M and Myrabo L N 2005 *Beamed Energy Propulsion: Third International Symposium on Beamed Energy Propulsion, AIP Conference Proceedings* **766** 166–177
[13] Libeau M, Myrabo L N, Filippelli M and McInerney J 2002 AIAA Paper 2002–3781
[14] Kenoyer D A, Anderson K S and Myrabo L N 2008 *Beamed Energy Propulsion: Fifth International Symposium on Beamed Energy Propulsion, AIP Conference Proceedings* **997** 325–337
[15] Takahashi M and Ohnishi N 2011 *Beamed Energy Propulsion: Seventh International Symposium on Beamed Energy Propulsion, AIP Conference Proceedings* **1402** 132–141
[16] Takahashi M and Ohnishi N 2012 *AIAA Journal* **50** 2600–2608
[17] Takahashi M and Ohnishi N 2012 AIAA Paper 2012–3302
[18] Wada Y and Liou M S 2006 AIAA Paper 2006–1358
[19] Van Leer B 1979 *Journal of Computational Physics* **32** 101–136
[20] Toro E F 2010 *Riemann Solvers and Numerical Methods for Fluid Dynamics, 3rd ed.* (New York: Springer Press)
[21] Toro E F and Clarke J F 1998 *Numerical Methods for Wave Propagation* (London: Kluwer Academic Publishers)
[22] Hirsch C 1984 *Numerical Computation of Internal and External Flows, Vol. 1* (New York: John Wiley and Sons)
[23] Spalart P R and Allmaras S R 1992 AIAA Paper 92–0439
[24] Skullestad A 1999 *Modeling, Identification and Control* **20** 3–25
[25] Hirsch C 1984 *Numerical Computation of Internal and External Flows, Vol. 2* (New York: John Wiley and Sons)
[26] Born M and Wolf E 1959 *Principles of Optics, 1st ed.* (New York: Pergamon Press)

Journal of Physics: Conference Series **454** (2013) 012005 doi:10.1088/1742-6596/454/1/012005

Global Vlasov simulation on magnetospheres of astronomical objects

Takayuki Umeda[1], Yosuke Ito[1] and Keiichiro Fukazawa[2]
[1]Solar-Terrestrial Environment Laboratory, Nagoya University, Nagoya 464-8601, Japan
[2]Research Institute for Information Technology, Kyushu University, Fukuoka 812-8581, Japan

E-mail: umeda@stelab.nagoya-u.ac.jp

Abstract. Space plasma is a collisionless, multi-scale, and highly nonlinear medium. There are various types of self-consistent computer simulations that treat space plasma according to various approximations. We develop numerical schemes for solving the Vlasov (collisionless Boltzmann) equation, which is the first-principle kinetic equation for collisionless plasma. The weak-scaling benchmark test shows that our parallel Vlasov code achieves a high performance and a high scalability. Currently, we use more than 1000 cores for parallel computations and apply the present parallel Vlasov code to various cross-scale processes in space plasma, such as a global simulation on the interaction between solar/stellar wind and magnetospheres of astronomical objects.

1. Introduction

No less than 99.9% of the matter in the visible Universe is in the plasma state. The plasma is a gas in which a certain portion of the particles are ionized, which is considered to be the "fourth state" of the matter. The Universe is filled with plasma particles ejected from the upper atmosphere of stars. The stream of plasma is called the stellar wind, which also carries the intrinsic magnetic field of the stars. Our solar system is also filled with plasma particles from the Sun. Neutral gases in the upper atmosphere of the Earth are also ionized by a photoelectric effect due to absorption of energy from sunlight.

Computer simulation has now become an essential approach in studies of space plasma. Since the number density of plasma particles in space is low and the mean-free path (average distance between collisions of plasma particles) is large, the word "space plasma" is generally equivalent to collisionless plasma. For an example, the number density far above the Earth's ionosphere is $\sim 100\mathrm{cm}^{-3}$ or much less, and a typical mean-free path of solar-wind plasma is about 1AU (Astronomical Unit: the distance from the Sun to the Earth; 1AU$\sim$ 150,000,000km). In the case of long collision length, the scale of plasma variability is set by the Debye length, which varies from a few millimeters at the low-Earth orbit to a few kilometer or so in tenuous plasma. The plasma behaves as a dielectric medium with strong nonlinear interactions between plasma particles and electromagnetic fields. That is, the motion of plasma results in an electric current, electric currents modify the surrounding electromagnetic fields, and electromagnetic fields modify the motion of plasma, which results in a new electric current.

There are numerous types of self-consistent computer simulations that treat space plasma according to various approximations. The "global"-scale processes, such as structures and dynamics of solar/stellar winds, solar/stellar flares, and magnetospheres of stars, planets and satellites, are commonly described by magneto-hydro-dynamic (MHD), Hall-MHD and multi-fluid models. On the other hand, electron-scale processes, such as acceleration and heating of electrons, are described by the full kinetic model, i.e., the Maxwell equations and either the Newton-Lorentz equation or the Vlasov (collisionless Boltzmann) equation for both electron and ion particles. Hybrid methods treat ions as particles and electrons as a fluid for ion-scale processes.

Conventionally, MHD simulations have been widely used for numerical modeling of global-scale problems such as magnetospheres of stars and planets. The MHD or fluid equations are derived by taking the zeroth, first, and second moments of the kinetic Vlasov equations, with the zeroth, first, and second moments being the conservation laws of the density, momentum, and energy, respectively. Thus, the MHD simulations need diffusion coefficients, which are essentially due to kinetic processes that are eliminated in the framework of the MHD approximation. These coefficients are essentially due to first-principle kinetic processes that are eliminated in the framework of the fluid approximations. Recent high-resolution in-situ observations have also suggested that fluid scale and kinetic scale in space plasma are strongly coupled with each other, which is called cross-scale coupling. To understand the cross-scale coupling in space plasma, it is important to include full kinetics in global-scale simulations, which is a final goal of space plasma physics.

We develop numerical schemes for Vlasov simulations for practical use on currently-existing supercomputer systems. The Vlasov model treats "hyper"-dimensional (>3D) distribution functions in position and velocity spaces. It is still difficult to treat the 3P3V (three dimensions for position and three dimensions for velocity) phase space even with the recent supercomputers, because a computation with 40^6 grid points requires 160GB memory. With the help of recent development of new Vlasov solvers, however, there are several successful attempts of 2P2V and 2P3V hyper-dimensional Vlasov simulations for studying the Kelvin-Helmholtz instability[1] and the magnetic reconnection [2, 3], which are "local" processes taking place at boundary layers of magnetospheres. In the present study, on the other hand, we aim to perform a "global" Vlasov simulation of the magnetosphere of a small astronomical object.

2. Overview of Numerical Schemes

The Vlasov model solves the kinetics equations of space plasma, i.e., the Maxwell equations (1) and the Vlasov (collisionless Boltzmann) equation (2),

$$\left.\begin{aligned} \nabla \times \mathbf{B} &= \mu_0 \mathbf{J} + \frac{1}{c^2}\frac{\partial \mathbf{E}}{\partial t} \\ \nabla \times \mathbf{E} &= -\frac{\partial \mathbf{B}}{\partial t} \end{aligned}\right\} \tag{1}$$

$$\frac{\partial f_s}{\partial t} + \mathbf{v}\frac{\partial f_s}{\partial \mathbf{r}} + \frac{q_s}{m_s}\left[\mathbf{E} + \mathbf{v} \times \mathbf{B}\right]\frac{\partial f_s}{\partial \mathbf{v}} = 0 \tag{2}$$

where $\mathbf{E}$, $\mathbf{B}$, $\mathbf{J}$, ρ, μ_0, ϵ_0 and c represent electric field, magnetic field, current density, charge density, magnetic permeability, dielectric constant and light speed, respectively. The Vlasov equation (2) describes the development of the distribution functions by the electromagnetic (Lorentz) force, with the collision term in the right hand side set to be zero. The distribution function $f_s(\mathbf{r}, \mathbf{v}, t)$ is defined in position-velocity phase space with the subscript s being the species of singly-charged particles (e.g., $s = i, e$ for ions and electrons, respectively). The Maxwell equations and the Vlasov equation are coupled with each other via the current density $\mathbf{J}$ that satisfies the continuity equation for charge

$$\frac{\partial \rho}{\partial t} + \nabla \cdot \mathbf{J} = 0 \tag{3}$$

These equations are regarded as the "first principle" of the collisionless plasma.

It is not easy to solve hyper-dimensional Vlasov equation numerically, in terms of both computational resources and computational accuracy. The Vlasov equation (2) consists of two advection equations with a constant advection velocity and a rotation equation by a centripetal force without diffusion terms. To simplify the numerical time-integration of the Vlasov equation, we adopt a modified version of the operator splitting [4],

$$\frac{\partial f_s}{\partial t} + \mathbf{v}\frac{\partial f_s}{\partial \mathbf{r}} = 0 \tag{4}$$

$$\frac{\partial f_s}{\partial t} + \frac{q_s}{m_s}\mathbf{E}\frac{\partial f_s}{\partial \mathbf{v}} = 0 \tag{5}$$

$$\frac{\partial f_s}{\partial t} + \frac{q_s}{m_s}\left[\mathbf{v}\times\mathbf{B}\right]\frac{\partial f_s}{\partial \mathbf{v}} = 0 \tag{6}$$

Equations (4) and (5) are scalar (linear) advection equations in which $\mathbf{v}$ and $\mathbf{E}$ are independent of $\mathbf{r}$ and $\mathbf{v}$, respectively. We adopt a multidimensional conservative semi-Lagrangian scheme [4] for solving the multidimensional advection equations. In the full electromagnetic method, it is essential to use conservative schemes for satisfying the continuity equation for charge. With the multidimensional conservative semi-Lagrangian scheme, the continuity equation for charge (3) is exactly satisfied. In the present study, we compute the numerical flux by using the multi-dimensional advection scheme [4] with a positive, non-oscillatory and conservative limiter [5, 6] for stable time-integration of advection equations. Equation (6), on the other hand, is a multi-dimensional rotation equation which follows a circular motion of a profile at constant speed by a centripetal force. For stable rotation of the profile on the Cartesian grid system, the "back-substitution" technique [7] is applied. In addition, Maxwell's equations are solved by the implicit Finite Difference Time Domain (FDTD) method.

The velocity distribution function has both configuration-space and velocity-space dimensions, and defined as a hyper-dimensional (>3D) array, There are some additional communications overhead in parallelizing over the velocity-space dimensions since a reduction operation is required to compute the charge and current densities (the zeroth and first moments) at a given point in configuration space. Thus we adopt the "domain decomposition" only in configuration space, where the distribution functions and electromagnetic fields are decomposed over the configuration-space dimensions. This involves the exchange of ghost values for the distribution function and electromagnetic field data along boundaries of each procession element. The non-oscillatory and conservative scheme [5, 6] uses six grids for numerical interpolation, and three ghost grids are exchanged by using the "`Mpi_Sendrecv()`" subroutine in the standard message passing interface (MPI) library for simplicity and portability [8]. We also use the "`Mpi_Allreduce()`" subroutine for the convergence check on each iteration of the implicit FDTD method. Note that the code allows thread parallelization over the velocity-space dimensions via OpenMP.

3. Performance Evaluation

We conduct the performance measurement test of our parallel Vlasov code with a phase-space grid of $(N_{v_x}, N_{v_y}, N_{v_z}, N_x, N_y) = (30, 30, 30, 40, 20)$ on one core, which corresponds to a weak-scaling test with 1GB/core.

We use the following three systems for the benchmark test. The Hitachi HA8000 at the University of Tokyo has four AMD Opteron 8356 (four-core 2.3GHz) processors and 64GB memory per node. A total of 512 nodes are connected each other with four links of the Myrinet-10G interconnect. The Fujitsu FX1 at Nagoya University and JAXA has one SPARC64 VII (four-core 2.5GHz) processor and 32GB/16GB memory per node. A total of 768/3008 nodes are

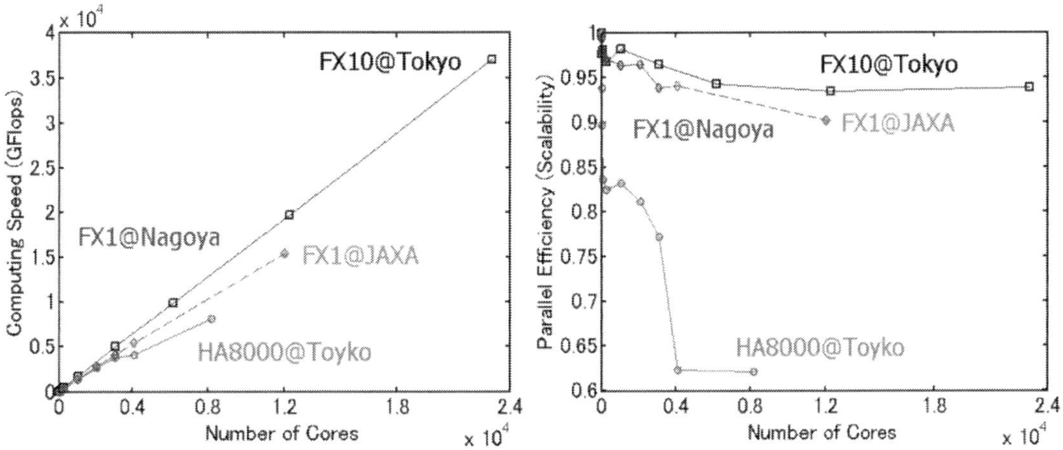

Figure 1. Performance of different supercomputer systems with scalar processors. (left) Computational speed and (right) scalability as a function of the number of cores. The circle, diamond, and square marks represent the results on the HA8000, FX1, and FX10, respectively.

connected each other with one link of the DDR InfiniBand. The Fujitsu FX10 at the University of Tokyo has one SPARC64 IXfx (sixteen-core 1.848GHz) processor and 32GB memory per node. A total of 4800 nodes are connected each other with the Tofu 6D mesh/torus interconnect.

Figure 1 shows the inter-node parallel performance on these systems. Note that the computational performance is measured by using the hardware counter installed on the Fujitsu FX1 and FX10 systems, and the performances on the Hitachi HA8000 are estimated based on the result on the FX1 and FX10. The peak performances (performance efficiencies) of the parallel Vlasov code are 8.19TFlops on the HA8000 with 8192 cores, 4.10TFlops/15.43TFlops on the FX1 with 3072/12,032 cores, and 37.05TFlops on the FX10 with 23,040 cores. We obtained a high scalability of ~90% on the FX1 and FX10 even with more than 10,000 cores. The Myrinet-10G internode-connection device of the HA8000 system also gives a high scalability of ~80% up to 2048 cores. However, the scalability becomes worse with more than 4096 cores because of the network bandwidth capacity.

4. Global Vlasov Simulation

4.1. Simulation Setup

We perform a two-and-half-dimensional (2.5D) Cartesian system in which spatial grids are taken in the two dimensional (2D) $x - y$ plane and velocity grids are taken in the three dimensional (3D) $v_x - v_y - v_z$ space (2P3V). There exists a dielectric circle (cylinder) as a small astronomical object at $r \equiv (x, y) = (0, 0)$, on the surface of which the electric charge accumulates by absorption of plasma particles. The radius of the object is set to be $R_S = 2r_i$, where r_i is the gyro radius of solar-wind ions. The simulation box is taken to be $-8R_S \leq L_x \leq 8R_S$ and $-12R_S \leq L_y \leq 12R_S$, and a total of $N_x \times N_y = 160 \times 240$ grid points are used. Thus the grid size is $\Delta \equiv \Delta x = \Delta y = 0.1R_S = 0.05r_i$. At all edges and corners of the simulation box, we applied so-called absorbing boundary conditions to suppress non-physical reflection of electromagnetic waves.

There exists a uniform plasma flow as the solar wind in the simulation box at the initial state. The plasma flow is directed in the x direction with a flow velocity $V_s = 5V_{ti}$. The uniform plasma flow is continuously injected from the left boundary, i.e., $x = -8R_S$, to the right. In order to save computational costs, electrons are assumed to be much heavier than the reality with the mass ratio $m_i/m_e = 25$. Assuming that the ions and electrons have the same temperature ($T_i/T_e = 1.0$), and thus the thermal velocity of electrons becomes $V_{te} = 5V_{ti}$. For the velocity space, we use $N_{v_x} \times N_{v_y} \times N_{v_z} = 60 \times 60 \times 60$ gird points. The range of the velocity space is taken to be $-15V_{ti} \le v_i \le 15V_{ti}$ and $-10V_{te} \le v_e \le 10V_{te}$ for all v_x, v_y and v_z directions, where V_t denotes the thermal velocity. Thus a total of 400 GB memory is used for the entire computation.

The solar wind carries an Inter-planetary Magnetic Field (IMF), which is the intrinsic magnetic field of the Sun. In the present study, an ambient magnetic field as the IMF is taken in the $x-z$ plane in order to include both ion gyro motion and magnetic convection in the $x-y$ plane. The magnitude of the ambient magnetic field is given by $B_0 \equiv \sqrt{B_{x0}^2 + B_{z0}^2} = m_i/q_i\omega_{ci}$ with $\omega_{ci}/\omega_{pi} = 0.02$, where ω_c and ω_p are the cyclotron and plasma angular frequency, respectively. Thus the ion inertial length and electron Debye length become $l_i = 2r_i$ and $\lambda_D = 0.02r_i$, respectively. The IMF is imposed in the 45° with respect to the x axis with $B_{x0} = B_{z0} = B_0/\sqrt{2}$. To carry the ambient magnetic field in the direction of plasma flow (x direction), we need a uniform electromotive (Faraday's inductive) force. A uniform external electric field is applied in the y direction with the magnitude of $E_{y0} = V_S B_{z0}$. The ion beta and Alfven Mach number are $\beta_i = 0.5$ and $M_A = 2.5$, respectively.

There exists an inner boundary at $r^2 = x^2 + y^2 = R_S$. We assume that the electric conductivity of the object is small, like an insulator. Numerically, the spatial numerical flux is set to be zero for $x^2 + y^2 < R_S$. Then, spatial advection in the object is not allowed and the electric charge of plasma particles onto the object accumulates on its surface. Note that we do not use any special treatment for electromagnetic waves at the inner boundary. That is, electromagnetic waves propagate freely into the object without conductance.

Suppose a typical velocity of the solar wind and the thermal velocity of ions are $V_S = 350$km/s and $V_{ti} = 70$km/s (i.e., about 50 eV), respectively, and a typical ion density is $N_i = 5\text{cm}^{-3}$, the ion plasma frequency and Debye length become about $f_{pi} = 468$Hz and $\lambda_{Di} = 24$m, respectively. Thus the radius of the object becomes about $R_S = 2.4$km, which corresponds to the typical size of an asteroid. The ion cyclotron frequency and the gyro radius become about $f_{ci} = 9.4$Hz and $r_i = 1.2$km, respectively.

In the present study, we performed two simulation runs without and with an intrinsic magnetic field of the object for Run A and Run B, respectively. In Run B, we assume a 2D dipole magnetic field as an intrinsic magnetic field,

$$\mathbf{B}_d = \frac{1}{2\pi}\left(2\mathbf{d}\frac{\mathbf{M}\cdot\mathbf{d}}{|\mathbf{d}|^4} - \frac{\mathbf{M}}{|\mathbf{d}|^2}\right) \tag{7}$$

where $\mathbf{M}$ denotes the dipole moment and $\mathbf{d} \equiv \mathbf{r} - \mathbf{d}_0$ denotes the distance from the center of the dipole ($\mathbf{d}_0$). In the present study, we use $\mathbf{d}_0 = (-0.6R_s, -0.3R_s)$ and $M_y = -6.4\pi B_0$ for Run B.

4.2. Simulation Result

Figure 2 shows the temporal development of the ion density together with magnetic field lines as a result of the interaction between the solar wind and the small dielectric object. It is found that a low-density region is formed on the nightside of the object in both runs. The result indicates the formation of wake tail by the absorption of solar-wind plasma on the surface of the body [9, 10]. As seen in figure 2, there are closed field lines at the dayside in Run B. A high-density region is formed at the dayside of the body, indicating the formation of bow shock by the interaction between the intrinsic magnetic field of the object and the solar wind with IMF. The structure of the wake tail at the nightside in Run B is modified from that in Run A, although the magnetic field lines at the nightside are mostly open.

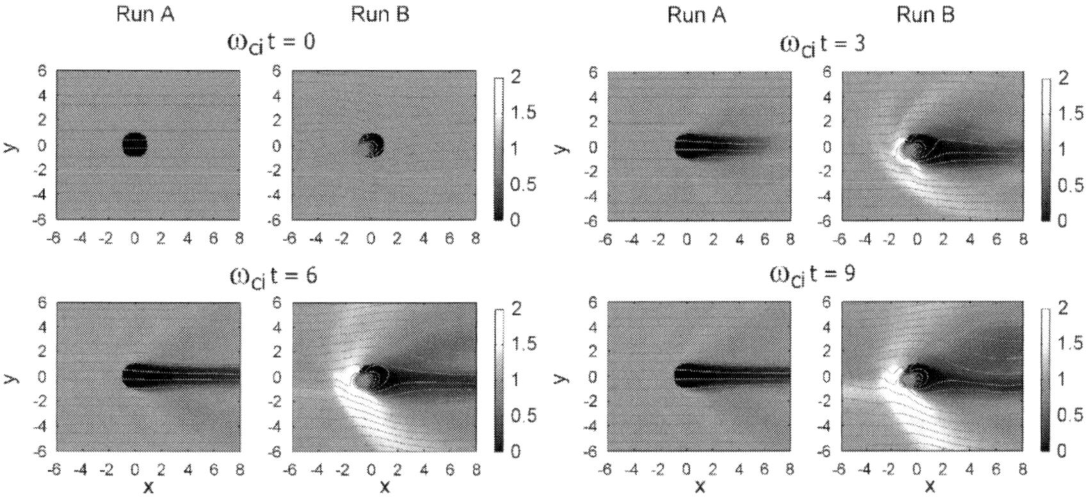

Figure 2. Spatial profiles of ion density together with magnetic field lines at different times for Runs A and B.

Figure 3 shows the reduced velocity distribution functions of ions, $f_i(y, v_x)$, $f_i(y, v_y)$, and ion density N_i at $x/R_S = 1.1$ and $\omega_{ci}t = 9$. In Run A, the ion density in the deep wake tail is < 0.01 of that of solar-wind ions. The $f_i(y, v_y)$ velocity distribution function shows that ions enter into the wake tail from both $+y$ and $-y$ directions. The entry of ions into the wake tail from the $+y$ side is due to the ion gyro motion. Since the thermal gyro radius of the solar-wind ions is $r_i = V_{ti}/\omega_{ci} = 0.5R_S$, only a small part (high-energy thermal tail) of solar-wind ions can enter into the deep wake tail. On the other hand, the entry of ions from the $-y$ side is due to the magnetic convection ($\mathbf{E} \times \mathbf{B}$ drift) [10]. At the nightside of the body, there appears a strong electric field directed toward the body ($-E_x$) due to the accumulation of electrons on the nightside surface of the body [11, 9]. This electric field is not Debye-shielded because of a low plasma density in the wake tail and becomes a motional (convective) electric field which moves magnetic fields in the $+y$ direction.

In Run B, on the other hand, the ion density in the deep wake tail is ~ 0.03 of that of solar-wind ions. The velocity distribution functions show that there exists a nonthermal ion component in a wide area of the downstream (tail) region. The nonthermal ions have a negative velocity in v_x and widely spread over v_y. We expect that origins of the nonthermal tail are the dayside bow shock and the dayside magnetopause where solar-wind ions are reflected backward by the magnetic pressure gradient. The reflected ions are picked up and carried downstream by IMF. Note that the pickup ions have a larger gyro radius than solar-wind ions and can enter into the deep wake tail by a trochoidal motion. Detailed analysis on these nonthermal ions is left as a future study. Note that the enhancement of ion flux in the wake tail of an unmagnetized small object is observed by the Kaguya spacecraft at the Moon [12].

5. Conclusion

In this paper, we have made performance measurements of a new Vlasov-Maxwell code on massively-parallel scalar computer systems, which are the Hitachi HA8000, the Fujitsu FX1 and the Fujitsu FX10 supercomputer systems. The weak-scaling benchmark test shows that our parallel Vlasov code achieves a high scalability (>90%) on more than 10,000 cores with recent inter-node connection devices.

We have also performed the "first" global Vlasov simulation of a magnetosphere of a small

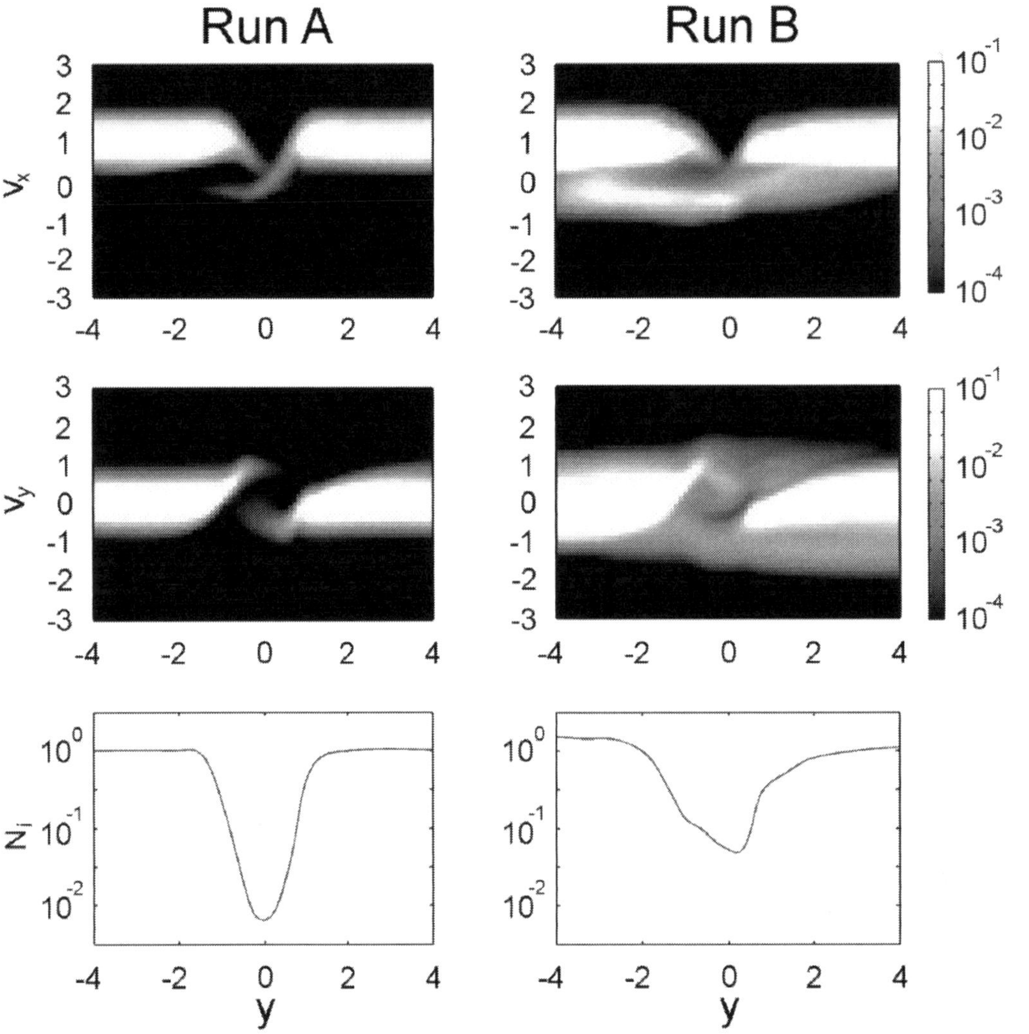

Figure 3. Reduced velocity distribution functions of ions, $f_i(y, v_x)$, $f_i(y, v_y)$, and ion density N_i at $x/R_S = 1.1$ and $\omega_{ci}t = 9$.

astronomical object with a spatial scale of ion gyro radius. When the spatial scale of the object and its magnetosphere is the same, the structures of the magnetosphere at the nightside may not be affected by the intrinsic magnetic field because most of magnetic fields at the nightside are open. However, the result indicates that the structure of the nightside wake tail is strongly modified by a weak intrinsic magnetic field at the dayside.

Finally, it should be noted that a 6D simulation (with both three spatial and three velocity dimensions: 3P3V) is essential for full understanding on the global structure and dynamics of magnetospheres. However this requires numerous computing resources with more than 50TB memory, and is left as a (far) future study.

Acknowledgments
The authors are grateful to Yasuhiro Nariyuki and Tatsuki Ogino for discussions. This work was supported by MEXT/JSPS under Grant-in-Aid for Young Scientists (B) No.23740367. The computational resource at Nagoya University is provided as a Nagoya University HPC program and a JHPCN program. The computational resource at JAXA is provided by Iku Shinohara. The computational resources at the University of Tokyo are provided as Large-Scale HPC Challenge programs and a JHPCN program. The global Vlasov simulation was performed on the DELL PowerEdge R815 at Solar-Terrestrial Environment Laboratory (STEL) in Nagoya University as a STEL computational joint program.

References
[1] Umeda T, Miwa J, Matsumoto Y, Nakamura T K M, Togano K, Fukazawa K and Shinohara I 2010 *Phys. Plasmas* **17** 052311
[2] Schmitz H and Grauer R 2006 *Phys. Plasmas* **13** 092309
[3] Umeda T, Togano K and Ogino T 2010 *Phys. Plasmas* **17** 052103
[4] Umeda T, Togano K and Ogino T 2009 *Comput. Phys. Commun.* **180** 365
[5] Umeda T 2008 *Earth Planets Space* **60** 773
[6] Umeda T, Nariyuki Y and Kariya D 2012 *Comput. Phys. Commun.* **183** 1094
[7] Schmitz H and Grauer R 2006 *Comput. Phys. Commun.* **175** 86
[8] Umeda T, Fukazawa K, Nariyuki Y and Ogino T 2012 *IEEE Trans. Plasma Sci.* **40** 1421
[9] Umeda T, Kimura T, Togano K, Fukazawa K, Matsumoto Y, Miyoshi T, Terada N, Nakamura T K M and Ogino T 2011 *Phys. Plasmas* **18** 012908
[10] Umeda T 2012 *Earth Planets Space* **64** 231
[11] Kimura S and Nakagawa T 2008 *Earth Planets Space* **60** 591
[12] Nishino M, Fujimoto M, Maezawa K, Saito Y, Yokota S, Asamura K, Tanaka T, Tsunakawa H, Matsushima M, Takahashi F, Terasawa T, Shibuya H and Shimizu H 2009 *Geophys. Res. Lett.* **36** L16103

Modelling migration in multilayer systems by a finite difference method: the spherical symmetry case

C I Hojbotă[1], V Toşa[1] and P V Mercea[2]

[1]National Institute for Research and Development of Isotopic and Molecular Technologies, 65-103 Donath, 400293 Cluj-Napoca, Romania

[2]Fabes Forschungs, 35 Schrangenhofstrasse, 80992 München, Germany

E-mail: valer.tosa@itim-cj.ro

Abstract. We present a numerical model based on finite differences to solve the problem of chemical impurity migration within a multilayer spherical system. Migration here means diffusion of chemical species in conditions of concentration partitioning at layer interfaces due to different solubilities of the migrant in different layers. We detail here the numerical model and discuss the results of its implementation. To validate the method we compare it with cases where an analytic solution exists. We also present an application of our model to a practical problem in which we compute the migration of caprolactam from the packaging multilayer foil into the food.

1. Introduction

Migration of substances from packaging to foods is an important issue for the industry due to the consumer protection regulations enforced by many industrialized countries (for example in the EU Regulation 10/2011 or the FDA regulations in the USA). The main concern is the existence of a dangerous chemical agents in the multi-layer packaging which migrate from one of the layers into the food affecting consumers' health. Migration can be determined experimentally and/or estimated by using appropriate theoretical models which allow the calculation of the migrant concentration in the food under specific contact conditions with the packaging. The development of such models allows estimating the migrant concentration in the product instead of performing costly laboratory tests. Modeling also helps laboratories to understand the physical and chemical processes behind the measured data and to estimate material parameters like diffusion and partition coefficients.

The migration model is essentially based on the diffusion equation assisted by partitioning: the ratio of migrant concentrations in two adjacent layers is imposed by the ratio of the migrant solubility in the two layers. Finite difference, finite element, finite volume, boundary element and semi-analytical methods have been developed to handle this problem (see [1] for a review). In particular finite difference methods proved to be simple enough to be suitable for planar geometries [2-4] and ware used to model packaging with plastic foils as well as for cylindrical geometries which ware developed to model cylindrical packaging [5] and water pipe cases [6].

A different problem is related to the study of the so-called master-batch additives. In this case the plastic foil which will subsequently used to pack the product is treated with chemical agents with the aim of modifying some of its properties (such as UV resistance, stiffness, scent, etc.) or color. The migrant is initially incorporated into tiny granules which are then added to the final plastic product. The substance starts to migrate from the spheres and slowly diffuses into the plastic until it reaches a uniform concentration.

In this paper we apply a finite difference method to solve the migration problem for spherical geometry in a multi-layer configuration. We will formulate the problem and present the method used for solving it in section 2, while in section 3 we validate the procedure as we compare it to the analytical form. In the last section we draw the conclusions and suggest further developments and tests for our method.

2. The model

The problem presentation starts from Fick's law of diffusion: for a concentration whose spatial profile is known at $t = 0$, the equation gives the concentration at time t given the diffusion coefficient D.

$$\frac{\partial c}{\partial t} = D\nabla \cdot (\nabla c) \tag{1}$$

Transforming into spherical coordinates and assuming isotropic medium for all directions of r we get the equation for our model :

$$\frac{\partial c}{\partial t} = D\left(\frac{\partial^2 c}{\partial r^2} + \frac{2}{r} \cdot \frac{\partial c}{\partial r}\right) \tag{2}$$

To solve the diffusion equation numerically we perform the following steps : (i) we define a non-uniform numerical grid (ii) we use Lagrange polynomials to approximate the differences at mesh points; (iii) we discretize the diffusion equation using finite differences; (iv) eventually we solve the resultant system of algebraic equations, having the following form:

$$-\theta g_{i-1} c_{i-1}^{n+1} + (1-\theta g_i) c_i^{n+1} - \theta g_{i+1} c_{i+1}^{n+1} = -\bar{\theta} g_{i-1} c_{i-1}^{n} + (1-\bar{\theta} g_i) c_i^{n} - \bar{\theta} g_{i+1} c_{i+1}^{n}, \tag{3}$$

with c_i^n representing the concentration of the migrant for the mesh point i and time $n\delta t$. The g_i coefficients are obtained from the Lagrange polynomials.

Discretization also implies deriving the equations at the interface between adjacent layers (dependent on the partition factor), as well as for boundary conditions. The partitioning between two adjacent layers i and j impose that the concentration c at the interface obeys the condition $k_{ij} = c_i/c_j$. The algebraic equations for these grid points were obtained by applying the fictitious point method as described in [7]. The model was implemented on a computer program and compared with the analytical solution given by [8] for a particular case of initial and boundary conditions. The validation is presented in the next section.

3. Testing and Results

3.1. *Validation*

The next step was to validate the numerical method used to solve the diffusion problem. To do this we found cases in which the problem is analytically tractable, for example in Crank's book [8] for different initial and boundary conditions. We choose the case of migration from a well-stirred solution into a medium which surrounds the sphere uniformly and symmetrically. The case corresponds to a migration in a two-layered configuration with a partition coefficient K between them. We consider the sphere with a uniform distribution of concentration C_0 and the surrounding medium free of migrant.

The expression for the amount of migrant at time t, as a fraction of the mass found at in infinitely long period of time M_∞ is given by the following equation:

$$\frac{M_t}{M_\infty} = 1 - \sum_{n=1}^{\infty} \frac{6\alpha(\alpha+1)exp[-Dq_n^2 t/a^2]}{9+9\alpha+q_n^2\alpha^2}, \tag{4}$$

where q_n are the the non-zero roots of the equation

$$\tan(q_n) = \frac{3q_n}{3+\alpha q_n^2}. \tag{5}$$

Although (4-5) require numerical methods to find the roots and to sum up the series, in the following we will to refer to it as the analytical solution.

The parameter $\alpha = V_{medium}/(K \cdot V_{sphere})$ represents the ratio of the volumes of the medium and the sphere taking into account α. The partition coefficient$K$ has the significance that the concentration of the migrant found in the surrounding medium at the interface with the sphere, is K times the concentration found in the sphere at the interface between the two layers.

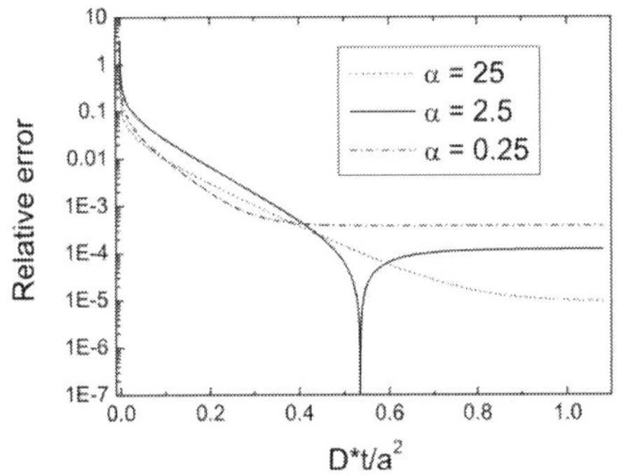

Figure 1. Relative error as a function of Dt/a^2, for different values of α up to large period of time.

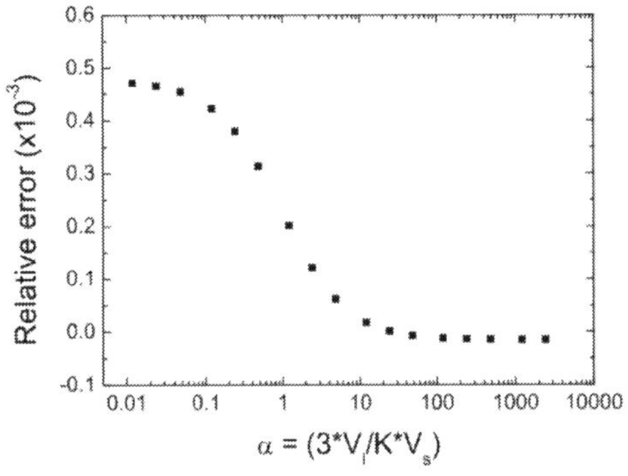

Figure 2. Relative error between analytical and numerical results, as a function of the parameter α, computed for large values of t.

We calculated numerically and analytically the total quantity of migrant found in the surrounding medium as a function of time t. The other parameters are a=1 cm (the radius of the sphere), C_0=10 ppm, D=$10^{-9}cm^2/s$. In general we obtained a very good agreement between the numerical and analytical results although the earlier the time the larger the difference in concentration. This aspect is illustrated in more detail in figure 1 where we represent the relative error (in absolute values) between the two sets of data for different values of α. One can note a decrease of the relative error towards a stable value as the time increases to very large values. One can see a cusp in the case where we chose the partition coefficient $\alpha = 25$ which occurs due to a change of sign of the error.

Since figure 1 contains the curves for different values of α we can plot directly the dependency of the error on α at a given time. We illustrate this in figure 2 where it can be clearly seen how the value of α affects the relative error of the two cases. For α values up to 10 the error values are of order 10^{-3} while as $\alpha > 10^2$ the relative error becomes insignificant, with values very close to zero. The plot is made for time value chosen as $t = 3 \cdot 10^5$ hours.

3.2. Application to food packaging

The migration module was integrated into the FP7 program "Flavorings, Additives and Food Contact Materials Exposure Task" (FACET) with the role of computing the quantity of migrant that ends up in food from food contact materials, especially packaging. The presented model development is dedicated to spherical symmetry which can be found in various food packagings (cheese, chocolate, etc.)

As an example we present here the migration of caprolactam from a 3-layered polymer packaging composed of polyamide (PA), glue and polyethylene (PE) into the food sample (in this case, Gouda cheese).

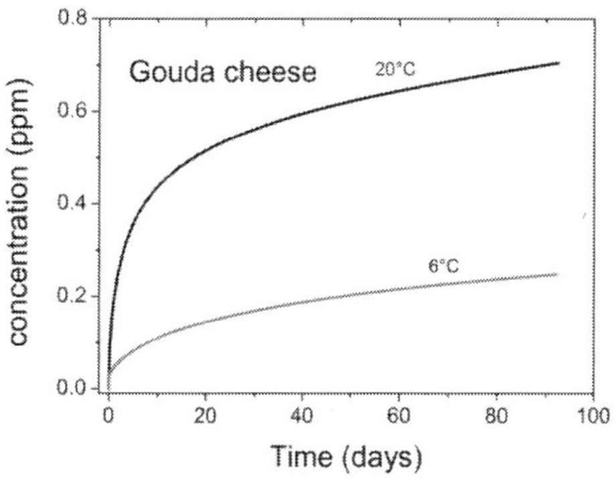

Figure 3. Amount of caprolactam found in cheese at different periods of time, for 6°C (shelf temperature) and 20°C (room temperature).

Caprolactam is a chemical compound having the chemical formula $(CH_2)_5C(O)NH$. It is used as an additive in plastics and polymers, and in the practical case it is initially found in the external layer made of polyamide. The maximum allowed concentration of caprolactam in food, as indicated by the European directive 2002/72/EC, is 15 mg/kg in food.

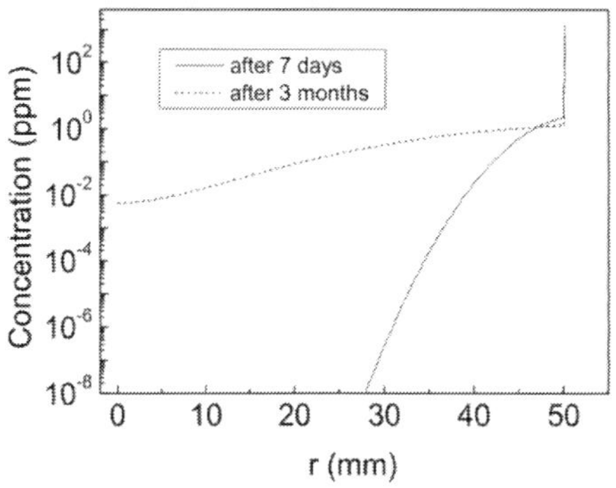

Figure 4. Spatial profile of the concentration in cheese and packaging foil after 90 days.

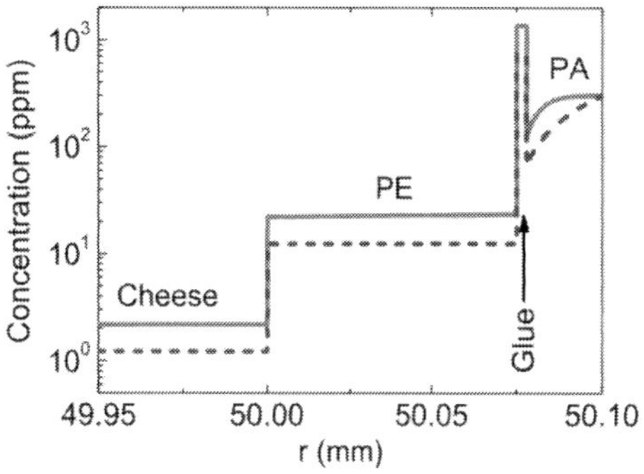

Figure 5. Detailed spatial profile of the concentration for the cheese/packaging interface.

The values for partition and diffusion coefficients were estimated from measurements or, when not available, from the Piringer formula [7]. We estimated the migration at 6°C (shelf temperature) and 20°C (room temperature). The results are illustrated in figure 3, showing the variation in time of the caprolactam concentration in food. We see that for both temperatures the final concentration after 90 days is lower than the specific migration limit (SML).

If we analyze the spatial profile of caprolactam concentration in food, see figure 4 and figure 5 we note that for regions in contact with the packaging, SML in cheese is exceeded, reaching more than 20 *ppm* after 7 days of migration. This result shows the utility of the numerical modeling as one can estimate the amount of food in which SML is exceeded while the laboratory experiments give only the average amount of caprolactam extracted from the food.

4. Conclusions

In this paper we presented a numerical model based on finite differences to solve the problem of chemical impurity migration within a multilayer spherical system. A practical application of this model is in the case of chemical species migration from polymer packaging into food products.

In conclusion, we can state that the model is well suited to simulate the migration from spherical objects in multilayer systems, as it can be seen from the error plots and the scenario for foil migration. Due to the difficulty and limited possibilities of the analytical treatment, this numerical model is more flexible regarding the parameters of the real system such as diffusion and partition coefficients or layer thickness, and it has the advantage that it can be generalized to an arbitrary number of layers.

The model can be further extended to incorporate time and space dependent diffusion coefficients as well as time dependent partition coefficients. For the planar geometry it was integrated in the FACET software [9] which calculates the exposure of the European customer to the chemicals found in food contact materials, in particular packaging. Including cylindrical and spherical geometries is also intended.

Acknowledgments

We gratefully acknowledge partial financing from FP7 Project *Flavorings, Additives and Food Contact Materials Exposure Task* (FACET) and from project 1/2012/PNII financed by Romanian Authority for Scientific Research (ANCS).

References

[1] Helmroth E, Rijk R, Dekker M and Jongen W 2002 Trends in Food Science and Technology **13** 102
[2] Dai W, Li Q, Nassar R and Shen L 2004 *International Journal for Numerical Methods in Engineering* **59** 493
[3] Tosa V, Kovacs K, Mercea P and Piringer O 2008 *AIP Conference Proceedings* **1048** 802
[4] Tosa V and Kovacs K 2009 *Journal of Physics: Conference Series* **182** 012042
[5] Pennarun P Y, Ngono Y, Dole P and Feigenbaum A 2003 *Journal of Applied Polymer Science* **92** 2859
[6] Denberg M, Denberg M, Arvin E and Hassager O 2007 *Journal of Water Supply: Research and TechnologyAQUA* **56** 435
[7] Piringer O, Baner A 2008 *Plastic Packaging* (Wiley-VCH, Weinheim)
[8] Crank J 1975 *The Mathematics of Diffusion* (New York: Oxford University Press)
[9] http://www.ucd.ie/facet

Journal of Physics: Conference Series **454** (2013) 012007 doi:10.1088/1742-6596/454/1/012007

Kinetic energy spectrum of low-Reynolds-number turbulence with polymer additives

Takeshi Watanabe and Toshiyuki Gotoh

Graduate School of Engineering, Department of Scientific and Engineering Simulations, Nagoya Institute of Technology, Gokiso, Showa-ku, Nagoya 466-8555, Japan

E-mail: watanabe@nitech.ac.jp, gotoh.toshiyuki@nitech.ac.jp

Abstract. The effects of polymer additives on the behavior of the kinetic energy spectrum in isotropic decaying turbulence were numerically investigated by hybrid Eulerian-Lagrangian simulations making full use of large-scale parallel computation. The kinetic energy spectrum was found to obey the power-law $E(k) \sim k^{-\alpha}$ in the range below the Kolmogorov length l_K when the turbulence decayed. The exponent α satisfied $4 < \alpha < 5$ and decreased with the increase in the Weissenberg number W_i. The value of α obtained by the largest W_i run was close to the values obtained in previous experimental and numerical studies on elastic turbulence, which is characterized by a larger W_i and a Reynolds number of less than unity. The relationship between the results of the present study and the results obtained for elastic turbulence is also discussed.

Small amounts of polymers added to a fluid significantly affect large-scale flow structures. Turbulence drag reduction [1, 2] is one of the most important phenomena related to polymer solution flows because this phenomenon has a broad range of industrial applications involving turbulent transport. On the other hand, the polymer solution flow characterized by both a smaller Reynolds number and higher elasticity indicates chaotic fluctuations in space and time. This phenomenon is referred to as elastic turbulence [3]. It is vital to examine the meso-scale dynamics of polymers in order to clarify the peculiar nature of polymer solution flows because the origin of the turbulence modifications or the origin of the appearance of random motions for low-Reynolds-number flow are due to the strong elasticity of polymers in flows.

Dilute polymer solution flows have been extensively investigated using constitutive equations, such as the Oldroyd-B or FENE-P models [4]. These are the evolution equations for the tensor field representing the polymer conformation and are constructed based on a simple polymer model. The constitutive equations are widely used in numerical studies of the above-mentioned phenomena [2] because of the ease of handling the polymer effects on fluid motions for a reasonable computational cost.

Another method by which to numerically investigate the polymer solution flow is a hybrid approach,*e.g.* [5]: the fluid motion is computed using the Navier–Stokes (NS) equations, while the polymer dynamics are determined by molecular or Brownian dynamics simulations (BDSs) using an appropriate polymer model [6]. In a previous study [7], we developed a two-way coupled simulation method using this approach, in which a BDS for a dumbbell model coupled with a direct numerical simulation (DNS) of turbulent flow was performed using large-scale parallel computations. A number of dumbbells, on the order of ten billion ($O(10^{10})$), were dispersed in a turbulent flow, and their advection and deformation were tracked during the time evolution of the system. We then examined the modification of the decaying turbulence for various polymer concentration and Weissenberg number, $W_i = \tau/\tau_K$, which is the ratio of the polymer relaxation time τ to the Kolmogorov time $\tau_K \equiv (\nu_s/\varepsilon)^{1/2}$ for the smallest eddy in turbulence (ν_s and ε are respectively the kinematic viscosity of the solvent fluid and the average rate of the energy dissipation of turbulence).

Fundamental quantity characterizing the fluctuations of the solvent fluid is the kinetic energy spectrum which is defined by

$$E(k,t) = \sum_{\boldsymbol{k}}{}' \frac{1}{2}|\hat{\boldsymbol{u}}(\boldsymbol{k},t)|^2 \tag{1}$$

in terms of the Fourier amplitude $\hat{\boldsymbol{u}}(\boldsymbol{k},t)$ where $\sum'_{\boldsymbol{k}}$ is the summation taken over the spherical shell within $k - \Delta k/2 < |\boldsymbol{k}| \leq k + \Delta k/2$ in the wavenumber space. One of the most remarkable results reported in [7] is the fact that when turbulence sufficiently decays and $W_i > 1$ the kinetic energy spectrum obeys the power-law decay of $E(k,t) \sim k^{-\alpha}$ with $\alpha = 4.7$ at wavenumbers higher than the Kolmogorov wavenumber $1/l_K \equiv (\varepsilon/\nu_s^3)^{1/4}$. On the other hand, previous studies on elastic turbulence revealed experimentally that $E(k) \sim k^{-3.5}$ [3] and by the DNS of the Oldroyd-B model under a two-dimensional Kolmogorov flow that $E(k) \sim k^{-3.8}$ [8]. Thus, the power-law exponent obtained in a previous study (approximately 4.7) is larger than the exponents obtained for elastic turbulence, although this result is consistent with the theoretical prediction using the simplified viscoelastic model, where the scaling exponent α must be greater than 3 [9].

The intrinsic difference between the previous study [7] and studies on elastic turbulence [3, 8] is that the former was performed with a smaller W_i and a larger Reynolds number, on the order of $R_\lambda \simeq 3.5$, than in [3, 8], where R_λ is the Taylor microscale Reynolds number defined by (9). This raises the question of how the scaling behavior of the kinetic energy spectrum varies with the variation of parameters W_i and R_λ in the hybrid simulation.

The purpose of the present study is to examine the scaling behavior of the kinetic energy spectrum $E(k,t)$ in isotropic decaying turbulence with polymer additives for a wider parameter space. We investigate the power-law behavior of $E(k,t)$ for cases of larger W_i values than those in our previous study [7]. This means that the system more closely resembles elastic turbulence.

The dumbbell model used in the present study is given by

$$\frac{d\boldsymbol{R}^{(n)}}{dt} = \boldsymbol{u}_1^{(n)} - \boldsymbol{u}_2^{(n)} - \frac{1}{2\tau} f\left(\frac{|\boldsymbol{R}^{(n)}|}{L_{max}}\right)\boldsymbol{R}^{(n)} + \frac{r_{eq}}{\sqrt{2\tau}}\left(\boldsymbol{W}_1^{(n)} - \boldsymbol{W}_2^{(n)}\right), \tag{2}$$

$$\frac{d\boldsymbol{r}_g^{(n)}}{dt} = \frac{1}{2}\left(\boldsymbol{u}_1^{(n)} + \boldsymbol{u}_2^{(n)}\right) + \frac{r_{eq}}{\sqrt{8\tau}}\left(\boldsymbol{W}_1^{(n)} + \boldsymbol{W}_2^{(n)}\right), \quad \boldsymbol{u}_\alpha^{(n)} \equiv \boldsymbol{u}(\boldsymbol{x}_\alpha^{(n)}(t), t), \tag{3}$$

where $\boldsymbol{R}^{(n)}(t)$ and $\boldsymbol{r}_g^{(n)}(t)$ are, respectively, the end-to-end vector and the center-of-mass vector of the n-th dumbbell. We adopt the finitely extensible nonlinear elastic (FENE) model $f(z) = 1/(1-z^2)$ for the elastic force of a dumbbell. In equation (2), L_{max} is the maximum extension length of the dumbbell because $f(z \to 1) = \infty$. The term $\boldsymbol{W}_{1,2}^{(n)}(t)$ indicates a random force representing the Brownian motion of particles in the solvent fluid, which obeys Gaussian statistics with a white-in-time correlation of $\langle W_{\alpha,i}^{(n)}(t)\rangle = 0$ and $\langle W_{\alpha,i}^{(m)}(t) W_{\beta,j}^{(n)}(s)\rangle = \delta_{\alpha\beta}\delta_{ij}\delta_{mn}\delta(t-s)$, where $\langle\cdots\rangle$ denotes the ensemble average. The subscripts α, β, i, j, n, and m take the values $(\alpha,\beta) = 1$ or 2, $(i,j) = 1$, 2, and 3, and $(n,m) = 1, 2, \cdots, N_t$, respectively. Moreover, δ_{ij} denotes the Kronecker delta, and $\delta(t)$ is the Dirac delta function. The constants $\tau \equiv \zeta/4k$ and $r_{eq} \equiv \sqrt{k_B T/k}$ are, respectively, the relaxation time and the equilibrium length of the dumbbell under $\boldsymbol{u}(\boldsymbol{x},t) = \boldsymbol{0}$. Here, k is the spring constant, and $\zeta \equiv 6\pi\nu_s\rho_s a$ (ρ_s is the density of the solvent fluid). Also, k_B and T are the Boltzmann constant and temperature, respectively.

The turbulent velocity field obeys the continuity equation for an incompressible fluid and the NS equations

$$\nabla \cdot \boldsymbol{u} = 0, \quad \frac{\partial \boldsymbol{u}}{\partial t} + \boldsymbol{u} \cdot \nabla \boldsymbol{u} = -\nabla p + \nu_s \nabla^2 \boldsymbol{u} + \nabla : \boldsymbol{T}^p, \tag{4}$$

where $p(\boldsymbol{x},t)$ is the pressure field. Here, ρ_s is set to unity and is equal to the density of bead ρ_p representing the polymer, and $\boldsymbol{T}^p(\boldsymbol{x},t)$ is the polymer stress tensor due to the force acting on the fluid from the dispersed dumbbells and is defined by

$$T^p_{ij}(\boldsymbol{x},t) = \frac{\nu_s\eta}{\tau}\left(\frac{L^3_{box}}{N_t}\right)\sum_{n=1}^{N_t}\left[\frac{R_i^{(n)}R_j^{(n)}}{r_{eq}^2} f\left(\frac{|\boldsymbol{R}^{(n)}|}{L_{max}}\right) - \delta_{ij}\right]\delta(\boldsymbol{x} - \boldsymbol{r}_g^{(n)}), \tag{5}$$

where N_t is the total number of polymers required to realize the experimental situation, $\eta \equiv (3r_{eq}/4a)^2\Phi_V$ is the zero-shear viscosity ratio of the polymer ν_p to the solution viscosity ($\eta = \nu_p/\nu_s$), and $\Phi_V \equiv (8\pi N_t/3)(a/L_{box})^3$ represents the volume fraction of the ensemble of dumbbells.

The numerical simulations of equations (2) through (5) are performed in a periodic box with periodicity $L_{box} = 2\pi$ using the pseudo-spectral method in space and the second-order Runge–Kutta method in time. A total of 128^3 grid points are set to solve the NS equations. The total number of dumbbells in computation is $N_{comp} = N_t/b = 5.04 \times 10^8$, where b is the artificial parameter representing the number of replica dumbbells. This reduces the computational cost of the computation of $N_t = O(10^{13})$ dumbbells [7]. The initial velocity field is given by the random solenoidal field obeying Gaussian statistics, with an energy spectrum

$$E(k,0) = 16\sqrt{\frac{2}{\pi}}\left(\frac{u_0^2}{k_0}\right)\left(\frac{k}{k_0}\right)^4 \exp\left(-2\left(\frac{k}{k_0}\right)^2\right) \quad (u_0 = 1, k_0 = 2). \tag{6}$$

In this setting, the initial value of R_λ is $R_\lambda(0) = 52$. The dumbbells are uniformly and randomly distributed over the computational domain. The initial configuration of each dumbbell is set according to $\boldsymbol{R}^{(m)}(0) = \sqrt{3}r_{eq}\boldsymbol{n}^{(m)}$, where $\boldsymbol{n}^{(m)}$ is a random unit vector, which is isotropically distributed.

Parallel computations are performed in order to evaluate the convection and deformation of the dispersed dumbbells. The program is parallelized using Message Passing Interface (MPI). The maximum number of MPI processors is 64, and the total number of dumbbells in computation, N_{comp}, is divided into 63 groups, with each group assigned one processor that computes the temporal evolution of the group. One processor is also assigned to perform the DNS of the solvent fluid. Figure 1 shows the schematic diagram of the parallel computation in the hybrid simulations. The field data of the fluid velocity $\boldsymbol{u}(\boldsymbol{x},t)$ and $\boldsymbol{u}(\boldsymbol{x} + \Delta\boldsymbol{x}/2, t)$ from the process dedicated to the time integration of the turbulence is transferred to all of the other 63 processes for the computation of the dumbbells. The fluid velocity at the bead positions is then interpolated using the TS13 scheme [10], whereas the polymer stress term is evaluated as follows:

(i) The polymer stress field $\boldsymbol{T}^p_m(\boldsymbol{x})$ ($m = 1, \cdots, 63$) is computed according to (5) for each processor.
(ii) The results are gathered by the processor ($m = 0$) for the turbulence DNS and are summed as $\boldsymbol{T}^p(\boldsymbol{x}) = \sum_{m=1}^{63}\boldsymbol{T}^p_m(\boldsymbol{x})$.
(iii) The obtained $\boldsymbol{T}^p(\boldsymbol{x})$ is incorporated into the DNS of the NS equations.

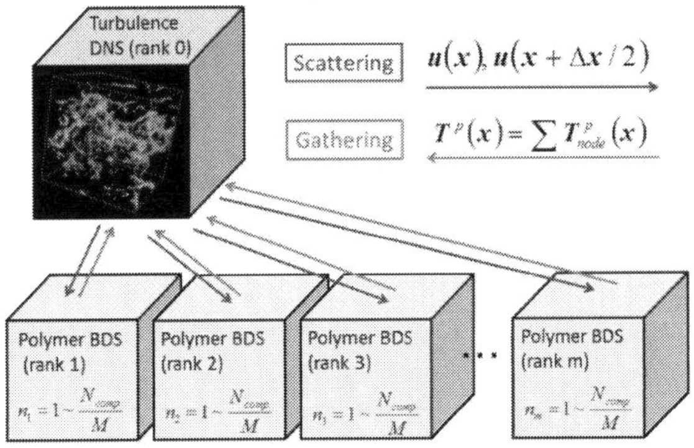

Figure 1. Schematic diagram of the parallelized method used in the hybrid simulation. The total number of process $M+1$ is set to be $M+1=64$ in the present study.

In step (i), since the center-of-mass vector $\boldsymbol{r}_g^{(n)}$ for each of the dumbbells is not on the grid points for the DNS computation, we need an approximate expression instead of (5). The delta function in (5) is approximated by the weight function $\delta_\Delta(\boldsymbol{x}-\boldsymbol{r}_g^{(n)})$ used for the tri-linear interpolation scheme [11]. The other parameter settings are described in detail in [7]. We examine three cases of decaying turbulence for three values of W_i while fixing the other parameters. The values of the numerical parameters are listed in Table 1.

Table 1. Parameters for the hybrid simulations of decaying turbulence and Brownian dynamics for dispersed dumbbells. The values of τ listed below are determined using the values of W_i given in the table and the minimum value of $\tau_K(t)$ during the simulation without the polymer additives (one-way coupling simulation).

	N_{comp} ($\times 10^9$)	b ($\times 10^5$)	$\Phi_V (\times 10^{-4})$	η	W_i	τ
Run D1	0.504	0.9	1.01	0.1045	25	4.903
Run D2	0.504	0.9	1.01	0.1045	50	9.806
Run D3	0.504	0.9	1.01	0.1045	100	19.612

Figure 2 shows the temporal evolutions of the kinetic energy of fluid motion $E(t)$ and of the potential energy for the ensemble of dumbbells $U(t)$, which are defined, respectively, as follows:

$$E(t) = \frac{1}{2}\langle \boldsymbol{u}(\boldsymbol{x},t)^2 \rangle_V \tag{7}$$

and

$$U(t) = -\frac{\nu_s \eta}{2\tau}\left(\frac{L_{max}}{r_{eq}}\right)^2 \frac{1}{N_{comp}} \sum_{n=1}^{N_{comp}} \ln\left[1-\left(\frac{|\boldsymbol{R}^{(n)}(t)|}{L_{max}}\right)^2\right] \tag{8}$$

for all of the runs, where $\langle\cdots\rangle_V$ represents the volume average over the computational domain $V = L_{box}^3$. Hereinafter, figures are plotted using non-dimensional time $t^* \equiv k_0 u_0 t$. In figure 2 (a), $E(t)$ decays monotonically with time, and the curves collapse onto a single curve irrespective of W_i. The values of $E(t)$ at the end of the simulations are much smaller than their initial values, meaning that the turbulence adequately decayed at $t^* = 20$. On the other hand, $U(t)$ increases with time for $t^* < 7$, which indicates that the turbulence kinetic energy is transferred

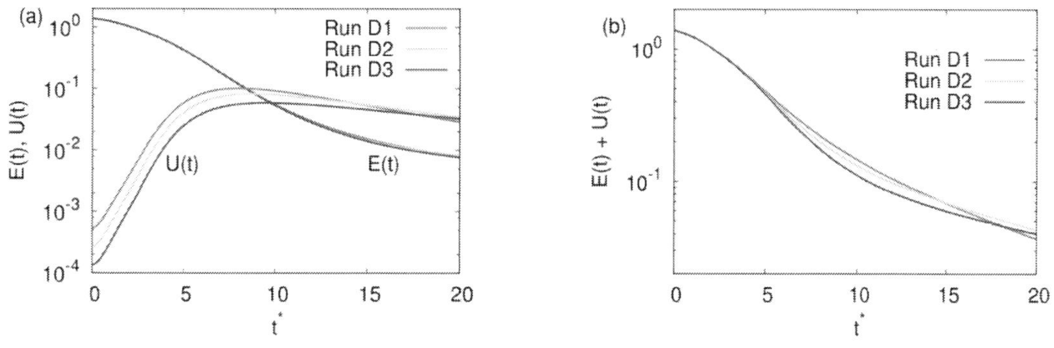

Figure 2. Comparison of the temporal evolutions of (a) the kinetic energy of fluid motion $E(t)$ and the potential energy for the ensemble of dumbbells $U(t)$ and of (b) the total energy $E(t) + U(t)$ for Runs D1, D2, and D3.

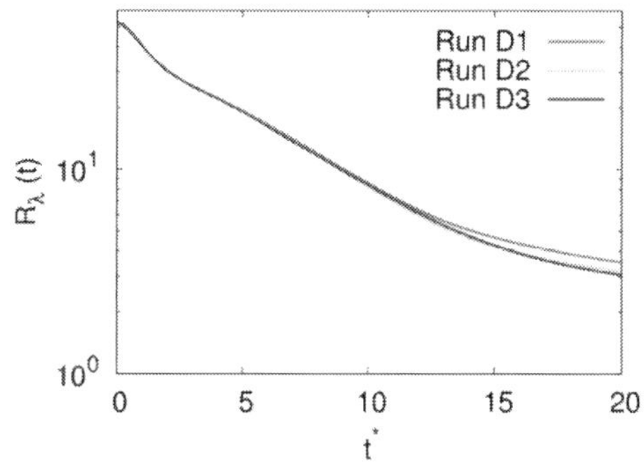

Figure 3. Comparison of the temporal evolutions of the Taylor microscale Reynolds number $R_\lambda(t)$ for Runs D1, D2, and D3.

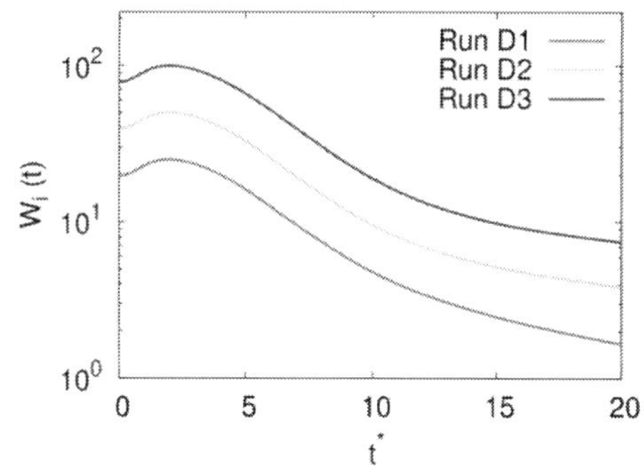

Figure 4. Comparison of the temporal evolutions of the local Weissenberg number $W_i(t)$ for Runs D1, D2, and D3.

to the potential energy of the ensemble of dumbbells in addition to the heat. Moreover, $U(t)$ starts to decay at around $t^* = 7$, indicating that the portion of restored potential energy is also back-transferred to turbulence and is dissipated due to the viscosity. Note that $U(t)$ is larger than $E(t)$ in the range of $t^* > 10$ for all of the runs. Figure 2(b) shows the temporal variation of the total energy $E(t) + U(t)$ based on the data of figure 2(a). We confirmed that the total energy also decays monotonically with time.

The temporal evolutions of the Taylor microscale Reynolds number $R_\lambda(t)$, which is defined by using $E(t)$ (7) and $\varepsilon(t) = \nu_s \langle (\nabla \boldsymbol{u})^2 \rangle_V$ as follows:

$$R_\lambda(t) = \sqrt{\frac{20}{3\nu_s \varepsilon(t)}} E(t), \tag{9}$$

are shown in figure 3. The curves almost collapse onto a single curve for $t^* < 10$, and the curve for Run D1 slightly deviates upward from the other curves for $t^* > 10$. Thus, R_λ is almost insensitive to the variation of W_i, and the values at $t^* = 20$ are $R_\lambda = 3.5, 3.2$, and 3.1 for Runs D1, D2, and D3, respectively.

Figure 4 compares the temporal evolutions of the local Weissenberg number $W_i(t) = \tau/\tau_K(t)$

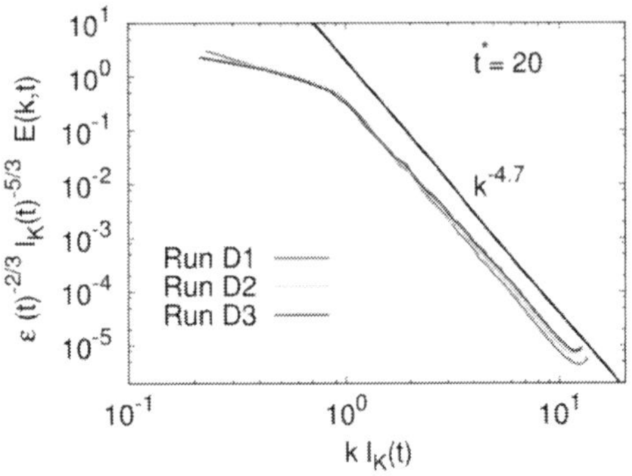

Figure 5. Comparison of the kinetic energy spectra normalized using the Kolmogorov scaling of (10) obtained at $t^* = 20$ for Runs D1, D2, and D3. The reference line shows the power law scaling of $k^{-4.7}$

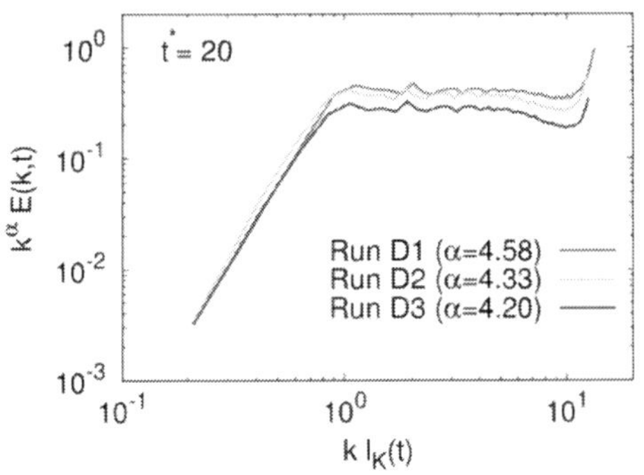

Figure 6. Compensated kinetic energy spectra $k^\alpha E(k,t)$ obtained at $t^* = 20$ for Runs D1, D2, and D3. The value of α is evaluated by a least squares fit in the range of $2 \le kl_K(t) \le 5$. Plateaus are clearly confirmed in the range $kl_K(t) > 1$.

for all runs. Based on figure 4, we confirm that the curves reach their maximum values at approximately $t^* = 2$, and $W_i(t)$ decays with time for $t^* > 2$. The value obtained for Run D3 is the greatest among the three values in the entire time region. At the end of simulation ($t^* = 20$), we have $W_i = 1.7, 3.8$, and 7.5 for Runs D1, D2, and D3, respectively, which indicates that numerous dumbbells remain in the stretched configuration for Runs D2 and D3 because the values of W_i are just above the critical value $W_i^c = 3-4$ at which the coil-stretch transition occurs [12].

Since the turbulence adequately decays for $t^* = 20$, where W_i is larger than unity for all of the runs, we investigate the scaling behavior of $E(k,t)$ obtained at $t^* = 20$. Figure 5 compares the kinetic energy spectra obtained for all of the runs, where we plot the dimensionless form of spectra $\hat{E}(kl_K(t),t)$ defined using the Kolmogorov scaling as

$$\hat{E}(kl_K(t),t) = \frac{E(k,t)}{\varepsilon(t)^{2/3} l_K(t)^{5/3}}. \tag{10}$$

As shown in figure 5, the normalized spectra generally collapse in the range of $kl_K(t) < 1$, and exhibit a power-law decay similar to $k^{-4.7}$ over one decade for $kl_K(t) > 1$. However, the spectrum becomes less steep with the increase of W_i. To see this point in more details, we make the compensated plot $k^\alpha E(k,t)$ in figure 6, where the exponent α is evaluated by using a least squares fit in the range $2 \le kl_k(t) \le 5$. This figure clearly indicates that the value of α decreases from 4.58 to 4.20 when W_i is increased. For the greatest values of W_i, $\alpha = 4.20$, which is close to the values obtained for elastic turbulence [3, 8].

Next, we will comment on the difference between the hybrid DNS using FENE dumbbells and the DNS of the elastic turbulence using the Ordroyd-B model [8]. The FENE model has the upper limit of the dumbbell extension length, L_{max}. This means that many dumbbells tend to remain at around L_{max} when W_i is larger. Figure 7 shows the probability density function (PDF) for the end-to-end distance of dumbbell $r = |\boldsymbol{R}^{(n)}|/L_{max}$ at $t^* = 20$ for all of the runs. This figure indicates that there are more stretched dumbbells for larger W_i, even when R_λ is sufficiently small. This also implies that the contribution of each dumbbell to the polymer stress tensor $\boldsymbol{T}^p$ becomes larger as W_i increases because $f(z) \to \infty$ as $z \to 1$. In contrast, the Oldroyd-B model is a linear equation for the conformation tensor of the polymer, which corresponds to

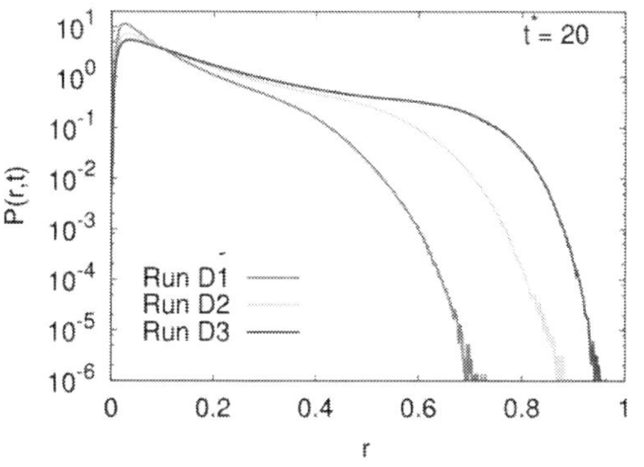

Figure 7. Probability density function for the end-to-end distance of dumbbells $P(r,t)$ normalized by the maximum extension length L_{max} obtained at $t^* = 20$.

the case in which the linear spring model is adapted to (2) by setting $f(z) = 1$. In this case, the stretched dumbbells can exceed L_{max}, leading to a greater contribution to the increase in $\boldsymbol{T}^p$ due to an increase in $|\boldsymbol{R}^{(n)}|$. Thus, the mechanism for increasing the amplitude of $\boldsymbol{T}^p$, which strongly affects the power-law behavior of $E(k,t)$ through the NS equations, differs significantly between the FENE model and the Oldroyd-B model.

Another important point is the nature of the isotropy of the flow field. In the present case, the flow field is regarded as nearly isotropic, so that the effect of anisotropy on the scaling behavior of $E(k,t)$ is negligible. In contrast, the flow field is strongly anisotropic for the DNS study under the Kolmogorov flow [8] and for the experimental viscoelastic flow within the rotating disks [3]. The persistent anisotropy leads to the modification of the spectral dynamics at small scales, so that we must carefully compare the present results to the previous results for elastic turbulence.

The above considerations raise the question of how the power-law form of the kinetic energy spectrum is universal for the changes in the mechanical properties of the chain polymer model or the degree of anisotropy. Moreover, since the flow field is nearly isotropic in the present study, it is appropriate to explore whether the power-law exponent is universal for $W_i \gg 1$ and $R_\lambda \ll 1$. We may need more computational resources to examine this problem because the case in which $W_i \gg 1$ requires much finer spatial resolution for computing the NS equation and much finer time steps for computing the dumbbell deformations in the Lagrangian frame with reasonable computation accuracy. This is perhaps the greatest computational challenge in examining the ultimate state of elastic turbulence by large-scale simulations.

We investigated the scaling behavior of the kinetic energy spectrum in decaying isotropic turbulence with polymer additives for various values of W_i. We obtained the power-law form of $E(k) \sim k^{-\alpha}$ in the range $kl_K > 1$, and α was approximately 4.2–4.6 and decreased with the increase in W_i. The value of 4.20 obtained for Run D3, which has the greatest value of W_i in the present study, was similar to the values obtained in the study on elastic turbulence ($\alpha = 3.5$ experimentally [3] and $\alpha = 3.8$ by DNS of the constitutive equation [8]).

T. W. and T. G. were supported in part by Grants-in-Aid for Scientific Research Nos. 23760156 and 24360068, respectively, from the Ministry of Education, Culture, Sports, Science, and Technology of Japan. T. W. would like to thank S. Uno and D. Sugimoto for their assistance in writing the parallelized code. The authors would like to thank the Theory and Computer Simulation Center of the National Institute for Fusion Science and JHPCN and HPC at the Information Technology Center of Nagoya University for providing the computational resources.

References

[1] Lumley J L 1973 *J. Polymer Sci.: Macromolecular Reviews.* **7** 263
[2] Proccacia I, L'vov V S and Benzi R 2008 *Rev. Mod. Phys.* **80** 225
[3] Groisman A and Steinberg V 2000 *Nature.* **405** 53

[4] Bird R B, Curtiss C F, Amstrong R C and Hassager O 1987 *Dynamics of Polymetric Liquids, Vol.2 Kinetic Theory, 2nd ed.*. Wiley, New York
[5] Laso M and Ottinger H C 1993 *J. Non-Newtonian Fluid Mech.* **47** 1
[6] Doi M and Edwards S F 1986 *The Theory of Polymer Dynamics.* Oxford Univ. Press, New York
[7] Watanabe T and Gotoh T 2013 *J. Fluid. Mech.* **717** 535
[8] Berti S, Bistagnino A, Boffetta G, Celani A and Musacchio S 2008 *Phys. Rev. E.* **77** 055306
[9] Fouxon A and Lebedev V 2003 *Phys. Fluids.* **15** 2060
[10] Yeung P K and Pope S B 1988 *J. Comp. Phys.* **79** 373
[11] Prosperetti A and Tryggvason G 2007 *Computational Methods for Multiphase Flow.* Cambridge University Press, Cambridge
[12] Watanabe T and Gotoh T 2010 *Phys. Rev. E.* **81** 066301

Journal of Physics: Conference Series **454** (2013) 012008 doi:10.1088/1742-6596/454/1/012008

Radiation hydrodynamics simulation of high-Z doped ICF targets

Takashi Shiroto[1], **Naofumi Ohnishi**[1], **Atsushi Sunahara**[2] **and Shinsuke Fujioka**[3]

[1]Department of Aerospace Engineering, Tohoku University, Sendai 980-8579, Japan
[2]Institute for Laser Technology, Osaka 550-0044, Japan
[3]Institute of Laser Engineering, Osaka University, Suita 565-0871, Japan

E-mail: tshiroto@rhd.mech.tohoku.ac.jp

Abstract. We conducted a comparative study among the Bi-CG family with incomplete LU factorization employed for solving diffusion-type radiative transfer equation in radiation hydrodynamics (RHD) simulations for laser-produced plasmas. The result suggests that Bi-CGSTAB is the most suitable method thanks to its high convergence stability and low computational cost. RHD simulations were performed for Rayleigh–Taylor instability experiments of an inertial confinement fusion target. We found that higher harmonics of static pressure perturbation become important at an intermediate wavelength.

1. Introduction

Suppression of Rayleigh-Taylor (RT) instability [1] at the ablation surface of a fuel shell is still a critical issue for inertial confinement fusion (ICF) [2] despite the fact that continuous efforts have been devoted to perform it for a long time. On a frame of the ablation surface, the low density plasma corona sustains massive fuel plasma in a gravitational field. Hence the surface becomes RT unstable while the ablated material flows through it. In classical RT instability which occurs at a material interface of an incompressible fluid, an analytical growth rate of the interface γ_{cl} is obtained by linear theory:

$$\gamma_{cl} = \sqrt{\frac{\rho_h - \rho_l}{\rho_h + \rho_l} kg}, \tag{1}$$

where $\rho_h(\rho_l)$ is the density of higher (lower) material, k is the wavenumber of the given perturbation, and g is the acceleration. Perturbation growth in an ICF target is suppressed due to the ablative flow through the unstable surface. The growth rate γ can be expressed by the modified Takabe formula [3, 4]

$$\gamma = \alpha\sqrt{\frac{kg}{1+kL}} - \beta k v_a, \tag{2}$$

where α and β are the constants determined by the ablation structure, L is the density scale length at the ablation surface, and v_a is the ablation velocity.

A concept of indirect-direct hybrid drive is preferable for achieving both target stability and high energy coupling efficiency. A high-Z doped target was proposed in the context of the hybrid drive by utilizing the double ablation structure formed by self-emission of high-Z dopant [5]. Radiation hydrodynamics (RHD) simulations demonstrated that perturbation growth in the high-Z doped target is suppressed because the radiative ablation (RA) surface is stabilized by its high ablation rate. However, the detailed property of such a target should be assessed for practical use; in particular the distance between the RA front and the electron-conduction ablation (EA) one may govern the behavior of a specific wavelength [6].

Although analysis of the perturbation growth can be performed by a multi-dimensional RHD code, reliable simulations should be conducted with a large number of computational cells for resolving the ablation structure ($< \mu$m) in the acceleration area ($\sim$ mm) and so they have a large computational costs. Since the radiative transfer solver is the most time-consuming module in the RHD code, it is important to make it efficient. If an optically thick flowfield can be assumed, the radiative transfer equation is simplified to a diffusion-type equation. We usually employ the bi-conjugate gradient (Bi-CG) method [7] for obtaining a solution of such an equation because the coefficient matrix that appears in implicit differentiation with 9–point stencil becomes asymmetric. In this study, the coefficient matrix becomes symmetric since we use Cartesian grids with isotropic radiation. However, we investigate the convergence properties of the Bi-CG family to develop a versatile code.

In this paper, we first conducted a comparative study among methods of the Bi-CG family. The Bi-CG method, conjugate gradient squared (CGS) method, bi-conjugate gradient stabilized (Bi-CGSTAB) method, and Bi-CGSAFE method are examined for the comparison with test matrices obtained from radiation transfer problems of a practical RHD simulation. Then, we performed a set of RHD simulations corresponding to planar ICF target experiments and analyzed the flowfield to discuss a significant suppression appearing in intermediate wavelengths.

2. Numerical methods and conditions

Numerical simulations were carried out by a two-dimensional Eulerian RHD code, RAICHO [8]. The code solves the two-dimensional Euler equations including some energy sources with realistic equations of state (EOS) for electrons and ions. The EOS for ions is based on the Cowan model and that for electrons is based on the Thomas–Fermi model. In the case of low-density and high-temperature plasma, the electrons and ions are in thermal non-equilibrium. Hence the energy equation is divided into two energy equations for electrons and ions. One of the source term is estimated by solving the radiative transfer equation which is approximated in a diffusion-type form.

The governing equations were discretized in a cell-centered finite volume fashion. The AUSM-DV method [9] was employed for the numerical flux with the second-order MUSCL method [10]. The dependent variables were integrated with a first-order explicit Euler method with inviscid flux and with an implicit method with the source terms in an operator splitting manner.

The laser is assumed to be normally irradiated on the planar target, and the laser absorption of inverse–bremsstrahlung is calculated with one-dimensional ray-tracing. The thermal conduction is formulated by a diffusion approximation with the Spitzer-Härm conductivity and a flux limiter of 0.1 [11]. In this study, the number of energy groups is 32 for the radiative transfer equation under a multi-group approximation.

Bromine-doped plastic ($C_{50}H_{47}Br_3$, hereafter CHBr) targets and the surrounding vacuum region were represented with uniform grids of 1537×49 in figure 1. The target is irradiated by a foot pulse and a main pulse as depicted in figure 2. The foot pulse is a Gaussian pulse with maximum intensity of 1.0×10^{13} W/cm^2, 530-nm wavelength, and 1.3-ns full width half maximum (FWHM). It has a sinusoidal spatial distribution to imprint a perturbation on the target. The main pulse is flat-topped with maximum intensity of 1.0×10^{14} W/cm^2, 530-nm

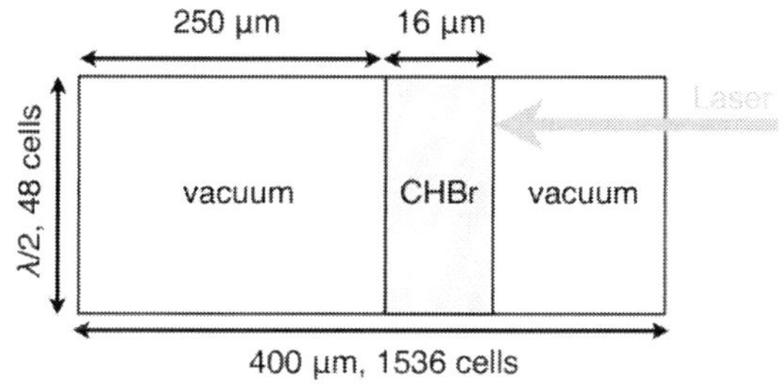

Figure 1. Simulation domain.

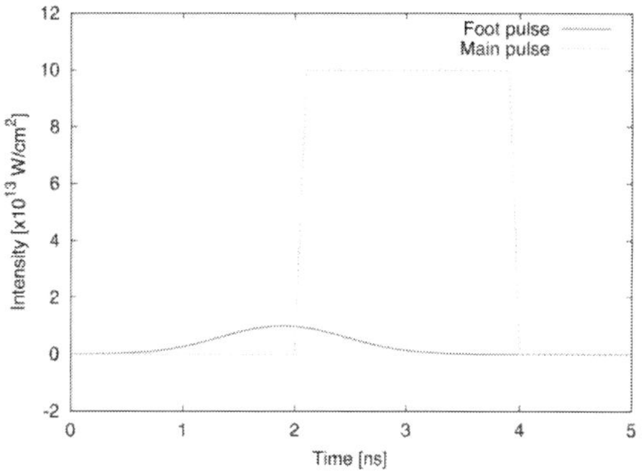

Figure 2. Temporal profile of input laser pulse.

wavelength, and 1.9-ns FWHM. The initial target density is 1.26 g/cm^3, and its thickness is set to 16 μm for all the simulations. The critical density of the target plasma is 3.98×10^{21} cm^{-3}.

3. Convergence property in radiative transfer solver

Figures 3 and 4 show examples of time history of relative residual norm for the Bi-CG, CGS, Bi-CGSTAB, and Bi-CGSAFE for two cases, in the radiation transfer computation. All methods are preconditioned by incomplete LU factorization vectorized in a hyperplane manner. Test matrices of the diffusion coefficient were obtained from the radiation transfer process of the practical RHD simulation. Case-A is a typical example for the early stage of laser incidence at which the radiation field drastically changes due to initial ionization. Case-B is for a matrix when the input laser becomes constant and the ablation structure is in a quasi-steady state. In Case-A, the matrix becomes stiffer than in Case-B, so a larger number of iterations is required to obtain a solution. Table 1 shows numbers of iterations and CPU time for solving Case-A and Case-B with each method. The numbers of iterations for the Bi-CGSTAB and the Bi-CGSAFE are much smaller than that of the others because of their high convergence stability. Although the Bi-CGSAFE is more state-of-the-art than the Bi-CGSTAB, the former needs more CPU time than the latter. This difference originates from the low computational cost of the Bi-CGSTAB.

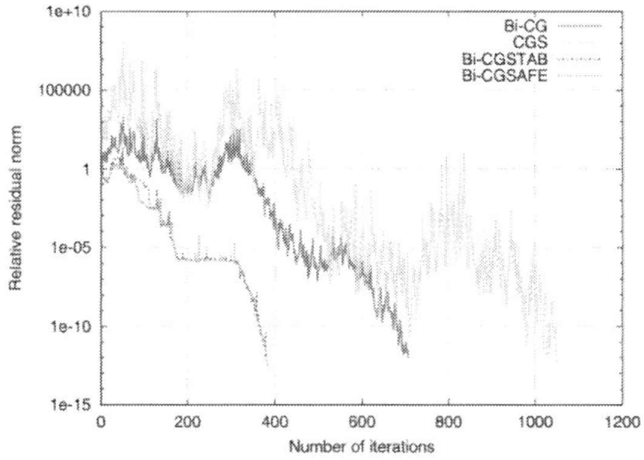

Figure 3. Convergence history (Case-A).

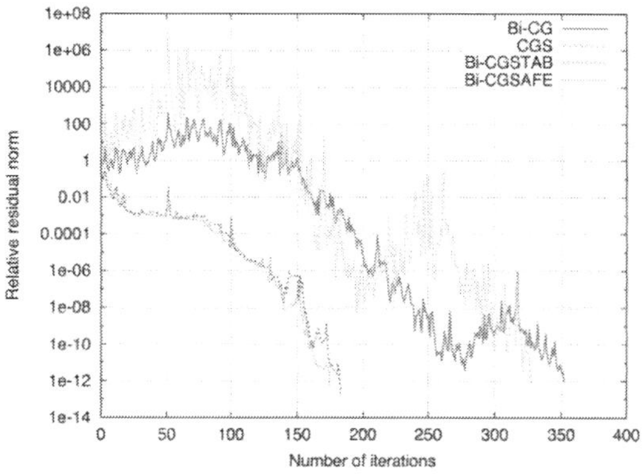

Figure 4. Convergence history (Case-B).

Hence the Bi-CGSTAB is the most suitable method among the Bi-CG family especially for the radiative transfer problem.

Table 1. Comparison of performances in the Bi-CG family.

Scheme	Case	♯ of iterations	CPU time [ms]
Bi-CG	A	706	1380
	B	352	688
CGS	A	1050	1955
	B	327	630
Bi-CGSTAB	A	379	666
	B	183	323
Bi-CGSAFE	A	382	948
	B	175	436

4. Perturbation growth

Figure 5 shows that a significant suppression of the RT growth rate in the intermediate wavelength was found as predicted in [6]. In order to identify the cause of such a suppression, we defined static pressure perturbation between two ablation fronts in an acceleration frame of the EA surface as follows:

$$\delta p(x, y) = p(x, y) - \frac{1}{\lambda} \int_{-\frac{\lambda}{2}}^{\frac{\lambda}{2}} p(x, y) dy, \tag{3}$$

where $p(x, y)$ is the static pressure in the EA acceleration frame, and λ is the perturbation wavelength. Thus the static pressure perturbation means the difference between static pressure

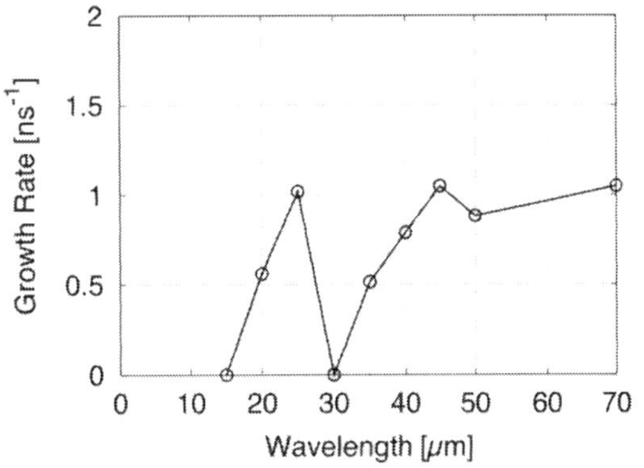

Figure 5. Dispersion relation of RT growth at RA front.

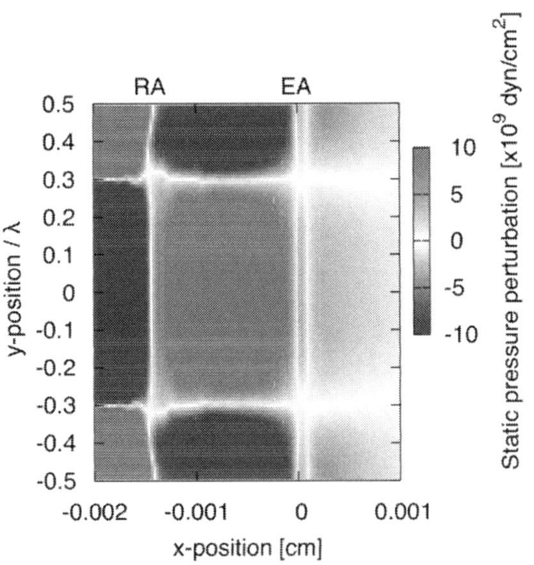

Figure 6. Static pressure perturbation for $\lambda = 25$ μm at 3.9 ns.

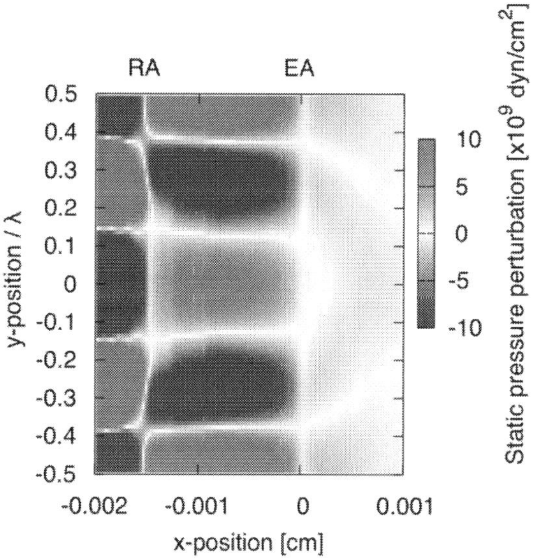

Figure 7. Static pressure perturbation for $\lambda = 30$ μm at 3.9 ns.

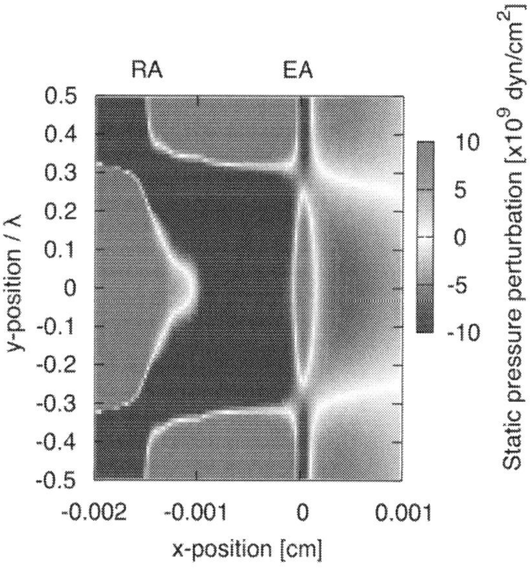

Figure 8. Static pressure perturbation for $\lambda = 35$ μm at 3.9 ns.

and its average value along y-axis at a given x-position in this paper. In the case of $\lambda = 30$ μm, at which the perturbation growth is significantly suppressed, higher harmonics of the static pressure perturbation appear between the RA front and EA front (figures 6–8). The fundamental mode is damped for this wavelength and then cannot grow at the RA front deformation. There is some possibility that a suppression effect emerges via pressure waves from the EA front to the RA front.

5. Conclusions

We conducted a comparative study among methods of family of conjugate gradient. The results suggest that the Bi-CGSTAB is the most suitable method among the Bi-CG family on RHD simulations. We also investigated the dynamics of ablation fronts in a high-Z doped ICF target using a two-dimensional RHD code. We found that higher harmonics of the static pressure perturbation between two ablation fronts develop at intermediate wavelengths, at which a significant suppression appears.

Acknowledgments

Computations were carried out on NEC SX-8R at Institute of Laser Engineering and Cybermedia Center, Osaka University.

References

[1] Chandrasekhar S 1968 *Hydrodynamic and Hydromagnetic Stability* Oxford University Press
[2] Nuckolls J, Wood L, Thiessen A and Zimmerman G 1972 *Nature* **239** 139
[3] Takabe H and Mima K 1985 *Phys. Fluids* **28** 3676
[4] Budil K S, Remington B A, Peyser T A, Mikaelian K O, Woolsey N C, Wood-Vasey W M and Rubenchik A M 1996 *Phys. Rev. Lett.* **76** 4536
[5] Fujioka S *et al* 2004 *Phys. Rev. Lett.* **92** 195001
[6] Yañez C, Sanz J, Olsazabal-Loumié M and Ibañez L F 2012 *Phys. Plasmas* **19** 062705

[7] Barrett R, Berry M, Chan T F, Demmel J, Donato J, Dongarra J, Eijkhout V, Pozo R, Romine C and Van Der Vorst H 1987 *Templates for the Solution of Linear Systems: Building Blocks for Iterative Methods* SIAM
[8] Ohnishi N, Nishikino M and Sasaki A 2006 *J. Phys. IV* France **133** 1193
[9] Wada Y and Liou M S 1994 *AIAA Paper* 94-0083
[10] Leer B V 1979 *J. Comp. Phys.* **32** 101
[11] Spitzer L and Härm R 1953 *Phys. Rev.* **89** 977

Fluid simulation of plume head-on collision dynamics during pulsed laser ablation

R Fujii[1,3], K Doi[1], Y Yokoyama[1], H Fukuoka[2], A Sugimura[1], N Tominaga[1], H Susa[1] and I Umezu[1]

[1]Department of Physics, Konan University, Kobe 658-8501, Japan

[2]Department of Mechanical Engineering, Nara National College of Technology, Nara 639-1080, Japan

[3]Department of Earth and Planetary Sciences, Kobe University, Kobe 657-8501, Japan

E-mail: umezu@center.konan-u.ac.jp

Abstract. Expansion dynamics of plume after irradiation of the target material is essential to prepare nanoparticles by pulsed laser ablation and it can be modified by collision of two plumes. In the present paper, effect of head-on collision on the expansion dynamics is discussed by numerical simulation based on the fluid dynamics and compared with the experimental results of plume emission. Suppression of plumes by collision with counter plume observed by experiment is reproduced by numerical simulation. Results of the numerical calculation indicate that shockwave induced by the irradiation of the opposite target suppress vapor expansion. The vapors do not mix around the center of the targets when the two targets are irradiated simultaneously and unstable flow is seen when delay between laser pulses was applied for irradiation of two targets. The results of the numerical simulation suggest that formation of combined and alloy nanoparticles are expected for former and latter cases.

1. Introduction

Formation of nanoparticles during pulsed laser ablation (PLA) in background gas is reported for various materials. Strong pulsed laser irradiation of target induces high-density, -pressure and -temperature plasma, plume, and it expands to the background gas. The analysis of point explosion well describes dynamic process of the plume expansion [1–3]. Nano-droplets are thought to be formed during the free expansion of the plume by the adiabatic cooling followed by nano-crystallization on further cooling [4, 5]. The control of nanostructure is expected by varying dynamic plume expansion processes. The dynamic expansion process can be modified by the background gas pressure and laser fluence, and correlation between the preparation condition and structure of nanoparticles has been discussed so far [6, 7]. Introducing synchronized two laser beams enable us to collide two plumes. The collision of the two plumes modifies plume expansion. For example, a layer of stagnated plasma is formed at the collision front when two plumes collide under appropriate conditions [8, 9]. One of the practical applications of colliding two plumes is formation of high quality droplet free films [10]. Recently, a method to form novel nanoparticles by colliding two plumes is proposed [11, 12].

Not only for the material processing, PLA has many similarities with astrophysical phenomena such as point source explosion, expansion of plasma, formation of shockwave and so on. The effect of collision of two plasmas is also important to reveal the evolution of structures within supernova remnants [13, 14]. In the present paper, we focused on the effect of plume collision on the nanoparticle formation. Irradiation of a target by a pulsed laser and that of two targets by two pulsed lasers are referred to as "single pulsed laser ablation (SPLA)" and "double pulsed laser ablation (DPLA)", respectively. Effect of collision on the expansion dynamics is discussed by comparing numerical and experimental methods and a way to control nanoparticle formation is proposed.

2. Experiment

Figure 1 shows experimental setup of DPLA. Si and Ge targets were placed parallel to each other at distance of 9 mm. The targets were irradiated by individual two YAG lasers. The wavelength, pulse width, repetition rate and fluence were 355 nm, 7 ns, 10 Hz and 1.6 J/cm^2 for ablation of the Si target and 266 nm, 7 ns, 10 Hz and 1.9 J/cm^2 for that of the Ge target, respectively. Helium gas was introduced as a background gas and kept constant value during the ablation. The plume expansion dynamics was monitored by gated ICCD equipped with monochrometer. The ICCD camera is positioned at the front side of the sheet in figure 1. The details of experimental set up are shown in Ref. 11.

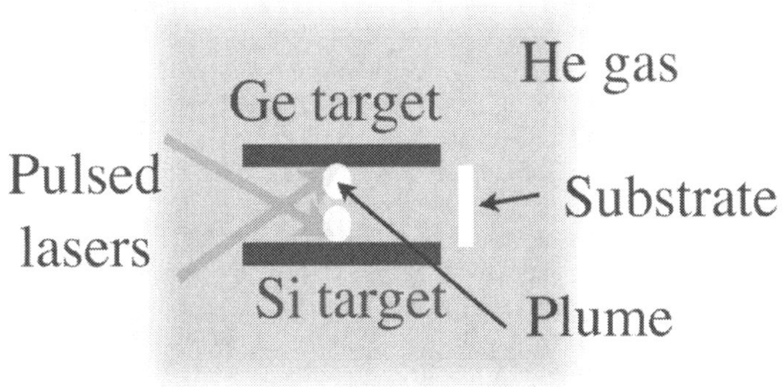

Figure 1. Experimental setup of DPLA. Targets are placed parallel and plumes are collided head-on condition.

3. Numerical method and initial conditions

We perform two-dimensional hydrodynamics calculations in the cylindrical coordinate system. We use following ordinary set of equations:

$$\frac{\partial \rho_i}{\partial t} = -\nabla(\rho_i \cdot \vec{v}) \qquad (\rho = \rho_{He} + \rho_{Ge} + \rho_{Si}), \tag{1}$$

$$\rho\left(\frac{\partial \vec{v}}{\partial t} + (\vec{v} \cdot \nabla \vec{v})\right) = -\nabla(P+q), \tag{2}$$

$$\frac{\partial e}{\partial t} = -\nabla((e+P)P \cdot \vec{v}), \tag{3}$$

$$e = \frac{1}{\gamma - 1}\frac{P}{\rho}. \tag{4}$$

Here we solve continuity equations for three species (suffix i in (1) denotes He, Ge and Si). As shown in (1), ρ represents the total density of the fluid, while $\vec{v}$ is the velocity, P and e denote the thermal pressure and the internal energy, respectively. q denotes the additional pressure term originated from the von-Neumann type artificial viscosity. In practice, we combine (3) and (4) to solve the energy equation. Substituting (4) into (3), we have

$$\frac{\partial P}{\partial t}+\vec{v}\cdot\nabla P=-\left(\gamma P+(\gamma-1)q\right)(\nabla\cdot\vec{v}). \quad (5)$$

We solve (5) instead of (3). Even though the plume of PLA is plasma state, neutral gas is assumed here for simplicity and electromagnetic effects are not taken into account. From this reason, we use the term 'vapor' instead of 'plume' for the result of calculation.

Here, we adopt an equation of state of ideal gas. We assumed the initial densities, pressure, temperature to reproduce the laser fluence and ejected mass per laser shot from experimental results [7]. The adopted parameters are summarized in table 1.

Table 1. Initial conditions used in the calculation.

	Density (kg/m^3)	Pressure (Pa)	Temperature (K)
Si	0.038	6.3×10^7	5.2×10^6
Ge	0.038	7.4×10^7	1.4×10^7
He	0.042	260	300

In order to treat a shock wave, we applied Constrained Interpolation Profile (CIP) scheme in which advection and non-advection terms in the equations are separately solved. We also employ von-Neumann type artificial viscosity in order to capture the shock front with a given grid size [15-17]. The additional pressure term q in (2) and (5) is given as

$$q=\alpha\left(-\rho C_s(\nabla\cdot\vec{v})\lambda+\frac{\gamma+1}{2}(\nabla\cdot\vec{v})^2\lambda^2\right) \quad \text{if} \quad \nabla\cdot\vec{v}<0, \quad (6)$$

$$q=0 \quad \text{if} \quad \nabla\cdot\vec{v}>0, \quad (7)$$

where C_S is the sound velocity of the fluid, λ is the assumed thickness of the shock front which is set to be the grid size in the present simulation. We adopt α=0.75 being appropriate for shock waves.

4. Results and discussion

The images of the plume expansion dynamics observed by the gated ICCD camera when two targets are irradiated simultaneously are shown in figure 2. Contour plot images of the total gas pressure and densities of the Si and Ge vapors obtained by numerical simulation are shown in figures. 3a and 3b, respectively. The profiles of densities of the Si and Ge vapors and temperature along symmetric axis under background gas pressure of 270 Pa at 100 ns after the laser irradiation are shown in figure 4. The high-density region roughly corresponds to the high temperature region. Slight increase in the temperature around the center of targets is observed by collision. However the temperature around the center of targets is about one order smaller than the highest temperature, it increases to considerable value by selecting initial condition.

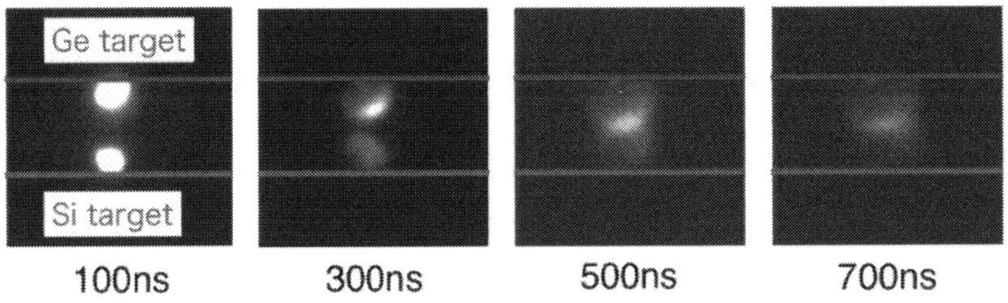

Figure 2. The images of plume emission under background gas of 270 Pa at 100, 300, 500 and 700 ns. The red and blue lines in the figures show the positions of the Si and Ge targets, respectively. The distance between targets is 9 mm.

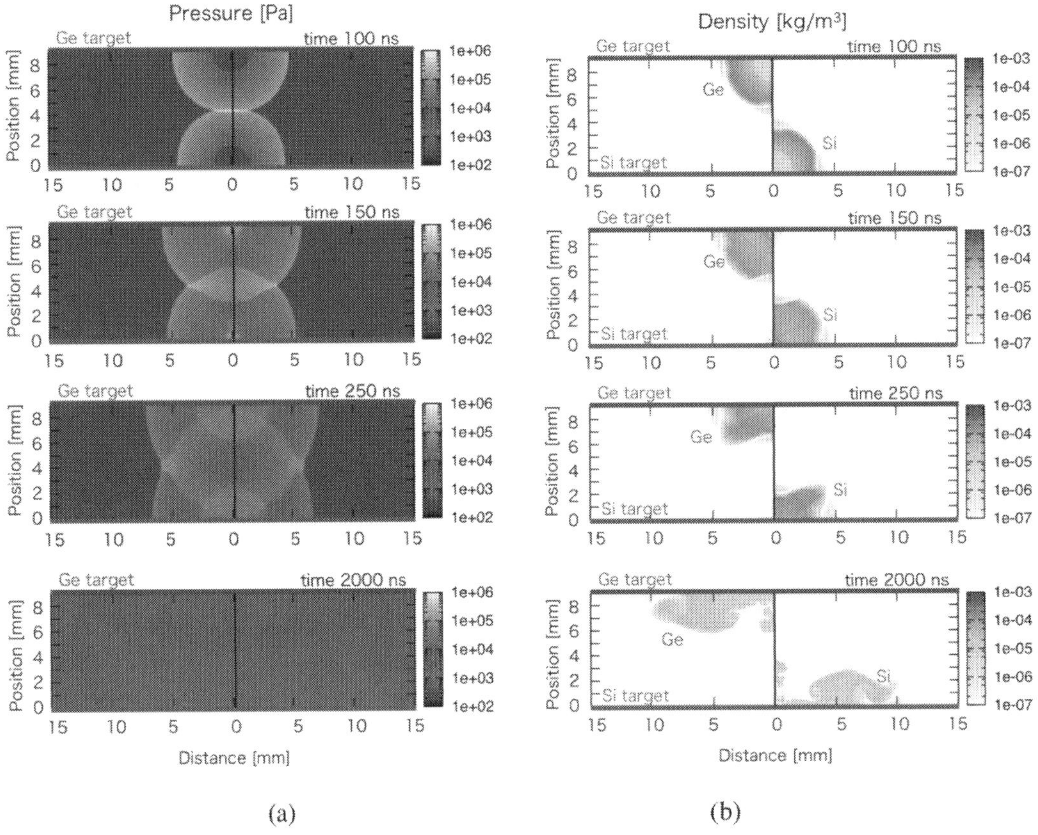

Figure 3. (a) Contour plots of total pressure and (b) Contour plots of Si and Ge vapor densities at 100, 150, 250 and 2000 ns after targets are irradiated simultaneously. The Si and Ge targets are at 0 and 9 mm, lower and upper side of the figure.

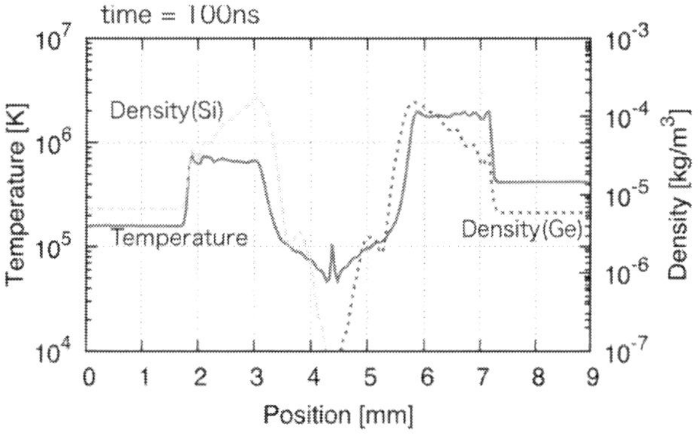

Figure 4. The densities of the Si and Ge vapors and temperature under background gas pressure of 270 Pa at 100 ns after the laser irradiation.

We define front edge of vapors at the position where density decreases to $1\text{x}10^6$ kg/m^3. The position of vapor front-edges is plotted in figure 5 versus time after the pulsed irradiation. Positions of the Si and Ge targets are 0 and 9 mm, respectively, on vertical axis of these figures. Even though the agreement of absolute value of time scale is not good, results of calculation are qualitatively similar to those of experiments [12]. The effect of DPLA is remarkable above 140 ns. The vapor front-edge moves backward at this time. The positions of the shockwave and vapor front-edge can be estimated from figures 3a and 3b, respectively. The shockwave induced by irradiation of the opposite target reaches the vapor front-edge at around 150 ns as shown in figure 3. Since this value is close to 140 ns, the backward motion observed in figure 5 should be the effect of counter shockwave. As we can see in the figure 3, the Ge and Si vapors are spatially separated even at 2000 ns after irradiation. This suggests formations of the Si and Ge nanoparticles in the spatially separated Ge and Si vapors individually. Formation of aggregates of combined Si and Ge nanoparticles is expected at later stage.

The results of calculation are suggestive also for position of substrate. The gas flow perpendicular to the symmetric axis around the center of targets is observed by calculation. This suggests that collection of nanoparticles is possible by placing substrate perpendicular to the substrate apart from symmetric axis as shown in figure 1.

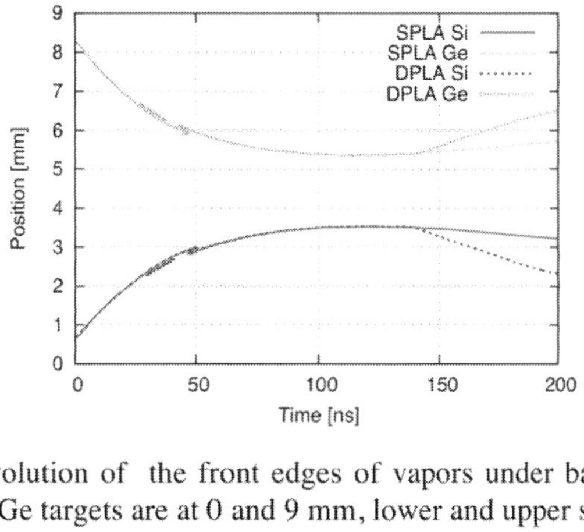

Figure 5. Time evolution of the front edges of vapors under background gas of 270Pa. The Si and Ge targets are at 0 and 9 mm, lower and upper side of the figure.

Delay-time between the two laser pulses is easily controlled parameter both for the experiment and calculation. Effect of the delay-time on the vapor collision is shown in figure 6. This is a result when the Ge target is irradiated following the irradiation of the Si target and the delay-time is set at 700 ns. The extensive unstable flow of the vapors begin to seen after 1000 ns of Si target irradiation, which is after 300 ns of Ge target irradiation, and clearly seen at 2000 ns as shown in figure 6. This time corresponds to the time when the Ge vapor reaches the Si target. The position of Ge vapor front as a function of the time after irradiation of Ge target when delay time is applied is shown in figure 7. However the effect of the counter vapor seen in figure 5 is observed at delay-time of 100 ns, such effect is not seen at delay-time of 700 ns or higher. The results of vapor expansion profile indicate that the delay time of 700 ns corresponds to the time when the Si vapor front reaches the Ge target. The very unstable feature is found at delay-time of 700 ns as shown in figure 6. A similar feature is observed at higher delay-time. One of the possible reasons of the very unstable feature is that the density of the Si vapor becomes lower with time, and the more compact dense Ge vapor penetrates the Si vapor like a bullet at higher delay-time. As a result, strong shear flow arise at the contact surface of Si and Ge vapors, and Kelvin-Helmholtz instability becomes prominent. Unfortunately, this unstable flow is not observed in the experimental plume emission. It is presumably because unstable flow grows at later stage of the plume expansion and the plume emission vanishes at this stage. We can expect mixture of two plumes by the extensive unstable flow shown here and formation of Si-Ge alloy nanoparticles if the temperature is not below the nanocrystallization at this time scale.

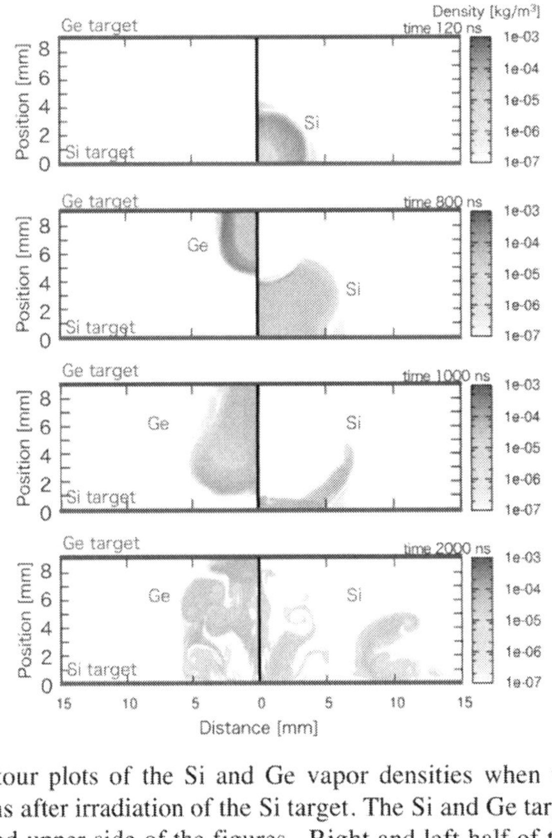

Figure 6. Contour plots of the Si and Ge vapor densities when the Ge target is irradiated 700 ns after irradiation of the Si target. The Si and Ge targets are at 0 and 9 mm, lower and upper side of the figures. Right and left half of the figures show densities of Si and Ge vapors, respectively.

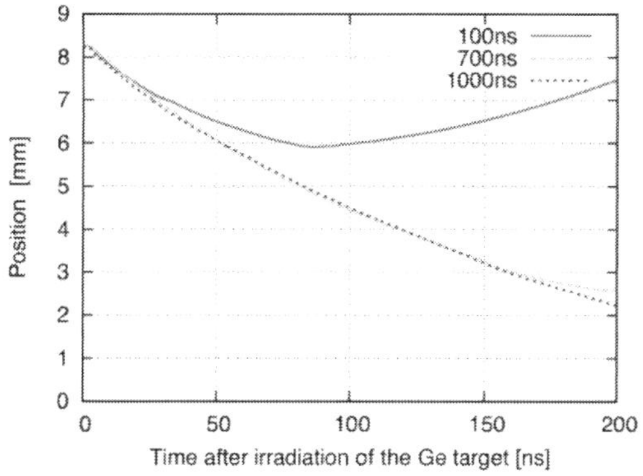

Figure 7. Time evolution of the front edge of the Ge vapor under background gas of 270 Pa. The Ge target is irradiated at 100, 700 and 1000 ns after the Si target is irradiated. The Si and Ge targets are at 0 and 9 mm, lower and upper side of the figure.

5. Conclusion

We show effects of collision of two vapors by numerical simulation. The vapor expansion dynamics is qualitatively similar to experimental plume expansion. The effects of shockwave on the plume expansion dynamics are discussed from the simulation. Results of numerical simulation show possibility to control nanostructure by controlling the delay-time between the laser pulses. The shockwave induced by irradiation of the opposite target suppress expansion of vapors. The two vapors are confined by counter shockwave and mixture of the vapor is not seen when the targets are irradiated simultaneously. The extensive unstable flow of the vapors is observed when the delay was applied between the laser pulses. These results give valuable information for design of the hybrid nanoparticles generation.

Acknowledgement

This study was partially supported by a Grant-in Aid for Scientific Research from the Japan Society for the Promotion of Science.

References

[1] Sedov L I 1959 Similarity and Dimensional Methods in Mechanics (London: Academic Press)
[2] Zeldovich Y B and Raizer Y P 1966 *Physics of Shock waves and High-temperature Hydrodynamic Phenomena* (New York: Academic Press)
[3] Anisimov S I, Bauerle D and Luk'yanchuk B S 1993 *Phys. Rev.* B **48** 12076
[4] Luk'yanchuk B S, Marine W and Anisimov S I 1998 *Laser Phys.* **8** 291
[5] Marine W, Patrone L, Luk'yanchuk B and Sentis M 2000 *Appl. Surf. Sci.* **154-155** 345
[6] Bäuerle D 2000 *Laser Processing and Chemistry* (Berlin: Springer-Verlag)
[7] Umezu I, Sugimura A, Inada M, Makino T, Matsumoto K and Takata M 2007 *Phys. Rev.* B **76** 045328
[8] Rambo P W and Denavit J 1994 *Phys. Plasmas* **1** 4050–60
[9] Hough, P Hayden P, Fallon C, Kelly T J, McLoughin C, Yeates P, Mosnier J P, Kennedy E T, Harilal S S and Costello J T 2011 *J. Phys. D* **44** 355203
[10] Irissou E, Vidal F, Johnston T, Chaker M, Guay D and Ryabinin A N 2006 *J. Appl. Phys.* **99** 034904

[11] Umezu I, Yamamoto S and Sugimura A 2010 *Appl. Phys. A* **101** 133
[12] Umezu I, Sakamoto N, Fukuoka H, Yokoyama Y, Nobuzawa K and Sugimura A 2012 *Appl. Phys. A* **110** 629
[13] Smith R A, Lazarus J, Hohenberger M, Moore A S, Robinson J S, Gumbrell E T and Dunne M 2007 *Astrophys. Space Sci.* **307** 131
[14] Gregory C D, Howe J, Loupias B, Myers S, Notley M M, Sakawa Y, Oya A, Kodama R, Koenig M and Woolsey N C 2008 *Astrophys. J.* **676** 420
[15] Takewaki H, Nishiguchi A and Yabe T 1985 *J. Comput. Phys.* **61** 261
[16] Yabe T and Aoki A 1991 *Comput. Phys. Commun.* **66** 219
[17] Ogata Y and Yabe T 1999 *Comput. Phys. Commun.* **119** 179

Finite difference simulations of seismic wave propagation for understanding earthquake physics and predicting ground motions: Advances and challenges

Hideo Aochi[1], Thomas Ulrich, Ariane Ducellier, Fabrice Dupros and David Michea

Bureau de Recherches Géologiques et Minières, 3 Avenue Claude Guillemin, 45100 Orléans, France

E-mail: h.aochi@brgm.fr

Abstract. Seismic waves radiated from an earthquake propagate in the Earth and the ground shaking is felt and recorded at (or near) the ground surface. Understanding the wave propagation with respect to the Earth's structure and the earthquake mechanisms is one of the main objectives of seismology, and predicting the strong ground shaking for moderate and large earthquakes is essential for quantitative seismic hazard assessment. The finite difference scheme for solving the wave propagation problem in elastic (sometimes anelastic) media has been more widely used since the 1970s than any other numerical methods, because of its simple formulation and implementation, and its easy scalability to large computations. This paper briefly overviews the advances in finite difference simulations, focusing particularly on earthquake mechanics and the resultant wave radiation in the near field. As the finite difference formulation is simple (interpolation is smooth), an easy coupling with other approaches is one of its advantages. A coupling with a boundary integral equation method (BIEM) allows us to simulate complex earthquake source processes.

1. Overview of finite difference approach in seismology

Seismic waves radiated from an earthquake propagate in the Earth, which is often considered as an elastic medium although the waves attenuate due to some anelasticity. The theory of wave propagation in an elastic medium was well established before the 20th century. In the early years of seismology, researchers tackled the relevant equations to obtain analytical (or semi-analytical) solutions [1]. These solutions are usually limited to some simple structures (e.g. 1D-layered model without any lateral medium heterogeneity). These solutions, however, are still useful as seismologists need a simple solution to rapidly analyse earthquakes. The advances in computing environments allowed the direct evaluation of the differential equation in the 1970s [2]. Since the mid-1990s, seismologists have been able to routinely carry out huge computations for various applications [3].

Let us consider a 3D isotropic elastic medium, characterized by two elastic (Lamé) coefficients λ and μ and density ρ. We denote displacement, velocity and stress as u, v, and τ, respectively. Taking a Cartesian coordinate (x, y, z), the equations of motion are written as

[1] Sabbatical leave during 2012 to the Disaster Prevention Research Institute of Kyoto University, Japan.

$$\rho\frac{\partial^2}{\partial t^2}u_i = \frac{\partial}{\partial x_j}\tau_{ij} + F_i$$

where indices represent each component or direction in the coordinates, and the summation convention about the same index is assumed. F is a body force (usually zero except for a particular source). The strain is defined as

$$\varepsilon_{ij} = \frac{1}{2}\left(\frac{\partial u_i}{\partial x_j} + \frac{\partial u_j}{\partial x_i}\right).$$

The constitutive relation between the stress and the strain is

$$\tau_{ij} = c_{ijkl}\varepsilon_{kl}$$

and

$$c_{ijkl} = \lambda\delta_{ij}\delta_{kl} + \mu(\delta_{ik}\delta_{jl} + \delta_{il}\delta_{jk})$$

where $\delta_{ij} = 1$ if $i = j$, otherwise 0. The differential equations of the seismic wave propagation consist of

$$\rho\frac{\partial}{\partial t}v_x = \frac{\partial}{\partial x}\tau_{xx} + \frac{\partial}{\partial y}\tau_{xy} + \frac{\partial}{\partial z}\tau_{xz} + F_x \text{ and two other equations,}$$

and

$$\frac{\partial}{\partial t}\tau_{xx} = \lambda\left(\frac{\partial}{\partial x}\mu_x + \frac{\partial}{\partial y}v_y + \frac{\partial}{\partial z}v_z\right) + 2\ \frac{\partial}{\partial x}v_x \text{ and five other equations.}$$

This is called the stress-velocity formulation, for which it is easy to discretize the differential terms. For reasons of space, we have shown only one equation for each component of velocity and stress.

As for many other differential equations, various approximations of these equations are possible. However, the staggered-grid discretization in space and time is common for good stability [4]. Figure 1 shows the spatial definition of each component of velocity and stress, noting that each component is separated from the others by half a grid step (except for the diagonal component of stress). Stress and velocity are evaluated at different time steps (by half a time step). Recently partially-staggered grids were proposed [5] to combine all the stress components at one point and all the velocity components at another point separated by half a grid step. Such a modified method gives a better estimation of the boundary conditions, but additional operations are necessary. In many cases, the 4th order approximation on the spatial differential operator is adopted because of its efficiency;

$$\frac{\partial}{\partial x}f(x) = \frac{1}{h}\left(\frac{9}{8}\left\{f(x+\frac{h}{2}) - f(x-\frac{h}{2})\right\} - \frac{1}{24}\left\{f(x+\frac{3}{2}h) - f(x-\frac{3}{2}h)\right\}\right).$$

Note that a time step is taken as $\Delta t \le c\cdot\Delta s / V_{max}$ and the upper frequency limit is conventionally $f_{max} = V_{min}/(5\ \Delta s)$, where V_{max} and V_{min} are the maximum and minimum wave velocities in the modeled structure, respectively and Δt and Δs are the time step and grid spacing. The constant c is often taken to be about 0.5.

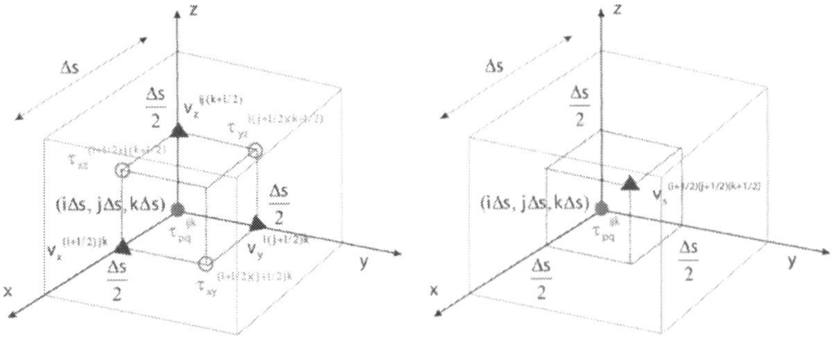

UNIT STAGGERED-GRID PARTIALLY-STAGGERED-GRID

Figure 1. Illustration of standard staggered grid and partially staggered grid. The denoted components are defined half a grid apart from the others. For simplicity, equal grid size is taken for all the directions.

Seismologists often treat a finite volume extracted from the Earth for a regional problem (e.g. a cube of 100 x 100 x 50 km from a sphere of radius 6400 km). Therefore, we should not have any artificial reflections from the given model boundaries, which should continue propagating outwards. For this purpose, an absorbing condition is required as a boundary to eliminate the reflected waves or a boundary zone absorbing the seismic waves [6]. Appropriate absorbing boundary conditions over a wide range of frequencies allow proper calculation not only of the wave propagation phase but also the residual displacement field. This feature becomes important when seismologists are interested in geodetic deformation of very long periods as well as short-period ground motions.

Figure 2 shows a simulation example of the wave propagation during the M9 2011 Tohoku-oki, Japan, earthquake, which was extremely large (a few hundred kilometers of fault length and more than a hundred seconds of rupture duration). High-sampling (1 Hz) continuous GPS observed the deformation process in displacement. We aim to reproduce such deformations without losing information on the residual absolute displacement, which is difficult to obtain from accelerometric data because of sensor response and low-frequency noise. In figure 2, we show two finite difference simulations low-pass filtered (cut-off at 50 s); one is properly calibrated to use an absorbing boundary condition (FDM1), while the other, using the same code, is not (FDM2). We compare with the reference solution synthetically calculated with the wave-number method in the same 1D structure. We observe that FDM1 provides a reasonable solution compared to the reference solution including the residual displacement, while FDM2 has some bias. We note that both finite difference simulations are consistent in a band-pass filter range (1 to 10 s), so that the difference comes from the response at long periods. In fact, FDM1 is calibrated in its absorbing condition so as to work better at around a period of 100 s, while FDM2 is our conventional simulation of wave propagation adopted for a period of 1 s. This discrepancy is important to understand when applying finite difference schemes for long-period deformation processes. The simulations do not perfectly match the observations because of the use of a simple source model with very few parameters, but these simulations are good enough as a first-order approximation.

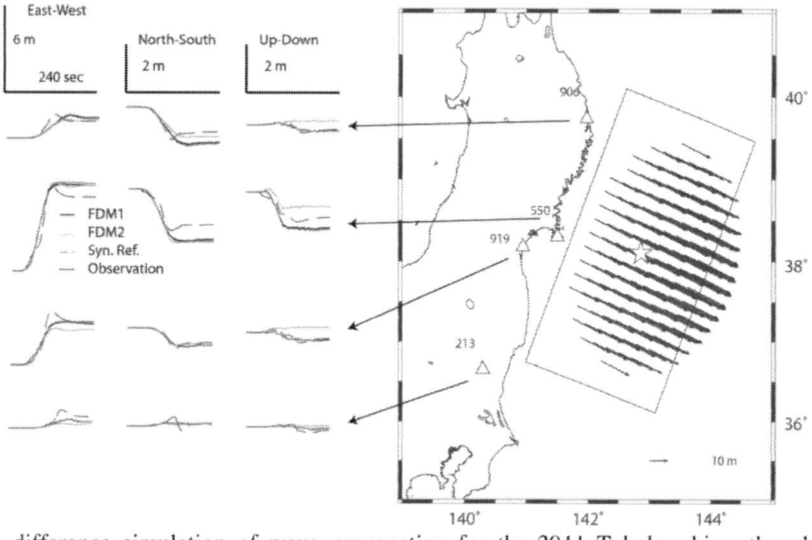

Figure 2. Finite difference simulation of wave propagation for the 2011 Tohoku-oki earthquake. A simple source model (a huge ellipse with Gaussian slip distribution and a constant rupture speed) is assumed (Madariaga, pers. comm. 2012). The waveforms (deformation) applied with a low-pass filter at 50 s are shown at four GPS stations. FDM1: Absorbing boundary condition is adapted for a period of 100 s, FDM2: the same code in the same condition except for the absorbing boundary condition calibrated for a conventional seismological frequency range of 1 s. Syn. Ref.: synthetic reference solution calculated with a discrete wavenumber method in the same 1D layered structure. Observation: displacement recorded by high-sampling (1Hz) continuous GPS network of the Geospatial Information Authority of Japan.

Journal of Physics: Conference Series **454** (2013) 012010 doi:10.1088/1742-6596/454/1/012010

Other aspects, such as implementation of earthquake sources, treatment of the ground surface and material contrast are found in various studies [7]. The extensions to visco-elastic medium are given in [8]. The finite difference scheme has been generally developed for equally-spaced structural grids, so that it is rather easy to manage memory allocation in parallel computing. However it is possible to mix different grids or even combine with finite element grids [9]. From their early era of development, finite difference methods are used to apply a Message Passing Interface (MPI). This is necessary to reduce the unbalances between CPUs. New approaches using graphic card as well as the usual CPU (GPU-CPU) have been recently developed [10].

2. Earthquake physics and ground motion

2.1. Coupling with a boundary integral equation method

The finite difference scheme always has difficulty in treating dislocations (earthquake faulting) in the medium, although some improvement is possible [11]. The reason why this is difficult is that the dislocation problem is a discontinuous phenomenon in the continuum medium, governed by nonlinearity (fracture mechanics). Nevertheless, there is no problem in calculating wave propagation for any known sources within the proper frequency range. The convenient and practical solution is to combine it with other methods, in particular with a boundary integral equation (boundary element) method specialized for the dislocation [12]. Figure 3 shows the schematic illustration of the simulation procedure. In the following session, we present a study on the rupture dynamics and wave propagation in the very-near-field of an earthquake.

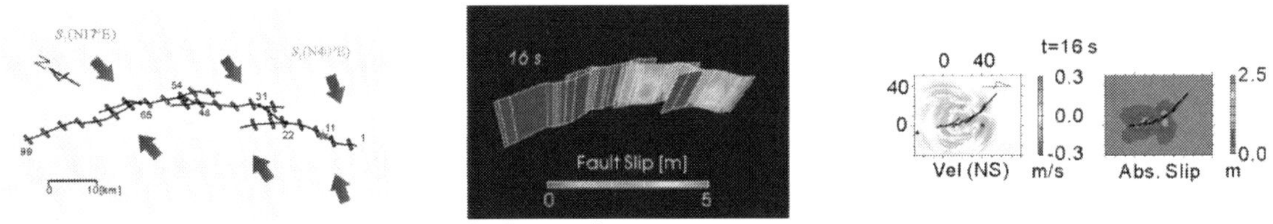

A: Initial Condition

B: Dynamic rupture process (boundary integral equation)

C: Near-field ground motion (finite difference simulation)

Figure 3. Simulation strategy of earthquake dynamics to wave propagation starting from the physical and geological initial condition (A), using a boundary integral equation (B) and a finite difference method (C). The shown example is the case of the Mw7.3 1992 Landers earthquake [13].

2.2. Very-near-field ground motion

2.2.1. *Unresolved scientific problems.*

Nowadays, seismologists are able to describe the earthquake rupture phenomena by kinematics in order to explain the regional or teleseismic data set (kinematic inversion). In other words, the regional seismic wave propagation (say at more than 10 km in a low frequency range below 0.5 – 1 Hz) is reasonably comprehensible. However, some observations very close to the causal fault (those leading to significant structural damage) can be somewhat unexpected. Although it is considered that these mechanisms are related to the dynamic rupture process of the causal fault, each case is not yet well resolved. The 2008 Mw6.9 Iwate-Miyagi, Japan, earthquake is remembered for the maximum recorded acceleration of more than 3800 cm/s^2 at the surface sensors of the station IWTH25 (Kik-net, NIED) located above the ruptured thrust fault (figure 4). An acceleration of more than 1000 cm/s^2 is observed even at the borehole sensor (280 m depth) at the same station. The PGV also reaches about 100 cm/s. We focus on the fact that an extreme ground motion is found even at depth, and this should have originated from the source process. In particular, the PGV is found at the very beginning of the ground motions, about 2 seconds after the arrival of S wave, as the first main pulse.

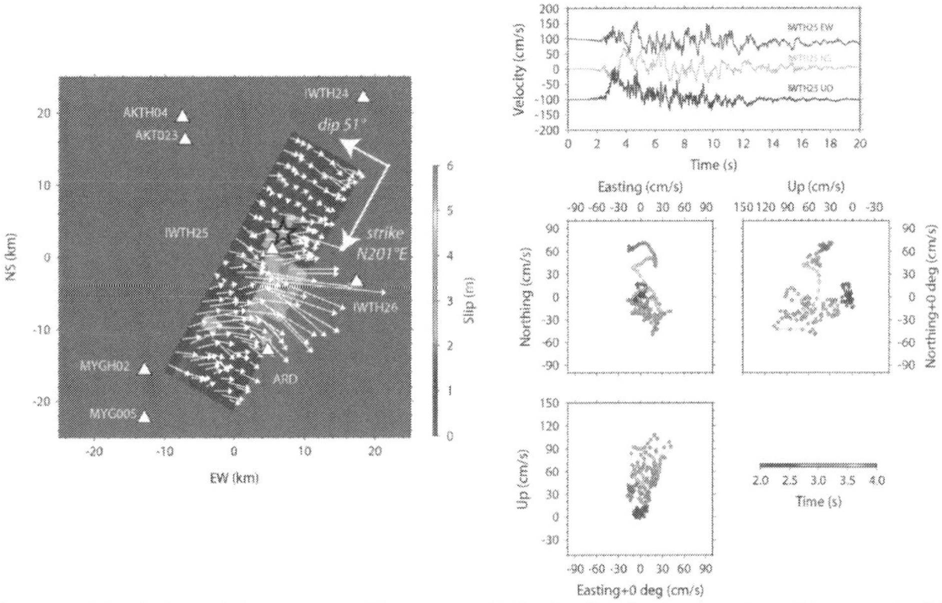

Figure 4. Map of the fault model (courtesy of Dr. Asano [14], slip distribution in color with arrows) of the 2008 Miyagi-Iwate earthquake in the left panel. Epicenter is shown as a star. A seismograph station, IWTH25, is located above the fault plane and close to the estimated epicenter, and this station recorded the very strong ground motions (100 cm/s) as shown in the right panel. The three components recorded at borehole (260 m depth) and the trajectory between 2 and 4 seconds, the arrival of the strong pulse, are shown. Very strong vertical motion is recorded at around 3.5 second.

2.2.2. *Kinematic and dynamic source process and wave radiation.*

We compute the wave propagation from the published kinematic source models using a finite difference method. The 1D structure is assumed. We aim to calculate up to 2 Hz. We first model the ground motion based on the rupture scenario obtained from the kinematic inversion [14] (Fault strike = N201°E, dip = 51°, focal depth = 7.77 km). As already pointed out [15], mainly because of the resolution limit of the inversion model (subfaults of 2 km), the strong first pulse is hardly reproduced in terms of its huge amplitude and short duration. This infers that a heterogeneity equivalent or smaller than this scale may be important. Then we try to simulate the beginning of the rupture process (about the first 3 s). We investigate kinematically the effect of a small circular patch (radius = 1 km) on the ground motion at IWTH25 by shifting its location. For the given geometrical situation, it is found that a patch not located beneath the station, but slightly shallower (2 km away from the hypocentral depth) can give a relatively significant fault parallel (e.g. significant vertical) component in the ground motion, but no model leads to a perfect match to the observations.

We then try to model the rupture process dynamically supposing not the kinematic scenario but the physical parameters (stress field and rupture criterion). We assume a uniform shear stress field (5 MPa) except for a patch representing heterogeneity (if any) as demonstrated in figure 2. A simple slip-weakening law is assumed (peak strength = 10 MPa, residual stress = 0 MPa, critical slip displacement = 5cm). Introducing a patch with higher initial stress generates heterogeneity in the rupture process. On the patch, the rupture passes locally faster than in other locations and leads to larger slip. We calculate the ground motion from this source model again, as shown in figure 5, in which, for reference, the case of uniform dynamic rupture (without the patch) and the kinematic source model are also shown. The synthetic seismograms still have a smoother waveform regardless of the heterogeneity. The ground motion is a convolution over the fault plane, so the rupture propagation on the rest of the fault plane other than the given small patch affects it. We encounter two difficulties. The station is unfavorably located with respect to the S-wave radiation from the assumed fault plane, and it is difficult to give the dynamic model a strong heterogeneity at the beginning without significantly affecting the later rupture process. As recent geodetic studies imply [16], the true fault geometry may not be as simple as a single fault.

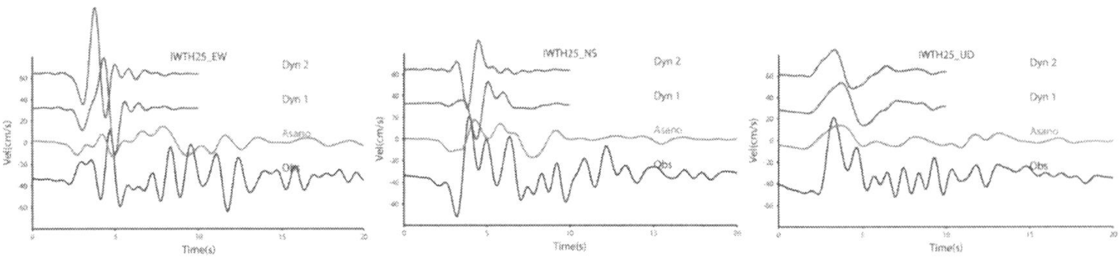

Figure 5. Comparison of seismograms (low-pass filtered at 1 Hz) at IWTH25 at depth for various source models. The calculation is made by the finite difference method using a 1D structure model beneath IWTH25. From top to down, dynamic rupture model presented with heterogeneity, dynamic rupture model under the homogeneous condition, kinematic rupture model of Asano and Iwata (2008), and the observation. For the three synthetic models, the origin time is set at t=0. Because of uncertainty in the real origin time, the observation may be shifted in time. The dynamic rupture model provides a shaper pulse, but no model can explain the pulse of the first two seconds.

2.2.3. *Discussion – do we really know the earthquake process?*

In fact, we have been trying to simulate the recent earthquakes according to the known parameters (source parameters and structure) [17]. We think that such reproduction process is important, because predicting the ground motion in any given situation is the objective of the quantitative seismic hazard assessment. From our several experiences, we find that the "known" parameters are not always satisfactory to explain the phenomena. This is because the earthquake process is too complex to determine all the numerous parameters necessary to construct the models. Any parameter is obtained under some hypothesis, e.g., fixing the other parameters. This is also why inversion analyses sometimes provide different models even using the same data set.

The seismology (and also the other domains of geoscience) is always in a knowledge gap; we try to model more precisely the phenomena using any numerical sophisticated method and intensive calculation (forward modeling), while the model parameters are not acquired precisely enough regardless of the increase of high-quality various data (inverse modeling). The sustainable efforts to fill in this gap are required, since this cannot be done easily. The intensive numerical simulation will be able to be a powerful tool towards this objective.

3. Perspectives and conclusion

Nowadays, when the accuracy of the calculation is aimed, higher-order finite element methods (spectral element or discontinuous Galerkin) could be better regardless of their costs [18]. Each method always has its merits and limits so that it is waited that various methods are developed and utilized. Such numerical codes have become also popular for engineering/industrial purpose such as the quantitative seismic hazard/risk assessments [19], however guaranteeing the numerical performance on parallel computers and obtaining the physically reasonable solutions are difficult tasks. Since 2000, various international benchmark and cross-checking tests have been also performed [20]. This is necessary for us to advance beyond.

The finite difference formulation for the seismic wave propagation is the most primary scheme. However, this always shows a good (computing) performance and this is often satisfactory for tackling the unsolved physical problems, as shown briefly in this paper. The basic formulation of finite difference is very well established and we can relatively easily handle the intensive calculation. Thus making full use of it, we are now able to explore the earthquake dynamics and the consequent ground motion. Coupling with a boundary integral equation is a practical option to take advantage of both methods. At the end, we precise that the finite difference scheme is classical, but the formulation always evolves for improving solutions and developing for other utilities [21].

Acknowledgments

We thank Dr John Douglas and an anonymous reviewer for their comments to improve the manuscript. This paper is a contribution to the French national project ANR DYNTOHOKU (2011-JAPN-009). Most of the recent numerical simulations are carried out at the French national computing center GENCI-CINES (grant 2012-46700), and some other results realised in previous projects are also included (BRGM, ANR, and CARNOT). H.A. also thanks DPRI – Kyoto University for his sabbatical stay in 2012. The data recorded by National Institute of Earth Science and Disaster Prevention, Japan (strong ground motion) and Geospatial Information Authority of Japan (GPS) are used.

References

[1] Aki K and Richards P 2002 *Quantitative seismology, 2nd edition* (University Science Books)

[2] Boore D M 1972 *Finite difference methods for seismic wave propagation in heterogeneous materials, in Methods in Computational Physics* (Academic Press) pp 1-37

Madariaga R 1976 Dynamics of an expanding circular fault *Bull. Seism. Soc.* **66**, 639-666

[3] Olsen K B, Archuleta R J and Matarese J. Three-dimensional simulation of a magnitude 7.75 earthquake on the San Andreas fault *Science* **270** 1628-1632

Furumura T and Chen L. 2004 Large scale parallel simulation and visualization of 3D seismic wavefield using the Earth Simulator *Comp. Model. Eng. Sci.* **6** 153-168

[4] Virieux J and Madariaga R. 1982 Dynamic faulting studied by a finite difference method *Bull. Seism. Soc. Am.* **72** 345-369

[5] Saenger E H, Gold N and Shapiro S A 2000 Modeling the propagation of elastic wave waves using a modified finite-difference grid *Wave Motion* **31** 77-92

[6] Clayton R and Engquist B 1977 Absorbing boundary conditions for acoustic and elastic wave equations *Bull. Seism. Soc. Am.* **67** 1529-1540

Collino F and Tsogka C 2001 Application of the perfectly matched absorbing layer model to the linear elastodynamic problem in anisotropic heterogeneous media *Geophysics* **66** 294-307

Komatitsch D and Martin R 2007 An unsplit convolutional perfectly matched layer improved at grazing incidence for the seismic wave equation *Geophysics* **72** SM155-167

[7] Graves R 1996 Simulating seismic wave propagation in 3D elastic media using staggered-grid finite differences *Bull. Seism. Soc. Am.* **86** 1091-1106

Mozco P, Kristek J, Galis M, Pazak P and Balazovjech M 2007 The finite-difference and finite-element modeling of seismic wave propagation and earthquake motion *Acta Physica Slovaca* **57** 177-406

[8] Day S M and Bradley C 2001 Memory-efficient simulation of anelastic wave propagation *Bull. Seism. Soc. Am.* **91** 520-531

Kristek J and Moczo P 2003 Seismic-wave propagation in viscoelastic media with materialdiscontinuities: A 3D fourth-order staggered-grid finite-difference modeling *Bull. Seism. Soc. Am.* **93** 2273-2280

[9] Aoi S and Fujiwara H 1999 3-D finite-difference method using discontinuous grids *Bull. Seism. Soc. Am.* **89** 918-930

Ducellier A and Aochi H 2012 Effects of the interaction of topographic irregularities on seismic ground motion, investigated by the hybrid FD-FE method *Bull. Earthq. Eng.* **10** 773-792

[10] Dupros F, Aochi H, Ducellier A, Komatitsch D and Roman J 2008 Exploiting intensive multithreading for efficient simulaton of seismic wave propagation *11th Int. Conf. Computational Science and Engineering* 253-260 doi:10.1109/CSE.2008.51

Michéa D and Komatitsch D 2010 Accelerating a 3D finite-difference wave propagation code using GPU graphics cards *Geophys. J. Int.* **182** 389-402

[11] Cruz-Atienza V M, Virieux J and Aochi H 2007 3D finite-difference dynamic-rupture modeling along nonplanar faults *Geophysics* **72** SM123-137

Dalguer L A and Day S M 2007 Staggered-grid split-node method for spontaneous rupture simulation *J. Geophys. Res.* **112** B02302 doi:10.1029/2006JB004467

[12] Aochi H and Madariaga R 2003 The 1999 Izmit, Turkey, earthquake: Nonplanar fault structure, dynamic rupture process, and strong ground motion *Bull. Seism. Soc. Am.* **93** 1249-1266

Fukuyama E, Ando R, Hashimoto C, Aoi S and Matsu'ura M 2009 Physics-based simulation of the 2003 Tokachi-oki, Japan, earthquake to predict strong ground motions *Bull. Seism. Soc. Am.* **99** 3150-3171

[13] Aochi H, Fukuyama E and Madariaga R 2003 Constraints of fault constitutive parameters inferred from non-planar fault modeling, *Geochemistry, Geophysics, Geosystems* **4** doi:10.1029/2001GC000207

[14] Asano K and Iwata T 2008 Kinematic source rupture process of the 2008 Iwate-Miyagi Nairiku earthquake, a Mw6.9 thrust earthquake in northeast Japan, using Strong Motion Data *EOS Trans. Am. Geophys. Union* **89**(53) Fall Meeting Suppl. Abstract S23B-1890.

[15] Suzuki W, Aoi S and Sekiguchi H 2010 Rupture process of the 2008 Iwate-Miyagi Nairiku, Japan, earthquake derived from near-source strong-motion records *Bull. Seism. Soc. Am.* **100** 256-266

[16] Takada Y, Kobayashi T, Furuya M and Murakami M 2011 Post-seismic surface deformation following the 2008 Iwate-Miyagi Nairiku earthquake detected by ALOS/PALSAR *J. Geod. Soc. Jpn.* **57** 181-193

[17] Aochi H, Durand V and Douglas J 2011 Influence of super-shear earthquake rupture models on simulated near-source ground motion from the 1999 Izmit (Turkey) earthquake *Bull. Seism. Soc. Am.* **101** 726-741

Aochi H, Ducellier A, Dupros F, Delatre M, Ulrich T, de Martin F and Yoshimi M 2013 Finite difference simulations of the seismic wave propagation for the 2007 Mw6.6 Niigata-ken Chuetsu-oki earthquake : Validity of models and reliable input ground motion in the near field, *Pure appl. Geophys.* **170** 43-64.

[18] Komatitsch D and Vilotte J P 1998 The spectral-element method: An efficient tool to simulate the seismic response of 2D and 3D geological structures *Bull. Seism. Soc. Am.* **88** 368-392

Chaljub E, Komatitsch D, Vilotte J, Capdeville Y, Valette B, and Festa G 2007 Spectral element analysis in seismology, *Advances in wave propagation in heterogeneous media, Advances in Geophysics* (Elsevier Academic Press) 365-419

Dumbser M and Käser M 2006 An arbitrary high-order discontinuous Galerkin method for elastic waves on unstructured meshes – II. The three-dimensional isotropic case *Geophys. J. Int.* **167** 319-336

Etienne V, Chaljub E, Virieux J and Glinsky N 2010 An hp-adaptive Galerkin finite-element method for 3D elastic wave modelling *Geophys. J. Int.* **183** 941-962

De Martin F 2011 Verification of a spectral-element method code for the Southern California Earthquake Center LOH.3 Viscoeastic case *Bull. Seism. Soc. Am.* **101** 2855-2865

[19] Douglas J and Aochi H 2008 A survey of techniques for predicting earthquake ground motions for engineering purposes *Survey in Geophysics* **29** 187-220

[20] Day S M, Bielak J, Dreger D, Graves R, Larsen S, Olsen K B and Pitarka A 2001 Tests of 3D elastodynamic odes: final report for lifelines project 1A01 *report to Pacific Earthquake Engineering Research Center*

Chaljub E, Moczo P, Tsuno S, Bard P, Kristek J, Käser M, Stupazzini M and Kristekova M 2010 Quantitative comparison of four numerical predicsions of 3D ground motion in the Grenoble Valley, France *Bull. Seism. Soc. Am.* **100** 1428-1455

Harris R A, Barell M, Andrews D J, Duan B, Ma S, Dunham E M, Gabriel A, Kaneko Y, Kase Y, Aagaard B T, Oglesby D D, Ampuero J, Hanks T C and Abrahamson N 2011 Verifying a compuational method for predicting extreme ground motion *Seis Res Lett* **82** 638-644

[21] Kozdon J E, Dunham E M and Nordstrom J 2012 Simulation of dynamic earthquake ruptures in complex geometries using high-order finite difference methods *J. Scientific Computing* doi:10.1007/s10915-012-9624-5

Zhang W, Zhang Z and Chen X 2012 Three-dimensional elastic wave numerical modelling in the presence of surface topography by a collocated-grid finite-difference method on curvilinear grids *Geophys. J. Int.* **190** 358-378

Journal of Physics: Conference Series **454** (2013) 012011 doi:10.1088/1742-6596/454/1/012011

Finite difference time domain method for calculating the band structure of a 2D photonic crystal and simulating the lensing effect

S Rafiee Dastjerdi, M Ghanaatshoar

Laser and Plasma Research Institute, Shahid Beheshti University, G.C., Evin 1983963113, Tehran, Iran

Email: m-ghanaat@sbu.ac.ir

Abstract. A finite difference time domain method based on regular Yee's algorithm in an orthogonal coordinate system is utilized to calculate the band structure of a two-dimensional square-lattice photonic crystal comprising dielectric cylinders in air background and to simulate the image formation of mentioned structure incorporating the perfectly matched layer boundary condition. By analyzing the photonic band diagram of this system, we find that the frequency region of effective negative refraction exists in the second band in near-infrared domain. In this case, electromagnetic wave propagates with a negative phase velocity and the evanescent waves can be supported to perform higher image resolution.

1. Introduction

Photonic crystals (PCs) as human-made structures are able to modify the flow of light by means of periodic spatial modulation of the refractive index [1, 2]. These structures exhibit the frequency band gap which eliminates the electromagnetic wave propagation [3]. Currently PCs in the form of meta-materials have attracted a lot of attention due to anomalous behaviour that has potential to develop biomedical imaging, optical communications and optoelectronic technologies. These meta- materials are able to restore both the phase of propagating waves and amplitude of evanescent waves to make perfect lens and as a result can overcome the diffraction limit of conventional convex lenses [4, 5].

Because of strong diffraction and multiple scattering in periodic structure of PCs, Maxwell's equations need to be numerically solved. Finite difference time domain method (FDTD) has been widely applied in simulating and modelling photonic band gap materials. This method linearly scales with the simulation dimension and can be implemented in a parallel computing environment very efficiently [6].

In this paper, FDTD method based on common Yee's lattice is implemented for calculating the dispersion band diagram of a 2D photonic crystal made from partitioned cylinders. We use the standard central-difference equations in an orthogonal grid scheme. By analyzing the band structure we show that this model system can be regarded as a superlensing device which can work at near-infrared regime. We also employ the FDTD method with perfectly matched layer (PML) boundary condition for simulating the field pattern of the point source and its image.

2. FDTD method for calculating the band structure

The FDTD method imitates the electromagnetic wave by volumetric sampling of unknown fields [7, 8]. In photonic band structure materials where the dielectric constant is periodically modulated, the electric and magnetic fields of the electromagnetic waves, $\vec{E}(\vec{r})$ and $\vec{H}(\vec{r})$, through the structure can be described by band index n and a wave vector k in irreducible Brillouin zone due to Bloch's theorem:

$$\begin{aligned} \vec{E}_{n,k}(\vec{r}) &= u_{n,k}(\vec{r})\exp(i\vec{k}.\vec{r}), \quad u_{n,k}(\vec{r}+\vec{R}) = u_{n,k}(\vec{r}) \\ \vec{H}_{n,k}(\vec{r}) &= v_{n,k}(\vec{r})\exp(i\vec{k}.\vec{r}), \quad v_{n,k}(\vec{r}+\vec{R}) = v_{n,k}(\vec{r}) \end{aligned} \tag{1}$$

where $\vec{R}$ is the lattice constant and $u_{n,k}(\vec{r})$ and $v_{n,k}(\vec{r})$ are periodic functions related to the Bloch waves. From (1), one can find eigenvalues of the periodic structure for a given periodic boundary condition. In order to compute the eigenmodes of the structure by means of FDTD method, random initial condition is used. Also we need sufficient amount of time to catch enough accuracy. For attaining eigenmodes we choose various low-symmetry locations in the unit cell as probes, in order to record peaks of the Fourier transform of the complex field components in the time domain for a given propagation constant. We should notice that, probes at high-symmetry locations of the unit cell are not able to detect all eigenmodes.

The PC proposed here is a 2D square lattice consisting of periodic arrays of long dielectric partitioned rods of Silica (SiO_2) and Silicon (Si) immersed in air background. It is assumed that the length of cylinders is long enough compare with the lattice constant. Furthermore, the width of the slab in the x direction is limited while in the y direction is unrestricted. Dielectric constants of two substances are $\varepsilon_{SiO_2} = 2.1$ and $\varepsilon_{Si}=12.1$ at $\lambda=1.55$ μm and the radius of cylinders is $R=0.45a$, where a, is the lattice constant of the PC structure. Here the lattice constant a, is set to 460 nm and the TM-polarized electromagnetic wave is considered for calculating the band structure and simulating the electromagnetic wave propagation. Figure 1 displays the band diagram of proposed configuration.

It should be noted that the partitioned cylinders create a complete band gap between second and third bands while there is no band gap within the first five bands of the PC that includes just silicon cylinders.

To achieve the frequency with the negative refractive effect, we should pay attention to some important conditions. First of all, the frequency should be below $0.5\times 2\pi c/a$ to avoid higher order Bragg diffractions (c is the speed of light in vacuum); secondly, the equifrequency surface (EFS) of the PC around the center of Brillouin zone (Γ point), should have an inward gradient leading to negative group velocity; and finally, the PC's EFS should contain the EFS of the background [9, 10]. We choose the frequency of the intersection point of PC and air band diagrams to evaluate the imaging properties of this structure. The intersection normalized frequency, $\omega a/2\pi c$, is around 0.295. It is apparent from Figure 1, that there is an inward gradient around the Γ point. Moreover, for this normalized frequency, the PC can be represented as a nearly isotropic medium and the normalized frequency can be managed by the refractive index of the materials incorporated in the PC [11].

3. Near-infrared imaging

We consider a $4a$-thick PC slab and locate a point source working at 0.295 normalized frequency at distance "a" from the left side of the PC structure. Then we utilize FDTD method with PML boundary condition to simulate the propagation of TM-polarized electromagnetic wave. Figure 2 shows the electric field distribution of the point source and its image. Two lines show the boundaries of the PC slab.

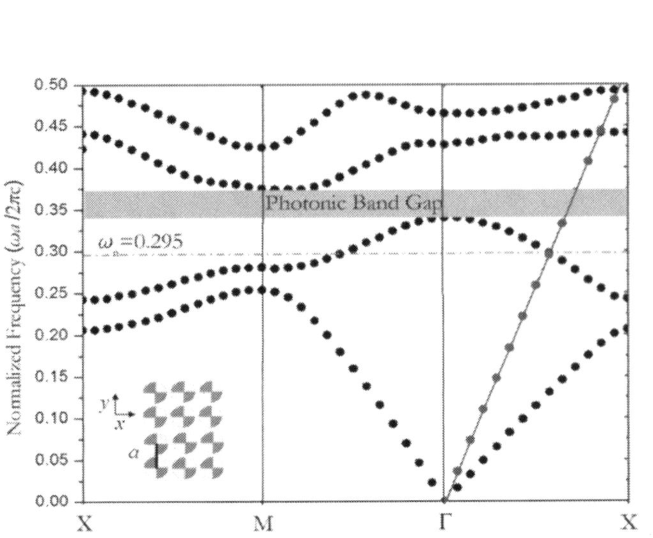

Figure1. Band diagram of corresponding structure composed of Si and SiO_2 cylinders in air with the radius of R=0.45a. Inset: partitioned cylinder PC made from Si (light parts) and SiO_2 (dark parts) with lattice constant a.

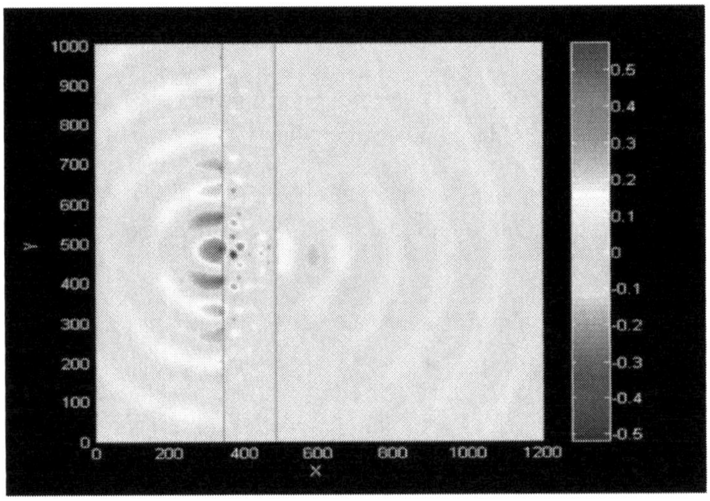

Figure2. Electric field distribution of the point source located at the distance "a" from the left side of the PC slab and its image in the x-y plane.

According to Figure 2 we can easily see the point image of the point source on the right side of the PC slab and far from it. Narrowing behavior of the wave front of electromagnetic wave when transmits through the slab, implies that negative refraction effect is the dominant phenomena in image formation.

4. Conclusion

In this report, we introduced a novel 2D photonic crystal made by Si and SiO_2 partitioned cylinders to investigate the image formation characteristics. We used FDTD method for band diagram calculations and imaging simulation of this system. We found that the band diagram of this structure opens a gap between the second and third bands and satisfies three conditions required to create a PC superlens.

References

[1] Yablonovitch E 1987 *Phys. Rev. Lett.* **58** 2059.
[2] Joannopoulos J D, Villeneuve P R and Fan S 1997 *Nature* **386** 143.
[3] Joannopoulos J D, Mead R D and Winn J N 1995 *Photonic Crystals: Molding the flow of light* (Princeton University Press, Princeton, NJ).
[4] Veselago V G 1968 *Sov. Phys. Usp.* **10** 509.
[5] Pendry J B 2000 *Phys. Rev. Lett.* **85** 3966.
[6] Kuang W, Kim W J and O'Brien J D 2007 *J.lightwave Tech.* **25** 2612.
[7] Taflove A and Hagness S C 2000 *Computational Electrodynamics: The Finite-Difference Time Domain Method* (Norwood, MA: Artech House).
[8] Yee K S 1966 *IEEE Trans. Antennas Propag.* **AP-14** 302.
[9] Notomy M 2000 *Phys. Rev. B* **62** 10696.
[10] Luo C, Johnson S G, Joannopoulos J D and Pendry J B 2003 *Phys. Rev. B* **68** 045115.
[11] Wang X, Ren Z F and Kempa K 2004 *Opt. Exp.* **12** 2919.

Resolving the outer density profile of dark matter halo in Andromeda galaxy

Takanobu Kirihara, Yohei Miki and Masao Mori
University of Tsukuba, Tennodai 1-1-1, Tsukuba, Ibaraki, Japan
E-mail: kirihara@ccs.tsukuba.ac.jp

Abstract. Large-scale faint structure detected by the recent observations in the halo of the Andromeda galaxy (M31) provides an attractive window to explore the structure of outer cold dark matter (CDM) halo in M31. Using an N-body simulation of the interaction between an accreting satellite galaxy and M31, we investigate the mass-density profile of the CDM halo. We find the sufficient condition of the outer density profile of CDM halo in M31 to reproduce the Andromeda giant stream and the shells at the east and west sides of M31. The result indicates that the density profile of the outer dark matter halo of M31 is a steeper than the prediction of the theory of the structure formation based on the CDM model.

1. Introduction

The cold dark matter (CDM) model, which is the standard paradigm for cosmological structure formation, predicts that galaxies have experienced many mergers with less massive galaxies and have grown larger. Cosmological simulations of structure formation suggest that dark matter halos associated with galaxies have some universal density profile under an assumption of spherical symmetric halo model such as Navarro-Frenk-White (NFW) profile [1] or Fukushige-Makino-Moore (FMM) profile [2] [3]. In these profiles, much controversy exists over the mass-density distribution of the inner dark matter halo. With respect to the density distribution of the outermost region, however, the mass density always decreases with the cube of the distance from the galactic center in these profiles. That is to say the outer region of dark matter halo is the excellent laboratory to examine the prediction of CDM model. However, it is extremely difficult to measure the mass distribution of the outer region of a galaxy because the stellar and/or gas density is too low to detect even by the latest instruments. Consequently, observational verifications of the theoretical prediction are not yet enough so far.

Only recently, a deep and wide-field photometric survey of the Andromeda galaxy (M31), which is the nearest large galaxy, revealed that the outer region of M31 showed a wealth of substructures such as the Andromeda giant stream and the shells at the east and west sides of M31[4] [5]. The Andromeda giant stream extends about 120 kpc further away along the line of sight from M31 [6], and the distribution of the radial velocity has already been observed [7]. These large-scale structures have been considered the evidence of a galaxy collision occurred about 1 Gyr ago [8] [9] [10]. The N-body simulations in earlier studies examined the orbits if an infalling satellite dwarf galaxy which reproduce all of these spatial structures [8] [11] [12]. In addition, Mori & Rich [9] showed that the collision of M31 and the infalling satellite with the total mass ranging from $5 \times 10^8 M_\odot$ to $5 \times 10^9 M_\odot$ successfully reproduced the Andromeda giant stream and the shells by the N-body simulations.

Such a large-scale structure spreads far beyond the galactic disk and is suitable for the study of the actual density profile of the dark matter halo in M31. Nevertheless, all of the earlier simulations always assumed that the mass density of the outer dark matter halo of M31 decreased with the cube of the distance from the galactic center in accordance with the prediction of the CDM model. These situations motivate us to test the prediction of the CDM model on the density profile of the dark matter halo, using the N-body simulation for the formation of the Andromeda giant stream in the dark matter halo with the different density profiles. In this paper, we introduce our numerical model in §2. In §3, we present simulation results, and a brief summary is given in §4.

2. Numerical model

In this study, we simulate the interaction between an accreting satellite dwarf galaxy and M31 using N-body simulations, which especially concentrated on the density profile of the dark matter halo in M31. It should be noted that Mori & Rich (2008) [9] studied the self-gravitating response of the disk, bulge, and dark matter halo of M31 to an accreting satellite and concluded that satellites less massive than $5 \times 10^9 M_\odot$ had a negligible effect on the gravitational potential of M31. Consequently, in this study, we treated M31 as the source of a fixed gravitational potential composed of a disk, a bulge, and a dark matter halo. For bulge and disk component, we adopt the model which reproduce the profile of surface brightness of M31 disk and bulge, rotation velocity curve of disk and velocity dispersion of bulge [8] [13].

The NFW model that is widely accepted density profile of CDM halos was empirically derived from cosmological N-body simulation [1]. The resultant empirical profiles of density distribution is approximately fitted by

$$\rho_{\rm NFW}(r) = \frac{\rho_s}{(r/r_s)(1+r/r_s)^2}, \tag{1}$$

where r_s and ρ_s are the scale radius and the scale density, respectively. We focus on making a diagnosis of the density profile in the outer CDM halo using the Andromeda giant stream and the N-body experiments. In this purpose, we introduce the index a in the equation of the density distribution as described below:

$$\rho_{\rm DMhalo}(r) = \frac{\rho_{s,\,a}}{(r/r_{s,\,a})(1+r/r_{s,\,a})^a}, \tag{2}$$

where $r_{s,\,a}$ and $\rho_{s,\,a}$ are the scale radius and the scale density, respectively. They are determined in such a way that the enclosed mass at 7.63 kpc and 125 kpc are identical for all models to reproduce the observed rotation velocity curve of M31. We vary the parameter a ranging from $1.5 < a < 4.0$ in this paper. In this case, NFW profile corresponds to $a = 2$.

To investigate the dynamical response of the orbiting satellite, we modeled the satellite dwarf galaxy by a self-consistent N-body realization of stars under the influence of an external force provided by M31. The satellite dwarf galaxy is adopted as a Plummer's sphere [14] with 49,152 particles. The total mass is $2.2 \times 10^9 M_\odot$, and the scale radius is 1.03 kpc. The initial position vector and velocity vector for the standard coordinates centered on M31 are taken from Fardal et al. (2007) [8]. Computation has been done on the FIRST cluster and Blade-GRAPE on FIRST in calculating self-gravity of satellite particles at Center for Computational Science (CCS), University of Tsukuba.

3. Results

The result of simulations indicates that the first pericentric passage occurred about 0.8 Gyr ago, and the satellite collides almost head-on with M31. Then, the distribution of satellite particles is distorted and is spread out significantly. A large fraction of the satellite particles acquires a high velocity relative to the center of M31. This debris expands to a great distance, remaining collimated and giving rise to the Andromeda giant stream. Stellar particles that initially constituted the satellite start to form a clear shell structure at the north east area of M31. After the formation of the eastern shell, stars move to the west area and produce the western shell. Figure 1 shows the surface mass density of the disrupted satellite galaxy at the present-day epoch. Each panel corresponds to the case for $a = 2.0$ (upper left), the case for $a = 2.5$ (upper right), the case for $a = 3.0$ (lower left), and the case for $a = 3.5$ (lower right). The positional coordinates ξ and η are the eastern and the northern direction on the sky, respectively. They are centered on the center of M31 and 1° in angle corresponds to 13.6 kpc. Black circles show the edges of the observed shells taken from Table 1 of Fardal et al. (2007) [8], and blue circles corresponds to the observed position of the Andromeda giant stream taken from Font et al. (2006) [12].

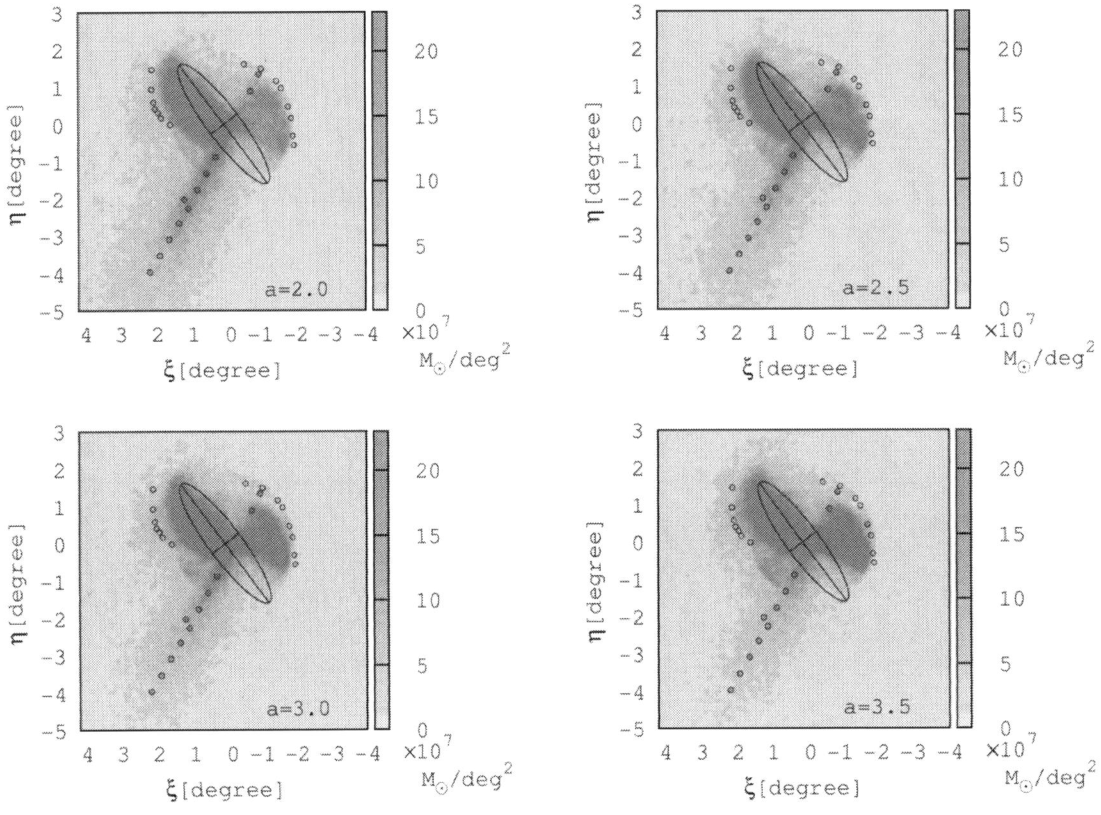

Figure 1. Surface mass density of the satellite galaxy at the present-day epoch. Each panel corresponds to the case for $a = 2.0$ (upper left), the case for $a = 2.5$ (upper right), the case for $a = 3.0$ (lower left), and the case for $a = 3.5$ (lower right). The positional coordinates ξ and η are the eastern and the northern direction on the sky, respectively. They are centered on the center of M31 and 1° in angle corresponds to 13.6 kpc.

When the parameter a becomes greater, the surface density of the Andromeda giant stream becomes lower. This difference mainly comes from the difference of enclosed mass of dark matter halo around the Andromeda giant stream. The larger enclosed mass corresponds to the shorter free-fall time. In other words, the model with a large a further accelerates the dynamical

Journal of Physics: Conference Series **454** (2013) 012012 doi:10.1088/1742-6596/454/1/012012

evolution of the whole system. Since stellar particles move quickly to the eastern shell after the formation of the Andromeda giant stream, the Andromeda giant stream has low surface density in case of a large a.

Figure 2 shows the radial velocity distribution of the Andromeda giant stream. The blue symbols in the figures are observed radial velocity distribution [15]. Red points show the results of N-body simulations. It clearly shows that the radial velocity distribution of all simulations is consistent with that of observations.

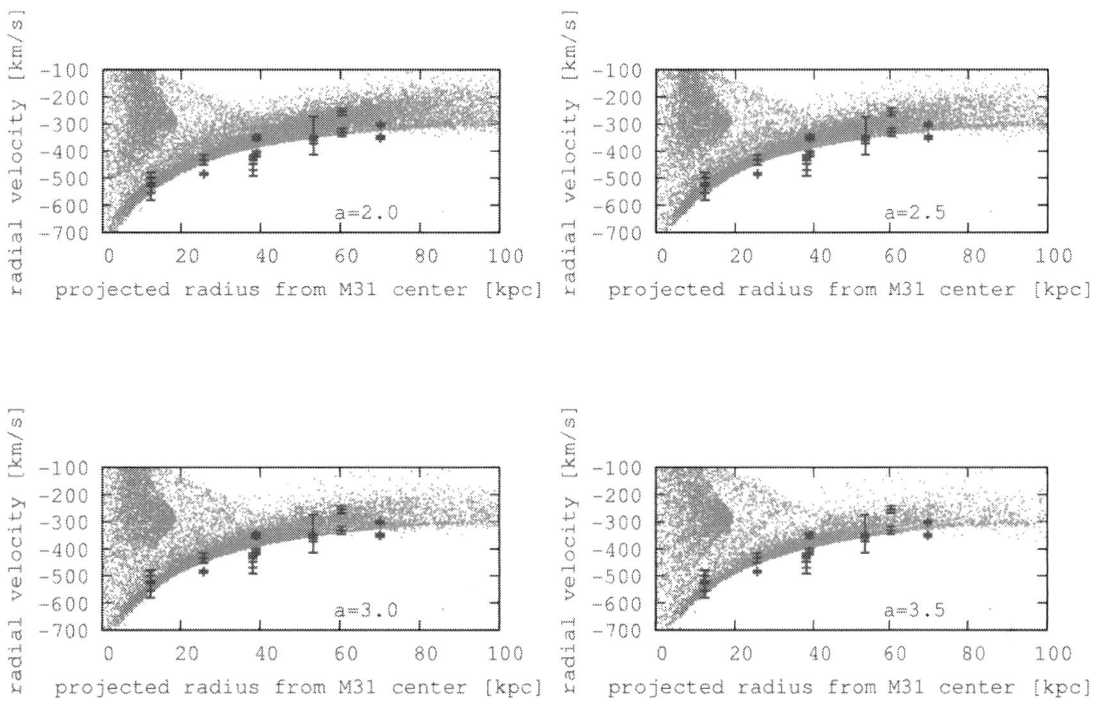

Figure 2. Radial velocity distribution of the Andromeda giant stream. Blue symbols show the observational data of [15]. Red points show the results of N-body simulations in the stream region ($0° \le \xi \le 8°$ and $-10° \le \eta \le 0°$). The upper-left panel shows the case of $a = 2.0$, the upper-right panel shows the case of $a = 2.5$, the lower-left panel shows the case of $a = 3.0$ and the lower-right shows the case of $a = 3.5$.

For the purpose of quantitative comparison between the observation and the simulation, we use the surface density ratio among the Andromeda giant stream and the two shells. Figure 3 shows the result of χ^2 analysis about the mutual combination of the density ratio among the Andromeda giant stream and the two shells. Here, we show the levels of $\chi^2_\nu = 2.3, 6.2$ and 11.8, corresponding to $68, 95$ and 99.7 percent confidence intervals in the case of three parameters.

We must note that the NFW model ($a = 2$) fails to reproduce the apparent Andromeda giant stream and shell structures. Only for the outer dark matter halo with a density power-law slope of $2 < a < 3.5$, the results of simulations successfully reproduce the observed Andromeda giant stream and shell structures. The blue line, corresponds to the NFW model, which is the standard model suggested by cosmological N-body simulations.

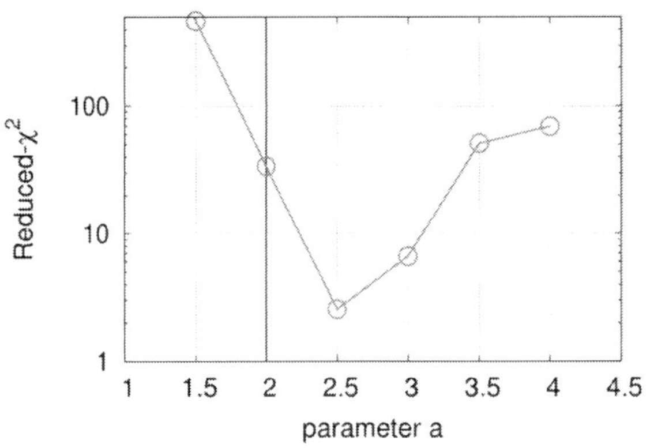

Figure 3. Reduced-χ^2 of the surface density ratio of the Andromeda giant stream, the eastern shell and the western shell as a function of density power-law slope a. The blue vertical line, $a = 2$, is for the prediction by the cosmological N-body simulations based on the cold dark matter model.

4. Summary

We examined the mass-density profile of the dark matter halo of M31 using N-body simulation of the galaxy collision. As a result, the density profile of the dark matter halo suggested by the cosmological N-body simulation based on the CDM model fails to reproduce the Andromeda giant stream and the shell structures. These structures are reproduced for $2 < a < 3.5$, and the best-fit parameter is $a = 2.5$. This result indicates that the mass-density profile of dark matter halo in M31 is steeper than that of the prediction of the CDM model. This discrepancy may be attributed to the triaxiality of the CDM halo, the tidal effect of M33 and Milky Way, and/or the morphology of the infalling satellite galaxy. Such models will be addressed in a forthcoming study.

Acknowledgments

We thank M. Umemura for use of the FIRST cluster at CCS, the University of Tsukuba. This work was partially supported by the program of the Pre-Strategic Initiatives, University of Tsukuba, the FIRST project based on Grants-in-Aid for Specially Promoted Research by MEXT (16002003), and the Grant-in-Aid for Scientific Research (S)(2024002), (A)(21244013), and (C)(25400222).

5. References

[1] Navarro J F, Frenk C S and White S D M 1996 The structure of cold dark matter halos *ApJ* **462** 563
[2] Fukushige T and Makino J 1997 On the origin of cusps in dark matter halos *ApJ* **477** 9
[3] Moore B, Governato F, Quinn T, Stadel J and Lake G 1998 Resolving the structure of cold dark matter halos *ApJ* **499** 5
[4] Ibata R, Irwin M, Lewis G, Ferguson A and Tanvir N 2001 A giant stream of metal-rich stars in the halo of the galaxy M31 *Nature* **412** 49
[5] Irwin M J, Ferguson A M N, Ibata R A, Lewis G F and Tanvir N R 2005 A minor-axis surface brightness profile for M31 *ApJ* **628** 105
[6] McConnachie A W, Irwin M J, Ibata R A, Ferguson A M N, Lewis G F and Tanvir N 2003 The three-dimensional structure of the giant stellar stream in andromeda *MNRAS* **343** 1335
[7] Ibata R, Chapman S, Ferguson A M N, Irwin M, Lewis G and McConnachie A 2004 Taking measure of the andromeda halo: a kinematic analysis of the giant stream surrounding M31 *MNRAS* **351** 117

[8] Fardal M, Guhathakurta P, Babul A and McConnachie A W 2007 Investigating the Andromeda stream - III. A young shell system in M31 *MNRAS* **380** 15
[9] Mori M and Rich M 2008 The once and future andromeda stream *ApJ* **674** 77
[10] Miki Y, Mori M and Rich M in prep
[11] Fardal M A, Babul A, Geehan J J and Guhathakurta P 2006 Investigating the Andromeda stream - II. Orbital fits and properties of the progenitor *MNRAS* **366** 1012
[12] Font A S, Johnston K V, Guhathakurta P, Majewski S R and Rich M 2006 Dynamics and stellar content of the giant southern stream *AJ* **131** 1436
[13] Geehan J J, Fardal M A, Babul A and Guhathakurta P 2006 Investigating the andromeda stream - I. Simple analytic bulge-disc-halo model for M31 *MNRAS* **366** 996
[14] Aarseth S J, Hénon M and Wielen R 1974 A comparison of numerical methods for the study of star cluster dynamics *A&A* **37** 183
[15] Trethewey D, Chapman S, Irwin M, McConnachie A, Ibata R, Ferguson A, Lewis G and Tanvir N in prep

24th IUPAP Conference on Computational Physics (IUPAP-CCP 2012) IOP Publishing
Journal of Physics: Conference Series **454** (2013) 012013 doi:10.1088/1742-6596/454/1/012013

Hunting a wandering black hole in M31 halo using GPU cluster

Yohei Miki, Masao Mori and Toshihiro Kawaguchi
University of Tsukuba, 1-1-1 Tennodai, Tsukuba, Ibaraki 305-8571, Japan
E-mail: ymiki@ccs.tsukuba.ac.jp

Abstract. In the hierarchical structure formation scenario, galaxies have experienced many mergers with less massive galaxies and have grown larger and larger. On the other hand, the observations indicate that almost all galaxies have a central massive black hole (MBH) whose mass is $\sim 10^{-3}$ of its spheroidal component. Consequently, MBHs of satellite galaxies are expected to be moving in the halo of their host galaxy after a galaxy collision, although we have not found such MBHs yet. We investigate the current-plausible position of an MBH of the infalling galaxy in the halo of the Andromeda galaxy (M31). Many substructures are found in the M31 halo, and some of them are shown to be remnants of a minor merger about 1 Gyr ago based on theoretical studies using N-body simulations. We calculate possible orbits of the MBH within the progenitor dwarf galaxy using N-body simulations. Our results show that the MBH is within the halo, about 30 kpc away from the center of M31.

In addition, further simulations are necessary to restrict the area in which the MBH exists, and hence to determine the observational field for the future observational detection. The most uncertainty of the current MBH position is caused by uncertainty about the infalling orbit of the progenitor dwarf galaxy. Therefore, we have performed a large (a few 10^4 realizations) set of parameter study to constrain the orbit in the six-dimensional phase space. For such purpose, we have already investigated in detail a few ten thousand orbit models using HA-PACS, a recently installed GPU cluster in University of Tsukuba.

1. Astrophysical Background and Motivation

In the context of hierarchical structure formation scenario under the Cold Dark Matter (CDM) universe, large galaxies, such as Milky Way or Andromeda galaxy (M31), have likely experienced many mergers with less massive galaxies and have grown larger and larger. Furthermore, a well known observed correlation between the mass of spheroidal component of galaxies and the mass of central massive black holes (MBHs) in their central region, so called the Magorrian relation [1, 2], suggests the coevolution of galaxies and their central MBHs. However, little is known how the coevolution of galaxies and MBHs proceeds. In the hierarchical structure formation scenario, galaxies collide and merge with each other and subsequently less massive galaxies and their central MBHs drift around the halo region of their host galaxy. In other words, MBHs move in the halo of their host galaxy after galaxy merging events, and finally they sink towards the central region of the host galaxy due to dynamical friction. Therefore, MBHs also locate outside the nucleus of their host galaxy, not only in the central region of galaxies like active galactic nuclei. However, we have not found such MBHs yet. Thus, searching for such MBHs is very hot issue recently [3, 4]. In this study, we investigate the probable position of such an MBH theoretically.

Many cosmological N-body simulations of the hierarchical structure formation exhibit a wealth of merger remnants around host galaxies [5]. To test the current cosmology by verifying such theoretical predictions from the CDM scenario, many observational studies have been examined to investigate merger remnants [6, 7]. In such a context, the giant stellar stream was discovered in the south region of the M31 halo [8], which is our neighbor galaxy. Further photometric and spectroscopic observations of spatial distribution [9, 10, 11, 12, 13, 14, 15], radial velocity distribution of red giant stars [16, 17, 18, 19, 20, 21, 22] and metallicity distribution [17, 18, 12, 20, 21, 22] clearly show other substructures near M31. Calculations of the motion of test particles under the gravitational potential of M31 [16, 23] and N-body simulations on the interaction between the progenitor of the stream and M31 [24, 25, 26, 27] suggest that the stream, the northeast shell and the west shell are the tidal debris formed in the last pericentric passage of a satellite on a radial orbit. These models reproduce the observed features and successfully constrain the orbit and properties of the progenitor.

The Magorrian relation suggests that the progenitor dwarf galaxy has an MBH whose mass $M_{\rm BH}$ is about 10^{-3} of the mass of their host galaxy's spheroidal component $M_{\rm sph}$ [1, 2]. The similar relation between $M_{\rm BH}$ and host galaxy's velocity dispersion σ, $M_{\rm BH}-\sigma$ relation, is held down to $M_{\rm BH} \sim 10^5 M_\odot$ [28, 29]. Therefore, the relation between MBHs and their host galaxies is held down to $M_{\rm sph} \sim 10^8 M_\odot$. Since the dynamical mass of the progenitor is estimated to be an order of $10^9 M_\odot$ [25, 26, 27], the progenitor likely has an MBH whose mass is up to an order of $10^6 M_\odot$ if the progenitor consists of spheroidal stellar component alone. If this is true, then an MBH should be now moving in the merger remnants. Finding such an MBH will provide us a hint for understanding the coevolution process of galaxies and MBHs. Thus, we investigate the current position of the MBH using N-body simulations, in view of future observational detections.

2. Numerical Modeling of the Interaction between M31 and the Infalling Satellite

The basic equation of N-body simulation by direct summation is Newton's equation of motion expressed as

$$\boldsymbol{a}_i = \sum_{j=0, j\neq i}^{N-1} \frac{G m_j \left(\boldsymbol{x}_j - \boldsymbol{x}_i\right)}{\left(\left|\boldsymbol{x}_j - \boldsymbol{x}_i\right|^2 + \epsilon^2\right)^{3/2}}, \tag{1}$$

where G is the gravitational constant, m_i, $\boldsymbol{x}_i$ and $\boldsymbol{a}_i$ are mass, position, and acceleration of i-th particle out of N particles, respectively. The gravitational softening parameter ϵ, introduced to avoid divergence due to division by zero, eliminates self interaction when calculating gravitational force. We use the word i-particles, and j-particles to denote particles feel gravitational force, and particles cause gravitational force, respectively.

We assume a fixed potential model (Hernquist bulge [30], exponential disk, and NFW halo [31]) for M31 [32, 25], because Mori & Rich [26] analytically and numerically showed the dynamical response of M31's disk against the collision with the progenitor is negligible. They represented the progenitor and M31 using N-body particles and focused on the thickness of the M31's disk due to the disk heating by dynamical friction after the collision. Their results showed that the effects on the disk thickness and the disk kinematics are negligibly small as far as the dynamical mass of the progenitor is less than $5 \times 10^9 M_\odot$. In this study, we assume a King sphere of $M = 3 \times 10^9 M_\odot$, $c = 0.7$, $r_{\rm t} = 4.5$ kpc, the best fit model derived by [27], as the progenitor dwarf galaxy.

2.1. *Finding the Current-Plausible Position of the MBH*

To investigate the current-plausible position of the MBH in the former satellite, we calculate orbital evolution of an MBH particle whose mass is $3 \times 10^6 M_\odot$ and $524,288$ N-body particles which represent the satellite. We set the gravitational softening length ϵ to be 13 pc. Following

[25], we adopt an initial position and velocity vector of (−34.75, 19.37, −13.99) kpc and (67.34, −26.12, 13.50) km s^{-1}, respectively. We calculate the self gravity of N-body particles using Blade-GRAPE on FIRST simulator at CCS, the University of Tsukuba. We use 2nd-order Runge-Kutta integrator and shared, adaptive time step.

2.2. Constraining Uncertainty for the Current Probable Position of the MBH

Further simulations are necessary to restrict the area in which the MBH exists, and hence to determine the observational field for the future observational detection. The most uncertainty of the current MBH position is caused by uncertainty about the infalling orbit of the progenitor dwarf galaxy. Therefore, we have performed a large set of parameter study to constrain the orbit of the infalling satellite which reproduce the observed structures in the six-dimensional phase space. Since the number of dimension for the parameter space, six, is too large to sweep the whole parameter space, the number of dimension is reduced as follows. First, we fix the initial distance of the infalling satellite as 7.63 kpc away from the center of the M31 (corresponds to scale radius of DM halo [25]). In addition, M31 is modeled as an axisymmetric system in this study. Therefore, the resultant number of dimension becomes four; however, the parameter space is still large.

To study such a wide parameter space, we have performed a large set of parameter survey utilizing GPU cluster. In this parameter survey, we represent the infalling satellite with 65536 particles and set the gravitational softening length ϵ as 50 pc. Numerical simulations of parameter study are performed on HA-PACS at CCS, the University of Tsukuba. We use 2nd-order leap-frog integrator and shared, fixed time step. Algorithm and implementation of our code is explained in Section 4.

3. General Purpose computing on Graphics Processing Unit and HA-PACS

Hence performing four dimensional parameter study is a challenging task, we need to accelerate calculation and sweep wide parameter space concurrently to complete the parameter study in realistic time. In recent days, GPU (Graphics Processing Unit) becomes one of the most attractive accelerator due to development of GPGPU (General Purpose computing on GPU). A C/C++ based programming environment named CUDA (Compute Unified Device Architecture) provided by NVIDIA enables programmers to quite easily implementing GPU codes run on NVIDIA's GPUs [33]. Furthermore, many GPU clusters exhibit on the TOP 500 list [34], such as Titan, Tianhe-1A, Nebulae, TSUBAME 2.0, and HA-PACS. Rapid performance increase of GPUs and development of such GPU clusters support accelerating numerical simulations. This preferred feature of GPU computing strongly encourages us performing four dimensional parameter study about the infalling orbit of the satellite on GPU cluster.

In this study, we have used HA-PACS (Highly Accelerated Parallel Advanced system for Computational Sciences), a newly installed GPU cluster at University of Tsukuba [35]. HA-PACS is equipped with the high-end GPUs and CPUs connected by PCI-express generation 3.0. Each node of HA-PACS consists of two sockets of Intel Sandy Bridge-EP and four boards of NVIDIA Tesla M2090, and the CPUs support full bandwidth connection of the GPUs without any performance bottleneck. The peak performance of HA-PACS is 1.604 PFLOPS in single precision, due to high performance GPUs of 1.427 PFLOPS in single precision. The Table 1 lists other detailed information of HA-PACS.

4. Implementation and Optimization of Single GPGPU Code

Many earlier studies reported that massive parallelization about i-particles achieves high performance of GPU [36, 37, 38, 39, 40]. Implementation and optimization technique in this work is based on CUDA SDK [37] and our previous work [40] except for one additional optimization. Implementation of CUDA SDK, based on [37], and our previous work [40] achieve

Table 1. Detailed information of HA-PACS

Number of nodes	268
CPU	Intel Xeon E5-2670
	16 cores per node, 2.6GHz
RAM	128 GB (DDR 3, 1,600MHz)
GPU	NVIDIA Tesla M2090
	512 CUDA cores, 1.3GHz
	4 boards per node
Video RAM	6 GB (GDDR 5, ECC on) per GPU
C Compiler	icc 12.1.0
MPI Library	Intel MPI 4.0.3.008
CUDA toolkit	nvcc 4.1, CUDA SDK 4.1.28
Interconnection	Infiniband QDR ×2 rails

Listing 1. Difference between CUDA SDK and our previous work

```
/* Implementation of CUDA SDK */
float r2 = rji.x * rji.x + rji.y * rji.y + rji.z * rji.z;
r2 += eps2;
/* Our Previous Work */
float r2 = eps2 + rji.x * rji.x + rji.y * rji.y + rji.z * rji.z;
```

high performance of 930 and 991 GFLOPS in single precision on a board of NVIDIA Tesla M2090.

In both implementation, a block contains 256 threads and the shared memory stores position data of 256 j-particles to minimize access time to the global memory within the innermost loop. Differences between [37] and [40] are the number of unroll count for the innermost loop, cache configuration, and the number of operations to calculate gravitational interaction. In our implementation, the number of unroll count for the innermost loop is 128 against for 32 of CUDA SDK, and we set "L1 cache preferred" since experiments exhibit slight performance increase compared with "shared memory preferred" in most cases. The last difference, the most influential one, is due to calculation process of $\mathrm{r}_{ji}{}^2 + \epsilon^2$. In both implementation, a `float3` type variable `rji`, a `float` type variable `eps2`, and a `float` type variable `r2` store the displacement vector $\mathrm{r}_{ji} \equiv \mathrm{x}_j - \mathrm{x}_i$, ϵ^2, and result of $\mathrm{r}_{ji}{}^2 + \epsilon^2$ calculated as Listing 1, respectively. The source codes shown in the Listing 1 look like almost the same, however, generated instruction sets are quite different. For implementation of CUDA SDK, one multiplication and two fused multiply-add (FMA) operations are performed at first, and one addition follows at the next step. Thus, its computational cost corresponds to 4 clock cycles according to CUDA C Programming Guide [33]. On the other hand, only three FMA operations are performed using 3 clock cycles in our implementation. Therefore, our implementation would be faster than CUDA SDK. The most influential point of this optimization is that the innermost loop includes the calculation of `r2`; thus, this small care directly increase performance.

Furthermore, we have implemented one additional optimization to hide memory access to the global memory. Since the shared memory stores the position data of j-particles, synchronization of whole threads within a block is necessary before and after updating information of j-particles. As far as a streaming multiprocessor contains multiple blocks, memory access time of a block can be hidden by overlapping with calculation of other blocks. However, such overlapping might not be occur since the CUDA schedulers determine which calculation of executable warp proceeds. Therefore, maximizing probability to occur the overlapping between calculation and memory access can help to achieve high performance. For the purpose, we have separated the load instruction from the global memory and the store instruction to the shared memory by using the two `__syncthreads()`. By this careful treatment, overlapping of memory access and calculation becomes more effective within a block, and the peak performance of our implementation reaches 1004 GFLOPS in single precision.

5. Results: Where is the Wandering MBH?

Results of Section 2.1 and Section 2.2 are shown in Section 5.1 and Section 5.2, respectively.

5.1. Current-Plausible Position of the MBH

The result of the N-body simulation to investigate the current plausible-position of the MBH when the infalling satellite follows Fardal's orbit [25] is shown in Figure 1. The red curve on the mass distribution map shows the orbit of the MBH from 910 Myr ago to 320 Myr future. The position of the MBH with 99.7% confidence level is shown by a blue curve. The black circle indicates the most probable current position of the MBH, where it is approaching the Milky Way with a (heliocentric) line-of-sight velocity of -170 km s^{-1} (the combination of a receding velocity of $+130$ km s^{-1} in the M31 rest frame plus the approaching heliocentric velocity of M31 of -300 km s^{-1} [41]). The MBH is close to the apocenter, where it is closer to the Milky Way than the M31 center, which means that the velocity of the MBH is relatively slow, and the uncertainty of the current position is smaller than any other position such as pericenter. The distance of the MBH from the center of M31 is about 30 kpc, so it is far away from the disk of M31.

5.2. Results of Parameter Study

The first results for 34,000 runs (among $\sim 10^5$ in total) of the parameter study to investigate the infalling orbit which reproduce observed structures well are shown in Figure 2. Black circles represent results reproduce the stream structure, shell structure and contrast among the stream and the two shells. On the other hand, crosses are runs which failed to reproduce observed structures. The infalling orbit assumed in earlier studies ([25, 26, 27]) is infalling velocity of -430 km s^{-1}, specific angular momentum of 660 kpc km s^{-1} (corresponds to the result shown in Figure 1). The Figure 1 shows that the maximum initial infalling velocity is about -300 km s^{-1}. Furthermore, the escape velocity of the infalling satellite is about -560 km s^{-1} at 7.63 kpc away from the center of the M31. Therefore, the minimum initial infalling velocity is expected to be around -560 km s^{-1}.

By completing this parameter survey, the plausible area where the wandering MBH exists will be restricted more tightly.

Figure 1. Mass distribution (column density) maps of the debris of the dwarf galaxy in standard coordinates centered on M31. Global distribution of N-body particles are shown as color image while white curves and lines show the disk of M31. White open squares represent observed points of the giant stellar stream [23], and white filled circles show edge of the observed shells [25]. Black circle shows the most probable current position of the MBH. Magenta and blue curve show the position of the MBH when the observed shells are reproduced with 95.4% and 99.7% confidence level, respectively. The red curve shows the orbit of the MBH from 910 Myr ago to 320 Myr future.

6. Summary

We investigate the current position of an MBH moving in the M31 halo. The current-plausible position of the MBH is within the halo, 30 kpc away from the center of M31. To determine the observational field for the future observational detection, we study uncertainty of the infalling orbit of the satellite by performing parameter study. Hence the required parameter space is large, we develop a highly optimized collisionless N-body code runs on GPU cluster, and we have performed a parameter study about the infalling orbit of the progenitor galaxy to determine the observation field to detect the MBH. The tentative results of the parameter study begin to restrict the possible parameter space of the infalling orbit which reproduce the observed structures well.

Acknowledgments

Numerical Simulations have been performed with FIRST and HA-PACS at the Center for Computational Sciences, University of Tsukuba. We appreciate HA-PACS project team for providing an opportunity to use a constructing super computer and giving many technical helps. This work was partially supported by the program of the Pre-Strategic Initiatives, University of Tsukuba, the FIRST project based on Grants-in-Aid for Specially Promoted Research by MEXT (16002003), and the Grant-in-Aid for Scientific Research (S)(20224002), (A)(21244013), (C)(18540242), and (C)(25400222).

References

[1] Magorrian J, Tremaine S, Richstone D, Bender R, Bower G, Dressler A, Faber S M, Gebhardt K, Green R,

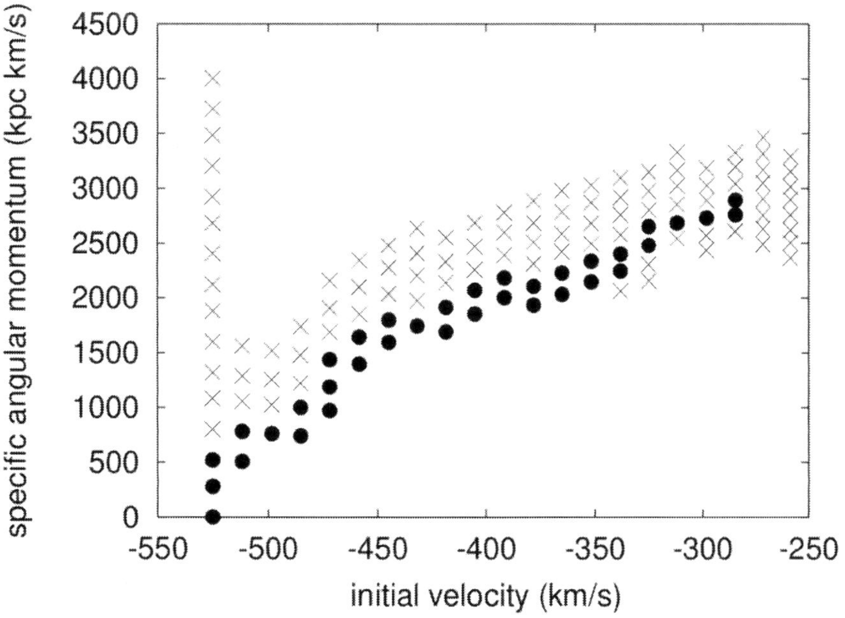

Figure 2. Preliminary results for 34,000 runs of parameter study for the infalling orbit of the satellite. The horizontal axis is the infalling velocity of the satellite at 7.63 kpc away from the center of the M31, and the vertical axis represents the specific angular momentum of the infalling satellite. Filled circles corresponds to results reproduce the stream structure, shell structure and contrast among the stream and the two shells. Crosses represent results which failed to reproduce observed structures.

Grillmair C, Kormendy J and Lauer T 1998 *Astronomical Journal* **115** 2285–2305
[2] Marconi A and Hunt L K 2003 *Astrophysical Journal* **589** L21–L24
[3] Farrell S A, Webb N A, Barret D, Godet O and Rodrigues J M 2009 *Nature* **460** 73–75
[4] Wiersema K, Farrell S A, Webb N A, Servillat M, Maccarone T J, Barret D and Godet O 2010 *Astrophysical Journal* **721** L102–L106
[5] Bullock J S and Johnston K V 2005 *Astrophysical Journal* **635** 931–949
[6] Chiba M, Minezaki T, Kashikawa N, Kataza H and Inoue K T 2005 *Astrophysical Journal* **627** 53–61
[7] Minezaki T, Chiba M, Kashikawa N, Inoue K T and Kataza H 2009 *Astrophysical Journal* **697** 610–618
[8] Ibata R, Irwin M, Lewis G, Ferguson A M N and Tanvir N 2001 *Nature* **412** 49–52
[9] Ferguson A M N, Irwin M J, Ibata R A, Lewis G F and Tanvir N R 2002 *Astronomical Journal* **124** 1452–1463
[10] McConnachie A W, Irwin M J, Ibata R A, Ferguson A M N, Lewis G F and Tanvir N 2003 *Monthly Notices of the Royal Astronomical Society* **343** 1335–1340
[11] Irwin M J, Ferguson A M N, Ibata R A, Lewis G F and Tanvir N R 2005 *Astrophysical Journal* **628** L105–L108
[12] Ibata R, Martin N F, Irwin M, Chapman S, Ferguson A M N, Lewis G F and McConnachie A W 2007 *Astrophysical Journal* **671** 1591–1623
[13] McConnachie A W, Irwin M J, Ibata R A, Dubinski J, Widrow L M, Martin N F, Côté P, Dotter A L, Navarro J F, Ferguson A M N, Puzia T H, Lewis G F, Babul A, Barmby P, Bienaymé O, Chapman S C, Cockcroft R, Collins M L M, Fardal M A, Harris W E, Huxor A, Mackey A D, Peñarrubia J, Rich R M, Richer H B, Siebert A, Tanvir N, Valls-Gabaud D and Venn K A 2009 *Nature* **461** 66–69
[14] Tanaka M, Chiba M, Komiyama Y, Guhathakurta P, Kalirai J S and Iye M 2010 *Astrophysical Journal* **708** 1168–1203
[15] Richardson J C, Irwin M J, McConnachie A W, Martin N F, Dotter A L, Ferguson A M N, Ibata R A, Chapman S C, Lewis G F, Tanvir N R and Rich R M 2011 *Astrophysical Journal* **732** 76
[16] Ibata R, Chapman S, Ferguson A M N, Irwin M, Lewis G and McConnachie A 2004 *Monthly Notices of the Royal Astronomical Society* **351** 117–124
[17] Kalirai J S, Guhathakurta P, Gilbert K M, Reitzel D B, Majewski S R, Rich R M and Cooper M C 2006 *Astrophysical Journal* **641** 268–280
[18] Kalirai J S, Gilbert K M, Guhathakurta P, Majewski S R, Ostheimer J C, Rich R M, Cooper M C, Reitzel D B and Patterson R J 2006 *Astrophysical Journal* **648** 389–404
[19] Guhathakurta P, Rich R M, Reitzel D B, Cooper M C, Gilbert K M, Majewski S R, Ostheimer J C, Geha M C, Johnston K V and Patterson R J 2006 *Astronomical Journal* **131** 2497–2513
[20] Gilbert K M, Fardal M, Kalirai J S, Guhathakurta P, Geha M C, Isler J, Majewski S R, Ostheimer J C, Patterson R J, Reitzel D B, Kirby E and Cooper M C 2007 *Astrophysical Journal* **668** 245–267
[21] Koch A, Rich R M, Reitzel D B, Martin N F, Ibata R A, Chapman S C, Majewski S R, Mori M, Loh Y S, Ostheimer J C and Tanaka M 2008 *Astrophysical Journal* **689** 958–982
[22] Gilbert K M, Guhathakurta P, Kollipara P, Beaton R L, Geha M C, Kalirai J S, Kirby E N, Majewski S R and Patterson R J 2009 *Astrophysical Journal* **705** 1275–1297
[23] Font A S, Johnston K V, Guhathakurta P, Majewski S R and Rich R M 2006 *Astronomical Journal* **131** 1436–1444
[24] Fardal M A, Babul A, Geehan J J and Guhathakurta P 2006 *Monthly Notices of the Royal Astronomical Society* **366** 1012–1028
[25] Fardal M A, Guhathakurta P, Babul A and McConnachie A W 2007 *Monthly Notices of the Royal Astronomical Society* **380** 15–32
[26] Mori M and Rich R M 2008 *Astrophysical Journal* **674** L77–L80
[27] Miki Y, Mori M and Rich R M in preparation
[28] Barth A J, Greene J E and Ho L C 2005 *Astrophysical Journal* **619** L151–L154
[29] Xiao T, Barth A J, Greene J E, Ho L C, Bentz M C, Ludwig R R and Jiang Y 2011 *Astrophysical Journal* **739** 28
[30] Hernquist L 1990 *Astrophysical Journal* **356** 359–364
[31] Navarro J F, Frenk C S and White S D M 1996 *Astrophysical Journal* **462** 563
[32] Geehan J J, Fardal M A, Babul A and Guhathakurta P 2006 *Monthly Notices of the Royal Astronomical Society* **366** 996–1011
[33] Nvidia 2011 *NVIDIA CUDA C Programming Guide Version 4.1*
[34] Top500 list URL `http://www.top500.org/`
[35] HA-PACS Project URL `http://www.ccs.tsukuba.ac.jp/CCS/eng/research-activities/projects/ha-pacs`
[36] Hamada T and Iitaka T 2007 *ArXiv Astrophysics e-prints* (*Preprint* `arXiv:astro-ph/0703100`)
[37] Nyland L, Harris M and Prins J 2007 Fast N-Body Simulation with CUDA

[38] Hamada T, Narumi T, Yokota R, Yasuoka K, Nitadori K and Taiji M 2009 42 TFlops hierarchical N-body simulations on GPUs with applications in both astrophysics and turbulence *Proceedings of the Conference on High Performance Computing Networking, Storage and Analysis* SC '09 (New York, NY, USA: ACM) pp 62:1–62:12 ISBN 978-1-60558-744-8

[39] Hamada T and Nitadori K 2010 190 TFlops Astrophysical N-body Simulation on a Cluster of GPUs *Proceedings of the 2010 ACM/IEEE International Conference for High Performance Computing, Networking, Storage and Analysis* SC '10 (Washington, DC, USA: IEEE Computer Society) pp 1–9 ISBN 978-1-4244-7559-9

[40] Miki Y, Takahashi D and Mori M 2012 Procedia Computer Science **9** 96 – 105 ISSN 1877-0509 proceedings of the International Conference on Computational Science, ICCS 2012 URL `http://www.sciencedirect.com/science/article/pii/S1877050912001329`

[41] de Vaucouleurs G, de Vaucouleurs A, Corwin Jr H G, Buta R J, Paturel G and Fouqué P 1991 *Third Reference Catalogue of Bright Galaxies. Volume I: Explanations and references. Volume II: Data for galaxies between 0^h and 12^h. Volume III: Data for galaxies between 12^h and 24^h.*

Journal of Physics: Conference Series **454** (2013) 012014 doi:10.1088/1742-6596/454/1/012014

Studying the core-cusp problem in cold dark matter halos using N-body simulations on GPU clusters

Go Ogiya[1], Masao Mori[1], Yohei Miki[1], Taisuke Boku[1] and Naohito Nakasato[2]

[1] University of Tsukuba, Tsukuba, Ibaraki, Japan
[2] University of Aizu, Aizu-Wakamatsu, Fukushima, Japan

E-mail: ogiya@ccs.tsukuba.ac.jp

Abstract. The discrepancy in the mass–density profile of dark matter halos between simulations and observations, the core–cusp problem, is a long–standing open question in the standard paradigm of cold dark matter cosmology. Here, we study the dynamical response of dark matter halos to oscillations of the galactic potential which are induced by a cycle of gas expansion and contraction in galaxies driven by supernova feedback. We developed a fast tree–code for PC clusters with GPU which displays high performance and high scalability. We perform large scale N–body simulations to follow the dynamical evolution of dark matter halos under the effect of oscillating gravitational potential. Furthermore, we compare the results of simulations with an analytical model of the resonance between particles and density waves to understand the physical mechanism of the cusp-core transition.

1. Introduction

The core–cusp problem remains one of the main unresolved contradictions between observation and theory predicted by the standard paradigm of the cold dark matter (CDM) cosmology. ΛCDM–cosmological N–body simulations have always predicted a steep power-law mass–density profile at the center of CDM halos [1], [2], [3]. However, recent observations of lower mass galaxies have revealed that the density profile of the dark matter (DM) halo is nearly constant around the center [4], [5], [6].

Mashchenko et al. [7] proposed that periodic variations in the galactic potential driven by recurrent star burst events may flatten the central DM density profile. Bursts of star formation induce a large–scale outflow from the galactic center. In time, the mass–loss leads to gas depletion and temporarily terminates star formation. Subsequently, the expelled gas falls back toward the galactic center, loses a large amount of internal energy by radiative cooling, and, once sufficient cold gas has accumulated, a starburst arises again. This cycle of expansion and contraction of the interstellar gas could lead to a cyclic change in the gravitational potential around the center of galaxies and have an impact on the density profile of DM halos. Some cosmological hydrodynamic simulations exhibit this starburst–outflow cycle and also observed a cusp–core transition at the center of CDM halos [7], [8], [9]. However, the physical connection between these two phenomena is still unclear.

We investigate the dynamical response of DM halos with a central cusp to a recurring change of the baryon gravitational potential with collisionless N–body simulations. To this end, we have developed a fast tree–code on PC clusters with GPU, which displays high performance and high scalability parallel processing. In section 2, we briefly describe the tree method. In section 3, we devise a data distribution procedure in memory space and propose an expedient method to reduce the frequency of warp branches. This implementation speeds up the kernel function more than three times compared with the methods proposed in previous work.

The results of our DM simulations presented in section 4 show that the cycle of expansion and contraction of the interstellar gas is an effective mechanism for flattening the central cusp in the mass density profile of the CDM halo. The recurrence frequency of star formation is one of the important factors in determining the dynamical response of DM halos. In section 5, we compare these results with an analytical model of the resonance between DM particles and the density waves of the interstellar gas that induces the recurring change of the gravitational potential. This analysis indicates that the resonance is effective in the cusp to core transition of the CDM halos and that this theory can resolve the core–cusp problem.

2. The tree method for collisionless gravitational systems

DM halos are collisionless self–gravitational systems in which the gravity of the whole system is more important than encounters of nearby particles. To follow the dynamical evolution of such systems, N–body simulations are a suitable numerical method. N–body simulations with large number of particles, N, must be performed to avoid artificial two–body relaxation and to achieve a sufficiently high resolution that ensures the softening length is much smaller than the resolution limit of observations of density profiles of DM halos.

The most straightforward and accurate algorithm to compute gravity among particles is the "direct" method. The computational cost is $O(N^2)$, since we would have to compute gravitational acceleration between all pairs of particles. Therefore this algorithm is unfavorable to solve problems with large N. Hereafter, we call the particles for which the gravitational acceleration is computed the "i–particles" and call the particles that interact with i–particles the "j–particles". The tree method reduces the cost to $O(N \log N)$ by treating a group of j–particles that are sufficiently far from the i–particle as a single heavy j–particle [10].

The tree method consists of 2 parts, tree–construction and tree–traversal. In the first part, the tree–construction, we construct an oct–tree of particles. We split the cube containing all particles recursively into 8 (in 3D) smaller cubes (figure 1), and we link all cubes to their daughter– and sister–cubes (figure 2). We continue the recursion until every cube contains less than or equal to $N_{\rm crit}$ particles. We set the parameter $N_{\rm crit} = 4$. After the linking operation, we compute the mass and position of each cube. The mass of the cube is the sum of that of the contained particles, and its position is the center of mass of the contained cubes. In the second part, the tree–traversal, we walk the tree constructed in the previous part for each i–particle. A cube is sufficiently far from an i–particle if its angular size, θ, is smaller than a critical size $\theta_{\rm c}$ =0.6.

When we reach a cube for which $theta > theta_{\rm c}$, we walk to a daughter–cube. If we reach a cube for which $theta < theta_{\rm c}$ we walk to a sister–cube. In the latter case, j–particles contained in the cube are treated as a single heavy j–particle, reducing the cost of computation.

The tree method to solve collisionless systems reduces the computational cost to $O(N \log(N))$ and produces results of comparable accuracy to the direct method if $\theta_{\rm c}$ is sufficiently small. N–body simulations are manageable because they are inherently boundaryless and scale–free. Thus, N–body simulations utilizing the tree method are commonly used in computational astrophysics.

3. Implementation and performance of our tree code on GPU clusters

3.1. *Overview of the code*

We speed–up our tree code by optimizing it for GPU clusters. Following previous work, we let CPU cores compute the tree–construction and let GPUs compute the tree–traversal [11], [12]. We order particle data along a Peano–Hilbert curve, a type of space–filling curve, in

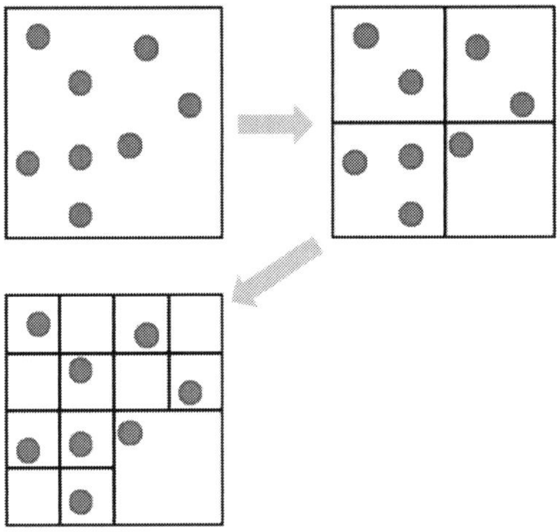

Figure 1. Recursive splitting of cubes into 8 (in 3D) smaller cubes until all cubes contain less than $N_{\rm crit}$ particles (blue dots). In this schematic diagram, $N_{\rm crit}$ is 1, whereas in our simulations $N_{\rm crit} = 4$.

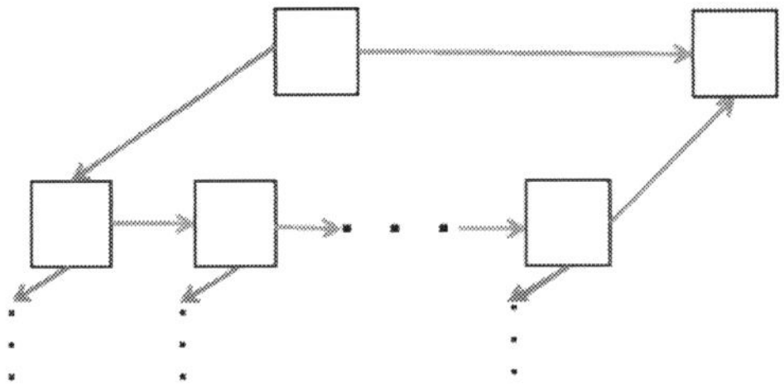

Figure 2. Links between cubes. Red and green arrows indicate links to daughter– and sister–cube, respectively.

memory space superior to Morton ordering, to achieve high cache hit rate. Furthermore, we have parallelized our code using MPI. Each MPI process becomes the host of a GPU, owns partial particle data, constructs a partial tree composed of its i–particles, and communicates necessary tree data with other processes.

"Warp branches" are an important issue in GPU computing. We explain the problem for the case of the NVIDIA Fermi architecture. In this architecture of GPUs, 32 cores work concurrently with the same instruction stream. In other words, GPU cores work in the manner of SIMD. This cluster of 32 cores is called a warp. When two or more different instructions emerge among the 32 cores, a large overhead arises because the instruction can only be processed sequentially within a warp. Such warp branches occur during the tree–traversal part whenever we step to a daughter– or sister–cube. Warp branches become less frequent if we bundle routes for given i–particles in the tree structure.

To achieve this, we propose a method in 2 steps: The first step is a "vectorization" step. We vectorize the operations of route selection in the tree structure and that of calculating gravitational acceleration for V i–particles, where V is the vector length or the number of particles per GPU core. We show a conceptual diagram of this step in figure 3. We bundle routes of V i–particles in a GPU core by measuring the distance to the cube from the nearest particle in a core, which is circled red in figure 3. We then select the next step in the tree structure, a daughter– or the sister–cube, by using the distance to the present cube from the nearest particle among the bundled ones. If $\theta > \theta_c$ for the nearest particle, we take a step to a daughter–cube even if other particles are sufficiently far from the present cube. If $\theta < \theta_c$ for the nearest particle, we compute gravitational acceleration from that cube to the i–particles by using the respective distance between each i-particle and the cube. After that, the walk continues to a sister–cube. The frequency of warp branches drops with increasing vector length, V. However, the amount of work per GPU core is proportional to V. Thus it is not advisable to set V very large.

The second step is a "grouping" step, which alleviates the problem of vectorization described above to some degree. We bundle routes for i–particles among several cores in the same fashion as we bundle the routes for i–particle in a core (figure 4). Again, the nearest particle in the group is circled red. We select the next step in the tree-traversal for i–particles by the value of the distance to a cube from the nearest particle. By grouping, we can increase the number of bundled i–particles without increasing the number of particles owned by a GPU core. The number of bundled particles becomes $V \times G$, where G is the number of group members.

Vectorizing and grouping will decrease the frequency of warp branches, but will increase the amount of computation. This is because, by choosing the particle in a vector or group nearest to the cube, will reduce the number of particles for which $\theta < \theta_c$ and for which we could treat the cube as a single heavy j–particle. Thus we expect that there exist an optimal pair of values for the vector length and number of group members. The Peano–Hilbert ordering of particles is very important in our method since bundling routes for spatially distant particles results in a large amount of unnecessary computations.

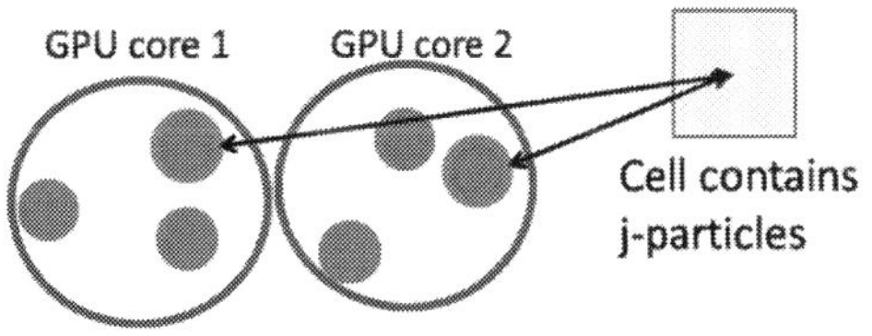

Figure 3. Conceptual diagram of vectorization. Blue dots represent i–particles. The red circled particles are the closest ones to the cube of j–particles. In this case, each GPU core holds 3 i–particles ($V = 3$).

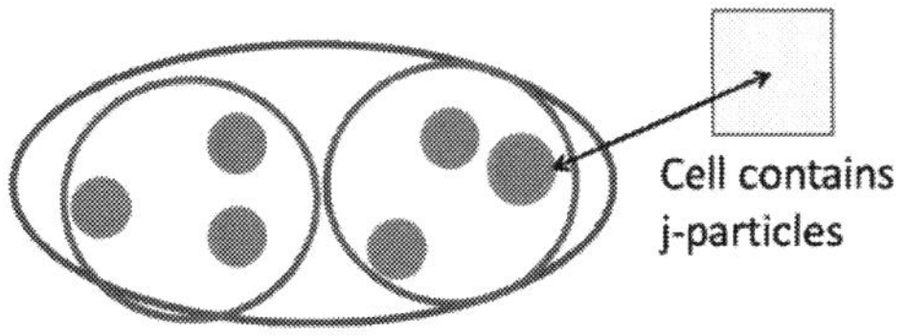

Figure 4. Conceptual diagram of grouping. The same i–particles are shown as figure 3. In this case, the number of group member, G, is 2, and 6 i–particles have been bundled into a group.

3.2. Computing environment for performance measurements

We measure the performance of our code on the GPU cluster HA-PACS, which was installed at the Center for Computational Sciences of the University of Tsukuba in 2012 [13]. HA–PACS has

4 NVIDIA Tesla M2090 GPU cards per node. As described above, each MPI process is the host of a GPU, and we launch up to four MPI processes per node. The specifications of HA–PACS are shown in Table 1.

Table 1. Compendium of HA–PACS

CPU	Intel(R) Xeon(R) CPU E5-2670 2.60GHz (8 cores/socket × 2 sockets = 16 cores/node)
GPU	NVIDIA Tesla M2090 (4 GPUs / node)
Main Memory	128 GB, DDR3 1600MHz, 4 channel / socket, 102.8 GB/s/node
OS	CentOS release 6.1 (Final)
CPU Compiler	GCC 4.4.5-6
GPU Toolkit	CUDA 4.0.17
Interconnection	Infiniband QDR × 2 rails
MPI	MVAPICH2 1.8

3.3. Performance of the code

In figure 5, we showed the results of performance measurements for the GPU part of the kernel function, which is essentially the tree–traversal part. In this test, we simply follow the evolution of particles initially distributed with the Navarro–Frenk–White (NFW) profile, which is often used for CDM halos [1] and a random velocity field. We set the number of particles, $N = 2^{23} \approx$ 8 million, and the tolerance parameter of the tree method, $\theta_c = 0.6$. We study the effects of vectorization and grouping by comparing runs with different combinations of V and G. Compared with the case of no vectorization or grouping, $(V, G) = (1, 1)$, tree–traversal may become more than 3 times faster. As expected, there exists an optimal pair of values for the vector length and the number of group members. In this test, the optimal pair is $(V, G) = (4, 4)$. In general, the optimal pair values depends on the N–body problem and the GPU architecture.

We have parallelized the tree–construction and tree–traversal operations with MPI. N particles are sorted according to the Peano–Hilbert space–filling curve, and N/N_p particles are assigned to each MPI process, where N_p is the number of MPI processes or GPU cards.

We now check the performance of the whole code. The tree–construction, tree–traversal, and the construction of the minimum partial trees that are communicated to other processes (Locally Essential Tree; LET) [14] take up most of the computational time.

Figure 6 shows the performance for strong scaling as a function of N_p. We set the total number of particles, $N = 2^{25} \approx 32$ million. When $N_p < 10$, the scaling is good. However, the performance saturates for $N_p > 10$. We suspect the reason for the performance saturation of the kernel function (blue line) is the worsening load balance between MPI processes. Because we have sorted and assigned N/N_p particles to each process, some processes are assigned particles from the central part of the computational domain while others are assigned from the outskirts of the domain. The number of cubes encountered in the tree–traversal for these processes can differ in this test by a factor ~ 5. To improve the load balance, we need to change the way particles are assigned to processes. The time to construct the LET (magenta) is nearly constant, and becomes dominant when $N_p > 10$. Another issue is that, in the current implementation, some CPU cores remain idle. The idle CPUs could be used to accelerate the operation of construction of LET.

We show the performance for weak scaling in figure 7. Each process holds $N_1 = 2^{24} \sim 16$M particles. The total number of particles is $N = N_1 \times N_p$. The time to compute gravitational

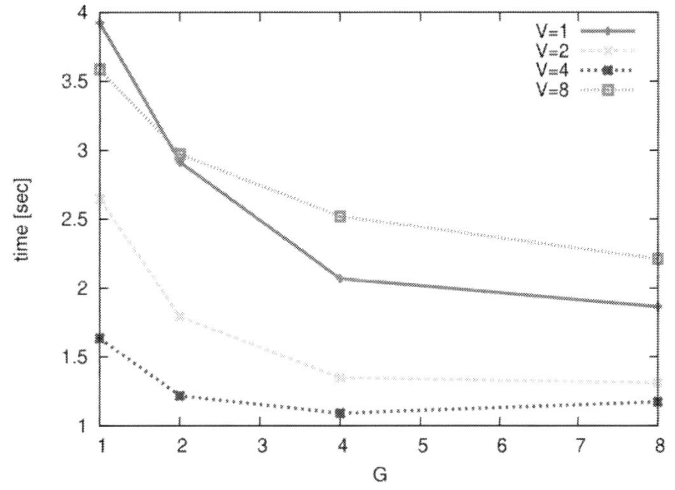

Figure 5. The dependence of the computational time of the kernel function on the vector length, V, and group size, G. Each line is the result for a different V. The best performance for this test calculations is obtained for $(V, G) = (4, 4)$.

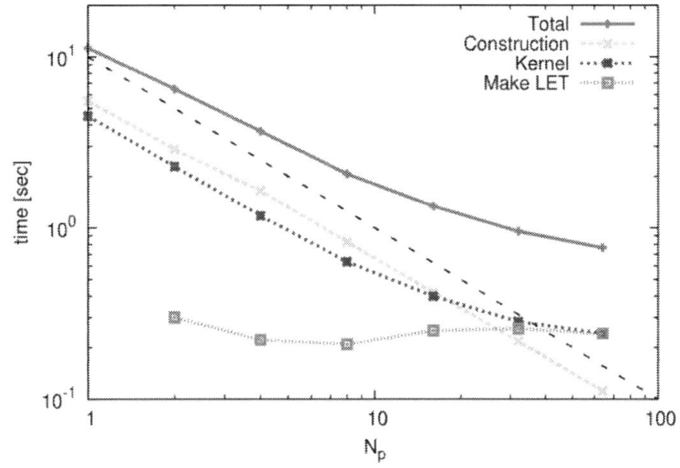

Figure 6. Strong scaling of different components of our code. Horizontal and vertical axes are number of MPI processes, $N_{\rm p}$ and computational time in the unit of second. Each line represents the time for tree–construction (green), kernel function (blue), and making LET (magenta). The red line is the total time. The black dashed one shows a scaling $\propto N_{\rm p}^{-1}$.

acceleration (red) is proportional to $\log N_{\rm p}$. This can be explained as follows: The computational cost of the tree method is $O(N \log N) = O(N_1 N_{\rm p} \log N_1 N_{\rm p})$. Through parallelization, it becomes $O(N_1 \log N_1 N_{\rm p})$. Since N_1 is a constant, the computational cost $\propto \log N_{\rm p}$. In other words, the depth of the tree structure is proportional to $\log N \propto \log N_{\rm p}$. In future work, we will enable the computation of the kernel function and the construction of the LET to occur in parallel through MPI communication to improve performance.

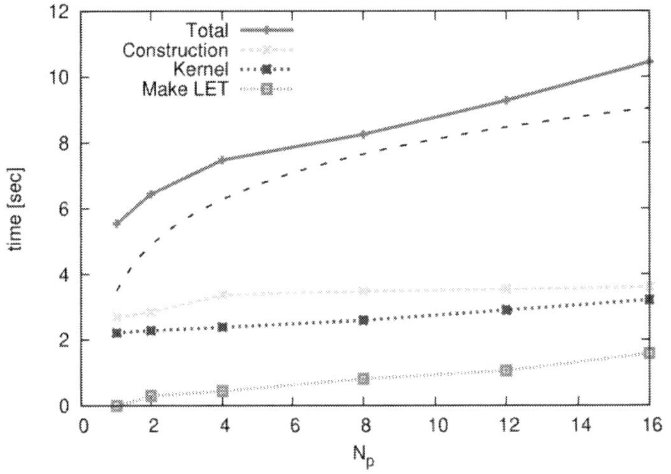

Figure 7. Weak scaling of different component of our code. The black line represents a scaling $\propto \log N_\mathrm{p}$. Other lines and axes have the same meaning as in figure 6.

4. Results of N–body simulations with oscillating galactic potential

We investigate the dynamical response of DM halos to oscillations of the galactic potential using N–body simulations.

4.1. Set up

The N–body system represents a DM halo. Initially, the halo profile follows the NFW model in virial equilibrium state. The NFW profile has a central cusp. To express the baryon component, we add an external Hernquist potential [15] to the N–body system and change its scale length in an oscillating manner to represent the gas contraction and expansion cycle driven by periodic starbursts. We choose a scale length typical for dwarf galaxies, 1kpc and an oscillation amplitude of 1kpc, that represent massive outflows. Fundamental parameters adopted in our simulations are listed in Table 2. We investigate three different values for the oscillation period of the external potential, T, which are chosen to reflect the star formation histories of observed local dwarf galaxies [16] [17] [18].

Table 2. Fixed parameters in our N–body simulation.

Number of particles	$2^{24} \sim 16$M, $2^{27} \sim 128$M
Softening (resolution)	0.004 kpc
Mass of DM halo	$10^9 M_\odot$
Scale length of DM halo	2 kpc
Baryonic mass	$1.7 \times 10^8 M_\odot$

4.2. Results

Figure 8 demonstrates the resultant density profile of DM halos after 10 oscillation periods. The red, blue, and magenta lines show the results for $T = 10$, 30 and 100 Myr, respectively. The yellow line is the result of a high resolution run for $T = 30$ Myr, in which we used 8 times the number of particles, which demonstrates good convergence of the solution. The black dashed line represents the initial NFW model, which has a central cusp. It is clear that the cusp has transitioned to a core, and that the resultant core scale depends on the oscillation period of the external potential, T. We have confirmed that the resultant core structure is stable during at least the first 100 oscillation periods.

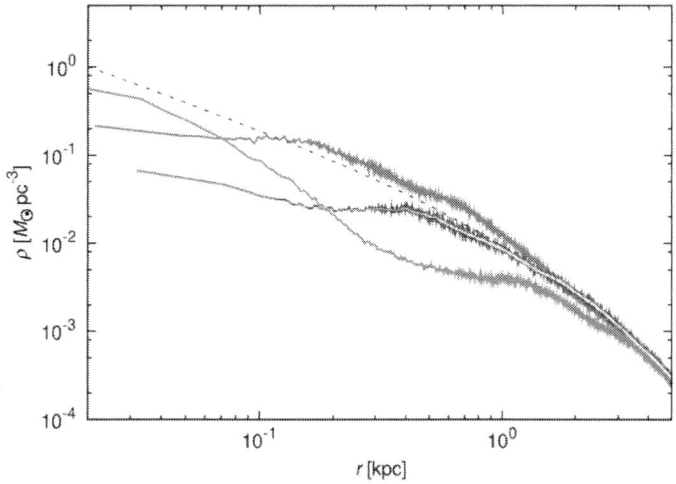

Figure 8. Density profiles of DM halos after 10 oscillation periods. The horizontal and vertical axes are the distance r from the galactic center and mass-density of the DM halo, respectively. The red, blue, and magenta lines show the results for $T = 10$, 30 and 100 Myr, respectively. Vertical lines are the predictions of the core scale by our analytical model (see section 5).

We also analyze the temporal Fourier spectrum of radial velocities of the modified DM halos. In figure 9, we show the Fourier spectrum of the radial velocity of the system, $\hat{v}_{\rm r}$, as a function of r for the Fourier frequencies, $\omega = 2\pi/T$. Comparing the location of the peaks of the spectra in this figure with the core length scale in figure 8, we find that the two match very well. We now turn to the physics of the mechanism that accelerates the DM particles on core length scales and creates the core structure from an initially cuspy profile.

5. Linear analysis of resonance

What is the physical reason for the transition from DM cusp to core in our simulations? To investigate this, we performed a linear analysis of the resonance between the density waves and the particles. We approximate the particle system by a fluid and consider the situation in which the equilibrium system, labeled 0, is perturbed by the external force (subscript ex), inducing a change in some physical quantities (subscript ind). We focus on a particular group of particles of some constant density, ρ_0, and velocity, v_0, in the equilibrium state. We assume the external force has a sinusoidal form,

$$-\frac{\partial \Phi_{\rm ex}}{\partial x} = A \sin{(kx - \Omega t)}, \qquad (1)$$

where A, k, and Ω are the oscillation strength, which is a positive constant, the wavenumber, and the frequency of the external force, respectively.

We solve the linearized Euler's equation and the equation of continuity, and derive the following solutions for the induced velocity and induced density enhancement;

$$v_{\rm ind}(t, r) = \frac{A}{\Omega - kv_0} \cos{(kr - \Omega t)}, \qquad (2)$$

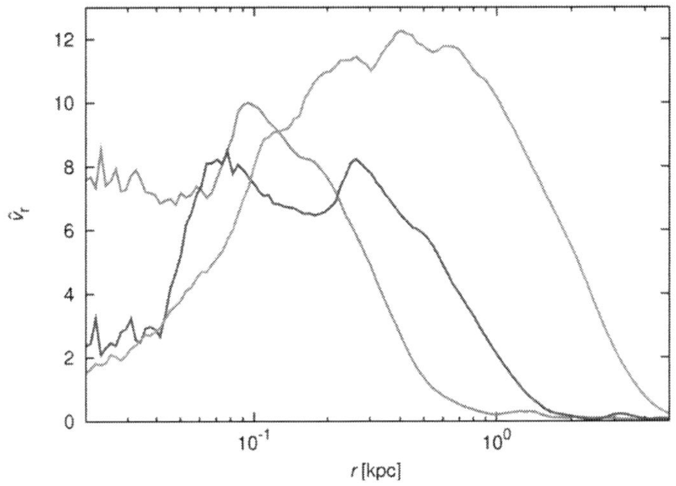

Figure 9. Fourier spectrum of the radial velocity of the system, $\hat{v}_r$ as a function of r. Each line represents the spectrum of same run as figure 8.

$$\rho_{\rm ind}(t, r) \quad = \quad \frac{A\rho_0 k}{(\Omega - kv_0)^2} \cos{(kr - \Omega t)}. \tag{3}$$

We find that the coefficients will diverge when the frequency of the external force $\Omega \equiv 2\pi/T$ equals kv_0. This is analogous to the resonance in the forced oscillation problem of harmonic oscillators. Thus, we conclude that DM halos are strongly affected by resonance between DM particles and density waves, and a core structure can form this way.

We may rewrite the resonance condition, $kv_0 \sim \Omega$, as

$$t_{\rm d}(r) \sim T, \tag{4}$$

where $t_{\rm d}(r)$ is the local dynamical time of the DM halo measured at r. Using this condition, we can predict the core scale created by resonance. The vertical lines in figure 8 are the predictions of this analytical model. They match well with the scales of resultant core.

6. Summary

To resolve the core-cusp problem of DM halos, we have studied the dynamical response of DM halos to oscillations of the gravitational potential. We have performed collisionless N-body simulations utilizing the tree algorithm. We speed-up our tree code by optimizing it for GPU clusters. We have proposed a method to reduce the frequency of Warp branches, and, as a result, the tree-traversal operation has been accelerated. From our simulations, we have arrived at the conclusion that resonance between DM particles and gas oscillation may play an important role to flatten the central cusp. Our analytical model predicts the core scales seen in the simulations quite accurately.

Acknowledgments

We greatly thank Alexander Wagner for checking our manuscript. Numerical simulations were performed with HA-PACS at the Center for Computational Sciences, University of Tsukuba. This work was supported by the Grant-in-Aid for Scientific Research (A)(21244013), and (C)(25400222).

Journal of Physics: Conference Series **454** (2013) 012014 doi:10.1088/1742-6596/454/1/012014

References

[1] Navarro J F, Frenk C S and White S D M 1997 *The Astrophysical Journal* **490** 493
[2] Moore B, Quinn T, Governato F, Stadel J and Lake G 1999 *Monthly Notices of the Royal Astronomical Society* **310** 1147
[3] Ishiyama T, Makino J, Portegies Zwart S *et al.* 2011 *Preprint* astro-ph/1101.2020
[4] Moore B 1994 *Nature* **370** 629
[5] Burkert A 1995 *The Astrophysical Journal Lett.* **447** L25
[6] Oh S-H, de Blok W J G, Brinks E, Walter F and Kennicutt R C Jr. 2011 *The Astronomical Journal* **141** 193
[7] Mashchenko S, Wadsley J and Couchman H M P 2008 *Science* **319** 174
[8] Governato F, Brook C, Mayer L *et al.* 2010 *Nature* **463** 203
[9] Pontzen A and Governato F 2012 *Monthly Notices of the Royal Astronomical Society* **421** 3464
[10] Barnes J and Hut P 1986 *Nature* **324** 446
[11] Nakasato N 2012 *Journal of Comp. Sci.* Volume 3 **3** 132141
[12] Nakasato N, Ogiya G, Miki Y, Mori M and Nomoto K 2012 *Preprint* astro-ph/1206.1199
[13] HA-PACS, http://www.ccs.tsukuba.ac.jp/CCS/eng/research-activities/projects/ha-pacs/base-cluster
[14] Warren M and Salmon J 1993 *ACM/IEEE Conference on Supercomputing* (New York: ACM.) pp 1221
[15] Hernquist L 1990 *The Astrophysical Journal* **356** 359
[16] Mateo M L 1998 *Annual Review of Astronomy and Astrophysics* **36** 435
[17] Tolstoy E, Hill V and Tosi M 2009 *Annual Review of Astronomy and Astrophysics* **47** 371
[18] McQuinn K B W, Skillman E D, Cannon J M *et al.* 2010 *The Astrophysical Journal* **724** 49

Journal of Physics: Conference Series **454** (2013) 012015 doi:10.1088/1742-6596/454/1/012015

Exploiting parallelism in many-core architectures: Lattice Boltzmann models as a test case

F Mantovani[1], M Pivanti[2], S F Schifano[3] and R Tripiccione[4]

[1] Department of Physics, Univesität Regensburg, Germany
[2] Department of Physics, Università di Roma *La Sapienza*, Italy
[3] Department of Mathematics and Informatics, Università di Ferrara and INFN, Italy
[4] Department of Physics and CMCS, Università di Ferrara and INFN, Italy

E-mail: `filippo.mantovani@regensburg.de`, `pivanti@fe.infn.it`, `schifano@fe.infn.it`, `tripiccione@fe.infn.it`

Abstract. In this paper we address the problem of identifying and exploiting techniques that optimize the performance of large scale scientific codes on many-core processors. We consider as a test-bed a state-of-the-art Lattice Boltzmann (LB) model, that accurately reproduces the thermo-hydrodynamics of a 2D-fluid obeying the equations of state of a perfect gas. The regular structure of Lattice Boltzmann algorithms makes it relatively easy to identify a large degree of available parallelism; the challenge is that of mapping this parallelism onto processors whose architecture is becoming more and more complex, both in terms of an increasing number of independent cores and – within each core – of vector instructions on longer and longer data words. We take as an example the Intel *Sandy Bridge* micro-architecture, that supports AVX instructions operating on 256-bit vectors; we address the problem of efficiently implementing the key computational kernels of LB codes – streaming and collision – on this family of processors; we introduce several successive optimization steps and quantitatively assess the impact of each of them on performance. Our final result is a production-ready code already in use for large scale simulations of the Rayleigh-Taylor instability. We analyze both raw performance and scaling figures, and compare with GPU-based implementations of similar codes.

1. Introduction

For several years, the main path to performance in large scale scientific computations has been that of introducing more parallelism in the associated algorithms and in their implementations. This trend reflects a matching evolution in system architectures that – over the years – have delivered more computing power by increasing the number of processing elements: today, high-end systems routinely have tens of thousands of interconnected cores.

Following this trend, application development has focused on identifying and exploiting a larger fraction of parallelism from the underlying algorithms and on mapping applications on a larger number of processing threads; several programming and runtime environments, such as MPI or openMP, are now available to support this trend. For some algorithms, whose parallelizable fraction is very large, the limits set by Amdahl's law have not prevented to reach good *strong* scaling figures. In many other cases, increased parallel computing resources have been used to tackle problems of larger scale, and, as summarized by Gustafson's law [1], at least good *weak* scaling figures have been obtained.

Recent developments in processor architecture introduce yet a new feature: they try to exploit several *different* parallelization options within the *same* processor: not only is the number of processors of one processing element and the number of cores available inside one processor steadily increasing, but data-paths within each core have now more parallel features, such as SIMD or vector extensions. At the extreme edge of this spectrum one finds GP-GPU architectures, that can be seen as assemblies of several hundred "slim" cores, so typically one core processes one element of the data-set of the program.

Memory access has also become more complex: most current processing elements support a common addressing space for all their cores, and at least some level of cache is kept coherent across the node, but memory bandwidth and latency – as seen by one specific core – is not uniform across the addressing space, as it depends strongly on how "close" a given memory location is to the processing core.

Performance optimization for these recent architectures is becoming more complex, as some difficult balance between conflicting requirements has to be found: the problem has to be partitioned over many independent threads, but – within each thread – one has to identify data elements with no data-dependencies that can be operated upon by vector instructions; at the same time data allocation and data sharing policies may have a strong impact on cache performance and on memory bandwidth.

In this paper, we try to identify several optimization strategies to tackle and solve the problems described above, and to assess their impact on performance. We focus on one specific test-case, a state-of-the-art Lattice Boltzmann (LB) algorithm in 2 dimensions that has a large and easily identified available parallelism and a very regular algorithmic structure; we also limit ourselves to latest generation multi-core Intel architectures, that assemble a large number of "fat" cores within one processor.

While our main motivation for this work has been that of developing a very fast code for large scale fluid-dynamics simulations, here we offer to the reader an analysis of the problems that will be encountered in most large scale scientific applications and that must be overcome to sustain performance on present (and possibly future) processors. Most of our optimization steps match specific architectural features of these processors; this allows a close assessment of the potential for performance of each architectural feature, but on the other side makes programming difficult and strongly machine dependent; it seems obvious to us that the development of efficient (semi-)automatic tools able to carry out at least part of this work starting from high level programs is an important problem, that should be extensively addressed.

This paper is structured as follows: in section 2 we discuss the architecture of typical large scale HPC commodity systems, and general software optimization strategies suitable for recent processors; section 3 briefly describes our Lattice Boltzmann algorithm, while section 4 presents in detail all optimizations steps that we have developed for our LB code. Section 5 summarizes our performance results and draws some comparisons with different architectures, such as GPUs; the paper ends with our concluding remarks. This paper draws on and expands upon earlier results for a previous generation of Intel processors based on the *Nehalem* micro-architecture [2, 3], and on GP-GPU [9, 10] systems based on the Fermi processor.

2. Multi-core CPU Commodity Systems

The typical HPC system for scientific computing today is a large cluster of nodes based on commodity multi-core CPUs, and interconnected by a high-speed switched network, such as Infiniband.

In recent years multi-core architecture has become the standard approach to develop high performance processors. Several cores are integrated on a single chip device; this has allowed peak performance at node level to scale according to Moore's law. In many cases, the cores are x86-CPUs, usually including two levels of caches, and a floating-point unit able to execute vector instructions; vector size increases steadily as newer processor versions are released. Today, 256 bit vectors have been adopted in the Intel Sandy Bridge micro-architecture, while the forthcoming Intel MIC architectures foresee 512 bits. Cores within a device share a large L3-cache and external memory.

Many system vendors provide so-called multi-socket platforms, based on several multi-core CPUs. Typically, two, four or eight sockets can be installed on a mother-board; each socket has its own memory-bank and is connected to the other sockets by dedicated busses. Appropriate protocols allow to share the global memory space and keep the content of caches coherent.

From a programming point of view these systems operate as *Symmetric Multiprocessors* (SMP), which means that they can be programmed as a single processor with N_c cores (N_c equal to the sum of cores of all sockets) sharing the full memory space and controlled by a single instance of the *Linux* operating system. Performances heavily rely on the ability of programmers, compilers and run-time support to carefully exploit parallelism on *all* available hardware features. Relevant optimizations for performances are:

- core parallelism: the code should allow all cores to work in parallel exploiting MIMD or SPMD multi-task parallelism; in the first case the application is decomposed in several sub-tasks and each one is executed by a different core; in the latter case, that can be also combined with the previous one, the data-set is typically partitioned among the cores, each one executing the same task.
- vector programming: each core has to process the data-set of the application using vector instructions and exploiting streaming-parallelism (SIMD); the number of data-items that can be processed by vector instructions depends on the architecture of the CPU; on the latest Sandy Bridge architecture the vector size is 256-bit, that is up to 8 single-precision or 4 double-precision floating point numbers.
- efficient use of caches: memory hierarchies of commodity systems are based on the concept of cache to minimize over-heads associated to accessing main memories. The application code has to exploit cache-reuse in order to save time in memory access; this may have a serious impact on the organization of the data-layout for the application.
- NUMA control: multi-socket platforms are *Non Uniform Memory Access* (NUMA) systems; this means that the time to access data in memory is related to the allocation of the threads that perform the memory access; the time is shorter to access data stored onto memory attached to the socket where the thread is running.

3. Lattice Boltzmann

The Lattice Boltzmann method is a computational approach that describes fluid dynamics by lattice discretization in position and momentum space [4]. On a discrete lattice (in $D = 2$ or 3 dimensions), the actual dynamics is replaced by a synthetic one, based on a set of lattice populations ($f_l(x,t)$); each population has a given fixed lattice velocity $\boldsymbol{c}_l$, associated to the fact that, at each time step, it drifts towards a nearby lattice site; populations evolve in (discrete) time according to the following equation:

$$f_l(\boldsymbol{x} + \boldsymbol{c}_l\Delta t, t + \Delta t) - f_l(\boldsymbol{x}, t) = -\frac{\Delta t}{\tau}\left(f_l(\boldsymbol{x}, t) - f_l^{(eq)}\right) \quad (1)$$

Macroscopic observables (density ρ, velocity $\boldsymbol{u}$ and temperature T) are defined in terms of the $f_l(x,t)$:

$$\rho = \sum_l f_l, \quad (2)$$

$$\rho \boldsymbol{u} = \sum_l \boldsymbol{c}_l f_l, \tag{3}$$

$$D\rho T = \sum_l |\boldsymbol{c}_l - \boldsymbol{u}|^2 f_l, \tag{4}$$

and the equilibrium distributions ($f_l^{(eq)}$) are themselves function of these macroscopic quantities [4].

LB models are now available in many different versions, characterized by the dimension of the lattice, the number of populations that they use (they are usually labelled as DnQm, where n is the number of dimensions and m is the number of populations) and by the explicit functional form of the equilibrium distributions $f_l^{(eq)}$. Starting from early models (e.g., D2Q9 or D3Q19), able to describe an incompressible fluid at constant temperature, recent LB algorithms describe the thermo-hydrodynamics of multiple-phases or correctly treat compressible fluids with realistic equations of state.

In the following we focus on a recently proposed D2Q37 model that correctly reproduces the equation of state of the fluid, regarded as a perfect gas ($p = \rho T$); see [5, 6] for full details. For this model, one shows that, after appropriate shift and re-normalization of the velocity and temperature fields, one recovers, via a Taylor expansion in Δt, the following set of thermo-hydrodynamical equations:

$$D_t \rho = -\rho \partial_i u_i^{(H)} \tag{5}$$

$$\rho D_t u_i^{(H)} = -\partial_i p - \rho g \delta_{i,2} + \nu \partial_{jj} u_i^{(H)} \tag{6}$$

$$\rho c_v D_t T^{(H)} + p \partial_i u_i^{(H)} = k \partial_{ii} T^{(H)}; \tag{7}$$

where superscripts H flag renormalized (physical) quantities, $D_t = \partial_t + u_j^{(H)} \partial_j$ is the material derivative and we neglect viscous heating; c_v is the specific heat at constant volume for an ideal gas, $p = \rho T^{(H)}$, and ν and k are the transport coefficients; g is the acceleration of gravity. In this model, population velocities are associated to moves on the lattice that reach points up to three lattice points away, see figure 3.

¿From the point of view of the present paper, LB algorithms are a nice test case, as they offer a huge degree of available parallelism, that can be immediately and easily identified: defining $\boldsymbol{y} = \boldsymbol{x} + \boldsymbol{c}_l \Delta t$, one rewrites the main evolution equation as:

$$f_l(\boldsymbol{y}, t + \Delta t) = f_l(\boldsymbol{y} - \boldsymbol{c}_l \Delta t, t) - \frac{\Delta t}{\tau}\left(f_l(\boldsymbol{y} - \boldsymbol{c}_l \Delta t, t) - f_l^{(eq)}\right) \tag{8}$$

It is now easy to identify the overall structure of the computation that evolves the system by one time step Δt; for each point $\boldsymbol{y}$ in the discrete grid one:

(i) gathers from neighboring sites the values of the fields f_l corresponding to populations that drift towards $\boldsymbol{y}$ with velocity $\boldsymbol{c}_l$ and then

(ii) performs all mathematical processing needed to compute (*in a completely local fashion*) the quantities appearing in the equation above. This step is slightly more complex if one wants to take into account reactive effects (combustion); indeed, in that case the divergence of the velocity field has to be explicitly computed. This means that a further gather operation must be performed midway in this (otherwise local) compute intensive step.

The key remark is that both steps above are completely uncorrelated for different points of the grid, so they can be parallelized according to any convenient schedule, as long as one makes sure that, for each and all grid points, step 1 is performed before step 2. In principle, the available

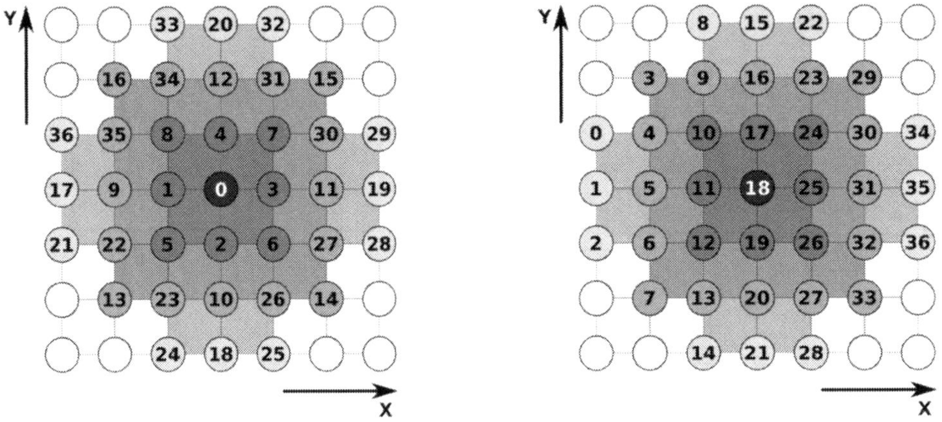

Figure 1. Velocity vectors for the LB populations in the D2Q37 model. Population labelling is arbitrary; populations are stored at consecutive memory locations, in the order given by their labels: at left our first labelling choice; at right, a different labeling , that improves cache reuse, as discussed in detail in section 4.

parallelism is as large as the number of sites of the lattice (that easily grows to hundreds of millions of sites).

A reference implementation of the LB algorithm repeatedly evolves the lattice-cells of the system for one time step. For each point in the grid, data needed to compute the new value of each population at each site in the grid is gathered from nearby sites, and then a fully local processing step is performed.

Reflecting this overall organization, the class of codes that we have developed processes every grid-point by applying in order the following three main computational phases:

- stream(): this phase gathers for each site 37 populations from neighbors according to the scheme of figure 3. This process does not make any floating-point computation but only accesses sparse blocks of memory locations. It collects at each site all the populations that will interact at the next computational phase, collide(). In this step, each site accesses the populations of neighbor cells at distance up to 3 in the physical grid.
- bc(): this phase adjusts values of the cells at the top and bottom edges of the lattice to enforce appropriate boundary conditions (e.g., a constant given temperature and zero velocity). At the right and left boundaries, we apply periodic boundary conditions. This is most easily done by allocating additional storage where copies of an appropriate number (3 in our case) of the rightmost and leftmost columns of the lattice are placed before the stream() step. Points at the right/left boundaries can then be processed as those in the bulk. If needed, boundary conditions could of course be enforced in the same way as it is done for the top and bottom edges.
- collide(): this phase performs all the mathematical steps associated to equation 8 and needed to compute the population values at each lattice site at the new time step (this is called "collision", in LB jargon). Input data for this phase are the populations gathered by the previous stream() phase. This step is the floating point most intensive section of the code; it uses only the population members of the site on which it operates, making the processing of different sites fully uncorrelated.

In the following section we describe in detail a family of implementations of these kernels that try to match at best the available parallelism with the architectural structure of a recent many-core CPU.

4. D2Q37 Implementation and Optimization

Our codes are targeted for multi-socket systems based on the Intel Sandy Bridge micro-architecture. All performance results shown below refer to a pre-production system based on dual Intel Xeon E5-2680 sockets (2-SB, in short). Each socket is an eight-core processor operating at 2.7 GHz, with 20 MB of L3-cache and 32 GB of RAM. Each core supports the execution of the *Advanced Vector Extensions* instruction set, using 256-bit register operands, on which 8 single-precision or 4 double-precision numbers can be packed; it completes two AVX double precision operations – one add and one multiply – at each clock cycle, delivering a peak performance of 21.6 double precision Gflops per core. All in all, the system has a peak memory bandwidth of 85.3 GB/s, and a peak double precision performance of 345.6 Gflops.

Data organization schemes for LB codes are mainly of two kinds, namely array-of-structures (AoS) and structure-of-arrays (SoA). For cache-based CPUs the AoS scheme is preferable as it keeps all populations of each lattice-site close together in memory; this improves the locality of populations associated to each lattice point, and better suits the cache hierarchy of the processor. This is relevant for performance of the collision kernel, the most compute intensive part of the code, since all computation involved at each lattice site only depends on the populations associated to the site itself. In the AoS scheme each lattice-cell is stored as an array of 37 double-precision floating-point values representing the populations of the D2Q37 LB model. In our implementation we store the lattice in column-major order, and we keep in memory two copies of the lattice; each step reads inputs from one copy and writes results to the other. This solution is more memory expensive than having a single copy, but it makes the implementation of the code much easier, and on current systems memory size is not a limiting resource for this problem. Our LB code runs one thread per core, and a lattice of size $L_x \times L_y$ is split in *sub-lattices* along the X dimension, each associated to a different thread. The threads are topologically arranged in a ring, and each sub-lattice includes three left and three right *ghost columns*; at the beginning of each loop, threads copy the three leftmost and rightmost columns from the sub-lattice of the previous and next thread onto the ghost columns of its own sub-lattice. Within each core, K sites of the sub-lattice are processed in parallel, exploiting data-parallelism using vector instructions; in the case of Sandy Bridge architectures, four double precision number can be packet into an AVX vector, so $K = 4$.

Parallelism at core level is managed through the standard-POSIX Pthreads library, the closer approach to the *Linux* kernel, to avoid any overheads associated to high-level libraries such as openMP or openMPI.

We have developed two versions of the code. The first version (V1) keeps the computation of the stream and collide kernels separate; this is necessary if reactive dynamics is enabled because we have to compute the divergence of the velocity field which requires to access all lattice points and has to be performed by an additional step between stream and collide. Version V2 is used if reactive dynamics is not enabled; in this we can merge computation of stream and collide in just one step, applied in sequence to all cells of the lattice. This optimization step is well known in literature, see for example [7].

Code version V1 is organized in several computational steps, each synchronized through the use of a pthread-barrier:

(i) two threads (in our case threads 0 and 1) copy the left and right borders onto the appropriate frames. This operation is simply performed by memory-copies; after receiving data from the neighbor threads, both threads apply stream kernel to the three lattice columns close to the frame that they have just copied.

(ii) while step (i) is executed, the remaining threads compute in parallel the stream kernel on all columns far enough from their Y-edges.

(iii) at the end of step (i) and (ii) two threads compute the bc kernel on the cells close to the top and the bottom of the lattice.

(iv) after step (iii), all threads start computing the collide phase, each on a different portion of the sub-lattice.

The organization of code version V2 is not fully straightforward, as the correct enforcement of the boundary conditions (execution of the bc kernel) has to be done after stream but before collide. In this case the computation is organized in three main steps, each followed by a synchronization barrier:

step 1: two threads copy the border-columns from neighbor threads, while two other threads execute stream for the upper and lower rows of the grid, leaving out the first and the last 3 columns; for these columns stream will be computed by threads performing the copy of border columns, as soon as they have finished to copy data into the ghost columns.

step 2: two threads adjust the boundary conditions for the cells located at the top and bottom rows of the grid. After one step of synchronization, all threads start computing the collide kernel for the same cells.

step 3: all threads compute stream and collide for the remaining cells of grid (the ones in the bulk).

The organization of the codes described above allows us to exploit core parallelism. A further optimization aims to exploit data parallelism, using the AVX vector instructions available on the Sandy Bridge processor, and processing four cells of the lattice at the same time. Data parallelism can be exploited either by the compiler, or by using *intrinsic* functions. In the first case the program is a scalar code, and is compiled enabling the *auto-vectorization* mode of the compiler. This lets the compiler introduce vector instructions, e.g. unrolling a loop and processing two or more independent iterations (according to the type of variables involved) using SIMD AVX instructions. Alternatively, the program may be handcrafted to use vector variables which are processed using *intrinsic* functions mapped by the compiler directly onto vector instructions. For example a double precision vector sum can be performed by the code line `c = _mm256_add_pd (a, b)` where `a, b, c` are vector variables of type `__mm256d`. This approach is allowed both by the GCC and ICC compilers. A third, more elegant and transparent way to generate vector code is that of using custom data vector definitions, which is allowed by the GCC compiler. GCC allows to define a vector data-type where two or more primitive-types (int, float, double, ...) can be packed in a vector; this can be done using the vector_size attribute in the type definition. Operations on vector data-type variables can be performed using the usual infix notation, and are then translated by the compiler generating assembly code that uses streaming instructions for the micro-architecture selected by the appropriate compilation flag. We have used this approach for all our codes, and we have used the GCC compiler throughout; we have also verified in a few selected cases that the ICC compiler reaches similar performances ($\pm 10\%$). We use this feature to define a new vector data type called fourD_t, see figure 2 left, to pack four double variables in an AVX vector. Each element of the population array then stores populations of four lattice cells. In our case we pack four lattice cells at distance $L_y/4$, as shown in figure 2 right.

Figure 3 shows performances for the collide (left) and stream (right) benchmarks as a function of the numbers of threads, running on a dual-socket eight-core Sandy Bridge system. We compare performance of the scalar code (GCC autovec) compiled enabling the auto-vectorization mode of the compiler, with that obtained by the vector code using vector variables (GCC vec). Performance of the collide kernel scales well with the number of threads; the version using vector variables is a factor $2\times$ better than the scalar code with auto-vectorization, and a factor $3.5\times$

[t]

```
typedef double fourD_t
        __attribute__ ((vector_size(32)));

typedef struct {
  fourD_t p[37];        // populations array
} v_pop_t;
```

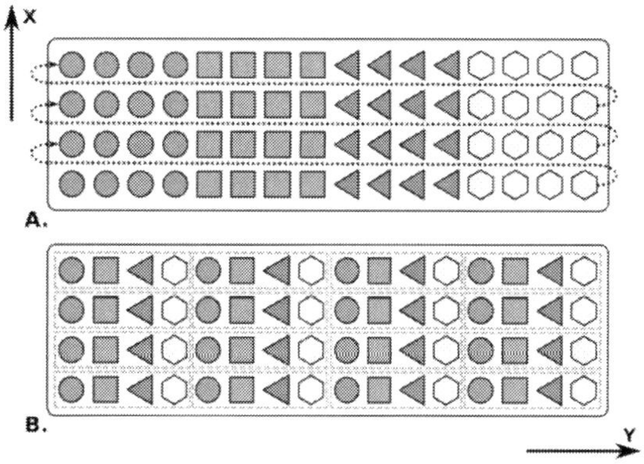

Figure 2. Left: definition of the structure holding four lattice cells at distance $NY/4$, that we use to exploit the AVX vector instructions of the Sandy Bridge processors. Right: visualization of how cells of a sample lattice of size 4×16 (A) have been packed to be processed by AVX vector instructions (B).

better than the scalar code compiled without auto-vectorization. Using 16 threads the kernel delivers 221 Gflops corresponding to 64% of peak.

Concerning the performance of the stream kernel, the compiler is not able to vectorize the code, so the compiled code executes loads and stores of scalar registers. Since loads and stores are actually made to cache, and eventually full cache lines are moved to/from main memory, the impact on performance should not be too large. However, a significant part of the workload associated to stream goes into the quite complex evaluation of addresses. In the vectorized version, this burden is reduced by a factor four, so one expects much better performance in this case. Indeed, the code version using vector variables – which already includes some of the optimizations that will be discussed later – is a factor approximately 1.7× better than the scalar code, and the memory bandwidth, saturates at 35 GB/s, ≈ 39% of peak.

We have introduced further optimizations to improve memory and cache efficiency; these optimizations are mainly relevant for the performance of the stream kernel. We remind that, in our implementation the cells of the lattice are stored in column major order; moreover, the

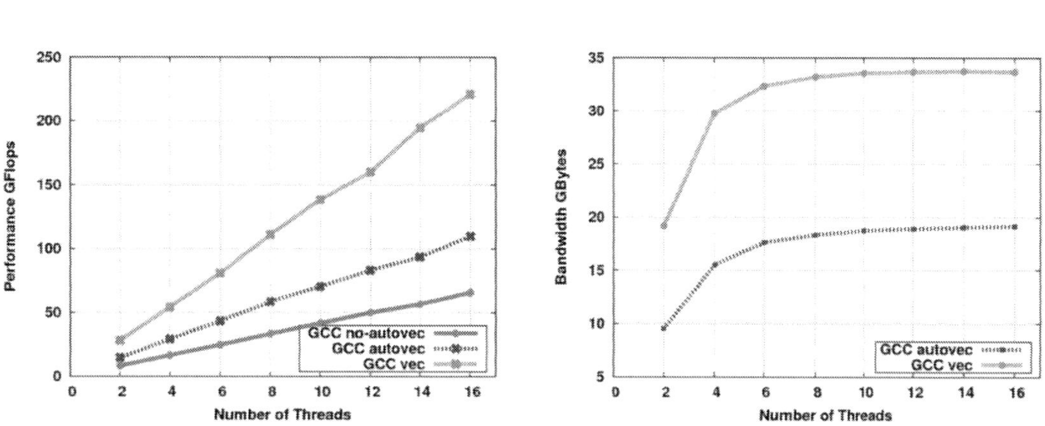

Figure 3. Performance figures for collide (left) and stream (right) kernels on a lattice of 1680×16000 points. For both kernels we show the performances obtained by the code compiled enabling the auto-vectorization mode of the compiler, and by the code that use intrinsic AVX instructions.

Journal of Physics: Conference Series **454** (2013) 012015 doi:10.1088/1742-6596/454/1/012015

Table 1. Performance of the stream benchmark running on 16 threads, for several sizes of the lattice. We show the effective bandwidth, as we add more and more optimizations steps. Left to right, basic code, NUMA control (+NUMA), new population labelling (+New-Labelling), cache-blocking (+Cache-Blocking), non-temporal instructions (+NT).

Stream Bandwidth (GB/s))						
$L_x \times L_y$	Size (GB)	Base	+NUMA	+New-Labelling	+Cache-Blocking	+NT
256 × 32000	2.26	11.7	25.4	33.9	44.0	58.5
480 × 32000	4.23	12.8	25.5	33.7	44.3	58.2
1680 × 16000	7.41	11.6	25.4	33.7	43.8	58.5

stream() step performs scattered accesses in local memory to gather populations at distance 1, 2 and 3 in the grid; one expects that performances may depend on details of cache and memory-allocation policies, as described in [8]. To improve cache performance, we have first chosen an appropriate labeling for the populations, as shown in figure 3 right; this new labeling maximizes the number of population values loaded into cache when processing point (x, y) in the lattice, and still available in cache when considering points $(x, y+1)$ and $(x, y+2)$. Moreover, to improve cache re-use, we have also implemented *cache-blocking*; each core executes the stream operation on a block of $B_x \times B_y$ cells such that the data size associates to each block fits into the L3 cache. We combine cache-blocking with the use of *non-temporal* store instructions; non-temporal stores do not require a prior cache-line read for ownership (RFO), but write data to memory directly, improving memory bandwidth. Finally, we also managed memory and thread allocation using the Linux NUMA library. The lattice array allocated on each node is split into sub-arrays, and each one is mapped onto the memory-bank close to the CPU of the thread processing it. This avoid conflicts due to a thread running on a CPU and accessing memory locations allocated on the banks physically attached to a different CPU. For a detailed analysis of this allocation style, see [2, 3].

Table 1 lists the effective bandwidth obtained by the stream benchmark running on the dual socket Sandy Bridge system, as we introduce more and more optimization steps. The version including all optimizations improves by a factor ≈ 10 with respect to the Base version; it delivers a peak bandwidth of ≈ 58 GB/s corresponding to approximately 68% of the theoretical peak.

5. Results and Conclusions

In this section we summarize the results of our implementation, and compare performances of our full codes (versions V1 and V2) with that obtained on a GP-GPU system.

Table 2 lists performance figures of production-ready code versions V1 and V2, running on the 2-SB system; we compare performances with that of our previous implementation [11] developed for a NVIDIA C2050 card based on the Fermi processor. Code version V1 delivers ≈ 166 Gflops on the 2-SB system, corresponding to 48% of the peak, while Code version V2 delivers ≈ 216 Gflops on the the same system, faster than the corresponding version for the GP-GPU; this corresponds to 62% of peak.

Our LB code optimized for multi-core CPUs has been extensively used to study several features of the Rayleigh-Taylor instability in 2 dimension; physics results will be published elsewhere; in picture 4 we show the temperature maps from a simulation of a small Rayleigh-Taylor cell of size 512 × 1024, at several stages during time evolution; the simulation as been performed using code version V2 running on the 2-SB system.

In this paper, we have described several optimization-steps that we have applied to our LB

Table 2. Performances of production-ready code versions V1 (left) and V2 (right), on a C2050 board based on a NVIDIA Fermi GP-GPU, and on a dual-socket Sandy Bridge system. For each code version we show the execution time, and the corresponding performance figures; we also list the sustained performance P of the full code, its efficiency, and two metrics relevant for the users: the time to update one site and its inverse (MLUps).

	C2050	2-SB
propagate (ms)	29.11	42.12
collide (ms)	154.10	146.00
propagate (GB/s)	84	60
collide (GF/s)	205.4	220
T/site (ns)	44	46
MLUps	22	21.7
P (GF/s)	172	166
R_{max}	33%	48%

	C2050	2-SB
propagateCollide (ms)	167.2	144.0
propagateCollide (GF/s)	190	224
T/site (ns)	40	35
MLUps	25	28.2
P (GF/s)	188	216
R_{max}	36%	62%

code to boost performances for a class of high-end commodity multi-core processors widely used for scientific computing. Our optimizations have implied a rather complex restructuring of the original algorithm, in order to exploit all available parallelism on the target architecture. Several successive code transformation have been applied: we have carefully controlled data allocation in main memory (including careful consideration of NUMA issues), we have split the computational task in many threads trying to maximize concurrent execution of threads, we have organized the computation to exploit SIMD features of the processor, and cache data reuse to improve memory bandwidth.

In summary, we can say that rather complex applications can sustain high performance if carefully implemented; on the other hand, no single optimization step is able to exploit a major fraction of the theoretically possible performance of the system; rather, many steps have to be implemented, and a substantial restructuring of the code is needed to carefully match the hardware structure of the processor.

Obviously we hope that automatic optimization tools will be able to perform this job in the

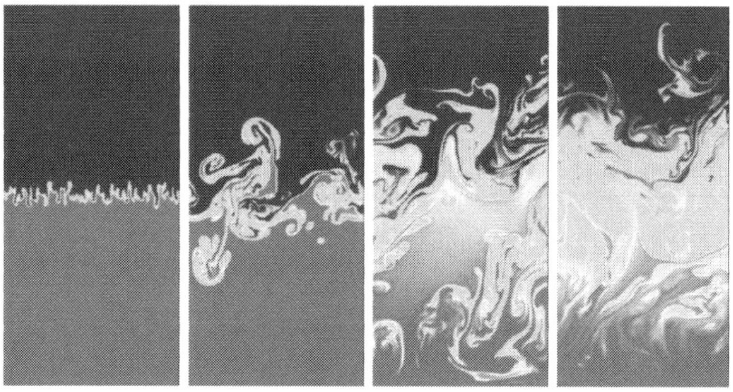

Figure 4. Temperature maps of a simulation of a Rayleigh-Taylor cell on a sample lattice size of 512×1024 points at several stages of its time evolution.

(not too far) future; we think that this is possible in principle, but it is hard to reach this goal starting from an implementation written in "low-level" languages such as C, where parallelism is not clearly expressed; a more high level description of the algorithm is probably necessary to help tools identify the available parallelism and adopt strategies to expose it.

Acknowledgments

This work has been done in the framework of the COKA and SUMA projects of INFN. We would like to thank INFN-CNAF and CINECA for providing access to their Sandy Bridge prototypes.

References

[1] Gustafson J L Reevaluating Amdahl's Law 2008 *Comm. of the ACM* **31** 532-3
[2] Biferale L et al 2011 Lattice Boltzmann Method Simulations on Massively Parallel Multi-core Architectures *Proc. of High Performance Computing Symposium* Boston MA
[3] Biferale L et al 2011 Optimization of Multi-Phase Compressible Lattice Boltzmann Codes on Massively Parallel Multi-Core Systems *Procedia Computer Science* **4** 994:1003
[4] Succi S 2001 *The Lattice Boltzmann Equation for Fluid Dynamics and Beyond* (Oxford:University Press)
[5] Sbragaglia M et al 2009 Lattice Boltzmann method with self-consistent thermo-hydrodynamic equilibria *J. Fluid Mech.* **628** 299
[6] Scagliarini A et al 2010 Lattice Boltzmann methods for thermal flows: Continuum limit and applications to compressible Rayleigh-Taylor systems *Phys. Fluids* **22** 055101
[7] Pohl T et al 2003 Optimization and Profiling of the Cache Performance of Parallel Lattice Boltzmann Codes *Parallel Processing Letters* **13** 549
[8] Wellein G et al 2006 On the Single Processor Performance of Simple Lattice Boltzmann Kernels *Computers and Fluids* **35** 910
[9] Biferale L et al 2012 A multi-GPU implementation of a D2Q37 Lattice Boltzmann Code *Lecture Notes in Computer Science* vol 7203 (Heidelberg:Springer) pp 640-650
[10] Bertazzo A et al 2012 Implementation and Optimization of a Thermal Lattice Boltzmann Algorithm on a multi-GPU cluster *Proc. of Innovative Parallel Computing* San Jose CA
[11] Biferale L et al 2012 An Optimized D2Q37 Lattice Boltzmann Code on GP-GPUs *Computers and Fluids* special issue on computer and fluids

24th IUPAP Conference on Computational Physics (IUPAP-CCP 2012) IOP Publishing
Journal of Physics: Conference Series **454** (2013) 012016 doi:10.1088/1742-6596/454/1/012016

Plasma particle-in-cell simulations with QED reactions for pair production experiments using a high-Z solid target

Toseo Moritaka[1], Luca Baiotti[1,2], An Lin[1,2], Li Weiwu[1,2], Youichi Sakawa[1,2], Yasuhiro Kuramitsu[1], Taichi Morita[1] and Hideaki Takabe[1,2]

[1]Institute of Laser Engineering, Osaka University, 2-6 Yamadaoka, Suita, Osaka, Japan
[2]Graduate School of Science, Osaka University, Machikaneyama, Toyonaka, Osaka, Japan

E-mail: moritaka-t@ile.osaka-u.ac.jp

Abstract. A new simulation scheme has been developed to investigate high-energy plasma phenomena including pair production and Bremsstrahlung in a high-Z nuclear field. The simulation scheme consists of (1) a Particle-in-Cell scheme for relativistic plasma dynamics, (2) a conservative semi-Lagrangian scheme for hard photon transport and (3) a Monte-Carlo scheme for the considered QED reactions. The developed scheme is applied to a test simulation relevant to recent experiments of positron production from a thin gold target irradiated by a high intensity laser. The simulation results successfully show a basic process leading to the positron ejection from the target. The process involves electron acceleration due to laser-plasma interaction at the target front, hard photon emission and pair production inside the target, and electrostatic positron acceleration at the target rear side.

1. Introduction

Recent progress in high-intensity laser technology is making possible a new experimental approach to high-energy plasma dynamics coupled with quantum electrodynamics (QED) processes. Such plasma dynamics play an important role in high-energy astronomical phenomena in neutron star magnetospheres [1], accretion discs around black holes [2] and so on. Theoretical estimates indicate that laser-plasma interaction under the dominant influence of QED effects can be obtained by using high-intensity lasers with $I \sim 10^{24-25}\mathrm{W/cm^2}$ [3,4], which will be realized at facilities currently under project [5]. In this interaction regime, active pair production would take place mediated by hard photon emission from accelerated electrons in the strong laser field [6-13].

Pair production could be attained also with the available laser intensities of $I \sim 10^{20-21}\mathrm{W/cm^2}$ by using a high-Z solid target. In this case, the strong nuclear field of high-Z ions enhances hard photon emission and subsequently pair production [14,15]. Such experiments have already been performed; for example energy spectrum measurements have been carried out for positrons ejected from a thin gold target irradiated by a high intensity laser [16-18]. The experimental results are, however, not fully described by current numerical simulations because traditional simulation methods, such as plasma particle-in-cell (PIC) [19] and Monte-Carlo particle transport methods [20], can address only one side of the laser-target interaction, i.e., either the collective plasma dynamics or the quantum stochastic process of positron production.

In order to approach a full numerical description of pair production processes in high-Z targets, we have developed a new simulation scheme that includes both QED processes and plasma dynamics. The simulation scheme is based on the PIC model commonly used to study kinetic phenomena in laser-plasma interaction. Additional computation procedures are integrated into the PIC simulation scheme to evaluate the time evolution of hard-photons and stochastic QED reactions.

The considered QED reactions, Bremsstrahlung and pair production [14,21,22], are characterized by two features: a rather weak transition rate (typically $\lambda \sim (10 - 100) \times L$) and an enormous number of potential seed particles (typically $N \sim 10^{21}[1/\mathrm{cm}^3]$), where λ, L and N denote interaction mean free path, system size and plasma number density, respectively. These features are relevant to the experimental result that the detected number of positrons is statistically considerable but much less than that of electrons [16,18]. In order to capture the products of Bremsstrahlung, we employ a fluid description for hard photons in multi-dimensional phase space. Since one PIC super-particle contains a lot of real particles, the super-particle emits a certain number of hard photons in one time step interval. Electron super-particles lose a part of their momentum every time step, and corresponding fractions of photon momentum are distributed into the discrete photon momentum space. In this way, low photon density can be captured as a continuous quantity. Pair production is also regarded as a continuous process that takes place every few time steps. Generated electron-positron pairs are given by test particles with continuous small weights estimated from the cross section of pair production and the photon momentum distribution.

Another difficulty arises from the treatment of the nuclear field producing the QED reactions. PIC simulations cannot resolve the nuclear field explicitly because its spatial scale is much smaller than the Debye length characterizing the grid separation. The Bremsstrahlung process is described as momentum and energy exchange processes among randomly selected electron-ion pairs and emitted photons, and is evaluated by an extension of the Monte-Carlo binary collision model [23]. Photon momentum and emission angle are computed from random numbers distributed according to the cross sections for relative momenta of each particle pair.

As a test, the scheme is applied to high-intensity laser interaction with a thin solid target. Simulation results successfully show a series of plasma and QED processes leading to positron ejection from the target. Seed particles for the QED reaction are high-energy electrons accelerated by the pulse laser in the low-density pre-plasma region at the target front. These electrons penetrate into the target with higher density and result in pair production mediated by hard-photon emission. It is also confirmed that positron acceleration due to the electrostatic field at the target's rear side dominantly characterizes the positron energy distribution.

This paper is organized as follows. In section 2, we introduce the details of the simulation scheme. Results of the test simulation are shown in section 3. Finally, the present study is summarized in section 4.

2. Numerical Scheme

The numerical scheme is based on a plasma PIC model [19]. PIC simulations describe collective plasma dynamics and have been employed widely for numerical studies on kinetic plasma phenomena in space and laboratories. We integrate two additional procedures relevant to pair production into the PIC simulation model. One evaluates the time evolution of the hard photons that mediate pair production. The other estimates stochastic QED reactions by Monte-Carlo calculation using random numbers.

Figure 1 shows the global picture of the simulation scheme. Physical objects considered in the simulation are classified into two categories - photons and particles (left part of figure 1).

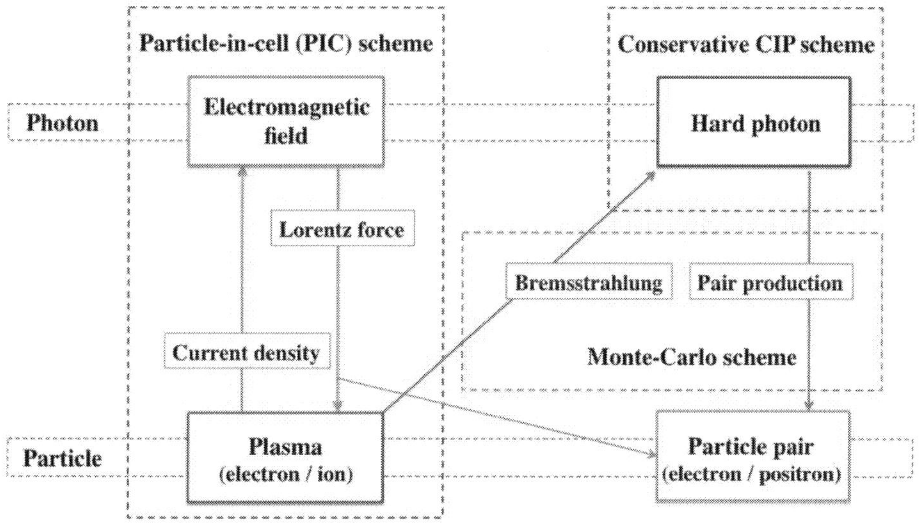

Figure 1. Schematic diagram of the simulation scheme. Physical objects considered in the simulation are surrounded with a color solid line. Arrowed lines represent the connections among these objects. Boxes surrounded by color dashed lines refer to processes computed with a given numerical schemes.

"Photons" include the electromagnetic field and hard photons (top boxes in figure 1). Their time evolutions are computed separately by different numerical schemes. In the present simulation, electron-positron pairs produced during the evolution are treated separately from the initial electron-ion plasma (bottom boxes in figure 1). Details of the numerical treatment for these physical objects are explained in the following subsections.

2.1. Time evolution of electromagnetic field and plasma

Time evolutions of the plasma and the electromagnetic field are determined by Newton-Lorentz and Maxwell equations in the framework of the PIC simulation (left side of figure 1). The electromagnetic field is discretized on a uniform spatial grid and the Maxwell equations are solved by a leapfrog finite difference method. The plasma dynamics are given by the motions of super-particles with a finite volume comparable to the Debye shell. One super-particle represents a macroscopic number of real particles. This treatment enables us to examine various kinetic properties of the plasma by using a reasonable number of computational particles. At each time step, the displacements of super-particles are combined to estimate the current density term in the Maxwell equations. The calculation cycle is closed through the computation of the Lorentz force and current density, as shown in the left side of figure 1.

2.2. Time evolution of hard photons

Time evolution of hard photons is computed to examine the pair production process that is mediated by hard photons. The energy of a hard photon should exceed twice the electron rest mass energy. The corresponding frequency is much higher than the plasma frequency in our case. Therefore hard photons can be clearly separated from the electromagnetic field evaluated in the PIC simulation. Photons with intermediate frequency contribute to neither plasma dynamics nor pair production and are ignored in the present simulation.

The spatial grid employed in the PIC simulation cannot resolve the hard photons because their wavelength is much smaller than the grid size. Instead, we employ a multi-dimensional phase space for hard photons. Hard photons with different momenta are distinguished by their position in the additional momentum space. In phase space, time evolution of hard photons is governed by the photon transport equation,

$$\frac{\partial f(r,p)}{c\partial t} = \hat{n} \cdot (\nabla f(r,p)) = \delta f(r,p), \tag{1}$$

where c, $\hat{n}$ and $f(r,p)$ denote light speed, propagation direction and photon distribution in phase space (r,p). Here, left-hand side terms indicate photon advection and the right-hand side term gives the photon absorption associated with photon-matter interaction.

A conservative Constrained Interpolation Profile scheme (CIP-CSL2) is employed to simulate photon advection. The CIP scheme is a semi-Lagrangian method and has been widely used in fluid simulations [24]. The total number of fluid elements is exactly conserved in this scheme. Photon advection in real space is calculated individually with computational cells in momentum space. Accordingly, total momentum and energy are conserved in the photon advection process.

The photon-matter interaction depends on the photon energy and the kind of material. For gold, the dominant interactions of hard photons are pair production ($\geq$ 5MeV) and Compton scattering ($\leq$ 5MeV) [22]. Total cross sections of these interactions have relatively weak dependency on photon energy, which is roughly characterized by the logarithm of photon energy. Therefore non-uniform logarithmic coordinates would be effective to represent photon momentum space within a limited computational memory. Figure 2 shows the cylindrical coordinate used for photon momentum space. Logarithmic coordinates are employed for radial distance in the x-y plane P_{xy} and for the distance in the z-direction P_z.

2.3. Treatment of QED reactions

The dominant QED reaction leading to pair production is the interaction between hard photons and the high-Z nuclear field [17,22]. Bremsstrahlung from electrons accelerating in the high-Z nuclear field mainly contributes the hard photon emission [14,21]. Therefore these two QED reactions are considered in the present simulation. Their positions in the whole simulation procedure can be seen in the right side of figure 1.

The spatial grid used in the PIC simulation does not resolve Bremsstrahlung processes, which have characteristic scale of atomic size. Thus, we employ a coarse-graining approach statistically

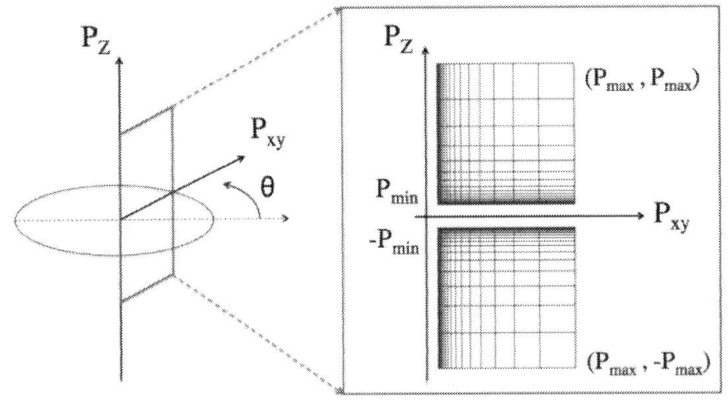

Figure 2. Cylindrical coordinates for photon momentum space. The photon momentum is given by three independent parameters, the angle θ and the absolute value of P_{xy} in the $x-y$ plane and the $z-$component P_z. Logarithmic coordinates are employed for P_{xy} and P_z.

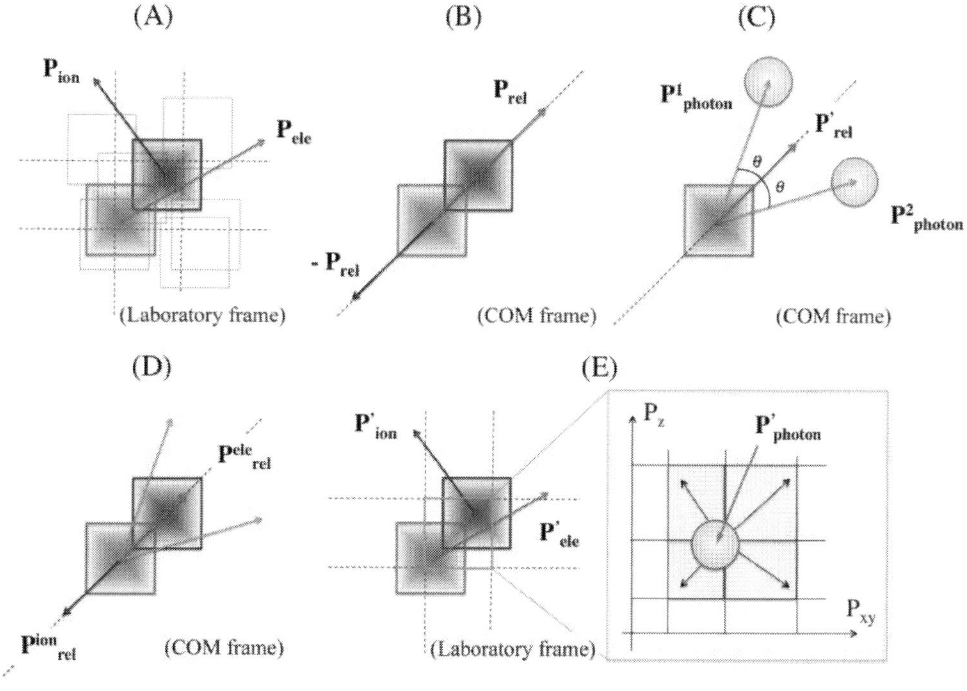

Figure 3. Numerical procedures for Bremsstrahlung. Red and blue boxes indicate ion and electron super-particles, respectively. Yellow circles indicate temporary photons with individual momenta. These photons are distributed into the photon momentum space in the last procedure (E).

consistent with theoretical cross sections. This approach is similar to the Monte-Carlo binary collision model with particle pairing [23]. In this model, binary collisions are described as a momentum exchange process between a pair of super-particles selected within a common spatial grid. This approach is extended to the momentum exchange among electrons, ions and photons consistently with Bremsstrahlung cross sections.

Numerical procedures associated with the Bremsstrahlung are summarized in figure 3. The following items A)∼ E) relate to the panels (A)∼(E) in the figure.

A) An electron-ion pair is randomly selected from super-particles (gray boxes) that belong to a common spatial grid. The red and blue boxes indicate the selected electron-ion pair. The symbols P_{ele} and P_{ion} denote the electron and ion momenta, respectively.

B) A Lorentz transformation is applied to move into the center-of-mass (COM) frame of the electron-ion pair. The electron and ion have symmetric momenta P_{rel} and $-P_{rel}$ in the COM frame.

C) Bremsstrahlung photons (yellow circles) are considered. The Bremsstrahlung cross sections are estimated from the relative momentum by using a prepared data table. Electron energy loss is determined by the total cross section. P'_{rel} gives the resulting electron momentum. The photon momenta P_{photon} and emission angles θ are computed from random numbers statistically distributed according to the differential cross sections, and the photon number is determined so as to satisfy the energy conservation. The angle components are chosen axisymmetrically with respect to the direction of relative momentum.

D) Total photon momentum is directed parallel to the electron-ion relative momentum. However, its absolute value is not consistent with the electron momentum loss. This inconsistency is resolved by re-distributing the momentum from electron to ion. The electron and ion momenta are corrected as $P'_{rel} \rightarrow P^{ele}_{rel}$ and $-P_{rel} \rightarrow P^{ion}_{rel}$, respectively.

E) The Lorentz transform is applied inversely. Electron, ion and photon momenta are updated in the laboratory frame (P'_{ele}, P'_{ion} and P'_{photon}, respectively). Finally, the photon momenta are distributed to the nearest cells in the momentum space as shown in the right panel.

The procedures described above are repeated for each spatial grid until all electron (or ion) super particles are once processed.

Next, we explain the numerical procedures for pair production. The reduction rate of photons is determined by the total cross section of pair production based on the photon energy. In principle, the computation of pair production should happen in each cell in momentum space according to the individual photon cross section. However, this approach requires a large number of computational particles if pair production is processed every few time steps. In order to avoid this problem, we define a few representative photons for each spatial grid consistently with the expected cross section in momentum space. Pair production is estimated for these representative photons and randomly selected ions. This approach limits the increase of computational particles.

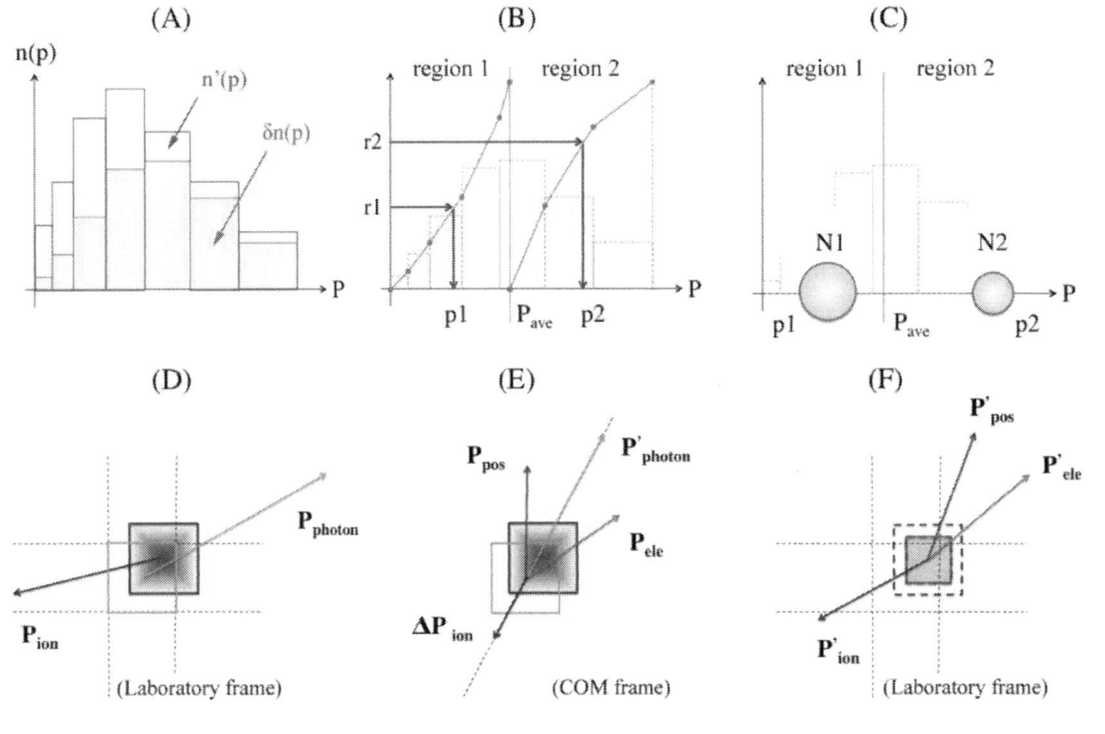

Figure 4. Numerical procedures for pair production. Top panels represent statistical treatments for photon distribution in momentum space. Yellow circles indicate representative photons with individual momentum and the number of contained photons. Bottom panels represent the interaction of each representative photon with a randomly selected ion. Blue and red boxes represent an ion super-particle and an electron-positron pair. Initially, electron and positron are placed at the same position.

Numerical procedures associated with pair production are summarized in figure 4. The following items A) ∼ F) relate to the panels (A)∼(F) in the figure.

A) The momentum-space distribution of the photons contributing pair production, $\delta n(p)$, is estimated for each cell in momentum space according to the total cross section. The histogram in the figure indicates the profile of photon number in momentum space. Photon momentum distribution is updated with the remaining photon number $n'(p) = n(p) - \delta n(p)$, where $n(p)$ denotes initial photon distribution.

B) Linearly interpolated cumulative distribution functions (red line with dots) are computed from $\delta n(p)$ for two momentum spaces, region 1 and region 2. Here, the momentum space is divided at the average momentum in $\delta n(p)$, namely p_{ave}. The momenta of the two representative photons, $p_1 (< p_{ave})$ and $p_2 (> p_{ave})$, are defined from random numbers, r_1 and r_2, by using the cumulative distribution functions.

C) Total number of photon reduction ($N = \sum_p \delta n(p)$) is distributed into these representative photons, N_1 and N_2 ($N = N_1 + N_2$), so that the total reduction of photon momentum is equal to the representative photon momentum ($Np_{ave} = N_1 p_1 + N_2 p_2$). In the figure, the yellow circles indicate the representative photons. Positive solutions N_1, N_2 can be obtained if $p_1 < p_{ave} < p_2$ is fulfilled.

D) We consider the interaction between the representative photon and ion. The ion (blue box) is randomly selected from the ion super-particles belonging to the same spatial grid (orange square).

E) In the ion rest frame, the momenta of the generated particle pair, p_{ele} and p_{pos}, are computed according to the photon momentum p'_{photon} and the cross section of pair production. The number of particle pairs is equal to the photon number contained in the representative photon. The inconsistency between photon momentum and total momentum of the particle pair is resolved by re-distributing the momenta between ion and particle pair. At the end, the ion has a finite momentum Δp_{ion} in this frame.

F) Finally, a Lorentz transform into the laboratory frame is applied. One obtains electron, positron and ion momenta (p'_{ele}, p'_{pos} and p'_{ion}) consistent with the photon reduction.

The generated particle pair is described as two test particles with continuous weights. The particle weight represents the number of contained real particles. The test particles move according to the Lorentz force but do not contribute current density and recursive pair production (see the bottom right part of figure 1). This treatment is valid for the present situation because the number density of particle pairs is negligibly small compared to that of the initial plasma. Numerical cost due to the increment of particle pairs is reduced dramatically by this treatment. Additionally, Bremsstrahlung procedures would become more complicated without this treatment because random pairing cannot be applied directly for super-particles with different weights.

3. Test simulation

3.1. Simulation model

The developed simulation scheme is applied to a test simulation of positron production in a solid gold target. Figure 5 shows the initial setup of the simulation. A two-dimensional simulation domain is considered with an initial laser pulse and a high-density target. The initial laser pulse is placed at the left side of the simulation domain and propagates rightward ($+x$ direction). A low-density plasma region (pre-plasma region) is located at the target front [10.5 μm < x < 19.5 μm in figure 5(A)]. The laser frequency ω_L is set so as to fulfill the condition $\omega_{pe}^p < \omega_L < \omega_{pe}^t$, where ω_{pe}^p and ω_{pe}^t are the plasma frequencies at the pre-plasma and the target, respectively. Under this condition, the laser pulse basically penetrates into the pre-plasma but is reflected in the

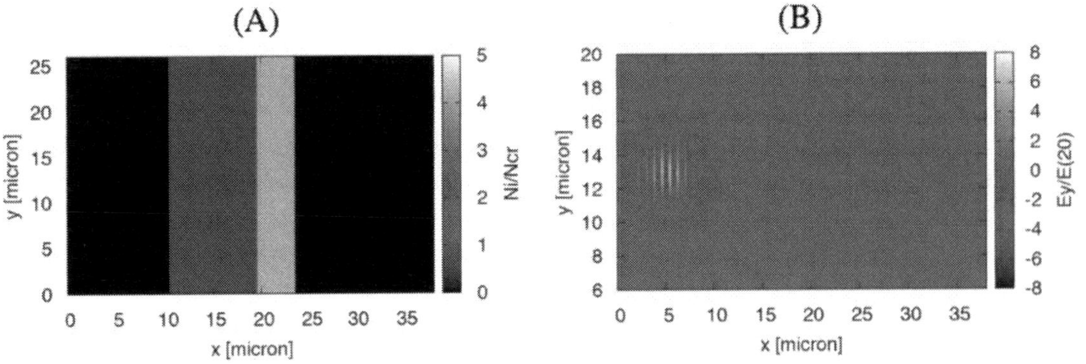

Figure 5. Initial condition for the test simulation. (A) Spatial profile of initial ion density normalized to the critical density N_{cr}. The initial density profile of electrons is the same as that of ions. (B) Spatial profile of initial electric field E_y normalized by the unit electric field $E(20)$ that corresponds to the focusing intensity $10^{20}\mathrm{W/cm^2}$.

target. In the experiment, the pre-plasma is generated because of the ionization of the solid target by the preceding pulse of the high intensity laser. A similar model has been employed for some numerical studies on ion acceleration using high-intensity laser and solid target by means of traditional PIC simulations [25].

The principal simulation parameters are set as follows. The simulation domain is implemented on a (1536×1024) grid. Initial number of super-particle per spatial grid is 67 at the pre-plasma and 480 at the target for each particle species. Physical parameters are estimated for reference plasma density $10^{21}\mathrm{cm^{-3}}$ at the pre-plasma. Laser intensity corresponding to the initial pulse laser is $7 \times 10^{20}\mathrm{W/cm^2}$. Pulse and spot sizes are 4.9μm and 3.1μm, respectively. Lengths of pre-plasma and target regions are 9.0μm and 3.9μm, respectively. Target density is $7.2 \times 10^{21}\mathrm{cm^{-3}}$, while the critical density for the initial laser pulse is estimated to be $N_{cr} = 1.7 \times 10^{21}\mathrm{cm^{-3}}$. Radiation and periodic boundary conditions are employed for the $x-$ and $y-$directions, respectively.

3.2. Simulation results

Figures 6(A) and (B) show the electric field and electron density profiles at the early phase ($ct = 12.8$ μm) when the laser pulse penetrates into the under-dense pre-plasma region [see figure 6(A)]. Vertical lines in the figures indicate the initial locations of the pre-plasma region and the target. The laser pulse pulls out the plasma from the laser transit region by exerting ponderomotive force [26]. In consequence, a wake structure is formed in the pre-plasma region as shown in figure 6(B).

Figures 6(C) and (D) show the phase-space plots for electron momenta in the x- and y-directions, respectively. One can see the electron acceleration in the $+x$-direction as well as the oscillation in the $\pm y$-directions. The momentum of some electrons reaches up to $p_x \sim 100m_ec$. The high-energy electron component comes from resonance acceleration in the oscillating plasma wake structure, while the electron motion in the $\pm y$-direction originates from a direct response to the laser fields. The basic acceleration mechanism in such non-linear laser-plasma interaction is referred to as "plasma wake-field acceleration" [27].

Figure 7(A) shows the electron phase-space plot at a later phase ($ct = 20.3$ μm) after the laser pulse reaches the target surface. The accelerated electrons pass through the high-density target region in contrast to the laser field that dissipates in the target. These electrons play a critical

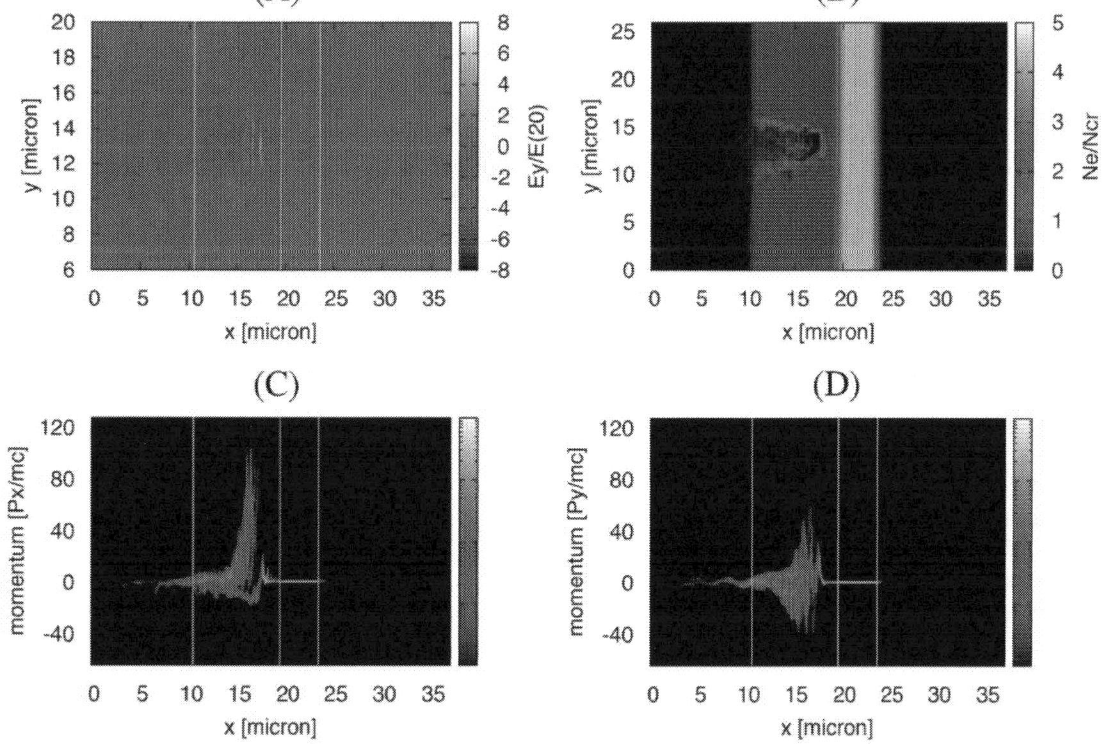

Figure 6. Laser-plasma interaction in the pre-plasma region at $ct = 12.8$ μm. (A) Spatial profile of the electric field E_y normalized to the unit electric field $E(20)$. (B) Spatial profile of the electron density N_e normalized to the critical density N_{cr}. (C) Phase-space plot for electron momentum in the x-direction P_x. (D) Phase-space plot for electron momentum in the y-direction P_y. Vertical lines indicate the initial locations of the pre-plasma region and the target.

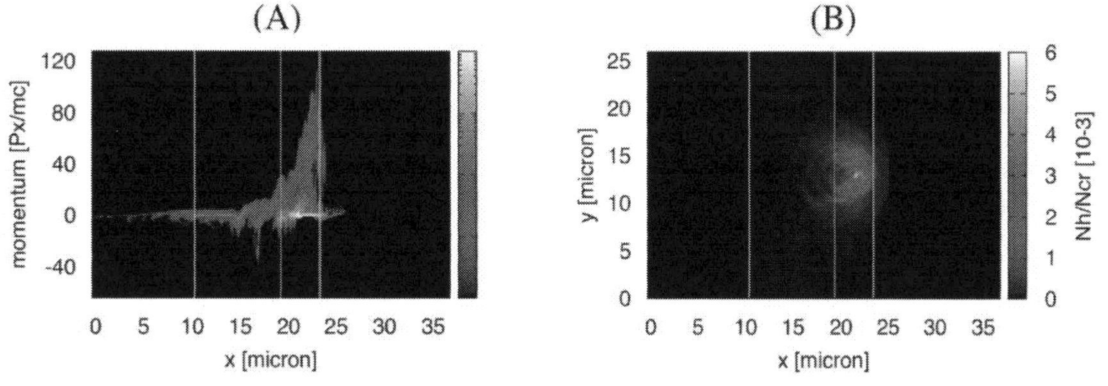

Figure 7. Photon emission from accelerated electrons inside the target at $ct = 20.3$ μm. (A) Phase-space plot for electron momentum in x-direction, P_x. (B) Spatial profile of photon number N_h. Hard photons with energy $E_p > 2m_ec^2$ are counted here and the photon number is normalized to $10^{-3}N_{cr}$. Vertical lines indicate the initial locations of the pre-plasma region and the target.

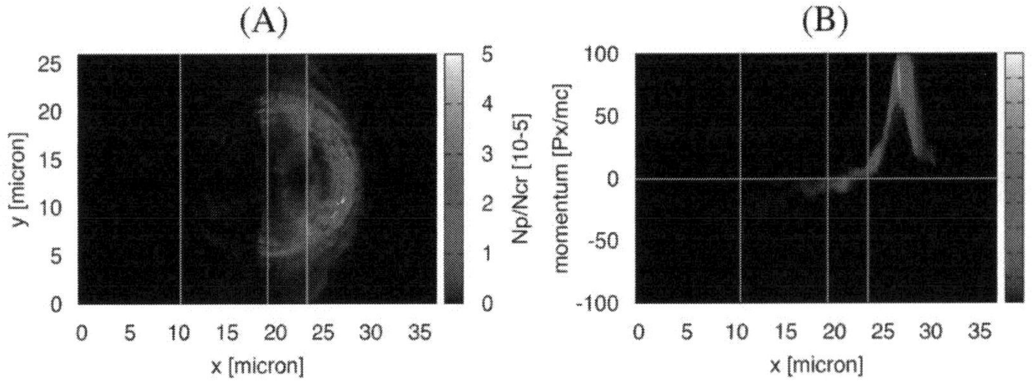

Figure 8. Positron generation and acceleration at $ct = 25.4$ μm. (A) Spatial profile of positron density, N_p, normalized to $10^{-5}N_{cr}$. (B) Phase-space plot for positron momentum in the x-direction, P_x. Vertical lines indicate the initial locations of pre-plasma region and target.

role as seed particles in subsequent photon emission and pair production because Bremsstrahlung is enhanced for higher electron energy and higher density of high-Z nuclei. Figure 7(B) shows the spatial profile of a hard photon with energy larger than $2m_ec^2$. Hard photons are found dominantly at the vicinity of the accelerated electrons passing through the target.

Figure 8(A) shows the positron density profile normalized to $10^{-5}N_{cr}$ at $ct = 25.4$ μm. Positrons are generated as hard photons propagate inside the target. As a result, the positron distribution follows the arch distribution of spreading hard photons. Figure 8(B) shows the phase-space plot for positron momentum in the x-direction, p_x. The distribution is sampled from the positron test particles in the central region of the simulation domain, 12.2 μm < y < 13.8 μm. Inside the target, positron momentum stays at a lower level compared to that of the seed electrons accelerated in the pre-plasma. This is because the differential cross section of Bremsstrahlung is strongly biased toward low-energy photon emission.

By contrast, positrons with considerable momentum are found behind the target ($x > 23$ μm). These positrons are accelerated originally by the sheath electric field induced at the rear side surface of the target. Induction of the sheath electric field comes from the violation of charge neutrality of the target due to electron penetration from pre-plasma. The resulting negative charge propagates through the free electrons in the target, and finally an electrostatic field is induced at the rear side in the direction perpendicular to the target surface ($+x$ direction). While electrons tend to be confined inside the target under the influence of this electrostatic field, positrons with positive charge tend to accelerate in the rearward direction.

The acceleration mechanism due to sheath electric field has been referred to as the origin of effective ion acceleration in the laser-solid interaction. The simulation results indicate that the positron energy spectrum is dominantly determined by the acceleration in the sheath electric field. This mechanism might elucidate the discrepancy of positron energy spectrum between Monte-Carlo simulation and experimental results [16,18].

4. Conclusion

Recent progress in high intensity lasers is allowing the experimental investigation of high-energy plasma phenomena under the influence of quantum electrodynamics (QED) effects. One available example is the positron measurement carried out on a thin gold target with a high-intensity laser [16-18]. Aiming at the numerical investigation of such experiments, we have developed a new simulation scheme including both collective plasma dynamics and stochastic QED reactions.

The developed scheme is based on the plasma particle-in-cell (PIC) method. Additional procedures associated with QED reactions are integrated into the PIC framework. Considered QED reactions are Bremsstrahlung and pair production in a high-Z nuclear field. Under experimental conditions, these reactions are characterized by an extremely low transition rate and a macroscopic number of seed particles. A fluid description of hard photons is employed in a multi-dimensional phase space to capture the small fraction of photon emission as a continuous quantity. Additional procedures include a conservative Constrained Interpolation Profile (CIP) scheme for hard-photon transport and Monte-Carlo schemes for QED reactions. The QED reactions are described as momentum and energy exchange processes between plasma (super-particles), hard photons (fluid) and particle pairs (test particles with continuous weights). The Monte-Carlo schemes are designed so as to satisfy conservation laws and statistical consistency with theoretical cross sections.

A test simulation of positron production in a thin solid target is performed by means of the developed simulation scheme. The simulation results successfully show that a series of basic physical processes leads to positron ejection from the target. Considered physical processes involve electron acceleration due to laser-plasma interaction in the pre-plasma region, pair production mediated by hard photon emission inside the target and positron acceleration due to sheath electric field at the target rear side.

The developed simulation scheme is a promising computational approach for full-scale simulation studies of pair production experiments using high-Z targets. Key themes for future work would be, for example, quantitative understanding of the controlling mechanism of the positron energy spectrum and ejection angle distribution and of the dependency of laser-to-positron energy conversion rate on experimental setup, and experimental optimization for effective positron production.

Acknowledgments

The authors acknowledge Professor A Titov (Joint Institute for Nuclear Research, Russia) and Professor H Ruhl (Universität München, Germany) for useful discussions. The calculations were performed on the ILE computer system at the Institute of Laser Engineering in Osaka University and on the Plasma Simulator at the National Institute for Fusion Science. Part of the simulation code is developed by using the K computer at the RIKEN Advanced Institute for Computational Science (Proposal number hp120267). This work was partially supported by the General Coordinated Research at National Institute for Fusion Science (NIFS13KNSS039).

References

[1] Timokhin A N 2010 *Monthly Notices of the Royal Astronomical Society* **408** 2092
[2] Ruffini R, Vereshchagin G and Xue S 2010 *Physics Reports* **487** 1
[3] Bulanov S V *et al* 2011 *Nuclear Instruments and Methods in Physics Research* **A 660** 31
[4] Sokolov I V, Naumova N M and Nees J A 2001 *Phys. Plasmas* **18** 093109
[5] THE EXTREME LIGHT INFRASTRUCTURE (http://www.extreme-light-infrastructure.eu/)
[6] Bell A R and Kirk J G 2008 *Phys. Rev. Lett.* **101** 200403
[7] Kirk J G, Bell A R and Akra I 2009 *Plasma Phys. Controlled Fusion* **51** 085008
[8] Nerush E N, Kostyukov I Yu, Fedotov A M, Narozhny N B, Elkina N V and Ruhl H 2011 *Phys. Rev. Lett* **106** 035001
[9] Elkina N V, Fedotov A M, Kostyukov I Yu, Legkov M V, Narozhny N B, Nerush E N and Ruhl H 2011 *Physical Review Special Topics - Accelerator and Beams* **14** 054401
[10] Ridgers C P, Brady C S, Duclous R, Kirk J G, Bennett K, Arber T D, Robinson A P L and Bell A R 2012 *Phys. Rev. Lett* **108** 165006

[11] Erber T 1966 *Reviews of Modern Physics* **15** 626
[12] Ritus V I 1985 *Journal of Soviet Laser Research* **6** 497
[13] Titov A I, Takabe H, Kampfer B and Hosaka A 2012 *Phys. Rev. Lett* **108** 240406
[14] Heitler W 1954 *The Quantum Theory of Radiation* (The Oxfort University Press)
[15] Koch H W and Motz J W 1959 *Reviews of Modern Physics* **31** 920
[16] Cowan T E *et al* 1999 *Laser and Particle Beams* **17** 773
[17] Nakashima K and Takabe H 2002 *Phys. Plasmas* **9** 1505
[18] Chen H *et al* 2011 *High Energy Density Physics* **7** 225
[19] Birdsall C K and Langdon A 1991 *Plasma Physics via Computer Simulation* (Adam Hilger)
[20] For example, The Electron Gamma Shower (EGS) (http://rcwww.kek.jp/research/egs/)
[21] ESTAR: Stopping-Power and Range Tables (http://www.nist.gov/pml/data/star/index.cfm)
[22] XCOM: Photon Cross Section Database (http://www.nist.gov/pml/data/xcom/index.cfm)
[23] Sentoku Y, Mima K, Kishimoto Y and Honda M 1998 *Journal of the Physical Society of Japan* **67** 4084
[24] Yabe T, Tanaka R, Nakamura T and Xiao F 2001 *Mon. Wea. Rev.* **129** 332
[25] Fourkal E, Shahine B, Ding M, Li J S, Tajima T and Ma C.-M 2002 *Medical Physics* **29** 2788
[26] Landau L D and Lifshitz E M 1975 *The Classical Theory of Fields*, 4th ed., (Pergamon Press, Oxford, New York)
[27] Tajima T and Dawson J 1979 *Phys. Rev. Lett.* **43** 267

Electron behavior in ion beam neutralization in electric propulsion: full particle-in-cell simulation

Hideyuki Usui, Akihiko Hashimoto and Yohei Miyake

Graduate School of System Informatics, Kobe University, 1-1 Rokkoudai, Nada, Kobe 657-8501, Japan

E-mail: h-usui@port.kobe-u.ac.jp

Abstract. By performing full Particle-In-Cell simulations, we examined the transient response of electrons released for the charge neutralization of a local ion beam emitted from an ion engine which is one of the electric propulsion systems. In the vicinity of the engine, the mixing process of electrons in the ion beam region is not so obvious because of large difference of dynamics between electrons and ions. A heavy ion beam emitted from a spacecraft propagates away from the engine and forms a positive potential region with respect to the background. Meanwhile electrons emitted for a neutralizer located near the ion engine are electrically attracted or accelerated to the core of the ion beam. Some electrons with the energy lower than the ion beam potential are trapped in the beam region and move along with the ion beam propagation with a multi-streaming structure in the beam potential region. Since the locations of the neutralizer and the ion beam exit are different, the above-mentioned bouncing motion of electrons is also observed in the direction of the beam diameter.

1. Introduction

As an interplanetary flight system, electric propulsion has been used for many spacecraft. Ion engine which is one of the typical electric propulsion systems gains the thrust by emitting heavy ions such as Xenon and Argon as an ion beam accelerated through grids in the electrostatic manner. To avoid the space-charge effect by excess emission of ions near the engine [1] and mitigate the spacecraft charging, thermal electrons are simultaneously released from a neutralizer attached in the vicinity of the ion engine location.

Ion beam emission and its neutralization by thermal electrons in electric propulsion is one of the most fundamental problems in spacecraft-plasma interactions. By emitting the same amount of electrons as the beam ions, the charge neutralization can be basically achieved in the spacecraft environment. However, because of large difference between electron and ion mass and dynamics, the charge neutralization between the two species does not simply occur outside the engine and the understanding of the neutralization process still remains at a rather primitive level. Although the previous works handled uniform ion beam which has an infinite cross section [2,3], we consider a situation in which an ion beam is emitted with a finite radius from a spacecraft [4-8] and simultaneously thermal electrons for the charge neutralization are released from a different position from the ion emitter [4]. Our focus is on the transient process of electron mixing to the ion beam in the very vicinity of the ion engine. To examine the transient process of the electron response in detail, we performed Particle-In-Cell [9,10] simulations in which both electrons and ions are treated as macro-particles and their dynamics are coupled with the electromagnetic field through electric current. In the current study, we considered the electrostatic fields only in the simulations. In section 2, we introduce our simulation model and parameters used in the simulations. In section 3, we show some results obtained in the simulation in terms of electron behavior in the ion beam region. Section 4 contains the conclusion.

2. Simulation model

To examine the detail of electron behaviors in the ion beam region in the vicinity of a spacecraft, we performed three-dimensional full Particle-In-Cell (PIC) simulations by using our original software called "EMSES"[11], which we developed for the analysis of spacecraft-plasma interactions. Like other conventional PIC simulation codes, EMSES basically solves Maxwell's equation to update the electromagnetic fields defined at spatial grid points by using the Finite-Difference Time-Domain (FDTD) method. In addition to the field components, dynamics of a large number of macro-particles representing electrons and ions which are distributed in the simulation space are updated by solving the equation of motion for each particle at each time step [9,10]. One of the unique features of EMSES is that it can define some internal conduction boundaries which represent to spacecraft conducting surfaces and solve the interactions with the surrounding plasma such as sheath formation near the surface and the spacecraft charging.

Figure 1 shows the three-dimensional simulation model used for the current study. An ion beam emitter and a neutralizer which releases thermal electrons are separately located on the spacecraft surface which is indicated in the figure as the injection plane. The location of the electron emitting neutralizer is set at the upper side of the ion emitter which is located in the middle of the injection plane. To shorten the calculation time in the simulation, we assumed protons as ions which are much lighter than the actual propellant such as Xenon. To keep the charge neutrality inside the spacecraft, total fluxes of the beam ions and thermal electrons are set to be the same to each other. We provide the beam velocity for the ions only and the electrons are thermally released from the neutralizer. We set the ratio of the electron thermal velocity v_{te} to the ion beam velocity v_{ibeam} as 1.25. The ion thermal velocity v_{ti} is much smaller than v_{ibeam}. Since the density of the background space plasma is small enough in comparison with that of the ion beam, we assumed no background plasma for simplicity. No interplanetary magnetic field is concerned in the current simulation. Since our aim is to examine

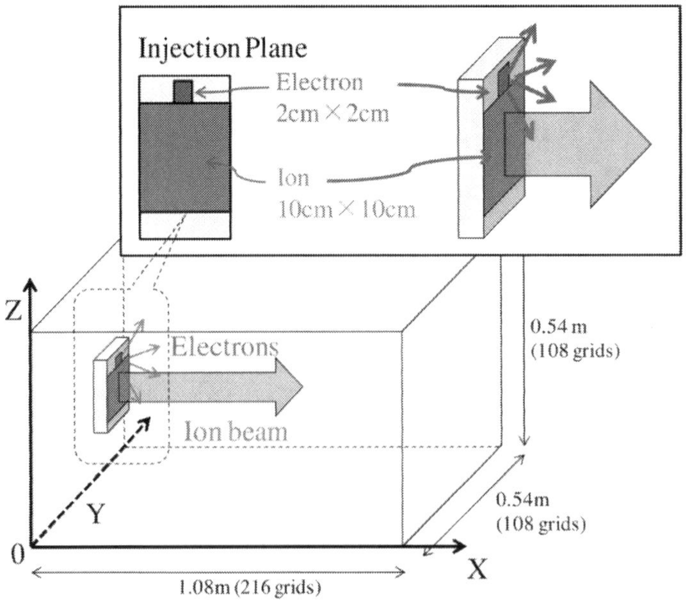

Figure 1. Simulation model.

electron response to the ion beam, we used EMSES in the electrostatic regime in which Poisson's equation is solved to update the electrostatic fields defined at the spatial grid points in the simulation space. The potential of the injection plane is treated as floating while the potential of the boundary surrounding the simulation space is fixed. To avoid the influence of the boundary potential to the plasma dynamics, we elongated the simulation space along the beam direction. We performed simulations up to the time when the beam reaches the simulation boundary.

3. Transient Response of Electrons

Figure 2 shows a bird's-eye view of spatial distribution of beam ions and thermal electrons measured at $\omega_{pe}t$ =400 where ω_{pe} denotes the electron plasma frequency at the neutralizer. The ions and electrons emitted from the injection plane located near the left boundary of the simulation space are indicated in a large number of red and blue dots, respectively. Although the neutralizer which releases the thermal electrons is located at the upper side of the plane with respect to the ion beam emitter, the electrons are overall distributed in the ion beam region shown in red. If one carefully sees the electron distribution shown in blue, however, the distribution is not uniform particularly at the beam front as well as near the beam emitter.

In figure 3, we show temporal evolution of the ion beam and the thermal electrons in terms of density contour maps measured on an x-z plane including the center axis of the ion beam. Spatial scales and density values in the figure are normalized to the local Debye length λ_D and the maximum density n_{max} measured at the exit of the electron neutralizer, respectively. As clearly shown in the figure, the ion beam simply propagates away from the emitter as time elapsed. Although some diffusion of ions is seen at the beam front as in the bottom panel, the signature of the ion beam propagation is very straightforward because of ion's large inertia.

On the contrary, the emitted electrons show very complex signatures as shown in the right panels of figure 3. Thermal electrons are emitted from the upper side of the yellow body representing an ion engine. Note that the highest density is shown in blue in the right panels. It seems that the emitted

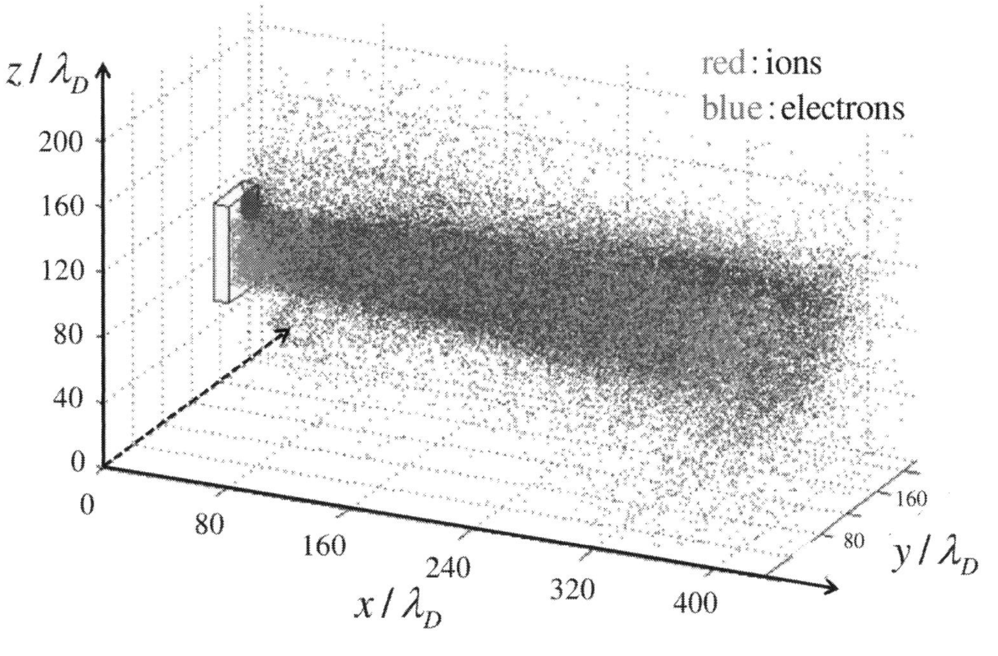

Figure 2. Spatial distribution of beam ions and thermal electrons emitted from a neutralizer measured at $\omega_{pe}t$ =400.

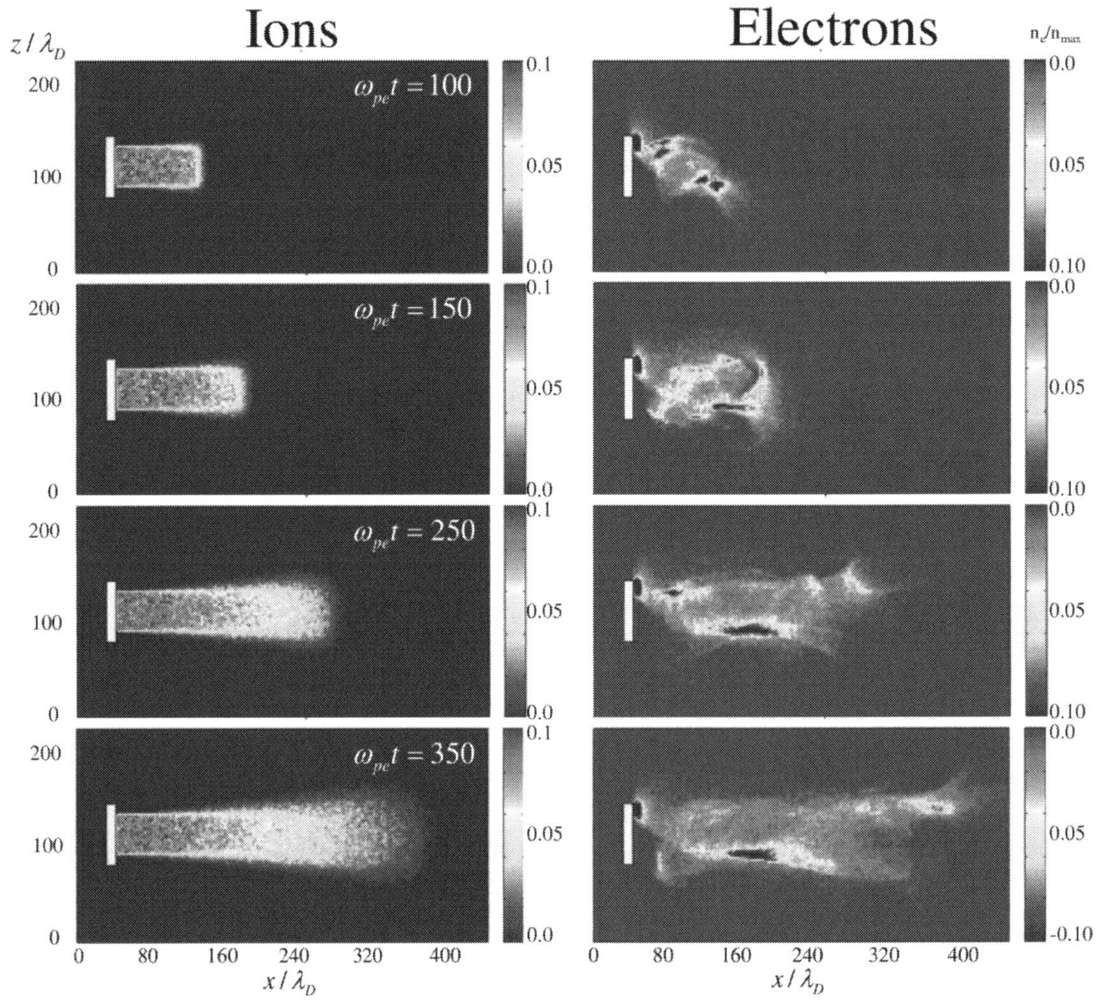

Figure 3. Contour maps for densities of ion beam and thermal electrons measured on an x-z plane including the center axis of the ion beam at different times.

electrons are overall attracted to the ion beam which has positive charges and they propagate along with the ion beam in the x direction to neutralize the positive ion charges. However, the electrons are not uniformly distributed in the beam region and the profile at each time seems asymmetric with respect to the ion beam direction. As shown in the third and fourth panels from the top, the electron density values become locally high at the surface surrounding the ion beam in the radial direction. The highest density region is constantly indicated at the location of the neutralizer attached at the upper side of the injection plane. In the beam region, however, the electron densities drastically change in time.

Figure 4 also shows density contour maps for the thermal electrons in different planes perpendicular to the ion beam propagation. Panels show the maps measured at x/λ_D=80, 160, 240, and 320, respectively. As previously shown in the profiles along the beam direction, the electrons are overall distributed in the beam region also in the radial direction. The profiles are symmetric with respect to the center line of the y-axis. However, they are asymmetric between the upper and lower regions with respect to the center line of the z-axis. The maximum density is found at the lower side of the beam surface at x/λ_D=160 while it is around the top surface of the beam at x/λ_D=320 as shown in

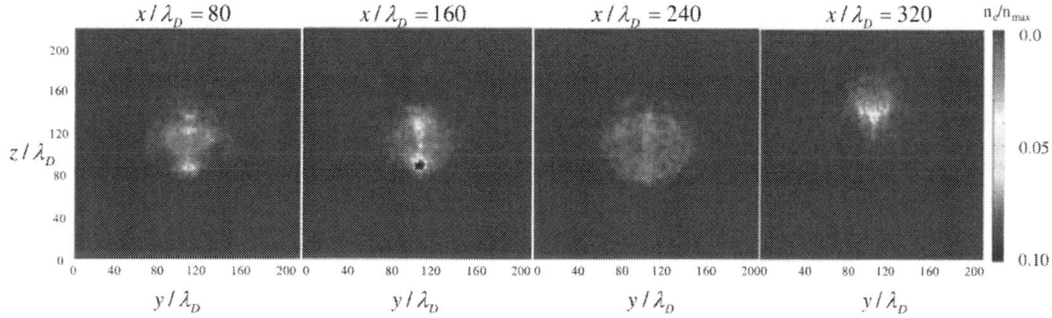

Figure 4. Contour maps for the electron densities measured on a *y*-*z* plane at the different locations along the *x* direction.

the most right panel. From these snapshots of the density profiles, it can be speculated that the thermal electrons released from the neutralizer make a kind of meandering or bouncing motion in the radial direction of the beam.

To understand the complex dynamics of the electrons shown in the previous figures, we examined the relation of the electrons to the electrostatic potential. In figure 5, we show a contour map of normalized electrostatic potential and the corresponding electron dynamics in an x-V_x phase space. The upper panel shows a snapshot of electrostatic potential measured on a *x*-*z* plane including the center axis of the ion beam measured at $\omega_{pe}t$ =300. The potential values are color-coded and the high potential with respect to that of the vacuum region is shown in red. The potential values are plotted in terms of energy and they are normalized to the initial thermal energy of electrons. Because of large thermal velocity of the electrons the ion beam region is not perfectly neutralized and a positive potential region remains in the beam region as shown in the upper panel. Since the potential energy in the beam region is much larger than the initial electron thermal energy, most of the electrons released from the neutralizer are electrostatically trapped in the positive potential region.

In the lower panel of figure 5, we focus on the electron dynamics along the beam direction by showing an x-V_x phase diagram in which electrons are plotted in blue dots. The electrons emitted from the neutralizer are quickly accelerated along the beam direction and some fast electrons escape from the beam front, which is shown beyond x/λ_D=320 in the panel. However, the most electrons are quickly decelerated at the beam front and are reflected back to the opposite direction to the beam propagation with negative velocities. They move toward the injection plane and are reflected back again toward the ion beam propagation with positive velocities. This bouncing motion of the electrons seems to be repeated multiple times in the beam region along the propagation direction. Although not displayed, as time elapses, the bouncing regions are elongated along the *x* direction in accordance with the ion beam propagation.

The bouncing period of the trapped electrons changes as the ion beam propagates away from the spacecraft. According to the conventional linear theory on the dynamics of electrons trapped in a potential well, the bouncing frequency ω_0 is approximately given as $\sqrt{e\Phi k^2/m}$ where e, m, ϕ, and k denote the electron charge, mass, the electrostatic potential and its wavenumber, respectively. Since the potential region is elongated as the ion beam propagates, k becomes smaller and smaller as time elapses. Then ω_0 becomes correspondingly small and eventually the bouncing period becomes long. ϕ is another factor which determines ω_0. ϕ is basically determined by the ion beam profile and may be slightly modified by the electron thermalization in the potential well. As shown in figure 5, the electron thermalization is caused by multiple bouncing motions in the potential well. These heated electrons may affect the potential structure as well as the ion beam propagation. However, as far as the current simulation results are concerned, no influence on the ion beam propagation is observed. It may be because the simulation space and time is so limited and we could just focus on the initial stage of the ion beam emission.

Since the electron bouncing motions occur in the three-dimensional potential structure, which is unlike the conventional one-dimensional ideal situation, the bouncing motions are not so simple.

The potential region is also finite in the direction of the beam diameter and the potential has a peak at the core of the ion beam. Therefore, the electrons are also trapped in the radial direction of the beam. Non-uniform density profiles of electrons as shown in figure 3 and figure 4 are caused by the dynamic electron response to the ion beam potential. As soon as electrons are emitted from the neutralizer they are quickly attracted to the core of the ion beam and are initially accelerated to the negative z direction. Then most of them penetrate the core part of the ion beam with the maximum velocity. However, when the electrons pass through the beam core, they start to be decelerated and finally stagnate at the other side of beam surface in the radial direction. This stagnation can cause relatively high density found near the surface of the ion beam as shown in figure 3 and 4. The electrons again start to be accelerated back to the beam core by the electric field and this process seems to be repeated multiple times before the electrons reach the beam front.

Back to the electron motion along the beam propagation, we display a profile of velocity

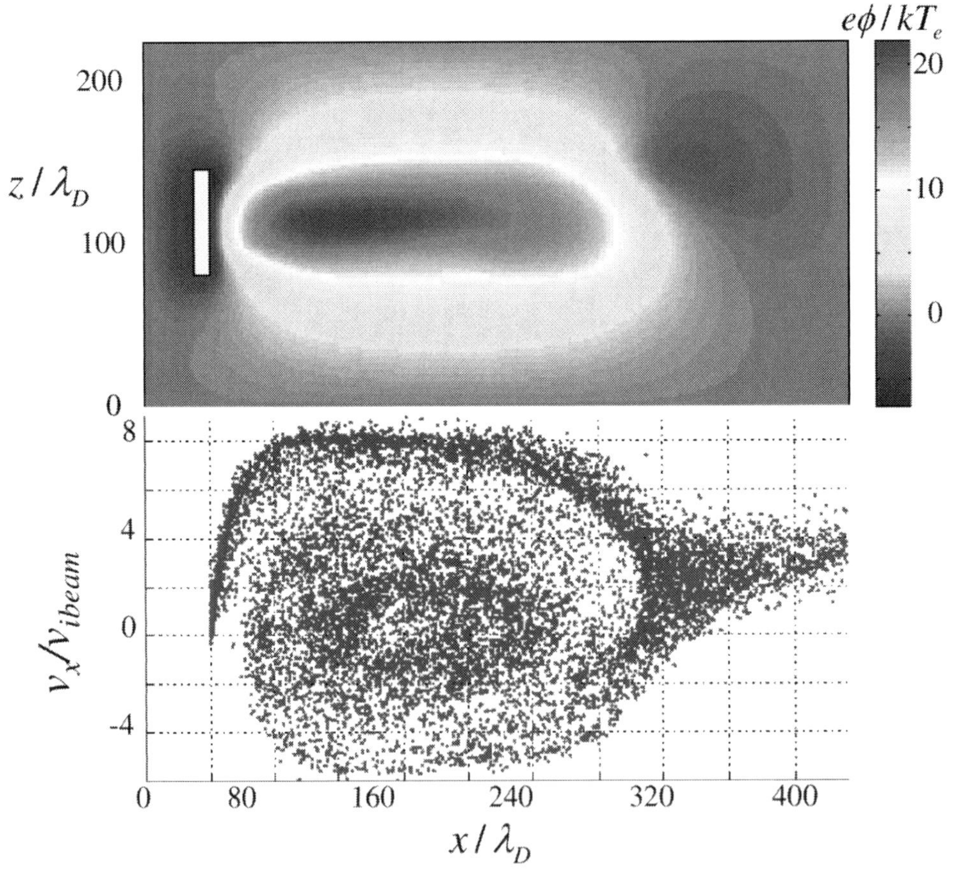

Figure 5. A contour map for electrostatic potential measured on a x-z plane including the ion beam propagation (upper panel) at $\omega_{pe}t$ =300 and the corresponding electron dynamics in a x-V_x phase space (lower panel).

distribution function of the electrons measured near the ion beam front in figure 6. As clearly shown, the function does not consist of a single Maxwellian but of two major components. One component has a peak which is located at positive velocity three times larger than v_{ibeam}. The other component which has a peak of negative velocity is owing to the electron reflection at the ion beam front. The velocity difference between the two peaks seems larger than the width of each distribution function. In such a situation, electron two-stream type instability may occur near the beam front and causes the local enhancement of the electric field which eventually heats electrons. Moreover, there may be some current-driven interactions between a cold ion beam and the accelerated electrons stated above. However, since the trapped electrons basically move back and forth along the beam direction in the potential well multiple times, the electron distribution overall tends to be smeared out and the clear velocity function which has two peaks can be seen only near the beam front as shown in figure 6. In other words, as shown in the lower panel of figure 5, the electron velocity distribution is not spatially uniform and rather varies much depending on the location along the beam direction. In such a situation, it may be difficult to observe electron two-stream instabilities because, according to the linear theory, the most unstable electrostatic mode has a wavelength which is much longer than the local region where the above velocity distribution is formed. This result basically agrees with the previous work [8] treating a local and finite beam emission. Meanwhile, the possibility of current-driven instabilities due to the difference between the cold ion beam and the accelerated electron component is not examined yet, which is left as a future work.

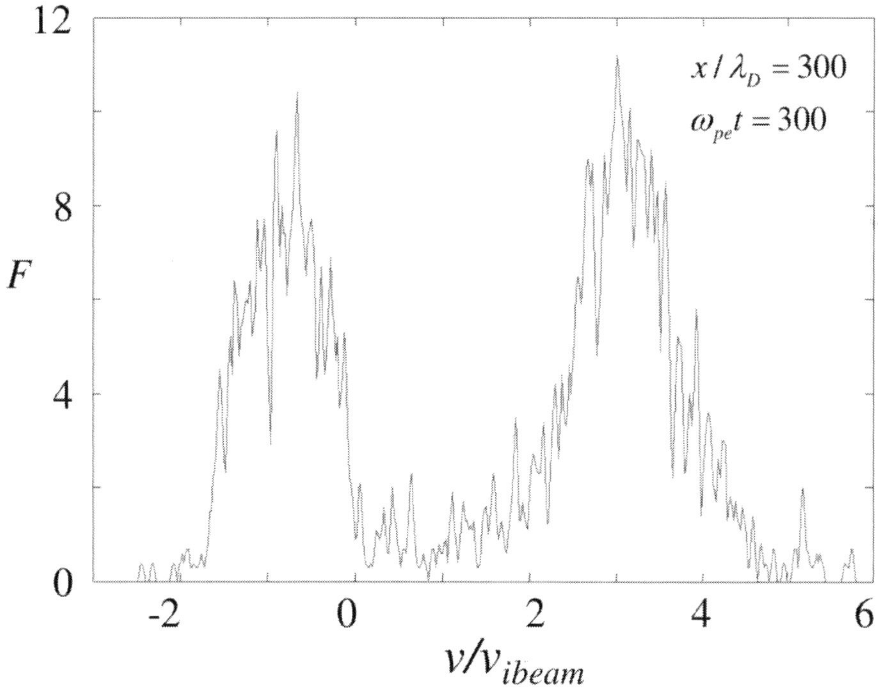

Figure 6. A distribution function of electron velocity along the ion beam propagation. F in the vertical axis is a normalized distribution function and the velocity v is normalized to v_{ibeam}.

4. Conclusion

We have examined the transient behavior of electrons released for ion beam neutralization in electric propulsion by performing full PIC simulations with EMSES. Unlike simple signature of ion beam propagation, we found that the behavior of thermal electrons emitted from the neutralizer is very complex. They basically propagate along with the ion beam to neutralize the positive charges of the ions. The electrons, however, are electrostatically trapped in a positive potential structure formed by the ion beam. The potential energy in the ion beam region is much larger than the initial thermal energy of electrons. The electrons released from the neutralizer are accelerated at the boundary of the potential well and move back and forth in the both directions parallel and perpendicular to the direction of the beam propagation. Although two major electron beam components are found particularly near the ion beam front, they are not uniform enough to cause electron two-stream instabilities. The electrostatic heating of electrons or the increase of the electron temperature is rather caused by the macroscopic trapping in the potential structure created in the whole region of the ion beam, not by the local beam instabilities. To examine the influence of electron trappings on the ion beam propagation, we need a much longer simulation space and time, which is left as a future work.

Acknowledgements

The computation in the present study was performed with the KDK system of Research Institute for Sustainable Human sphere (RISH) at Kyoto University.

References

[1] Wang J and Lai S 1997 Virtual anode in ion beam emissions in space: Numerical simulations *J. Spacecraft Rockets* **34** pp 829-836

[2] Buneman O and Kooyers G 1963 Computer Simulation of the Electron Mixing Mechanism in Ion Propulsion *J. AIAA(Am. Inst. Aeron. Astronaut.)* **1**(11) pp 2525-2528

[3] Wadhwa R P, Buneman O, and Brauch D F 1965 Two-Dimensional Computer Experiments on Ion Beam Neutralization *J. AIAA(Am. Inst. Aeron. Astronaut.)* **3**(6) pp 1076-1081

[4] Othmer C, Glassmeier K H, Schule J and Motschmann U 2000 Three dimensional simulations of ion thruster beam neutralization *Physics of Plasmas*, **7**(12) pp 5242-5251

[5] Othmer C, Glassmeier K H, Motschmann U, Schule J and Frick Ch 2002 Numerical simulation of ion thruster-induced plasma dynamics. the model and initial results *Advances in Space Research* **29** pp 1357-1362.

[6] Wheelock A, Gatsonis N and Cooke D 2003 Ion Beam Neutralization Processes for Electric Micropropulsion Applications *AIAA* 2003-5148

[7] Brieda L and Wang J 2005 Modeling Ion Thruster Beam Neutralization Using a Fully Kinetic ES-PIC Code *AIAA* 2005-4045

[8] Wang J, Chang O and Cao Y 2012 Electron-Ion Coupling in Mesothermal Plasma Beam Emission: Full Particle PIC Simulations *IEEE Transactions on Plasma Science* **40** pp 230 - 236

[9] Birdsall C K and Langdon A B 1985 P*lasma Physics via Computer Simulation* McGraw-Hill Inc. New York

[10] Omura Y and Matsumoto H 1993 *KEMPO1: Technical Guide to one-Dimensional Electro-magnetic Particle Code, Computer Space Plasma Physics: Simulation Techniques and Softwares* ed. By Matsumoto H and Omura Y Terra Scientific, Tokyo pp 21-65

[11] Miyake Y and Usui H 2009 A new electromagnetic particle code for the analysis of conductive spacecraft-plasma interactions *Physics of Plasmas* **16** 062904

Journal of Physics: Conference Series **454** (2013) 012018 doi:10.1088/1742-6596/454/1/012018

High performance computing for drug development on K computer

Hideaki Fujitani, Keiko Shinoda, Takefumi Yamashita and Tatsuhiko Kodama

Laboratory for Systems Biology and Medicine, Research Center for Advanced Science and Technology, The University of Tokyo, 4-6-1 Komaba Meguro-ku Tokyo 153-8904 Japan

E-mail: fjtani@lsbm.org

Abstract. Massively parallel computations (MP-CAFEE) ware developed to calculate absolute binding free energies of small molecules bound to a protein by all-atom molecular dynamics. It uses the nonequilibrium work measurement and Bennett acceptance ratio methods to calculate the free energy difference between the bound and unbound states. The FUJI force field was developed in order to assign force field parameters to arbitrary organic molecules in a unified manner including proteins and nucleic acids. Its dihedral parameters agree with the torsion energy profiles calculated by high-level ab initio molecular orbital theory for the model systems of protein backbone. Comparing with various force fields it agrees well with recent observations by vibrational spectroscopy on Ramachandran angle's population of alanine dipeptide in water. MP-CAFEE with FUJI force field has an efficient parallel algorithm and enough accuracy for computer aided drug design.

1. Introduction

One of the biggest challenges in drug development is to design small molecules of high affinity to a pharmaceutical target protein based on physical principles for all-atom molecular dynamics in water. In biology and pharmacology, a ligand means a substance (usually, a small molecule) bound to a protein which has a biological function. The binding occurs by intermolecular forces, caused by Coulomb and van der Waals interactions between all atoms. The binding of the protein and ligand is usually reversible without forming covalent bond. The Gibbs free energy difference between the bound and unbound states characterizes the binding affinity of the protein and ligand. The stronger the ligand binds, the more effectively the biological function works. Therefore, the binding affinity is an important criterion to design a new ligand to the target protein.

In order to explore dynamical behaviors of protein and ligand in water, it is popular to use molecular dynamics simulations with classical force field like AMBER, CHARMM, OPLS-AA, and so on. Based on the molecular force field many free energy calculation methods have been developed [1], but it was hard to predict the binding affinity with enough accuracy for drug design. In 2005 we reported the massively parallel computation of absolute binding free energy (MP-CAFEE) for FKBP and ligand systems using AMBER99 force field for the FKBP protein and general AMBER force field (GAFF) and Austin Model 1 bond charge correction (AM1BCC) charges for the ligands [2]. However, our calculated values were shifted by about 3.2 kcal/mol from experimental absolute binding free energies [3].

2. FUJI force field

The binding affinity of the protein and ligand is affected by thermal fluctuations. In order to rank molecules in computer aided drug design, we must predict the binding affinity within an error of 1 kcal/mol, while the thermal energy of the room temperature is 0.6 kcal/mol. From our early stage experiences in MP-CAFEE calculations, we realized that we should refine the force field itself, instead of sticking to the sampling issue on the free energy calculation. Using a unified force field for proteins and organic molecules, we got better absolute binding free energies for the FKBP and ligands [3].

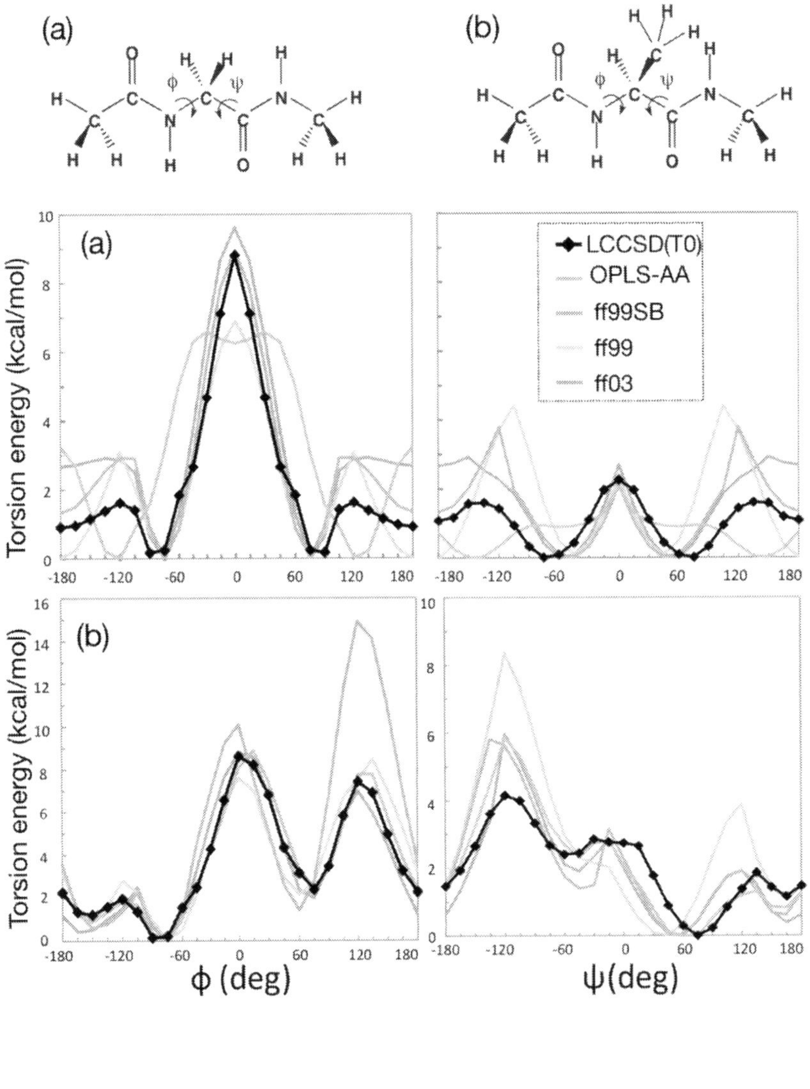

Figure 1. The torsional energy profiles of ϕ and ψ calculated with OPLS-AA and three AMBER force field variants (ff99SB, ff99, ff03) for (a) glycine dipeptide and (b) alanine dipeptide [4, 5, 6]. The filled squares with a black line show the torsional energy profile of ϕ and ψ calculated by MOLPRO quantum chemistry package with DF-LCCSD(T0)/Aug-cc-pVTZ//DF-LMP2/Aug-cc-pVTZ level.

In the 1990s Kollman's group developed the AMBER force field. The dihedral torsion parameters of protein backbone were determined to agree with quantum mechanical calculations for the model systems of glycine and alanine dipeptides [7]. But it was difficult to determine accurate torsion parameters by quantum mechanics, because of the shallow minima and very flat regions of potential energy surfaces with respect to the Ramachandran angles ϕ and ψ. We intensively investigated accuracy of various quantum mechanical theory levels to determine the appropriate theory level for the torsional energy profiles. The filled squares with black lines in figure 1 show the torsional energy profiles of glycine and alanine dipeptides calculated by the high level theory of DF-LCCSD(T0)/Aug-cc-pVTZ//DF-LMP2/Aug-cc-pVTZ. The colored lines show torsional energy profiles obtained by various molecular mechanical force fields, which all deviate from the ab initio torsional energy profile. We refined the relevant molecular mechanical dihedral parameters to fit the ab initio torsional energy profiles [8].

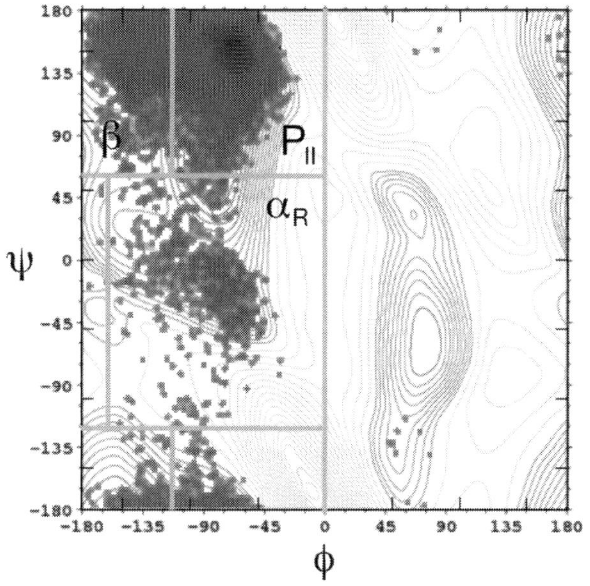

Figure 2. Ramachandran angle's population of alanine dipeptide in TIP3P water. The buried contour shows an energy landscape of alanine dipeptide in vacuum.

Grdadolnik *et al.* measured a short-lived population of Ramachandran angles of alanine dipeptide in aqueous solution using vibrational spectroscopy (infrared and Raman spectra) [9]. In order to compare with their experimental population we performed molecular dynamics simulation of alanine dipeptide in TIP3P water using the FUJI force field. Figure 2 shows an example of the Ramachandran angle's population in a 100 ns simulation at room temperature. The purple area in P_{II} region refers to dense population. The buried contour shows an energy landscape of alanine dipeptide in vacuum [8]. It has a metastable state around ($\phi = 60, \psi = -60$), but the alanine dipeptide in water rarely takes such conformation.

Seabra *et al.* calculated the Ramachandran angle's population of alanine dipeptide in water by the replica exchange method using various semiempirical quantum mechanics and classical force field molecular mechanics. Their results were in contrast with the computational chemist's intuitive idea that the more expensive a method the better its accuracy. Actually, they found

Table 1. Ramachandran angle's population of alanine dipeptide in water. The top two rows (infrared and Raman) are experiments and others are theoretical results.

method	α_R	β	$\mathrm{P_{II}}$	other	reference
infrared	0.11	0.29	0.60	-	[10]
Raman	0.09	0.29	0.62	-	[10]
ff94	0.84	0.04	0.11	0.01	[11]
ff99SB	0.32	0.24	0.40	0.04	[11]
ff03	0.45	0.19	0.35	0.01	[11]
PM3	0.14	0.51	0.33	0.02	[11]
CHARM27	0.53	0.07	0.40	0.00	[12]
FUJI	0.05	0.17	0.76	-	[13]
FUJI	0.06	0.23	0.71	0.00	this work

the ff99SB results were more accurate than most of the semiempirical quantum mechanical methods. Table 1 shows the Ramachandran angle's population in the three regions (α_R, β, $\mathrm{P_{II}}$). Our population by the FUJI force field was derived from three 100 ns trajectories and agrees with the metadynamics analysis by Vymětal and Vondrášek [13].

Even taking into account some ambiguities in the experimental measurement and in the theoretical boundaries of the three regions, we can conclude that only the FUJI force field agrees with the experimental Ramachandran angle's population of alanine dipeptide in water. This conclusion is extremely significant for molecular dynamics simulations with classical force field, because the same dihedral parameters are used for the protein backbone of all amino acids.

3. MP-CAFEE

There are important requirements for accurate absolute binding free energy calculations. The first requirement is a well-equilibrated bound structure including the conformational change of the protein induced by the binding of the small molecule. The second one is the convergence of the work distribution with a sufficient number of trajectories and dense spacing of the coupling constant λ between the small molecule and the rest of the system. Finally, the most important requirement is an accurate force field for the protein and small molecule. Our calculations were performed by GROMACS [14].

Using the FUJI force field we performed absolute binding free energy calculations for the major urinary protein (MUP) and two ligands: 2-methoxy-3-isopropylpyrazine (IPMP) and 2-methoxy-3-isobutylpyrazine (IBMP). MUP has a unique binding structure such that the whole ligand is wrapped in the protein. The binding site cannot be seen from outside (figure 3). Homans's group performed excellent works on this system including the binding free energy measurements by isothermal titration calorimetry [15]. It is a good test system for the absolute binding free energy calculations by all atom molecular dynamics.

Table 2. The calculated binding free energy $\Delta G_{\mathrm{bind}} = \Delta G_{\mathrm{complex}} - \Delta G_{\mathrm{sol}}$ and experimental binding free energy ΔG_{exp} in cal/mol.

ligand	ΔG_{sol}	$\Delta G_{\mathrm{complex}}$	ΔG_{bind}	ΔG_{exp}
IPMP	-1.6	-9.9	-8.3	-8.1
IBMP	-1.9	-11.3	-9.4	-9.2

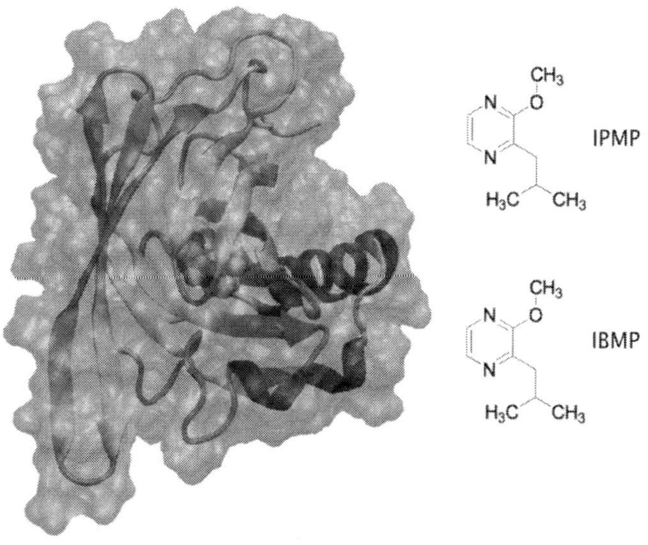

Figure 3. Binding structure of IPMP to MUP and two chemical formulas of IPMP and IBMP.

We used the FUJI force field for MUP and ligands. The point charges of the ligands were determined by the restrained electrostatic potential fit (RESP) using two conformations for each ligand [16]. After 20 ns equilibration from X-ray binding structures (PDB: 1QY1, 1QY2) we performed free energy calculations using 32 λ points and 12 trajectories for each λ point. The details of the MP-CAFEE procedure are explained in [3]. Calculated free energies and experimental binding free energies are listed in table 2. They show excellent agreement between theory and experiment.

In order to examine how much the force field difference affects the binding free energy, we performed the same binding free energy calculation using the ff99SB-ildn force field which was improved by ff99SB by D. E. Shaw's research [17]. The difference is only the force field for MUP. The ligand force field and other calculation details are the same as the FUJI force field calculation. The ff99SB-ildn gave 1.9 kcal/mol stronger binding free energy than the FUJI force field (figure 4). It seems reasonable because ff99SB has larger torsional energy barrier between the lowest energy position ($\phi = -80$) and the second lowest energy position ($\phi = -180$) in figure 1(a). But its deviation from the experimental value is too large for the computer aided drug design.

4. Drug Design on K computer

We believe that the most important issue for computational physics is to prepare correct input data. "Garbage in, garbage out" is always true on computers. The force field is vital to all atom molecular dynamics on biomolecules. As the FUJI force field seems enough accurate for computer aided drug design, we start to design small molecules for pharmaceutical target proteins with MP-CAFEE.

K computer is one of the most powerful computers; it has 88,128 CPU nodes. We calculated the binding free energies of 40 small molecules to a pharmaceutical target protein in one parallel job using 15,360 CPU nodes of K computer. Some of the 40 molecules were synthesized and their calculated binding free energies agreed with experimental measurements. Now is the time to utilize the MP-CAFEE with FUJI force field in order to accelerate the drug development by K computer.

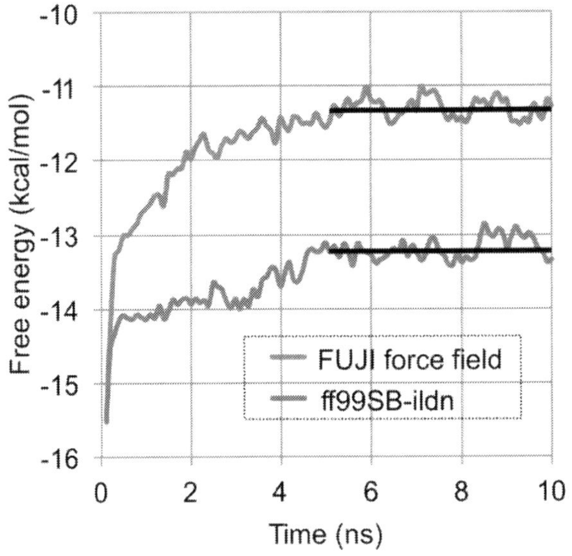

Figure 4. Binding free energy convergence of IBMP to MUP calculated with two different force fields: FUJI force field and ff99SB-ildn.

Acknowledgments

We like to thank all our collaborators on the technology development for the computer aided drug design. This research has been funded by MEXT SPIRE Supercomputational Life Science and by FIRST Kodama project in Japan.

References

[1] Kollman P A 1993 *Chem. Rev.* **93** 2395–2417
[2] Fujitani H, Tanida Y, Ito M, Jayachandran G, Snow C D, Shirts M R, Sorin E J and Pande V S 2005 *J. Chem. Phys.* **123** 084108
[3] Fujitani H, Tanida Y and Matsuura A 2009 *Phys. Rev. E* **79** 021914
[4] Hornak V, Abel R, Okur A, Strockbine B, Roitberg A and Simmerling C 2006 *Proteins: Struct. Funct. Bioinf.* **65** 712–725
[5] Wang J, Cieplak P and Kollman P A 2000 *J. Comput. Chem.* **21** 1049–1074
[6] Duan Y, Wu C, Chowdhury S, Lee M C, Xiong G, Zhang W, Yang R, Cieplak P, Luo R, Lee T, Caldwell J, Wang J and Kollman P 2003 **24** 1999–2012
[7] Cornell W D, Cieplak P, Bayly C I, Gould I R, Merz, Jr K M, Ferguson D M, Spellmeyer D C, Fox T, Caldwell J W and Kollman P A 1995 *J. Am. Chem. Soc.* **117** 5179–5197
[8] Fujitani H, Matsuura A, Sakai S, Sato H and Tanida Y 2009 *J. Chem. Theory Comput.* **5** 1155–1165
[9] Grdadolnik J, Grdadolnik S G and Avbelj F 2008 *J. Phys. Chem. B* **112** 2712–2718
[10] Grdadolnik J, Mohacek-Grosev V, Baldwin R L and Avbelj F 2011 *Proc. Natl. Acad. Sci.* 1794–1798
[11] de M Seabra G, Walker R C and Roitberg A E 2009 *J. Phys. Chem. A* **113** 11938–11948
[12] Vymětal J and Vondrášek J 2010 *J. Phys. Chem. B* **114** 5632–5642
[13] Vymětal J and Vondrášek J 2011 *Chem. Phys. Lett.* **503** 301–304
[14] Hess B, Kutzner C, van der Spoel D and Lindahl E 2008 *J. Chem. Theory Comput.* **4** 435–447
[15] Bingham R J, Findlay J B C, Hsieh S Y, Kalverda A P, Kjellberg A, Perazzolo C, Phillips S E V, Seshadri K, Trinh C H, Turnbull W B, Bodenhausen G and Homans S W 2004 *J. Am. Chem. Soc.* **126** 1675–1681
[16] Bayly C I, Cieplak P, Cornell W D and Kollman P A 1993 *J. Phys. Chem.* **97** 10269
[17] Lindorff-Larsen K, Piana S, Palmo K, Maragakis P, Klepeis J L, Dror R O and Shaw D E 2010 *Proteins: Struct. Funct. Bioinf.* **78** 1950–1958

Journal of Physics: Conference Series **454** (2013) 012019 doi:10.1088/1742-6596/454/1/012019

Development of a first-principles code based on the screened KKR method for large super-cells

S Doi, M Ogura and H Akai

Department of Physics, Graduate School of Science, Osaka University, 1-1 Machikaneyama, Toyonaka, Osaka 560-0043, Japan

E-mail: doi@presto.phys.sci.osaka-u.ac.jp

Abstract. The procedures of performing first-principles electronic structure calculation using the Korringa–Kohn–Rostoker (KKR) and the screened KKR methods are reviewed with an emphasis put on their numerical efficiency. It is shown that an iterative matrix inversion combined with a suitable preconditioning greatly improves the computational time of screened KKR method. The method is well parallelized and also has an $O(N)$ scaling property.

1. Introduction

First-principles electronic structure calculations based on the density functional theory (DFT) have been widely used and are now indispensable for both basic condensed matter physics and applications. Microscopic information obtained by first-principles calculations not only helps us to understand the physics behind experimental results but also makes it possible for us to predict properties of materials that have not been synthesized yet. This even opens up a possibility of computational materials design (CMD). However, in order for CMD to be a reality, it is necessary to execute first-principles calculations for larger scale systems than ever. In a first-principles calculation based on the DFT, the ground state electronic structure $\rho(\mathbf{r})$ (or the corresponding one body effective potential $V_{\text{eff}}(\mathbf{r})$) which satisfies the variational principle of energy functional $\delta E[\rho(\mathbf{r})] = 0$ is calculated by solving the so-called Kohn-Sham equations self-consistently [1, 2]. In each Kohn-Sham iteration cycle, one-body Schrödinger-like equation must be solved. As long as the usual band structure calculation algorithms are used, the numerical costs are proportional to the third power of the system size. Due to this property, large scale electronic structure calculations using a traditional method need large amount of numerical resources and computational time. To overcome these difficulties, algorithms and codes which have a property of $O(N)$ have been developed for many years. $O(N)$ means that the total steps of the calculation is proportional to the system size (i.e., the number of atoms N in the system).

Among many other methods solving the Kohn-Sham equations, the KKR Green's function method is one of the most accurate and powerful methods. Moreover, screened KKR, an extension of the method to $O(N)$, was introduced in 1995 [3]. The most time consuming part in the KKR Green's function method is the process of inverting a matrix having the dimension proportional to the number of atoms in the system. In the screened KKR, by choosing a suitable reference system such that the matrix pattern becomes sparse, and by combining the matrix sparsity with an efficient matrix solver, the total computational work is greatly reduced without sacrificing the accuracy. The screened KKR method has been mainly applied to multilayered systems [4]. For multilayered systems, assuming 2D periodicity, the unit cell is extended to the layer growth direction. It was shown that in this case the total step needed for the calculation is exactly proportional to the number of layers N. Then, the electronic structure calculations for a system containing layers larger than ten thousands become possible on even a single CPU of present-day desktop PCs. This is due to the fact that in such layered systems, the problem is reduced to a matrix inversion of a simple block tridiagonal sparse matrix. Beyond multilayered structures, if the unit cell is extended to x, y, and z directions, the matrix pattern becomes general block sparse. As long as a direct method like LU factorization, which is used in traditional KKR and screened KKR for multilayered systems, is used, it is very difficult to obtain sufficient efficiency for the matrix inversion of these general block sparse matrices. Recently, high efficiency in using an iterative method, which has been widely used to solve simultaneous linear equations with sparse coefficient matrix, was proven for these large super-cell systems [5]. The electronic structure calculation code, which uses the iterative matrix inversion technique, for super-cell systems with massively parallel computing `'KKRnano'` was developed by the Jülich group [6, 7]. Referring to their work, we have developed a first-principles electronic structure calculation code for large super-cell systems by using the screened KKR method, based on the full-potential KKR (FP-KKR) code that was developed in our group [8].

We briefly outline the KKR Green's function method in section 2. Section 3 describes the concept of screened KKR and explains how $O(N)$ scaling properties are obtained for both multilayered systems and general large super-cell systems. Implementation of parallel computing using the iterative method as a sparse matrix solver is also discussed. In section 4, results of some test calculations are shown. The efficiency of the iterative method when combined with a suitable preconditioning is confirmed. Section 5 is devoted for summary.

2. KKR Green's function method

2.1. KKR Green's function method

In contrast to usual schemes of first-principles methods, in the KKR Green's function method, instead of solving eigenvalue problems, the corresponding one-body Green's function $G(\mathbf{r}, \mathbf{r}', E)$ which satisfies

$$[-\Delta + V_{\text{eff}}(\mathbf{r}) - E]G(\mathbf{r}, \mathbf{r}', E) = -\delta(\mathbf{r} - \mathbf{r}') \tag{1}$$

is calculated. The electron density $\rho(\mathbf{r})$ is constructed directly from the energy integration of the Green's function up to the Fermi level E_{F} [9]:

$$\rho(\mathbf{r}) = -\frac{1}{\pi}\Im \int_{-\infty}^{E_{\text{F}}} dEG(\mathbf{r}, \mathbf{r}, E) \tag{2}$$

$G(\mathbf{r}, \mathbf{r}, E)$ has a very complex structure near the real energy axis, and evaluation of the integral (2) along real energy axis requires many energy sampling points. However, off the real energy axis, $G(\mathbf{r}, \mathbf{r}, E)$ becomes structureless without any singularity on the upper half of complex energy plane. Therefore by an analytic continuation, the integral (2) can be evaluated by numerical integration with a few tens of energy sampling points accurately. Practically, core electrons can be considered as well localized and they can be determined by atomic calculation for given effective potentials. In this case the complex energy contour can start from E_{B} (between the bottom of valence band and the top of core level) and approaches to E_{F} through the upper half of complex energy plane. This procedure is important not only for the numerical integration but also for the matrix inversion in the iterative method to obtain the rapid convergences.

2.2. Calculation of crystal Green's function

The one-body Green's function in a crystal is obtained by multiple scattering theory. The system is divided into cells around each lattice point and local effective potential $V^n(\mathbf{r})$ for the n-th cell

Journal of Physics: Conference Series **454** (2013) 012019 doi:10.1088/1742-6596/454/1/012019

is defined within the cell. Green's function can be expressed in a cell-centered form:

$$G(\mathbf{r}+\mathbf{R}^n, \mathbf{r}'+\mathbf{R}^{n'}, E) = \delta_{nn'} G_S^n(\mathbf{r}, \mathbf{r}', E) + \sum_{LL'} R_L^n(\mathbf{r}, E) G_{LL'}^{nn'}(E) R_{L'}^{n'}(\mathbf{r}', E) , \tag{3}$$

where $\mathbf{R}^n$ denotes the center of the n-th cell, $L \equiv (l, m)$ the combined indices of the orbital angular momentum l and the magnetic quantum number m, $R_{L'}^n(\mathbf{r}, E)$ the regular solution of the single-site Schrödinger equation for the local potential. The single-site Green's function G_S^n is defined as

$$G_S^n(\mathbf{r}, \mathbf{r}', E) = \sqrt{E} \sum_L R_L^n(\mathbf{r}_<, E) H_L^n(\mathbf{r}_>, E) , \tag{4}$$

where $\mathbf{r}_<(\mathbf{r}_>)$ means the smaller (larger) of $\mathbf{r}$ and $\mathbf{r}'$, $H_L^n(\mathbf{r}, E)$ irregular solution of the single-site Schrödinger equation for the local potential. In the original KKR formulation, the potential shape is assumed to be muffin-tin potentials. However, an FP extension is possible if all regions are Voronoi cells and the potentials are expanded into real spherical harmonics $X_L(\hat{\mathbf{r}})$ around the cell center [8, 10]. In both potential models, the existence of the local potential around each cell can be described by the scattering t-matrix. The scattering t-matrix $t_{LL'}^n(E)$ for the n-th cell is defined as

$$t_{LL'}^n(E) = \int_{\Omega_n} d\mathbf{r} j_l(\sqrt{E} r) X_L(\hat{\mathbf{r}}) V^n(\mathbf{r}) R_{L'}^n(\mathbf{r}, E) , \tag{5}$$

where Ω_n is the volume of the cell, and j_l is the spherical Bessel function. Calculation of the first term in (3) is rather straightforward. The most time consuming part in KKR is the determination of the expansion coefficients $G_{LL'}^{nn'}(E)$ in the second term in (3). $G_{LL'}^{nn'}(E)$ expresses the effect of the multiple scattering due to the existence of the t-matrices on the lattice points in the system and is connected to the expansion coefficient of an arbitrary reference system $G_{LL'}^{\mathrm{r},nn'}(E)$ through the Dyson equation:

$$G_{LL'}^{nn'}(E) = G_{LL'}^{\mathrm{r},nn'}(E) + \sum_{n'',L''} G_{LL''}^{\mathrm{r},nn''}(E) \sum_{L'''} \left[t_{L''L'''}^{n''}(E) - t_{L''L'''}^{\mathrm{r},n''}(E) \right] G_{L'''L'}^{n''n'}(E) . \tag{6}$$

Here, the reference system is defined as the system which has the same lattice structure as the crystal, but has a different kind of potential scatterers, described by a set of t-matrices $t_{LL'}^{\mathrm{r},n}(E)$ on the lattice positions. Therefore, if $G_{LL'}^{\mathrm{r},nn'}(E)$ is known, we can calculate $G_{LL'}^{nn'}(E)$ by solving the Dyson equation (6).

2.3. Standard KKR Green's function method

In the standard KKR, the free space, which has no potential scatterers, is chosen as a reference system. For the free space reference, an analytical expression of the expansion coefficient $g_{LL'}^{nn'}(E)$ (called *structure constant*) is known:

$$g_{LL'}^{nn'}(E) = -4\pi i \sqrt{E} (1 - \delta_{nn'}) \sum_{L''} i^{l+l''-l'} C_{LL'L''} h_{l''}(\sqrt{E} |\mathbf{R}^n - \mathbf{R}^{n'}|) X_{L''}(\widehat{\mathbf{R}^n - \mathbf{R}^{n'}}) , \tag{7}$$

where $C_{LL'L''} \equiv \int d\hat{\mathbf{r}} X_L(\hat{\mathbf{r}}) X_{L'}(\hat{\mathbf{r}}) X_{L''}(\hat{\mathbf{r}})$ is the Gaunt number and h_l the spherical Hunkel function. The Dyson equation for the free space reference is now given by

$$G_{LL'}^{nn'}(E) = g_{LL'}^{nn'}(E) + \sum_{n'',L''} g_{LL''}^{nn''}(E) \sum_{L'''} t_{L''L'''}^{n''}(E) G_{L'''L'}^{n''n'}(E) . \tag{8}$$

For periodic crystals, by applying the 3D Fourier transform

$$G_{LL'}^{\mu\mu'}(\mathbf{k},E) = \sum_{n'} \exp(i\mathbf{k}\cdot(\mathbf{R}^n - \mathbf{R}^{n'}))G_{LL'}^{nn',\mu\mu'}(E)\ , \tag{9}$$

the Dyson equation for each $\mathbf{k}$ point in the irreducible Brillouin zone (BZ) is obtained:

$$G_{LL'}^{\mu\mu'}(\mathbf{k},E) = g_{LL'}^{\mu\mu'}(\mathbf{k},E) + \sum_{L'',\mu''} g_{LL''}^{\mu\mu''}(\mathbf{k},E)t_{L''}^{\mu''}(E)G_{L''L'}^{\mu''\mu'}(\mathbf{k},E)\ . \tag{10}$$

Here, $\mu(=1,\cdots,N)$ indicates an atom in the unit cell of periodic crystal (N is the number of atoms in the unit cell). In the matrix notation of combined indices of μ and L, $g(\mathbf{k},E) = \left\{g_{LL'}^{\mu\mu'}(E)\right\}, G(\mathbf{k},E) = \left\{g_{LL'}^{\mu\mu'}(E)\right\}$, and $t(E) = \{\delta_{\mu\mu'}t_{LL'}^{\mu}(E)\}$, with matrix dimension $(l_{\max}+1)^2 \times N$, (10) can be rewritten as

$$G(\mathbf{k},E) \quad = t(E)^{-1}\left[g(\mathbf{k},E) - t(E)^{-1}\right]^{-1} t(E)^{-1} - t(E)^{-1}\ . \tag{11}$$

Then, the problem to solve Dyson equation (10) reduced to the problem of calculating the inverse of the matrix $M(\mathbf{k},E) \equiv g(\mathbf{k},E) - t(E)^{-1}$. Matrix inversion is usually done by the *LU* factorization; then the number of steps needed increases as $O(N^3)$. Therefore, if the system size becomes large, the computational time explodes rapidly. To overcome this difficulty, the concept of screened KKR described in the next section was proposed.

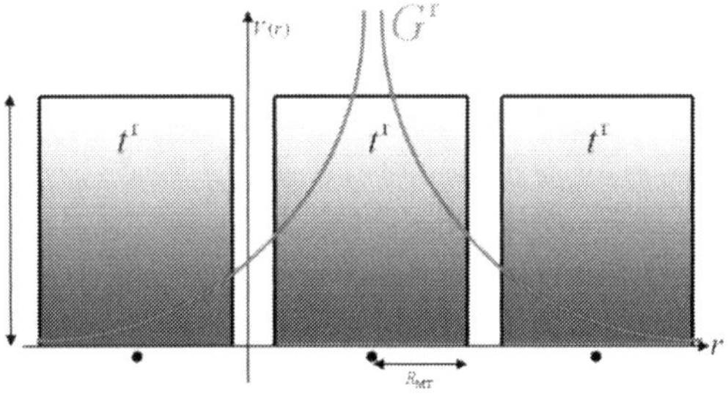

Figure 1. Repulsive potential reference system. $G_{LL'}^{\mathrm{r},nn'}(E)$ decays exponentially in real space.

Figure 2. Real space cluster to obtain $G_{LL'}^{\mathrm{r},nn'}(E)$ in solving Dyson equation (12) with the number of repulsive potential arrays N_{cl}

3. Screened KKR method

3.1. The concept of screened KKR method

For the free space reference system, Green's function has the analytical form $G^{\mathrm{r}}(\mathbf{r},\mathbf{r}',E) \equiv g(\mathbf{r},\mathbf{r}',E) = -\exp(i\sqrt{E}|\mathbf{r}-\mathbf{r}'|)/4\pi|\mathbf{r}-\mathbf{r}'|$. From this expression, we can see that $g(\mathbf{r},\mathbf{r}',E)$ decays exponentially in real space when E is negative. In other words, $g(\mathbf{r},\mathbf{r}',E)$ decays exponentially in real space if energy region is lower than the continuous eigenstates region of free space.

As suggested by the above fact, in screened KKR, a reference system with constant repulsive potential arrays of muffin-tin radius R_{MT} for each lattice position (figure 1) is adopted [3]. For these repulsive potential systems, continuous eigenstates are pushed up to a energy region much higher than E_{F} if the height of repulsive potential is sufficiently high. Then the Green's function of the reference system decays exponentially in real space over the energy region used in evaluating (2). In this case, it is justified to neglect electron propagation longer than some threshold distance without sacrificing the accuracy. Correspondingly, the matrix pattern appearing in the Dyson equation becomes sparse. By combining this feature with an efficient sparse matrix solver, it is possible to reduce the computational time drastically while keeping the accuracy.

Structure constants $G^{\mathrm{r},nn'}_{LL'}(E)$ of the reference system with repulsive potentials are easily calculated by solving a Dyson equation

$$G^{\mathrm{r},nn'}_{LL'}(E) = g^{nn'}_{LL'}(E) + \sum_{n'',L''} g^{nn''}_{LL''}(E) \sum_{L'''} t^{\mathrm{r},n''}_{L''L'''}(E) G^{\mathrm{r},n''n'}_{L'''L'}(E) \tag{12}$$

in real space within small cluster (with the number of repulsive potential N_{cl} within the cluster) as shown in figure 2, due to exponential decay of $G^{\mathrm{r},nn'}_{LL'}(E)$. After obtaining $G^{\mathrm{r},nn'}_{LL'}(E)$, we can calculate the crystal Green's function through (6). In short, the reference system with repulsive potentials is inserted as an intermediate step, between initial free space and final crystal system in the screened KKR.

3.2. Screened KKR method for multilayered systems

If the screened KKR is applied to a multilayered system which is extended to the layer growth direction (figure 3), assuming 2D periodicity for the rest two directions, the Dyson equation can be solved for each $\mathbf{k}_{||}$ in the 2D BZ, and the matrix pattern becomes just block tridiagonal [4, 10]. For a block tridiagonal matrix, the site-diagonal elements of its inverse, which are necessary for the construction of charge density in Kohn-Sham iteration, can be calculated in $O(N)$ (N is the number of layer) using the LU factorization. This technique have been widely and successfully applied to layered systems since the screened KKR was first proposed.

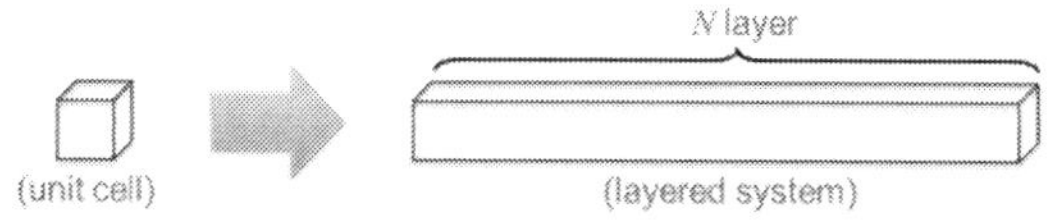

Figure 3. Multilayered system.

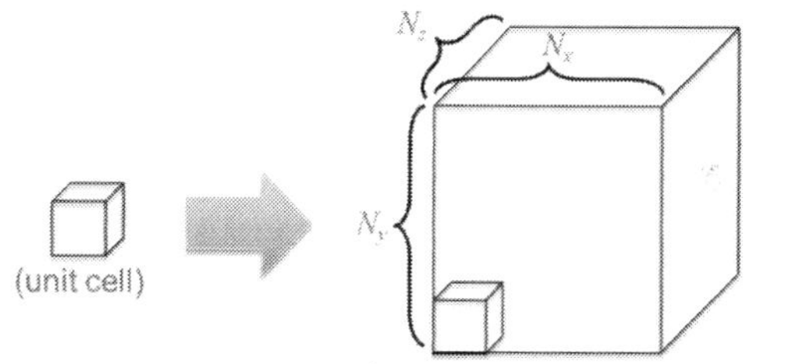

Figure 4. Super-cell system which is extended to x, y, and z directions.

3.3. Screened KKR method for super-cell systems

If we consider systems which are extended to x, y, and z directions (figure 4), the sparse matrix pattern becomes general block sparse form, differently from the case of multilayered systems. For a general block sparse matrix, there is no known $O(N)$ direct algorithm to calculate its inverse such as the LU factorization for block tridiagonal matrices. If LU factorization is applied, so called *fill-in* elements appear which require much larger memory storage than that of the original sparse matrix. Now, the problem is, for given matrix A, a calculation of matrix A^{-1} which satisfies $AA^{-1} = I$ (I is the unit matrix). And for the i-th column vector x_i of A^{-1}, it satisfies $Ax_i = I_i$ (I_i : the i-th column of I). It means that the problem can be considered as a simultaneous linear equation for each x_i sharing a sparse coefficient-matrix A.

Then, an iterative method for simultaneous linear equations is a good candidate as an efficient sparse matrix solver. For sparse coefficient-matrices of large dimension, nonstationary iterative methods categorized as the Krylov subspace methods have been widely used. In the Krylov subspace method, to solve an equation $Ax = b$, the true solution which minimizes the norm of the residual vector $||r|| \equiv ||x - b||$ is searched in the Krylov subspace defined as $K_n(c, A) = \mathrm{span}(c, Ac, A^2c, \ldots, A^{n-1}c)$ by iterations. Here c is an arbitrary vector. Starting from the initial guess x_0 and, normally, using $r_0 \equiv x_0 - b$ as c, an approximate solution $x_n \in x_0 + K_n(r_0, A)$ is updated in the n-th iteration. For each iteration, only (sparse matrix)-by-(vector) manipulations are required and no additional memory larger than A is needed, in contrast to the case of a direct method. If the iteration converges rapidly, the method has an advantage over direct method from the view point of both the memory usage and the computational time. Since non-Hermitian general sparse matrices appear in the present cases, the transpose-free quasi-minimal residual (TFQMR) algorithm [11] is suitable as an iterative method [5, 12]. Further, as long as iterative methods are used, the calculation for each column vector is independent on the other columns. Therefore, parallelization with respect to the number of atoms is easily realized.

Usually, iterative methods are used in combination with a preconditioning of the coefficient matrix. For a linear equation $Ax = b$, the preconditioning matrix is defined as

$$A'(M_2x) = M_1^{-1}b\,, \tag{13}$$

where

$$A' = M_1^{-1}AM_2^{-1}\,. \tag{14}$$

M_1 is a left preconditioning matrix and M_2 a right preconditioning matrix. As a general characteristics of the Krylov subspace method, the iteration convergence is fast when a distribution of eigenvalues of A is dense near a non-zero point. Therefore, the acceleration in the convergence is expected if A' approaches to I through preconditioning. The efficiency of the preconditioning is shown by test calculations in the next section.

4. Test calculations and preconditioning

Following previous studies [12, 5, 6, 7], we have developed a code for first-principles electronic structure calculation of large super-cell systems using the screened KKR with the TFQMR algorithm. The modification of TFQMR routine available from QMRPACK in Netlib [13] is implemented, where the sparse matrix vector manipulation for each iteration is executed for $(l_{\mathrm{max}} + 1)^2$ column vectors at the same time, in contrast to the original routine where each column is treated separately. This elaboration is implemented so as to take an advantage of cache access as mentioned in [6, 7]. MPI parallelization with respect to atoms in the super-cell is implemented. To check the performance and validity of the code, intensive test calculations were made. As a system for the test calculations, we took a single Ni impurity in a large Cu super-cell of the face-centered-cubic (fcc) structure. The primitive cubic unit cell consists of four atoms forming a fcc lattice with the lattice constant $a = 6.83$ a.u. is considered. Repeating this unit cell along three directions eight times, we construct super-cell containing $2048 (= 4\times8\times8\times8)$ Cu atoms of a g structure. Then, one of 2048 Cu atom is replaced by a single Ni atom (figure 5).

The local electronic structure of a Ni impurity site in the super-cell is calculated using a single core of Intel Core2 Quad Z6700 2.66GHz. As a minimal setting, $l_{\mathrm{max}} = 2$ and only nearest neighbor propagations of screened structure constants are considered ($N_{\mathrm{cl}} = 13$), the Dyson equation is solved only at Γ-point in the BZ of the super-cell to construct the charge density. The convergence of iterative method is strongly dependent on the electronic structure at the sampling points used for the integration on the complex energy plane. Especially at a

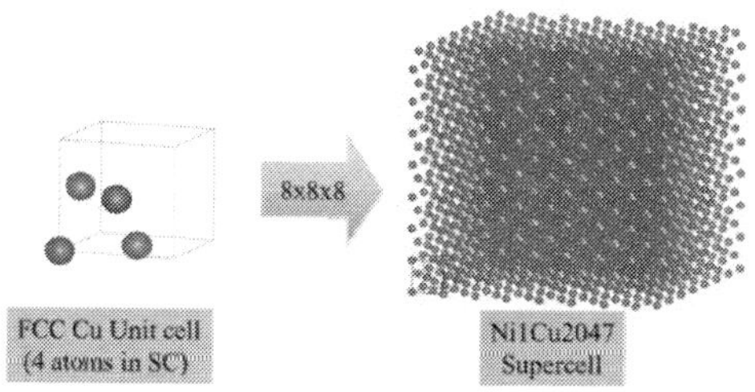

Figure 5. Ni1Cu2047 super-cell system for test calculation. One atom in Cu super-cell is substituted for Ni impurity.

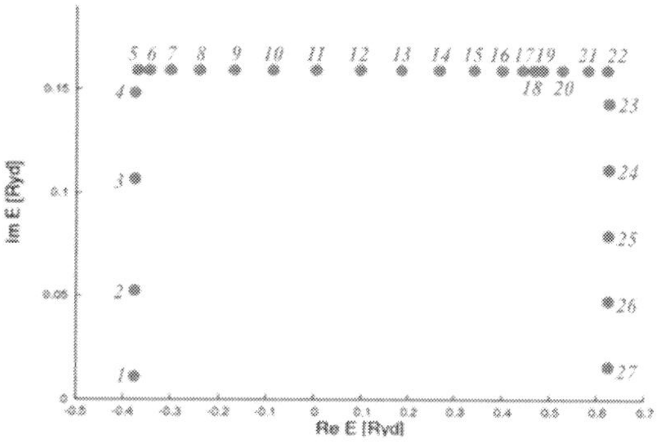

Figure 6. Sampling points on complex energy plane. Numbers with blue character in this figure distinguish energy sampling points.

sampling point near the real energy axis with continuous eigenstates, the convergence is worse. It means that calculations at the energy sampling points near E_F could be a bottleneck in the case of metallic systems. Therefore, it is desirable to use energy sampling points as apart from the real axis as possible. For this, the finite-temperature formalism obtained by introducing the Fermi–Dirac distribution function $f(E, E_F, T) = (\exp(1 + (E - E_F)/k_B T))^{-1}$ in (2), is useful [14]. In the finite-temperature formalism, the nearest sampling point from E_F is the first "Matsubara frequency" with a vertical distance $\pi k_B T$ from the real axis. In the test calculations, 27 sampling points, including five Matsubara frequencies with $T = 800$ K, are used in complex energy integration (figure 6). All the test calculations were started with the initial guess of $x_0 = (0, \ldots, 0)^T$ for Ni impurity. The TFQMR iterations were repeated until relative residual norm approaches to $||r||/||r_0|| = 10^{-6}$.

4.1. TFQMR without preconditioning
The number of the iteration and CPU time needed in one self-consistent field (SCF) calculation for each energy sampling point is shown in figure 7 with red line and points. As can been seen from this figure, calculations at a few Matsubara energies near E_F are the bottlenecks.

4.2. TFQMR with the block Jacobi (diagonal scaling) preconditioning
The block Jacobi (BJCB) (or called as diagonal scaling) preconditioning, which uses diagonal blocks of sparse matrix as left preconditioning matrix M_1 ($M_2 = I$) such that the diagonal blocks of A' become unit matrices, is one of the simplest preconditionings. Results using BJCB left preconditioning are shown in figure 7 with green line and points. Even a simple preconditioning such as BJCB improves the convergence of TFQMR considerably. However, the bottlenecks near E_F still exist.

4.3. TFQMR with the block-circulant matrix preconditioning
An efficient algorithm, named block-circulant matrix (BCM) preconditioning, suitable for super-cell systems, was proposed in [15]. Generally, a circulant matrix $C \in \mathbb{C}^{N \times N}$

$$C = \begin{pmatrix} c_0 & c_1 & \cdots & c_{N-1} \\ c_{N-1} & c_0 & \cdots & c_{N-2} \\ \vdots & \vdots & \ddots & \vdots \\ c_1 & c_2 & \cdots & c_0 \end{pmatrix} \tag{15}$$

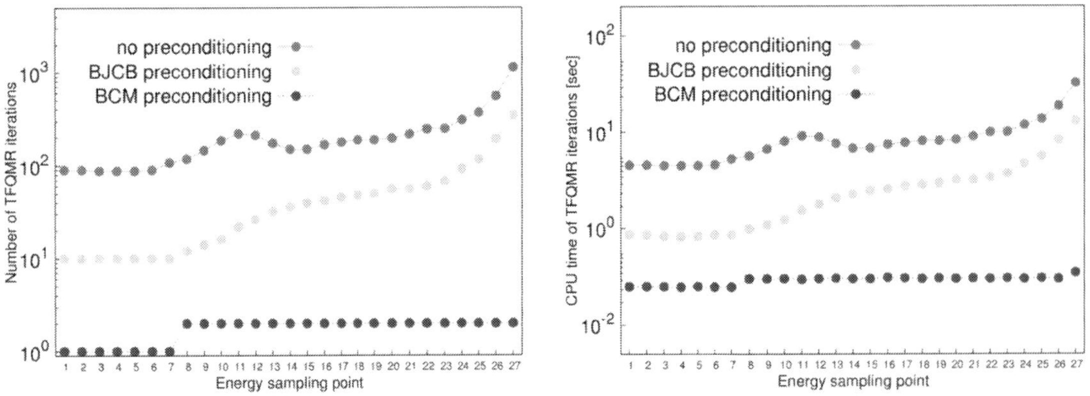

Figure 7. The numbers and CPU time of iteration for each sampling point in one SCF with/without using preconditioning. Numbers on the horizontal axis distinguish energy sampling points (see figure 6).

can be diagonalized by a Fourier matrix F_N defined as $(F_N)_{j,k} = \exp(2\pi i \cdot jk/N)$. Once the matrix is diagonalized as $C = F_N \text{diag}(\lambda_1, \ldots, \lambda_N) F_N^{-1}$, where λ_i $(i = 1, \ldots, N)$ are the eigenvalues of C, the inverse operation is easily done as $C^{-1} = F_N \text{diag}(\lambda_1^{-1}, \ldots, \lambda_N^{-1}) F_N^{-1}$. Also, a d-level circulant matrix (d-level means that the circulant matrix pattern has d-fold multiperiod, and hence C appearing in (15) is a 1-level circulant matrix) can be diagonalized by a d-dimensional Fourier-transform. For the super-cell system consisting of a repeated unit cell structure without substitutional atoms, the sparse matrix pattern becomes just a 3-level block-circulant. Then, the Dyson equation for the super-cell of a pure system keeping the ideal structure of the unit cell without any substitutional impurities can be solved by using the fast-Fourier-transform (FFT) if the repetition of the primitive unit cell along x, y, and z directions in the unit super-cell are powers of two. If substitutional impurities or lattice distortions appear, the periodicity will be lost and the matrix pattern is no longer circulant. However, even in these cases, BCM can be constructed through the averaging over the corresponding blocks (figure 8). Once BCM is constructed, diagonalization is easy, and multiplying the inverse of BCM to an arbitrary column vector can be done within ~ 0.1 sec in the present calculation of setting. Therefore, BCM can be used as right preconditioning matrix M_2 $(M_1 = I)$. Using $M_2 \neq I$ requires a $M_2^{-1} \times$ (vector) operation for each iteration, but from the view point of the parallelization and memory usage this procedure is favorable with respect to the case of $M_1 \neq I$ if the preconditioning matrix is not simple like BJCB. In contrast to the left preconditioning which requires memory resource to store A' itself, as long as the right preconditioning is used, it is sufficient to store only the result of $M_2^{-1} \times$ (vector) operations.

Results using BCM preconditioning are shown in figure 7 with blue line and points. As can be seen from this figure, linear equations are already virtually solved in the first step of the iteration for simple systems where a few impurities exist in a large super-cell such as the case of this test calculation. The needed number of iterations and CPU time are drastically reduced over all the sampling points. Even at the first Matsubara frequency, the TFQMR iteration converges after a few iterations, and the dependence of numerical effort on the position of sampling points is entirely smoothened. Now, the CPU time per single SCF cycle is reduced to ~ 13 sec, which is surprisingly fast for an all-electron level calculation.

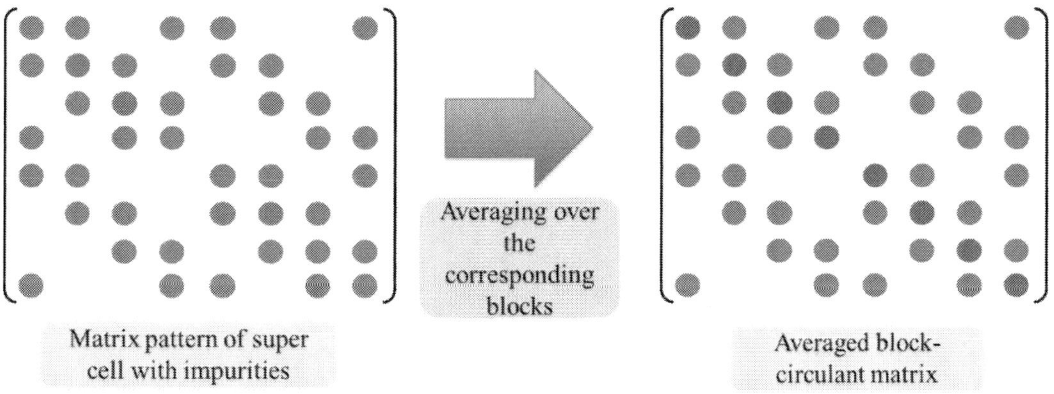

Figure 8. Construction of BCM by averaging over corresponding blocks. If atoms in pure super-cell are substituted for impurities, diagonal blocks are substituted for that of impurities (shown as blue circle in the figure) due to the difference of t-matrix. Averaging over diagonal blocks (shown as purple circle), BCM can be constructed again.

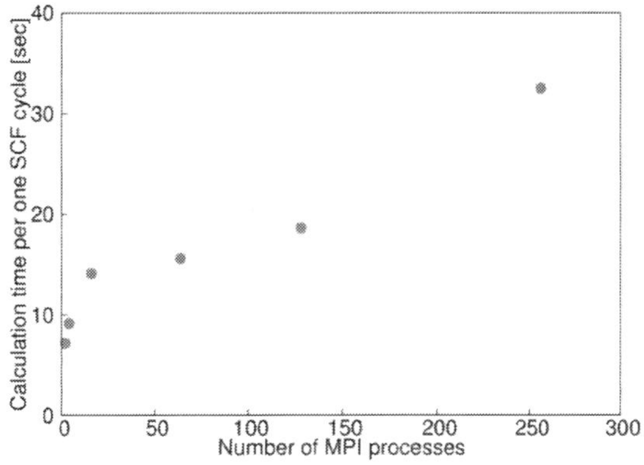

Figure 9. Calculation time per one SCF cycle as a function of the number of MPI processes.

4.4. Parallel efficiency

The parallel efficiency of our code is explained in this subsection. Figure 9 shows the calculation time per one SCF cycle as a function of the number of MPI processes. The calculated super-cell is the same as in former test calculations. In contrast to the former test calculations, the electronic structure of an impurity atom as well as other atoms is calculated using many MPI processes. The calculation time is estimated on the calculation server which consists of Intel Xeon E5-2690 2.90GHz nodes. Note that the maximum number of MPI processes is limited to 256 because of a limitation of numerical resources which we can use (2048 MPI processes are required to determine the full electronic structure of super-cell with 2048 atoms). From this figure, one can confirm that the calculation time increases with the number of MPI processes. Ideally, the dependence of the calculation time with respect to the number of MPI process should be constant. This deviation from the ideal situation is due to the load of MPI communications which are required to construct sparse coefficient matrices from each piece of information of each MPI process. But the calculation time is still tens of seconds with 256 MPI processes, which makes us expect that the calculation time per one SCF cycle for the full determination of electronic structure of the whole super-cell is several tens of seconds with thousands of MPI processes.

5. Summary

We have developed a code for electronic structure calculations of large super-cell systems. It uses the screened KKR and TFQMR algorithms combined with FP-KKR and is MPI parallelized with respect to the number of atoms. The efficiency of TFQMR combined with the BJCB and BCM preconditioning for super-cell systems is confirmed by several test calculations. The calculations shown here is rather limited for a single atom in a unit super-cell with a single MPI process. Due to the perfect MPI parallelization with respect to the number of atoms, the full determination of the electronic structure of the whole super-cell containing thousands of atoms is expected to be possible in an all-electron level within a reasonable CPU time comparable with that of the calculation that uses a single MPI process as is shown in the test calculations. The BCM preconditioning is efficient for super-cells with substitutional impurities, if ideal geometry of the unit structure or only with a slight distortion from ideal position is assumed. The TFQMR algorithm with a suitable preconditioning seems to surprisingly broaden the range of the application of the KKR Green's function method.

Acknowledgments

This research was supported by a grant for Osaka University Global COE Program "Core Research and Engineering of Advanced Materials-Interdisciplinary Education Center for Materials Science". Theoretical calculations were partly performed using Research Center for Computational Science, Okazaki, Japan.

References

[1] Hohenberg P and Kohn W 1964 *Phys. Rev.* **136** B864
[2] Kohn W and Sham L J 1965 *Phys. Rev.* **140** A1133-A1138
[3] Zeller R, Dederichs P H,Újfalussy B, Szunyogh L and Weinberger P 1995 *Phys. Rev.* B **52** 8807
[4] Wildberger K, Zeller R and Dederichs P H 1997 *Phys. Rev.* B **55** 10074
[5] Zeller R 2008 *J. Phys: Condens. Matter* **20** 294215
[6] Thieß A 2011 *Dissertation* Rheinisch-Westfälische Technische Hochschule Aachen
[7] Thieß A, Zeller R, Bolten M, Dederichs P H and Blügel S 2012 *Phys. Rev.* B **85** 235103
[8] Ogura M and Akai H 2005 *J. Phys: Condens. Matter* **17** 5741-5755
[9] Akai H 2000 *Korringa-Kohn-Rostoker Method* (`http://kkr.phys.sci.osaka-u.ac.jp/pdf/kkrnote.pdf`)
[10] Papanikolaou N, Zeller R and Dederichs P H 2002 *J. Phys: Condens. Matter* **14** 2799-2823
[11] Freund R W 1993 *SIAM J. Sci. Comput.* **14** 470
[12] Smirnov A V and Johnson D D 2002 *Comput. Phys. Commun.* **148** 74
[13] Freund R W, Nachtigal N M and Reeb J C 1994 *QMRPACK User's Guide, ORNL Technical Report* ORNL/TM-12807; and (`http:/www.netlib.org/linalg/qmr`)
[14] Wildberger K, Lang P,Zeller R and Dederichs P H 1995 *Phys. Rev.* B **52** 11502
[15] Bolten M, Thiess A, Yavneh I and Zeller 2012 *Linear Algebra Appl.* **436** 436

Journal of Physics: Conference Series **454** (2013) 012020 doi:10.1088/1742-6596/454/1/012020

Optimization of the Jastrow factor in the correlated wave function of electrons using the first-principles transcorrelated method for solid-state calculations

Masayuki Ochi[1] **and Shinji Tsuneyuki**[1,2]

[1] Department of Physics, The University of Tokyo, Bunkyo-ku, Tokyo 113-0033, Japan
[2] Institute for Solid State Physics, Kashiwa, Chiba 277-8581, Japan

E-mail: ochi@cms.phys.s.u-tokyo.ac.jp

Abstract. The transcorrelated (TC) method is one of the wave-function-based methods used for first-principles electronic structure calculations, and in terms of the computational cost is applicable to solid-state calculation. In this method, a many-body wave function of electrons is approximated as a product of the Jastrow factor and the Slater determinant, and the first-principles Hamiltonian is similarity-transformed by the Jastrow factor. The Schrödinger equation is rewritten as an eigenvalue problem for this similarity-transformed Hamiltonian, from which one obtains a self-consistent field (SCF) equation for optimizing one-electron orbitals in the Slater determinant at low computational cost. In contrast, optimization of the Jastrow factor is computationally much more expensive and has not been performed for solid-state calculation of the TC method before. In this study, we develop a new method for optimizing the Jastrow factor at a reasonable computational cost using the random-phase approximation (RPA) and pseudo-variance minimization. We apply this method to some simple solids, and find that the band gap of a wide-band-gap insulator is much improved by RPA.

1. Introduction

First-principles electronic structure calculation has played an active and important role in condensed matter physics. In particular, density functional theory (DFT) [1, 2] has produced prominent successes because simple approximations for the exchange-correlation functional of DFT, such as local density approximation (LDA) or generalized gradient approximation (GGA), provide satisfactory results for many systems at relatively low computational cost. However, it is well known that such simple approximations have several problems in accuracy, e.g., the band gap is much underestimated, and van der Waals interaction is not correctly described. In addition, improving accuracy is difficult to achieve systematically for DFT-based methods because, in DFT, we have to handle the exchange-correlation functional, the exact form of which is quite nontrivial. On the other hand, wave-function theory, in which one explicitly handles the many-body wave function, has a systematic way of improving accuracy, hence it is expected to be a good candidate for high-accuracy calculation. Unfortunately, however, the Hartree-Fock (HF) method, which is the starting point for most of wave-function theories, is known to be highly inaccurate for solid-state calculations because it neglects all electron correlation effects such as the screening effect in solids, whereas other sophisticated wave-function theories are often too expensive for solid-state calculations.

The transcorrelated (TC) method [3, 4, 5, 6, 7, 8] is one of the promising wave-function theories, which is applicable to solid-state calculation in terms of the computational cost [9, 10]. In the TC method, a many-body wave function is approximated as a product of the Jastrow factor and the Slater determinant, and the first-principles Hamiltonian is similarity-transformed by the Jastrow factor. Using this similarity-transformed Hamiltonian, a self-consistent field (SCF) equation is derived just like the HF method, and then one-electron orbitals in the Slater determinant can be optimized by solving the SCF equation with reasonable computational cost. The TC method has been applied to various systems, such as atoms or molecules [3, 4, 8, 11, 12], the homogeneous electron gas [13, 14, 15, 16], solids [9, 10], and the Hubbard model [17]. Because the TC method takes into account the electron correlation effect through the Jastrow factor, its accuracy is much improved over the HF method: for example, calculated band gaps show good agreement with the experimental values for some semiconductors and insulators [9, 10]. However, for solid-state calculations, optimization of the Jastrow factor is very computationally intensive, so instead we use a parameter-free Jastrow function determined from an analysis based on the random phase approximation (RPA) for the homogeneous electron gas [18] and the cusp condition [19]. Although this simple Jastrow function yields accurate results as mentioned above, this function is unsatisfactory for an accurate description of solids: e.g., the screening effect for large-gap insulators should not be correctly described by using the analysis of the homogeneous electron gas.

For optimization of the Jastrow factor, the most popular method is the variational Monte Carlo (VMC) [20], which has been widely applied to solid-state calculation and known to provide accurate results. However, VMC is computationally intensive, needing to perform $3N$-dimensional integrations for calculations of the total energy for a system of N electrons. A few researchers have proposed alternative optimization methods and applied these to atomic or molecular systems [8, 12], but for solid-state calculations these are also computationally expensive.

In this study, we propose a new method for optimizing the Jastrow factor aiming for a low-cost computation. The long-range behavior of the Jastrow factor is determined using the static dielectric constant calculated using the RPA; the rest of the Jastrow factor is optimized by minimizing the pseudo-variance, which is defined for the similarity-transformed Hamiltonian of the TC method and can be calculated at reasonable computational cost. In section 2, we introduce a basic idea of the TC method. Our new method for optimizing the Jastrow factor is described in section 3. Some results from applications of our method to some simple solids are presented in section 4, and the final section is devoted to the conclusion of this study.

2. Theory: the TC method

In the TC method, we rewrite a many-body wave function Ψ_0 in the form $\Psi_0 = F\Phi_0$, where F is the Jastrow factor, a symmetric product of two-body positive function, with Φ_0 a many-body function formally defined as $\Phi_0 = \Psi_0/F$. Then the Schrödinger equation,

$$\mathcal{H}F\Phi_0 = E_0 F\Phi_0, \tag{1}$$

is equivalent to a similarity-transformed eigenvalue equation,

$$\mathcal{H}_{TC}\Phi_0 = E_0\Phi_0 \quad (\mathcal{H}_{TC} = F^{-1}\mathcal{H}F). \tag{2}$$

For the first-principles Hamiltonian of many-electron systems,

$$\mathcal{H} = \sum_{i=1}^{N}\left(-\frac{1}{2}\nabla_i^2 + v_{\text{ext}}(x_i)\right) + \frac{1}{2}\sum_{i=1}^{N}\sum_{j=1(\neq i)}^{N}\frac{1}{|\mathbf{r_i}-\mathbf{r_j}|}, \tag{3}$$

the TC Hamiltonian $\mathcal{H}_{TC}$ can be expressed in the form,

$$\mathcal{H}_{TC} = \mathcal{H} + \frac{1}{2}\sum_{i=1}^{N}\sum_{j=1(\neq i)}^{N} (\nabla_i^2 u(x_i, x_j) - (\nabla_i u(x_i, x_j))^2 + 2\nabla_i u(x_i, x_j) \cdot \nabla_i)$$
$$-\frac{1}{2}\sum_{i=1}^{N}\sum_{j=1(\neq i)}^{N}\sum_{k=1(\neq i,j)}^{N} \nabla_i u(x_i, x_j) \cdot \nabla_i u(x_i, x_k), \quad (4)$$

where the Jastrow factor used in our calculation is

$$F = \exp\left(-\frac{1}{2}\sum_{i=1}^{N}\sum_{j=1(\neq i)}^{N} u(x_i, x_j)\right), \quad (5)$$

with the usual notation that x_i represents position $\mathbf{r_i}$ and spin σ_i of the i-th electron. Without loss of generality, we can assume the Jastrow function $u(x_i, x_j)$ to be symmetric.

Because the electron correlation effect is partially taken into account through F, it is expected that we can obtain a relatively accurate result with a simple approximation for Φ_0. In this paper, we assume that Φ_0 is a single Slater determinant:

$$\Phi_0 = \frac{1}{\sqrt{N!}}\det[\phi_{1,\cdots,N}(x_{1,\cdots,N})]$$
$$\equiv \frac{1}{\sqrt{N!}}\det\begin{pmatrix} \phi_1(x_1) & \phi_1(x_2) & \cdots & \phi_1(x_N) \\ \phi_2(x_1) & \ddots & \ddots & \vdots \\ \vdots & \ddots & \ddots & \vdots \\ \phi_N(x_1) & \cdots & \cdots & \phi_N(x_N) \end{pmatrix}. \quad (6)$$

This approximation is equivalent to the so-called Slater-Jastrow ansatz.

We can derive an SCF equation like the HF method for the similarity-transformed Hamiltonian,

$$\left(-\frac{1}{2}\nabla_1^2 + v_{\text{ext}}(x_1)\right)\phi_i(x_1) + \sum_{j=1}^{N}\int dx_2\, \phi_j^*(x_2)v_{\text{2body}}(x_1, x_2)\det[\phi_{i,j}(x_{1,2})]$$
$$-\frac{1}{2}\sum_{j=1}^{N}\sum_{k=1}^{N}\int dx_2 dx_3\, \phi_j^*(x_2)\phi_k^*(x_3)v_{\text{3body}}(x_1, x_2, x_3)\det[\phi_{i,j,k}(x_{1,2,3})] = \sum_{j=1}^{N}\epsilon_{ij}\phi_j(x_1), \quad (7)$$

where one-electron orbitals $\phi(x)$ are orthonormalized, v_{2body} and v_{3body} are defined as,

$$v_{\text{2body}}(x_1, x_2) \equiv \frac{1}{|\mathbf{r_1} - \mathbf{r_2}|} + \frac{1}{2}(\nabla_1^2 u(x_1, x_2) + \nabla_2^2 u(x_1, x_2)) - (\nabla_1 u(x_1, x_2))^2 - (\nabla_2 u(x_1, x_2))^2)$$
$$+\nabla_1 u(x_1, x_2) \cdot \nabla_1 + \nabla_2 u(x_1, x_2) \cdot \nabla_2, \quad (8)$$

and

$$v_{\text{3body}}(x_1, x_2, x_3) \equiv \nabla_1 u(x_1, x_2) \cdot \nabla_1 u(x_1, x_3) + \nabla_2 u(x_2, x_1) \cdot \nabla_2 u(x_2, x_3)$$
$$+\nabla_3 u(x_3, x_1) \cdot \nabla_3 u(x_3, x_2). \quad (9)$$

We can then optimize the one-electron orbitals of the Slater determinant and obtain the band structure while using an explicitly correlated many-body wave function. This is one advantage of the TC method. The total energy of the ground state is evaluated as $E_0 = \text{Re}[\langle\Phi_0|\mathcal{H}_{TC}|\Phi_0\rangle]$. In the next section, we introduce our new method for optimizing the Jastrow factor.

3. Optimization of the Jastrow factor

Because optimizing the Jastrow factor is computationally intensive, we have used a very simple Jastrow function formerly [9, 10],

$$u(x,x') = \frac{A_0}{|\mathbf{r}-\mathbf{r}'|}\left(1-\exp\left(-|\mathbf{r}-\mathbf{r}'|/C_{0;\sigma,\sigma'}\right)\right), \tag{10}$$

where $A_0 = (4\pi N/V)^{-1/2}$ (N: the number of electrons in the simulation cell, V: the volume of the simulation cell) and $C_{0;\sigma,\sigma'} = \sqrt{2A_0}$ (para : $\sigma = \sigma'$), $\sqrt{A_0}$ (anti : $\sigma \neq \sigma'$). The former condition for A_0 is derived by the RPA analysis of the uniform electron gas [18], and the latter condition for C_0 by the cusp condition [19]. Thus we can take the screening effect in solids into account to some extent and have a many-body wave function satisfying the cusp condition even though a simple Jastrow function is used. However, we can see some unfavorable features. First, we only use the conditions for $|\mathbf{r}-\mathbf{r}'| \to \infty$ (RPA) and $|\mathbf{r}-\mathbf{r}'| \to 0$ (cusp condition), and say nothing of and have no degrees of freedom for the intermediate region. Second, RPA is applied not to a target system but to the uniform electron gas with the same valence-electron-density on average as the target system. This causes too much screening of the electron-electron interaction, especially for large-gap insulators.

To resolve these problems, we added additional terms to the Jastrow factor, and optimized parameters therein:

$$\begin{aligned} u(x,x') &= \frac{A}{|\mathbf{r}-\mathbf{r}'|}\left(1-\exp\left(-|\mathbf{r}-\mathbf{r}'|/C_{\sigma,\sigma'}\right)\right) \\ &+ \left(\sum_{m=0}^{M-1} c_{m;\sigma,\sigma'}|\mathbf{r}-\mathbf{r}'|^m\right)\left(\frac{|\mathbf{r}-\mathbf{r}'|}{L}-1\right)^3\Theta\left(\frac{|\mathbf{r}-\mathbf{r}'|}{L}-1\right), \end{aligned} \tag{11}$$

where $\Theta(x)$ is the Heaviside step function defined as $\Theta(x) = 0$ $(x < 0)$, 1 $(x > 0)$. In this paper, we call the first term of (11), $(A/|\mathbf{r}-\mathbf{r}'|)(1-\exp(-|\mathbf{r}-\mathbf{r}'|/C_{\sigma,\sigma'}))$, the long-range term because it describes the asymptotic behavior at infinity ($|\mathbf{r}-\mathbf{r}'| \to \infty$); the second term we call the short-range polynomials as these involve a cutoff length, L. No cutoff length is required for the long-range term because a special treatment exists for $1/r$-type long-range functions, which was originally developed for the HF method by Gygi and Baldereschi [21]. The short-range polynomials have the same form as those in [22]. In this reference, the performance of the damping factor $(|\mathbf{r}-\mathbf{r}'|/L-1)^P$ before the Heaviside step function was checked for $P = 2$ and 3, and it was found that the results for $P = 3$ are more accurate than those for $P = 2$; we adopt $P = 3$ here.

We proceed by determining the value of A using the dielectric static constant ε such that $A = A_0\sqrt{1-(1/\varepsilon)}$, which reproduces the macroscopic screening of the electron-electron interaction in insulators: $1/r \to 1/(\varepsilon r)$. This is based on the fact that the screening effect is caused by the three-body terms $\nabla u \cdot \nabla u$ [10] and its strength is proportional to A^2 in the long-range region. The dielectric static constant is calculated by an RPA relation [23] using LDA orbitals ϕ^{LDA} and LDA orbital energies ϵ^{LDA},

$$\varepsilon = 1 + \frac{8\pi}{V}\lim_{\mathbf{q}\to 0}\frac{1}{|\mathbf{q}|^2}\sum_{\mathbf{k},\sigma}\sum_{i}^{\text{vale.}}\sum_{a}^{\text{cond.}}\frac{|\langle\phi^{\mathrm{LDA}}_{\mathbf{k}+\mathbf{q},\sigma,a}|e^{i\mathbf{q}\cdot\mathbf{r}}|\phi^{\mathrm{LDA}}_{\mathbf{k},\sigma,i}\rangle|^2}{\epsilon^{\mathrm{LDA}}_{\mathbf{k}+\mathbf{q},\sigma,a}-\epsilon^{\mathrm{LDA}}_{\mathbf{k},\sigma,i}}, \tag{12}$$

where $\mathbf{k}$ and $\mathbf{q}$ are wave-vectors in the k-space, and i and a are indices of the valence and conduction bands respectively. RPA was also used to optimize the Jastrow factor in the previous study by other researchers [24], but they adopted a more complex form of the Jastrow function and used RPA in a different manner from ours. Here, we use this simple way to achieve a low-cost computation.

Next, we determine the values of $C_{\sigma,\sigma'}$ from $C_{\sigma,\sigma'} = \sqrt{2A}$ (para : $\sigma = \sigma'$), $\sqrt{A}$ (anti : $\sigma \neq \sigma'$), to satisfy the cusp condition. The value of L is determined from $(4/3)\pi L^3 = V$, where we use a large value for L to obtain an adequate optimization within the size of the simulation cell for which the periodic boundary condition is imposed.

Finally, with the cusp condition being imposed as a constraint, we optimize the values of $c_{m;\sigma,\sigma'}$ by the following method. For this purpose, we cannot use minimization of the expectation value, $\langle\Phi_0|\mathcal{H}_{TC}|\Phi_0\rangle$, of $\mathcal{H}_{TC}$, because $\mathcal{H}_{TC}$ is non-Hermitian and the variational principle does not hold. On the other hand, minimization of the variance, $\sigma^2_{TC} = \langle\Phi_0|(\mathcal{H}^\dagger_{TC} - E_0)(\mathcal{H}_{TC} - E_0)|\Phi_0\rangle$, is still applicable to this non-Hermitian eigenvalue problem and was used in [8] for atomic and molecular systems. However, it involves six-body terms, and is not easy to apply to solid-state calculations. Instead, we define a new variable, called the pseudo-variance here, for arbitrary many-body ($3N$-dimensional) functions Φ_p:

$$\sigma^2_{PS} = \sum_p^{\text{finite}} |\langle\Phi_p|(\mathcal{H}_{TC} - E_0)|\Phi_0\rangle|^2, \tag{13}$$

and we use its minimization as the guiding principle for optimizing the parameters $c_{m;\sigma,\sigma'}$ in the Jastrow function. Pseudo-variance has several favorable features: (i) it is a non-negative real number, (ii) it is easy to calculate because it does not involve $3N$-dimensional integrations nor evaluations of six-body terms such as the variance of $\mathcal{H}_{TC}$, (iii) it equals zero if $\mathcal{H}_{TC}\Phi_0 = E_0\Phi_0$ exactly holds, and (iv) if the number of Φ_p is infinite and the completeness relation, $\sum_p |\Phi_p\rangle\langle\Phi_p| = \text{id.}$, holds, it equals the variance of $\mathcal{H}_{TC}$. Additionally, in (13), we can see that the length of the vector $(\mathcal{H}_{TC} - E_0)\Phi_0$, which is zero for the exact ground state, is measured with projection to each direction Φ_p, so it is expected that the pseudo-variance minimization does work for optimizing the Jastrow function even if the number of Φ_p is small. This is a similar idea to [12] or the variance minimization of VMC calculation [25].

Because $c_{m;\sigma,\sigma'}$ are linear parameters in the Jastrow function, the pseudo-variance is written as a polynomial of $c_{m;\sigma,\sigma'}$. Optimization of these parameters is carried out with low computational cost because the coefficients of such a polynomial are unchanged during the optimization process, and so the calculation is required only once. In principle, also the 'A' parameter in the Jastrow function can be optimized with pseudo-variance minimization, but we do not take this measure, aiming to perform an efficient, i.e., low-cost computation.

We should mention the choice of Φ_p. In this study, we used doubly excited configurations $\Phi^{a,b}_{i,j} \equiv (1/\sqrt{N!})\det[\phi_{1,2,\ldots,\hat{i},\ldots,\hat{j},\ldots,N-1,N,a,b}(x_{1,\cdots,N})]$, where electrons in the i-th and j-th occupied states of the ground state are excited to a-th and b-th unoccupied states. It is noteworthy that $\langle\Phi_p|\mathcal{H}_{TC}|\Phi_0\rangle$ equals zero for a singly excited configuration because of the Brillouin's theorem, proved in the Appendix, and also zero for an excited configuration involving four or more excited electrons because the similarity-transformed Hamiltonian consists of up to three-body interactions. The contribution from the triply excited configuration is non-zero but its calculation is too intensive, hence not considered here.

In fact, a very similar method to pseudo-variance minimization was already proposed and applied to a neon atom by Boys and Handy, who originally proposed the TC method [4, 5]. They used $\sum_p^{\text{finite}} |\langle\tilde{\Phi}_p|\mathcal{H}_{TC} - E_0|\Phi_0\rangle|^2/\Delta E$ with $\Delta E = \langle\tilde{\Phi}_p|\mathcal{H}_{TC}|\tilde{\Phi}_p\rangle - \langle\Phi_0|\mathcal{H}_{TC}|\Phi_0\rangle$, instead of the pseudo-variance, where $\tilde{\Phi}_p$ are determinants with one or two orbitals in the Slater determinant of the ground state being replaced with some basis functions. $(\Delta E)^{-1}$ is called a weighting factor, which suppresses contribution from $\tilde{\Phi}_p$ of high energy. This weighting factor was introduced to realize feasible computational cost at that time, but there is no physical reason to choose this form of the weighting factor. Different choices may produce different results. Moreover, minimization of this guiding variable can in principle yield unstable behavior, because it is not necessarily positive if we change the values of the Jastrow parameters with Φ_0 and $\tilde{\Phi}_p$ being fixed. Therefore, for our study, we used the pseudo-variance, which does not introduce a weighting factor.

4. Results

4.1. Procedure and some conditions of our calculations

Our calculation was carried out as follows: (i) perform an LDA calculation, (ii) calculate the static dielectric constant with the RPA relation (12), (iii) determine the values of A and $C_{\sigma,\sigma'}$ in the Jastrow function using the static dielectric constant, (iv) solve the SCF equation (7) to optimize one-electron orbitals of the TC method with the Jastrow function without the short-ranged polynomial terms, (v) use the Jastrow function of (11) hereafter, and optimize $c_{m;\sigma,\sigma'}$ by pseudo-variance optimization, (vi) obtain physical quantities by solving the SCF equation (7), and (vii) repeat steps (v) and (vi) until convergence is achieved. Initial guesses for the values of $c_{m;\sigma,\sigma'}$ in step (v) are all zero. Step (vii) was always unnecessary for the calculations leading to results presented in this paper because our results, such as the total energy and the band gap, changed very little.

The LDA [26] calculations were done with TAPP code [27, 28] using a plane-wave basis, and the HF and TC calculations with TC++ code [9, 10]. For this paper, one-electron orbitals in the HF and TC method are expanded as linear combinations of LDA orbitals. Convergence was achieved with respect to the number of LDA orbitals. Non-local norm-conserving pseudopotentials generated by the method developed by Troullier and Martins [29] were used for the LDA, HF, and TC calculations. For the HF and TC calculations, singularities in the electron-electron Coulomb interaction and the Jastrow function in the k-space were handled with a method proposed by Gygi and Baldereschi [21], where we used an auxiliary function of the same form as that proposed by Massidda *et al.* [30]. The Message Passing Interface (MPI) system was used for parallelization of the calculations.

4.2. Results: determination of the long-range parameter using the RPA

First, we use the Jastrow function without the short-ranged polynomials, with values for A and $C_{\sigma,\sigma'}$ determined by the RPA, as described above. In other words, we will show results for step (iv) of our procedure described in the previous subsection.

The results for the band-gap calculations applied to bulk silicon and lithium fluoride are listed in the Table 4.2. The LDA calculations needed to evaluate the static dielectric constant were performed using a $16 \times 16 \times 16$ to a $32 \times 32 \times 32$ k-point mesh, and the following TC calculations were performed using a $4 \times 4 \times 4$ k-point mesh. The cutoff energy for the plane waves for silicon and lithium fluoride were 36 and 144 Ry, respectively. While the band gap obtained for silicon did not change very much, that for lithium fluoride was much improved by RPA. This is reasonable because the discrepancy between the electronic structure of solids and that of the homogeneous electron gas is much larger for a large-gap insulator like lithium fluoride than for silicon.

4.3. Results: pseudo-variance minimization for the short-range parameters

Next, we will present some results for the pseudo-variance minimization using a Jastrow function with the short-range polynomials, (11), i.e., results obtained by performing the whole procedure of our calculations described in section 4.1.

The results for bulk silicon are listed in Table 4.3. We used four valence bands (i.e., all valence bands) and 48 conduction bands for doubly excited configurations Φ_p for calculation of the pseudo-variance, (13). It should be noted that we used a $4 \times 4 \times 4$ k-point mesh, but

Table 1. Band gap [eV] for bulk silicon and lithium fluoride calculated with several methods. "TC (previous)" means TC calculations using the previous simple Jastrow function, (10), and "TC (RPA)" does that using the same form of the Jastrow function, but the values of A and $C_{\sigma,\sigma'}$ are determined with the RPA.

	Si	LiF
LDA	0.5[a]	9.9
TC (previous)	1.7[a]	10.5
TC (RPA)	2.0	14.0
HF	6.7[a]	21.7
Experiment	1.2[b]	14.2[c]

[a] See [10]
[b] See [31]
[c] See [32]

electron excitations were restricted on a $2 \times 2 \times 2$ k-point mesh; a similar idea is used in [33] applied to MP2 calculations. This restriction reduces the computational cost by a factor of several dozen. The cutoff energy for FFT was 36 Ry. With M denoting the number of the short-range polynomial terms, $M = 0$ corresponds to the Jastrow function without any short-range polynomials. Because the cusp condition makes $c_{0;\sigma,\sigma'}$ zero when $M = 1$, results for $M = 1$ is not included here; the Jastrow function for $M = 1$ is completely equivalent to that for $M = 0$. We can see that the pseudo-variance is actually decreased with increasing M within a reasonable computation time. However, the band gap and valence bandwidth change very little with short-range polynomials added. It is an open question whether the position-dependent Jastrow function like $\chi(\mathbf{r})u(|\mathbf{r}-\mathbf{r}'|)\chi(\mathbf{r}')$ or three-body function including the positions of nuclei might improve the band gap. Also the application of the pseudo-variance minimization to other materials is an important future problem to check whether the same trends as described in this section can be found for other materials.

Table 2. Results for pseudo-variance minimization applied to bulk silicon. M is the number of the short-range polynomial terms defined in (11). $M = 0$ corresponds to the Jastrow function without the short-range polynomials. Note that the degrees of freedom for $c_{m;\sigma,\sigma'}$ are $2(M-1)$ because of the spin degrees of freedom and the cusp condition imposed as a constraint.

M	0	2	3	4	Experiment
Band gap [eV]	2.0	2.1	2.1	2.1	1.2[a]
Valence bandwidth [eV]	15.1	15.1	15.1	15.2	12.5(6)[a]
Computation time* [hh:mm]	-	0:52	1:37	2:54	-
Pseudo-variance σ_{PS}^2 [(hartree)^2]	0.0254	0.0185	0.0184	0.0179	-

[a] See [10]
* Computation time was measured only for the pseudo-variance minimization; 64 MPI processes are used for parallelization on the K-computer, Kobe, Japan.

5. Conclusion

In this study, we developed a new method for optimizing the Jastrow factor aimed at achieving low-cost computation. In our method, we use the RPA and minimization of the pseudo-variance, the latter of which is defined for the similarity-transformed Hamiltonian and can be calculated at reasonable computational cost. We have applied this method to some simple solids, and found that the band gap of lithium fluoride, one of the large-gap insulators, is much improved using the RPA. In addition, we found that the optimization of the Jastrow factor by pseudo-variance minimization can be performed at low computational cost, although the short-range polynomials of the electron-electron distance did not improve the band gap of silicon.

Acknowledgments

This work was partly supported by a Grant-in-Aid for JSPS Fellows as well as a Grant-in-Aid project from MEXT, Scientific Research on Innovative Areas, "Materials Design through Computics: Complex Correlation and Non-equilibrium Dynamics", and the Computational Materials Science Initiative, Japan. A part of the computation was performed at the supercomputer center of the Institute for Solid State Physics, The University of Tokyo, Tokyo, Japan, and K-computer, Kobe, Japan.

Appendix: Brillouin's theorem for the TC method

We provide a proof of the Brillouin's theorem for the TC method,

$$\langle \Phi_i^a | \mathcal{H}_{TC} | \Phi_0 \rangle = 0, \tag{14}$$

for any singly excited configuration Φ_i^a defined as $\Phi_i^a \equiv (1/\sqrt{N!})\mathrm{det}[\phi_{1,2,\dots,\hat{i},\dots,N-1,N,a}(x_{1,\cdots,N})]$, where an electron of the i-th occupied state is excited to the a-th unoccupied state.

This relation is easily verified by explicitly calculating the left-hand side of (14) as follows:

$$\begin{aligned}
&\text{LHS of Eq. (14)} \\
&= \int \mathrm{d}x_1\ \phi_a^*(x_1)\Big(-\frac{1}{2}\nabla_1^2 + v_{\mathrm{ext}}(x_1)\Big)\phi_i(x_1) \\
&\quad + \sum_{j=1}^{N} \int \mathrm{d}x_1 \mathrm{d}x_2\ \phi_a^*(x_1)\phi_j^*(x_2) v_{\mathrm{2body}}(x_1, x_2)\mathrm{det}[\phi_{i,j}(x_{1,2})] \\
&\quad - \frac{1}{2}\sum_{j=1}^{N}\sum_{k=1}^{N} \int \mathrm{d}x_1 \mathrm{d}x_2 \mathrm{d}x_3\ \phi_a^*(x_1)\phi_j^*(x_2)\phi_k^*(x_3) v_{\mathrm{3body}}(x_1, x_2, x_3)\mathrm{det}[\phi_{i,j,k}(x_{1,2,3})] \\
&= \int \mathrm{d}x_1\ \phi_a^*(x_1)\sum_{j=1}^{N}\epsilon_{ij}\phi_j(x_1) \quad (\because \text{Eq. (7)}) \\
&= 0 \quad (\because \phi_a \text{ is orthogonal to every } \phi_j.)
\end{aligned} \tag{15}$$

It is noteworthy that $\langle \Phi_0 | \mathcal{H}_{TC} | \Phi_i^a \rangle \neq 0$ caused by the non-Hermiticity of the TC Hamiltonian. However, if we use a biorthogonal formulation of the TC method, we can obtain the Brillouin's theorem for both sides. [34]

References

[1] Hohenberg P and Kohn W 1964 *Phys. Rev.* **136** B864
[2] Kohn W and Sham L J 1965 *Phys. Rev.* **140** A1133
[3] Boys S F and Handy N C 1969 *Proc. R. Soc. London Ser. A* **309** 209
[4] Boys S F and Handy N C 1969 *Proc. R. Soc. London Ser. A* **310** 43

[5] Boys S F and Handy N C 1969 *Proc. R. Soc. London Ser. A* **310** 63
[6] Boys S F and Handy N C 1969 *Proc. R. Soc. London Ser. A* **311** 309
[7] Handy N C 1971 *Mol. Phys.* **21** 817
[8] Umezawa N and Tsuneyuki S 2003 *J. Chem. Phys.* **119** 10015
[9] Sakuma R and Tsuneyuki S 2006 *J. Phys. Soc. Jpn.* **119** 103705
[10] Ochi M, Sodeyama K, Sakuma R and Tsuneyuki S 2012 *J. Chem. Phys.* **136** 094108
[11] Ten-no S 2000 *Chem. Phys. Lett.* **330** 169
[12] Luo H, Hackbusch W and Flad H J 2010 *Mol. Phys.* **108** 425
[13] Dharma-wardana M W C and Grimaldi F 1976 *Phys. Rev. A* **13** 1702
[14] Amour E A G 1980 *J. Phys. C: Solid St. Phys.* **13** 343
[15] Umezawa N and Tsuneyuki S 2004 *Phys. Rev. B* **69** 165102
[16] Luo H 2012 *J. Chem. Phys.* **136** 224111
[17] Tsuneyuki S 2008 *Prog. Theor. Phys. Supplement* **176** 134
[18] Bohm D and Pines D 1953 *Phys. Rev.* **92** 609
[19] Kato T 1957 *Commun. Pure Appl. Math.* **10** 151
[20] Foulkes W M C, Mitas L, Needs R J and Rajagopal G 2001 *Rev. Mod. Phys.* **73** 33
[21] Gygi F and Baldereschi A 1986 *Phys. Rev. B* **34** 4405
[22] Drummond N D, Towler M D and Needs R J 2004 *Phys. Rev. B* **70** 235119
[23] Gajdoš M, Hummer K and Kresse G 2006 *Phys. Rev. B* **73** 045112
[24] Gaudoin R, Nekovee M, Foulkes W M C, Needs R J and Rajagopal G 2001 *Phys. Rev. B* **63** 115115
[25] Umrigar C J, Wilson K G and Wilkins J W 1988 *Phys. Rev. Lett.* **60** 1719
[26] Perdew J P and Zunger A 1981 *Phys. Rev. B* **23** 5048
[27] Yamauchi J, Tsukada M, Watanabe S and Sugino O 1996 *Phys. Rev. B* **54** 5586
[28] Sugino O and Oshiyama A 1992 *Phys. Rev. Lett.* **68** 1858
[29] Troullier N and Martins J L 1991 *Phys. Rev. B* **43** 1993
[30] Massidda S, Posternak M and Baldereschi A 1993 *Phys. Rev. B* **48** 5058
[31] CKittel 1986 *Introduction to Solid State Physics, 6th ed.* (Wiley, New York)
[32] Piacentini M, Lynch D W and Olson C G 1976 *Phys. Rev. B* **13** 5330
[33] Shimazaki T and Hirata S 2009 *Int. J. Quant. Chem.* **13** 2953
[34] Hino O, Tanimura Y and Ten-no S 2001 *J. Chem. Phys.* **115** 17

Journal of Physics: Conference Series **454** (2013) 012021 doi:10.1088/1742-6596/454/1/012021

First-principles band structure and FLEX approach to the pressure effect on T_c of the cuprate superconductors

Hirofumi Sakakibara[1], Katsuhiro Suzuki[1], Hidetomo Usui[1], Kazuhiko Kuroki[1], Ryotaro Arita[2,5], Douglas J Scalapino[3] and Hideo Aoki[4]

[1]Department of Engineering Science, The University of Electro-Communications, Chofu, Tokyo 182-8585, Japan
[2]Department of Applied Physics, The University of Tokyo, Hongo, Tokyo 113-8656, Japan
[3]Physics Department, University of California, Santa Barbara, California 93106-9530, USA
[4]Department of Physics, The University of Tokyo, Hongo, Tokyo 113-0033, Japan
[5]JST, PRESTO, Kawaguchi, Saitama 332-0012, Japan

E-mail: hiro_rebirth@vivace.e-one.uec.ac.jp

Abstract. High-temperature cuprate superconductors have been known to exhibit significant pressure effects. In order to fathom the origin of why and how T_c is affected by pressure, we have recently studied the pressure effects on T_c adopting a model that contains two copper d-orbitals derived from first-principles band calculations, where the d_{z^2} orbital is considered on top of the usually considered $d_{x^2-y^2}$ orbital. In that paper, we have identified two origins for the T_c enhancement under hydrostatic pressure: (i) while at ambient pressure the smaller the hybridization of other orbital components the higher the T_c, an application of pressure acts to reduce the multiorbital mixing on the Fermi surface, which we call the orbital distillation effect, and (ii) the increase of the band width with pressure also contributes to the enhancement. In the present paper, we further elaborate the two points. As for point (i), while the reduction of the apical oxygen height under pressure tends to increase the d_{z^2} mixture, hence to lower T_c, here we show that this effect is strongly reduced in *bi-layer* materials due to the pyramidal coordination of oxygen atoms. As for point (ii), we show that the enhancement of T_c due to the increase in the band width is caused by the effect that the many-body renormalization arising from the self-energy is reduced.

1. Introduction

Although many kinds of superconductors have been discovered, the superconducting transition temperature T_c of the cuprate superconductors still remains to be the highest, and the possibility of further enhancing T_c still attracts much attention. To achieve higher T_c, it is important to understand the key parameters that control T_c, and from this viewpoint, there have been studies that have focused on the correlation between T_c and the lattice parameters such as the in-plain bond length(l)[1, 2] or the apical oxygen height(h_O) [3, 4, 5, 6, 7, 8, 9, 10, 11] measured from the CuO_2 planes. Theoretically, we have studied the material dependence of T_c in [10, 11], and introduced a two-orbital model that takes into account the $d_{x^2-y^2}$ and d_{z^2} Wannier orbitals. In most of the theories of the cuprates, only the $d_{x^2-y^2}$ (and the hybridized oxygen p) orbital is considered, but actually it has been noticed from the early days that in $(La,Sr)_2CuO_4$, which has small h_O and relatively low T_c, there is a strong mixture of the d_{z^2} orbital component near the Fermi level[12, 13, 14]. In fact, nowadays there have been more studies that focus on the d_{z^2} orbital or the related apical oxygen [7, 8, 15, 16, 17, 18]. In [10, 11], we have shown that this mixture of the d_{z^2} orbital component near the Fermi level works destructively against $d-$wave superconductivity, and hence this is the main reason of the material dependence of T_c.

Studying the pressure effect on T_c is an *in situ* way of attacking the problem of the correlation between T_c and the lattice structure. It is well known that in most of the cuprates, application of pressure enhances T_c[19, 20]. On the other hand, it has been revealed that the pressure effect exhibits strong anisotropy[21, 22, 23]. In our recent theoretical study on the pressure effect in the cuprates, it has been revealed that in addition to the d_{z^2} effect, the roundness of the Fermi surface, which is controlled by Cu-4s-$d_{x^2-y^2}$ hybridization, and also the band width are important parameters that governs T_c under pressure[24].

In the present paper, after briefly reviewing the main results obtained in [24], we study more closely the effect of the band width by applying the fluctuation exchange approximation(FLEX)[25, 26, 27] to the two-orbital model. Secondly, we will explain the difference of the d_{z^2} orbital effect between the multi-layer and single-layer systems under hydrostatic pressure.

2. Calculation Method

2.1. Determination of the Crystal Structure Under Pressure

We first obtain the crystal structure of the single-layer cuprates La_2CuO_4 and Hg_2BaCuO_4 under ambient pressure. Namely, we calculate the total energy by first principles calculation[28] varying the lattice constants, and fit the result by the standard Burch-Marnaghan formula[29] to obtain the structure at the most stable point. From such calculation, we can obtain the crystal structure within one percent discrepancy from the lattice constants determined experimentally[30, 31]. To obtain the lattice structure under hydrostatic pressure, we optimize the Poisson's ratio and atomic position by first principles reducing the cell volume to 95% or 90% of the lattice structure under ambient pressure. Since the compressibility in the cuprates is known to be about ~ 0.01 GPa^{-1}[32], $V = 0.9V_0$ corresponds to a pressure of about 10 GPa.

2.2. Construction of the two-orbital model and FLEX approximation

Using the obtained crystal structure under ambient or hydrostatic pressure, we construct maximally localized Wannier orbitals[33, 34] to extract the hopping parameters of the $d_{x^2-y^2}$-d_{z^2} two-orbital model[10]. As for the electron-electron interactions, we consider the on-site intraorbital Coulomb repulsion U, interorbital repulsion U', the Hund's coupling J and the pair-hopping J'. Here we also keep the orbital SU(2) requirement, $U - U' = 2J$. We set $U = 3.0$ eV, $U' = 2.4$ eV and $J = J' = 0.3$ eV unless mentioned otherwise. Estimates of U for the cuprates is $7 - 10t$, $t \simeq 0.45$eV (namely, U is about $3 \simeq 4.5$ eV), and $J(J') \simeq 0.1U$, so the values adopted here are within the widely accepted range.

We apply FLEX to this model to obtain the Green's function renormalized by the many-body self-energy correction. In FLEX, we define the spin and charge susceptibilities as follows;

$$\hat{\chi}_s(q) = \frac{\hat{\chi}^0(q)}{1 - \hat{S}\hat{\chi}^0(q)}, \tag{1}$$

$$\hat{\chi}_c(q) = \frac{\hat{\chi}^0(q)}{1 + \hat{C}\hat{\chi}^0(q)}, \tag{2}$$

where $q \equiv (\vec{q}, i\omega_n)$, the irreducible susceptibility is

$$\chi^0_{l_1,l_2,l_3,l_4}(q) = \sum_q G_{l_1 l_3}(k+q) G_{l_4 l_2}(k) \tag{3}$$

with the interaction matrices

$$S_{l_1l_2,l_3l_4} = \begin{cases} U, & l_1 = l_2 = l_3 = l_4 \\ U', & l_1 = l_3 \neq l_2 = l_4 \\ J, & l_1 = l_2 \neq l_3 = l_4 \\ J', & l_1 = l_4 \neq l_2 = l_3, \end{cases} \tag{4}$$

$$C_{l_1l_2,l_3l_4} = \begin{cases} U & l_1 = l_2 = l_3 = l_4 \\ -U' + J & l_1 = l_3 \neq l_2 = l_4 \\ 2U' - J, & l_1 = l_2 \neq l_3 = l_4 \\ J' & l_1 = l_4 \neq l_2 = l_3, \end{cases} \tag{5}$$

here, l_1, l_2 are orbital indices. Considering the self-energy correction originating from these susceptibilities, we solve the Dyson equation in a self-consistent manner. Then the renormalized Green's function is substituted to the linearized Eliashberg equation for superconductivity,

$$\lambda \Delta_{ll'}(k) = -\frac{T}{N} \sum_q \sum_{l_1l_2l_3l_4} V_{ll_1l_2l'}(q) G_{l_1l_3}(k-q) \Delta_{l_3l_4}(k-q) G_{l_2l_4}(q-k). \tag{6}$$

The maximum eigenvalue λ in the above equation reaches unity at the superconducting transition temperature $T = T_c$, so that λ calculated at a fixed temperature can be used as a measure for T_c. Here, we calculate λ at $T = 0.01$ eV for La and $T = 0.03$ eV for Hg. The reason for this is because $T_c \sim 100$ K of Hg is about three times larger than for La with $T_c \sim 40$ K[35].

3. Results and Discussions: effect of band width

3.1. Orbital Distillation

In figure 3.1, λ is depicted as a function of the unit cell volume. λ increases in La and in Hg cuprates with hydrostatic pressure, and this agrees with the well-known experimental result that T_c goes up monotonically under pressure in the cuprates[19, 20]. To understand this result, we have introduced three important parameters, the level offset ΔE, the roundness of Fermi surface ($r_{x^2-y^2}$ defined below) and the band width W[24]. To decompose the pressure effect into the contribution from the variation of these parameters, we vary each parameter "by hand" from the original value at $V = V_0$ to the value at $V = 0.9V_0$ separately, and calculate the variation in λ. This result is also depicted in figure 3.1 as the length of arrows. ΔE is the on-site energy difference between $d_{x^2-y^2}$ and d_{z^2} Wannier orbitals in the two-orbital model, so this is a measure of the d_{z^2} orbital effect, which has been found to work destructively against d-wave superconductivity in our previous study[10]. The effect of ΔE is dominant in La compound because La originally has small ΔE thereby suppressing T_c, so its increase under pressure is effective for the T_c enhancement. Note that in La system, ΔE increases under pressure because the absolute magnitude of the crystal field increases in spite of the reduction of h_O/l. On the other hand, in the Hg compound the pressure effect through ΔE is negligible because ΔE is intrinsically large. Instead, the other two parameters are effective. $r_{x^2-y^2}$ is defined as $r_{x^2-y^2} \equiv (|t_2| + |t_3|)/|t_1|$, where t_i is the i-th neighbor hopping within the $d_{x^2-y^2}$ orbital. The roundness of Fermi surface is enhanced by the increase of this value, and t_2 and t_3 are mostly mediated by the Cu-4s orbital (which is effectively included in the two Wannier orbitals in the present model) with the path of $d_{x^2-y^2} \to 4s \to d_{x^2-y^2}$[4]. It is known that the roundness of the Fermi surface works against the spin-fluctuation-mediated superconductivity[36], so the reduction of the 4s effect enhances T_c. The 4s orbital effect is reduced by hydrostatic pressure because the energy level offset between the $d_{x^2-y^2}$ and the 4s orbital is enhanced when the oxygen ligands approach Cu. The effect of d_{z^2} and 4s orbitals put together, we can say that T_c increases when the main band has more pure $d_{x^2-y^2}$ component, namely, when the "orbital distillation" takes place.

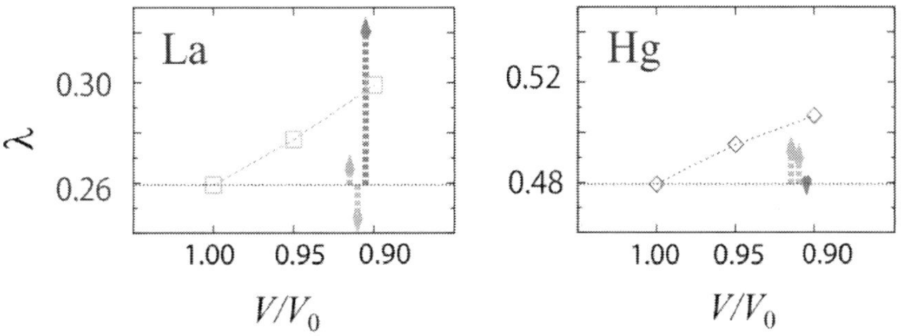

Figure 1. Eigenvalue λ of the Eliashberg equation plotted against the volume compression V/V_0. Arrows indicate the contribution to the λ variance from the parameters $r_{x^2-y^2}$(leftmost), W(center) and ΔE(rightmost)(see text).

3.2. The Effect of Band Width W

Now, we turn to the first main topic of this paper, i.e., the effect of the band width W. The band width W is defined as the energy range between the top and the bottom of the main band. It is evident that the band width is controlled by the in-plain bond length l, so that hydrostatic pressure enhances W. In figure 3.1, we can see that the increase in W results in an enhancement of λ in Hg, while the opposite occurs for La. The reason for this can be understood as follows. Let us first start with the U dependence of T_c for a fixed band width. In the top of figure 3.2, we show the absolute value of the renormalized Green's function squared $|G|^2$ in the Hg compound at $(\vec{k}, i\omega) = (\pi, 0, i\pi k_B T)$ for several values of $2 < U < 5$ eV. The value of $|G|^2$ monotonically decreases with larger U because the self-energy, which increases with U, suppresses $|G|^2$. On the other hand, the pairing interaction shown in the middle increases with U because the spin fluctuations develop monotonically. Consequently, $V|G|^2$ shown on the right exhibits a maximum at a certain U. Since $V|G|^2$ can be considered as a rough measure of the eigenvalue of the Eliashberg equation for $d-$wave superconductivity, λ (and thus T_c) is expected to be *maximized* around a certain U_{optim}.

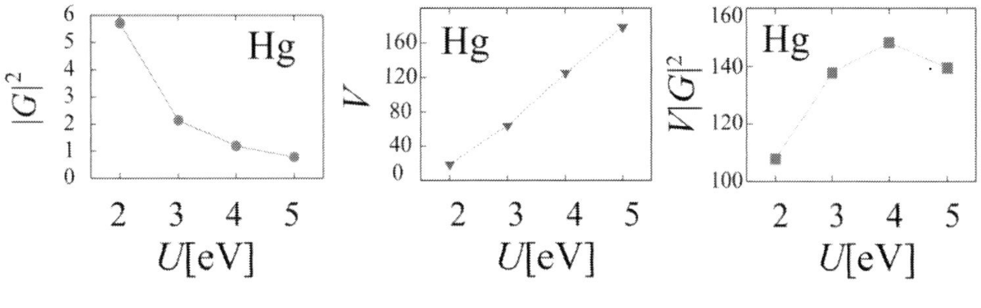

Figure 2. U dependence of the FLEX calculation result for the Hg compound. (left) the absolute value of the renormalized Green's function squared $|G|^2$ at $(\vec{k}, i\omega) = (\pi, 0, i\pi k_B T)$, (middle) the effective pairing interaction V at $(\vec{q}, i\omega) = (\pi, \pi, 0)$ and (right) the product $V|G|^2$.

In the left panels of figure 3.2, we show the U dependence of λ for $3 < U < 5$ eV for the two compounds with $V = V_0$ and $0.9V_0$. We can see that $U < 5$ eV lies on the left side of U_{optim} for La, while 3 eV$< U$ is on the right side of U_{optim} in Hg. This means that larger values of

U is necessary for La to be in the "strongly correlated regime". This is because electrons can avoid the strong intraorbital repulsion within $d_{x^2-y^2}$ orbitals by using the d_{z^2} orbital degrees of freedom in materials with small ΔE.

In the right panel of figure 3.2, we show the U dependence of the increment $\Delta\lambda$ of the eigenvalue λ induced by the increase of W under pressure. In this calculation, W is increased "by hand" up to its value at $V = 0.9V_0$, while the other two parameters are fixed at their original values. From this figure, we can see that the increase in W always enhances T_c in the Hg cuprate within the realistic U range. For the La cuprate on the other hand, $\Delta\lambda$ is negative for small values of U, and this is the reason why W affects superconductivity in opposite ways between La and Hg in figure 3.1. For larger values of U, however, $\Delta\lambda$ becomes positive even for La.

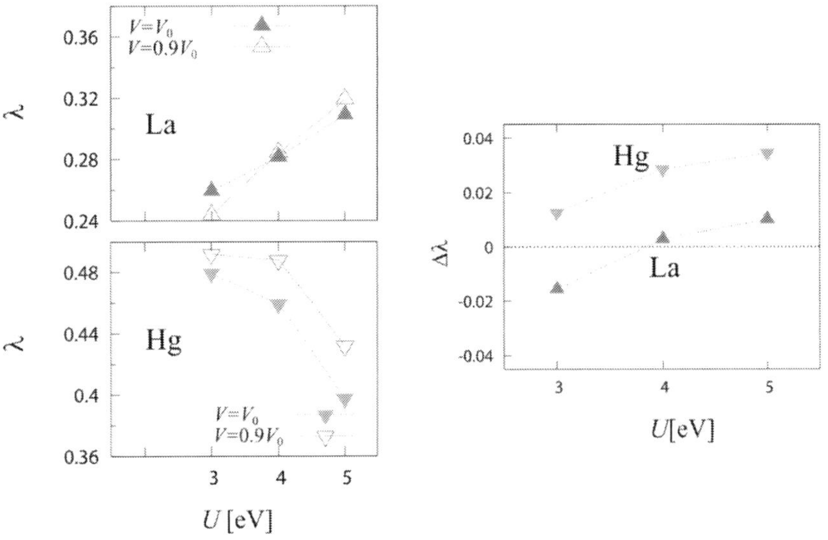

Figure 3. (Left) λ against U for La (top) and Hg (bottom). Filled (opened) symbols displays the result for $V = V_0 (V = 0.9V_0)$. (Right) $\Delta\lambda$ (increment in λ when W is increased up to its value at $V = 0.9V_0$) against U.

To see the effect of the band width more clearly, we provide in figure 3.2 a schematic band width dependence of the U vs T_c plot. Namely, T_c should depend on U and W essentially in the form $Wf(U/W)$, where f is a certain function that gives the overall dependence of T_c against the electron correlation strength. Therefore, as the pressure is applied, W increases so that $U_{\rm optim}$ (peak position of the curve) also increases accordingly keeping $U_{\rm optim}/W$ constant. At the same time, the absolute value of the maximized T_c is enhanced by the application of the pressure because the entire energy scale increases (i.e., both $U_{\rm optim}$ and W are enhanced).

4. The d_{z^2} Orbital Effect: Multi-layer vs Single-layer Systems

So far, we have concentrated on single-layer systems. In this section, we will discuss the difference of the pressure effect through ΔE between single and multi-layer systems. Here, we focus on the comparison between single- and bi-layer La, Hg, Tl, Bi, and Y cuprates. We consider ΔE_d, the energy difference between the $d_{x^2-y^2}$ and the d_{z^2} orbital in the d-p model which considers all of the Cu 3d and O 2p orbitals explicitly. In the d-p model, the basis functions for the hopping part are close to the atomic orbitals, so ΔE_d can be considered as the energy difference between the $d_{x^2-y^2}$ and the d_{z^2} atomic orbitals. ΔE is the energy difference between the $d_{x^2-y^2}$ and the

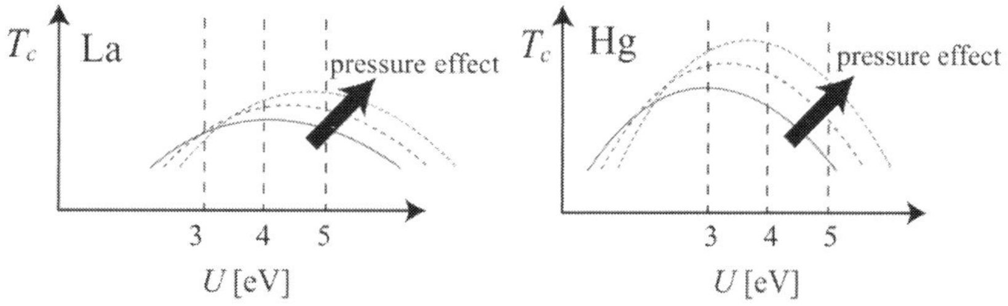

Figure 4. Schematic figure of T_c variation induced by applied pressure. Each line is a U vs. T_c plot for fixed W, and W is larger for solid green → dashed blue → dotted red.

d_{z^2} Wannier orbitals which effectively take the O2p orbitals into account, so ΔE_d and ΔE are positively correlated[11].

In figure 4, we show the relationship between the apical oxygen height $h_{\rm O}$ and ΔE_d. We can see that, while ΔE_d is positively correlated with the apical oxygen height in both single- and bi-layer systems as expected, ΔE_d is overall significantly greater in the bi-layer systems than in the single-layer systems. This is because bi-layer cuprates take pyramidal coordination of the oxygen ligands, while the single-layer cuprates take an octahedral one. Hence the effect of the apical oxygen should be weaker in the bi-layer systems, so that the "effective $h_{\rm O}$" is larger. As we have seen in the previous section, ΔE (and hence the d_{z^2} orbital) plays minor role in the T_c enhancement under pressure in materials with large ΔE, so the effect of ΔE should become less relevant as the number of the layers increases. This can be considered as one reason why T_c increases in spite of somewhat larger reduction in $h_{\rm O}$ in bi-layer system than in single-layer system under hydrostatic pressure[1].

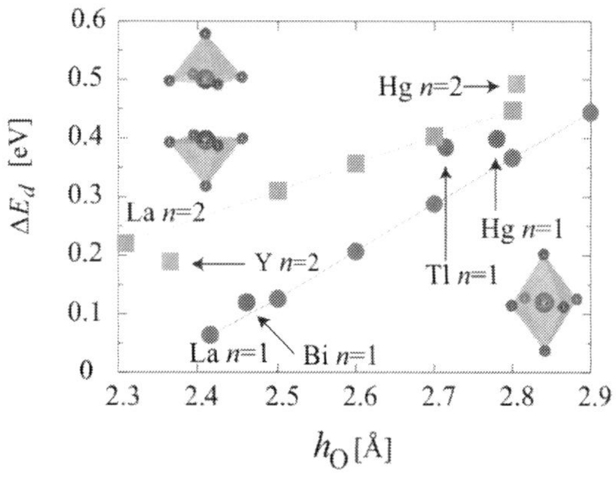

Figure 5. Relationship between the apical oxygen height $h_{\rm O}$[Å] and the level difference ΔE_d [eV]. The squares(circles) show the value of ΔE_d in each bi-layer(single-layer) system(the materials are indicated by arrows). The dashed(solid) line presents the calculation result when $h_{\rm O}$ is varied hypothetically in bi-layer (single-layer) La system by hand.

Journal of Physics: Conference Series **454** (2013) 012021 doi:10.1088/1742-6596/454/1/012021

5. Conclusion

In conclusion, we have performed first-principle calculation+FLEX study to analyze the pressure effect on T_c of the cuprates. To explain the T_c enhancement induced by hydrostatic pressure, we introduce three important parameters, level offset ΔE, roundness of the Fermi surface $r_{x^2-y^2}$ and the band width W. ΔE and $r_{x^2-y^2}$ are the measure of "orbital purity" in the main band, and T_c is enhanced by "orbital distillation". We have shown that hydrostatic pressure enhances the distillation, thereby enhancing T_c. We have also analyzed the effect of the band width W in detail. It has been shown that λ (and hence T_c) is maximized around a certain U, which depends on the band width as well as ΔE. For materials with large ΔE such as the Hg compound, the increase of W under pressure leads to an enhancement of λ for realistic values of U.

We also discuss the difference between bi- or single-layer systems from the view point of ΔE. Even when h_{O} is suppressed under pressure in bi-layer systems, the effect on T_c should be small because of the pyramidal coordination of the oxygen ligands.

6. Acknowledgments

The numerical calculations were performed at the Supercomputer Center, ISSP, University of Tokyo. This study has been supported by Grants-in-Aid for Scientific Research from JSPS (Grants No. 23340095, RA; No. 23009446, HS; No. 21008306, HU; and No. 22340093, KK and HA). RA acknowledges financial support from JST-PRESTO. DJS acknowledges support from the Center for Nanophase Material Science at Oak Ridge National Laboratory.

References

[1] Jorgensen J D, Hinks D G, Chmaissem O, Argyiou D N, Mitchell J F and Dabrowski B 1996 *Lecture Notes in Physics* vol 475, (Berlin: Springer) p 1; Jorgensen J D, Hinks D G, Chmaissem O, Argyiou D N, Mitchell J F and Dabrowski B 1995 *Recent Developments in High Temperature Superconductivity*, ed Klamut J *et al.* (Berlin: Springer) pp 1-15

[2] Bianconi A, Bianconi G, Caprara S, Di Castro D, Oyanagi H and Saini N L 2000 *J. Phys.: Condens. Matter* **12** 10655; Bianconi A, Agrestini S, Bianconi G, Di Castro D, and Saini N L 2001 *J. Alloys Compd.* **317-318** 537; Poccia N, Ricci A and Bianconi A 2010 *Adv. Condens. Matter Phys.* **2010** 261849

[3] Maekawa S, Inoue J and Tohyama T 1992 *The Physics and Chemistry of Oxide Superconductors* vol 60, ed Iye Y and Yasuoka H (Berlin: Springer) pp 105-115

[4] Andersen O K, Liechtenstein A I, Jepsen O and Paulsen F 1995 *J. Phys. Chem. Solids* **56** 1573

[5] Feiner L F, Jefferson J H and Raimondi R 1996 *Phys. Rev. Lett.* **76** 4939

[6] Pavarini E, Dasgupta I, Saha-Dasgupta T, Jepsen O and Andersen O K 2001 *Phys. Rev. Lett.* **87** 047003

[7] Weber C, Haule K and Kotliar G 2010 *Phys. Rev.* B **82** 125107

[8] Weber C, Yee C -H, Haule K and Kotliar G 2012 *Eur. Phys. Lett.* **100** 37001

[9] Takimoto T, Hotta T and Ueda K 2004 *Phys. Rev.* B **69** 104504

[10] Sakakibara H, Usui H, Kuroki K, Arita R and Aoki H 2010 *Phys. Rev. Lett.* **105** 057003

[11] Sakakibara H, Usui H, Kuroki K, Arita R and Aoki H 2012 *Phys. Rev.* B **85** 064501

[12] Shiraishi K, Oshiyama A, Shima N, Nakayama T and Kamimura H, 1988 *Solid State Commun.* **66** 629

[13] Kamimura H and Eto M 1990 *J. Phys. Soc. Jpn.* **59** 3053; Eto M and Kamimura H 1991 *J. Phys. Soc. Jpn.* **60** 2311

[14] Freeman A J and Yu J 1988 *Physica* B **150** 50

[15] Wang X, Dang H T and Millis A J 2011 *Phys. Rev.* B **84** 014530

[16] Hozoi L, Siurakshina L, Fulde P and van den Brink J 2011 *Sci. Rep.* **1** 65

[17] Uebelacker S and Honerkamp C 2012 *Phys. Rev.* B **85** 155122

[18] Mori M, Khaliullin G, Tohyama T and Maekawa S 2008 *Phys. Rev. Lett.* **101** 247003

[19] Klehe A -K, Gangopadhyay A K, Diederichs J and Schilling J S 1993 *Physica* C **213** 266; Klehe A -K, Schilling J S, Wagner J L and Hinks D G 1994 *Physica* C **223** 313

[20] Gao L, Xue Y Y, Chen F, Xiong Q, Meng R L, Ramirez D, Chu C W, Eggert J H and Mao H K 1994 *Phys. Rev.* B **50** 4260

[21] Hardy F, Hillier N J, Meingast C, Colson D, Li Y, Barisic N, Yu G, Zhao X, Greven M and Schilling J S 2010 *Phys. Rev. Lett.* **105** 167002

[22] Gugenberger F, Meingast C, Roth G, Grube K, Breit V, Weber T, Wuhl H, Uchida S and Nakamura Y 1994 *Phys. Rev.* B **49** 13137

[23] Meingast C, Junod A and Walker E 1996 *Physica* C **272** 106
[24] Sakakibara H, Suzuki K, Usui H, Kuroki K, Arita R, Scalapino D J and Aoki H 2012 *Phys. Rev.* B **86** 134520
[25] Bickers N E, Scalapino D J and White S R 1989 *Phys. Rev. Lett.* **62** 961
[26] Dahm T and Tewordt L 1995 *Phys. Rev. Lett.* **74** 793
[27] Yada K and Kontani H 2005 *J. Phys. Soc. Jpn.* **74** 2161
[28] Blaha P, Schwarz K, Madsen G K H, Kvasnicka D and Luitz J 2001 *Wien2k: An Augmented Plane Wave + Local Orbitals Program for Calculating Crystal Properties* (Wien: Vienna University of Technology)
[29] Birch F 1947 *Phys. Rev.* B **71** 809
[30] Jorgensen J D, Schuttler H -B, Hinks D G, Capone D W, Zhang K, Brodsky M B and Scalapino D J 1987 *Phys. Rev. Lett.* **58** 1024
[31] Wagner J L, Radaelli P G, Hinks D G, Jorgensen J D, Mitchell J F, Dabrowski B, Knapp G S and Beno M A 1993 *Physica* C **210** 447
[32] Balagurov A M, Sheptyakov D V, Aksenov V L, Antipov E V, Putilin S N, Radaelli P G and Marezio M 1999 *Phys. Rev.* B **59** 7209
[33] Kunes J, Arita R, Wissgott P, Toschi A, Ikeda H and Held K 2010 *Comp. Phys. Commun.* **181** 1888
[34] Marzari N and Vanderbilt D 1997 *Phys. Rev.* B **56** 12847; Souza I, Marzari N and Vanderbilt D 2001 *Phys. Rev.* B **65** 035109 ; The Wannier functions are generated by the code developed by Mostofi A A, Yates J R, Marzari N, Souza I and Vanderbilt D http://www.wannier.org/
[35] Eisaki H, Kaneko N, Feng D L, Damascelli A, Mang P L, Shen K M, Shen Z -X and Greven M 2004 *Phys. Rev.* B **69** 064512
[36] Scalapino D J 2007 *Handbook of High Temperature Superconductivity*, ed Schrieffer J R and Brooks J S (New York: Springer) Chapter 13

Journal of Physics: Conference Series **454** (2013) 012022 doi:10.1088/1742-6596/454/1/012022

Ab initio study of dissociation reaction of ethylene molecules on Ni cluster

K Shimamura[1], **T Oguri**[2], **Y Shibuta**[2], **S Ohmura**[3], **F Shimojo**[1] **and S Yamaguchi**[2]

[1]Department of Physics, Kumamoto University, Kumamoto 860-8555, Japan
[2]Department of Materials Engineering, The University of Tokyo, Tokyo 113-8656, Japan
[3]Department of Physics, Kyoto University, Kyoto 606-8502, Japan

E-mail: shimojo@kumamoto-u.ac.jp

Abstract. The dissociation reaction of ethylene molecules on the Ni cluster surface is investigated by *ab initio* molecular dynamics simulations. We observe that hydrogen atoms are generated from ethylene molecules at a rate of about 20 ps^{-1}. The activation energy for the dissociation of a hydrogen atom is estimated to be about 0.52 eV, which corresponds to a rate of only about 0.1 ps^{-1}. We find that the adsorption energy of an ethylene molecule on the Ni cluster is more than 1.5 eV, which is three times greater than the activation energy for the hydrogen dissociation. It is, therefore, suggested that the adsorption energy is responsible for the increase of the rate of the dissociation reaction. Based on these results, we discuss the microscopic process of the reaction of ethylene molecules on the Ni cluster in detail.

1. Introduction

The formation mechanism of carbon nanotubes (CNTs) [1, 2] via a catalytic chemical vapor deposition (CCVD) technique [3–5] has been widely studied since the CCVD is developed as a promising technique for large quantity synthesis of CNTs. It is well known that catalytic metals are essential to synthesize single-walled carbon nanotubes (SWNTs), whereas multi-walled carbon nanotubes (MWNTs) can be formed without catalytic metals [6]. Therefore, the role of catalytic metal in the growth of SWNTs has been widely discussed [7,8]. In 2003, Shibuta and Maruyama [9] first reported a continuous simulation of metal-catalyzed growth of initial cap structure of the SWNT by classical molecular dynamics (MD) simulation. In the simulation, graphite networks precipitated after isolated carbon atoms dissolved into a Ni cluster, and then a cap structure was formed on the surface of the Ni cluster. The metal-catalyzed growth model was then followed by a lot of numerical works [10–30]. In addition, in situ environmental transmission electron microscopy [31, 32] directly captured the nucleation process of the initial cap structure of SWNTs, which validates the metal-catalyzed growth model predicted by above numerical works. Therefore, there has been a broad consensus up to now that the cap structure is nucleated on the surface of the metal nanoparticle as an initial step of SWNT growth in the CCVD process.

On the other hand, reaction processes of carbon source molecules and additives on the catalytic metal surface are poorly understood since these reactions are complexly intertwined with diffusion and dissociation processes. However, it is not straightforward to treat dissociation process of carbon source molecules and subsequent formation process of graphite network at the

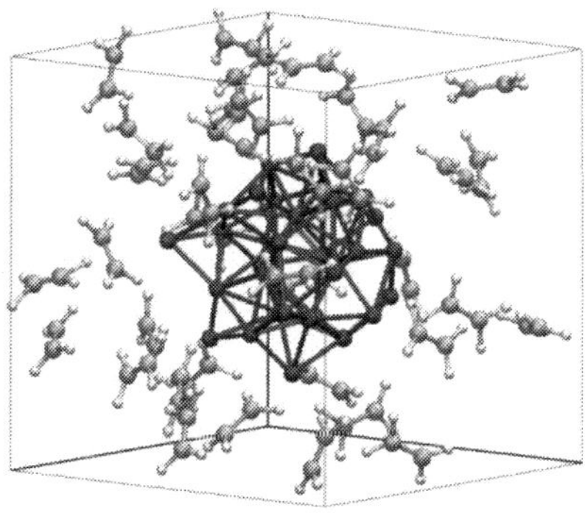

Figure 1. Snapshot of the initial atomic configuration of a Ni cluster, which consists of 32 nickel atoms, and 37 ethylene molecules in a box of dimensions 15×15×15 Å^3.

same time in numerical simulation due to that there is a discrepancy in timescale between dissociation and subsequent graphite formation processes. Therefore, when a formation of the nanotube cap structure on the metal surface is focused on, it has been compelled to ignore the initial dissociation process as in the most of numerical studies introduced above [8–30].

Under such circumstance, we investigate a continuous simulation starting from dissociation of carbon source molecules onto the surface of a metal cluster by *ab initio* MD simulations in this study. Specifically, dissociation of ethylene molecules onto the Ni cluster is examined, which is one of the typical combinations for the CCVD synthesis of SWNTs.

2. Method of calculation

The electronic states are calculated using the projector-augmented-wave (PAW) method [33], which is an all-electron electronic-structure-calculation method within the frozen-core approximation. In the framework of density functional theory, the generalized gradient approximation [34], is used for the exchange-correlation energy. The plane-wave-cutoff energies are 30 and 250 Ry for the electronic pseudo-wave functions and the pseudo-charge density, respectively. The energy functional is minimized iteratively using a preconditioned conjugate-gradient method. The gamma point is used for Brillouin zone sampling. Projector functions are generated for the $3d$, $4s$ and $4p$ states of nickel atom, the $2s$ and $2p$ states of carbon atom, and the $1s$ state of hydrogen atom. MD simulations are carried out at a temperature of 1500 K in the canonical ensemble using the Nosé-Hoover thermostat technique [35]. The equations of motion are integrated numerically using an explicit reversible integrator [36] with a time step of 0.242 fs. All computation time is 2.42 ps. The system studied in our MD simulations consists of a Ni_{32} cluster and 37 ethylene molecules (in total of 254 atoms) in a box of dimensions 15×15×15 Å^3 under periodic boundary conditions. In the initial configuration, the ethylene molecules are arranged around the Ni cluster, which is annealed at 1500 K without ethylene molecules beforehand, as shown in figure 1. The number density of the ethylene molecules corresponds to that of the liquid state. We use our own computer code in the present work.

To quantify the change in the bonding properties of atoms associated with the dissociation reaction of hydrogen atoms, we use a population analysis [37] by expanding the electronic wave functions in an atomic-orbital basis set. Based on the formulation generalized to the PAW method [38], we obtain the gross population for each atom and the bond-overlap population for each atomic pair. From the gross population, we estimate the charge of atoms, and the bond-overlap population gives a semi-quantitative estimate of the strength of covalent bonding between atoms.

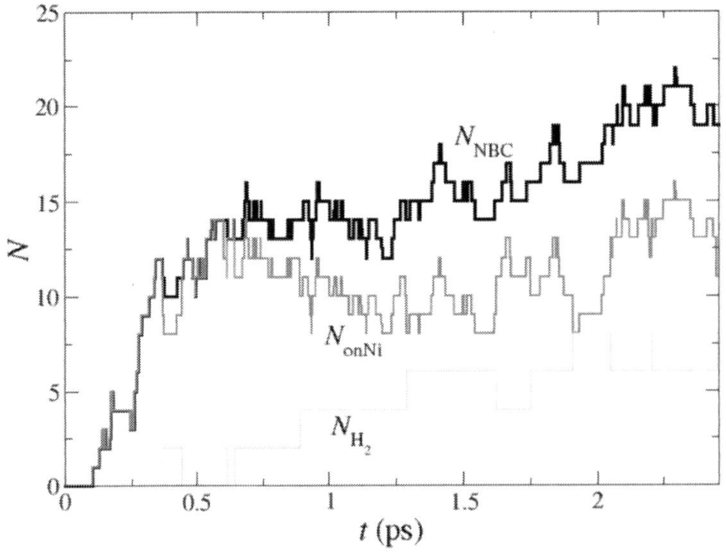

Figure 2. Time evolution of the number of hydrogen atoms. N_{NBC} shows the number of hydrogen atoms which are not bonded to carbon atoms. N_{onNi} shows the number of hydrogen atoms which exist as a hydrogen atom around the Ni cluster. N_{H2} is the number of hydrogen atoms constituting hydrogen molecules.

3. Results and Discussion

3.1. The number of dissociated hydrogen atoms

We observe a lot of dissociation reactions of hydrogen atoms from ethylene molecules on the Ni cluster. The reaction occurs more than 10 times during 400 fs in the beginning of our simulation, which corresponds to a reaction rate of about 20 ps^{-1}. The number of hydrogen atoms, which are not bonded to carbon atoms, N_{NBC} increases with time as shown in figure 2. More than half of them exist as a hydrogen atom around the Ni cluster within 2.42 ps, the number of which is N_{onNi} in figure 2. The rest form hydrogen molecules. N_{H2} in figure 2 is the number of the hydrogen atoms constituting hydrogen molecules. The averaged value of N_{onNi} is nearly constant ~11 after 0.5 ps. The number of hydrogen atoms that can exist around the Ni cluster is limited because the Ni cluster has a finite volume. Therefore, hydrogen molecules are generated when the number of the hydrogen atoms exceeds the limitation. Since it is considered that the hydrogen atoms that exist around the catalysts are responsible for inhibiting the synthesis of the CNTs, an efficient process of removing the dissociated hydrogen atoms may be important for a smooth synthesis.

3.2. Dissociation process of a hydrogen atom from an ethylene molecule

Figure 3 shows a typical example of the dissociation process of a hydrogen atom from an ethylene molecule on the Ni cluster. The snapshot at 100 fs (figure 3(a)) represents an atomic configuration just after the adsorption on the Ni cluster. In this configuration, the C-C bond is a single bond because the carbon atoms are bonded to the nickel atoms. After the formation of a bonding state spreading over nickel, carbon and hydrogen atoms labeled "Ni1", "C1" and "H1", respectively, at around 137 fs (figure 3(b)), the dissociation reaction of H1 occurs at about 155 fs (figure 3(c)). On the other hand, the C-C bond is a little bit strengthened during the reaction of the dissociation. Actually, the dissociation of C-C bonds does not occur even once in our simulations.

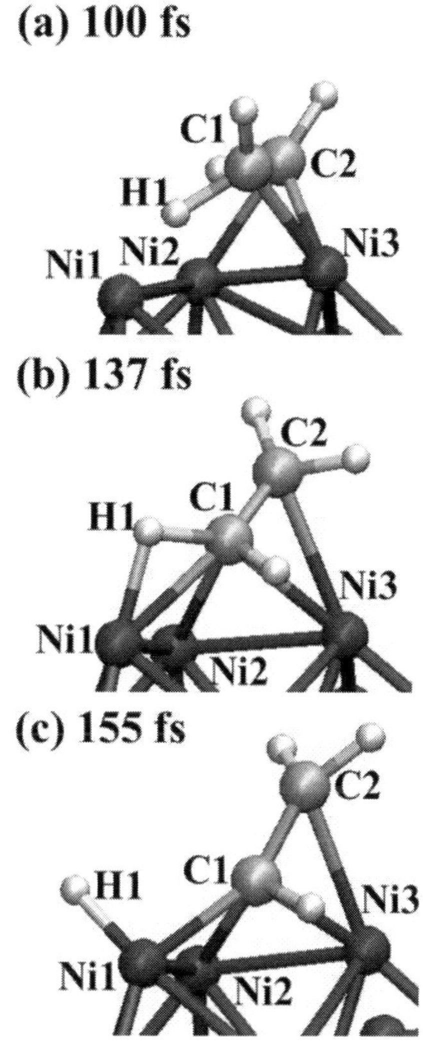

Figure 3. A typical example of the dissociation process of a hydrogen atom from an ethylene molecule on the Ni cluster. Atomic configurations at 100, 137 and 155 fs are shown.

3.3. Snapshot of atomic configuration

Figure 4 shows a snapshot of atomic configuration at 2.42 ps in our simulations. 17 ethylene molecules are adsorbed on Ni cluster, some of which dissociate hydrogen atoms. The number of the carbon atoms which dissociate one hydrogen atom (labeled "C_{H1}" in figure 4) are 7, and the number of the carbon atoms which dissociate two hydrogen atoms (labeled "C_{H0}" in figure 4) are 5. The dissociated hydrogen atoms exist around the Ni cluster (labeled "H" in figure 4), or as hydrogen molecules (labeled "H_2" in figure 4) as described in the section 3.1. Since the C-C bonds are never broken, it is considered that pairs of carbon atoms are the smallest components for the synthesis of CNTs rather than single carbon atoms in the case of the ethylene molecule as a carbon source molecule. We observe actually a carbon chain formed from two pairs of carbon atoms as shown in the circle of figure 4.

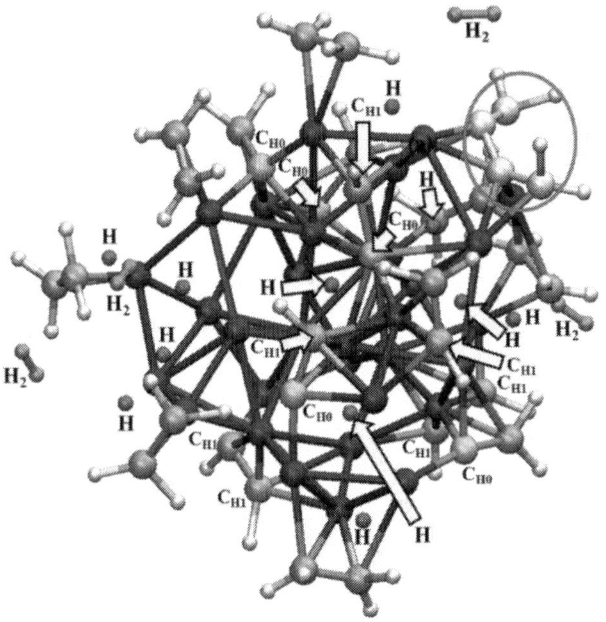

Figure 4. Snapshot of atomic configuration at 2.42 ps in MD simulations. Spheres labeled "C_{H1}" are carbon atoms that are bonded to one hydrogen atom. Spheres labeled "C_{H0}" are carbon atoms that are not bonded to hydrogen atoms. Spheres labeled "H" are dissociated hydrogen atoms. Pairs of spheres labeled "H_2" are hydrogen molecules. The numbers of C_{H1}, C_{H0}, H and H_2 are 7, 5, 11 and 4, respectively. The circle shows a carbon chain which consists of four carbon atoms.

3.4. Rate of the dissociation reaction

To find the minimum energy path of the dissociation reaction of a hydrogen atom from an ethylene molecule on the Ni cluster, we adopt the nudged elastic band (NEB) method [39]. As a discrete representation of a path from the reactant configuration $\mathbf{R_0}$ to the product configuration $\mathbf{R}_M$, $M-1$ replicas of the system are created and connected together with springs. The images are then relaxed toward the minimum energy path. In this paper, we use $M = 10$. The reactant and product configurations are prepared as follows. Firstly, we pick up the atomic coordinates of the Ni cluster and the dissociated ethylene molecule just before and after the dissociation reaction reported in the section 3.2. Secondly, we perform the structural optimization for the two systems to obtain the initial and final states for the NEB calculation. The activation energy for the reaction is estimated to be about 0.52 eV, which corresponds to a reaction rate of only about 0.1 ps^{-1} according to the transition state theory [40]. However, the dissociation reaction occurs actually at a much faster rate of about 20 ps^{-1} in our simulations as mentioned in the section 3.1. Obviously, it is much faster than the reaction rate estimated from the activation energy.

In order to explain the large difference between these reaction rates, we investigate adsorption energies of the ethylene molecule on the Ni cluster. We calculate the adsorption energy as follows. Firstly, we calculate the sum of potential energies of two isolated systems, an isolated Ni cluster and an isolated ethylene molecule. Secondly, we calculate the potential energy of a system in which an ethylene molecule is put on the Ni cluster. Thirdly, the latter energy is subtracted from the former to obtain the adsorption energy. Figure 5 shows one of the adsorption sites on the Ni cluster, which has the adsorption energy of 1.573 eV. It is clarified that the ethylene

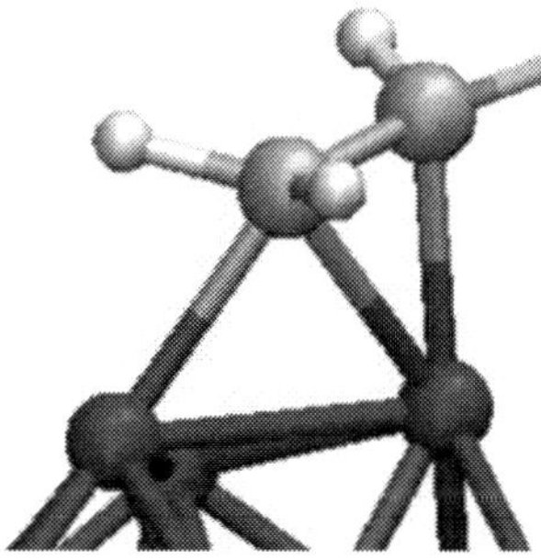

1.573 eV

Figure 5. Atomic configuration of an adsorption site of an ethylene molecule on the Ni cluster. The value of the adsorption energy is shown.

molecule can obtain more than 1.5 eV when it is adsorbed on the Ni cluster. This value is about three times larger than the activation energy of 0.52 eV. Therefore, the adsorption energy is responsible for accelerating the reaction rate.

4. Summary

To investigate the microscopic process of the dissociation reaction of the ethylene molecules on the Ni cluster in detail, we perform *ab initio* molecular dynamics simulations. We observe a lot of dissociation reactions of hydrogen atoms from ethylene molecules on the Ni cluster at a fast rate of about 20 ps^{-1}. While more than half of the dissociated hydrogen atoms exist around the Ni cluster, the rest form hydrogen molecules. Since there is a limit for the number of hydrogen atoms around the Ni cluster, hydrogen molecules are generated when the number of hydrogen atoms exceeds the limitation. We also calculate the activation energy and the reaction rate for the dissociation reaction. The reason why the reaction rate of the dissociation reaction is so fast is that the ethylene molecules obtain the adsorption energy of at least 1.5 eV, which is about three times larger than the activation energy of 0.52 eV for the dissociation reaction.

Acknowledgments

This work was supported by KAKENHI [Grant-in-Aid for Young Scientists (A) No. 24686026]. The authors furthermore thank the Research Institute for Information Technology, Kyushu University, for the use of its facilities. The computations were also carried out using the computer facilities at the Supercomputer Center, Institute for Solid State Physics, The University of Tokyo.

References

[1] Iijima S 1991 *Nature* **354** 56
[2] Iijima S and Ichihashi T 1993 *Nature* **363** 603
[3] Dai H, Rinzler A G, Nikolaev P, Thess A, Colbert D T and Smalley R E 1996 *Chem. Phys. Lett.* **260** 471
[4] Nikolaev P, Bronikowski M J, Bradley R K, Rohmund F, Colbert D T, Smith K A and Smalley R E 1999 *Chem. Phys. Lett.* **313** 91
[5] Maruyama S, Kojima R, Miyauchi Y, Chiashi S and Kohno M 2002 *Chem. Phys. Lett.* **360** 229
[6] Dresselhaus M S, Dresselhaus G and Avouris P 2001 *Carbon nanotubes: synthesis, structure, properties, and applications* (Topics in Applied Physics vol 80) (Berlin: Springer-Verlag)
[7] Banhart F 2009 *Nanoscale* **1** 201
[8] Shibuta Y 2011 *Diamond Relat. Mater.* **20** 334
[9] Shibuta Y and Maruyama S 2003 *Chem. Phys. Lett.* **382** 381

[10] Ding F, Rosén A and Bolton K 2004 *J. Chem. Phys.* **121** 2775
[11] Ding F, Bolton K and Rosén A 2004 *J. Phys. Chem. B* **108** 17369
[12] Ding F, Bolton K and Rosén A 2006 *Comput. Mater. Sci.* **35** 243
[13] Zhao J, Martinez-Limia A and Balbuena P B 2005 *Nanotechnology* **16** S575
[14] Gómez-Gualdrón D A, McKenzie G D, Alvarado J F J and Balbuena P B 2012 *ACS Nano* **6** 720
[15] Shibuta Y and Maruyama S 2007 *Chem. Phys. Lett.* **437** 218
[16] Shibuta Y and Maruyama S 2007 *Comput. Mater. Sci.* **39** 842
[17] Shibuta Y and Elliott J A 2006 *Chem. Phys. Lett.* **427** 365
[18] Shibuta Y and Elliott J A 2009 *Chem. Phys. Lett.* **472** 200
[19] Neyts E C, Shibuta Y, Van Duin A C T and Bogaerts A 2010 *ACS Nano* **4** 6665
[20] Neyts E C, Van Duin A C T and Bogaerts A 2011 *J. Am. Chem. Soc.* **133** 17225
[21] Neyts E C, Van Duin A C T and Bogaerts A 2012 *J. Am. Chem. Soc.* **134** 1256
[22] Amara H, Bichara C and Ducastelle F 2006 *Phys. Rev. B* **73** 113404
[23] Amara H, Bichara C and Ducastelle F 2008 *Surf. Sci.* **602** 77
[24] Amara H, Bichara C and Ducastelle F 2008 *Phys. Rev. Lett.* **100** 056105
[25] Amara H, Roussel J M, Bichara C, Gaspard J P and Ducastelle F 2009 *Phys. Rev. B* **79** 014109
[26] Ohta Y, Okamoto Y, Irle S and Morokuma K 2008 *ACS Nano* **2** 1437
[27] Ohta Y, Okamoto Y, Irle S and Morokuma K 2009 *J. Phys. Chem. C* **113** 159
[28] Page A J, Irle S and Morokuma K 2010 *J. Phys. Chem. C* **114** 8206
[29] Page A J, Minami S, Ohta Y, Irle S and Morokuma K 2010 *Carbon* **48** 3014
[30] Page A J, Yamane H, Ohta Y, Irle S and Morokuma K 2010 *J. Am. Chem. Soc.* **132** 15699
[31] Hofmann S, Sharma R, Ducati C, Du G, Mattevi C, Cepek C, Cantoro M, Pisana S, Parvez A, Cervantes-Sodi F, Ferrari A C, Dunin-Borkowski R, Lizzit S, Petaccia L, Goldoni A and Robertson J 2007 *Nano Lett.* **7** 602
[32] Yoshida H, Takeda S, Uchiyama T, Kohno H and Homma Y 2008 *Nano Lett.* **8** 2082
[33] Blöchl P E 1994 *Phys. Rev. B* **50** 17953
[34] Perdew J P, Burke K and Ernzerhof M 1996 *Phys. Rev. Lett.* **77** 3865
[35] Hoover W G 1985 *Phys. Rev. B* **31** 1695
[36] Tuckerman M, Berne B J and Martyna G J 1992 *J. Chem. Phys.* **97** 1990
[37] Mulliken R S 1955 *J. Chem. Phys.* **23** 1833
[38] Shimojo F, Nakano A, Kalia R K and Vashishta P 2008 *Phys. Rev. E* **77** 066103
[39] Jónsson H, Mills G and Jacobsen K W 1998 *Nudged elastic band method for finding minimum energy paths of transitions* Classical and Quantum Dynamics in Condensed Phase Simulations (Singapore: World Scientific)
[40] Truhlar D G, Garrett B C and Klippenstein S J 1996 *J. Phys. Chem. C* **100** 12771

Variational path integral molecular dynamics study of a water molecule

Shinichi Miura

School of Mathematics and Physics, Kanazawa University,
Kakuma, Kanazawa 920-1192, Japan.

E-mail: smiura@mail.kanazawa-u.ac.jp

Abstract. In the present study, a variational path integral molecular dynamics method developed by the author [*Chem. Phys. Lett.* **482**, 165 (2009)] is applied to a water molecule on the adiabatic potential energy surface. The method numerically generates an exact wavefunction using a trial wavefunction of the target system. It has been shown that even if a poor trial wavefunction is employed, the exact quantum distribution is numerically extracted, demonstrating the robustness of the variational path integral method.

1. Introduction

Quantum Monte Carlo (QMC) methods provide computational tools for accurately calculating ground state properties of many body systems [1, 2, 3, 4]. The Variational Monte Carlo (VMC) method [5], for example, is used to calculate expectation values of physical quantities using a trial wavefunction of the target system. The more sophisticated diffusion Monte Carlo (DMC) method [6, 7] is a projector approach in which a stochastic imaginary time evolution is used to improve a starting trial wavefunction. The QMC methods including the VMC and DMC methods have successfully been applied to various quantum systems ranging from quantum liquids like helium to electronic structure of atoms and molecules [1, 4].

Recently, we have developed a molecular dynamics algorithm for a variational path integral method [1, 8] that is closely related to the diffusion Monte Carlo method [9, 10, 11, 12, 13]. The variational path integral method [1], which is also called path integral ground state [8], is a method to numerically generate exact ground state of many body systems. We have constructed the molecular dynamics algorithm to carry out the variational path integral calculations on the basis of path integral molecular dynamics method developed for finite temperature quantum systems [14, 15]. We call it a variational path integral molecular dynamics (VPIMD) method. In the present study, we apply the VPIMD method to vibrational fluctuation of a water molecule on an adiabatic potential energy surface.

2. Methods

2.1. Variational molecular dynamics

We start to consider a system consisting of N particles whose coordinates are collectively represented to be R. The ground state of the system is described using a trial wavefunction $\Phi_T(R, \mathbf{A})$ where $\mathbf{A}$ denotes a set of variational parameters. Expectation value of a Hamiltonian

$\hat{H}$ using the trial wavefunction reads [1, 4]

$$\begin{aligned} \langle H \rangle &= \frac{\int dR \Phi_T^*(R) \hat{H} \Phi_T(R)}{\int dR |\Phi_T(R)|^2} \\ &= \frac{\int dR |\Phi_T(R)|^2 E_L(R)}{\int dR |\Phi_T(R)|^2}, \end{aligned} \tag{1}$$

where $E_L(R)$ represents a local energy defined by

$$E_L(R) = \Phi_T^{-1}(R) \hat{H} \Phi_T(R). \tag{2}$$

The variational Monte Carlo method is designed so as to generate coordinates R according to a distribution function $\rho(R) \propto |\Phi_T(R)|^2$. Then, the expectation value of the Hamiltonian (1) can be calculated by the local energy averaged along a Monte Carlo trajectory. Here, we consider a molecular dynamics method to generate the distribution function $\rho(R)$. First, we define an effective potential $V_{\rm VMC}(R)$ using the distribution function $\rho(R)$:

$$\rho(R) \equiv e^{-\beta V_{\rm VMC}(R)}, \tag{3}$$

where a parameter β is a fictitious inverse temperature. The distribution function $\rho(R)$ can be regarded as a canonical distribution at the fictitious β. To sample the distribution, we introduce the following classical Hamiltonian:

$$H_{\rm VMC} = \sum_{i=1}^{N} \frac{\mathbf{p}_i^2}{2m'_i} + V_{\rm VMC}(R), \tag{4}$$

where $\mathbf{p}_i$ is a fictitious momentum of an i-th particle and m'_i is an associated fictitious mass. Then, we can obtain equations of motion based on the Hamilton equation. In the present study, we attach a single Nosé-Hoover chain thermostat [16] to each degrees of freedom for generating the canonical distribution compatible with (3). The resulting thermostatted equations of motion are basic equations of the variational molecular dynamics method [9]. Then, as in the variational Monte Carlo method, we can obtain quantum mechanical expectation values of various physical quantities using a molecular dynamics trajectory.

2.2. Variational path integral molecular dynamics

In this subsection, a method to systematically improve the description of the system is considered. The exact ground state of the system, $|\Psi_0\rangle$, can be obtained using a trial wavefunction $|\Phi_T\rangle$ by the following relation: [1, 4]

$$|\Psi_0\rangle = \lim_{\beta \to \infty} e^{-\frac{\beta}{2}\hat{H}} |\Phi_T\rangle, \tag{5}$$

where β is an imaginary time [1]. Here, we refer to a scalar product of the above exact wavefunction as a pseudo partition function Z_0 [17], which plays a central role to construct a variational path integral:

$$\begin{aligned} Z_0 &= \langle \Psi_0 | \Psi_0 \rangle = \langle \Phi_T | e^{-\beta \hat{H}} | \Phi_T \rangle \\ &= \int \int dR dR' \langle \Phi_T | R \rangle \langle R | e^{-\beta \hat{H}} | R' \rangle \langle R' | \Phi_T \rangle, \end{aligned} \tag{6}$$

where we use the closure relation for the coordinate basis: $\int dR|R\rangle\langle R| = 1$. A matrix element $\langle R|e^{-\beta\hat{H}}|R'\rangle$ in (6) is found to be the same as a density matrix at the inverse temperature β, $\rho(R, R'; \beta)$. The density matrix can be written on the basis of the discretized path integral as [1]

$$\begin{aligned}\rho(R, R'; \beta) &= \langle R|[e^{-\Delta\tau\hat{H}}]^M|R'\rangle \qquad (7)\\ &= \int\cdots\int\{\prod_{s=1}^{M-1} dR^{(s)}\}\prod_{s=0}^{M-1}\langle R^{(s)}|e^{-\Delta\tau\hat{H}}|R^{(s+1)}\rangle\\ &\propto \int\cdots\int\{\prod_{s=1}^{M-1} dR^{(s)}\}e^{-S(\{R^{(s)}\};\Delta\tau)},\end{aligned}$$

where $\Delta\tau = \beta/M$ and $S(\{R^{(s)}\}; \Delta\tau)$ is a discretized imaginary time action. Explicit expression of the action is dependent on an approximation on a short time propagator $\langle R|e^{-\Delta\tau\hat{H}}|R'\rangle$. In the present study, the primitive approximation [1] is adopted. Then, the pseudo partition function can be written by

$$Z_0 \propto \int\cdots\int\{\prod_{s=0}^{M} dR^{(s)}\}\Phi_T(R^{(0)})e^{-S(\{R^{(s)}\};\Delta\tau)}\Phi_T(R^{(M)}). \qquad (8)$$

As in the standard path integral method for finite temperature systems [18], the pseudo partition function can be regarded as a configurational integral of classical polymers. However, in the variational path integral, the classical isomorphic systems consist of open chain polymers. Furthermore, distributions of end-point coordinates at $s = 0$ and M are affected by the trial wavefunction $\Phi_T(R^{(0)})$ and $\Phi_T(R^{(M)})$, respectively. Here, we consider a molecular dynamics method to sample configurations of the above isomorphic polymers. First, we define the following classical Hamiltonian:

$$H_{\mathrm{VPIMD}} = \sum_{s=0}^{M}\sum_{i=1}^{N}\frac{(\mathbf{p}_i^{(s)})^2}{2m_i'} + \frac{S(\{R^{(s)}\})}{\beta} - \frac{\ln\Phi_T(R^{(0)})}{\beta} - \frac{\ln\Phi_T(R^{(M)})}{\beta}, \qquad (9)$$

where $\mathbf{p}_i^{(s)}$ denotes a fictitious momentum of an i-th particle at an s-th time slice and m_i' is a fictitious mass of the i-th particle. Using the above Hamiltonian, we can derive equations of motion based on the Hamilton equation. Then, to generate the distribution compatible with (8), we attach a single Nosé-Hoover chain thermostat to each degree of freedom. The resulting equations of motion are basic equations for the variational path integral molecular dynamics (VPIMD) method [9]. In the present study, we use staging coordinates [14] to describe the polymer configurations for enhancing sampling efficiency. The standard definition [15, 19] on the staging variables and associated staging masses $m^{(s)}$ are adopted.

3. Results

In the present study, a trial wavefunction is constructed using a local mode S_ν representing the distance between a pair of atoms in a molecule [20]. The local modes are defined for a molecule consisting of N atoms by

$$S_\nu = |\mathbf{r}_i - \mathbf{r}_j|, \quad 1 \le i < j \le N, \quad 1 \le \nu \le N_{\mathrm{pair}} = \frac{N(N-1)}{2} \qquad (10)$$

where $\mathbf{r}_i$ is an i-th atomic coordinate. Then, the trial wavefunction is written by [20]

$$\Phi_T(R) = \exp\left\{\sum_{\mu=1}^{N_{\mathrm{pair}}}\sum_{\nu=1}^{N_{\mathrm{pair}}}\Delta S_\mu A_{\mu\nu}\Delta S_\nu\right\} \qquad (11)$$

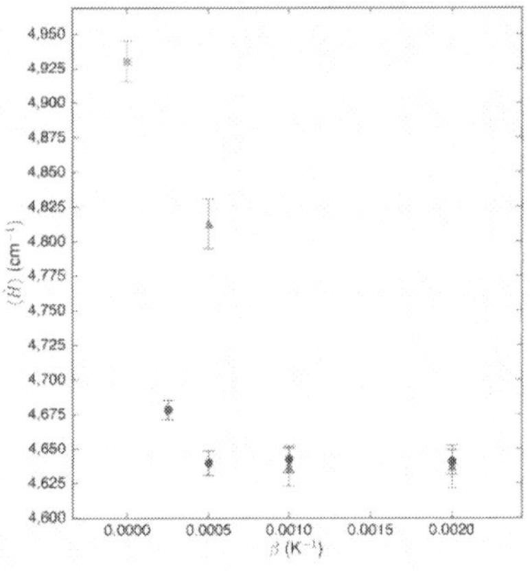

Figure 1. Total energy of a water molecule by the variational PIMD calculations. The energy is shown as a function of the projection time β. Blue filled circles are the results with the variational parameter set 1, and red filled triangles with the set 2. A green filled square indicates the energy by the variational MD with the set 1. The error bar is expressed at 95 % confidence level.

with

$$\Delta S_\nu = S_\nu - S_\nu^0. \tag{12}$$

Here, S_ν^0 indicates the value of a ν-th mode at the minimum of the potential energy. The symmetric matrix $\mathbf{A}$ defines variational parameters. For a water molecule, four parameters are needed for the trial wavefunction (11). Let us define $S_1 = r_{\mathrm{OH}_1}, S_2 = r_{\mathrm{OH}_2}$, and $S_3 = r_{\mathrm{H}_1\mathrm{H}_2}$. For the water molecule, the matrix $\mathbf{A}$ has the following symmetry: $A_{ij} = A_{ji}, A_{11} = A_{22}$, and $A_{13} = A_{23}$. Then, the following four parameters are independent: A_{11}, A_{12}, A_{13}, and A_{33}. In the present study, two sets of the variational parameters have been tested. The set 1 is given by $A_{11} = -19.61, A_{12} = -6.11, A_{13} = 4.32$, and $A_{33} = -8.60$ in units of Bohr^{-2} [21], and the set 2 by $A_{11} = -19.61, A_{12} = 12.20, A_{13} = 8.63$, and $A_{33} = -8.60$ in units of Å^{-2} [20].

Regarding the interatomic interaction for the water molecule, we have used a potential function derived from experimental data by Carney, Curtiss and Langhoff [22]. It involves an expansion in Simons-Parr-Finlan coordinates about the equilibrium geometry and consists 19 terms. Then, the variational molecular dynamics and the variational path integral molecular dynamics calculations for the H_2O molecule have been performed. Here, we comment that the variational MD results presented in the paper are all with the parameter set 2, since the trial wave function with the parameter set 1 causes unphysical behavior of the water molecule; actually, the water molecule dissociated in the course of the VMD calculation. The variational PIMD calculations have been performed for various β; both sets of the parameters, 1 and 2 have been tested. The imaginary time increment is fixed to be $\Delta\tau = 2.5\times10^{-5}$ K^{-1} for all the VPIMD calculations. Here and hereafter, the inverse temperature β and the associated $\Delta\tau$ is represented to be in units of K^{-1}, corresponding to the β multiplied by the Boltzmann constant.

In Fig. 1, we show the projection time β dependence of the total energy. The total energy was calculated using the mixed estimator derived by the following relation [4]:

$$E_0 = \frac{\langle \Phi_T | \hat{H} e^{-\beta\hat{H}} | \Phi_T \rangle}{\langle \Phi_T | e^{-\beta\hat{H}} | \Phi_T \rangle}, \tag{13}$$

where E_0 is the ground state energy. We first discuss the results by the parameter set 1. Using the variational MD method, the total energy by the trial wave function with the set 1 was calculated to be 4930 $\pm$ 15 cm^{-1}. With increasing β, the variational PIMD energy decreases quickly and reaches the exact ground state energy. For the parameter set 1, $\beta = 5.0 \times 10^{-4}$ K^{-1} is found to be large enough to obtain the exact ground state energy. On the other hand, the energy using the set 2 also decreases with increasing β. As seen in the figure, we need twice larger β to obtain the exact ground state energy. This is due to the fact that the set 1 is

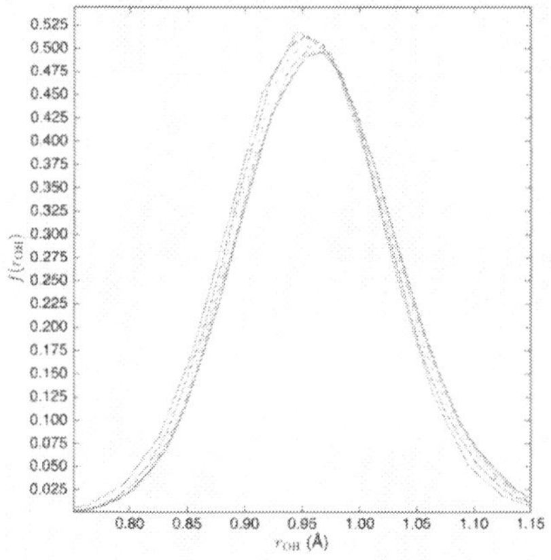

Figure 2. O-H bond length distribution $f(r_{\rm OH})$ by the variational PIMD calculations. Vermilion dashed line, green twin dot line, orange dot dashed line, and pink solid line indicate the results with the projection time $\beta = 2.5, 5.0, 10$, and 20×10^{-4} K^{-1}, respectively. Blue dotted line indicates the variational MD result. All the calculations are performed using the trial wavefunction with the parameter set 1.

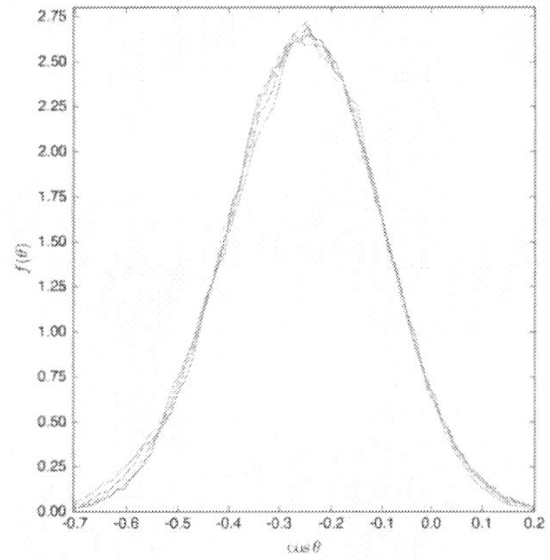

Figure 3. H-O-H bond angle distribution $f(\theta)$ by the variational PIMD calculations. Vermilion dashed line, green twin dot line, orange dot dashed line, and pink solid line indicate the results with the projection time $\beta = 2.5, 5.0, 10$, and 20×10^{-4} K^{-1}, respectively. Blue dotted line indicates the variational MD result. All the calculations are performed using the trial wavefunction with the parameter set 1.

much better variational parameters than the set 2. As mentioned above, we could not obtain the total energy by the VMD, since the water molecule was dissociated in the course of the simulation. Similarly, the VPIMD calculations could not been performed stably in the range of $\beta < 5.0 \times 10^{-4}$ K^{-1}. For such a poor set of the parameters, we could obtain the exact ground state energy, demonstrating the robustness of the variational path integral method.

Figure 2 shows the structural fluctuation of the water molecule at the ground state. Quantum expectation values of observables diagonal in the coordinate space can straightforwardly be obtained using the variational path integral, since the system coordinates at the time slices around $\tau = \beta/2$ are distributed according to the exact wavefunction with β large enough. In Fig. 2, O-H bond length and H-O-H bond angle distributions are presented. For each VPIMD calculation, the distributions are evaluated using the coordinates at $\tau = \beta/2$. In both distributions, trial wavefunction describes well the structural correlation in the exact wavefunction. This is due to the trial wavefunction with the set 1 provides the good trial wavefunction for the water molecule. With increasing β, the structural correlations approach those for the exact wavefunction.

4. Concluding remarks

In the present study, the variational path integral molecular dynamics (VPIMD) method has been applied to quantum vibrational fluctuation of a water molecule on the adiabatic potential energy surface. To perform the VPIMD calculations, we need to prepare a trial wavefunction of the water molecule. The VPIMD method numerically extracts the exact ground state from the trial wavefunction. In the present study, we have tested two sets of variational parameters for a vibrational wavefunction. It is demonstrated that the VPIMD method generates the same results for the different sets of the parameters, when the projection time is long enough. It is worthwhile to remark that one parameter set is so poor that the bound state of the water molecule could not been reproduced; even for such a poor set of parameters, the VPIMD method is found to generate the exact ground state, demonstrating the robustness of the variational path integral.

An important direction on further methodological development regarding the VPIMD is on systematically improving the description of adiabatic potential energy surfaces of atoms and molecules. This can be achieved by combining the variational path integral with electronic structure calculations as in the case of the finite temperature path integral molecular dynamics [23, 24]. Since this method needs to perform electronic structure calculations at each VPIMD step, computational cost is expected to be rather expensive. As demonstrated in this paper, we can reduce the projection time β directly related with the computational cost if we adopt a good trial wavefunction. This indicates that the variational parameter optimization is crucial for efficient VPIMD calculations combined with *ab initio* electronic structure theory. These issues will be addressed in the near future.

Acknowledgments
This work was partially supported by the Grant-in-Aid for Scientific Research (C) (No. 23550011) from the Japan Society for the Promotion of Science and by the Strategic Programs for Innovative Research (SPIRE), MEXT, and the Computational Materials Science Initiative (CMSI), Japan.

References

[1] Ceperley D M 1995 *Rev. Mod. Phys.* **67** 279
[2] Lester W A ed 1997 *Recent Advances in Quantum Monte Carlo Methods* (Singapore: World Scientific)
[3] Nightingale M P and Umrigar C J ed 1999 *Quantum Monte Carlo Methods in Physics and Chemistry* (Dordrecht: Kluwer Academic)
[4] Foulkes W M C, Mitas L, Needs R J and Rajagopal G 2001 *Rev. Mod. Phys.* **73** 33
[5] McMillan W L 1965 *Phys. Rev.* **138** A442
[6] Anderson J B 1975 *J. Chem. Phys.* **63** 1499
[7] Anderson J B 1976 *J. Chem. Phys.* **65** 4121
[8] Sarsa A, Schmidt K E and Magro W R 2000 *J. Chem. Phys.* **113** 1366
[9] Miura S 2009 *Chem. Phys. Lett.* **482** 165
[10] Miura S 2011 *Comp. Phys. Commu.* **182** 274
[11] Miura S 2011 *GAKUTO International Series — Mathematical Sciences and Applications* **34** 129
[12] Miura S 2012 *Mol. Sim.* **38** 378
[13] S. Miura 2012 *Advances in Quantum Monte Carlo* ed Tanaka S, Rothstein S and Lester W (Washington DC: American Chemical Society) pp 177–186
[14] Tuckerman M, Berne B J, Martyna G J and Klein M L 1993 *J. Chem. Phys.* **99** 2796
[15] Tuckerman M E and Hughes A 1999 *Classical and Quantum Dynamics in Condensed Phase Simulations* ed Berne B J, Ciccotti G and Coker D F (Singapore: World Scientific) p 311
[16] Martyna G J, Klein M L and Tuckerman M 1993 *J. Chem. Phys.* **98** 2796
[17] Baroni S and Moroni S 1999 *Phys. Rev. Lett.* **82** 4745
[18] Chandler D and Wolynes P G 1981 *J. Chem. Phys.* **74** 4078
[19] Miura S and Tanaka J 2004 *J. Chem. Phys.* **120** 2161
[20] Bernu B, Ceperley D M and Lester Jr. W A 1990 *J. Chem. Phys.* **93** 552
[21] Typographical errors regarding values of variational parameters in [20] have been corrected in the following paper: Bernu B, Ceperley D M, and Lester Jr. W A 1991 *J. Chem. Phys.* **95** 7782
[22] Carney G D, Curtiss L A and Langhoff S R 1976 *J. Mol. Spectrosc.* **61** 371
[23] Marx D and Parrinello M 1996 *J. Chem. Phys.* **104** 4077
[24] Shiga M, Tachikawa M and Miura S 2001 *J. Chem. Phys.* **115** 9149

Shape transition of micelles in amphiphilic solution: A molecular dynamics study

Susumu Fujiwara[1], Masato Hashimoto[1], Takashi Itoh[1], Hiroaki Nakamura[2] and Yuichi Tamura[3]

[1]Department of Macromolecular Science and Engineering, Kyoto Institute of Technology, Matsugasaki, Sakyo-ku, Kyoto 606-8585, Japan
[2]Fundamental Physics Simulation Research Division, National Institute for Fusion Science, 322-6 Oroshi-cho, Toki 509-5292, Japan
[3]Department of Information Science and Systems Engineering, Konan University, 8-9-1 Okamoto, Higashinada-ku, Kobe 658-8501, Japan

E-mail: fujiwara@kit.ac.jp

Abstract. The shape transition of micelles in an amphiphilic solution is studied by a molecular dynamics simulation of coarse-grained rigid amphiphilic molecules with explicit solvent molecules. Our simulations show that the micellar shape changes from a disc into a cylinder, and then into a sphere as the hydrophilic interaction increases. We find that the potential energy decreases monotonically even during the micellar shape transition as the hydrophilic interaction increases. In contrast, it is ascertained that there exists a wide coexistence region in the intensity of the hydrophilic interaction between a cylinder and a sphere.

1. Introduction

Amphiphilic molecules such as lipids and surfactants contain both a hydrophilic part and a hydrophobic part. In aqueous or organic solvents, these molecules often spontaneously self-assemble into various structures such as micelles, bilayer membranes and bicontinuous structures [1, 2, 3]. Shape transitions of micelles play an important role in cellular physiology. The sphere-to-rod transitions of surfactant micelles in aqueous solution have recently been investigated by coarse-grained molecular dynamics (MD) simulations [4, 5]. Although numerous computer simulation studies have so far been carried out on micellar systems in amphiphilic solution [6], the detailed mechanisms of micellar shape transition in amphiphilic solution have not been fully understood at the molecular level. The purpose of this study is to clarify the molecular mechanism of shape transition of micelles in an amphiphilic solution. With a view to investigating the effect of the hydrophilicity on the micellar shapes in amphiphilic solution at the molecular level, we perform the MD simulations of coarse-grained rigid amphiphilic molecules with explicit solvent molecules and analyze the micellar shape transitions systematically.

2. Simulation model and method

The computational model is the same as that used in the previous work [7, 8, 9, 10]. An amphiphilic molecule is a rigid rod which consists of one hydrophilic particle and two hydrophobic particles. A solvent molecule is modeled as a hydrophilic particle. The interaction between a hydrophilic particle and a hydrophobic particle is modeled by a repulsive soft core potential

$$U_{\mathrm{SC}}(r) = 4\varepsilon_{\mathrm{SC}} \left(\frac{\sigma_{\mathrm{SC}}}{r}\right)^9, \tag{1}$$

the interaction between a hydrophilic head particle and a solvent particle is modeled by a Lennard-Jones (LJ) potential

$$U_{\mathrm{LJ-hs}}(r) = 4\varepsilon_{\mathrm{hs}} \left[\left(\frac{\sigma}{r}\right)^{12} - \left(\frac{\sigma}{r}\right)^6\right], \tag{2}$$

and all other interactions are modeled by an LJ potential

$$U_{\mathrm{LJ}}(r) = 4\varepsilon \left[\left(\frac{\sigma}{r}\right)^{12} - \left(\frac{\sigma}{r}\right)^6\right]. \tag{3}$$

Here, r is the interparticle distance, $\varepsilon_{\mathrm{SC}}$ is an interaction parameter for the intensity of the hydrophobic interaction and $\varepsilon_{\mathrm{hs}}$ is that for the intensity of the hydrophilic interaction. The parameter σ_{SC} is set to $\sigma_{\mathrm{SC}} = 1.05\sigma$ as in [11]. To avoid discontinuities in both the potential energy and the force due to the potential energy cutoff, we use the shifted force variant of these nonbonded potentials

$$V_X(r) = U_X(r) - U_X(r_c) - \left.\frac{\partial U_X}{\partial r}\right|_{r=r_c} (r - r_c), \tag{4}$$

where X = SC, LJ-hs or LJ, and r_c is the cutoff distance, which is set to $r_c = 3.0\sigma$ here. In what follows, we represent dimensionless quantities by an asterisk, e.g., number density $\rho^* = \rho\sigma^3$, time $t^* = t\sqrt{\varepsilon/m\sigma^2}$ and temperature $T^* = k_B T/\varepsilon$, where k_B is the Boltzmann constant.

The equations of motion for all particles are solved numerically by using the leap-frog algorithm at constant temperature with a time step of $\Delta t^* = 0.0025$ and the temperature is controlled at every 10 time steps by *ad hoc* velocity scaling [12]. We apply the periodic boundary conditions and the number density is set to $\rho^* = 0.75$.

Initially, we prepare an isolated micelle of 97 amphiphilic molecules with ($\varepsilon^*_{\mathrm{SC}}$, $\varepsilon^*_{\mathrm{hs}}$)=(1.0, 1.0) in solution at $T^* = 1.3$. The number of solvent particles is 5541, which leads to the amphiphilic concentration of 0.05. The intensity of the hydrophilic interaction $\varepsilon^*_{\mathrm{hs}}$ is then changed to various values ($\varepsilon^*_{\mathrm{hs}}$=1.2, 1.4, ..., 4.0) and MD simulations of $t^* = 2.0 \times 10^4$ (8.0×10^6 time steps) are carried out for each simulation run.

3. Simulation results and discussion

3.1. Micellar shape

We show, in figure 1, snapshots of micelles formed by amphiphilic molecules at $\varepsilon^*_{\mathrm{hs}}$=1.0, 2.0 and 4.0. Isosurfaces of the density of the tail particles, which are calculated by Gaussian splatting techniques, are depicted in this figure. Figure 1 tells us that the micellar shape changes from a disc into a cylinder, and then into a sphere as the intensity of the hydrophilic interaction $\varepsilon^*_{\mathrm{hs}}$ increases.

3.2. Potential energy

In figure 2, the total potential energy E^*_{pot} is shown as a function of the intensity of the hydrophilic interaction $\varepsilon^*_{\mathrm{hs}}$. This figure indicates that the total potential energy changes monotonically with the intensity of the hydrophilic interaction even during the micellar shape transition.

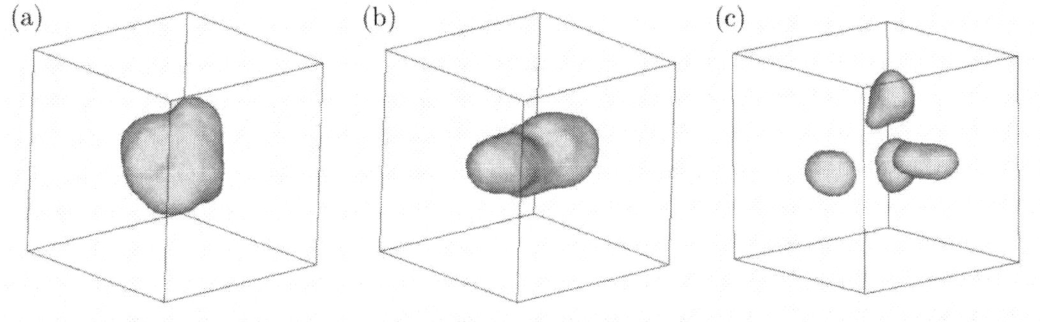

Figure 1. Snapshots of micelles formed by amphiphilic molecules for various values of the hydrophilic interaction parameter $\varepsilon^*_{\rm hs}$. (a) A disc micelle for $\varepsilon^*_{\rm hs} = 1.0$, (b) a cylindrical micelle for $\varepsilon^*_{\rm hs} = 2.0$ and (c) spherical micelles for $\varepsilon^*_{\rm hs} = 4.0$. Isosurfaces of the density of the tail particles, which are calculated by Gaussian splatting techniques, are depicted to show the micellar shape clearly. Note that solvent molecules are not displayed for clarity.

3.3. Fraction of micellar shapes

Here, we examine the micellar shape distribution in order to quantitatively investigate how the micellar shape changes with the hydrophilicity. As in our previous papers [8, 9, 10], we use the orientational order parameters as indices to characterize the micellar shape. We introduce a coordinate system that uses the three principal axes of inertia of the micelle. The origin is located at the center-of-mass position of the micelle, the x-axis is the principal axis with the largest moment of inertia and the z-axis is the principal axis with the smallest moment of inertia. The orientational order parameters p_x, p_y and p_z are defined by

$$p_i = \left\langle \frac{3\cos^2\theta_i - 1}{2} \right\rangle \qquad (i = x, y, z), \tag{5}$$

where θ_i is the angle between the end-to-end vector of an amphiphilic molecule and the i-axis ($i = x, y, z$), and $\langle \cdots \rangle$ denotes the average over of all amphiphilic molecules. Note that the average is taken for the amphiphilic molecules in the vicinity of the center-of-mass position of the micelle, that is, those in the region $-\Delta r < x, y, z < \Delta r$. We set $\Delta r = 2.5\sigma$ in the calculation of p_i. Ideally, the orientational order parameters take the following values: $(p_x, p_y, p_z) = (1, -0.5, -0.5)$ for a disc, $(p_x, p_y, p_z) = (0, 0, -0.5)$ for a cylinder and $(p_x, p_y, p_z) = (0, 0, 0)$ for a sphere. Detailed analysis of the distribution functions of these orientational order parameters showed that, in practice, three types of micellar shapes are clearly distinguishable by the orientational order parameters: $0.5 < p_x < 1.0$ and $-0.5 < p_y, p_z < -0.25$ for a disc micelle, $-0.25 < p_x, p_y < 0.5$ and $-0.5 < p_z < -0.25$ for a cylindrical micelle, and $-0.25 < p_x, p_y, p_z < 0.5$ for a spherical micelle [8, 9, 10]. We calculate the fractions of the micellar shapes on the basis of these orientational order parameters.

Figure 3 shows the fraction of various micellar shapes as a function of the intensity of the hydrophilic interaction $\varepsilon^*_{\rm hs}$. We find from this figure that the dominant micellar shape is a disc for $\varepsilon^*_{\rm hs} \leq 1.2$, a cylinder for $\varepsilon^*_{\rm hs} = 2.0$ and 2.2, and a sphere for $\varepsilon^*_{\rm hs} \geq 3.6$. It is also found that there exists a wide coexistence region in the intensity of the hydrophilic interaction between a cylinder and a sphere for $2.4 \leq \varepsilon^*_{\rm hs} \leq 3.4$. The micellar shapes are visibly distinguishable even in the coexistence region, that is, a cylinder and a sphere coexist dynamically.

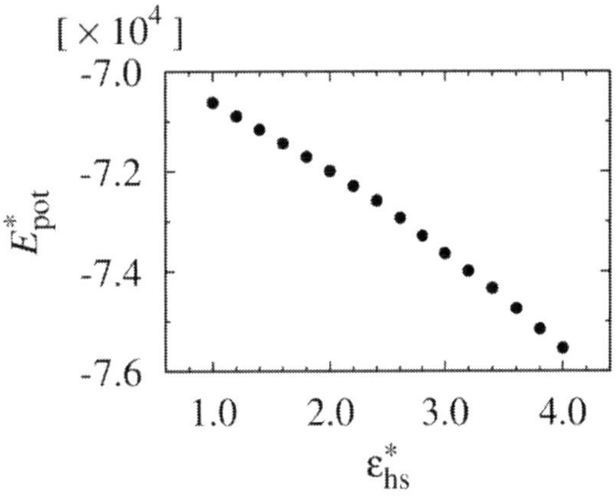

Figure 2. The total potential energy $E^*_{\rm pot}$ versus the hydrophilic interaction parameter $\varepsilon^*_{\rm hs}$.

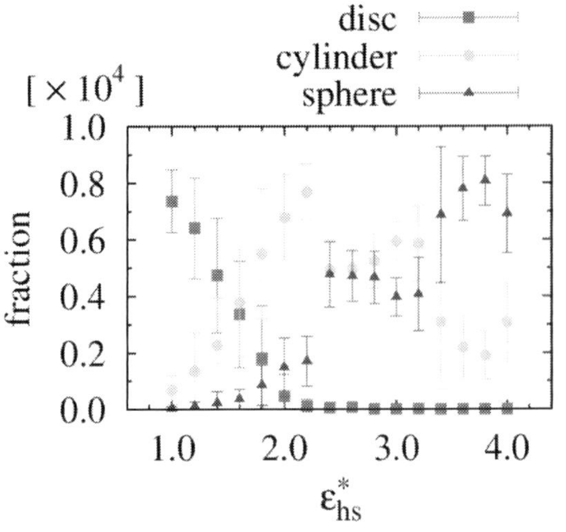

Figure 3. The fraction of various micellar shapes versus the hydrophilic interaction parameter $\varepsilon^*_{\rm hs}$.

4. Conclusions

We have obtained the following new results by performing MD simulations of coarse-grained rigid amphiphilic molecules with explicit solvent molecules. (1) As the intensity of the hydrophilic interaction, $\varepsilon^*_{\rm hs}$, increases, the micellar shape changes from a disc into a cylinder, and then into a sphere. (2) The potential energy decreases monotonically even during the shape transition of micelles as $\varepsilon^*_{\rm hs}$ increases. (3) There exists a wide coexistence region in the intensity of the hydrophilic interaction between a cylinder and a sphere.

Acknowledgments

This study was partially supported by JSPS KAKENHI Grant Number 22540419 and was partially performed with the support and under the auspices of the NIFS Collaborative Research Program (NIFS12KNTS016). The authors thank the Supercomputer Center, Institute for Solid State Physics, University of Tokyo for the use of the facilities.

References

[1] Gelbart W M, Ben-Shaul A and Roux D (eds) 1994 *Micelles, Membranes, Microemulsions, and Monolayers* (Springer-Verlag, New York) pp 1 – 104
[2] Israelachvili J N 1992 *Intermolecular and Surface Forces* (Academic Press, London, 2nd ed.)
[3] Hamley I W 2007 *Introduction to Soft Matter* (J. Wiley, Chichester, Rev. ed.)
[4] Sangwai A V and Sureshkumar R 2011 *Langmuir* **27** 6628
[5] Velinova M, Sengupta D, Tadjer A V and Marrink S J 2011 *Langmuir* **27** 14071
[6] Brodskaya E N 2012 *Colloid J.* **74** 154
[7] Fujiwara S, Hashimoto M and Itoh T 2006 *J. Plasma Phys.* **72** 1011
[8] Fujiwara S, Itoh T, Hashimoto M and Tamura Y 2007 *Mol. Simul.* **33** 115
[9] Fujiwara S, Itoh T, Hashimoto M and Horiuchi R 2009 *J. Chem. Phys.* **130** 144901
[10] Fujiwara S, Hashimoto M, Itoh T and Horiuchi R 2012 *Chem. Lett.* **41** 1038
[11] Goetz R and Lipowsky R 1998 *J. Chem. Phys.* **108** 7397
[12] Allen M P and Tildesley D J S 1987 *Computer Simulation of Liquids* (Clarendon Press, Oxford)

Journal of Physics: Conference Series **454** (2013) 012025 doi:10.1088/1742-6596/454/1/012025

A graph theoretical approach to fluctuating networks in glass-forming liquids: A molecular dynamics study with applications of the pebble game algorithm

Yasushi Takéuchi

Hinode Science Center, National Astronomical Observatory of Japan, Osawa 2-21-1, Mitaka, Tokyo, Japan

E-mail: takeuchi.yasushi@nao.ac.jp

Abstract. The network structures fluctuating in space and time are studied in glass-forming liquids of silica (SiO_2) and silicate (Mg_2SiO_4), by carrying out molecular dynamics (MD) simulations and then applying the graph theoretical algorithm of a "pebble game". The pebble game algorithm was developed in Thorpe and coworkers' studies on the percolation of rigidity to form amorphous solids and, although its exactness is proved only in two dimensions, it has been demonstrated to apply to three-dimensional networks. In liquid silica and silicates, the network connections are extended with amounts of network-forming Si, O atoms, and infinitely-percolating clusters emerge over an extensive range of compositions. The search along a network for free "pebbles", attached virtually to the constituent atoms to represent their degrees of freedom of motions, shows that one infinitely-percolating cluster contains some rigid clusters. These rigid clusters in liquid states fluctuate in space and time because of the alteration of connections at high temperatures, in contrast to those solidified below glass transition temperatures, and these fluctuations are responsible for the slow structural relaxations. The pebble game analyses thus give insights into the internal structures in infinitely-spanning networks, beyond surveys by the conventional theory of percolation, and reveal the behaviors of rigid clusters playing the dominant role in the slow structural relaxations in glass-forming liquids. We discuss these results especially by taking focus on varieties in network structures, such as the composition-dependent degrees of connectivity and structural transformation under high pressures.

1. Introduction

The high abilities to form glasses of various inorganic materials, *e.g.* silica (SiO_2), have their origins in network structures. Three-dimensional networks are infinitely spanned by Si-O connections in liquid or vitreous silica. The connections impose constraints on motions of the network-forming atoms and, correspondingly, fluctuations to relax strains in networks are constrained. Viscous flow relaxations in network liquids are especially suppressed, because these relaxations undergo the thermal activation of altering connections and rearranging atoms. In lowering temperatures, fluctuations are converged to infinitesimal motions and the random configurations of atoms under constraints are readily "frozen in." An intuitive understanding thus ascribes glass transitions to stiffening of liquids, or, inhibiting fluctuations for fluid relaxations at low temperatures [1].

The degrees of connectivity in network structures depend strongly on chemical compositions of glass-forming liquids. If one- or two-valent cations are added to silica to form silicates such as $MgSiO_3$ and Mg_2SiO_4, these cations are coordinated to dangling O atoms and intercalated amongst networks for the local charge neutrality. The degrees of connectivity are therefore reduced with amounts of intercalated cations. The decreasing degrees of connectivity induce the decrease in shear viscosities by several orders of magnitude. Long shear relaxation times, tens nanoseconds or longer in highly-connected network liquids, correspondingly decrease in ill-connected network liquids and approach normal relaxation times of the order of 0.1 picosecond. The transitions from "strong", highly-connected networks to "fragile", ill-connected networks are responsible for the distinct temperature-dependences of shear viscosities, Arrhenian in highly-connected network liquids and non-Arrhenian in ill-connected network liquids [1]. The network connectivities thus play the dominant roles in slow structural relaxations, and consequently in glass-forming abilities.

In addition to depending strongly on chemical compositions, network structures undergo transformations under high pressures, and thereby their structural relaxations are affected. Highly-connected network liquids especially exhibit enigmatic behaviors, *viz.* decreasing shear viscosities under high pressures [2, 3]. High-pressure transformations enhance structural relaxations in highly-connected networks because some fractions of regularly network-forming, 4-fold Si atoms are compacted to irregularly network-forming, 5-fold atoms and catalyze the alteration of Si-O connections [3, 4]. These spatial and temporal fluctuations of network connections are intrinsic in glass-forming liquids and determine the behaviors of their structural relaxations.

The time-dependent configurations of atoms in network liquids can be obtained from molecular dynamics (MD) simulations. The network structures are readily identified from these configurations and, if their degrees of connectivity exceed the percolation thresholds, major parts of these networks are occupied by singly-connected clusters which percolate infinitely over three-dimensional space [5]. Besides their single connectedness, these infinitely-percolating networks are comprised of finite domains having different flexibilities or rigidities in structural relaxations, because of spatially and temporally fluctuating connections. These domains are ill identified, however, in a conventional framework of the percolation theory. The connectedness of flexible or rigid domains can rather be explored by employing the theory of rigidity percolation, which was addressed pioneeringly in Thorpe's studies on ascribing glass transitions to the percolation of rigid domains under the constraints of covalent bondings [6]. To identify the topology of constraints determining the rigidity percolation, the protocols are specified by the algorithm of a "pebble game" [7, 8, 9, 10, 11, 12, 13, 14] in a framework of graph theory.

This article reports the preliminary results of a graph theoretical approach applying the pebble game algorithm to network structures in glass-forming liquids. MD simulations were carried out at first to generate highly-connected network structures in liquid SiO_2 (silica) and ill-connected network structures in liquid Mg_2SiO_4 (Mg-olivine) at high temperatures and high pressures. The methodology of these simulations is described in Section 2. The pebble games are then played on these network structures to identify their flexibilities or rigidities under the constraints of Si-O connections. Section 3 is devoted to describing the pebble game algorithm, which specifies the feasible protocols to identify the edges (or, connections) as the constraints on motions of domains in a network, and thereby to enumerate the degrees of freedom of domain motions. This algorithm was exploited originally to apply to two-dimensional networks [7, 8, 9, 10], according to its rigorous proof in two dimensions by graph theory, whilst recent studies have empirically verified its applicability in three dimensions [11, 12, 13, 14]. In Section 4, the applications of the pebble game algorithm to network liquids are discussed. The results of these applications reveal that, in network liquids, the rigidities differ by the degrees of connectivity, and that these different rigidities are consistent with the different behaviors in slow structural relaxations, *viz.* large and small shear viscosities of highly- and ill-connected network liquids, respectively, and their decrease and increase under high pressures.

2. Molecular dynamics simulations

The NVE (or, microcanonical) ensembles of network structures were generated by carrying out MD simulations on liquid SiO_2 and Mg_2SiO_4 at high temperatures $T \simeq$ 2200-3300 K and high pressures $P \simeq$ 0.1-10 GPa. Temperatures $T \simeq 1.1 \times T_m$ were elevated along the elevation of melting temperatures T_m under high pressures. The finite numbers of the constituent atoms (512 Si, 1024 O of SiO_2, and 512 Mg, 256 Si, 1024 O of Mg_2SiO_4) were confined to the cubic boxes with the periodic boundaries, and neither external degrees of freedom nor constraints were imposed to the systems conserving the numbers N of atoms, volumes V, and energies E. The box sizes $L \simeq$ 24-28 Å were fixed to serve the densities under the relevant temperatures and pressures. The constituent atoms were assumed to be the rigid ions (*i.e.* the firm couplings of cores and valence electrons) interacting in a framework of classical mechanics by the Born-Mayer-Huggins pairwise potentials with Matsui's empirical parameters [15]. The Ewald method [16] was applied to sum the long-range Coulombic interactions over all pairs of atoms in the fundamental box and its replicas, whilst the short-range interactions of Born-Mayer repulsion and van der Waals dispersion were truncated at some separations of atoms. The equations of motion of atoms were integrated numerically by Verlet's central difference algorithm [17] with the time-step increment $\Delta t = 1$ fs over 102400 time-steps, corresponding to the uppermost time $t_{max} = 102.4$ ps, and their positions and velocities were accumulated at intervals $\Delta t' = 100\Delta t = 0.1$ ps. Prior to thus achieving equilibria, temperatures were controlled by scaling the velocities of atoms. The time series of the atomic configurations fluctuating in equilibria were generated consequently and, according to the ergodic hypothesis, these time series are equivalent to NVE ensembles.

3. The pebble game algorithm

An intuitive understanding speculates that a larger number of edges cause the larger rigidity in a network whilst, if edges are overabundant, not all of them independently impose the constraints on motions of incident vertices and thereby some of them are redundant. This speculation is verified in a framework of graph theory by Laman's theorem [18]. We assume generic networks hereafter, which have no special symmetry such as parallel edges or collinearly connected edges [7, 8, 9, 10]. Laman's theorem states that a generic network of two dimensions ($d = 2$) contains no redundant edge if, and only if, none of its subgraphs violates $b \leq 2n - 3$ (n, b are the numbers of vertices and edges of a subgraph), or equivalently, if and only if no subgraph contains larger numbers of edges than the rest of subtracting the degrees of freedom of trivial motions of a rigid body ($d(d+1)/2 = 3$) from those of all vertices in a subgraph ($dn = 2n$). If overall subgraphs are naively enumerated, however, their number increases divergingly with the size of their parent network and, inevitably, the search over them is not tractable in large networks.

To implement Laman's theorem by circumventing the diverging numbers of combinatorial countings, Jacobs and Hendrickson proposed the algorithm of a "pebble game" [9, 19]. Their algorithm deals with the virtual objects, pebbles, by which the degrees of freedom are visualized in a two-dimensional generic network. Two ($= d$) pebbles are attached to each vertex and represent its inherent degrees of freedom of two-dimensional unconstrained motions. Free pebbles are located on vertices, and one free pebble contributes one degree of freedom to those of motions of the distinct domain containing its attached vertex. Those other than free pebbles are, on the other hand, assigned to independent constraints and have no contribution to the degrees of freedom of domain motions. One independent edge is covered by one pebble, by consuming one free pebble complementarily from one of its incident vertices. Each domain contains at least three ($= d(d+1)/2$) free pebbles of a rigid body. The domains containing more than three free pebbles have larger degrees of freedom of motions and are, therefore, more flexible than rigid bodies. The flexible domains are thus defined by the underconstrained domains consisting of n vertices and less than $2n-3$ edges. The domains containing only three free pebbles are rigid, on the other hand, and each rigid domain contains $2n-3$ independent edges. The rigid domains containing just $2n-3$ edges are isostatic, and all of their edges are independent constraints. If more than $2n-3$ edges are contained in a rigid domain, its excess edges other than $2n-3$ independent constraints are redundant constraints. The rigid domains containing redundant edges are overconstrained. The redundant edges are not covered by pebbles, because all pebbles are either free on vertices or assigned to independent edges. The underconstrained, isostatic, and overconstrained domains under the independent and redundant constraints are thus identified throughout a two-dimensional generic network. The protocols to search over all edges and to rearrange pebbles do not consume so huge computational efforts as combinatorial countings of subgraphs, and consequently, along these protocols, the computations amenable rigorously to Laman's theorem are feasible.

Besides its exactness in two dimensions, no rigorous extension of the pebble game algorithm is proven in three dimensions. This is because the sufficient condition in Laman's theorem (or, the part of "only if" in its statement) is broken in three dimensions [19]. Jacobs conjectured that the pebble game algorithm applies to three-dimensional bond-bending networks, because the angular constraints on bond-bending rule out the "double banana graph", an infamous counterexample to Laman's theorem in three dimensions [11]. His conjecture is analogous to the molecular framework conjecture [20] on applying Laman's theorem to the flexibilities of framework molecules. Some works have, in fact, applied the pebble game algorithm successfully to macromolecules such as proteins [12, 13]. Furthermore, no failure has been reported in its tentative applications to both bond-bending and non-bond-bending networks in three dimensions [14].

The pebble game algorithm specifies the same protocols in three dimensions ($d=3$) as in two dimensions, except for attaching three ($=d$) pebbles to each vertex and fixing six ($=d(d+1)/2$) free pebbles of a rigid body in each domain [11, 12, 13, 14]. A three-dimensional pebble game is executed along the following protocols:

(i) Initially, locate all vertices without edges, and attach three free pebbles to each vertex.

(ii) Add one edge, e_{ab} connecting vertices v_a and v_b, and try to cover e_{ab} by one pebble;

  (a) Collect three free pebbles to each of v_a and v_b, and fix these six free pebbles on v_a and v_b temporarily.

  (b) Search for a seventh free pebble to be located on one of the vertices which are connected to v_a or v_b by the edges added in the preceding protocols, without rearranging the six free pebbles on v_a and v_b.

    - If a seventh free pebble is found and is located on a vertex connected to v_a (or v_b), then identify e_{ab} as an independent constraint, and cover e_{ab} by consuming one of the three free pebbles on v_a (or v_b).
    - Otherwise, identify e_{ab} as a redundant constraint. Furthermore, identify v_a, v_b, e_{ab}, and the vertices and edges which were trailed in the failed search for a seventh free pebble, as some parts of an overconstrained domain.

(iii) Iterate Protocol (ii) until all edges are added.

In these protocols, the search for a free pebble which can be located on a vertex v_x is executed by rearranging the pebbles in the sequences of connected edges starting from v_x, under the definitions of attaching three pebbles to each vertex and covering each independent edge by one pebble. ¿From a pebble game along these protocols, the flexible and rigid domains are identified in a three-dimensional network, and their degrees of freedom under the independent and redundant constraints are obtained.

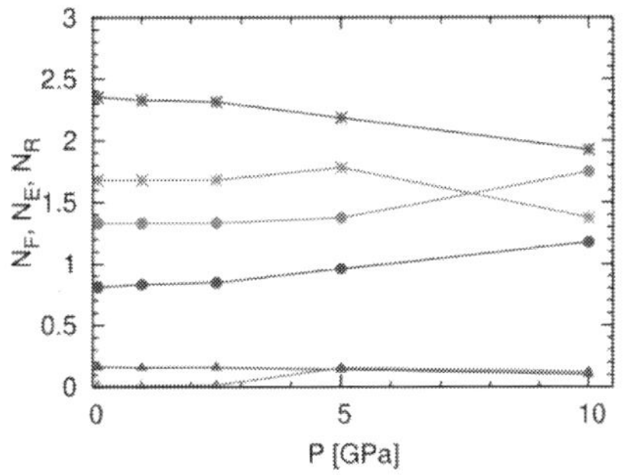

Figure 1. The average numbers N_F of floppy modes (asterisks), N_E of edges (circles), and N_R of redundant constraints (triangles) in liquid SiO_2 (red lines) and Mg_2SiO_4 (blue lines) at various pressures P.

4. Rigidity in network liquids

The network structures in liquid SiO_2 and Mg_2SiO_4 consist of the vertices and edges which are defined by the network-forming Si, O atoms and the Si-O connections, respectively. Playing the pebble games on these networks, the flexible and rigid domains are identified with their degrees of freedom of unconstrained motions, or the numbers of "floppy" modes which in glassy states are excited at no (or negligible) energy cost [6, 7, 8]. The time averages of the fluctuating numbers of floppy modes, N_F, of a network-forming atom are shown in figure 1. The average numbers of edges N_E and redundant constraints N_R are additionally shown.

The results in figure 1 show the smaller values of N_F in highly-connected networks of liquid SiO_2 than those in ill-connected networks of liquid Mg_2SiO_4, and therefore indicate the larger rigidities in highly-connected networks than in ill-connected networks. These results further show that N_F in liquid SiO_2 is insensitive to pressure whereas N_F in liquid Mg_2SiO_4 decreases under high pressures, and these behaviors are consistent with the decreasing and increasing shear viscosities respectively of liquid SiO_2 and Mg_2SiO_4 under high pressures, as predicted in Takéuchi's recent studies [2].

As a consequence, the pebble games in network liquids reveal that their rigidities differ by the degrees of connectivity, and that these different rigidities reflect the different behaviors in structural relaxations. Takéuchi's future works will advance from these preliminary results, by exploring the dynamics of flexible and rigid network domains to determine their behaviors in slow structural relaxations, especially under high pressures: This will be discussed elsewhere.

Acknowledgments

The author thanks the CCP2012 committee for offering good opportunities of useful discussions in this vigorous conference. The author also thanks an anonymous referee for his or her kindly and useful comments.

References

[1] Angell C A, Ngai K L, McKenna G B, McMillan P F and Martin S W 2000 *J. Appl. Phys.* **88** 3113
[2] Takéuchi Y 2009 *Prog. Theor. Phys. Suppl.* **178** 181
[3] McMillan P F and Wilding M C 2009 *J. Non-Cryst. Solids* **355** 722
[4] Tsuneyuki S and Matsui Y 1995 *Phys. Rev. Lett.* **74** 3197
[5] Stauffer D and Aharony A 1994 *Introduction to Percolation Theory* 2nd Ed. (London: Taylor & Francis)
[6] Thorpe M F 1983 *J. Non-Cryst. Solids* **57** 355
[7] Jacobs D J and Thorpe M F 1995 *Phys. Rev. Lett.* **75** 4051
[8] Jacobs D J and Thorpe M F 1996 *Phys. Rev.* E **53** 3682

[9] Jacobs D J and Hendrickson B 1997 *J. Comput. Phys.* **137** 346
[10] Thorpe M F, Jacobs D J, Chubynsky N V and Rader A J 1999 *Rigidity Theory and Applications*, ed M F Thorpe and P M Duxbury (New York: Plenum) p 239
[11] Jacobs D J 1998 *J. Phys.* A **31** 6653
[12] Jacobs D J, Kuhn L A and Thorpe M F 1999 *Rigidity Theory and Applications*, ed M F Thorpe and P M Duxbury (New York: Plenum) p 357
[13] Jacobs D J, Rader A J, Kuhn L A and Thorpe M F 2001 *Proteins* **44** 150
[14] Chubynsky M V and Thorpe M F 2007 *Phys. Rev.* E **76** 041135
[15] Matsui M 1998 *Properties of Earth and Planetary Materials at High Pressure and Temperature*, ed M H Manghnani and T Yagi (Washington, D.C.: American Geophysical Union) p 145
[16] Sangster M J L and Dixon M 1976 *Adv. Phys.* **25** 247
[17] Verlet L 1967 *Phys. Rev.* **159** 98
[18] Laman G 1970 *J. Eng. Math.* **4** 331
[19] Hendrickson B 1992 *SIAM J. Comput.* **21** 65
[20] Whiteley W 2005 *Phys. Biol.* **2** S116

Journal of Physics: Conference Series **454** (2013) 012026 doi:10.1088/1742-6596/454/1/012026

Order parameter by instantaneous normal mode analysis for melting behaviour of cluster $Ag_{17}Cu_2$

Pin-Han Tang[1], **Ten-Ming Wu**[1] **and S K Lai**[2]

[1] Institute of Physics, National Chiao-Tung University, Hsinchu 300, Taiwan
[2] Complex Liquid Laboratory, Department of Physics, National Central University, Chungli 320, Taiwan

Email: tmw@faculty.nctu.edu.tw; sklai@coll.phy.ncu.edu.tw

Abstract. We perform isothermal Brownian-type molecular dynamics simulations with the Gupta potential for bimetallic cluster $Ag_{17}Cu_2$ from T=0 to 1500K, across the temperature range associated with the melting behaviour determined by the specific heat of the cluster. We also use the instantaneous normal mode (INM) analysis to dissect dynamics of the cluster. In terms of the projected density of states of the vibrational INMs, we propose a new order parameter that specifically describes the melting behaviour of the cluster. The calculated result of our order parameter is consistent with the information inferred from the specific heat data.

1. Introduction

Melting of clusters, which is associated with a transition from the solid-like to the liquid-like cluster structures, is more complicated than that of bulk systems [1]. Due to the finite sizes of clusters, the melting transition of a cluster occurs over a temperature range, rather than at a well-defined temperature [2,3,4]. Instead of a discontinuous jump as in the specific heat of a bulk system at the melting temperature, the specific heat C_V of a cluster changes continuously with temperature and generally exhibits a single broader peak or a main peak with a prepeak at a lower temperature [3,4,5]; the melting temperature T_m is usually defined as the temperature at the single peak or the main peak.

Physically, the melting phenomena are described by some appropriated chosen order parameters. The melting of bulk systems at a given pressure is a first-order transition in thermodynamics and commonly characterized by a discontinuous jump in the order parameter, such as entropy or volume, at the melting temperature. For clusters, other order parameters have been used to characterize the transition between the solid-like and liquid-like cluster structures [2]. Theoretically, the bimodality in the probability distribution of the short-time averaged temperature in microcanonical ensemble provides an evidence for the coexistence of the solid-like and liquid-like cluster isomers within a temperature range [6]. By the microcanonical ensemble simulations, the solid-like and liquid-like isomers of a cluster are found to be separated by a barrier in Landau free energy and, therefore, potential energy serves as the order parameter of the free energy [7]. Also based on the short-time average, some geometric order parameters are used to address the transition between two stable isomers and their coexistence [8]. Continuously used in computer simulations [5,9,10], the root-mean-square bond length fluctuation constant δ acts as a Lindemann-like order parameter, which signals some kinds of structural or phase transformation. However, the information of the transition deduced from δ and C_V is not always consistent. Recently, by analyzing the cluster dynamics of $Ag_{17}Cu_2$ with the instantaneous normal mode (INM) theory, we propose a new order parameter, which describes the melting behavior of the cluster in consistence with the information inferred from its specific heat [11]. The main purpose of this proceeding is to address this new order parameter.

2. Instantaneous Normal Mode Theory for Clusters

Originally developed for studying the short-time dynamics of liquids from the viewpoint of potential energy landscape [12,13], the INM theory has been a useful analytic tool for investigating microscopic dynamics in condensed matter and fruitful results of bulk systems have been presented. A recent review of the INM theory is given in Ref. [14]. Before applied for bulk systems, the INM analysis has been used to study cluster dynamics in a series of papers in 1990 [15,16,17]. Despite some insight provided by those studies, the strategy to exploit the cluster dynamics with the INM analysis still demands further exploration. The major impediment lies in the particle number. Even for a cluster containing particles less than a hundred, the potential energy landscape of the cluster is extremely complicated in a multi-dimensional space. Differing from the particle motions in bulk systems, whose particle numbers are infinite, a cluster with finite particle number has rotational motions of a whole about its principal axes. This feature causes the INM analysis for clusters in some aspects different from that for bulk systems and results in a new order parameter to monitor the melting behavior of the cluster.

The INMs of a cluster configuration are referred to as the eigenmodes of the Hessian matrix evaluated at the configuration, where the Hessian matrix is defined as the second derivatives of the potential energy function with respect to particle displacements [18]. For a cluster of N particles with mass m_j, for $j=1,\dots, N$, which may be of different species, the $3N$ INMs generally contain three translational (T_x, T_y, T_z), three rotational (R_x, R_y and R_z) and $3N$-6 vibrational ($V_1,\dots,V_{3N-6}$) modes. The normalized eigenvector of an INM, indexed with α, is denoted as $\vec{e}_j^{\,\alpha}$, where $\vec{e}_j^{\,\alpha} \equiv (e_{jx}^{\alpha},\ e_{jy}^{\alpha},\ e_{jz}^{\alpha})$ is the three-dimensional displacement of particle j in this INM.

The three translational INMs are easily identified with zero eigenvalues. Because of the vibrational-rotational couplings, the three rotational modes of a cluster can be separated out from the rest of vibrational modes by two methods. As the vibrational-rotational couplings are extremely weak, the clusters are close to rigid so that the rotational and vibrational degrees of freedom are assumed to be decoupled and three purely rotational modes of the cluster can be separated out by the standard projection technique [19], which is referred to *Method I* in this proceeding. On the other hand, as the vibrational-rotational couplings are significant, the rotational and vibrational degrees of freedom of a cluster mix to some extent so that the purely rotational modes are not well defined. However, under the INM approximation, a cluster at an instant is considered as a rigid body with the instantaneous structure and its instantaneous total angular momentum is regarded as a constant. Thus, by an alternative method [11,17], which is referred as *Method II*, the approximated rotational modes of a cluster are identified as the INMs that have the three largest angular momenta, whose magnitudes are evaluated by the following formula

$$L_\alpha = c \left| \sum_{j=1}^{N} \sqrt{m_j} (\vec{r}_j \times \vec{e}_j^{\,\alpha}) \right|$$

where $\vec{r}_j$ is the position of particle j and c is the proportionality constant between the instantaneous particle velocity and the normalized INM eigenvector. Note that the constant c given here is different from the one given in Ref. [17], in which the particles of a cluster are of the same species.

After excluding the translational and rotational modes, the remaining $3N$-6 INMs of a cluster behave vibrationally and are indexed with the INM frequencies ω_α, which are the square roots of the INM eigenvalues. The normalized vibrational density of states (DOS) of a cluster is expressed as

$$D_{Vib}(\omega) = \frac{1}{3N-6} \left\langle \sum_{\alpha=1}^{3N-6} \delta(\omega - \omega_\alpha) \right\rangle$$

where the brackets denote an ensemble average over cluster configurations. Generally, $D_{Vib}(\omega)$ consists of two lobes, $D_{Vib}^{(s)}(\omega)$ and $D_{Vib}^{(u)}(\lambda)$, associated with the INMs of real frequencies ω_α and imaginary frequencies $\omega_\alpha = i\lambda_\alpha$ respectively. The normalization of $D_{Vib}(\omega)$ yields a unit area under the curves of $D_{Vib}^{(s)}(\omega)$ and $D_{Vib}^{(u)}(\lambda)$, which are usually plotted in the positive and negative axes of ω, respectively [10,11].

In the literatures, with the INM eigenvector, the projection operator of particle j is defined as [18,20]

$$P_j^\alpha = |\vec{e}_j^{\,\alpha}|^2 = \sum_{\mu=x,y,z} e_{j\mu}^\alpha \cdot e_{j\mu}^\alpha .$$

In terms of the projection operator, the contribution of particle j to the vibrational DOS is expressed as

$$D_j(\omega) = \frac{1}{3N-6}\left\langle \sum_{\alpha=1}^{3N-6} P_j^\alpha \delta(\omega-\omega_\alpha)\right\rangle .$$

The integration of $D_j(\omega)$ over frequency is given as

$$I_j \equiv \int_0^\infty D_j^{(s)}(\omega)d\omega + \int_0^\infty D_j^{(u)}(\lambda)d\lambda = \frac{1}{3N-6}\left\langle \sum_{\alpha=1}^{3N-6} P_j^\alpha \right\rangle .$$

Thus, I_j is also realized as an ensemble average of all vibrational projection operators associated with particle j. According to the results given in the following section, we emphasize that the temperature variations of all I_j quantities of a species in a cluster shed light on the melting behaviour of the cluster.

3. Simulation and Melting Behaviour for $Ag_{17}Cu_2$

With the Gupta potentials of Cu-Cu, Ag-Ag and Cu-Ag, we generate the configurations of $Ag_{17}Cu_2$ by Brownian-type isothermal MD simulations with a time step of 1 *fs* [11]. We start the simulation at T=0 and, then, heat up to 1500 K with a step of 10 K. Following a previous work [5], the specific heat of the cluster is calculated by

$$C_V = \frac{\langle E_{tot}^2\rangle_t - \langle E_{tot}\rangle_t^2}{k_B T^2},$$

where E_{tot} includes the kinetic energies of all atoms and the Gupta potential energies of every atomic pairs. At each T, the variation of E_{tot} is averaged over a long period in a range of $(1-3)\times 10^{-7} s$.

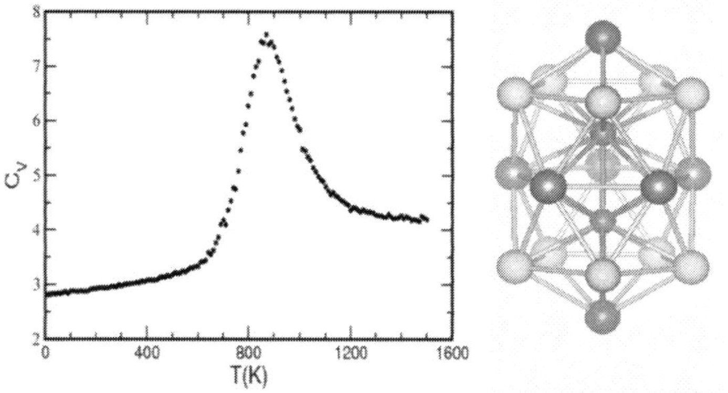

Figure 1. Specific heat C_V (left) and ground-state structure (right) of $Ag_{17}Cu_2$. C_V is in unit of k_B. The atoms in the ground-state structure are classified into four subsets, denoted as $Cu^{(2)}$ (red), $Ag^{(2)}$ (violet), $Ag^{(5)}$ (brown) and $Ag^{(10)}$ (green), as in Ref. [11].

In the left part of Fig. 1, the specific heat of $Ag_{17}Cu_2$ shows a maximum located at $T_m \approx 890$ K. Shown in the right part of Fig. 1 is the ground-state structure of $Ag_{17}Cu_2$ (at T=0), which has the D_{5h} symmetry [21] with the principal axis going through the two centrally located Cu atoms: a five-fold rotational symmetry about the principal axis, a reflection symmetry about the plane containing the middle pentagonal ring and another reflection symmetry about the plane containing the principal axis and anyone of Ag atoms residing in the middle pentagonal ring. Owing to the D_{5h} point group, we classify the atoms in the ground-state structure presented in Fig. 1 into four subsets of atoms: two Cu atoms, top and bottom Ag atoms, five Ag atoms in the middle pentagonal ring and ten Ag atoms in the upper and lower pentagonal rings, which are denoted as $Cu^{(2)}$, $Ag^{(2)}$, $Ag^{(5)}$ and $Ag^{(10)}$, respectively [11].

Averaged over 10^6 cluster configurations at each temperature, the INM DOS of $Ag_{17}Cu_2$ are depicted in Fig. 2. After removing the contributions of the rotational INMs, which are selected by anyone of the two methods described above, the vibrational INM DOS is obtained. Since the vibrational INM DOS obtained by the two methods are found to be only slight different [11], we only present in this proceedings the results by *Method II*. At T=100K, the INM spectra with and without the rotational contributions exhibit structural characteristics and all vibrational INMs possess real frequencies, displaying a solid-like behaviour, since the cluster configurations are around the ground-state structure corresponding to the lowest energy of the potential energy landscape. At T=500K, the INM spectra are generally smoothed out and the amount of the imaginary-frequency vibrational INMs are noticeable, which suggests that the cluster has the possibility to surmount the energy barriers to higher-energy excited states. As T increases further to 900K or 1300K, the INM DOS exhibits a large portion of imaginary-frequency vibrational INMs, indicating a liquid-like behaviour, in which a cluster at sufficiently high temperatures would go around everywhere the potential energy landscape.

The INM analysis is applied to the $Ag_{17}Cu_2$ cluster configurations at finite temperatures and at the ground-state structure (T=0). At finite T, the atoms in $Ag_{17}Cu_2$ are also classified into four subsets according to the following procedure: By tracing the atom labels of the subsets in the ground-state structure, the subsets of atoms at a finite temperature can be obtained in a simulation, in which the cluster temperature is gradually raised from zero. At low temperatures, one can recognize the four subsets of atoms by the cluster structures, in which the relative positions of atoms are similar as those in the ground-state structure. On the contrary, at temperatures where the cluster is realized as the liquid-like behaviour, the atom positions of the same species become completely random within the range of cluster size so that the subsets of atoms of $Ag^{(2)}$, $Ag^{(5)}$ and $Ag^{(10)}$ lose the identities they had at T=0. As a result, atoms in the cluster at high temperatures are only distinguished by species of different masses.

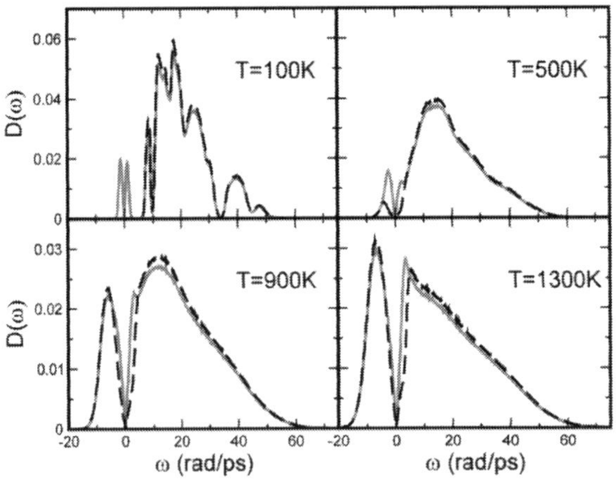

Figure 2. Normalized INM DOS of $Ag_{17}Cu_2$ at four temperatures. The red-solid lines are the spectra with the contributions of rotational and vibrational INMs and the black-dashed lines are the spectra by removing the contributions of the rotational INMs selected by *Method II* described in Ref. [11].

Because the atoms in a subset play the same role in dynamics, we average their $D_j(\omega)$ given as

$$D_X(\omega) = \frac{1}{N_X}\sum_{j=1}^{N_X} D_j(\omega),$$

where the subscript X=$Cu^{(2)}$, $Ag^{(2)}$, $Ag^{(5)}$ and $Ag^{(10)}$ and N_X is the total number of atoms in subset X [11]. Therefore, $D_X(\omega)$ is the averaged contribution of an atom in subset X to all vibrational INMs with ω. Shown in Fig. 3 are $D_X(\omega)$ for the four subsets of atoms of $Ag_{17}Cu_2$ at T=0, 100 and 500K. At zero temperature, $D_X(\omega)$ is a bunch of straight lines and, for any subset of atoms, there is no line falling in the range between 30 and 37 rap/ps. The vibrational INMs with ω above 37 rap/ps are dominated by the Cu atoms and the INMs with ω below 30 rap/ps are dominated by the Ag atoms. At T=100K, the temperature effect smears the straight line of $D_X(\omega)$ at T=0 into a continuous function and the vibrational-frequency gap between 30 and 37 rap/ps is generally filled out. At T=500K, with the appearance of the imaginary-frequency lobe, $D_X(\omega)$ of the three subsets of Ag atoms become similar in spectral shape but are distinguishable from that of the subset $Cu^{(2)}$.

4. Order Parameter by the INM Analysis

By integrating $D_j(\omega)$ of the vibrational INMs obtained by *Method II* [11], the I_j values of atoms in the four subsets of $Ag_{17}Cu_2$ are plotted in Fig. 4(a) with a temperature resolution of 10*K*. Due to the difference in masses of the Ag and Cu atoms, the I_j of the Cu atoms are well separated from those of the Ag atoms at all *T*. Below 450K, the I_j of the three subsets of Ag atoms split into three branches and as *T* approaches to zero the three I_j branches converge to the values corresponding to the subsets of Ag atoms in the ground-state structure. The three branches indicate that the atoms in the cluster in this temperature range vibrate with small amplitudes around the grounds-state structure and the vibrations result in a mild increase of the specific heat with temperature. From 450 to 600K, the three I_j branches are still visibly separated but somewhat defile by exchanges between two branches, physically corresponding to the site permutations among Ag atoms in different subsets or scatter due to the cluster in the excited states. In this temperature range, the cluster structures mostly remain similar as the ground-state configuration but, with some probability, the cluster is driven into the configurations of the first and

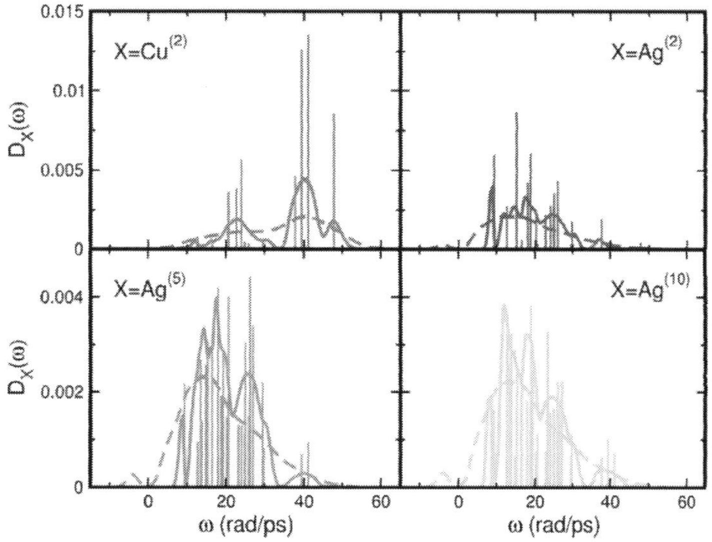

Figure 3. Vibrational $D_X(\omega)$ spectra for the four subsets of atoms of $Ag_{17}Cu_2$. As in Ref. [11], the straight lines and the solid curves are for the cluster at the ground-state structure and 100K, respectively. The dashed curves are for the cluster at T=500*K*.

second lowest excited states by thermal activation over the energy barriers in the potential energy landscape [11]. This thermal activation causes the behavior of the specific heat with increasing temperature different from that at low temperatures. Between 600 and 900K, the I_j of the three subsets of Ag atoms start to coalesce and the branches are no longer distinguishable, indicating that the ground-state structure is being melted and the original subsets of Ag atoms gradually lose their identities. With enough thermal energy, the cluster surmounts energy barriers with a large probability into even higher excited states [11], which are far from the ground state in the configuration space, so that the C_V within this temperature range increases dramatically and reaches a maximum at T=900K. After 900K, the I_j values of all Ag atoms merge into one, indicating that the ground-state structure could have melted and the Ag atoms are completely random. Corresponding to a sharp drop in the C_V within this range, the randomness among the Ag atoms is no longer enhanced by increasing further the cluster temperature so that we can safely say that the $Ag_{17}Cu_2$ cluster is in the liquid-like behaviour.

In terms of the I_j values of the Ag atoms, we propose a new order parameter $\tau(T)$ for describing the melting transition of $Ag_{17}Cu_2$. The variation of the I_j values for the 17 Ag atoms is calculated as

$$\sigma_I^2(T) = \frac{1}{N_{Ag}} \sum_{j=1}^{N_{Ag}} I_j^2 - \left(\frac{1}{N_{Ag}} \sum_{j=1}^{N_{Ag}} I_j \right)^2,$$

where N_{Ag} =17. The order parameter $\tau(T)$ is defined as the standard deviation $\sigma_I(T)$ normalized with $\sigma_I(T=0)$, i.e.

$$\tau(T) = \frac{\sigma_I(T)}{\sigma_I(T=0)}.$$

The results of $\tau(T)$ calculated for $Ag_{17}Cu_2$ are shown in Fig. 4(b). Below 300K, $\tau(T)$ is almost one, signaling that the ground-state structure is well preserved. From 300 to 600K, $\tau(T)$ declines from one to 0.85, indicating again that the ground-state structure is generally maintained. $\tau(T)$ starts to drop sharply around T=600*K* and approaches asymptotically to a small value around 890*K*, which is T_m defined by the peak position in the C_V curve shown in Fig. 1. This sharp decay in $\tau(T)$ marks a melting shake-up of the ground-state structure, with the cluster being thermally activated into the higher-energy excited states. Beyond T_m, $\tau(T)$ stays at a non-zero constant, which is due to the finite size of the cluster. Physically, the constancy in $\tau(T)$ manifests no further distinction in the atomic distribution under an ensemble average among the Ag atoms by further increasing temperature of the cluster.

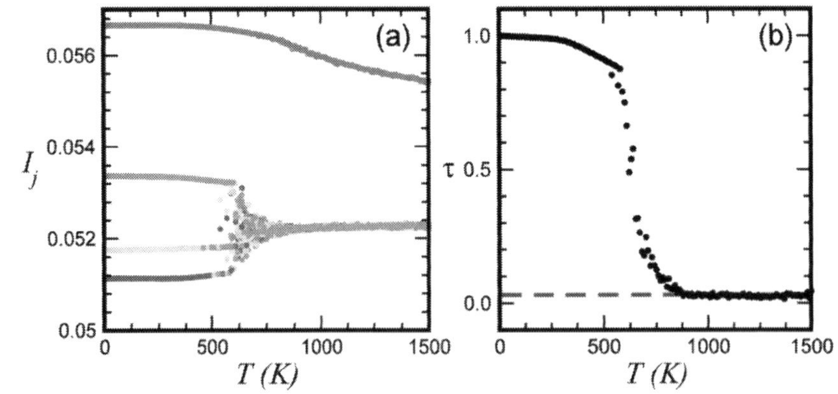

Figure 4. (a) Temperature variations of I_j for the four subsets of atoms (b) Order parameter $\tau(T)$ *vs* *T*. The dash line is the high-temperature asymptote of $\tau(T)$. The figures are referred from Ref. [11].

5. Conclusions

We have performed isothermal Brownian-type MD simulations with the Gupta potentials for $Ag_{17}Cu_2$ cluster. With the simulations, the melting behavior of the cluster is obtained by the specific heat data. On the other hand, the dynamic properties of $Ag_{17}Cu_2$ are analyzed by the INM approach. With the projection operators defined by the INM eigenvectors, the vibrational INM DOS, $D_j(\omega)$, of each atom is given. With the integral value I_j of $D_j(\omega)$ over frequency, we define a new order parameter $\tau(T)$ as the normalized standard deviation of the I_j value for the Ag atoms to describe the melting behavior of $Ag_{17}Cu_2$. In terms of $\tau(T)$, the melting behavior of the cluster is distinguished into three temperature regimes: (I) At low temperatures, where $\tau(T)$ is almost or close to one, the ground-state structure is generally preserved; (II) At intermediate temperatures, where $\tau(T)$ drops sharply from one to a non-zero small value, the cluster is generally found in the ground-state structure but, with some probability, in the higher-energy excited states; (III) At high temperatures, where $\tau(T)$ stays at a non-zero value, the cluster rarely resides in the ground-state structure and is almost found in the higher-energy excited states, which are significantly deviated from the ground state in the configuration space. According to our results for the $Ag_{17}Cu_2$ cluster, the three regimes characterized by $\tau(T)$ are consistent with the melting behavior obtained by the temperature variation of the specific heat. More recently, we have examined the new order parameter for Ag_{14} cluster, which specific heat shows both a main peak and a pre-peak at a lower temperature, and the order parameter also works well for Ag_{14} cluster [22].

Acknowledgement

TMW acknowledge supports from National Science Council, Taiwan under Grant No. NSC 101-2112-M-009-007.

References

[1] Wales D J 2003 *Energy Landscapes, With Applications to Clusters, Biomolecules and Glasses* (Cambridge University Press, Cambridge)
[2] Berry R S and Wales D J 1989 *Phys. Rev. Lett.* **63** 1156
[3] Schmidt M, Kusche R, Kronmüller W, von Issendorff B and Haberland H 1997 *Phys. Rev. Lett.* **79** 99
[4] Schmidt M, Kusche R, von Issendorff B and Haberland H 1998 *Nature* **393** 238
[5] Yen T W, Lai S K, Jakse N and Bretonnet J L 2007 *Phys. Rev. B* **75** 165420
[6] Jellinek J, Beck T L and Berry R S 1986 *J. Chem. Phys.* **84** 2783
[7] Lynden-Bell R M and Wales D J 1994 *J. Chem. Phys.* **101** 1460
[8] Doye J P K and Wales D J 1995 *J. Chem. Phys.* **102** 9673
[9] Calvo F and Spiegelmann F 2000 *J. Chem. Phys.* **112** 2888
[10] Hsu P J, Luo J S, Lai S K, Wax J F and Bretonnet J L 2008 *J. Chem. Phys.* **129** 194302
[11] Tang P H, Wu T M, Yen T W, Lai S K and Hsu P J 2011 *J. Chem. Phys.* **135** 094302
[12] Stratt R M 1995 *Acc. Chem. Res.* **28** 201
[13] Keyes T 1997 *J. Phys. Chem. A* **101** 2921
[14] Stanley H E, Buldyrev S V, Giovambattista N, La Nave E, Mossa S, Scala A, Sciortino F, Starr F W and Yamada M 2003 *J. Stat. Phys.* **110** 1039
[15] Adams J E and Stratt R M 1990 *J. Chem. Phys.* **93** 1332
[16] Adams J E and Stratt R M 1990 *J. Chem. Phys.* **93** 1358
[17] Adams J E and Stratt R M 1990 *J. Chem. Phys.* **93** 1632
[18] Larsen R E, Goodyear G and Stratt R M 1996 *J. Chem. Phys.* **104** 2987
[19] Page M and McIver Jr. J W 1988 *J. Chem. Phys.* **88** 922
[20] Buchner M, Ladanyi B M and Stratt R M 1992 *J. Chem. Phys.* **97**, 8522
[21] Carter R L 1998 *Molecular Symmetry and Group Theory* (Wiley, New York)
[22] Tang P H, Wu T M, Hsu P J and Lai S K 2012 *J. Chem. Phys.* **137** 244304

Molecular dynamics with atomic transitions and nuclear reactions

Richard More[1] and Feilu Wang[2]

[1] Lawrence Berkeley National Laboratory, Berkeley, California, USA
[2] National Astronomical Observatory, Chinese Academy of Sciences, Beijing, China

E-mail: rmore2262@sbcglobal.net

Abstract. We describe molecular dynamics particle simulations with particles that have internal structure and/or undergo reactions. These calculations give an atomic-scale description of hot plasma based on well-established microphysics and test kinetic theories used to calculate energy exchange and transport in plasma hydrodynamic simulations. The computer experiments can be given detailed diagnostics, without the usual limits of resolution of Laboratory equipment. Typical applications are non-equilibrium atomic kinetics for emission or absorption of X-ray laser radiation by solid targets and/or atomic-scale simulation of hot plasma with fusion reactions, as in inertial fusion ignition experiments.

1. Introduction

Molecular Dynamics (MD) particle simulation using large digital computers was introduced by Alder and Wainwright in 1957 [1]. MD studies a system of particles which represent atoms, molecules or ions. Each particle has classical coordinates $R_i(t)$, $V_i(t)$, and moves according to Newton's laws governed by forces from the neighbor atoms. For many years the particles were structureless hard spheres or point charges and the goal was to study particle correlations and phase transitions. Early calculations showed that computer particles rapidly relax to an equilibrium Maxwell distribution as a result of binary collisions [2]. In 1966, Brush, Sahlin and Teller applied the complementary technique of Monte Carlo (MC) simulation for the equilibrium statistical mechanics of point charges in dense plasmas [3]. The MC method is also widely used for dynamical processes.

Here we consider particle simulation *in which the particles can change state*. For example, if the particles are atoms or ions with excited states, we can study atomic kinetics and emission and absorption of radiation. If the particles are low-Z hydrogen isotopes, we can study fusion reactions and fusion ignition [4]. The particle motion in our simulation is classical Molecular Dynamics while the quantum transitions are treated by a Monte Carlo method (see Appendix A, Appendix B). The composite simulation can give new insights for a wide class of important scientific questions.

A fundamental scientific challenge of the calculations is the interface between classical and quantum descriptions: how can we include quantum effects, as completely and accurately as possible, in a basically classical particle simulation? Can we treat the particle motion by classical mechanics, turn the calculation over to quantum mechanics for an atomic transition or reaction, and then revert to classical dynamics for the reaction outputs? What is the best way to handle these abrupt changes of description (see Appendix B)?

A second fundamental challenge concerns computational efficiency. What new computational methods are needed to simulate atomic-scale many-body systems with reactions? We need to add or subtract particles created or destroyed in nuclear reactions; we want to use smaller time-steps for fast-moving parts of the system, and would like to use importance-weighted sampling for efficient computation (see Appendix A). A strategy is needed to handle many atoms when each atom can make many possible transitions between many excited states.

Fusion simulation must include ion-ion collisions, electron-ion and electron-radiation coupling, fusion alpha energy-loss and the production of energetic "knock-on" recoil ions, and the fusion reactions should be calculated for the actual ion velocity-distribution when this is not a simple Maxwellian. In section 3 we describe calculations of fusion ignition at inertial fusion target conditions and illustrate some scientific conclusions we can draw from these calculations.

2. MD code structure

The codes described here have the usual structure of Molecular Dynamics particle simulations [5]. They treat a small volume inside an assumed infinite medium, imposing periodic boundary conditions that keep the particles inside the simulation box. Sub-boxes are used to efficiently locate other atoms near any given atom. For our calculations, the simulation box is divided into $3375 = (15)^3$ sub-boxes. Atoms move freely between the sub-boxes.

Pointers (ion--> box, box --> neighbor boxes, box --> ions) serve to locate the atoms. For each atom one pointer tells which sub-box it is in, a second pointer identifies the 26 neighbor sub-boxes, and a third pointer tells which ion is in each sub-box. The pointers are updated during each time-step. A simple calculation gives the *periodic distance* between two particles, i.e., the distance calculated the shortest way in the toroidal topology of the periodic boundary conditions. Neighbor boxes are defined consistent with periodic boundary conditions.

Fast-moving particles take smaller time-steps than normal particles and those subcycle time steps are synchronized to the main time-steps. For example, in fusion calculations each fusion alpha particle should take 10 time steps during one time-step for the majority D,T ions because its velocity is ~ 10 times larger. The motion of slow particles is interpolated to supply position data for the subcycle time-steps.

In the usual kinetic theory of gases, the internal state of atoms is assumed to be separate from or uncoupled to the motion in the gas. The MD simulation is able to study the effects of coupling between particle position, velocity and internal state.

The internal state of atoms or ions is coupled to position and velocity $\mathbf{R}_i$, $\mathbf{V}_i$ in various ways: A fast-moving atom has more chances for collision than a slow-moving atom, and the cross-sections for reactions depend on the relative velocities $|\mathbf{V}_1 - \mathbf{V}_2|$ of the reacting particles. The Doppler shift of emission lines depends on the atom's velocity. In a plasma or ionized gas, the electric field $\mathbf{E}(\mathbf{r})$ of nearby ions causes a Stark-effect splitting of spectral lines and the Stark profile depends on the atomic position $\mathbf{R}_i$. Laser gain profiles and escape factors depend on the statistical distribution of velocities and line-widths in the whole plasma. There are simple theories for most of these couplings, but the theories typically treat one process at a time and/or assume a Maxwell distribution of velocities. Molecular Dynamics can relax these assumptions and test the theories.

3. Atomic kinetics for MD

We compare three methods to compute the atomic kinetics. These methods aim to calculate the statistical distribution of excited states, which is summarized by

$$N_{ij} = \text{probability that atom i is in state j} \quad (j = 0, 1, 2, ...) \tag{1}$$

Method 1.) Schroedinger Equation.

It is possible to solve a numerical Schroedinger equation for each atom i:

$$i\hbar \frac{\partial}{\partial t} \hat{U}_i(t) = \hat{H}_i(t) \circ \hat{U}_i(t) \tag{2}$$

Here H_i is a one-atom Hamiltonian function. $U_i(t)$ is time-evolution operator which tells how the wave-function $\Psi_i(t)$ evolves. Projected onto ψ_j, which is the eigenstate for the excited state j, this gives the population

$$N_{i,j}(t) = \left| \left\langle \psi_j \middle| U_i(t) \middle| \Psi_i(0) \right\rangle \right|^2 \tag{3}$$

In recent years the numerical Schroedinger eqn. has been used to study harmonic generation by high-intensity laser radiation. Reference [6] describes a one-atom Schroedinger code used to study mixed Stark and Zeeman effects for a hydrogen atom subject to high-frequency microwave radiation. For electron cyclotron resonance (ECR) heating in magnetic fusion plasma experiments, the microwave frequency is nearly resonant with the atomic Zeeman splitting.

The Schroedinger equation describes time-resolved atomic kinetics, i.e., it describes the time evolution of radiation emission or electron impact excitation. It would be possible to modify the Schroedinger equation to describe rapid transitions that are not time-resolved by introducing an effective Hamiltonian operator.

Using the Schroedinger equation, it is possible to study certain phenomena not encompassed by the other methods, for example: time-dependent Stark splitting [6] in the local atomic environment, or ion-ion collisions during emission of light and the consequent modification of Doppler line-width [7].

Method 2.) CR Method.

It is also possible to solve many copies of the Collisional-Radiative (CR) atomic kinetics model

$$\frac{dN_{ij}}{dt} = -N_{ij} \sum_k T_{j \to k} + \sum_k N_{ik} T_{k \to j} \tag{4}$$

The transition rate-matrix is schematically indicated by

$$T_{j \to k} = n_e \left\langle \sigma_{jk} v_e \right\rangle + A_{jk} \left(n_{h\nu} + 1 \right) \tag{5}$$

where σ_{jk} is the electron collision cross-section to change the atomic state j --> k, A_{jk} is a radiative rate (written for emission), and $n_{h\nu}$ is the number of photons per mode in the radiation environment. The notation is schematic because there are many additional processes [8]:

Electron collisional excitation	Collisional de-excitation
Electron-impact ionization	3-Body recombination
Absorption of radiation (excitation)	Emission
Photo-ionization	Radiative recombination
Autoionization	Dielectronic recombination
Stark ionization	Stark recombination

For several of these processes, simulations with classical point electrons have difficulty to obey detailed balance, and this is an argument in favor of the *fluid description* of free electrons which we adopt in this work.

The one-atom Collisional-Radiative model is a well-established computational method and has been used to analyze spectra from astrophysics and magnetic fusion plasmas for almost 50 years with extensive comparison to experiments [8]. The applications have required detailed knowledge of atomic processes, and useful collections of atomic data are available to accurately describe specific atoms and ions [9].

The CR rate equation describes atomic processes as instantaneous transitions; this is the opposite extreme from the Schroedinger equation. We should distinguish the *rate* of transitions (which depends on density) from the *duration* of the transition event. The actual duration of transitions depends on the atom (ion) considered, but roughly speaking,

e-i collision	i-i collision	radiation
$\sim 10^{-17}$ sec	$\sim 10^{-15}$ sec	$\sim 10^{-12}$ sec

We underline that the transition duration depends on the ion and the temperature; for example, radiation emission takes as long as 10 nsec for hydrogen (Z = 1) but is much quicker for high-Z ions, while "forbidden" transitions have lifetimes as long as 1 sec and are only seen from atoms which are not often disturbed by collisions in low-density astrophysical plasmas.

Method 3.) "Markov" Method.

In this method, each atom i is assumed to be in a definite excited state j(i) at each time. The state can change during each time-step, governed by probabilities; this is a mixed MC/MD method. The probability of the transition j --> k during dt is taken to be

$$P_{jk} = T_{j\text{-->}k}\, dt \tag{6}$$

The time-step dt must be small enough so that the sum (over k) of the probabilities is less than unity for all initial states j. In fact, for most excited states, on most time-steps, there is no transition; for efficiency, the MC test considers this most-likely case first.

Since this method appears to assume the atomic state is *constantly being measured*, there's a natural question whether it gives the same answer as the other methods. In general, we expect the Markov method to be entirely equivalent to the CR equations but to neglect quantum interference effects that are included in the Schroedinger method. An example of such a quantum effect is the mixing of 2s and 2p states that occurs when an atom is in a local electric field (Stark effect) [6]; this mixing effect has no representation in the CR or Markov descriptions.

3.1 Test of the Markov method

We describe a simple test of the Markov method for atomic kinetics by comparison with the usual CR model.

We consider neutral hydrogen, and describe the atomic dynamics by the populations of the 1,012 excited states having principal quantum number n = 1 to n = 11. The CR calculation follows the evolution of one atom, and finds the probabilities for the excited states. The atom begins in the n = 1 groundstate. The transition rates obey detailed balance consistent with a plasma temperature of $T_e = T_r = 10$ eV. The excited state populations relax to thermal equilibrium (with the expected Boltzmann distribution) in one nsec. (This is not complete equilibration because there is no coupling between electronic excited states and the ion kinetic energy, so the ion temperature remains zero in this test.)

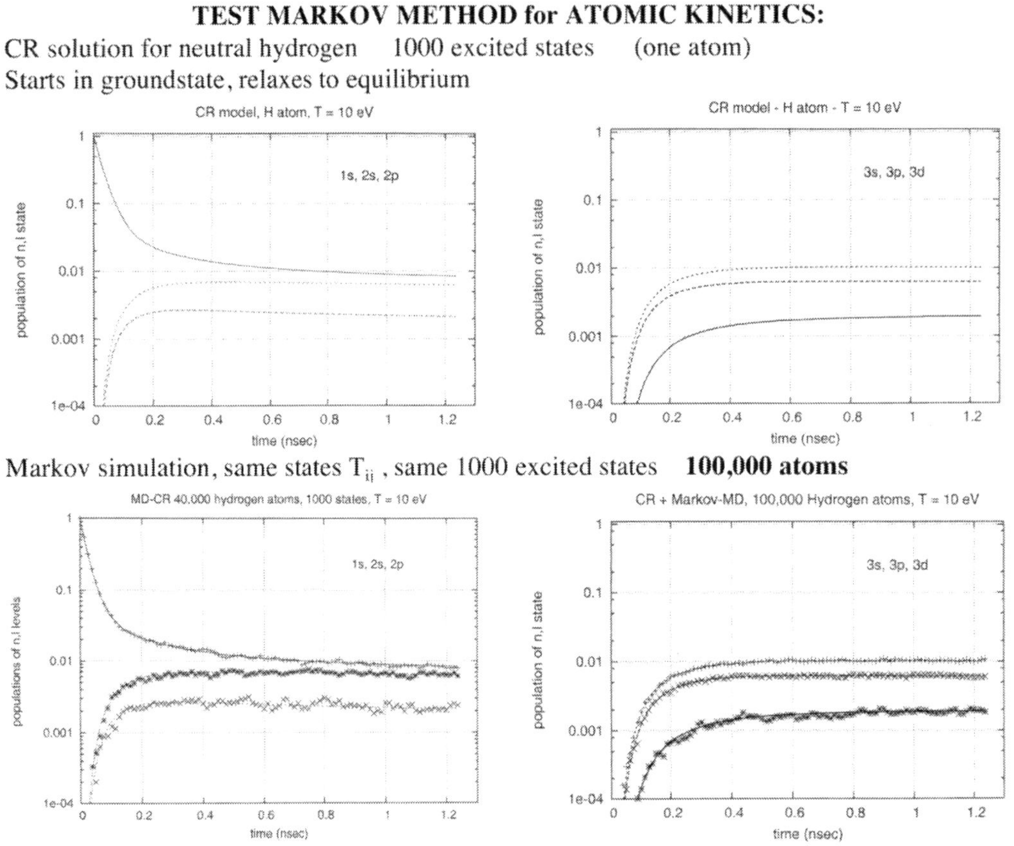

Figure 1. Comparison of one-atom collisional-radiative (CR) model for hydrogen with the Markov MD model described in the text. For this comparison the atoms are not moving and all have the same local environment. The 1012 states used and transition rates are the same in the two calculations; the purpose is to test the Markov MD method to calculate atomic kinetics. Agreement with the CR method is excellent, but similar calculations with smaller numbers of atoms showed larger effect of fluctuations, especially for excited states with small populations. The most interesting conclusion is that even 1000 excited states and 100,000 atoms represent a feasible calculation on a desktop workstation.

We repeat the same calculation by the Markov method, using the same set of excited states and the same rates T_{jk}. It was possible to perform the calculation on a desktop computer for a plasma containing 100,000 atoms. For this test, all the atoms are identical so the result should be the same as the CR calculation. The agreement is very good (Figure 1). If the atoms were moving or in different environments, the MD calculation could include those effects and presumably would be more accurate.

The most interesting result is the observation that, using the Markov method, the MD+CR calculation *is feasible*.

3.2 Practical limits

First, treating electrons as classical particles would require a very small time-step dt ~ 10^{-21} sec, because without this time-resolution the calculation of the strongly curved orbit of electron scattering by a nucleus will not be accurate enough to conserve energy. Because such short time-steps severely limit the calculation, we adopt a fluid description of free electrons, described below as a *Langevin method*.

The Schroedinger equation description also requires a small time step,

$$dt \ll \hbar / |E_j| \sim (5\ 10^{-17}\ \text{sec})/Z^2$$

where E_j is the binding energy of an atomic state. With such a small time-step, the plasma simulation can only be followed for relatively short times. The minimum storage required for the Schroedinger eqn. calculation is something like

$$\sim (6\ N_{atoms})\ X\ (2\ N_{states})$$

i.e., six position-velocity variables for each atom and the real and imaginary parts of the probability amplitude for each excited state. In fact, most MD codes have several copies ("old" and "new") for each of the variables, so the storage is *proportional* to the number indicated above.

For the collisional-radiative model, the time-step must at least resolve the transition rates,

$$dt \ll 1 / T_{jk} \sim 10^{-12}\ \text{sec}$$

The time-step could be further restricted by the MD atomic motion, depending on the density; ions should not be able to cross a sub-box during one time-step to avoid mis-counting their collisions. The storage needed is proportional to $\sim (6\ N_{atoms})\ X\ (N_{states})$, an estimate which includes atomic position-velocity variables and the probability N_{ij} for the excited states on each atom.

In contrast, the Markov method stores only one integer for each atom, as if each atom is instantaneously "in" one state. The time step $dt \ll 1 / T_{jk} \sim$ psec is similar to the CR model but the required storage is much less:

$$\text{Storage} \sim 7\ N_{atoms}$$

However with this method one needs to average over many atoms to reduce the fluctuations, which are especially visible for excited states with small populations.

3.3 Langevin Description of free electrons

For practical reasons mentioned above, we seek to replace particle electrons by an effective fluid which will represent the behavior of the electron gas over longer time-scales. The fluctuations in this fluid produce important electron-ion energy-transfer effects, so we must keep track of the fluctuations. For the present we assume the electrons have equilibrium Maxwellian distribution at the current electron temperature $T_e(t)$, and work out the fluctuations from that assumption, using statistical methods described by Chandrasekhar [10]. Electrons equilibrate much more rapidly than ions, so it is possible to assume Maxwellian electrons even while allowing non-Maxwellian ions.

We describe the Langevin treatment of ion coupling to electrons. During one ion time-step dt there are many electron-ion collisions. The momentum transfers from these collisions combine as a random walk to produce a total momentum change $d\mathbf{p}_{tot} = M_i\, d\mathbf{v}$. During time dt, the probability that the ion changes velocity by $d\mathbf{v}$ as a result of the electron-ion collisions is

$$P(\vec{v}_i, d\vec{v}, dt) = \left(\frac{M_i \tau}{2\pi k T_e dt}\right)^{3/2} e^{-\frac{M_i}{2kT_e}\frac{\tau}{dt}(d\vec{v}+\gamma\vec{v}_i)^2} \tag{7}$$

Here the parameter $\gamma = dt / 2\tau$ where dt is the time-step and τ is the collision time given below. The average (over this probability distribution) of the change of ion velocity is

$$< dv > = -\gamma v_i = -(v_i / 2\tau)\, dt \tag{8}$$

Eq. (8) describes the ion energy-loss dE/dx to electrons. If the electrons are hotter than the ions, the random transfers $d\mathbf{v}$ will heat the ions. Eqn. (7) has been written for the case in which the ion velocity is not large compared to the

electron thermal velocity. The electron-ion collision time τ that results from the statistical calculation is:

$$\frac{1}{\tau} = \frac{8\sqrt{2\pi}}{3} \frac{Z^2 e^4 n_e}{M_i kT_e} \sqrt{\frac{m_e}{kT_e}} \left(\frac{m_e}{m_r}\right) \log\Lambda \tag{9}$$

This well-known collision time goes under the names of Landau and Spitzer.

The Langevin method gives a real benefit for the time-step in an MD code; in our fusion ignition simulations of multi-keV plasmas we can use a time step of 10^{-17} sec which makes it possible to carry a calculation for as long as 10-20 psec.

The time-step is a concern even for the largest parallel supercomputers, because while the supercomputer can do many calculations at the same time, the largest number of sequential time-steps that can be chained together (on today's machines) is only of the order of 10^6 or 10^7.

3.4 Applications of the MD + CR method

Most of the applications will be studies of dynamical or non-equilibrium phenomena. We can offer a list of likely prospects:

X-ray laser interaction with cold matter [11]
Gas laser gain profiles; gain narrowing
Laser cooling; coherent emission by neighbor atoms (super-radiance)
Tokamak edge plasmas; collisional modification of Doppler profiles
Astrophysical plasmas, interstellar medium
Combustion Science (H_2, O_2 combustion)
Fusion ignition and novel fusion schemes

Of these possibilities, only the fusion ignition particle simulation has been worked out in detail.

4. Fusion Ignition Simulations

Fusion ignition is a suitable application for molecular dynamics, because the atomic-scale physics of fusion involves a small number of essential species: deuterium and tritium ions, electrons, radiation and fusion reaction product alphas and neutrons [12, 13, 14].

We describe fusion simulations performed with a hybrid MD/MC computer code named *OK* (for Okehazama, a historic spot in Aichi prefecture). Table I gives a brief overview of the physical processes included in the code.

Ten thousand D,T ions are described by their positions and velocities $\{\mathbf{R}_j(t), \mathbf{V}_j(t)\}$. At keV temperatures the DT is fully ionized and there are no impurities other than fusion output alpha particles. The velocity distribution is constrained only by collision processes in the code. An "ion temperature" $T_i(t)$ is calculated as 2/3 of the instantaneous average ion kinetic energy.

Special computational techniques used in the *OK* code include:

1.) Fast-moving α particles are subcycled (synchronized small time-steps)
2.) Electrons and ions are coupled by the Langevin method described above.
3.) Electrons emit and absorb bremsstrahlung radiation, calculated with Kramers' cross-sections [15]
4.) DT ion pair-collisions are treated by an analytic method
5.) Fusion reactions occur based on a Monte Carlo test imposed during each DT collision.
6.) The Monte Carlo tests are "guided" to reduce fluctuations (see Appendix A).

Journal of Physics: Conference Series **454** (2013) 012027 doi:10.1088/1742-6596/454/1/012027

Table 1. Physical processes included in the *OK* code simulations described in the text.

OK Code PHYSICS

IONS: N_i = 10,000 DT IONS Described by classical { $R_i(t)$ }, { $V_i(t)$ }

ELECTRONS $N_e = N_i$, $T_e(t)$ The electron velocity distribution is assumed to be instantaneously Maxwellian

RADIATION { I_ν } Isotropic, homogeneous radiation is described by a 200 group photon frequency spectrum. Radiation is emitted and absorbed during electron-ion collisions; Kramers' bremsstrahlung cross-sections are used.

ALPHAS Variable number $N_\alpha(t)$ { $R_j(t)$ }, { $V_j(t)$ } Move in plasma with subcycle time-steps $dt_\alpha = dt_{DT} / 10$.

NEUTRONS Collected in output spectrum P(v)

Processes:
Ion-ion Collisions
DT fusion reactions
Electron-ion collisions
Emission, absorption of radiation

7.) The plasma composition changes with nuclear reactions. A flag or pointer vector identifies whether each ion is currently "active" or "consumed". When DT ions react they are promptly rendered inactive. The ion list is not rearranged, but rather the "active ion" flag is switched and that ion is skipped for all further processes and audits. Reacting D and T ions are replaced by a fusion alpha and neutron having high kinetic energies.

The Monte Carlo method to test for fusion is described in Appendix B.

4.1 Ignition simulations

We describe simulations of inertial fusion ignition using assumed inertial fusion target hot-spot parameters,

DT fuel density = 100 g/cm^3 = 2.41 10^{25} ions/cm^3
Temperatures begin near 10 keV (electrons and ions) and 1 keV (radiation)

Laboratory experiments performed to date do not quite achieve these desired initial conditions [16, 17].

It is useful to describe the physical conditions in such a hot dense plasma in order to provide an intuitive idea about which physical processes are most important.

At density of 100 g/cm^3, there are ~ 2.4 10^{25} ions/cm^3. The Debye length is ~ 0.1 nm and the (usual) Coulomb logarithm ~ 4.5. At the plasma ion density N_i considered here, the ion-sphere radius (~ half the distance between neighbor ions) is $R_o = (3/4\pi N_i)^{1/3}$ ~ 0.02 nm. For $kT \geq 10$ keV, the ion coupling parameter $\Gamma = Z^2e^2/R_okT$ is not large:

$$\Gamma = Z^2e^2/R_okT \leq 0.007$$

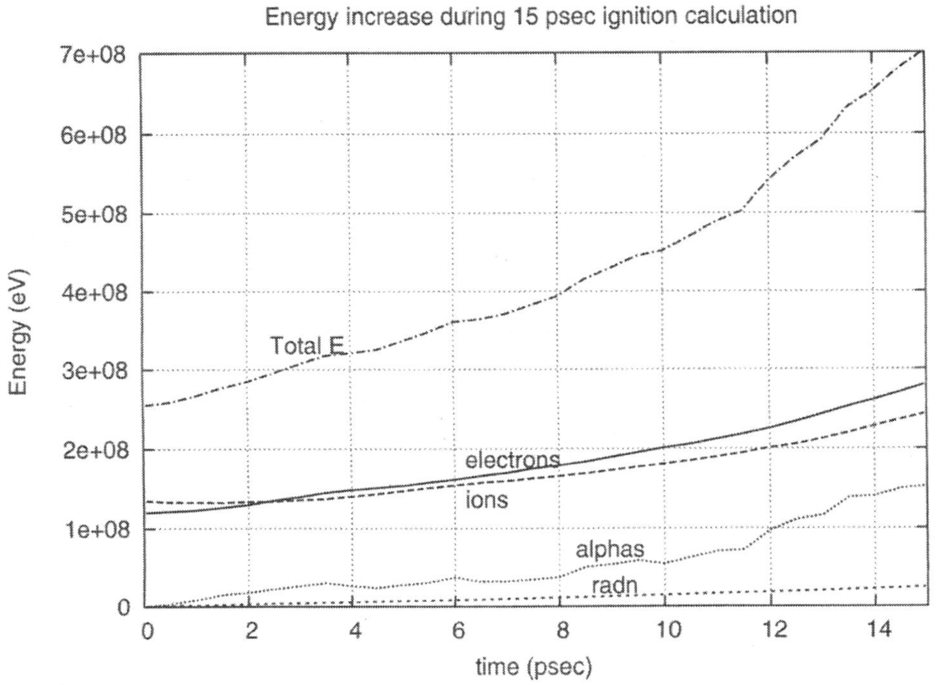

Figure 2. Energy changes during the ignition calculation reported as Case A in the text. The total energies of 10,000 DT ions (and electrons, radiation and fusion alphas) are reported in electron volts (eV). Careful inspection shows that while the alpha energy increases after each fusion reaction, it decreases during the intervals between fusions due to energy deposition in the plasma.

This inequality shows that equilibrium ion pair-correlations due to the Coulomb potential, which would be important at lower temperature, do not have a large effect on plasma at fusion ignition conditions. The Fermi energy ~ 300 eV << kT so the plasma is entirely non-degenerate. Strong ion-coupling and degeneracy effects are more important during assembly (implosion) of the fuel.

The computational box size is L = 0.75 nanometers. With DT ion velocities $v_D \sim v_T \sim 10^8$ cm/sec, the time taken to cross a sub-box (L/15) is about 5 10^{-17} sec, i.e., five time-steps.

Ion distribution becomes approximately Maxwellian within ~ .1 - .2 psec due to ion-ion collisions, but the high-energy tail of the distribution may not be Maxwellian.

Fusion alpha energy-deposition requires ~ 2 - 5 psec. Because of this delay, fusion ignition requires ~ 10-20 psec.

We describe results from 4 ignition self-heating calculations which start from slightly different initial temperatures and run for different times. (Figure 2) All these calculations assume a constant DT fuel density 100 g/cm^3. In each case the radiation begins as a 1 keV blackbody and heats to ~ 5 keV but is not a Planck spectrum.

case A Initial temperatures (T_e^0 = 8 keV, T_i^0 = 9 keV)
123 fusions during 15 psec clock time = 84 h
Final temperatures (T_e^f = 18.7 keV, T_i^f = 16.6 keB)

case B Initial temperatures (T_e^0 = 9 keV, T_i^0 = 11 keV)
100 fusions during 10 psec clock time = 60 h
Final temperatures (T_e^f = 17.3 keV, T_i^f = 15.7 keV)

Journal of Physics: Conference Series **454** (2013) 012027 doi:10.1088/1742-6596/454/1/012027

case C Initial temperatures (T_e^0 = 10 keV, T_i^0 = 7 keV)
45 fusions during 10 psec clock time = 48 h
Final temperatures (T_e^f = 12.5 keV, T_i^f = 10.4 keV)

case D Initial temperatures (T_e^0 = 11 keV, T_i^0 = 9 keV)
316 fusions during 20 psec clock time = 89 h
Final temperatures (T_e^f = 32.2 keV, T_i^f = 31.9 keV)

These calculations were repeated with different random-number seeds using the fluctuation-suppression algorithm described in Appendix A. The results were the same to 1 or 2 %.

4.2 Fusion "ignition attractor" phenomenon
In comparing several ignition calculations (like cases A-D) with similar starting conditions, it is found that the plasma rapidly relaxes to a definite ignition heating trajectory and self-heats along this well-defined path in the space of T_e, T_i, T_r. The heating path can be considered to be an *attractor* in the parameter space [4].

4.3 Non-Maxwell ion distribution
Most fusion events involve an ion that has energy >> kT. While the bulk of the ion distribution is quite close to a Maxwellian (at a changing temperature), the most important ions are a small minority in the high-energy *tail* of the distribution, which we consider to be the range from 2 kT_i to perhaps 10 kT_i. Molecular dynamics is able to calculate this minority population with no *a priori* assumption aside from the geometry and boundary conditions.

A non-Maxwell ion distribution f(E) can result from

1.) *Knock-on* DT ions at 20-100 keV produced by Coulomb collisions with fast alpha particles. As the fuel temperature rises and energy loss dE/dx to electrons decreases, this process becomes more important.
2.) Ion beam heating (fast-ion fast-ignition). Comparisons have found that the knock-on DT ions assist simple plasma heating to reach a self-sustaining reaction.
3.) One might expect that if the plasma temperature would rise too fast, there would not be enough time for tail-filling, leading to a depleted tail. In fact the MD calculations do not seem to support this expectation, probably because the heating rate is itself limited by the same couplings that relax the ion velocity distribution.
4.) In real fusion targets, the heated material has a finite ρR. In that case, spatial loss of high-energy DT ions depletes the active tail of the DT distribution and can significantly reduce the fusion yield. This mechanism was not included in the calculations described here but can readily be included in future work. (See Petschek & Henderson [18], Nishikawa, Takabe & Mima [19], and Molvig, Hoffman et al. [20])

4.4 Neutron TOF signal
The clearest experimental signature of non-Maxwellian ion distributions would be in output neutron velocity spectra (time of flight spectra). The MD calculations described here predict a small pulse of energetic neutrons that move faster than the tail of a Maxwellian curve-fit [4].

The MD predicts one in 10^6 of the neutrons will have velocity

$$\delta v \sim 5\ 10^7 \text{ cm/sec}$$

faster than the neutrons in the tail of a Gaussian curve-fit to the overall neutron velocity spectrum. This is only a small fraction of the neutrons but may be easy to detect because they arrive early. At a distance 10 meter away from the fusion capsule, these fast neutrons would arrive ~ 2 nsec early.

4.5 Code tests

As the *OK* code was constructed, the physical processes were tested one by one to verify that they give reasonable rates and properly obey detailed balance. The results of these tests are briefly summarized:

I. Ion-ion collisions relax an initial non-Maxwellian ion distribution to a Maxwellian in about the expected time. This comparison is an important test of the small-ball method for ion-ion collisions. For the test, a delta-function distribution f(E) is evolved by a calculation which includes only ion-ion collisions. The relaxation is examined two ways: the overall ion velocity distribution (plotted as a histogram) is compared to the expected Maxwellian and three moments of the distribution are calculated and seen to approach the expected Maxwellian limits. The time to evolve from a delta-function distribution to a good Maxwellian is about 0.12 psec when the ion temperature is 5 keV (density = 100 g/cm^3), and ~ .35 psec when the temperature is 10 keV.

II. The electron-ion (Langevin) coupling is tested by a special calculation in which the electron temperature is held fixed, ion-ion collisions are suppressed, and the only coupling is the Langevin coupling between electrons and ions. In this calculation the ion distribution relaxes to a Maxwellian at the electron temperature. This relaxation takes as long as 5 psec for T_e = 5 keV, significantly slower than the ion-ion thermalization.

III. The radiation emission/absorption formulas used here obey detailed balance and would eventually relax the photon distribution to an equilibrium Planck distribution [15]. Radiation equilibration is much slower than ion-ion or electron-ion equilibration and radiation is far from equilibrium throughout the ignition calculations.

IV. Energy conservation is an important test. The code conserves energy to a few % even on long calculations, approximately one part in 10^8 per time-step.

4.6 Limitations

The *OK* code has several limitations: The ion-ion collision algorithm assumes weak ion coupling and uniform electrons. This assumption is appropriate for fusion ignition conditions, where $\Gamma << 1$, but would not be justified at lower temperature (implosion conditions). For this reason the code is suitable above $T_{e,i}$ = 2 or 3 keV, but would need modification for lower temperatures.

There are statistical fluctuations ~ $1/N^{1/2}$. With 10^4 ions, the fluctuations in the main plasma properties are not too large. For example, when we plot separate temperatures for D and T ions, these show fluctuations ~ 1-2 %. However there are only about 100 fusions in a typical calculation and this number shows fluctuations ~ 10 %, as verified by repeating runs with various random-number seeds. A technique to reduce these fluctuations is described in Appendix A.

4.7 Verification of hydrodynamic simulation codes

Large-scale inertial fusion experiments have been designed and are modeled by elaborate "multiphysics" hydrodynamic simulation codes. These codes represent a remarkable investment of scientific expertise and have been tested over many years on a large variety of laser (and other) experiments [21].

However, fusion experiments performed on the NIF laser do not (today) agree with the predictions of "multiphysics" hydrodynamic simulations [16,17]. There could be several reasons for the disagreements, some involving the calculations, others involving the laser itself or the target fabrication.

It should be clear that the independent calculation of ignition in simple homogeneous fuel by the MD simulation does not directly give much information about target experiments, but could be useful to help identify aspects of the hydrodynamic simulations that need improvement.

Journal of Physics: Conference Series **454** (2013) 012027 doi:10.1088/1742-6596/454/1/012027

5. X-Ray Laser Interaction with cold matter

H. Yoneda has recently described experiments on the SPring-8 X-Ray Free Electron Laser in which a high-intensity short pulse of X-rays interacts with an iron target [11]. The remarkable parameters of the experiment are X-ray energy $h\nu \sim 7.1$ keV, pulse-length 10 fsec pulse, focused intensity $I_\nu \sim 2\ 10^{19}$ Watts/cm^2. Simple estimates show that during the pulse there are several hundred x-rays per atom in the target.

The handbook cross-section for K-shell absorption by neutral Fe is

$$\sigma_K \sim 2.47\ 10^{-20}\ \text{cm}^2$$

and the lifetime of a K-shell hole is $\tau \sim 1$ fsec. This short lifetime is limited by both X-ray (K --> L) emission and the Auger effect.

During the pulse, approximately 40 % of the Fe atoms should have K-shell holes:

$$\frac{I_\nu}{h\nu} \cdot \sigma_K \cdot \tau \approx 0.39 \tag{10}$$

The K-shell absorption edge rises by 600 eV when the same atom has a K-shell hole, so it cannot easily absorb a second photon from the laser source. The absorption edge broadens and shifts when one or more near neighbor atom has a K-shell hole, and such a broadening is visible in the preliminary experimental absorption spectra. Free electrons are also rapidly heated (maybe too rapidly to remain in thermal equilibrium) and this leads to a *blurring of the Fermi surface* which also permits absorption at lower photon energies.

A many-atom atomic model is the best way to describe this type of experiment because different target atoms are in different ionization/excitation states and different environments during the short pulse of X-rays. A calculation treating the target as having some "temperature" would not capture the inhomogeneity and transient dynamics.

6. Summary and Conclusions

This paper has described a type of atomic-scale simulation which can be called *Multi-physics Molecular Dynamics*. The method can give atomic-scale models for fusion and other atomic processes.

A key challenge for describing realistic atomic-scale dynamics is to incorporate quantum phenomena (especially abrupt transitions) within a largely classical simulation framework. An interesting method to describe these transitions in treatments of chemical reactions during atomic and/or molecular collisions was proposed by Tully and Preston, [22]. In this work the Landau-Zener approximation was used to justify treating the transition as happening at a specific curve-crossing point in the classical motion.

We would like to find a more general method, which can decide how to change the description from classical (to or from a quantum description) at a predetermined position or time without assuming there is a curve-crossing or other physical transition simultaneous with the change of description. The "small-ball" method summarized in Appendix B gives only a hint of the desired general method.

This method of multiphysics molecular dynamics could be especially useful for preliminary evaluation of new ideas for exotic phenomena. For example, heavy-ion fusion, fusion with spin-polarized fuel, or "ion-quiver" fusion have been proposed but are not being studied experimentally for economic reasons. Particle simulation can offer a first test of these ideas which would help identify candidates for further exploration.

The molecular dynamics simulations can also test the theory of plasma processes by examining situations in which many processes occur simultaneously. The physical models assumed in hydrodynamic simulation codes can only benefit as a result of this type of comparison and testing.

Acknowledgement
The authors acknowledge useful conversations with many scientists especially including H. Takabe, H. Yoneda, K. Mima, F. Graziani, J. Glosli, J. Barnard, A. Friedman and G. Faussurier. The work was partially supported by the Office of Fusion Energy Sciences, of the U.S. Department of Energy under Contract No. DE-AC02-05CH11231 and partially supported by the Lawrence Livermore National Laboratory under contract DE-AC52-07NA27344 through the LDRD Program at LLNL under project code 09-SI-011.

Appendix A. "Guiding" the Monte Carlo

In the fusion calculations, the fusion probability per DT collision is very small, ~ 10^{-6}. When the Monte Carlo test is passed, the resulting *real fusions* are implemented but are at most a few hundred events in 10-20 psec. The fluctuations in this number are about ± 10 %.

We also calculate *virtual fusions* using an enhanced cross-section F_{enh} x σ_{DT}. The enhancement factor F_{enh} is typically 100. These virtual fusions have the same statistical distribution as a function of particle energy because the enhanced cross-section has the same energy-dependence as the real cross-section. If we implemented these fusions and allowed them to produce energy or deplete the fuel, it would dramatically alter the ignition dynamics. However we can use them to improve the statistics of the calculation because they have smaller fluctuations (~ 1 %).

We use the virtual fusions to predict the output neutron time-of-flight spectrum and obtain a smooth spectrum consistent with the spectrum from the real fusions. However, the ratio

$$\text{(Number of virtual fusions)}/F_{enh}$$

also predicts the number of real fusions *with less fluctuation*.

It is possible to dynamically adjust the fusion conditional probability used in the MC test to guide the number of real fusions toward the expectation based on the virtual fusions. When we do this we can obtain million-particle accuracy from a 10,000-particle calculation. We used this method in re-runs of some of our reference cases and find the algorithm has entirely satisfactory behavior.

Appendix B. "Small-ball" method

In the *OK* code, we treat ion-ion Coulomb collisions by analytic solution rather than by numerical integration. This is combined with a MC test for QM processes such as fusion (or radiation emission in previous versions of the code).

Some advantages of this *small-ball* method [15, 23, 24] are: The method makes it easy to calculate radiation emission and absorption consistent with detailed balance. The method gives a way to calculate fusion reactions and quantum transitions in atomic kinetics. The calculation assumes a pure Coulomb potential inside the small ball, with no need for a pseudopotential. The small-ball method allows use of large time-steps for the MD calculation.

Some difficulties of the method are, in brief: it must be integrated into MD code and this raises issues about the time-steps, synchronization, labeling and storage issues. The small-ball method neglects some forces between ions for short times (while they're small), and there is a rare possibility of multiple ions being inside the small ball at the same time.

For fusion reactions, the small-ball approximation assigns a *conditional probability* of reaction

$$P_{Cond} = \frac{\sigma_{reaction}}{\pi R_{sb}^2} \tag{B-1}$$

to be tested each time that a D and T ion collide. Since the arriving flux is proportional to πR_{sb}^2, the reaction rate is approximately independent of R_{sb}

As a more general question, we can argue that when a particle crosses a predefined surface (such as the surface of a small sphere), we want to change description from a classical description with position-momentum coordinates **r, p** to a quantum description with a wave-function ψ_j. This change should be governed by a conditional probability $P_{\mathbf{rp}}(\psi_j)$ to decide which wave function is chosen in each individual case.

How should this conditional probability be chosen to obtain the most realistic description of the particle dynamics? Conversely, when the particle leaves the sphere, it should again select the outgoing coordinates **r'**, **p'** using another conditional probability $P_{\psi j}(\mathbf{r}', \mathbf{p}')$. What is the best way to select these conditional probabilities?

We can propose use of Bayes theorem and the Wigner distribution to guide these choices. At least in a near-equilibrium system, we have *a priori* probabilities $P(\mathbf{r}, \mathbf{p})$ and $P(\psi_j)$ for the classical and quantum probabilities; these can be obtained from Maxwell or Fermi-Dirac statistics, depending on the system. The conditional probabilities should then obey Bayes' theorem,

$$P(\mathbf{r}, \mathbf{p})\, P_{\mathbf{rp}}(\psi_j) \;=\; P(\psi_j)\, P_{\psi j}(\mathbf{r}, \mathbf{p}) \qquad \text{(B-2)}$$

In this equation we can use the Wigner distribution,

$$P_{\psi_j}(r_o, p_o) = \langle \psi_j |\, \delta(r - r_o)\; \delta(p - p_o)\, | \psi_j \rangle \qquad \text{(B-3)}$$

to determine the conditional probability for $\mathbf{r}_o$, $\mathbf{p}_o$ given the wave-function ψ_j. (This expression should be symmetrized by taking the real part.) If we know the wave function in both position-space $\psi_j(\mathbf{r}_o)$ and momentum-space $\phi_j(\mathbf{p}_o)$, then the Wigner distribution can be explicitly evaluated as

$$\langle \psi_j |\, \delta(r - r_o)\; \delta(p - p_o)\, | \psi_j \rangle = \psi_j^*(r_o) \frac{e^{i p_o r_o / \hbar}}{(2\pi)^{3/2}} \varphi_j(p_o) \qquad \text{(B-4)}$$

This method has the well-known imperfection that the Wigner distribution assigns unphysical negative probabilities over a small part of the **r, p** phase-space.

References

[1] Alder B and Wainwright T 1975 J. *Chem. Phys.* **27** 1208; 1959 J. *Chem. Phys.* **31** 459
[2] Kittel C 1958 *Elementary Statistical Mechanics* (J. Wiley and Sons)
[3] Brush S, Sahlin H and Teller E 1966 *J. Chem. Phys.* **45** 2102
[4] More R 2103 "Particle Simulation Code for Fusion Ignition" to appear in the *Proceedings of the Heavy-Ion Fusion Conference September 2012* (Berkeley CA)
[5] Frenkel D and Smit B 2002 *Molecular Simulation* (San Diego: Academic Press); Hansen J P 1973 *Phys. Rev. A* **8** 3096; Hansen J P and MacDonald I R 1976 *Theory of Simple Liquids* (London: Academic Press)
[6] Fujimoto T and A Iwamae A 2008 *Plasma Polarization Spectroscopy* (Springer)
[7] Burgess D, Everett D and Lee R 1979 *J. Phys. B* **12** L755
[8] Griem H 1997 *Principles of Plasma Spectroscopy* (Cambridge); Salzmann D 1998 *Atomic Physics in Hot Plasmas* (Oxford); Fujimoto T 2004 *Plasma Spectroscopy* (Oxford); More Kato Libby and Faussurier 2001 *J. Quant. Spectrosc. Radiat. Transfer* **71**
[9] Atomic Data Base (Oroshi, Toki Gifu Japan: National Institute for Fusion Science)
[10] Chandrasekhar S 1943 *Rev. Mod. Phys.* **15** 1
[11] Yoneda H, *private communication*
[12] Lindl J D 1998 *Inertial Confinement Fusion* (New York: AIP Press-Springer-Verlag)
[13] Atzeni S and Meyer-ter-Vehn J 2004 *Physics of Inertial Fusion* (Oxford: Clarendon Press)
[14] Fraley G, Linnebur E, Mason R and Morse R 1974 *Phys. Fluids* **17** 474
[15] More R, Graziani F, Glosli J and Surh M 2010 *High Energy Density Physics* **6** 29

[16] Moses E Plenary Talk at the 24th IAEA Meeting on Fusion Science San Diego CA October 8-14 2012

[17] U.S. Department of Energy Report to Congress December 2012 NNSA Path Forward to Achieving Ignition in the Inertial Confinement Fusion Program ed. W Goldstein, Report LLNL-TR-570412 Science of Fusion Ignition on NIF May 2012

[18] Petschek A G and Henderson D B 1979 *Nucl. Fusion* **19** 1678

[19] Nishikawa T, Takabe H and Mima K 1989 *Japan. J. Appl. Phys.* **28** 2004

[20] Molvig K, Hoffman N, Albright B and Webster R 2012 *Knudsen Layer Reduction of Fusion Reactivity* preprint LANL

[21] Marinak M, this meeting

[22] Tully J and Preston R 1971 *J. Chem. Phys.* **55** 562

[23] Glosli J, Graziani F, More R, Murillo M, Streitz F and Surh M 2009 *J. Phys. A* **42** 214030

[24] More R, Goto M, Graziani F, Ni P and Yoneda H 2010 *Plasma and Fusion Research* **5** S2008; More R, Graziani F, Glosli J and Surh M 2010 *High Energy Density Physics* **6** 29

Journal of Physics: Conference Series **454** (2013) 012028 doi:10.1088/1742-6596/454/1/012028

Evaluations of the conformational search accuracy of CAMDAS using experimental three-dimensional structures of protein-ligand complexes

A Oda[1,2,3], N Yamaotsu[4], S Hirono[4], Y Takano[3], S Fukuyoshi[1], R Nakagaki[1] and O Takahashi[2]

[1] Institute of Medical, Pharmaceutical and Health Sciences, Kanazawa University, Kakuma-machi, Kanazawa, Ishikawa 920-1192, Japan

[2] Faculty of Pharmaceutical Sciences, Tohoku Pharmaceutical University, 4-4-1 Komatsushima, Aoba-ku, Sendai, Miyagi 981-8558, Japan

[3] Institute for Protein Research, Osaka University, 3-2 Yamadaoka, Suita, Osaka 565-0871, Japan

[4] School of Pharmacy, Kitasato University, 5-9-1 Shirokane, Minato-ku, Tokyo 108-8641, Japan

E-mail:oda@p.kanazawa-u.ac.jp

Abstract. CAMDAS is a conformational search program, through which high temperature molecular dynamics (MD) calculations are carried out. In this study, the conformational search ability of CAMDAS was evaluated using structurally known 281 protein-ligand complexes as a test set. For the test, the influences of initial settings and initial conformations on search results were validated. By using the CAMDAS program, reasonable conformations whose root mean square deviations (RMSDs) in comparison with crystal structures were less than 2.0 Å could be obtained from 96% of the test set even though the worst initial settings were used. The success rate was comparable to those of OMEGA, and the errors of CAMDAS were less than those of OMEGA. Based on the results obtained using CAMDAS, the worst RMSD was around 2.5 Å, although the worst value obtained was around 4.0 Å using OMEGA. The results indicated that CAMDAS is a robust and versatile conformational search method and that it can be used for a wide variety of small molecules. In addition, the accuracy of a conformational search in relation to this study was improved by longer MD calculations and multiple MD simulations.

1. Introduction

In computer-aided drug design (CADD) trials, a technique to construct accurate three-dimensional (3D) molecular structures is important [1]. A wide variety of compounds has to be considered for drug design and development, but their experimental 3D structures, such as X-ray crystallographic structures and NMR structures, are not always obtained. Thus, computational approaches to determine the 3D structures of small molecules are widely used. Although structural optimizations are carried out to procure the most stable structures of isolated molecules, the optimized structures of isolated small molecules are occasionally inappropriate for drug design. In drug design and development trials, the molecular structures of small ligands in biopolymer-ligand complexes play important roles instead of isolated ligands. Because of this, the interactions between drug target biopolymers and drug candidates (small molecule) attracted attention. The structures of small molecules in the complexes are frequently

Journal of Physics: Conference Series **454** (2013) 012028 doi:10.1088/1742-6596/454/1/012028

different from the most stable state structures of isolated molecules [2]. Thus, many types of conformers should be generated for small ligand molecules, and the appropriate conformer is then selected from the generated conformer set. The appropriate conformer which can be used as the ligand structure in biopolymer-ligand complex has to be included in the generated conformer set. Sequentially, the appropriate conformer is selected from the conformer set by using computational methods, such as structure-based drug design (SBDD) and/or ligand-based drug design (LBDD) techniques. The selected conformer is eventually used for the next steps of drug design and development trials. Because the conformational differences cause the differences in physical and chemical properties of molecules [3], the computational method for generating a conformer set should be carefully considered. Several software programs have been developed for conformational searches, such as CAMDAS (Conformational Analyzer with Molecular Dynamics And Sampling) [4] which is the molecular-dynamics-based method and OMEGA [5] which is the knowledge-based method. The conformers obtained are mostly used for several CADD calculations, i.e. computational docking and virtual screening [6] in SBDD or molecular superposition [7] and 3D-quantitative structure-activity relationship (QSAR) study [8] in LBDD. This attests that conformational searching is one of the key steps in CADD.

As mentioned above, the constructions of a wide variety of conformers are required for conformational searches, and several computational methods have been developed for this purpose. Computational techniques such as systematic search, random search, and distance geometry methods have been applied to conformational searches [1]. Moreover, methods using molecular dynamics (MD) calculations are also frequently used. Although some conformational search methods have to include anomalistic procedures for particular molecules such as macrocycle compounds, a conformational search method using MD can be carried out without any pre-processing. To generate a wide variety of conformers by using MD calculations, simulations are generally carried out in an extremely high temperature because the energy barriers among conformers need to be climbed over. The effectiveness of high temperature MD in searching conformational space has already been evaluated. For example, replica exchange MD (REMD) simulations which include the high temperature MD of replica and simulated annealing using MD have been applied to several molecular systems [9-10]. Through these procedures, efficient sampling can be performed. In REMD, a temperature above 500 K is frequently used. In fact, quadruple-digit temperatures are sometimes adopted. Because MD programs only solve Newton's equation of motion, particular treatments such as "conformation dictionary" are unnecessary for any compounds. In addition, an unusual conformational change, e.g. torsional rotations of unsaturated bonds and ring flips, can occur because of high temperature. Thus, high performance conformational searches can be carried out by using high temperature MD calculations.

Previously, we have developed a computational program CAMDAS in which high temperature MD calculations are carried out for the conformational search of molecules [4]. In this software, clustering of conformers generated by high temperature MD is performed, where RMSDs are used as distances, and the representative conformers are outputs. The current version of CAMDAS 2.1 can use several reliable molecular force fields, i.e. MM2 [11], MMFF [12], and AMBER force field [13], and several file formats (PDB, mol2, and AMBER parameter file formats) can be treated. In addition to user-friendliness, improvements for its computational accuracy are implemented. The multi-copy procedure, in which multiple initial conformers are generated from input coordinates by using distance geometry method and multiple conformational search trials are simultaneously carried out, has been implemented. Previously, we have reported the conformational searching ability of CAMDAS by using *N*-acetylalanine-*N'*-methylamide and cyclodecane as a test set [4]. Recently, CAMDAS has been used in several studies related to drug design and development. For example, in some studies, CAMDAS was used together with SBDD techniques for the investigations of ligand recognition mechanisms of cytochrome P450 2C9 and 2C19 [14] and for molecular modelling of a human acidic mammalian chitinase in complex with argifin which is a cyclopentapeptide chitinase inhibitor with a flexible ring structure [15]. By using CAMDAS not only with SBDD but also with LBDD procedures, useful results were obtained for the design and synthesis of novel delta opioid receptor agonists [16], for the estimation of the 3D pharmacophore of ligands for rat multidrug-resistance-associated protein 2 [17], and for the identification of the 3D pharmacophore of κ-opioid receptor agonists [18]. CAMDAS is also useful for a wide variety of molecular modelling trials other than drug design, such as the determination of the stereostructure of luminamicin [19]. On the other hand, exhaustive tests using a large test set for evaluating the search ability of CAMDAS were not carried out, and the influences of computational settings on search results were not clearly investigated.

In this study, the ability of CAMDAS for conformational search was evaluated by using structurally known protein-ligand complexes. The results indicated the usefulness of high-temperature MD for the conformational search and revealed the desirable settings for CAMDAS.

2. Methods

2.1. Test set

The test set consisted of 260 systems which were extracted from 281 protein-ligand complexes in a Glide test set [20]. From the Glide test set, only the systems with rotatable bonds of ligand molecules more than or equal to 1 were used as tests. The numbers of rotatable bonds were counted by using OMEGA [5]. The test systems are listed in table 1.

Table 1. Test set. Rigid molecules are described in italics, and flexible ligands are underlined.

121p	1aaq	1acm	*1aco*	*1add*	1adf	1ake	1apt	1apu	1apv
1apw	1atl	*1avd*	*1azm*	1b6j	1b6k	1b6l	1b6m	1baf	1bbp
1bkm	1bma	*1bra*	1byb	1c1b	1c3i	*1c5p*	*1c83*	*1c84*	*1c86*
1c87	*1c88*	*1c8k*	*1cbs*	*1cbx*	1cde	*1cdg*	*1cil*	1cnx	*1com*
1cps	*1ctr*	*1ctt*	1d3d	1d3p	1d7x	1d8f	*1dbb*	1dbm	1dd6
1dds	1dhf	*1did*	*1die*	1dih	*1dm2*	*1dog*	*1dr1*	*1dwb*	1dwc
1dwd	*1e5i*	1eap	*1ebg*	*1ecv*	1eed	1ejn	1ela	1elb	1elc
1eld	1ele	*1epb*	*1eta*	1etr	1ets	1ett	1ezq	*1f0r*	*1f0s*
1f0t	1f0u	*1fen*	*1fh8*	*1fh9*	*1fhd*	1fjs	1fkg	1fq5	1frp
1fvt	*1g45*	*1g46*	*1g48*	*1g4j*	*1g4o*	*1g52*	*1g53*	*1g54*	*1ghb*
1glp	1glq	*1gsp*	1hbv	1hdc	1hef	1hfc	1hgg	1hgh	1hgi
1hgj	1hih	1hps	1hpv	1hpx	1hri	1hsg	*1hsl*	1hte	1htf
1hti	1hvr	*1hyt*	1icn	1ida	*1igj*	*1imb*	*1ivb*	*1ivc*	*1ivd*
1ive	1ivf	*1lah*	*1lcp*	*1ldm*	1lic	1lmo	1lna	*1lst*	1mcr
1mdr	1mfe	*1mld*	1mmq	1mnc	*1mrk*	*1mup*	1nco	*1nis*	1nnb
1nsc	1nsd	1odw	*1okl*	*1pbd*	1pgp	1pha	*1phd*	*1phf*	*1phg*
1poc	1ppc	1pph	1ppi	1ppk	1ppl	1ppm	1pro	1pso	*1rbp*
1rds	*1rhl*	*1rls*	1rne	*1rnt*	*1rob*	1sbg	1slt	1snc	*1sre*
1srj	*1stp*	*1tdb*	*1thy*	1tka	1tlp	1tmn	*1tng*	*1tnh*	*1tni*
1tnj	*1tnk*	*1tnl*	*1tph*	1trk	*1tyl*	*1ukz*	*1wap*	*1xid*	*1xie*
2ack	*2ada*	*2ak3*	2cgr	*2cht*	*2cmd*	*2ctc*	2dbl	*2gbp*	2ifb
2lgs	*2mcp*	*2phh*	*2pk4*	2plv	2r04	2r07	2sim	*2tmn*	2tpi
2upj	*2xis*	*2yhx*	*2ypi*	3cla	3cpa	3dfr	*3hvt*	*3mth*	*3ptb*
3tpi	*4aah*	*4cts*	4dfr	*4fab*	*4fbp*	4fxn	4hmg	4phv	*4tim*
4tln	4tmn	4tpi	*4ts1*	*5abp*	5cts	5p2p	5tln	5tmn	6cpa
6rnt	*6tim*	6tmn	7cpa	*7tim*	*8abp*	8atc	8gch	*9abp*	9hvp

For the CAMDAS calculations, only the ligand structures were extracted from the complex structures registered in the Protein Data Bank (PDB) [21]. Hydrogen atoms were added for the ligand structures by using an SYBYL6.9 [22] software. The complemented ligand structures were optimized through AM1 method using MOPAC2007 [23]. The addition of hydrogen is a generally required process for a conformational search. In fact, it is indispensable to using both CAMDAS and OMEGA. The procedure of filling the valence can be performed not only by means of SYBYL but also by means of other programs such as Maestro and ChemBio 3D. Meanwhile, the influence of the procedures is removed through structural optimizations with the use AM1 method. The optimized PDB structures are defined as "test set **i**."

Because the conformers included in set **i** were constructed from crystal structures and only structural optimizations were performed, the conformations of set **i** were similar to the structures of crystal. The ability of conformational search program may be overestimated by using set **i** as initial structures, because the conformational search tests using initial conformers which are similar to experimental structures may result in undue success. Thus, the preliminary conformational search was carried out before the main tests in order to remove the influences of crystal structures. The generated conformations through the preliminary search were used as initial structures of the main tests. The flow chart of the tests is shown in figure 1. The preliminary and main conformational searches were performed using CAMDAS. The generated conformers using preliminary searches were optimized through AM1 method, and the lowest energy conformers (set **ii**), the conformers with the smallest RMSDs in comparison with experimental structures (set **iii**), and the conformers with the largest RMSDs (set **iv**) in comparison with crystal structures were selected. That is to say that in relation to experimental structures, conformers iii were the nearest and conformers iv were the farthest. Conformer sets **ii**, **iii**, and **iv** were used as initial structures of the main tests as well as set **i**.

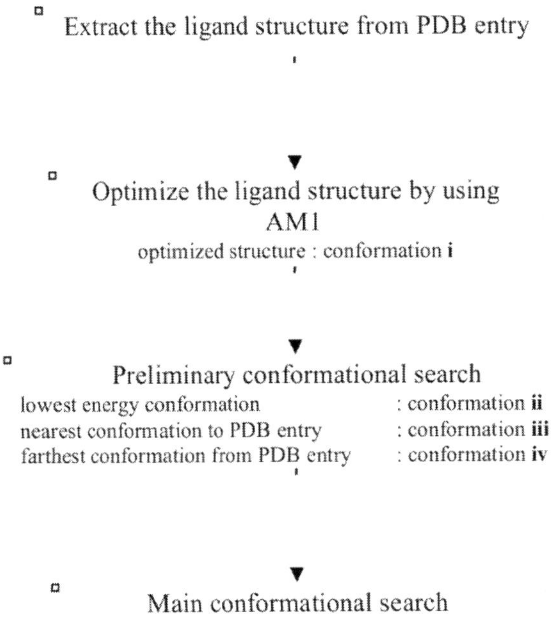

Figure 1. Flowchart of conformational search tests.

2.2. Evaluations for search results

The results of the main tests were compared with the experimental structures, and the accuracy of search results were evaluated using the RMSDs between the generated conformers and the experimental structures. In drug design and development studies, protein-ligand complex structures play important roles. Therefore, conformational search programs are required to reproduce ligand conformers in protein-ligand complexes. In this study, "reasonable" conformers were defined as the conformers whose RMSDs from the experimental structure were smaller than or equal to 2.0 Å. When at least one reasonable conformer is obtained, the conformational search trial for the test system is defined as successful. The numbers of successful systems were compared among search trials using several initials (sets **i**, **ii**, **iii**, and **iv**) and several settings. Thus, the dependencies of the results of conformational searches on initial structures and settings were investigated. In addition to CAMDAS tests, conformational searches using OMEGA [5], in which the molecules are divided into fragments and conformers are fast constructed by assembling the fragments, were also performed for comparison.

2.3. Computational setting

For both the preliminary and the main conformational searches using CAMDAS, the same settings described below were used. The general AMBER force field (GAFF) [24] was adopted. The assignment of atom types and the atomic charge calculations were carried out by using the antechamber module [25] of AMBER9 [26]. Although the molecular files for CAMDAS calculations can be created by using other software programs such as SYBYL, the antechamber module, which is currently included in the free software of AmberTools, is easier to obtain. For CAMDAS calculations, the dielectric constant was set to 80, and the conformational searches using 1000000 steps of MD calculations under 1200 K were performed. 1200 K is sometimes used as the temperature for REMD and annealing [10]. The conformational sampling was conducted every 100 steps during MD trajectory. Meanwhile, the maximum number of sampled conformers was set to 10000. After the conformational sampling, structural optimizations with maximum 1000000 cycles were carried out for the sampled conformers. The clustering of optimized conformers was carried out after conformer generation steps. In the clustering, two conformers were placed in the same cluster when the RMSD between them is less than 1.5 Å. After the clustering, the representative conformers were extracted from the obtained clusters. This procedure (1000000-steps MD and 1.5 Å threshold) was defined as procedure **I**. The OMEGA calculations for the comparative study were performed using default settings. Because the conformational search results were unaffected by the initial conformers in the default settings of OMEGA calculations, only conformer set **ii** was used as initial conformers. MOPAC2007 was used for AM1 calculations, and an implicit water model by using COSMO was adopted.

2.4. Additional settings

In addition to procedure **I**, different settings of CAMDAS calculations were tested. In procedure **I**, 1,000,000-steps MD was carried out, and clustering threshold was set to RMSD < 1.5 Å. On the other hand, the procedure with 1,000,000-steps MD and a threshold of 0.5 Å (procedure **II**), the procedure with 5,000,000-steps MD and a threshold of 1.0 Å (procedure **III**), the procedure with 5,000,000-steps MD and a threshold of 0.5 Å (procedure **IV**), and the procedure with multicopy MD calculations using the same settings as procedure I (procedure **V**) were carried out. In the multicopy MD of procedure **V**, ten types of initial conformations were generated and ten MD calculations were simultaneously carried out. For procedures **II** to **V**, only conformer set **ii** was used. The versions of the conformational search programs were CAMDAS 2.1 and OMEGA 2.2.1.

3. Results and discussion

3.1. Dependencies on initial conformations

In table 2, the results of conformational searches using four types of initial conformers with procedure **I** are shown. The results of OMEGA calculations are also shown for comparison. The success rate p was calculated as follows:

$$p = n_{\text{success}} / n_{\text{all}}$$

where, n_{success} is the number of systems for which reasonable conformers were obtained by the conformational search, and n_{all} is the total number of test systems. "Reasonable conformer" is defined as the conformer whose RMSD against an experimental structure is less than or equal to 2.0 Å. When at least one reasonable conformer is obtained for one test system, the test is defined as "success". In the test, the n_{all} is equal to 260 as shown in table 1. In addition to success rate, the PDB ID of the test system, where the RMSD of the best conformer is the largest of all 260 test systems, and the best RMSD of the system (the largest, best RMSD) are described in table 2. For example, in the conformational search for 1ake using initial conformer **i**, the RMSD of the best conformer obtained from the search was 2.310 Å. RMSDs of other conformers were worse than 2.310 Å; nevertheless, "2.310 Å" was the best RMSD for 1ake. For all test systems, the best RMSD was less than or equal to 2.310 Å. In this case, we can say that 1ake was "the system where the RMSD of the best conformer is the largest of all 260 test systems" and that 2.310 Å was the "largest, best RMSD." In the table, the averages of generated conformers are also shown.

As shown in table 2, CAMDAS can generate reasonable conformers for almost all the test systems regardless of initial conformers. The success rates for initial conformers **i** and **iii** were better than those for conformers **ii** and **iv**. Conformer **i** was generated from the crystal structure using simple optimizations only, and conformer **iii** had the most similar structure to the experimental structure. However, the difference of success rates between them was only around 3% despite the differences of initial conformers. The results indicated that the conformational searches by CAMDAS were robust for the difference of initial structures. Although the structural optimization using MOPAC were carried out to generate initial conformers, the robustness suggested that such pre-processing not always have to be carried out for CAMDAS. In addition, the success rates of CAMDAS were comparable to that of OMEGA, which is one of the most widely used and the most reliable conformational search programs [27-28]. The results suggested that CAMDAS is useful for the conformational searches of small molecules.

Table 2. Results of conformational searches using several initial conformations.

conformer	i	ii	iii	iv	OMEGA
$n_{success}$	256	248	255	249	252
success rate	0.985	0.953	0.981	0.958	0.969
largest best RMSD/Å	2.310	2.474	2.477	2.403	3.862
(PDB ID)[a)]	(1ake)	(1adf)	(5cts)	(5cts)	(1ake)
num. of conformers	69.67	68.65	68.47	67.74	127.55

Although the success rate of CAMDAS calculations were comparable with that of OMEGA as mentioned above, the results of CAMDAS were better than those of OMEGA in terms of the calculations for the largest best RMSD. The system where the RMSD of the best conformation is the largest of all CAMDAS test was 5cts using conformer **iii**. The best RMSD of the test of 5cts with initial conformer **iii** using CAMDAS was only 2.477 Å. On the other hand, the system where the RMSD of the best conformation is the largest in OMEGA test was 1ake, and the best RMSD of 1ake using OMEGA was 3.862 Å. Thus, CAMDAS can generate conformations with an RMSD < 2.5 Å even for the systems where reasonable conformations cannot be generated. Because the numbers of rotatable bonds are 23 and 16 for the ligands of 5cts and 1ake, respectively, their flexibilities are relatively high (in general, the number of rotatable bonds for drug-like molecules are less than or equal to ten [29]). The results indicated that CAMDAS can be effectively used for flexible molecules. The best conformations are shown in figures 2 and 3. Figure 2 is the best conformation generated by CAMDAS for 5cts using initial conformer **iii**, while figure 3 is the best conformation calculated by OMEGA for 1ake. For comparison, both conformations were superposed onto crystal structures. As seen in figures 2 and 3, the structures colored by elements are generated conformation, and the structures in light blue are crystal structures. The hydrogen atoms were removed from this figure. In figure 2, although the conformation of the flexible chain moiety was different between the generated conformation and the crystal structure, many functional groups such as carboxyl, phosphate, and adenine groups were well-overlapped. Because the interactions between functional groups of the drug target and the ligand molecule are important for drug design trials, the generated conformations that have functional groups with locations similar to crystal structures seem to be useful for practical use despite an RMSD > 2.0 Å. The conformation shown in figure 2 was the worst case of the "best conformations" in this study, because the RMSD of the best solution for 5cts using conformer **iii** was 2.477 Å. Moreover, it was the largest RMSD of all "best conformations." Therefore, CAMDAS can generate conformations where, even in the worst case, the functional groups were located in appropriate positions. On the other hand, for the result of 1ake obtained by using OMEGA which is shown in figure 3, the distance between two adenine moieties in the computational result was shorter than that in the crystal structure. Furthermore, both adenine moieties did not overlap the crystal structure. Because the locations of adenine moieties and phosphate groups were different from the crystal structure, the calculated locations of functional groups which are potentially-important for protein-ligand interactions seemed to be inappropriate when using OMEGA. As shown in table 2 and figures 2 and 3, although the success rates were comparable between CAMDAS and OMEGA, the effectiveness for failed systems where reasonable conformations cannot be generated was different. CAMDAS can generate the reasonable positions of conformers in which the functional groups are located. Additionally, CAMDAS seems to be a more robust and fail-safe method for the conformational searches of wide varieties of small molecules. In general, ligand conformations in the protein-ligand complexes are not always the same as the conformations of isolated ligand molecules which are located in the minima of potential surface of isolated ligand [2]. Thus, conformational search methods such as OMEGA sometimes fail to search the ligand conformations in protein-ligand complexes. Even for such cases, CAMDAS can generate conformations close to the ligand structures in the protein-ligand complexes. A knowledge-based approach is adopted by OMEGA. In contrast, CAMDAS uses high temperature MD simulations. Therefore, CAMDAS can search the conformers without any previous knowledge, and it may be the reason behind the robustness of CAMDAS. Although the inaccuracy of force field parameters is considered a cause of the difference between the calculated conformation and the ligand structure in protein-ligand complexes [2], the results indicated that GAFF is appropriate for conformational searches of small ligands.

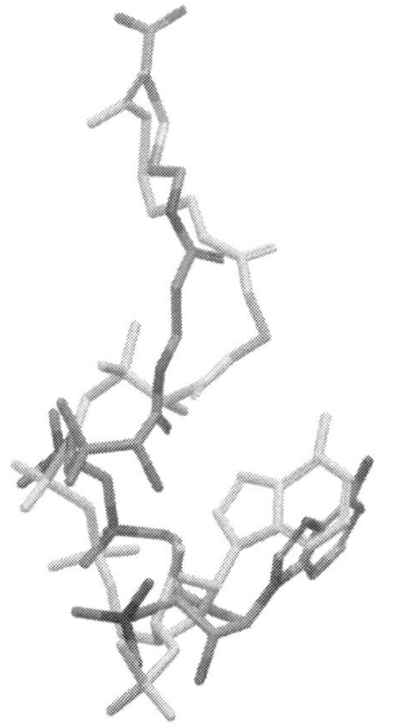

Figure 2. The best conformation with largest RMSD obtained by using CAMDAS. The molecules colored light blue are crystal structures, and the molecules colored by elements are generated conformers.

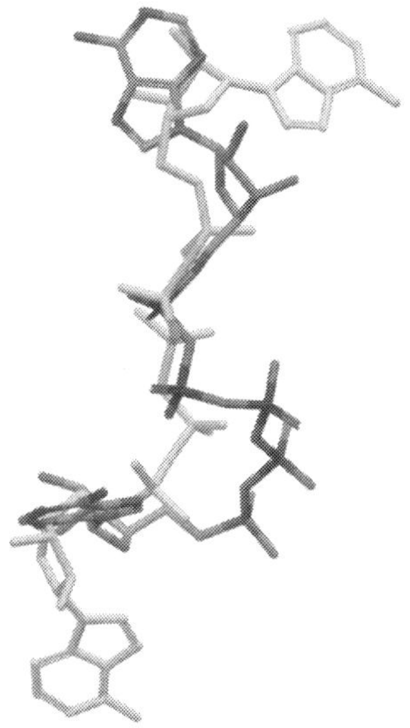

Figure 3. The best conformation with largest RMSD obtained by using OMEGA. The molecules colored light blue are crystal structures, and the molecules colored by elements are generated conformers.

The numbers of generated conformers are also shown in table 2. The number of generated conformers by using OMEGA is around twice as many as that by CAMDAS. For molecular superposition studies such as pharmacophore search and 3D-QSAR studies, all conformers have to be tested, and it is $O(n^2)$ problem. That is, the increase of the number of conformers to be tested is directory linked to the increase of computational costs. Thus, less number of generated conformations is better, when the similar success rates can be achieved. Because of this point of view, CAMDAS seems to be more useful than OMEGA. In OMEGA, the fragment decomposition and the reconstruction of small molecules are carried out for conformational searches, in contrast to CAMDAS in which high temperature MD calculations of complete small molecules are performed. Because CAMDAS generates the conformations by using MD trajectory in accordance with the physical law without artificial decomposition of molecules, only natural structures are generated. The generations of natural conformations may be the reason of efficiency of CAMDAS by which high success rates can be obtained despite less number of generated conformations than by OMEGA. OMEGA gives an output with a large number of conformers even for small and rigid molecules; however, structural optimizations and clustering might reduce the number of generated conformations similar to CAMDAS.

3.2. Results of conformational searches using additional settings

In table 3, the results by using different settings for conformational searches are shown. The detailed searches, which are frequently required for the ligand-based drug design trials of large flexible molecules, can be carried out using these settings. For the test calculations of this study, conformer **ii** was used as a set of initial conformations. As shown in the table, these settings can be used to perform very accurate conformational searches, even though conformer **ii** which was constructed without reference to crystal structures was used as initials. For example, the number of failed systems were only two using method **IV** (1hbv and 5cts), and the best RMSDs of the failed systems were 2.125 Å (1hbv) and 2.013 Å (5cts). Using method **V**, conformational searches for eight systems (1eed, 1ejn, 1fq5, 1ida, 1ppi, 1pso, 1rne, 9hvp) could not be carried out and no conformations were generated, because multicopy method could not generate 10 different initial conformers. For other test systems, the number of failed systems was only three (1ake, 1hvr, 5cts) using method **V**. The best RMSDs were 2.147 Å, 2.024 Å, and 2.377 Å for 1ake, 1hvr, and 5cts, respectively. The conformational search trials with no solutions could be easily decided as "failure", and recalculations could be carried out immediately. Thus, the search trials with no solutions were "desirable failure" in comparison with the search trials in which many solutions were obtained but reasonable solutions were not generated. The result indicated the usefulness of the multicopy method through which multiple MD simulations were simultaneously carried out.

Table 3. Results of conformational searches using additional settings.

setting	I	II	III	IV	V
$n_{success}$	248	254	256	258	249[a)]
success rate	0.953	0.977	0.985	0.992	0.988[a)]
largest best RMSD/Å	2.474	2.613	2.251	2.125	2.377
(PDB ID)	(1adf)	(1ake)	(1aaq)	(1hbv)	(5cts)
num. of conformers	68.65	380.04	563.23	1606.63	502.84

[a)] CAMDAS calculations with setting **V** could not be carried out for eight molecules, and n_{all} was equal to 252 for setting **V**.

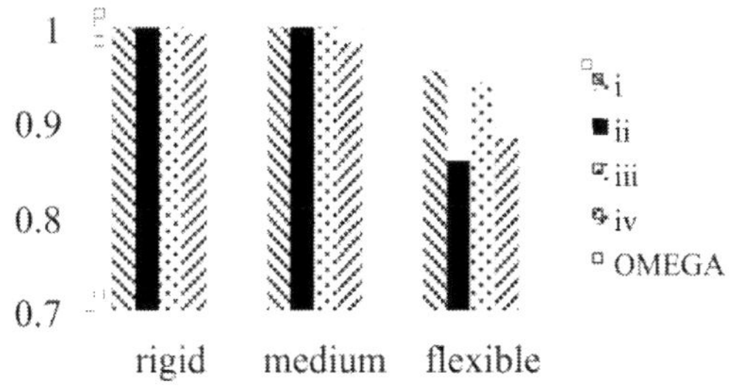

Figure 4. The success rates for several types of molecules.

3.3. Influences of molecular flexibilities

The relationship between molecular flexibilities and the success rates are shown in figure 4. In the figure, the results by setting **I** and OMEGA are illustrated. The molecules in which the numbers of rotors are less than or equal to 5 are defined as "rigid", and the molecules with 6 ~ 10 rotors are defined as "medium", and the molecules with 11 ~ rotors are "flexible". In Veber's drug-likeness criterion [29], the drug-like molecules have ten or less rotatable bonds. Thus, "rigid" and "medium" meet the Veber's criterion in terms of the number of rotatable bonds. As shown in the figure, the success rates for "flexible" are the worst of all molecular groups. Thus, more detailed procedures such as settings **IV** and **V** may be useful for these flexible molecules. The success rates for "flexible" group were 0.976 and 0.974 by using setting **IV** and **V**, respectively (data not shown). These success rates were larger than those of setting **I** and OMEGA. On the other hand, CAMDAS could generate reasonable conformations for all "rigid" and "medium" molecules, except for the tests using conformer **iv** as initials. Because conformer **iv** was the farthest conformer from the crystal structure, the results suggested that CAMDAS could generate reasonable conformations for drug-like molecules unless the unnatural initial conformations are used. On the other hand, OMEGA failed the conformational searches even for rigid molecules. Because it is difficult to prepare the countermeasure for unanticipated failures, CAMDAS seems to be more suitable for the conformational searches of drug-like molecules than OMEGA.

4. Conclusion

The conformational search abilities of CAMDAS were evaluated by using various initial conformers and settings. The success rates of conformational searches by CAMDAS were comparable to those by OMEGA, which is one of the most efficient and widely used conformational search methods. Although the unanticipated failures rarely occurred when OMEGA was used, only "manageable failures" occurred in the case of using CAMDAS. For example, OMEGA rarely failed for the conformational searches of rigid molecules, and all the generated conformations were rarely far from the crystal structures. On the other hand, the reasonable conformations could be obtained for all rigid molecules through CAMDAS, and the conformations whose RMSDs were around 2.5 Å (or less) could be obtained even for the failed tests. CAMDAS searched the conformations in accordance with the physical law, and any artificial treatments were not adopted. The manageability of CAMDAS may have been caused by the procedure without artificial treatments. The results indicated that CAMDAS is a useful and robust tool for conformational searches of small molecules.

Acknowledgments

This work was supported by a grant-in-aid for scientific research (23790137) from Japan Society for the Promotion of Science.

References

[1] Leach A R 2001 *Molecular Modelling*, 2nd edition (Essex: Pearson Education Limited) pp 457–508.
[2] Perola E and Charifson P S 2004 *J. Med. Chem.* **47** 2499
[3] Barton D H R 1950 *Experientia* **6** 316
[4] Tsujishita H and Hirono S 1997 *J. Comput.-Aided Mol. Des.* **11** 305
[5] *Omega 2.2.1* (Boston, MA: OpenEye Scientific Software, Inc)
[6] Kitchen D B, Decornez H, Furr J R and Bajorath J 2004 *Nat. Rev. Drug. Discovery* **3** 935
[7] Iwase K and Hirono S 1999 *J. Comput.-Aided Mol. Des.* **13** 499
[8] Cramer R D III, Patterson D E and Bunce J D 1988 *J. Am. Chem. Soc.* **110** 5959
[9] Sugita Y and Okamoto Y 1999 *Chem. Phys. Lett.* **314** 261
[10] Kalurachchi K, Uma K, Zimmermann R A and Nikonowicz E P 1997 *Proc. Natl. Acad. Sci. USA* **94** 2139
[11] Allinger N L 1977 *J. Am. Chem. Soc.* **99** 8127
[12] Halgren T A 1996 *J. Comput. Chem.* **17** 490
[13] Cornell W D, Cieplak P, Bayly C I, Gould I R, Merz K M, Ferguson D M, Spellmeyer D C, Fox T, Caldwell J W and Kollman P A 1995 *J. Am. Chem. Soc.* **117** 5179
[14] Oda A, Yamaotsu N and Hirono S 2004 *Pharm. Res.* **21** 2270
[15] Gouda H, Terashima S, Iguchi K, Sugawara A, Saito Y, Yamamoto T, Hirose T, Shiomi K, Sunazuka T, Ōmura S and Hirono S 2009 *Bioorg. Med. Chem.* **17** 6270
[16] Nagase H, Osa Y, Nemoto T, Fujii H, Imai M, Nakamura T, Kanemasa T, Kato A, Gouda H and Hirono S 2009 *Bioorg. Med. Chem. Lett.* **19** 2792
[17] Hirono S, Nakagome I, Imai R, Maeda K, Kusuhara H and Sugiyama Y 2005 *Pharm. Res.* **22** 260
[18] Yamaotsu N, Fujii H, Nagase H and Hirono S 2010 *Bioorg. Med. Chem.* **18** 4446
[19] Gouda H, Sunazuka T, Ui H, Handa M, Sakoh Y, Iwai Y, Hirono S and Ōmura S 2005 *Pro. Natl. Acad. Sci. USA* **102** 18286
[20] Friesner R A, Banks J L, Murphy R B, Halgren T A, Klicic J J, Mainz D T, Repasky M P, Knoll E H, Shelley M, Perry J K, Shaw D E, Francis P and Shenkin P S 2004 *J. Med. Chem.* **47** 1739
[21] Berman H M, Westbrook J, Feng Z, Gilliland G, Bhat T N, Weissig H, Shindyalov I N and Bourne P E 2000 *Nucleic Acids Res.* **28** 235
[22] *SYBYL 6.9* (St. Louis: Tripos Inc.)
[23] Stewart J J P 2007 *MOPAC2007* (Colorado Springs, CO: Stewart Computational Chemistry)
[24] Wang J, Wolf R M, Caldwell J W, Kollamn P A and Case D A 2004 *J. Comput. Chem.* **25** 1157
[25] Wang J, Wang W, Kollman P A and Case D A 2006 *J. Mol. Graph. Model.* **25** 247
[26] Case D A, Darden T A, Cheatham T E III, Simmerling C L, Wang J, Duke R E, Luo R, Merz K M, Pearlman D A, Crowley M,Walker R C, Zhang W, Wang B, Hayik S, Roitberg A, Seabra G, Wong K F, Paesani F, Wu X, Brozell S, Tsui V, Gohlke H, Yang L, Tan C, Mongan J, Hornak V, Cui G, Beroza P, Mathews D H, Schafmeister C, Ross W S and Kollman P A 2006 *AMBER9* (San Francisco, CA: University of California, San Francisco)
[27] Boström J, Greenwood J R and Gottfries J 2003 *J. Mol. Graph. Model.* **21** 449
[28] Kirchmair J, Wolber G, Laggner C and Langer T 2006 *J. Chem. Inf. Model.* **46** 1848
[29] Veber D F, Johnson S R, Cheng H, Smith B R, Ward K W and Kopple K D 2002 *J. Med. Chem.* **45** 2615

Journal of Physics: Conference Series **454** (2013) 012029 doi:10.1088/1742-6596/454/1/012029

Photoexcited electron and hole dynamics in semiconductor quantum dots: phonon-induced relaxation, multiple exciton generation and recombination

Kim Hyeon-Deuk

Department of Chemistry, Kyoto University, Kyoto, 606-8502, Japan

kim@kuchem.kyoto-u.ac.jp

Abstract. Photoexcited dynamics of electrons and holes in semiconductor quantum dots (QD), including phonon-induced relaxation, multiple exciton generation, and recombination (MEG and MER), were simulated by combining *ab initio* time-dependent density functional theory and non-adiabatic (NA) molecular dynamics. These nonequilibrium phenomena govern the optical properties and photoexcited dynamics of QDs, determining the branching between electronic processes and thermal energy losses. Our approach accounts for QD size and shape as well as defects, core–shell distribution, surface ligands and charge trapping, which significantly influence the properties of photoexcited QDs. The method creates an explicit time-domain representation of photoinduced processes and describes various kinetic regimes owing to the non-perturbative treatment of NA couplings in the quantum dynamics. QDs of different sizes and materials, with and without ligands, are considered. The simulations provide direct evidence that the high-frequency ligand modes on the QD surface play a pivotal role in the electron–phonon relaxation, MEG, and MER. The insights reported here suggest novel routes for controlling the photoinduced processes in semiconductor QDs and lead to new design principles for increasing efficiencies of photovoltaic devices.

1. Time-domain *ab initio* simulation of excited-state dynamics in quantum dots

Confinement of charge carriers in quantum dots (QDs) of sizes smaller than the Bohr exciton radius of the corresponding bulk material determine the electronic properties of the QD, leading to a variety of applications. High cross-sections for light absorption make QDs excellent materials for photovoltaic devices. Similarly to molecular systems, spatial confinement of electronic states in a QD creates relatively strong charge-phonon couplings, and efficiencies of QD photovoltaic devices and lasers depend upon charge-phonon relaxation rates. The phonon-mediated relaxation of electrons and holes in QDs is the focus of many experimental and theoretical investigations.

In our study, state-of-the-art time domain density functional theory (DFT) and non-adiabatic (NA) molecular dynamics simulations are applied to study phonon-induced relaxation of photo-excited electrons and holes in Ge and Si QDs.[1,2] Our *ab initio* calculations showed that the quantum confinement makes the electron and hole density of states (DOS) more symmetric in the Ge and Si QDs compared to the bulk semiconductors as figure 1(a) shows. Figure 1(b) demonstrates that, in spite of the symmetric DOS, the electrons decay faster than the holes, and so the electron and hole relaxations are quite asymmetric. Such excited-state dynamics cannot be predicted only by the static band structure calculations, suggesting importance of the real-time NA dynamics simulation including the phonon dynamics. This asymmetry arises due to the stronger NA couplings caused by the phonon dynamics in the conduction band (CB) than in the valence band (VB). The stronger NA couplings of the electrons compared to the holes was rationalized by the more delocalized electron density over the hydrogen surface ligands and thus by the larger contribution of the high-frequency Ge–H and Si–H surface bonds to the CB. Our results demonstrate that surface passivation is another option to control excited electron and hole dynamics in nano materials. Figure 1(b) also shows a linear relationship between the electron and hole relaxation rates and the CB and VB DOS. This can be regarded as a new type of Fermi's golden rule since the rates are now proportional not to the square of the static Coulomb couplings but to the square of the dynamical NA couplings.

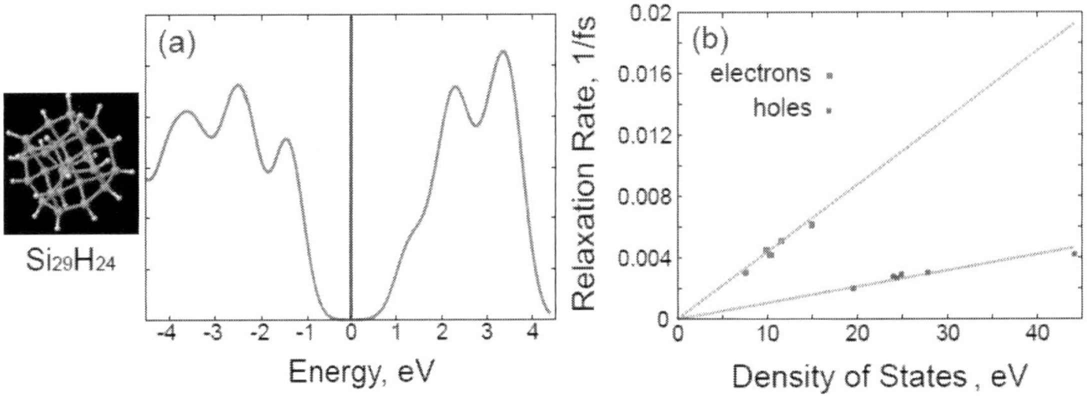

Figure 1: (a) Density of electron and hole states in the Si QD. The electron and hole DOS are quite symmetric due to the quantum confinement. The CB is about twice as high as the VB in the corresponding bulk DOS. The energy gap E_g is calculated as 2.1 eV. (b) Asymmetric relaxation of the electron and hole in the symmetric DOS. The rates scale linearly with the DOS, leading to Fermi's golden rule.

2. Multiple exciton generation and recombination dynamics in semiconductor quantum dots

Further advantages of QDs in photovoltaic devices are due to the possibility of generating multiple electron-hole pairs (excitons) upon absorption of a single photon as in figure 2. This is called multiple exciton generation (MEG), and provides great potential for increasing solar energy conversion efficiencies relative to those in bulk systems since the MEG provides new mechanisms for utilizing excess photon energy and avoiding energy loss to heat. We recently developed a real-time atomistic simulation method for studying the MEG and its inverse process, multiple exciton recombination (MER).[2,3,4] Our atomistic simulation is performed by directly solving the time-dependent Schrödinger equation based on multiple exciton bases with NA coupling and band energies obtained by the time-domain DFT calculation on nano materials. Our method calculates various real-time dynamics of the MEG and MER treating NA phonon couplings non-perturbatively, and the MEG and MER are allowed to occur simultaneously. Our atomistic simulation method can take into account QD size, shape, defects, core-shell distribution, surface ligands, and charge trapping, which significantly influence the photoexcited dynamics in nano materials. These are all advantages of our method which static electronic calculations of band structures and perturbative rate theories such as Fermi's golden rule cannot account for.

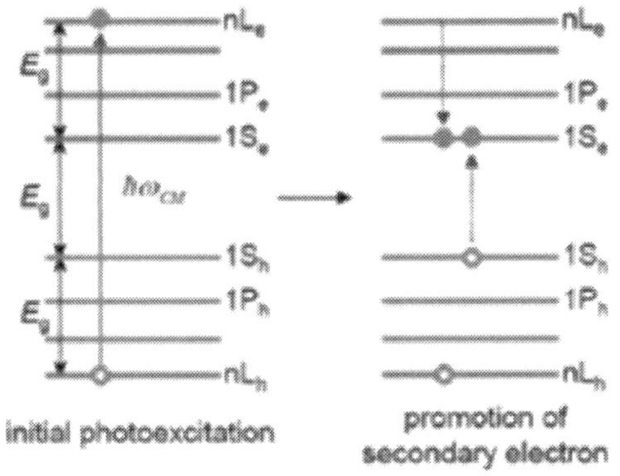

Figure 2: Schematic depiction of the MEG

As a result, we found the following important insights. (1) The MEG rapidly accelerates with initial excitation energy, reflecting strong energy dependence of double exciton (DE) density of states.[2,3,4] (2) At early times, the MEG exhibits Gaussian relaxation rather than exponential decay because the initial dynamics involves fewer states and is not statistically enough. The exponential dynamics, assumed in the standard rate theories, starts at a later time and becomes more important in larger QDs where a number of states are involved in the MEG dynamics. [2,3] (3) Phonon-assisted MEG is observed at energies below a purely electronic threshold due to presence of high-frequency ligand vibrations. The lack of the energy to generate DEs is compensated by the high-frequency phonon modes.[2,3] (4) Dissipation by phonons is essential for the MER since lower-energy DEs can be a main gateway to recombine into single excitons (SEs) due to the larger SE/DE DOS ratio around the lower-energy DEs.[4] (5) The MER simulated starting from a DE is significantly slower than the MER involving an optical excitation of a SE, followed by the MEG and then the MER (See figure 3). The latter time scale agrees well with the experiments [5-8], emphasizing the importance of superpositions of many DEs prepared by the preceding MEG for the efficient MER. [4]

The above detailed descriptions of the interplay between the MEG and MER coupled to phonon modes provide important insights for the excited-state dynamics in semiconductor QDs and in other nano materials, and also for improving efficiency of solar energy conversion using them.

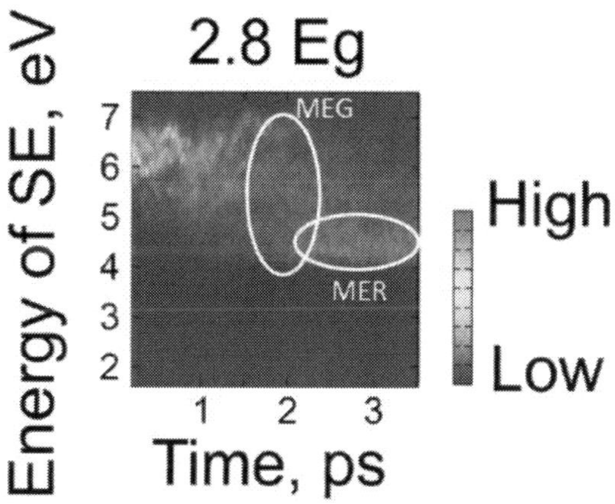

Figure 3: SE population dynamics accompanying the DE dynamics. The initially excited SE evolves into other SE and DE states. The decrease in the SE population seen at the intermediate time corresponds to the MEG. The recursion of the SE population at the later time corresponds to the MER, demonstrating that our method accounts for the MEG and MER simultaneously. The MER appears around 2 E_g, supporting the Auger recombination picture proposed in the experiments.

References

[1] K. Hyeon-Deuk, A. B. Madrid and O. V. Prezhdo, Dalton Transaction, **45**, 10069-10077 (2009)
[2] K. Hyeon-Deuk and O. V. Prezhdo, Journal of Physics: Condensed Matter, **24**, 363201 (2012)
[3] K. Hyeon-Deuk and O. V. Prezhdo, Nano Letters, **11**, 1845-1850 (2011)
[4] K. Hyeon-Deuk and O. V. Prezhdo, ACS Nano, **6**, 1239 -1250 (2012)
[5] R. D. Schaller, J. M. Pietryga and V. I. Klimov, Nano Letters, **7**, 3469–3476 (2007)
[6] I. Robel, R. Gresback, U. Kortshagen, R. D. Schaller and V. I. Klimov, Physical Review Letters, **102**, 177404 (2009).
[7] Y. Kobayashi, L. Pan and N. Tamai, Journal of Physical Chemistry C, **113**, 11783–11789 (2009)
[8] Y. Kobayashi, T. Nishimura, H. Yamaguchi and N. Tamai, Journal of Physical Chemistry Letters, **2**, 1051–1055 (2011)

24th IUPAP Conference on Computational Physics (IUPAP-CCP 2012) IOP Publishing
Journal of Physics: Conference Series **454** (2013) 012030 doi:10.1088/1742-6596/454/1/012030

An efficient computational method for the implementation of a semi-classical instanton approach using discretized path integrals

T Kawatsu[1,2] and S Miura[2]
[1] Institute for Molecular Science, National Institute of Natural Science, 38 Nishigonaka, Myodaiji, Okazaki 222-8585, Japan
[2] Graduate School of Natural Science and Technology, Kanazawa University, Kakuma, Kanazawa 920-1192, Japan

E-mail: kawatsu@fukui.kyoto-u.ac.jp, smiura@mail.kanazawa-u.ac.jp

Abstract. In the present paper, a numerical method for a semi-classical instanton method was examined. We implemented the instanton approach using discretized path integrals. The computational accuracy of the method is controlled by the following two parameters: the imaginary time duration (τ) and the time increment ($\Delta\tau$), which represents the discretized path integral. To obtain accurate results, a long τ must be used in combination with a short $\Delta\tau$; however, because the computational cost is virtually proportional to $\tau/\Delta\tau$, the instanton calculations were computationally expensive under these conditions. In the present study, we propose a method that reduces the computational cost and represents long τ instanton trajectories by employing an extended instanton trajectory from calculations based on a short τ. We applied the method to calculate tunnel splitting in a HO_2 hydrogen transfer reaction using the double many-body extension IV potential as a validation.

1. Introduction

A useful method of analyzing chemical reactions is the intrinsic reaction coordinate (IRC), which is a minimum energy path connecting an initial state to a final state in the configurational space of target systems. When quantum mechanical fluctuations are important for describing molecular processes, the description of chemical processes must be improved from a simple IRC. For example, a reaction path Hamiltonian method provides a theoretical tool that describes quantum mechanical processes on the basis of the IRC [1]. Nevertheless, Carrington Jr. and Miller have reported that the reaction pathway of intra-molecular hydrogen transfer in a malonaldehyde molecule deviates from the IRC due to tunneling effects [2]. To represent molecular processes affected by tunneling, the aforementioned authors developed a reaction surface Hamiltonian approach where two reaction coordinates are adopted to describe large amplitude motions that cannot be represented by the reaction path Hamiltonian approach. The reaction surface Hamiltonian approach provides useful tools and a rigid theoretical basis; however, the reaction coordinates must be pre-defined, which may be difficult for multi-dimensional systems. In such cases, path integral methods can be effectively applied because all of the degrees of freedom can be treated equally. Approaches based on path integral Monte Carlo [3] and molecular dynamics [4] can yield numerically exact results for molecular processes without pre-defined reaction coordinates. However, these methods require massive samplings to calculate highly quantitative tunnel effects, and many tunneling pathways must be included. As a result, the computational cost of these methods is rather expensive. In contrast, the instanton approach [5-10] in path integral theory, which is classified as a Wentzel-Kramers-Brillouin type approximation in standard quantum mechanics, is based on a semi-classical method and can be applied to a path integral representing tunneling effects. In the instanton approach, the instanton trajectory is determined to minimize the action for the duration of the imaginary time. Harmonic fluctuations along the instanton trajectory are included analytically. Katsnelson et al. proposed a method to determine the instanton trajectory by directly solving the imaginary time Newtonian equation [7]. In contrast, Mil'nikov and Nakamura determined the instanton trajectory by optimizing the action of the path [8,9]. Following the work of Mil'nikov and Nakamura, Richardson and Althorpe implemented the instanton approach using a discretized path integral representation [10]. Because the discretized path integral representation method has been used extensively in path integral Monte Carlo and molecular dynamics, the well-tested algorisms used in the aforementioned calculations can be used to develop the code for instanton calculations. Here, we briefly describe the Richardson-Althorpe instanton approach [10].

In the path integral theory, a propagator between the initial ($q^{(1)}$) and final ($q^{(M+1)}$) states is written as Equation (1), using a second order Suzuki-Trotter formula [10]:

$$G\left(q^{(1)}, q^{(M+1)}; \tau\right) \equiv \left\langle q^{(1)} \middle| e^{-\frac{\hat{H}\tau}{\hbar}} \middle| q^{(M+1)} \right\rangle = \left(\frac{1}{2\pi\hbar\Delta\tau}\right)^{\frac{3NM}{2}} \int dq^{(2)} \cdots dq^{(M)} e^{-\frac{1}{\hbar}S\left(q^{(1)},\cdots,q^{(M+1)}\right)}, \tag{1}$$

and

$$S\left(q^{(1)}, \cdots, q^{(M+1)}\right) = \sum_{n=1}^{M} \frac{\Delta\tau}{2}\left(\frac{q^{(n)} - q^{(n+1)}}{\Delta\tau}\right)^2 + \sum_{n=2}^{M} \Delta\tau V\left(q^{(n)}\right) + \frac{\Delta\tau}{2}\left(V\left(q^{(1)}\right) + V\left(q^{(M+1)}\right)\right), \tag{2}$$

where τ is an imaginary time duration from $q^{(1)}$ to $q^{(M+1)}$, $\Delta\tau$ is an interval between adjacent imaginary time slices, and N is the number of atoms in the original system. The total number of time slices is expressed as $M \equiv \tau/\Delta\tau$, $\hat{H}$ is the Hamiltonian, and $q^{(n)}$ is the set of mass-weighted Cartesian coordinates of the original system at imaginary time slice n. Equations (1) and (2) can be interpreted as a configurational integral of classical polymers with a potential function $S/\tau\hbar$, where each polymer consists of N interaction sites or beads. Snapshots or replicas of the system at adjacent imaginary time slices are connected by harmonic springs. Interatomic interaction at each time slice is expressed as $V\left(q^{(n)}\right)$ (n=2,..., M) and $V\left(q^{(n)}\right)/2$ (n=1, M+1). In the instanton approach, instead of integrating the entire $\{q^{(n)}\}$ space (= integrating all paths connecting from $q^{(1)}$ to $q^{(M+1)}$), we evaluate a stationary phase contribution by applying the following conditions: $\partial S/\partial q^{(n)} = 0$, which gives the instanton trajectory or optimized configuration of the polymer and includes its surrounding harmonic fluctuation. Path integral theory suggests that such a path (namely, the classical solution) has more possibilities than other paths. This approximation significantly reduces the computational cost compared to sampling approaches such as path integral Monte Carlo or molecular dynamics simulations yet still includes the harmonic component of quantum fluctuations along the instanton trajectory. Although an instanton with a single transition is determined, the pathway that goes back and forth many times also provides a stationary phase contribution, which can be included analytically by neglecting correlations among transitions. The action along the instanton trajectory is denoted as S_{inst}. Tunnel splitting ($\Delta\epsilon$) is then expressed in accordance with the results described in the literature [10]:

$$\Delta\epsilon = \lim_{\tau\to\infty} 2\hbar K e^{-S_{inst}/\hbar}, \tag{3}$$

$$K \equiv \sqrt{\frac{S_{inst}}{2\pi\hbar}}\Phi^{-1}, \tag{4}$$

and

$$\Phi \equiv \left(\frac{\det' J_{inst}}{\det J_0}\right)^{\frac{1}{2}}, \tag{5}$$

where K is a coefficient that counts contributions from harmonic fluctuations around the instanton trajectory, J_{inst} and J_0 are $3MN \times 3MN$ Hessian matrices of the instanton trajectory and the trajectory that all time slices remain on in either the initial or final state, respectively. The Hessian J_{inst} has one zero eigenvalue associated with the translation of the instanton trajectory in imaginary time, and $\det'$ indicates the determinant without a zero eigenvalue mode.

In the present study, we adopted the Richardson-Althorpe approach for instanton calculations. Possible improvements to their computational methods were also examined. First, we investigated the effect of $\Delta\tau$ on the evaluation of S_{inst} and Φ for a given value of τ. Next, we determined the dependence on τ in the tunnel splitting calculation. We then proposed an efficient method of extending τ to reduce the computational cost and maintain accuracy. Section 2 includes the computational results and a brief discussion. In subsection 2.1, we show a computational result of the instanton trajectory and tunnel splitting for hydrogen transfer in a HO_2 molecule using a double many-body extension (DMBE) IV potential [11]. In subsection 2.2, the parameter dependences of systematic errors on the instanton action and tunnel splitting are discussed. In section 2.3, we propose a method to efficiently extend τ. Section 3 includes a summary of our results.

2. Results and discussions

2.1. HO_2 hydrogen transfer pathway

To examine the usefulness of the method, we applied the instanton approach to hydrogen transfer in a molecule of HO_2. This system has been well studied and is considered a benchmark molecular system [8-10]. In accordance with prior research [8-10], we employed the DMBE IV empirical potential field [11]. The instanton trajectory was obtained by optimizing action Equation (2) using the L-BFGS method [12]. To diagonalize Hessians, LAPACKE in Intel MKL library was employed. In the course of the optimization, both ends of the instanton trajectory were considered to be movable to exclude the overall rotation and translation between initial and final coordinates. In figure 1, the computed instanton trajectory and IRC trajectory are drawn on the potential surface. The angle γ indicates the progress of hydrogen transfer from one oxygen to the other. The O-O distance (r_{OO}) represents collective heavy atomic motion (r_{cH}), which is the distance between the hydrogen atom and the center of the oxygen pair, and helps to represent hydrogenic motion with an angle of γ. Figure 1a shows that the instanton trajectory represented by r_{OO} and γ deviates from the IRC. The characteristic difference between the instanton and the IRC is the motion of oxygen. In the IRC, the O-O bond stretches when the system approaches the potential barrier, which leads to a relatively low potential barrier. In contrast to the IRC, the instanton pathway is primarily constructed based on the motion of hydrogen. Because the instanton corresponds to tunnel transfer, heavy atoms are much less active than light atoms. The instanton goes over a higher potential barrier than the IRC due to the tunneling effect. Figure 1b primarily shows hydrogenic motion because the center of the oxygen pair remains nearly constant during the hydrogen transfer reaction. Subsequently, both the IRC and instanton trajectories draw curves that are close to each other on a plane by the coordinates r_{CH} and γ. The computed value of tunnel splitting was $\Delta\epsilon = 7.0 \times 10^{-9}$ cm^{-1}, which is in agreement with Richardson's prior calculation [10].

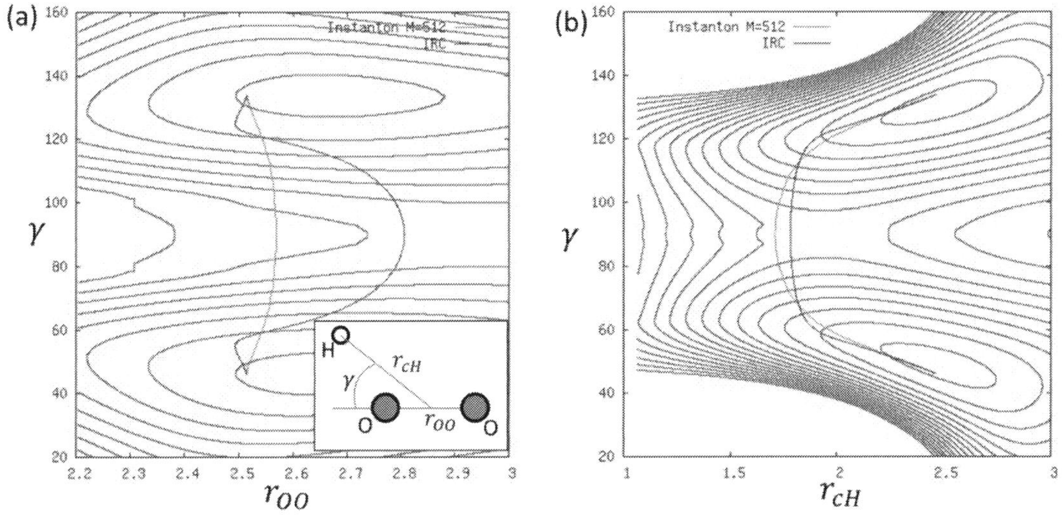

Figure 1. Instanton and IRC trajectories for HO_2 hydrogen transfer. Green and blue curves represent the instanton and IRC, respectively. Red curves are contour lines of the potential surface. The parameters τ=3600 and M=512 were applied. Axis notations are described in the inset of figure 1a.

2.2. Parameter dependences of systematic errors

We first present the systematic errors on the numerical values of S_{inst} and Φ that arise from the discretized approximation of the path integral. Due to the second order Suzuki-Trotter decomposition, the systematic error of the action of S_{inst} and Φ will be proportional to $\Delta\tau^2$ for a given τ. Because the convergence of S_{inst} and Φ with respect to $\Delta\tau$ also depends on τ, we must use a sufficiently large value of τ, as shown in Equation (3). Figure 2a shows the corresponding proportions for various values of τ. As shown in the figure, when τ is greater than or equal to 1800 au, a single line is observed. However, when τ=1200 and 1800 au, the values are present in different series. This disagreement arises from the fact that the imaginary time duration is not large enough to represent the instanton trajectory connecting the initial and final configurations. A representative reaction coordinates γ at each imaginary time slice $n\Delta\tau$ are presented in figure 3a. The reaction coordinates γ with τ=1200, 2400, 4800, and 9600 au are over-written in the figure. In the case of τ=1200 au, the configurations at both ends of the imaginary time duration τ do not reach the initial and final configurations, respectively. In figure 3a, the time for the transition from the initial to final configuration is estimated to be approximately 2000 au, which is the minimum duration of time (τ) required to converge S_{inst} in the present system. Thereafter, we adopted only results with τ greater than 1800 au to avoid the aforementioned problem. Figure 2b presents $\Delta\tau^2$ dependences for errors of Φ, which is another factor that affects tunnel splitting. The error of Φ was also proportional to $\Delta\tau^2$; however, the calculated values converge more slowly on τ compared to S_{inst}. The Φ value is determined by the ratio of eigenvalues for J_{inst} and J_0 in Equation (5). When τ lengthens, the eigenvalues can change even though the transition region from the initial to the final configurations is identical. Because the values of Φ obtained from τ=14400 and 19200 au were similar, τ must be set to at least 14000 au to obtain an exact value of $\Delta\epsilon$. This value of τ is much longer than that used to represent S_{inst}, and the computational cost is high.

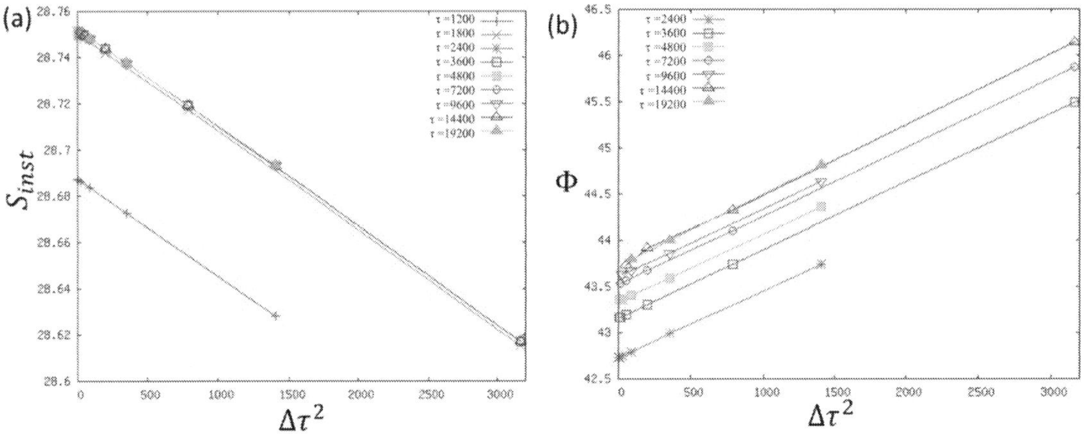

Figure 2. $\Delta\tau^2$ dependences of systematic errors on (a) S_{inst} and (b) Φ with various values of τ.

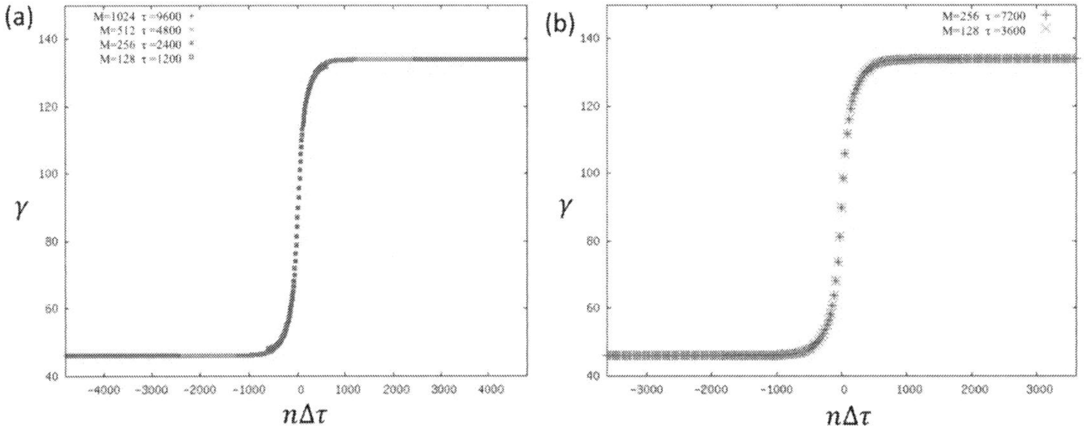

Figure 3. Representative reaction coordinates (γ) at each imaginary time slice $n\Delta\tau$ in the instanton trajectory. (a) $\Delta\tau$=9.375 with short to long τ parameters: τ=1200 (purple), 2400 (blue), 4600 (green), and 9600 (red). (b) $\Delta\tau$=28.125 with two medium τ parameters: τ=3600 (green) and 7200 (red). The horizontal axis γ is described in figure 1.

2.3. Extension of the polymer length τ

In figure 3b, the reaction coordinates γ at each imaginary time slice $n\Delta\tau$ were obtained with parameters τ=3600 or 7200 au and $\Delta\tau$=28.125 au. Comparing these two calculations, the reaction coordinates γ at each time slice on the instanton are identical. Therefore, the differences caused by τ come from the difference in the number of time slices at the initial and final configurations. These results suggest that we can add time slices at the initial and final configurations to both ends of the instanton trajectory to extend τ without re-calculating the stationary phase for high values of τ. When the configurations at both ends of the imaginary time duration τ reach the initial and final configurations, respectively, Equation (2) predicts that the τ extension does not affect S_{inst}. However, Φ value depends on the τ extension of the trajectory. The convergence for τ value requests at least 14000 au as estimated in previous section. For example, if we can calculate Φ value at τ=19200 au using the instanton trajectory optimized at τ=3600, the size of the optimized system is four times smaller than fully optimized trajectory, because using the same $\Delta\tau$, the number of imaginary time slices M are proportional to τ. The computational cost for the optimization reduces four times or more.

In addition, we can also reduce the computational cost to calculate interatomic potential for added imaginary time slices by the τ extension because these slices are the same as the initial or final geometries.

Figure 4 shows the computed values of (a) S_{inst} and (b) Φ using the fully optimized calculations (horizontal axis) and the extension of τ without re-optimization of the instanton trajectory (vertical axis). Superscript * indicates the τ extension results. S^*_{inst} is in agreement with the S_{inst} values shown in figure 4a. In figure 4b, the τ extension results (Φ^*) also agree with the original Φ values, except when τ=2400 au. These results demonstrate that we can extend τ without re-optimizing the instanton trajectory obtained with a short τ. We remark that τ must not be too short. Namely, τ=2400 au is not long enough to provide a correct instanton trajectory, and the τ extension process may expand the error. Φ is computed using Hessians, and such calculations appear to be more sensitive to the accuracy of computational conditions. As shown in figure 4b, numerical convergence on Φ is achieved for τ equal to or greater than 3600 au, which is approximately twice as long as the transition time estimated above. Thus, to obtain sufficient numerical accuracy with the τ extension method, the optimized result for the instanton trajectory for the imaginary time duration had to be two times greater than the transition time in this system. Although several small errors are observed in figure 4b, these errors are caused by the instanton trajectory optimization error for determining the stationary phase in the current calculation level.

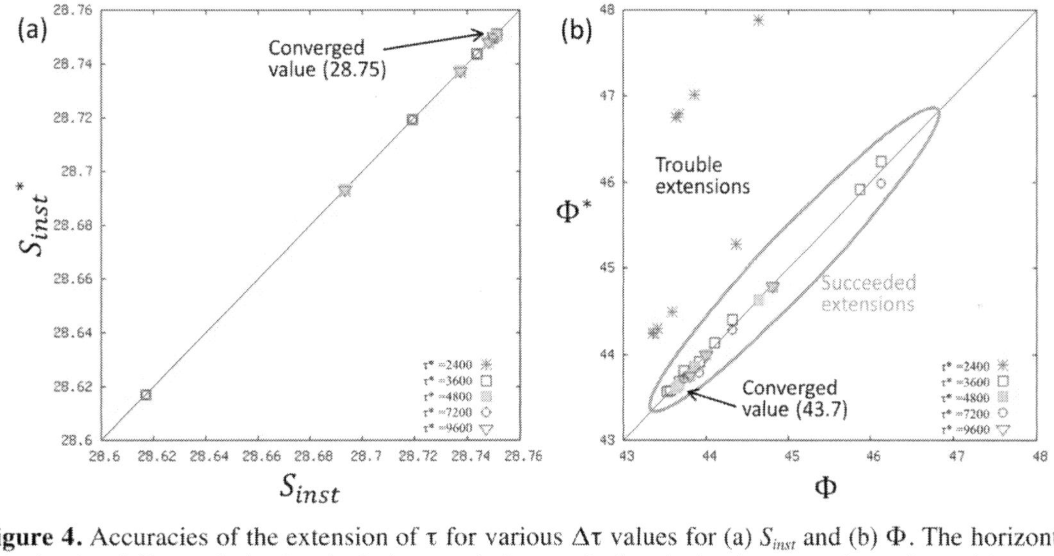

Figure 4. Accuracies of the extension of τ for various $\Delta\tau$ values for (a) S_{inst} and (b) Φ. The horizontal axis is the fully optimized calculation, and the vertical axis is the extension of τ without re-optimization. τ^* indicates τ before extensions. Pre-extended calculations are M=64 (τ^*=2400, 3600), 128 (τ^*=2400-7200), 256 (τ^*=2400-7200), 512 (all τ^*), and 1024 (all τ^*) and all of these are extended to up to either M=2048 or τ=14400 or 19200. The converged values are calculated by the least square fitting using results τ=19200 in figure 2.

3. Concluding remarks

The instanton approach with the discretized path integral method can be directly applied to multi-dimensional systems. In the present paper, we proposed a method for efficient instanton calculations. The method reduces the computational cost associated with taking the long limit of τ. In our method, the imaginary time duration τ was extended by recycling the result of the converged instanton trajectory for a relatively short τ. The numerical results presented in this paper indicated that our method produces accurate numerical values.

Similar to path integral Monte Carlo and molecular dynamics, interatomic interaction at each imaginary time slice can be calculated independently, and the instanton method can be easily implemented in highly parallel computational conditions with great efficiency. Additionally, the method presented in this paper can reduce computational costs; thus, we can invest these reduced computational costs to calculate the interatomic potential at each imaginary time slice using accurate quantum chemical calculations. These issues will be addressed in the near future.

Acknowledgments
This work was partially supported by a Grant-in-Aid for Scientific Research (C) (No. 23550011) from the Japan Society for the Promotion of Science and by the Strategic Programs for Innovative Research (SPIRE), MEXT, and the Computational Materials Science Initiative (CMSI), Japan.

References

[1] Miller W H, Handy N C and Adams J E 1980 *J Chem Phys* **72** 99
[2] Carrington Jr. T and Miller W H 1986 *J Chem Phys* **84** 4364
[3] Tuckerman E M and Hughes A 1999 *Classical and Quantum Dynamics in Condensed Phase Simulations* ed G Ciccotti and D F Coker (Singapore: World Scientific) p 311
[4] Ceperley D M 1995 *Rev Mod Phys* **67** 279
[5] Coleman S 1985 *Aspects of Symmetry* (New York, Melbourne: Cambridge University Press)
[6] Altland A and Simons B 2006 *Condensed Matter Field Theory* (New York: Cambridge University Press)
[7] Katsnelson M I, van Schilfgaarde M, Antropov V P and Harmon B N 1996 *Phys Rev A* **54** 4802
[8] Mil'nikov G and Nakamura H 2008 *Phys Chem Chem Phys* **10** 1374
[9] Mil'nikov G V and Nakamura H 2001 *J Chem Phys* **115** 6881
[10] Richardson J O and Althorpe S C 2011 *J Chem Phys* **134** 054109
[11] Pastrana M R, Quintales L A M, Brandão J and Varandas A J C 1990 *J Phys Chem-Us* **94** 8073
[12] Nocedal J 1980 *Math Comput* **35** 773

Journal of Physics: Conference Series **454** (2013) 012031 doi:10.1088/1742-6596/454/1/012031

Sampling from a polytope and hard-disk Monte Carlo

Sebastian C Kapfer and Werner Krauth

Laboratoire de Physique Statistique, Ecole Normale Supérieure, UPMC, CNRS
24 rue Lhomond, 75231 Paris Cedex 05, France

E-mail: `sebastian.kapfer@ens.fr, werner.krauth@ens.fr`

Abstract. The hard-disk problem, the statics and the dynamics of equal two-dimensional hard spheres in a periodic box, has had a profound influence on statistical and computational physics. Markov-chain Monte Carlo and molecular dynamics were first discussed for this model. Here we reformulate hard-disk Monte Carlo algorithms in terms of another classic problem, namely the sampling from a polytope. Local Markov-chain Monte Carlo, as proposed by Metropolis et al. in 1953, appears as a sequence of random walks in high-dimensional polytopes, while the moves of the more powerful event-chain algorithm correspond to molecular dynamics evolution. We determine the convergence properties of Monte Carlo methods in a special invariant polytope associated with hard-disk configurations, and the implications for convergence of hard-disk sampling. Finally, we discuss parallelization strategies for event-chain Monte Carlo and present results for a multicore implementation.

1. Introduction

The hard-disk system is a fundamental model of statistical and computational physics. For more than a century, the model and its generalization to d-dimensional spheres have been central to many advances in physics. The virial expansion is an example: Boltzmann's early calculations of the fourth virial coefficient [1] ultimately led to Lebowitz and Onsager's proof of the convergence of the virial expansion up to finite densities [2] for all d and to the general and systematic study of virial coefficients. The theory of phase transitions provides another example for the lasting influence of the hard-disk model and its generalizations. Kirkwood and Monroe [3] first hinted at the possibility of a liquid–solid transition in three-dimensional hard spheres. This prediction was surprising because of the absence of attractive interactions in this system. The depletion mechanism responsible for the effective-medium attraction was also first studied in hard spheres, by Asakura and Oosawa [4]. In two dimensions, the liquid–solid phase transition was first evidenced by Alder and Wainwright [5]. It lead to far-reaching theoretical [6], computational [7, 8] and experimental [9] work towards the understanding of 2D melting. In mathematics, hard disks and hard spheres have also been at the center of attention [10]. A rigorous existence proof of the melting transition in hard spheres is still lacking, but the ergodicity of the molecular dynamics evolution of this system has now been established rigorously [11, 12].

Arguably the most important role for the hard-disk model has been in the development of numerical simulation methods. Molecular dynamics [13, 14] and Markov-chain Monte Carlo [15] were first formulated for hard disks. The early algorithms have continued to be refined: Within the molecular dynamics framework, this has lead to highly efficient event-scheduling strategies [14, 16] and, for Monte Carlo, to the development of cluster algorithms [17, 18, 19].

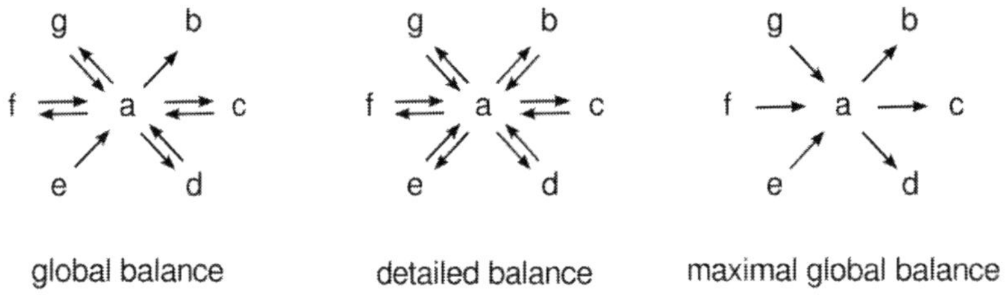

Figure 1. Monte Carlo balance conditions: Arrows represent probability flows $P_{a\to b} = \pi_a\, p_{a\to b}$ between configurations a and b (each arrow stands for a probability flow of same magnitude). *Left*: Global balance, as required for Markov-chain Monte Carlo. The total flow $\sum_c P_{c\to a}$ into the configuration a equals the flow $\sum_c P_{a\to c}$ out of it. *Center*: Detailed balance: the net flow between any two configurations is zero, $P_{a\to b} = P_{b\to a}$. *Right*: Another special case of global balance: maximal global balance at a ($P_{a\to b} > 0 \implies P_{b\to a} = 0$).

Even the modern simulation algorithm remain slow, however, and revolutions like the cluster algorithms for spin systems [20, 21] have failed to appear. Moreover, rigorous mathematical bounds for the correlation time (mixing time) of Monte Carlo algorithms were obtained in the thermodynamic limit only for small densities [22, 23, 24], which are far inside the liquid phase. At higher densities, close to the liquid–solid transition, many numerical calculations have suffered from insufficient simulation times until recently [7, 8].

In the present article, we discuss computational aspects of the hard-disk model, starting with an introduction (Section 2). In particular, we reinterpret hard-sphere Monte Carlo in terms of the sampling of points from high-dimensional polytopes (Section 3). Local Monte Carlo amounts to random walks in a sequence of such polytopes, while event-chain Monte Carlo is equivalent to molecular dynamics evolutions with particular initial conditions for the velocities. We analyze the convergence properties of the algorithms in these polytopes for the hard-disk case. Parallel event-chain algorithms emerge naturally as molecular dynamics with more general initial conditions (Section 4). We describe several parallelization strategies and report on implementations.

2. Local Monte Carlo and event-chain Monte Carlo

We consider N equal hard disks of unit radius $\sigma = 1$ in a square box of size $L \times L$. In the following, we assume throughout periodic boundary conditions for positions and pair distances. The statistical weights π_a are equal to unity for configurations a without overlaps (all pair distances larger than 2) and zero for illegal configurations (with overlaps). The phase diagram of the system depends only on the packing fraction $\eta := N\pi\sigma^2/L^2$. The letters i, j, k number disks and a, b, c, …, label hard-disk configurations of N disks, given by the coordinates of the disk centers $\boldsymbol{r}_i$.

2.1. Balance conditions

Markov-chain Monte Carlo algorithms are governed by balance conditions for the flows $P_{a\to b} := \pi_a\, p_{a\to b}$ from configuration a to configuration b (see Fig. 1); $p_{a\to b}$ is the conditional probability to move from a to b, given that the system is in a. To converge towards the stationary distribution π_a, the *global* balance condition must be satisfied: The total flow onto configuration a must equal the total flow out of a,

$$\sum_c P_{c\to a} = \sum_c P_{a\to c}. \tag{1}$$

The local Monte Carlo algorithm, introduced by Metropolis et al. in 1953 [15] (see Fig. 2), uses the more restrictive *detailed* balance condition $P_{a\to b} = P_{b\to a}$ for which the net flow between each pair of configurations a and b is zero. Moving from configuration $a = \{\boldsymbol{r}_1, \dots \boldsymbol{r}_i, \dots, \boldsymbol{r}_N\}$ to $b = \{\boldsymbol{r}_1, \dots \boldsymbol{r}_i + \boldsymbol{\delta}, \dots, \boldsymbol{r}_N\}$ involves sampling the disk i to be displaced and the displacement $\boldsymbol{\delta}$. For detailed balance, the probability to sample $\boldsymbol{\delta}$ at $\boldsymbol{r}_i$ must equal the probability to sample $-\boldsymbol{\delta}$ at position $\boldsymbol{r}_i'$. In order to be ergodic, the displacements $\boldsymbol{\delta}$ are chosen such that each disk can eventually reach any position in the system.

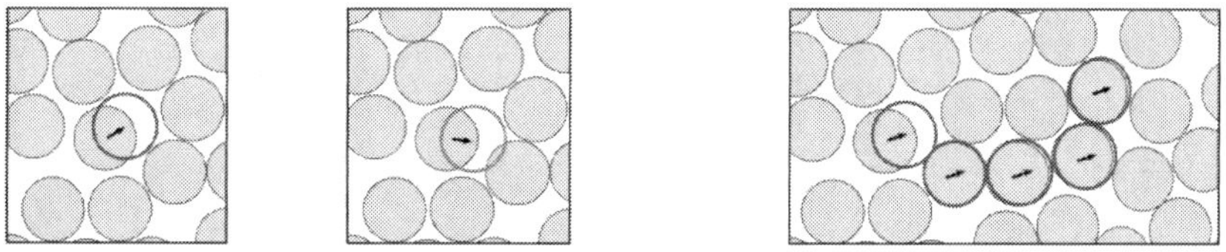

Figure 2. Monte Carlo moves for hard disks. *Left*: Accepted and rejected Metropolis moves. *Right*: Event-chain move. The sum of individual displacements equals a predefined value ℓ. With periodic boundary conditions, the event-chain move is rejection-free.

Unlike the local Monte Carlo algorithm, a single move of the event-chain algorithm [19] may displace several disks. An event-chain move is parameterized by a total displacement ℓ and a direction $\boldsymbol{e}_\ell$, which together form a vector $\boldsymbol{\ell} := \ell\, \boldsymbol{e}_\ell$. The move starts by sampling a disk i and "sliding" it in the $\boldsymbol{e}_\ell$ direction until it hits another disk j, or at most for the distance ℓ. The disk j is then displaced in its turn, also in the $\boldsymbol{e}_\ell$ direction, see Fig. 2. This process continues until the displacements of the individual disks sum up to ℓ. After this, a new disk and possibly a new direction are sampled for the next move. With periodic boundary conditions, no rejections occur in this algorithm. For a given displacement vector $\boldsymbol{\ell}$, any disk configuration a can reach N other configurations, using each of the N disks to start an event chain. Likewise, a can be reached from N other configurations which may be reconstructed by event chains with displacement vector $-\boldsymbol{\ell}$. This implies that the event-chain satisfies the global balance condition (1). If the vectors $\pm\boldsymbol{\ell}$ are equally likely, it also satisfies detailed balance. In order to be ergodic, the displacements $\boldsymbol{\ell}$ must span space: By choosing $\boldsymbol{e}_\ell \in \{\boldsymbol{e}_x, \boldsymbol{e}_y\}$, the event-chain algorithm realizes the maximal global balance (see Fig. 1), where flow between two configurations is possible only in one direction. This version is more efficient than detailed balance versions (for example, $\pm\boldsymbol{e}_x$ and $\pm\boldsymbol{e}_y$) [19]. It is again possible to alternate repeated moves in the $\boldsymbol{e}_x$ direction with repeated moves in $\boldsymbol{e}_y$ without destroying the correctness of the algorithm. For displacements ℓ smaller than the mean free path l_{mfp}, the event-chain algorithm is roughly equivalent to the local Monte Carlo algorithm. It accelerates for increasing ℓ, and for ℓ much larger than the mean free path, it is about two orders of magnitude faster than the local Monte Carlo method, and about ten times faster than the best current implementations [16] of event-driven molecular dynamics (see [25]).

2.2. *Correlation times and orientational order*

The characteristic challenge of numerical simulations for the hard-disk model resides in the extremely long correlation time. This is illustrated in Fig. 3 using snapshots of configurations

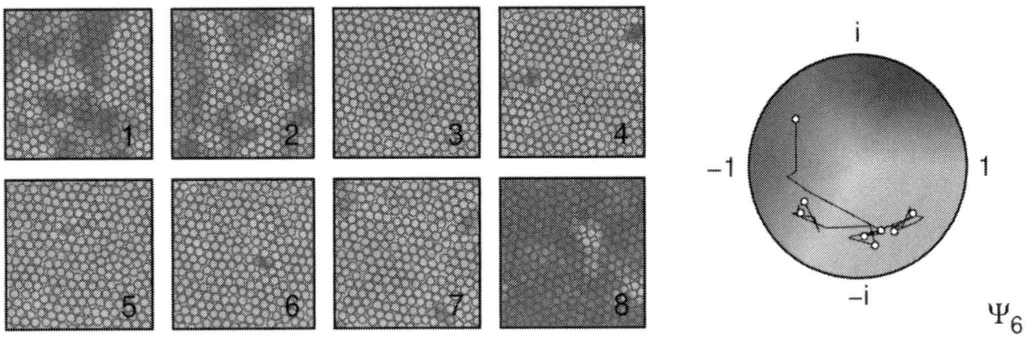

Figure 3. Local Monte Carlo evolution of 16^2 disks in a square box with periodic boundary conditions at packing fraction $\eta = 0.707$. *Left*: Disk configurations and their local orientational field ψ_j for one simulation run. Frames are separated by $100,000N$ iterations (10^5 sweeps) of local Monte Carlo. The slow decorrelation of the orientation is manifest. *Right*: For the same simulation run, the evolution of the global orientational order parameter Ψ_6, defined by Equation (3), in the complex plane.

obtained during a long simulation run. The system is quite small and not extremely dense, yet correlations in the orientation of the system persist over millions of Monte Carlo moves. To quantify the orientations and their correlations, we consider the local orientational field

$$\psi_j := \sum_{k=1}^{N_j} w_{j,k} \exp(6\mathrm{i}\phi_{j,k}), \tag{2}$$

where N_j is the number of Voronoi neighbors of disk j. The $w_{j,k}$ (with $\sum_k w_{j,k} = 1$) are normalized weights according to the length of the Voronoi interface between disks j and k, and $\phi_{j,k}$ is the angle of the vector between the disk centers [26]. The average of (2) over all disks yields the global orientational order parameter,

$$\Psi_6 := \frac{1}{N} \sum_j \psi_j. \tag{3}$$

In a square box, the mean value of Ψ_6 is zero because of the $\phi_{j,k} \to \phi_{j,k} + \pi$ symmetry, and its correlation function

$$C_6(\Delta t) := \frac{\langle \Psi_6(t)\Psi_6^*(t+\Delta t)\rangle_t}{\langle |\Psi_6(t)|^2\rangle_t}. \tag{4}$$

decays to zero for infinite times Δt. We conjecture that Ψ_6 is the slowest observable in the system. For large times, global orientational correlations decay exponentially, $C_6(\Delta t) \propto \exp(-\Delta t/\tau)$, and we obtain the empirical correlation time τ from an exponential fit to C_6.

3. Polytope representation of event-chain moves

Event-chain moves along a single direction $\boldsymbol{e}_\ell = \boldsymbol{\ell}/\ell$ sample a restricted configuration space. For the remainder of this section, we take the chains to move in the positive x direction, unless specified otherwise, to simplify the notation. Since all y coordinates are fixed, two disks whose

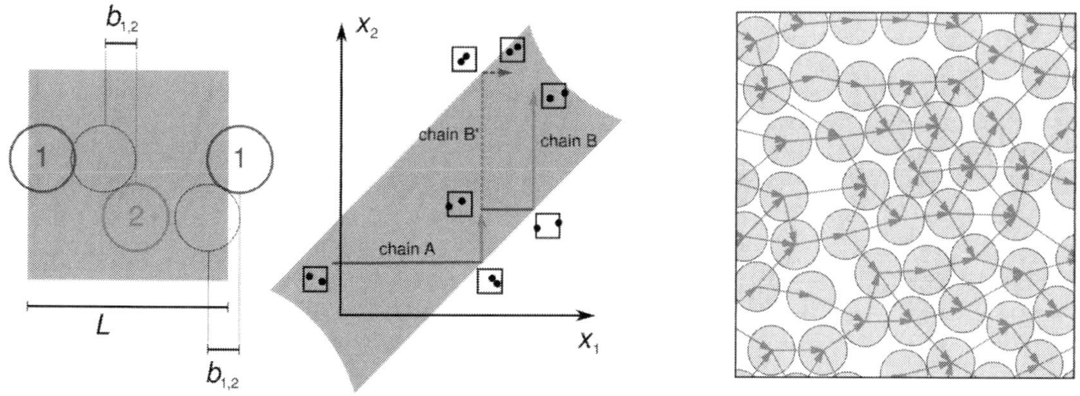

Figure 4. Event-chain move and polytope representation. *Left*: Two disks in a periodic box. The constraints (5) are $x_1 \leq x_2 - b_{1,2}$ and $x_2 \leq x_1 + L - b_{1,2} = x_1 - b_{2,1}$. *Center*: Molecular dynamics evolution in the polytope corresponding to two event chains with moves of disk 1 (blue segments) and disk 2 (red segments). The trajectory begins with disk 1, and it depends on the choice of the starting disk (1 or 2) for the second chain. Periodic boundary conditions are ignored for clarity. Snapshots of the configuration are sketched along the trajectory. *Right*: Hard-disk configuration with its constraint graph for motion along the x axis. Each node has at most three forward and three backward links. This graph is invariant under event-chain moves in the x direction.

y coordinates differ by less than 2 radii cannot slide across each other, and their relative order is fixed. Furthermore, while in x collision mode, any disk can collide with not more than six other disks, at most three in the forward direction, and at most three in backward direction (see Fig. 4). The collision partners of a disk may include itself, because of boundary conditions. The relations among disks constitute a *constraint graph*, which expresses the partial order between them (see Fig. 4). This graph remains invariant while performing event-chain moves in the $\pm \boldsymbol{e}_x$ direction. Each directed edge from i to k corresponds to a linear inequality for the x coordinates of the disks i and k:

$$x_i \leq x_k - b_{i,k}, \tag{5}$$

with $b_{i,k} := \sqrt{4 - (y_i - y_k)^2}$. The constant $b_{i,k}$ can be adjusted to also account for periodic boundary conditions in the x direction. The inequalities (5) imply that no more than three forward collision partners can be present[1].

The system of linear inequalities (5) delimit a subset of the N-dimensional space of x coordinates $X = (x_1, \ldots, x_N)$, an N-dimensional polytope, bounded by at most $3N$ hyperplanes. This convex object is easier to analyze than the highly intricate $2N$-dimensional configuration space of the full hard-disk problem. The polytope is unbounded in the $(1, 1, \ldots, 1)$ direction in consequence of the periodic boundary conditions, since uniform translation of all the disks is always permitted. Also, since the constraint graph is invariant under event-chain moves in the

[1] A superset of the actual constraint graph can be computed from efficient local criteria. This superset contains redundant inequalities, but describes the same polytope; it is, in particular, useful for the practical implementation: If i collides forward with j, and j collides forward with k, we have $x_i \leq x_k - b_{i,j} - b_{j,k}$; if now $b_{i,j} + b_{j,k} > b_{i,k}$, the disks i and k can never come into contact; disk j *covers* disk k. Applying this rule iteratively, the disks in the forward direction can be reduced to at most three: at most one each with $y \in (y_i - 1, y_i + 1)$, with $y \in [y_i + 1, y_i + 2)$ and with $y \in (y_i - 2, y_i - 1]$. Thus, each node in the constraint graph has at most degree six.

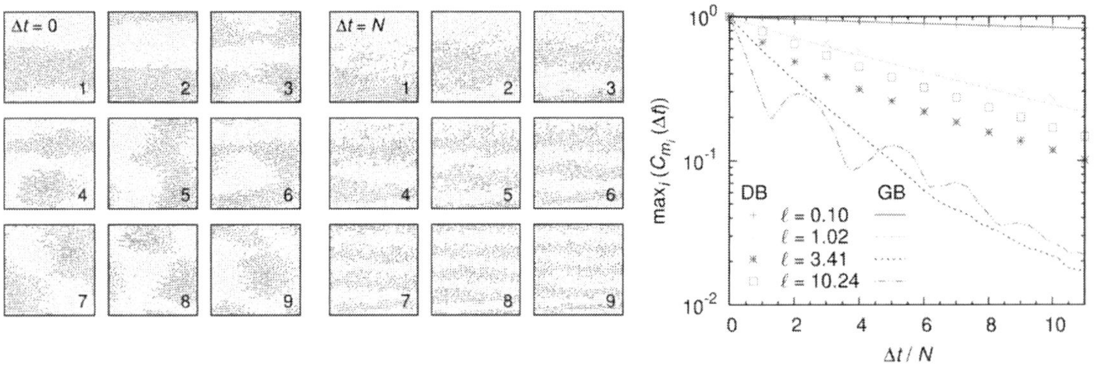

Figure 5. Relaxation dynamics in the invariant polytope, for a given initial configuration of the 32^2 hard-disk system at packing fraction $\eta = 0.698$. *Left*: Slowly decaying modes. Configurations are shown with red disks moving in the $+x$ direction and green disks in $-x$. The modes shown are the eigenvectors of the largest eigenvalues of $U(0)$. *Center*: Remaining correlation after $\Delta t = N$ of event-chain moves in the horizontal direction. These are the largest eigenvectors of $U(N)$. Correlations in the horizontal direction have all but disappeared. *Right*: Decay of the slowest modes, for the event-chain simulations with various total displacements ℓ, and for both the global (GB) and the detailed (DB) balance version.

x direction, so is the polytope. However, the polytope becomes bounded by taking a section orthogonal to $(1, 1, \ldots, 1)$.

In the invariant polytope, an event chain of total displacement ℓ corresponds to a molecular dynamics evolution of duration ℓ: Displacing the i-th disk corresponds to the "particle" X moving in the i-th coordinate direction, and each collision event (the transfer of momentum from one disk to another) to a right-angle reflection at the facets of the polytope (see Fig. 4). The construction of the event-chain move is finished at time ℓ. The next move involves the sampling of a new starting disk and possibly of one of the $\pm \boldsymbol{e}_\ell$ directions. In the invariant polytope, this is the choice of new velocities. Local Monte Carlo on the other hand, if restricted to moves in x direction, implements diffusive motion in the invariant polytope[2].

The invariant constraint graph allows for fast lookup of possible collision partners, and may even replace the customary cell grids (see, for example, Section 2.4 of [28]). While computation of the actual constraint graph requires depth search, a superset sufficient for practical computations can be computed efficiently, see the footnote on page 5.

3.1. *Correlation functions in the invariant polytope*

Although the sampling problem from the invariant polytope concerns a convex body, it is notoriously nontrivial [27]. The inequalities (5) essentially amount to a system of coupled one-dimensional hard-disk problems. To study the relaxation behavior effected by the event-chain algorithm in the polytope, we consider the cross-covariance of the disk coordinates,

$$U_{ij}(\Delta t) := \left\langle \tilde{x}_i(t + \Delta t) \cdot \tilde{x}_j(t) \right\rangle_t, \tag{6}$$

[2] To preserve the polytope, a "sliding" version of local Monte Carlo must be considered: The move $\boldsymbol{r}_i \to \boldsymbol{r}_i + \boldsymbol{\delta}$ is valid only if all intermediate positions $\boldsymbol{r}_i + \alpha\boldsymbol{\delta}$ with $\alpha \in [0, 1]$ yield legal hard-sphere configurations.

where $\tilde{x}_i(t)$ is the x coordinate of the disk i, compensated for the overall translation of the system due to the event-chain moves,

$$\tilde{x}_i(t) := x_i(t) - \frac{\alpha t}{N} - \left\langle x_i(t) - \frac{\alpha t}{N} \right\rangle_t . \tag{7}$$

Here, α is 1 for the global balance version of the event-chain algorithm (chains only in $+x$ direction), and 0 for the detailed balance version ($\pm x$). The eigenvectors of $U(0)$ are the polytope's normal modes m_i, $i = 1, \dots, N$, in the sense of principal component analysis. The nature of the modes m_i depends on the structure of the invariant polytope and captures the relative order of colliding disks and their frozen-in y coordinates. The normal modes to the largest eigenvalues are large-scale cooperative rearrangements of the disks (see Fig. 5). They are the slowest modes to decay under both local and event-chain Monte Carlo and govern the global decorrelation of the disk configuration. In particular, two modes dominated by antiparallel flow bands are very slow to decay (mode 1 and 2 in Fig. 5).

At delay times $\Delta t > 0$, the cross-covariance $U_{ij}(\Delta t)$ captures residual correlations among the disk coordinates. The event-chain moves couple more efficiently to the longitudinal modes of the system, and we find that after $\Delta t \approx N$, the event-chain algorithm has virtually erased longitudinal correlations. The most prominent residual correlations carry a transverse band structure (see Fig. 5). The result is a substantial decrease in efficiency of the algorithm for simulated duration in a single direction larger than $\approx N$.

To estimate the convergence time, we study the projection of the system's evolution $X(t)$ onto a single mode, $X(t) \cdot m_i$. The autocorrelation function

$$C_{m_i}(\Delta t) := \frac{\langle (X(t) \cdot m_i)(X(t + \Delta t) \cdot m_i) \rangle_t}{\langle (X(t) \cdot m_i)^2 \rangle_t} \tag{8}$$

is, for short chain lengths ℓ, monotonously decaying. Larger chain lengths accelerate the decay, as the coupling to large-scale modes is improved (Fig. 5). For chains spanning several times the box, however, the autocorrelation functions C_{m_i} develop oscillations with very weak damping, offsetting the benefits of longer chains. The detailed balance version of event-chain Monte Carlo is generally slower and less prone to oscillations. For optimal performance, the global balance version should thus be used with ℓ larger, but on the order of $l_{\text{mfp}}\sqrt{N}$, and for times $\theta \approx N$ (see Fig. 5). For disk configurations larger than the correlation length, ℓ can be reduced appropriately.

3.2. Convergence of the full hard-disk problem

The invariant polytope representation allows us to interpret the convergence of the full hard disk sampling problem. The conceptually simplest Monte Carlo algorithm for hard disks consists entirely in polytope sampling: One iteration amounts to direct sampling a new configuration a_{n+1} from the invariant polytope of the starting configuration a_n, and exchanging the x and y coordinates of all the disks. This Markov-chain algorithm satisfies detailed balance. In our experiments, the timescale τ, measured in iterations, for relaxation to equilibrium increases only as $N^{1/4}$ for large systems, implying that most of the complexity of the hard-disk sampling problem resides in the polytope sampling.

Since direct sampling is a hard problem for high-dimensional polytopes (see Section 3.3), we replace it by Markov chains of a fixed number of event-chain moves, in effect performing molecular dynamics in the invariant polytopes for fixed duration θ:

$$\left\{ \begin{array}{l} \text{polytope MD} \\ \text{in } x \text{ direction} \\ \text{for duration } \theta \end{array} \right\} \to \left\{ \begin{array}{l} \text{polytope MD} \\ \text{in } y \text{ direction} \\ \text{for duration } \theta \end{array} \right\} \to \cdots . \tag{9}$$

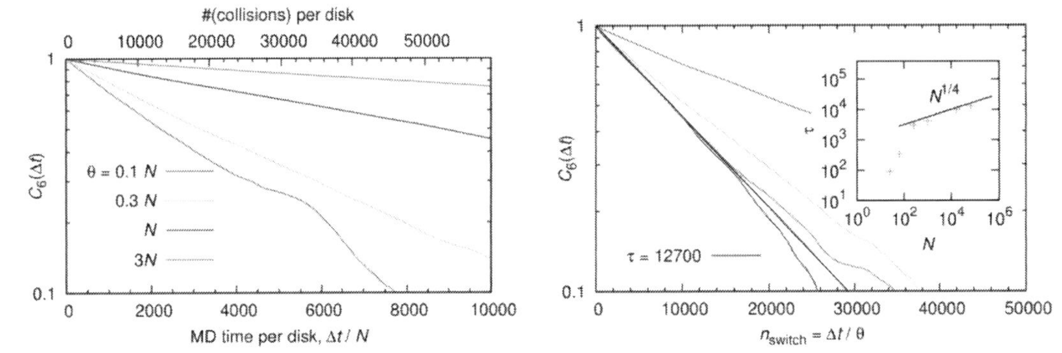

Figure 6. *Left*: Decay of the Ψ_6 autocorrelation function $C_6(\Delta t)$ for several switching intervals θ, as a function of the simulated MD time (*bottom axis*), or alternatively, the number of collisions per disk (*top axis*). *Right*: Decay of C_6 as a function of the number of x/y switches n_{switch}. As θ approaches N, the curves approach the limit of direct sampling from the polytope, with a mixing time of τ cycles. All curves were averaged from systems of $N = 256^2$ disks at packing fraction $\eta = 0.698$; the chain length was $\ell = 6.5 \cdot 10^3$. *Inset*: The mixing time τ first increases rapidly with system size, but only grows as $N^{1/4}$ for larger systems (also $\eta = 0.698$).

This algorithm satisfies detailed or global balance depending on the version of the event-chain algorithm that is used for polytope sampling.

We study the influence of the switching interval θ on convergence properties. In Fig. 6, the autocorrelation function C_6 of the complex order parameter Ψ_6 is plotted vs. cumulative molecular dynamics time. C_6 decays most quickly when the switching interval θ is small, but the decay speed deteriorates very slowly with θ. Only at $\theta \approx N$ (corresponding to about 6-7 collisions per disk at these densities), the algorithm becomes notably less efficient. The efficiency drop thus follows the decay of longitudinal (in x direction) correlations in the invariant polytope, and is to be expected from the results in Section 3.

In the limit $\theta \to \infty$, the event-chain algorithm realizes direct sampling in the invariant polytope. The approach to this limit is illustrated in Fig. 6 by plotting C_6 against the number of x/y switching cycles n_{switch}. As the switching interval θ increases, the autocorrelation functions approach an asymptotic curve $\propto \exp(-n_{\text{switch}}/\tau)$, where τ is the correlation time of the direct sampling algorithm. We find that for practical purposes, event-chain Monte Carlo reaches the asymptotic regime for $\theta \approx N$, and thus samples an approximately independent point in the invariant polytope in $O(N)$ operations. Importantly, the correlation time $\tau(N)$ increases rapidly only for small system size N. After the system size surpasses the correlation length, τ grows only as $N^{1/4}$.

3.3. Application to general polytopes

The invariant polytope is bounded by hyperplanes which are normal to $N-2$ coordinate axes and have unit derivative along the remaining axes. By choice of the $\boldsymbol{\ell}$, the molecular dynamics evolution is aligned with the coordinate axes at all times, and computations of intersections are of complexity $O(1)$. As shown in Section 3.2, the event-chain algorithm seems to achieve an effective mixing time of $O(N)$ collision events, so that the cost of sampling the hard-disk polytope appears as $O(N)$.

The event-chain algorithm also allows to sample general polytopes. Direct sampling from polytopes is straightforward only in low dimensions N, especially in $N = 2$: A two-dimensional polytope with n edges (a convex n-sided polygon), can be decomposed into n triangles, using an interior point. Triangles may then be sampled according to their areas, and a random point may be sampled inside the sampled triangle (see, e. g. chap. 6.2 of [28]). In higher dimensions N, triangulation by simplices generalizes this decomposition. Since polytopes such as the invariant hard-disk polytope have an exponential number of facets, direct sampling algorithms are no longer practical. Markov-chain sampling [29, 30, 31, 32] achieves mixing times of $O(MN)$ steps, where M is the number of bounding hyperplanes ($M \leq 3N$ for hard disks), and where each move may be implemented in $O(M)$ steps. It will be interesting to see how event-chain polytope sampling compares with existing polytope sampling methods, in particular the 'hit-and-run' algorithms.

4. Parallel Monte Carlo algorithms for hard disks

In view of the long running times of Monte Carlo simulations and of the current standstill in computer clock speeds, it is essential to develop *parallel Monte Carlo methods* which distribute the work load among several *threads* performing independent computation with as few communication as possible. Such methods will allow to study not only the standard hard disk ensemble, but also related systems such as soft disks and polydisperse disk packings. However, parallel Monte Carlo algorithms for continuum systems pose many more problems than for lattice models, for example the Ising spins, where straightforward parallel application of local Metropolis updates converges to the Boltzmann distribution [33].

4.1. Parallel implementation of local Monte Carlo

A massively parallel implementation of the local Monte Carlo algorithm was applied recently to the hard-disk melting problem [34, 25]. It sets up square cells according to a four-color checkerboard pattern. Disks in same-color cells can be updated simultaneously, but moves across cell boundaries are rejected. To ensure ergodicity, a new cell grid must be sampled periodically. Massive parallelism of $\sim$ 1500 threads on a graphics card offsets the slowness of local Monte Carlo compared to event-chain algorithm Monte Carlo [25]. These calculations confirmed the first-order liquid-hexatic phase transition in hard disks [8].

4.2. Parallel implementations of event-chain Monte Carlo

For parallel implementations of event-chain Monte Carlo, we consider only parallel threads that run chains in the same direction $\pm \boldsymbol{e}_\ell$. This minimizes the chance that two chains cross each other and move the same disks. It also allows us to apply the invariant polytope framework of Section 3. It is instructive to realize that the effects of an event-chain move can be summarized in the difference vector of the new and old x coordinates: $\Delta X = X(\ell) - X(0)$, with $X(t) = (x_1(t), \ldots, x_N(t))$. Moreover, if two chains are *independent*, meaning their sets of disks touched are disjoint, the net effect of running both chains is the sum of their individual difference vectors, $\Delta X^{\text{net}} = \Delta X^{(1)} + \Delta X^{(2)}$. If, however, any disk is touched by both chains, the chain reaching this disk earlier in MD time has precedence, the later chain sees a modified environment, and consequently takes a different evolution. Thus, interdependent chains cannot be added arithmetically[3]. The primary obstacle in parallelizing event-chain Monte Carlo consists in preserving the correct causal relations between subsequent chains, as required for the convergence to correct equilibrium distribution.

In the following, we discuss three strategies to parallelize event-chain Monte Carlo. The *predict/execute algorithm* distributes work among threads for a model of chains that follow

[3] Note, however, that due to the convexity of the accessible configuration space, the arithmetic average of two moves, $(\Delta X^{(1)} + \Delta X^{(2)})/2$, is always admissible.

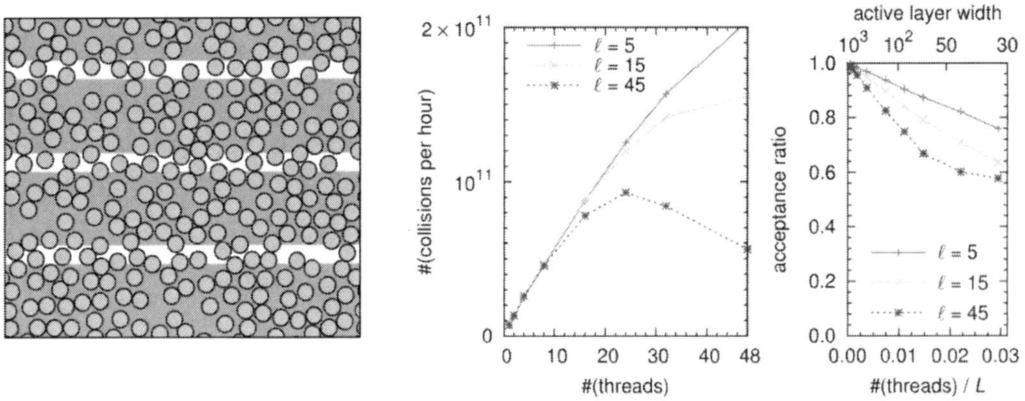

Figure 7. *Left*: Two-color stripe scheme with active (green backdrop) and isolation layers for the event-chain algorithm. Chains may run simultaneously if they are located in different stripes. *Center*: Scaling of the isolation layer algorithm on a shared-memory machine (Opteron 6276, 2.3 GHz), for a $N = 4096^2$ disk packing at $\eta = 0.698$. We plot the number of collisions in accepted chains per hour of computation. On the same machine, the serial version has a performance of about $8.5 \cdot 10^9$ collisions per hour. The event-chain routine is the same in both programs. *Right*: The acceptance ratio for chains depends on the thickness of the active layers (which is decreases as more threads are added) and the total displacement ℓ of the chains.

each other chronologically. The effects ΔX of several chains are predicted in advance from the current disk configuration. The effects of the chains are then applied to the system state in the chronological order in which the starting disks were sampled. To detect conflicts, it is sufficient to compute the intersection of the set of disks touched by the current chain and of the chains that ran since the beginning of planning; if this intersection is not empty, the chain has to be recomputed from the updated state of the disk configuration. Planning and execution of chains can proceed in parallel on a shared-memory machine. Using lock-free data structures, we attain collision rates in excess of 10^{11} per hour in x collision mode on a four-processor machine. Due to its serial nature, this algorithm does not scale well beyond a few threads, however; with too many chains predicted in advance, the probability for recomputations rises. Moreover, switching between x and y collision modes requires reinitialization of the data structures and is rather expensive.

A variation of the four-color scheme adapted to the event-chain algorithm partitions the system in horizontal stripes, separated by *frozen isolation layers* of thickness ≥ 2 disk radii (see Fig. 7). Disks with their centers in the isolation layers are kept fixed, and thus guarantee the independence of chains running in neighboring stripes. To preserve the isolation layers, chains colliding with a frozen disk are rejected. As there are rejected moves, the global balance condition is no longer guaranteed: the number of accepted forward chains can be different from the number of accepted backward chains (see Section 2.1). When allowing chains in both the $\pm \boldsymbol{e}_\ell$ directions, however, the isolation layer algorithm satisfies detailed balance. Furthermore, in order to limit the rejection rate, the per-chain total displacement ℓ has to be kept lower than in the serial algorithm. In view of the discussion of Section 3, these necessities reduce somewhat the efficiency of the method. Due to the isolation layers, the accessible configuration space is restricted, and for ergodicity, the layer boundaries have to be resampled periodically, as in the four-color version of local Monte Carlo.

We have implemented the isolation layer algorithm in parallel on a shared-memory machine. Using several cores in parallel, it is possible to achieve effective collision rates which are 10–

30 times the single-core performance (see Fig. 7), for systems of sufficient size. For systems too small, less threads can be used without shrinking the active strips to a point where the acceptance ratio becomes a limiting factor. Systems of physical interest, however, are on the order of $N = 1024^2$, and allow to use 10–20 cores with moderate ℓ. At this time, the algorithm is not bound by rejection rates, but by communication between threads.

Finally, the event-chain scheme is not fundamentally limited to a single moving disk at any time. We may indeed launch multiple *concurrent chains*, which run at the same simulated MD time, and interact with each other. This is different from the parallel simulation of chains which interact in sequential manner. In the invariant polytope picture, multiple concurrent chains correspond to choosing more general initial conditions, where more than one disk is given an initial velocity of 1. After time ℓ, multiple chains have executed, and possibly interacted with each other; there is no rejection in this algorithm. The problem has some resemblance with event-driven molecular dynamics, because the scheduling of collisions must be foreseen, but there are several simplifications: all velocities are in the same direction and of magnitude 0 or 1. As a consequence, two moving disks cannot collide with one other; however, the faithful simulation of chains close by and possibly interacting requires careful synchronization among threads. In our experiments, this limits the speedup by parallelization. Our most efficient method at this point is the isolation layer algorithm.

5. Conclusion

We have reached in this paper a better understanding of the event-chain Monte Carlo algorithm for the hard-disk sampling problem. By restricting the algorithm to chains in a single direction, a connection appears to the well-known problem of sampling random points from a polytope: A move of the event-chain algorithm consists in performing a finite-time molecular dynamics simulation in the invariant polytope of the disk configuration. This connection offers new strategies to solve the hard-disk sampling problem in terms of polytope sampling; it also suggests to investigate the utility of event-chain methods for the sampling of general polytopes. Finally, it will be interesting to study the combinatorial structure of the typical invariant polytope, and its dependence on thermodynamical parameters.

By the study of correlation functions, we have shown that the Monte Carlo relaxation process in the invariant polytope separates into two phases: A rapid longitudinal relaxation, followed by a much slower relaxation of the transverse degrees of freedom. We have given recommendations for the parameters of the algorithm based on these results. Finally, we have discussed several strategies for parallelizing Monte Carlo algorithms for hard disks, alleviating the problem of the long simulation times in hard disk Monte Carlo. The parallelization of the hard-disk ensemble remains challenging due to its unique combination of very little actual computation and long correlation times. Efficient methods to tackle the hard-disk ensemble are, however, crucial in order to treat related systems such as soft disks with the same level of success as the hard disks. New concepts such as the link to polytope sampling will be essential in this effort.

Acknowledgment

We thank P. Diaconis, E. P. Bernard, S. Leitmann and M. Hoffmann for fruitful discussions.

References

[1] Boltzmann L 1986 Sitzungsberichte der Akad. Wiss. Wien Math-Naturw. Kl. **105-2a** 695.
[2] Lebowitz JL and Penrose O 1964 *J. Math. Phys.* **5** 841.
[3] Kirkwood JG and Monroe E 1940 *J. Chem. Phys.* **8** 845.
[4] Asakura S and Oosawa F 1954 *J. Chem. Phys.* **22** 1255.
[5] Alder BJ and Wainwright TE 1962 *Phys. Rev.* **127**, 359.
[6] Kosterlitz JM and Thouless DJ 1972 *J. Phys. C: Solid State Phys.* **6** 1181; Halperin BO and Nelson DR 1978 *Phys. Rev. Lett.* **41** 121.

[7] Mak CH 2006 *Phys. Rev. E* **73** 065104(R).
[8] Bernard EP and Krauth W 2011 *Phys. Rev. Lett.* **107**, 155704.
[9] Zahn K, Lenke R and Maret G 1999 *Phys. Rev. Lett.* **82** 2721.
[10] Diaconis P 2011 *J. Stat. Phys.* **144** 445.
[11] Sinai YG 1970 *Russian Mathematical Surveys* **25**, 137.
[12] Simanyi N 2003 *Inventiones Mathematicae* **154**, 123.
[13] Alder BJ and Wainwright TE 1957 *J. Chem. Phys.* **27**, 1208 (1957).
[14] Rapaport DC 1980 *J. Comp. Phys.* **34**, 184.
[15] Metropolis N, Rosenbluth AW, Rosenbluth MN, Teller AH and Teller E 1953 *J. Chem. Phys.* **21** 1087.
[16] Isobe M 1999 *Int. J. Mod. Phys. C* **10**, 1281.
[17] Dress C and Krauth W 1995 *J. Phys. A, Math. Gen.* **28** L597.
[18] Jaster A 1999 *Phys. Rev. E* **59**, 2594.
[19] Bernard EP, Krauth W and Wilson DB 2009 *Phys. Rev. E* **80**, 056704.
[20] Swendsen RH and Wang JS 1987 *Phys. Rev. Lett.* **58** 86.
[21] Wolff U 1989 *Phys. Rev. Lett.* **62** 361.
[22] Kannan R, Mahoney MW and Montenegro R 2003 *Algorithms and Computation, 14th International Symposium, ISAAC 2003. Proceedings. Lecture Notes in Computer Science 2906* ed Ibaraki T, Katoh N and Ono H (Springer).
[23] Wilson DB 2000 *Random Struct. Algorithms* **16**, 85.
[24] Chanal C and Krauth W 2010 *Phys. Rev. E* **81** 016705.
[25] Engel M, Anderson JA, Glotzer SC, Isobe M, Bernard EP and Krauth W 2013 *Phys. Rev. E* **87**, 042134.
[26] Mickel W, Kapfer SC, Schröder-Turk GE and Mecke K 2013 *J. Chem. Phys.* **138** 044501; arXiv:1209.6180.
[27] Dyer ME and Frieze AM 1988 *SIAM J. Comput.* **17** 967; Kaibel V and Pfetsch ME 2003 *Algebra, Geometry and Software Systems*, ed Joswig M and Takayama N (Springer); arXiv:math/0202204.
[28] Krauth W 2006 *Statistical Mechanics: Algorithms and Computations* (Oxford University Press).
[29] Smith RL 1984 *Operations Research* **32**, 1296.
[30] Rubin PA 1984 *Communications in Statistics - Simulation and Computation* **13**, 375.
[31] Dyer ME, Frieze A and Kannan R 1991 *J. ACM* **38**, 1.
[32] Kannan R and Narayaran H 2012 *Mathematics of Operations Research* **37**, 1.
[33] Berg BA 2004 *Markov Chain Monte Carlo Simulations And Their Statistical Analysis* (World Scientific).
[34] Anderson JA, Jankowski E, Grubb TL, Engel M and Glotzer SC 2012 arXiv:1211.1646.

Fourier Monte Carlo simulation of crystalline membranes in the flat phase

Andreas Tröster

Vienna University of Technology, Wiedner Hauptstrasse 8-10/136, A-1040 Wien, Austria

E-mail: andreas.troester@tuwien.ac.at

Abstract. Stimulated by the recent interest in graphene, the elastic behavior of crystalline membranes continues to be under debate. In their flat phase, one observes scaling of the correlation functions of in-plane and out-of-plane deformations $\boldsymbol{u}(\boldsymbol{x})$ and $f(\boldsymbol{x})$ at long wavelengths with respect to a given reference plane governed by a single universal exponent η. The purpose of the present article is to explain the ideas and techniques underlying our Fourier Monte Carlo simulation approach to the numerical determination of η in much greater detail than was possible in a recent letter that is currently under review. Our simulations are based on an effective Hamiltonian first derived by Nelson and Peliti formulated exclusively in terms of the Fourier amplitudes $\tilde{f}(\boldsymbol{q})$ of the field $f(\boldsymbol{x})$, and we calculate the out-of-plane correlation function $\langle|\tilde{f}(\boldsymbol{q})|^2\rangle = \tilde{G}(\boldsymbol{q})$ and their related mean squared displacement $\langle(\Delta f)^2\rangle$. The key to the progress reported in this work is the observation that on tuning the Monte Carlo acceptance rates separately for each wave vector, we are able to eliminate critical slowing down and thus achieve unprecedented statistical accuracy. A finite size scaling analysis for $\langle(\Delta f)^2\rangle$ gives $\eta = 0.795(10)$. In the alternative approach, where we study the scaling of $\tilde{G}(\boldsymbol{q})$, we observe an unexpected anisotropic finite size effect at small wave vectors which hampers a similarly accurate numerical analysis.

1. Introduction

Membranes are two-dimensional surfaces embedded in three-dimensional space, and like their one-dimensional cousins, the linear polymers, their phase diagram is determined by the competition between energy loss and entropy gain accompanying the membrane's exploration of possible deformed configurations. Sophisticated theoretical and computational approaches have been developed to classify the resulting wealth of phases and determine their properties. A full survey of all the corresponding past achievements is well beyond the scope of this article and we may refer the novice reader to the excellent source [1] instead. Instead, we only choose to focus on those aspects of the statistical mechanics of membranes that are needed in order to understand the context of our present work.

In classifying the various phases of fluctuating phases, most pertinent theories aim at computing the analytic properties (i.e. decay length scales and/or scaling powers) of the out-of-plane deformations (OPDs) and the surface unit normals and their correlation functions, since the membrane in-plane deformations (IPDs) are usually harder to assess experimentally. As a matter of fact, the variety of possible conformational phases of membranes is much richer than that of polymers, and it is worth noting that the corresponding universality classes of asymptotic long wavelength conformations also depend crucially on microscopic structural features of the membrane, in contrast to what one might naively expect from arguments based on the standard renormalization group (RG) folklore. The explanation of this seemingly paradoxical statement is that distinct microstructures support different types of elastic interactions, which may induce long-range effective interactions between the OPDs. For instance, a membrane may have the microscopic characteristics of a two-dimensional liquid, with breaking and recombining bonds between its molecules. Familiar examples of such liquid membranes from soft matter physics are lipid bilayers [2]. Such a liquid membrane has a non-vanishing bending rigidity, but exhibits no static resistivity to shear, and its energy can thus only depend on its shape. It turns out that if self-avoidance effects are ignored, in which case one technically speaks of a "phantom" membrane model, liquid membranes are predicted to be always in a rotationally invariant "crumpled" phase, i.e. the correlations of their unit normals decay exponentially on a scale given by the so-called de Gennes-Taupin persistence length [3]. If only short-range interactions were at work, spontaneous breaking of this continuous symmetry would be ruled out by the Mermin-Wagner-Hohenberg theorem [4]. However, in [5] it was noted that the presence of (effective) long-range interactions between the unit normals may stiffen the membrane, and a so-called crumpling transition to a low temperature "flat" phase, whose long wavelength undulations define a new nontrivial universality class, may take place. In crystalline membranes, which support a nonzero shear modulus μ, such long-range forces are supplied by the shear-mediated coupling of the the IPDs to the OPDs. This is the case that we will concentrate on below, graphene being the most prominent, but certainly not the only example (see, e.g., [6, 7]). Relaxing the requirement of a regular lattice structure to that of a microscopic network of at the given temperature thermally unbreakable bonds defines so-called tethered membranes. Biological cell walls, which consist of lipid bilayers interacting with a shear-resistant protein network, the so-called cytoskeleton, are certainly the most important example of this species [8]. Inclusion of anisotropy in the elastic energy of a membrane may lead to tubular phases which appear flat in one and crumpled in the other direction. At finite temperature, a crystalline membrane may also undergo a dislocation unbinding transition to an hexatic phase [9], in which, even though μ vanishes, a residual directional bond stiffness remains, leading to an algebraic instead of an exponential decay of the correlations between the unit normals in this so-called crinkled phase for a phantom membrane. At present, there is good evidence but, at least to the author's best knowledge, no rigorous proof that when passing from phantom to physical membranes by inclusion of an energy penalty enforcing self-avoidance, the crumpled phases become unstable while self-avoidance remains irrelevant (in the RG sense) within the flat phases [10].

2. Theory

In the present work, we consider a continuum elastic energy model for an isotropic crystalline phantom membrane. In the Monge parameterization of the flat phase, membrane deformations with respect to a given two-dimensional reference plane with coordinates $\boldsymbol{x} = (x_1, x_2)$ are parameterized by a deformation vector field $(\boldsymbol{u}(\boldsymbol{x}), f(\boldsymbol{x}))$, where $\boldsymbol{u}$ is a two-dimensional vector of IPDs, while the "height function" f parameterizes the OPDs. As the membrane free energy is invariant under homogeneous translations, it can depend on deformations only via deformation gradients. Following [11, 12], an effective Hamiltonian containing all relevant couplings for describing the flat phase is given by

$$\tilde{\mathcal{H}}[f, \boldsymbol{u}] = \tilde{\mathcal{H}}_b[f] + \tilde{\mathcal{H}}_s[f, \boldsymbol{u}] \tag{1}$$

where the bending energy

$$\tilde{\mathcal{H}}_b[f] = \frac{\kappa_0}{2} \int d^2x \, (\Delta f)^2 \, (\boldsymbol{x}) \tag{2}$$

is augmented by an harmonic stretching part

$$\tilde{\mathcal{H}}_s[f, \boldsymbol{u}] = \frac{1}{2} \sum_{ij} \int d^2x \left(2\mu_0 u_{ij}^2(\boldsymbol{x}) + \lambda_0 u_{ii}(\boldsymbol{x}) u_{jj}(\boldsymbol{x})\right) \tag{3}$$

Here κ_0 denotes the (bare) bending rigidity, $\Delta = \frac{\partial^2}{\partial x_1} + \frac{\partial^2}{\partial x_2}$ is the two-dimensional Laplacian and λ_0 and μ_0 are the bare Lamé constants. Moreover, in the Lagrangian strain tensor

$$u_{ij}(\boldsymbol{x}) = \epsilon_{ij}(\boldsymbol{x}) + f_{ij}(\boldsymbol{x}) \tag{4}$$

all nonlinear terms beyond the infinitesimal strain tensor

$$\epsilon_{ij}(\boldsymbol{x}) = \frac{1}{2}\left(\frac{\partial u_i(\boldsymbol{x})}{\partial x_j} + \frac{\partial u_j(\boldsymbol{x})}{\partial x_i}\right) \tag{5}$$

can be dropped for the IPDs, while one needs to keep a quadratic term symmetric "tensor potential"

$$f_{ij}(\boldsymbol{x}) = \frac{1}{2} \frac{\partial f(\boldsymbol{x})}{\partial x_i} \frac{\partial f(\boldsymbol{x})}{\partial x_j} \tag{6}$$

in the OPDs. In the continuum formalism presented in [12], which we sketch here for completeness, a transverse projector$P_{ij}^T = \delta_{ij} - \partial_i \partial_j / \Delta$ is employed to split $f_{ij}(\boldsymbol{x})$ into longitudinal and transverse parts $f_{ij}^L(\boldsymbol{x})$ and $f_{ij}^T(\boldsymbol{x})$, respectively. In the calculation of the partition function, the fact that $\epsilon_{ij}(\boldsymbol{x})$ is purely longitudinal allows $f_{ij}^L(\boldsymbol{x})$ to be "gauged away", i.e. absorbed into redefined path integration variables $\epsilon_{ij}(\boldsymbol{x}) \to \epsilon'_{ij}(\boldsymbol{x})$. Gaussian integration over these shifted variables produces a mere scalar factor, leaving a new effective Hamiltonian [11]

$$\mathcal{H}[f] = \frac{\kappa_0}{2} \int d^2x \, (\Delta f)^2 + \frac{K_0}{2} \int d^2x \left[P_{ij}^T f_{ij}(\boldsymbol{x})\right]^2 \tag{7}$$

for the OPDs which only involves the remaining transverse part $f_{ij}^T(\boldsymbol{x})$, where

$$K_0 = \frac{4\mu(\mu_0 + \lambda_0)}{2\mu_0 + \lambda_0} \tag{8}$$

is the bare 2d Young modulus [13]. This Hamiltonian resembles a field theory with a non-standard $\boldsymbol{q}^4$ type of dispersion, and a non-local fourth order contribution that, as promised above, resembles the effective long-range interaction of OPD's mediated by the eliminated IPD's. We introduce the continuous Fourier amplitudes

$$f(\boldsymbol{x}) = \int_{|\boldsymbol{q}|>(2\pi/L)}^{|\boldsymbol{q}|<\Lambda} \frac{d^2q}{(2\pi)^2} \widehat{f}(\boldsymbol{q}) e^{i\boldsymbol{q}\boldsymbol{x}} \tag{9}$$

where $\Lambda^{-1} \sim a$ denotes some microscopic length for which the continuum approximation ceases to be valid, and consider the correlations

$$\widehat{G}(\boldsymbol{p})\delta^2(\boldsymbol{p} + \boldsymbol{q}) \equiv \langle \widehat{f}(\boldsymbol{p}) \widehat{f}(\boldsymbol{q}) \rangle \tag{10}$$

of the OPD's. For a fluid membrane, μ_0 vanishes identically, which implies that $K_0 \equiv 0$ vanishes as well and the equipartition theorem applied to the remaining harmonic bending contribution yields

$$\widehat{G}(\boldsymbol{q}) = \frac{1}{\kappa_0 q^4} \tag{11}$$

In the case of a solid membrane it is tempting to try to do the same, i.e. ignore the term $\propto K_0$ as a first guess, which effectively amounts to a harmonic approximation. However, such a strategy fails to capture the essential physics of crystalline membrane elasticity, which, as we have stressed above, is caused by the effective long-range interaction encoded in the anharmonic part of (7). Technically, the coupling K_0 is relevant in the sense of the renormalization group and effectively leads to a renormalization of the bare bending rigidity κ_0. More precisely, one can show that the resulting renormalized bending rigidity acquires a nontrivial $\boldsymbol{q}$-dependence, such that actually

$$\widehat{G}(\boldsymbol{q}) = \frac{1}{\kappa(\boldsymbol{q}) q^4} \tag{12}$$

At this stage the critical exponent η is introduced as

$$\kappa(\boldsymbol{q}) \sim q^{-\eta} \tag{13}$$

The exponent η is indeed the central quantity governing the elastic behavior of the flat phase at long wavelengths, as other observables of interest like the $\boldsymbol{q}$-dependent scaling of the Lamé coefficients, the mean square height fluctuations (see below) and the correlation function of the unit normals are governed by scaling laws, whose critical exponents are related to η by simple algebraic scaling relations [13].

Due to the complicated structure of the anharmonic nonlocal fourth order term in (7), it is difficult to obtain precise estimates of η. Thus, the literature exhibits a remarkable dispersion of published numerical results for η, which are summarized in [8, 14]. Specifically, an early one-loop perturbative calculation by Nelson and Peliti [11] gave $\eta = 1$, whereas Le Doussal and Radzihovsky [15] obtained $\eta = 4/(1+\sqrt{15}) \approx 0.821$ from self-consistent RPA. RG calculations based on an ϵ-expansion [16] and a large-d expansion [17] have been carries out, but their numerical extrapolation to the physically relevant case of a two-dimensional membrane in three-dimensional space is not straightforward. Recently, the result $\eta = 0.849$ was derived from a nonperturbative RG approach [18]. Due to the difficulties encountered in these analytical approaches, extensive simulations have also been carried out. For instance, in [19] large-scale Monte Carlo simulations were performed based on a simple model with Gaussian spring pair potentials, and the estimates $\eta = 0.750(5)$ and $\eta = 0.72(4)$ were obtained from monitoring the mean squared fluctuations of IP and OP phonons (see below) and making use of the appropriate scaling relations. Motivated by the enormous interest in the properties of graphene, the problem was recently re-investigated with atomistic MC and molecular dynamics simulations based on a realistic many-body potential for carbon. While preliminary [20] conclusions based on the analysis of the observed normal-normal correlation function $\widehat{G}(\boldsymbol{q})$ of graphene membranes, which seemed to indicate a completely unexpected "preferred wavelength" for elastic fluctuations in graphene turned out to be based on artifacts caused by lack of sufficient convergence, subsequent work [21] bases on a simple pseudo-harmonic model finally yielded a result $\eta \approx 0.85$ in reasonable agreement with the bulk of the existing literature. These simulations are also very instructive, because, as is explained at length in the authoritative reference [13], they illustrate the crossover of $\widehat{G}(\boldsymbol{q})$ from mean-field (MF) ($\sim q^{-4}$) to critical ($\sim q^{-4+\eta}$) behavior, which roughly sets in at a "Ginzburg wave vector"

$$q_G \approx \sqrt{\frac{3k_B T K_0}{8\pi\kappa_0^2}} \tag{14}$$

analytically defined by the breakdown of the harmonic approximation $K_0 \approx 0$ to the effective Hamiltonian (7). For graphene, using the parameters $T = 300\text{K}$, $K_0 = 12.4\text{Å}^{-1}$ and $\kappa_0 = 1.1\text{eV}$ yields $q_G \approx 0.18\text{Å}^{-1}$ [22, 23]. The relevance of such simulations for determining critical behavior depends, of course, on how far we are able to penetrate the scaling region. Obviously, a convenient measure for this ability is thus the ratio $q_G/(2\pi/L)$ of q_G to the smallest accessible wave vector component $2\pi/L$. Unfortunately, for the large scale atomistic MC simulations presented in [21] this ratio turns out to be only approximately 9. In other words, in each linear direction there are only 9 data points that may carry relevant information on the scaling region. For the purpose of determining η with sufficient precision it is clearly desirable to generate a larger fraction of relevant data.

3. Simulation

Common sense indicates that it may not be extremely efficient to base simulations of universal properties on atomistic potential models anyway. In calculating long wavelength characteristics, one should get rid of all irrelevant microscopic details that only make simulations more expensive, ideally ending up with an effective Hamiltonian like (7), which also allows to choose the coupling parameters κ_0, K_0 independently in a convenient way. Unfortunately, as mentioned above, the presence of the projector P_{ij}^T in (7) makes the fourth order contribution to (7) nonlocal, reflecting the long-range effective interaction of the OPD's induced by the IPD's. Thus, an attempt to base a real-space MC simulation directly on it is a rather hopeless enterprise. To clarify its nonlocal structure, let us rewrite (7) in terms of the Fourier amplitudes (9). Abbreviating $\hat{\boldsymbol{Q}} := \boldsymbol{Q}/|\boldsymbol{Q}|$, one finds [13]

$$\mathcal{H}_\Lambda[f] = \frac{\kappa_\Lambda}{2}\int \frac{d^2q}{(2\pi)^2} q^4 |\widehat{f}(\boldsymbol{q})|^2 + \frac{K_\Lambda}{8}\int \frac{d^2Q}{(2\pi)^2}|\widehat{\mathcal{F}}(\boldsymbol{Q})|^2 \tag{15}$$

where

$$\widehat{\mathcal{F}}(\boldsymbol{Q}) := \int \frac{d^2q}{(2\pi)^2} R(\boldsymbol{Q},\boldsymbol{q})\widehat{f}(\boldsymbol{q})\tilde{f}(\boldsymbol{Q}-\boldsymbol{q}) \tag{16}$$

and in the definition of the function

$$R(\boldsymbol{Q},\boldsymbol{q}) := (\hat{\boldsymbol{Q}} \times \boldsymbol{q})^2 \tag{17}$$

the vectors $\hat{\boldsymbol{Q}}$ and $\boldsymbol{q}$ are formally embedded in 3-dim. space for the sake of a compact and intuitive notation. Here it is important to keep in mind that (15) represents an effective field-theoretic Hamiltonian, in which a wave vector cutoff $|\boldsymbol{q}| < \Lambda$ for the modes $\widehat{f}(\boldsymbol{q})$ is implicit, and, of course, the numerical value of its parameters will also depend on the choice of Λ, which our notation reflects by introducing the subscript Λ in $\mathcal{H}_\Lambda, \kappa_\Lambda$ and K_Λ.

Hypothetically replacing $R(\boldsymbol{Q},\boldsymbol{q})$ by 1, $\widehat{\mathcal{F}}(\boldsymbol{Q})$ reduces to the Fourier transform of the squared OPD field $f^2(\boldsymbol{x})$. In this sense, we may say that $\widehat{\mathcal{F}}(\boldsymbol{q})$ has the structure of a "generalized square" [24]. Furthermore, the above effective Hamiltonian actually resembles a formal continuum limit of a discrete lattice model with lattice constant a, and in writing down (7) and (15), infrared and ultraviolet cutoffs $2\pi/ \leq |q_i| \leq \pi/a$ are implicitly understood. Thus, the above effective Hamiltonian for the flat phase of a crystalline model is accessible to our recently developed [25]

Fourier Monte Carlo (FMC) algorithm. In fact, the present situation is reminiscent of that of compressible spin models [26, 27], in which elimination of the phonons by Gaussian integration induces a long-range effective interaction for the spin degrees of freedom. For a detailed account on the nuts and bolts of FMC we refer to [28, 29, 24]. Here we content ourselves with the following brief description.

In applying the FMC algorithm to the present problem, universality authorizes us to consider the crystalline structure as a square lattice Γ of volume $V = La_0 \times La_0$ with $N_0 = L \times L$ sites and lattice constant a_0. In the present work, we use the the discrete Fourier conventions

$$f(\boldsymbol{x}) = \frac{1}{N_0}\sum_{\boldsymbol{q}\in\tilde{\Gamma}} \tilde{f}(\boldsymbol{q})e^{i\boldsymbol{q}\boldsymbol{x}}\ , \qquad \tilde{f}(\boldsymbol{q}) = \sum_{\boldsymbol{x}\in\Gamma} f(\boldsymbol{x})e^{-i\boldsymbol{q}\boldsymbol{x}} \tag{18}$$

for which the discrete Fourier amplitudes $\tilde{f}(\boldsymbol{q})$ are extensive quantities with the same physical dimension as the discretized field $f(\boldsymbol{x})$. These conventions, although asymmetric, are very convenient since, when combined with the rescaling

$$\widehat{f}(\boldsymbol{q}) \equiv a_0^2 \tilde{f}(\boldsymbol{q}) \tag{19}$$

they allow for easy translation of continuum formulae following the rules

$$\sum_{\boldsymbol{q}\in\widehat{\Gamma}} \leftrightarrow V\int_{\widehat{V}} \frac{d^2q}{(2\pi)^2},\ \sum_{\boldsymbol{x}} \leftrightarrow \int \frac{d^2x}{a_0^2} \tag{20}$$

FMC employs these complex Fourier amplitudes $\tilde{f}(\boldsymbol{q})$ as basic MC variables. MC moves correspond to shifts

$$\tilde{f}(\boldsymbol{q}) \to \tilde{f}(\boldsymbol{q}) + \epsilon\delta_{\boldsymbol{q},\boldsymbol{q}_0} + \epsilon^*\delta_{\boldsymbol{q},-\boldsymbol{q}_0} \tag{21}$$

where ϵ is a random complex number with bounded modulus, and the wave vector $\boldsymbol{q}_0 \in \tilde{\Gamma}$ is chosen at random, but no reference to the direct lattice is made whatsoever. In the vast majority of situations encountered, this algorithm may be quite inefficient compared to real-space based ones, but for problems like coarse graining [25] and RG transformations [30, 31, 32] or, like in the present case, the computation of critical exponents [27], in which the subset of $\boldsymbol{q}$-vectors parameterizing the nonzero Fourier amplitudes can be confined to a narrow region bounded by wave vector cutoffs while the effect of the remaining modes is accounted for by the use of an effective Hamiltonian, it is ideally suited. This is even more so, if long-range interactions are involved [30, 33]. In the simulation of critical phenomena, it is essential to gather as much information as possible on the behavior of the system at "small" wave vectors, i.e. those around the critical wave vector ($\boldsymbol{q} = \boldsymbol{0}$ in the present case). Real space simulations necessarily take place in finite systems, of, say, linear size L_0a_0, whose smallest accessible nonzero wave vector components are $2\pi/L_0a_0$, but also have to deal with all other, increasingly unimportant modes up to the boundary of the Brillouin zone $|q_i| \leq \pi/a_0$ determined by the lattice constant a_0. However, if we are able to choose a wave vector cutoff $\Lambda \equiv \Lambda_0/b$, $b > 1$, and replace the original Hamiltonian by an effective one involving only modes with wave vectors inside Λ, then this corresponds to working with the original Hamiltonian at a system of lattice constant $a = ba_0$ and with $L = L_0/b$ unit cells in each direction, thus holding $N = N/b^2$ unit cells altogether. In turn this, implies that our FMC simulation effort with effective Hamiltonian and cutoff Λ resembles that of a direct lattice simulation with atomistic Hamiltonian on a system whose volume is b^2 times larger. In what follows we set a_0 to 1 and keep in mind that working at a cutoff Λ implies an effectively larger lattice constant $a \sim \pi/\Lambda$.

We now enter the discussion of the specific details of the present simulations. Since the effective Hamiltonians (1), (7) both involve spatial derivatives, we must first find a suitable discretization scheme. In a continuum Fourier transform, a partial derivative $\partial_j g(\boldsymbol{x})$ of a function $g(\boldsymbol{x})$ translates into multiplying its Fourier transform $\widehat{g}(\boldsymbol{q})$ by a factor $\widehat{\partial}(\boldsymbol{q}) = iq_j$, and taking the Laplacian gives a factor $\widehat{\Delta}(\boldsymbol{q}) = -\sum_{j=1}^{2} q_j^2 = -q^2$. On the lattice we may replace derivatives by Fourier transforms of nearest neighbor finite difference operators, which yields

$$\tilde{\partial}_j(\boldsymbol{q}) = \frac{i}{a}\sin aq_j,\ \tilde{\Delta}(\boldsymbol{q}) = -\frac{4}{a^2}\sum_{i=1}^{2}\sin^2\frac{aq_j}{2} \tag{22}$$

Of course, for $aq \to 0$, these operators asymptotically approach their continuum limits. However, working on a finite lattice with finite a, deviations from linearity in $\tilde{\partial}_j(\boldsymbol{q})$ and from cubic isotropy in $\tilde{\Delta}(\boldsymbol{q})$ become noticeable for large enough $\boldsymbol{q}$. In order to extract the exponent η from the $\boldsymbol{q}$-dependence of $\tilde{G}(\boldsymbol{q})$, it is essential to reduce these finite size effects, and in particular the accompanying cubic anisotropy. However, at the same time we would still like to maintaining overall periodicity in reciprocal space. For $\partial_j f(\boldsymbol{x})$, one may replace the above simple 2-point discretization by a four-point scheme (cf. [34], p. 15), which on Fourier transformation translates into

$$\tilde{\partial}_j(\boldsymbol{q}) = \frac{i}{a}\frac{(4-\cos aq_j)\sin aq_j}{3} \tag{23}$$

Let us take this approach one step further by considering a corresponding improved version of the lattice Laplacian $\tilde{\Delta}(\boldsymbol{q})$ derived from taking contributions from lattice neighbors beyond nearest neighborhood into account. Technically, in two dimensions this works as follows. One introduces polar coordinates $\boldsymbol{q} \to (q, \phi)$ and Taylor-expands an ansatz of type

$$\begin{aligned} -a^2\tilde{\Delta}(\boldsymbol{q}) \;\equiv\;& b + c(\cos aq_1 + \cos aq_2) + d[\cos a(q_1+q_2) + \cos a(q_1-q_2)] \\ & + e(\cos 2aq_1 + \cos 2aq_2) \\ & + f\Big[\cos a(q_1-2q_2) + \cos a(q_1+2q_2) + \cos a(2q_1-q_2) + \cos a(2q_1+q_2)\Big] + \ldots \end{aligned} \tag{24}$$

in q, successively choosing the coefficients b, c, d, e, f to eliminate low order ϕ-dependence, which by cubic symmetry must appear through terms proportional to $\cos(4\phi), \cos(8\phi), \ldots$ and unwanted low order powers of q except the leading q^2 one. Choosing $b = \frac{21}{5}, c = -\frac{26}{15}, d = -\frac{8}{15}, e = \frac{1}{30}$ and $f = \frac{1}{15}$, we obtain

$$-a^2\tilde{\Delta}(\boldsymbol{q}) = q^2 - \frac{q^6}{90} + O\left(q^8, \cos(4\phi), \cos(8\pi)\right) \tag{25}$$

such that cubic anisotropy is minimal for small enough cutoff Λ.

Since the notion of transverseness for a discretized system may not be well-defined independently of a chosen underlying discretization scheme, the safest way to discretize the membrane problem should be based on the original effective Hamiltonian (1) rather than (7). Nevertheless, even without resorting to the trick of splitting $f_{ij}(\boldsymbol{x})$ into transverse and longitudinal part, elimination of the IPD's by discrete Fourier transformation of all fields and Gaussian integration over the homogeneous and inhomogeneous contributions to $\tilde{\epsilon}_{ij}(\boldsymbol{q})$ is in principle straightforward, albeit tedious. The result is

$$\mathcal{H}[f] = \frac{\kappa_\Lambda}{2N}\sum_{\boldsymbol{q}\neq\boldsymbol{0}}\tilde{\Delta}(\boldsymbol{q})|\tilde{f}(\boldsymbol{q})|^2 + \frac{K_\Lambda}{2N}\sum_{\boldsymbol{Q}\neq\boldsymbol{0}}|\mathcal{F}(\boldsymbol{Q})|^2 \tag{26}$$

Journal of Physics: Conference Series **454** (2013) 012032 doi:10.1088/1742-6596/454/1/012032

and

$$\mathcal{F}(\boldsymbol{Q}) = \frac{1}{2N} \sum_{\boldsymbol{q}\neq 0,\boldsymbol{Q}} R(\boldsymbol{Q},\boldsymbol{p}) \tilde{f}(\boldsymbol{q}) \tilde{f}(\boldsymbol{Q}-\boldsymbol{q}) \tag{27}$$

The function $R(\boldsymbol{Q},\boldsymbol{p})$ is now defined as follows. We introduce the abbreviation $\tilde{\partial}(\boldsymbol{q}) \equiv i\tilde{\boldsymbol{q}}$ and form the 2×2 real-valued matrix

$$\mathcal{M}(\boldsymbol{Q}) = \frac{1}{\tilde{Q}_1^2 + \tilde{Q}_2^2} \begin{pmatrix} -\tilde{Q}_2^2 & \tilde{Q}_1\tilde{Q}_2 \\ \tilde{Q}_1\tilde{Q}_2 & -\tilde{Q}_1^2 \end{pmatrix} \tag{28}$$

Then it is not hard to see that in the long wavelength limit the real function

$$R(\boldsymbol{Q},\boldsymbol{q}) := \rho(\boldsymbol{q}|\boldsymbol{Q}|\boldsymbol{Q}-\boldsymbol{q}) \tag{29}$$

where

$$\rho(\boldsymbol{p}|\boldsymbol{Q}|\boldsymbol{q}) := \sum_{ij} \tilde{p}_i \mathcal{M}_{ij}(\boldsymbol{Q}) \tilde{q}_j \tag{30}$$

indeed reduces to the continuum version encountered in equation (17):

$$R(\boldsymbol{Q},\boldsymbol{q}) \to (\hat{\boldsymbol{Q}} \times \boldsymbol{q})^2 \quad \text{for} \quad \boldsymbol{Q},\boldsymbol{q} \to \boldsymbol{0} \tag{31}$$

This completes the description of our discretization scheme.

4. Optimization of FMC

In our present FMC simulations we monitored the set of squared moduli $|\tilde{f}(\boldsymbol{q})|^2$ for wave vectors $\boldsymbol{q}$ inside a suitably chosen cutoff Λ, together with their sum (33) and the total energy of the system. From the resulting separate time series of these MC data, which we also analyzed to ensure the complete equilibration of our simulations before MC measurements, we calculated estimates of the corresponding statistical error using the standard jackknife procedure [35, 36]. These estimates were cross-checked using the standard analysis of integrated autocorrelation times [37, 35] and the blocking method [38, 36] and we report nice numerical agreement of the different error estimates derived from these methods. An addition, we also introduce the following refinement to FMC. In the basic algorithm outlines by (21), the complex number ϵ is chosen at random within a circle of fixed radius r_ϵ in the complex plane. In the start-up phase of the simulation, one can incrementally adjust r_ϵ as to achieve a decent overall acceptance rate of FMC moves (usually somewhere between $30-60\%$). Still, as could have been anticipated from the well-known phenomenon of critical slowing down [36], one observes a dramatic growth of the integrated autocorrelation times $\tau_{\text{int}}(\boldsymbol{q})$ for $|\boldsymbol{q}| \to 0$, which in turn implies that, similar to the case of a real space algorithm based on non-collective moves, efficiency of plain FMC breaks down where we need it the most. Looking at the individual acceptance rates for moves of type (21) for different wave vectors $\boldsymbol{q}_0$, one realizes that this behavior results from the fact that for a uniform choice of r_ϵ, the acceptance rates for "small" $\boldsymbol{q}_0$-vectors are close to 100%, while they may fall well below 30% for large ones, which, however, make up the vast majority of $\boldsymbol{q}$-vectors, as the number of wave vectors of length q grows as q^{d-1} in d dimensions (cf. inset of figure 1). Since the equilibrium values of $|\tilde{f}(\boldsymbol{q}_0)|^2$ are dramatically larger for smaller than for larger wave vectors, this still implies that, since they all move on the same scale r_ϵ, the modes close to criticality make no headway in comparison to the noncritical ones. At this point, the following idea comes to rescue. In principle, (21) can be regarded as a different type of MC move for each vector $\boldsymbol{q}_0$. Thus, without violating detailed balance, we can in principle pick a different value $r_\epsilon(\boldsymbol{q}_0)$ separately for each $\boldsymbol{q}_0$, offering the possibility to also tune the corresponding values $r_\epsilon(\boldsymbol{q}_0)$ separately. Of course, this tuning is not a trivial matter, since changing $r_\epsilon(\boldsymbol{q}_0)$ for one $\boldsymbol{q}_0$ will influence all other $r_\epsilon(\boldsymbol{q})$ in a possibly nonlinear way. Nevertheless, it is straightforward to implement a corresponding individual optimization for each $r_\epsilon(\boldsymbol{q}_0)$ during the start-up period of the simulation. And once this is done, we indeed observe an almost uniform common value of the integrated autocorrelation times $\tau_{\rm int}(\boldsymbol{q})$ independent of the wave vector under consideration. In other words, on tuning $r_\epsilon(\boldsymbol{q}_0)$ individually, our FMC algorithm becomes free of critical slowing down! Actually, considering our previous work [29, 24], the underlying basic idea of this Optimized Fourier Monte Carlo (OFMC) algorithm is embarrassingly self-evident and should have come to mind much earlier. In fact, after completing this work we realized that even the authors of [21], who certainly were aware of our work, had come up with a similar idea. However, they had only considered occasional "wave vector moves" applied to a simple "pseudo-harmonic" model on top of a local MC move set defined in real space. Thus it is understandable that, despite of the resulting improvements that allowed them to correct the errors contained in their previous work [20], they obviously were not able to observe the dramatic reduction of autocorrelation times we report here.

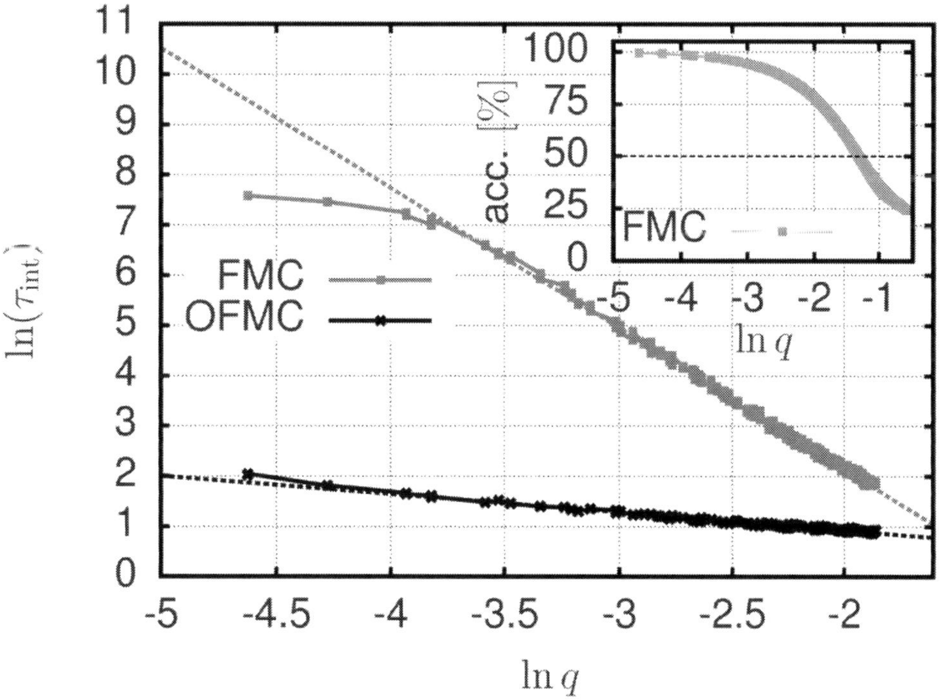

Figure 1. Main plot: Dramatic reduction of integrated autocorrelation times $\tau_{\rm int}(\boldsymbol{q})$ for moduli $|\tilde{f}(\boldsymbol{q})|^2$ for our new OFMC algorithm in comparison to plain FMC for system size $L = 640$ and cutoff $\Lambda = \pi/8$. Deviations from linearity are due to the finite MC run time of 2^{20} MC steps (plain FMC) and finite tolerance $\pm5\%$ for a 50% target acceptance (OFMC). Inset: q-dependent acceptance rates as measured for plain FMC.

5. Results

From the FMC point of view, the observables that suggest themselves naturally for the purpose of extracting η are, of course, $\tilde{G}(\boldsymbol{q}) = \langle|\tilde{f}(\boldsymbol{q})|^2\rangle$, and the mean square displacement $\langle(\Delta f)^2\rangle$. Let us first consider the latter.

5.1. Mean squared displacement $\langle(\Delta f)^2\rangle\rangle$

In continuum formalism $\langle(\Delta f)^2\rangle$ is calculated from $\widehat{G}(\boldsymbol{q})$ as

$$\langle(\Delta f)^2\rangle = G(\mathbf{0}) = \int \frac{d^2q}{(2\pi)^2}\widehat{G}(\boldsymbol{q}) \tag{32}$$

For $L \to \infty$, $\langle(\Delta f)^2\rangle$ diverges like $\langle(\Delta f)^2\rangle \sim L^{2\zeta}$, where ζ is known as the roughness exponent [8]. Explicitly calculating the integral (32) using the infrared cutoff $|q_i| \geq 2\pi/L$ and the scaling form (12) with (13), one easily finds the scaling relation $\zeta = 1 - \eta/2$. On our discrete lattice, the integral in (32) is, of course, replaced by the sum

$$(\Delta f)^2 \sim \sum_{\boldsymbol{q}} |\tilde{f}(\boldsymbol{q})|^2 \tag{33}$$

In contrast to $\tilde{G}(\boldsymbol{q})$, the observable $\langle(\Delta f)^2\rangle\rangle$ directly lends itself to a simple FSS approach. Trying to take possible sub-leading FSS corrections into account, we employ the ad hoc FSS ansatz

$$\langle(\Delta f)^2\rangle = \delta + \alpha L^{2-\eta}(1 + \beta/L + \gamma/L^2) \tag{34}$$

In providing proper data for an accurate FSS analysis, it is important to use a cubic cutoff geometry, since on a finite lattice the fraction of $\boldsymbol{q}$-vectors inside and outside of a fixed spherical cutoff generally does not scale with system size in a perfectly linear way. For the FSS analysis, we performed FMC simulations using the effective Hamiltonian (15) with renormalized parameters $\kappa_\Lambda = 0.1$, $K_\Lambda = 0.1$, in which a factor $(k_BT)^{-1}$ has been absorbed, and investigated a collection of systems with parameters

$$L = 32n, \ \ n = 1, 2, \ldots, 20, \quad \Lambda = \pi/8 \tag{35}$$

The above choice of parameters, which was used in the production runs, was fixed during the preparing test runs and was motivated by the heuristic principle to have an approximate balance of the harmonic and anharmonic contributions to the total average energy changes recorded during the simulation. For this choice, the largest considered system with $(L, \Lambda) = (640, 40)$ yields a value of $q_G/(2\pi/L) \approx 33$. In other words, we are able to evaluate a factor of ≈ 13 times more data describing the scaling region than those reported in [21]. Moreover, the statistical quality of these data is not affected by critical slowing down. And in fact, a standard least squares fit of (34) to our data works amazingly well (cf. figure 2). Numerically it results in the estimate

$$\eta = 0.795(10) \tag{36}$$

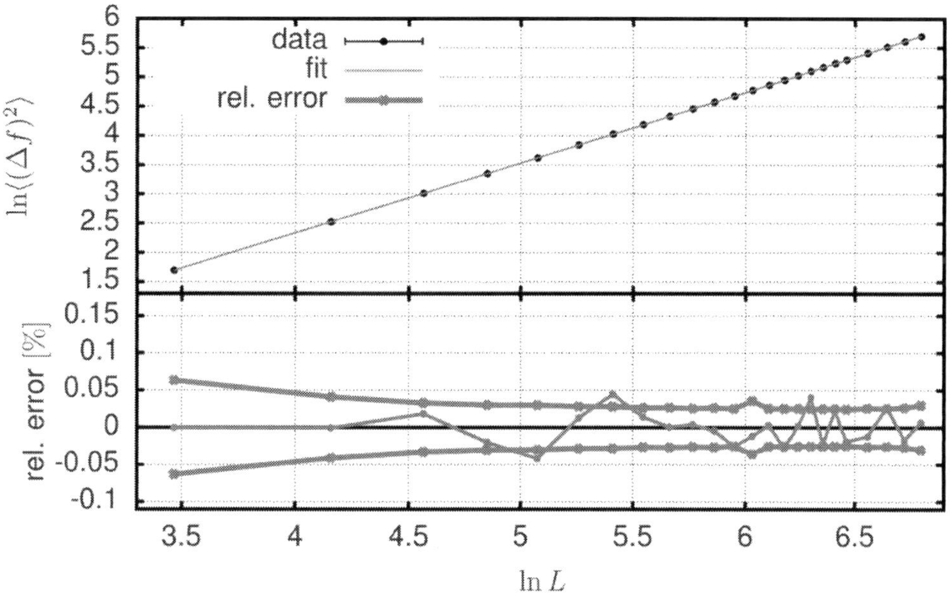

Figure 2. Top: fit of FSS ansatz (34) to simulation results for $\langle(\Delta f)^2\rangle$ obtained for the collection of systems (35). Error bars are smaller than symbol size. Bottom: Relative deviations of simulation data from the fit. Red lines are a guide to the eye. Gray lines indicate the relative statistical errors of the simulation data in percent.

5.2. Correlation function $\tilde{G}(\boldsymbol{q})$

As we now show, it is much harder to extract a reliable result for η from the critical scaling relation $\tilde{G}(\boldsymbol{q}) \sim k^{\eta-4}$. In view of (11-13), we actually should expect to see a crossover from MF behavior $\tilde{G}(\boldsymbol{q}) \sim q^{-4}$ for $q \gg q_G$ to critical scaling for $q \ll q_G$. However, it is difficult to judge the corresponding width of the crossover region. Thus, to reliably isolate η from a fit of the $\tilde{G}(\boldsymbol{q})$ data, an ansatz which allows for enough degrees of freedom to smoothly follow such a crossover behavior is required. In [13], Katsnelson considers a straightforward combination of the type

$$\tilde{G}^{-1}(\boldsymbol{q}) = \frac{\kappa_\Lambda q^4}{k_B T}\left[1 + \frac{1}{\alpha}\left(\frac{q_\Lambda}{q}\right)^\eta\right] \tag{37}$$

with a single free mixing parameter α (apart from the sought-after exponent η) that implicitly parameterizes the location of the crossover region in terms of a wave number $q_\Lambda = \sqrt{\frac{K_\Lambda}{\kappa_\Lambda}}$ introduced merely for dimensional reasons. Taking the logarithmic derivative of this expression, such an ansatz yields an effective exponent

$$\eta_{\text{eff}}(q) = \frac{\eta}{1 + \alpha\left(\frac{q}{q_\Lambda}\right)^\eta} \tag{38}$$

In [39], another type of crossover function is considered, which would translate into

$$\eta_{\text{eff}}(q) = \frac{\eta}{1 + \left(\frac{q}{q_\Lambda}\right)^\sigma} \tag{39}$$

in our present setting. Compared to Katsnelson's crossover function (38), this functional form rigidly fixes the location of the crossover on the q-axis, since it imposes $\alpha \equiv 1$. On the other hand, if we regard the parameter σ, which is fixed as $\sigma = (4-d)/2$ by the spatial dimension d in [39], as a free parameter in the present setting, it represents an additional degree of freedom that also allows to tune the width of the crossover region. A comparison of (38) and (39) thus suggests to introduce a more general crossover function with two free parameters α, σ, as

$$\eta_{\text{eff}}(q) \equiv \frac{\eta}{1+\alpha\left(\frac{q}{q_\Lambda}\right)^{\sigma}} \tag{40}$$

If we reintegrate this formula and nail down the unknown integration constant by comparing its asymptotics at $q \to \infty$ to the required MF behavior $\tilde{G}^{-1}(\boldsymbol{q}) \sim q^4$, we obtain

$$\tilde{G}^{-1}(\boldsymbol{q}) = \frac{\kappa_\Lambda q^4}{k_B T} \cdot \left[1 + \frac{1}{\alpha}\left(\frac{q_\Lambda}{q}\right)^{\sigma}\right]^{\eta/\sigma} \tag{41}$$

which is the general fitting function with parameters η, σ and α we used to describe our data.

As a further refinement, we made an effort to correct for residual traces of the underlying discretization (i.e. spurious higher powers of q and cubic anisotropy effects implicitly caused by the replacements $\boldsymbol{q} \to \tilde{\boldsymbol{q}}$ and $q^2 \to -\tilde{\Delta}(\boldsymbol{q})$). In polar coordinates $\boldsymbol{q} \to (q, \phi)$, such corrections should be of the general form

$$|\boldsymbol{q}| \to |\boldsymbol{q}|_{\text{corr}} = q + (a_3 + b_3 \cos 4\phi)q^3 + (a_5 + b_5 \cos 4\phi + c_5 \cos 8\phi)q^5 + \dots \tag{42}$$

In fitting (41) to the $\tilde{G}(\boldsymbol{k})$ data this replacement was made, treating a_3, b_3, a_5, b_5 and c_5 as additional fit parameters. However, this correction turned out to be of minor importance. Another concern was that since the number of $\boldsymbol{q}$-vectors of length q in two dimensions grows linearly with q, the data points in the scaling region might be underrepresented in a straightforward least squares fit. Thus, fits were performed using a minimization procedure for a sum of squared differences, weighted not only by the inverse variances of the data points, but in addition also by $q^{-\omega}$ for values $\omega = 0.0$, 0.5 and 1.0. To calculate proper statistical error bars for this hand-made weighted fit procedure, we employed a standard bootstrap approach [40] using 100 bootstrap samples. However, it turned out that for all values of ω the results were statistically compatible with each other, the smallest error resulting for $\omega = 0$.

Figure 3 shows the typical results of such a fit obtained for a large system of linear size $L = 1800$ and cutoff $\Lambda = \pi/15$ (corresponding, as explained above, to a real space system of linear size $L_0 = 27000$) from a parallel version of our FMC code which will be described elsewhere. At first glance, the fit looks quite acceptable, producing a value of $\eta = 0.751 \pm 0.017$. However, monitoring the relative deviations of individual data from the fit as compared to their relative statistical errors, we notice that while most of the deviations remain in the range of the statistical noise of the data, for the smallest $\boldsymbol{q}$-vectors $(1,0)\cdot 2\pi/L$ and $(0,1)\cdot 2\pi/L$ that we can study on the cubic lattice the deviation exceed 10%. Following next in size are the vectors $(1,1)\cdot 2\pi/L$ and $(1,-1)\cdot 2\pi/L$, who also show a noticeable deviation, albeit much smaller and opposite in its tendency. This behavior is in violent conflict with the supposed scaling behavior, but our error bars are much too small to put the blame on statistical inaccuracies. Clearly, a similar behavior will also be seen for any other reasonable choice of fitting function.

Let m, n denote two arbitrary integers. In the following we will denote the set of all wave vectors $(m,n)\cdot 2\pi/L$ together with all vectors which are equivalent to these modulo cubic symmetry as as $(m,n)_c$ and refer to this set as a (cubic) symmetric direction. A separate analysis of the deviations in various such symmetric directions (cf. figure 4) reveals a certain systematic behavior detectable on top of the statistical noise for vectors along $(1,0)_c$ and $(1,1)_c$,

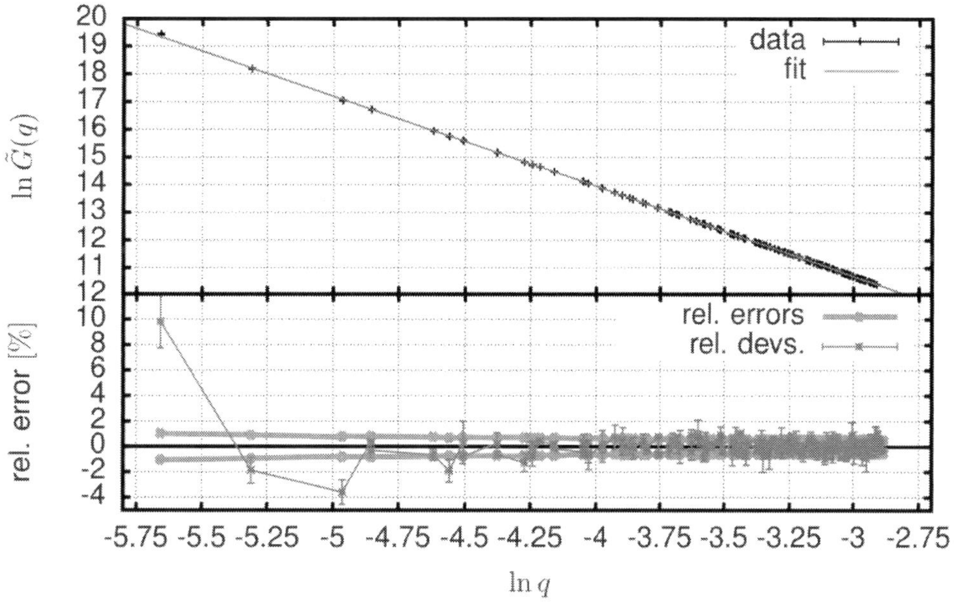

Figure 3. Top: fit of crossover ansatz (41) to simulation results for $\tilde{G}(\boldsymbol{q})$ at linear system size $L = 1800$ and cutoff $\Lambda = \pi/15$. Error bars are smaller than symbol size. Bottom: Relative deviations of simulation data from the fit. Red lines are a guide to the eye. Gray lines indicate the relative statistical errors of the simulation data in percent.

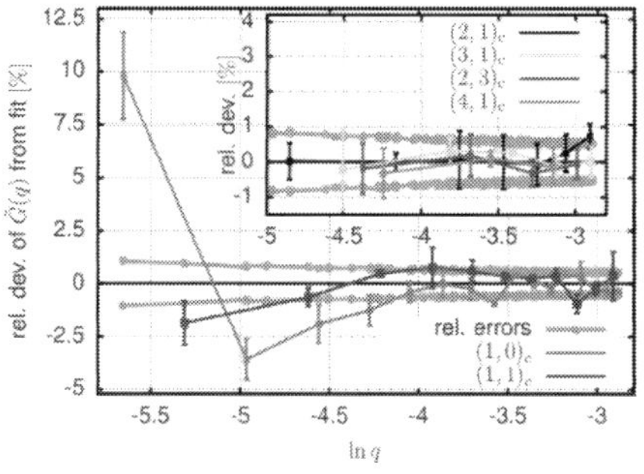

Figure 4. Separate comparison of relative deviations of simulation data for $\tilde{G}(\boldsymbol{q})$ for wave vectors $\boldsymbol{q}$ along various symmetric directions as indicated. Lines are a guide to the eye. Gray lines indicate the relative statistical errors of the simulation data in percent. Main plot: symmetric directions of type $(1,0)_c, (1,1)_c$. Inset: various other directions.

which decreases rapidly with the length of the corresponding wave vector, while the remaining modes reveal no such behavior within statistical errors. Initially suspecting a bug in our code, we introduced a number of modifications to our original program to investigate this phenomenon further. To eliminate possible anisotropy effects originating form a cubic cutoff geometry, we replaced the cubic cutoff $|q_i| \leq \Lambda$ by a spherical one with $|\boldsymbol{q}| \leq \Lambda$. We also performed additional simulations using modified discretization schemes for $\tilde{\partial}(\boldsymbol{q})$ and $\tilde{\Delta}(\boldsymbol{q})$. Since we recall infrequent reports on spurious oscillations arising from the use of sharp cutoffs in the literature on the exact renormalization group [41], we even tried to replace our sharp cutoff by a soft one, i.e. we introduced a modification of the dispersion function $\tilde{\Delta}(\boldsymbol{q})$, which rises smoothly near the boundary of the cutoff and finally diverges beyond Λ. None of these modifications had a noticeable effect on the observed special behavior of the $(1,0)_c$ and $(1,1)_c$ modes. However, switching off anharmonicity by hand by putting $K = 0$ immediately eliminates the problem, and so does the brute force modification $R(\boldsymbol{Q}, \boldsymbol{q}) \equiv 1$. Finally, we chose to investigate the following toy model. Consider two wave vectors $\boldsymbol{k}, \boldsymbol{p}$, and let $\boldsymbol{Q} = \boldsymbol{k} + \boldsymbol{p}$. Taking the angular average of the continuum limit formula (17)

$$R(\boldsymbol{k}+\boldsymbol{p}, \boldsymbol{k}) = ((\widehat{\boldsymbol{k}+\boldsymbol{p}}) \times \boldsymbol{p})^2 = \frac{k^2p^2 + (\boldsymbol{kp})^2}{(\boldsymbol{k}+\boldsymbol{p})^2} \tag{43}$$

yields

$$\langle R(\boldsymbol{k}+\boldsymbol{p}, \boldsymbol{k})\rangle_\phi = \frac{k^2 + p^2 + (p+q)|p+q|}{4} \tag{44}$$

Plugging this into our algorithm, we again observe that any possible residual pathological deviations of the $(1,0)_c$ and $(1,1)_c$ modes are dissolved in the statistical noise of the data. Although these ad hoc replacements of $R(\boldsymbol{Q}, \boldsymbol{q})$ are, of course, physically unacceptable, they nevertheless demonstrate that the behavior reported above is very unlikely caused by a mere bug in our code, but seems to be related to the angle-dependent structure between $\boldsymbol{k}$ and $\boldsymbol{p}$ of encoded in (43) in combination with the finite size of the system. We currently believe that it represents an, albeit completely unexpected and somewhat exotic, finite size effect resulting from the interplay of the effective Hamiltonian (26) with the topology of underlying set of admissible wave vectors, which induces a kind of "surface effect in reciprocal space". Indeed, the subset of wave vectors of the Brillouin zone taking part in the simulation is bounded by the cutoff Λ for large wave vectors. In a transition with critical wave vector $\boldsymbol{q}_c = \boldsymbol{0}$, the modes with large $\boldsymbol{q}$-vectors just below the cutoff Λ are the least critical and are not expected to contribute significantly to the critical asymptotics. However, in the special care of the current membrane problem, the Brillouin zone of a finite membrane represents a lattice, in which a hole is pinched at its center $\boldsymbol{q} = \boldsymbol{0}$, inducing an additional "boundary" besides the one defined by Λ, and the resulting cubic anisotropy is felt by the small wave vector modes in its neighborhood. Intuitively, the effect of this hole at the center of the Brillouin zone may be viewed as a breaking of anharmonic "reciprocal space bonds" that are mediated by the strongly direction-dependent coupling (43). And in contrast to the large wave vector cutoff, modes with small q do fluctuate violently, while (43) reveals that the couplings to all other modes are bounded by q and thus small. For the smallest admissible wave vectors $(2\pi/L, 0)$ and $(0, 2\pi/L)$, we may thus expect a certain deviatoric tendency of $\tilde{G}(\boldsymbol{q})$ towards MF behavior (and thus "up"), which is indeed identified in the data as the large positive deviation of $\tilde{G}(\boldsymbol{q})$ at this wave vector in figure 4. For larger vectors, however, we find it difficult to predict even the qualitative behavior of the resulting anisotropy.

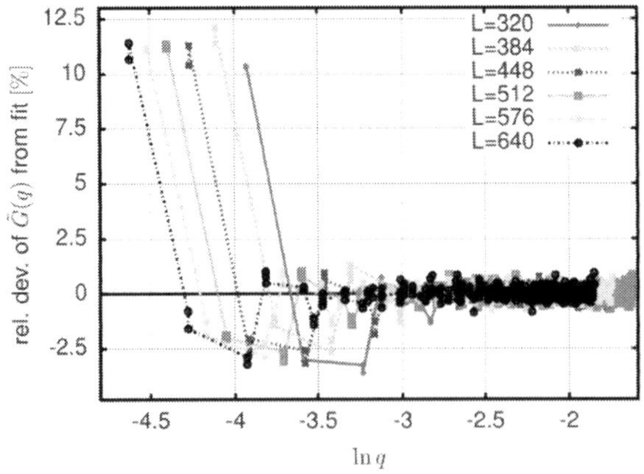

Figure 5. Comparison of relative deviations of simulation data for $\tilde{G}(\boldsymbol{q})$ as observed for the collection of systems (35).

Based on the hypothesis that what we see at small wave vectors is a finite size effect, we decided to monitor the corresponding systematic trends observed with growing system size. As figure 5 demonstrates, the pathological effects observed in the deviations of the $\tilde{G}(\boldsymbol{k})$ data from the collective trends found in a least squares fit of all modes are indeed qualitatively completely similar for all considered system sizes, but for growing system size L are confined to a successively shrinking region around $\boldsymbol{q} = \boldsymbol{0}$. It is thus reasonable to expect that these irregularities may indeed asymptotically disappear in the thermodynamic limit.

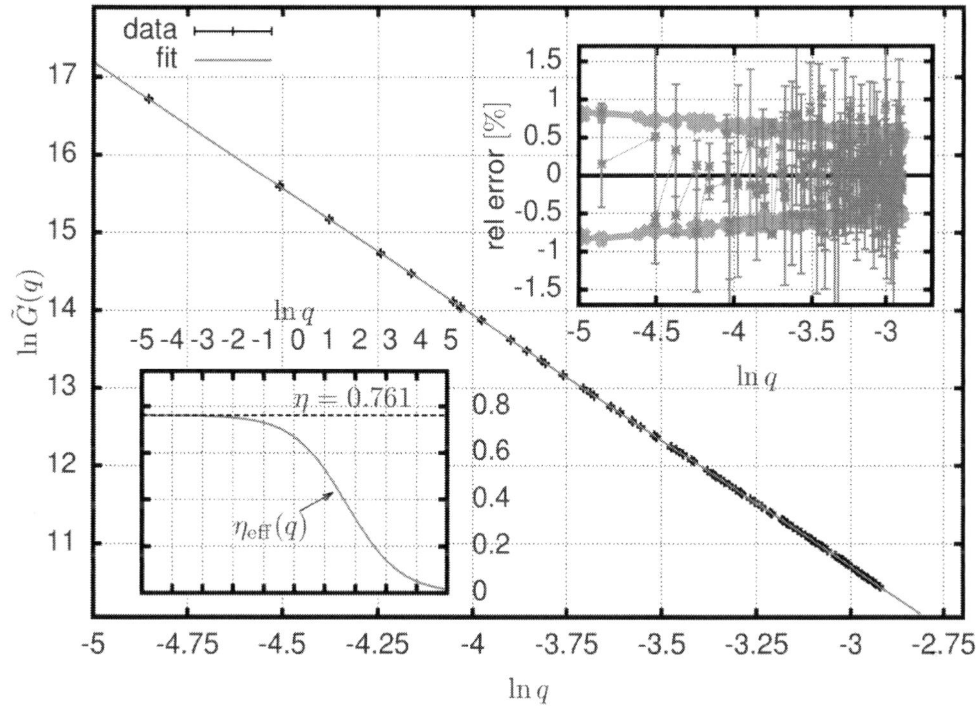

Figure 6. Main plot: fit of crossover ansatz (41) to simulation results for $\tilde{G}(\boldsymbol{q})$ with all contributions for $\boldsymbol{q}$-vectors of type $(1,0)_c$ and $(1,1)_c$ omitted. Error bars are smaller than symbol size. Right upper inset: relative deviations of simulation data from the fit. Red lines are a guide to the eye. Gray lines indicate the relative statistical errors of the simulation data in percent. Left lower inset: crossover function (40) as determined by the fit.

As a finite size effect, the presence of the above described deviations should be properly covered by our FSS ansatz (34). In addition, note from figure 4 that the tendencies observed in the small wave vector deviations along $(1,0)_c$ and $(1,1)_c$ are different, and therefore tend to at least partially compensate themselves in the total sum (33). Thus, we are still confident that the estimate (36) is correct.

To be able to squeeze out a reasonable numerical estimate from $\tilde{G}(\boldsymbol{q})$ for comparison, we decided to omit the most deviatoric $(1,0)_c$ and $(1,1)_c$ data in performing our fits. As figure 6 shows, the quality of the resulting fit is excellent, producing the result

$$\eta = 0.761 \pm 0.008 \tag{45}$$

However, it is not straightforward to combine this estimate with our previous one. On the one hand, the difficulties encountered in deriving (45) may be the source of a systematic error superimposed on the observed statistical uncertainties. On the other hand, our guess (40) for the true crossover function $\eta_{\rm eff}(k)$ may still be less than perfect, introducing an additional systematic error in the result. Last but not least, a small conventional finite size effect may still remain. Reviewing these arguments we cannot help falling back on our previous FSS result (36), which is not plagued by these difficulties.

6. Conclusion

In summary, we have presented a thorough discussion of our OFMC approach to calculating the critical exponent η governing the long wavelength scaling of elastic deformations of crystalline membranes in the flat phase. The great efficiency and excellent statistical accuracy of OFMC is basically due to two reasons. On the one hand, by working with the Fourier amplitudes of deformations as the basic MC variables and using an effective Hamiltonian at a wave vector cutoff, we manage to project out all uninteresting microscopic details and to focus only on the long wavelength degrees of freedom that determine the universal critical behavior. On the other hand, we are using the collective MC move set (21), for which we manage to overcome critical slowing down by tuning the MC acceptance rates separately for each $\boldsymbol{q}$ vector, which yields a huge effective speedup of our simulations.

In future work, we plan to investigate the behavior of crystalline membranes close to a substrate and/or under uniaxial tension. We also expect a similar approach to work for hexatic membranes [12]. However, the OFMC approach is, of course, not limited to membranes. Further applications to e.g. compressible spin systems and other long-range lattice spin models are under way.

Note added in proof. A short account on the results presented above has now been published in [42].

Acknowledgments

Major parts of our computations were performed on the Vienna Scientific Cluster (VSC2). We are grateful for support by the Austrian Science Fund (FWF) Project P22087-N16 and would like to thank K. Binder, C. Dellago, G. Kahl, M. Katsnelson and U. Pedersen for valuable discussions.

References

[1] Nelson D, Piran T and Weinberg S (eds) 2004 *Statistical Mechanics of Membranes and Surfaces, Second Edition* (Singapore: World Scientific)
[2] Leibler S 2004 *Statistical Mechanics of Membranes and Surfaces, Second Edition* ed Nelson D et al (Singapore: World Scientific) chapter 3 pp 49–101
[3] De Gennes P and Taupin C 1982 *J. Phys. Chem.* **86** 2294
[4] Mermin N and Wagner H 1966 *Phys. Rev. Lett.* **17** 1133
[5] Peliti L and Leibler S 1985 *Phys. Rev. Lett.* **54**(15) 1690–1693
[6] Novoselov K S, Jiang D, Schedin F, Booth T J, Khotkevich V V, Morozov S V and Geim A K 2005 *Proceedings of the National Academy of Sciences of the United States of America* **102** 10451–10453
[7] Pisana S, Lazzeri M, Casiraghi C, Novoselov K, Geim A, Ferrari A and Mauri F 2007 *Nature Mat.* **6** 198
[8] Bowick M J 2004 *Statistical Mechanics of Membranes and Surfaces, Second Edition* ed Nelson D et al (Singapore: World Scientific) chapter 11 p 323
[9] Chaikin P M and Lubensky T 1995 *Principles of Condensed Matter Physics* (Cambridge, UK: Cambridge University Press)
[10] Bowick M and Travesset A 2001 *Phys. Rept.* **344** 255
[11] Nelson D and Peliti L 1987 *J. Phys. (Paris)* **48** 1085
[12] Nelson D 2004 *Statistical Mechanics of Membranes and Surfaces, Second Edition* ed Nelson et al (Singapore: World Scientific) chapter 6 p 131
[13] Katsnelson M I 2012 *Graphene: Carbon in Two Dimensions* (Cambridge, UK: Cambridge University Press)
[14] Gompper G and Kroll D M 2004 *Statistical Mechanics of Membranes and Surfaces, Second Edition* ed Nelson D et al (Singapore: World Scientific) chapter 12 pp 359–426

[15] Le Doussal P and Radzihovsky L 1992 *Phys. Rev. Lett.* **69** 1209
[16] Aronovitz J A and Lubensky T 1988 *Phys. Rev. Lett.* **60** 2634
[17] David F and Guitter E 1988 *Europhys. Lett.* **5** 709
[18] Kownacki J P and Mouhanna D 2009 *Phys. Rev. E* **79**(4) 040101
[19] Bowick M J, Catterall S M, Falcioni M, Thorleifsson G and Anagnostopoulos K N 1996 *J. Phys. I France* **6** 1321
[20] Fasolino A, Los J and Katsnelson M I 2007 *Nature Materials* **6** 858
[21] Los J H, Katsnelson M I, Yazyev O V, Zakharchenko K V and Fasolino A 2009 *Phys. Rev. B* **80**(12) 121405(R)
[22] Zakharchenko K V, Los J H, Katsnelson M I and Fasolino A 2010 *Phys. Rev. B* **81**(23) 235439
[23] Zakharchenko K V, Roldán R, Fasolino A and Katsnelson M I 2010 *Phys. Rev. B* **82**(12) 125435
[24] Tröster A and Dellago C 2010, Physics Procedia **6**, pp 106–116
[25] Tröster A 2007 *Phys. Rev. B* **76** 012402
[26] Bergman D and Halperin B 1976 *Phys. Rev. B* **13** 2145
[27] Tröster A 2008 *Phys. Rev. Lett.* **100** 140602
[28] Tröster A and Dellago C 2007 *Ferroelectrics* **354** 225
[29] Tröster A 2008 *Comput. Phys. Comm.* **179** 30
[30] Tröster A 2009 *Phys. Rev. E* **79** 036707
[31] Tröster A 2010 *Phys. Rev. B* **81** 125135
[32] Tröster A 2011 *Computer Physics Communications* **182** 1837–1841
[33] Tröster A 2010 *Phys. Rev. B* **81** 012406
[34] Moin P 2010 *Fundamentals of Engineering Numerical Analysis* (Cambridge, UK: Cambridge University Press)
[35] Berg B A 2004 *Markov Chain Monte Carlo Simulations and Their Statistical Analysis* (Singapore: World Scientific)
[36] Amit D and Martin-Mayor V 2005 *Field Theory, the Renormalization Group, and Critical Phenomena: Graphs to Computers (3rd Edition)* (Singapore: World Scientific)
[37] Binder K and Heermann D W 2010 *Monte Carlo Simulations in Statistical Physics* 5th ed Graduate Texts in Physics (Berlin: Springer)
[38] Flyvbjerg H and Petersen H G 1989 *The Journal of Chemical Physics* **91** 461–466
[39] Luijten E and Binder K 1998 *Phys. Rev. E* **58** R4060
[40] Efron B and Tibshirani R 1994 *An Introduction to the Bootstrap* (Boca Raton: Chapman and Hall)
[41] Morris T R and Tighe J F 1999 *Journal of High Energy Physics* **1999** 007
[42] Tröster A 2013 *Phys. Rev. B* **87** 104112

Validity of the Rayleigh hypothesis for two-dimensional randomly rough metal surfaces

T Nordam, PA Letnes, and I Simonsen

Department of Physics, Norwegian University of Science and Technology (NTNU), NO-7491 Trondheim, Norway

E-mail: tor.nordam@gmail.com

Abstract. The Rayleigh hypothesis is the assumption that the field in the region above (below) a rough surface can be expressed as a weighted sum of upwards (downwards) propagating scattered (transmitted) modes, and that these expressions can be used to satisfy the boundary conditions on the fields at the surface. This hypothesis is expected to be valid for surfaces of sufficiently small slopes. For one-dimensional sinusoidal surfaces, the region of validity is known analytically, while for randomly rough surfaces in one and two dimensions, the limits of validity of the Rayleigh hypothesis are not known. In this paper, we perform a numerical study of the validity of the Rayleigh hypothesis for two-dimensionally rough metal and perfectly conducting surfaces by considering the conservation of energy. It is found for a perfect electric conductor that the region of validity is defined by the ratio of the root-mean-square roughness, δ, over the correlation length, a, being less than about 0.2, while for silver we find $\delta/a \lesssim 0.08$ for an incident wavelength $\lambda = 457.9$ nm. Limitations in our simulations made us unable to check the Rayleigh hypothesis for roughness where $\delta \gtrsim 0.13\lambda$.

1. Introduction

Randomly rough surfaces are abundant in nature and such structures may influence surface processes and phenomena in numerous ways. Lord Rayleigh is credited for being the first to study how wave scattering from surfaces is influenced by their roughness. In particular, at the end of the 19th century, he studied how waves are scattered by sinusoidal surfaces [1,2]. To this end, he assumed that the fields in the *surface region*, defined as the spatial region between the maximum and minimum points of the surface, can be written in a form similar to the asymptotic fields. This means that the reflected (transmitted) fields can be expressed as a weighted sum of upward (downward) propagating plane waves [see (4) and Ref. 3 for details].

This way of writing the fields in the surface region is an approximation which is expected to be valid if the surface structures have sufficiently small local slopes. Today this approximation is known as the *Rayleigh hypothesis* in honor of its inventor [1–4], and most perturbation theoretical treatments of the scattering problem can be constructed based on the assumption that the Rayleigh hypothesis is valid [3]. Moreover, by applying the Rayleigh hypothesis and imposing the boundary conditions satisfied by the fields at the surface, the so-called *reduced Rayleigh equation* for reflection (transmission) can be derived from the resulting coupled set of integral equations after eliminating the transmission (reflection) amplitudes [3,5]. The reduced Rayleigh equation, which contains either the unknown reflection or transmission amplitudes, has recently been solved by direct numerical means to produce non-perturbative results [3,6–8].

To take full advantage of an approximation, it is imperative to know its region of validity. About 75 years after its introduction, it was proven mathematically that the Rayleigh hypothesis for a sinusoidal surface[1], i.e., a surface which deviates from its mean by $x_3 = \zeta_0 \sin(\Lambda x_1)$, is formally valid if $\zeta_0 \Lambda \leq 0.448$ [9, 10]. However, for a randomly rough surface, the precise region of validity is still not formally known. For rough surfaces that are constant along one direction in the mean plane, i.e., their height can be written as $x_3 = \zeta(x_1)$, there have been attempts to use a direct numerical solution of the one-dimensional reduced Rayleigh equation to evaluate the validity of the Rayleigh hypothesis [11].

To the best of our knowledge, no attempt at a numerical study of the validity of the Rayleigh hypothesis has been reported in the literature for surfaces that are randomly rough along both directions of the mean plane. The lack of such a study in the literature is probably due to the significant computational cost. The purpose of this paper is to fill this gap in the literature by reporting some preliminary results for the validity of the Rayleigh hypothesis applied to electromagnetic wave scattering from two-dimensional randomly rough metal and perfectly conducting surfaces. Since this problem has no known analytic solution, we approach it by using a purely numerical solution of the reduced Rayleigh equation, which was recently implemented and tested [7].

For scattering from non-absorbing metallic substrates in vacuum, one must require that the power incident upon the surface equals the power reflected from it, since no energy can be transmitted through the metal. Hence, by numerically studying the conservation of energy in the simulated scattering process over the surface roughness parameter space, one may indirectly get information about the validity of the Rayleigh hypothesis which underlies the simulation approach being used. It should be stressed that the conservation of energy is a necessary, but not sufficient condition for correct results. Hence, the region of validity of the Rayleigh hypothesis for the systems we study here may in fact be even more restricted than what we report.

This paper is organized as follows: In Section 2, we define the geometry and some conventions used throughout the paper. Next, in Section 3, we describe the asymptotic field expansions, and write down the reduced Rayleigh equation for reflection from a perfectly conducting substrate, as well as the expression we will use to evaluate the conservation of energy, and we briefly describe how we proceed to solve the reduced Rayleigh equation. In Section 4, we present our results. We also include a discussion of these results, and what they can tell us about the limits on the validity of the Rayleigh hypothesis, as well as some discussion of other potentially limiting steps in our approach. Finally, we present some concluding remarks in Section 5.

2. Scattering geometry

The system we study consists of vacuum ($\varepsilon_1 = 1$) in the region $x_3 > \zeta\left(\boldsymbol{x}_\parallel\right)$, where $\boldsymbol{x}_\parallel = (x_1, x_2, 0)$ and a perfect electric conductor (PEC) or a metal in the region $x_3 < \zeta\left(\boldsymbol{x}_\parallel\right)$. The surface profile function $\zeta\left(\boldsymbol{x}_\parallel\right)$ is assumed to be a single-valued function of $\boldsymbol{x}_\parallel$ that is differentiable with respect to x_1 and x_2, and constitutes a zero-mean, stationary, isotropic, Gaussian random process defined by

$$\left\langle \zeta(\boldsymbol{x}_\parallel)\zeta(\boldsymbol{x}'_\parallel) \right\rangle = \delta^2 W\left(\left| \boldsymbol{x}_\parallel - \boldsymbol{x}'_\parallel \right| \right). \tag{1}$$

Here the angular brackets denote an average over the ensemble of realizations of the surface profile function. In writing (1), we have defined the root-mean-square height of the surface, $\delta = \left\langle \zeta^2(\boldsymbol{x}_\parallel) \right\rangle^{1/2}$, and $W(\boldsymbol{x}_\parallel - \boldsymbol{x}'_\parallel)$ is the height-height auto-correlation function of the surface,

[1] The coordinate system used is defined in Fig. 1.

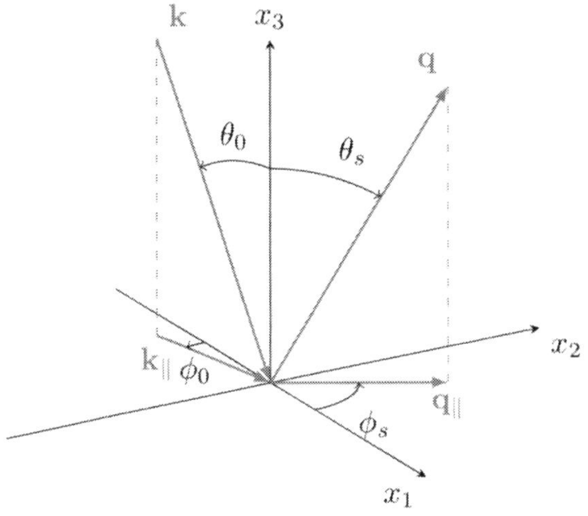

Figure 1. A sketch of the scattering geometry assumed in this work. The figure also shows the coordinate system used, angles of incidence (θ_0, ϕ_0) and scattering (θ_s, ϕ_s), and the corresponding lateral wavevectors $\boldsymbol{k}_\parallel$ and $\boldsymbol{q}_\parallel$, respectively.

normalized so that $W(\mathbf{0}) = 1$ [3]. The correlation function that will be considered in this work is isotropic and of the Gaussian form, given by

$$W\left(|\boldsymbol{x}_\parallel|\right) = \exp\left(-x_\parallel^2/a^2\right), \tag{2}$$

where a denotes the lateral correlation length.

3. Scattering theory

The electric field in the vacuum $[x_3 > \zeta(\boldsymbol{x}_\parallel)]$ is the sum of an incident field and a scattered field, $\boldsymbol{E}(\boldsymbol{x};t) = [\boldsymbol{E}(\boldsymbol{x}|\omega)_{\mathrm{inc}} + \boldsymbol{E}(\boldsymbol{x}|\omega)_{\mathrm{sc}}]\exp(-\mathrm{i}\omega t)$. The Rayleigh hypothesis consists of the assumption that the scattered field can be written as a weighted sum of upwards propagating modes, which allows us to write

$$\begin{aligned}
\boldsymbol{E}(\boldsymbol{x}|\omega)_{\mathrm{inc}} &= \left\{\frac{c}{\omega}\left[\hat{\boldsymbol{k}}_\parallel \alpha_1(k_\parallel) + \hat{\boldsymbol{x}}_3 k_\parallel\right] B_{\mathrm{p}}(\boldsymbol{k}_\parallel) + \left(\hat{\boldsymbol{x}}_3 \times \hat{\boldsymbol{k}}_\parallel\right) B_{\mathrm{s}}(\boldsymbol{k}_\parallel)\right\} \exp\left(\mathrm{i}\boldsymbol{k}_\parallel \cdot \boldsymbol{x}_\parallel - \mathrm{i}\alpha_1(k_\parallel)x_3\right) \\
\boldsymbol{E}(\boldsymbol{x}|\omega)_{\mathrm{sc}} &= \int \frac{\mathrm{d}^2 q_\parallel}{(2\pi)^2}\left\{\frac{c}{\omega}\left[\hat{\boldsymbol{q}}_\parallel \alpha_1(q_\parallel) - \hat{\boldsymbol{x}}_3 q_\parallel\right] A_{\mathrm{p}}(\boldsymbol{q}_\parallel) + \left(\hat{\boldsymbol{x}}_3 \times \hat{\boldsymbol{q}}_\parallel\right) A_{\mathrm{s}}(\boldsymbol{q}_\parallel)\right\} \\
&\quad \times \exp\left(\mathrm{i}\boldsymbol{q}_\parallel \cdot \boldsymbol{x}_\parallel + \mathrm{i}\alpha_1(q_\parallel)x_3\right).
\end{aligned} \tag{3}$$

Here the subscripts p and s denote the p-polarized and s-polarized components of these fields with respect to the local planes of incidence and scattering. The expressions in front of the amplitudes A_{p} and B_{p} are the unit polarization vectors for p polarized light, and similarly the expressions in front of the amplitudes A_{s} and B_{s} are the unit polarization vectors for s polarized light. A caret over a vector indicates that it is a unit vector. The third component of the wave vector $\boldsymbol{q}$ in vacuum, $\alpha_1(q_\parallel)$, is given by

$$\alpha_1(q_\parallel) = \left[\left(\frac{\omega}{c}\right)^2 - q_\parallel^2\right]^{1/2}, \qquad \mathrm{Re}\,\alpha_1(q_\parallel) > 0,\ \mathrm{Im}\,\alpha_1(q_\parallel) > 0. \tag{4}$$

We assume that a linear relation exists between the amplitudes $A_\alpha(\boldsymbol{q}_\parallel)$ and $B_\beta(\boldsymbol{k}_\parallel)$ $(\alpha, \beta = \mathrm{p}, \mathrm{s})$, which we write as

$$A_\alpha(\boldsymbol{q}_\parallel) = \sum_\beta R_{\alpha\beta}(\boldsymbol{q}_\parallel|\boldsymbol{k}_\parallel) B_\beta(\boldsymbol{k}_\parallel).$$

It then can be shown that the scattering amplitudes, organized into the following 2×2 matrix

$$\boldsymbol{R}(\boldsymbol{q}_\parallel|\boldsymbol{x}_\parallel) = \begin{pmatrix} R_{\mathrm{pp}}(\boldsymbol{q}_\parallel|\boldsymbol{k}_\parallel) & R_{\mathrm{ps}}(\boldsymbol{q}_\parallel|\boldsymbol{k}_\parallel) \\ R_{\mathrm{sp}}(\boldsymbol{q}_\parallel|\boldsymbol{k}_\parallel) & R_{\mathrm{ss}}(\boldsymbol{q}_\parallel|\boldsymbol{k}_\parallel) \end{pmatrix}, \tag{5}$$

satisfy the matrix integral equation [12]

$$\int \frac{\mathrm{d}^2 q_\parallel}{(2\pi)^2} \boldsymbol{M}^+(\boldsymbol{p}_\parallel|\boldsymbol{q}_\parallel) \boldsymbol{R}(\boldsymbol{q}_\parallel|\boldsymbol{k}_\parallel) = -\boldsymbol{M}^-(\boldsymbol{p}_\parallel|\boldsymbol{k}_\parallel). \tag{6a}$$

This equation is known as the *reduced* Rayleigh equation since it is an equation for the scattered field in the medium of incidence only. The elements of the matrices $\boldsymbol{M}^\pm$ for a perfect electric conductor take the form

$$\boldsymbol{M}^\pm(\boldsymbol{p}_\parallel|\boldsymbol{q}_\parallel) = I(\pm\alpha_1(q_\parallel)|\boldsymbol{p}_\parallel - \boldsymbol{q}_\parallel) \begin{pmatrix} \pm\frac{c}{\omega}\frac{p_\parallel q_\parallel - (\omega/c)^2 \hat{\boldsymbol{p}}_\parallel \cdot \hat{\boldsymbol{q}}_\parallel}{\alpha_1(q_\parallel)} & [\hat{\boldsymbol{p}}_\parallel \times \hat{\boldsymbol{q}}_\parallel]_3 \\ \pm\frac{\omega}{c}\frac{[\hat{\boldsymbol{p}}_\parallel \times \hat{\boldsymbol{q}}_\parallel]_3}{\alpha_1(q_\parallel)} & \hat{\boldsymbol{p}}_\parallel \cdot \hat{\boldsymbol{q}}_\parallel \end{pmatrix}, \tag{6b}$$

where

$$I\left(\gamma|\boldsymbol{Q}_\parallel\right) = \int \mathrm{d}^2 x_\parallel \exp\left(-\mathrm{i}\boldsymbol{Q}_\parallel \cdot \boldsymbol{x}_\parallel\right) \exp\left[-\mathrm{i}\gamma\zeta\left(\boldsymbol{x}_\parallel\right)\right]. \tag{6c}$$

For a penetrable metal or dielectric surface, the corresponding matrix elements are given explicitly in Ref. 7, and will not be repeated here.

3.1. Mean Differential Reflection Coefficient

The solution of the reduced Rayleigh equation determines the scattering amplitudes $R_{\alpha\beta}(\boldsymbol{q}_\parallel|\boldsymbol{k}_\parallel)$, where α and β signify the polarization of the reflected and incident light, respectively. When the incident field is known, this quantity completely specifies the scattered field in the region above the maximum point of the surface. However, $R_{\alpha\beta}(\boldsymbol{q}_\parallel|\boldsymbol{k}_\parallel)$ is not directly measurable in experiments. A more useful quantity is the mean differential reflection coefficient (DRC), which is defined as the fraction of the time-averaged incident power which is scattered by the surface into the solid angle $\mathrm{d}\Omega_s$ about the scattering direction $\boldsymbol{q}$ defined by the angles of scattering (θ_s, ϕ_s) (see Fig. 1). The mean DRC is defined as [13]

$$\left\langle \frac{\partial R_{\alpha\beta}}{\partial \Omega_s} \right\rangle = \frac{1}{L^2} \frac{\omega^2}{4\pi^2 c^2} \frac{\cos^2\theta_s}{\cos\theta_0} \left\langle \left| R_{\alpha\beta}(\boldsymbol{q}_\parallel|\boldsymbol{k}_\parallel) \right|^2 \right\rangle, \tag{7}$$

where L^2 is the area of the plane $x_3 = 0$ covered by the rough surface, and θ_0 is the angle of incidence (see Fig. 1).

3.2. Conservation of Energy

To test whether the Rayleigh hypothesis is fulfilled, we use the conservation of energy as a criterion. Considering a perfectly conducting substrate, all the incident power must be reflected, and we can check whether our calculations conserve energy from the reflected field alone. The fraction of the time-averaged incident power in the form of light of polarization β, scattered by the surface into light of polarization α, is given by the integral of the corresponding mean DRC over the upper hemisphere:

$$\mathcal{U}_{\alpha\beta} = \int \mathrm{d}\Omega_s \left\langle \frac{\partial R_{\alpha\beta}}{\partial \Omega_s} \right\rangle . \tag{8}$$

If the substrate is non-absorbing, and the incident light has polarization β, one should have

$$\mathcal{U}_{\beta} = \sum_{\alpha} \mathcal{U}_{\alpha\beta} = 1, \tag{9}$$

since energy is conserved in the scattering process. To improve statistics, we define the quantity

$$\mathcal{U} = \frac{\mathcal{U}_{\mathrm{p}} + \mathcal{U}_{\mathrm{s}}}{2}, \tag{10}$$

which we find by separately calculating both $\mathcal{U}_{\mathrm{p}}$ and $\mathcal{U}_{\mathrm{s}}$ for each surface realisation, and taking their average. This is the quantity we will use when testing the validity of the Rayleigh hypothesis.

While the conservation of energy is a necessary, but not sufficient, condition for correct results, we have previously made successful direct comparisons of results obtained by the numerical solution of the reduced Rayleigh equation for weakly rough surfaces against both experimental data, and simulation results obtained by the rigorous surface integral method [14–16]. We are thus confident that our approach works well for weakly rough surfaces. Furthermore, if we hold the transverse correlation length of the surface roughness fixed, while increasing the rms height of the surface, we typically see that at some point, $\mathcal{U}$ starts increasing past 1. While $\mathcal{U} > 1$ is clearly unphysical, and thus evidence that the technique is not working, $\mathcal{U} = 1$ is not by itself proof that the results are correct (for a non-absorbing substrate). Still, our comparisons with both experimental data and other numerical data indicate that the conservation of energy is a useful guide to the validity of our approach.

Thus, in this paper we use $\mathcal{U} \approx 1$ as a criterion for when our approach is valid when considering a perfect electric conductor, and $\mathcal{U} \lesssim 1$ when considering a penetrable metal. Note that this criterion does not directly test the Rayleigh hypothesis, but rather our complete method for solving the scattering problem, which among other things relies on the Rayleigh hypothesis. There could be other reasons why our approach fails for strongly rough surfaces, and this will be discussed in more detail when considering our results.

3.3. Solving the reduced Rayleigh equation

To solve the reduced Rayleigh equation, we begin by numerically generating a realization of the surface profile function on a grid of N_x^2 points within a square region of the x_1x_2 plane of edge L, so that the in-plane sampling interval is $\Delta x = L/N_x$. A two-dimensional version of the filtering method used in Refs. [7, 16, 17] is used to generate the surface realizations from a given correlation function.

The next step is to evaluate the integrals $I(\gamma|\boldsymbol{Q}_{\parallel})$ defined in (6c). These integrals are so-called Fourier integrals and care should be taken when evaluating them due to the oscillating integrands [18]. Using direct numerical integration routines for their evaluation will typically result in inaccurate results. Instead, a (fast) Fourier transform technique with end point corrections may be adapted for their evaluation (the details of the method is outlined in Ref. [18]). However, these calculations are time consuming. In solving for $\boldsymbol{R}(\boldsymbol{q}_\parallel|\boldsymbol{k}_\parallel)$, we will discretize both $\boldsymbol{p}_\parallel$, $\boldsymbol{q}_\parallel$ and $\boldsymbol{k}_\parallel$, and $I(\gamma|\boldsymbol{Q}_\parallel)$ must be evaluated for all relevant combinations of the arguments $\gamma = \pm\alpha_1(p_\parallel)$ and $\boldsymbol{Q}_\parallel = \boldsymbol{p}_\parallel - \boldsymbol{q}_\parallel$ and $\boldsymbol{Q}_\parallel = \boldsymbol{p}_\parallel - \boldsymbol{k}_\parallel$. For the calculations used to generate the results presented in this paper, this would amount to evaluating $I(\gamma|\boldsymbol{Q}_\parallel)$ on the order of 10^9 times per surface realization.

A computationally more efficient way of evaluating $I(\gamma|\boldsymbol{Q}_\parallel)$ is to assume that the exponential function $\exp\left[-\mathrm{i}\gamma\zeta(\boldsymbol{x}_\parallel)\right]$, present in the definition of $I(\gamma|\boldsymbol{Q}_\parallel)$, can be expanded in powers of its argument, and then evaluating the resulting expression term-by-term by employing the Fourier transform. This gives

$$I(\gamma|\boldsymbol{Q}_\parallel) = \sum_{n=0}^{\infty} \frac{(-\mathrm{i}\gamma)^n}{n!} \tilde{\zeta}^{(n)}(\boldsymbol{Q}_\parallel), \tag{11a}$$

where $\tilde{\zeta}^{(n)}(\boldsymbol{Q}_\parallel)$ denotes the Fourier transform of the nth power of the surface profile function, i.e.,

$$\tilde{\zeta}^{(n)}(\boldsymbol{Q}_\parallel) = \int \mathrm{d}^2x_\parallel \zeta^n(\boldsymbol{x}_\parallel) \exp\left(-\mathrm{i}\boldsymbol{Q}_\parallel \cdot \boldsymbol{x}_\parallel\right). \tag{11b}$$

In practice, the sum in (11a) will be truncated at a finite value of n. The Fourier transforms are calculated using a fast Fourier transform (FFT) algorithm.

The primary advantage of using Eqs. (11) for calculating $I(\gamma|\boldsymbol{Q}_\parallel)$ is that the rewrite in (11a) moves γ outside the integral, and calculating the integral in (11b) by FFT gives us the value of the integral as a function of $\boldsymbol{Q}_\parallel$. By using the proper discretization, we can make sure that the values of $\boldsymbol{Q}_\parallel$ resolved by the FFT are precisely those we need when setting up a linear equation system to solve for $R_{\alpha\beta}(\boldsymbol{q}_\parallel|\boldsymbol{k}_\parallel)$. This alternative way of evaluating the I-integrals results in a significant reduction in computational time. The same method has previously been applied successfully to the numerical solution of the one-dimensional reduced Rayleigh equation [6,19,20].

It should be noted that the series expansion used to arrive at (11a) requires $\left|\gamma\zeta(\boldsymbol{x}_\parallel)\right| \ll 1$ to converge reasonably fast. Even though the series expansion of the exponential function is always convergent in theory, this procedure might not be numerically stable for large values of $\left|\gamma\zeta(\boldsymbol{x}_\parallel)\right|$, in particular due to the oscillatory nature of the series. Thus, limited numerical precision means that using the expansion presented in (11a) places additional constraints on the amplitude of the surface roughness which may be more restrictive than those introduced by the Rayleigh hypothesis. In particular, the Rayleigh hypothesis places a constraint on the maximum *slope* of the surface roughness, while the numerical procedure used to evaluate the I-integrals, needed to solve the reduced Rayleigh equation, also limits the *amplitude* of the surface roughness. Hence, surfaces exist for which the Rayleigh hypothesis is satisfied, but the above expansion method will not converge numerically, and the much more time-consuming approach of Ref. 18 will have to be applied.

Finally, in evaluating the integral in (6a) over $\boldsymbol{q}_\parallel$, the integration limits were truncated to the circular region defined by $\left(q_1^2 + q_2^2\right)^{1/2} \leq Q/2$. The Nyquist sampling theorem requires that $|q_1|$ and $|q_2|$ be smaller than $Q_c = \pi/\Delta x$ [21, p. 605]. The components of the vector $\boldsymbol{p}_\parallel - \boldsymbol{q}_\parallel$ entering $I(\gamma|\boldsymbol{p}_\parallel - \boldsymbol{q}_\parallel)$ lie in the interval $[-Q, Q]$, so we have chosen $Q = Q_c$. A quadratic grid with grid constant $\Delta q = 2\pi/L$ was constructed within the circular region of the q_1q_2 plane where $\left(q_1^2 + q_2^2\right)^{1/2} \leq Q/2$. The integral over this region in (6a) was carried out by a two-dimensional version of the extended midpoint method [21, p. 161] and the values of $R_{\alpha\beta}(\boldsymbol{q}_\parallel|\boldsymbol{k}_\parallel)$ were calculated for values of $\boldsymbol{q}_\parallel$ on the grid for a given value of $\boldsymbol{k}_\parallel$ (or equivalently, a given angle of incidence), which was also a point on the grid. The resulting matrix equations were solved

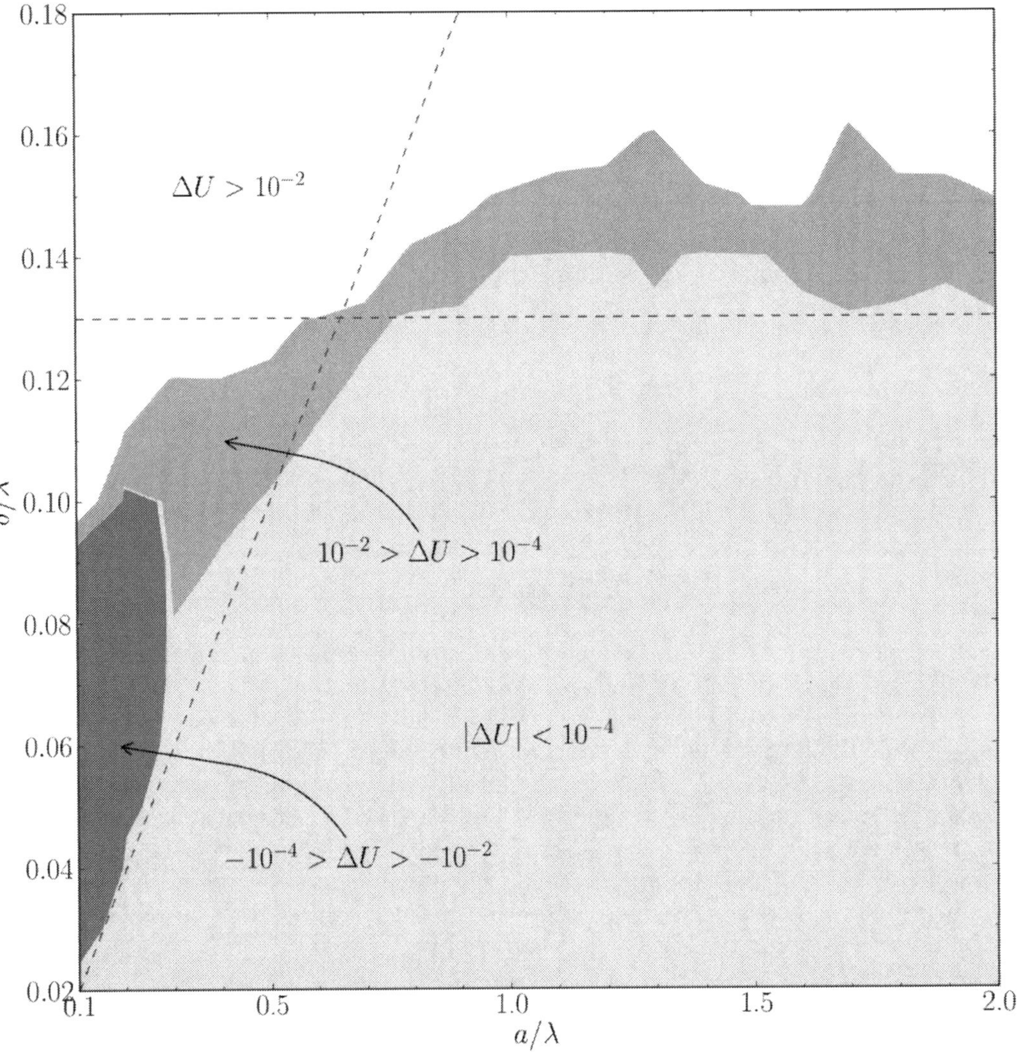

Figure 2. Contour plot of $\Delta\mathcal{U} \equiv \mathcal{U} - 1$, i.e., the deviation from unity of the fraction of incident power reflected from randomly rough surfaces on perfectly conducting substrates. The surface roughness is characterized by the correlation function given in (2), with the correlation length, a, given on the horizontal axis and the rms height of the surface, δ, on the vertical axis. The slope of the diagonal dashed line is given by $\delta/a = 0.2$, which corresponds to an rms slope of the rough surface of 0.28.

by LU decomposition and back substitution [21, p. 48]. In this way we obtain $R_{\alpha\beta}(\boldsymbol{q}_\parallel|\boldsymbol{k}_\parallel)$, from which the mean DRC can be calculated [see (7)]. For a more detailed description of how we solve the reduced Rayleigh equation, we refer the interested reader to Ref. 7.

4. Results

The direct numerical solution method of the reduced Rayleigh equation from Sec. 3 has been used to calculate the scattering amplitudes $R_{\alpha\beta}(\boldsymbol{q}_\parallel|\boldsymbol{k}_\parallel)$ for randomly rough surfaces of edges $L \times L$, where $L = 15\lambda$. The surface profile function $\zeta(\boldsymbol{x}_\parallel)$ was discretized to $N_x \times N_x$ points, with

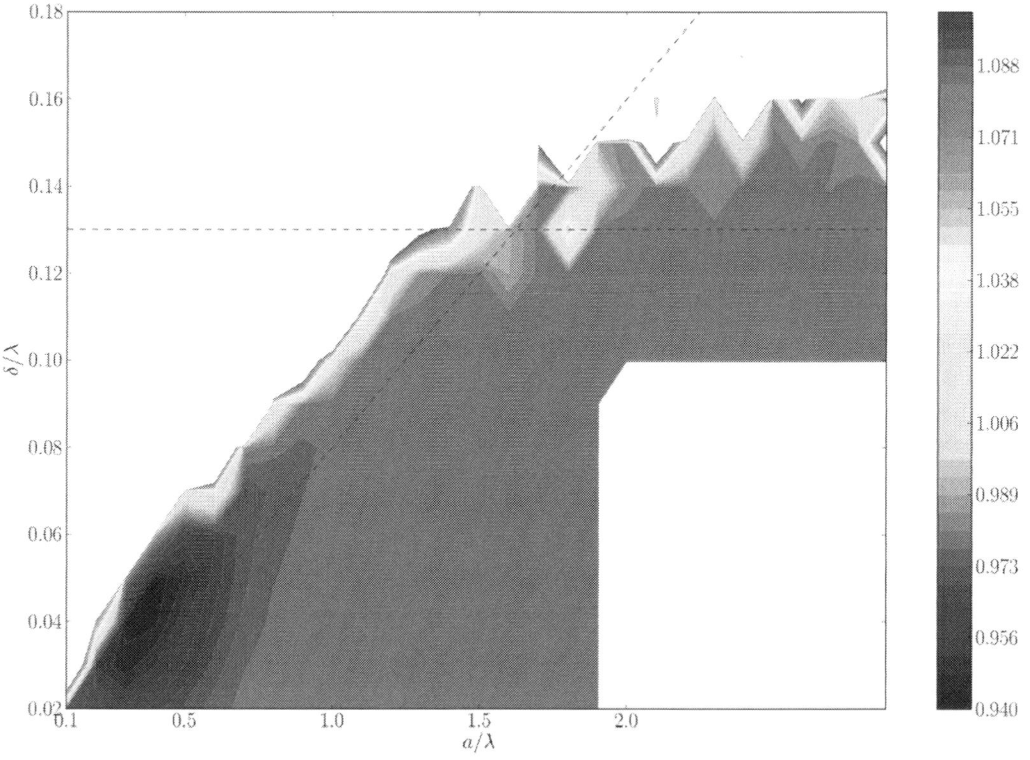

Figure 3. Contour plot of $\mathcal{U}$, the fraction of incident power reflected from randomly rough surfaces on silver substrates. The incident wavelength is 457.9 nm, which corresponds to a dielectric constant of $\epsilon_2 = -7.5+0.24i$. The surface roughness is characterized by the correlation function given in (2), with the correlation length, a, given on the horizontal axis and the rms height, δ, on the vertical axis. The white area in the bottom right corner represents parameters for which we have no data, but where we are confident the value of $\mathcal{U}$ is the same as in the surrounding region. The white area in the upper left represents values of $\mathcal{U}$ greater than 1.1. The slope of the diagonal dashed line is given by $\delta/a = 0.08$, which corresponds to an rms slope of the rough surface of 0.11.

$N_x = 319$, and the roughness was characterized by the height-height autocorrelation function given by (2). The sum in (11a) was truncated at $n = 20$. Solving the reduced Rayleigh equation for one surface realization with these parameters takes approximately 17 minutes on a machine with two six-core AMD Opteron 2.1 GHz processors. The majority (around 90%) of this time is spent performing the LU-decomposition of the linear equation system.

The first set of results we present is for a perfectly conducting substrate. For this system, Fig. 2 depicts a contour plot of $\Delta\mathcal{U} \equiv \mathcal{U} - 1$, i.e., the deviation from unity of the fraction of reflected power, as a function of the transverse correlation length, a, and the rms height, δ, of the rough surface. The contours are defined by $-10^{-2} < \Delta\mathcal{U} < -10^{-4}$, $-10^{-4} < \Delta\mathcal{U} < 10^{-4}$, $10^{-4} < \Delta\mathcal{U} < 10^{-2}$, and $10^{-2} < \Delta\mathcal{U}$. The results of Fig. 2 were obtained from calculations on a grid of points in parameter space defined by $\delta \in [0.02\lambda, 0.18\lambda]$ and $a \in [0.1\lambda, 2.0\lambda]$ with spacing $\Delta\delta = 0.01\lambda$ and $\Delta a = 0.1\lambda$. For each pair of surface parameters (a, δ), we calculated the fraction of reflected power, $\mathcal{U}$, and averaged the solution over 20 surface realizations. This was found to be sufficient to get convergent results, at least for the surface type and parameters considered in this work.

In the wave scattering from a perfectly conducting substrate, one should formally always have $\mathcal{U} \equiv 1$ independent of the level of surface roughness. However, in numerical calculations this condition may, or may not, be satisfied. The latter case may indicate that the numerical method is being applied outside its range of validity, and that the results cannot be trusted. From Fig. 2 we find that for weakly rough surfaces $\mathcal{U}$ deviates from 1 by less than 10^{-4}, and that the boundary between the regions defined by $|\Delta\mathcal{U}| < 10^{-4}$ and $|\Delta\mathcal{U}| > 10^{-4}$ is approximately given by $\delta = 0.2a$ (indicated by the diagonal dashed line in Fig. 2) for sufficiently small δ/λ.

For a Gaussian correlation function, which was used to produce the results of Fig. 2, the rms slope of the surface is given by $\sqrt{2}\delta/a$ [17]. Thus, the limit of the validity of the Rayleigh hypothesis being given by a linear relationship between δ and a is consistent with the local slope of the surface being the critical parameter. Since the Rayleigh hypothesis is a *small slope* approximation, our finding is not unexpected [4]. For perfectly conducting substrates, we find the Rayleigh hypothesis to be valid for surfaces of rms slope less than about 0.28.

The Rayleigh hypothesis is not expected to introduce any additional restrictions, for example on the amplitude of the rough surface. However, from Fig. 2 we observe that regardless of the correlation length, $\mathcal{U}$ deviates from 1 when $\delta \gtrsim 0.13$. Since this is not an expected consequence of the Rayleigh hypothesis, one may wonder about the cause of this behaviour.

As was discussed in Sec. 3.3, with increasing amplitudes of $\zeta(\boldsymbol{x}_\parallel)$ we no longer expect to be able to accurately calculate the integrals $I(\gamma|\boldsymbol{Q}_\parallel)$. Preliminary tests indicate that this problem can be alleviated somewhat by increasing the order up to which the sum in (11a) is carried out, but at some point limited numerical precision, leading to numerical cancellations, becomes an issue. Hence, we concluded that the upper finite roughness level for which $\mathcal{U}$ starts deviating significantly from 1 is not a consequence of the Rayleigh hypothesis. Rather, it is a numerical artifact of how the integrals encoding the surface roughness are calculated. We speculate that using an alternative way of calculating $I(\gamma|\boldsymbol{Q}_\parallel)$ which is not based on a series expansion may resolve this issue, but at the cost of significantly increased computational time. A more promising approach might be to employ software libraries which allow calculations to be carried out with arbitrary precision. Tests we have performed, as well as the results of Ref. 11, indicate that higher precision allows the accurate calculation of the integrals $I(\gamma|\boldsymbol{Q}_\parallel)$, though again this comes with a price in terms of computational time.

The second set of results that will be presented are for silver substrates. Here the wavelength of the incident light was assumed to be 457.9 nm, for which the dielectric function of silver is $\epsilon_2 = -7.5 + 0.24\mathrm{i}$. In Fig. 3, we present the dependence over parameter space of $\mathcal{U}$, i.e., the fraction of the incident power reflected from randomly rough silver surfaces. Since the silver substrate is absorbing, $\mathcal{U}$ will now in general be less than 1, whereas $\mathcal{U} > 1$ still represents an unphysical situation. Again, the values of the rms height, δ, and transverse correlation length, a, were varied, ranging from 0.02 to 0.18λ in 0.01λ increments for δ, and from 0.1λ to 2.9λ in 0.1λ increments for a. For each pair of values, (a, δ), we calculated $\mathcal{U}$, and averaged the solution over 10 surface realizations. The white area in the upper left corner of Fig. 3 represents values where $\mathcal{U} > 1.1$. The large white area in the bottom right corner of Fig. 3 represents surface parameters for which we did not perform any calculations. This was done to reduce computational time, since we are rather confident that for these parameters we will find $\mathcal{U} < 1$.

For a silver substrate, the region of validity of the Rayleigh hypothesis is less pronounced than for the same geometry but with a perfectly conduction substrate. Unlike in a perfect conductor, light can penetrate into the silver and so-called surface modes can be excited. Changing roughness allows light to couple more or less strongly into intermediate surface modes before being scattered back into the vacuum, and the coupling to surface modes is not only dependent on the amplitude of the surface, but also on the length scale of the surface structures.

Still, with these physical differences between a perfectly conducting and a silver substrate, we observe by comparing Figs. 2 and 3 that the structure of the region of validity of the Rayleigh hypothesis appears roughly similar. For a silver substrate, it follows from an inspection of Fig. 3 that the region of validity is defined by the intersection between the following two regions: (*i*) $\delta \lesssim 0.08a$ corresponding to a critical rms slope of 0.11; and (*ii*) $\delta \lesssim 0.13\lambda$. Note that the critical rms slope is less for a silver (0.11) than a PEC substrate (0.28), while the second region is the same for the same two substrates.

5. Conclusions

We have presented preliminary results for the numerical investigation of the validity of the Rayleigh hypothesis obtained by studying the fraction of the incident power which is reflected from rough surfaces. In this way it is found that the Rayleigh hypothesis is valid for perfectly conducting substrates when $\delta \lesssim 0.2a$ and for silver substrates when $\delta \lesssim 0.08a$, where δ denotes the rms height and a the correlation length of the surfaces roughness. These regions correspond to critical rms slopes of 0.28 (PEC) and 0.11 (silver), respectively. Both regions are bounded upwards by an rms height of approximately $\delta \approx 0.13\lambda$. We argue that this is not an inherent limitation of the Rayleigh hypothesis, which is a *small-slope* approximation, not a *small-amplitude* approximation, but rather a consequence of the way we calculate the integrals $I(\gamma|\boldsymbol{Q}_{\parallel})$ [see (11)] present in the reduced Rayleigh equation. It may be possible to get around this limitation by using increased numerical precision, or another algorithm for calculating $I(\gamma|\boldsymbol{Q}_{\parallel})$, but at the cost of increased computational time.

Acknowledgments

We would like to thank A.A. Maradudin for fruitful discussions. The work of T.N. was supported by Nordic Center of Excellence in PV, funded by Nordic Energy Research (contract no. 06-Renew-C31). This work was partially carried out under the HPC-EUROPA2 project (project number: 228398) with support of the European Commission – Capacities Area – Research Infrastructures, and also received support from NTNU by the allocation of computer time.

References

[1] Rayleigh J W S 1896 *The Theory of Sound* 2nd ed vol II (London: MacMillan) pp. 89, 297–311.
[2] Rayleigh J W S 1907 *Proc. R. Soc. Lon. Ser. A* **79** 399
[3] Simonsen I 2010 *Eur. Phys. J.-Spec. Top.* **181** 1
[4] Voronovich A 2007 *Light Scattering and Nanoscale Surface Roughness* Nanostructure Science and Technology ed Maradudin A (Springer US) pp 93–105
[5] Soubret A, Berginc G and Bourrely C 2001 *Phys. Rev. B* **63** 245411
[6] Simonsen I and Maradudin A 1999 *Opt. Commun.* **162** 99
[7] Nordam T, Letnes P A and Simonsen I *arXiv:1204.4984*
[8] Nordam T, Letnes P A, Simonsen I and Maradudin A A 2012 *Opt. Express* **20**(10) 11336
[9] Millar R F 1969 *Math. Proc. Cambridge* **65** 773
[10] Millar R 1971 *Proc. Camb. Phil. Soc.* **69** 217
[11] Tishchenko A V 2009 *Opt. Express* **17** 17102
[12] Nordam T, Letnes P A, Simonsen I and Maradudin A A 2012 Numerical solution of the Rayleigh equations for the scattering of light from a two-dimensional randomly rough perfectly conducting surface (unpublished)
[13] McGurn A R and Maradudin A A 1996 *Wave Random Media* **6** 251
[14] Simonsen I, Maradudin A A and Leskova T A 2010 *Phys. Rev. A* **81** 013806
[15] Simonsen I, Maradudin A A and Leskova T A 2010 *Phys. Rev. Lett.* **104** 223904
[16] Simonsen I, Kryvi J B, Maradudin A A and Leskova T A 2011 *Comp. Phys. Commun.* **182** 1904
[17] Maradudin A A, Michel T, McGurn A R and Méndez E R 1990 *Ann. Phys. (NY)* **203** 255
[18] Press W H, Teukolsky S A, Vetterling W T and Flannery B P 1992 *Numerical Recipes in Fortran. The Art of Scientific Computing* 2nd ed (Cambridge: Cambridge University Press)
[19] Madrazo A and Maradudin A A 1997 *Opt. Commun.* **134**
[20] Simonsen I 2010 *Phys. Status Solidi B* **247** 2075
[21] Press W, Flannery B, Teukolsky S and Vetterling W 2007 *Numerical recipes* 3rd ed (Cambridge: Cambridge University Press)

Multicanonical simulation of coupled folding and binding of intrinsically disordered protein using an Ising-like protein model

Katsuyoshi Matsushita[1,2] **and Macoto Kikuchi**[1,3,4]

[1] Cybermedia Center, Osaka University, Toyonaka 560-0032, Japan
[2] Institute for Protein Research, Osaka University, Suita, 565-0871, Japan
[3] Graduate School of Science, Osaka University, Toyonaka 643-0043, Japan
[4] Graduate School of Frontier Biosciences, Osaka University, Suita, 565-0871, Japan

E-mail: `kmatsu@cp.cmc.osaka-u.ac.jp`

Abstract. An intrinsically disordered protein is one that does not spontaneously fold in physiological conditions but only folds when it binds to a target protein. Computer simulation of this coupled folding and binding is one of the central subjects of computational biophysics. Computing the free energy landscape is helpful in understanding coupled folding and binding. For this reason, we developed an Ising-like protein model and a multicanonical simulation in an energy-entropy space. The calculated free energy landscape indicates that coupled folding and binding induces rapid structural switching of the bound target protein.

1. Introduction

The multicanonical ensemble method [1] for model proteins has shed light on protein folding and binding. In particular, multicanonical simulations [2, 3, 4, 5, 6, 7, 8, 9, 10, 11, 12] provide us with a tool to investigate the free energy landscape, which is a key to understanding protein folding and binding. Recently, Higo *et al.* [13] applied a multicanonical simulation to the coupled folding and binding [14, 15, 16, 17, 18, 19, 20, 21] of intrinsically disordered (ID) proteins [22, 23, 24, 25]. An ID protein is one that is partially or wholly unstructured in physiological conditions, and coupled folding and binding are a folding transition driven by binding between the ID protein and a target protein. Multicanonical simulations succeeded in sampling various low-energy binding structures of an ID protein. However, information only from low-energy structures is not sufficient to achieve a deeper understanding of coupled folding and binding. This is because a protein is a finite size system, and the folding and binding occur at physiological temperatures. We have to clarify the structure of the free energy landscape. Multicanonical simulations to calculate the free energy landscape are difficult not only for elaborate atomic scale models [26, 27] but also for coarse-grained (Gō) models [28, 29], because such simulations require a very large amount of computational power to achieve a thermal equilibrium state. Therefore a simple model expressing the fundamental characteristics of the coupled folding and binding is necessary.

In the present work, we introduce a simple Ising-like protein model which is a variant of the Wako-Saitō-Muñoz-Eaton (WSME) model [30, 31, 32, 33] The model consists only of binary

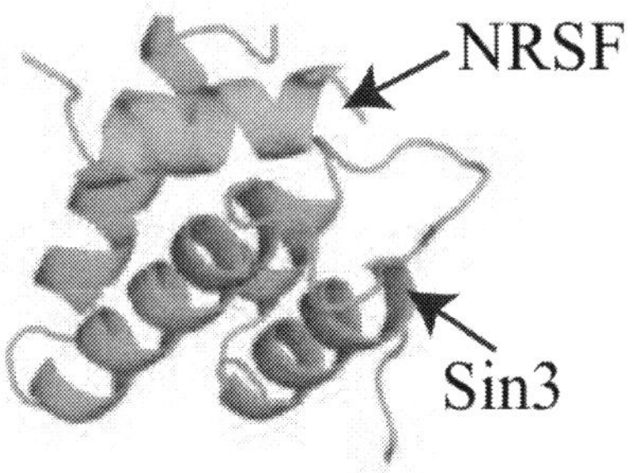

Figure 1. The binding structure of the binding part of NRSF on that of Sin3 (PDB:2czy).

variables and therefore enables us to easily carry out simulations in thermal equilibrium. We apply a multicanonical simulation method to the model. With this model and method, we investigate the coupled folding and binding of the neuron-restrictive silencer factor (NRSF) [34, 35] and corepresser Sin3. NRSF is an ID protein, and Sin3 is a NRSF-binding target protein. We obtain the free energy landscape representing the coupled folding and binding of NRSF. By comparing it with the free energy landscape of a fictitious binding part which folds in physiological conditions, we show that helical domains of Sin3 are prevented from forming non-cooperatively by the coupled folding and binding. As a result, cooperative forming of the helical domains of Sin3 is promoted. This result implies that the ID protein controls interprotein interaction with the target through the promotion of cooperative folding of the target protein.

2. Model and Method

For simplicity, we consider only the binding part of NRSF and that of Sin3, instead of full length proteins. The binding structure of the binding parts was obtained through a NMR experiment, as shown in figure 1 [36]. In the structure, NRSF forms a helix and bridges 4 helices of Sin3. In contrast, the binding part of NRSF without Sin3 is intrinsically disordered. Namely, the binding part has no specific native structure in physiological conditions. Since the structure is not obtained by NMR, it will be natural to consider that the binding part of Sin3 without NRSF is partially structured: the binding part consists of 4 α-helical domains which form independently.

To further gain information for constructing our model, we review a previous numerical study of low energy structures in the ID state of NRSF. The binding part of NRSF consists of 15 residues corresponding to the 43th-57th residues from the N-terminal. The size of the binding part is a tractable size on an atomic scale simulation. In fact, low energy structures were examined by a multicanonical simulation [13]. In the simulation, the binding part of NRSF without Sin3 predominantly adopts hairpin-like structures and adopts an α-helix structure with a low probability. Furthermore a simulation of binding with Sin3 was also carried out and showed that the binding part of NRSF adopts various binding structures. We note that the structure of the binding part of Sin3 in the simulation was artificially restrained. This is because the number of residues of the binding part, 77, which corresponds to the 31st-107th residues from the N-terminal, is too large.

On the basis of the above observations of low energy structures, we consider the following simple model in order to illustrate the free energy landscape of the coupled folding and binding. We express a state of the binding parts using a set of binary variables, m_i and $c_{i,j}$. m_i represents the structure of the i-th residue and takes unity in an α-helix and zero otherwise. $c_{i,j}$ represents a contact between the i-th residue and j-th one and takes unity in hairpin structures and zero otherwise. The energy of a state expressed by $\{m_i\}$ and $\{c_{i,j}\}$ is defined by

$$\mathcal{H}\left(\{m_i\},\{c_{i,j}\}\right) = \varepsilon\mathcal{H}_{\mathrm{helix}}\left(\{m_i\}\right) + \eta\mathcal{H}_{\mathrm{hairpin}}\left(\{m_i\},\{c_{i,j}\}\right). \quad (1)$$

The first term in r. h. s. represents the energy stabilizing the helix structures of NRSF and Sin3. The explicit form of $\mathcal{H}_{\text{helix}}$ is

$$\mathcal{H}_{\text{helix}}\left(\{m_i\}\right) = -\sum_{\langle i,j\rangle}\prod_{k=i}^{j} m_k, \tag{2}$$

where the summation is taken over all residue pairs natively contacting. We adopt the residue pairs from the helix domains of the binding structure obtained by NMR [36]. In this case, as the criterion of the inter-residue contact, we employ $d_{i,j} < 0.65$nm and $|i-j| > 3$, where $d_{i,j}$ is distance between the i-th residue and j-th one.

The second term in r. h. s. of (1) is introduced to energetically stabilize hairpin structures of NRSF, which are observed as low-energy structures in the full atomic simulation [13]. Here $\mathcal{H}_{\text{hairpin}}$ is

$$\mathcal{H}_{\text{hairpin}}\left(\{m_i\},\{c_{i,j}\}\right) = -\sum_{\langle i,j\rangle} c_{i,j}\prod_{k=i}^{j}(1-m_k) \tag{3}$$

where the summation is taken over the 2nd-14th, 5th-11th and 7th-9th residue pairs in NRSF. Here the residues are renumbered from N-terminal of the NRSF binding part. ε and η denote the coupling constant for inter-residue contacts in the helix structure and that in the hairpin-like structures, respectively. η is an important parameter for controlling disorder in the binding part of NRSF. In fact $\eta = 0$ corresponds to a fictitious binding part that folds; and $\eta = 2\varepsilon$ corresponds to the ID binding part in our simulation.

In this model, the state of the helix structure, namely $m_i = 1$, is unique and hence has no entropy. In contrast, the state of the hairpin structures or that of the others, namely $m_i = 0$, consists of many structures. Therefore we consider their entropy with the following term:

$$\mathcal{S}\left(\{m_i\},\{c_{i,j}\},\{b_{i,j}\}\right) = \mathcal{S}_{\text{otherwise}}\left(\{m_i\},\{b_{i,j}\}\right) + \mathcal{S}_{\text{hairpin}}\left(\{m_i\},\{c_{i,j}\},\{b_{i,j}\}\right), \tag{4}$$

where $\mathcal{S}_{\text{hairpin}}$ represents the entropy for hairpin structures of NRSF, and $\mathcal{S}_{\text{otherwise}}$ represents that for the other structures of NRSF and Sin3. $b_{i,j}$ represents a binding state between the i-th residue in NRSF and the j-th residue in Sin3 and adopts unity in binding and zero otherwise.

The entropy terms explicitly take the forms

$$\mathcal{S}_{\text{otherwise}}\left(\{m_i\},\{b_{i,j}\}\right) = s\sum_{i}(1-\tau_i\left(\{c_{i,j}\},\{b_{i,j}\}\right))(1-m_i), \tag{5}$$

$$\mathcal{S}_{\text{hairpin}}\left(\{m_i\},\{c_{i,j}\},\{b_{i,j}\}\right) = (1-\mu)s\sum_{i}\tau_i\left(\{c_{i,j}\},\{b_{i,j}\}\right)(1-m_i), \tag{6}$$

where $\tau_i\left(\{c_{i,j}\},\{b_{i,j}\}\right) = \max_{\langle j,k\rangle\langle p,q;r,s\rangle}\{\{c_{j,k}|j<i<k\},\{b_{p,q}b_{r,s}|p<i<r\}\}$ for the i-th residue of NRSF and $\tau_j\left(\{c_{i,j}\},\{b_{i,j}\}\right) = \max_{\langle p,q;r,s\rangle}\{b_{p,q}b_{r,s}|r<j<s\}$ for the j-th residue of Sin3. In the maximization in τ_i, $\langle p,q;r,s\rangle$ is taken over all neighboring binding pair. The neighboring binding pair $\langle p,q;r,s\rangle$ indicates that there is no other binding site between the p-th residue and the r-th one or between the q-th residue and the s-th residue. For reproducing the intrinsically disordered state of NRSF, we take $s = 1$ and $\mu = 0.25$. The value of μ indicates that the binding part of NRSF loses 25% of the entropy at a cost of forming the hairpin structures. In (5) and (6), we introduce the dependence of τ_i on $b_{i,j}$ in order to consider an entropy effect coming from protein binding. For simplicity, we deal with the entropy effect of the binding by the same way with the entropy loss for the state forming hairpin structures.

We note that $\varepsilon\mathcal{H}_{\rm helix}$, $\eta\mathcal{H}_{\rm hairpin}$, $\mathcal{S}_{\rm hairpin}$, and $\mathcal{S}_{\rm otherwise}$ described so far are adopted from the WSME model [30, 31, 32, 33, 37, 38] with some modifications for expressing the intrinsic disorder.

We also take into account the binding chemical potential defined on binding sites in NRSF and those in Sin3, $h\mathcal{G}(\{m_i\},\{b_{i,j}\})$, to investigate binding-driven folding.

$$\mathcal{G}(\{m_i\},\{b_{i,j}\}) = \sum_{\langle i,j\rangle} b_{i,j} m_i m_j \gamma_{i,j}(\{b_{k,l}\},\{m_p\}), \tag{7}$$

where the summation is taken over all binding site pairs between NRSF and Sin3 and h denotes the chemical potential gain for each binding $b_{i,j}$. $\gamma_{i,j}(\{b_{k,l}\},\{m_p\}) = \prod_{\langle k,l\rangle}\{b_{k,l}\left(\prod_{p=i}^{k} m_p \prod_{q=j}^{l} m_q - 1\right)+1\}$ and the product over $\langle k,l\rangle$'s is taken over the set of the neighboring binding of $\langle i,j\rangle$. The binding sites are adopted from the NMR structure [36] which satisfies the same criterion for the native contacts in $\mathcal{H}_{\rm helix}$. The pairs of binding sites are the pair of the 29th residue from the N-terminal in Sin3 binding part and the 11th one from the N-terminal in NRSF binding part, 32nd-8th, 33rd-8th, 36th-5th and 66th-3rd. We introduces $\gamma_{i,j}$ in order to express cooperativity between the neighboring binding. In $\mathcal{G}$, $b_{i,j}$ is coupled with a set of m_i and thereby gives rise to chemical potential gain only in the native binding structure ($m_i = 1$). Therefore the coupling leads to the binding-driven folding.

The probability density of the state $(\{m_i\}, \{c_{i,j}\}, \{b_{i,j}\})$ is given by

$$P(\{m_i\},\{c_{i,j}\},\{b_{i,j}\}) \propto \exp\{-\beta\left[\mathcal{H}(\{m_i\}) + h\mathcal{G}(\{m_i\},\{b_{i,j}\})\right] + \mathcal{S}(\{m_i\},\{c_{i,j}\},\{b_{i,j}\})\}. \tag{8}$$

β denotes the inverse temperature, $1/k_{\rm B}T$, where T and $k_{\rm B}$ denote a temperature and the Boltzmann constant, respectively.

In order to obtain the free energy landscape, we calculate the density of states

$$\begin{aligned} n(C_{\rm NRSF}, C_{\rm Sin3}, E, S) = & \sum_{\{m_i\},\{c_{i,j}\},\{b_{i,j}\}} \delta\left(\mathcal{H}(\{m_i\}) + h\mathcal{G}(\{m_i\},\{b_{i,j}\}) - E\right) \\ & \times \delta\left(\mathcal{S}(\{m_i\},\{c_{i,j}\},\{b_{i,j}\}) - S\right)\delta\left(C_{\rm NRSF}(\{m_i\}) - C_{\rm NRSF}\right)\delta\left(C_{\rm Sin3}(\{m_i\}) - C_{\rm Sin3}\right), \end{aligned}$$

where $C_{\rm NRSF}(\{m_i\})$ denotes an order parameter for the α-helix structure of NRSF, and $C_{\rm Sin3}(\{m_i\})$ denotes that of Sin3. They are defined by

$$C_p(\{m_i\}) = \frac{1}{N_p}\sum_{i\in\Omega_p} m_i, \tag{9}$$

where Ω_p and N_p denote the set and the number of all residues of the binding part of protein p, respectively. From the density of states, we obtain the free energy landscape

$$\beta F(C_{\rm NRSF}, C_{\rm Sin3}) = \ln\sum_{E,S} n(C_{\rm NRSF}, C_{\rm Sin3}, E, S)P(E,S), \tag{10}$$

where $P(E,S) = \exp\left[-\beta E + S\right]$. We calculate $n(C_{\rm NRSF}, C_{\rm Sin3}, E, S)$ by using the Wang-Landau method [39] for the two dimensional space of E and S.

3. Results and Discussions

Let us compare the free energy landscapes between the case of a fictitious binding part ($\eta = 0$) and that of the ID binding part ($\eta = 2\varepsilon$) in order to examine the effect of the coupled folding and binding of the ID binding part. In figures 2(a)-2(f), we show the free energy landscapes for

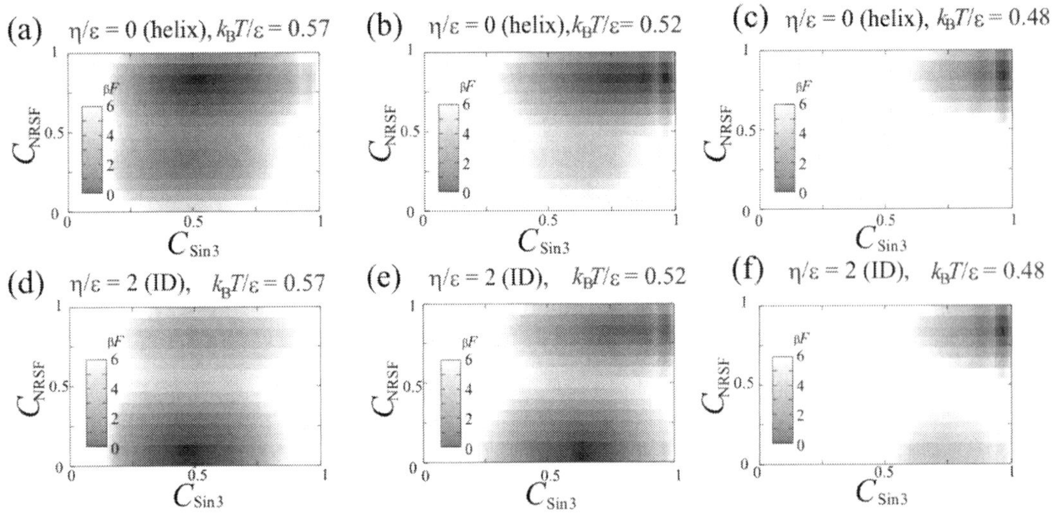

Figure 2. Density plots of the free energy landscape $\beta F(C_{\rm NRSF}, C_{\rm Sin3})$. The data are plotted for (a) $k_{\rm B}T = 0.57\varepsilon$ and $\eta = 0$, (b) $k_{\rm B}T = 0.52\varepsilon$ and $\eta = 0$, (c) $k_{\rm B}T = 0.48\varepsilon$ and $\eta = 0$, (d) $k_{\rm B}T = 0.57\varepsilon$ and $\eta = 2\varepsilon$, (e) $k_{\rm B}T = 0.52\varepsilon$ and $\eta = 2\varepsilon$ and (f) $k_{\rm B}T = 0.48\varepsilon$ and $\eta = 2\varepsilon$. Panels (a)-(c) correspond to the binding part folded at $k_{\rm B}T = 0.52\varepsilon$ and panels (e)-(f) correspond to the ID binding part at $k_{\rm B}T = 0.52\varepsilon$.

$h = \varepsilon$ at temperatures around $k_{\rm B}T = 0.52\varepsilon$, which corresponds to the folding temperature of NRSF for $\eta = 0$. At this temperature NRSF is in intrinsic disorder and exhibits binding-driven folding for $h = 2\varepsilon$.

In the case of $\eta = 0$, as shown in figure 2(a), there are two free energy minima corresponding to the folded state ($C_{\rm NRSF} = 0.25$) and the unfolded one ($C_{\rm NRSF} = 0.75$) of NRSF at high temperatures. In this case, helical domains of Sin3 independently fold and therefore $C_{\rm Sin3}$ takes values around 0.5. In the case of $k_{\rm B}T = 0.52\varepsilon$, the unfolded NRSF becomes unstable, as shown in figure 2(b). Hence folding of NRSF is promoted with decreasing temperature. In this case, $C_{\rm Sin3}$ of the free energy minimum at $C_{\rm NRSF} = 0.75$ increases. Therefore the folding of NRSF promotes that of Sin3. However, Sin3 does not perfectly fold, as indicated by the $C_{\rm NRSF}$, being widely distributed, from 0.5 to 1.0. At low temperatures, Sin3 almost perfectly folds as shown in figure 2(c). These free energy landscapes indicate that non-cooperative helical domain formation of Sin3 is promoted by the binding of NRSF for $\eta = 0$.

In the case of the ID binding part, namely, $\eta = 2\varepsilon$, the free energy landscape in figure 2(d) has two minima similar to those observed for $\eta = 0$. At $k_{\rm B}T = 0.52\varepsilon$, the intrinsically disordered state of NRSF, corresponding to the minimum for $C_{\rm NRSF} \leq 0.25$, is predominantly remained as shown in figure 2(e), in contrast to the case of $\eta = 0$. As a result, the partial folding of Sin3, $(C_{\rm NRSF}, C_{\rm Sin3}) = (0.75, 0.75)$, is not promoted by the binding of NRSF. Namely the partially-folded state of Sin3 is not predominant even at low temperatures. At $k_{\rm B}T = 0.48\varepsilon$ only the completely-folded state of Sin3 is realized, as show in figure 2(f). Thus, at the binding-driven folding transition, the ID binding part does not promote the non-cooperative formation of the helical domains of Sin3, and thereby only helps their cooperative formation.

To see the effects of cooperative helical domain formation on the structural switching of Sin3 in binding with NRSF, we examine the structural order parameter of Sin3, $C_{\rm Sin3}$, as a function of h. We firstly identify the characteristic chemical potential, $h_{2\varepsilon}$, where binding-driven folding of NRSF occurs for $\eta = 2\varepsilon$. At $h_{2\varepsilon}$, the binding susceptibility of $C_{\rm b}$ shows a peak due to the

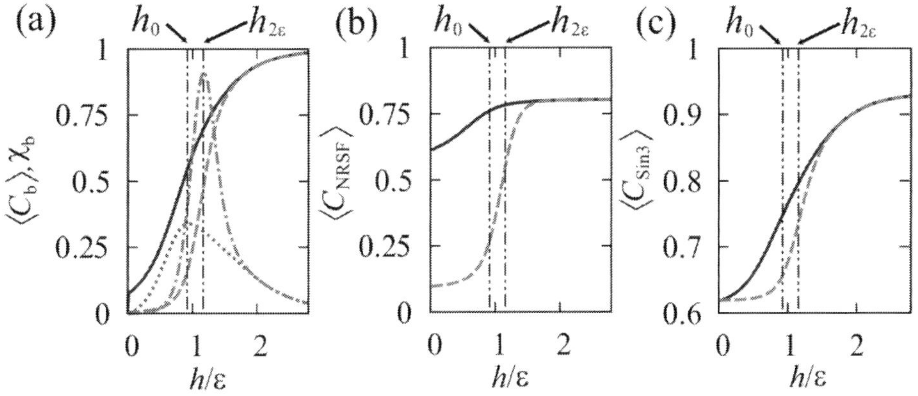

Figure 3. (a) $\langle C_b \rangle$ for $\eta = 0$ (solid line) and 2ε (dashed line) and its susceptibilities for $\eta = 0$ (dotted line) and 2ε (dashed-dotted line). (b) $\langle C_{NRSF} \rangle$ for $\eta = 0$ (solid line) and 2ε (dashed line). (c) $\langle C_{Sin3} \rangle$ for $\eta = 0$ (solid line) and 2ε (dashed line). The dashed double-dotted lines denote binding chemical potential h_0 for the folded binding part and $h_{2\varepsilon}$ for ID binding part.

binding. Here, C_b is a binding order parameter defined by

$$C_b(\{m_i\}, \{b_{i,j}\}) = \frac{\mathcal{G}(\{m_i = 0\}, \{b_{i,j} = 0\}) - \mathcal{G}(\{m_i\}, \{b_{i,j}\})}{\mathcal{G}(\{m_i = 0\}, \{b_{i,j} = 0\}) - \mathcal{G}(\{m_i = 1\}, \{b_{i,j} = 1\})}. \tag{11}$$

We also denote h with the susceptibility peak for $\eta = 0$ by h_0. We show $\langle C_b \rangle$ and susceptibilities as a function of h for $\eta = 0$ and 2ε in figure 3(a), where $\langle \cdots \rangle$ denotes an ensemble average. h_0 and $h_{2\varepsilon}$ are indicated by the dashed double-dotted lines. $\langle C_{NRSF} \rangle$ for $\eta = 2\varepsilon$ exhibits rapid switching around $h_{2\varepsilon}$, as shown in figure 3(b). Namely, a binding-driven folding transition actually occurs at $h_{2\varepsilon}$. Lastly, we show $\langle C_{Sin3} \rangle$ as a function of h in figure 3(c). $\langle C_{Sin3} \rangle$ for $\eta = 2\varepsilon$ around $h_{2\varepsilon}$ increases rapidly as compared to that for $\eta = 0$ at h_0. Thus the coupled folding and binding of the ID binding part enhances sensitivity in the folding response of Sin3 to the change of chemical potential, namely a small change in h induces a large change in $\langle C_{Sin3} \rangle$.

Finally, we consider the implication of the sensitivity. The chemical potential correlates to the concentration of NRSF and Sin3. The change in chemical potential corresponds to the change in concentration of NRSF and Sin3. Therefore, the high sensitivity is efficient in sensing the concentration change. In addition, the folding of Sin3 controls the interaction between NRSF and Sin3 [40]. Thus ID, which induces high sensitivity, is important in the control of the interaction between NRSF and Sin3.

4. References

[1] Berg B A and Neuhaus T 1991 *Physics Letters B* **267** 249–253
[2] Hansmann U H E and Okamoto Y 1993 *J. Comp. Chem* **14** 1333–1338
[3] Hao M H and Scheraga H A 1994 *J. Chem. Phys* **98** 4940–4948
[4] Y Okamoto U H E H 1995 *J. Phys. Chem* **99** 11276–11287
[5] Kolinski A, Galazka W and Skolnick J 1996 *Proteins* **26** 271–287
[6] Urakami N and Takasu M 1996 *J. Phys. Soc. Jpn* **65** 2694–2699
[7] Kumar S, Payne P and Vásquez M 1996 *J. Comp. Chem.* **17** 1269–1275
[8] Nakajima N, Nakamura H and Kidera A 1997 *J. Phys. Chem.* **101** 817–824
[9] Noguchi H and Yoshikawa K 1997 *Chem. Phys. Lett.* **278** 184–188
[10] Higo J, Nakajima N, Shirai H, Kidera A and Nakamura H 1997 *J. Comp. Chem.* **18** 2086–2092
[11] Bartels C and Karplus M 1998 *J. Phys. Chem. B* **102** 865–880

[12] Iba Y, Chikenji G and Kikuchi M 1998 *J. Phys. Soc. Jpn* **67** 3327–3330
[13] Higo J, Nishimura Y and Nakamura H 2011 *J. Am. Chem. Soc.* **133** 10448–10458
[14] Dunker A K, Garner E, Guilliot S, Romero P, Albrecht K, Hart J, Obradovic Z, Kissinger C and Villafranca J E 1998 *Pac. Symp. Biocomput.* **3** 473–484
[15] Wright P E and Dyson H J 1999 *J. Mol. Biol.* **293** 321–331
[16] Verkhiviker G M, Bouzida D, Gehlhaar D K, Rejto P A, Freer S T and Rose P W 2003 *Proc. Natl. Acad. Sci. USA* **100** 5148–5153
[17] Gupta N and Irbäck A 2004 *J. Chem. Phys.* **120** 3983
[18] Hilser V H and Tompson E B 2007 *Proc. Natl. Acad. Sci. USA* **104** 8311–8315
[19] Turjanski A D, Gutkind J S, Best R B and Hummer G 2008 *PLoS one* **4** e1000060
[20] Chen H F 2009 *PLoS one* **4** e6516
[21] Houng Y and Liu Z 2009 *J. Mol. Bio.* **393** 1143–1159
[22] Dyson H J and Wright P E 2002 *Curr Opin Struct Biol.* **12** 54–60
[23] Papoian G A and Wolynes P G 2003 *Biopolymers* **68** 333–349
[24] Sugase K, Dyson H J and Wright P E 2007 *Nature* **447** 1021–1025
[25] Wright P E and Dyson H J 2009 *Curr Opin Struct Biol.* **19** 31–38
[26] McCammon J A 1976 Molecular dynamics study of the bovine pancreatic trypsin inhibitor *Models for Protein Dynamics* (CECAM, Orsay, France) p 137
[27] McCammon J A, Gelin B R and Karplus M 1977 *Nature* **267** 585–590
[28] Clementi C, Nymeyer H and Onuchic J N 2000 *J. Mol. Biol* **298** 937–953
[29] Koga N and Takada S 2001 *J. Mol. Biol.* **313** 171–180
[30] Wako H and Saitô N 1978 *J. Phys. Soc. Jpn.* **44** 1931–1938
[31] Wako H and Saitô N 1978 *J. Phys. Soc. Jpn.* **44** 1939–1945
[32] Muñoz V, Henry E R, Hofrichter J and Eaton W A 1998 *Proc. Natl. Acad. Sci. USA* **95** 5872–5879
[33] Muñoz V and Eaton W A 1999 *Proc. Natl. Acad. Sci. USA* **96** 11311–11316
[34] Chong J A, Tapia-Ramirez J, Kim S, Toledo-Aral J J, Yingcong Zheng M C B, Altshuller Y M, Frohman M A, Kraner S D and Mandel G 1995 *Cell* **80** 949–957
[35] Schoenherr C J and Anderson D J 1995 *Science* **267** 1360–1363
[36] Nomura M, Uda-Tochio H, Murai K, Mori N and Nishimura Y 2005 *J. Mol. Bio.* **354** 903–915
[37] Bruscolini P and Pelizzola A 2002 *Phys. Rev. Lett.* **88** 258101
[38] Itoh K and Sasai M 2010 *Proc. Natl. Acad. Sci. USA* **107** 7775–7780
[39] Wang F and Landau D P 2001 *Phys. Rev. Lett.* **86** 2050–2053
[40] Fischer E 1894 *Ber. Dt. Chem. Ges.* **27** 2985–2993

Stochastic and fractal properties of silicon and porous silicon rough surfaces

S Hosseinabadi [1] and M Rajabi [2]
[1] Department of Physics, East Tehran Branch, Islamic Azad University, P. O. Box 163-33955, Tehran, Iran
[2] Iranian Research Organization for Science and Technology(IROST), P. O. Box 33535-111, Tehran, Iran

E-mail: hosseinabadi.s@gmail.com

Abstract. In this paper, we investigate the stochastic properties and fractal behavior of Si and porous silicon (PS) rough surfaces to characterize the complexity of their morphology. To this end, height fluctuations of these rough surfaces are determined by Atomic Force Microscopy (AFM) and then roughness and correlation length of the surfaces are calculated. The generalized Hurst exponent, $h(q)$ and singularity spectrum, $f(\alpha)$ are obtained by using two dimensional MF-DFA method for both rough surfaces; Our results show that both mentioned surfaces are multifractal and have different scaling exponents. To investigate the reason of the observed multifractality behavior, we determine height distribution, skewness and kurtosis measures and show that the deviation from the Gaussian distribution for the height fluctuations of the surfaces can be a reason for the observed multifractality behavior.

1. Introduction

In the last decades, exploring and characterization of stochastic surface morphology has been one of the interesting fields of study, because surface roughness has an enormous influence on many important physical phenomena such as optical and optoelectronics properties, surface conductivity, reflection and scattering [1, 2, 3]. One can characterize many natural phenomena by a degree of stochasticity; Seismic recordings, cosmic background radiation[4, 5], turbulent flows [6, 7, 8, 9], stock market [10, 11], earthquakes[12] and the surface roughness during growing of many materials and corrosion[13, 14, 15] are examples of such phenomena and processes . Many random processes in nature generate fractal structures; Fractal structures or scaling properties have been observed in space fluctuations of rough surfaces and such interfaces show self-similar or self-affine properties [16, 17, 18, 19, 20]. Fractal properties have been also studied in the other topic such as biology[21, 22], bacterial colony growth [23], diffusion-limited aggregation [24] and climate indicators [25]. When the structure is uniform and regular, we have a monofractal structure and the system is described only by one scaling exponent named Hurst exponent, H; any irregularity in the system can reduce to the multifractality behavior. A multifractal system is combination of different monofractal subsets [17, 18] and various regions of the system have different scaling exponents, named generalized Hurst exponent, $h(q)$; In other words, multifractals should be described by many different numbers of scaling exponents $h(q)$, where q is a real number. Because of infinite scaling exponents , the theoretical and the numerical study of the multifractal surfaces is more complicated than those of monofractal ones. Changing

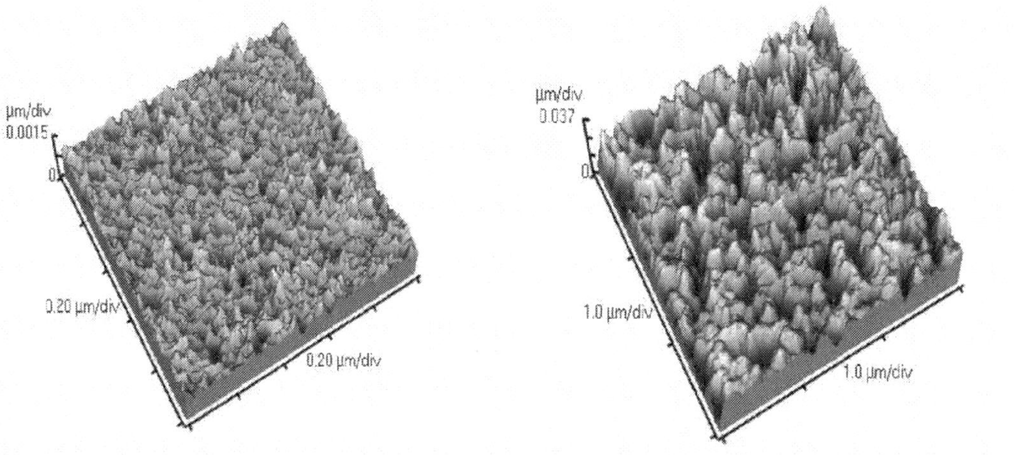

Figure 1. Left: AFM image of Si rough surface; Right: AFM image of PS rough surface.

one of the $h(q)$'s can lead to different feature in the system. Another important feature of the multifractal systems, is the presence of the singularity spectrum, $f(\alpha)$, where α is the Hölder exponent. Recently there has been an increasing interest in the notion of multifractality because of its extensive applications in different areas such as complex systems, industrial and natural phenomena. In this paper, we study the statistical and fractal properties of silicon surface and porous silicon (PS) rough surface produced during electrochemical etching of Si . Because of band gap broadening of PS compared to silicon and room temperature visible photoluminescence, more studies make it a promising material for Si based optoelectronic devices [26, 27, 28]. In this study, we determine the height fluctuations of Si and PS surfaces with Atomic Force Microscopy (AFM); Then we investigate the complexity of the height fluctuations and determine the fractal properties of them by two dimensional multifractal detrended fluctuation analysis. We show that the studied rough surfaces are multifractal and investigate the reason of the observed multifractality behavior.

2. Experiment

In this study we use (100) oriented, boron doped p-type silicon films with thickness of $525 \pm 20 \mu m$. Electrochemical etching of Si surfaces are performed at a constant current density of $20mA/cm^2$ for $20min$ in solution of $32\% HF(HF : C_2H_5OH = 4:1)$ and porous silicon (PS) rough surfaces are produced. The surface topography of Si and PS surfaces are characterized by Atomic Force Microscope (AFM; Veeco-Termo Microscopes) in contact mode. Figure 1 shows the AFM images of the Si and PS rough surfaces.

3. Statistical and Fractal Analysis

3.1. Roughness and correlation function

A rough surface is usually studied in terms of its deviation from a smooth surface or roughness, σ as $\sigma = \langle [h(\vec{r}) - \bar{h}]^2 \rangle^{1/2}$ in which $h(\vec{r})$ is the height at position $\vec{r}$ and $\bar{h}$ is the mean height [13]. The surfaces with equal σ may have different shapes and morphology; Therefore, the correlation function $C(\vec{r}, \vec{r'})$ or the correlation length scale , over which the correlation function reaches $1/e$ of its maximum should be investigated. The correlation function is defined by

$$C(\vec{r}, \vec{r'}) = \langle [h(\vec{r}) - \bar{h}][h(\vec{r'}) - \bar{h}] \rangle \quad (1)$$

For an isotropic surface, we can define the normalized correlation function as follows:

$$C(|\vec{r}-\vec{r'}|) = \frac{\langle [h(\vec{r}+\vec{l}) - \bar{h}][h(\vec{r}) - \bar{h}] \rangle}{\sigma^2} \tag{2}$$

where $l = |\vec{r} - \vec{r'}|$.

Our calculations shows that the roughness of PS rough surface is greater than Si one ($\sigma_{Si} = 7.3 \pm 0.1nm$ and $\sigma_{PS} = 14.200 \pm 0.001\mu m$); Also the height fluctuations of PS surface are more correlated than Si one. Figure 2 shows the correlation function, $C(l)$ with respect to the separation distance, l; This diagram shows that the correlation length of PS is greater than Si or in other words, the correlation length increases when the Si thin film is etched by HF solution.

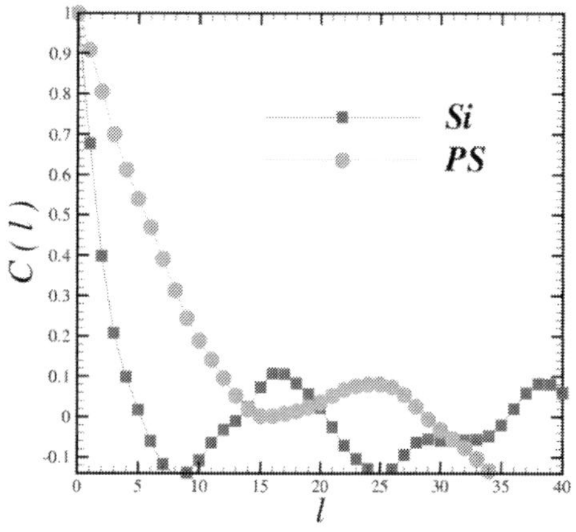

Figure 2. Correlation function $C(l)$ for Si and PS surfaces.

3.2. Fractal properties

There are different methods to determine the fractal properties of random data sets; These methods are such as spectral analysis [29], fluctuation analysis [30], detrended fluctuation analysis (DFA) [22, 32, 33], wavelet transform module maxima (WTMM) [34, 35] and discrete wavelets [36, 37]. For real and experimental data sets and in the presence of noise, the multifractal DFA (MF-DFA) method gives very reliable results [31, 32]. This algorithm is simpler than WTMM but involves a bit more effort in programming. To analyse the fractal behavior of the height fluctuations of our rough surfaces, we use the two dimensional MF-DFA algorithm and determine the spectrum of the generalized Hurst exponent, $h(q)$. Two dimensional MF-DFA algorithm is as follows: In this method, height fluctuation of the rough surface is represented by $Z(\vec{r})$ at coordinate $\vec{r} = (i, j)$ and has the following steps [31]:

Step1: Consider a two dimensional array $Z(i, j)$ where $i = 1, 2, ..., M$ and $j = 1, 2, ..., N$. Divide the $h(i, j)$ into $M_s \times N_s$ non-overlapping square segments of equal sizes $s \times s$, where $M_s = [\frac{M}{s}]$ and $N_s = [\frac{N}{s}]$. Each square segment can be denoted by $Z_{\nu,w}$ such that $u_{\nu,w}(i, j) = Z(l_1 + i, l_2 + j)$ for $1 \leq i, j \leq s$, where $l_1 = (\nu - 1)s$ and $l_2 = (w - 1)s$.

Step 2: The cumulative sum in each segment is calculated by:

$$u_{\nu,w}(i, j) = \sum_{k_1=1}^{i} \sum_{k_2=1}^{j} Z_{\nu,w}(k_1, k_2); \tag{3}$$

where $1 \leq i, j \leq s$.

Step 3: Local trend in each segments is determined by fitting procedure as follows:

$$\tilde{u}_{\nu,w}(i,j) = ai + bj + c. \tag{4}$$

Variance for each segment is determined as follows:

$$\delta_{\nu,w}(i,j) = u_{\nu,w}(i,j) - \tilde{u}_{\nu,w}(i,j), \tag{5}$$

$$F^2_{\nu,w}(s) = \frac{1}{s^2}\sum_{i=1}^{s}\sum_{j=1}^{s}\delta^2_{\nu,w}(i,j). \tag{6}$$

Step 4: Fluctuation function in q'th order is obtained by averaging over all segments:

$$F_q(s) = \Big(\frac{1}{M_s \times N_s}\sum_{\nu=1}^{M_s}\sum_{w=1}^{N_s}\big[F^2_{\nu,w}(s)\big]^{q/2}\Big)^{1/q}, \tag{7}$$

$F_q(s)$ increases with increasing s and depends on the order q. In the above relation, q can take any real value except zero. For $q = 0$, equation (7) becomes

$$F_0(s) = \exp\Big(\frac{1}{2M_s \times N_s}\sum_{\nu=1}^{M_s}\sum_{w=1}^{N_s}\ln F^2_{\nu,w}(s)\Big). \tag{8}$$

Step 5: Finally, the scaling behavior of the fluctuation functions should be investigated by analyzing log-log plots of $F_q(s)$ versus s for each moment of q,

$$F(s) \sim s^{h(q)}. \tag{9}$$

For $q = 2$ the Hurst exponent is obtained by:

$$H \equiv h(q = 2) - 1. \tag{10}$$

If $h(q)$ with respect to q is a constant value, the system will be monofractal and the constant value named Hurst exponent; in other words, a monofractal system can be characterized by a single scaling law with one scaling exponent (Hurst exponent) in all scales [17, 18]. In multifractal structures, scaling behavior of the system is changing with q and should be described with the whole spectrum of scaling exponents $h(q)$.

Figure 3 shows the generalized Hurst exponent, $h(q)$ as a function of q for Si and PS rough surfaces. The q dependence of the $h(q)$ spectrum represents that these rough surfaces have a multifractal structure. Using standard multifractal formalism [32] we have

$$\tau(q) = qh(q) - 2. \tag{11}$$

Figure 4 shows the $\tau(q)$ as a function of q for Si and PS surfaces. Slope of this diagram is changing with q and shows the multifractality behavior in the height fluctuations of the studied rough surfaces. The singularity spectrum, $f(\alpha)$ of a multifractal rough surface is given by the Legendre transformation of $\tau(q)$ as

$$f(\alpha) = q\alpha - \tau(q), \tag{12}$$

where $\alpha = \frac{\partial\tau(q)}{\partial q}$. For a multifractal surface, various parts of the height fluctuations are characterized by different values of α, causing a set of Hölder exponents instead of a single α. Figure 5 shows $f(\alpha)$ versus α and confirms the multifractality nature of studied rough surfaces.

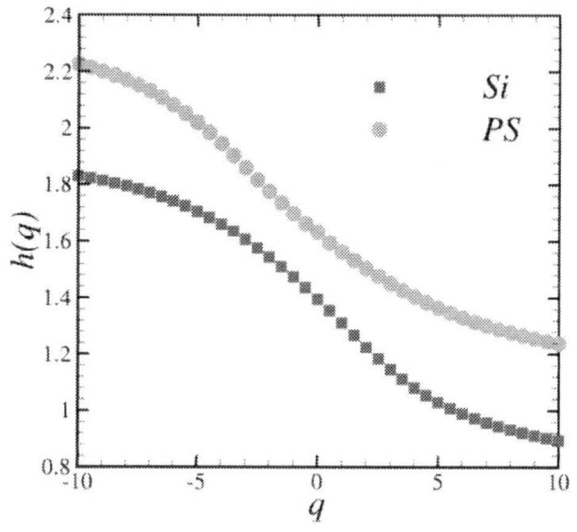

Figure 3. Diagram of $h(q)$ for Si and PS surfaces.

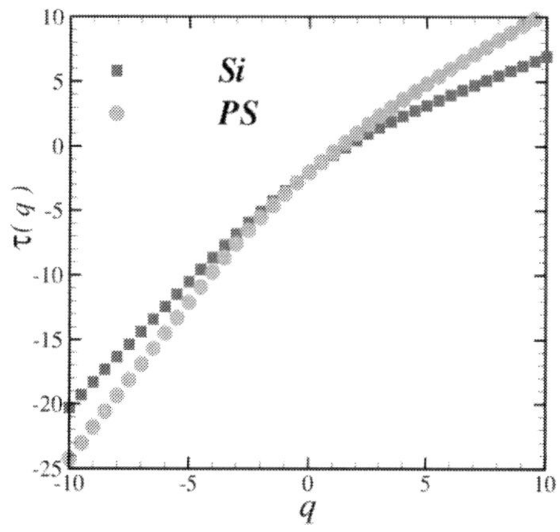

Figure 4. Diagram of $\tau(q)$ for Si and PS surfaces.

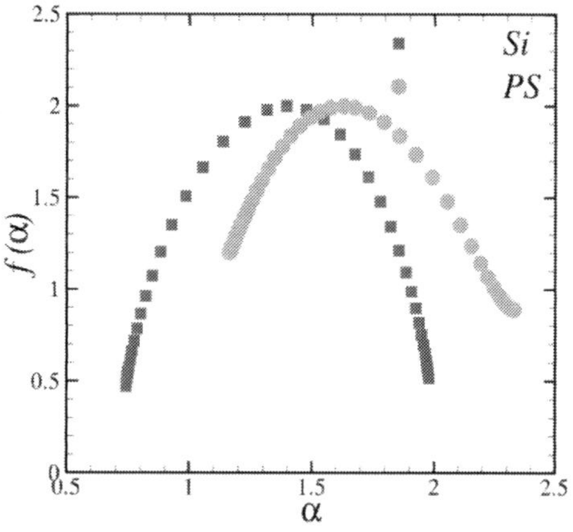

Figure 5. Diagram of $f(\alpha)$ for Si and PS surfaces.

4. Height distribution

The multifractality behavior in a system can have different reasons such as the appearance of different correlations in the system or data distribution deviation from the Gaussian one. To investigate the cause of multifractality in the Si and PS rough surfaces, we study the height distribution and higher moments of height fluctuations in these surfaces.

To analyze symmetry of the height under the transformation $h \to -h$, we can determine the skewness measure as follows:

$$S \equiv \frac{\langle (h-\overline{h})^3 \rangle}{\langle (h-\overline{h})^2 \rangle^{3/2}}. \tag{13}$$

This parameter is a measure of the asymmetry of the distribution. Skewness can come in the form of positive skewness or negative skewness, depending on whether data points are skewed to the right or to the left of the data average. Another measure which is based on the fourth moment of the data is kurtosis. This measure compares the weight of distribution tails with

Gaussian and is a measure of the fatness of the probability distribution compared to the Gaussian distribution as follows:

$$Q \equiv \frac{\langle (h-\overline{h})^4 \rangle}{\langle (h-\overline{h})^2 \rangle^2} - 3. \tag{14}$$

For the Gaussian distribution this measure is zero. In Figure 6, the normalized height distributions $P(h)$ for Si and PS rough surfaces are shown. Regarding to these diagrams, we can see the height distributions of the rough surfaces have been deviated from the Gaussian distribution. Utilizing 13 and 14, we calculated the skewness and kurtosis measures for our surfaces; For the Si, skewness is positive, $S = +7.58 \times 10^{-4}$ and kurtosis is $Q = 6.91 \pm 0.02$ which indicates a peaked distribution compared to the Gaussian (as seen in the Figure 6) and for the PS rough surfaces, skewness is negative, $S = -1.12 \times 10^{-2}$ and $Q = 0.71 \pm 0.02$. As our results show, the height distribution of the Si and PS rough surfaces are deviated from the Gaussian one and it can be a reason for the multifractality behavior in these rough surfaces.

5. Conclusion

In this paper, we have studied the stochastic and fractal properties of silicon and porous silicon (PS) rough surfaces. To this end, we determined the roughness parameter and correlation length of the mentioned surfaces and showed that the height fluctuations in PS rough surface are more correlated than Si one. Also to investigate the fractal nature of the surfaces, we determined the $h(q)$, $\tau(q)$ and $f(\alpha)$ spectra, utilizing the two-dimensional MF-DFA method and showed that the surfaces are multifractal. To find out the observed multifractality behavior reason, we studied the distribution function of the height fluctuations, skewness and kurtosis measures; Our results showed that the height distribution deviation from the Gaussian one can be a reason for the observed multifractality behavior in the Si and etched Si (PS) rough surfaces.

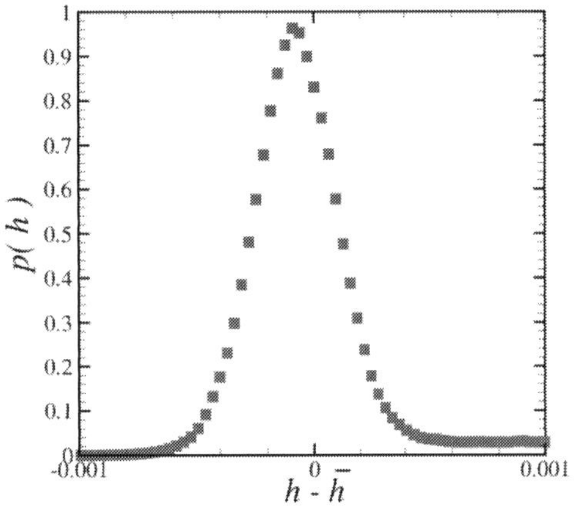

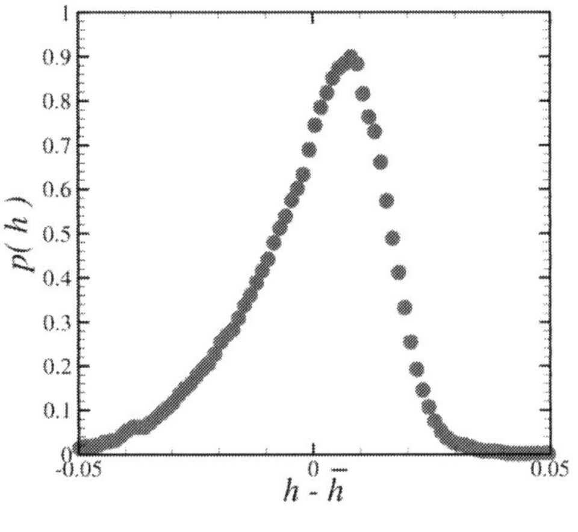

Figure 6. Normalized height distributions $P(h)$ for Si (left) and PS (right) surfaces.

6. Acknowledgments

S. Hosseinabadi is grateful to M. Sadegh Movahed for his supervision on the MF-DFA programming. This work was supported in part by East Tehran branch, Islamic Azad University.

7. References

[1] Chatterjee S 2003 *Phys. Scr.* **67** 234
[2] Laybourn R, Dakin JP, Gambling WA 1970 *Opt. Quantum Electron.* **2(1)** 36
[3] Beckman P, Spizzichino A 1987 *The Scattering of Electromagnetic Wave from Rough Surface* (Artech: Norwood)
[4] Ghasemi F, Bahraminasab A, Sadegh Movahed M, Rahvar S, Sreenivasan KR, Reza Rahimi Tabar M 2006 *J. Stat. Mech.* **84** 68
[5] Movahed M S, Ghasemi F, Rahvar S, Rahimi Tabar M R 2011 *Phys Rev. E* **84** 021103
[6] Davoudi J, Rahimi Tabar M R 2000 *Phys. Rev. E* **61** 6563
[7] Rahimi Tabar M R, Rouhani S 1997 *Phys. Lett.* A **224** 331
[8] Zhou W X, Sornette D 2002 *Physica D* **165** 94
[9] Meneveau C and Sreenivasan K R 1987 *Phys. Rev. Lett.* **59** 1424
[10] Jafari G R, Movahed M S, Fazeli S M, Reza Rahimi Tabar M , Masoudi S F 2006 *J. Stat.Mech.* 06008
[11] Johansen A, Sornette D, Ledoit O 1999 *J. Risk* **1** 5
[12] Stauer D, Sornette D 1998 *Physica A* **252** 271
[13] Barabasi A L, Stanley H E 1995 *Fractal Concepts in Surface Growth* (Cambridge University Press: Cambridge)
[14] Hosseinabadi S , Mortezaali A and Masoudi A A 2008 *Surf. Interface Anal.* **40** 71
[15] Hosseinabadi S , Masoudi A A, Movahed M S 2010 *Physica B: Condensed Matter* **405** 2072
[16] Sornette D 2000 *Critical Phenomena in Natural Sciences: Chaos, Fractals, Self Organization and Disorder: Concepts and Tools* (Heidelberg, Germany: Springer-Verlag)
[17] Stanley H E and Meakin P 1988 *Nature* **335** p 405
[18] Ausloos M and Berman D H 1985 *Proc. R. Soc. London* A **400** 331
[19] Hosseinabadi S, Rajabpour M A , Movahed M Sand Vaez Allaei S M 2012 *Phys.Rev. E* **85** 031113
[20] Feder J, 1988 *Fractals* (Plenum Press New York and London)
[21] Ivanov P C, Nunes Amaral L A, Goldberger A L, Havlin S, Rosenblum M G, and H. Eugene Stanley, 1999 *Nature* **399** 461
[22] Peng C K , Buldyrev S V, Havlin S, Simons M, Stanley H E and Goldberger A L 1994 *Phys. Rev. E* **49** 1685
[23] Ben-Jacob E, Shochet O, Tenenbaum A, Cohen I, Czirok A and Vicsek T 1994 it Fractals **2** 15
[24] Vicsek T 1990 *Physica A* **168** 490
[25] Lovejoy S 1982 *Science* **216** 185 ; Cahalan R F 1989 *Advances in Remote Sensing and Retrieval Methods* ed Deepak A, Fleming H , Theon J (Deepak Publishing, Hampton) p 371
[26] Canham L T 1990 *Appl. Phys. Lett.* **57** 1046
[27] Bisi O, Ossicini S and Pavesi L 2000 *Surface Science Reports* **38** 1
[28] Ben Rabha M, Dimassi W, Gaid M, Ezzaouia H and Bessais B 2011 *Physica Status Solidi (c)* **8** 1874
[29] Hurst H E 1951 *Trans. Am. Soc. Civ. Eng.* **116** 770
[30] Peng C K, Buldyrev S , Goldberger A , Havlin S, Sciortino F, Simons M and Stanley H E 1992 *Nature* **356** 168
[31] Gao-Feng Gu and Wei-Xing Zhou 2006 *Phys. Rev. E* **74** 061104
[32] Kantelhardt J W, Zschiegner S A, Koscielny-Bunde E , Havlin S, Bunde A, Stanley H E, 2002 *Physica A* **316** 87
[33] Hu K, Ivanov P C, Chen Z, Carpena P and Stanley H E 2001 *Phys. Rev. E* **64** 011114
[34] Struzik Z R, Siebes A P J M 2002 *Physica A* **309** 388
[35] Arrault J, Arneodo A , Davis A and Marshak A 1997 *Phys. Rev. Lett.* **79** 75
[36] Kantelhardt J W, Roman H E and Greiner M 1995 *Physica A* **220** 219
[37] Roman H E, Kantelhardt J W and Greiner M 1996 *Europhys. Lett.* **35** 641

Monte Carlo simulation of InAs HEMTs considering strain and quantum confinement effects

Akira Endoh[1,2], Issei Watanabe[1], Akifumi Kasamatsu[1] and Takashi Mimura[1,2]

[1] National Institute of Information and Communications Technology (NICT)
4-2-1 Nukui-Kitamachi, Koganei, Tokyo 184-8795, Japan
[2] Fujitsu Laboratories Ltd., 10-1 Morinosato-Wakamiya, Atsugi, Kanagawa 243-0197, Japan

E-mail: aendoh@nict.go.jp

Abstract. We carried out Monte Carlo simulation of $In_{0.53}Ga_{0.47}As$/strained-InAs/$In_{0.53}Ga_{0.47}As$ composite channel high electron mobility transistors (HEMTs) considering strain and quantum confinement effects in very thin InAs layer. We calculated the unstrained and the strained band structures of InAs. We also considered the self-consistent analysis of 2-dimensional electron gas (2DEG) by solving Schrödinger and Poisson equations. With considering the effect of 2DEG, the drain-source current I_{ds} decreases. However, the negative threshold voltage shift due to the short-channel effects is not affected by considering 2DEG. The threshold voltage shift occurs in the region $L_g/d < \sim 5$ (L_g: gate length, d: sum of the barrier and channel layer thicknesses). We also obtained the cutoff frequency f_T values. At a gate length L_g of 20 nm, the calculated f_T values were 943 GHz without 2DEG and 813 GHz with 2DEG. The trend of the f_T values with L_g reflects that of the electron velocities mainly. The present simulation results indicate that the record f_T might be obtained by reducing L_g to around 20 nm for $In_{0.53}Ga_{0.47}As$/strained-InAs/$In_{0.53}Ga_{0.47}As$ channel HEMTs.

1. Introduction

InP-based $In_{0.52}Al_{0.48}As/In_xGa_{1-x}As$ ($x \geq 0.53$) high electron mobility transistors (HEMTs) are one of the most promising devices for future ultrahigh-speed applications because of their high electron mobilities, high electron velocities and high sheet-electron densities. At present, the record cutoff frequency f_T of field effect transistors (FETs) is 688 GHz for a 40-nm-gate $In_{0.7}Ga_{0.3}As$ channel HEMT [1]. On the other hand, the record f_T value for any type of transistor is 765 GHz for a pseudomorphic InP/InGaAs heterojunction bipolar transistor (PHBT) [2]. To achieve higher-speed operations of HEMTs, using an InGaAs/strained-InAs/InGaAs composite layer as a channel is one of the most effective methods [3]. In our previous work [4], we reported Monte Carlo (MC) simulation results of InGaAs/InAs/InGaAs composite channel HEMTs without considering strain and quantum confinement effects in very thin InAs layer.

In this work, we calculated the band structures of the unstrained and the strained InAs and carried out MC simulation of $In_{0.53}Ga_{0.47}As$/strained-InAs/$In_{0.53}Ga_{0.47}As$ composite channel HEMTs considering 2-dimensional electron gas (2DEG) self-consistent analysis by solving Schrödinger and Poisson equations. For $In_{0.53}Ga_{0.47}As$/strained-InAs/$In_{0.53}Ga_{0.47}As$ channel HEMTs, very thin InAs layer must be used to prevent degradation of crystal quality due to the difference in the lattice constant between InAs and $In_{0.53}Ga_{0.47}As$ (about 3%). Therefore, strain and quantum confinement effects are indispensable to theoretical analyses such as MC simulation for InAs layer. We examined the effect of the gate length L_g on DC and RF performances in the HEMTs with and without considering the effect of 2DEG.

2. Band structures of unstrained and strained InAs

The unstrained and the strained band structures of InAs were calculated by using all-electron full-potential linearized augmented-plane-wave (FLAPW) method in the local density approximation (LDA). The computational code ABCAP (All-electron Band-structure CAlculation Package) used was originally developed by Kodama, Hamada and Yanase [5]. Figure 1 shows the conduction band structures of the unstrained and the strained InAs. By applying -3% biaxial compressive strain to [100] and [010] directions in *x*-*y* plane, the increase of the electron effective mass in the Γ-valley and the decrease of the Γ-L and the Γ-X valley energy separations were observed. We obtained the electron effective mass in the Γ-valley along the [100] ([010]) direction to be $0.032m_0$ for the unstrained InAs and $0.038m_0$ for the -3% (compressive) strained InAs with considering nonparabolicity, where m_0 is electron rest mass. The electron effective mass in the Γ-valley for the unstrained InAs is almost same as that calculated by the empirical pseudopotential method ($0.031m_0$) [6, 7]. On the other hand, the electron effective mass for the -3% (compressive) strained InAs is lighter than that by the pseudopotential method ($0.043m_0$) [7]. Note that the electron effective mass in the unstrained n-InAs deduced from magnetophonon resonance is $0.0219m_0$ at 250 K [8]. The details of the calculation results will be described elsewhere [9].

3. Monte Carlo simulation

MC simulations were carried out by using the program "COSMOS," developed by Mizuho Information & Research Institute, Inc. [10]. Figure 2 shows a schematic cross-sectional model structure of the composite channel HEMT used in the present MC simulations. We used a three-valley model (Γ, L, X) with nonparabolicity for the conduction band structures of InAs, $In_{0.53}Ga_{0.47}As$, $In_{0.52}Al_{0.48}As$ and InP layers. $In_{0.53}Ga_{0.47}As$ and $In_{0.52}Al_{0.48}As$ layers are lattice-matched to InP layer. InAs layer is compressively strained. We used the unstrained and the strained band structures of InAs obtained in section 2. The band parameters of AlAs, GaAs and InP were taken from the literature [6,

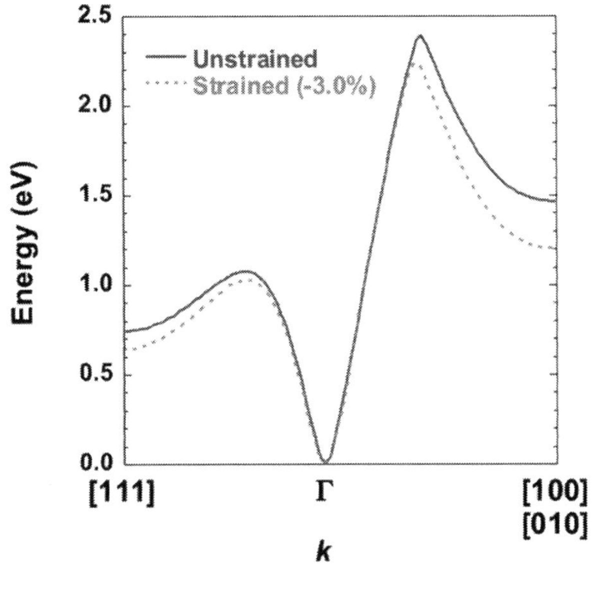

Figure 1. Conduction band structures of unstrained and strained InAs.

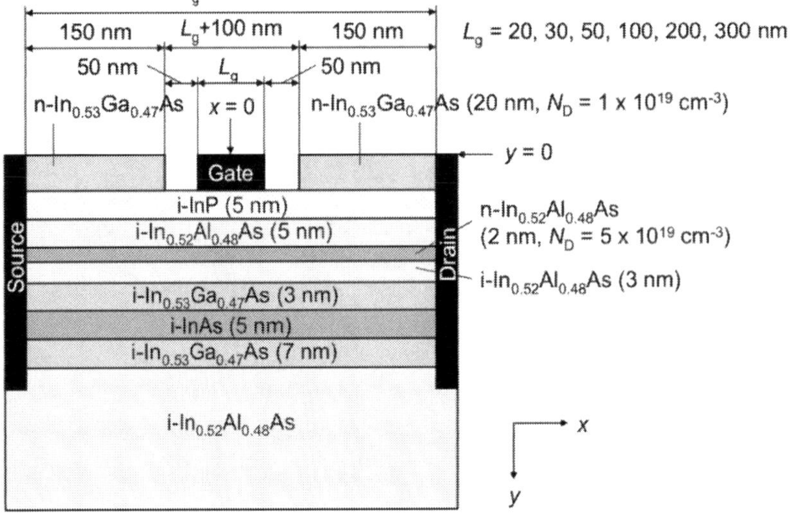

Figure 2. Schematic cross-sectional model structure of InGaAs/strained-InAs/InGaAs composite channel HEMT.

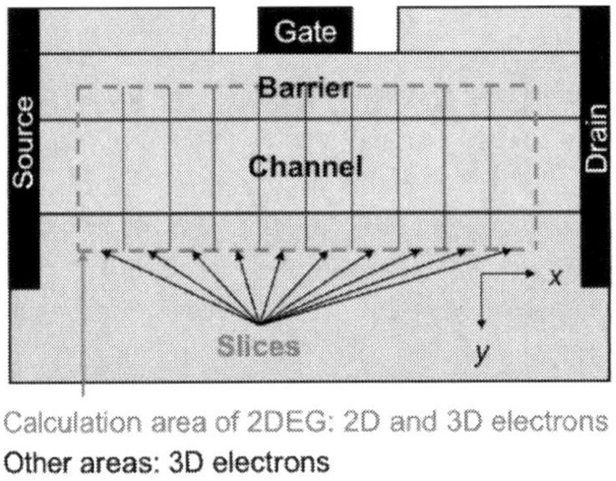

Figure 3. Schematic description of the calculation area of 2DEG.

11]. Figure 3 shows a schematic description of the calculation area of 2DEG. We considered the 2DEG in and near the channel layer by solving the 1-dimensional Schrödinger equation along y-axis at various x-positions [12]. The calculation slices in figure 3 were dense in the region with high electric field. We considered the lowest three quantum levels in the channel. The electrons above the highest quantum level energy were treated as 3-dimensional electrons. The electron scattering mechanisms [13, 14] considered were polar optical phonon scattering, non-polar optical phonon scattering, acoustic phonon scattering, inter-valley phonon scattering and ionized impurity scattering. We did not consider impact ionization in the present simulations since the previous MC simulation results [15] show that the influence of impact ionization is relatively small for the InGaAs/strained-InAs/InGaAs channel HEMT in the drain-source voltage V_{ds} range between 0 and 0.8 V. We did not consider electron-electron interactions either. Sub-band scattering was considered for 2DEG. Dirichlet boundary conditions were applied to all metal-semiconductor interfaces, and Neumann boundary conditions (the zero normal derivative of the potential) were applied to other surfaces. The potential was calculated by the finite difference method. The time step was set to 0.5 fs. The gate length L_g was varied from 20 to 300 nm. All simulations were carried out at a lattice temperature of 300 K.

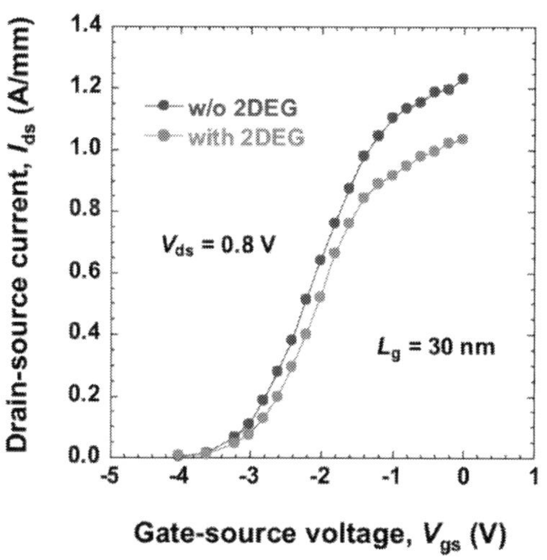

Figure 4. Drain-source current vs. gate-source voltage (I_{ds}-V_{gs}) characteristics of 30-nm-gate HEMTs with and without 2DEG. The drain-source voltage V_{ds} is 0.8 V.

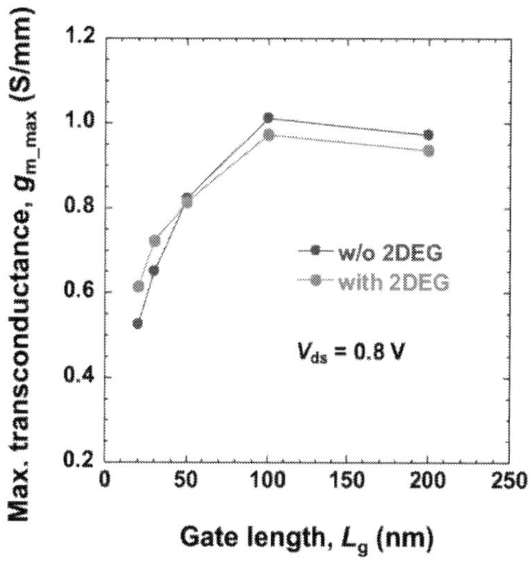

Figure 5. Gate length L_g dependence of the maximum transconductance g_{m_max} under a drain-source voltage V_{ds} of 0.8 V.

4. Results and discussion

4.1. DC performance

Figure 4 shows the drain-source current vs. gate-source voltage (I_{ds}-V_{gs}) characteristics of a 30-nm-gate HEMT with and without considering the effect of 2DEG under a drain-source voltage V_{ds} of 0.8 V. The I_{ds} with 2DEG is lower than that without 2DEG. With considering the effect of 2DEG, the reductions of electron velocity and electron density in the channel occur due to the quantum levels in the channel. Figure 5 shows the gate length L_g dependence of the maximum transconductance g_{m_max} under a V_{ds} of 0.8 V. The g_{m_max} increases with decreasing L_g in the region $L_g > \sim 100$nm, and decreases in the region $L_g < \sim 100$ nm. The decrease of g_{m_max} in the region $L_g < \sim 100$ nm results from the "so-called" short-channel effects [16-18].

To clarify the short-channel effects on performances further, we obtained the threshold voltage shift in the I_{ds}-V_{gs} curves. Figure 6 shows the channel aspect ratio L_g/d dependence of the threshold voltage shift. In this work, d was defined as the sum of the barrier and channel layer thicknesses. The threshold voltage shift was defined as the difference from the threshold voltage V_{th} at $L_g/d = 10$ (L_g = 300 nm). The negative threshold voltage shift due to the short-channel effects occurs in the region $L_g/d < \sim 5$. This phenomenon is similar to those for AlGaAs/GaAs [16], InAlAs/InGaAs [17] and AlGaN/GaN [18] HEMTs. The difference between with and without 2DEG is very small. Thus, the same scaling rule is valid for the InGaAs/strained-InAs/InGaAs composite channel HEMTs.

4.2. Cutoff frequency

The cutoff frequency f_T values were calculated by the method presented by Kwon and Pavlidis [19]. Figure 7 shows the gate length L_g dependence of the cutoff frequency f_T under a V_{ds} of 0.8 V. The f_T increases with decreasing L_g in the present simulation range. The f_T's without 2DEG are higher than those with 2DEG. At an L_g of 20 nm, f_T without 2DEG is 943 GHz. On the other hand, f_T with 2DEG is 813 GHz. Thus, the record f_T might be obtained by reducing L_g to around 20 nm for InGaAs/strained-InAs/InGaAs channel HEMTs [2].

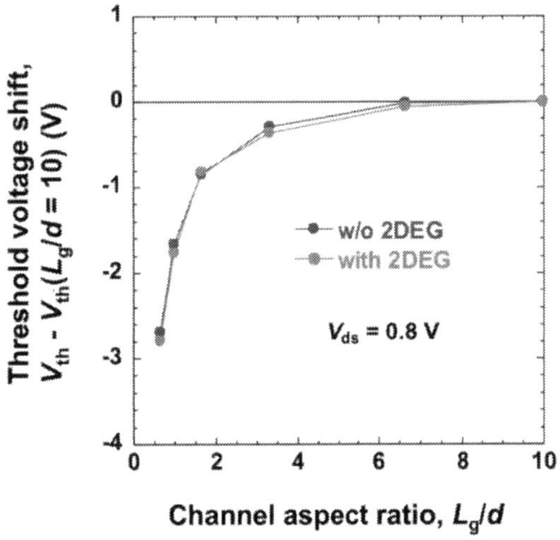

Figure 6. Channel aspect ratio L_g/d dependence of the threshold voltage shift under a drain-source voltage V_{ds} of 0.8 V.

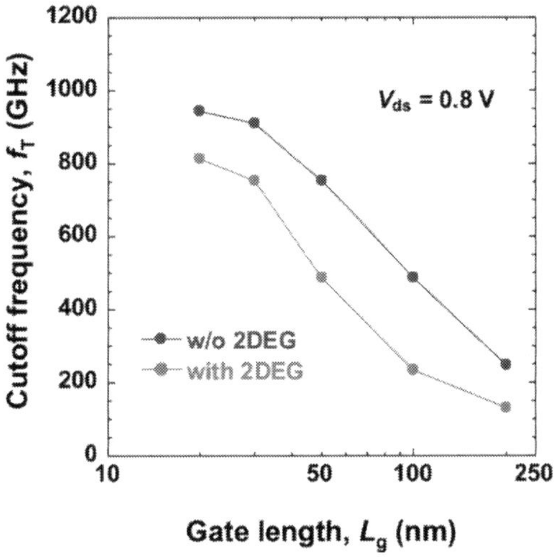

Figure 7. Gate length L_g dependence of the cutoff frequency f_T with and without 2DEG under a drain-source voltage V_{ds} of 0.8 V.

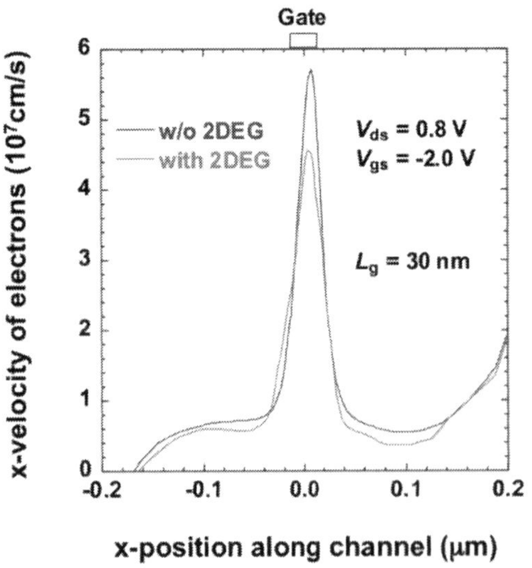

Figure 8. Electron velocity profiles in the channel layer for 30-nm-gate HEMTs with and without 2DEG. The drain-source voltage V_{ds} is 0.8 V, and the gate-source voltage V_{gs} is -2.0 V.

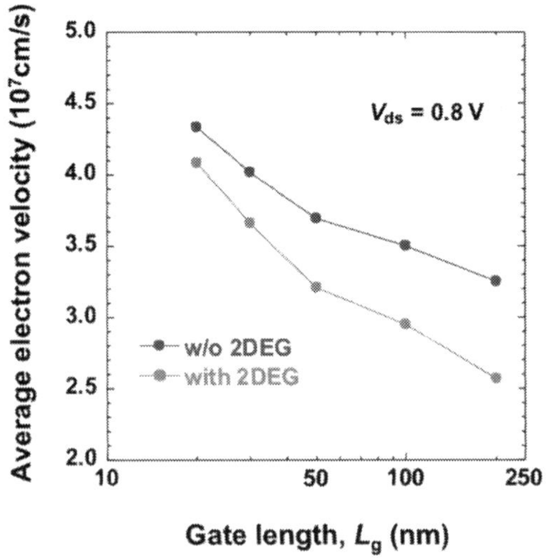

Figure 9. Gate length L_g dependence of the average electron velocity in the channel layer under the gate electrode. The drain-source voltage V_{ds} is 0.8 V.

To understand the trend of the cutoff frequency f_T in detail, we obtained the electron velocity in the channel layer. The electron velocity was obtained by taking the average for the whole channel depth. Figure 8 compares the electron velocity profiles in the channel layer of the 30-nm-gate HEMTs with and without considering the effect of 2DEG under a V_{ds} of 0.8 V and a V_{gs} of -2.0 V. There is a velocity overshoot under the gate electrode. In the almost whole region, the electron velocity without 2DEG is higher than that with 2DEG. The peak electron velocities are about 5.7×10^7 cm/s without 2DEG and about 4.6×10^7 cm/s with 2DEG. Figure 9 shows the gate length L_g dependence of the average electron velocity in the channel layer under the gate electrode under a V_{ds} of 0.8 V. The V_{gs} values are the voltages with the g_{m_max} values. The average electron velocity increases with decreasing L_g in the present simulation range. The increase of average electron velocity with decreasing L_g results from the increase of the electric field under the gate electrode. On the other hand, the average velocity without 2DEG is higher than that with 2DEG. The trend of the f_T values with L_g reflects that of the electron velocities mainly.

Figure 10 shows the potential profiles of the 30-nm-gate HEMTs without and with 2DEG under a V_{ds} of 0.8 V and a V_{gs} of -2.0 V. There is a negligible difference in the potential profiles between with and without 2DEG. Therefore, the difference in electron velocity between with and without 2DEG does not result from the difference in electric field. Figure 11 shows the electron density distribution along y-axis at $x = 0$, i.e. the centre of the gate electrode (See figure 2), of the 30-nm-gate HEMTs. As clearly seen from figure 11, the electron density in the InAs layer without 2DEG is higher than that with 2DEG. The occupancy of electrons in the InAs layer is about 80% without 2DEG. On the other hand, the occupancy is about 65% with 2DEG. Therefore, the higher electron velocity without 2DEG results from the higher occupancy of electrons in the InAs layer in which the electron effective mass is lighter.

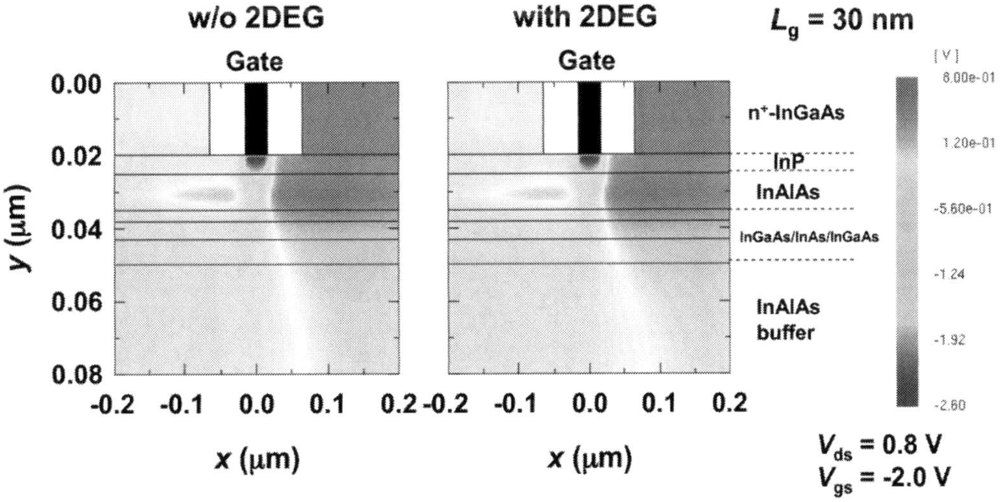

Figure 10. Potential profiles for 30-nm-gate HEMTs with and without 2DEG.

4.3. Comments on strain effects

So far, we discussed the quantum confinement effects on the performances in the InGaAs/strained-InAs/InGaAs composite channel HEMTs. In our MC simulations, the compressive strain is applied to the InAs layer. For the InGaAs/InAs/InGaAs channel HEMTs with and without the strain of the InAs layer, Machida *et al.* [15] reported that the maximum transconductance g_{m_max} without strain is higher than that with strain for 30-nm-gate HEMTs. The cutoff frequency f_T can be calculated by the equation, $f_T = g_m/(2\pi C_g)$, where C_g is the gate capacitance. At each gate length, the C_g value is approximately constant. Therefore, the cutoff frequency f_T without strain might be higher than that with strain. The MC simulation results of strain effects will be described elsewhere [20].

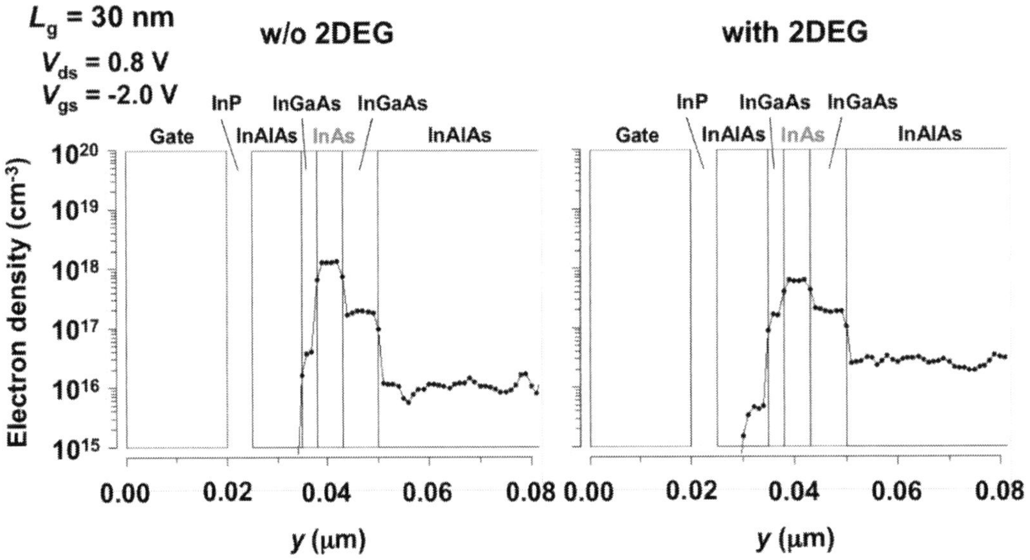

Figure 11. Electron density along y-axis at x = 0 for 30-nm-gate HEMTs with and without 2DEG.

5. Summary

In summary, we carried out MC simulation of $In_{0.53}Ga_{0.47}As$/strained-InAs/$In_{0.53}Ga_{0.47}As$ composite channel HEMTs considering strain and quantum confinement effects in very thin InAs layer to examine the effect of L_g on DC and RF performances. We calculated the unstrained and the strained band structures of InAs by using all-electron FLAPW method in the LDA. We also considered the 2DEG self-consistent analysis by solving Schrödinger and Poisson equations. The I_{ds} with 2DEG is lower than that without 2DEG due to the reductions of electron velocity and electron density. The reduction of the g_{m_max} due to the short-channel effects was observed in the region L_g < ~100 nm. To clarify the short-channel effects, we obtained the threshold voltage shift in the I_{ds}-V_{gs} curves. The negative threshold voltage shift due to the short-channel effects is not affected by considering 2DEG. The trend of the threshold voltage shift is similar to those of AlGaAs/GaAs, InAlAs/InGaAs and AlGaN/GaN HEMTs, i.e. the threshold voltage shift occurs in the channel aspect region L_g/d < ~5. We also obtained the cutoff frequency f_T values by the method presented by Kwon and Pavlidis. The f_T values at L_g = 20 nm are 943 GHz without 2DEG and 813 GHz with 2DEG. The trend of the f_T values with L_g reflects that of the electron velocities mainly. Thus, the record f_T might be obtained by reducing L_g to around 20 nm for $In_{0.53}Ga_{0.47}As$/strained-InAs/$In_{0.53}Ga_{0.47}As$ channel HEMTs.

Acknowledgment

This work was supported in part by "The research and development project for the expansion of radio spectrum resources" of the Ministry of Internal Affairs and Communications, Japan.

References

[1] Kim D-H, Brar B and del Alamo J A 2011 f_T = 688 GHz and f_{max} = 800 GHz in L_g = 40 nm $In_{0.7}Ga_{0.3}As$ MHEMTs with g_{m_max} > 2.7 mS/μm *IEDM Tech. Dig.* p 319

[2] Snodgrass W, Hafez W, Harff N and Feng M 2006 Pseudomorphic InP/InGaAs Heterojunction Bipolar Transistors (PHBTs) Experimentally Demonstrating f_T = 765 GHz at 25°C Increasing to f_T = 845 GHz at -55°C *IEDM Tech. Dig.* 22.1

[3] Kim D-H and del Alamo J A 2010 30-nm InAs PHEMTs With f_T = 644 GHz and f_{max} = 681 GHz *IEEE Electron Device Lett.* **31** 806

[4] Endoh A, Watanabe I, Hirose N, Mimura T and Matsui T 2010 Monte Carlo Simulations of Electron Transport in Nanogate $In_{0.7}Ga_{0.3}As/InAs/In_{0.7}Ga_{0.3}As$ Composite Channel HEMTs *Abst. 37th Int. Symp. Comp. Semicon.* p 30

[5] Umekawa M, Hamada N, Kodama A and Moritomo Y 2004 Electronic Structure of $RbMnFe(CN)_6$: Ground State *J. Phys. Soc. Jpn.* **73** 430
The code MIZUHO/ABCAP is supplied from Mizuho Information & Research Institute, Inc.

[6] Fischetti M V 1991 Monte Carlo Simulation of Transport in Technologically Significant Semiconductors of the Diamond and Zinc-Blende Structures — Part I: Homogeneous Transport *IEEE Trans. Electron Devices* **38** 634

[7] Nishino H, Kawahira I, Machida F, Hara S and Fujishiro H I 2010 Monte Carlo Study of Strain Effect on High Field Electron Transport in InAs and InSb *Proc. 22nd Int. Conf. Indium Phosphide Related Mat.* p 156

[8] Takayama J, Shimomae K and Hamaguchi C 1981 Temperature Dependence of the Band-Edge Effective Mass in n-InAs Deduced from Magnetophonon Resonance *Jpn. J. Appl. Phys.* **20** 1265

[9] Endoh A, Watanabe I, Kasamatsu A and Mimura T, to be published

[10] http://www.mizuho-ir.co.jp/solution/research/semiconductor/devicemeister/montecarlo/

[11] Madelung O 1996 *Semiconductor — Basic Data*, 2nd ed. (Berlin: Springer)

[12] Fukada T, Tanino N, Nakazato M and Kohyama Y 1997 A Self-Consistent Analysis of Two-Dimensional Electron Gas in HEMT Structure Using Monte Carlo Simulation *Technical Report of IEICE* **97**(269) 1 [in Japanese]

[13] Tomizawa K 1996 *Semiconductor Device Simulation — Visualization of the Operation of Submicron Semiconductor Devices by Simulation* — (Tokyo: Corona) [in Japanese]

[14] Jacoboni C and Lugli P 1989 *The Monte Carlo Method for Semiconductor Device Simulation* (Heidelberg: Springer)

[15] Machida F, Nishino H, Sato J, Watanabe H, Hara S and Fujishiro H I 2011 Strain Effects on Performances in InAs HEMTs *Proc. 23rd Int. Conf. Indium Phosphide Related Mat.* p 437

[16] Awano Y, Kosugi M, Kosemura K, Mimura T and Abe M 1989 Short-Channel Effects in Subquarter-Micrometer-Gate HEMT's: Simulation and Experiment *IEEE Trans. Electron Devices* **36** 2260

[17] Endoh A, Yamashita Y, Shinohara K, Higashiwaki M, Hikosaka K, Mimura T, Hiyamizu S and Matsui T 2002 Fabrication Technology and Device Performance of Sub-50-nm-Gate InP-Based High Electron Mobility Transistors *Jpn. J. Appl. Phys.* **41** 1094

[18] Endoh A, Yamashita Y, Hikosaka K, Matsui T, Hiyamizu S and Mimura T 2006 Threshold voltage shifts in decananometre-gate AlGaN/GaN HEMTs *Electron. Lett.* **42** 490

[19] Kwon Y and Pavlidis D 1996 Delay Time Analysis of Submicron InP-Based HEMT's *IEEE Trans. Electron Devices* **43** 228

[20] Endoh A, Watanabe I, Kasamatsu A and Mimura T, to be published

Conformational analysis investigation into the influence of nano-porosity of ultra-permeable ultra-selective polyimides on its diffusivity as potential membranes for use in the "green" separation of natural gases

Tarek M Madkour
The American University in Cairo (AUC), School of Sciences and Engineering, Department of Chemistry, AUC Avenue, P.O. Box 74, New Cairo 11835, Egypt

E-mail: tarek.madkour@aucegypt.edu

Abstract. Nano-porous polymers of intrinsic microporosity, PIM, have exhibited excellent permeability and selectivity characteristics that could be utilized in an environmentally friendly gas separation process. A full understanding of the mechanism through which these membranes effectively and selectively allow for the permeation of specific gases will lead to further development of these membranes. Three factors obviously influenced the conformational behavior of these polymers, which are the presence of electronegative atoms, the presence of non-linearity in the polymeric backbones (backbone kinks) and the presence of bulky side groups on the polymeric chains. The dipole moment increased sharply with the presence of backbone kinks more than any other factor. Replacing the fluorine atoms with bulky alkyl groups didn't influence the dipole moment greatly indicating that the size of the side chains had much less dramatic influence on the dipole moment than having a bent backbone. Similarly, the presence of the backbone kinks in the polymeric chains influenced the polymeric chains to assume less extended configuration causing the torsional angles around the interconnecting bonds unable to cross the high potential energy barriers. The presence of the bulky side groups also caused the energy barriers of the cis-configurations to increase dramatically, which prevented the polymeric segments from experiencing full rotation about the connecting bonds. For these polymers, it was clear that the fully extended configurations are the preferred configurations in the absence of strong electronegative atoms, backbones kinks or bulky side groups. The addition of any of these factors to the polymeric structures resulted in the polymeric chains being forced to assume less extended configurations. Rather interestingly, the length or bulkiness of the side groups didn't affect the end-to-end distance distribution to a great deal since the presence of quite large bulky side chain such as the pentyl group has caused the polymeric chains to revert back to the fully extended configurations possibly due to the quite high potential energy barriers that the chains have to cross to reach the less extended configurational states.

1. Introduction

Nanoporous materials are characterized with small pore sizes whose diameters can be adjusted and therefore can be used for gas mixture separations depending on their size exclusion [1]. It was also mentioned that the lower energy and cost needed for membrane-based gas separation techniques than traditional processes made the membrane methods more preferred [2,3]. It was illustrated that the membranes must have certain properties such as high permeability, high selectivity, mechanical stability and low cost to be useful for gas mixture separation such as hydrogen and nitrogen gas mixture and the purification of natural gas through the removal of carbon dioxide [4]. Silicon rubber, cellulose acetate, polysulfone materials are all known to be used in case of O_2/N_2 separation, however polyimides are used for both O_2/N_2 and CO_2/CH_4 separations. Both the diffusivity and solubility parameters control the diffusion mechanism at which the mobility of the molecules is reflected by its diffusivity. It was claimed that the increase in the molecular size of the gas leads to reduction in the diffusion coefficient, and increasing in the solubility coefficient [5]. Enhancement in the diffusivity or the solubility leads to a decrease in the membrane selectivity since increasing the permeability occurs on the expense of the selectivity of the membranes. Glassy polymers of microporosity nature achieve high permeability/selectivity ratios that are close to the upper bound in the separation processes of simple gases, at which they allow the diffusion of smaller molecules more rapidly than larger ones. It was claimed [5] that the high selectivity and good mechanical properties are the reasons for using membranes formed by glassy polymers in gas separations. Rubbery polymers, however, are shown to be more effective in isolating bulky organic molecules from smaller gas molecules since rubbery membranes have high permeability with low selectivity.

Production of polymeric membranes with both high permeability and selectivity while maintaining long-term stability is considered a challenge. Polyimides of intrinsic microporosity are such examples [6]. It was claimed that the rigid and contorted macrostructures that are not able to pack space efficiently are the reason for their microporosity and high gas permeability, at which the rigidity is obtained by fused ring structure and the sites of contortion are made by spirocentres. It was also reported that these nanoporous polymeric materials are similar to glassy polymers but with higher solubility coefficient [7] which subsequently allows for permeability enhancement without selectivity loss. It is believed that these materials are excellent candidates as separation membranes for their high permeability, good selectivity and long thermal stability.

Polymers of intrinsic microporosity (PIMs) that were first created by Mckeown and Budd [8] were synthesized from rigid and bent monomers which have a tetrahedral carbon atom known as a "site of contortion" responsible for the porosity in PIMs. This site of contortion led to the observed free volume between the polymer chains. They also illustrated the different ways by which PIMs can be synthesized. Dioxine-forming reaction between an orthodihydroxy monomer and an orthodihalide monomer can be used to produce linear polymer. Interestingly, a network can be also synthesized if one of the monomers has more than two pairs of halides. Early microporous PIM networks were based on phthalocyanines and porphyrins and enclosed metal ions or 2H in the cavities. More Recent PIM networks were synthesized from a hexaazatrinaphythylene monomer, which made metal binding possible and contributed to greater mass that will eventually lead to a decrease in the surface area. Other PIMs could be synthesized by incorporating bis(phenazyl)s, fused fluorenes and imides, which led to a variety of gas permeability characteristics.

PIMs synthesized from rigid twisted spirobiflurones through imide- and amide-forming reactions showed the highest surface area and rigidity. The surface areas were somewhat low when the imide-forming reactions formed PIMs with binaphthalene as the site of contortion. Advanced polyimide forming reactions including binaphthalene and 4-4`-(9-fluorenylidene) dianiline were done and when the polymers were used as films, they exhibited selective CO_2/H_2 gas uptake. Hyper cross-linked polymers (HCPs) such as cross-linked polystyrene otherwise known as Davankov resins were created by the same way resulting in highly rigid networks that are characterized by small pore sizes, high surface areas and high resistance to collapsing [9]. They can be formed by the polymerization of vinylbenzyl chloride with a small amount of divinylbenzene cross-linker to form a lightly cross-linked copolymer. The copolymer is then allowed to be "hyper cross-linked" via a Fidel-craft alkylation reaction using a Lewis acid such as iron (III) chloride after being swollen in an appropriate solvent [10]. Germain et al. [11] created hyper crosslinking of linear polyaniline and polypyrole precursors, without requiring the use of lithiating agents or huge amounts of Lewis acids and consequently, it does not produce large volumes of HCl gas [12]. Polyaniline can be cross-linked using Ullmann and Buchwald coupling reaction, leading to the formation of networks connected by aromatic groups. They also highlighted that thermodynamic compatible solvents used with the precursor led to faster desolvation during crosslinking. To determine the best solvent for a certain crosslinking process,

Table 1. Different molecular structures of the PIM-polyimides under investigation.

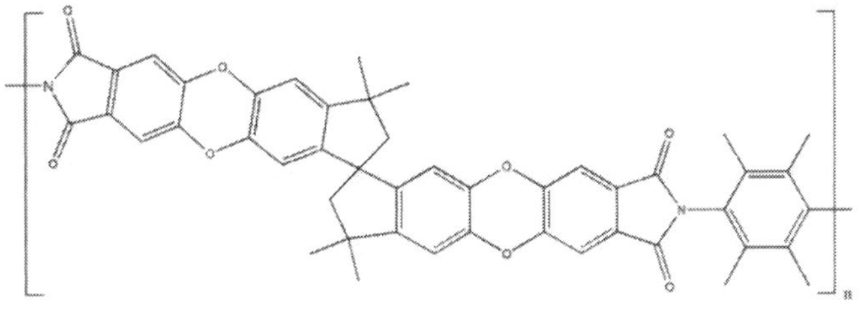

PIM-1

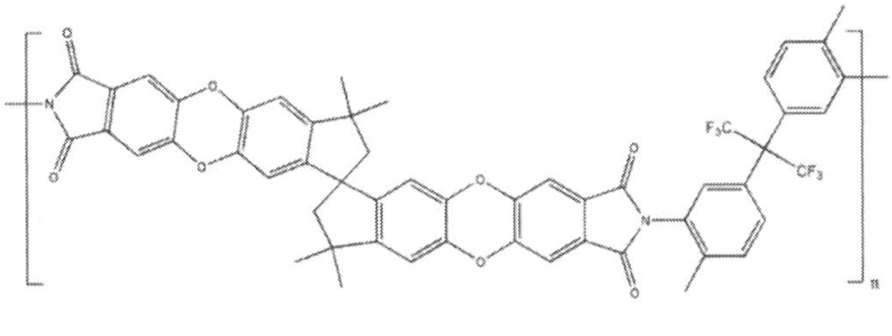

PIM-2

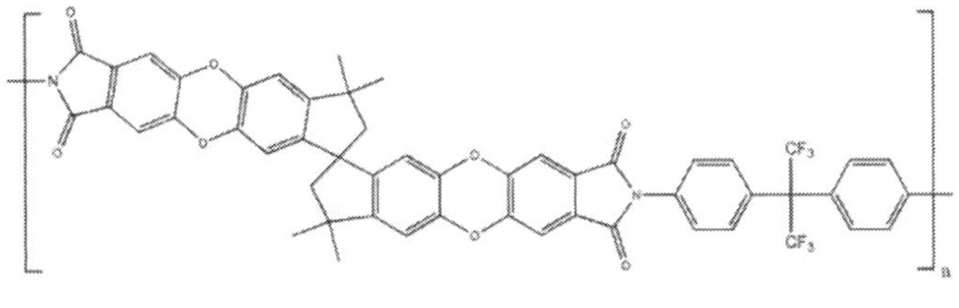

PIM-3

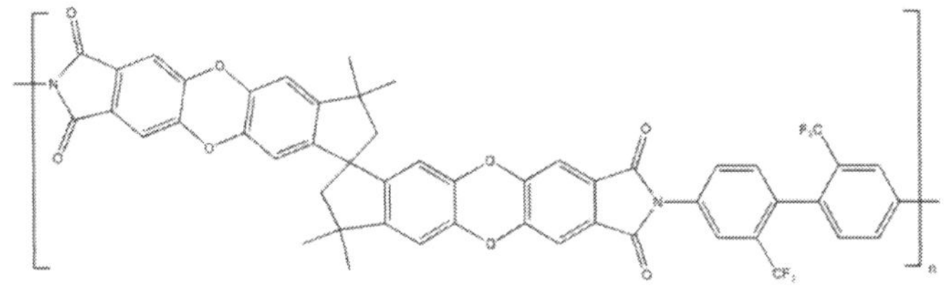

PIM-4

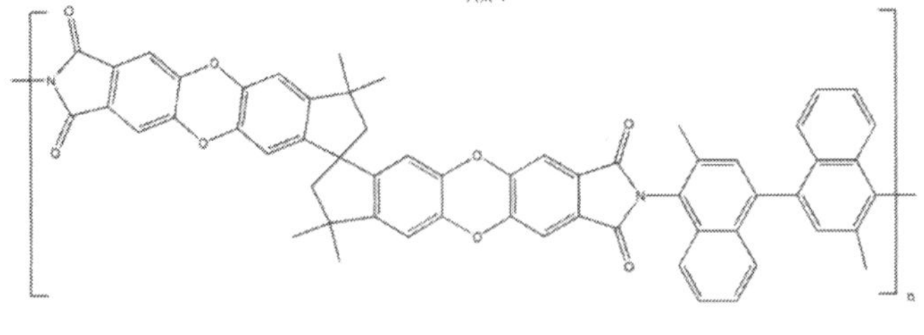

PIM-8

Hildebrand and Hansen parameters were used. Compatible monomer/solvent systems have similar values for the Hildebrand and Hansen parameters. The performed reaction should have fast chemical kinetics and results in the least amount of non-crosslinked by-products. Finally, there should be no chemical reaction between the reactant and the solvent since they may lead to the formation of two intermixed polymer networks [13].

2. Simulation methodology

2.1 The chemical structures of the investigated polymers

The polymeric structures simulated in this article are polyimides that are commonly prepared by a step polymerization involving a cycloimidization reaction between a bis(carboxylic anhydride) and a diamine. The investigated polyimides were synthesized and characterized by Ghanem et al [14]. The polymers showed excellent selectivity and permeability characteristics due to the high restriction for rotation around the imide linkage. According to the authors [14], to create even more permeable polyimides, the PIM concept was applied. A dianhydride was prepared that incorporated a spiro-centre, referred to as a centre of contortion. Table 1 illustrates the chemical structure of five different PIMs (PIM-1, PIM-2, PIM-3, PIM-4 and PIM-8) that were prepared and characterized experimentally by Ghanem et al [14] and chosen here to examine the influence of the polymeric chemical structure on the conformational behavior of the investigated polymer in relation to its experimentally determined permeability.

2.2 Choice of force field

All calculations were performed using Materials Studio® Package from Accelrys, inc. The quality of the forcefield, which consists of a set of analytical function terms and parameters, and provides a detailed description of the inter- and intramolecular interactions as a function of the molecular structures and their relative positions influences directly the reliability of the simulations to a great deal. The simulation of the different polymeric systems has been carried out using the COMPASS forcefield. The COMPASS forcefield is a CFF-type forcefield [15] constructed using exact ab-initio quantum mechanics calculations and developed to improve upon existing forcefields. In the forcefield approach, the potential energy of a system is expressed as a sum of valence and non-bonded interactions:

$$E_{total} = E_{valence} + E_{nonbond} = (E_{bond} + E_{angle} + E_{torsion} + E_{inv}) + (E_{vdW} + E_{Coulomb}) \quad (1)$$

The valence terms parameterize the short-range intramolecular interactions that determine the bond lengths (E_{bond}), bond angles (E_{angle}), torsion angles ($E_{torsion}$) and out-of-plane inversion angles (E_{inv}). The nonbonded interactions, which can be both intra- and intermolecular, are split into a dispersive van der Waals term (E_{vdW}), and electrostatic term ($E_{Coulomb}$). Other cross-terms are also important and well accounted for by the COMPASS forcefield.

2.3 Conformational Analysis

Various conformational properties of the investigated polymeric systems evaluated using RIS Metropolis Monte Carlo methods, RMMC, available through the Conformer module® as part of Materials Studio® Package from Accelrys, inc. [16-18]. Unlike traditional rotational isomeric states methods, RMMC methods allow torsion angles to vary continuously and therefore do not impose the assumption of discrete rotational states. The method is thus based on relative energies of the new and the old conformation. The bond-based cut-off method was used for the simulations of the single polymeric chains in the θ-condition. In this method, the parameters n_{min} and n_{max} affect the way van der Waals and Coulomb energies are evaluated in the RMMC simulation. The nonbond energies were not computed for atoms that are closer than n_{min} or further apart than n_{max}. The usual value for n_{min} was 3 whereas that of n_{max} was taken as 6 in order to include the interactions of the side chains of the neighbouring repeat units in the energy evaluation.

Table 2 lists the chemical structures of new compounds that have not been synthesized experimentally yet but envisaged to have good permeability characteristics and may assist in the further understanding through the conformational analysis of the origin of microporosity in these PIM-polyimides.

Table 2. Different molecular structures of new envisaged PIM-polyimides.

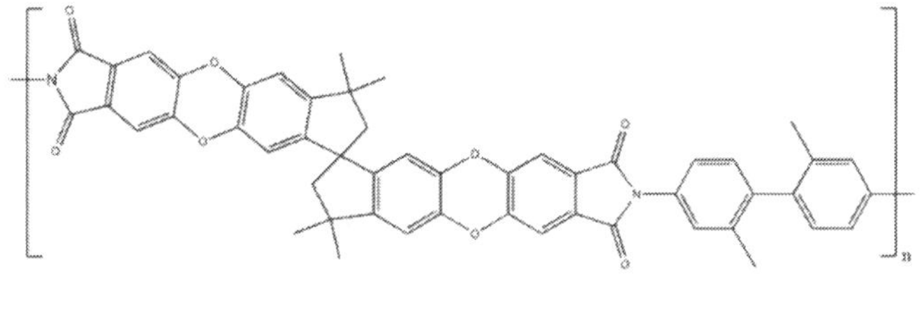

PIM-4-CH_3

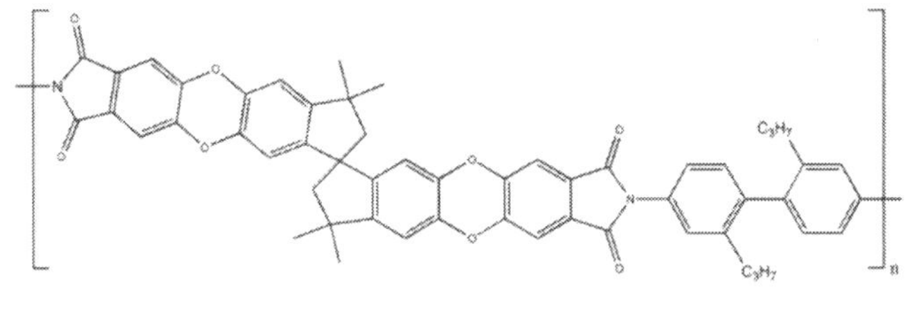

PIM-4-C_3H_7

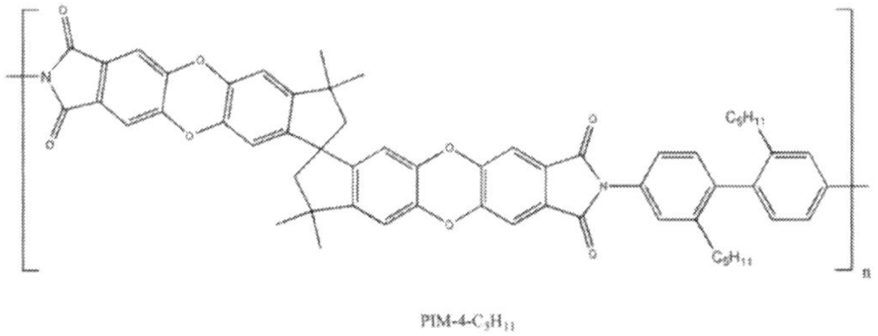

PIM-4-C_5H_{11}

RMMC simulations were done for polymers with 100 repeat units at room temperature. The number of the equilibration steps was 10,000 for each rotatable bond. This stage was necessary to bring the polymer chain from its initial conformation to a conformation typical for the chain in the melt at the desired temperature. The number of simulation steps was 50,000 steps for each rotatable bond. The block-averaged values of the mean squared end-to-end distance and energy were plotted as functions of the simulation step number in order to determine if the system was fully equilibrated. There was no evidence of any long-term drift in mean squared end-to-end distance and energy over the course of simulation, which indicated that the chains had reached a steady state. RMMC was used to calculate the conformational parameters such as the dipole moment, the end-to-end distances and the radii of gyration. The end-to-end distance, r, of a single chain is defined as:

$$r = \sum_{i=1}^{n} l_i \tag{2}$$

where l_i denotes the vector along backbone bond i, and n denotes the number of backbone bonds in the molecule. The radius of gyration, S, is defined as the root mean squared distance of the atoms in the molecule from their common centre of mass and is given by:

$$S^2 = \frac{\sum_{i=1}^{N} m_i S_i^2}{\sum_{i=1}^{N} m_i} \tag{3}$$

where S_i denotes the distance of atom i from the centre of mass, and N denotes the total number of atoms. The radial distribution function (RDF) gives a measure of the probability that, given the presence of an atom at the origin of an arbitrary reference frame, there will be an atom with its center located in a spherical shell of infinitesimal thickness at a distance r from the reference atom. The radial distribution function, $g_{\alpha\beta}(r)$, is calculated from the average of the static relationship of every given pair of particles, $\alpha\beta$, as:

$$C(m) = \frac{1}{n}\sum_{i=1}^{n} V(m+i)\times V(i) \tag{4}$$

where $\langle n_{\alpha\beta}(r)\rangle$ is the average number of atom pairs in the spherical shell between r and $r+\Delta r$, $\alpha\beta$ denotes the possibility that α and β particles may be of a different chemical nature, and $\rho_{\alpha\beta}$ is the density of atom pairs of type $\alpha\beta$. By investigating the radial distribution function, more information on the local packing influenced by specific interactions could be obtained. To evaluate the influence of contortion of these polymers on its conformational behavior, the probability distribution function of the angle of contortion around the spiro-centre is evaluated. Figure 1a shows a schematic drawing illustrating the specified angle of contortion (θ_c).

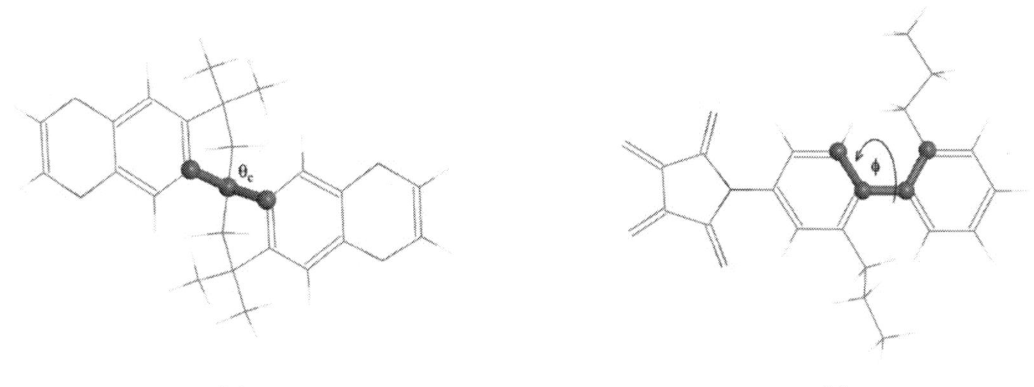

(a) (b)

Figure 1. Schematic drawing of a portion of PIM molecule illustrating (a) the angle of contortion, θ_c, and (b) the torsional angle, ϕ, evaluated for the various polymeric structures.

Furthermore the torsional angle (ϕ) about the backbone bonds connecting the aromatic rings of a PIM molecule, figure 1b, is also evaluated to test the influence of the side groups of the polymeric structure on the restriction of the rotation around the polymeric backbone axis.

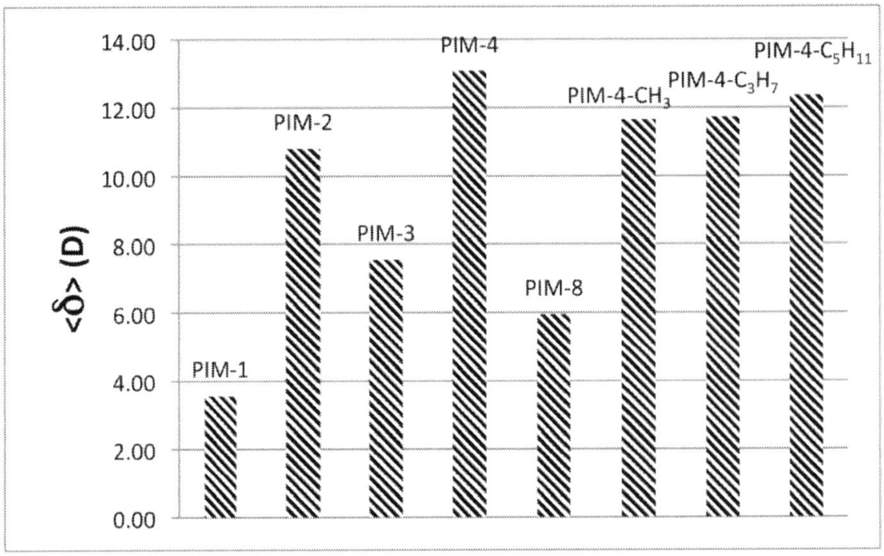

Figure 2. The influence of the molecular structure on the dipole moment of the PIM-polyimides.

3. Results and Discussion

The use of molecular simulation techniques allows for the modeling and investigation of the influences of the molecular structures on the conformational behavior of PIM-polyimides. Figure 2 illustrates the influence of the chemical structure of the various polymers listed in tables 1 and 2 on the dipole moment of these polymers.

It is obvious from the figure that the dipole moment of all the linear polymers, PIM-1, 3, 8 have low dipole moment despite the presence of the fluorine atoms in PIM-3, which obviously cancel each other to a large extent. Once the polymer exhibit non-linear backbone, the dipole moment increases sharply. Replacing in the fluorine atoms in PIM-4 with bulky alkyl groups didn't influence the dipole moment greatly indicating that the size of the side chains has much less dramatic influence than having a bent backbone such as in PIM-2 or the presence of alternating side chains on neighbouring aromatic rings.

Figure 3 illustrates the influence of the molecular structure on the radius of gyration of the PIM-polyimides. The figure clearly shows that all the PIMs have similar Rg values with the exception of PIM-2 due to the presence of the bent angle in the backbone which allows the polymeric chains to assume less extended configuration. PIM-3 also exhibited a low value for the radius of gyration, which can be explained on the basis of the presence of the tetrahedral carbon atom connecting the two phenyl group. Unlike, PIM-4 in which both phenyl rings lie on the same plane, the presence of the tetrahedral carbon atom between the two phenyl rings causes the bond angle of the C (ring 1) – C (tetrahedral atom) – C (ring 2) to be 109.5^{o} causing an out-of-plane kink and a decrease in the radius of gyration as in the case of PIM-2 but to a lesser extent.

Figure 4 illustrates the influence of the molecular structure on the probability distribution function of the torsional angle (ϕ) of PIM-1, 2, 3 and 8. It is clear from the figure that the bulkiness and the bent backbone configurations of PIM-2 and 8 has caused the torsional angle to be limited to a narrow range around a certain value and unable to cross a potential energy barrier, hence the zero-value of the probability function for these two structures. Simple and linear backbone structures such as the case of PIM-1 and 3 showed the ability of the torsional angles to visit various angle and to cross the various low potential energy values.

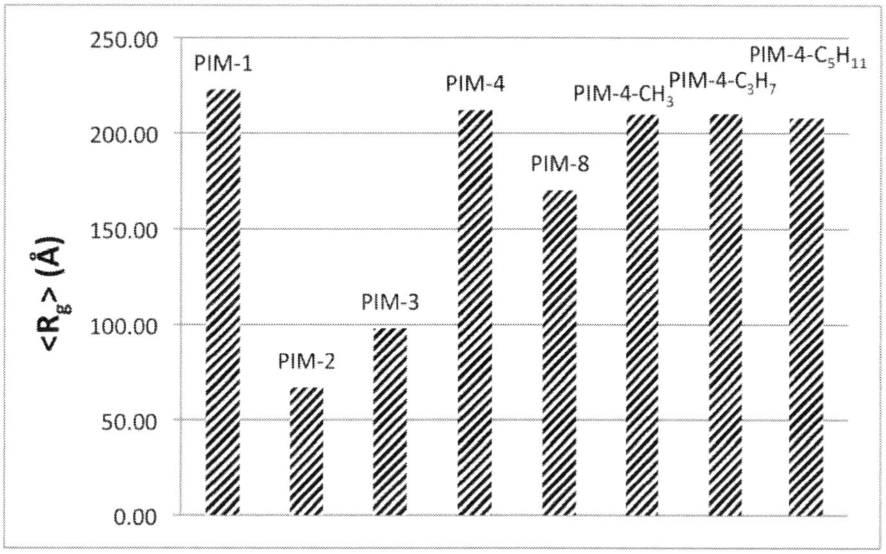

Figure 3. The influence of the molecular structure on the radius of gyration of the PIM-polyimides.

Figure 5 depicts the influence of the molecular structure on the probability distribution function of the torsional angle (ϕ) of PIM-4, 4-CH_3, 4-C_3H_7 and 4-C_5H_{11}. Similarly to the observation made in figure 4, the presence of the bulky side groups in PIM-4, 4-C_3H_7 and 4-C_5H_{11} has caused the energy barriers of cis-configurations to increase dramatically, which prevents the polymeric segments from experiencing full rotation about the connecting bonds. To a lesser extent, this was observed for PIM-4-CH_3 due to the smaller size of the methyl group.

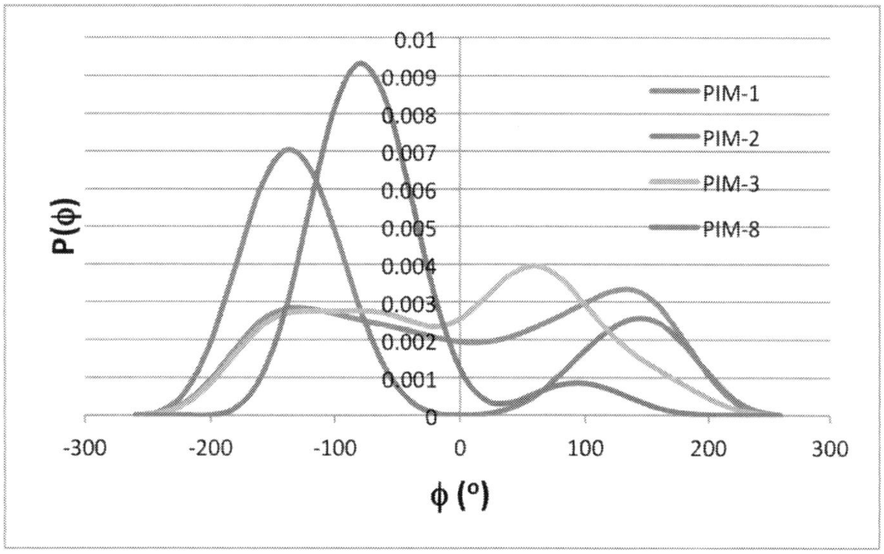

Figure 4. The influence of the molecular structure on the probability distribution function of the torsional angle (ϕ) of PIM-1, 2, 3 and 8.

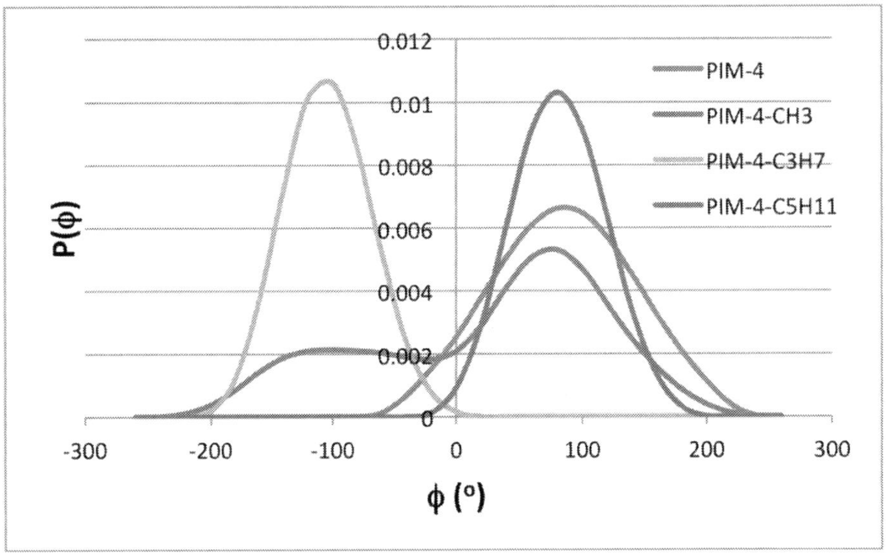

Figure 5. The influence of the molecular structure on the probability distribution function of the torsional angle (ϕ) of PIM-4, 4-CH_3.

Figure 6 shows the influence of the molecular structure on the probability distribution function of the angle of contortion (θ_c) of the PIM-polyimides. It is obvious from the figure that the contortion angle had a very limited range of bending due to the high steric hindrance around the sp^3 hybridized tertiary C-atom. Interestingly, the values of the probability distribution function itself didn't show much change with the variation of the molecular structure of the various polymers.

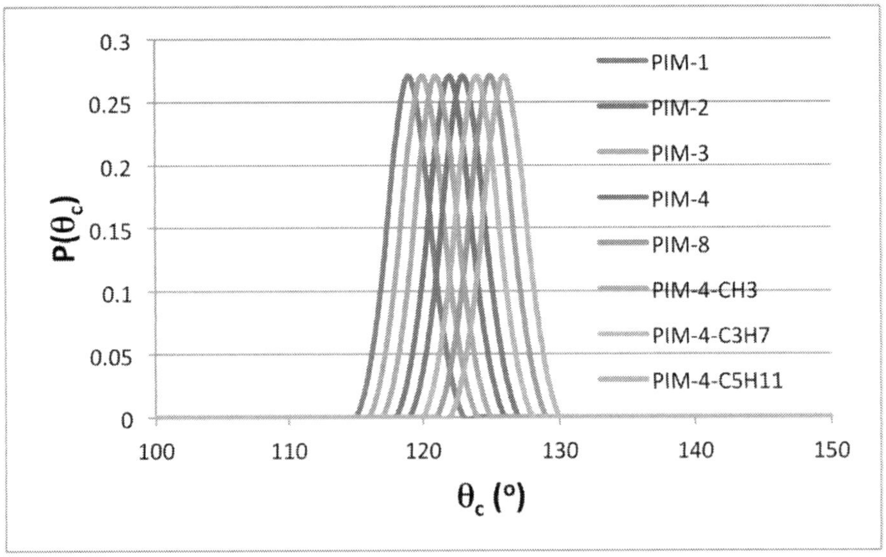

Figure 6. The influence of the molecular structure on the probability distribution function of the angle of contortion (θ_c) of the PIM-polyimides.

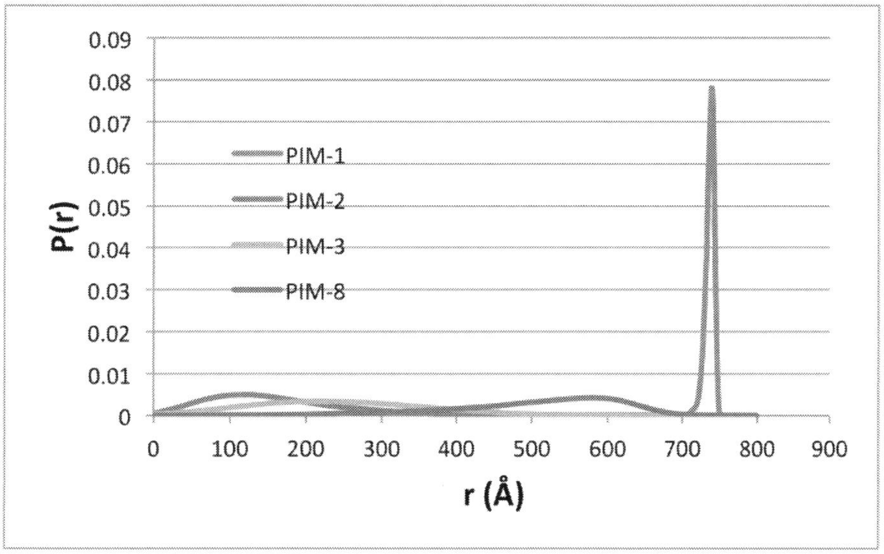

Figure 7. The influence of the molecular structure on the probability distribution function of the end-to-end distance (r) of PIM-1, 2, 3 and 8.

Figure 7 exhibits the influence of the molecular structure on the probability distribution function of the end-to-end distance (r) of PIM-1, 2, 3 and 8. It is rather interesting that the probability distribution functions showed a high value for PIM-1 and quite narrow variation in the value of the end-to-end distance. This indicates that fully extended configuration is the preferred configurations for these polymers in the absence of strong electronegative atoms, backbones kinks or bulky side groups. The addition of any of these factors to the polymeric structures results in the polymeric chains being forced to assume less extended configurations.

Figure 8 illustrates the influence of placing bulky side chains of different length on the probability distribution function of the end-to-end distance (r) of PIM-4, 4-CH_3, 4-C_3H_7 and 4-C_5H_{11}. Rather interstingly, the length or bulkiness of the side groups didn't affect the end-to-end distance distribution very much in clear agreement with the observations made earlier for the radius of gyration, figure 3.

Figure 9 exhimibts the influence of the molecular structure on the probability distribution function of radius of gyration of the PIM-polyimides. As discussed earlier, in the absence of side groups, electronegative atoms such as the fluorine atoms or backbone kinks, PIM-1 show the most preferred fully extended configuration with high value for both the end-to-end distance and the radius of gyration. The presence of these factors has caused the radius of gyration to assume various values in response to the polymeric chains assuming different configurations. Interesting, figure 9 predicts that the presence of quite large bulky side chain such as the pentyl group has caused the polymeric chains to revert back to the fully extended configurations possibly due to the quite high potential energy barriers that the chains have to cross to reach the less extended configurational states.

4. Conclusion

The use of molecular simulation techniques allows for the modeling and investigation of the influences of the molecular structures on the conformational behavior of PIM-polyimides of different molecular structures. Three factors obviously influenced the conformational behavior of these polymers, which are the presence of electronegative atoms, the presence of non-linearity in the polymeric backbones (backbone kinks) and the presence of bulky side groups on the polymeric chains. The dipole moment increased sharply with the presence of backbone kinks more than any other factor. Replacing the fluorine atoms in PIM-4 with bulky alkyl groups didn't influence the dipole moment greatly indicating that the size of the side chains had much less dramatic influence on the dipole moment than having a bent backbone. Similarly, the presence of the backbone kinks in the polymeric chains influenced the polymeric chains to assume less extended configuration causing the torsional angles around the interconnecting bonds unable to cross the high potential energy barriers. The presence of the bulky side groups also caused the energy barriers of the cis-configurations to increase dramatically, which prevented the polymeric segments from experiencing full rotation about the connecting bonds. For these polymers, it was clear that the fully extended configurations are the preferred configurations in the absence of strong electronegative atoms, backbones kinks or bulky side groups. The addition of any of these factors to the polymeric structures resulted in the polymeric chains being forced to assume less extended configurations. Rather interestingly, the length or bulkiness of the side groups didn't affect the end-to-end distance distribution to a great deal since the presence of quite large bulky side chain such as the pentyl group has caused the polymeric chains to revert back to the fully extended configurations possibly due to the quite high potential energy barriers that the chains have to cross to reach the less extended configurational states.

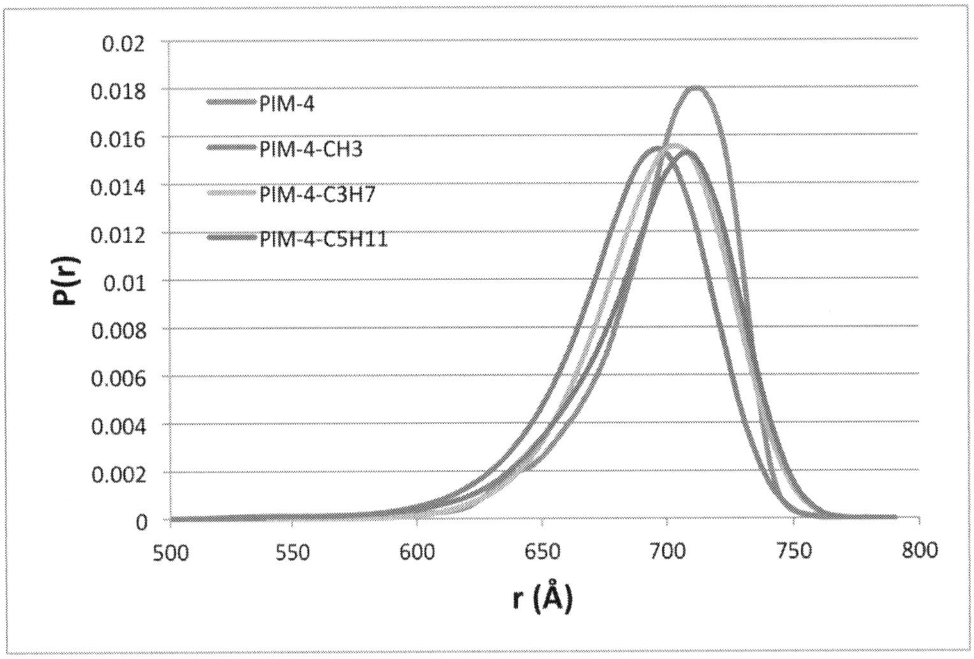

Figure 8. The influence of the molecular structure on the probability distribution function of the end-to-end distance (r) of PIM-4, 4-CH_3, 4-C_3H_7 and 4-C_5H_{11}.

References

[1] Lu G Q and Zhao X S 2004 *Nanoporous Materials: Science and Engineering* (London: Imperial College Press)

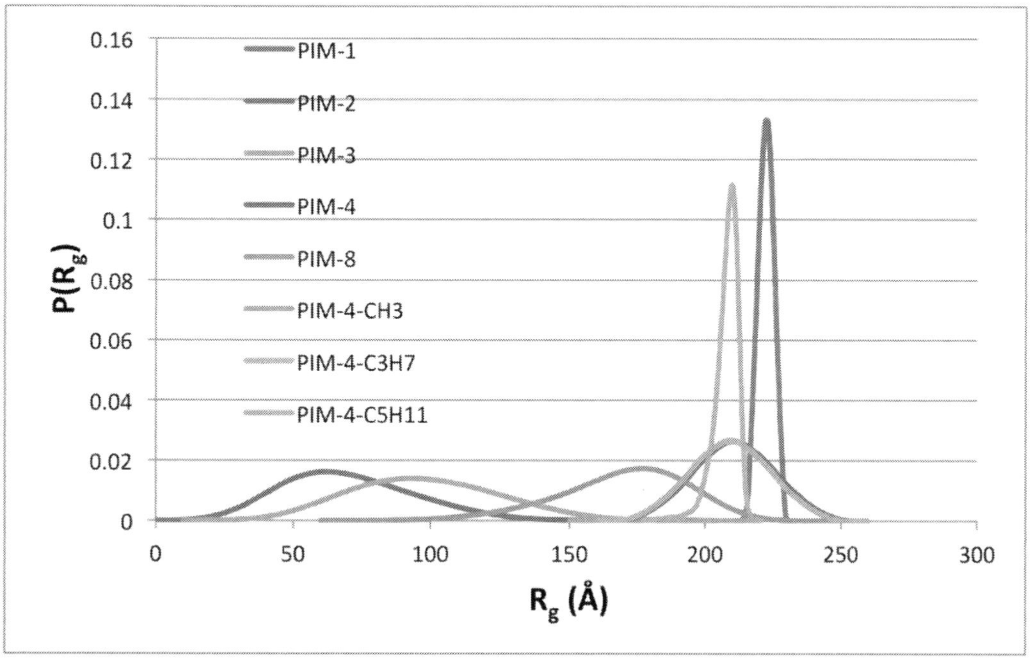

Figure 9. The influence of the molecular structure on the probability distribution functions of radius of gyration of the PIM-polyimides.

[2] Bae T H 2010 *Engineering Nanoporous Materials for Application in Gas Separation Membranes* (Ph.D. thesis: Georgia Institute of Technology)
[3] Madkour T and Mark J 1998 *Macromolecular Theory and Simulations* **7**, 69-77
[4] Budd P M, Msayib K J, Tattershall C E, Ghanem B S, Reynolds K J, McKeown N B and Fritsch D 2005 *Journal of Membrane Science* **251** 263
[5] Trewin A and Cooper A 2010 *Angew. Chem. Int. Ed.* **49** 1533
[6] Ghanem B S, Mckeown N B, Budd P M, Selbie J D and Fritsch D 2008 *Advanced Materials* **20** (14) 2766
[7] Madkour T and Mark J 2002 *J. Polym. Sci. Polym. Phys.* **40**(9) 840
[8] McKeown N B, Budd P M and Book D 2007 *Macromol. Rapid. Commun.* **28** 995
[9] Svec F, Germain J and Fréchet J M 2009 *Small* **5** 1098
[10] McKeown N B and Budd P M 2006 *Chem. Soc. Rev.* **35** 675
[11] Germain J and Fréchet J M 2008 *J. Chem. Mater.* **20** 7069
[12] Ghanem B S, McKeown N B, Budd P M, Al-Harbi N M, Fritsch D, Heinrich K, Starannikova L, Tokarev A and Yampolskii Y 2009 *Macromolecules* **42** 7881
[16] Madkour T and Mark J 1997 *Journal of Polymer Science Part B: Polymer Physics* **35(16)** 2757
[13] Germain J B 2008 *Nanoporous Polymers for Hydrogen Storage and Sulfur as a thin-film Etch Resist* (PhD thesis: University of California, Berkeley)
[14] Madkour T and Barakat A M 1997 *Computational and Theoretical Polymer Science* **7(1)** 35
[15] Du X, Sun Y L, Tan B E, Teng Q F, Yao X J, Su C Y and Wang W 2010 *Chem. Commun.* **46** 970
[17] Madkour T and Soldera A 2001 *European Polymer Journal* **37**(6), 1105
[18] Madkour T, Azzam R and Mark J 2006 *J. Polym. Sci. Polym. Phys.* **44** 2524

GPGPU simulations of 2D lattice neutral models in ecology

Takeshi Oura[1] **and Kei Tokita**[2]

[1]Cybermedia Center and Graduate School of Science, Osaka University, Toyonaka, Osaka 560-0043, Japan

[2]Cybermedia Center, Graduate School of Science and Graduate School of Frontier Biosciences, Osaka University, Toyonaka, Osaka 560-0043, Japan

E-mail: [1]`oura@cp.cmc.osaka-u.ac.jp`, [2]`tokita@cmc.osaka-u.ac.jp`

Abstract. 2D lattice neutral models in ecology are studied using general-purpose computing on graphic processing units (GPGPU). Processing times of GPGPU and CPU simulations are compared for various system sizes and it is found out that the larger the system size, the faster the GPGPU version, and the efficiency of GPGPU is maximally 263 times higher. Ecological significance of the GPGPU simulations and the lattice neutral model is also reported.

1. Introduction

General-purpose computing on graphic processing units (GPGPU) [1] is one of the recent hot topics not only in computer sciences but also in simulation sciences like computational physics. It is still fresh in our memory that the N-body simulations on GPUs for astrophysics and turbulence by the DEGIMA GPGPU cluster system in Nagasaki University won the Gordon Bell award in the Price/Performance category in 2009 and 2010 [2, 3], the GPU-accelerated TSUBAME 2.0 supercomputer at the Tokyo Institute of Technology ranked #5 on the TOP500 in 2010 [4] and the peta-scale phase field simulation for dendritic solidification on TSUBAME won the Gordon Bell award in the Special Achievement in Scalability and Time-to-Solution category in 2011 [5]. In the present study we demonstrate that GPGPU is also highly effective for simulations in biological physics and achieves maximally 263-fold speedup compared to CPU simulations.

We here study a 2D lattice version of the Hubbell's neutral model of biodiversity [6, 7] which appropriately predicts species abundance distributions (SAD) while there are some controversy on the validity of its "neutral hypothesis". SAD has been thoroughly surveyed by field ecologists and also recently theoretically led by statistical mechanics for population dynamics with complex interspecies interactions [8]. One of the major problems of many studies on the neutral model is that they are mainly based on the mean-field theory [9] for essentially infinite dimensional systems though real ecosystems of sedentary species such as tropical rain forests and coral reefs originally intended by the neutral theory are essentially of two dimensional and finite length of migration of plants' seeds and coral larvae is important. Because mathematical analysis for 2D models has not progressed very much we here study 2D lattice neutral model using GPGPU.

2. Model

We consider an ecological model of L^2 individuals of species occupying $L \times L$ 2D lattice sites, being born, dying, migrating and evolving. Every lattice site is at anytime occupied by an individual of some species, that is, no vacant site occurs and the total number of individual is constant ($= L^2$) at all times, which is modeled on a "saturated" community like reefs with no space between corals and rain forests covered by tree crowns with no vacant space.

The model has five parameters: the death rate d, the birth rate b, the migration distance r, the speciation rate ν and the system size $L \times L$. The neutral theory assumes that the values of these parameters are constants for each given species and independent from those of other species, but they are different from one species to the other. The migration area A is a set of the neighborhood of (i, j) and in the present study we assume the Moore neighborhood except at the center ($A = M(i,j) = \{(i', j') : |i - i'| \leq r, |j - j'| \leq r, (i', j') \neq (i, j)\}$). The migration distance r is one of the essential parameters of the present model. The simulation starts from an initial state in which random natural numbers $s\,(s = 1, 2, 3, \cdots)$ are distributed onto $L \times L$ lattice sites. The number $s(i, j)$ on the site (i, j) denotes the species number. In one Monte Carlo step the following procedures are executed and repeated $L \times L$ times.

3. Algorithm

3.1. Original algorithm

(1) Choose randomly one site (i, j) and kill the individual on it with probability $d(\leq 1)$. Here we set $d = 1$ without loss of generality because d changes only the time scale of the dynamics.

(2) With probability $\nu(\leq 1)$ assign a new species $s' \notin \{s(i,j)|i,j = 1, 2, \cdots, L)\}$ onto (i, j). This process denotes evolution of a new species or migration of a new species from outside the system. In the present study we use a same value of $\nu = 0.001$.

(3) If the new species is not assigned in (2), choose one individual (s'') in the migration area A with probability in proportion to $b\rho$, where ρ is the density of species s'' in A, and copy s'' onto (i, j). The migration-distance r is one of the essential parameters of the present model. This process corresponds to fixation of pollens of plants or coral larvae from parental individuals in the migration area. We here set $b = 1$ without loss of generality because b is independent of species number under the neutral hypothesis.

The above Monte Carlo steps are iterated until the system leads to a stationary state at a time which depends on the parameters. We use periodic boundary conditions. Simple implementation by pseudocode of the original algorithm is given in Algorithm 1.

Algorithm 1 Original algorithm

```
for time t < end do
  i ⇐ integer random number generated by U(L)
  j ⇐ integer random number generated by U(L)
  x ⇐ random number generated by U(1)
  if x < ν then
    s(i, j) ⇐ new species
  else
    i' ⇐ integer random number generated by U(i − r, i + r)
    j' ⇐ integer random number generated by U(j − r, j + r)
    s(i, j) ⇐ s(i', j')
  end if
end for
```

Table 1. Sites with same number updated at the same time.

1	**2**	1	2	1	2	1	2
3	**4**	3	4	3	4	3	4
1	2	1	2	1	2	1	2
3	4	3	4	3	4	3	4
1	2	1	2	1	2	1	2
3	4	3	4	3	4	3	4
1	2	1	2	1	2	1	2
3	4	3	4	3	4	3	4

(a) $r = 1$, $L = 8$

1	**2**	**3**	1	2	3
4	**5**	**6**	4	5	6
7	**8**	**9**	7	8	9
1	2	3	1	2	3
4	5	6	4	5	6
7	8	9	7	8	9

(b) $r = 2$, $L = 6$

1	**2**	**3**	**4**	1	2	3	4
5	**6**	**7**	**8**	5	6	7	8
9	**a**	**b**	**c**	9	a	b	c
d	**e**	**f**	**0**	d	e	f	0
1	2	3	4	1	2	3	4
5	6	7	8	5	6	7	8
9	a	b	c	9	a	b	c
d	e	f	0	d	e	f	0

(c) $r = 3$, $L = 8$

3.2. Modified algorithm for parallel processing

We modified the algorithm for parallel processing using the checker-board decomposition[10]. Sites which have no influence on each other are able to be updated at the same time. These sites are calculated in parallel. Table 1 is a scheme to construct such sites. For example, in the case of Table 1(a), the sites with same number ("1", "2", "3" and "4") are updated in parallel and the one Monte Calro step (MCS) is processed in four parallel steps for "1", "2", "3" and "4" in principle.

Sites processed at the same time are at the interval of r. The smaller r, the larger the number of sites is, the better performance is. As the number of the groups updated in parallel A is $(r+1)^2$ shown in Table 1, the number of the sites which can be updated in parallel is $L^2/(r+1)^2$. If the number of available cores is sufficiently larger than the number of parallel tasks, processing times are proportional to $(r+1)^2$. Simple implementation by pseudocode of the modified algorithm for the parallel processing is given in Algorithm 2.

Algorithm 2 Modified algorithm

```
for time t < end do
  for set ∈ {set'|{i, j|each s(i, j) have no influence }} do
    for each (i, j) ∈ set in parallel do
      x ⇐ random number generated by U(1)
      if x < ν then
        s(i, j) ⇐ new species
      else
        i' ⇐ integer random number generated by U(i − r, i + r)
        j' ⇐ integer random number generated by U(j − r, j + r)
        s(i, j) ⇐ s(i', j')
      end if
    end for
  end for
end for
```

4. Results

4.1. Settings

We compared processing times of simulations by GPUs and multi-core environments of CPUs. For GPGPU the calculation is executed by nVidia Tesla C2050 and C2050×2(+openMP). As a multi-CPUs environment we use Intel Xeon W3350 4 cores for openMP calculation and Xeon 5160 4∼100 cores for MPI. One node has 2 Xeon 5160 (= 4 cores). In the following processing times by Xeon W3350 1 core is used as a baseline for the performance.

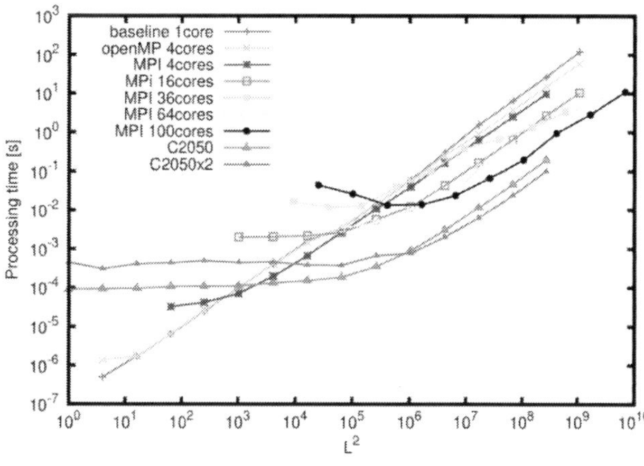

Figure 1. Processing times per 1MCS vs. system size L^2 ($r = 1$).

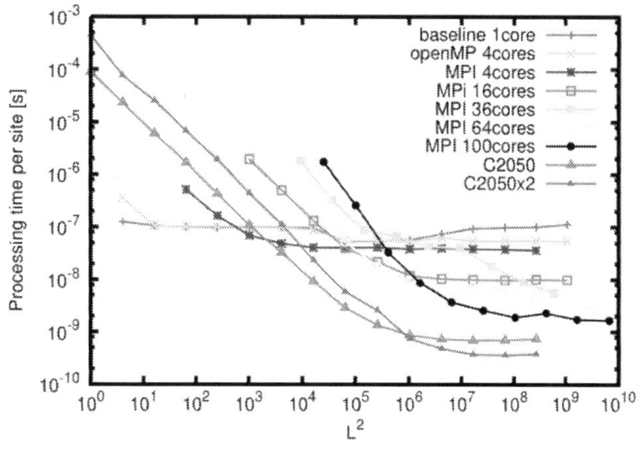

Figure 2. Processing times per site vs. system size L^2 ($r = 1$).

4.2. System-size (L^2) dependency of processing times

Processing times at various system size L^2 are mesuared to clarify size dependency. Figure 1 shows observed processing times per 1 MCS by GPUs and multi-core environments of CPUs. In a wide range of L^2, C2050 and C2050×2 are faster than multi-CPUs. In MPI calculations the more the cores used, the worse the processing times at small L. In addition system-size (L) dependency of MPI processing with higher degree of parallel (> 36 cores) fluctuates in comparison with C2050's. Figure 2 represents processing times per sites. At $L^2 > 10^7$ effects of overhead on C2050 are ignorable. In contrast effects of overhead on MPI are significant. One weak point of C2050 is that C2050 can calculate up to $L \sim 16384$ because C2050's memory size is only 2 GByte. Peak performance of GPUs and CPUs is summarized in Table 2. From figure 2 and table 2 for the system $L \times L = 16384 \times 16384 = 268435456$ the GPGPU calculation is 136 and 263 times more efficient than the CPUs for C2050 and C2050×2 respectively.

Table 2. Peak performance of the simulations by single core, multicore (MPI) and many cores of GPGPU.

	Processing times per site [s]	Scale factor $\left(\frac{\text{Baseline time}}{\text{Processing time}}\right)$
Xeon W3350 1core (baseline)	1×10^{-7}	1
Xeon W3350×4cores (openMP)	5.42×10^{-8}	1.84
Xeon 5160	1.92×10^{-7}	0.52
Xeon 5160×4cores (MPI)	3.63×10^{-8}	2.756
Xeon 5160×16cores (MPI)	1.07×10^{-8}	9.35
Xeon 5160×36cores (MPI)	6.08×10^{-9}	16.45
Xeon 5160×64cores (MPI)	3.71×10^{-9}	27.06
Xeon 5160×100cores (MPI)	1.65×10^{-9}	44.38
Tesla C2050	7.35×10^{-10}	136
Tesla C2050×2	3.79×10^{-10}	263

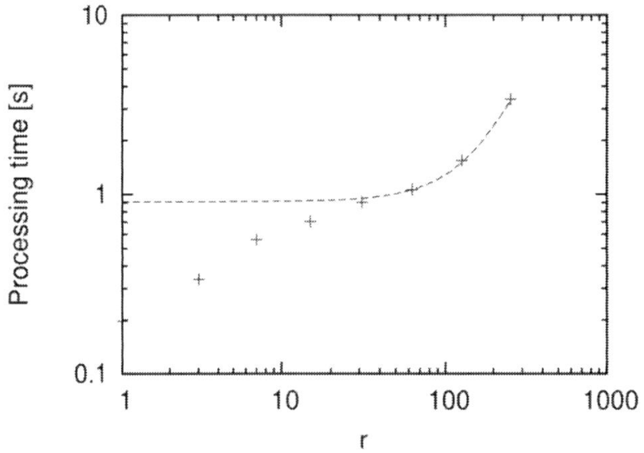

Figure 3. Processing times per 1MCS vs. migration distance r ($L = 16384$)

4.3. Migration-distance (r) dependency of processing times

In Figure 3 cross symbols represent processing times for various r at 1MCS. The dashed line is proportional to $(r+1)^2$. As noted previously, processing times are proportional to $(r+1)^2$. But at small r the number of sites processed simultaneously is large, thus processing times increase against $(r+1)^2$. For large r, processing times become close to $(r+1)^2$ asymptotically. At $r > 60$ processing times almost coincide with $(r+1)^2$.

4.4. Rank abundance relations
While we saw the advantage of GPGPU against multi-core environments in the previous section, we here consider the biological significance of the present study. In the context of the study of the neutral model one of the most important observables is the rank abundance relations (RAR) (the relations between abundance and the specie's rank, which is essentially equivalent to SAD). The migration-distance (r) dependency of RAR is depicted in figure 4, which clearly displays that migration-distance changes RAR. This migration-distance (r) dependency was essentially not clarified by the mean field theory of the neutral model. As the theoretical treatment of the spatial explicit model has not made progress, the present study verifies that large scale GPGPU simulations are effective. As the derivative of the inverse function of RAR is SAD, we observe that for larger r the width of SAD becomes larger, which means that abundant and rare species coexist in the system.

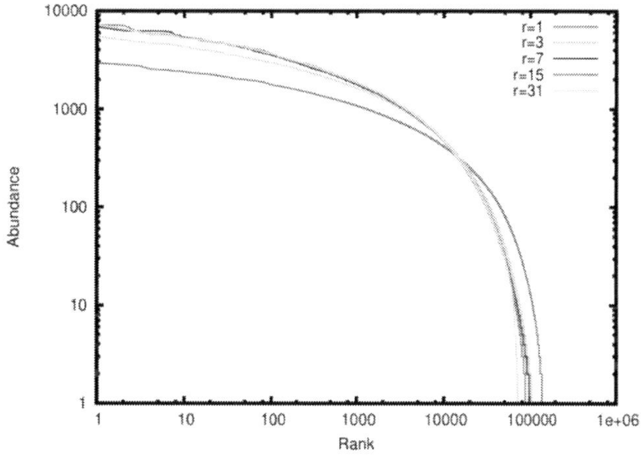

Figure 4. Rank abundance relations for various r ($L = 4096$).

5. Summary
Simulations are executed for various system sizes and each processing time of simulations using GPGPU (nVIDIA Tesla C2050) and CPU (Intel Xeon W3530 2.80GHz and Intel Xeon 5160 3.00 GHz) are compared. In the calculations of the present model the performance of one board of C2050 is comparative to a PC cluster with hundreds of cores. In this sense, GPGPU provides much better cost performance. As the present model is a variant of a lattice model with mesoscopic interactions often studied in physics, the scalability and the high cost-performance of the GPGPU simulation is promising for other studies of computational physics.

6. Acknowledgements
The authors thank Mr. Sugahara for illuminating discussions. This work was supported in part by the Global COE Program (Core Research and Engineering of Advanced Materials-Interdisciplinary Education Center for Materials Science), MEXT, Japan. This work was also supported in part by the initiative-based project E0-5 "Creation and Sustainable Governance of New Commons through Formation of Integrated Local Environmental Knowledge", Research Institute for Humanity and Nature (RIHN). K.T. acknowledges support by grants-in-aid from MEXT, Japan (Grants NO.24570099). MPI calculations are carried out on the PC clusters at Cybermedia Center, Osaka University.

References

[1] *GPGPU*, Wikipedia, `http://en.wikipedia.org/wiki/GPGPU`
[2] Hamada T *et al.* 2009 42 TFlops hierarchical N-body simulations on GPUs with applications in both astrophysics and turbulence,*IEEE/ACM SC09*, Portland, USA, ACM Gordon Bell prize price/performance, November 2009
[3] Hamada T and Nitadori K 2010 190 TFlops Astrophysical N-body Simulation on a Cluster of GPUs,*IEEE/ACM SC10*, New Orleans, USA, ACM Gordon Bell prize honorable mention price/performance, November 2010
[4] *TOP500* `http://www.top500.org/lists/2010/11`
[5] Shimokawabe T *et al.* 2011 Peta-scale Phase-Field Simulation for Dendritic Solidification on the TSUBAME 2.0 Supercomputer, *IEEE/ACM SC11*, Seattle, USA, ACM Gordon Bell prize special achievements in scalability and time-to-solution, November 2011
[6] Hubbell S P 2001 *The Unified Neutral Theory of Biodiversity and Biogeography* (Princeton: Princeton University Press)
[7] Rosindell J, Hubbell S P and Etienne R S 2011 *TREE* **26** 340–348
[8] Tokita K 2004 *Phys. Rev. Lett.* **93** 178102
[9] Etienne R and Alonso D 2007 *J. Stat. Phys.* **128** 485–510
[10] Landau D P and Binder K 2000 *A Guide to Monte Carlo Simulations in Statistical Physics* (Cambridge University Press)

Multicanonical simulation of the Domb-Joyce model and the Gō model: new enumeration methods for self-avoiding walks

Nobu C Shirai[1,2] **and Macoto Kikuchi**[2,1,3]

[1] Graduate School of Science, Osaka University, Toyonaka, Osaka 560-0043, Japan
[2] Cybermedia Center, Osaka University, Toyonaka, Osaka 560-0043, Japan
[3] Graduate School of Frontier Biosciences, Osaka University, Suita, Osaka 565-0871, Japan

E-mail: `shirai@cp.cmc.osaka-u.ac.jp`

Abstract. We develop statistical enumeration methods for self-avoiding walks using a powerful sampling technique called the multicanonical Monte Carlo method. Using these methods, we estimate the numbers of the two dimensional N-step self-avoiding walks up to $N = 256$ with statistical errors. The developed methods are based on statistical mechanical models of paths which include self-avoiding walks. The criterion for selecting a suitable model for enumerating self-avoiding walks is whether or not the configuration space of the model includes a set for which the number of the elements can be exactly counted. We call this set a scale fixing set. We selected the following two models which satisfy the criterion: the Gō model for lattice proteins and the Domb-Joyce model for generalized random walks. There is a contrast between these two models in the structures of the configuration space. The configuration space of the Gō model is defined as the universal set of self-avoiding walks, and the set of the ground state conformation provides a scale fixing set. On the other hand, the configuration space of the Domb-Joyce model is defined as the universal set of random walks which can be used as a scale fixing set, and the set of the ground state conformation is the same as the universal set of self-avoiding walks. From the perspective of enumeration performance, we conclude that the Domb-Joyce model is the better of the two. The reason for the performance difference is partly explained by the existence of the first-order phase transition of the Gō model.

1. Introduction

A self-avoiding walk (SAW) is a path on a lattice with the spatial restriction that the path must not visit the same site more than once [1]. A SAW is used as a model of chain polymers, because the above restriction represents the excluded volume effect. Lattice protein models such as the Gō model [4, 5, 6] and the HP model [7] have played important roles in the construction of the funnel picture of the energy landscape [6, 8, 9], which forms the basic theoretical understanding of protein folding.

Many sophisticated methods have been developed to analyze the thermodynamic properties of a lattice protein. The multi-self-overlap ensemble (MSOE) [14, 15], which is an extended version of the multicanonical Monte Carlo method [10, 11], is one of the best methods to find a ground state and calculate thermodynamic properties of a long lattice protein. The main idea of the MSOE is that the self-avoiding condition is relaxed so that some intersections of a chain are allowed. In the present work, we apply the idea to the enumeration problem of finite-step SAWs.

Besides having many applications like polymers, the SAW itself has many interesting asymptotic behaviors of infinite steps and has been studied by physicists, chemists, and mathematicians for a long time. In spite of great efforts, major parts of proposed asymptotic behaviors have not been solved in the rigorous mathematical sense and remain as conjectures. Enumeration of N-step SAWs is a famous unsolved problem and the exact number of two-dimensional SAWs is known up to only 71 steps [2]. The total number of N-step SAWs is written as c_N, and c_{71} is counted as

$$c_{71} = 4\ 190\ 893\ 020\ 903\ 935\ 054\ 619\ 120\ 005\ 916 \simeq 4.1909 \times 10^{30},$$

which is larger than Avogadro's constant ($\simeq 6.0221 \times 10^{23}$). Since c_N increases exponentially with N, it will soon be impossible to deal with all components of N-step SAWs on a computer and we have to use statistical methods for approximate enumeration of c_N with large N.

In our study, in order to estimate c_N with large N accurately, we developed new enumeration methods using multicanonical simulations of the following two kinds of statistical mechanical models: the Domb-Joyce model [3] for generalized random walks and the Gō model [4, 5, 6] for lattice proteins. By using our methods, we were able to estimate the number of SAWs on a square lattice up to 256 steps with error estimation.

2. Models and Methods

2.1. Self-avoiding walk (SAW)

We denote points on a d-dimensional cubic lattice by $\omega(i) \in \mathbb{Z}^d$ $(i = 0, 1, 2, \cdots)$ and a set of points by a path ω. An N-step random walk (RW) is defined by $\omega = (\omega(0), \omega(1), \cdots, \omega(N))$ starting from the origin of the lattice with the constraint $|\omega(i+1) - \omega(i)| = 1$ $(i = 0, 1, \cdots, N-1)$. We denote the universal set of RWs as $\{\omega\}_{\text{RW}}$. Since each site has $2d$ nearest neighbors, the total number of N-step RWs is exactly $(2d)^N$. In the case of a SAW, a further constraint imposed by $\omega(i) \neq \omega(j)$ for all $i \neq j$ and we denote the universal set of SAWs as $\{\omega\}_{\text{SAW}}$. This constraint makes it difficult to exactly count the total number of N-step SAWs c_N.

2.2. Multicanonical Monte Carlo method

Two enumeration methods proposed in this paper are based on the same sampling technique called the multicanonical Monte Carlo method [10, 11]. If we define Hamiltonian $\mathcal{H}$ as a function which maps a path on a d-dimensional lattice to energy, $\mathcal{H} : \omega_a \to E_a$, we can introduce an energy structure into $\{\omega\}_{\text{RW}}$ and $\{\omega\}_{\text{SAW}}$. Using the multicanonical method, we can estimate the number of states $\Omega(E)$ accurately over a wide range of energy, as we will explain in the following sections.

Introducing a weight $W(E)$ as a function of E into the Markov chain Monte Carlo, we define the transition probability from a path ω_a to another path ω_b

$$p(\omega_a \to \omega_b) = \min\left[\frac{W(E_b)}{W(E_a)}, 1\right], \tag{1}$$

where E_a and E_b are energy of ω_a and ω_b, respectively. To calculate thermodynamic values at specific inverse temperature β, we can use the Metropolis' method with $W(E) \propto \exp(-\beta E)$. In the multicanonical method, the energy space of $W(E)$ is divided into n bins and inverse temperature β_i $(i = 1, 2, \cdots, n)$ is introduced to each bin. $W(E)$ of the ith bin is set to be proportional to $\exp(-\beta_i E)$, and joint parameters α_i $(i = 1, 2, \cdots, n-1)$ are introduced to connect $W(E)$ continuously at the boundaries of the bins. By modifying β_i $(i = 1, 2, \cdots, n)$ and α_i $(i = 1, 2, \cdots, n-1)$, we determine $W(E)$ to be approximately proportional to the inverse

Journal of Physics: Conference Series **454** (2013) 012039 doi:10.1088/1742-6596/454/1/012039

of the number of states $1/\Omega(E)$. If the number of energy levels n_E is finite and n is equal to n_E, which is true throughout our study, the multicanonical method is identical to the entropic sampling [16].

In our study, we built up $W(E)$ using the Wang-Landau method [12, 13], the detailed procedure of which is given in the Appendix. If we give a transition probability using this $W(E)$, the Markov chain Monte Carlo produces a flat histogram $H(E)$. We can obtain $H(E)/W(E) \propto \Omega(E)$ with high accuracy over a wide range of the energy scale. In order to estimate the number of states, $\Omega(E)$, we need to introduce an absolute scale. In section 3 and 4, we introduce two different absolute scales.

2.3. Multi-self-overlap ensemble

To efficiently sample SAWs from $\{\omega\}_{\text{SAW}}$, we can use the multi-self-overlap ensemble (MSOE) [14, 15], which is an extended version of the multicanonical method. Relaxing the self-avoiding conditions, the MSOE makes it possible to explore the configuration space faster than the case that the self-avoiding conditions are strictly kept despite the extra configuration space. The reason for the fast exploration is that transition paths from one SAW to another increase via configurations with one intersection or more. Let V be the number of the overlaps of a path. If k points of a path ($k \geq 3$) are on the same site, we define that there are $k-1$ overlaps on that site. Note that, in the original paper, V is defined as penalty for overlaps, and they used $(k-1)^2$ instead of $k-1$. Using V and the prescribed cutoff, V_{cut}, we denote the original configuration space and the expanded configuration space as $\{\omega\}_{\text{RW}}^{V=0}$ and $\{\omega\}_{\text{RW}}^{V\leq V_{\text{cut}}}$, respectively.

In the same way as with the multicanonical method, we build up the weight function $W(E,V)$. By using the transition probability

$$p(\omega_a \to \omega_b) = \min\left[\frac{W(E_b, V_b)}{W(E_a, V_b)}, 1\right], \tag{2}$$

we obtain a two-dimensional flat histogram, $H(E,V)$. The number of $\{\omega\}_{\text{SAW}}$ can be obtained from the equation $\Omega(E) \propto H(E,0)/W(E,0)$. We used the MSOE in the first enumeration method using the Gō model. The idea of using the number of overlaps as a variable is shared by the second enumeration method using the Domb-Joyce model, which has the fully expanded configuration space $\{\omega\}_{\text{RW}}$.

3. Multicanonical simulation of the Gō model

3.1. Gō model

In the first enumeration method, we estimate c_N by multicanonical simulations of the Gō model [4, 5, 6], which was introduced originally to investigate the protein folding problem theoretically. The Gō model is defined as a SAW with specialized interactions that gives one conformation ω_{native}, called the native structure, as a unique ground state apart from the trivial spatial symmetry. Using two indices of points of ω, i and j, a native contact pair is defined by a pair of i and j ($j > i+1$) which satisfies the condition $|\omega_{\text{native}}(i) - \omega_{\text{native}}(j)| = 1$. The Hamiltonian of the intrachain interactions can be written as

$$\mathcal{H}(\omega) = \sum_{(i,j)\in\{\text{native contact pairs}\}} -\varepsilon\ \delta(|\omega(i) - \omega(j)|, 1). \tag{3}$$

If we define n_{ncp} as the total number of native contact pairs, the ground state energy E_{GS} is written as $-\varepsilon n_{\text{ncp}}$.

3.2. Designing a native state

We calculate $g(E) = H(E,0)/W(E,0)$ of the Gō model using the MSOE. In order to estimate $\Omega(E)$ from $g(E)$, we need to know the number of states of at least one bin of $\Omega(E)$. We design a ω_{native} that is suitable for that purpose, leaving aside the protein folding problem.

Selecting a compact structure which can maximize the number of the native contact pairs for ω_{native}, we can easily count the number of ground states. In the case of $d = 2$ and $N = 24$, for example, if we choose the configuration shown in figure 1-(a) ("roll" shape) or (b) ("beta" shape) as a native structure, the number of ground states is 8 for both cases considering spatial symmetry. Roll-shaped or beta-shaped native states of any N are made from the following rules.

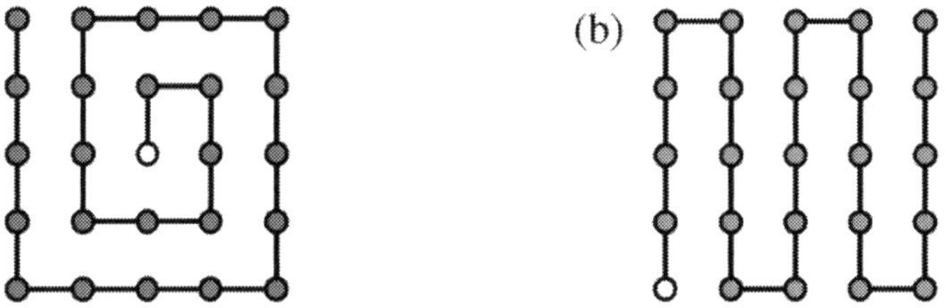

Figure 1. Two types of native states of the Gō model ($N = 24$): (a) roll and (b) beta shape. The start point, $\omega(0)$, of each chain is shown as a white circle.

I. Find L ($\in \mathbb{N}$) which satisfies the condition $\sqrt{N} < L \leq \sqrt{N} + 1$.
II. Draw the $L \times L$ compact structure using a roll or beta shape (figure 2 for $L = 2, 3, 4, 5$).
III. Select N steps from the start point $\omega(0)$. This can be used as ω_{native} for the N-step Gō model.
IV. The number of ground states are determined as below.

In the case of a roll shape,

$$\Omega(E_{\text{GS}}) = \begin{cases} 16 & \text{when } N = (\lfloor\sqrt{N}\rfloor)(\lceil\sqrt{N}\rceil) \\ 8 & \text{otherwise,} \end{cases}$$

and in the case of a beta shape,

$$\Omega(E_{\text{GS}}) = \begin{cases} 16 & \text{when } N = L^2 \\ 8 & \text{otherwise,} \end{cases}$$

where $\lfloor\ \rfloor$ is the floor function and $\lceil\ \rceil$ is the ceiling function. In the cases of $\Omega(E_{\text{GS}}) = 16$, there is an extra double degeneracy at the end of the paths. The roll-shaped and beta-shaped native states for $N \leq 24$ are illustrated in figure 2-(a) and (b), respectively.

3.3. Scale fixing by the number of ground states

If we design ω_{native} with specified $\Omega(E_{\text{GS}})$, like a roll or beta shape, the estimated number of states, $\Omega^*(E)$ (we denote estimated values calculated by simulations with asterisk $(*)$), can be calculated from $g(E)$ as

$$\Omega^*(E) = \frac{\Omega(E_{\text{GS}})}{g(E_{\text{GS}})} g(E) = \lambda_{\text{Gō}}\, g(E). \tag{4}$$

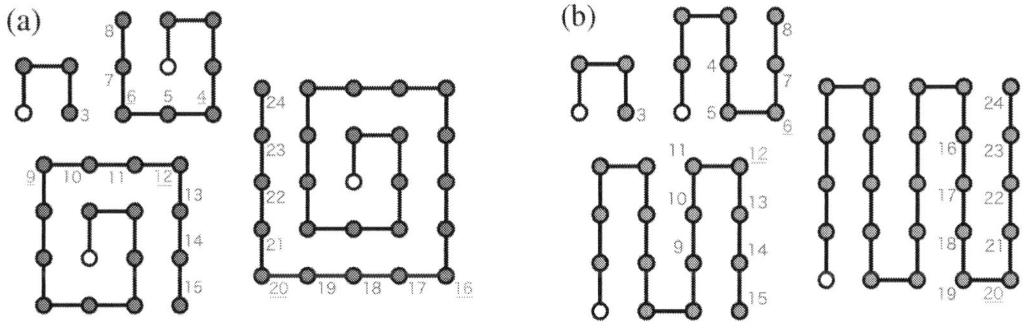

Figure 2. (a) Roll- and (b) beta-shaped native states for the Gō models with $N = 3 - 24$. The start point, $\omega(0)$, of each chain is shown as a white circle. The steps with blue numbers have 8 ground states and the steps with underlined red numbers have 16 ground states.

Here we defined the scale fixing factor, $\lambda_{\text{Gō}}$, by $\Omega(E_{\text{GS}})/g(E_{\text{GS}})$. Since the N-step Gō model has the same configuration space as $\{\omega\}_{\text{SAW}}$, the whole sum of $\Omega(E)$ is equal to c_N, and the estimated number of N-step SAWs is given by

$$c_N^* = \sum_E \Omega^*(E). \tag{5}$$

The scale of $\Omega(E)$ is determined by the set of the ground states $\{\omega\}_{\text{GS}}$, which we call the scale fixing set. The idea of the scale fixing with the Gō model is illustrated in figure 3. This model includes $\{\omega\}_{\text{SAW}}$ as the whole of its configuration space and, if it has a well-defined native state, it also includes a scale fixing set $\{\omega\}_{\text{GS}}$ as a part.

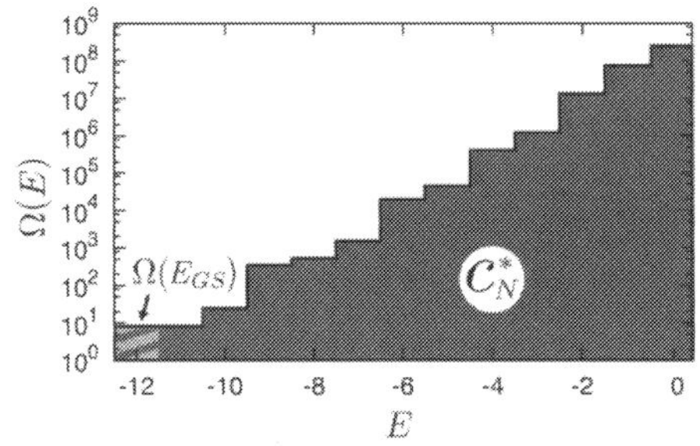

Figure 3. The idea of scale fixing demonstrated with a roll-shaped ω_{native} of the Gō model ($N = 19$). This model includes $\{\omega\}_{\text{SAW}}$ as the whole of its configuration space and also includes a scale fixing set $\{\omega\}_{\text{GS}}$ as a part.

4. Multicanonical simulation of the Domb-Joyce model

4.1. Domb-Joyce model

In the second enumeration method, we estimate c_N by multicanonical simulations of the Domb-Joyce model [3]. The configuration space of the Domb-Joyce model is $\{\omega\}_{\text{RW}}$. In the Domb-Joyce model, each path is weighted with the factor

$$\prod_{i=0}^{N-2} \prod_{j=i+2}^{N} (1 - u\delta_{ij}), \tag{6}$$

where u is a parameter of the model. If we replace u by $1 - \exp(-\beta J)$, each weight is identical to the Boltzmann factor of the Hamiltonian, which is written as

$$\mathcal{H}(\omega) = \sum_{i<j} J\, \delta(\omega(i), \omega(j)) = J V_{\mathrm{DJ}}, \tag{7}$$

where V_{DJ} is the number of intersections of a path. In the discussion below, we assume $J > 0$, which means a pair of crossed steps as a repulsive interaction of the strength J. In the limit of $\beta \to 0$ ($w \to 0$), all the configurations are equally weighted and the Domb-Joyce model corresponds to N-step RWs. In the opposite limit, $\beta \to \infty$ ($w \to 1$), the Domb-Joyce model reduces to N-step SAWs.

4.2. Modified Domb-Joyce model

In the actual simulation, we used a slightly modified model of the Domb-Joyce model in order to narrow the range of energy. In the modified model, we use V in section 2.3 instead of V_{DJ}. We explain their difference. If we introduce a number of intersections of each site on a lattice as $V(x, y)$, we calculate V_{DJ} or V by

$$\sum_{x=-N}^{N} \sum_{y=-N}^{N} V(x, y). \tag{8}$$

where $V(x, y) = 0$ at sites without points of a path. In the original Domb-Joyce model, if there are k points ($k \geq 2$) on the same site (x_s, y_s), $V(x_s, y_s)$ is given by $V(x_s, y_s) = k(k-1)/2$. We, however, use the definition given by $V(x_s, y_s) = k - 1$. This modification does not change the properties in $\beta \to 0$ and $\beta \to \infty$ of the Domb-Joyce model.

4.3. Scale fixing by the total number of N-step random walks

From the multicanonical simulations of the modified Domb-Joyce model, we can obtain $g(E) = H(E)/W(E)$ ($E = JV$) of the conformation space $\{\omega\}_{\mathrm{RW}}$. Since the total number of $\{\omega\}_{\mathrm{RW}}$ is exactly known as $(2d)^N$ in the d-dimensional cubic lattice, the universal set of the modified Domb-Joyce model $\{\omega\}_{\mathrm{RW}}$ can be used as a scale fixing set. Using a scale fixing factor $\lambda_{\mathrm{DJ}} = \sum_E \Omega(E) / \sum_E g(E)$, we can calculate the estimated number of states $\Omega^*(E)$ as

$$\Omega^*(E) = \lambda_{\mathrm{DJ}}\, g(E). \tag{9}$$

Then, c_N^* is obtained by

$$c_N^* = \Omega^*(0). \tag{10}$$

The idea of the scale fixing with the modified Domb-Joyce model is illustrated in figure 4. In contrast with the Gō model, the modified Domb-Joyce model includes $\{\omega\}_{\mathrm{SAW}}$ as a part of its configuration space and also includes a scale fixing set $\{\omega\}_{\mathrm{RW}}$ as a whole.

5. Results and Discussions

We estimated c_N up to 143 steps with the Gō model and up to 256 steps with the modified Domb-Joyce model with the same amount of computational effort. Thus, we conclude that the modified Domb-Joyce model is more efficient than the Gō model for enumerating SAWs. The results are shown in table 1 along with the known exact numbers [2] up to 71 steps. c_N^* agree with the exact numbers within the statistical error. It should be noted that the present methods are statistically unbiased. These enumeration methods are highly efficient and successfully counted large numbers up to 10^{108}. $c_N^{1/N}$ is shown in figure 5.

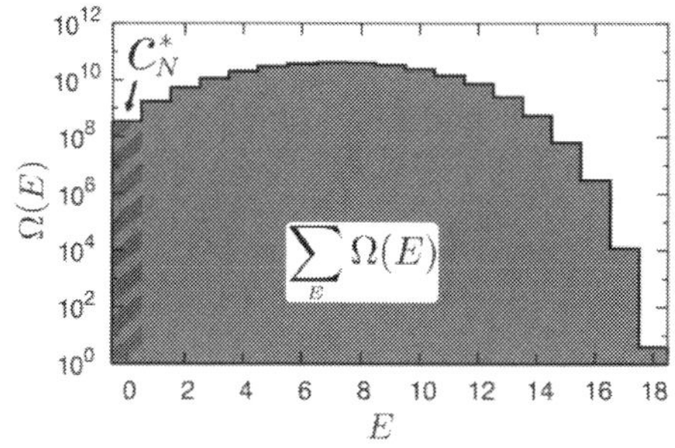

Figure 4. The idea of scale fixing demonstrated with the modified Domb-Joyce model ($N = 19$). This model includes $\{\omega\}_{\rm SAW}$ as a part of its configuration space and also includes a scale fixing set $\{\omega\}_{\rm RW}$ as a whole.

Table 1. A table of the estimated number of N-step SAWs c_N^* with the N-step exact number of SAWs c_N. The exact and estimated numbers are rounded off to 5 digits or less. The numbers in parentheses are the statistical errors for the last digits of the estimated numbers.

N	L	c_N Exact	c_N^* Gō (roll)	c_N^* Gō (beta)	c_N^* Domb-Joyce
3	2	36	$3.602(1) \times 10^1$	$3.60(2) \times 10^1$	$3.60(3) \times 10^1$
8	3	5916	$5.92(1) \times 10^3$	$5.92(3) \times 10^3$	$5.92(5) \times 10^3$
15	4	6.4166×10^6	$6.42(3) \times 10^6$	$6.44(4) \times 10^6$	$6.42(6) \times 10^6$
24	5	4.6146×10^{10}	$4.62(3) \times 10^{10}$	$4.62(2) \times 10^{10}$	$4.61(4) \times 10^{10}$
35	6	2.2525×10^{15}	$2.24(2) \times 10^{15}$	$2.26(1) \times 10^{15}$	$2.25(2) \times 10^{15}$
48	7	7.5014×10^{20}	$7.48(8) \times 10^{20}$	$7.55(7) \times 10^{20}$	$7.51(7) \times 10^{20}$
63	8	1.7155×10^{27}	$1.72(3) \times 10^{27}$	$1.72(3) \times 10^{27}$	$1.74(2) \times 10^{27}$
71	8	4.1909×10^{30}	$4.3(2) \times 10^{30}$	$4.24(8) \times 10^{30}$	$4.20(5) \times 10^{30}$
80	9		$2.74(8) \times 10^{34}$	$2.71(6) \times 10^{34}$	$2.68(3) \times 10^{34}$
99	10		$2.9(3) \times 10^{42}$	$3.0(1) \times 10^{42}$	$2.91(4) \times 10^{42}$
120	11		$2.5(2) \times 10^{51}$	$2.16(9) \times 10^{51}$	$2.17(4) \times 10^{51}$
143	12		$1.4(4) \times 10^{61}$	$1.19(8) \times 10^{61}$	$1.12(3) \times 10^{61}$
168	13				$4.1(1) \times 10^{71}$
195	14				$9.9(4) \times 10^{82}$
224	15				$1.77(8) \times 10^{95}$
255	16				$2.1(1) \times 10^{108}$
256	16				$6.2(4) \times 10^{108}$

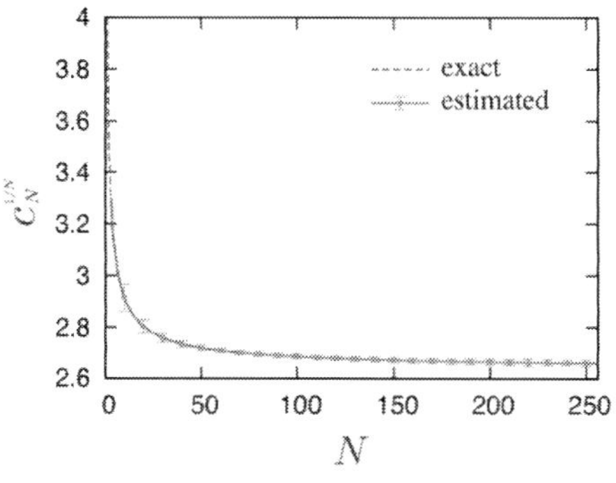

Figure 5. A plot of $(c_N^*)^{1/N}$ (pink) and $c_N^{1/N}$ (blue). The error bars were shown every 10 steps and their values were magnified by a factor of 20.

Using $g(E)$ calculated from the multicanonical simulations, we calculated specific heats of the two models (see figure 6 and 7). Based on the size dependence of the specific heats of the Gō model, we expect that the model exhibits a first-order phase transition at the long chain limit. We consider that this transition can partly explain the reason for the performance difference between the two developed methods, and we suggest that the performance of the statistical

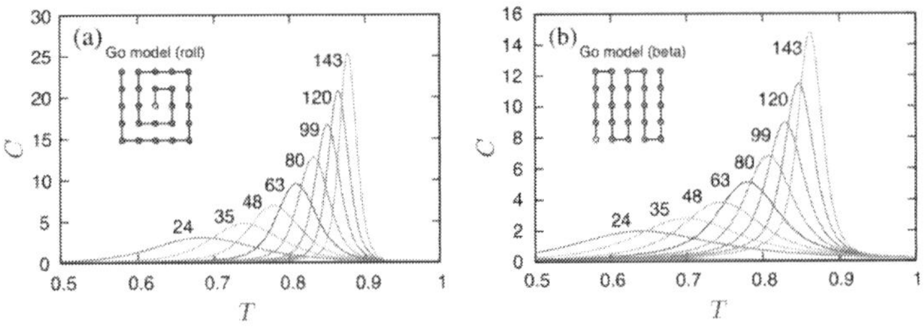

Figure 6. Temperature dependence of the specific heats of the Gō model with (a) roll- and (b) beta-shaped native states.

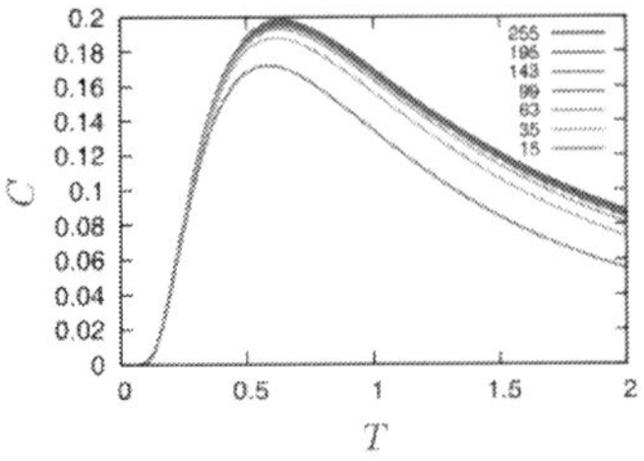

Figure 7. Temperature dependence of the specific heats of the modified Domb-Joyce model.

enumeration can be evaluated by the thermodynamic behavior of the models.

Acknowledgments
This work was supported by the Global COE Program Core Research and Engineering of Advanced Materials-Interdisciplinary Education Center for Materials Science, MEXT, Japan.

Appendix. Algorithm of the developed enumeration methods for N-step SAWs

A pseudo-code of the Wang-Landau method and the multicanonical method with a flow chart of the whole procedure to calculate c_N^* with statistical errors.

Algorithm 1. The Wang-Landau method and the multicanonical method (lines with WL were skipped in the latter algorithm)

Initialization

```
for the whole range of $E$ do
    $W(E) \leftarrow 1$
    $H(E) \leftarrow 0$
end for
make the first conformation $\omega_0$ and calculate its energy $E_0$
$f \leftarrow e$    ▷ $f$ is a modification factor of the weight function $W(E)$.
$T \leftarrow 10^7$ ▷ $T$ is a cycle of judgment for updating $f$.
$\lambda_{\mathrm{flat}} \leftarrow 0.7 \sim 0.95$
        ▷ $\lambda_{\mathrm{flat}}$ is a flatness parameter of judgment for updating $f$.
        ▷ These values of the initial settings are typical values.
```

Functions

```
function MOVE($\omega$)
    make a candidate state $\omega'$ from $\omega$
    returning $\omega'$
end function

function ENERGY($\omega$)
    calculate energy $E$ of $\omega$ from $\mathcal{H}$
    returning $E$
end function

function JUDGE($E_a$,$E_b$,$W(E)$)
    generate a uniform random number $r \in [0,1)$ $(r \in \mathbb{R})$
    if $r < \frac{W(E_b)}{W(E_a)}$ then
        returning 1
    else
        returning 0
    end if
end function

function UPDATE($H(E)$)
    $U \leftarrow 1$
    calculate the average of the histogram $H(E)$, $\overline{H}$
    for the whole range of $E$ do
            ▷ In some cases, the range of $E$ is limited by hand
        if $H(E) < \overline{H} \cdot \lambda_{\mathrm{flat}}$ then
            $U \leftarrow 0$
        end if
    end for
end function

function INITIALIZE($H(E)$)
    for the whole range of $E$ do
        $H(E) \leftarrow 0$
    end for
end function
```

Main part

```
while $f - 1 > 10^{-8}$ do
    for $t := 0$ to $T - 1$ do
        $\omega_t' \leftarrow$ MOVE($\omega_t$)
        $E_t' \leftarrow$ ENERGY($\omega_t'$)
        $J \leftarrow$ JUDGE($E_t$,$E_t'$,$W(E)$)
        if $J = 1$ then
            $\omega_{t+1} \leftarrow \omega_t'$
            $E_{t+1} \leftarrow E_t'$
        else
            $\omega_{t+1} \leftarrow \omega_t$
            $E_{t+1} \leftarrow E_t$
        end if
        $W(E_{t+1}) \leftarrow W(E_{t+1})/f$                ▷ WL
        $H(E_{t+1}) \leftarrow H(E_{t+1}) + 1$
    end for
    $U \leftarrow$ UPDATE($H(E)$)                           ▷ WL
    if $U = 1$ then                                          ▷ WL
        $f \leftarrow f^{\frac{1}{2}}$                       ▷ WL
    end if                                                   ▷ WL
    INITIALIZE($H(E)$)                                       ▷ WL
end while
```

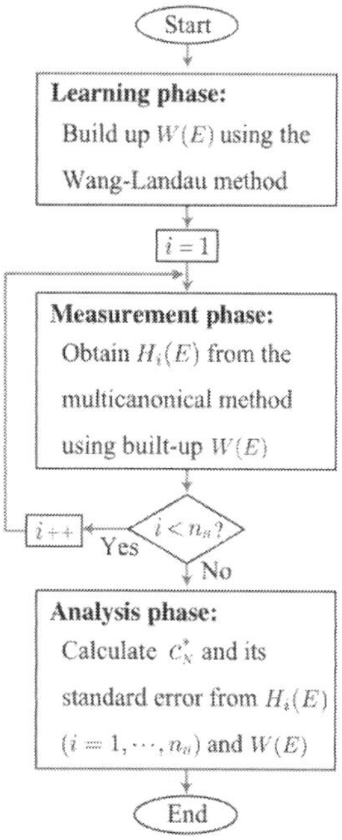

Figure 8. Flow chart of the developed enumeration methods for N-step SAWs. n_H is the number of Histograms.

References

[1] Madras N and Slade G 1993 *The Self-Avoiding Walk* (Boston, MA: Birkhäuser)
[2] Jensen I 2004 *J. Phys. A: Math. Gen.* **37** 5503
[3] Domb C and Joyce G S 1972 *J. Phys. C: Solid State Phys.* **5** 956
[4] Taketomi H, Ueda Y and Gō N 1975 *Int. J. Peptide Protein Res.* **7** 445
[5] Gō N and Taketomi H 1978 *Proc. Natl. Acad. Sci. USA* **75** 559
[6] Gō N 1983 *Ann. Rev. Biophy. Bioeng.* **12** 183
[7] Lau K F and Dill K A 1989 *Macromolecules* **22** 3986
[8] Onuchic J N, Luthey-Schulten Z and Wolynes P G 1997 *Annu. Rev. Phys. Chem.* **48**, 545
[9] Dill K A and Chan H S 1997 *Nat. Struct. Mol. Biol.* **4** 10
[10] Berg B A and Neuhaus T 1991 *Phys. Lett. B* **267** 249
[11] Berg B A and Neuhaus T 1992 *Phys. Rev. Lett.* **68** 9
[12] Wang F and Landau D P 2001 *Phys. Rev. Lett.* **86** 2050
[13] Wang F and Landau D P 2001 *Phys. Rev. E* **64** 056101
[14] Iba Y, Chikenji G and Kikuchi M 1998 *J. Phys. Soc. Jan* **67** 3327
[15] Chikenji G, Kikuchi M and Iba Y 1999 *Phys. Rev. Lett.* **83** 1886
[16] Lee J 1993 *Phys. Rev. Lett.* **71** 211

Journal of Physics: Conference Series **454** (2013) 012040 doi:10.1088/1742-6596/454/1/012040

Empirical study of the GARCH model with rational errors

Ting Ting Chen[1] **and Tetsuya Takaishi**[2]

[1]Faculty of Integrated Arts and Sciences, Hiroshima University, Higashi-Hiroshima 739-8521, Japan

[2]Hiroshima University of Economics, Hiroshima 731-0192, JAPAN

E-mail: [1]`d102355@hiroshima-u.ac.jp`

E-mail: [2]`tt-taka@hue.ac.jp`

Abstract. We use the GARCH model with a fat-tailed error distribution described by a rational function and apply it to stock price data on the Tokyo Stock Exchange. To determine the model parameters we perform Bayesian inference to the model. Bayesian inference is implemented by the Metropolis-Hastings algorithm with an adaptive multi-dimensional Student's t-proposal density. In order to compare our model with the GARCH model with the standard normal errors, we calculate the information criteria AIC and DIC, and find that both criteria favor the GARCH model with a rational error distribution. We also calculate the accuracy of the volatility by using the realized volatility and find that a good accuracy is obtained for the GARCH model with a rational error distribution. Thus we conclude that the GARCH model with a rational error distribution is superior to the GARCH model with the normal errors and it can be used as an alternative GARCH model to those with other fat-tailed distributions.

1. Introduction

In finance volatility plays a central role for risk management such as derivative price estimation and portfolio allocation for which it is important to measure a reliable volatility from the data observed in the financial markets. Since volatility is not a direct observable in the financial markets we have to rely on a certain estimation technique. Parametric volatility models designed to capture asset return and volatility properties are often used in empirical finance. The most popular and successful model is the GARCH model[1], which is a generalized version of the ARCH model[2].

In the GARCH model asset returns r_t at time t are expressed as $r_t = \sigma_t \epsilon_t$ where σ_t^2 is the time-changing volatility which is given by a function of past returns and past volatilities. In the original GARCH model the standard normal distribution, i.e. $N(0,1)$ was used for ϵ_t errors. It is known that the GARCH model well captures relevant properties of asset returns such as the fat-tailed behavior of the return distribution and the volatility clustering that are classified as stylized facts[3]. On the other hand in empirical studies it is often observed that the GARCH model does not sufficiently account for the leptokurtosis of the return distributions. To circumvent this it is advocated to apply a different distribution having a fatter tail than that of the normal distribution for the ϵ_t error. Several distributional forms having a fatter tail than the normal distribution such as Student's t-distribution[4] and the generalized error distribution

(GED)[5] are applied for the ϵ_t error term. By using the Student's t-distributions or GED for ϵ_t errors, usually one gets a better goodness-of-fit to the financial return data. However the Student's t-distributions and GED are not necessarily the optimal solution for the ϵ_t error term of the GARCH model and one could also choose other fat-tailed distributions.

In this study we apply Padé approximants described by a rational function for the ϵ_t error term. The Padé approximants are flexible to approximate a function in a certain domain. In finance Padé approximants are used to describe the interest rate return distributions[6, 7], where the parameters of rational functions are obtained by fitting to the interest rate return distributions. Here we apply a rational function for the ϵ_t error of the GARCH model. In [8] the GARCH model with rational errors was investigated by using USD/JPN exchange rate returns and the goodness-of-fit by the Akaike information criterion (AIC)[9] and deviance information criterion (DIC)[10] showed that the GARCH model with rational errors is superior to the GARCH mode with normal errors.

We further investigate the effectiveness of the GARCH model with rational errors by using stock return data on the Tokyo Stock Exchange. In this study in order to clarify the model-effectiveness, in addition to AIC and DIC, we utilize realized volatility which is a model-free estimate of the integrated volatility. Using realized volatility as a proxy of the true volatility we calculate the accuracy of the volatility by a loss function for both models and using the loss function we determine which model is more effective.

2. GARCH model with normal error distribution

Bollerslev introduced the GARCH(p,q) model[1] which is a generalized version of the ARCH model[2]. The GARCH(p,q) model is expressed as

$$y_t = \sigma_t \epsilon_t, \tag{1}$$

and

$$\sigma_t^2 = \omega + \sum_{i=1}^{q} \alpha_i y_{t-i}^2 + \sum_{i=1}^{p} \beta_i \sigma_{t-i}^2, \tag{2}$$

where α_i, β_i and ω are parameters of the GARCH model. These parameters are determined so that the model matches the return data. Since the volatility σ_t^2 should be positive the GARCH parameters are restricted to $\omega > 0$, $\alpha_i > 0$ and $\beta_i > 0$ to ensure a positive volatility. ϵ_t is an independent normal error following $N(0,1)$ and the return time series is given by y_t. In this study we focus on the GARCH(1,1) model, i.e. $p = 1$ and $q = 1$, where the volatility process is given by

$$\sigma_t^2 = \omega + \alpha y_{t-1}^2 + \beta \sigma_{t-1}^2, \tag{3}$$

and hereafter the GARCH model simply denotes the GARCH(1,1) model. Moreover for the GARCH model with normal errors we call it the GARCH-N model.

3. GARCH model with rational error distribution

In general a rational function of the Padé approximants is expressed by two polynomial functions $T_M(x)$ and $B_N(x)$ as

$$P_{M,N}(x) = \frac{T_M(x)}{B_N(x)}, \tag{4}$$

where M and N stand for the degrees of the polynomial $T_M(x)$ and $B_N(x)$ respectively. In order to consider $P_{M,N}(x)$ as a probability distribution we have to impose conditions that it must be positive and normalized to 1. Furthermore similar to the normal distribution we assume that $P_{M,N}(x)$ takes a maximum value at the origin and it is symmetric to the $x = 0$ axis, i.e. $P_{M,N}(x) = P_{M,N}(-x)$. In [6] possible normalizable distributions with finite variances are derived to approximate the interest rate distributions. The simplest normalized probability distribution with tunable parameters a_1 and a_2 is given by

$$P_{0,4}(x) = \frac{a_1}{\pi(1 + (a_1^2 + 2a_2)x^2 + a_2^2x^4)}. \tag{5}$$

The variance of this probability distribution is calculated to be $-1/a_2$. Since usually the variance of the GARCH error distribution is set to 1 we also set the variance of $P_{0,4}(x)$ to 1, i.e. $a_2 = -1$. Finally we obtain our rational error distribution for the GARCH model as

$$P(x) = \frac{a}{\pi(1 + (a^2 - 2)x^2 + x^4)}. \tag{6}$$

When we use the rational error distribution of (6) for ϵ_t of the GARCH model we call it the GARCH model with rational errors (GARCH-RE model)[8].

4. Realized volatility

Recent availability of high frequency financial data enables us to calculate the realized volatility constructed as a sum of squared intraday returns[11, 12, 13, 14], see also e.g.[15]. Let us assume that the logarithmic price process $\ln p(s)$ follows a continuous time stochastic diffusion,

$$d\ln p(s) = \tilde{\sigma}(s)dW(s), \tag{7}$$

where $W(s)$ stands for a standard Brownian motion and $\tilde{\sigma}(s)$ is a spot volatility at time s. Under this assumption the integrated volatility defined by

$$\sigma_h^2(t) = \int_t^{t+h} \tilde{\sigma}(s)^2 ds, \tag{8}$$

where h stands for the interval to be integrated. In empirical finance "daily volatility" is of primary importance and for the daily volatility h takes one day. Since $\tilde{\sigma}(s)$ is latent and not observed in the financial markets, (8) can not be evaluated analytically.

Let us define a sampling period Δ by $\Delta = h/n$, i.e. we sample n returns in the time interval of h. Then the $i-th$ intraday return on the day t with Δ sampling period is given by a log-price difference as

$$r_{t+i\Delta} = \ln P_{t+i\Delta} - \ln P_{t+(i-1)\Delta}, \tag{9}$$

where P_t is an asset price at time t. Using these intraday returns the realized volatility RV_t on the day t is given by a sum of squared intraday returns as

$$RV_t = \sum_{i=1}^{n} r_{t+i\Delta}^2. \tag{10}$$

Under ideal circumstance RV_t is proved to converge to the integrated volatility of (8) in the limit of $n \to \infty$. However in the real financial markets there exist several types of bias such as microstructure noise[16], and thus in the presence of the bias the convergence of RV_t to the integrated volatility is not guaranteed. Let us assume that the log-price observed in financial markets is contaminated with independent noise[17], i.e.

$$\ln P_t^* = \ln P_t + \xi_t, \tag{11}$$

where $\ln P_t^*$ is the observed log-price in the markets which consists of the true log-price $\ln P_t$ and noise ξ_t with mean 0 and variance ρ^2. Under this assumption the observed return r_t^* is given by

$$r_t^* = r_t + \eta_t, \tag{12}$$

where $\eta_t = \xi_t - \xi_{t-\Delta}$. Thus RV_t^* actually observed from the market data is obtained as a sum of the squared returns r_t^*,

$$RV_t^* = \sum_{i=1}^{n}(r_{t+i\Delta}^*)^2, \tag{13}$$

$$= RV_t + 2\sum_{i=1}^{n} r_{t+i\Delta}\eta_{t+i\Delta} + \sum_{i=1}^{n}\eta_{t+i\Delta}^2. \tag{14}$$

With these independent noises the bias appears as $\sum_{i=1}^{n}\eta_{t+i\Delta}^2$ which corresponds to $\sim 2n\rho^2$. Thus due to the bias the RV_t^* diverges as $n \to \infty$.

Practically in order to avoid the distortion from microstructure noise one needs to choose a good sampling period which reduces the microstructure noise bias and at the same time to maintain the accuracy of the realized volatility. The optimal sampling period is suggested to be around 5min[18]. One could also use kernel-based estimations which are designed to reduce the microstructure noise[17, 19, 20].

Another type of bias is due to "non-trading hours". Since stock markets are not open 24 hours the high-frequency data are only available for a part of 24 hours. At the Tokyo stock exchange market domestic stocks are traded in the two trading sessions: (a) morning trading session 9:00-11:00. (b) afternoon trading session 12:30-15:00. The daily realized volatility calculated without including intraday returns during the non-trading periods can be underestimated. When we consider volatility only in each trading session[21, 22] this bias problem does not arise. Otherwise we need to deal with this bias appropriately.

Hansen and Lunde[23] advocated an idea to circumvent the problem by introducing an adjustment factor which modifies the realized volatility so that the average of the realized volatility matches the variance of the daily returns. Let $(R_1, ..., R_N)$ be N daily returns constructed by close-close daily log-price difference. The adjustment factor c (HL adjustment factor) is given by

$$c = \frac{\sum_{t=1}^{N}(R_t - \bar{R})^2}{\sum_{t=1}^{N} RV_t}, \tag{15}$$

where $\bar{R}$ denotes the average of R_t. Then using this factor the daily realized volatility is modified to cRV_t. Although originally the HL adjustment factor is introduced to correct the bias of the non-trading hours it can also correct the microstructure noise bias effects to some extent.

5. Empirical results

In this study we analyze the stock price data of Panasonic Co. traded on the Tokyo Stock Exchange. This stock is listed in the Topix core 30 index which includes the 30 most liquid

Table 1. Results of the Bayesian inference for GARCH-RE and GARCH-N models. The values marked by $*$ show the autocorrelation time τ_{int} defined by $\tau_{int} = 1 + 2\sum_{t=1}^{\infty} ACF(t)$, where $ACF(t)$ stands for the autocorrelation function.

	GARCH-RE		GARCH-N	
α	0.132(38)	5.4(10)*	0.148(31)	2.7(3)*
β	0.858(41)	5.8(12)*	0.836(33)	2.8(2)*
ω	$2.8(1.2) \times 10^{-5}$	6.5(16)*	$1.3(5) \times 10^{-5}$	3.3(4)*
a	1.57(9)	4.2(6)*	–	–
AIC	-4151.29		-4148.35	
DIC	-4156.30		-4151.98	

and highly market capitalized stocks. Our data set begins June 3, 2006 and ends December 30, 2009. Figure 1 shows the daily return time series of Panasonic Co. We apply the GARCH-RE and GARCH-N models for the daily returns shown in Figure 1 and estimate the daily volatilities corresponding to those daily returns. The parameter estimation of the GARCH-RE and GARCH-N models is conducted by the Bayesian inference. A popular approach to perform the the Bayesian inference is the Markov Chain Monte Carlo (MCMC) methods. Since there exist a variety of MCMC methods we need to choose an adequate method for the Bayesian inference of the GARCH model. We perform the Bayesian inference by the Metropolis-Hastings algorithm[24, 25] with an adaptive multi-dimensional Student's t-proposal density (MHAS algorithm)[26, 27, 28, 29]. In the MH algorithm we need to specify the proposal density. In [30] the proposal densities constructed from an auxiliary process are used for the MH algorithm. References [31, 32] use a multi-dimensional Student's t-proposal density for which density parameters are determined by the maximum likelihood method. Here we use MHAS algorithm where density parameters of a multi-dimensional Student's t-proposal density are determined adaptively during the Monte Carlo simulations so that the multi-dimensional Student's t-proposal density matches the posterior distributions of the model. The MHAS algorithm has been shown to be very efficient for the Bayesian inference of the GARCH models[26, 27, 28, 29]. The implementation of the MHAS algorithm was done as follows. We discarded the first 6000 Monte Carlo updates by the MHAS algorithm. Then we accumulated 50000 Monte Carlo samples for analysis. Table 1 shows the values of the parameters averaged over the Monte Carlo samples. The values marked by $*$ show the autocorrelation time of the Monte Carlo data generated by the MHAS algorithm. We find that the values of the autocorrelation time are small which indicates that the MHAS algorithm generates effectively uncorrelated Monte Carlo samples.

In order to compare the goodness-of-fit of the models we utilize two information criteria: AIC[9] and DIC[10]. The AIC is defined by $AIC = -\ln L(\bar{\theta}) - 2k$ where k is the number of the parameters of the model and $L(\bar{\theta})$ is the likelihood function of the model at $\bar{\theta}$. θ stands for $\theta = (\alpha, \beta, \omega, a)$ for the GARCH-RE model and $\theta = (\alpha, \beta, \omega)$ for the GARCH-N model.

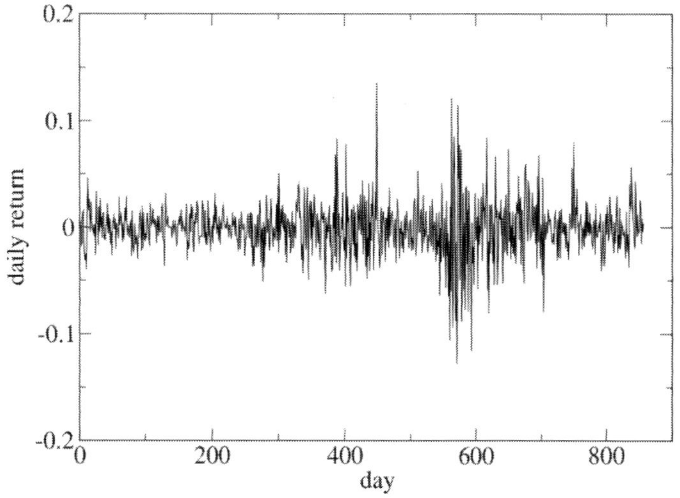

Figure 1. Daily return time series of Panasonic Co.

$\bar{\theta}$ stands for the parameters averaged over the Monte Carlo samples. The DIC is defined by $2[\ln L(\bar{\theta}) - 2E(\ln L(\theta))]$ where $E(\ln L(\theta))$ is the Monte Carlo average of $\ln L(\theta)$. For both AIC and DIC the model with the smallest value is chosen as the one which would best predict the time series. As seen in Table 1 both of AIC and DIC give smaller values for the GARCH-RE model. Thus we find that the GARCH-RE model is superior to the GARCH-N model.

Next we compare the GARCH-RE and GARCH-N models with the accuracy of the volatility. To do that we measure the difference between the volatility from the models and the true volatility. Since we do not know the value of the true volatility we use the realized volatility as a proxy of the true volatility. The realized volatility is constructed by a sum of squared intraday returns as (10). Figure 2 shows the average realized volatility as a function of sampling period, i.e. the realized volatility is averaged at each sampling period. Such plot is call "volatility signature plot" advocated in [33] to visualize the microstructure noise bias on the realized volatility. As expected in (14) we find that the realized volatility diverges at small sampling period (or at high sampling frequency). In this study we correct this bias with the HL adjustment factor which adjusts the average of the realized volatility to the variance of the daily return.

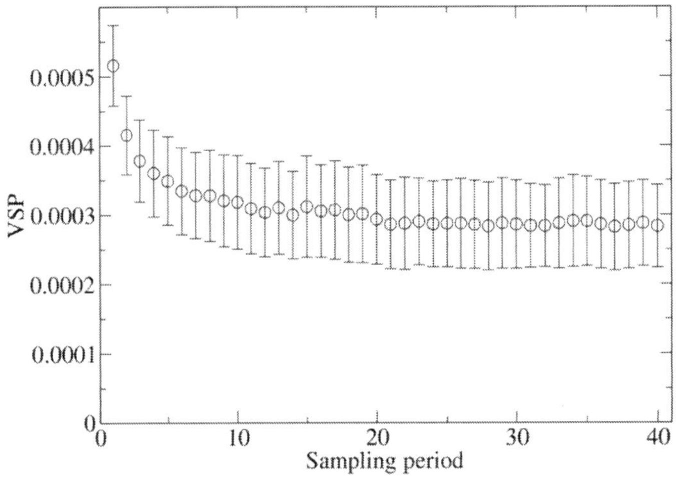

Figure 2. Volatility signature plot: Average realized volatility at each sampling period.

Figure 3 shows the HL adjustment factor as a function of sampling period. The HL adjustment factor decreases as the sampling period decreases. This decrease is explained by the microstructure noise bias which inflates the realized volatility at small sampling periods. As the sampling period increases the HL adjustment factor reaches a plateau around 2 where the microstructure noise bias effects are expected to be small. This factor of 2 means that the original realized volatility is underestimated due to non-trading hours and the size of the volatility during no-trading hours is about the same size as that during the trading hours. On the Tokyo Stock Exchange there are two non-trading periods: lunch break and night break. Since during the lunch break the size of the volatility is observed to be small[21] the dominant contribution to the factor of 2 comes from the night break.

Figure 4 compares volatilities from the GARCH-RE and GARCH-N models, and the realized volatility at 1-min sampling period. To quantify the accuracy of the volatility we measure a loss

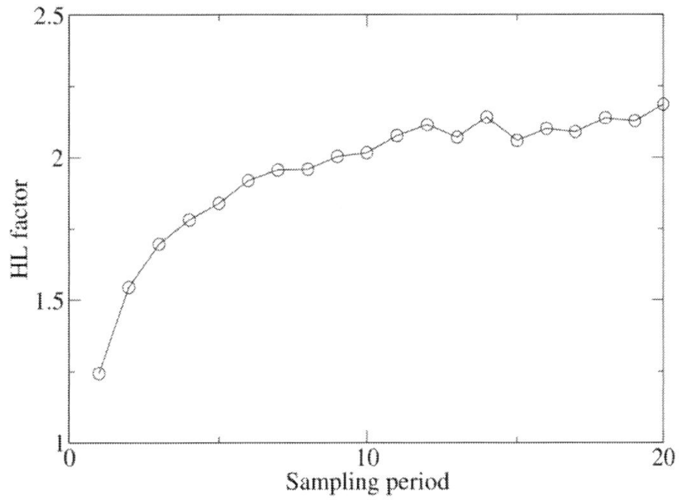

Figure 3. HL adjustment factor as a function of sampling period.

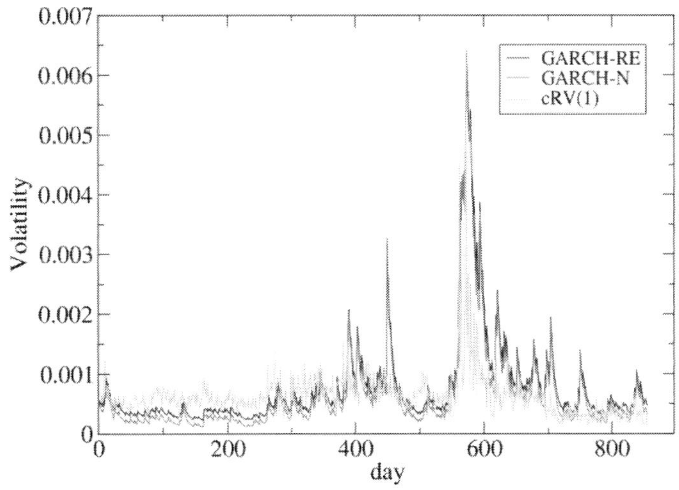

Figure 4. Volatility from GARCH-RE, GARCH-N models and the realized volatility at 1-min sampling period. The realized volatility is adjusted by the HL adjustment factor.

function of the root mean square percentage error (RMSPE) defined by

$$RMSPE = \left(\sum_{t=1}^{N} (\frac{\bar{\sigma_t^2} - cRV_t}{cRV_t})^2 \right)^{1/2}, \qquad (16)$$

where $\bar{\sigma_t^2}$ is the volatility estimated from the Bayesian inference of the GARCH-RE or GARCH-N models. $\bar{\sigma_t^2}$ is also adjusted so that the average of $\bar{\sigma_t^2}$, i.e. $\sum_{t=1}^{N} \bar{\sigma_t^2}/N$ matches the variance

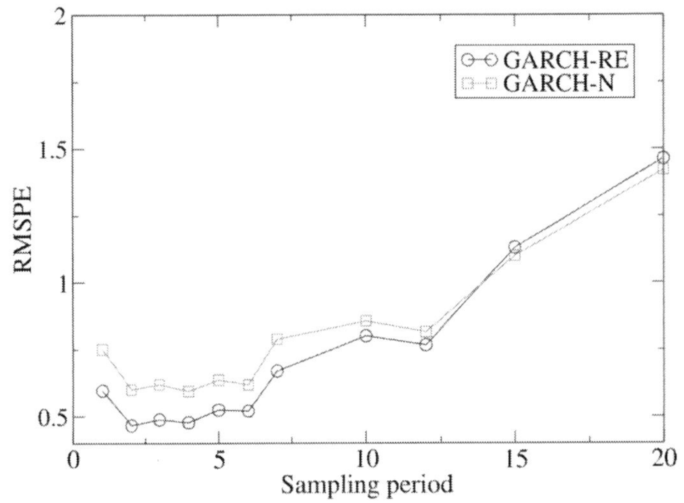

Figure 5. RMSPE of GARCH-RE and GARCH-N models as a function of sampling period.

of the daily returns. Figure 5 shows RMSPE of the GARCH-RE and GARCH-N models. We find that RMSPE takes a minimum around 1 to 6-min sampling periods where the GARCH-RE model gives smaller values. It is also noted that the sampling frequencies which take the minimum of the RMSPE are very similar to the optimum sampling frequencies obtained from the mean squared error of the realized volatility[18]. Our result of the RMSPE also indicates that the GARCH-RE model is more effective than the GARCH-N model.

6. Conclusions

We performed Bayesian inference of the GARCH-RE model and the GARCH-N model for the stock price data of Panasonic Co. on the Tokyo Stock Exchange. The Bayesian inference is implemented by the MHAS algorithm. In order to compare models we calculated the information criteria AIC and DIC, and find that both criteria favor the GARCH-RE model. We also calculated the accuracy of the volatility by the RMSPE and found that the smaller RMSPE is obtained for the GARCH-RE model. Thus we conclude that the GARCH-RE model is superior to the GARCH-N model and it can be used as an alternative GARCH model to those with other fat-tailed distributions.

Acknowledgments

Numerical calculations in this work were carried out at the Yukawa Institute Computer Facility and at the facilities of the Institute of Statistical Mathematics. This work was supported by Grant-in-Aid for Scientific Research (C) (No.22500267).

References

[1] Bollerslev T 1986 Generalized Autoregressive Conditional Heteroskedasticity *Journal of Econometrics* **31** 307-327

[2] Engle R F 1982 Autoregressive Conditional Heteroskedasticity with Estimates of the Variance of the United Kingdom inflation *Econometrica* **50** 987-1007

[3] Cont R 2001 Empirical Properties of Asset Returns: Stylized Facts and Statistical Issues *Quantitative Finance* **1** 223–236

[4] Bollerslev T 1987 A conditional heteroskedastic time series model for speculative prices and rates of returns *Review of Economics and Statistics* **69** 542-547
[5] Nelson D 1991 Conditional heteroskedasticity in asset returns: A new approach *Econometrica* **59** 347-370
[6] Nuyts J and Platten I 2001 Phenomenology of the term structure of interest rates with Pade Approximants *Physica A* **299** 528-546
[7] Alderweireld T and Nuyts J 2004 Detailed empirical study of the term structure of interest rates. Emergence of power laws and scaling laws *Physica A* **331** 602-616
[8] Takaishi T and Chen TT 2012 Bayesian Inference of the GARCH model with Rational Errors *International Proceedings of Economics Development and Research* **29** 303–307
[9] Akaike H 1973 Information theory and an extension of the maximum likelihood principle *Proceedings of the 2nd International Symposium on Information Theory* Petrov B. N. and Caski F (eds.), Akadimiai Kiado, Budapest: 267-281
[10] Spiegelhalter D J *et al.* 2002 Bayesian Measures of Model Complexity and Fit *Journal of the Royal Statistical Society* **64(4)** 583-639
[11] Andersen T G and Bollerslev T 1998 Answering the Skeptics: Yes, Standard Volatility Models do Provide Accurate Forecasts *International Economic Review* **39** 885–905
[12] Andersen T G, Bollerslev T, Diebold F X and Labys P 2000 Exchange Rate Returns Standardized by Realized Volatility are (Nealy) Gaussian *Multinational Finance Journal* **4** 159–179
[13] Andersen T G, Bollerslev T, Diebold F X and Labys P 2001 The distribution of realized exchange rate volatility *Journal of the American Statistical Association* **96** 42–55
[14] Andersen T G, Bollerslev T, Diebold F X and Ebens H 2001 The distribution of realized stock return volatility *Journal of Financial Economics* **61** 43–76
[15] McAleer M and Medeiros M C 2008 Realized Volatility: A Review *Econometric Reviews* **27** 10-45
[16] Campbell J Y, Lo A W and MacKinlay A C 1997 The Econometrics of Financial Markets Princeton University Press
[17] Zhou B 1996 High-frequency data and volatility in foreign-exchange rates *Journal of Business & Economics Statistics* **14** 45–52
[18] Bandi F M and Russell J R 2008 Microstructure noise, realized variance, and optimal sampling *The Review of Economic Studies* **75** 339-369
[19] Hansen P R and Lunde A 2006 Realized variance and market microstructure noise *Journal of Business and Economics Statistics* **24** 127-218
[20] Barndorff-Nielsen O E, Hansen P R, Lunde A and Shephard N 2008 Designing realized kernels to measure the ex-post variation of equity prices in the presence of noise *Econometrica* **76** 1481-1536
[21] Takaishi T, Chen TT and Zheng Z 2012 Analysis of Realized Volatility in Two Trading Sessions of the Japanese Stock Market *Prog. Theor. Phys. Supplement* **194** 43-54
[22] Takaishi T 2012 Finite-Sample Effects on the Standardized Returns of the Tokyo Stock Exchange *Procedia - Social and Behavioral Sciences* **65** 968–973
[23] Hansen P R and Lunde A 2005 A forecast comparison of volatility models: does anything beat a GARCH(1,1)? *Journal of Applied Econometrics* **20** 873-889
[24] Metropolis N *et al.* 1953 Equations of State Calculations by Fast Computing Machines *J. of Chem. Phys.* **21** 1087-1091
[25] Hastings W K 1970 Monte Carlo Sampling Methods Using Markov Chains and Their Applications *Biometrika* **57** 97-109
[26] Takaishi T 2009 An Adaptive Markov Chain Monte Carlo Method for GARCH Model *Lecture Notes of the Institute for Computer Sciences, Social Informatics and Telecommunications Engineering. Complex Sciences* Vol. **5** 1424
[27] Takaishi T 2009 Bayesian Estimation of GARCH Model with an Adaptive Proposal Density *New Advances in Intelligent Decision Technologies, Studies in Computational Intelligence* Vol. **199** 635-653
[28] Takaishi T 2009 Bayesian Inference on QGARCH Model Using the Adaptive Construction Scheme *Proceedings of 8th IEEE/ACIS International Conference on Computer and Information Science* 525-529 doi:10.1109/ICIS.2009.173
[29] Takaishi T 2010 Bayesian inference with an adaptive proposal density for GARCH models *J. Phys.: Conf. Ser.* **221** 012011
[30] Nakatsuma T 2000 Bayesian analysis of ARMA-GARCH models: Markov chain sampling approach *Journal of Econometrics* **95** 57–69
[31] Mitsui H and Watanabe T 2003 Bayesian analysis of GARCH option pricing models *J. Japan Statist. Soc. (Japanese Issue)* **33** 307–324
[32] Asai M 2006 Comparison of MCMC Methods for Estimating GARCH Models *J. Japan Statist. Soc.* **36** 199-212

[33] Andersen T G, Bollerslev T, Diebold F X and Labys P 2000 Great Realization, *Risk March* 105–108

Journal of Physics: Conference Series **454** (2013) 012041 doi:10.1088/1742-6596/454/1/012041

Analysis of Spin Financial Market by GARCH Model

Tetsuya Takaishi

Hiroshima University of Economics, Hiroshima 731-0192, Japan

E-mail: tt-taka@hue.ac.jp

Abstract. A spin model is used for simulations of financial markets. To determine return volatility in the spin financial market we use the GARCH model often used for volatility estimation in empirical finance. We apply the Bayesian inference performed by the Markov Chain Monte Carlo method to the parameter estimation of the GARCH model. It is found that volatility determined by the GARCH model exhibits "volatility clustering" also observed in the real financial markets. Using volatility determined by the GARCH model we examine the mixture-of-distribution hypothesis (MDH) suggested for the asset return dynamics. We find that the returns standardized by volatility are approximately standard normal random variables. Moreover we find that the absolute standardized returns show no significant autocorrelation. These findings are consistent with the view of the MDH for the return dynamics.

1. Introduction

Statistical properties of asset price returns have been extensively studied and some pronounced properties are found and classified as stylized facts [1]. One of the stylized facts known for many years is that the probability distributions of returns exhibit fat-tailed distributions. This evidence indicates that asset price dynamics is not a simple Gaussian random walk. A possible origin for the fat-tailed distributions has been explained by the mixture-of-distribution hypothesis (MDH) [2] where the price return dynamics is described by a Gaussian random process with time-varying volatility. Let $r(t)$ be a return at time t. Then under the MDH the return is described by $r(t) = \sigma_t \epsilon_t$, where σ_t is a standard deviation and ϵ_t is a Gaussian random number with variance 1 and mean 0. In finance σ_t^2 is also called "volatility". Using high-frequency financial data the MDH has been examined and evidence on the MDH has been found [3, 4, 5, 6, 7, 8, 9].

Empirically it is well known that the volatility changes with time and exhibits persistence of the same magnitude of the volatility. This feature of the volatility is called "volatility clustering" which is also one of the stylized facts. In order to forecast future volatility one need to use volatility models which possess the properties of return and volatility. The most successful models in empirical finance are the ARCH model [10] and its generalized version, the GARCH model [11]. Since the invention of the ARCH and GARCH models many extended models have been also proposed and applied for empirical finance. Some examples of those models are EGARCH [12], QGARCH [13, 14], GJR [15], APARCH [16] and GARCH-RE [17] models etc. See also, e.g., [18].

In physics to understand financial market dynamics a variety of agent-based models have been proposed and examined [19, 20, 21, 22, 23, 24, 25, 27, 28, 29, 30, 31, 32, 33]. It is found that to some extent those models are able to capture some of stylized facts such as fat-tailed distributions of returns and long autocorrelation times of the absolute returns.

In this study we perform simulations of financial markets in a three-state spin model [32] which is an extended version of the two-state Bornholdt model [27] and determine volatility of returns simulated in the model. In the real financial markets volatility itself is not directly observed. Therefore in order to infer the latent volatility one usually utilizes parametric models such as the GARCH model. Here we also take the same approach and use the GARCH model to determine the spin-market volatility. After determining the volatility we further examine the view of the MDH for the return dynamics of the financial spin model.

2. Three-state spin model

The model we use here is a Potts-like model [32] in which agents (or spins S_i) locate on one of lattice sites, having one of three states $(1,-1,0)$. The three states $(1,-1,0)$ are assigned to "buy", "sell" and "inactive" orders respectively. The model includes two interactions which conflict each other. One is the nearest neighbor interaction which causes the ferromagnetic order or the majority effect. In other words with this interaction the agents tend to imitate their neighbor agents and as a result belong to the majority group. In the real financial markets this imitation corresponds to the herding behavior. The other interaction is a global interaction proportional to the magnitude of the magnetization, which causes the anti-ferromagnetic order or the minority effect. When the magnitude of the magnetization is big, most of spins takes 1 (or -1). Such a state corresponds to "bubble" state in the real financial markets. When the bubble state appears agents should change their spins to the minority group. By doing so agents are able to avoid future possible loss when the "bubble" economy bursts. These two interactions conflicting each other cause a complicated dynamics. For instance we observe no single stable phase, but instead find ferromagnetic and anti-ferromagnetic phases repeatedly. We again come to this point below.

Based on the heat bath dynamics proposed by Bornholdt [27] we update spins according to the following probability.

$$P(S_i \to S_i') = \begin{cases} \exp\left(\lambda(h(i,S_i') - \mu S_i S_i' |M|)\right)/C & \text{for } S_i' = 1,-1 \\ \exp\left(\lambda(h(i,S_i') - \mu(\epsilon_i |M| - 2\gamma K))\right)/C & \text{for } S_i' = 0 \end{cases} \tag{1}$$

where

$$h(i,S_i') = \sum_{\langle i,j \rangle} \delta_{S_j,S_i'}, \tag{2}$$

$\langle i,j \rangle$ stands for the summation over the nearest neighbors j of the site i and ϵ_i is defined by $\epsilon_i = 2\delta_{S_i,0} - 1$ which takes 1 for $S_i = 0$ and -1 for $S_i \neq 0$. C is the normalization factor determined so that the following equation is satisfied.

$$\sum_{k=1,-1,0} P(S_i \to k) = 1. \tag{3}$$

M stands for the magnetization calculated by

$$M = \frac{1}{N}\sum_{i=1}^{N} S_i, \tag{4}$$

where N is the number of agents in the system and K is the inactivity rate given by

$$K = \frac{1}{N}\sum_{i=1}^{N} \delta_{S_i,0}. \tag{5}$$

The Hamiltonian of the model could be written as

$$H = -J\sum_i h(i, S_i) + \sum_i L_i, \tag{6}$$

where

$$L_i = \begin{cases} \mu\text{sign}(S_i)S_i|M| & \text{for } S_i = 1, -1 \\ \mu(\epsilon_i|M| - 2\gamma K) & \text{for } S_i = 0 \end{cases} \tag{7}$$

and $J = 1$ for a ferromagnetic coupling. The partition function $Z(\lambda, H)$ is given by

$$Z(\lambda, H) = \sum_{S_1,\ldots,S_N} \exp(-\lambda H). \tag{8}$$

To have an insight on the model let us consider two-state model (Ising-type model) for simplicity. For the two-state model the Hamiltonian can be written as

$$H = -J\sum_i h(i, S_i) + \sum_i \mu\text{sign}(S_i)S_i|M|. \tag{9}$$

With a redefined coupling $J = 2J'$ the nearest neighbor interaction term may be rewritten as

$$H = -J'\sum_i 2(h(i, S_i) - 2d) + \sum_i \mu\text{sign}(S_i)S_i|M|, \tag{10}$$

where d is the dimension of the model, i.e. $d = 2$ in this case. Equation (10) corresponds to the usual Ising model representation as

$$H = -J\sum_{\langle i,j\rangle} S_j S_i + \sum_i \mu\text{sign}(S_i)S_i|M|. \tag{11}$$

Let E_l be a term containing S_l and we obtain

$$E_l = -(J\sum_{\langle l,j\rangle} S_j - \mu\text{sign}(S_l)|M|)S_l. \tag{12}$$

Using the mean field $\phi = \langle S_l\rangle$ we obtain

$$E_l = -(Jz\phi - \mu\text{sign}(S_l)|M|)S_l, \tag{13}$$

where $z = 2d$ is the number of the neighbor sites of an agent S_l. Each agent may interact with some other agents. In the realistic situation the number of agents from which an agent can obtain information on the markets should be limited. Therefore a very big z for which the mean field approximation is appropriate may not be realistic and, certainly the infinite z is not. There also exists a warning that the mean field approximation should not be used for agent models [34]. Since we already know that simulations on $d = 2$ [27, 31] and on low dimensions [30] can successfully produce some similarity with the real financial markets such as the power-law return distribution, here we employ 2 dimensional lattice simulations.

Contrast to the original Ising model, a term $\mu\text{sign}(S_l)|M|$ behaves differently, i.e. it changes the sign depending on S_l and $|M|$ varies with time. These features cause a more complex behavior. Setting the temperature T ($\sim 1/\lambda$) below the critical temperature T_c and $\mu = 0$, the system of this model is in the ordered phase. In addition, assuming $|M|$ is a constant, [33] considered the phase diagram as a function of $|M|$ and found that the system undergoes the phase transition to the disordered phase at some critical $|M|_c$. In the disordered phase, M fluctuates strongly, that creates a bigger volatility. On the other hand in the ordered phase the fluctuation of M is small, which results in exhibiting a small volatility. When we let $|M|$ vary, the system can switch the phase to the other one repeatedly in simulations. The durations of the phases correspond to the appearance of the volatility clustering.

The introduction of γ in (1) gives a more flexibility to the model. By varying γ one can observe the exponential return distribution [32], in addition to the power-law distribution. Empirically the exponential return distributions have been also observed in the Indian market [35]. Since γ can change the number of agents who participate buy-or-sell transactions, γ might be related to tune the number of transactions or volume.

3. GARCH Model

The GARCH(p,q) model by Bollerslev [11] is given by

$$y_t = \sigma_t \epsilon_t, \tag{14}$$

$$\sigma_t^2 = \omega + \sum_{i=1}^{q} \alpha_i y_{t-i}^2 + \sum_{i=1}^{p} \beta_i \sigma_{t-i}^2, \tag{15}$$

where the GARCH parameters are restricted to $\omega > 0$, $\alpha_i > 0$ and $\beta_i > 0$ to ensure a positive volatility. ϵ_t is an independent normal error $\sim N(0,1)$ and returns are given by y_t.

We focus on the GARCH(1,1) model where the volatility process is given by

$$\sigma_t^2 = \omega + \alpha y_{t-1}^2 + \beta \sigma_{t-1}^2. \tag{16}$$

Although the GARCH(1,1) model is the simplest, in empirical studies the GARCH(1,1) model is often chosen as the best one by comparison of information criteria such as AIC [36]. In this study the GARCH(1,1) model is denoted simply by GARCH model.

The unconditional volatility $\bar{\sigma}^2$ can be found by substituting $\sigma_t^2 = \sigma_{t-1}^2 = \bar{\sigma}^2$ and $E[y_{t-1}^2] = \sigma_t^2$ to (16). For $(\alpha + \beta) < 1$, we obtain

$$\bar{\sigma}^2 = \frac{\omega}{1 - (\alpha + \beta)}. \tag{17}$$

The rate of convergence to the unconditional volatility can be measured by $\alpha + \beta$. When $\alpha + \beta$ is near 1, volatility persists very long. Empirically the value of $\alpha + \beta$ is often inferred to be close to 1.

The GARCH model includes three model parameters, α, β and ω. These parameters are determined so that the model matches the observed time series data. To determine the model parameters we employ the Bayesian inference performed by the Markov Chain Monte Carlo (MCMC) method. The MCMC method we use here is the Metropolis-Hastings method with an adaptive multi-dimensional student's t-distributions, which is shown to be very efficient for the Bayesian inference of the GARCH model [37, 38, 39, 40].

4. Simulations of three-state spin model

We use a 100×100 square lattice with the periodic boundary condition and start simulations on a random configuration. The spins are updated one by one in a random order. We define "one sweep" as 100^2 updates and use one sweep as unit time. We discard the first 3×10^4 sweeps as thermalization and then accumulate 10^4 sweeps for analysis. We make simulations for two parameter sets (a):$(\lambda, \mu, \gamma) = (4.0, 10.0, 1.0)$ and (b):$(\lambda, \mu, \gamma) = (4.0, 10.0, 0.05)$, i.e. λ and μ are fixed and γ is varied. It is found that the power-law return distribution is obtained for $\gamma = 1.0$ and the exponential return distribution for $\gamma = 0.05$ [32].

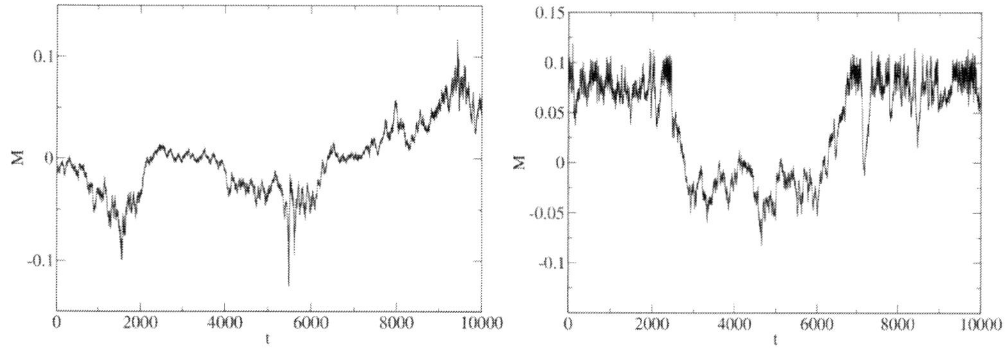

Figure 1. Magnetization $M(t)$ simulated at $\gamma = 1.0$ (left) and $\gamma = 0.05$ (right).

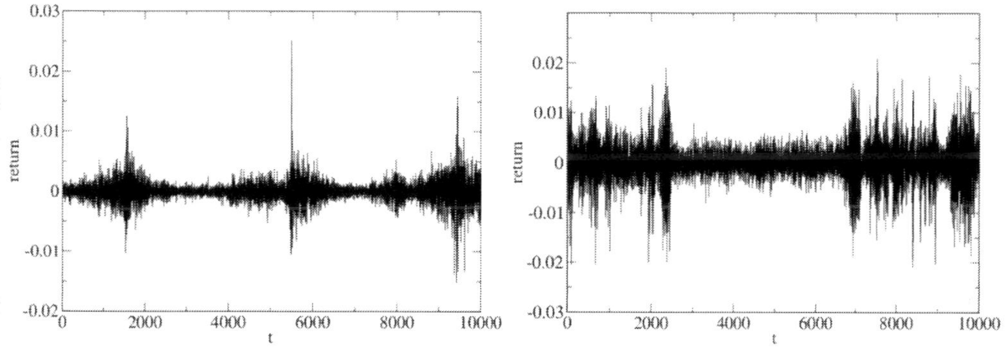

Figure 2. Return time series simulated at $\gamma = 1.0$ (left) and $\gamma = 0.05$ (right).

Figure 1 shows the magnetization $M(t)$ as a function of time (sweep). Following [31] we define the return $r(t)$ through the magnetization as

$$r(t) = M(t+1) - M(t), \tag{18}$$

where $M(t)$ is the magnetization given by (4). Figure 2 shows time series of return $r(t)$ where we observe the intermittency of returns. As mentioned in section 2 such intermittent behavior is expected to be caused by the phase change. Namely the ordered phase where $|M|$ and its fluctuation are small corresponds to low-volatility phase. Conversely the disordered phase corresponds to high-volatility phase. These properties can also be confirmed by comparing figures 1 and 2. Since values of volatility are not obtained directly from returns we estimate them by the GARCH model as used in the volatility estimation in empirical finance.

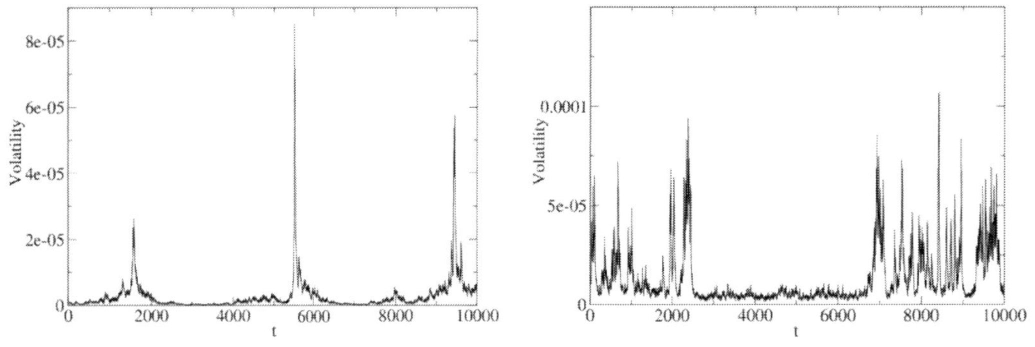

Figure 3. Volatility time series obtained by the GARCH model: $\gamma = 1.0$ (left) and $\gamma = 0.05$ (right).

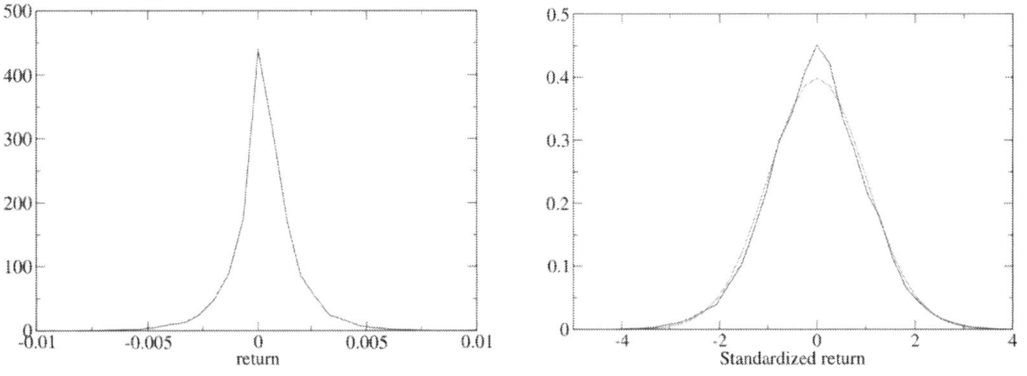

Figure 4. Return distribution (left) and standardized return distribution (right) simulated at $\gamma = 1.0$. The red line shows the standard normal distribution.

5. Volatility from the GARCH model

We apply the GARCH model for return data obtained from the financial spin model and estimate volatility corresponding to each return. The Bayesian inference of the GARCH model is performed by the Metropolis-Hastings algorithm with an adaptive multi-dimensional student's t-distributions [37, 38, 39, 40]. In total we make 4×10^4 Monte Carlo updates of GARCH parameters. The first 1.1×10^4 updates are discarded as thermalization and the 2.9×10^4 updates are used for analysis. At each update we also store the values of volatility and at the end we average the values of volatility for final output. The GARCH parameters determined by the Bayesian inference are listed in table 1. For both simulation parameters, values of $\alpha + \beta$ are found to be very close to 1, which means that the return time series obtained from those simulations have the strong persistency in volatility.

Figure 3 shows the time series of volatility obtained by the GARCH model. It is seen that there exist high-volatility and low-volatility periods. Such behavior of the volatility corresponds to "volatility clustering" which often observed in the real financial markets.

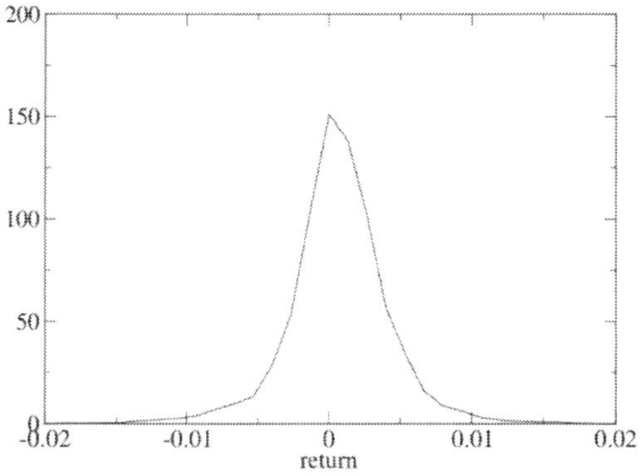

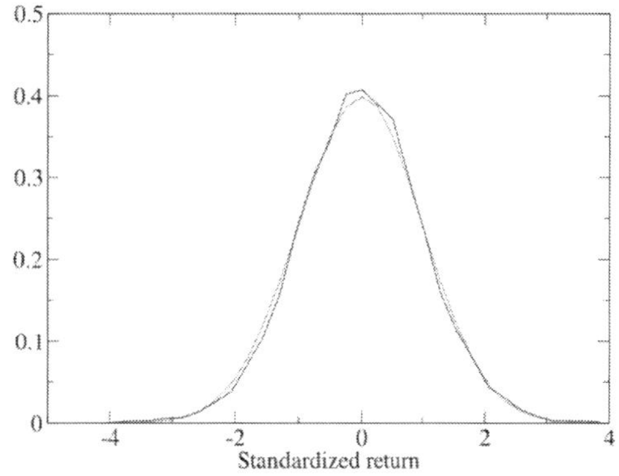

Figure 5. Return distribution (left) and standardized return distribution (right) simulated at $\gamma = 0.05$. The red line shows the standard normal distribution.

Table 1. Results of GARCH parameters. SD and SE stand for standard deviation and statistical error respectively. Statistical errors are estimated by the jackknife method. τ_{int} is the autocorrelation time defined by $\tau_{int} = 1 + 2\sum_{i=1}^{\infty} ACF(t)$ where $ACF(t)$ stands for the autocorrelation function.

	α	β	ω
$\gamma = 1.0$	0.0495	0.9508	2.1×10^{-9}
SD	0.0037	0.0013	6×10^{-10}
SE	0.00003	0.00003	4×10^{-12}
τ_{int}	1.8 ± 0.4	1.8 ± 0.3	1.8 ± 0.4
$\gamma = 0.05$	0.0836	0.9100	8.6×10^{-8}
SD	0.0055	0.0057	1.4×10^{-8}
SE	0.00003	0.00003	8×10^{-11}
τ_{int}	1.41 ± 0.05	1.40 ± 0.04	1.42 ± 0.05

Next we examine the view of the MDH for the return data from the financial spin model. In the MDH the return $r(t)$ is assumed to be given by $r(t) = \sigma_t \epsilon_t$, where σ_t^2 is volatility and ϵ_t is an independent Gaussian random variable with mean 0 and variance 1. Under this assumption the returns standardized by σ_t, i.e. $r(t)/\sigma_t$ should be standard normal random variables. Using volatility determined by the GARCH model as a proxy of the true volatility we standardize returns $r(t)$.

Figures 4 and 5 compare the distributions of the unstandardized and standardized returns. The unstandardized return distributions are very different from the standard normal distributions (variance 1 and mean 0). On the other hand the standardized return distributions come close to the standard normal distributions that are depicted in red. The normality of the standardized returns can be further confirmed by examining variance and kurtosis.

Table 2 lists variance and kurtosis of the unstandardized and standardized returns. The variance and kurtosis of the unstandardized returns differ from the values expected for standard normal random variables, i.e. variance = 1 and kurtosis = 3. On the other hand it is remarkable that the variances of standardized returns from both simulations at $\gamma = 1.0$ and 0.05 are consistent with 1. The values of kurtosis for standardized returns also come close to 3 although

Table 2. Variance and kurtosis of the unstandardized and standardized returns.

	$\gamma = 1.0$		$\gamma = 0.05$	
	variance	kurtosis	variance	kurtosis
unstandardized	$2.7 \times 10^{-6} \pm 9 \times 10^{-7}$	21 ± 9	$1.3 \times 10^{-5} \pm 2 \times 10^{-6}$	6.6 ± 0.8
standardized	0.997 ± 0.022	3.54 ± 0.09	0.999 ± 0.025	3.49 ± 0.10

their values are slightly higher than 3. This small disagreement might be understood by that the GARCH model we used here, i.e. the GARCH(1,1) model with normal errors, still does not capture completely the properties of the return data from the spin financial model. The similar disagreement has been observed in the real financial markets [6].

We further examine the MDH by calculating the autocorrelation function. As a stylized fact of asset returns it is known that the absolute return time series exhibits a very long correlation. However under the MDH the standardized returns are expected to be Gaussian random variables. Therefore if the MDH is hold we expect that the autocorrelation between the absolute standardized returns disappears. Figure 6 shows the autocorrelation functions of the absolute returns and absolute standardized returns. It is clear that the absolute returns show a very long correlation but the absolute standardized returns have no significant correlation. Thus this examination of the autocorrelation function also confirms the MDH for the return time series of the spin financial market. The similar behavior of the autocorrelation function on the absolute standardized returns is also observed in the real financial market [8].

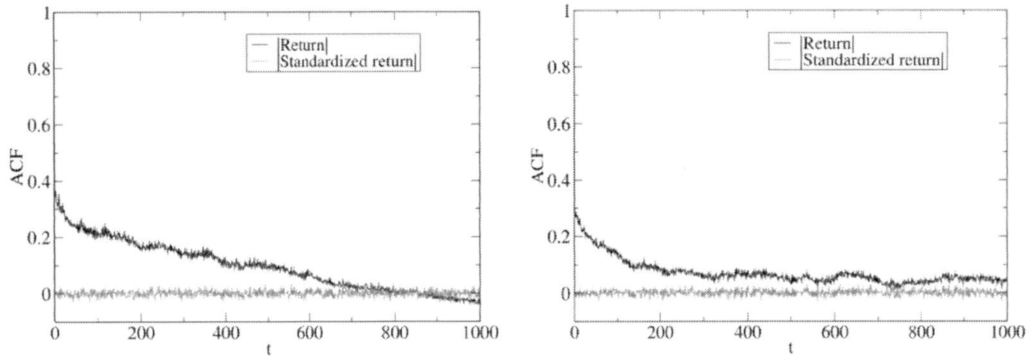

Figure 6. Autocorrelation functions of absolute returns and absolute standardized returns at $\gamma = 1.0$ (left) and $\gamma = 0.05$ (right).

6. Conclusions

The volatility of a financial spin model with three-states was investigated by the GARCH model. To determine volatility by the GARCH model the Bayesian inference by the MCMC method was applied. We find that volatility of the spin financial market exhibits "volatility clustering" which is often observed in the real financial markets. In order to examine the MDH for the spin financial market we analyzed returns standardized by the volatility obtained from the GARCH model. Under the MDH the normality is expected for the standardized returns. We find that variance and kurtosis of the standardized returns are very similar to those of standard normal random variables except for a slightly higher kurtosis. The autocorrelations of the absolute standardized returns were also investigated and no significant autocorrelation between absolute standardized returns was found, which can also be expected from the MDH. From these findings it is concluded that the spin financial market we simulated here is consistent with the view of the MDH for the return dynamics.

Acknowledgments

Numerical calculations in this work were carried out at the Yukawa Institute Computer Facility and the facilities of the Institute of Statistical Mathematics. This work was supported by Grant-in-Aid for Scientific Research (C) (No.22500267).

References

[1] Cont R 2001 Empirical Properties of Asset Returns: Stylized Facts and Statistical Issues *Quantitative Finance* **1** 223–236
[2] Clark P K 1973 A subordinated stochastic process model with finite variance for speculative prices *Econometrica* **41** 135-155
[3] Andersen T G, Bollerslev T, Diebold F X and Labys P 2000 Exchange Rate Returns Standardized by Realized Volatility are (Nealy) Gaussian *Multinational Finance Journal* **4** 159–179
[4] Andersen T G, Bollerslev T, Diebold F X and Ebens H 2001 The distribution of realized stock return volatility *Journal of Financial Economics* **61** 43–76
[5] Andersen T G, Bollerslev T and Dobrev, D 2007 No-arbitrage semi-martingale restrictions for continuous-time volatility models subject to leverage effects, jumps and i.i.d. noise: Theory and testable distributional implications *Journal of Econometrics* **138** 125-180
[6] Andersen T G, Bollerslev T, Frederiksen P and Nielsen M Ø 2010 Continuoustime models, realized volatilities, and testable distributional implications for daily stock returns *Journal of Applied Econometrics* **25** 233-261
[7] Fleming J and Paye B S 2011 High-frequency returns, jumps and the mixture of normals hypothesis *Journal of Econometrics* **160** 119-128
[8] Takaishi T, Chen TT and Zheng Z 2012 Analysis of Realized Volatility in Two Trading Sessions of the Japanese Stock Market *Prog. Theor. Phys. Supplement* **194** 43-54
[9] Takaishi T 2012 Finite-Sample Effects on the Standardized Returns of the Tokyo Stock Exchange *Procedia - Social and Behavioral Sciences* **65** 968–973
[10] Engle R F 1982 Autoregressive Conditional Heteroskedasticity with Estimates of the Variance of the United Kingdom inflation *Econometrica* **50** 987–1007
[11] Bollerslev T 1986 Generalized Autoregressive Conditional Heteroskedasticity *Journal of Econometrics* **31** 307–327
[12] Nelson D B 1991 Conditional Heteroskedasticity in Asset Returns: A New Approach *Econometrica* **59** 347-370
[13] Engle R F and Ng V 1993 Measuring and testing the impact of news on volatility *Journal of Finance* **48** 1749-1778
[14] Sentana E 1995 Quadratic ARCH models *Review of Economic Studies* **62** 639-661
[15] Glosten L R, Jaganathan R and Runkle D E 1993 On the Relation Between the Expected Value and the Volatility of the Nominal Excess on Stocks *Journal of Finance* **48** 1779-1801
[16] Ding Z, Granger C W J and Engle R F 1993 A long memory property of stock market returns and a new model *Journal of Empirical Finance* **1** 83–106
[17] Takaishi T and Chen TT 2012 Bayesian Inference of the GARCH model with Rational Errors *International Proceedings of Economics Development and Research* **29** 303–307
[18] Bollerslev T, Chou R Y and Kroner K F 1992 ARCH modeling in finance *Journal of Econometrics* **52** 5–59
[19] Cont R and Bouchaud J P 2000 Herd behaviour and aggregate fluctuations in financial markets *Macroecon. Dynamics* **4** 170-196
[20] Stauffer D and Penna T J P 1998 Crossover in the Cont-Bouchaud percolation model for market fluctuations *Physica A* **256** 284–290
[21] Sato A H and Takayasu H 1998 Dynamic numerical models of stock market price: from microscopic determinism ot macroscopic randomness *Physica A* **250** 231–252
[22] Lux T and Marchesi M 1999 Scaling and Criticality in a Stochastic Multi-Agent Model of a Financial Market *Nature* **397** 498–500
[23] Iori G 1999 Avalanche Dynamics and Trading Friction Effects on Stock Market Returns *Int. J. Mod. Phys. C* **10** 1149–1162

[24] da Silva L R and Stauffer D 2001 Ising-correlated clusters in the Cont-Bouchaud stock market model *Physics A* **294** 235–238

[25] Challet D, Marsili M and Zecchina R 2000 Statistical mechanics of system with heterogeneous agents: Minority games *Phys. Rev. Lett.* **84(8)** 1824–1827

[26] Raberto M, Cincotti S, Focardi S M and Marchesi M 2001 Agent-based Simulation of a Financial Market *Physics A* **299** 319–327

[27] Bornholdt S 2001 Expectation Bubbles in a Spin Model of Markets: Intermittency from Frustration across Scales *Int. J. Mod. Phys. C* **12** 667–674

[28] Sznajd-Weron K and Weron R 2002 A Simple Model of Price Formation *Int. J. Mod. Phys. C* **13** 115–123

[29] Sanchez J R 2002 A Simple Model for Stocks Markets *Int. J. Mod. Phys. C* **13** 639–644

[30] Yamano Y 2002 Bornholdt's Spin Model of a Market Dynamics in High Dimensions *Int. J. Mod. Phys. C* **13** 89–96

[31] Kaizoji T, Bornholdt S and Fujiwara Y 2002 Dynamics of Price and Trading Volume in a Spin Model of Stock Markets with Heterogeneous Agents *Physica A* **316** 441–452

[32] Takaishi T 2005 Simulations of Financial Markets in a Potts-like Model *Int. J. Mod. Phys. C* **13** 1311–1317

[33] Krause S M and Bornholdt S 2011 Spin models as microfoundation of macroscopic financial market models *arXiv:1103.5345*

[34] Stauffer D 2012 A Biased Review of Sociophysics *J. Stat. Phys.* **151** 9–20

[35] Matia K, Pal M, Salunkay H and Stanley H E 2004 Scale-dependent price fluctuations for the Indian stock market *Europhys. Lett.* **66** 909–914

[36] Akaike H 1973 Information theory and an extension of the maximum likelihood principle *Proceedings of the 2nd International Symposium on Information Theory* Petrov B N and Caski F (eds.), Akadimiai Kiado, Budapest: 267-281

[37] Takaishi T 2009 An Adaptive Markov Chain Monte Carlo Method for GARCH Model *Lecture Notes of the Institute for Computer Sciences, Social Informatics and Telecommunications Engineering. Complex Sciences* Vol. **5** 1424–1434

[38] Takaishi T 2009 Bayesian Estimation of GARCH Model with an Adaptive Proposal Density *New Advances in Intelligent Decision Technologies, Studies in Computational Intelligence* Vol. **199** 635–643

[39] Takaishi T 2009 Bayesian Inference on QGARCH Model Using the Adaptive Construction Scheme *Proceedings of 8th IEEE/ACIS International Conference on Computer and Information Science* 525–529 doi:10.1109/ICIS.2009.173

[40] Takaishi T 2010 Bayesian inference with an adaptive proposal density for GARCH models *J. Phys.: Conf. Ser.* **221** 012011

Journal of Physics: Conference Series **454** (2013) 012042 doi:10.1088/1742-6596/454/1/012042

Lattice computations for high energy and nuclear physics

K Jansen
NIC, DESY Zeuthen, Platanenallee 6, D-15738 Zeuthen, Germany
E-mail: Karl.Jansen@desy.de

Abstract. An overview is given on present lattice field theory computations. We demonstrate the progress obtained in the field due to algorithmic, conceptual and supercomputer advances. We discuss as particular examples Higgs boson mass bounds in lattice Higgs-Yukawa models and the baryon spectrum, the anomalous magnetic moment of the muon and nuclear physics for lattice QCD. We emphasize a number of major challenges lattice field theory is still facing and estimate the computational cost for simulations at physical values of the pion mass.

1. Introduction

When quantum chromodynamics as our theory of the strong interaction was developed, it became clear that one would have to deal with very non-perturbative phenomena. The interaction between the postulated quarks and gluons can not be addressed by perturbation theory anymore when distances of about 1fm or, correspondingly energy scales below 1GeV are considered. However, this is exactly the energy regime where the observed particles such as the proton and the neutron appear. This let K. Wilson to state in his 1977 Cargese lectures [1]

Unfortunately, it is not known yet whether the quarks in quantum chromodynamics actually form the required bound states. To establish whether these bound states exist one must solve a strong coupling problem and present methods for solving field theories don't work for strong coupling.

This observation then led to the development of lattice field theory as a non-perturbative tool to address the fundamental question raised by K. Wilson who then himself [2] suggested to reformulate the theories of elementary particle interactions on a 4-dimensional lattice in order to establish a well-defined and regulated formulation of field theories which are furthermore amenable to non-perturbative computations.

Whereas Wilson was thinking of strong coupling expansions to evaluate the theories [3], it was shown in [4, 5], by drawing analogy to statistical physics, that such lattice gauge theories can also be calculated numerically. This opened a much broader road to carry through ab-initio calculations of theories in high energy physics with techniques that can cover an energy range from large scales, where perturbation theory can be applied, up to very low scales where truly non-perturbative phenomena play the dominant role.

In this article I want to provide an overview how far lattice field theory has progressed in the understanding of the question "whether the quarks in quantum chromodynamics actually form the required bound states". We will see that there is in fact a tremendous development. There is substantial improvement of the used algorithms and a much better conceptual understanding of lattice field theories. In addition, there are new supercomputer architectures reaching now the multi-petaflop regime. All this furthered the field to a point where lattice computations make contact to and has an impact on experiments and phenomenology. In addition, we will provide an example that lattice field is not only a tool to address QCD to answer Wilson's original question, but also other field theories such as the Higgs-Yukawa model discussed here.

2. Lattice field theory

The principles of lattice field theory can already be explained at the example of the Schwinger model [6] which describes quantum electrodynamics in two dimensions. In the continuum, the model is defined through the path integral [7]

$$\mathcal{Z} = \int \mathcal{D}A_\mu \mathcal{D}\Psi \mathcal{D}\bar{\Psi} e^{-S_{\mathrm{gauge}} - S_{\mathrm{ferm}}} \tag{1}$$

which is taken here in Euclidean time. In this form, the path integral resembles a partition function in statistical mechanics and, in fact, this resemblance goes much deeper and the study of phase transitions, critical phenomena and also algorithmic developments are common to both fields. The path integral is determined through the actions S_{gauge} for the gauge field and S_{ferm} for the fermion fields, where S_{ferm} is given by

$$S_{\mathrm{ferm}} = \int d^2x \bar{\Psi}(x) \left[D_\mu + m\right] \Psi(x) \tag{2}$$

with m being the fermion mass and the gauge covariant derivative

$$D_\mu \Psi(x) \equiv \gamma_\mu(\partial_\mu - ig_0 A_\mu(x))\Psi(x) , \tag{3}$$

where Ψ, $\bar{\Psi}$ denote anti-commuting fermion fields, A_μ is the gauge potential and g_0 the gauge coupling. The $2 \otimes 2$ matrices γ_μ are the standard Pauli-matrices. The gauge field action is given as

$$S_{\mathrm{gauge}} = \int d^2x F_{\mu\nu} F_{\mu\nu} \ , \ F_{\mu\nu}(x) = \partial_\mu A_\nu(x) - \partial_\nu A_\mu(x) \ . \tag{4}$$

Taking the gauge field action S_{gauge} and deriving the equations of motion will give the Maxwell equations in Euclidean time. It is worth noting that this simple field theory has already a number of interesting properties such as a dynamical generation of a mass gap of bound states or the non-perturbative generation of a fermion condensate [8, 9].

Going to a lattice description of this model, one introduces a lattice spacing a and places the fermion fields on the integer valued lattice sites $x = (t, \mathbf{x})$. The lattice fermion action then assumes the form

$$S \to a^2 \sum_x \bar{\Psi} \left[\gamma_\mu \partial_\mu - \partial_\mu^2 + m\right] \Psi(x) \tag{5}$$

with the most obvious choice of the lattice derivative assuming nearest neighbour differences,

$$\partial_\mu = \frac{1}{2}\left[\nabla_\mu^* + \nabla_\mu\right] \tag{6}$$

with, neglecting for a moment the gauge fields,

$$\nabla_\mu \Psi(x) = \frac{1}{a}\left[\Psi(x + a\hat{\mu}) - \Psi(x)\right] \ , \quad \nabla_\mu^* \Psi(x) = \frac{1}{a}\left[\Psi(x) - \Psi(x - a\hat{\mu})\right] \tag{7}$$

which in the limit of vanishing lattice spacing will lead to the desired continuum derivative. The second order derivative term ∂_μ^2 in (5) appears in order to remove unwanted modes on the

Journal of Physics: Conference Series **454** (2013) 012042 doi:10.1088/1742-6596/454/1/012042

lattice, the so-called doubler modes. The conceptual drawback of this term is that it breaks a most important symmetry of the continuum action, namely continuum chiral symmetry[1], i.e. the invariance of the action under exchange of mass-less left- and right-handed fermions. Thus it seems that there is a price to pay when putting fermions on the lattice and, in fact, already at an early stage of lattice field theory a no-go theorem was established. It states that on the lattice either the fermion spectrum, the locality of the theory, chiral symmetry or the correct dispersion relation has to be altered compared to the continuum theory [12, 13, 14]. Only in the limit when the lattice spacing a vanishes, the continuum theory with all the desired properties listed above is expected to be recovered. However, this expectation needs, of course, to be explicitly verified. Therefore, it is very important to have different lattice fermion formulations since taking the continuum limit from these alternative theories, which break different continuum properties, physical results have to agree.

Let us now look at the gauge part of the Schwinger model. It was Wilson's essential observation that not the gauge potentials themselves, but the parallel transporters,

$$U(x,\mu) = e^{iaA_\mu(x)} \in U(1) \tag{8}$$

are the fundamental objects for a lattice formulation of gauge theories. Employing the parallel transporter, a lattice covariant derivative can be defined,

$$\nabla_\mu \Psi(x) = \frac{1}{a}\left[U(x,\mu)\Psi(x+\mu) - \Psi(x)\right] \tag{9}$$

which guarantees the gauge invariance of the lattice theory. The gauge interaction is then realized through a so-called plaquette variable,

$$U_{(x,p)} = U(x,\mu)U(x+\mu,\nu)U^\dagger(x+\nu,\mu)U^\dagger(x,\nu) \;\rightarrow F_{\mu\nu}F^{\mu\nu}(x) \text{ for } a \rightarrow 0\,. \tag{10}$$

As indicated in (10) the so defined plaquette variable assumes the continuum expression for the gauge field interaction. Using the plaquette variable, a lattice action for the Schwinger model can be formulated,

$$S = a^2 \sum_x \left\{ \beta\left[1 - \mathrm{Re}(U_{(x,p)})\right] + \bar{\psi}\left[m_0 + \frac{1}{2}\{\gamma_\mu(\nabla_\mu + \nabla_\mu^\star) - a\nabla_\mu^\star\nabla_\mu\}\right]\psi \right\}\,. \tag{11}$$

This action, with $\beta = 1/g_0^2$ and m_0 the fermion mass on the lattice, can be used to compute now expectation values of physical quantities,$\langle\mathcal{O}\rangle$ in close analogy to statistical mechanics by

$$\langle\mathcal{O}\rangle = \frac{1}{\mathcal{Z}}\int_{\text{fields}} \mathcal{O}e^{-S} \tag{12}$$

where $\mathcal{Z} = \int_{\text{fields}} e^{-S}$ is the path integral or partition function. Having now formulated the Schwinger model in Euclidean space-time on a discrete lattice, using a finite number of lattice points, the integral in (12) can be solved numerically. Since in the integrand the Boltzmann factor e^{-S} strongly favors only certain field configurations, Monte Carlo methods relying on importance sampling are used for this task; see [15, 16] for introductions of simulation algorithms used in lattice field theory. Having established the lattice theory of the Schwinger model, a corresponding action for quantum chromodynamics (QCD), our target theory for describing the strong interaction, can be given. Here, one has go to four dimensions and the $U(1)$ gauge fields of (8) are replaced by non-abelian SU(3) valued gauge fields which represent the gluon fields. The fermions, i.e. the quark fields carry now a colour charge through which they interact with

[1] It is possible to establish a modified but exact chiral symmetry on the lattice [10], based on the Ginsparg-Wilson relation [11]; see also section 4

the gluons and also a Lorentz index thus leading to the fact that they are represented as twelve component Grassman valued fields. We will not detail the lattice QCD action but refer the interested reader to [17, 18, 19, 20].

3. Evaluation of the fermion action

Having established an Euclidean space-time lattice action and correspondingly the pathintegral for lattice field theory, in principle numerical evaluations for calculating physical quantities become possible. However, there is a major and severe obstacle in computing the fermionic part of the pathintegral which leads to very computer time consuming calculations when fermions are employed such as in lattice QCD or also in lattice Higgs-Yukawa models as discussed below. To understand why this is so, let us have a look at the generic form of the fermion part of the partition function,

$$\mathcal{Z}_{\text{ferm}} = \int \mathcal{D}\bar{\psi}\mathcal{D}\psi e^{-\bar{\psi}\{D_{\text{ferm}}\}\psi} , \tag{13}$$

where D_{ferm} is a given, generic lattice Dirac operator. The key observation is that fermions, as Grassman variables, cannot be handled on a computer directly. However, since the action is quadratic in the fermion fields, the fermion degrees of freedom can be integrated out leading to the expression

$$\mathcal{Z}_{\text{ferm}} \propto \det[D_{\text{ferm}}] \tag{14}$$

and hence to the evaluation of a determinant. The difficulty arises from the fact that present days lattice computations are performed for lattices with a number of sites $48^3 \cdot 96$. Since, as discussed above, in QCD the fermion field per lattice site has 12 components we would have to compute a determinant of a huge matrix of size $[48^3 \cdot 96 \cdot 12] \otimes [48^3 \cdot 96 \cdot 12]$ which is, of course, prohibitive.

The way out is to represent the fermion determinant as an integral over bosonic degrees of freedom which can be handled on a computer,

$$\det[D_{\text{ferm}}] \propto \int \mathcal{D}\Phi^\dagger\mathcal{D}\Phi e^{-\Phi^\dagger\{D_{\text{ferm}}^{-1}\}\Phi} . \tag{15}$$

To evaluate this representation of the fermionic path integral, we therefore need to compute a vector $X = D_{\text{ferm}}^{-1}\Phi$, or equivalently, we have to solve a set of linear equations

$$D_{\text{ferm}}X = \Phi . \tag{16}$$

Still, given the huge size of the matrix D_{ferm} this looks like an unsurmountable task. What helps here is that the matrix D_{ferm} is a sparse matrix and hence well suited algorithms such as conjugate gradient can be used.

Nevertheless, in the algorithms used to simulate the fermion degrees of freedom, the set of linear equations of (16) needs to solved $O(100)$ times to generate just one of the gauge field configurations on which eventually physical measurements are to be performed. And, the goal is to reach finally a set of $O(1000 - 10000)$ of such gauge field configurations. If in addition, one keeps in mind that, say, in a conjugate gradient computation for the solution of (16) $O(100 - 1000)$ iterations are needed, one can easily estimate the extremely large number of matrix vector operations, scalar products and norms that is needed for just one simulation point.

All this led in the 20th century to the fact that lattice simulations were performed in the so-called quenched approximation where the above sketched determinant evaluation was completely neglected. It was then the CP-PACS collaboration situated in the home country of the Conference on Computational Physics 2012, i.e. Japan, which showed in a remarkable paper [21] that this quenched approximation must be dismissed to perform QCD calculations.

In this paper [21], for the first time, the complete low-lying hadron spectrum was computed. By fixing two input masses, all the rest of the baryon masses were "predicted" by lattice QCD. The precision reached in this calculation was so high that in fact significant deviations between the quenched lattice QCD computation and the experimentally determined baryon masses were detected. As a conclusion, it became clear that the quenched approximation is a much too strong truncation of the theory and that the quarks have to be taken as truly dynamical degrees of freedom into account.

The problem of taking the step to keep the determinant in the simulation became very clear at the Berlin lattice symposium in 2001 [22]. At this conference a number of collaborations, using their particular formulation of lattice QCD, showed results for the simulation costs when the pion mass was shifted from an unphysically large value of around 700GeV towards its physical value of 140MeV.

It was found that the simulation costs grow with an inverse power of the pion mass m_π as m_π^{-s} with $s = 1-2$. In addition, when decreasing the lattice spacing while keeping the physical extent of the box fixed, the number of lattice points to be simulated needs to be increased leading to a scaling of a^{-4}. When studying the actual lattice spacing dependence of the simulation cost, it was found that there is an additional dependence, leading to a total scaling of a^{-t} with $t = 6-7$. All this led to the conclusion that simulations close to the physical point, i.e. at physical values of the pion mass, seemed unreachable [23].

The hope at that time was then that chiral perturbation theory could be used to extrapolate results obtained at too large pion masses to the physical point. However, in the course of the years, it turned out that there are significant challenges in performing chiral extrapolations of baryon quantities; see e.g. [24] for a recent discussion of the subject.

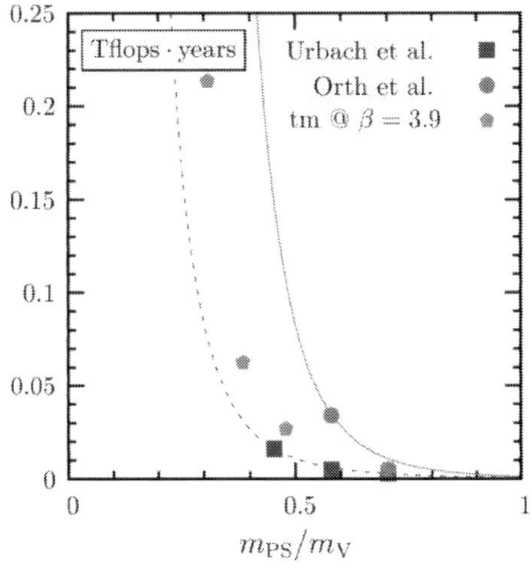

Figure 1. The simulation cost of lattice QCD calculations is shown. The ratio $m_{\rm PS}/m_{\rm V}$ in the graph assumes a physical value of about 0.18 and the graph shows how the computation costs, expressed in teraflop years increases when $m_{\rm PS}/m_{\rm V}$ is moved to the physical value. In the past, these costs were several hundred or even thousands of teraflop years. The figure demonstrates the improvement in developing algorithms since it demonstrates that nowadays an amount of teraflop years is necessary that can be reasonably realized on present supercomputer architectures.

In this somewhat desperate situation, around 2005, a number of publications appeared [25, 26, 27] which demonstrated that the standard Hybrid Monte algorithm [28] used for simulations of fermion system can be supplied with very substantial improvements. This led to a scaling with the pion mass as shown in figure 1, as one example. The graph shows the typical simulation cost in teraflop years to solve a problem of lattice QCD as a function of the ratio of the pion to the vector meson mass m_V with a physical value of this ratio to be $m_\pi/m_V = 0.18$. The finding of the Berlin 2001 lattice conference was that the simulation cost at the physical value of the this ratio becomes several hundred or even more teraflop years. In figure 1, with the mentioned algorithmic improvements implemented, the simulation costs stay at a reasonable level and given present days supercomputers it seems to become perfectly reasonable to simulate directly at the physical point, an enterprise that is in fact started by many collaborations world-wide. The status of such simulations of lattice QCD over the last years is summarized in the reviews of [29, 30, 31, 32].

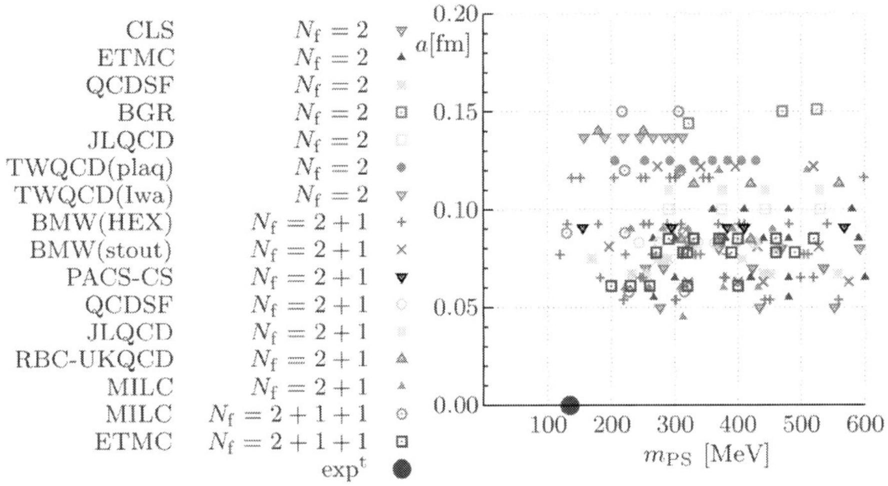

Figure 2. The simulation landscape of present lattice QCD calculations. In the graph, we show the values of the lattice spacing a and pion masses m_{PS} where present simulations are performed world-wide today. Note that already a number of simulations are carried out at the physical point where the pion mass assumes its experimentally measured value.

The progress in supercomputer architectures and, in particular, in the algorithmic improvements are documented in the "landscape" graph of figure 2. This figure shows the lattice spacing and the pion mass where simulations are performed. It can be clearly seen that nowadays simulations with small lattice spacings, $a \lesssim 0.1$fm and pion masses well below 500MeV are performed. Even more, there are already simulations at the physical value of the pion mass. As a result of these calculations a first ever computation of the baryon spectrum could be achieved treating the quarks as truly dynamical degrees of freedom [33], see figure 3. Such a calculation has by now been repeated by many lattice collaborations world-wide; see the review [34].

The calculation of the light hadron spectrum can certainly be considered as a breakthrough for lattice QCD computations since this was taken as a most important benchmark result. However, the lattice community has gone already significant steps forward. Following the theoretical basis developed in [35], it is now possible, to address also resonances on the lattice. In figure 4. we show an example for a calculation of the phase shift for the ρ-meson resonance [36]. For a comprehensive review of the status of resonance calculations from lattice field theory, see [37]. An alternative way to compute phase shifts on the lattice is by using a partial wave analysis of a hadron wave function computed on the lattice; see section 6. Even more, as the lattice

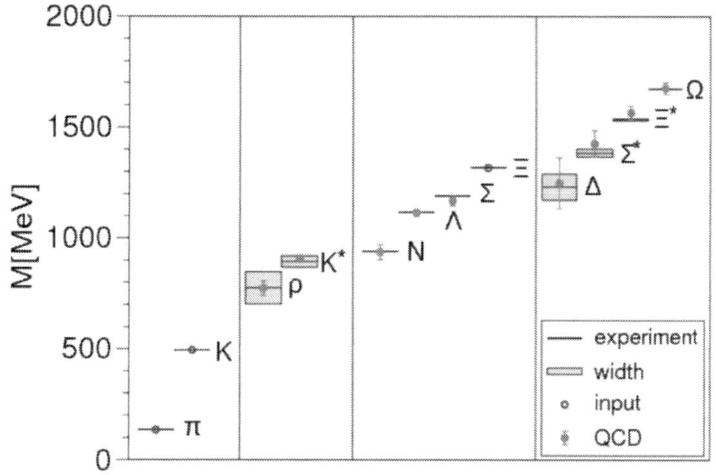

Figure 3. The hadron spectrum calculation of [33]. Using only two input masses, all the rest of the hadron masses are directly computed from lattice QCD. The nice agreement between the lattice calculation and experimental measurement is very reassuring that lattice QCD can provide non-perturbative results that can be directly compared to experiment.

calculations become more and more precise, groups have taken the effects of electromagnetism and an explicit breaking of isospin into account, [38, 39, 40]. In fact, it is now possible to even disentangle the effects of electromagnetism and isospin breaking on the mass splitting of hadrons (see [41]). This is a truly remarkable result since these effects are on the MeV level, while the hadron masses are on a scale of GeV.

In the discussion above, we have seen that there has been a tremendous progress in lattice field theory calculations. In the following, we will discuss two particular examples for such present computations. The first example describes results for bounds on the Higgs boson mass. These bounds can be directly related to the energy scale up to which the standard model of particle interaction can be valid. The second example discusses a computation of the leading order hadronic contribution to the muon anomalous magnetic moment.

After this discussion, we will emphasize some open demanding problems and challenges to lattice field theory. These problems and challenges will for sure necessitate the usage of todays supercomputers of the K-machine class.

4. Higgs-Yukawa

The Higgs-Yukawa sector of the Standard Model (SM) describes the generation of fermion masses via the non-vanishing vacuum expectation value (vev) acquired by the Higgs field which couples through a Yukawa coupling to the fermions. The essential element in this picture is that the coupling of the fermions to the Higgs field is chirally invariant which leads to the gauge invariant electroweak sector of the SM in the presence of gauge fields.

In principle, the involved couplings of the theory, the quartic self-interaction of the Higgs field and the Yukawa-coupling between the Higgs field and the fermions, can grow strong. This happens, when the involved masses are large and then perturbation theory might fail to analyze the theory. There are indeed examples where the applicability of perturbation theory is questionable. The first is the upper Higgs boson mass bound which is based on triviality arguments [42]. Here the Higgs boson mass can become large, resulting in a strong value of the quartic coupling such that perturbation theory may not work anymore. The second is the lower Higgs boson mass bound which is based on vacuum instability arguments [43, 44, 45, 46]. Here

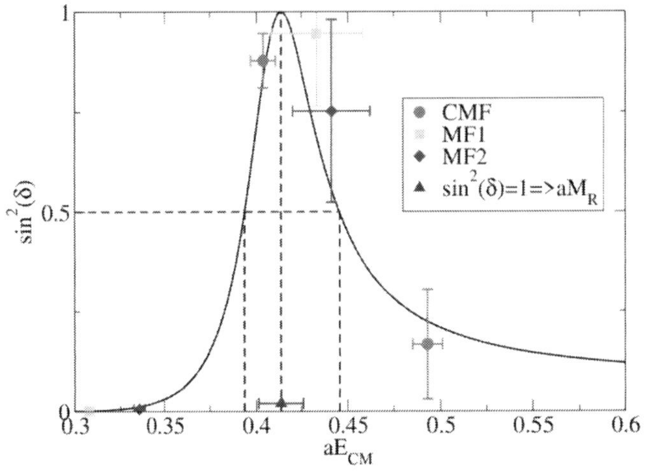

Figure 4. We show the scattering phases calculated in the center of mass frame (CMF), as well as in two moving frames (MF1 and MF2) together with the fits to an effective range formula; see [36]. At the position where the scattering phase passes $\pi/2$, the resonance mass m_ρ (denoted as aM_R in the graph) is determined. Through the fit, the coupling constant $g_{\rho\pi\pi}$ and the decay width Γ_ρ are also extracted.

it is unclear whether this instability is not an artifact of perturbation theory applied at large values of the Higgs field such that an expansion around the minimum of the effective potential is not justified anymore.

It is important to stress that both the lower and the upper Higgs boson mass bounds are intrinsically related to the cut-off of the theory. Thus, a calculation of the Higgs boson mass bounds can in turn be used to determine the cut-off up to which the SM is valid, once the SM Higgs boson mass has been determined. If, for example, the recent result for a scalar particle at the Large Hadron Collider (LHC) [47, 48] is confirmed as a SM Higgs boson with a mass of about 125 GeV, the SM could be valid up to very high energies before violating the Higgs boson mass bounds; see [49] for a recent analysis at next-to-next leading order of perturbation theory.

Another example where non-perturbative calculations are necessary is the possibility of a heavy fourth fermion generation [50, 51] which would lead to a large value of the corresponding Yukawa coupling.

Conceptually clean investigations of Higgs-Yukawa models on the lattice became possible when it was realised that –based on the Ginsparg-Wilson relation [11]– there exists a consistent formulation of an exact lattice chiral symmetry [52], which emulates the chiral character of the Higgs-fermion coupling structure of the SM. This triggered a number of lattice investigations of Higgs-Yukawa like models [53, 54, 55, 56, 57, 58, 59, 60, 61].

As an example, we show here the results of the Higgs boson mass bound calculations discussed in [61, 62]. In the left graph of figure 5 the upper and lower bounds were computed at several choices of the cut-off scale, with the fermion masses m_f at the physical top-quark mass. The right graph of this figure shows the fermion mass dependence of the lower and upper Higgs boson mass bounds at a fixed value of the cut-off of about 1.5TeV. It can be clearly seen that while the upper bound is relatively unaffected when using a heavy fermion mass, the lower bound increases substantially.

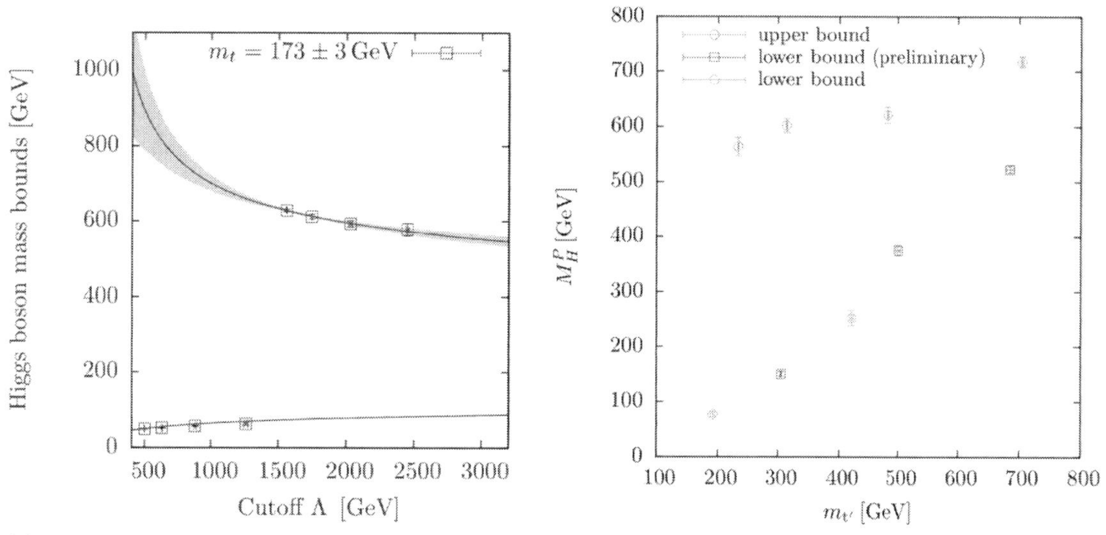

(a) The cut-off dependence of the upper and lower Higgs boson mass bounds for fermion a mass of about ∼ 173GeV.

(b) The mass dependence of the lower and upper Higgs boson mass bounds at a cut-off of about 1.5TeV.

Figure 5. Results for the Higgs boson mass bounds.

The graphs in figure 5 have a very interesting interpretation in the light of the SM and also the fourth fermion generation. Concerning the SM, a Higgs boson mass of about 125GeV just seems to escape the Higgs boson mass bounds leading the SM to be valid up to very high energies. On the other hand a fourth fermion generation seems to be ruled out for fermion masses larger than about 300GeV. Combining this with phenomenological analyses of allowed fermion masses to be larger than about 500GeV, this indicates that a simple extension of the SM with a fourth fermion generation is not compatible with the experimental finding of a possible 125GeV Higgs boson mass.

The discussion of the Higgs boson mass bounds here, as they are interesting by themselves, served as an example that lattice field theory can also provide important information for non-QCD like theories. We note in passing that for the Higgs-Yukawa model considered here, the Higgs boson has also been treated as a true resonance [63]. In addition, the model is studied at non-zero temperature and also at very strong values of the Yukawa-coupling to explore whether a generic non-perturbative interaction can take place that is different from standard model physics [62].

5. The muon anomalous magnetic moment

Let us now turn back to lattice QCD, the original target theory. Here we want to discuss one particular example, namely the anomalous magnetic moment of the muon.

In Dirac's theory the electron is described as a spin 1/2 particle with the positron as its anti-particle. Like in the classical theory the angular momentum, the spin $\vec{S}$ leads to a magnetic moment of the electron

$$\vec{\mu}_m = -g_e \mu_0 \vec{S} \tag{17}$$

with $\mu_0 = e/4m_e$, e being the electric charge and m_e the mass of the electron. The quantity g_e is the gyro-magnetic ratio of the electron and assumes in Dirac's theory a value of $g_e = 2$.

When a quantum field theory is considered, deviations from $g_e = 2$ appear, originating from a virtual photon exchange which leads to an *anomalous magnetic moment*

$$a_e = \frac{g_e - 2}{2} \ . \tag{18}$$

The most simple correction to g_e is just the one *virtual* photon exchange between the charged electrons. This correction, first considered by Schwinger [64], can be directly computed from quantum electrodynamics (QED), the relativistic theory of electromagnetic interaction. The seminal work by Schwinger revealed that the quantum correction $\delta(a_e)$ is

$$\delta(a_e) = \frac{\alpha}{2\pi} \ . \tag{19}$$

Experimental measurements [65] found a value for $g_e = 2.00238(10)$ and Schwinger's calculation yielded $g_e = 2.00232$. The agreement of these results established the theoretical and experimental fact that quantum corrections lead indeed to deviations from the classical value of $g_e = 2$. Subsequent more precise theoretical calculations and experimental measurement confirmed the agreement between QED predictions and observation (see [66]).

Today, the theoretical efforts have been driven to enormously complicated calculations involving the computation of thousands of diagrams. The calculations of [67] involve nowadays diagrams including four loops, we speak of a 4-loop order calculation and even a 5-loop order calculation is in sight. The theoretical computations are based on a systematic expansion of the theory in the fine structure constant α. Employing such high loop order quantum field theoretical calculations and extremely accurate and subtle experimental measurements, one finds an incredible agreement between experiment and and standard model predictions for the anomalous magnetic moment of the electron, i.e. (see e.g. [68])

$$\begin{aligned} a_e(\text{th}) &= 1159652201.1(2.1)(27.1) \cdot 10^{-12} \\ a_e(\text{exp}) &= 1159652188.4(4.3) \cdot 10^{-12} \ . \end{aligned} \tag{20}$$

This establishes the electron magnetic moment as one of the most precisely determined quantity in nature and the agreement between the two numbers in (20) achieved with such an amazing accuracy is one of the cornerstone results to make us confident that indeed quantum field theories are the correct method to describe particle interactions.

However, the story does not end here. Nature has decided, for an unknown reason, that it comprises three particle generations. Besides the electron, which belongs to the first generation, there are further so-called leptons, the muon with a mass of about 105.7MeV and the τ having a mass of 1777MeV. Comparing these values to the mass of the electron of 0.511MeV, we see that the masses of the leptons of the second (the muon) and the third (the τ) generations are much larger than the electron mass. This fact will play an important role in the discussion below. Being leptons, also the muon and τ carry spin, leading to magnetic moments which become, due to the discussed quantum corrections, anomalous. For example, a comparison for the case of the muon anomalous magnetic moment, a_μ, between experiment and theory yields [66]

$$\begin{aligned} a_\mu^{\text{exp}} &= 1.16592080(63) \times 10^{-3} \\ a_\mu^{\text{theory}} &= 1.16591790(65) \times 10^{-3} \ . \end{aligned} \tag{21}$$

Here the agreement between theory and experiment is not as nice as in the case of the electron and in fact, one finds $a_\mu^{\text{exp}} - a_\mu^{\text{th}} = 2.90(91) \times 10^{-9}$ which leads to a larger than 3σ level discrepancy.

However, this is a very interesting result. It means that either in the theoretical calculation something has been neglected or has not properly been included. Or, somewhat much more exciting, the discrepancy points to a breakdown of the standard model of particle interactions and the inconsistency stems from effects of some yet unknown new physics beyond the standard model.

Indeed, calculations show that these new physics effects would lead to a correction to the anomalous magnetic moment of size

$$\delta(a_l^{\rm newphysics}) = m_{\rm lepton}^2/M_{\rm newphysics}^2 \,. \tag{22}$$

Here $m_{\rm lepton}$ is the mass of one of the leptons and $M_{\rm newphysics}$ represents the mass (or scale) of a particle originating from the (unknown) new physics beyond the standard model. The formula in (22) shows that in the case of the muon anomalous magnetic moment the effect of new physics would show up about $(m_\mu/m_e)^2 \approx 4 \cdot 10^4$ times stronger than in the case of the electron. In principle, the τ-lepton would be even more suitable to detect these new physics effects, but unfortunately due to the very short lifetime of the τ lepton the experimental measurements of the anomalous magnetic moment of the τ are presently much too imprecise to unveil a possible new physics contribution. This leaves us then with the muon anomalous magnetic moment as the ideal place to look for new physics and indeed a large number of works has been devoted to explore this possibility (see [66]).

As we discussed for the case of the electron, theoretically the anomalous magnetic moment is computed in an expansion of the theory in the fine structure constant to a given loop order. In the loops, in principle all particles of the standard model can appear, if allowed by symmetries and quantum numbers. Therefore, besides leptons and photons, also the vector bosons, W and Z, of the weak interaction and the quarks and gluons of the strong interaction, described by quantum chromodynamics (QCD) can and will contribute. It can be shown that the contribution of the W and Z bosons are first of all rather small and can be, secondly, well controlled by perturbation theory [66].

However, the strong interaction of quarks and gluons in QCD lead to so-called hadronic contributions $a_\mu^{\rm had}$ which are intrinsically of non-perturbative nature. Taking these contributions into account by perturbation theory is therefore rather doubtful. Employing additional model assumptions to estimate $a_\mu^{\rm had}$ will not provide a fully controlled and reliable calculation of the hadronic contributions and hence an unambiguous and stringent test whether the standard model is correct or must be extended by some new physics cannot be performed.

It is exactly at this point where lattice field theory methods applied to quantum chromodynamics can help – at least in principle. Of course, the discretization itself induces a systematic error which must be removed by making the lattices finer and finer until the continuum of space time points is recovered by some suitable extrapolation process, a procedure which is called the *continuum limit*. In addition, the simulations necessarily demand a finite number of lattice points which can lead to *finite size effects* when the lattice is not large enough in physical units. Finally, often the simulations need to be performed at values of hadron masses that are larger than observed in nature. The reason is that, as discussed above, for smaller and smaller hadron masses the computational costs increase rapidly such that one is restricted to values of, say, pion masses that are a factor of about two larger than the ones observed in nature.

All these systematic effects that appear in lattice simulations need to be controlled in a quantitative way. For example, to reach physical values of the pion masses, an extrapolation to the physical point where the pion mass assumes its physical value needs to be performed. This appeared to be very problematic in the past. This is illustrated in figure 6(a) by the lowest lying points which were obtained in the past by the standard way of computing $a_\mu^{\rm had}$. The figure shows that these older lattice results for $a_\mu^{\rm had}$ are significantly below the experimental number. An extrapolation to the physical point, reconciling the lattice data with experiment becomes in this situation very difficult and even needs some additional model assumptions. This all has led to an error of $a_\mu^{\rm had}$ as obtained from lattice simulations that is about a factor of 10 larger than the phenomenological one [69]. The lattice community has been therefore rather sceptical in the past that lattice QCD can provide a significant contribution to our understanding of the discrepancy in $g_\mu - 2$.

One additional suspicion has been that so-called dis-connected (singlet) contributions could be substantial. In all existing lattice calculations these contributions were neglected, however. The reason is simply that these contributions are very noisy and therefore hard to compute reliably. In [70] a dedicated effort to calculate these contributions was undertaken for the first time. As a result, it could be established that the dis-connected contributions are in fact small and can be safely neglected. In addition, also the effects of non-zero values of the lattice spacing and the finite volume turned out to be small. Thus the difficulty to reconcile lattice data with the experimental result, shown in figure 6(a), is rather puzzling.

A resolution of this puzzle was only given in [70] where it was observed that by a suitable redefinition of the lattice observable needed to compute $a_\mu^{\rm had}$ a much smoother and much better controlled approach to the physical point can be achieved.

To illustrate the idea, let us give the definition of $a_\mu^{\rm had}$,

$$a_\mu^{\rm had} = \alpha^2 \int_0^\infty dQ^2 \frac{1}{Q^2}\omega(r)\Pi_R(Q^2)\ . \tag{23}$$

Here α is the electromagnetic coupling and $\Pi_R(Q^2)$ the renormalized vacuum polarization function, $\Pi_R(Q^2) = \Pi(Q^2) - \Pi(0)$. The functional form of $\omega(r)$ is analytically known and the argument r is given by $r = Q^2/m_\mu^2$ where m_μ denotes the mass of the muon and Q a generic momentum. The key observation is now that on the lattice there is a large freedom to choose a definition of r. The only requirement is that in the limit of reaching a physical pion mass the continuum definition of $r = Q^2/m_\mu^2$ is recovered. Hence, on may define

$$r_{\rm latt} = Q^2 \cdot \frac{H^{\rm phys}}{H} \tag{24}$$

with possible choices for H

$$\begin{aligned} r_1: &\quad H = 1 &\quad H^{\rm phys} &= 1/m_\mu^2 \\ r_2: &\quad H = m_V^2(m_{\rm PS}) &\quad H^{\rm phys} &= m_\rho^2/m_\mu^2 \\ r_3: &\quad H = f_V^2(m_{\rm PS}) &\quad H^{\rm phys} &= f_\rho^2/m_\mu^2\ . \end{aligned} \tag{25}$$

Here, $m_V(m_{\rm PS})$ is the mass of the ρ-meson and $f_V(m_{\rm PS})$ the ρ-meson decay constant as determined on the lattice at unphysical pion masses $m_{\rm PS}$. Furthermore, m_ρ and f_ρ denote the corresponding ρ-meson mass and decay constant at the physical point. All the definitions in (25) guarantee that indeed the desired definition of r is recovered in the limit of a physical pion mass since then by definition $m_V(m_{\rm PS})$ and $f_V(m_{\rm PS})$ assume their physical values. In figure 6(a) we show the results for $a_\mu^{\rm had}$ for all three definitions of r. Clearly, for the definitions r_2 and r_3 the behaviour of the lattice data towards physical pion masses is simply linear and allows for a controlled extrapolation to the physical point. As a result, one finds using definition r_2 in (25) values from the lattice computations and experiment

$$\begin{aligned} a_{\mu,N_f=2}^{\rm had,latt} &= 5.72\,(16)\cdot 10^{-8} \\ a_{\mu,N_f=2}^{\rm had,exp} &= 5.660\,(47)\cdot 10^{-8}\ . \end{aligned} \tag{26}$$

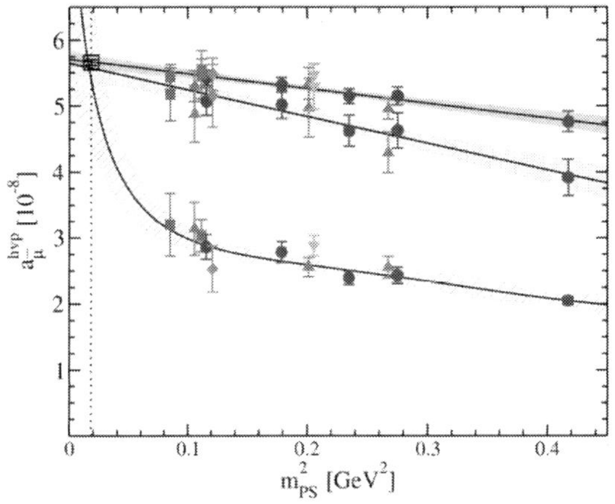

(a) We show a result for the hadronic contribution to the muon anomalous magnetic moment using different choices for r in (25). The curves correspond to the definitions r_1, r_2 and r_3 in (25) from bottom to top. Clearly, the improved definitions, r_2 and r_3, show a much milder pion mass dependence than the standard definition, r_1.

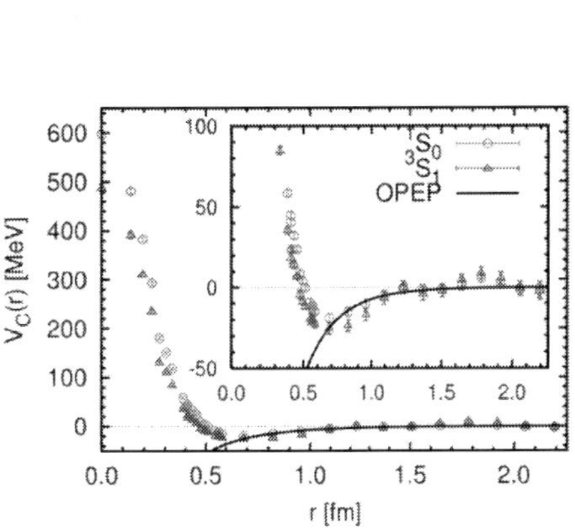

(b) The lattice QCD result of the effective central part of the NN potential $V_C(r)$ ($V_C^{\text{eff}}(r)$) in the $^1S_0$ ($^3S_1$) channel for $m_\pi/m_\rho = 0.595$. The inset shows its enlargement. The solid lines correspond to the one-pion exchange potential (OPEP).

Figure 6. The leading order hadronic contribution to the muon anomalous magnetic moment and an example of a nuclear potential computed in lattice QCD.

One can notice that the lattice result is not only consistent with the experimental determination, but also that the error of the lattice calculation assumes a comparable magnitude as the experimental one. This opens the very interesting possibility that lattice computations can significantly contribute to resolve the discrepancy between the standard model and experimental determinations of the anomalous magnetic moment of the muon.

The hear discussed results for a_μ^{had} have been achieved for the case of two mass-degenerate quark flavours. What is needed in the future is the inclusion of the strange and the charm quarks to allow for a direct comparison to the experimental results. In addition, newly planned experiments at Fermilab [71] and J-PARC [72] are aiming at an accuracy of below 0.5% for the hadronic contribution to the muon anomalous magnetic moment. To match this accuracy, lattice simulations have possibly to include explicit effects of isospin breaking and electromagnetism. All this is in principle reachable within lattice QCD, see section 3, but, it constitutes a real challenge for the lattice community. An even larger challenge are contributions to $g_\mu - 2$ that appear at higher order of the electromagnetic coupling, most notably the so-called light-by-light contributions (see also section 7.1). However, the lattice community is ready to attack these challenges and a large activity has been started to compute a_μ^{had} at leading order and beyond [73, 74, 75, 76, 77, 78, 79, 80, 81, 82, 70].

6. Nuclear physics

The examples discussed above were concerned with questions in high energy physics. However, there is also a significant activity in the lattice community to address problems in nuclear physics; see [83, 84] for reviews of different approaches.

Let me discuss here the method described in [85]. The starting point here is the Schrödinger equation,

$$[H_0 + V(x)]\,\phi_k(x) = \epsilon_k \phi(x)_k \qquad (27)$$

with the energy $\epsilon_k = \frac{k^2}{2\mu}$, the reduced mass $\mu = m_N/2$ and the free Hamiltonian $H_0 = \frac{-\nabla^2}{2\mu}$. $V(x)$ denotes the nuclear potential. The basic idea, pioneered in [85] is to revert this equation. If we would know the exact QCD wave function at wavenumber k, $\phi_k(x)$, for a certain physical situation, we can compute the corresponding potential by solving

$$[\epsilon_k - H_0]\,\phi_k(x) = \int d^3y V(x,y)\phi_k(y)\ . \tag{28}$$

Examples for such wave functions could be the nucleon wave function, three body wave functions etc. The basic idea is then to use the potential calculated in finite volume within lattice QCD to solve the Schrödinger equation in infinite volume obtaining in this way the wave function in infinite volume at a given wave number k.

The wave function can be computed on a lattice in a straightforward way. E.g., if we want the nucleon wave function, it can be obtained via

$$\phi_k(r) = \langle 0|N(x+r)N(x)|NNW_k\rangle \tag{29}$$

using solely lattice QCD. Here $W_k = \sqrt{k^2 + m_N^2}$, $N(x)$ a nucleon interpolating operator and $|NNW_k\rangle$ a 2-nucleon state with energy W_k. Knowing the wave function ϕ_k a partial wave analysis can be performed,

$$\phi_k^l \to A_l \frac{\sin(kt - l\pi/2 + \delta_l(k))}{kr}\ for\ r \to \infty \tag{30}$$

which eventually allows access to the scattering phase shifts $\delta_l(k)$. The strategy can be summarized with the following steps:

- compute the wave function within lattice QCD in *finite volume*
- determine the potential $U(x)$ in *finite volume*
- use this potential to solve the Schrödinger equation in *infinite volume*
- determine scattering phases or binding energies from the so obtained wave function and the partial wave analysis

Following these steps it then becomes, at least in principle, possible to address questions in nuclear physics, neutron stars, supernovae or other related areas. Note that the here described way of calculating phase shifts can be considered as an alternative way to the methods proposed in [35] which has by now extensively been used in lattice QCD computations; see [37] for a recent review.

In the work of [85] following the above sketched strategy it was demonstrated that it is possible to compute directly from lattice QCD the nucleon-nucleon potential (see figure 6(b)). This work, which has been selected as one of 21 papers in Nature Research Highlights 2007, opened the road for further investigations of nuclear physics from lattice QCD, following this approach. Demanding problems that are targeted now are in-elastic scattering, resonances and the extension to weak interactions; see [83] for a recent overview.

7. Challenges

In the previous sections we have discussed a number of topics where lattice field theory has been very successful and important results could be obtained that directly influence the interpretation of experiments or enter phenomenological analyses. However, there are still "big rocks on the road" that constitute major challenges for lattice field theory calculations. In this section, I only want to discuss a few of them.

7.1. *Anomalous magnetic moment of the muon*

As described above, the lattice can now provide very promising results for the leading order hadronic contribution to the anomalous magnetic moment of the muon. However, getting this contribution more and more precise, it becomes very important to also consider the next to leading order contributions. The dominant piece here is the so-called light-by-light contribution which is given by a 4-point function

$$\Pi_{\mu\nu\alpha\beta}(q_1, q_2, q_3) = \int_{xyz} e^{iq_1\cdot x + iq_2\cdot y + iq_3\cdot z} \langle j_\mu(0) j_\nu(x) j_\alpha(y) j_\beta(z)\rangle \tag{31}$$

with the electromagnetic quark current

$$j_\mu = \frac{2}{3}\bar{u}\gamma_\mu u - \frac{1}{3}\bar{d}\gamma_\mu d - \frac{1}{3}\bar{s}\gamma_\mu s + \frac{2}{3}\bar{c}\gamma_\mu c\,. \tag{32}$$

To compute $\Pi_{\mu\nu\alpha\beta}(q_1, q_2, q_3)$ at all momenta needed, one would have to solve the linear set of equation in (16) $O(V^3)$ times, with V the lattice volume. Having again lattices of size $V = 48^3{\cdot}96$ in mind, a straightforward calculation of this light-by-light contribution is clearly prohibitive. The lattice community is therefore very active to find better methods to obtain information on $\Pi_{\mu\nu\alpha\beta}(q_1, q_2, q_3)$ but, it is rather clear that this will be a very demanding task in any case, needing presumably supercomputer capabilities well beyond the K-computer class.

7.2. *Hadron structure*

One main activity of the lattice community is the exploration of the quark and gluon structure of a hadron [86, 87, 88, 89]. There is clearly a need from the lattice to provide inside into the hadron structure since this is, by nature of the problem, a non-perturbative task. There are two basic benchmark observables which are relatively easily accessible to lattice QCD calculations. The first is the axial charge of the nucleon and the second is the average momentum of the quark in a nucleon. Both quantities can be determined on the lattice rather precisely and they can be measured accurately experimentally or can be extracted from a global phenomenological analysis of deep inelastic scattering data.

Thus, similar to the hadron spectrum computation, it would be very important and reassuring if g_A and $\langle x\rangle$ as computed within lattice QCD would agree with experiment or phenomenology. However, presently this is not so, as can be seen in figures 7 and 8; see [87] for a more detailed discussion on this topic. The lattice results for g_A and $\langle x\rangle$ which are obtained at still unphysically large pion masses clearly deviate from the experimental or phenomenological values.

It needs to be stressed that the lattice community has undertaken a large effort to understand the discrepancies shown in figure 7 and figure 8. Many possible systematic effects were studied such as lattice spacing or finite volume artifacts, contaminations of excited states or the extrapolations to the physical point where the pion mass assumes its experimentally determined value (see the above listed reviews). None of these attempts could resolve the tension and it is presently a real puzzle what the origin of the discrepancy can be. To answer this question, most probably only a calculation directly at the physical point, on large enough lattices, several lattice spacings and with a sufficiently high accuracy would be needed.

7.3. *Nuclear physics*

The benchmark calculation for nuclear physics that has been discussed above has been done in the quenched approximation, see section 2, i.e. with infinitely heavy sea quark masses. Since the exploratory work in [85] was intended to provide a proof of concept, truncating the theory in such a way is justified as a first step. However, to make really contact to nuclear physics in the real world, also these computations need to be performed at the physical point with active quark degrees of freedom and preferably in large volume. Thus, there is also in this field of lattice research a very big challenge awaiting.

Journal of Physics: Conference Series **454** (2013) 012042 doi:10.1088/1742-6596/454/1/012042

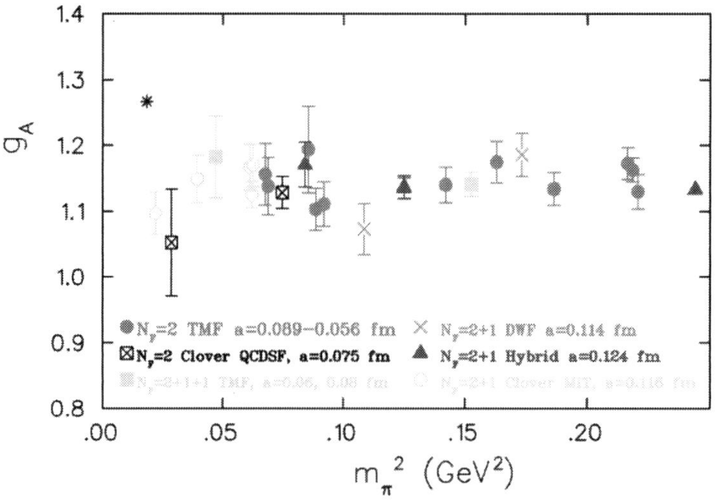

Figure 7. The nucleon axial charge g_A as determined by several lattice computations. Lattice results, taken at unphysical values of the pion mass undershoot the experimental value significantly. The graph is taken from [87].

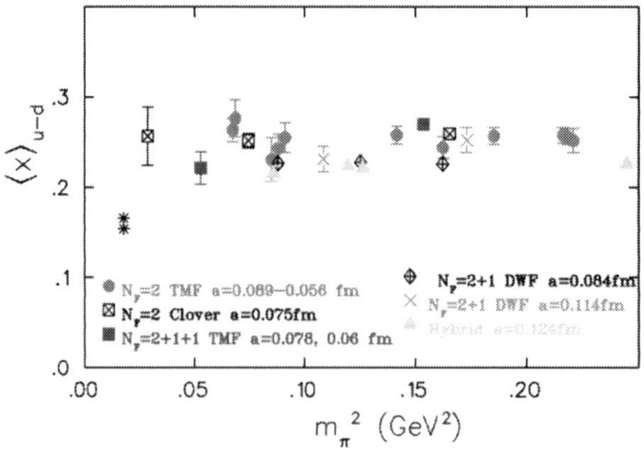

Figure 8. The average momentum of a quark in a nucleon. The lattice results in this case are systematically higher than the phenomenological value. The graph is taken from [87].

7.4. The physical point

As emphasized above, it would be highly desirable to have lattice calculations at the physical point for many physical quantities available to eliminate the systematic effect of working at unphysical values of the pion mass. From our knowledge today, it seems that the coarsest value of the lattice spacing should not exceed $a = 0.1$fm. At the same time, we want $L \cdot m_\pi > 5$ to suppress unwanted large finite volume effects. Combining this with the physical value of the pion mass of $m_\pi \approx 140$MeV, this leads to a linear extent of the box of $L \approx 5$fm. Thus to even start simulations at the physical point, a lattice with $48^3 \cdot 96$ lattice points is needed. However, this is typically already the largest lattice size that is used today.

In addition and even more demanding, one would like to perform a continuum limit and employ at least $a = 0.05$fm thus needing a box with $96^3 \cdot 192$ number of lattice points. Such a box size establishes challenges from the algorithmic point of view, the computational demand and also the storage requirement and builds a real challenge for the future.

Moreover, the lattice community has discovered that going to small values of the lattice spacing, the computational strategies used for coarser values cannot be simply taken over. The reason is that the system starts to freeze in certain topological charge sectors [90] and thus the configuration space is not fully explored and the simulations are not ergodic. Therefore the lattice community is presently exploring new strategies [91] for the computations which raises just another challenge for the future.

8. Concluding remarks

In a presentation like this one, necessarily a selection of topics addressed by the lattice community has to be made. I have left out such interesting directions as B-physics, the excited state spectrum of mesons and nucleons, computations of the quark masses and the strong coupling constant as fundamental parameters of QCD, decay constants or calculations at non-zero temperature and even efforts to look at QCD at non-vanishing chemical potential. The reader who is interested in these directions and others are referred to the annual lattice symposia which are published in [92, 93, 94, 95, 96] for the last five years. Especially the plenary presentations covering the different topics addressed by lattice field theory can provide good overviews about the status of lattice field theory computations in the different areas of modern research.

What I wanted to provide with these proceedings is to give an overview about he substantial progress that has been made in lattice field theory in the last years. At the examples of the baryon spectrum, Higgs boson mass bounds, the anomalous magnetic moment of the muon and efforts to cover nuclear physics I wanted to show the variety and quality of results as well as the accuracy modern lattice calculations can achieve today.

It is important for me to stress that the lattice does not only cover problems within QCD but also goes beyond. In these proceedings I have discussed the example of Higgs-Yukawa models, but there also large activities in the area of conformal field theories (see [97] for a review) and supersymmetry (see [98] for a review).

I hope, that I could convey that lattice field theory has developed into a mature field that has more and more an impact for experiments and phenomenological analyses. At the same time, it is becoming clear that there are major challenges for lattice computations on the way which can only be solved with new computational strategies, further improved algorithms and, last but not least, new supercomputer architectures towards the exaflops regime.

Acknowledgments

I wish to thank the organizers of this very pleasant conference for giving me the opportunity to report about lattice activities in the last years. I am grateful to G. Herdoiza, providing me with figure 2.

References

[1] Wilson K G 1977
[2] Wilson K G 1974 *Phys. Rev.* **D10** 2445–2459
[3] Wilson K G 2005 *Nucl. Phys. Proc. Suppl.* **140** 3–19 (*Preprint* hep-lat/0412043)
[4] Creutz M, Jacobs L and Rebbi C 1979 *Phys.Rev.* **D20** 1915
[5] Creutz M 1980 *Phys.Rev.* **D21** 2308–2315
[6] Schwinger J S 1962 *Phys.Rev.* **128** 2425–2429
[7] Feynman R P 1948 *Rev. Mod. Phys.* **20** 367–387

[8] Coleman S R 1976 *Annals Phys.* **101** 239
[9] Coleman S R, Jackiw R and Susskind L 1975 *Annals Phys.* **93** 267
[10] Lüscher M 1998 *Phys. Lett.* **B428** 342–345 (*Preprint* `hep-lat/9802011`)
[11] Ginsparg P H and Wilson K G 1982 *Phys.Rev.* **D25** 2649
[12] Nielsen H B and Ninomiya M 1981 *Nucl. Phys.* **B185** 20
[13] Nielsen H B and Ninomiya M 1981 *Phys. Lett.* **B105** 219
[14] Friedan D 1982 *Commun. Math. Phys.* **85** 481–490
[15] Luscher M 2010 331–399 (*Preprint* `1002.4232`)
[16] Kennedy A 2006 (*Preprint* `hep-lat/0607038`)
[17] Rothe H 1992 *World Sci.Lect.Notes Phys.* **43** 1–381
[18] Montvay I and Münster G 1994 *Quantum fields on a lattice* Cambridge Monographs on Mathematical Physics (Cambridge University Press)
[19] DeGrand T and Detar C E 2006
[20] Gattringer C and Lang C B 2010 *Lect.Notes Phys.* **788** 1–211
[21] Aoki S *et al.* (CP-PACS Collaboration) 2000 *Phys.Rev.Lett.* **84** 238–241 (*Preprint* `hep-lat/9904012`)
[22] Muller-Preussker M *et al.* Prepared for 19th International Symposium on Lattice Field Theory (Lattice 2001), Berlin, Germany, 19-24 Aug 2001
[23] Bernard C, Christ N, Gottlieb S A, Jansen K, Kenway R *et al.* 2002 *Nucl.Phys.Proc.Suppl.* **106** 199–205
[24] Walker-Loud A 2008 *PoS* **LATTICE2008** 005 (*Preprint* `0810.0663`)
[25] Luscher M 2005 *Comput.Phys.Commun.* **165** 199–220 (*Preprint* `hep-lat/0409106`)
[26] Urbach C, Jansen K, Shindler A and Wenger U 2006 *Comput. Phys. Commun.* **174** 87–98 (*Preprint* `hep-lat/0506011`)
[27] Clark M and Kennedy A 2007 *Phys.Rev.Lett.* **98** 051601 (*Preprint* `hep-lat/0608015`)
[28] Duane S, Kennedy A D, Pendleton B J and Roweth D 1987 *Phys. Lett.* **B195** 216–222
[29] Jansen K 2008 *PoS* **LATTICE2008** 010 (*Preprint* `0810.5634`)
[30] Jung C 2009 *PoS* **LAT2009** 002 (*Preprint* `1001.0941`)
[31] Scholz E E 2009 *PoS* **LAT2009** 005 (*Preprint* `0911.2191`)
[32] Hoelbling C 2010 *PoS* **LATTICE2010** 011 (*Preprint* `1102.0410`)
[33] Durr S, Fodor Z, Frison J, Hoelbling C, Hoffmann R *et al.* 2008 *Science* **322** 1224–1227 (*Preprint* `0906.3599`)
[34] Fodor Z and Hoelbling C 2012 *Rev.Mod.Phys.* **84** 449 (*Preprint* `1203.4789`)
[35] Luscher M 1991 *Nucl.Phys.* **B354** 531–578
[36] Feng X, Jansen K and Renner D B 2011 *Phys.Rev.* **D83** 094505 (*Preprint* `1011.5288`)
[37] Mohler D 2012 (*Preprint* `1211.6163`)
[38] Blum T, Zhou R, Doi T, Hayakawa M, Izubuchi T *et al.* 2010 *Phys.Rev.* **D82** 094508 (*Preprint* `1006.1311`)
[39] Horsley R *et al.* (QCDSF Collaboration, UKQCD Collaboration) 2012 *Phys.Rev.* **D86** 114511 (*Preprint* `1206.3156`)
[40] Portelli A, Durr S, Fodor Z, Frison J, Hoelbling C *et al.* 2011 *PoS* **LATTICE2011** 136 (*Preprint* `1201.2787`)
[41] Lellouch L 2012 1+ URL `http://www.confx.de/`
[42] Dashen R F and Neuberger H 1983 *Phys.Rev.Lett.* **50** 1897
[43] Lee T and Wick G 1974 *Phys.Rev.* **D9** 2291
[44] Linde A D 1976 *JETP Lett.* **23** 64–67
[45] Weinberg S 1976 *Phys.Rev.Lett.* **36** 294–296
[46] Sher M 1989 *Phys.Rept.* **179** 273–418
[47] Aad G *et al.* (The ATLAS Collaboration) 2012 (*Preprint* `1207.7214`)
[48] Chatrchyan S *et al.* (The CMS Collaboration) 2012 (*Preprint* `1207.7235`)
[49] Degrassi G, Di Vita S, Elias-Miro J, Espinosa J R, Giudice G F *et al.* 2012 (*Preprint* `1205.6497`)
[50] Carena M S, Megevand A, Quiros M and Wagner C E 2005 *Nucl.Phys.* **B716** 319–351 (*Preprint* `hep-ph/0410352`)
[51] Holdom B, Hou W, Hurth T, Mangano M, Sultansoy S *et al.* 2009 *PMC Phys.* **A3** 4 (*Preprint* `0904.4698`)
[52] Luscher M 1998 *Phys.Lett.* **B428** 342–345 (*Preprint* `hep-lat/9802011`)
[53] Bhattacharya T, Martin M R and Poppitz E 2006 *Phys.Rev.* **D74** 085028 (*Preprint* `hep-lat/0605003`)
[54] Giedt J and Poppitz E 2007 *JHEP* **0710** 076 (*Preprint* `hep-lat/0701004`)
[55] Poppitz E and Shang Y 2007 *JHEP* **0708** 081 (*Preprint* `0706.1043`)
[56] Gerhold P and Jansen K 2007 *JHEP* **0709** 041 (*Preprint* `0705.2539`)
[57] Gerhold P and Jansen K 2007 *JHEP* **0710** 001 (*Preprint* `0707.3849`)
[58] Fodor Z, Holland K, Kuti J, Nogradi D and Schroeder C 2007 *PoS* **LAT2007** 056 (*Preprint* `0710.3151`)
[59] Gerhold P and Jansen K 2009 *JHEP* **0907** 025 (*Preprint* `0902.4135`)
[60] Gerhold P (*Preprint* `1002.2569`)
[61] Gerhold P and Jansen K 2010 *JHEP* **1004** 094 (*Preprint* `1002.4336`)

[62] Bulava J, Gerhold P, Jansen K, Kallarackal J, Knippschild B *et al.* 2012 (*Preprint* 1210.1798)
[63] Gerhold P, Jansen K and Kallarackal J 2012 *Phys.Lett.* **B710** 697–702 (*Preprint* 1111.4789)
[64] Schwinger J S 1948 *Phys.Rev.* **73** 416–417
[65] Foley H and Kusch P 1948 *Phys.Rev.* **73** 412–412
[66] Jegerlehner F and Nyffeler A 2009 *Phys.Rept.* **477** 1–110 (*Preprint* 0902.3360)
[67] Aoyama T, Hayakawa M, Kinoshita T and Nio M 2008 *Phys.Rev.* **D77** 053012 (*Preprint* 0712.2607)
[68] Jegerlehner F 2008 *Lect.Notes Phys.* **745** 9–34
[69] Renner D B and Feng X 2008 *PoS* **LATTICE2008** 129 (*Preprint* 0902.2796)
[70] Feng X, Jansen K, Petschlies M and Renner D B 2011 (*Preprint* 1103.4818)
[71] Roberts B (Muon (g-2) Collaboration) 2004 *Nucl.Phys.Proc.Suppl.* **131** 157–161
[72] Sato S 2011 *AIP Conf.Proc.* **1336** 560–564
[73] Blum T 2003 *Phys.Rev.Lett.* **91** 052001 (*Preprint* hep-lat/0212018)
[74] Renner D B, Feng X, Jansen K and Petschlies M 2012 (*Preprint* 1206.3113)
[75] Gockeler M *et al.* (QCDSF Collaboration) 2004 *Nucl. Phys.* **B688** 135–164 (*Preprint* hep-lat/0312032)
[76] Aubin C and Blum T 2007 *Phys. Rev.* **D75** 114502 (*Preprint* hep-lat/0608011)
[77] Boyle P, Del Debbio L, Kerrane E and Zanotti J 2012 *Phys.Rev.* **D85** 074504 (*Preprint* 1107.1497)
[78] Hayakawa M, Blum T, Izubuchi T and Yamada N 2006 *PoS* **LAT2005** 353 (*Preprint* hep-lat/0509016)
[79] Della Morte M, Jager B, Juttner A and Wittig H 2012 *JHEP* **1203** 055 (*Preprint* 1112.2894)
[80] Aubin C, Blum T, Golterman M and Peris S 2012 *Phys.Rev.* **D86** 054509 (*Preprint* 1205.3695)
[81] Aubin C, Blum T, Golterman M and Peris S 2012 (*Preprint* 1210.7611)
[82] Feng X, Hotzel G, Jansen K, Petschlies M and Renner D B 2012 *PoS* **LATTICE2012** 174 (*Preprint* 1211.0828)
[83] Doi T and Collaboration f H Q 2012 (*Preprint* 1212.1572)
[84] Beane S, Detmold W, Orginos K and Savage M 2011 *Prog.Part.Nucl.Phys.* **66** 1–40 (*Preprint* 1004.2935)
[85] Ishii N, Aoki S and Hatsuda T 2007 *Phys.Rev.Lett.* **99** 022001 (*Preprint* nucl-th/0611096)
[86] Alexandrou C 2010 *PoS* **LATTICE2010** 001 (*Preprint* 1011.3660)
[87] Alexandrou C 2012 (*Preprint* 1208.5679)
[88] Renner D B 2012 *PoS* **QNP2012** 003 (*Preprint* 1207.2680)
[89] Lin H W Lattice 2012 **PoS** 013 (*Preprint* 1212.6849)
[90] Schaefer S, Sommer R and Virotta F (ALPHA Collaboration) 2011 *Nucl.Phys.* **B845** 93–119 (*Preprint* 1009.5228)
[91] Luscher M and Schaefer S 2011 *JHEP* **1107** 036 (*Preprint* 1105.4749)
[92] Orginos K *et al.* 2008 *Proceedings of Science* **66** 1+ URL http://pos.sissa.it/cgi-bin/reader/conf.cgi?confid=66
[93] Liu C *et al.* 2009 *Proceedings of Science* **91** 1+ URL http://pos.sissa.it/cgi-bin/reader/conf.cgi?confid=91
[94] Rossi G *et al.* 2010 *Proceedings of Science* **105** 1+ URL http://pos.sissa.it/cgi-bin/reader/conf.cgi?confid=105
[95] Vranas P *et al.* 2011 *Proceedings of Science* **139** 1+ URL http://pos.sissa.it/cgi-bin/reader/conf.cgi?confid=139
[96] Leinweber D *et al.* 2012 *Proceedings of Science* **164** 1+ URL http://pos.sissa.it/cgi-bin/reader/conf.cgi?confid=164
[97] Appelquist T 2010 *Prog.Theor.Phys.Suppl.* **180** 72–84
[98] Catterall S 2010 *PoS* **LATTICE2010** 002 (*Preprint* 1010.6224)

A first look at quasi-Monte Carlo for lattice field theory problems

K Jansen[1], H Leovey[2], A Nube[1,3], A Griewank[2] and M Mueller-Preussker[3]

[1] NIC, DESY Zeuthen, Platanenallee 6, D-15738 Zeuthen, Germany
[2] Institut für Mathematik, Humboldt Universität zu Berlin, Unter den Linden 6, D-10099 Berlin, Germany
[3] Institut für Physik, Humboldt-Universität zu Berlin, Newtonstr. 15, D-12489 Berlin, Germany

E-mail: griewank@mathematik.hu-berlin.de, Karl.Jansen@desy.de, leovey@mathematik.hu-berlin.de, mmp@physik.hu-berlin.de, Andreas.Nube@desy.de

Abstract. In this project we initiate an investigation of the applicability of Quasi-Monte Carlo methods to lattice field theories in order to improve the asymptotic error behavior of observables for such theories. In most cases the error of an observable calculated by averaging over random observations generated from an ordinary Monte Carlo simulation behaves like $N^{-1/2}$, where N is the number of observations. By means of Quasi-Monte Carlo methods it is possible to improve this behavior for certain problems to up to N^{-1}. We adapted and applied this approach to simple systems like the quantum harmonic and anharmonic oscillator and verified an improved error scaling.

1. Introduction

Four-dimensional quantum field theories play a crucial role in the mathematical description of the fundamental forces of nature. The path integral formalism developed by R.P. Feynman has been established since decades to quantize classical field theories and thus, to formulate quantum field theories.

A quantum field theory (QFT) describes the behavior of certain particles and their interaction. In the absence of an interaction the path integral of a QFT is usually trivial. It is mostly the interaction term that makes the path integral challenging to evaluate. In cases where the coefficient of the interaction term, the coupling constant, is small enough a perturbative treatment of the path integral is possible and often sufficient.

However many interesting quantities, like e.g. the hadron spectrum, decay constants, certain matrix elements and form factors, have to be calculated in a regime of a strong coupling(-constant), where a perturbative approach must fail. In such situations it is necessary to treat the path integral non-perturbatively. Because of the lack of closed form solutions the path integral can only be evaluated numerically. In order to do so, it is necessary to give the path integral a mathematically well defined meaning. A straightforward method is the discretization of space and time by introducing a lattice spacing a and a Euclidean space-time lattice with a fixed spacing between two neighboring lattice points. Restricting the system to a finite lattice extension (and applying certain boundary conditions) the infinite-dimensional path integral is converted to a finite-dimensional integral. (In principle, one still has to perform the transition to zero lattice spacing and therefore to infinitely many space-time points at the end.)

Such lattice path integrals can easily have dimensions of $O(10^9)$ (e.g. simulations of lattice-discretized quantum chromodynamics (QCD), the theory of strong interactions of elementary particles). The high dimensionality of the problem restricts the spectrum of applicable algorithms to Monte Carlo-based methods. Especially Markov chain-Monte Carlo (Mc-MC) methods have been successfully applied since the beginning of the study of lattice field theories. With the algorithms employed observables calculated from a Monte Carlo chain of N steps will obey a statistical error proportional to $1/\sqrt{N}$.

Recent developments in the field of Quasi-Monte Carlo (QMC) methods, discussed in more detail in sections 3 to 5, show that under certain conditions it is possible to construct sets of samples of integration points leading to much faster rates of convergence of an observable and a much better asymptotic error behavior of up to $1/N$.

Such an improved error behavior would decrease the number of samples necessary to achieve a certain error bound, resulting in a drastic reduction of runtime. Note that for present computations in field theory state-of-the-art supercomputers are used. It is unclear, whether QMC methods can be used for lattice field theory simulations. As a first step, to nevertheless investigate this possibility, we will focus in this work on the study of much simpler models, namely the quantum mechanical harmonic and anharmonic oscillator.

2. Quantum mechanical harmonic and anharmonic oscillator

In this section we will discuss the basic steps for the quantization of the theory in the path integral approach and the discretization on a time lattice. The first step is the construction of the Lagrangian (or the action) of the corresponding classical mechanical system for a given path $x(t)$ of a particle with mass M_0. For a numerically stable evaluation of the path integral it is essential to pass on to Euclidean time. In this case the Lagrangian L and the action S is given by:

$$L(x,t) = \frac{M_0}{2}\left(\frac{dx}{dt}\right)^2 + V(x) \tag{1}$$

$$S(x) = \int_0^T L(x,t)\, dt. \tag{2}$$

Depending on the scenario (harmonic or anharmonic oscillator) the potential $V(x)$ consists of two parts

$$V(x) = \underbrace{\frac{\mu^2}{2}x^2}_{\text{harmonic part}} + \underbrace{\lambda x^4}_{\text{anharmonic part}} , \tag{3}$$

such that the parameter λ controls the anharmonic part of the theory. It should also be mentioned that in the anharmonic case the parameter μ^2 can take on negative values, leading then to a double well potential.

The next step is to discretize time into equidistant time slices with a spacing of a. The path is then only defined on the time slices:

$$t \to t_i = (i-1)\cdot a \quad i = 1 \ldots d \tag{4}$$

$$x(t) \to x_i = x(t_i) . \tag{5}$$

On the lattice the derivative with respect to the time appearing in (1) (first term) will be replaced by the forward finite difference $\nabla x_i = \frac{1}{a}(x_{i+1} - x_i)$. The choice of the lattice derivative is not unique and requires special care, particularly if one considers more complicated models like lattice QCD. But in [1] it was shown that the lattice derivative chosen here permits a well defined continuum limit. Putting all the ingredients together, we can write down the lattice action for the (an)harmonic oscillator

$$S^{\mathrm{latt}}(x) = a \sum_{i=1}^{d} \frac{M_0}{2} (\nabla x_i)^2 + V(x_i) . \tag{6}$$

For the path a cyclic boundary condition $x_{d+1} = x_1$ can be assumed. In the following the superscript "latt" will be dropped, as we will only refer to the lattice action from now on. The expectation value of an observable O of the quantized theory expressed in terms of the path integral reads as follows:

$$\langle O(x) \rangle = \frac{\int_{\mathbb{R}^d} O(x) e^{-S(x)} dx_1 ... dx_d}{\int_{\mathbb{R}^d} e^{-S(x)} dx_1 ... dx_d} . \tag{7}$$

This expression is suitable for a numerical evaluation of certain quantities of the underlying theory. Up to now only Monte Carlo methods are known to give reliable results for dimensions $d \gg 10$. One type of such methods, often used in physics, is the Markov chain-Monte Carlo approach mostly applying the weight $\propto e^{-S(x)}$ for sampling paths $\{x_i\}$ (so-called "importance sampling"). Especially the Metropolis algorithm[2] is suitable and a straightforward solution of (7) (also described in [1]) and serves as a reference method for the QMC approach, which is much less intuitive. The theory of QMC methods is a purely mathematical topic. During the discussion of the key aspects of QMC, following in the next sections, we will stick to a rather mathematical language, being more adequate for the description of a mathematical issue.

3. Direct Monte Carlo and quasi–Monte Carlo methods

In many practical applications one is interested in calculating quotients of the form (7) where the action $S(.)$ and the observables $O(.)$ are usually smooth functions in high dimensions. In some special situations where one would like to deal with integrands of moderately high dimensions, one may consider an estimator for the integral I_1 in the numerator and I_2 in the denominator of (7) separately, and then take I_1/I_2 as an estimation of $\langle O(x) \rangle$. Another possibility is to take a joint estimator for the total quantity $\langle O(x) \rangle$ using a single direct sampling method. A well known approach based on direct sampling is the so called weighted uniform sampling (WUS) estimator, analyzed in [3]. We will show some characteristics of the WUS estimator in section 6, and we will refer from now on to these methods as *plain* or *direct* sampling methods for estimating (7). In many interesting examples, we encounter the case were the action $S(.)$ and the observable $O(.)$ lead to integrals I_1,I_2 of Gaussian type. Then the integrals I_1,I_2 can be written in the form

$$I_i = \frac{1}{(2\pi)^{d/2}\sqrt{\det(C)}} \int_{\mathbb{R}^d} g_i(\mathbf{x}) e^{-\frac{1}{2}\mathbf{x}^\top C^{-1} \mathbf{x}} d\mathbf{x}, \quad \mathbf{x} = (x_1, \ldots, x_d),\ i = 1, 2 \quad ,$$

where C denotes the covariance matrix of the Gaussian density function. A transformation to the unit cube in $\mathbb{R}^d$ can be applied such that the corresponding integrals take the form

$$I = \int_{[0,1]^d} g(A\mathbf{\Phi}^{-1}(\boldsymbol{z})) d\boldsymbol{z} = \int_{[0,1]^d} f(\boldsymbol{z}) d\boldsymbol{z} = I_{[0,1]^d}(f), \quad \boldsymbol{z} = (z_1, \ldots, z_d) . \tag{8}$$

Here $AA^\top = C$ is some symmetric factorization of the covariance matrix, and $\mathbf{\Phi}^{-1}(\boldsymbol{z}) := (\Phi^{-1}(z_1), \ldots, \Phi^{-1}(z_d))^\top$, where $\Phi^{-1}(\cdot)$ represents the inverse of the normal cumulative distribution function $\Phi(\cdot)$.

In the classical plain or direct Monte–Carlo (MC) approach one tries to estimate (8) by generating samples pseudo-randomly. One starts with a finite sequence of independent identically distributed (i.i.d.) samples $P_N = \{\boldsymbol{z}_1, \ldots, \boldsymbol{z}_N\}$, where the points $\boldsymbol{z}_j$, $1 \leq j \leq N$, have been generated from the uniform distribution in $[0,1]^d$. Then, the quadrature rule is fixed by taking the average of the function evaluations for f

$$Q_N := \frac{1}{N} \sum_{j=1}^{N} f(\boldsymbol{z}_j),$$

as an approximation of the desired integral $\int_{[0,1]^d} f(\boldsymbol{z})\, d\boldsymbol{z}$. The resulting estimator $\hat{Q}_N$ is unbiased. The integration error can be approximated via the central limit theorem, given that f belongs to $L_2([0,1]^d)$. The variance of the estimator $\hat{Q}_N$ is given by

$$\frac{\sigma^2}{N} = \frac{1}{N}\left(\int_{[0,1]^d} f^2(\boldsymbol{z})\, d\boldsymbol{z} - \left(\int_{[0,1]^d} f(\boldsymbol{z})\, d\boldsymbol{z}\right)^2\right).$$

As measured by its standard deviation from zero the integration error associated with the MC approach is then of order $O(N^{-\frac{1}{2}})$. The quality of the MC samples relies on the selected pseudo–random number generators of uniform samples, here we use the *Mersenne Twister* generator from Matsumoto and Nishimura (see [4]). MC is in general a very reliable tool in high–dimensional integration, but the order of convergence is in fact rather poor.

In contrast, quasi–Monte Carlo (QMC) methods generates deterministically point sets that are more regularly distributed than the pseudo–random points from MC (see [5], [6], [7], [8]). Typical examples of QMC are shifted lattice rules and low–discrepancy sequences. To explain what we mean by "regularly distributed", we define now the classical notion of discrepancy of a finite sequence of points P_N in $[0,1)^d$. Given $P_N = \{\boldsymbol{z}_1, \ldots, \boldsymbol{z}_N\}$ a set of points in $[0,1)^d$, and a nonempty family $\mathbb{I}$ of Lebesgue-measurable sets in $[0,1)^d$, we define the classical discrepancy function by

$$D(\mathbb{I}; P_N) := \sup_{B \in \mathbb{I}} \left| \frac{\sum_{i=1}^{N} c_B(\boldsymbol{z}_i)}{N} - \lambda_d(B) \right|,$$

where c_B is the characteristic function of B. This allows us to define the so called *star discrepancy.*

Definition 3.1 *We define the star discrepancy* $D^\star(P_N)$ *of the point set* P_N *by* $D^\star(P_N) := D(\mathbb{I}; P_N)$, *where* $\mathbb{I}$ *is the family of all sub-intervals of the form* $\prod_{i=1}^{d}[0, u_i)$, *with* $u_i \geq 0$, $1 \leq i \leq d$.

The *star discrepancy* can be considered as a measure of the worst difference between the uniform distribution and the sampled distribution in $[0,1)^d$ attributed to the point set P_N. The usual way to analyze QMC as a deterministic method is by choosing a class of integrand functions F, and a measure of discrepancy $D(P_N)$ for the point sets P_N. Then, the deterministic integration error is usually given in the form

$$|Q_N - \int_{[0,1]^d} f(\boldsymbol{z})\, d\boldsymbol{z}| \;\leq\; D(P_N) V(f),$$

where $V(f)$ measures a particular variation of the function $f \in F$. A classical particular error bound in this form is the famous Koksma–Hlawka inequality, where $D(P_N)$ is taken to be the *star discrepancy* of the point set P_N, and $V(f)$ is the variation in the sense of Hardy and Krause of f. In the context of QMC, a sequence of points in $[0,1)^d$ is called a low–discrepancy sequence if $D^\star(P_N) = O(N^{-1}(\log(N))^d)$ for all truncations of the sequence to its first N terms.

3.1. *Quasi-Monte Carlo errors and complexity*

There are certain reproducing kernel Hilbert spaces $\mathbb{F}_d$ of functions $f : [0,1]^d \to \mathbb{R}$ which are particularly useful for estimating the quadrature error of QMC methods (see [9]). Consider a kernel $K : [0,1]^d \times [0,1]^d \to \mathbb{R}$ satisfying $K(\cdot, \boldsymbol{y}) \in \mathbb{F}_d$ and $\langle f, K(\cdot, \boldsymbol{y})\rangle = f(\boldsymbol{y})$ for each $\boldsymbol{y} \in [0,1]^d$ and $f \in \mathbb{F}_d$. We denote now with $\langle \cdot, \cdot \rangle$ and $\|\cdot\|$ the inner product and norm in $\mathbb{F}_d$. If the integral

$$I(f) = \int_{[0,1]^d} f(\boldsymbol{z}) d\boldsymbol{z}$$

is a continuous functional on the space $\mathbb{F}_d$, then the worst case quadrature error $e_N(\mathbb{F}_d)$ for point sets $P_N = \{\boldsymbol{z}_1, \ldots, \boldsymbol{z}_N\}$ and quasi-Monte Carlo algorithms for the space $\mathbb{F}_d$ can be given by

$$e_N(\mathbb{F}_d) := \sup_{f \in \mathbb{F}_d, \|f\| \leq 1} |I(f) - Q_N(f)| = \sup_{\|f\| \leq 1} |\langle f, h_N \rangle| = \|h_N\|,$$

due to Riesz' representation theorem for linear bounded functionals. In this case, the *representer* $h_N \in \mathbb{F}_d$ of the quadrature error is given by

$$h_N(\boldsymbol{z}) = \int_{[0,1]^d} K(\boldsymbol{z}, \boldsymbol{y}) d\boldsymbol{y} - \frac{1}{N} \sum_{i=1}^{N} K(\boldsymbol{z}, \boldsymbol{z}_i) \quad (\forall \boldsymbol{z} \in [0,1]^d).$$

In QMC error analysis, one usually considers the weighted (anchored) tensor product Sobolev space introduced in [10]

$$\mathbb{F}_d = \mathcal{W}_{2,\text{mix}}^{(1,\ldots,1)}([0,1]^d) = \bigotimes_{i=1}^{d} W_2^1([0,1]) \,,$$

with the weighted norm $\|f\|_\gamma^2 = \langle f, f \rangle_\gamma$ and inner product

$$\langle f, g \rangle_\gamma = \sum_{u \subseteq \{1,\ldots,d\}} \prod_{j \in u} \gamma_j^{-1} \int_{[0,1]^{|u|}} \frac{\partial^{|u|}}{\partial \boldsymbol{z}_u} f(\boldsymbol{z}_u, \mathbf{1}) \frac{\partial^{|u|}}{\partial \boldsymbol{z}_u} g(\boldsymbol{z}_u, \mathbf{1}) d\boldsymbol{z}_u,$$

where for $u \subseteq \{1, \ldots, d\}$ we denote by $|u|$ its cardinality, and $(\boldsymbol{z}_u, \mathbf{1})$ denotes the vector containing the coordinates of $\boldsymbol{z}$ with indices in u, and the other coordinates set equal to 1.

The corresponding reproducing kernel is given by

$$K_{d,\gamma}(\boldsymbol{z}, \boldsymbol{y}) = \prod_{j=1}^{d} (1 + \gamma_j [1 - \max(z_j, y_j)]) \quad (\boldsymbol{z}, \boldsymbol{y} \in [0,1]^d).$$

There are several other examples considered for error analysis. For example, the weighted Walsh space consisting of Walsh series (see [7, Example 2.8] and [11]). The weighted tensor product Sobolev space allow for explicit QMC constructions deriving error estimates of the form

$$e_N(\mathbb{F}_d) \leq C(\delta) N^{-1+\delta} \quad (\delta \in (0, \tfrac{1}{2}]), \tag{9}$$

where the constant $C(\delta)$ is independent on the dimension d, given that the sequence of weights (γ_j) satisfies (see [12])

$$\sum_{j=1}^{\infty} \gamma_j^{\frac{1}{2(1-\delta)}} < \infty \,.$$

Journal of Physics: Conference Series **454** (2013) 012043 doi:10.1088/1742-6596/454/1/012043

Traditional unweighted function spaces considered for integration suffer the from the curse of dimensionality. Their weighted variants describe a setting where the variables or group of variables may vary in importance. Thus, they give a partial explanation of why some very high-dimensional spaces become tractable for QMC.

Explicit QMC constructions satisfying (9) are *shifted lattice rules* for weighted spaces. The rate (9) can be also obtained for Niederreiter and Sobol' sequences (see [13]).

The idea of "weighting" the norm of the spaces to obtain tractable results can be applied in fact to more general function spaces than smooth function spaces of tensor product form, and many integration examples can be found in [6]. In our numerical experiments, we used so far QMC algorithms based on a particular type of low–discrepancy sequences. Numerical experiments with shifted lattice rules will be carried out in the near future, following new techniques for fixing adequate weights introduced in [14].

4. Low–discrepancy (t, d)-sequences

The most well known type of low–discrepancy sequences are the so called (t, d)-sequences. To introduce how (t, m, d)-nets and (t, d)-sequences are defined, we consider first *elementary intervals* in a integer base $b \geq 2$. Let E be any sub-interval of $[0, 1)^d$ of the form $E = \prod_{i=1}^{d}[a_i b^{-c_i}, (a_i + 1)b^{-c_i})$ with a_i, $c_i \in \mathbb{N}, c_i \geq 0$, $0 \leq a_i < b^{-c_i}$ for $1 \leq i \leq d$. An interval of this form is called an elementary interval in base b.

Definition 4.1 *Let $0 \leq t \leq m$ be integers. A (t, m, d)-net in base b is a point set P_N of $N = b^m$ points in $[0, 1)^d$ such that every elementary interval E in base b with $\lambda_d(E) = \frac{b^t}{b^m}$ contains exactly b^t points.*

Definition 4.2 *Let $t \geq 0$ be an integer. A sequence $\mathbf{x}_1, \mathbf{x}_2, ...$ of points in $[0, 1)^d$ is a (t, d)-sequence in base b if for all integers $k \geq 0$ and $m > t$, the point set consisting of $N = b^m$ points $\mathbf{x}_i$ with $kb^m \leq i < (k + 1)b^m$, is a (t, m, d)-net in base b.*

The parameter t is called the *quality parameter* of the (t, d)–sequences. In [15], theorem 4.17, it is shown that (t, d)-sequences are in fact low–discrepancy sequences. We reproduce this result in the following

Theorem 4.3 *The star-discrepancy $D^\star$ of the first N terms P_N of a (t, d)-sequence in base b, satisfies*

$$ND^\star(P_N) \leq C(d, b)b^t(log(N))^d + O(b^t(log(N))^{d-1}),$$

where the implied constants depend only on b and d. If either $d = 2$ or $b = 2$, $d = 3, 4$, we have

$$C(d, b) = \frac{1}{d}\left(\frac{b-1}{2log(b)}\right)^d,$$

and otherwise

$$C(d, b) = \frac{1}{d!}\frac{b-1}{2\lfloor b/2 \rfloor}\left(\frac{\lfloor b/2 \rfloor}{log(b)}\right)^d.$$

Explicit constructions of (t, d)-sequences are available. Some of them are the generalized Faure, Sobol', Niederreiter and Niederreiter–Xing sequences. All these examples fall into the category of constructions called *digital sequences*. We refer to [7] for further reading on this topic.

5. Randomized QMC

There are some advantages in retaining the probabilistic properties of the sampling. There are practical hybrid methods permitting us to combine the good features of MC and QMC. Randomization is an important tool for QMC if we are interested for a practical error estimate of our sample quadrature Q_N to the desired integral. One goal is to randomize the deterministic point set P_N generated by QMC in a way that the estimator $\hat{Q}_N$ preserves unbiasedness. Another important goal is to preserve the better equidistribution properties of the deterministic construction.

The simplest form of randomization applied to *digital sequences* seems to be the technique called *digital b–ary shifting.* In this case, we add a random shift $\Delta \in [0,1)^d$ to each point of the deterministic set $P_N = \{\boldsymbol{z}_1, ..., \boldsymbol{z}_N\}$ using operations over the selected ring $\mathbb{F}_b$. The application of this randomization preserves in particular the t value of any projection of the point set (see [5] and references therein). The resulting estimator is unbiased.

The second randomization method we present is the one introduced by Art B. Owen ([16]) in 1995. He considered (t, m, d)-nets and (t, d)-sequences in base b and applied a randomization procedure based on permutations of the digits of the values of the coordinates of points in these nets and sequences. This can be interpreted as a random scrambling of the points of the given sequence in such a way that the net structure remains unaffected. We do not discuss here in detail Owen's randomization procedure, or from now on called *Owen's scrambling.* The main results of this randomization procedure can be stated in the following

Proposition 5.1 ***(Equidistribution)***
A randomized (t, m, d)-net in base b using Owen's scrambling is again a (t, m, d)-net in base b with probability 1. A randomized (t, d)-sequence in base b using Owen's scrambling is again a (t, d)-sequence in base b with probability 1.

Proposition 5.2 ***(Uniformity)***
Let $\tilde{\boldsymbol{z}}_i$ be the randomized version of a point $\boldsymbol{z}_i$ originally belonging to a (t, m, d)-net in base b or a (t, d)-sequence in base b, using Owen's scrambling. Then $\tilde{\boldsymbol{z}}_i$ has the uniform distribution in $[0,1)^d$, that is, for any Lebesgue measurable set $G \subseteq [0,1)^d$, $P(\tilde{\boldsymbol{z}}_i \in G) = \lambda_d(G)$, with λ_d the d-dimensional Lebesgue measure.

The last two propositions state that after *Owen's scrambling* of *digital sequences* we retain unaffected the low discrepancy properties of the constructions, and that after this randomization procedure we obtain random samples uniformly distributed in $[0,1)^s$.

The basic results about the variance of the randomized QMC estimator $\hat{Q}_N$ after applying *Owen's scrambling* to (t, m, d)-nets in base b (or of (t, d)-sequences in base b) can be found in [17]. We summarize these results in the following

Theorem 5.3 *Let $\tilde{\boldsymbol{z}}_i$, $1 \leq i \leq N$, be the points of a scrambled (t, m, d)-net in base b, and let f be a function on $[0,1)^d$ with integral I and variance $\sigma^2 = \int (f - I)^2 d\boldsymbol{z} < \infty$. Let $\hat{Q}_N = N^{-1} \sum_{i=1}^{N} f(\tilde{\boldsymbol{z}}_i)$, where $N = b^m$. Then for the variance $V(\hat{Q}_N)$ of the randomized QMC estimator it holds*

$$V(\hat{Q}_N) = o(1/N), \quad \text{as } N \to \infty, \quad \text{and} \quad V(\hat{Q}_N) \leq \frac{b^t}{N} \left(\frac{b+1}{b-1} \right)^d \sigma^2.$$

For $t = 0$ we have

$$V(\hat{Q}_N) \leq \frac{1}{N} \left(\frac{b}{b-1} \right)^{d-1} \sigma^2.$$

The above theorem says that the variance of scrambled $(0, m, d)$–nets is never more than 3 times the variance of the corresponding MC estimator. The bound of the theorem above can be improved (see theorem 13.9 in [7]) to show that the variance of scrambled $(0, m, d)$–nets are in fact always smaller than the variance of the MC estimator. If the integrand at hand is smooth enough, using *Owen's scrambling* it can be shown that one can obtain an improved asymptotic error estimate of order $O(N^{-\frac{3}{2}-\frac{1}{d}+\delta})$, for any $\delta > 0$, see [18]. Improved scrambling techniques have been developed in [19],[20].

6. Weighted uniform sampling

Weighted uniform sampling is a way of estimating a quotient of integrals of the form

$$R := \frac{\int_{[0,1]^d} f_1(\boldsymbol{z})d\boldsymbol{z}}{\int_{[0,1]^d} f_2(\boldsymbol{z})d\boldsymbol{z}}$$

by taking the estimator

$$\hat{R}_N := \frac{\sum_{j=1}^{N} f_1(\boldsymbol{z}_j)}{\sum_{j=1}^{N} f_2(\boldsymbol{z}_j)}\,, \tag{10}$$

where the points $\boldsymbol{z}_j$, $1 \leq j \leq N$, have been generated from the uniform distribution in $[0, 1]^d$. This estimator was analyzed in [3] and applications have been investigated for example in [21] and [22]. The bias and the root mean square error (RMSE) of this estimator satisfy

$$\begin{aligned} Bias(\hat{R}_N) &= \frac{R\,var(f_2)}{N} - \frac{cov(f_1, f_2)}{N} + O(N^{-\frac{3}{2}}) \\ RMSE(\hat{R}_N) &= \frac{\sqrt{var(f_1) + R^2 var(f_2) - 2R\,cov(f_1, f_2)}}{\sqrt{N}} + O(N^{-\frac{3}{4}})\,. \end{aligned}$$

The bias of the estimator is asymptotically negligible compared with the RMSE. One clear disadvantage of WUS against Mc-MC or Importance Sampling for problems with large regions of relative low values of the integrands is that with WUS we sample over the entire unit cube $[0, 1]^d$ uniformly, while Mc-MC and Importance Sampling based techniques try to concentrate in more characteristic or important regions of the integrands. These limitations where observed in our numerical experiments.

7. Numerical experiments

We consider for our numerical tests the *quantum mechanical harmonic and anharmonic oscillator* in the *path integral approach* as described in section 2. For definiteness we repeat here the expression for the action of the system:

$$S(x) = \frac{a}{2}\sum_{i=1}^{d} \frac{M_0}{a^2}(x_{i+1} - x_i)^2 + \mu^2 x_i^2 + 2\lambda x_i^4\,. \tag{11}$$

We investigate the two observable functions

$$O_1(x) = \frac{1}{d}\sum_{i=1}^{d} x_i^2\,,\; O_2(x) = \frac{1}{d}\sum_{i=1}^{d} x_i^4\,,$$

using the notation $\langle X^2\rangle$,$\langle X^4\rangle$ for $\langle O_1(x)\rangle$,$\langle O_2(x)\rangle$ in our tests.

7.1. Harmonic Oscillator

For the harmonic oscillator we can apply immediately the direct sampling approach described in sections 3 and 6 for calculating estimates of observables O by setting

$$f_1 = O(A\Phi^{-1}(\boldsymbol{z})) \ , \quad f_2 = 1$$

in (10). The matrix A is a square root of C, the covariance matrix of the variables x_i, appearing in the action if it is expressed as a bilinear form: $S = \frac{1}{2}x^T C^{-1} x$. Different factorizations, namely Cholesky and PCA (principle component analysis) have been tried out. The PCA based factorization seemed to perform better in our tests, which is the reason why we will only show results for this method. The PCA can be explicitly obtained for circulant Toeplitz matrices and the matrix–vector products can be efficiently computed by means of the fast Fourier transform. In the ordinary Mc-MC approximation, we used the Mersenne Twister[4] pseudo random number generator. For the QMC tests, we use randomly scrambled Sobol' sequences using the technique proposed by J. Matousěk[19]. The error of $\langle X^2 \rangle$ was obtained by scrambling 10 times the QMC sequence and making 10 runs of an Mc-MC simulation (with different seeds). This procedure is repeated 30 times in both cases to obtain the error of the error. From the results, shown in figure 1, we can see a scaling that agrees perfectly with the expected behavior, namely $N^{-0.5}$ for Mc-MC and N^{-1} for QMC, for large N. Although this example is trivial, it was our first successful application of the QMC approach in a physical model and motivated us to pass on to more complicated models.

7.2. Anharmonic Oscillator

The WUS approach was also used for this problem to estimate $\langle X^4 \rangle$ and $\langle X^2 \rangle$. With the anharmonic term in the action the distribution function of the variables x_i becomes very complicated. This makes it very hard to generate the samples directly from the PDF of the anharmonic oscillator. Instead of this, the anharmonic term and a part of the harmonic term is treated as part of the weight functions f_1 and f_2 in (10), leaving the sampling procedure of the x_i as it was for the harmonic oscillator, accept for a different factor μ_{sim}^2 in front of the harmonic term

$$f_1(\boldsymbol{z}) = O(A\Phi^{-1}(\boldsymbol{z}))f_2(\boldsymbol{z}) \ , \quad f_2(\boldsymbol{z}) = e^{-\sum a\left(\frac{\mu^2-\mu_{sim}^2}{2}\right)(A\Phi^{-1}(\boldsymbol{z}))_i^2 + a\lambda(A\Phi^{-1}(\boldsymbol{z}))_i^4} \ . \tag{12}$$

This procedure is necessary, because of $C = A^T A$ being positive definite only if $\mu_{sim}^2 > 0$, which is necessary for the existence of A^{-1} during the sampling procedure. Further, it is important to note that the PCA factorization during the generation of the Gaussian samples is essential for an efficient reduction of the effective dimension (see [23]) of the problem. For the parameters listed below, we estimated the effective dimensions of the functions 12 to be close to 20 (for a 99% variance concentration). On the other hand we found out that the effective dimension depends also very strong on the parameter $T = da$, the physical time extent of the system. For small T-values, say $T < 0.2$, the effective dimension is reduced sufficiently good like in the harmonic case, such that the QMC approach leads to a $1/N$ error scaling. The situation changes for $T = 1.5$, where the error behaves only like $1/N^{0.75}$, due to the increase of the effective dimension. The parameters were set to $M_0 = 0.5$, $\lambda = 1.0$, $\mu^2 = -16$. In the two tests the parameters a and μ_{sim}^2 had been adjusted such, that T was kept fixed. We set $a = 0.015$ and $\mu_{sim}^2 = 0.015$ for $d = 100$, whereas for $d = 1000$ $a = 0.0015$ and $\mu_{sim}^2 = 0.0015$ was chosen. The error analysis of $\langle X^2 \rangle$ and $\langle X^4 \rangle$ has been adopted from the harmonic oscillator test case described in the last subsection 7.1. The result is shown in figure 2. For reasons mentioned earlier, WUS shows its limitations for large T in our experiments. If $T \geq 5$ and $\mu^2 \leq -4$, then we observe poor results with the Mc-MC or RQMC direct WUS sampling method. For $T \in [1, 1.5]$ and $\mu^2 \in [-20, 10]$ the

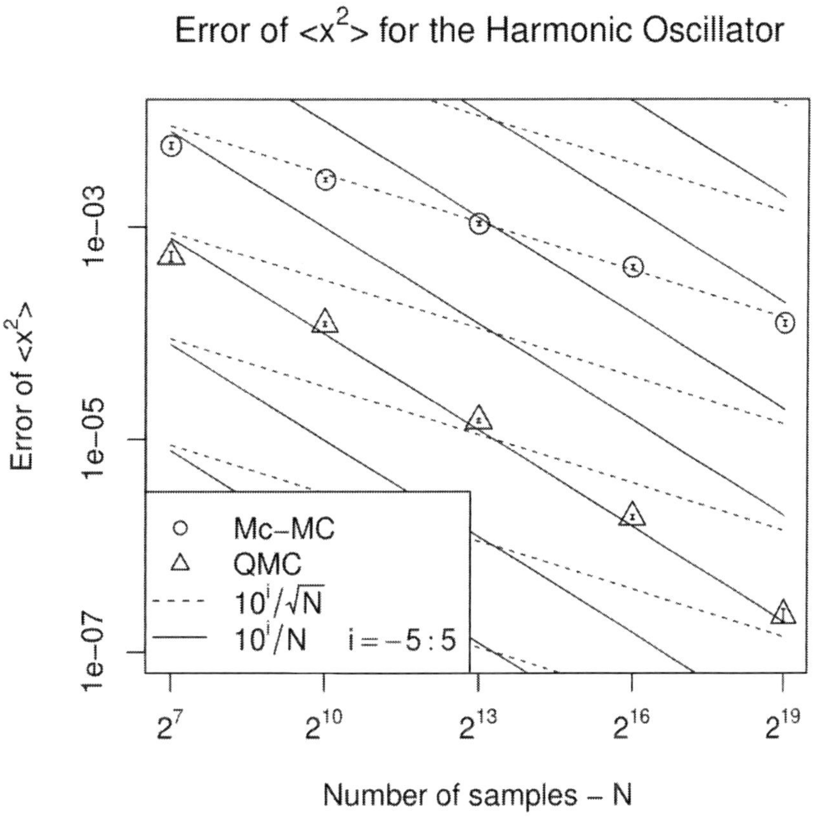

Figure 1. Error of $\langle X^2 \rangle$ in dependence of the number of samples N, $\lambda = 0$ (harmonic oscillator),$d = 51$, $M_0 = 0.5$ and $\mu^2 = 2.0$

PCA results for RQMC seem satisfactory. The resulting estimation of the ground state energy matches in at least two significant digits with the theoretical value, $E_0 = 3.863...$, calculated in [24], namely $\hat{E}_0 = 3.856 \pm 0.004$ for $d = 100$ and $\hat{E}_0 = 3.864 \pm 0.003$ for $d = 1000$.

8. Concluding Remarks

For the harmonic oscillator we found a large-N error behavior as expected for QMC ($\sim 1/N$) and Mc-MC ($\sim 1/\sqrt{N}$). Also for the anharmonic oscillator the estimation procedure leads to a significant improvement when employing the QMC approach. In this case, the error scaling is only of $O(N^{-0.75})$ instead of the theoretically best case of $O(N^{-1})$. Further, we found that the applicability of the WUS approach seems to be limited by the physical time extent $T = da$. Stable results could only by found for values $T \leq 1.5$. On the other hand, the choice of a does not seem to have any effect and the accessible range of T values gives already estimates of the ground state energy, compatible (within errors) with the theoretical prediction (valid in the limit $T \to \infty$ and $a \to 0$). For the case that the improved error scaling and the mild dependence on the lattice spacing a found here will also be present in more elaborate models, the QMC has the potential to become very valuable in the future.

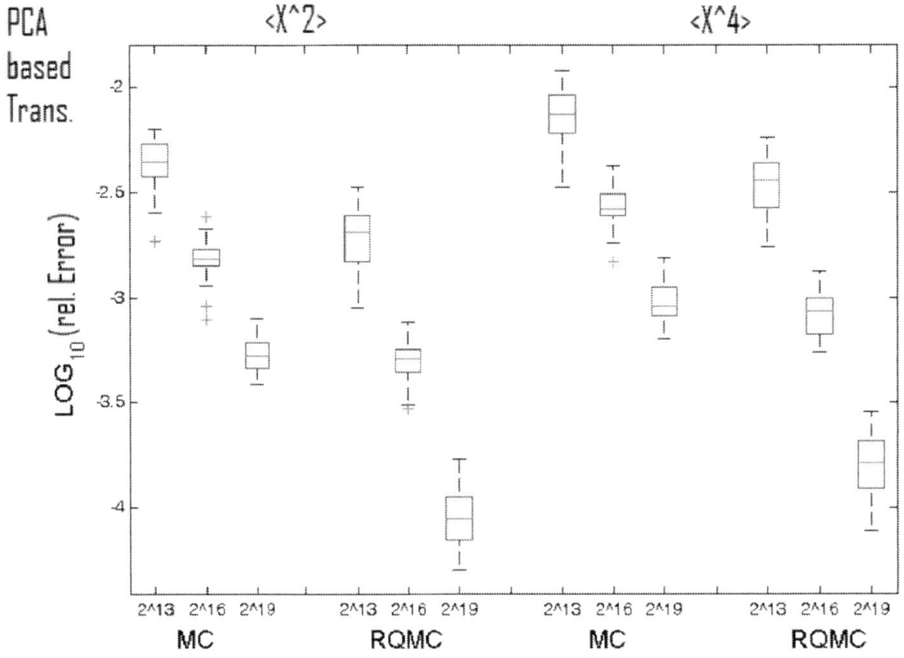

(a) $a = 0.015$, $d = 100$ ($T = 1.5$)

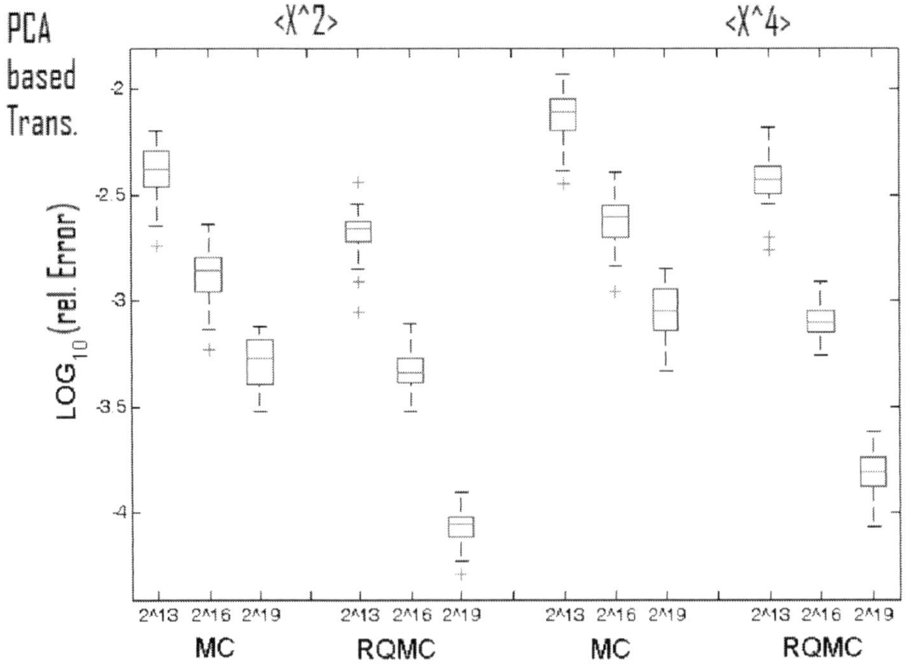

(b) $a = 0.0015$, $d = 1000$ ($T = 1.5$)

Figure 2. Shown is the log_{10}(relative error) as box plots with 30 repetitions of the experiment with $\lambda = 1.0$, $\mu^2 = -16$ and a and d as indicated. For the sample generation MC and randomly scrambled Sobol' (RQMC) was used with $2^{13}, 2^{16}$ and 2^{19} points. The approximate convergence rate is of $O(N^{-0.75})$ for RQMC.

Acknowledgments
The authors wish to express their gratitude to Alan Genz (Washington State University) and Frances Kuo (University of New South Wales, Sydney) for inspiring comments and conversations, which helped to develop the work in this report. Frances Kuo collaborated with us during her visit to the Humboldt-University Berlin in 2011. A.N., K.J. and M.M.-P. acknowledge financial support by the DFG-funded corroborative research center SFB/TR9.

References

[1] Creutz M and Freedman B A 1981 *Ann. Phys.* **132** 427–462
[2] Metropolis N, Rosenbluth A, Rosenbluth M, Teller A and Teller E 1953 *J. Chem. Phys.* **21** 1087–1092
[3] Powell M J D and Swann J 1966 *J.Inst.Maths Applics* **2** 228–236
[4] Matsumoto M and Nishimura T 1998 *ACM Trans. Model. Comput. Simul.* **8** 3–30 ISSN 1049-3301 URL `http://doi.acm.org/10.1145/272991.272995`
[5] L'Ecuyer P and Lemieux C 2005 *Modeling Uncertainty* (*International Series in Operations Research & Management Science* vol 46) ed Dror M, LEcuyer P and Szidarovszky F (Springer US) pp 419–474 ISBN 978-0-7923-7463-3 URL `http://dx.doi.org/10.1007/0-306-48102-2_20`
[6] Novak E and Woźniakowski H 2010 *Tractability of multivariate problems. Volume II: Standard information for functionals* (*EMS Tracts in Mathematics* vol 12) (European Mathematical Society (EMS), Zürich) ISBN 978-3-03719-084-5 URL `http://dx.doi.org/10.4171/084`
[7] Dick J and Pillichshammer F 2010 *Digital Nets and Sequences: Discrepancy Theory and Quasi-Monte Carlo Integration* (New York, NY, USA: Cambridge University Press) ISBN 0521191599, 9780521191593
[8] Kuo F, Schwab C and Sloan I 2012 *ANZIAM Journal* **53**
[9] Hickernell F J 1998 *Math. Comp* **67** 299–322
[10] Sloan I H and Wozniakowski H 1997 *J. Complexity* **14** 1–33
[11] Dick J 2008 *SIAM J. Numer. Anal.* **46** 1519–1553 ISSN 0036-1429 URL `http://dx.doi.org/10.1137/060666639`
[12] Kuo F Y 2003 *J. Complexity* **19** 301–320 ISSN 0885-064X numerical integration and its complexity (Oberwolfach, 2001) URL `http://dx.doi.org/10.1016/S0885-064X(03)00006-2`
[13] Wang X 2003 *Math. Comput.* **72** 823–838 ISSN 0025-5718 URL `http://dx.doi.org/10.1090/S0025-5718-02-01440-0`
[14] Griewank A, Lehmann L, Leovey H and Zilberman M 2012 *Math. Comp.*
[15] Niederreiter H 1992 *Random number generation and quasi-Monte Carlo methods* (*CBMS-NSF Regional Conference Series in Applied Mathematics* vol 63) (Philadelphia, PA: Society for Industrial and Applied Mathematics (SIAM)) ISBN 0-89871-295-5
[16] Owen A B 1995 *Monte Carlo and Quasi-Monte Carlo Methods in Scientific Computing* (*Lecture Notes in Statistics* vol 106) ed Niederreiter H and Shiue P J S (Springer-Verlag) pp 299–317
[17] Owen A B 1997 *SIAM J. Numer. Anal.* **34** 1884–1910 ISSN 0036-1429 URL `http://dx.doi.org/10.1137/S0036142994277468`
[18] Owen A B 2008 *Ann. Statist.* **36** 2319–2343
[19] Matoušek J 1998 *Journal of Complexity* **14** 527–556
[20] Tezuka S and Faure H 2003 *Journal of Complexity* **19** 744 – 757 ISSN 0885-064X URL `http://www.sciencedirect.com/science/article/pii/S0885064X03000359`
[21] Spanier J and Maize E H 1994 *SIAM Rev.* **36** 18–44 ISSN 0036-1445 URL `http://dx.doi.org/10.1137/1036002`
[22] Caflisch R E and Moskowitz B 1995 *Lecture Notes in Statistics 106* (Springer-Verlag) pp 1–16
[23] Caflisch R E, Morokoff W and Owen A 1997 *The Journal of Computational Finance* **1** 27–46
[24] Blankenbecler R, DeGrand T A and Sugar R 1980 *Phys.Rev.* **D21** 1055

Simulation of lattice QCD with domain-wall fermions

Ting-Wai Chiu (for the TWQCD Collaboration)
Physics Department, National Taiwan University, Taipei 106, Taiwan
Center for Quantum Science and Engineering, National Taiwan University, Taipei 106, Taiwan
E-mail: twchiu@phys.ntu.edu.tw

Abstract. Quantum Chromodynamics (QCD) is the fundamental theory for the interaction between quarks and gluons. It manifests as the short-range strong interaction inside the nucleus and plays an important role in the evolution of the early universe, from the quark-gluon phase to the hadron phase. To solve QCD is a grand challenge, since it requires very large-scale numerical simulations of the discretized action of QCD on the 4-dimensional space-time lattice. Moreover, since quarks are relativistic fermions, the fifth dimension is introduced so that massless quarks with exact chiral symmetry can be realized at finite lattice spacing, on the boundaries of the fifth dimension, the so-called domain-wall fermion (DWF). In this work, I discuss how to simulate lattice QCD with DWF so that the chiral symmetry can be preserved optimally with a finite extent in the fifth dimension. I also outline the simulations which have been performed by the TWQCD Collaboration and present some recent physical results.

1. Introduction

Quantum Chromodynamics (QCD) is the fundamental theory for the interaction between quarks and gluons. It provides the theoretical framework to understand the nuclear force/energy from the first principles. Moreover, QCD plays an important role in the evolution of the early universe, from the quark-gluon phase to the hadron phase. Since quarks are relativistic fermions, they possess chiral symmetry in the massless limit. Chiral symmetry forbids additive mass renormalization which causes the fine-tuning problem associated with the scalar field. In QCD, the chiral symmetry $[SU_L(N_f) \times SU_R(N_f)]$ of N_f massless quarks is spontaneously broken to $SU_V(N_f)$, due to the strong interaction between quarks and gluons. This gives the (nearly) massless Goldstone bosons (pions) and their specific interactions. To investigate the spontaneous chiral symmetry breaking as well as hadron physics from the first principles of QCD, it requires nonperturbative methods. So far, lattice QCD is the most promising approach. It consists in discretizing the continuum space-time on a 4-dimensional lattice [1] and computing physical observables by Monte Carlo simulations [2]. However, in lattice QCD, formulating lattice fermions with exact chiral symmetry at finite lattice spacing is rather nontrivial. This is realized through domain-wall fermions (DWF) on the (4+1)-dimensional lattice [3, 4] and overlap fermions on the 4-dimensional lattice [5, 6].

Lattice QCD with exact chiral symmetry [3, 5] is an ideal theoretical framework to study nonperturbative physics from the first principles of QCD. However, it is rather nontrivial to perform Monte Carlo simulations such that the chiral symmetry is preserved at high precision and all topological sectors are sampled ergodically.

Currently, there are three groups (RBC/UKQCD, JLQCD, TWQCD) around the world performing large-scale dynamical simulations of lattice QCD with exact chiral symmetry. Since the computational requirement for these dynamical simulations is 10 − 100 times that of their counterparts using traditional lattice fermions (e.g., Wilson fermions, staggered fermions, and their variants), they are often performed with state-of-the-art architectures. While RBC/UKQCD and JLQCD have been using IBM Blue Gene supercomputers, TWQCD has been using a GPU cluster since 2009 (currently consisting of 320 Nvidia GPUs, with sustained 100 Tflops/s).

The RBC/UKQCD Collaborations have been using conventional domain-wall fermions with the Shamir kernel [7, 8], which suffers from large chiral symmetry breaking (i.e., large residual mass), especially in the finite temperature QCD. On the other hand, the JLQCD Collaboration used overlap fermions in a fixed topology [9], which attains very good chiral symmetry but at the expense of sampling all topological sectors. To overcome the deficiencies of the above two approaches, the TWQCD Collaboration has been using optimal domain-wall fermions (ODWF) [10, 13, 14] to preserve the chiral symmetry, which not only attains a good chiral symmetry with a modest extension (e.g., $N_s = 16$) in the fifth dimension, but also samples all topological sectors ergodically. Recently, the JLQCD Collaboration has started a new project [15] to use conventional domain-wall fermions with the scaled Shamir kernel, performing dynamical simulations with more than 6 racks of the IBM Blue Gene/Q supercomputer. In other words, now all 3 groups (RBC/UKQCD, JLQCD, TWQCD) are using domain-wall fermions (DWF) to perform large-scale dynamical simulations of lattice QCD.

In this work, we discuss how to simulate lattice QCD with DWF such that the chiral symmetry can be preserved optimally with a finite extent in the fifth dimension. We also outline the simulations which have been performing by the TWQCD Collaboration and present some recent physical results.

Mathematically, optimal domain-wall fermions (ODWF) is a theoretical framework to preserve the chiral symmetry maximally with a set of analytical weights, $\{\omega_s, s = 1, \cdots, N_s\}$, one for each layer in the fifth dimension [10]. Thus the artifacts due to the chiral symmetry breaking with finite N_s can be reduced to the minimum, especially in the chiral regime. In general, the 4-dimensional effective Dirac operator of massless ODWF can be written as [16]

$$
\begin{aligned}
D &= \frac{1}{2r}[1 + \gamma_5 S_{opt}(H)], \qquad S_{opt}(H) = \frac{1 - \prod_{s=1}^{N_s} T_s}{1 + \prod_{s=1}^{N_s} T_s}, \\
T_s &= \frac{1 - \omega_s H}{1 + \omega_s H}, \qquad H = cH_w(1 + d\gamma_5 H_w)^{-1}, \qquad r = [2m_0(1 - dm_0)]^{-1},
\end{aligned}
\tag{1}
$$

where c and d are constants, and $H_w = \gamma_5 D_w$, with D_w the usual Wilson-Dirac operator plus a negative parameter $-m_0$ ($0 < m_0 < 2$). Here $S_{opt}(H) = HR_Z(H)$, where $R_Z(H)$ is the Zolotarev optimal rational approximation of $(H^2)^{-1/2}$ [17].

Recently we have demonstrated that it is feasible to perform a large-scale dynamical QCD simulation with ODWF, which not only preserves the chiral symmetry to a good precision, but also samples all topological sectors ergodically [18]. To recapitulate, we perform HMC simulations of 2 flavors QCD on a $16^3 \times 32$ lattice, with ODWF at $N_s = 16$ and plaquette gauge action at $\beta = 5.95$. Then we compute the low-lying eigenmodes of the overlap Dirac operator and use its index to obtain the topological charge of each gauge configuration, from which we compute the topological susceptibility for 8 sea-quark masses. Our result of the topological susceptibility agrees with the sea-quark mass dependence predicted by the NLO ChPT [19], and provides the first determination of both the pion decay constant (F_π = 92(12)(2) MeV) and the chiral condensate ($\Sigma^{\overline{\rm MS}}$(2 GeV) = [259(6)(7) MeV]^3) simultaneously from the topological susceptibility. Furthermore, our recent results of the mass and the decay constant of the pseudoscalar meson [20] are also in good agreement with the sea-quark mass dependence predicted by NLO ChPT [21], from which we obtain the low-energy constants F, Σ, $\bar{l}_3$ and $\bar{l}_4$. With the low-energy constants, we determine the average up and down quark mass ($m_{ud}^{\overline{\rm MS}}(2 \text{ GeV}) = 4.17(13)(19)$ MeV), and the chiral condensate ($\Sigma^{\overline{\rm MS}}(2 \text{ GeV}) = [230(4)(6) \text{ MeV}]^3$). Our results of the topological susceptibility together with the mass and decay constant of the pseudoscalar meson assert that the nonperturbative chiral dynamics of the sea-quarks are well under control in our HMC simulations.

Recently we have extended our simulations to two sets of larger lattices: (i) $20^3 \times 40 \times 16$ with plaquette gauge action at $\beta = 5.95$, for 6 sea-quark masses corresponding to pion masses in the range 230-450 MeV; (ii) $24^3 \times 48 \times 16$ with plaquette gauge action at $\beta = 5.95$, for 4 sea-quark masses corresponding to pion masses in the range 230-450 MeV. Now, after simulating 2-flavors QCD on various lattices ($16^3 \times 32$, $20^3 \times 40$, $24^3 \times 48$), we are ready to perform dynamical simulations of $(2+1)$-flavors and $(2+1+1)$-flavors QCD on the $32^3 \times 64 \times 16$ lattice, with pion mass close to the physical value. Our strategy will be outlined in the final section.

2. Theoretical aspects of domain-wall fermions

In general, for a given N_s (the number of sites in the fifth dimension), the (mathematically) maximal chiral symmetry can be attained by the optimal domain-wall fermion (ODWF) [10] with the operator

$$[\mathcal{D}(m_q)]_{xx';ss'} = (\rho_s D_w + 1)_{xx'} \delta_{ss'} + (\sigma_s D_w - 1)_{xx'} L_{ss'}, \tag{2}$$

where $\rho_s = c\omega_s + d$, $\sigma_s = c\omega_s - d$, and c, d are constants. The indices x and x' denote the sites on the 4-dimensional space-time lattice, while s and s' the layers in the fifth dimension. Here D_w is the standard Wilson Dirac operator plus a negative parameter $-m_0$ $(0 < m_0 < 2)$,

$$(D_w)_{xx'} = -\frac{1}{2} \sum_\mu \left[(1-\gamma_\mu) U_\mu(x) \delta_{x+\hat{\mu},x'} + (1+\gamma_\mu) U_\mu^\dagger(x') \delta_{x-\hat{\mu},x'} \right] + (4 - m_0), \tag{3}$$

where $U_\mu(x)$ denotes the link variable pointing from x to $x + \hat{\mu}$,

$$L = P_+ L_+ + P_- L_-, \quad P_\pm = (1 \pm \gamma_5)/2, \tag{4}$$

and

$$(L_+)_{ss'} = \begin{cases} -r m_q \delta_{N_s,s'}, & s = 1, \quad r \equiv 1/[2m_0(1-dm_0)] \\ \delta_{s-1,s'}, & 1 < s \leq N_s \end{cases}, \quad L_- = (L_+)^T. \tag{5}$$

The weights $\{\omega_s\}$ along the fifth dimension are fixed according to the formula derived in [10] such that the maximal chiral symmetry is attained. In general, for other DWFs without maximal chiral symmetry, the weights $\{\rho_s\}$ and $\{\sigma_s\}$ have different values. For example, for the conventional (Shamir) DWF, which has been used by the RBC/UKQCD Collaborations, $c = d = 1/2$ and $\omega_s = 1, \forall s$; for the scaled Shamir DWF (with scaling factor = 2), which is being used by the JLQCD Collaboration, $c = 1$, $d = 1/2$, and $\omega_s = 1, \forall s$.

The breaking of chiral symmetry due to finite N_s in the fifth dimension can be measured by the residual mass emerging in the axial Ward identity on the lattice. For the general DWF operator (2), the axial Ward identity is derived in [16], and a new formula for the residual mass is also obtained,

$$m_{res} = \left\langle \frac{\text{tr}(D_c + m_q)^{-1}_{0,0}}{\text{tr}[(D_c^\dagger + m_q)(D_c + m_q)]^{-1}_{0,0}} \right\rangle_{\{U\}} - m_q, \tag{6}$$

where $(D_c + m_q)^{-1}$ denotes the valence quark propagator with m_q equal to the sea-quark mass, tr denotes the trace running over the color and Dirac indices, and the subscript $\{U\}$ denotes averaging over an ensemble of gauge configurations. Formula (6) is useful in practice, since it immediately gives the residual mass once the 12 columns of the quark propagator are computed. Moreover, an upper-bound for the residual mass of ODWF is derived in [16]. It asserts that for N_s less than some threshold value ($\sim 16-18$), the residual mass is an exponentially decay function of N_s. This implies that ODWF can provide a viable way to preserve chiral symmetry to a good precision (e.g., $m_{res}a \sim 10^{-5}$) with a modest N_s (e.g., $N_s \simeq 16$).

3. HMC simulation of lattice QCD with domain-wall quarks

First of all, we perform the even-odd preconditioning on the DWF operator (2), which is essential for lowering the condition number as well as halving the memory consumption. Since D_w commutes with $(\rho)_{ss'} \equiv \rho_s \delta_{ss'}$ and $(\sigma)_{ss'} = \sigma_s \delta_{ss'}$, (2) becomes

$$\mathcal{D}(m_q) = D_w(\rho + \sigma L) + (1-L) = D_w[c\omega(1+L) + d(1-L)] + (1-L), \tag{7}$$

where $(\omega)_{s,s'} = \omega_s \delta_{s,s'}$ is a diagonal matrix in the fifth dimension. Now, separating the even and the odd sites on the 4D space-time lattice, (7) can be written as

$$\mathcal{D}(m_q) = \begin{pmatrix} 4-m_0 & D_w^{\rm EO} \\ D_w^{\rm OE} & 4-m_0 \end{pmatrix} [c\omega(1+L) + d(1-L)] + (1-L) = \begin{pmatrix} X & D_w^{\rm EO} Y \\ D_w^{\rm OE} Y & X \end{pmatrix}, \tag{8}$$

where

$$X \equiv (4-m_0)[c\omega(1+L) + d(1-L)] + (1-L), \quad Y \equiv c\omega(1+L) + d(1-L). \tag{9}$$

Now we further rewrite it in a more symmetric form by defining

$$M_5 \equiv \omega^{-1/2} Y X^{-1} \omega^{1/2} = \left\{ (4-m_0) + \omega^{-1/2} \left[c(1+L)(1-L)^{-1} + d\omega^{-1} \right]^{-1} \omega^{-1/2} \right\}^{-1}, \tag{10}$$

and

$$S_1 \equiv \omega^{-1/2} Y X^{-1} = M_5 \omega^{-1/2}, \quad S_2 \equiv Y^{-1} \omega^{1/2}. \tag{11}$$

Then (8) becomes

$$\mathcal{D}(m_q) = S_1^{-1} \begin{pmatrix} 1 & M_5 D_w^{\rm EO} \\ M_5 D_w^{\rm OE} & 1 \end{pmatrix} S_2^{-1} = S_1^{-1} \begin{pmatrix} 1 & 0 \\ M_5 D_w^{\rm OE} & 1 \end{pmatrix} \begin{pmatrix} 1 & 0 \\ 0 & C \end{pmatrix} \begin{pmatrix} 1 & M_5 D_w^{\rm EO} \\ 0 & 1 \end{pmatrix} S_2^{-1}, \tag{12}$$

where the Schur decomposition has been used in the last equality, with the Schur complement

$$C \equiv 1 - M_5 D_w^{\rm OE} M_5 D_w^{\rm EO}. \tag{13}$$

Note that the quark mass dependence ($m \equiv r m_q$) only resides in M_5 through L.

Since $\det \mathcal{D} = \det S_1^{-1} \cdot \det C \cdot \det S_2^{-1}$, and S_1 and S_2 do not depend on the gauge field, we can just use C in the Monte Carlo simulation. After including the Pauli-Villars fields (with $m_q = 1/r$), the pseudofermion action for 2 flavors QCD (in the isospin limit $m_u = m_d$) can be written as

$$S_{pf} = \phi^\dagger C_1^\dagger (C C^\dagger)^{-1} C_1 \phi, \quad C_1 \equiv C(m_q = 1/r), \tag{14}$$

where ϕ and $\phi^\dagger$ are complex scalar fields carrying the same quantum numbers (color, spin) of the quark fields. Including the gluon fields, the partition function for 2 flavors QCD can be written as

$$Z = \int [dU][d\phi^\dagger][d\phi] \exp\left(-S_g[U] - \phi^\dagger C_1^\dagger (C C^\dagger)^{-1} C_1 \phi \right), \tag{15}$$

where $S_g[U]$ is the lattice action for the gauge field, e.g., the Wilson plaquette action

$$S_g[U] = \beta \sum_{plaq.} \left\{ 1 - \frac{1}{3} \mathrm{Re}\,\mathrm{Tr}(U_p) \right\}, \quad \beta = \frac{6}{g^2}. \tag{16}$$

It is rather difficult to simulate (15) directly with the Metropolis algorithm. The conventional wisdom to handle this problem is to introduce a fictitious dynamics to guide the Monte Carlo simulation, i.e., the Hybrid Monte Carlo (HMC) [22]. Since the pseudofermion action (14) is positive-definite, ϕ can be generated by the heat-bath method with Gaussian noise η satisfying the Gaussian distribution $\exp(-\eta^\dagger \eta)$. That is, to solve the following equation with the conjugate gradient algorithm

$$C_1 \phi = C\eta \Leftrightarrow C_1^\dagger C_1 \phi = C_1^\dagger C \eta. \tag{17}$$

Then the fictitious molecular dynamics only involves the gauge fields $\{A_l\}$ and their conjugate momenta $\{P_l\}$, where $A_l = A_l^a t^a$ is the matrix-valued gauge field corresponding to the link variable $U_l = \exp(iA_l^a t^a)$. The Hamiltonian of the molecular dynamics is

$$\mathcal{H} = \frac{1}{2} \sum_{l,a} (P_l^a)^2 + S_g[U] + \phi^\dagger C_1^\dagger (CC^\dagger)^{-1} C_1 \phi, \tag{18}$$

and the partition function can be written as

$$\mathcal{Z} = \int [dU][dP][d\phi][d\phi^\dagger] \exp(-\mathcal{H}). \tag{19}$$

The Hamilton equations for the fictitious molecular dynamics are

$$\frac{dA_l^a(\tau)}{d\tau} = \frac{\partial \mathcal{H}}{\partial P_l^a(\tau)} = P_l^a(\tau) \Leftrightarrow \frac{dU_l(\tau)}{d\tau} = iP_l(\tau) U_l(\tau), \tag{20}$$

$$\frac{dP_l^a(\tau)}{d\tau} = -\frac{\partial \mathcal{H}}{\partial A_l^a(\tau)} = -\frac{\partial S_g}{\partial A_l^a(\tau)} - \frac{\partial S_{pf}}{\partial A_l^a(\tau)}. \tag{21}$$

These two equations together imply that $d\mathcal{H}/d\tau = 0$, which gives

$$P_l^a \frac{dP_l^a(\tau)}{d\tau} = -\frac{dS_g}{d\tau} - \frac{dS_{pf}}{d\tau}, \tag{22}$$

as an alternative form of (21).

The algorithm of HMC simulation can be outlined as follows:

(i) Choose an initial gauge configuration $\{U_l\}$.

(ii) Generate P_l^a with Gaussian weight $\exp(\{P_l^a\}^2/2)$.

(iii) Generate η with Gaussian weight $\exp(-\eta^\dagger \eta)$.

(iv) Compute ϕ according to (17).

(v) With $\{\phi\}$ held fixed, integrate (20) and (21) by an algorithm (leapfrog/Omelyan) which ensures exact reversibility and area-preserving map in the phase space for any $\delta\tau$.

(vi) Accept the new configuration $\{U_l'\}$ generated by the molecular dynamics with probability $min(1, e^{-\Delta\mathcal{H}})$, where $\Delta\mathcal{H} \equiv \mathcal{H}(U_l', P_l') - \mathcal{H}(U, P)$. This completes one HMC trajectory.

(vii) For the next trajectory, go to (ii).

To summarize, we first generate the random noise vector η with Gaussian distribution, $\exp(-\eta^\dagger\eta)$, then we obtain $\phi = C_1^{-1}C\eta$ using the conjugate gradient (CG). With fixed ϕ, the system is evolved under a fictitious Hamiltonian dynamics, the so-called molecular dynamics (MD). In MD, we use the Omelyan integrator [23, 24, 25], and the Sexton-Weingarten multiple-time scale method [26]. The most time-consuming part in the MD is to compute the fermion force $-\partial S_{pf}/\partial A_l^a(\tau)$ which is required for updating the conjugate momenta in (21), since it involves solving the linear system $(CC^\dagger)v = C_1\phi$ with CG. Here we take advantage of the remarkable floating-point capability of the Nvidia GPU and perform the CG with mixed precision [27]. Moreover, the computations of the gauge force and the update of the gauge field are also performed in GPUs. In other words, the entire HMC trajectory is computed by GPUs. Furthermore, we introduce an auxiliary heavy fermion field with mass m_H ($m_q \ll m_H < 1/r$) similar to the case of Wilson fermions [28]. For two-flavors QCD, the pseudofermion action (with $C_H \equiv C(m_H)$) becomes,

$$S_{pf}^H = \phi^\dagger C_H^\dagger (CC^\dagger)^{-1} C_H \phi + \phi_H^\dagger C_{PV}^\dagger (C_H C_H^\dagger)^{-1} C_{PV} \phi_H,$$

which gives exactly the same fermion determinant of (14). Nevertheless, the presence of the heavy fermion field plays a crucial role in reducing the light fermion force and its fluctuation, thus diminishes the change of the Hamiltonian in the MD trajectory, and enhances the acceptance rate.

In two-flavors QCD, only the quantum fluctuations (internal fermion loops) of the lightest u and d quarks (in the isospin limit $m_u = m_d$) are incorporated, while those of the heavier quarks (s, c, b and t) are neglected. To incorporate the dynamical s quark, one must use a pseudofermion action different from that of two-flavors QCD with degenerate masses (14). A straightforward approach is to take the inverse square root of the quark matrix (and the quarter-root of the Pauli-Villars matrix) of the two-flavors QCD, namely

$$S_{pf}^{N_f=1} = \phi^\dagger (C_1^\dagger C_1)^{1/4} (CC^\dagger)^{-1/2} (C_1^\dagger C_1)^{1/4} \phi.$$

However, the inverse square root and the quarter-root operations cannot be evaluated exactly since the cost of solving the eigenproblem of $CC^\dagger$ or $C_1^\dagger C_1$ is prohibitive. A way to resolve this problem is to use the optimal rational approximations for $(CC^\dagger)^{-1/2}$ and $(C_1^\dagger C_1)^{1/4}$ respectively, similarly to the rational hybrid Monte Carlo (RHMC) [29].

Recently, Chen and Chiu have constructed a novel DWF action for one flavor QCD [30], which is exact, without taking the square root, and amenable to HMC simulation. This novel pseudofermion action for one flavor QCD can be written as

$$\begin{aligned} S_{pf}^{N_f=1} &= \begin{pmatrix} 0 & \phi_1^\dagger \end{pmatrix} \left[I - c_- v_-^T \omega^{-1/2} \frac{1}{H(m)} \omega^{-1/2} v_- \right] \begin{pmatrix} 0 \\ \phi_1 \end{pmatrix} + \\ & \begin{pmatrix} \phi_2^\dagger & 0 \end{pmatrix} \left[I + c_+ v_+^T \omega^{-1/2} \frac{1}{H(1) - \Delta_+(m) P_+} \omega^{-1/2} v_+ \right] \begin{pmatrix} \phi_2 \\ 0 \end{pmatrix}, \end{aligned} \tag{23}$$

where ϕ_1 and ϕ_2 are complex scalar fields on the 4-dimensional space-time lattice, with color and 2-spinor indices, and

$$\begin{aligned} H(m) &= \gamma_5 R_5 \left[D_w + \omega^{-1/2} \left\{ \omega^{-1} d + c[1 + L(m)][1 - L(m)]^{-1} \right\}^{-1} \omega^{-1/2} \right] \\ &\equiv \gamma_5 R_5 \left[D_w + M_+ P_+ + M_- P_- \right], \quad (R_5)_{s,s'} = \delta_{s', N_s+1-s}, \\ \Delta_\pm(m) &= R_5 [M_\pm(1) - M_\pm(m)]. \end{aligned}$$

The coefficients $c_\pm$ and the vectors $v_\pm$ in (23) only depend on the parameters c, d, and $\{\omega_s, s = 1, \cdots, N_s\}$, and they will be shown explicitly in [30]. A detailed description of our HMC simulations will be presented in a forthcoming paper.

4. Gauge ensembles

For zero temperature QCD, we have been working on the following ensembles of two-flavors QCD with ODWF:

- $16^3 \times 32$ ($a \sim 0.1$ fm), with plaquette gauge action at $\beta = 5.95$, for 8 sea-quark masses corresponding to pion masses in the range 230-560 MeV [18, 20].
- $20^3 \times 40$ ($a \sim 0.1$ fm), with plaquette gauge action at $\beta = 5.95$, for 6 sea-quark masses corresponding to pion masses in the range 230-450 MeV [31].
- $24^3 \times 48$ ($a \sim 0.1$ fm) with plaquette gauge action at $\beta = 5.95$, for 4 sea-quark masses corresponding to pion masses in the range 230-450 MeV.

The first two lattice sizes ($16^3 \times 32$, and $20^3 \times 40$) have been completed, while the third lattice size ($24^3 \times 48$) is expected to be completed in the summer of 2013.

For finite temperature QCD, we have been working on the following ensembles of two-flavors QCD with ODWF:

- $16^3 \times 6$, with plaquette gauge action, for 30-40 β values in the range $[5.00, 5.95]$, each of 3-4 sea-quark masses.
- $24^3 \times 8$, with plaquette gauge action, for 30-40 β values in the range $[5.00, 5.95]$, each of 3-4 sea-quark masses.

The first lattice size ($16^3 \times 6$) has been completed, while the lattice size ($24^3 \times 8$) is still in progress.

For the quark part, we use ODWF with $c = 1$, $d = 0$ (i.e., $H = H_w$), $N_s = 16$, and $\lambda_{min}/\lambda_{max} = (0.01, 0.02, 0.05)/6.2$, where different values of λ_{min} have been used for different gauge ensembles. In general, for each sea-quark mass, we generate the initial 400-500 trajectories on a single GPU. After discarding 200-300 trajectories for thermalization, we sample one configuration every 5 trajectories, resulting in 20-32 "seed" configurations for each sea-quark mass. Then we use these seed configurations as the initial configurations for independent simulations on 20-32 GPUs. Each GPU generates 200-250 trajectories independently. Then we accumulate a total of 5000 trajectories for each sea-quark mass. From the saturation of the binning error of the plaquette, as well as the evolution of the topological charge, we estimate the autocorrelation time to be around 10 trajectories. Thus we sample one configuration every 10 trajectories, and obtain 500 configurations for each sea-quark mass.

For each configuration, we calculate the zero modes plus 80-180 conjugate pairs of the lowest-lying eigenmodes of the overlap Dirac operator. We outline our procedures as follows. First, we project 250 low-lying eigenmodes of H_w^2, using adaptive thick restart Lanczos algorithm (a-TRLan), where each eigenmode has a residual less than 10^{-12}. Then we approximate the sign function of the overlap operator by the Zolotarev optimal rational approximation with 64 poles, where the coefficients are fixed with $\lambda_{max}^2 = (6.4)^2$ and λ_{min}^2 equal to the maximum of the 250 projected eigenvalues of H_w^2. Then the sign function error is less than 10^{-14}. Using the 250 low-modes of H_w^2 and the Zolotarev approximation with 64 poles, we use the a-TRLan algorithm again to project the zero modes plus 80-180 conjugate pairs of the lowest-lying eigenmodes of the overlap operator, where each eigenmode has a residual less than 10^{-12}. We store all projected eigenmodes for the later use.

In figures 1-2, we present the essential characteristics of our HMC simulations, which are quite universal for any lattice sizes and sea-quark masses. In figure 1 (a), we plot the topological charge versus the trajectory in the HMC simulation of two-flavors QCD with ODWF. The lattice is $16^3 \times 32 \times 16$ with the spatial box $\sim (2 \text{ fm})^3$ and the sea-quark mass corresponding to $M_\pi \sim 360$ MeV. All these HMC trajectories are computed in one GPU, with a total of 543 trajectories, and the initial 200 trajectories are discarded for thermalization. We see that the topological charge has a short auto-correlation time, and each topological sector is sampled

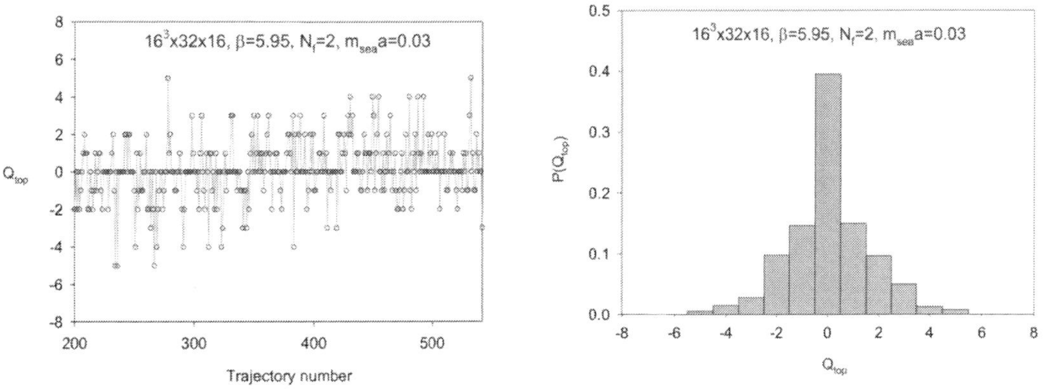

(a) (b)

Figure 1. (a) The topological charge versus the trajectory in the HMC simulation of two-flavors QCD with ODWF. The lattice is $16^3 \times 32 \times 16$ with the spatial box $\sim$ (2 fm)^3 and the sea-quark mass corresponding to $M_\pi \sim 360$ MeV. The line connecting the data points is only for guiding the eyes. (b) The histogram of the topological charge distribution in (a).

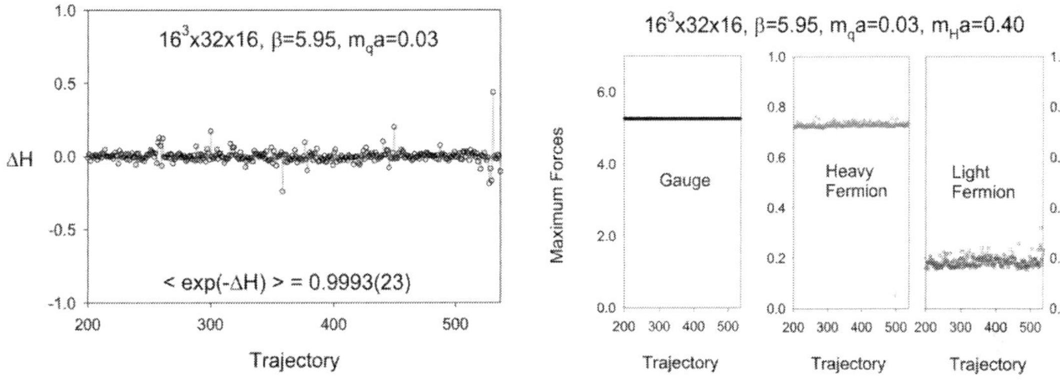

(a) (b)

Figure 2. (a) The $\Delta\mathcal{H}$ versus the trajectory in the HMC simulation of two-flavors QCD with ODWF. The lattice is $16^3 \times 32 \times 16$ with the spatial box $\sim$ (2 fm)^3, and the sea-quark mass corresponding to $M_\pi \sim 360$ MeV. The line connecting the data points is only for guiding the eyes. (b) The maximum forces of the gauge field, heavy fermion field, and light fermion field.

ergodically. Moreover, the topological charge distribution behaves like a Gaussian distribution with the mean close to zero, as shown in figure 1 (b). In figure 2 (a), we plot the change of the Hamiltonian of each trajectory, which is quite smooth, without any spikes in all trajectories. Using the measured value of $\langle \Delta\mathcal{H} \rangle = 0.00167(240)$, we estimate the theoretical acceptance rate $P_{\text{HMC}} = \text{erfc}\left(\sqrt{\langle \Delta\mathcal{H} \rangle/2}\right) = 0.977(10)$, which agrees with the measured acceptance rate 0.983(7). Moreover, the measured value of $\langle \exp(-\Delta\mathcal{H}) \rangle = 0.9993(23)$, in good agreement with the condition $\langle \exp(-\Delta\mathcal{H}) \rangle = 1$ which follows from the area-preserving property of the HMC simulation. In figure 2 (b), we plot the maximum force (averaged over all links) among all momentum updates in a trajectory, for the gauge field, the heavy fermion field, and the light

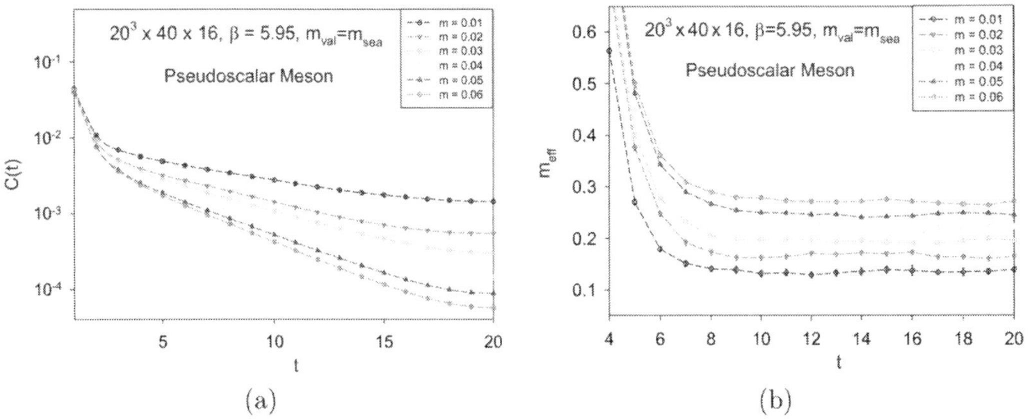

Figure 3. (a) The time-correlation function of the pseudoscalar meson for six sea-quark masses. (b) The effective mass of (a). The dashed line connecting the data points of the same sea-quark mass is for guiding the eyes.

fermion field respectively. The forces all behave smoothly for all trajectories.

5. Some recent results

To verify whether a dynamical simulation of lattice QCD captures the nonperturbative chiral dynamics of the sea-quarks, a prerequisite is to examine to what extent the pion mass (M_π) and decay constant (F_π) agree with the sea-quark mass dependence as predicted by the next-to-leading-order chiral perturbation theory (NLO ChPT) [21] and to check whether the resulting low-energy constants are reasonable or not. Furthermore, to check whether a dynamical simulation of lattice QCD samples all topological sectors ergodically, it is necessary to measure the topological susceptibility versus the sea-quark mass and to compare the results with the sea-quark mass dependence as predicted by the NLO ChPT [19]. We have performed these tests for the gauge ensembles on the $16^3 \times 32 \times 16$ lattice [18, 20].

In this section, we present preliminary results of these tests for the gauge ensembles on the $20^3 \times 40 \times 16$ lattice [31].

Using the valence quark propagator with quark mass equal to the sea-quark mass, we compute the time-correlation function of the pseudoscalar interpolator

$$C_\pi(t) = \sum_{\vec{x}} \mathrm{tr}\{\gamma_5 (D_c + m_q)^{-1}_{0,x} \gamma_5 (D_c + m_q)^{-1}_{x,0}\},$$

where the trace runs over the Dirac and color space. Then the ensemble average $\langle C_\pi(t) \rangle$ of each m_q is fitted to the formula $Z[e^{-M_\pi a t} + e^{-M_\pi a (T-t)}]/(2 M_\pi a)$ to extract the pion mass $M_\pi a$ and the decay constant $F_\pi a = m_q a \sqrt{2Z}/(M_\pi a)^2$.

In figure 4, we plot $(M_\pi a)^2/(m_q a)$ and $F_\pi a$ versus $m_q a$ respectively. Here we have made the correction for the finite volume effect using the estimate within ChPT calculated up to $\mathcal{O}(M_\pi^4/(4\pi F_\pi)^4)$ [32]. Taking into account of the correlation between M_π^2/m_q and F_π for the same sea-quark mass, we fit our data to the formulas of NLO ChPT,

$$\frac{M_\pi^2}{m_q} = \frac{2\Sigma}{F^2}\left[1 + \left(\frac{\Sigma m_q}{16\pi^2 F^4}\right)\ln\left(\frac{2\Sigma m_q}{F^2 \Lambda_3^2}\right)\right], \quad (24)$$

$$F_\pi = F\left[1 - \left(\frac{\Sigma m_q}{8\pi^2 F^4}\right)\ln\left(\frac{2\Sigma m_q}{F^2 \Lambda_4^2}\right)\right], \quad (25)$$

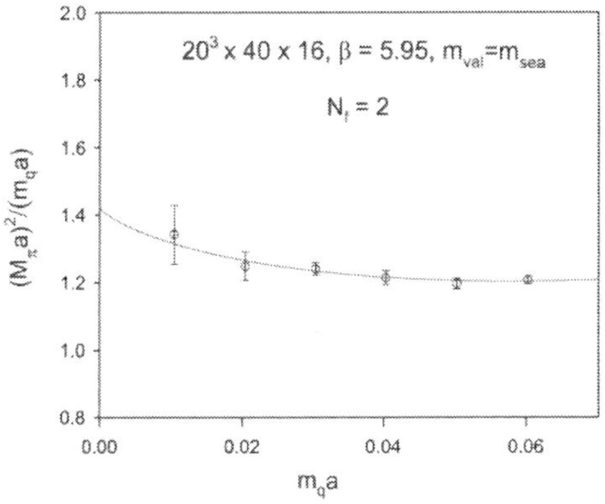

(a)

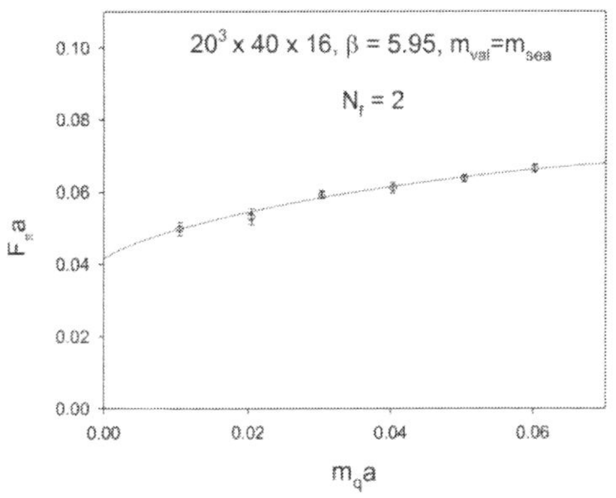

(b)

Figure 4. Pseudoscalar meson of 2 flavors QCD with ODWF (a) $(M_\pi a)^2/(m_q a)$ and (b) $F_\pi a$. The solid lines are the simultaneous fits to the NLO ChPT, for six sea-quark masses.

where Λ_3 and Λ_4 are related to the low energy constants $\bar{l}_3$ and $\bar{l}_4$ as follows.

$$\bar{l}_3 = \ln\left(\frac{\Lambda_3^2}{m_{\pi^\pm}^2}\right), \quad \bar{l}_4 = \ln\left(\frac{\Lambda_4^2}{m_{\pi^\pm}^2}\right), \quad m_{\pi^\pm} = 0.140 \text{ GeV}.$$

Our procedure of data fitting to extract the parameters (Σ, F, Λ_3 and Λ_4) has been outlined in [20]. For six sea-quark masses, our fit gives

$$\begin{array}{ll} \Sigma a^3 = 0.00122(6)(2), & Fa = 0.0414(10)(16), \\ \bar{l}_3 = 3.829(105)(43), & \bar{l}_4 = 4.755(93)(23), \end{array} \quad (26)$$

where the systematic errors are estimated by varying the number of data points from 6 to 4 ($m_q a \leq 0.04$).

With the fitted parameters, we use the physical ratio

$$\left(\frac{M_\pi}{F_\pi}\right)^{phys} = \frac{0.135 \text{ GeV}}{0.093 \text{ GeV}} \simeq 1.45$$

as the input, and solve the equation $M_\pi(m_q)/F_\pi(m_q) = 1.45$ to obtain the physical bare quark mass $m_q^{phys} a = 0.00254(10)(16)$. From (25) and the physical pion decay constant $F_\pi = 92.6$ MeV, we determine the inverse lattice spacing at the physical point,

$$1/a = 2.076(6)(5) \text{ GeV}.$$

From (24), we obtain the pion mass at the physical point, $M_\pi = 0.134(5)(3)$ GeV, which serves as a consistency check.

In order to convert the chiral condensate Σ and the average m_u and m_d to those in the $\overline{\text{MS}}$ scheme, we calculate the renormalization factor $Z_s^{\overline{\text{MS}}}(2 \text{ GeV})$ using the non-perturbative renormalization technique through the RI/MOM scheme [33], which gives $Z_s^{\overline{\text{MS}}}(2 \text{ GeV}) = 1.244(18)(39)$. Then the values of Σ and the average of m_u and m_d are transcribed to

$$\Sigma^{\overline{\text{MS}}}(2 \text{ GeV}) = [238(10)(6) \text{ MeV}]^3, \quad (27)$$

$$m_{ud}^{\overline{\text{MS}}}(2 \text{ GeV}) = 4.07(13)(12) \text{ MeV}. \quad (28)$$

Our results of the chiral condensate (27) and the average up and down quark mass (28) are in good agreement with our previous ones on the $16^3 \times 32$ lattice [20]. Since our calculation is done at a single lattice spacing the discretization error cannot be quantified reliably, but we do not expect much larger errors because our lattice action is free from $O(a)$ discretization effects.

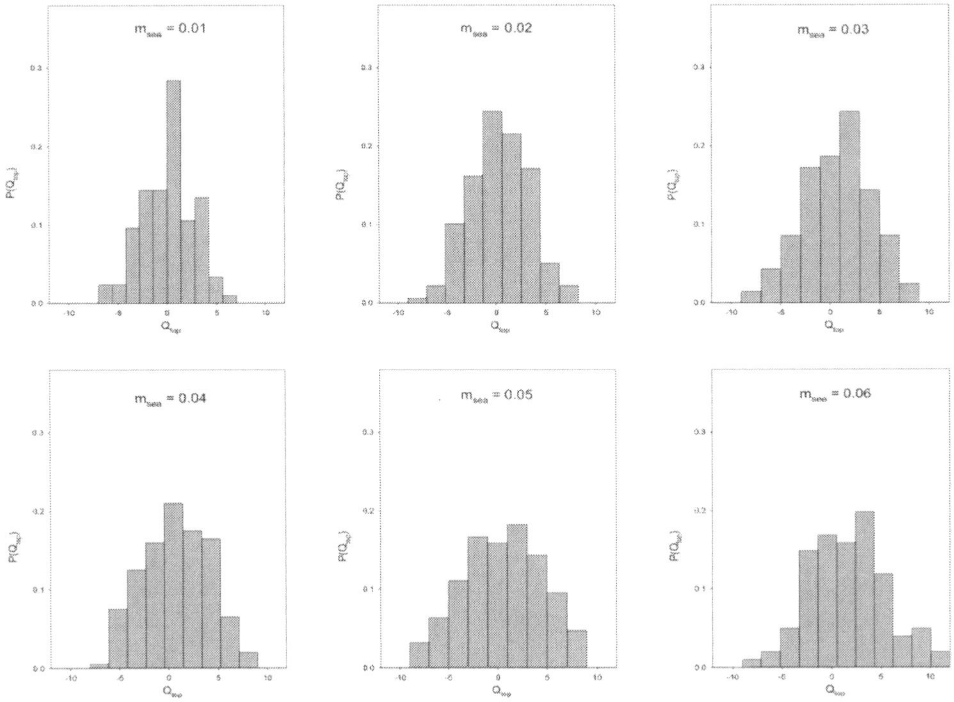

Figure 5. Histogram of topological charge distribution for six sea-quark masses (preliminary results with ~ 200 configurations for each ensemble).

For the projection of low-lying eigenmodes (zero modes plus 180 pairs of lowest-lying eigenmodes) of the overlap operator for each configuration, we have completed about half of the total ($500 \times 6 = 3000$) configurations. In figure 5, we plot the histogram of topological charge distribution for $m_q a = 0.01, 0.02, \cdots, 0.06$ respectively. Evidently, the probability distribution of Q_t for each sea-quark mass behaves like a Gaussian distribution, and it becomes more sharply peaked around $Q_t = 0$ as the sea-quark mass m_q gets smaller. We will measure the topological susceptibility and related physical quantities after the projections of all 3000 configurations are completed.

6. Conclusion and outlook

Chiral symmetry plays an important role in particle physics. It is one of the most salient features of the massless fermion field, a consequence of the principles of quantum mechanics and special relativity. Thus it is vital to preserve the chiral symmetry at finite lattice spacing. Theoretically, this is realized by the domain-wall fermions with infinite extent ($N_s \to \infty$) in the fifth dimension, or the overlap fermion in the four dimension. However, in practice, it is a rather challenging problem to perform dynamical simulations of lattice QCD with overlap or domain-wall fermions such that the chiral symmetry is preserved at a high precision and all topological sectors are sampled ergodically. Naively, one might expect that a high precision of chiral symmetry can be attained with a sufficiently large N_s. However, the relevant question is how the chiral symmetry violation (e.g., the residual mass) decreases with respect to N_s.

In this work, it has been shown that lattice QCD with ODWF provides a viable approach to perform large-scale simulations of unquenched QCD, which not only preserves the chiral symmetry to a good precision, but also samples all topological sectors ergodically. The upper-bound of the residual mass for lattice QCD with ODWF [16] asserts that for N_s less than some threshold value ($\sim 16-18$), the residual mass decays exponentially with respect to N_s. Thus, for lattice QCD with ODWF, chiral symmetry can be preserved to a good precision ($m_{res}a \sim 10^{-5}$) with a modest $N_s \sim 16$. On the other hand, for lattice QCD with the conventional DWF, the residual mass behaves like $N_s^{-\alpha}$ ($\alpha \sim 1-2$) for any N_s [34]. Thus it is difficult for the conventional DWF to attain $m_{res}a \sim 10^{-5}$, even for $N_s \sim 32$. Another approach to avoid large chiral symmetry violations is to use smeared links rather than thin links, since the number of low-lying eigenmodes of $H = cH_w(1 + d\gamma_5 H_w)^{-1}$ would be largely reduced for smeared links. However, it is unclear to what extent the short-distance physics would be affected even if the smeared links are only used in the lattice fermion operator.

Finally, we briefly outline the new project of the TWQCD Collaboration. Now, after simulating two-flavors QCD on various lattices ($16^3 \times 32$, $20^3 \times 40$, $24^3 \times 48$), we are ready to perform dynamical simulations of $(2+1)$-flavors and $(2+1+1)$-flavors QCD on the $32^3 \times 64$ lattice, with pion mass close to the physical value, and residual mass $m_{res}a \sim 10^{-5}$. We outline our strategy as follows. In order to compute the fermion force (by conjugate gradient) for lattice QCD with DWF on the $32^3 \times 64 \times 16$ lattice, at least 11 GB RAM are necessary and this exceeds the maximum memory (6 GB) currently available in a single GPU (Nvidia C2070/K20x/GTX-TITAN). In other words, we must use multiGPUs to meet the memory requirement, as well as to speed up the computation. Recently, we have developed efficient CUDA codes for the computation of entire HMC trajectories with multiGPUs. In table 1, we summarize our benchmark for various Nvidia GPUs [35]. This suggests that it is feasible to perform dynamical simulations of $(2+1)$-flavors and $(2+1+1)$-flavors QCD on the $32^3 \times 64 \times 16$ lattice, with a GPU cluster of Nvidia K20/C2075/GTX680/GTX-TITAN. Details will be presented in a forthcoming paper.

Table 1. Benchmark of Nvidia GPUs, using HMC simulation of lattice QCD with ODWF on the $32^3 \times 48 \times 16$ lattice. All numbers are in units of Gflops/sec.

2*C2070	2*K20c	4*GTX680	2*GTX-TITAN
340	535	945	774

Acknowledgments

I thank the members of the TWQCD Collaboration for many discussions, in particular, Yu-Chih Chen, Wen-Ping Chen, Han-Yi Chou, Tung-Han Hsieh, and Yao-Yuan Mao. Also, I would like to thank the Organizers of CCP2012 for invitation and kind hospitality. This work is supported in part by the National Science Council (Nos. NSC99-2112-M-002-012-MY3) and NTU-CQSE (No. 102R891404). I also thank NCHC and NTU-CC for providing facilities to perform part of the calculations.

References

[1] Wilson K G 1974 *Phys. Rev.* D **10** 2445
[2] Creutz M 1980 *Phys. Rev.* D **21** 2308

[3] Kaplan D B 1992 *Phys. Lett.* B **288** 342
[4] Kaplan D B 1993 *Nucl. Phys. Proc. Suppl.* **30** 597
[5] Neuberger H 1998 *Phys. Lett.* B **417** 141
[6] Narayanan R and Neuberger H 1995 *Nucl. Phys.* B **443** 305
[7] Shamir Y 1993 *Nucl. Phys.* B **406** 90
[8] Allton C *et al.* [RBC and UKQCD Collaboration] 2007 *Phys. Rev.* D **76** 014504
[9] Fukaya H *et al.* [JLQCD Collaboration] 2006 *Phys. Rev.* D **74** 094505
[10] Chiu T W 2003 *Phys. Rev. Lett.* **90** 071601
[11] Chiu T W 2003 *Phys. Lett.* B **552** 97
[12] Chiu T W 2012 *Phys. Lett.* B **716** 461
[13] Chiu T W 2004 *Nucl. Phys. Proc. Suppl.* **129** 135
[14] Chiu T W *et al.* [TWQCD Collaboration] 2009 *PoS* **LATTICE2009** 034
[15] JLQCD Collaboration, private communications.
[16] Chen Y C and Chiu T W [TWQCD Collaboration] 2012 *Phys. Rev.* D **86** 094508
[17] Chiu T W, Hsieh T H, Huang C H and Huang T R 2002 *Phys. Rev.* D **66** 114502
[18] Chiu T W, Hsieh T H and Mao Y Y [TWQCD Collaboration] 2011 *Phys. Lett.* B **702** 131
[19] Mao Y Y and Chiu T W [TWQCD Collaboration] 2009 *Phys. Rev.* D **80** 034502
[20] Chiu T W, Hsieh T H and Mao Y Y [TWQCD Collaboration] 2012 *Phys. Lett.* B **717** 420
[21] Gasser J and Leutwyler H 1985 *Nucl. Phys.* B **250** 465
[22] Duane S, Kennedy A D, Pendleton B J and Roweth D 1987 *Phys. Lett.* B **195** 216
[23] Omelyan I P, Mryglod I M, and Folk R 2001 *Phys. Rev. Lett.* **86** 898
[24] Omelyan I P, Mryglod I M, and Folk R 2002 *Phys. Rev.* E **65** 056706
[25] Omelyan I P, Mryglod I M, and Folk R 2003 *Comput. Phys. Commun.* **151** 272
[26] Sexton J C and Weingarten D H 1992 *Nucl. Phys.* B **380** 665
[27] Chiu T W *et al.* [TWQCD Collaboration] 2010, *PoS* **LATTICE2010** 030
[28] Hasenbusch M 2001 *Phys. Lett.* B **519** 177
[29] Clark M A and Kennedy A D 2007 *Phys. Rev. Lett.* **98** 051601
[30] Chen Y C and Chiu T W 2013, in preparation.
[31] Chiu T W *et al.* [TWQCD Collaboration] 2012 *PoS* **LATTICE2012**, 205
[32] Colangelo G, Durr S and Haefeli C 2005 *Nucl. Phys.* B **721**, 136
[33] Martinelli G, Pittori C, Sachrajda C T, Testa M and Vladikas A 1995 *Nucl. Phys.* B **445** 81
[34] Antonio D J *et al.* [RBC and UKQCD Collaborations] 2008 *Phys. Rev.* D **77** 014509
[35] Chiu T W and Chou H Y 2013, in preparation.

Quantum Monte Carlo study of the half-filled Hubbard model on the honeycomb lattice

Yuichi Otsuka[1], Seiji Yunoki[1,2,3] and Sandro Sorella[1,4,5]

[1]Computational Materials Science Research Team, RIKEN AICS, Kobe, Hyogo 650-0047, Japan,
[2]Computational Condensed Matter Physics Laboratory, RIKEN ASI, Saitama 351-0198, Japan,
[3]CREST, Japan Science and Technology, Kawaguchi, Saitama 332-0012, Japan,
[4]SISSA – International School for Advanced Studies, Via Bonomea 265, 34136 Trieste, Italy, and
[5]Democritos Simulation Center, CNR – IOM Instituto Officina dei Materiali, 34151 Trieste, Italy

E-mail: otsukay@riken.jp

Abstract. The half-filled Hubbard model on the honeycomb lattice is investigated by numerically exact large-scale quantum Monte Carlo simulations for lattice sizes up to 2592 sites. By performing careful finite-size scaling for the spin-spin correlation functions, calculated with a high degree of accuracy, we find that the ground state is antiferromagnetically long-range ordered at $U/t = 4$, where U is the on-site Hubbard interaction and t is the nearest neighbor hopping. Our result is in sharp contrast to a recent report [Meng *et al.*, *Nature* 464, 847 (2010)], where instead strong evidence of spin liquid behavior is found.

1. Introduction

The Hubbard model, consisting of the on-site repulsive interaction U and the nearest neighbor hopping t, has been extensively investigated as the canonical model for strongly correlated electron systems, especially since the discovery of high-T_c cuprate superconductors [1]. On the square lattice at half filling, the model exhibits the Mott transition from a paramagnetic metal to an antiferromagnetic insulator (AFMI) upon inclusion of an infinitely small repulsive interaction U. This is due to the perfect nesting of the non-interacting Fermi surface for the square lattice, as well as a Van Hove singularity characteristic in the density of states [2]. In the case of the honeycomb lattice, there are no such instabilities, though the lattice is still bipartite. Therefore, we expect the Mott transition to occur at a finite value of the interaction, U_c. Indeed, the Mott transition from a semi-metal (SM) to the AFMI at $U_c/t \approx 4.5$ was reported two decades ago [3].

Recently, the discovery of graphene [4] has triggered renewed interests in the half-filled Hubbard model on the honeycomb lattice [5]. Among many works, Meng *et al.* [6] have recently reported that a gapped spin liquid phase without any broken symmetry resides between SM and AFMI. Their finding is rather surprising because it is widely believed that only large quantum fluctuations, usually induced by geometrical frustration, can stabilize a true spin liquid in the ground state. Therefore, the possibility to have a true spin liquid in the simple honeycomb lattice deserves much more careful analysis.

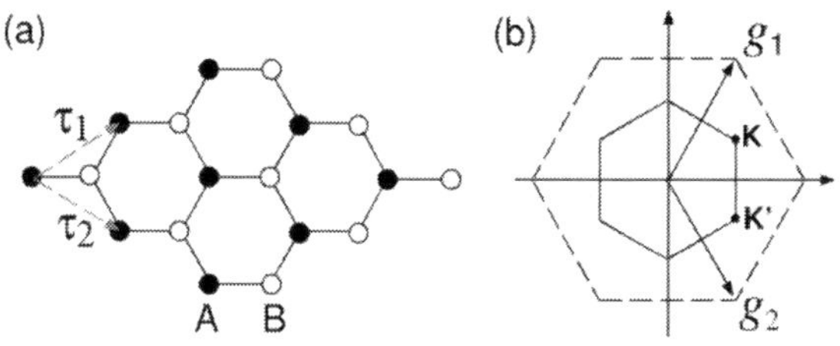

Figure 1. (a) Finite-size cluster of the honeycomb lattice with linear dimension $L = 3$. Closed (open) circles represent sites belonging to the A (B) sublattice. Red dashed lines are the primitive lattice vectors $\vec{\tau}_1$ and $\vec{\tau}_2$. (b) The first Brillouin zone denoted by solid lines in momentum space with the reciprocal lattice vectors $\vec{g}_1$ and $\vec{g}_2$. K and K′ are two inequivalent momenta, where the non-interacting dispersion is zero, i.e., two independent Dirac points appear.

In this paper, using a numerically exact quantum Monte Carlo method at zero temperature, we reexamine the Hubbard model on the honeycomb lattice with much larger clusters containing up to 2592 sites. We perform careful extrapolations of our numerical data to the thermodynamic limit (TDL) in order to estimate an antiferromagnetic (AF) order parameter which should have a strong finite-size effect near the phase boundary. Here, we focus on the case of $U/t = 4$, where the strongest evidence of the spin liquid behavior was previously found [6]. We show numerical evidence that the ground state is instead antiferromagnetically ordered.

2. Model and method

The Hubbard model on the honeycomb lattice is described by following Hamiltonian,

$$\hat{H} = -t \sum_{\langle i,j\rangle,\sigma} \left(c_{i\sigma}^{\dagger} c_{j\sigma} + c_{j\sigma}^{\dagger} c_{i\sigma} \right) + U \sum_{i} n_{i\uparrow} n_{i\downarrow}, \tag{1}$$

where $c_{i\sigma}^{\dagger}$ ($c_{i\sigma}$) creates (annihilates) an electron of spin $\sigma(=\uparrow,\downarrow)$ at site i, $\langle i,j\rangle$ runs over a pair of nearest neighbor sites i and j, and $n_{i\sigma} = c_{i\sigma}^{\dagger} c_{i\sigma}$. We consider finite-size clusters of $N = 2L^2$ sites with L being the linear dimension, as shown in figure 1(a) for $L = 3$. We assume periodic boundary conditions and that L is a multiple of 3, so that the two inequivalent Dirac points K and K′ [figure 1(b)], expected to be relevant for low-energy properties, are allowed in momentum space. Therefore, we use $L = 3, 6, 9, 12, 15, 18, 24$, and 36.

Since the honeycomb lattice is bipartite and the particle-hole symmetry holds at half filling, the auxiliary fields quantum Monte Carlo method [2] is the most powerful tool to calculate the exact ground state correlation functions of the model (1) for large L, without being vexed by the negative sign problem [3]. Here, we employ the ground-state auxiliary field quantum Monte Carlo method to evaluate the ground state expectation value $\langle O\rangle$ of any physical observable $\hat{O}$ by first calculating

$$O(\tau) = \frac{\langle\psi_{\mathrm{L}}|\, e^{-\frac{\tau}{2}\hat{H}}\, \hat{O}\, e^{-\frac{\tau}{2}\hat{H}}\, |\psi_{\mathrm{R}}\rangle}{\langle\psi_{\mathrm{L}}|\, e^{-\tau\hat{H}}\, |\psi_{\mathrm{R}}\rangle} \tag{2}$$

and then taking the limit $\tau \to \infty$, i.e., $\langle O\rangle = \lim_{\tau\to\infty} O(\tau)$. Here, $\langle\psi_{\mathrm{L}}|$ and $|\psi_{\mathrm{R}}\rangle$ are referred to the left and right trial (initial) wave functions, respectively, both of which are chosen so as not to be orthogonal to the exact ground state wave function. For the former, we take $|\psi_{\mathrm{L}}\rangle = \hat{P}_{\mathrm{G}}|\Phi_{\mathrm{MF}}\rangle$, where $\hat{P}_{\mathrm{G}} = \prod_i \exp\left(-g n_{i\uparrow} n_{i\downarrow}\right)$ is the partial Gutzwiller projection operator and $|\Phi_{\mathrm{MF}}\rangle$ is the

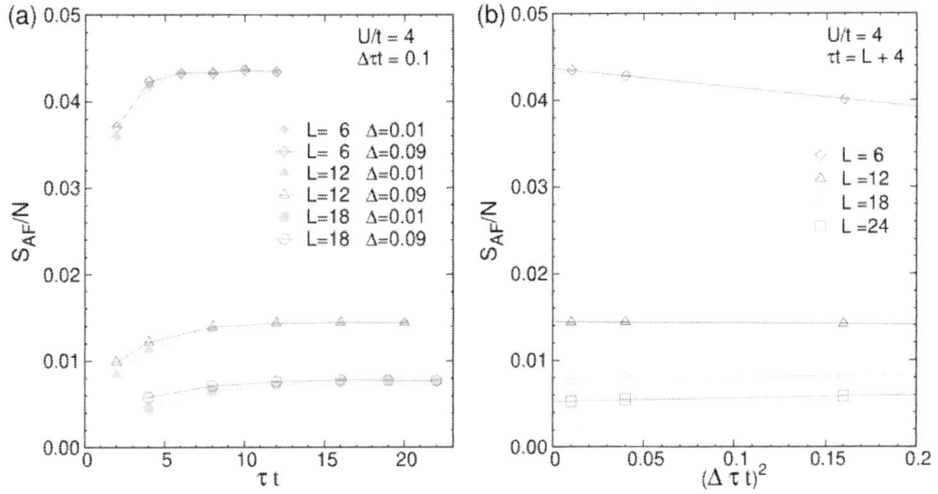

Figure 2. Spin structure factor $S_{\rm AF}/N$ as a function of τt (a) and $(\Delta\tau t)^2$ (b) for $U/t = 4$ and different system sizes L indicated in the figures. In (a), the AF order parameter Δ in $|\psi_{\rm L}\rangle$ is also indicated. In (b), $\tau t = L + 4$ is fixed and lines are linear fits in $(\Delta\tau t)^2$. The Gutzwiller parameter g in $|\psi_{\rm L}\rangle$ is 0.71.

mean-field solution of (1) with AF order parameter Δ (ordered along the x direction). For the latter, we use a simple Slater determinant of N electrons, the ground state for $U/t = 0$, with an additional small perturbation term to remove the degeneracy at momenta K and K′. It should be noted that these choices are crucial, because they allow fast convergences and stable numerical simulations [7]. The quantum mechanical expression in (2) is mapped onto the (2+1) classical system by means of the Suzuki-Trotter decomposition [8, 9] and by introducing discrete auxiliary fields [2] defined for each time slice and each site. The resulting classical partition function can be evaluated by the standard Monte Carlo method.

Apart for the statistical errors in the sampling, we have three types of systematic errors due to (i) finite projection time τ, (ii) finite width of the Trotter slice $\Delta\tau = \tau/N_\tau$ with N_τ being the Trotter number, and (iii) finite system size L. All of these should be eliminated by careful extrapolations to accurately estimate the order parameter in the TDL.

3. Results

In order to study the nonmagnetic to AF transition, we calculate the spin structure factor $S_{\rm AF} = \frac{1}{N}\langle[\sum_{\boldsymbol{r}}(\boldsymbol{S}_{\boldsymbol{r},{\rm A}} - \boldsymbol{S}_{\boldsymbol{r},{\rm B}})]^2\rangle$ and the spin-spin correlation $C_s(\boldsymbol{R}) = \langle \boldsymbol{S}_{\boldsymbol{r},{\rm A}} \cdot \boldsymbol{S}_{\boldsymbol{r}+\boldsymbol{R},{\rm A}}\rangle$ at the maximum distance $|\boldsymbol{R}| = L_{\rm max}$, where $\boldsymbol{S}_{\boldsymbol{r},{\rm A}}$ ($\boldsymbol{S}_{\boldsymbol{r},{\rm B}}$) is the spin operator at unit cell $\boldsymbol{r}$ on A (B) sublattice. In the TDL, $1/L \to 0$, both these quantities converge to the AF order parameter squared, $m_s^2 = S_{\rm AF}/N = C(L_{\rm max})$, which should be finite if the ground state is AF ordered.

First, let us study the effect of the finite projection time τ. The τ-dependences of $S_{\rm AF}/N$ and $C(L_{\rm max})$ are plotted in figure 2 (a) and figure 3(a), respectively. We can see in the figures that, within statistical errors, these quantities increase monotonically with increasing τ and converge to the same limit, regardless of the AF order parameter value Δ, defining the initial wave function $|\psi_{\rm L}\rangle$. Since the projection time τ required for convergence, significantly depends on the value of Δ, we determine its optimal value for small clusters and use it for large L. We can also find in figure 2 (a) and figure 3(a) that the projection time τ needed for convergence increases linearly with L. We thus take $\tau t = L + 4$ in our simulations. We check that this value

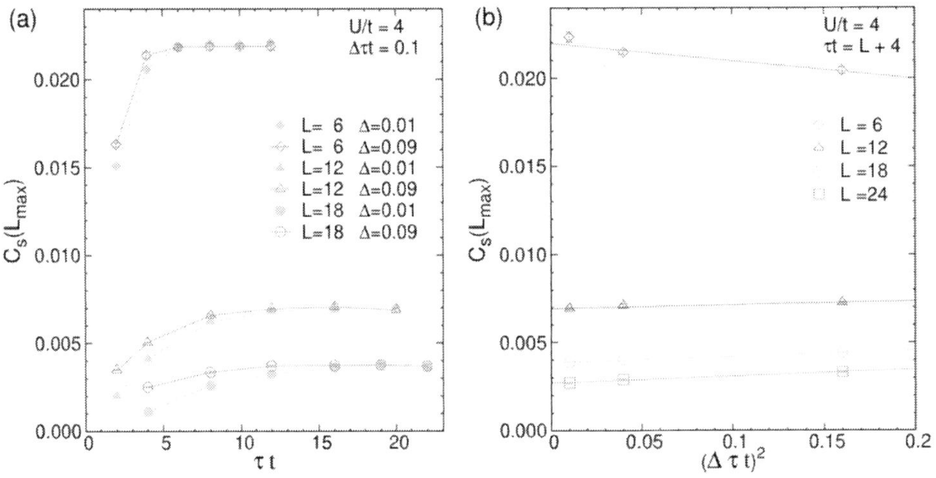

Figure 3. Spin-spin correlation function at the maximum distance $C(L_{max})$ as a function of τt (a) and $(\Delta\tau t)^2$ (b) for $U/t = 4$ and different system sizes L indicated in the figures. In (a), the AF order parameter Δ in $|\psi_{\rm L}\rangle$ is also indicated. In (b), $\tau t = L + 4$ is fixed and lines are linear fits in $(\Delta\tau t)^2$. The Gutzwiller parameter g in $|\psi_{\rm L}\rangle$ is 0.71.

of τ is large enough to obtain converged results in all cases studied. Notice also that the small τ tends to underestimate the AF order, which implies that our estimate of m_s should be at most a lower bound.

Next, we discuss the effect of the finite width of the Trotter slice $\Delta\tau$. Since we use the symmetric decomposition [8, 9], the systematic error introduced in the expectation value should be $\mathcal{O}(\Delta\tau^2)$. Indeed, as shown in figure 2 (b) and figure 3(b), the results for different values of $\Delta\tau$ are well fitted by a linear function of $(\Delta\tau)^2$. By extrapolating to $\Delta\tau \to 0$, we can in principle eliminate the systematic error due to a finite value of $\Delta\tau$ used in the simulations. However, we find that $\Delta\tau t = 0.1$ is already small enough compared to the statistical error. Since among the three types of the systematic errors, only this extrapolation process can result in reducing m_s, we analyze the finite size scaling for the results before and after $\Delta\tau \to 0$.

Finally, we extrapolate $S_{\rm AF}/N$ and $C(L_{\rm max})$ to the TDL by careful finite-size scaling. In figure 4, both $S_{\rm AF}/N$ and $C(L_{\rm max})$ are plotted as a function of $1/L$ [10] and fitted using cubic polynomials. We can see in figure 4 that the consistent AF order parameter m_s is obtained by the finite-size extrapolation of both quantities. This strongly supports that our finite-size extrapolation is reliable. The estimated $m_s^2 = \lim_{L\to\infty} S_{\rm AF}/N = \lim_{L\to\infty} C(L_{\rm max})$ is apparently small, which reflects that $U/t = 4$ is very close to $U_{\rm c}/t$. However, as clearly seen in the inset of figure 4, m_s^2 is finite.

To further confirm this result, we summarize in Tables 1 and 2 the results of m_s^2 obtained by several fitting procedures using different sets of data for $\Delta\tau t = 0.1, 0.2, 0.4$, and $\Delta\tau t \to 0$. As shown in Tables 1 and 2, there is a clear tendency that m_s^2 is larger when we use the more accurate (namely closer to the TDL) data, which includes the largest size simulations. This tendency is less clear in the results of $C(L_{\rm max})$, but here the finite-size effects are smaller and the statistical errors are much larger compared to those of $S_{\rm AF}/N$. Note, however, that the extrapolated values by $C(L_{\rm max})$ are systematically larger than those of $S_{\rm AF}/N$, which implies that our extrapolated value for m_s^2 based on $S_{\rm AF}/N$ can be considered to be a safe lower bound. We find that the most reliable fit is obtained when we use all data for L = 6–36, by judging from the values of the chi-square obtained in the extrapolations. It is also important to note

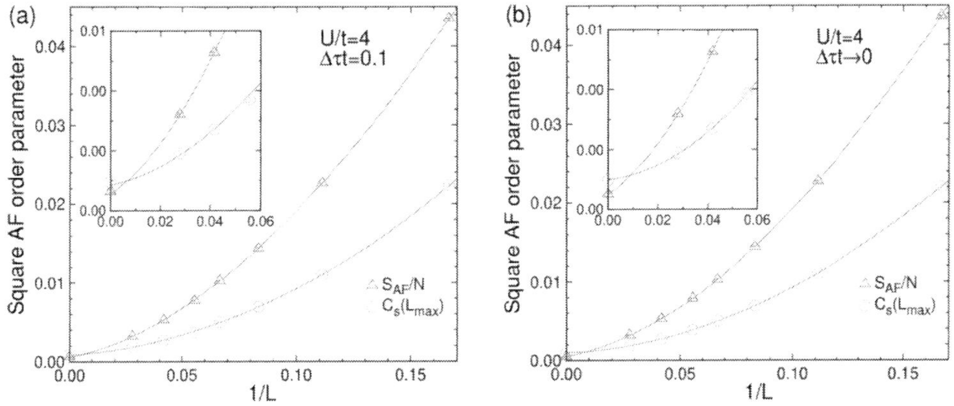

Figure 4. Finite-size scaling of the AF order parameter squared for $\Delta\tau t = 0.1$ (a) and $\Delta\tau t \to 0$ (b). Red triangle and green circle represents the spin structure factor $S_{\rm AF}/N$ and the spin-spin correlation function at the maximum distance $C(L_{\rm max})$, respectively. Data are extrapolated to the TDL by cubic polynomials in $1/L$ shown by solid curves. The extrapolated values are indicated at $1/L = 0$.

that the AF order parameters, estimated for all these different procedures, are finite.

4. Summary

The half-filled Hubbard model on the honeycomb lattice has been investigated by large-scale quantum Monte Carlo simulations. In order to examine whether the antiferromagnetic long-range order exists at $U/t = 4$, we have performed the simulations on lattices containing up to 2592 sites and carefully extrapolated data to the thermodynamic limit. In sharp contrast to the recent report by Meng et al. [6], our results support the existence of the antiferromagnetic long-range order, which suggests that the spin liquid phase is at least reduced to $U/t < 4$ if it ever exists. The results for other U/t and the ground state magnetic phase diagram will be published elsewhere [7].

Table 1. Extrapolated values in the TDL of the spin structure factor, $S_{\rm AF}/N$, (in unit of 10^{-4}) for $U/t = 4$ and $\Delta\tau t = 0.1, 0.2, 0.4$, and $\Delta\tau t \to 0$. The extrapolation is performed using different data sets (indicated in the first column) by cubic polynomials in $1/L$. Chi-square (χ^2) divided by the number of degrees of freedom (N_{DF}) is also shown. The numbers of data points used are 6, 7, and 8 for (i)-(iii), (iv) and (v), and (vi), respectively. The statistical errors are indicated by numbers in parentheses (corresponding to the last two digits).

		$\Delta\tau t = 0.4$		$\Delta\tau t = 0.2$		$\Delta\tau t = 0.1$		$\Delta\tau t \to 0$	
	L	$S_{\rm AF}/N$	$\chi^2/N_{\rm DF}$	$S_{\rm AF}/N$	$\chi^2/N_{\rm DF}$	$S_{\rm AF}/N$	$\chi^2/N_{\rm DF}$	$S_{\rm AF}/N$	$\chi^2/N_{\rm DF}$
(i)	3 - 18	8.6(14)	1.95	3.3(15)	1.66	3.0(14)	1.23	2.8(13)	2.44
(ii)	6 - 24	13.7(15)	1.46	8.9(17)	0.26	4.5(12)	0.72	3.6(11)	1.95
(iii)	9 - 36	16.1(17)	0.64	6.5(15)	0.32	6.3(16)	1.61	5.1(13)	3.27
(iv)	3 - 24	10.98(86)	2.82	5.54(93)	2.30	3.45(66)	0.87	2.76(62)	1.63
(v)	6 - 36	14.2(11)	1.07	7.61(83)	0.45	5.68(86)	1.15	4.97(74)	2.19
(vi)	3 - 36	11.97(64)	3.03	6.12(53)	1.89	4.34(52)	1.93	3.78(45)	2.83

Table 2. Extrapolated values in the TDL of the spin-spin correlation function at the maximum distance, $C(L_{\max})$, (in unit of 10^{-4}) for $U/t = 4$ and $\Delta\tau t = 0.1, 0.2, 0.4$, and $\Delta\tau t \to 0$. The extrapolation is performed using different data sets (indicated in the first column) by cubic polynomials in $1/L$. Chi-square (χ^2) divided by the number of degrees of freedom (N_{DF}) is also shown. The numbers of data points used are 6, 7, and 8 for (i)-(iii), (iv) and (v), and (vi), respectively. The statistical errors are indicated by numbers in parentheses (corresponding to the last two digits).

		$\Delta\tau t = 0.4$		$\Delta\tau t = 0.2$		$\Delta\tau t = 0.1$		$\Delta\tau t \to 0$	
	L	$C(L_{\max})$	$\chi^2/N_{\rm DF}$	$C(L_{\max})$	$\chi^2/N_{\rm DF}$	$C(L_{\max})$	$\chi^2/N_{\rm DF}$	$C(L_{\max})$	$\chi^2/N_{\rm DF}$
(i)	3 - 18	12.2(30)	0.45	4.2(33)	0.68	12.9(37)	0.74	8.9(28)	0.75
(ii)	6 - 24	14.8(43)	0.40	10.3(40)	0.91	6.6(32)	0.10	8.5(28)	0.75
(iii)	9 - 36	10.2(38)	0.26	11.5(48)	0.89	10.5(40)	0.37	10.6(36)	0.97
(iv)	3 - 24	13.0(19)	0.34	7.2(19)	0.88	10.1(19)	0.75	8.8(16)	0.50
(v)	6 - 36	12.5(25)	0.41	10.7(24)	0.61	8.6(23)	0.37	9.9(20)	0.67
(vi)	3 -36	12.4(13)	0.31	8.3(14)	0.84	10.5(15)	0.60	9.5(12)	0.52

Acknowledgments

We acknowledge E. Tosatti, F. Becca, and T. Li for useful discussions. We are also grateful to A. Muramatsu and F. F. Assaad for valuable comments. This work is supported by a PRACE grant 2010PA0447 and by MIUR-COFIN2012. Part of the results is obtained by K computer at RIKEN Advanced Institute for Computational Science.

References

[1] Imada M, Fujimori A and Tokura Y 1998 *Rev. Mod. Phys.* **70** 1039
[2] Hirsch J 1985 *Phys. Rev. B* **31** 4403
[3] Sorella S and Tosatti E 1992 *Europhys. Lett.* **19** 699
[4] Novoselov K S, Geim A K, Morozov S V, Jiang D, Zhang Y, Dubonos S V, Grigorieva I V and Firsov A A 2004 *Science* **306** 666
[5] Neto A H C, Guinea F, Peres N M R, Novoselov K S and Geim A K 2009 *Rev. Mod. Phys.* **81** 109
[6] Meng Z Y, Lang T C, Wessel S, Assaad F F and Muramatsu A 2010 *Nature* **464** 847
[7] Sorella S, Otsuka Y and Yunoki S 2012 Absence of a spin liquid phase in the hubbard model on the honeycomb lattice (*Preprint* `arXiv:1207.1783v1`)
[8] Trotter H F 1959 *Proc. Amer. Math. Soc.* **10** 545
[9] Suzuki M 1976 *Commun. Math. Phys.* **51** 183
[10] Reger J D, Riera J A and Young A P 1989 *J. Phys.: Cond. Matt.* **1** 1855

Improved multi-variable variational Monte Carlo method examined by high-precision calculations of one-dimensional Hubbard model

Ryui Kaneko, Satoshi Morita and Masatoshi Imada

Department of Applied Physics, The University of Tokyo, 7-3-1 Hongo, Bunkyo-ku, Tokyo 113-8656, Japan

E-mail: rkaneko@solis.t.u-tokyo.ac.jp

Abstract. We revisit the accuracy of the variational Monte Carlo (VMC) method by taking an example of ground state properties for the one-dimensional Hubbard model. We start from the variational wave functions with the Gutzwiller and long-range Jastrow factor introduced by Capello *et al.* [Phys. Rev. B **72**, 085121 (2005)] and further improve it by considering several quantum-number projections and a generalized one-body wave function. We find that the quantum spin projection and total momentum projection greatly improve the accuracy of the ground state energy within 0.5% error, for both small and large systems at half filling. Besides, the momentum distribution function $n(k)$ at quarter filling calculated up to 196 sites allows us direct estimate of the critical exponents of the charge correlations from the power-law behavior of $n(k)$ near the Fermi wave vector. Estimated critical exponents well reproduce those predicted by the Tomonaga-Luttinger theory.

1. Introduction

Strongly correlated electron systems have attracted much interest over the past decades. Although rigorous ground states are obtained analytically for some simple models, numerical methods offer powerful tools for understanding more realistic systems. There are several numerical methods for the purpose of calculating low-energy properties of the strongly correlated systems. At finite temperatures, the auxiliary-field quantum Monte Carlo (AFQMC) method [1, 2, 3] gives exact ground states within the statistical errors. However, for the geometrically frustrated systems, the errors become exponentially large, known as the negative sign problem. To overcome this difficulty, the Gaussian-basis Monte Carlo (GBMC) method [4, 5, 6] has been proposed as an alternative to the AFQMC method, although the tractable range of parameters are still under debate. The dynamical mean-field theory (DMFT) [7, 8] can also be applied to strongly correlated electron systems. However, when we wish to take into account sufficient spatial correlations to represent fluctuations around broken symmetry states, we need larger clusters beyond the present computational accessibility. At zero temperature, the exact diagonalization (ED) method gives exact ground states. This method, however, is not able to handle large system sizes. The density matrix renormalization group (DMRG) method [9] gives nearly exact ground states for relatively larger systems, though it is mainly applied to one-dimensional systems, or two-dimensional systems with cylindrical boundary conditions. The unbiased path-integral renormalization group (PIRG) method [10, 11, 12] has been applied to

Journal of Physics: Conference Series **454** (2013) 012046 doi:10.1088/1742-6596/454/1/012046

several strongly correlated electron systems, with essentially accurate ground states, even in the presence of geometrical frustrations. However, strong correlation regime requires a set of many basis functions for accurate estimates, sometimes beyond the computational accessibility. Among them, although a bias inevitably exists, the VMC method [13] is applicable to large systems in the presence of strong correlations and geometrical frustrations.

Accuracy of the calculated properties obtained by the VMC method relies on the choice of the trial wave functions. In most cases, the trial wave functions $|\psi\rangle$ are given by Slater determinants $|D\rangle$ supplemented by correlation factors $\mathcal{P}$ as $|\psi\rangle = \mathcal{P}|D\rangle$. One of the famous example of this trial function is a "projected Bardeen-Cooper-Schrieffer (BCS)" type wave function $|\psi\rangle = \mathcal{P}_{\mathrm{G}}^{\infty}|\mathrm{BCS}\rangle$, where $\mathcal{P}_{\mathrm{G}}^{\infty} = \prod_i(1 - n_{i\uparrow}n_{i\downarrow})$ denotes a Gutzwiller factor, which prohibits double occupancy of the sites, and $|\mathrm{BCS}\rangle = \prod_{\boldsymbol{k}}(u_{\boldsymbol{k}} + v_{\boldsymbol{k}}c_{\boldsymbol{k}\uparrow}^{\dagger}c_{-\boldsymbol{k}\downarrow}^{\dagger})|0\rangle$ denotes a BCS state, which is the ground state of the BCS mean-field Hamiltonian. This projected BCS wave function is equivalent to the so-called "resonating valence bond (RVB)" wave function, which is the superposition of all dimer coverings [14, 15]. The projected BCS and RVB wave functions provide highly accurate descriptions in two-dimensional strongly correlated electron systems [16, 17].

In principle, the trial wave functions based on Slater determinants become accurate for higher dimensions since the mean-field-like treatment becomes valid above the upper critical dimension. Although it seems difficult to capture the low-energy properties at lower spatial dimensions by using such a trial wave function, long-range correlation factors may restore the accuracy of the wave functions [18, 19]. Capello *et al.* have calculated the ground state of the one-dimensional Hubbard model by using the projected BCS type wave functions. They have shown that the long-range Jastrow factor is essential to describe the low-energy properties. They have also reproduced correct power-law behaviors of the spin and charge correlation functions.

Although Capello *et al.*'s scheme appears to substantially reproduce the low-energy properties of the one-dimensional model, the errors in the energy become larger even for small system sizes, typically more than 2% for $L = 18$ for intermediate interaction strength [19]. This is mainly because they have used the Slater determinants with a limited number of variational parameters. The Slater determinants in their wave function are just a Fermi sea of the tight binding model, or short-range BCS wave functions with nonzero amplitudes at most up to third-nearest-neighbors. In addition, in practical numerical calculations, since the accessible system sizes are limited, it is important to prepare the wave functions preserve the correct quantum numbers within its accessible range of sizes.

In this paper, to obtain more accurate wave functions of the low-dimensional systems for the smaller as well as for larger system sizes, we revisit the variational calculations of the one-dimensional Hubbard model, as a reference system. By considering the multi-variable Slater determinants and projections which restore the SU(2) spin rotational and spatially translational symmetries, we obtain more accurate ground states of the model than the previous studies. We also calculate the momentum distribution function $n(k)$ up to 196 sites, the first high-precision variational calculation of $n(k)$ to our knowledge, and directly estimate the critical exponents of the charge correlations. We compare them with the analysis by Kawakami *et al.* [20, 21] as well as Ogata *et al.* [22]. Estimated critical exponents well reproduce those obtained by the Tomonaga-Luttinger theory.

2. Model and method

We consider the one-dimensional Hubbard model. The Hamiltonian is given by

$$H = -t\sum_{\langle i,j\rangle,\sigma}\left(c_{i\sigma}^{\dagger}c_{j\sigma} + \mathrm{h.c.}\right) + U\sum_i n_{i\uparrow}n_{i\downarrow}, \tag{1}$$

where $c^\dagger_{i\sigma}$ ($c_{i\sigma}$), $n_{i\sigma}$, t and U denote the creation (annihilation) operator, the number operators, the nearest-neighbor hopping interaction, and the on-site Coulomb interaction, respectively.

Exact ground states of the one-dimensional Hubbard model are obtained by Lieb and Wu [23]. At half filling, the ground state is a metal for $U = 0$, and is an insulator for nonzero positive U. Away from half filling, the ground state is a metal for any $U \geq 0$. Unlike the normal Fermi liquid, the momentum distribution of this metal does not show a finite jump at the Fermi wave vector. Instead, it shows power-law decays near the Fermi wave vector. This type of the ground state, which appears in one-dimensional electron systems, is well known as the Tomonaga-Luttinger liquid [24, 25].

To obtain accurate variational description of the ground state of the model, we use a multi-variable variational Monte Carlo (mVMC) method [17]. We use the trial wave functions given by

$$|\psi\rangle = \mathcal{P}_{\rm G}\mathcal{P}_{\rm J}\mathcal{P}_{\rm dh}\mathcal{L}^{S=0}\mathcal{L}^{K=0}\,|\phi_{\rm pair}\rangle\,, \qquad (2)$$

where $|\phi_{\rm pair}\rangle$, $\mathcal{P}_{\rm G}$, $\mathcal{P}_{\rm J}$, $\mathcal{P}_{\rm dh}$, $\mathcal{L}^{S=0}$, and $\mathcal{L}^{K=0}$ denote the one-body part, the Gutzwiller factor, the Jastrow factor, the doublon-holon factor, the spin quantum-number projection on to $S = 0$ space, and the total momentum projection on to $K = 0$ space, respectively, as we detail later. We calculate the ground states of the model under the periodic boundary conditions at half filling ($n = 1$) for $N_{\rm s} = 4n + 2$ sites, and at quarter filling ($n = 0.5$) for $N_{\rm s} = 8n + 4$ sites, so that the systems obey the closed shell conditions.

For the one-body part $|\phi_{\rm pair}\rangle$, we use a generalized pairing wave function defined as

$$|\phi_{\rm pair}\rangle = \left(\sum_{i,j=1}^{N_{\rm s}} f_{ij}c^\dagger_{i\uparrow}c^\dagger_{j\downarrow}\right)^{N_{\rm e}/2} |0\rangle\,, \qquad (3)$$

where f_{ij}, $N_{\rm s}$, and $N_{\rm e}$ denote the variational parameters, the number of sites, and the number of electrons, respectively. When all f_{ij}'s are optimized simultaneously and independently, since the total number of f_{ij} is $N_{\rm s}^2$, the computation time for the optimization is highly demanding. To reduce the computation time to $\mathcal{O}(N_{\rm s})$, we assume f_{ij} to have a sublattice structure, namely $f_{ij} = f_{\sigma(i)}(R_i - R_j)$, where $\sigma(i)$ denotes a sublattice index at site i. Since optimized ground states by using f_{ij} with reflectional symmetries, namely $f_{ij} = f_{ji}$, have higher energies, we do not impose reflectional symmetries on f_{ij}. Before calculating the ground states of larger system sizes, we examine the accuracy of the wave functions using several sublattice structures of f_{ij} for smaller system sizes less than about 20 sites. We find that optimized energies for two-sublattice structure and $N_{\rm s}$-sublattice (fully optimized) structure are nearly the same. This suggests that the ground states are well-described by the wave functions with the two-sublattice structure for the one-dimensional Hubbard model. In the following, we show results for mainly two-sublattice structure, and one-sublattice translationally invariant structure.

To include correlation effects beyond the one-body approximation, we consider the correlation factors

$$\mathcal{P}_{\rm G} = \exp\left(-g\sum_{i=1}^{N_{\rm s}} n_{i\uparrow}n_{i\downarrow}\right), \qquad (4)$$

$$\mathcal{P}_{\rm J} = \exp\left(-\sum_{i<j} v_{ij}(n_{i\uparrow}+n_{i\downarrow})(n_{j\uparrow}+n_{j\downarrow})\right), \qquad (5)$$

and

$$\mathcal{P}_{\rm dh} = \exp\left(-\sum_{m=0}^{2}\sum_{l=1}^{N_{\rm s}}\alpha_{ml}\sum_{i=1}^{N_{\rm s}}\xi_{iml}\right), \qquad (6)$$

where g, v_{ij}, and α_{ml} denote variational parameters of the Gutzwiller, Jastrow, and the doublon-holon correlation factors, respectively. We take into account long-range part of the Jastrow factor, which is indispensable for reproducing the low-energy properties of the one-dimensional Hubbard model [18, 19], and the doublon-holon factor. A variable ξ_{iml} in equation (6) is a many-body operator, which gives one if a doublon (holon) exists at the i-th site and m holons (doublons) exist at l-th nearest neighbors, and gives zero otherwise. We assume the Jastrow factor to have translational symmetry, namely $v_{ij} = v(r_{ij})$, where $r_{ij} = |R_i - R_j|$ is the distance between the sites i and j. We note that the Gutzwiller factor and the doublon-holon factor have the translational symmetry by definition.

In addition to the one-body part and the correlation factors, we introduce a projection $\mathcal{L}^{S=0}$, which restores the SU(2) spin rotational symmetry, and a projection $\mathcal{L}^{K=0}$, which restores the translational symmetry. The spin quantum-number and total momentum projections are switched on at all optimization steps. Since the correlation factors have the SU(2) and translational symmetries, the quantum-number projections commute with the correlation factors.

To optimize a large number of these variational parameters, we use the stochastic reconfiguration (SR) method developed by Sorella [26]. We first optimize variational parameters f_{ij} for typically 1000 SR steps with fixed correlation factors. After confirming the energy convergence, we simultaneously optimize all the variational parameters (g, v_{ij}, α_{ml}, and f_{ij}) for more than 4000 SR steps.

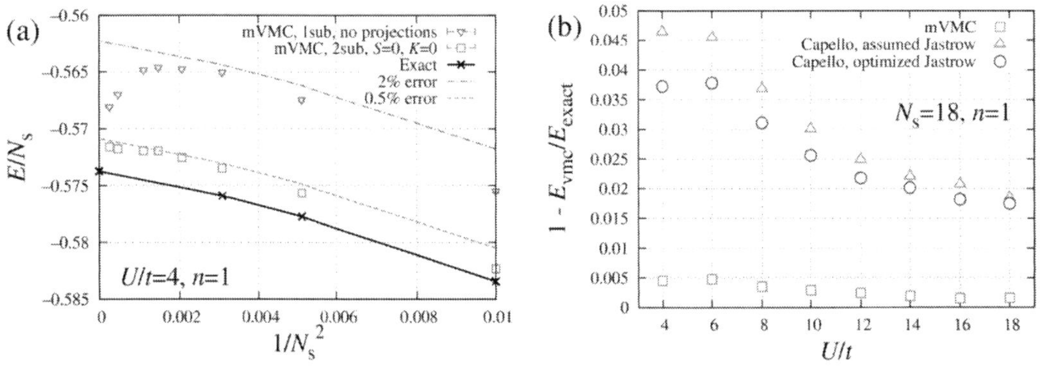

Figure 1. Calculated energy of Hubbard model at half filling. Statistical errors are smaller than the symbol sizes. (a) System size N_s dependence at $U/t = 4$. The black crosses denote the exact energies obtained by the ED method for finite sizes and analytically at the thermodynamic limit. The purple down-pointing triangles and red squares denote calculated energies by the mVMC method with the translationally invariant and two-sublattice f_{ij} trial wave functions, respectively. The former is obtained without the quantum-number projections and the latter is with the spin and momentum projections to the total singlet and zero total momentum. Two and 0.5 percent errors measured from the exact results are plotted by dash-dotted purple and dashed red curves, respectively. (b) Coulomb interaction U/t dependence of errors in calculated energies at half filling for $N_s = 18$. The red squares denote our mVMC results obtained by using the two-sublattice f_{ij} trial wave functions. The errors in the energy are much smaller than their results. The green triangles and blue circles denote the errors in the energy obtained by Capello *et al.* by using short-range BCS wave functions with the third-nearest-neighbor amplitudes [19]. The former is obtained with the assumed Jastrow factor and the latter is with the optimized Jastrow factor.

3. Results

We obtain the ground-state energy of the one-dimensional Hubbard model at half filling, as shown in figure 1. For the case of the two-sublattice f_{ij}, the errors in the energy are within approximately 0.5%, much smaller than the previous calculations by Capello *et al.* [19] The extended one-body part and quantum projections improve the accuracy of the ground state energy, even for the small system sizes. We show results of the two-sublattice f_{ij} condition, hereafter.

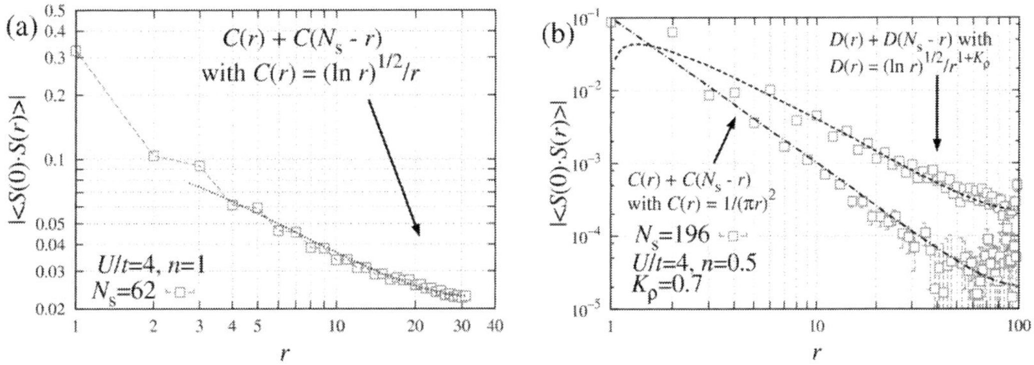

Figure 2. Calculated spin correlation functions of Hubbard model at $U/t = 4$, (a) at half filling and (b) at quarter filling, respectively. The red squares denote the mVMC results. These points are well described by the exact asymptotic expressions of the spin correlation functions, given by the dotted black lines.

In the gapless one-dimensional systems, spin and charge correlations show power-law decays [27, 28]. The asymptotic expressions of the spin correlation functions are given as

$$C(r) = \langle \boldsymbol{S}(0) \cdot \boldsymbol{S}(r) \rangle = -\frac{1}{\pi^2 r^2} + A\cos(2k_{\rm F}r)\frac{\sqrt{\ln r}}{r^{K_\rho+1}} + \cdots, \tag{7}$$

where A, $k_{\rm F}$, and K_ρ denote a constant, the Fermi wave vector, and a critical exponent, respectively. The Fermi wave vector $k_{\rm F}$ is written in terms of the filling n through the equation $k_{\rm F} = n\pi/2$. The critical exponent K_ρ is a function of the Coulomb interaction U and the filling n. At half filling, the charge degrees of freedom are gapped out with the exponential decay of correlations in distance, leaving only the gapless spin correlation in the form of equation (7) formally with $K_\rho = 0$ for nonzero repulsive U. Namely, the spin correlation functions decay as $C(r) \sim A(-1)^r\sqrt{\ln r}/r$ [27, 28, 29]. Away from half filling, the critical exponent satisfies $0 < K_\rho < 1$ for nonzero repulsive U. Thus the spin correlation functions decay as equation (7). To include finite size effects under the periodic boundary conditions, we compare the calculated correlation functions with $C(r) + C(N_{\rm s} - r)$ instead of the original correlation function $C(r)$ because of the reflection effects from the boundary. As shown in figure 2, calculated spin correlation functions of the Hubbard model reproduce $\sqrt{\ln r}/r$ decay at half filling, and $\sqrt{\ln r}/r^{1+K_\rho}$ and $1/r^2$ decays at quarter filling with the critical exponent $K_\rho \approx 0.7$ at $U/t = 4$, respectively. This critical exponent is nearly the same as the exact value, $K_\rho = 0.711$, predicted by the Tomonaga-Luttinger theory [18, 20, 21, 28].

In our calculations, the one-body part of the trial wave function includes long-range part in f_{ij} and v_{ij}. Although there should be no long-range order for the one-dimensional Hubbard model, our trial wave function could fallaciously stabilize the insulator with the antiferromagnetic long-range order. To examine whether our calculations correctly reproduce the absence of the

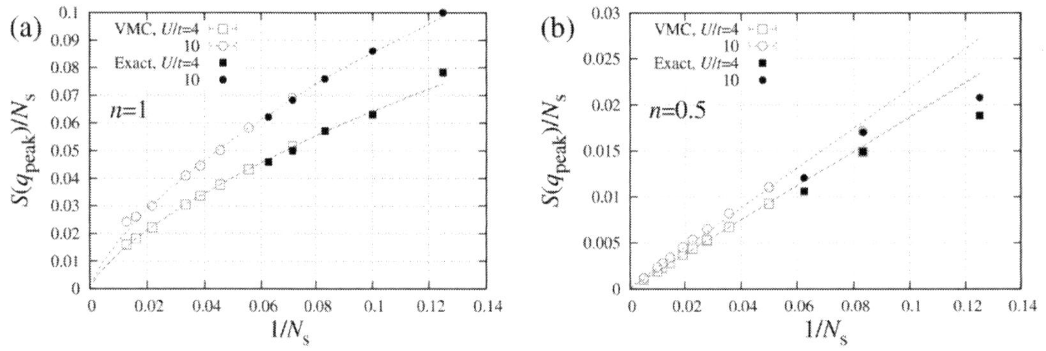

Figure 3. Peak values of spin structure factors of Hubbard model at $U/t = 4$ and $U/t = 10$ for each system size, (a) at half filling and (b) at quarter filling. Statistical errors are smaller than the symbol sizes. At half filling, the spin structure factor shows a peak at $q_{\text{peak}} = \pi$, while at quarter filling, it shows a peak at $q_{\text{peak}} = \pi/2$. The filled symbols denote results obtained by the ED method for the model under the periodic boundary conditions, while the open symbols denote the mVMC results. The red squares denote results at $U/t = 4$, and the green circles denote results at $U/t = 10$. The squared magnetizations, $S(q_{\text{peak}})/N_{\text{s}}$ converge to zero in the thermodynamic limit.

long-range order in the model, we calculate the peak values of the spin structure factors

$$S(q) = \frac{1}{3N_{\text{s}}} \sum_{i,j} e^{iq(R_i - R_j)} \langle \boldsymbol{S}_i \cdot \boldsymbol{S}_j \rangle \tag{8}$$

for each system size. For the insulator, size dependencies of the peak values of the spin structure factor are expected to follow the scaling

$$S(q_{\text{peak}}) = \frac{1}{3} \int dr\, e^{iq_{\text{peak}} r} \langle \boldsymbol{S}(0) \cdot \boldsymbol{S}(r) \rangle \sim \int_{\Lambda}^{N_{\text{s}}} dr \frac{\sqrt{\ln r}}{r} \sim (\ln N_{\text{s}})^{3/2}, \tag{9}$$

where Λ denotes the cutoff, while for the metal, since $1/r^{1+K_\rho} \gg 1/r^2$ for $r \gg 1$, we obtain

$$S(q_{\text{peak}}) = \frac{1}{3} \int dr\, e^{iq_{\text{peak}} r} \langle \boldsymbol{S}(0) \cdot \boldsymbol{S}(r) \rangle \sim \int_{\Lambda}^{N_{\text{s}}} dr \frac{1}{r^{1+K_\rho}} \sim \text{const.} + \mathcal{O}(N_{\text{s}}^{-K_\rho}). \tag{10}$$

By using these equations, we extrapolate the squared magnetization, the peak values of the spin structure factor $S(q_{\text{peak}})$ divided by the system size N_{s}, in the thermodynamic limit. As shown in figure 3, the squared magnetizations, $S(q_{\text{peak}})/N_{\text{s}}$ converge to zero in the thermodynamic limit. These results suggest that our trial wave functions with the long-range f_{ij} and symmetry-restoring projections $\mathcal{L}$ are able to reproduce ground states with no long-range magnetic order with power-law decay of correlations.

For the one-dimensional Hubbard model, the momentum distribution

$$n(k) = \frac{1}{2N_{\text{s}}} \sum_{i,j,\sigma} e^{ik(R_i - R_j)} \left\langle c_{i\sigma}^{\dagger} c_{j\sigma} \right\rangle \tag{11}$$

shows a power-law singularity

$$|n(k) - n(k_{\text{F}})| \propto |k - k_{\text{F}}|^{\alpha} \tag{12}$$

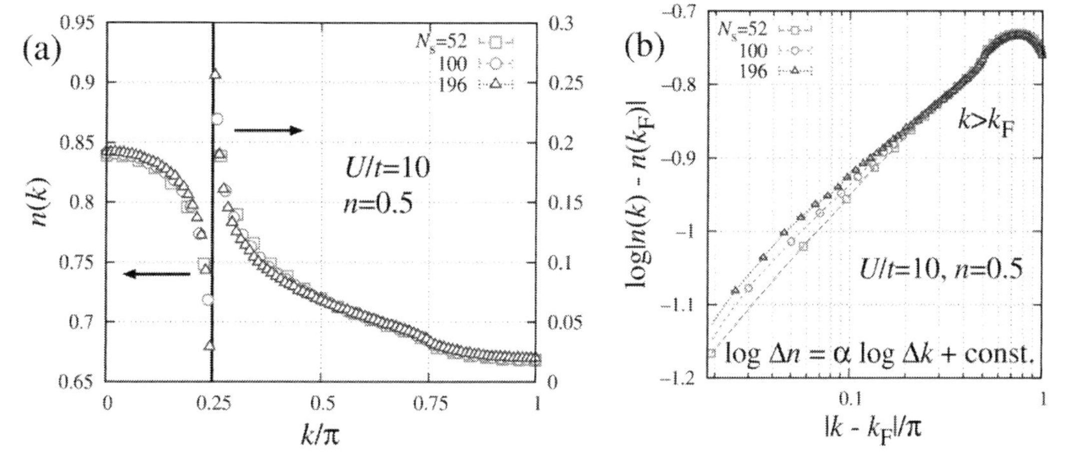

Figure 4. (a) Momentum distribution of Hubbard model at $U/t = 10$ for each system size, at quarter filling. Power-law singularities appear at the momenta $k_F = \pi/4$ and $3k_F = 3\pi/4$. (b) Logarithmic plot of momentum distribution near k_F for each system size. Statistical errors are smaller than the symbol sizes.

with a critical exponent α near the Fermi wave vector k_F, contrary to the ordinary Fermi liquid, which shows a jump at the Fermi wave vector [20, 21]. The critical exponent α is written in terms of K_ρ through the equation $\alpha = (K_\rho + 1/K_\rho - 2)/4$ [28, 30, 31]. For the infinite Coulomb interactions, Ogata *et al.* have calculated the momentum distribution $n(k)$ of the metal ground states by using the Bethe-ansatz wave functions, and have estimated the critical exponents α from equation (12) [22]. We calculate the momentum distribution of the Hubbard model at quarter filling up to 196 sites. As shown in figure 4(a), calculated momentum distributions show power-law behavior near the wave vector k_F. Another singularity also appears at $k = 3k_F$, as expected in the previous studies [22].

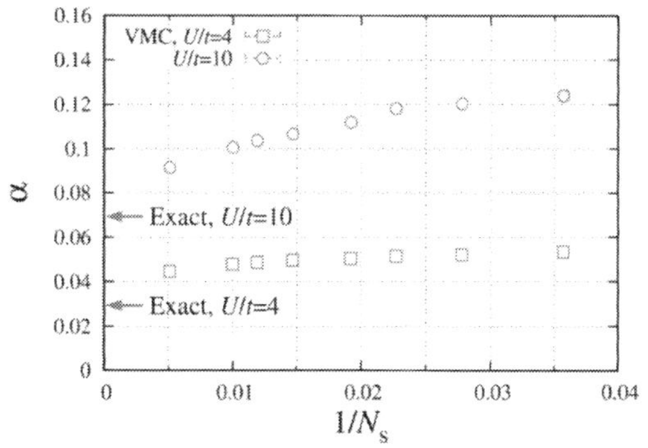

Figure 5. Estimated critical exponent α from relation $|n(k) - n(k_F)| \propto |k - k_F|^\alpha$ near k_F. Statistical errors are smaller than the symbol sizes. The red squares denote results at $U/t = 4$, and the green circles denote results at $U/t = 10$. These values are nearly the same as the exact results, $\alpha = 0.029$ ($U/t = 4$) and $\alpha = 0.069$ ($U/t = 10$), predicted by the Tomonaga-Luttinger theory [18, 20, 21, 28].

To estimate the critical exponent α, we plot $\ln |n(k) - n(k_F)|$ as a function of $|k - k_F|$ for each system size, as shown in figure 4(b), and extrapolate the slopes from data points by using equation (12). Since large finite-size effects appear near $|k-k_{\rm F}|/\pi \lesssim 0.1$, we use data that satisfy $|k-k_{\rm F}|/\pi \in [0.1, 0.4]$ for fitting. As shown in figure 5, as the system size $N_{\rm s}$ goes to infinity, the estimated critical exponent $\alpha(N_{\rm s})$ gradually converges to the exact value, $\alpha = 0.029$ at $U/t = 4$ and $\alpha = 0.069$ at $U/t = 10$, predicted by the Tomonaga-Luttinger theory [18, 20, 21, 28].

4. Summary

To improve the VMC method, we revisit the variational description of the ground states of the one-dimensional Hubbard model, by considering the multi-variable Slater determinants as well as the long-range correlation factors. We find that the quantum spin projection and total momentum projection greatly improve the accuracy of the ground state energy with less than 0.5% error, irrespective of the system size. By calculating the momentum distribution function $n(k)$ of the model at quarter filling up to 196 sites, we have shown that our method is able to reproduce power-law decays at the momenta $k_{\rm F}$ and $3k_{\rm F}$. By directly estimating the critical exponents of the charge correlations from the power-law behavior of $n(k)$ near $k_{\rm F}$, we have also shown that estimated critical exponents well reproduce those predicted by the Tomonaga-Luttinger theory.

Acknowledgments

The mVMC codes used for the present computation are based on that first developed by Daisuke Tahara. They thank Takahiro Misawa for fruitful discussions. This work was supported by a Grant-in-Aid for Scientific Research (Grant Nos. 22104010 and 22340090), from MEXT, Japan. A part of this research has been funded by the Strategic Programs for Innovative Research (SPIRE), MEXT, and the Computational Materials Science Initiative (CMSI), Japan.

References

[1] Blankenbecler R, Scalapino D J and Sugar R L 1981 *Phys. Rev.* D **24** 2278
[2] Sorella S, Baroni S, Car R and Parrinello M 1989 *Europhys. Lett.* **8** 663
[3] Imada M and Hatsugai Y 1989 *J. Phys. Soc. Jpn.* **58** 3752
[4] Corney J F and Drummond P D 2004 *Phys. Rev. Lett.* **93** 260401
[5] Assaad F F, Werner P, Corboz P, Gull E and Troyer M 2005 *Phys. Rev.* B **72** 224518
[6] Aimi T and Imada M 2007 *J. Phys. Soc. Jpn.* **76** 084709
[7] Metzner W and Vollhardt D 1989 *Phys. Rev. Lett.* **62** 324
[8] Georges A, Kotliar G, Krauth W and Rozenberg M J 1996 *Rev. Mod. Phys.* **68** 13
[9] White S R 1993 *Phys. Rev.* B **48** 10345
[10] Imada M and Kashima T 2000 *J. Phys. Soc. Jpn.* **69** 2723
[11] Kashima T and Imada M 2001 *J. Phys. Soc. Jpn.* **70** 2287
[12] Mizusaki T and Imada M 2004 *Phys. Rev.* B **69** 125110
[13] Ceperley D, Chester G V and Kalos M H 1977 *Phys. Rev.* B **16** 3081
[14] Anderson P W 1973 *Mat. Res. Bull.* **8** 153
[15] Anderson P W 1987 *Science* **235** 1196
[16] Liang S, Doucot B and Anderson P W 1988 *Phys. Rev. Lett.* **61** 365
[17] Tahara D and Imada M 2008 *J. Phys. Soc. Jpn.* **77** 114701
[18] Capello M, Becca F, Yunoki S, Fabrizio M and Sorella S 2005 *Phys. Rev.* B **72** 085121
[19] Capello M, Becca F, Fabrizio M, Sorella S and Tosatti E 2005 *Phys. Rev. Lett.* **94** 026406
[20] Kawakami N and Yang S K 1990 *Phys. Lett.* A **148** 359
[21] Kawakami N and Yang S K 1991 *J. Phys.: Cond. Matt.* **3** 5983
[22] Ogata M and Shiba H 1990 *Phys. Rev.* B **41** 2326
[23] Lieb E H and Wu F Y 1968 *Phys. Rev. Lett.* **20** 1445
[24] Tomonaga S 1950 *Prog. Theor. Phys.* **5** 544
[25] Luttinger J M 1963 *J. Math. Phys.* **4** 1154
[26] Sorella S 2001 *Phys. Rev.* B **64** 024512
[27] Giamarchi T and Schulz H J 1989 *Phys. Rev.* B **39** 4620
[28] Schulz H J 1990 *Phys. Rev. Lett.* **64** 2831
[29] Hallberg K A, Horsch P and Martínez G 1995 *Phys. Rev.* B **52** R719

[30] Parola A and Sorella S 1990 *Phys. Rev. Lett.* **64** 1831
[31] Sorella S, Parola A, Parrinello M and Tosatti E 1990 *Europhys. Lett.* **12** 721

Journal of Physics: Conference Series **454** (2013) 012047 doi:10.1088/1742-6596/454/1/012047

Variational Monte Carlo study for the insulating mechanism of Sr_2IrO_4: from the viewpoint of energy gain

Hiroshi Watanabe[1,2,3], **Tomonori Shirakawa**[1,2,3] **and Seiji Yunoki**[1,2,3]

[1]Computational Condensed Matter Physics Laboratory, RIKEN ASI, Saitama 351-0198, Japan
[2]CREST, Japan Science and Technology Agency (JST), Saitama 332-0012, Japan
[3]Computational Materials Science Research Team, RIKEN AICS, Kobe, Hyogo 650-0047, Japan

E-mail: h-watanabe@riken.jp

Abstract. To examine the insulating mechanism of a novel $5d$-electron system Sr_2IrO_4, we study the ground state properties of a three-orbital Hubbard model using a variational Monte Carlo method. We find that the insulating state in the ground state phase diagram shows crossover behavior from a weakly-correlated to a strongly-correlated antiferromagnetic state. This crossover is characterized by the different mechanisms of the insulating state, i.e., changing from an interaction-energy driven insulator to a band-energy driven insulator with increasing the interaction. We discuss that Sr_2IrO_4 is located around this crossover region and displays an anomalous behavior.

1. Introduction

Recently, a $5d$ transition metal oxide Sr_2IrO_4 [1] has attracted much attention. In this material, three t_{2g} orbitals of Ir atoms are hybridized with each other by the large spin-orbit coupling (SOC) of $5d$ electrons, and nominally five $5d$ electrons occupy these orbitals. Because of this quantum entanglement of spin and orbital degrees of freedom, an effective total angular momentum $J_{\rm eff}=|\boldsymbol{L}-\boldsymbol{S}|=1/2$ state is stable locally at each Ir atom. As a result, many interesting properties are experimentally observed, which include a novel $J_{\rm eff}=1/2$ insulating state [2, 3, 4, 5]. The origin of this insulating state is still under debate: Although the Mott-Hubbard type mechanism (i.e., a strongly-correlated insulator) is originally proposed for Sr_2IrO_4 [2, 3], there are several reports suggesting the Slater type mechanism (i.e., a weakly-correlated insulator) [6, 7, 8]. It should be noted, however, that experimentally the temperature dependence of the resistivity is found insulating and no significant change is observed at the Néel temperature [9], strongly suggesting that Sr_2IrO_4 is a Mott-Hubbard type insulator. In this paper, we study the ground state properties of a three-orbital Hubbard model using a variational Monte Carlo (VMC) method and examine the insulating mechanism of Sr_2IrO_4 from the viewpoint of energy gain.

2. Model and method

We consider a three-orbital Hubbard model on a two-dimensional square lattice defined by the following Hamiltonian $H=H_{\rm kin}+H_{\rm SO}+H_{\rm I}$, where $H_{\rm kin}=\sum_{\boldsymbol{k}\alpha\sigma}\varepsilon_\alpha(\boldsymbol{k})c^\dagger_{\boldsymbol{k}\alpha\sigma}c_{\boldsymbol{k}\alpha\sigma}$ is the kinetic term, $H_{\rm SO} = \lambda \sum_i \boldsymbol{L}_i \cdot \boldsymbol{S}_i$ is the SOC term with a coupling constant λ, and

$$\begin{aligned} H_{\rm I} = & \, U \sum_{i,\alpha} n_{i\alpha\uparrow} n_{i\alpha\downarrow}, \\ & + \sum_{i,\alpha<\beta,\sigma} \left[U' n_{i\alpha\sigma} n_{i\beta\bar{\sigma}} + (U' - J) n_{i\alpha\sigma} n_{i\beta\sigma} \right] \\ & + J \sum_{i,\alpha<\beta} \left(c^\dagger_{i\alpha\uparrow} c^\dagger_{i\beta\downarrow} c_{i\alpha\downarrow} c_{i\beta\uparrow} + c^\dagger_{i\alpha\uparrow} c^\dagger_{i\alpha\downarrow} c_{i\beta\downarrow} c_{i\beta\uparrow} + \text{H.c.} \right) \end{aligned} \tag{1}$$

is the Coulomb interactions including intraorbital (U), interorbital (U'), and spin-flip and pair-hopping (J) interactions. Here, $c^\dagger_{i\alpha\sigma}$ is a creation operator of electron at site i, spin $\sigma(=\uparrow,\downarrow)$, and orbital $\alpha(= yz, zx, xy)$ corresponding three t_{2g} orbitals.

The kinetic and SOC terms can be represented ($H_0 = H_{\rm kin} + H_{\rm SO}$) in the matrix form,

$$\begin{aligned} H_0(t_i, \mu_3, \lambda) &= \sum_{\boldsymbol{k}\sigma} \left(c^\dagger_{\boldsymbol{k}yz\sigma}, c^\dagger_{\boldsymbol{k}zx\sigma}, c^\dagger_{\boldsymbol{k}xy\bar{\sigma}} \right) \begin{pmatrix} \varepsilon_{yz}(\boldsymbol{k}) & {\rm i}\sigma\lambda/2 & -\sigma\lambda/2 \\ -{\rm i}\sigma\lambda/2 & \varepsilon_{zx}(\boldsymbol{k}) & {\rm i}\lambda/2 \\ -\sigma\lambda/2 & -{\rm i}\lambda/2 & \varepsilon_{xy}(\boldsymbol{k}) \end{pmatrix} \begin{pmatrix} c_{\boldsymbol{k}yz\sigma} \\ c_{\boldsymbol{k}zx\sigma} \\ c_{\boldsymbol{k}xy\bar{\sigma}} \end{pmatrix} \\ &= \sum_{\boldsymbol{k}ms} E_m(\boldsymbol{k}) a^\dagger_{\boldsymbol{k}ms} a_{\boldsymbol{k}ms}, \end{aligned} \tag{2}$$

where $c^\dagger_{\boldsymbol{k}\alpha\sigma}$ is the Fourier transformation of $c^\dagger_{i\alpha\sigma}$. Notice that the SOC mixes the three orbitals and up and down spins, and the new quasiparticles, obtained by diagonalizing H_0, are characterized by the band index m and pseudospin s with a creation operator $a^\dagger_{\boldsymbol{k}ms}$. In the atomic limit ($\varepsilon_{yz}(\boldsymbol{k}) = \varepsilon_{zx}(\boldsymbol{k}) = \varepsilon_{xy}(\boldsymbol{k}) = 0$), the sixfold degenerate states are split into doubly degenerate $J_{\rm eff} = 1/2$ states and fourfold degenerate $J_{\rm eff} = 3/2$ states [2].

First, the non-interacting tight-binding energy band is constructed so as to reproduce the result of LDA+SO (spin-orbit) calculation [2] for Sr_2IrO_4. The energy dispersion for three t_{2g} orbitals are defined as $\varepsilon_{yz}(\boldsymbol{k}) = -2t_5 \cos k_x - 2t_4 \cos k_y$, $\varepsilon_{zx}(\boldsymbol{k}) = -2t_4 \cos k_x - 2t_5 \cos k_y$, and $\varepsilon_{xy}(\boldsymbol{k}) = -2t_1(\cos k_x + \cos k_y) - 4t_2 \cos k_x \cos k_y - 2t_3(\cos 2k_x + \cos 2k_y) + \mu_3$ [4]. Here, we choose a set of parameters $(t_1, t_2, t_3, t_4, t_5, \mu_3, \lambda)$=(0.36, 0.18, 0.09, 0.37, 0.06, -0.36, 0.50) eV. In the following, we use t_1 as an energy unit.

Next, we introduce the the following Gutzwiller-Jastrow type trial wave function: $|\Psi\rangle = P_{\rm J_c} P_{\rm G} |\Phi\rangle$. $|\Phi\rangle$ is the one-body part of the wave function, constructed from the ground state of N_e number of electrons for $\tilde{H}_0(\tilde{t}_i, \tilde{\mu}_3, \tilde{\lambda}_{\alpha\beta})$, where $\tilde{t}_i$, $\tilde{\mu}_3$, and $\tilde{\lambda}_{\alpha\beta}$ are variational parameters. As an effect of the many-body Coulomb interaction, the "effective" coupling constant of the SOC has orbital dependence: $\lambda \to \tilde{\lambda}_{\alpha\beta}$. To consider magnetically ordered states, a term with magnetic order parameters is added to $\tilde{H}_0$. Here, we consider the commensurate antiferromagnetic (AF) state with an ordering vector $\boldsymbol{Q} = (\pi, \pi)$. Two states considered are i) an out-of-plane AF order (along z axis) and ii) an in-plane AF order (along x axis), described by $\sum_{i,m} M^z_m {\rm e}^{{\rm i}\boldsymbol{Q}\cdot\boldsymbol{r}_i} (a^\dagger_{im\uparrow} a_{im\uparrow} - a^\dagger_{im\downarrow} a_{im\downarrow})$ and $\sum_{i,m} M^x_m {\rm e}^{{\rm i}\boldsymbol{Q}\cdot\boldsymbol{r}_i} (a^\dagger_{im\uparrow} a_{im\downarrow} + a^\dagger_{im\downarrow} a_{im\uparrow})$, respectively. These two AF states have different energies in general since the rotational symmetry in a spin space is broken due to the SOC and the Hund's coupling J. However, here we consider only the case of $J = 0$ and thus they are energetically degenerated [4].

For the trial wave function of superconducting states, we consider a BCS-type one in a pseudospin representation,

$$|\Phi\rangle \propto \left(\sum_{\boldsymbol{k}mn} f_{\boldsymbol{k}mn} a^\dagger_{\boldsymbol{k}m\uparrow} a^\dagger_{\boldsymbol{k}n\downarrow} \right)^{N_e/2} |0\rangle . \tag{3}$$

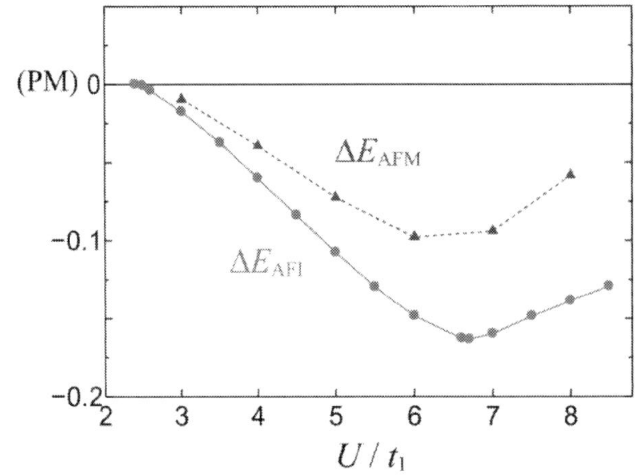

Figure 1. (color online) The condensation energy $\Delta E(U)$ defined as the energy difference between the AF and the paramagnetic states [see (4)]. $\Delta E_{\rm AFI}$ and $\Delta E_{\rm AFM}$ denote the condensation energies of AF insulating and metallic states. Here, $n = 5, \lambda/t_1 = 1.4$, and $J/U = 0$ are used. Statistical errors are within the width of the symbols.

Here, $f_{\boldsymbol{k}mn}$ is obtained by diagonalizing BCS-type Hamiltonian with three energy bands (6×6 matrix) and the explicit form is given in [10].

$P_{\rm G} = \prod_{i,\gamma} [1 - (1 - g_\gamma) |\gamma\rangle \langle\gamma|_i]$ in $|\Psi\rangle$ is a Gutzwiller factor extended to the three-orbital system [4]. i represents a site index and γ runs over all possible electron configurations at each site. For the three-orbital system, there are $4^3 = 64$ electron configurations, namely, $|0\rangle = |0\,0\,0\rangle$, $|1\rangle = |0\,0\uparrow\rangle$, $\cdots$, $|63\rangle = |\uparrow\downarrow\uparrow\downarrow\uparrow\downarrow\rangle$. g_γ is a weight of each electron configuration, and it takes the value from 0 to 1, controlling the local electron correlations. In this study, we classify the possible 64 patterns into 12 groups by the Coulomb interaction energy, and g_γ's in the same group are set to be the same.

$P_{\rm J_c} = \exp\left[-\sum_{i\neq j} v_{ij} n_i n_j\right]$ in $|\Psi\rangle$ is a charge Jastrow factor that controls the long-range charge correlations. Here, we assume that v_{ij} depends only on the distances, $v_{ij} = v(|\boldsymbol{r}_i - \boldsymbol{r}_j|)$ and consider all independent v_{ij}. For example, there are 41 independent v_{ij}'s for 16×16, taking into account up to the 41st-neighbor correlation.

Using the trial wave functions described above, we optimize the variational parameters to minimize the variational energy using the stochastic reconfiguration method [11]. Comparing the variational energy for paramagnetic, AF, and superconducting states, we determine the ground state phase diagram. For large size calculations, parallel computing is indispensable. The VMC method can be efficiently parallelized by distributing the independent Monte Carlo samplings to different processors. We parallelize the simulations using 32–128 processors and the total number of Monte Carlo samples taken is about 10^5–10^6, for example, $10,000 \times 64$ samples using 64 processors with $10,000$ independent samples for each processor.

3. Results

We study the ground state properties of the three-orbital Hubbard model on a 16×16 square lattice with electron density $n = 5$. In this paper, we consider only $J = 0$ and we impose the condition $U = U' + 2J$ [12], thus $U = U'$.

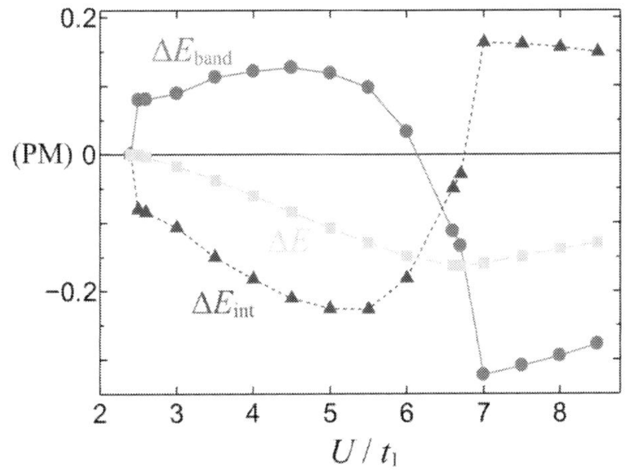

Figure 2. (color online) The condensation energy $\Delta E(U) = \Delta E_{\mathrm{band}}(U) + \Delta E_{\mathrm{int}}(U)$ defined as the energy difference between the AF insulating and the paramagnetic states. $\Delta E_{\mathrm{band}}(U)$ and $\Delta E_{\mathrm{int}}(U)$ are two separate contribution from the band energy and the interaction energy, respectively. Here, $n = 5, \lambda/t_1 = 1.4$, and $J/U = 0$ are used. Statistical errors are within the width of the symbols.

First, we calculate the condensation energy of the AF state, which is defined by

$$\Delta E(U) = E_{\mathrm{AF}} - E_{\mathrm{PM}}, \tag{4}$$

with varying U. Here, E_{AF} (E_{PM}) is the optimized variational energy of the AF (paramagnetic metallic) state for a given value of U. The critical U_{AF} for the AF order is where $\Delta E(U = U_{\mathrm{AF}}) = 0$. We find that with increasing U, the AF transition occurs at around $U_{\mathrm{AF}}/t_1 \sim 2.5$. Furthermore, we find that a metal-insulator transition occurs simultaneously and the system becomes an AF insulator. Namely, an AF metallic state has always larger variational energy than the AF insulating state, as shown in figure 1, and thus it is unstable. We also calculate the variational energies of the possible superconducting states and find that the one with $d_{x^2-y^2}$ symmetry has the smallest variational energy. However, its variational energy is always much larger than the AF insulating state and therefore the superconducting state is not the ground state for this electron density.

Next, to study the character of the AF insulating state, we examine the origin of this negative condensation energy $\Delta E(U)$ in the AF insulating state. Let us divide $\Delta E(U)$ into two parts,

$$\begin{aligned} \Delta E(U) &= \Delta E_{\mathrm{kin}} + \Delta E_{\mathrm{SO}} + \Delta E_U + \Delta E_{U'} + \Delta E_J + \Delta E_{J'} \\ &= \Delta E_{\mathrm{band}}(U) + \Delta E_{\mathrm{int}}(U), \end{aligned} \tag{5}$$

where $\Delta E_{\mathrm{band}}(U) = \Delta E_{\mathrm{kin}} + \Delta E_{\mathrm{SO}}$ is the contribution from the band (i.e., kinetic and SOC) energy and $\Delta E_{\mathrm{int}}(U) = \Delta E_U + \Delta E_{U'} + \Delta E_J + \Delta E_{J'}$ is the contribution from the interaction (i.e., Coulomb) energy. Figure 2 shows the U dependences of $\Delta E(U)$, $\Delta E_{\mathrm{band}}(U)$, and $\Delta E_{\mathrm{int}}(U)$. For small U, the energy gain of the AF insulating state is due to the interaction energy, i.e., $\Delta E_{\mathrm{int}}(U) < 0$ but $\Delta E_{\mathrm{band}}(U) > 0$, indicating that this AF insulator is interaction-energy driven. Instead, for large U, the AF insulating state is stabilized due to the band energy gain, i.e., $\Delta E_{\mathrm{band}}(U) < 0$ but $\Delta E_{\mathrm{int}}(U) > 0$, strongly suggesting that this AF insulator is band-energy driven. It is also clear in figure 2 that there is a crossover region $U_{\mathrm{co}}/t_1 \sim 6.5$ where

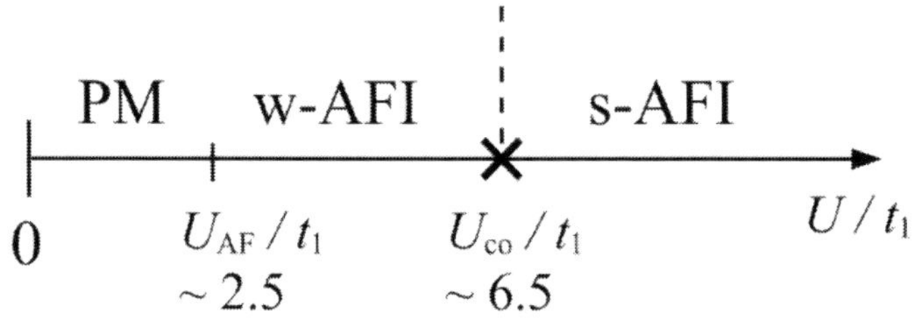

Figure 3. The ground state phase diagram for $n = 5, \lambda/t_1 = 1.4$, and $J/U = 0$, obtained from figure 2. PM, w-AFI, and s-AFI denote a paramagnetic metallic, a weakly-correlated AF insulating, and a strongly-correlated AF insulating states, respectively.

$\Delta E_{\rm band}(U)$ and $\Delta E_{\rm int}(U)$ are both negative and intersect. Note that this crossover is not a phase transition because both insulating states are described by the same AF insulating wave function. However, this crossover should be considered to be a boundary separating a weakly-correlated and a strongly-correlated regions. The ground state phase diagram is summarized in figure 3. The same kind of crossover for a AF and a superconducting states is also discussed in a single-orbital Hubbard model [13, 14, 15, 16] and the results are consistent with ours for the three-orbital Hubbard model.

There are several reports estimating the values of the Coulomb interactions for Ir oxides [7, 17, 18]. According to the constrained RPA calculation [7], the value of U is estimated as large as ~ 2.3 eV, corresponding to $U/t_1 \sim 6.5$ in our model. This value is very close to $U_{\rm co}/t_1 \sim 6.5$ as seen in figure 3. We also found that $U_{\rm co}$ depends on the values of λ/t_1 and J/U (a smaller λ/t_1 or a larger J/U increases $U_{\rm co}$), however, these effects are only quantitative and overall behaviors of $\Delta E_{\rm band}(U)$, $\Delta E_{\rm int}(U)$, and $\Delta E(U)$ are unchanged. Our result strongly indicates that Sr_2IrO_4 should be located at around the crossover region separating the Slater type and the Mott-Hubbard type insulators. Therefore, it is very natural to expect that Sr_2IrO_4 can exhibit dual characteristics of weakly-correlated and strongly-correlated natures.

4. Summary

In this paper, using the variational Monte Carlo method, we have studied the ground state properties of the three-orbital Hubbard model for Sr_2IrO_4. We have found that the insulating state in the ground state phase diagram shows a crossover from the weakly-correlated AF insulating state to the strongly-correlated AF insulating state, i.e., from the interaction-energy driven to the band-energy driven AF insulating state. Our result suggests that Sr_2IrO_4 is located at around this crossover region and can display an anomalous behavior.

Acknowledgments

The authors would like to thank S. Biermann and R. Arita for fruitful discussions. The computation has been done using the RIKEN Cluster of Clusters (RICC) facility and the facilities of the Supercomputer Center, Institute for Solid State Physics, University of Tokyo. This work has been supported by Grant-in-Aid for Scientific Research from MEXT Japan under the grant numbers 24740251 and 24740269.

References

[1] Randall J J *et al.* 1957 *J. Am. Chem. Soc.* **79** 266
[2] Kim B J *et al.* 2008 *Phys. Rev. Lett.* **101** 076402
[3] Kim B J, Ohsumi H, Komesu T, Sakai S, Morita T, Takagi H, Arima T 2009 *Science* **323** 1329

[4] Watanabe H, Shirakawa T and Yunoki S 2010 *Phys. Rev. Lett.* **105** 216410
[5] Shirakawa T, Watanabe H and Yunoki S 2011 *J. Phys. Soc. Jpn.* **80** SB010
[6] Kini N S, Strydom A M, Jeevan H S, Geibel C and Ramakrishnan S 2006 *J. Phys.: Condens. Matter* **18** 8205
[7] Arita R, Kuneš J, Kozhevnikov A V, Eguiluz A G and Imada M 2012 *Phys. Rev. Lett.* **108** 086403
[8] Hsieh D, Mahmood F, Torchinsky D H, Cao G and Gedik N 2012 *Phys. Rev. B* **86** 035128
[9] Chikara S, Korneta O, Crummett W P, DeLong L E, Schlottmann P and Cao G 2009 *Phys. Rev. B* **80** 140407(R)
[10] Watanabe H, Shirakawa T and Yunoki S 2013 *Phys. Rev. Lett.* **110** 027002
[11] Sorella S 2001 *Phys. Rev. B* **64** 024512; Casula M and Sorella S 2003 *J. Chem. Phys.* **119** 6500; Casula M, Attaccalite C and Sorella S 2004 *J. Chem. Phys.* **121** 7110; Yunoki S and Sorella S 2006 *Phys. Rev. B* **74** 014408; Sorella S, Casula M and Rocca D, *J. Chem. Phys.* **127** 014105
[12] Kanamori J, 1963 *Prog. Theor. Phys.* **30** 275
[13] Yokoyama H, Tanaka Y, Ogata M and Tsuchiura H 2004 *J. Phys. Soc. Jpn.* **73** 1119
[14] Maier Th A, Jarrell M, Macridin A and Slezak C 2004 *Phys. Rev. Lett.* **92** 027005
[15] Yanase Y and Ogata M 2005 *J. Phys. Soc. Jpn.* **74** 1534
[16] Yokoyama H, Ogata M, Tanaka Y, Kobayashi K and Tsuchiura H 2013 *J. Phys. Soc. Jpn.* **82** 014707
[17] Martins C, Aichhorn M, Vaugier L and Biermann S 2011 *Phys. Rev. Lett.* **107** 266404
[18] Comin R *et al.* 2012 *Phys. Rev. Lett.* **109** 266406; Kim B H, Khaliullin G and Min B I 2012 *Phys. Rev. Lett.* **109** 167205

Journal of Physics: Conference Series **454** (2013) 012048 doi:10.1088/1742-6596/454/1/012048

Fermion-induced decoherence of bosons in optical lattices

Akiko Masaki[1] **and Hiroyuki Mori**[2]

[1] Institute for Solid State Physics, University of Tokyo, Kashiwa, Chiba 277-8581, Japan

[2] Department of Physics, Tokyo Metropolitan University, Hachiohji-shi, Tokyo 192-0397, Japan

E-mail: amasaki@issp.u-tokyo.ac.jp

Abstract. Bose-Fermi mixtures with attractive Bose-Fermi interactions in one-dimensional optical lattices are studied by using Quantum Monte Carlo simulations of an extended Bose-Fermi Hubbard model. We first derived the extended Hubbard model with the hopping terms of each species modified to include interaction effects. Bosonic Mott transition induced by introducing fermions into bosons is demonstrated by the simulations.

1. Introduction

Ultracold atoms in optical lattices presented a new way of analyzing particles on periodic potentials [1]. Various phenomena have been realized in the experiments. For example, superfluid-Mott insulator transition [2], shell structure formation in real-space distribution under a confining potential [3], and localizations [4, 5, 6] were observed in bosonic systems. These phenomena can be well described with conventional Hubbard models.

Bose-Fermi mixtures are also an interesting target of ultracold atom experiments and physical properties of the mixtures are well described with a Bose-Fermi-Hubbard model [7]. However, phenomena that this model cannot explain have been observed in experiments of Bose-Fermi mixtures in optical lattices [8, 9, 10]. In these experiments, bosons lost their coherence and underwent a Mott transition as fermions, which attractively interacted with the bosons, were introduced to the system. The analysis of ordinary Bose-Fermi-Hubbard model with a confining potential however showed an opposite result that the bosons maintained the superfluid coherency even in the presence of the fermions [11, 12].

Lühmann et al. [13] discussed a self-trapping effect caused by the attractive boson-fermion interactions to explain the disappearance of the superfluid coherency. When there are some bosons in a well of the lattice potential, the wave function of the fermion in the same well would be squeezed through the boson-fermion interactions. Then the squeezed wave function would produce a narrower and deeper effective potential for the bosons, which makes the bosonic hopping parameter smaller and induces a bosonic Mott transition.

We propose another scenario about the change of the hopping parameters with focus on the terms that are usually ignored in the Hubbard model. In the followings it will be shown that the hopping parameter of the bosons/fermions could locally changes depending on the presence/absence of the fermions/bosons nearby.

For the observation of physical properties of our extended Hubbard model, numerical simulations were conducted and the fermion-induced Mott transition of the bosons was successfully demonstrated. In the simulations the superfluid density and the visibility were calculated as functions of the number of the fermions. It was then confirmed that the superfluid coherency of the bosons vanished with the introduction of the fermions when the attractive boson-fermion interactions were large enough.

The paper is organized as follows. We derive an extended Hubbard model in section 2. Section 3 presents the results of the simulations. A brief summary and discussions are given in section 4.

2. Extended Bose-Fermi Hubbard model

We introduce an effective model of bosons and fermion strongly correlating with each other. One-dimensional bosons and (spin-polarized) fermions in an optical lattice with contact interactions can be generally written as follows.

$$H = \int dx \psi_B^\dagger(x) \left[\frac{p_x^2}{2m_B} + V(x)\right] \psi_B(x) + \int dx \psi_F^\dagger(x) \left[\frac{p_x^2}{2m_F} + V(x)\right] \psi_F(x) \tag{1}$$
$$+\frac{g_{BB}}{2} \int dx \psi_B^\dagger(x)\psi_B^\dagger(x)\psi_B(x)\psi_B(x) + g_{BF} \int dx \psi_B^\dagger(x)\psi_F^\dagger(x)\psi_F(x)\psi_B(x),$$

where $V(x)$ presents the optical lattice potential and $\psi_B(x)$ and $\psi_F(x)$ are the boson and fermion field operators respectively. The coupling constants are given by $g_{BB} = \frac{4\pi a_{BB}\hbar^2}{m_B}$ and $\frac{2\pi a_{BF}\hbar^2}{M}$ where $M = \frac{2m_B m_F}{m_B + m_F}$ is the reduced mass, and a_{BB} and a_{BF} are boson-boson and boson-fermion s-wave scattering lengths respectively.

In the periodic optical lattice, the field operators can be expanded by Wannier functions as $\psi_B(x) = \sum_i b_i w_B(x - x_i)$ and $\psi_F(x) = \sum_i f_i w_F(x - x_i)$. Here, b_i and f_i are annihilation operators of the bosons and the fermions respectively at the i-th well of the potential and $w(x)$ is the Wannier function. Since we are only interested in low energy behaviors of the system, we only take account of the lowest energy Wannier function $w(x)$. Ordinary Bose-Fermi Hubbard Hamiltonian can be derived from the Hamiltonian H by using the expansion. In the derivation, the terms of relatively small overlap integrals are all discarded. For example, the following gives largest contribution to the boson-fermion interaction term.

$$g_{BF} \sum_i \int dx \, |w_B(x - x_i)|^2 \, |w_F(x - x_i)|^2 \, b_i{}^\dagger b_i f_i{}^\dagger f_i. \tag{2}$$

On the other hand, the terms of the following type contain smaller overlap integrals and are usually omitted in Bose-Fermi Hubbard model.

$$g_{BF} \sum_{\langle i,j \rangle} \int dx \, |w_F(x - x_i)|^2 \, w_B^*(x - x_i) w_B(x - x_j) f_i{}^\dagger f_i b_i{}^\dagger b_j. \tag{3}$$

However we consider that this kind of terms would be a key for the explanation of the experimentally observed fermion-induced Mott transition. Including (3), the hopping term of the bosons can be expressed as

$$H_B^{hop} = -\sum_{\langle i,j \rangle} \left[t_B^0 - \Delta_B(f_i{}^\dagger f_i + f_j{}^\dagger f_j)\right] b_i{}^\dagger b_j + h.c., \tag{4}$$

where t_B^0 is ordinary hopping parameter in the tight binding approximation and $\Delta_B = -(|g_{BF}|/2)\int dx\,|w_F(x-x_i)|^2 w_B^*(x-x_i)w_B(x-x_j)$. Here we assumed attractive boson-fermion interactions, $g_{BF} < 0$. t_F^0 and Δ_F are defined in a similar manner. With this hopping term, our extended Hamiltonian can be written in the form

$$H = -\sum_{\langle i,j\rangle}\left[t_B^0 - \Delta_B(n_{F,i}+n_{F,j})\right] b_i{}^\dagger b_j - \sum_{\langle i,j\rangle}\left[t_F^0 - \Delta_F(n_{B,i}+n_{B,j})\right] f_i{}^\dagger f_j + h.c. \qquad (5)$$
$$+\frac{U_{BB}}{2}\sum_i n_{Bi}(n_{Bi}-1) + U_{BF}\sum_i n_{Bi}n_{Fi},$$

where n_{Bi} (n_{Fi}) is the number operator of the bosons (fermions) at the site i, and U_{BB} and U_{BF} present on-site boson-boson and boson-fermion interactions respectively.

Note that, in this model, the hopping energy of the bosons (fermions) changes depending on the presence of the fermions (bosons) at each site. In a pure boson system with no fermions, the hopping parameter of the bosons is always t_B^0 at any site. With the introduction of fermions to the system, the hopping parameter of the bosons becomes $t_B^0 - \Delta_B$ or $t_B^0 - 2\Delta$ at the sites where the fermions are present. Figure 1 illustrates these three cases.

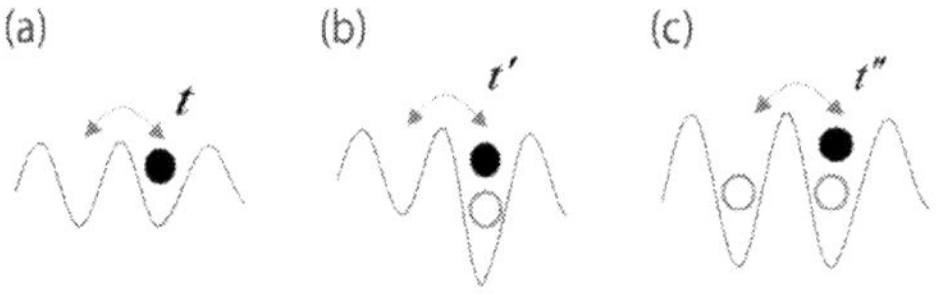

Figure 1. (a), (b) and (c) illustrate effective potential wells of the bosons (filled black circle). In (a) with no fermions, the boson hops to the next site with the transfer energy $t(= t_B^0)$. In (b) with a single fermion (open red circle) the boson hops with $t'(= t_B^0 - \Delta_B)$. In (c) with two fermions, the boson hops with $t''(= t_B^0 - 2\Delta_B)$.

In a mean-field sense, the average hopping energy of the bosons would decrease as the number of the fermions increases. We could therefore expect that the bosons would undergo a Mott transition when a sufficient number of the fermions are introduced to the system and when other conditions of e.g. the number of the bosons and the boson-boson interactions are met.

3. Numerical results

To confirm that the above scenario is correct, we performed Quantum Monte Carlo Simulation(QMC) of the extended Hubbard Hamiltonian (6) that we derived. The QMC is based on a stochastic Green's function method in the canonical ensemble [14, 15]. Since fermions in one dimension can be mapped on hard-core bosons exactly through the Jordan-Wigner transformation, we employ the hard-core bosons in the simulations to avoid a negative sign problem but have to take account of the transformation effect in the calculation of physical quantities of the fermions. We simulated 64 bosons on 64 sites and changed the number of the fermions N_f from 0 to 64. To obtain a qualitative view of the transition we set $t_B^0 = t_F^0 = 1$ for simplicity. The boson-boson interaction was set to $U_{bb} = 4$ with which the bosons are in a superfluid state in the absence of the fermions. The inverse temperature and the boson-fermion interaction were set to $\beta = 64$ and $U_{BF} = -4$ respectively.

Calculation result of the superfluid density ρ_b is shown in figure 2 as a function of N_f with different Δ_B. The superfluid density is calculated by $\rho_b = \frac{\langle W^2\rangle}{2t_B^0\beta L^{-1}}$ [16] where W is the winding number and L is the system size. As explained above, in the absence of fermions, the hopping parameter of the bosons is t_B^0 at any site and the bosons are in the superfluid state as we chose relatively small boson-boson interactions. In the presence of 64 fermions on 64 lattice sites, on the other hand, the hopping energy t_B is given by $t_B^0 - 2\Delta_B$ for any sites. While with $\Delta_B > 0.2$ the ratio U_{BB}/t_B becomes large enough to form a bosonic Mott state. Figure 2 shows that the superfluid density vanishes at around $N_f = 45$ with $\Delta_B = 0.3$ and at around $N_f = 35$ with $\Delta_B = 0.4$. When the attractive interaction U_{BF} is sufficiently large, the superfluid density decreases as the fermions are introduced, forming a bosonic Mott state. We confirmed that this fermion-induced bosonic decoherence could also be observed in a Bose-Fermi mixture with a confinement potential.

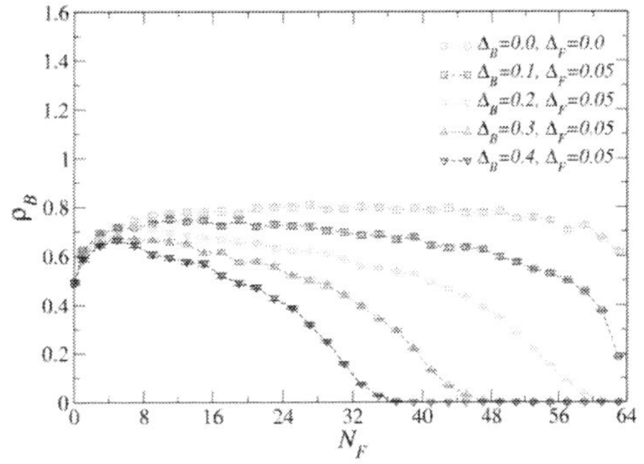

Figure 2. Superfluid density as a function of N_F with different Δ_B and Δ_F.

4. Conclusions

We numerically studied Bose-Fermi mixtures with attractive Bose-Fermi interactions on optical lattices using an extended Bose-Fermi-Hubbard model, which was derived to take account of the effective change of particle hopping energies due to the Bose-Fermi interactions. It was observed in the simulations that the introduction of the fermions induced a bosonic Mott transition as found in the experiments. More detailed calculation results will be given elsewhere.

References

[1] Jaksch D, Bruder C, Cirac J I, Gardiner C W and Zoller P 1998 *Phys. Rev. Lett.* **81** 3108
[2] Greiner M, Mandel O, Esslinger T, Hansch T, and Bloch I 2002 *Nature* **415** 39
[3] Campbell G K, Mun J, Boyd M, Medley P, Leanhardt A E, Marcassa L G, Pritchard D E and Ketterle W 2006 *Science* 313 649
[4] Billy J, Josse V, Zuo Z, Bernard A, Hambrecht B, Lugan P, Clement D, Sanchez-Palencia L, Bouyer P and Aspect A 2008 *Nature* **453** 891
[5] Roati G, D'Errico C, Fallani L, Fattori M, Fort C, Zaccanti M, Modugno G, Modugno M and Inguscio M 2008 *Nature* **453** 895
[6] Fallani L, Lye J E, Guarrera V, Fort C, and Inguscio M 2007 *Phys. Rev. Lett.* **98** 130404
[7] Sugawa S, Inaba K, Taie S, Yamazaki R, Yamashita M and Takahashi Y 2011 *Nature Phys.* **7**642
[8] Günter K, Stöferle T, Moritz H, Köhl M, and Esslinger T 2006 *Phys. Rev. Lett.* **96** 180402
[9] Ospelkaus S, Ospelkaus C, Wille O, Succo M, Ernst P, Sengstock K, and Bongs K 2006 *Phys. Rev. Lett.* **96** 180403
[10] Best Th, Will S, Schneider U, Hackermuller L, van Oosten D, and Bloch I 2009 *Phys. Rev. Lett.* **102** 030408
[11] Pollet L, Kollath C, Schollwock U and Troyer M 2008 *Phys. Rev. A* **77** 023608
[12] Varney C N, Rousseau V G, and Scalettar R T, 2008 *Phys. Rev. A* **77**, 041608
[13] Lühmann D-S, Bongs K, Sengstock K, and Pfannkuche D 2008 *Phys. Rev. Lett.* **101** 050402
[14] Rousseau V G 2008 *Phys. Rev. E* **77** 056705
[15] Rousseau V G 2008 *Phys. Rev. E* **78** 056707
[16] Pollock E L and Ceperley D M 1987 *Phys. Rev. B* **36** 8343

A study of parallelizing O(N) Green-function-based Monte Carlo method for many fermions coupled with classical degrees of freedom

Shixun Zhang[1], **Shinichi Yamagia**[2] **and Seiji Yunoki**[3,4,5]

[1]Department of Computer Science, University of Tsukuba, Tsukuba, Ibaraki 305-8573, Japan

[2]Faculty of Engineering, Information and Systems University of Tsukuba / JST PRESTO, 1-1-1 Tennodai, Tsukuba, Ibaraki 305-8573, Japan

[3]Computational Condensed Matter Physics Laboratory, RIKEN ASI, Wako, Saitama, 351-0198 Japan

[4]Computational Materials Science Research Team, RIKEN AICS, Kobe, Hyogo, 650-0047 Japan

[5]CREST, Japan Science and Technology Agency, Kawaguchi, Saitama 332-0012, Japan

E-mail: sxzhang@cs.tsukuba.ac.jp

Abstract. Models of fermions interacting with classical degrees of freedom are applied to a large variety of systems in condensed matter physics. For this class of models, Weiße [Phys. Rev. Lett. **102**, 150604 (2009)] has recently proposed a very efficient numerical method, called O(N) Green-Function-Based Monte Carlo (GFMC) method, where a kernel polynomial expansion technique is used to avoid the full numerical diagonalization of the fermion Hamiltonian matrix of size N, which usually costs O(N^3) computational complexity. Motivated by this background, in this paper we apply the GFMC method to the double exchange model in three spatial dimensions. We mainly focus on the implementation of GFMC method using both MPI on a CPU-based cluster and Nvidia's Compute Unified Device Architecture (CUDA) programming techniques on a GPU-based (Graphics Processing Unit based) cluster. The time complexity of the algorithm and the parallel implementation details on the clusters are discussed. We also show the performance scaling for increasing Hamiltonian matrix size and increasing number of nodes, respectively. The performance evaluation indicates that for a 32^3 Hamiltonian a single GPU shows higher performance equivalent to more than 30 CPU cores parallelized using MPI.

1. Introduction

In solid state materials, we observe quite different physical properties such as superconductivity and magnetism. These different properties are explained by different behavior of electrons, which in turn is caused mostly by Coulomb interactions among electrons or/and interactions between electrons and other degrees of freedom such as phonon. Therefore, theoretical understanding of these interacting electrons is crucial to, e.g., design new materials with rich functionalities.

However, precisely due to this many-body nature of these interactions, it is usually very difficult to treat these systems analytically and even numerically.

Among such interacting electron systems, models of electrons interacting with classical degrees of freedom can be applied to a large variety of systems. One of the most studied systems is the double exchange model for colossal magnetoresistive manganese oxides [1, 2, 3, 4]. In the double exchange model, electrons can move around a lattice site under the influence of the localized classical spins via Hund's rule coupling, the Hamiltonian being given in section 2. To solve this model, the Monte Carlo method is commonly used to carry out the importance sampling for the classical degrees of freedom after treating electron degrees of freedom by numerically diagonalizing the electron Hamiltonian of dimension N (where N is a system size) [1, 2, 3]. However, since at every Monte Carlo step one must fully diagonalize the Hamiltonian matrix to evaluate the partition function for a given configuration of the classical spins, it usually costs $O(N^3)$ numerical complexity. Because of this $O(N^3)$ scaling of the calculations, the accessible system size is strictly limited to up to hundreds to thousands sites, preventing us from studying important properties such as the Curie temperature of a ferromagnetic ordering [1, 2, 3].

In order to reduce the numerical complexity, Kernel Polynomial Method (KPM) is very advantageous [5]. Indeed, Motome and Furukawa have applied a Chebyshev expansion of the electron density of states for evaluating the partition function, which can reduce the numerical complexity down to $O(N^2)$ [6]. They have also proposed that the numerical complexity can be further reduced to $O(N)$ by truncating the moment calculations for the Chebyshev expansion [7]. Very recently, Weiße has pointed out that the same numerical complexity of $O(N)$ can be achieved by using a Green-function-based Monte Carlo (GFMC) method, instead of directly evaluating the partition function for a given spin configuration [8]. The GFMC method dramatically decreases the computational time by using only a few local Green's functions estimated by the Chebyshev expansion to calculate the change of the electron density of states. Moreover, this $O(N)$ GFMC method is more attractive than the one for directly evaluating the partition function with the same complexity of $O(N)$ [7] because the GFMC method achieve the $O(N)$ complexity without any truncation [8].

Motivated by these recent progress, here we apply the $O(N)$ GFMC method to the double exchange model. In this paper, we mainly focus on the implementation of the GFMC method using both MPI on a Central Processing Unit (CPU) cluster and Nvidia's Compute Unified Device Architecture (CUDA) programming techniques on a Graphics Processing Unit (GPU) cluster [9]. The time complexity of the algorithm is estimated. The implementation details on a GPU cluster is also described since programming on a GPU cluster requires more attention on hardware aspects such as memory copy and communication between CPU and GPU. The performance evaluation indicates that the GPU program can archive over 10× speedup as compared to the MPI parallelization on a CPU cluster.

This paper is organized as follows: section 2 explains the double exchange model and formulates the GFMC method, section 3 describes the implementation details and the parallelization techniques, which are tested in section 4, section 5 shows the performance on a single CPU processor as well as the whole cluster, and finally, section 6 will give the conclusions.

2. Model and Method Formulation

2.1. Double Exchange Model

The double exchange (DE) model describes electrons interacting the classical spins via Hund's rule coupling. The Hamiltonian is given by

$$H = -t \sum_{\langle i,j \rangle} \left(c^{\dagger}_{i\sigma} c_{j\sigma} + \mathrm{H.c.} \right) + J_H \sum_{i,\alpha\beta} \vec{S}_i \cdot c^{\dagger}_{i\alpha} \vec{\sigma}_{\alpha\beta} c_{i\beta}, \tag{1}$$

where $c^\dagger_{i\sigma}$ is a creation operator of electron at site i and with spin $\sigma = (\uparrow, \downarrow)$, $\langle i,j\rangle$ runs over a pair of nearest neighbor sites i and j, $\vec{S}_i$ is the classical spin at site i with its normalization $|\vec{S}_i| = 1$, and J_H denotes the Hund's rule coupling with $J_H > 0$. In the limit of infinite J_H ($J_H \to \infty$), the Hamiltonian becomes

$$H = -\sum_{\langle i,j\rangle} t_{ij}\left(c^\dagger_{i\sigma}c_{j\sigma} + \text{H.c.}\right) \tag{2}$$

with the hopping amplitude

$$t_{ij} = \cos\frac{\theta_i - \theta_j}{2}\cos\frac{\phi_i - \phi_j}{2} + i\cos\frac{\theta_i + \theta_j}{2}\sin\frac{\phi_i - \phi_j}{2}, \tag{3}$$

$$\tag{4}$$

where θ_i and ϕ_j denote the polar and the azimuthal angles of the classical spin $\vec{S}_i$, respectively.

The grand partition function of the DE model is written as

$$Z = \prod_i^N \int d^3\vec{S}_i \text{Tr}_\text{e}\left[\exp(-\beta H(\{\vec{S}_i\}) - \mu N\right], \tag{5}$$

where $\beta = 1/T$ is inverse of temperature T, μ is the chemical potential, N is the total number operator of electrons, and $\text{Tr}_\text{e}[\cdots]$ indicates the trace over the electron degrees of freedom in the Fock space. The trace over classical spin degrees of freedom, i.e., $\prod_i^N \int d^3\vec{S}_i P(\{\vec{S}_i\})$, is evaluated by the Monte Carlo importance sampling with its weight $P(\{\vec{S}_i\})$ for a given spin configuration $\{\vec{S}_i\}$,

$$P(\{\vec{S}_i\}) = \text{Tr}_\text{e}\left[\exp(-\beta H(\{\vec{S}_i\} - \mu N)\right] = \exp[-\text{S}_\text{eff}(\{\vec{S}_i\})], \tag{6}$$

where

$$S_\text{eff}(\{\vec{S}_i\}) = -\sum_i^N \log(1 + e^{-\beta[\epsilon_i(\{\vec{S}_i\}) - \mu]}) \tag{7}$$

$$= -\int \log(1 + e^{-\beta[E-\mu]})\rho(E)dE, \tag{8}$$

ϵ_i is the i-th eigenvalue of Hamiltonian $H(\{\vec{S}_i\})$, and $\rho(E)$ is the electron density of state (DOS) for a given spin configuration $\{\vec{S}_i\}$. Using the Metropolis algorithm, the possibility $P(\{\vec{S}_i\} \to \{\vec{S}'_i\})$ of accepting a new spin configuration $\{\vec{S}'_i\}$ is given by

$$P(\{\vec{S}_i\} \to \{\vec{S}'_i\}) = \frac{P(\{\vec{S}'_i\})}{P(\{\vec{S}_i\})} = e^{-\text{S}_\text{eff}(\{\vec{S}'_i\}) + \text{S}_\text{eff}(\{\vec{S}_i\})}.$$

Therefore, the quantity Δ_seff define by

$$\Delta_\text{seff} = \text{S}_\text{eff}(\{\vec{S}'_i\}) - \text{S}_\text{eff}(\{\vec{S}_i\}) = -\int \log(1 + e^{-\beta[E-\mu]})(\rho'(E) - \rho(E))dE \tag{9}$$

is all we need to carry out the Monte Carlo calculation. Here, $\rho'(E)$ is DOS for a given spin configuration $\{\vec{S}'_i\}$. As mentioned in section 1, we can exactly diagonalize $H(\{\vec{S}_i\})$ with $\text{O}(N^3)$

complexity to evaluate $\Delta_{\rm seff}$ [1, 2, 3], or we can use Kernel Polynomial Method (KPM) with $O(N^2)$ or $O(N)$ complexity to estimate the DOS directly [6, 7]. However, the latter one with $O(N)$ complexity must truncate the moment calculations. Here, we use the recently proposed GFMC method [8], as will be explained below for completeness.

2.2. Green-function-based Monte Carlo (GFMC) method

In this $O(N)$ GFMC method, we need to evaluate only several elements of the Green's function to calculate $\Delta_{\rm seff}$ given in (9), provided that the change of spin configuration $\{\vec{S}_i\} \to \{\vec{S}'_i\}$ is local, i.e, only a single spin at site i is changed with the rest of spins unaltered. We first give the definition of the Green's function $G(z)$ in the complex plane:

$$G(z) = \frac{1}{H - z}, \; z = E + i\epsilon, \tag{10}$$

where E is real and ϵ is a very small positive real. The Hamiltonian for a given spin configuration $\{\vec{S}_i\}$ ($\{\vec{S}'_i\}$) is denoted by H (H'), and the Hamiltonian difference Δ is thus $\Delta = H' - H$.

The determinate of $G(z)(H' - z)$, i.e.,

$$d(z) := \mathrm{Det}\left[G(z)(H' - z)\right] = \mathrm{Det}\left[G(z)(H - z) + G(z)\Delta\right] = \mathrm{Det}\left[\mathbb{1} + G(z)\Delta\right] \tag{11}$$

$$= \mathrm{Det}\left[G(z)\right]\left[(H' - z)\right] = \prod_i^N \frac{1}{\epsilon_i - z} \prod_i^N \epsilon'_i - z, \tag{12}$$

has a special role in the GFMC method, where ϵ_i and ϵ'_i are the i-th eigenvalues of H and H', respectively. It is readily shown that

$$\begin{aligned} \frac{1}{\pi} \lim_{\epsilon \to 0} \mathrm{Im} \frac{\mathrm{d}\log(d(z))}{dz} &= \frac{1}{\pi} \lim_{\epsilon \to 0} \mathrm{Im} \left(\sum_i^N \frac{1}{\epsilon_i - z} - \sum_i^N \frac{1}{\epsilon'_i - z} \right) \\ &= \sum_i^N \delta(E - \epsilon'_i) - \sum_i^N \delta(E - \epsilon_i) \\ &= \rho(E) - \rho'(E). \end{aligned} \tag{13}$$

Thus, this equation in the left hand side can be used in (9), and $\Delta_{\rm seff}$ is now described using $d(z)$:

$$\begin{aligned} \Delta_{\rm seff} &= \frac{1}{\pi} \lim_{\epsilon \to 0} \mathrm{Im} \int \log(1 + e^{-\beta[E-\mu]}) \frac{\mathrm{d}\log(d(z))}{dz} dE \\ &= \frac{\beta}{\pi} \int \frac{1}{1 + e^{\beta(E-\mu)}} \lim_{\epsilon \to 0} \mathrm{Im} \log(d(z)) dE. \end{aligned} \tag{14}$$

It is important to notice that we need only several elements of $G(z)$ to evaluate $d(z) = \mathrm{Det}\left[\mathbb{1} + G(z)\Delta\right]$. As an example, here we consider the simple cubic lattice with the nearest neighbor hopping, and we assume that only one spin at site o is changed: $\vec{S}_o \to \vec{S}'_o$. In the cubic lattice, site o has 6 nearest neighbors (NN) denoted by $\{n, e, s, w, t, b\}$. Then, the Δ matrix has a very simple form of

$$\Delta = \begin{bmatrix} 0 & 0 & 0 & \Delta_{n,o} & 0 & 0 & 0 \\ 0 & 0 & 0 & \Delta_{e,o} & 0 & 0 & 0 \\ 0 & 0 & 0 & \Delta_{s,o} & 0 & 0 & 0 \\ \Delta_{o,n} & \Delta_{o,e} & \Delta_{o,s} & 0 & \Delta_{o,w} & \Delta_{o,t} & \Delta_{o,b} \\ 0 & 0 & 0 & \Delta_{w,o} & 0 & 0 & 0 \\ 0 & 0 & 0 & \Delta_{t,o} & 0 & 0 & 0 \\ 0 & 0 & 0 & \Delta_{b,o} & 0 & 0 & 0 \end{bmatrix}. \tag{15}$$

Therefore, to evaluate $d(z)$, only the following 7×7 Green's functions have to be calculated

$$G = \begin{bmatrix} G_{n,n} & G_{n,e} & G_{n,s} & G_{n,o} & G_{n,w} & G_{n,t} & G_{n,b} \\ G_{e,n} & G_{e,e} & G_{e,s} & G_{e,o} & G_{e,w} & G_{e,t} & G_{e,b} \\ G_{s,n} & G_{s,e} & G_{s,s} & G_{s,o} & G_{s,w} & G_{s,t} & G_{s,b} \\ G_{o,n} & G_{o,e} & G_{o,s} & G_{o,o} & G_{o,w} & G_{o,t} & G_{o,b} \\ G_{w,n} & G_{w,e} & G_{w,s} & G_{w,o} & G_{w,w} & G_{w,t} & G_{w,b} \\ G_{t,n} & G_{t,e} & G_{t,s} & G_{t,o} & G_{t,w} & G_{t,t} & G_{t,b} \\ G_{b,n} & G_{b,e} & G_{b,s} & G_{b,o} & G_{b,w} & G_{b,t} & G_{b,b} \end{bmatrix}. \tag{16}$$

The computation can be further simplified by expanding $d(z)$ as the following:

$$\begin{aligned} d(z) &= \det(\mathbb{1} + \mathrm{G(z)}\Delta) && (17) \\ &= [1 + \sum_{j\in NN} \Delta_{jo} G_{oj}(z)][1 + \sum_{j \in NN} \Delta_{oj} G_{jo}(z)] && (18) \\ &- G_{oo}[\sum_{j,k\in NN} \Delta_{jo}\Delta_{ok} G_{kj}(z)], && (19) \end{aligned}$$

where

$$\Delta_{jo} = \langle j|\Delta|o\rangle, G_{oj} = \langle o|G|j\rangle. \tag{20}$$

Moreover, using the following state $|v\rangle$:

$$|v\rangle = \Delta|o\rangle = \sum_{j\in NN} \Delta_{jo}|j\rangle, \tag{21}$$

$d(z)$ can be compactly expressed as

$$d(z) = [1 + G_{ov}(z)][1 + G_{vo}(z)] - G_{oo}(z)G_{vv}(z). \tag{22}$$

Notice that we now need only a 2×2 Green's function to evaluate $d(z)$.

Now, a question is how to calculate efficiently the local Green's functions $G(z)$. For this purpose, we use the KPM [5], which can be efficiently implemented in a GPU cluster [10]. Using two types of Chebyshev polynomials (m: integer),

$$\begin{cases} T_m(x) = \cos\left[m \arccos(x)\right] \\ U_m(x) = \dfrac{\sin\left[(m+1)\arccos(x)\right]}{\sin\left[\arccos(x)\right]}, \end{cases} \tag{23}$$

the diagonal elements of the Green's function are expanded as

$$\begin{aligned} G_{ii}(\tilde{w}+i\epsilon) &= \frac{i}{\sqrt{1-\tilde{w}^2}}\left[\tilde{\mu}_0 + 2\sum_{m=1}^{M-1}\tilde{\mu}_m \mathrm{T_m}(\tilde{w})\right] + 2\sum_{m=1}^{M-1}\tilde{\mu}_m \mathrm{U_m}(\tilde{w}) \\ &= \frac{i}{\sqrt{1-\tilde{w}^2}}\left[\tilde{\mu}_0 + 2\sum_{m=1}^{M-1}\tilde{\mu}_m \exp\left[-im\arccos(\tilde{w})\right]\right]. \end{aligned} \tag{24}$$

Since the Chebyshev polynomials $T_m(x)$ and $U_m(x)$ requires that the argument x should be within $[-1, 1]$, we must renormalize the energy spectrum E to $\tilde{w}$. The $\tilde{\mu}_m$ represents the m-th moment (defined below) after applying a kernel function, $\tilde{\mu}_m = \mu_m g_m$, where g_m is the kernel function to eliminate Gibbs oscillations [5]. Here, we apply Lorenz kernel function which is defined as $g_m = \sinh[\lambda(1 - m/M)]/\sinh(\lambda)$ with appropriate choice of λ [5]. The m-th moment μ_m is defined as

$$\begin{aligned} \mu_m &= \frac{1}{\pi}\lim_{\epsilon\to 0}\mathrm{Im}\int G_{ii}(\tilde{w}+i\epsilon)T_m(\tilde{w})d\tilde{w} \\ &= \int \sum_n \langle i|n\rangle\langle n|i\rangle\delta(\tilde{w}-\tilde{\epsilon}_n)T_m(\tilde{w})d\tilde{w} \\ &= \sum_n \langle i|n\rangle\langle n|i\rangle T_m(\tilde{\epsilon}_n)dE \\ &= \sum_n \langle i|T_m(\tilde{H})|n\rangle\langle n|i\rangle \\ &= \langle i|T_m(\tilde{H})|i\rangle, \end{aligned} \tag{25}$$

where $\tilde{H}$ is the renormalized Hamiltonian to fit the spectra within $[-1, 1]$.

Defining $|a_m\rangle = T_m(\tilde{H})|i\rangle$ and thus $\mu_m = \langle i|a_m\rangle$, the Chebyshev series in (23) yields the recursive relation of these vectors $|a_m\rangle$, namely,

$$\begin{cases} |a_0\rangle = |i\rangle \\ |a_1\rangle = \tilde{H}|a_0\rangle \\ |a_m\rangle = 2\tilde{H}|a_{m-1}\rangle - |a_{m-2}\rangle \end{cases} . \tag{26}$$

In addition, the number M of coefficients is obtained with $M/2$ iterations if the following equation is used,

$$T_{2m-i} = 2T_{m-i}T_m - T_i,\ i = 0, 1 \tag{27}$$

$$\mu_{2m-i} = 2\langle r_{m-i}|r_m\rangle - \mu_i. \tag{28}$$

When the coefficients μ_m are all calculated, we can evaluate the Green's function using fast Fourier transform [5]. In (24), if we choose

$$\tilde{w} = \cos\frac{\pi(k+\frac{1}{2})}{M},\ (k = 0, 1, ..., M-1)$$

the Green's function becomes

$$G_{ii}(\tilde{w}+i\epsilon) = \frac{2i}{\sqrt{1-\tilde{w}^2}}\sum_{m=0}^{M-1}\mu'_m \exp\left[\frac{-im\pi(k+\frac{1}{2})}{M}\right], \tag{29}$$

where

$$\mu'_m = \begin{cases} \tilde{\mu}_0/2, \ m = 0 \\ \tilde{\mu}_m, \ m > 0 \end{cases}. \tag{30}$$

Let us now denote the summation part in (29) by χ_k:

$$\chi_k = \sum_{m=0}^{M-1} \mu'_m \exp\left[\frac{-im\pi(k+\frac{1}{2})}{M}\right].$$

It should be recalled that the following expression is required for the fast Fourier transformation [11]:

$$\gamma_n = \sum_{m=0}^{M-1} c_m \exp\left[\frac{-i2m\pi n}{M}\right], (m = 0, 1, \ldots, M-1) \tag{31}$$

where

$$c_m = \mu'_m \exp\left[\frac{-im\pi}{2M}\right] \tag{32}$$

Using the following correspondence between χ_j and γ_j

$$\begin{cases} \chi_{2j} = \gamma_j, \\ \chi_{2j+1} = \gamma^*_{M-j-1} \end{cases} \quad j = 0, 1, ..., M/2 - 1,$$

χ_j can be evaluated using the fast Fourier transformation, and thus the time complexity of calculating (29) reduces to $O(M\log(M))$, where M is the number of moments kept.

The off diagonal elements of the Green's function, G_{ov} and G_{vo}, can be evaluated similarly if we use the follow mixed elements of the Green's function (note that i below is the imaginary unit):

$$G_{o+v,o+v} = (\langle v| + \langle o|)G(|v\rangle + |o\rangle) = G_{oo} + G_{vv} + G_{ov} + G_{vo} \tag{33}$$
$$G_{o+iv,o+iv} = (-i\langle v| + \langle o|)G(|o\rangle + i|v\rangle) = G_{oo} + G_{vv} + iG_{ov} - iG_{vo}, \tag{34}$$

G_{ov} and G_{vo} are now expressed by the diagonal elements of the Green's functions:

$$G_{ov} = \frac{1}{2}[G_{o+v,o+v} - G_{oo} - G_{vv} + i(G_{oo} + G_{vv} - G_{o+iv,o+iv})], \tag{35}$$
$$G_{vo} = \frac{1}{2}[G_{o+v,o+v} - G_{oo} - G_{vv} + i(G_{o+iv,o+iv} - G_{oo} - G_{vv})]. \tag{36}$$

Using these four elements of the 2×2 Green's function, we can readily calculate $d(z)$ and thus Δ_{seff} can be evaluated.

Finally, it should be noted that if Hamiltonian matrix H is stored in a compression format, the time complexity of (26) is $O(NM)$, where N is the dimension of H. As mentioned above, the time complexity to calculate (24) is $O(M\log(M))$, and thus the total time complexity is expected to be $O(NM) + O(M\log(M))$. When M is fixed, the time complexity should scale linearly with the dimension of H. Indeed, we find, as shown in figure 1, that the execution time is approximately proportional to the Hamiltonian size N.

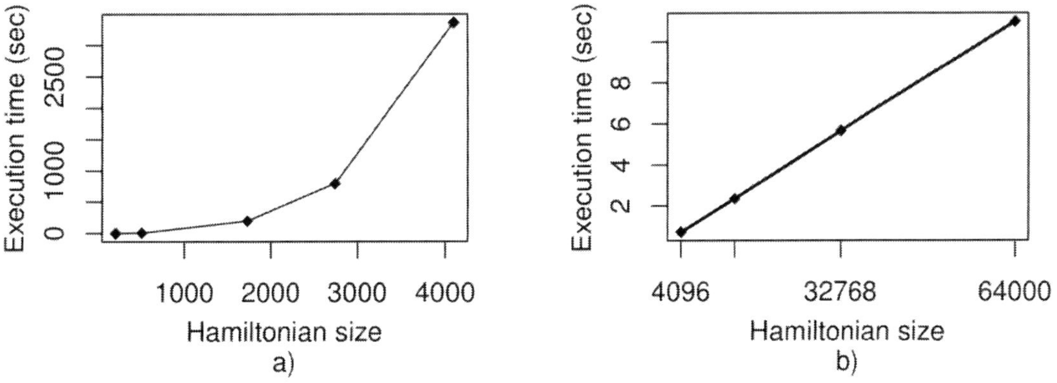

Figure 1. Hamiltonian size dependence of the execution time for (a) the exact diagonalization method [1, 2, 3] and (b)the GFMC method. Here, for simplicity, we performed only 10 Monte Carlo trial flips of spins, and M is fixed for different Hamiltonian sizes. We can clearly see that the time consumption for the exact diagonalization method roughly follow the complexity of $O(N^3)$, but for the GFMC method it is almost linear.

3. Implementation and Parallelization Schemes

3.1. Algorithms

For a given temperature T, we use Algorithm 1 to calculate the magnetization M for the classical spins through the average over S Monte Carlo sweeps (the loop from line 1), where each sweep corresponds to N spin trail flips (the loop from line 2). Since the direction of the magnetization is trivial, the magnetization M is defined here as the length of the total spin vector.

In the implementation, we apply the Metropolis method (line 10 in the Algorithm 1) to determine whether a trail flip is accepted by comparing with a random number between 0 and 1.

Section 2 has illustrated how to calculate the Δ_{seff} (from line 4 to 9) using the GFMC method. Especially, the calculation of the expansion coefficients μ_m plays a crucial role (line 6). This part occupies most of the execution time since the recursion [denoted by (26)] involves intensive matrix-vector multiplication (MVM) with complexity of $O(N)$. Note, however, that GPU is an ideal platform to parallelize MVM, because the multiplications between the rows and the vector could be distributed to hundreds of streaming processors. Therefore, we focus on a GPU implementation with the highly parallelism and expect a large speedup factor as compared with the CPU one.

3.2. Parallelization Frameworks

The magnetization M as a function of temperature T is obtained through the average over a large number of Monte Carlo sweeps. If the system stays in the thermal equilibrium, the Monte Carlo sweeps are independent with each other. Therefore, we could divide the outside loop S (line 1) into many threads. However, the warm up sweeps, which are necessary to achieve the thermal equilibrium, should be abandoned for taking the average, and this condition may prevent us from dividing S to too many threads because some threads may have too few sweeps to reach thermal equilibrium.

In addition to this parallelism for the Monte Carlo sampling, the intensive matrix-vector and vector-vector multiplications in line 6 can be effectively parallelized into multi-core CPU processors using OpenMP or many-core GPU processors using CUDA. In this paper, we focus on the implementation based on GPU to archive higher parallelism than multi-core CPU.

In our algorithm, we combine two parallelizations to achieve high efficiency, i.e., the Monte Carlo sweeps are divided into several MPI threads while in each thread we employ GPU to calculate matrix-vector operations.

Algorithm 1 Calculate the magnetization as a function of temperature

Require: Integer S to represent number of Monte Carlo sampling sweeps
Require: Hamiltonian matrix H of dimension $N \times N$
Require: Scalar M to store accumulated magnetization

```
for i = 1 → S do
  for j = 1 → N do
    Randomly choose site o and change the spin randomly
    Calculate the modification matrix Δ ← H − H' using (3)
    Get vector v⃗ ← Δ o⃗
    Calculate the coefficients μ̃_m^(o), μ̃_m^(v), μ̃_m^(o+iv), μ̃_m^(o−iv) applied Lorentz kernel function g_m,
    where μ̃_m^(V) = ⟨V|T_m(H̃)|V⟩ g_m
    Calculate 4 elements of the Green's function G_oo, G_vv, G_{o+iv,o+iv} and G_{o−iv,o−iv} using
    (24)
    Calculate d(z) using (22)
    Calculate Δ_seff using (14)
    if e^{−Δ_seff} > rand() then
      Accept the new spin configuration.
      Update H : H ← H + Δ
    end if
  end for
  M = M + |Σ_i^N S⃗_i| / N
end for
Magnetization ← M/S
```

3.3. Implementation on GPU

3.3.1. Compression Format of Hamiltonian Matrix Because the Hamiltonian matrix may be very large, e.g. $20^3 \times 20^3$, an effective compression technique must be introduced to reduce not only the memory consumption but also unnecessary memory access. The compression storage format of the Hamiltonian matrix may varies with different lattices and different physics models. For the simple cubic lattice used in this paper, each site has 6 nearest neighbors that lead to 6 non-zero entries in each row. This case is very suitable for ELLPack format [12] because no extra data need to be padded to the end of the rows. ELLPack format needs two matrices to store the non-zero entries and column indices, respectively. On GPU device, the two matrices are stored in column major to trigger coalesced memory access.

3.3.2. Arrangement of CUDA Kernel Functions One important issue that should be well addressed is the communication between GPU and CPU. Since the Hamiltonian matrix and the spin configurations may be updated after each iteration, if the Hamiltonian matrix are stored in CPU memory, the data transfer between CPU and GPU may significantly decrease the performance. This is especially true when the lattice size is small because the execution time for numerical calculation occupies a low percentage of the overall time consumption. According to our test, the simulation of $6 \times 6 \times 6$ cubic lattice on a 8x PCIe Tesla C2050 is about 30% slower than on a 16x PCIe C2050. Therefore, the data transfer may not only decrease the performance, but also make the program to be more dependent on the bandwidth between the CPU and GPU.

To avoid the memory transfer between CPU and GPU as much as possible, in our implementation the matrix H and the spin configurations are kept in GPU memory.

With these considerations, we propose a multi-kernel GPU implementation shown in the figure 2, in which the kernel functions are indicated by "<<<>>>" and arranged as the

```
for i = 1 → S
  for j = 1 → N
      // choose a spin randomly and flip a spin with a random direction
      // and calculate matrix Δ
      FlipSpinAndCalculateDelta <<<>>> ();
      // prepare vectors |o⟩, |v⟩, |o⟩ + i|v⟩, |o⟩ − i|v⟩
      PrepareVector <<<>>> ();
      // calculate the expansion coefficients for each Green's function
      for each vector in { |o⟩, |v⟩, |o⟩ + i|v⟩, |o⟩ − i|v⟩ }
          for m=0 to M/2
              // calculate 2 coefficients for each iteration
              CalculateCoefficient<<<>>> ();
          end for
      end for
      // transform (24) to meet the requirement of first Fourier transformation
      PrepareForFFT<<<>>> ();
      // perform first Fourier transformation to obtain 4 elements of Green's function

      PerformsFFT<<<>>> ();
      // Calculate Δseff
      CalculateDSEFF<<<>>> ();
      // If a trial spin flip is to be accepted, update the status
      UpdateStatus<<<>>> ();
      // Calculate the magnetization
      MeasureSpin<<<>>> ();
  end for
end for
```

Figure 2. Pseudocode of GPU implementation

following:

Function "FlipSpinAndCalculateDelta" flips a spin with a random direction and calculates the Hamiltonian matrix difference Δ, which is used in the function "PrepareVector" to create four vectors $\{|o\rangle, |v\rangle, |o\rangle + i|v\rangle, |o\rangle - i|v\rangle\}$. For each vector, the function "CalculateCoefficient" calculates two expansion coefficients μ_m in every iteration. After all the coefficients are ready, function "PrepareForFFT" prepares the data for fast Fourier transformation (FFT), including applying Lorentz kernel function and transform (24) to meet the requirement of FFT. After performing FFT by function "PerformsFFT", we obtain Δ_{seff} using function "CalculateDSEFF". If the flip is to be accepted, function "UpdateStatus" will update the spin configuration and the Hamiltonian matrix. After one MC sweep finishes, function "MeasureSpin" accumulates the spins to calculate the magnetization.

3.3.3. Implementation of the Recursion The most important kernel function is "CalculateCoefficient", which involves the most intensive computation due to the recursive matrix-vector operations. Figure 3 demonstrates the implementation of this CUDA kernel function to calculate the coefficients μ_{2m}. In CUDA architecture, the parallel running threads are divided into several thread groups, called "Blocks". For example, in figure 3 there are p thread blocks and each block has 4 threads, indexed from 0 to 3. Each thread performs the multiplication between

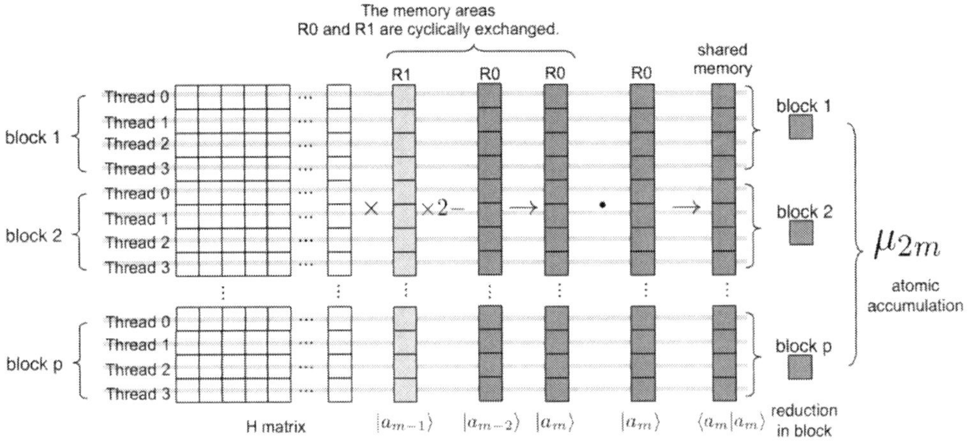

Figure 3. Implementation of function "CalculateCoefficient" to calculate the moments μ_{2m}

a row and the vector $|a_{m-1}\rangle$ that is stored in a 1D array R1. Since the produced vector $|a_m\rangle$ could be placed on R0 overwriting the vector $|a_{m-2}\rangle$, we only need two 1D memory arrays, R0 and R1, to perform calculations by exchanging their pointers after each iteration.

After the vector $|a_m\rangle$ is ready, a production $\langle a_m|a_m\rangle$ is performed, shared memory is applied to store the product result. $\langle a_m|a_m\rangle$ is a dot product, for CUDA, it is a reduction problem, which will be carried out in two steps, first, the product result in each block is accumulated to produce a scalar and then the scalar in each block is further accumulated to get coefficient μ_{2m} using (28).

This implementation could eliminate the most data transfers during the calculation. The bandwidth test shows the performance difference between a 8x and 16x PCIe based Tesla C2050 is very small (less than 6%) while calculating a 6^3 cubic lattice.

4. Validation

In order to check the availability of our implementation, The DE model on the simple cubic lattice is simulated. The magnetization M as a function of temperature T is examined and the results are shown in figure 4. The simulations are performed with the following parameters: Chebyshev truncation number is 256, chemical potential μ is 0, the average is taken over 5000 MC sweeps.

5. Performance evaluation

5.1. Experimental Setup

The performance evaluation is performed on a cluster system that has 16 nodes of the NEC LX series connected with 40GByte/sec Infiniband network via the switch. Each node has two NVIDIA Tesla M2050 GPUs connected to the separate PCI Express buses and 12 CPU cores (Xeon E5645 2.4GHz). The OS of the node is CentOS 6 and the graphics driver with CUDA 4.0 is installed. The compiler for CPU is gcc-4.4.4. Two CPUs are tightly coupled with sharing the memory bus (totally 12 CPU cores in a node). Amount of total memory is 12GBytes per node. All experiments performed below use the compiler option "-O3" for CPU and GPU programs. A head node of the cluster is also connected to the cluster via the Infiniband network to serve the home file system in order to share the program resources.

We evaluate the performance in two ways. The first is to measure the performance of a single CPU core and a single GPU with different Hamiltonian matrix sizes. The second is to evaluate

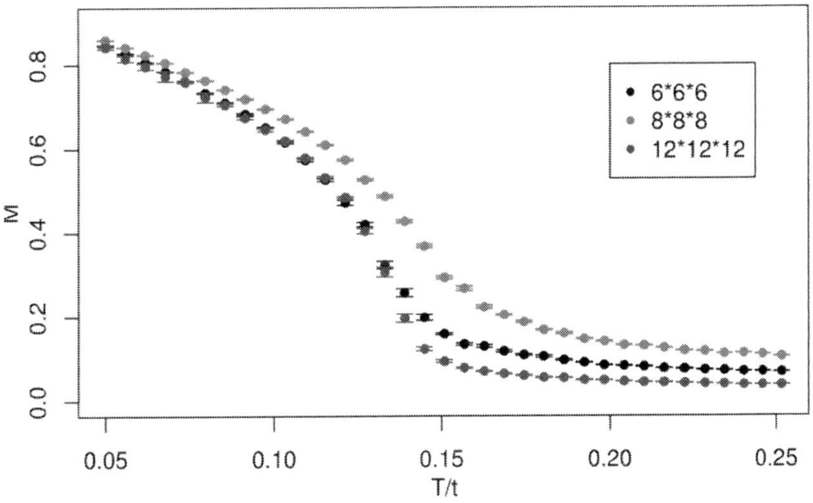

Figure 4. Magnetization M as a function of temperature T for different sizes (indicated in the figure) of the simple cubic lattice.

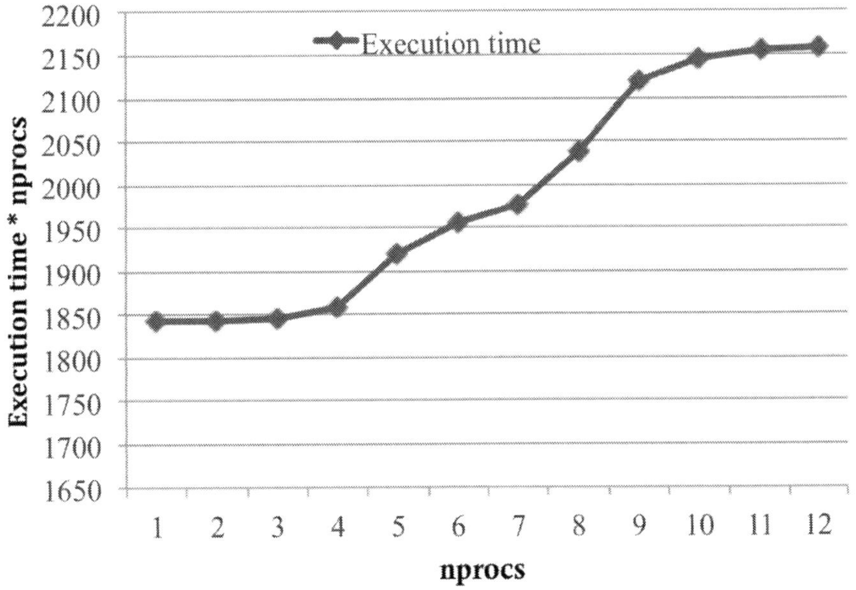

Figure 5. The performance scaling on multiple CPU cores within a node

the overall performance comparison between all CPU cores and all GPUs.

5.2. Performance Scaling on Multi-core CPU

Section 3.1 has explained that the process of calculating coefficients $\tilde{\mu}$ involves intensive MVM operations. The performance of MVM largely depends on the memory access speed. Therefore, the memory bandwidth plays an essential role in the overall performance. This conclusion is consistent with our multi-core CPU performance scaling test shown in figure 5.

This test is performed on a single node that has 2 CPUs, totally 12 cores and 12GB DDR2 800

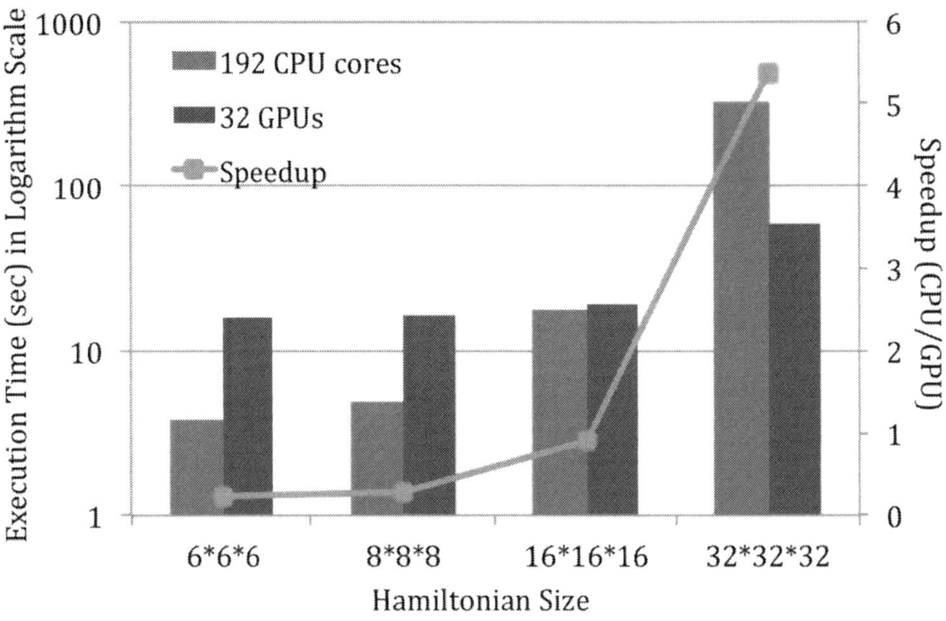

Figure 6. Performance scaling with Hamiltonian matrix size on cluster, the result is obtained with 38400 trail flips calculated and Chebyshev truncation number kept as 256

memory with a peak bandwidth of 12.5GB/s. In order to evaluate the parallelization efficiency, we examined the quantity expressed by $y = execution\ time \times nprocs$, in which $nprocs$ represents the number of MPI threads. As the outside loop [from line 1] in Algorithm 1 is divided into multiple MPI threads and there is very little communications among these threads, in an ideal case of parallelization, y should be a constant. However, we have noticed that when the threads number becomes larger than 4, y begins to increase fast, suggesting that the efficiency become to decrease rapidly.

It indicates that the memory bandwidth becomes saturated when $nprocs$ grows to be larger than 4. Therefore, we can conclude that the bottleneck of this method is memory bandwidth. In this paper we resort to solve the problem using Tesla C2050 GPU with maximum bandwidth of 144GB/s [9].

5.3. Performance Scaling for Increasing Hamiltonian Size on Cluster

Figure 6 shows the performance scaling on the cluster while the Hamiltonian size is increasing from 6^3 to 32^3. 38400 Monte Carlo trail flips are calculated while the number of Chebyshev moments are kept as 256. It can be observed that when the H matrix is small, e.g. 6^3 or 8^3, 192 CPU cores are much faster than 32 GPUs. To explain the reason, let us divide the total time consumption of GPU program into two parts: a) the time for numerical calculation and b) the time for other works such as initializing GPU device, copying memory between CPU and GPU, and MPI communication, etc. Usually the latter part consumes very little time, but when the matrix size is small, comparing with the first part, time consumption of the latter part can not be neglected and significantly decrease the performance. However, the speedup increases rapidly with increasing H matrix size and in the best case, i.e. 32^3, 32GPU is about over than 5 times faster than 192 CPU cores. If we define the speedup factor as the equivalent number of CPU cores comparing with one GPU. For Hamiltonian with size of 32^3, this factor is about 30 (192/32 * 5).

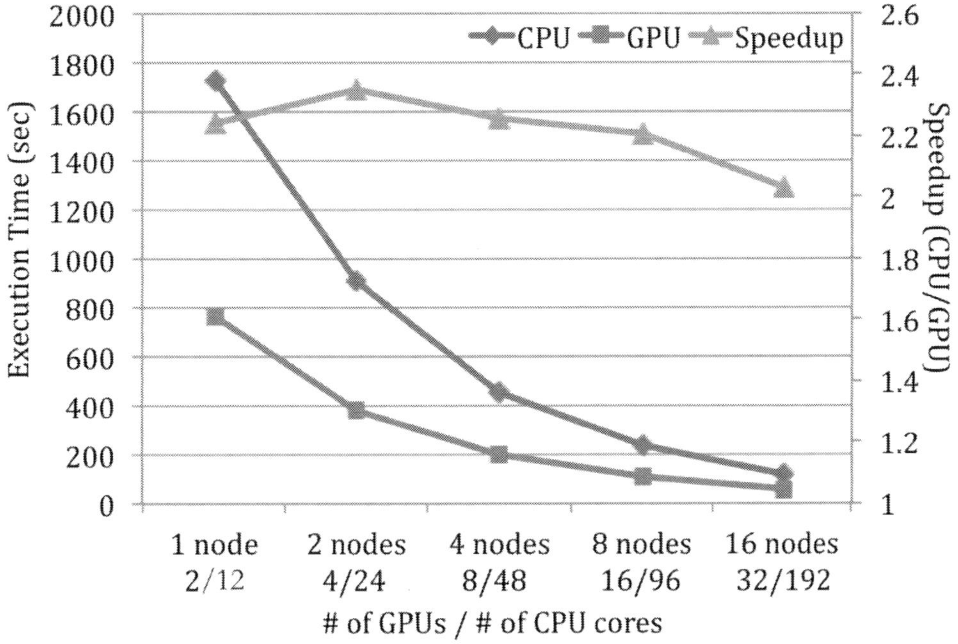

Figure 7. Performance scaling for increasing number of nodes, each node has 2 Tesla C2050 and 12 Xeon 2.4GHz cores. 8000 Monte Carlo trail flips of a 20^3 cubic lattice Hamiltonian are calculated with the Chebyshev truncation number kept as 256. The blue/green lines represent time consumption of CPU cores/GPUs

5.4. Performance Scaling for Increasing Number of Nodes

The performance scaling for increasing number of nodes is also very important reference for implementing the algorithm on a large cluster or even supercomputer. Here 8000 trail flips of a 20^3 cubic lattice Hamiltonian is calculated on our cluster that has 2 Tesla C2050 and 12 Xeon 2.4GHz cores in each node. The performance evaluation is shown in figure 7. As there is very little communication among MPI threads, it can be noticed that for both CPU and GPU implementation, the time consumption always decreases by half when the number of nodes is doubled , suggesting an ideal performance scaling with number of nodes.

5.5. Performance Considerations

As shown in figure 5, the bottleneck of this algorithm is the memory bandwidth, GPU could solve the memory bandwidth limitation very well and shows high speedup ratio. Due to the nature that there is little communication among MPI threads, this algorithm is also capable to run on many compute nodes while keeping good parallelization efficiency.

However, due to the thermal equilibrium problem explained in section 3.2, the of number of MPI threads is limited, in this case, a combination of parallelization techniques such as MPI, OpenMP and CUDA should be introduced. For example, the workload is firstly distributed into different nodes using MPI, and then in each node we use OpenMP to calculate the MVM, if the CUDA is capable, the OpenMP thread may employ GPU to calculate the MVM. The combination can increase the parallelism and gain high performance. However, it should be pointed out that the process to combine these techniques may not be very straightforward since we have to take trade-offs in many occasions.

6. Conclusion and Future Work

In this paper, we have provided the detailed formulation of the GFMC method for the DE model. We have proposed implementation methods to parallelize the GFMC method using MPI on CPU as well as CUDA on GPU. In order to eliminate the data transfer between CPU and GPU as much as possible, the program is implemented on GPU using several kernel functions. The performance evaluation indicates that the GPU implementation could effectively overcome the CPU memory bandwidth limitation and shows more than 30 speedup factor when Hamiltonian size is 32^3. The performance scaling test for increasing number of processors indicates that the MPI parallelization is very effective for this algorithm. Finally we discussed more considerations on the performance for future optimizations.

For the future work, we will implement this method as a GPU based library that can be used for other lattice models.

Acknowledgment

This work is partially supported by the Japan Science Technology Agency (JST) PRESTO program. And also this work is partially supported by KAKENHI (24300020) Grant-in-Aid for Scientific Research (B). We also would like to thank Dr. Alexander Weisse for insightful comments and suggestions to help implement the GFMC method.

References

[1] Yunoki S, Hu J, Malvezzi A L, Moreo A, Furukawa N and Dagotto E 1998 *Phys. Rev. Lett.* **80**(4) 845–848

[2] Yunoki S and Moreo A 1998 *Phys. Rev. B* **58**(10) 6403–6413

[3] Dagotto E, Yunoki S, Malvezzi A L, Moreo A, Hu J, Capponi S, Poilblanc D and Furukawa N 1998 *Phys. Rev. B* **58**(10) 6414–6427

[4] Moreo A, Yunoki S and Dagotto E 1999 *Science* **283** 2034–2040

[5] Weiße A, Wellein G, Alvermann A and Fehske H 2006 *Review of Modern Physics* **78** 275–306

[6] Motome Y and Furukawa N 1999 *Journal of the Physical Society of Japan* **68** 3853

[7] Furukawa N and Motome Y 2004 *Journal of the Physical Society of Japan* **73** 1482

[8] Weiß e A 2009 *Physical Review Letters* **102** 17–20 ISSN 0031-9007

[9] NVIDIA Corporation URL `http://developer.nvidia.com/cuda`

[10] Zhang S, Yamagiwa S, Okumura M and Yunoki S 2012 *International Journal of Parallel Programming* 1–30 ISSN 0885-7458

[11] Duhamel P and Vetterli M 1990 *Signal Processing* **19** 259 – 299 ISSN 0165-1684

[12] Grimes R, Kincaid D and Young D 1979 *ITPACK 2.0 User's Guide* (Technical Report CNA-150, Center for Numerical Analysis, University of Texas)

24th IUPAP Conference on Computational Physics (IUPAP-CCP 2012) IOP Publishing
Journal of Physics: Conference Series **454** (2013) 012050 doi:10.1088/1742-6596/454/1/012050

Cluster structure in Monte Carlo shell model

T Yoshida[1], **N Shimizu**[1], **T Abe**[2] **and T Otsuka**[1,2,3]

[1]Center for Nuclear Study, University of Tokyo, 7-3-1 Hongo, Bunkyo, Tokyo 113-0033, Japan.
[2]Department of Physics, University of Tokyo, 7-3-1 Hongo, Bunkyo, Tokyo, 113-0033, Japan.
[3]National Superconducting Cyclotron Laboratory, Michigan State University, East Lansing, Michigan, USA

E-mail: yoshida@cns.s.u-tokyo.ac.jp

Abstract. The wave function of ^{8}Be, which is obtained from the Monte Carlo shell model (MCSM), is discussed. A method to define an intrinsic state in the MCSM is proposed. The appearance of two-α-cluster structures in ^{8}Be is demonstrated.

1. Introduction

One of the purposes of theoretical nuclear physics is to understand the physical observables of nuclei in the large region of the nuclear chart. So far, nuclear density functional theory is the only method able to describe them across the whole nuclear chart. If we go beyond the theory in terms of the accurate description for the excited states and the reliability for nuclear interactions, we need to select a specific model to understand the properties of nuclei at each region. The shell model calculation is usually suited for the study of the medium mass region. In the large region of the nuclear chart, the shell model picture is essential to understand the nuclear magic numbers related to the single-particle motion of a nucleon.

The shell model calculation enables us to take all relevant degrees of freedom of nucleon motion into account when the particles are set to be in the valence shell on top of the core nucleus. For the lightest part of the nuclear chart (mass number lower than $A = 14$), shell model calculations without assuming an inert core can be performed. The model space without the core is important for the accurate prediction of the properties of light nuclei.

On the other hand, α cluster configuration exists in light nuclei, which has an important role at the α-breakup-threshold-energy region. In this picture, the four nucleons (spin-up proton, spin-down proton, spin-up neutron and spin-down neutron) become an ingredient of the subsystem, ^{4}He (α particle). The subsystem is so stable that one α cluster has possibility to exist separately from the remaining nucleus especially for the excited states near α-breakup thresholds.

The appearance of α cluster structure has been studied for a few decades. In the early stage of the study, phenomenological (*e.g.* α cluster model) calculations have been performed [1]. Recently, *ab initio* calculations have been started for this region. One example is the Green's function Monte Carlo (GFMC) [2] approach and another is the lattice effective field theory [3, 4]. In their analyses of $A = 8$ and 12 nuclei, the two and three-α-cluster shapes emerge in *a priori*. It is important to confirm the existence of the cluster configuration by calculating the density distribution of ^{8}Be from shell-model side. However, it is difficult to discuss α-cluster configuration in view of intrinsic shape utilizing shell model calculations. This is because the single-particle orbit of the shell model is expanded around the one center of the harmonic oscillator potential.

Recently, the size of the model space which is tractable in the shell model calculation has been increased due to the development of parallel computers and methodology of shell model calculations. Therefore, it is natural to expect that α-cluster configurations are included in the single-particle orbit of the shell model if the model space is large enough. In fact, the deformation and cluster-like shape for Li isotopes have been investigated by using the no core full configuration (NCFC) approach [5].

Another approach is the Monte Carlo shell model (MCSM) [6] method which has been applied to the various nuclei by using modern parallel computers with sophisticated energy-minimum search where the conjugate gradient method and energy-variance extrapolation [7, 8, 9] are combined. The MCSM is expected to be applicable toward the large region of the nuclear chart both with and without the inert core (no-core MCSM). In this work, we show the density profiles of the wave function obtained in the no-core-MCSM calculation. For this purpose, we define the intrinsic state from the no-core-MCSM wave function in the following section. We focus on the ground state of ^{8}Be which is expected to have the two-α-cluster configuration.

2. Formulation

In this section, we show how to obtain the intrinsic structure for ^{8}Be. In order to investigate the cluster-like state, it is important to extract the intrinsic state from the wave function whose angular momentum is a good quantum number in the no-core MCSM.

In the no-core-MCSM calculation, the wave function is projected to the state which has a good angular momentum quantum, J. We use a Slater determinant, $|\phi_n\rangle$, to describe a basis state which is defined as

$$|\phi_n\rangle = \prod_{\alpha=1}^{N_p} \sum_{i=1}^{N_{sp}} c_i^\dagger D_{i\alpha}^{(n)} |-\rangle, \tag{1}$$

where $|-\rangle$ is a vacuum state and $c_i^\dagger$'s are creation operators of the single-particle orbit, i. It is specified by coefficients of the harmonic oscillator state. The resultant wave function is given by a superposition of the angular-momentum-projected Slater determinants,

$$|\Psi\rangle = P_{MK}^{J\pi}|\Phi\rangle, \quad |\Phi\rangle = \sum_{n=1}^{N_b} f_n|\phi_n\rangle. \tag{2}$$

In this study, we use the values $\pi = +$, $J = 0$, $M = 0$ and therefore K-quantum number $K = 0$. The coefficients f_n's are determined by diagonalizing the Hamiltonian matrix, $\langle\phi_i|H|\phi_j\rangle$. In the no-core-MCSM procedure, we tune the value of $D^{(n)}$ for each Slater determinant by the conjugate gradient method in order to minimize the expectation value of the Hamiltonian with J-projected wave functions. Here, we use the JISP16 NN interaction which reproduces NN scattering data and properties of deuteron and other light nuclei [10].

In this J-projected wave function, the cluster-like structure is difficult to be visualized. To analyse the wave function in detail, we go back to the wave function before the angular momentum projection, $|\Phi\rangle$, which is the linear combination of the unprojected basis states. However, this state is not considered as an intrinsic state because the principal axis of a basis state, $|\phi_i\rangle$, is not necessarily in the same direction with that of other basis states. The directions are rather randomly distributed after the energy-minimum search in the MCSM procedure. As a result, intrinsic structures such as cluster configurations may be hidden.

In order to extract the wave function of the intrinsic state, we rotate each basis state so that its diagonalized Q-moment is aligned along the z-axis, following the concept of the intrinsic state in the GFMC [2]. The intrinsic wave function $|\Phi^{\rm intr}\rangle$ is defined as

$$|\Phi^{\rm intr}\rangle \equiv \sum_n f_n R(\Omega_n)|\phi_n\rangle = \sum_n f_n|\phi_n^R\rangle, \quad (3)$$

with the transformed basis state $|\phi_n^R\rangle$. Here, $R(\Omega_n)$ is the rotation operator with an Euler angle Ω_n, which is determined by the condition, $\langle\phi_n^R|Q_{i,j}|\phi_n^R\rangle = \delta_{i,j}q_i$, where $Q_{i,j(=x,y,z)}$ are the operators of Q-moment. We note that this state, $|\Phi^{\rm intr}\rangle$, exactly has the same energy after the angular momentum projection. After that, it is probable that these aligned basis states have large overlap with each other and make a distinct principal axis toward the z-axis when the condition $q_z > q_y > q_x$ is set. The intrinsic density is obtained by the expectation value of the one-body-density operator,

$$\rho^{\rm intr}(r) = \langle\Phi^{\rm intr}|\sum_i \delta(r-r_i)|\Phi^{\rm intr}\rangle, \quad (4)$$

where r_i denotes the position of the i-th nucleon.

3. Results

Figure 1 shows the proton densities for the 0^+ ground state of ^{8}Be. Here, the densities of the wave function Φ (before alignment) are shown in the left panel. The intrinsic densities of the wave function $\Phi^{\rm intr}$ are shown in the right panel. As shown in the results of $N_b = 1$, a clear deformation and its neck structure to be called dumbbell shape appear. We can see that as the number of the Slater determinants (N_b) increases, the densities before alignment become vaguer because of the mixture of different directions of principal axes of the basis states. On the other

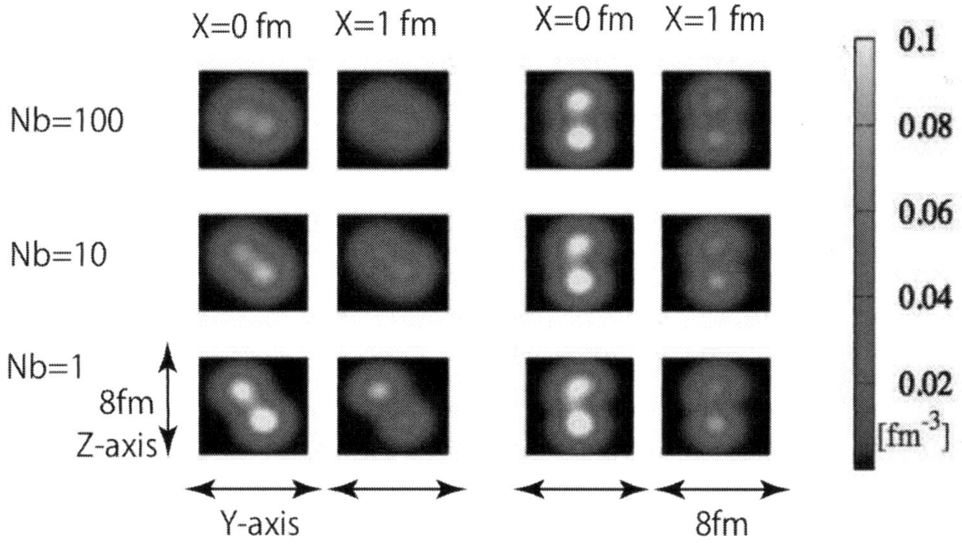

Figure 1. ^{8}Be proton densities before alignment of principal axes of unprojected basis states, $|\Phi\rangle$ (left panel) and intrinsic state, $|\Phi^{\rm intr}\rangle$ (right panel) for various N_b and slices along yz plane. The numbers of basis states are $N_b = 1, 10$ and 100 for the lower, middle and upper figures, respectively. Slices along yz plane are $x = 0$ fm plane (left) or $x = 1$ fm plane (right) for each panel. The size of each box is 8 fm $\times$ 8 fm.

hand, the intrinsic densities have clearer dumbbell-like structure for each N_b. In addition, the density distribution of the intrinsic state is almost unchanged with respect to N_b. This result indicates the appearance of cluster structures. We see that the neck of the dumbbell shape is more enhanced in $N_{shell} = 4$ than in $N_{shell} = 3$. Since the number of particles appeared in the density distribution for opposite sides of the principal axis are almost the same, this state can be considered as two α clusters. This result is similar to that of the GFMC [2].

In summary, we analyse the density distribution of the 0^+ ground state of ^{8}Be. We use a method by which one can extract the intrinsic structure from the wave function in the no-core MCSM. We show that the neck shape indicating the two α clusters emerges within a relatively small number of the harmonic oscillator quanta. We notice that the shape which appears in the no-core MCSM has the similar property in the GFMC in terms of the density. However, it is necessary to clarify - in future studies - whether the present analysis is valid with larger model spaces, with Coulomb interaction and without contamination of spurious center of mass motion.

Acknowledgments
This work has been supported by the SPIRE Field 5 from MEXT and the CNS-RIKEN joint project for large-scale nuclear structure calculations. The numerical calculations were performed on the T2K Open Supercomputers at the University of Tokyo and Tsukuba University.

References
[1] Fujiwara Y, Horiuchi H, Ikeda K, Kamimura M, Katō K, Suzuki Y and Uegaki E 1980 Prog. Theor. Phys. Suppl. **68** 29
[2] Wiringa R B, Pieper S C, Carlson J and Pandharipande V R 2000 Phys. Rev. C **62** 014001
[3] Epelbaum E, Krebs H, Lee D and Meißner U 2011 Phys. Rev. Lett **106** 192501
[4] Epelbaum E, Krebs H, Lahde T A, Lee D and Meißner U 2012 Phys. Rev. Lett **109** 252501
[5] Cockrell C, Vary J P and Maris P 2012 Phys. Rev. C **86** 034325
[6] Abe T, Maris P, Otsuka T, Shimizu N, Utsuno Y and Vary J P 2012 Phys. Rev. C **86** 054301
[7] Otsuka T, Honma M, Mizusaki T, Shimizu N and Utsuno Y 2001 Prog. Part. Nucl. Phys. **47** 319
[8] Shimizu N, Abe T, Tsunoda Y, Utsuno Y, Yoshida T, Mizusaki T, Honma M and Otsuka T 2012 Prog. Theor. Exp. Phys. **2012** 01A205
[9] Abe T, Maris P, Otsuka T, Shimizu N, Tsunoda Y, Utsuno Y, Vary J P and Yoshida T contribution to the CCP2012 proceedings
[10] Shirokov A M, Vary J P, Mazur A I, Zaytsev S A and Weber T A 2005 Phys. Lett. B **621** 96

Computational particle physics for event generators and data analysis

Denis Perret-Gallix

LAPP (IN2P3/CNRS), France

E-mail: denis.perret-gallix@in2p3.fr

Abstract. High-energy physics data analysis relies heavily on the comparison between experimental and simulated data as stressed lately by the Higgs search at LHC and the recent identification of a Higgs-like new boson. The first link in the full simulation chain is the event generation both for background and for expected signals. Nowadays event generators are based on the automatic computation of matrix element or amplitude for each process of interest.

Moreover, recent analysis techniques based on the matrix element likelihood method assign probabilities for every event to belong to any of a given set of possible processes. This method originally used for the top mass measurement, although computing intensive, has shown its efficiency at LHC to extract the new boson signal from the background.

Serving both needs, the automatic calculation of matrix element is therefore more than ever of prime importance for particle physics. Initiated in the 80's, the techniques have matured for the lowest order calculations (tree-level), but become complex and CPU time consuming when higher order calculations involving loop diagrams are necessary like for QCD processes at LHC. New calculation techniques for next-to-leading order (NLO) have surfaced making possible the generation of processes with many final state particles (up to 6). If NLO calculations are in many cases under control, although not yet fully automatic, even higher precision calculations involving processes at 2-loops or more remain a big challenge.

After a short introduction to particle physics and to the related theoretical framework, we will review some of the computing techniques that have been developed to make these calculations automatic. The main available packages and some of the most important applications for simulation and data analysis, in particular at LHC will also be summarized (see CCP2012 slides [1]).

1. Particle Physics goals and means

Particle physics targets the study of the ultimate basic elements of matter and of the fundamental forces generated by or acting on them; the smaller the element is, the smaller the probe wavelength must be and, therefore, the larger the interaction energy. Actually, particle physics or more precisely its high-energy physics branch tends to reproduce on earth the range of energies that was prevalent when matter did start to form, at a very early time of the Universe history (figure 1), namely some 10^{-10} s after the Big Bang. There is a profound duality between the today state of matter probed by high-energy colliders and the state of the Universe shortly after its creation, 13.7 billion years ago.

The highest man-made particle colliding energy reaches, today, 8 TeV (Tera Electron Volt ~ thousand times the proton mass) on the way to the nominal 14 TeV expected at the large hadron collider (LHC [2]) at CERN [3], Geneva (Switzerland). This energy, however, remains many orders of magnitude smaller than the typical interaction energy occurring in the Universe when quarks, gluons and electrons were moving in a confinement free "hot soup". From this even earlier Universe energy we only have a glimpse, today, through the very high-energy cosmic rays (more than 10^8 TeV) hitting the upper earth atmosphere. Producing huge particle showers they are detected by large area detectors on earth surface like at the 3000 km^2 Pierre Auger Observatory [4] or observed by satellite surveying the earth as proposed by the GEM-Euso Collaboration [5].

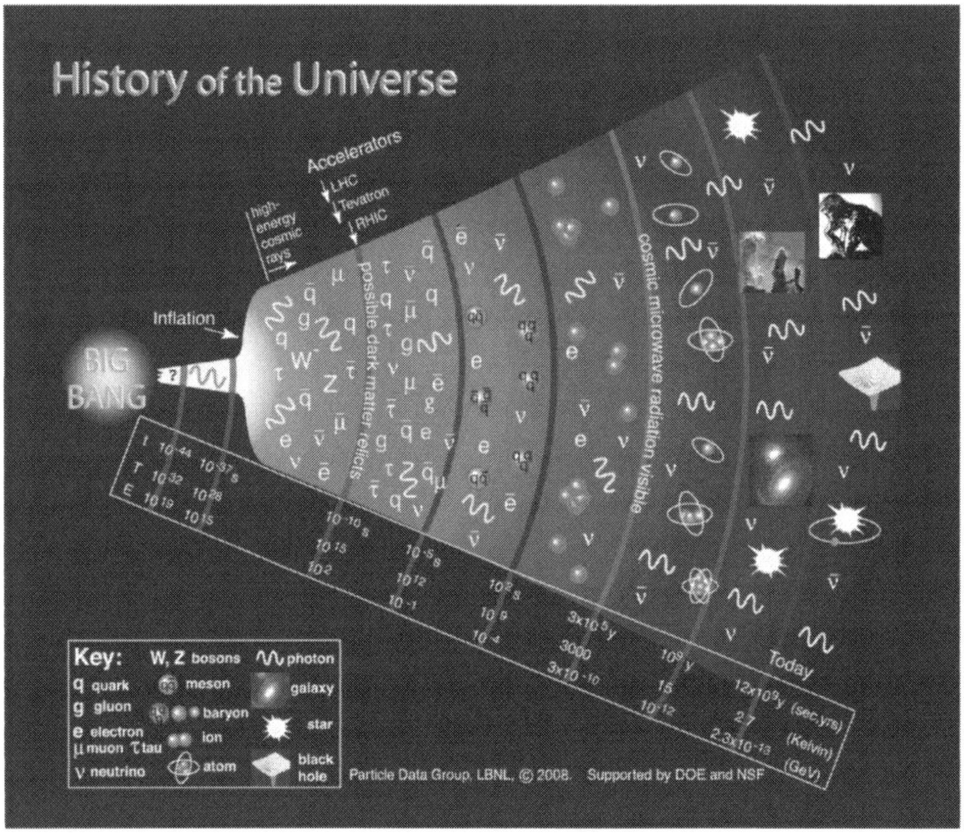

Figure 1. History of the Universe.

Located at CERN, the LHC (figure 2) is the latest built proton-proton collider. It is housed in a 27 km circular, 100 m deep underground tunnel in the Geneva region. The LHC benefits from a large accelerator infrastructure progressively installed at CERN since it was founded back in the 50's. The proton bunches produced from a hydrogen source are accelerated by the Linac and Booster. Then, the PS and SPS increase the energy up to 450 GeV before entering the LHC. Four experiments are located in the LHC ring: two general purpose ATLAS [6] and CMS [7] and two dedicated LHCb [8] for the meson B physics and ALICE [9] for hadronic physics when protons are replaced by heavy ions (e.g. lead-lead collision) in the LHC.

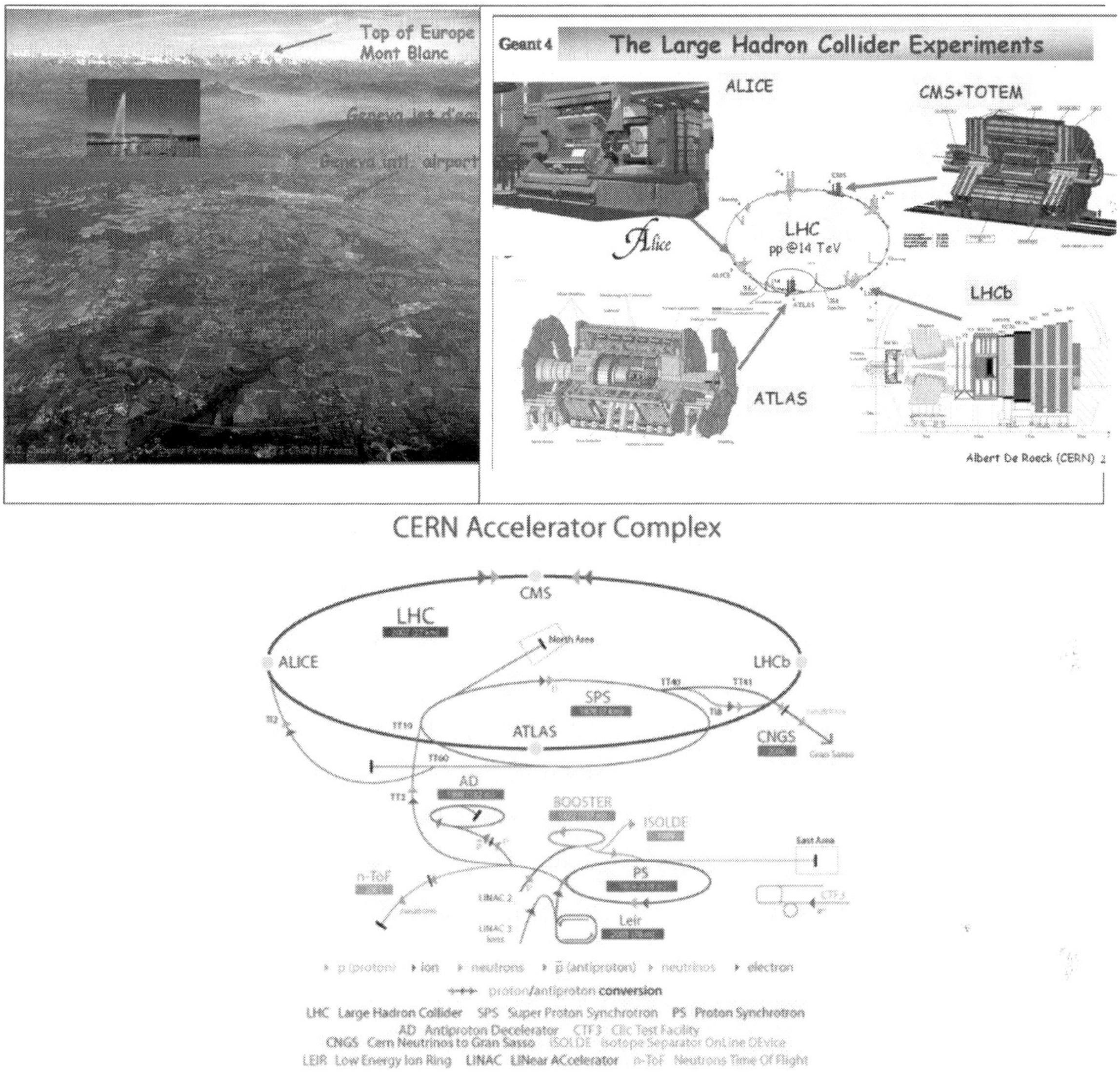

Figure 2. LHC and experiments at CERN.

The particle physics overall picture is now embedded in a very solid theory that, shyly enough, we keep calling a Model. The Standard Model (SM) is maybe one of the most developed and rugged theory of any physics fields left alone other research domains. This is probably due to the fact that dealing with fundamental processes, generally, does not lead to the complexity of n-body problems generally intractable, although supercomputers are opening new hopes, at least for small n.

But that does not mean that the theory is simple when it comes to numerical evaluations. Fortunately methods have been developed based on the underlying quantum field theory to make precise computations. Until now, the experimental data have always either confirmed the theoretical predictions or brought new insight reinforcing the general view.

This success story seems to continue with the recent discovery of a "predicted" boson at the CERN LHC, last July (ATLAS [10], CMS [11]). Nobody yet dares calling it "Higgs", the last missing ingredient of the SM at current scales. We will soon learn from the experiments whether or not this new boson is the actual SM Higgs proposed as a consequence of a mechanism giving mass to elementary particles [12] [13] [14] [15].

This discovery is materialized by the small bump popping out of the recombined $\gamma\gamma$ mass histograms (figure 3) for both ATLAS [10] and CMS [11]. These simple plots summarize the extensive computational undertaking behind high-energy physics data analysis. For example, in the CMS experiment, the dots and the error bars represent the observed number of events in a given mass bin. The numbers come from a long chain of data analysis beginning with the trigger event selection, followed by the event reconstruction, the event filtering and analysis for the millions of events recorded by the experiments.

The yellow/green bands represent the best estimation of the so-called background events. They are based on a fit of the data points away from the bump. This fit is also compared to simulated events belonging to background processes. In red, the sum of background and simulated Higgs signal is shown. It is a clear demonstration of the interplay between experimental and simulated data.

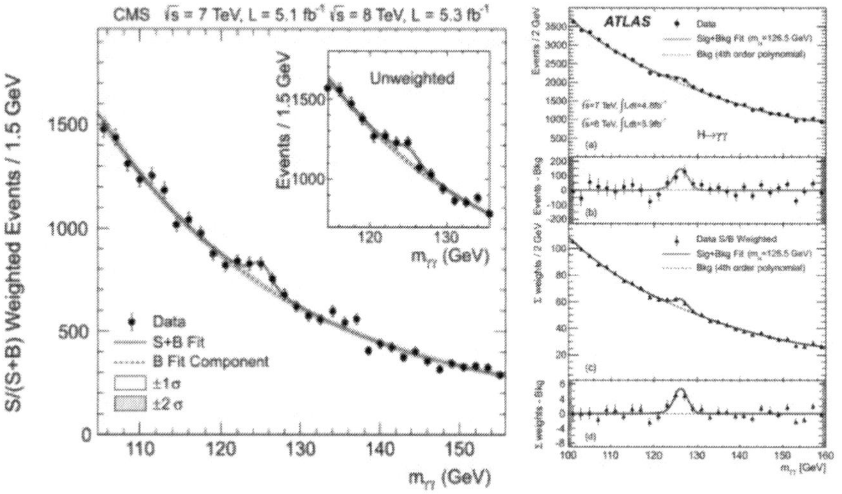

Figure 3. Higgs-like boson signal from CMS and ATLAS (CERN).

Most of the data simulation and analysis is performed on a distributed data grid system called the WLCG (Worldwide LHC Computing Grid) [16]. This infrastructure based on tiers computing centres has been installed all around the world to absorb the heavy computing load of the high LHC statistics. The tier 0 is located at the data production centre (CERN) and absorbs 20% of the load, 11 tiers 1 connected through direct high bandwidth fibre links (10 Gb/s) are scattered across the main participating countries/regions and ~ 140 tiers 2 are providing the specific computing support necessary to research teams. The WLCG provides more than 250,000 processors, and close to 150 PB of disk storage from over 150 sites in 34 countries, producing a massive distributed computing infrastructure that support the needs of more than 8,000 physicists.

In section 2, we will describe the general theoretical framework. Section 3 is a short presentation of the perturbative calculus and of the recent higher precision calculation breakthroughs. Then, in the following sections, like Russian dolls, we will go deeper in the experimental simulation chain (section 4): with the description the event formation phases (section 5), of the various steps leading to the building of event generator (section 6), including the automatic calculation of matrix elements (section 7) and their multi-dimensional integration giving the cross-sections (section 8). Section 9 describes the more recent usage of the same matrix element expressions for a more complete event analysis. Finally, section 10 lists some of the computational techniques that are used and shared with other research fields.

2. Perturbative and not perturbative? That is the ... question.

Let us first replace this computational activity in the general framework of particle physics and its Standard Model. There are basically two domains of calculation depending on the actual value of the

QCD[1] coupling constant α_s. This constant is actually "running" from small values close to zero in the "asymptotic freedom" regime, namely at high energy when probing the deep inside of hadrons, to large values ($\alpha_s \sim 1$) in the "confinement" region where the partons (quarks and gluons) are kept inside the hadrons.

If α_s is small when, for example, 2 partons, almost free inside the protons, interact, the solutions of the Schrödinger equation which represents the time evolution of the quantum state can be approximated by a series expansion in power of α_s. This is the so-called perturbative domain. Each term of the series corresponds to a level of precision, the higher the number of terms the more precise the calculation. We will see later also that each term of the series can be graphically represented by a set of "Feynman" diagrams [17], all having similar characteristics. This domain describes the dynamics of QCD interactions. This is the branch of computational particle physics that will be discussed in this paper.

Conversely, if α_s is large ~ 1, the hadron quantum system is probed at the confinement limit. Quarks and gluons are all strongly coupled and cannot escape the hadron boundary. Here the interest is more focused on the hadron as a whole. Global values like hadron masses are computed. The perturbative approximation does not apply though and the Schrödinger equation must be solved exactly. However, this has proved to be analytically unfeasible despite many years of trials until K.G. Wilson [18] proposed to replace the continuum phase space by a discrete approximation, the size of the lattice introducing a cut-off stabilizing the solutions. It is only these recent years that this approach became successful with the tremendous increase in computational power of the super-computers and new breakthrough in the basic algorithms. This technique is known as the QCD calculus on lattice or lattice QCD. Please refer to the excellent status report by Karl Jansen in these proceedings.

3. From tree level to higher orders: the NLO Revolution

Perturbative field theory allows level-by-level calculation of particle interactions. The technique was developed by R.P. Feynman [19] in the 50's. Each precision level can be matched to a set of similar diagrams. Let's take a simple example, in QED[2] (figure 4), at the lowest level or tree level in the coupling constant α, the simple $e^+e^- \rightarrow e^+e^-$ process involves only 2 diagrams: a s-channel graph when the electron and the positron annihilate to form a virtual photon, also called propagator, which then decay into an electron and a positron and a t-channel where a photon is exchanged between the electron and the positron. In QED, the coupling constant is small enough (~ 1/137, although also running) so that in most cases first order calculations provide good enough precisions matching experimental accuracies. The input and output particles are called "legs".

But in some cases, experimental errors are much smaller than tree level computations like at the precision e^+e^-colliders. They can even be orders of magnitude smaller when, for example measuring the muon anomalous magnetic moment factor $(g-2)/2$ at 0.14 ppm precision [20]. The theoretical precision should be, at least, as good and computing the next terms in the expansion series becomes necessary both for electroweak and hadronic corrections.

The next level in precision is the so-called "Next-to-Leading Order" (NLO). It essentially implies adding 1-loop corrections. A photon can be exchanged between the initial state electron and positron, an e^+e^- loop can be inserted in the propagator or even (not shown); a second photon is exchanged in the t-channel (creating box diagrams). But the virtual exchanged particle can also be real and diagrams with an external photon must be included. If one sticks precisely to the same final state, the additional photon should not be observed namely, it should be either soft or almost collinear with the emitted electron. Loops and/or legs can be put on all elementary parts of the tree-level diagrams and the number of diagrams increases dramatically. Next-to-Next-Orders calculations follow the same trend, adding more and more diagrams and therefore becoming more and more computational intensive.

[1] QCD: Quantum ChromoDynamics, model of the strong interaction particularly probed at protons colliders like

[2] QED: Quantum ElectroDynamics, quantum theory for the electromagnetic interactions.

At LHC, QCD plays a major role and things get more complicated. First the proton is not an elementary particle, but a complex system of 3 valence quarks in a gluon fog and a sea of quark-antiquark pairs. At LHC energy, the interest is on the hard scattering of proton constituents, namely gluon-gluon, quark-gluon or quark-quark. Therefore many different initial states must be taken into account. The selection of the initial partons within the proton is made through the use of the so-called parton distribution functions (PDF) which describe the parton content of the proton. These functions are mainly parameterizations based on data from the e-p experiments at HERA (DESY [21]) or in $p\bar{p}$ experiments at D0/CDF (Tevatron [22], Fermilab) and now at LHC.

Second, the final states may involve many particles, i.e. many legs, like shown in the experimental wish list discussed at the "Les Houches" meetings [23]. This is a huge step in complexity compared to the 2→4 channels studied at the electron-positron collider (LEP II [24]).

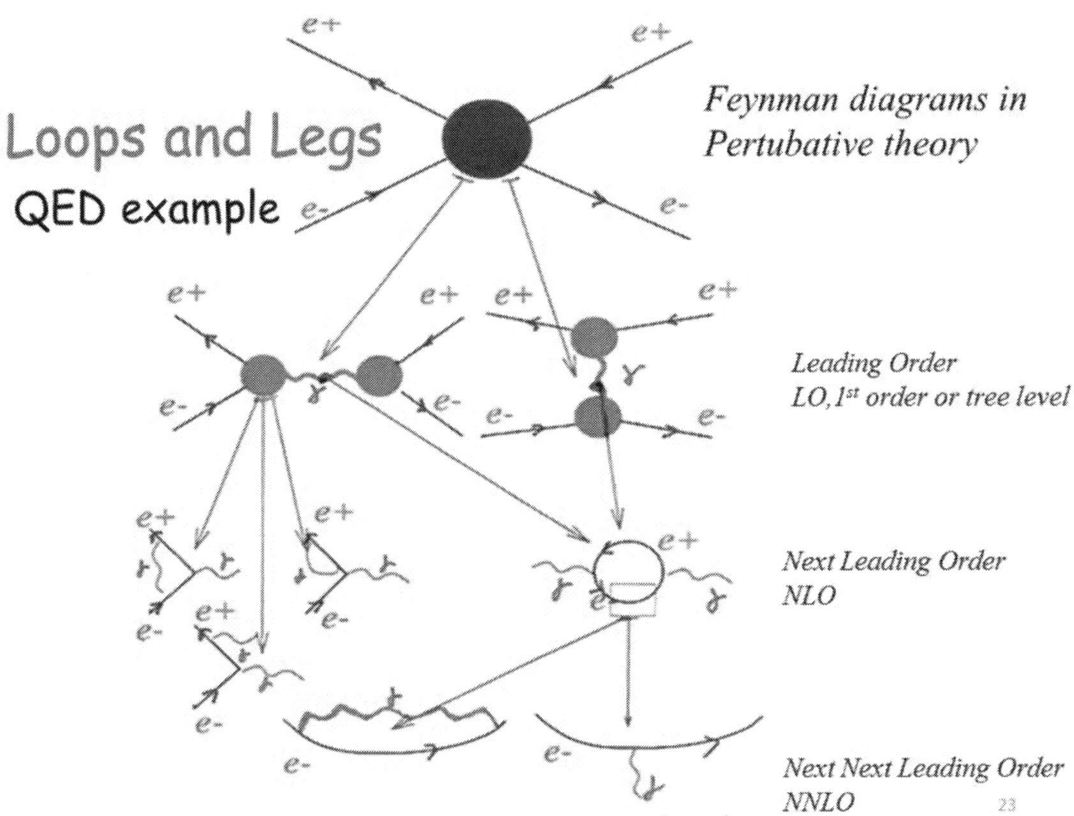

Figure 4. Loops and legs in QED.

Finally, contrary to QED where tree-level calculations are often a good approximation, QCD estimations must go beyond LO which depend too much on arbitrary scales. The renormalization scale μ_R enters in the cross-section as well as in the value of α_s. Renormalization is a necessary operation to cancel infinities arising in the calculation of the cross-section due to UV (Ultra-Violet) divergences. The factorization scale μ_F which also enters in the cross-section as well as in the parton distribution function is related to the soft and collinear divergences, namely in the IR (Infra-Red) region. These scales are arbitrary and induce large normalization errors (10-20% or even more). Higher order calculations reduce largely the sensitivity to these arbitrary scales and this trend is even more pronounced when the number of legs is large.

QCD NLO calculation of many legs processes may lead to more than $10^{5\text{-}6}$ diagrams and for each, the amplitude or matrix element can be so large in the usual Feynman representation that computer memory sizes become an issue. Clearly the Feynman diagrammatic approach was not quite suited for this kind of calculation.

Although this conventional method is still in use and even being improved, new approaches originated from the string/twistor theory have recently been developed for the calculation of one-loop processes. The on-shell unitarity method [25] to the expense of rewriting the quantum field theory in terms of invariant gauge on-shell intermediate expression simply proposes to cut the loop diagrams in one or many places decomposing the initial one-loop calculation to a set of tree level amplitudes, easily handled by the conventional tree-level diagram calculators. Following these ideas many developments have been undertaken (see for a summary until 2008 [26]) and have led to automatic calculation of 2→4 or 2→5 NLO processes (figure 5).

Even if these techniques for the most complex calculations have not reached full automation, they have made the computation possible with the current computers. Event generators at NLO based on SHERPA [27] and BlackHat [28], MadGraph [29] and others have been heavily used by the LHC experiments.

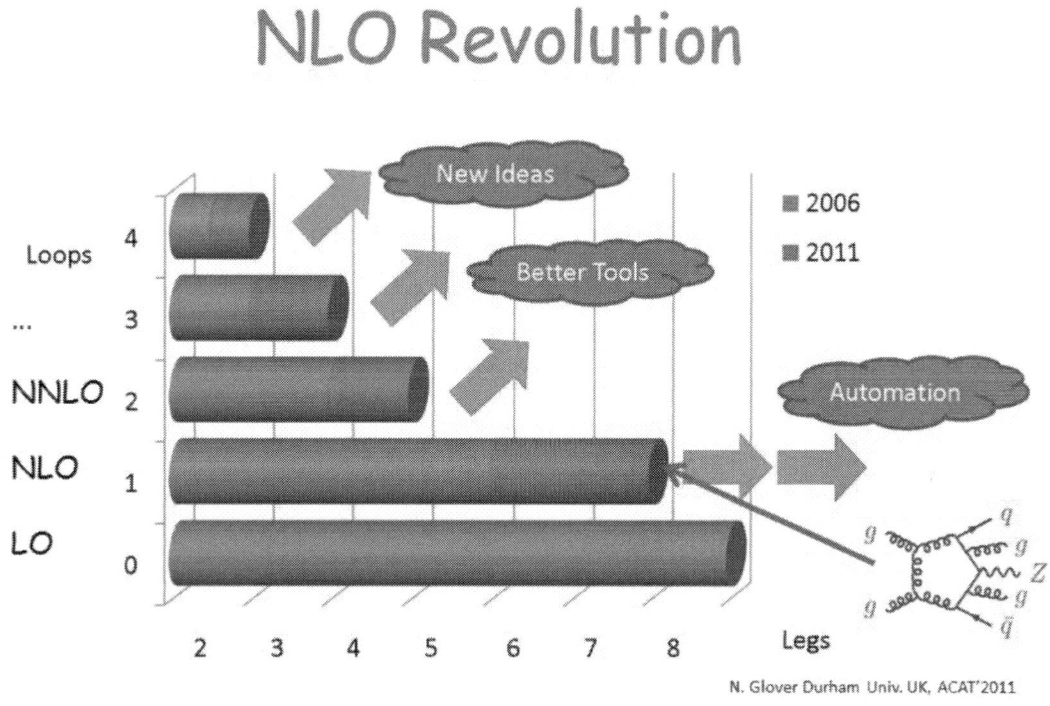

Figure 5. NLO Revolution [30].

4. Event Simulation and data analysis

It is impossible to calculate analytically the detector responses to an event produced in the quite complex LHC experiments. One therefore relies on Monte-Carlo simulations. Any process of interest can be actually generated and simulated in a virtual representation of the detectors. The simulation is supposed to mimic in detail what is detected when beams collide.

Figure 6 presents schematically the various steps of the simulation chain. On the top, the "physics model" describes the framework in which the calculation is performed. The basic model is the Standard Model (SM). But extension or beyond SM models are also implemented like for example the supersymmetric (SUSY) model. This model suggests that any known fermion (boson) has a partner boson (fermion) yet unobserved due to its high mass. Clearly beyond standard model (BSM) theories are made up to cure problems arising in the SM beyond current energy, in particular, the divergent contributions to the Higgs mass.

The actual process is, then, selected by defining the input and output particles as well as the order of calculation. The matrix element or the expression of the process probability is then automatically prepared based on the physics model data.

After integrating of the matrix element over the parameters phase space, the code for an event generator is constructed. It will generate random samples of energy-momentum four-vectors for all final state physical particles.

Each generated particle is then propagated into a model representing the experiment built by a detector simulation package. GEANT [31], initiated at CERN and developed by an international collaboration is the main package used by our community. Version 4 is an extremely detailed description of the detector components down to nuts and bolts and read-out cables. All physics particle-matter interactions have been implemented. Each particle, step by step is tracked in the material forming the structures as well as the detection sensitive elements. The charged particles may produce primary gas ionization, scintillation or Cherenkov light depending on the materials. Photons/gammas create electromagnetic showers producing hundred to thousand low energy particles which are all tracked down to final absorption. This information is collected by sensitive elements: wires, silicon strips/pixels or photo-detectors. All particles are followed until stopping in material or escaping the experimental set-up. The huge amount of information produced by the detector simulation should in principle represent exactly what a real event in a real detector would produce. Obviously, confidence on the detector simulation accuracy has been obtained after thousands of validation measurements for each type of particles, materials and detectors.

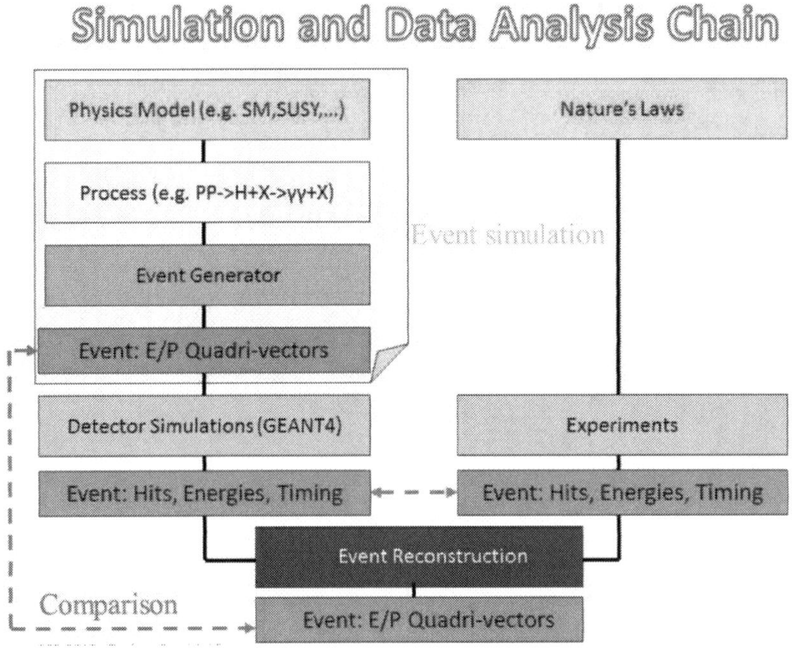

Figure 6. Monte-Carlo simulation and the data analysis chain.

At this level, one has now simulated i.e. theory "generated" event data that can be compared to the experimental data produced by Nature in the physical detector. Both simulation and real data are then reconstructed by the same reconstruction programme to get the initial set of energy-momentum four-vectors representing the actual simulated or observed event. Comparing the reconstructed simulated four-momenta to those produced by the event generator is a test of self-consistency of the whole system although departure between the simulated and the experimental data may give hints to new physics discovery.

The final analysis is generally done with the Root data analysis toolkit [32], developed at CERN. Events data are put in large ntuples (sort of expandable spread sheets) on which filters and fits can be applied. Estimators can then be computed and placed in histograms. This is how plots like those shown in figure 3 have been produced. Would the experimental data have been precisely fitted by the green-yellow background band, we would had concluded to a no Higgs discovery and higher limits on its production would had been set. But hopefully some excess was found, signalling the existence of something beyond the expected background, a new particle, maybe the expected Higgs.

5. Event sequential formation stages

Let's now focus on the physics of an event formation. In a high-energy *pp* scattering, the generation of a single event (figure 7) takes several steps. First, 2 partons (flavour and colour) are selected in each of the colliding protons. Before interacting, partons are allowed to emit gluons (initial state radiations). Then the actual hard-scattering interaction occurs and 2 or more particles are produced. These particles may then decay like in heavy quark/leptons or bosons (Z, W) producing several daughters. Finally, the last remaining quark and gluon undergo partonic showers before the final hadronization step occurs, namely the recombination of end partons in hadrons (baryons or mesons).

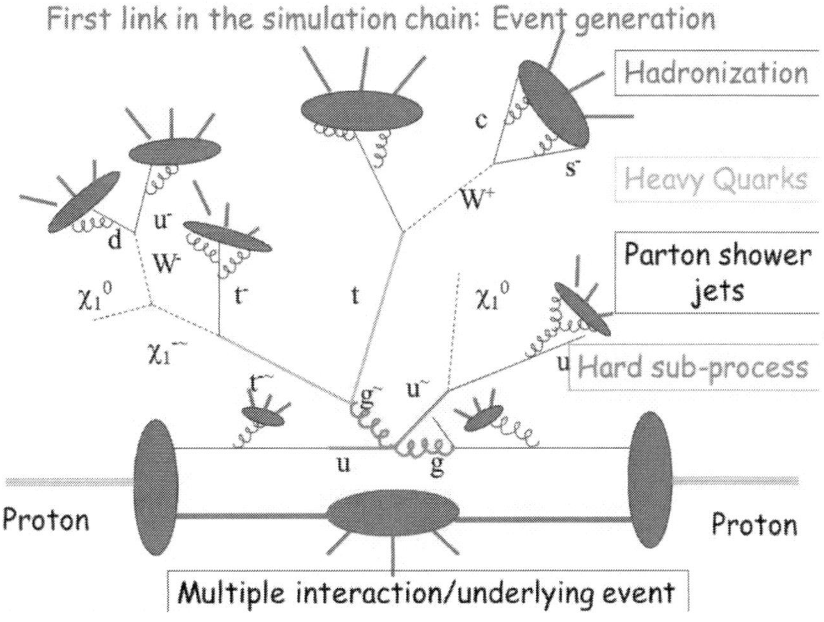

Figure 7. Sequential steps in the formation of an event. The event presented here is a SUSY event with a hard process: u-quark gluon →u-squark gluino, the u-squark being the superpartner of a u-quark and the gluino the gluon counterpart.

In addition, several interactions between different partons can occur in the same proton collision leading to multiple parton interactions (MPI). Finally, in the collision of proton bunches more than one collision can occur and multiple vertices are created, this is called the underlying events (UE). The odds that several interactions producing a hard scattering each leaving, in the detectors, high momentum transfer particles is quite small, but these extra collisions produce many additional tracks and energy depositions in the detector. These parasitic effects make the analysis more difficult as each track or energy deposition should be associated to the right vertex and, therefore, the whole reconstruction process takes more CPU resources and becomes more error prone.

6. Process calculation

Now that the physics of an event formation has been described, let us see how this is implemented in the global process calculation (figure 8). A flurry of packages has been developed to carry out the full event simulation.

Based on the model Lagrangian or, more generally, on the group symmetries and fields definitions, symbolic manipulation languages like Reduce, Maple, Mathematica or Form [33] are used to encode in a model file the various couplings and the propagators expressions either on an analytic form or as a set of numerical libraries. The Form language has been developed to handle very large expressions buffered on disk files. Specific programs like LanHEP [34], FeynRules [35] or SARAH [36,37] (SUSY) have made this derivation automatic.

Then a specific process can be selected. The initial particles undergoing the hard collision are generated after initial state radiation and beamstrahlung (CIRCE [38]) for electrons or parton extraction from the parton distribution functions (PDF) (LHApdf [39], CTEQ [40], MRST [41],) for the protons has been applied. One of the automatic amplitude/matrix element generators can be put in action to produce the global expression of the process integrand.

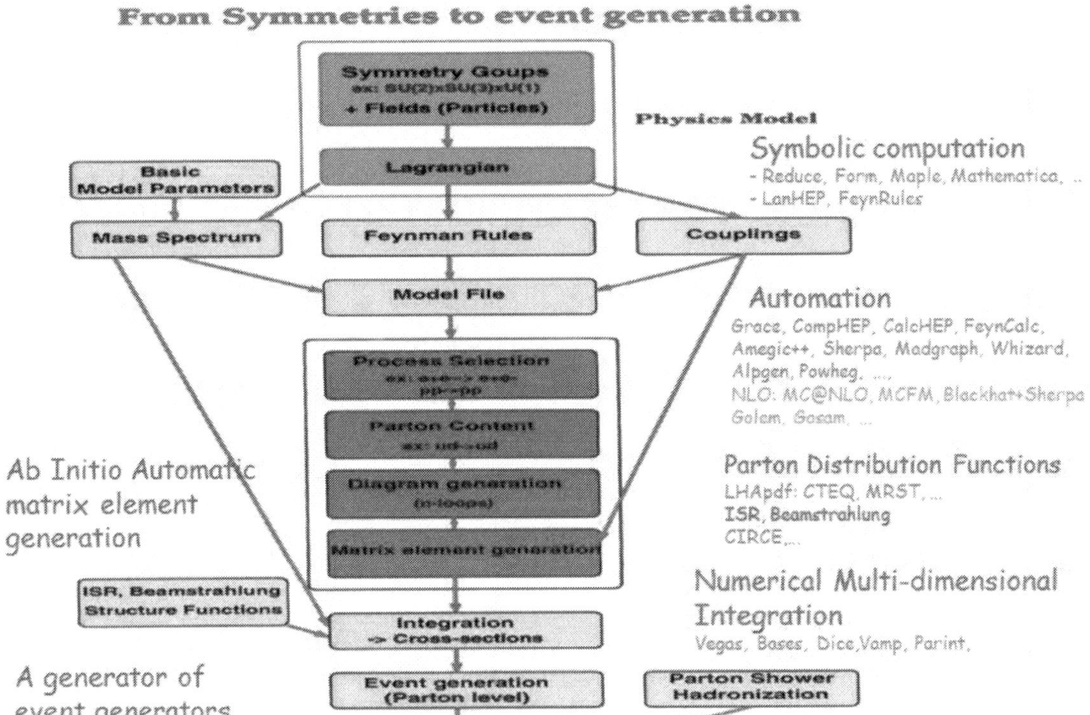

Figure 8. From symmetries to Event Generation.

This expression must then be integrated over the full phase space, namely over all possible legs momenta/energy, only constrained by the energy momentum conservation and the initial conditions. If loops are involved, integration over the loop propagators must also be performed. The multi-dimensional integration provides the total cross-section and the probability function for the generation of the events.

The event generator can, now, be constructed. It will randomly generate events at the parton level. Parton shower and hadronization programs like Pythia [42] or Herwig [43] will take over to produce the final full-fledged event ready to enter the detector simulation stage.

7. Matrix element automatic calculations

The matrix element calculation it-self has been for many years performed manually by talented theorists. Back in the 80's, a group in Japan from the KEK laboratory (Grace) [44] and a Russian group from Moscow State University (CompHEP) [45] [46] started to develop packages to do these lengthy and error prone calculations automatically, that was the beginning of the automatic process calculation endeavour.

The automation was initially seen as a toy project, at best to train students. But the experimental requests for new process calculations went exploding as well as the complexity of the process calculation and, finally, the automatic approach became the baseline. The first event generator with an automated computed matrix element was probably grc4f $e^+e^- \rightarrow$ 4 fermions [47] which was used for the LEP2 analysis. For the first time, spin correlations and mass effects were included. Nowadays most calculations are made by computer tools in a combination of symbolic and numerical operations.

The various steps of the automatic calculation flowchart, at tree level can be summarized as follows:

1. Enumeration and description of the set of diagrams entering the calculation of a given process matrix element (e.g. standalone QGRAF [48] or included in the automatic packages e.g. [44], [45])
2. Preparation of the matrix element expression for each diagram by various calls to the vertex and propagator subroutine library or by building a global analytical expression.
3. Multi-dimensional integration over the full phase space (including the PDF variables) to produce 1) the total cross-section 2) the probability function describing how likely an event with a given set of input and output four-vector will occur.
4. Preparation of the event generator code. According to the probability function, random set of the process four vectors will be generated.
5. Inclusion of the final state radiation calls as well as the parton showering and hadronization steps based on Pythia [42] or Herwig [43].

For reference here follows some tree level matrix element generators: GRACE [44], CompHEP [45], CalcHEP [49], FeynCalc [50], MadGraph [29], AMEGIC++ [51] as well as multi-channel event generators HELAC-PHEGAS [52], ALPGEN [53], GR@PPA [54], O'Mega [55], Sherpa [27], Whizard [56], Pythia [42], Herwig [43].

In practice there are, yet, no complete and fully automatic matrix element generators at NLO, but many packages are being developed in this direction. Recent reports can be found in [57,58,59] and the main contenders are: MCFM [60], NLOJET++ [61], BlackHat [62], Rocket [63], CutTools [64], MadLoop [65], OpenLoops [66], GOLEM [67], POWHEG [68], aMC@NLO [69,70], Sherpa+BlackHat [71], MadGOLEM [72], GoSam [73], HELAC-NLO [74].

8. Matrix element multi-dimensional Monte-Carlo integration

Unfortunately matrix elements tend to have singularities in some parts of the phase space making their integration difficult: for example, in the collinear region where two particles (e.g. an electron and a photon) get a small angular separation. This corresponds to the quantum physical situation where an electron alone cannot be distinguished from an electron closely accompanied by a soft photon. In that case one gets a singularity that can be tamed by introducing a cut on the photon-electron angle smaller than the actual experimental angular resolution.

Anyway singularities occur in the expression of the integrand and the skill of the event generator developers are still needed to craft the most efficient variable transformations so that singularities are avoided or minimized. The mapping between integration and physical variables is coded into the "kinematics subroutines". They are selectively plugged into the integrand code according to its sensitivity to specific divergences.

For the less acute variations, different techniques automatically adapting the sampling grid to the shape of the integrand have been implemented in the main multi-dimensional integration packages like Vegas [75,76], Bases/Spring [77], Dice [78], Foam [79], Vamp [80], Mint [81], ParInt [82]:

- Importance sampling: the grid intervals or bins get smaller and therefore the sampling is denser in regions of high integrand values as this phase space regions will contribute most to the total integral.
- Stratified sampling: the bins get smaller in the region of fast variation of the integrand so that from one interval to the next the function varies slowly and smoothly.
- Multi-channel sampling: several mappings, namely grid definitions, can be selected depending on the phase space region. The method is to always select the mapping that suits best the behavior of the integrand at the particular sampling point.

For processes up to 2→4, at tree level the full automation is now usual business. For more particles in the final state (e.g. 2→6), the integration stage may take a lot of CPU time and the calculation may become intractable if the kinematics mappings are not adequate.

For NLO calculations the number of integration dimensions increases due to the additional loop four-momenta and the integration difficulties become even more stringent. In addition, the size of the integrant can be enormous and compiler and computer memory limitations become an issue. Fortunately after the NLO revolution, some computations have become feasible, but automation has not yet been fully implemented.

When the integrand becomes too large, it can be cut in smaller parts (e.g. each sub-diagram) distributed over many nodes. This is one direction for parallelization. A second one to accelerate the computation is to share the whole phase space again over many nodes. So that, in the end, each node will compute a part of the integrand for a part of the phase space. However, the integration algorithm requires the full value of one or several integrand calculations over the whole phase space, before executing the next iteration. This is the serial part of the integration algorithm and load balancing between the nodes becomes an issue that has to be addressed [83,84].

Recently, it has been shown that the computation of loop integrals or tree level processes on GPUs have led to dramatic acceleration [85,86]. Supercomputers will certainly benefit these calculations.

9. Matrix element method for data analysis.

As we have seen, the process matrix element is at the heart of the event generator and its calculation can be made automatic in most cases, at least, at tree level. But a new application of the matrix element is looming up for the data analysis it-self.

The usual way of extracting new insights from the experiments is by comparing the simulated and experimental event distributions. This is the conventional approach. A more elaborate, but similar technique called the template analysis has been developed along these lines and involving likelihood maximization. But a new approach is nowadays preferred. It involves the use of the matrix element expression in the event selection algorithm it-self. This is the matrix-element method [87] or also called the matrix element likelihood algorithm (MELA) in CMS.

9.1 Template method

Let us assume that a process parameter is to be estimated from experimental data. Let's take for example, the mass of the top quark. The t-quark has a very short lifetime and decays, even before it gets hadronized, in various channels including in lepton + jet. From these decay products the top quark mass can be reconstructed. The template method requires the simulation of several similar distributions assuming different top masses. Then, a likelihood expression is built and its maximization provides the best fit between the experimental and simulated distributions. This method is straight forward, but only a small part of the full event information is used, namely the reconstructed top mass.

The rest of the event data like the distributions of remaining particles and their correlations does not enter this optimization.

9.2 Matrix-element method

From the same selected sample, a probability is built for each event taking into account all final state particle kinematics to estimate to which category it is most likely part of. The following probability P_{sig} is therefore estimated event per event:

$$P_{sig}(x, m_t) = \frac{1}{N} \int_{y,q_1,q_2} d\Phi(y)\ |M_{t\bar{t}}(y, m_t)|^2\ W(x, y)\ f_{pdf}(q_1)\ f_{pdf}(q_2)\ d\,q_1 dq_2$$

where:

- x represents the measured kinematic variables for all final state particles
- m_t is the parameter to be estimated, here the top mass
- $|M_{t\bar{t}}|^2$ is the square of the matrix element associated to the signal process under study, here the production of a pair of top quarks.
- $W(x, y)$ is the transfer function mapping the measured four-vectors and the matrix element theoretical parameters.
- $f_{pdf}(q)$ is the parton distribution function described above giving the probability that the initial colliding partons with q_1 and q_2 four-momenta are produced by the interacting protons.

This expression is integrated over the phase space and normalized to the observed cross-section N. Similar probabilities $P_{bkg1...n}$ are formed assuming that the same event belongs to background processes, namely processes having a similar final state but not through the decay of top quarks.

If the $f_{sig,bkg1...n}$ are the fractional parts of the total sample (signal or backgrounds), a global probability for one event with the kinematics x to form a mass m_t whether from signal or from background can then be computed:

$$P_{evt}(x, m_t) = f_{sig}\ P_{sig} + \ f_{bkg1}\ P_{bkg1} + \cdots + \ f_{bkgn}\ P_{bkgn}\ .$$

Finally, a sample level likelihood is computed as the product of the global probability of all events and the following inverse log is minimized to get the most probable m_t:

$$\mathcal{L}(m_t) = \ -ln(\textstyle\prod_{evt} P_{evt}(x, m_t))\ .$$

In this method implemented in the MadWeight [88] package, the whole event information is incorporated in the parameter estimation. In addition the separation between the experimental data and the theory input is well identified.

But this is a very CPU time consuming procedure as for each event several multi-dimensional integrations are necessary. To help solving this problem, the use of GPU has been implemented successfully [89].

The template (figure 9) and the matrix element method (figure10) have been used with great success for establishing the value of the top mass at the Tevatron by the D0 [90] and CDF [90] experiments.

A similar technique has also been used at LHC to estimate the Higgs mass. A global expression MELA is computed from the likelihood, namely MELA is 0 (1) if the probability of the event to be a background (signal) is maximal. Each event is presented on a scatter plot with the estimated Higgs mass on the x-axis and the MELA value on y-axis. Setting the threshold on MELA above 0.5 selects events more likely to be a signal than a background which greatly enhances the signal over noise ratio of the Higgs mass measurement (figure 11).

Matrix element methods incorporating NLO corrections have also been studied [91], but have not yet been implemented for experimental data analyses.

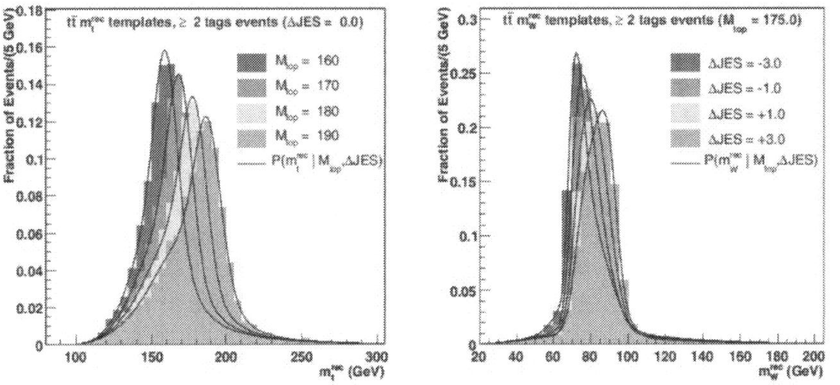

Figure 9. The template distributions for reconstructed top mass (left) and mw (right) for CDF [90]

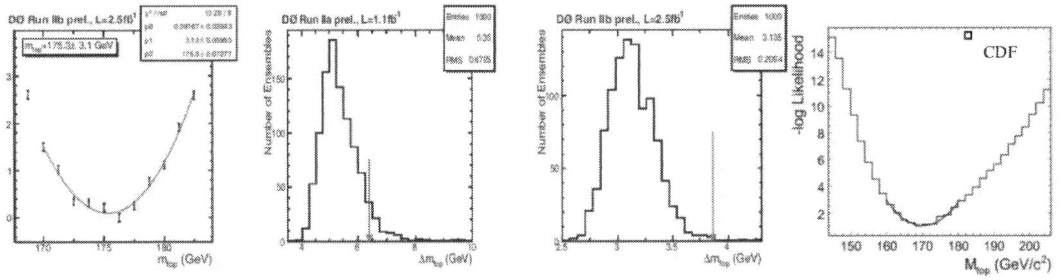

Figure 10. Measured likelihood for the top mass at D0 and CDF [90,92].

10. Cross-disciplinary fertilization and conclusions

Computational particle physics covers two main branches, one dealing with the non-perturbative calculations, lattice QCD, which is virtually intractable without the use of large parallel super computers. Recent developments show that thanks to a new algorithm and a drastic increase in computing power, the hadronic mass spectrum and other global parameters are now well reproduced.

The other branch is related to the perturbative region probing the inner part of the nucleon as well as its high-energy dynamics. These computations, for the highest available energy, have also become feasible thanks to new NLO calculation methods and, again, to the availability of a large Grid computing infrastructure. The automation is now on track for the most complex calculations. Experimental design, physics analysis and theoretical interpretation at current colliders deeply rely on the development of these "new" tools providing matrix elements and event generators.

In fact, perturbative calculations both benefit from and motivate the development of techniques and methods common to other domains in computational science including:

- Symbolic calculation: the traduction of symmetries and fields to Feynman rules and couplings has been automated using algebraic languages like Reduce, Maple or Mathematica. Often created to support these calculations, they are now multi-purpose languages and are heavily used in many other research fields. However, the handling of very large expression of high complexity has triggered the implementation of expression disk caching and parallel techniques like in the language Form.

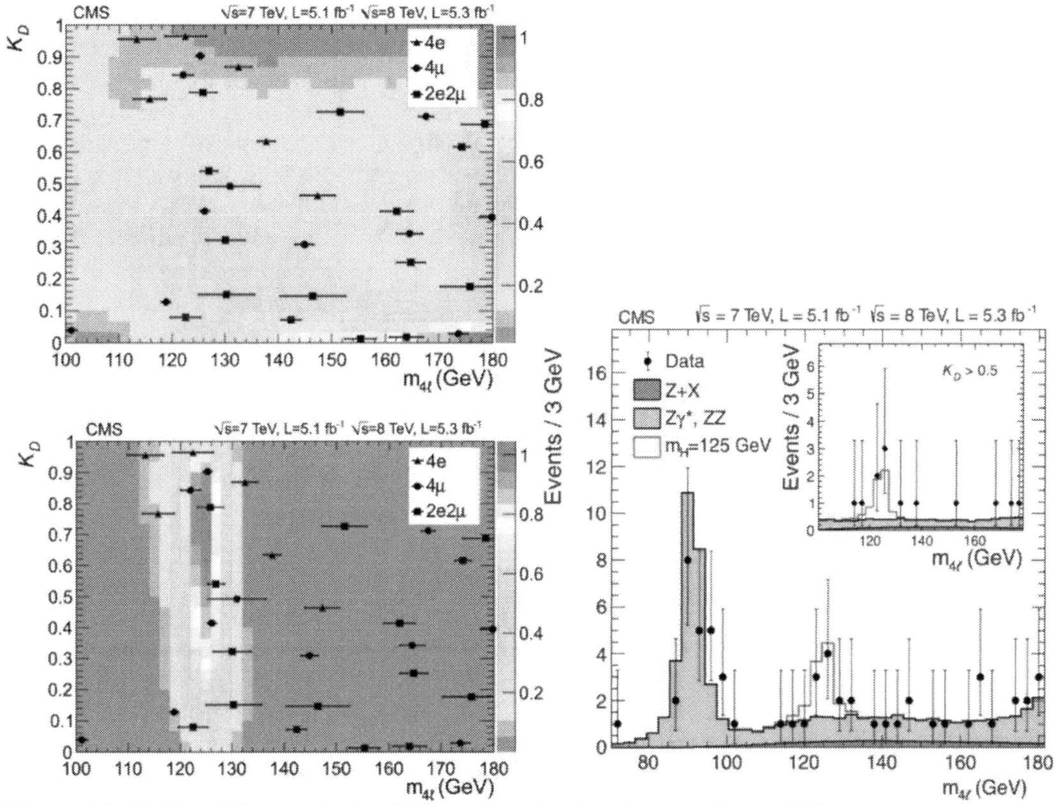

Figure 11. CMS: MELA analysis of Higgs decaying in 4 leptons (from [11]).

- Parallelism:
  - Numerical multi-dimensional integration: one of the main issues in perturbative computation is the integration over many dimensions of very large singular expressions. Some of the integration packages have been made parallel and able to run on GPU or supercomputers.
  - Process calculation factorial growth: the number of models, of processes and sub-processes is sky rocketing prompting for the use of massively parallel supercomputers or large clusters.
- High precision floating point computation: often double precision is not sufficient. Quadruple or octuple precision becomes necessary to achieve convergence for the most singular integrands leading to order of magnitude increase in computing time [93]. Specific libraries and hardware developments based on co-processors or GPUs [94] are in progress.
- Large computing infrastructures:
  - The internationally distributed Data Grid architecture, mostly created to cover the collider experiments needs, has fulfilled the expectations. It gave birth to the "Cloud" technology, a new commercial endeavor.
  - Although most of the current data analysis and simulations programs are serial, both the many-core/GPU long term industrial trends and the need for higher precision theoretical predictions are incentives to implement multi-threading and parallel execution techniques (e.g. for analysis [95] and simulation [96]). Certainly not an easy task in object oriented environments, but it will open access to the huge power of supercomputers.
- General data analysis toolkits (e.g. Root) implementing advanced statistical methods developed for particle physics has found applications in other environment like biology, finance and medicine.

Finally, the initial dream of automatic perturbative physics calculations has triggered the development of a flurry of packages and specific tools. Some are at the heart of today high-energy colliders and astroparticle experiment data analysis. But these packages have usually very little in common. Despite great attempts to standardize some aspects like the output files format (StdHep [97], HepMC [98]) or the interface with NLO calculations (BLHA [99]), they still have no common input syntax, no general parameter definition format, a poor modularity, no universal submission scheme, no common visualization and analysis tools, and so on. It is time to address these issues to facilitate their necessary detailed comparisons and their embedding in the large experimental simulation packages.

References

[1] Perret-Gallix D 2012 *CCP 2012: Computational particle physics for event generators and data analysis (slides)* http://acpp-coll.in2p3.fr/cgi-bin/twiki.source/pub/ACPP/PresentationsNotes/ccp_2012-3.pptx

[2] LHC: Large Hadron Collider http://public.web.cern.ch/public/en/LHC/LHC-en.html

[3] CERN: The European Center for Nuclear Research http://www.cern.ch

[4] Pierre Auger Observatory http://www.auger.org/

[5] GEM-EUSO Proposal http://jemeuso.riken.jp/en/index.html

[6] ATLAS experiment at CERN LHC http://atlas.ch/

[7] CMS experiment at CERN LHC http://cms.web.cern.ch/

[8] LHC-B experiment at CERN LHC http://lhcb.web.cern.ch/lhcb/

[9] ALICE experiment at CERN LHC http://aliceinfo.cern.ch/

[10] ATLAS Collaboration 2012 Observation of a new particle in the search for the Standard Model Higgs boson with the ATLAS detector at the LHC *Phys. Lett. B*, **716** p 1-29

[11] CMS Collaboration 2012 Observation of a new boson at a mass of 125 GeV with the CMS experiment at the LHC *Phys. Lett. B* **716** no.1 p 30-61

[12] CERN web site. The search for the Higgs boson http://public.web.cern.ch/public/en/science/higgs-en.html

[13] Brout R and Englert F, Broken Symmetry and the Mass of Gauge Vector Mesons 1964 *Phys. Rev. Lett.* 13 no. 9 p 321-323

[14] Higgs P 1964 Broken Symmetries and the Masses of Gauge Bosons *Phys. Rev. Lett.* **13** 508–509

[15] Guralnik G S, Hagen C R and Kibble T W B 1964 Global Conservation Laws and Massless Particles *Phys. Rev. Lett.* **13** 585–587

[16] WLCG: Worldwide LHC Computing GRID http://wlcg.web.cern.ch/

[17] Kaiser D, Physics and Feynman's Diagrams 2005 *American Scientist* **93** 156 http://web.mit.edu/dikaiser/www/FdsAmSci.pdf

[18] Wilson K G 1974 Confinement of quarks *Phys. Rev. D* **10** 2445

[19] Feynman R P 1949 Space-time approach to quantum electrodunamics *Physics Review* **76** 769-789

[20] E-898, Fermilab, The New (g – 2) Experiment: A proposal to measure the Muon Anomalous Magnetic moment to +- 14 ppm precision http://gm2.fnal.gov/public_docs/proposals/Proposal-APR5-Final.pdf

[21] HERA at DESY (Hamburg) http://www.desy.de/research/facilities/hera_experiments/index_eng.html

[22] TEVATRON at FERMILAB http://www.fnal.gov/pub/science/experiments/energy/tevatron/

[23] Les Houches Working group, The SM and NLO MultiLeg and SM MC Working Groups 2010

arXiv:1203.6803

[24] LEP at CERN (Geneva) http://public.web.cern.ch/public/en/research/lep-en.html

[25] Bern Z, Dixon L J, Dunbar D C and Kosower D A 1995 Fusing gauge theory tree amplitudes into loop amplitudes *Nucl. Phys. B* **435** 59–101

[26] Les Houches Working group, The NLO MultiLeg Working Group: Summary Report 2008 arXiv:0803.0494

[27] SHERPA Collaboration, SHERPA http://sherpa.hepforge.org/trac/wiki

[28] Bern Z, Diana G, Dixon L, Febres Cordero F, Hoeche S et al., Four-Jet Production at the Large Hadron Collider at Next-to-Leading Order in QCD 2012 *Phys. Rev. Lett.* **109** 042001

[29] Alwall J, Herquet M, Maltoni F, Mattelaer O and Stelzer T 2011 MadGraph 5: Going Beyond *JHEP* **1106** 128

[30] Glover N and Heinrich G 2011 Computations in Particle Physics: in *ACAT' 2011 Uxbridge, London UK (Advanced Computing and Analysis technologies)* http://indico.cern.ch/getFile.py/access?contribId=124&sessionId=15&resId=0&materialId=slides &confId=93877

[31] GEANT 4 Collaboration. Geant 4 http://geant4.cern.ch/

[32] Root Collaboration. Root http://root.cern.ch/drupal/

[33] Kuipers J, Ueda T, Vermaseren J A M and Vollinga J, FORM version 4.0. arXiv:1203.6543

[34] Semenov A 2010 LanHEP - a package for automatic generation of Feynman rules from the Lagrangian. Updated version 3.1 arXiv:1005.1909

[35] Christensen N D and Duhr C 2009 FeynRules - s - Feynman rules made easy *Comp. Phys. Comm.* **180** 1614-1641

[36] Staub F 2012 SARAH 3.2: Dirac Gauginos, UFO output, and more *Comp. Phys. Comm.* **184** 1792

[37] Staub F, Ohl T, Porod W and Speckner Ch 2012 A tool box for implementing supersymmetric models *Comp. Phys. Comm.* **183** 2165 http://sarah.hepforge.org/Toolbox.html

[38] Ohl T Beam spectra: CIRCE, Luminous, Pandora http://theorie.physik.uni-wuerzburg.de/~ohl/lc/generators-sections013.html#circe

[39] Whalley M and Buckley A LHAPDF http://lhapdf.hepforge.org/

[40] CTEQ Collaboration. The Coordinated Theoretical-Experimental Project on QCD http://www.phys.psu.edu/~cteq/

[41] Martin A, Stirling J, Thorne R and Watt G MRS/MRST/MSTW Parton distributions http://durpdg.dur.ac.uk/hepdata/mrs.html

[42] Sjöstrand T et al. Pythia http://home.thep.lu.se/~torbjorn/Pythia.html

[43] Webber B et al. HERWIG http://www.hep.phy.cam.ac.uk/theory/webber/Herwig/

[44] Kaneko T, Kawabata S and Shimizu Y 1997 Automatic generation of Feynman graphs and amplitudes in QED *Comp. Phys. Comm.* **43** 279-295 http://minami-home.kek.jp/

[45] Boos E E, Dubinin M N, Edneral V F, Ilyin V A, Kryukov A P, Pukhov A E, Rodionov A Ya, Savrin V I, Slavnov D A and Taranov A Yu 1989 CompHEP: Computer system for calculations of particle collision characteristics at High Energy Moscou State University: Institute of Nuclear Physics, Preprint 89-63/140 http://comphep.sinp.msu.ru/ http://ccdb5fs.kek.jp/cgi-bin/img/allpdf?199003342

[46] Boos E E and Dubinin M N 2010 Problems of automatic calculation for collider physics *Physics Uspekhi* **53** 1039-1051

[47] Fujimoto J, Ishikawa T, Kaneko T, Kato K, Kawabata S, Kurihara K, Munehisa T, Perret-Gallix

D, Shimizu Y and Tanaka H 1997 Grc4f v1.1: A Four fermion event generator for e+ e- collisions *Comp. Phys.Comm.* **100** 128-156

[48] Nogueira P, Automatic Feynman graph generation 1993 *Journal of Computational Physics* **105** 279-289 http://cfif.ist.utl.pt/~paulo/qgraf.html

[49] Puhkov A CalcHE http://theory.sinp.msu.ru/~pukhov/calchep.html

[50] Mertig R FeynCalc http://www.feyncalc.org/

[51] Krauss F, Kuhn R and Soff G 2002 AMEGIC++ 1.0, A Matrix Element Generator In C++ *JHEP* **0202** *044*

[52] Kanaki A and Papadopoulos C G, HELAC-PHEGAS: automatic computation of helicity amplitudes and cross sections http://arxiv.org/abs/hep-ph/0012004

[53] Mangano M L, Moretti M, Piccinini F, Pittau R and Polosa A 2000 ALPGEN V2.14: A collection of codes for the generation of multi-parton processes in hadronic collisions in *Workshop on Computer Particle Physics: (CCP 2001) Automatic Calculation for Future Colliders* arXiv:hep-ph/0012004

[54] Odaka S and Kurihara Y 2012 GR@PPA 2.8: initial-state jet matching for weak boson production processes at hadron collisions *Comp. Phys. Comm.* **183** 1014

[55] Moretti M, Ohl T and Reuter J 2001 O'Mega: An Optimizing matrix element generator in *2nd ECFA/DESY Study 1998-2001 1981-2009 [hep-ph/0102195]* http://theorie.physik.uni-wuerzburg.de/~ohl/omega/

[56] Kilian W, Ohl T and Reuter J 2011 WHIZARD: Simulating Multi-Particle Processes at LHC and ILC *Eur. Phys. J.*, no. C **71** 1742 http://whizard.hepforge.org/

[57] Binoth T 2008 LHC phenomenology at next-to-leading order QCD: theoretical progress and new results in *12th International Workshop on Advanced Computing and Analysis Techniques in Physics Research (ACAT 2008)* arXiv:0903.1876 POS (ACAT08) 011

[58] Maitre D 2010 Automation of Multi-leg One-loop virtual Amplitudes ,POS (ACAT2010)008 http://pos.sissa.it/archive/conferences/093/008/ACAT2010_008.pdf

[59] Tramontano F 2012 Progress in automated Next-to-Leading-Order calculations in *ACAT 2011: Advanced Computing and Analysis Technologies for Research, J. Phys. COnf. Ser.* **368** 012058

[60] Campbell J, Ellis K and Williams C. MCFM - Monte Carlo for FeMtobarn processes http://mcfm.fnal.gov/

[61] Nagy Z 2003 Next-to-leading order calculation of three jet observables in hadron hadron collision *Phys.Rev.* D **68** 094002 http://www.desy.de/~znagy/Site/NLOJet++.html

[62] Berger C F, Bern Z, Dixon L, Febres Cordero F, Forde D, Ita H, Kosower D A and Maitre D 2008 One-Loop Calculations with BlackHat *Nucl. Phys. Proc. Suppl.* **183** 313

[63] Ellis R K, Giele W T, Kunszt Z, Melnikov K and Zanderighi G 2009 One-loop amplitudes for W + 3 jet production in hadron collisions *JHEP* **0901** 012

[64] Ossola G, Papadopoulos C G and Pittau R 2008 *JHEP* **0803** 042 http://www.ugr.es/~pittau/CutTools/

[65] Hirschi V, Frederix R, Frixione S, Garzelli M V, Maltoni F and Pittau R 2011 Automation of one-loop QCD corrections *JHEP* **1105** 044

[66] Cascioli F, Maierhoefer P and Pozzorini S 2012 Scattering Amplitudes with Open Loops *Phys. Rev. Lett.* **108** 111601

[67] Cullen G et al. 2010 Recent Progress in the Golem Project *Nucl. Phys. Proc. Suppl.* **205-206** 67

[68] Oleari C 2010 The PowHeg box *Nucl. Phys. Proc. Suppl.* **205-206** 34

[69] Frixione S and Webber B R 2007 Matching NLO QCD computations with parton shower

simulations: the POWHEG method, *JHEP* **0711** 070

[70] Alwall J et al. aMC@NLO http://amcatnlo.web.cern.ch/amcatnlo/

[71] Bern Z, Ozeren K, Dixon L, Hoeche S, Febres Cordero F, Ita H, Kosower D and Maître D 2012 High multiplicity processes at NLO with BlackHat and Sherpa in *Proceedings of Loops and Legs 2012* http://arxiv.org/pdf/1210.6684.pdf

[72] Lopez-Val D, Goncalves-Netto D, Mawatari K, Plehn T, Wigmore I 2012 MadGolem: automating NLO calculations for New Physics http://arxiv.org/abs/1209.2797

[73] Cullen G, Greiner N, Heinrich G, Luisoni G, Mastrolia P, Ossola G, Reiter T and Tramontano F 2012 Automated One-Loop Calculations with GoSam *Eur. Phys. J.*, no. C **72** 1889 http://gosam.hepforge.org/

[74] Bevilacqua G, Czakon M, Garzelli M V, van Hameren A, Kardos A, Papadopoulos C G, Pittau R and Worek M 2011 HELAC-NLO *Comp. Phys. Comm.* **184** 986

[75] Lepage G P 1978 A New Algorithm for Adaptive Multidimensional Integration *Journal of Computational Physics* **27** 192–203

[76] VEGAS - GNU Scientific Library -- Reference Manual http://www.gnu.org/software/gsl/manual/html_node/VEGAS.html

[77] Kawabata S 1995 A new version of the multi-dimensional integration and event *Comp. Phys. Comm.* **88** 309-326

[78] Yuasa F, Tobimatsu K and Kawabata S 2003 Parallelization of the multidimensional integration *Nuclear Instruments and Methods in Physics Research A* **502** 599–601

[79] Jadach S 2000 Foam: Multi-Dimensional General Purpose Monte Carlo Generator With Self-Adapting Simplical Grid *Comp. Phys. Comm.* **130** 244-259

[80] Ohl T 1999 VAMP, Version 1.0: Vegas AMPlified: Anisotropy, Multi-channel sampling and Parallelization http://whizard.hepforge.org/vamp.pdf

[81] Nason P 2007 MINT: a Computer Program for Adaptive Monte Carlo Integration and Generation of Unweighted Distributions http://xxx.lanl.gov/abs/arXiv:0709.2085

[82] de Doncker E et al. Parint: Parallel Integration Package http://www.cs.wmich.edu/parint/

[83] Yuasa F, Ishikawa T, Kawabata S, Perret-Gallix D, Itakura K, Hotta Y and Okuda M 2000 Hybrid parallel computation of integration in GRACE in *6th International Workshop on New Computing Techniques in Physics Research (AIHEP99)* http://arxiv.org/abs/hep-ph/0006268

[84] Laccetti G, Lapegna M, Mele V, Romano D and Murli A 2012 A Double Adaptive Algorithm for Multidimensional Integration on Multicore Based HPC Systems *International Journal of Parallel Programming* **40** 397-409 http://rd.springer.com/article/10.1007%2Fs10766-011-0191-4

[85] Yuasa F, Ishikawa T, Hamaguchi N, Koike T and Nakasato N 2012 Acceleration of Feynman loop integrals in high-energy physics on many core GPUs in *Conference on Computational Physics (Kobe)*, 2012 http://acpp-coll.in2p3.fr/cgi-bin/twiki.source/pub/ACPP/PresentationsNotes/ccp2012-poster-v2.pdf

[86] Kanzaki J 2012 Monte Carlo integration on GPU *Eur. Phys. J. C* **71** 1559

[87] Kondo K 1991 Dynamical likelihood method for reconstruction of events with missing momentum. 2: Mass spectra for 2 → ; 2 processes *J. Phys. Soc. Ja.* **60** 836–844

[88] Artoisenet P, Lemaître V, Maltoni F and Mattelaer O 2010 Automation of the matrix element reweighting method *JHEP* **1012** 068, https://cp3.irmp.ucl.ac.be/projects/madgraph/wiki/MadWeight

[89] Harrington R, Optimization of Matrix element methods using GPUs in *Workshop on Future Computing, University of Edinburgh*, 2011 http://indico.cern.ch/getFile.py/access?contribId=10&sessionId=3&resId=0&materialId=slides&

confId=141309

[90] Brandt O, Top Quark mass mesurement at Tevatron 2010 *Il Nuovo Cimiento* **33** 73-80

[91] Campbell J M, Giele W T and Williams C 2012 Extending the matrix element method to next-to-leading order FERMILAB-CONF-12-176-T, FERMILAB-CONF-12-176-T http://arxiv.org/pdf/1205.3434.pdf

[92] CDF Collaboration 2012 Measurements of the Top-quark Mass and the tt-bar Cross Section in the Hadronic tau + Jets Decay Channel at sqrt(s)=1.96 TeV *Phys. Rev. Lett.* **109** 192001

[93] Fujimoto J, Ishikawa T and Perret-Gallix D 2005 High precision numerical computations: A case for an HAPPY design ACPP IRG note: ACPP-N-1 2005 http://emc2.in2p3.fr/cgi-bin/twiki.source/pub/ACAT/PresentationsNotes/Highprecisionnumericalcomputatio3.pdf

[94] Lu M, He B and Luo Q, Supporting Extended Precision on Graphics Processors in *DaMoN '10 Proceedings of the Sixth International Workshop on Data Management on New Hardware*, p 19-26 http://www.ntu.edu.sg/home/bshe/precision_damon10.pdf

[95] Jarp S, Lazzaro A, Nowak A and Valsan L 2012 Comparison of Software Technologies for Vectorization and Parallelization CERN openlab http://openlab.web.cern.ch/sites/openlab.web.cern.ch/files/technical_documents/Evaluation_of_Parallel_Technology.pdf

[96] Dong X, Cooperman G and Apostolakis J, Multithreaded Geant4: Semi-Automatic Transformation into Scalable Thread-Parallel Software College of Computer Science, Northeastern University, Boston and PH/SFT, CERN (Switzerland), http://citeseerx.ist.psu.edu/viewdoc/download?doi=10.1.1.174.8538&rep=rep1&type=pdf

[97] Garren L and Lebrun P. StdHep: StdHep provides a common output format for Monte Carlo events http://cepa.fnal.gov/psm/stdhep/

[98] Dobbs M and Hansen J B 2001 HepMC - a C++ Event Record for Monte Carlo Generators *Comp. Phys. Comm.* **134** 41 https://savannah.cern.ch/projects/hepmc/

[99] Binoth T, Boudjema F, Dissertori G, Lazopoulos A, Denner A, Dittmaier S, Frederix R, Greiner N, Hoche S, Giele W, Skands P, Winter J, Gleisberg T, Archibald J, Heinrich G, Krauss F, Maitre D, Huber M, Huston J, Kauer N, et al. 2010 A proposal for a standard interface between Monte Carlo tools and one-loop programs *Comp. Phys. Comm.* **181** 1612

Journal of Physics: Conference Series **454** (2013) 012052 doi:10.1088/1742-6596/454/1/012052

Time evolution of Heisenberg operators of nuclei and electrons of QED system based on field theory

Masato Senami, Toshihide Miyazato, Soujirou Takada, Yuji Ikeda and Akitomo Tachibana

Department of Micro Engineering, Kyoto University, Kyoto 615-8540, Japan

E-mail: akitomo@scl.kyoto-u.ac.jp

Abstract. The present status of the development of our code, QEDynamics, is reported. In this code, the time evolution simulation is carried out for Heisenberg operators based on quantum electrodynamics in a manner of quantum field theory. For this simulation, the treatment of the photon is important, particularly infrared and ultra-violet photons. As a result, renormalization is one of essential ingredients, since quantum field theory is adopted.

1. Introduction

For the description of the nature, QED (quantum electrodynamics) is the most celebrated theory. For example, the prediction of the anomalous magnetic dipole moment of electron by perturbative calculations shows extreme accordance with experimental observations [1, 2, 3]. Although the standard model of particle physics is now established completely by the discovery of a Higgs-like particle, the precision of this model is not so improved compared with QED due to the uncertainties of parameters. Hence, QED is suitable enough for the precise description of the interaction between photon and matter.

QED is extremely successful in perturbative calculations for free particles. The application to bound states is intensely studied, while the sophisticated calculation method has not been established due to its nonperturbativity. QED is based on quantum field theory, where all phenomena are essentially described by field operators. In computational simulations, operators are difficult to handle. The exact simulation of field theory is performed by lattice calculation, such as lattice QCD. However, QED is not suitable for the lattice computation, since the way to take the continuum limit has not been established. Hence, the computation of QED bound states has been developed as approximation in most studies. Some treatments, Bethe-Salpeter formalism or NRQED [4], are frequently discussed. In quantum chemistry, the QED correction to Hamiltonian for ab initio calculation is often used [5]. In most studies, the photon is treated as potential, and however this is not suitable for quantum system studies, for example systems with single photon.

In this work, we propose a new computation method of the time evolution of quantum systems based on QED and report the present status of the development of our program code, QEDynamics [6]. In our method, field operators are treated in Heisenberg picture. The evolution of the operators is calculated as the transition between modes prepared by the method of quantum chemistry. Since the aim of our work is the study of bound states, such as molecules, nuclei as well as electrons are treated as quantum fields. The QED including nuclei is formulated with the gauge invariance as Rigged QED [7, 8, 9].

2. Theory and computational details

In this paper, electron fields are treated as two-component Schrödinger fields. This treatment is called as primary Rigged QED [10]. Since the four-component relativistic treatment requires much larger computation resources, this approximation supports longer time evolution calculations and larger system calculations. It is noted that even a two-component treatment retrieves Lorentz symmetry by the addition of all expansion terms of the relativistic Lagrangian to the two-component Lagrangian [4, 11].

The electron and nucleus fields are expanded as

$$\hat{\Psi}(t,\vec{x}) = \sum_m \hat{e}_m(t)\psi_m(\vec{x}), \quad (1)$$

$$\hat{\chi}_a(t,\vec{x}) = \sum_m \hat{f}_{am}(t)\chi_{am}(\vec{x}), \quad (2)$$

where $\hat{e}_m(\hat{f}_{am})$ is the annihilation operator of the electron (nucleus) and $\psi_m(\chi_{am})$ is its expansion function. The photon field is given by [?, 7, 8]

$$\hat{A}^\mu(t,\vec{x}) = \hat{A}^\mu_{\mathrm{rad}}(t,\vec{x}) + \hat{A}^\mu_A(t,\vec{x}) + \hat{A}^\mu_M(t,\vec{x}). \quad (3)$$

Here radiation photon, $\hat{A}^\mu_{\mathrm{rad}}$, is derived by the quantization for a free field, while $\hat{A}^\mu_A$ and $\hat{A}^\mu_M$ are photons produced by dynamical effects of the system and environment, respectively. These terms are particularly important for the description of infrared photons.

The definition of $\hat{A}^\mu_{\mathrm{rad}}$ is given by

$$\hat{A}^\mu_{\mathrm{rad}}(t,\vec{x}) \equiv \frac{\sqrt{c}}{\pi\sqrt{2\hbar}}\sum_\sigma \int_k d^3\vec{k}\frac{1}{\sqrt{2k_0}}\hat{A}^\mu_r(t,\vec{x},\vec{k},\sigma), \quad (4)$$

$$\hat{A}^\mu_r(t,\vec{x},\vec{k},\sigma) = \hat{a}(\vec{k},\sigma)\epsilon^\mu(\vec{k},\sigma)\exp[-ik_\nu x^\nu/\hbar] + \hat{a}^\dagger(\vec{k},\sigma)\epsilon^{*\mu}(\vec{k},\sigma)\exp[ik_\nu x^\nu/\hbar], \quad (5)$$

where $\hat{a}^\dagger$ and $\hat{a}$ are the creation and annihilation operators of the photon, and $\vec{\epsilon}$ is its polarization. The integration in this definition is reduced to a summation for the implementation of this computation.

On the other hand, $\hat{A}^\mu_{A,M}$ is given with creation-annihilation operators of matter fields. These photons are divided into two parts, scalar one, $\hat{A}^0_{A,M}$, and vector one, $\hat{\vec{A}}_{A,M}$. The scalar part is given by

$$\hat{A}_{0_{A,M}}(t,\vec{x}) = Z_e e \int_{A,M} d^3\vec{s}\frac{\hat{\Psi}^\dagger(t,\vec{s})\hat{\Psi}(t,\vec{s})}{|\vec{x}-\vec{s}|} + \sum_a Z_a e\int_{A,M} d^3\vec{s}\frac{\hat{\chi}^\dagger_a(t,\vec{s})\hat{\chi}_a(t,\vec{s})}{|\vec{x}-\vec{s}|}. \quad (6)$$

The vector part is induced by the transverse component of currents,

$$\hat{\vec{A}}_{A,M}(t,\vec{x}) = \frac{1}{c}\int_{A,M} d^3\vec{s}\frac{\hat{\vec{j}}_{eT}(u,\vec{s})}{|\vec{x}-\vec{s}|} + \frac{1}{c}\int_{A,M} d^3\vec{s}\frac{\hat{\vec{j}}_{NT}(u,\vec{s})}{|\vec{x}-\vec{s}|}, \quad (7)$$

where $u = t - |\vec{r}-\vec{s}|/c$ is introduced for the retardation effect.

Next, the description of the evolution of the fields is introduced. For the photon field, the evolution is already included partially in (3)-(7). Further effects of the time evolution of operators beyond this approximation is not considered in this article. These effects are discussed in our next study. The equations of motion of the electron and nucleus fields are given by

$$i\hbar\frac{d}{dt}\hat{\Psi} = \left[\frac{1}{2m_e}\left(-i\hbar\partial_i - \frac{Z_e e}{c}\hat{A}^i\right)^2 + Z_e e\hat{A}_0\right]\hat{\Psi}, \tag{8}$$

$$i\hbar\frac{d}{dt}\hat{\chi}_a = \left[\frac{1}{2m_a}\left(-i\hbar\partial_i - \frac{Z_a e}{c}\hat{A}^i\right)^2 + Z_a e\hat{A}_0\right]\hat{\chi}_a. \tag{9}$$

By these equations, the equations of motion for annihilation operators are derived as

$$i\hbar\frac{d}{dt}\hat{e}_m(t) = \sum_n \int d^3\vec{x}\psi_m^\dagger(\vec{x})\left[\frac{1}{2m_e}\left(-i\hbar\partial_i - \frac{Z_e e}{c}\hat{A}^i\right)^2 + Z_e e\hat{A}_0\right]\psi_n(\vec{x})\hat{e}_n(t), \tag{10}$$

$$i\hbar\frac{d}{dt}\hat{f}_{am}(t) = \sum_n \int d^3\vec{x}\chi_{am}^\dagger(\vec{x})\left[\frac{1}{2m_a}\left(-i\hbar\partial_i - \frac{Z_a e}{c}\hat{A}^i\right)^2 + Z_a e\hat{A}_0\right]\chi_{an}(\vec{x})\hat{f}_{an}(t). \tag{11}$$

The creation operators at t can be derived as Hermitian conjugate operators. These equations are computed in a numerical manner. In the manner, the continuum time is discretized by finite time step, δt. The operators at t are calculated as an expansion by the operators at $t = 0$. This expansion should not be limited only at the first order of creation-annihilation operators. For example, the operator, $\hat{A}_0$, contains electron annihilation and creation operators, and $\hat{A}^i_{\rm rad}$ in $\hat{A}^i$ includes the creation-annihilation operators of the photon. Hence, the operators, $\hat{e}_m$ and $\hat{f}_{am}$, should be expanded as the polynomials of creation-annihilation operators of the electron, nucleus and photon. Accordingly, the coefficients of these polynomials are computed by our numerical code.

In the time evolution computation in this expansion procedure, the unitary operator, $\hat{U} = \exp(-i\hat{H}t)$, is impossible to use, since the Hamiltonian is not a number but an operator with creation-annihilation operators. As a result, we solve the equation of motion by linear approximation,

$$\hat{e}_m(t_{i+1}) \simeq \hat{e}_m(t_i) + \delta t\frac{d\hat{e}_m}{dt}(t_i), \tag{12}$$

$$\hat{f}_{am}(t_{i+1}) \simeq \hat{f}_{am}(t_i) + \delta t\frac{d\hat{f}_{am}}{dt}(t_i), \tag{13}$$

where $t_{i+1} = t_i + \delta t$. The value of δt should be small enough to satisfy the condition that this linear expansion is reliable. The order of the polynomials of the operators at $t = 0$ increases exponentially in an instant. In addition, $\hat{\vec{A}}_{A,M}$ is defined with the operators at past steps in order to represent retardation effects. To calculate these terms exactly, all expansion coefficients of the all past operators are required to store, and storing this data is not realistic due to their huge size. Hence, the order of operator polynomials is restricted to some order. In the present work, we adopt the minimal order expansion as a trial computation. For the electron field, the creation-annihilation operator of the electron at $t = 0$ is limited to third order, that of the photon is up to the second order, and those of nucleus fields are restricted to the zeroth order. This nucleus contribution to the electron field is given by using the expectation value of nucleus fields. In other words, this nucleus contribution is given by not an operator but a number. The third order electron operator terms are required to represent the exchange interaction, and hence this is considered to be minimal. For nucleus fields, the limitation of the operator order is chosen as the similar criterion. The same nucleus field is limited to the third order, and other species and the electron field are replaced by the expectation values.

The calculation of $\hat{\vec{A}}_{A,M}$ is tremendous task, since this term is difficult to compute in an analytic form. Hence, we use the expectation value of this term by a numerical integration and a difference method. (One approach to compute this term in a smart form is mentioned in [10].) The effects of $\hat{\vec{A}}_{A,M}$ are quantitatively small in spite of their qualitative importance. Hence we calculate these terms once fifties time steps and until $\hat{\vec{A}}_{A,M}$ is updated the same value of $\hat{\vec{A}}_{A,M}$ is used. The transverse component of current in equation (7) is computed as [8]

$$\hat{\vec{j}}_{eT}(t,\vec{x}) = \hat{\vec{j}}_e(t,\vec{x}) - \hat{\vec{j}}_{eL}(t,\vec{x}), \quad (14)$$

$$\hat{\vec{j}}_e = \frac{1}{2m_e}\left[i\hbar\hat{\Psi}^\dagger\hat{\vec{D}}\hat{\Psi} - i\hbar\left(\hat{\vec{D}}\hat{\Psi}\right)^\dagger\hat{\Psi}\right], \quad (15)$$

$$\hat{\vec{j}}_{eL} = \frac{1}{4\pi}\mathrm{grad}\frac{\partial}{\partial t}\hat{A}_0, \quad (16)$$

where the nuclear current has the same form. The difficulty of the analytic calculation of $\hat{\vec{A}}_{A,M}$ resides in $\mathrm{grad}\frac{\partial}{\partial t}\hat{A}_0$. The values of $\langle\hat{\vec{j}}_{eT}\rangle$ are calculated on grid points and stored. The numerical integration in $\hat{\vec{A}}_{A,M}$ is performed based on this stored data. The study about $\hat{\vec{A}}_{A,M}$ will be reported in our future work in detail.

After the evolution of operators is calculated, we can calculate physical quantities, such as charge density and energy density. Of course, these quantities depends on states, i.e. ket vector. Ket vector is defined with creation operators and the vacuum state $|0\rangle$. For the photon, coherent states are chosen in the present version of our code to suppress data size. For matter fields, the choice of ψ_m and χ_{am} is important for efficient and accurate computations. In this work, we simply use Hartree-Fock orbitals as ψ_m for electrons. The extension to post Hartree-Fock calculation including correlation effects is not so difficult, and we implement this in the near future. For nuclei, we use wave packets as χ_{am} so that the orthogonality, $\int \chi^\dagger_{am}\chi_{an}d^3x = \delta_{mn}$, is satisfied.

Before concluding this section, we mention the renormalization of fields. In the exact form of our method, all quantum effects of the system are included, and hence fields must face divergence. This divergence problem should be solved by the renormalization as ordinary quantum field theory. However, the ordinary renormalization scheme assumes the existence of asymptotic fields. We do not know the asymptotic state for a bound state in quantum field theory. We must adopt a novel prescription of renormalization for our method. For this purpose, we use conservation laws, such as energy and charge. For each time step, we calculate quantities, which must conserve, and if the values are deviated from the true values the operators are renormalized by the common normalization factor. We have coded the renormalization routine using particle number conservation. However other conservation quantities, total energy, momentum, and angular momentum, have not been coded yet. Hence for the calculation shown in this article we adopt the prescription used in NRQED for bound states [12]. In this prescription, cutoff, for the ultra-violet divergence, is used, since the nonrelativistic Lagrangian is incorrect for very high energy, for example higher energy than the electron mass. For higher energy than the cutoff scale, the renormalization for a free particle is adopted, since binding energy is negligible for such a high energy. Therefore effects by virtual ultra-violet photons from the second order operator expansion are appropriately renormalized.

3. Results

In this article, we show our results for hydrogen atom as a demonstration. The purpose of the usage of this model is to show the validity of the interaction by radiation photons. We can of course calculate larger models, and however the behavior of operators is slightly worse due to

Journal of Physics: Conference Series **454** (2013) 012052 doi:10.1088/1742-6596/454/1/012052

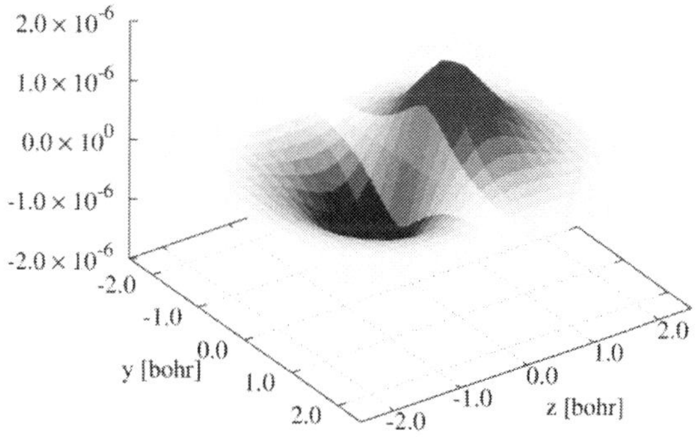

Figure 1. The electron density distribution on $y-z$ plane ($x=0$) at $t=2.0$ a.u. The density is depicted as the difference from that at $t=0$. The units are atomic units.

the insufficient renormalization of the present version of our code, since interactions among the system are much stronger. The results of molecules will be reported in the near future.

In our computation, the initial state of the hydrogen atom is prepared by our own self-consistent field simulation code which can calculate both electronic and nuclear wave functions. The initial state is prepared in a Hartree-Fock calculation. The basis set of the electron is cc-pVDZ [13]. For the nuclear one, we use our (3s1p) basis, where one exponent of Gaussian (1s1p) is chosen as the same in [14] and other two exponents are chosen to have larger and smaller values. Incident photons are assumed to move toward the positive y direction with kinetic energy, 10 hartree. The ket vector of the photon is the coherent state whose eigenvalue is 0.01. The time step width is chosen as $\delta t = 10^{-5}$ a.u. (1 a.u.$= 2.4 \times 10^{-17}$ s) and the cutoff scale is chosen as 20 a.u. for virtual photons. This cutoff scale is much lower than $m_e c^2$ for the sake of reliable computation.

In figure 1, the electron density distribution at $t=2.0$ a.u. is shown on $y-z$ plane ($x=0$). The density evolution is shown as the difference from the density at $t=0$. The response of the electron is correctly transverse to the photon momentum, and its distribution pattern is plausible. In figure 2, the evolution of the electron density is shown as a function of time. The electron density at $(x,y,z) = (0,0,0.8)$ bohr is depicted. Two time scale can be seen in this figure. The shorter time scale, whose period is about 0.6 a.u., corresponds to the photon oscillation of $\exp(\pm i E_{\rm photon} t)$. The longer time scale, whose period is about 5 a.u., is attributed to the energy difference between the ground and excited states. This oscillation behavior is seen to be disturbed slightly. We consider that this may be due to the effects of the nuclear motion and $\vec{A}_A$.

Finally, we should mention the thermalization of Hamiltonian. The initial state we prepared is not the QED initial state but the state which is derived by variational calculations in quantum mechanics. As the real QED initial state, we should calculate it in our QED program. For the calculation without external incident photon, we can derive the QED state, where the space is filled by real and virtual photons, after long enough time steps are computed. For this computation, an appropriate definition is required for the condition of sufficient thermalization. In addition, infinite real and virtual photon modes, which are described by both $\hat{A}_{\rm rad}$ and $\hat{A}_A$,

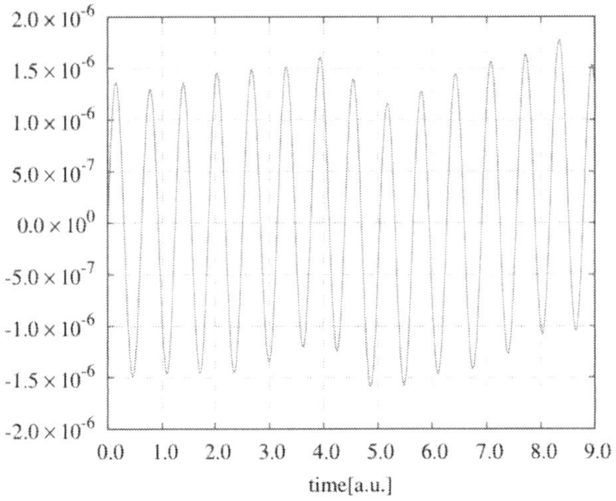

Figure 2. The evolution of the electron density as a function of time. The electron density at $(x, y, z) = (0, 0, 0.8)$ bohr is depicted. The units are atomic units.

are required to be treated. For these issues, we are now trying some ideas. Hence, the results shown in this article are not for the QED initial state.

4. Summary

We have proposed our computation method of the time evolution of quantum systems based on Rigged QED and reported the present status of the development of our program code, QEDynamics [6]. In our method, the evolution of Heisenberg operators of nuclei and electrons is calculated as coefficients of polynomials of creation-annihilation operators of electrons, nuclei, and photons. For the simulation based on QED, the treatment of the photon is one of the most important points, particularly infrared and ultra-violet photons. For the treatment of the ultra-violet photon, we have used the cutoff. In this article, the results for the hydrogen atom is shown as a demonstration. For larger atoms and molecules, where internal interactions are much stronger, the renormalization is extremely important in viewpoints of quantum field theory. The renormalization is only partially implemented in this version of our code. We improve the treatment of the renormalization method in the near future

Acknowledgments

This work was partially supported by a Grant-in-Aid for Scientific Research from the Japan Society for the Promotion of Science (JSPS) (22550011). M. S. is supported by a Grant-in-Aid for Young Scientists (B) from JSPS (24760028) and Mizuho Foundation for the Promotion of Sciences. Y. I. is supported by the Sasakawa Scientific Research Grant from the Japan Science Society.

References

[1] Aoyama T, Hayakawa M, Kinoshita T and Nio M 2012 Tenth-order qed contribution to the electron g-2 and an improved value of the fine structure constant, and references therein (*Preprint* `arXiv:1205.5368 [hep-ph]`)
[2] Hanneke D, Hoogerheide S F and Gabrielse G 2011 *Phys. Rev. A* **83** 052122
[3] et al [Particle Data Group Collaboration] J B 2012 *Phys. Rev. D* **86** 010001
[4] Caswell W E and Lepage G P *Phys. Lett.* **167**

[5] Dyal K G and Fægri K 2007 *Introduction to Relativistic Quantum Chemistry* (New York: Oxford University Press)
[6] Senami M, Ichikawa K and Tachibana A *QEDynamics, http://www.tachibana.kues.kyoto-u.ac.jp/qed*
[7] Tachibana A 2005 *J. Mol. Modeling* **11** 301
[8] Tachibana A 2010 *J. Mol. Struct.: THEOCHEM* **943** 138
[9] Tachibana A 2012 *J. Math. Chem.* **50** 669
[10] Tachibana A *Electronic Stress with Spin Vorticity, in Frontiers in Theoretical Chemistry: Concepts and Methods: A tribute to Professor B. M. Deb* ed Ghosh S K and Chattaraj P K
[11] Iliaš M and Sauc T 2007 *J. Chem. Phys.* **126** 064102
[12] Jentschura U D and Nandori I 2002 *Phys. Rev. A* **66** 022114
[13] T H Dunning J 1989 *J. Chem. Phys.* **90** 1007
[14] Nakai H 2002 *Int. J. Quant. Chem.* **86** 511

Journal of Physics: Conference Series **454** (2013) 012053 doi:10.1088/1742-6596/454/1/012053

Coupled perturbed Hartree-Fock method for non-Hermitian Hamiltonians

Yuji Ikeda, Masato Senami and Akitomo Tachibana*

Department of Micro Engineering, Kyoto University, Kyoto 615-8540, Japan

*E-mail: akitomo@scl.kyoto-u.ac.jp

Abstract. In order to investigate conductive properties of the system composed of a channel molecule and electrodes, a coupled perturbed Hartree-Fock method for non-Hermitian perturbation Hamiltonians is derived. A non-Hermitian perturbation Hamiltonian which describes interactions between the channel molecule and an electrode is given to represent inflow and outflow at the interface. The effects of electrodes includes the states of the density of state of the channel molecule and the statistical distribution of electrons in electrodes. Benzenedithiol and dihydro-benzenedithiol are studied in this paper, and the first order perturbations of molecular orbital energies of them are obtained.

1. Introduction

Conductive properties of single molecules have been discussed in many articles [1, 2, 3, 4, 5, 6, 7, 8]. The conductive properties of the molecules are often analyzed by using methods based on Green's functions [9, 10, 11, 12, 13, 14]. For these methods, the effects of electrodes are often described as non-Hermitian self-energy matrices [15]. This method has succeeded to estimate currents and conductances through the molecules. Previously, Shimazaki et al. calculated the transmission probabilities of benzenedithiol (BDT) and its derivatives based on the Hartree-Fock (HF) method with the self-energy matrices [13]. In their paper, an approximate Green's function for a channel molecule is calculated by using the first order perturbation energies of molecular orbitals (MOs), $\varepsilon_p^{(1)}$, where the subscript p indicates the pth MO. $\varepsilon_p^{(1)}$ is calculated as,

$$\varepsilon_p^{(1)} = \langle p^{(0)} | \Sigma_L + \Sigma_R | p^{(0)} \rangle, \tag{1}$$

where $\Sigma_{L,R}$ are the self-energies of the left and right electrodes, respectively. The obtained predictions of currents and conductances based on Green's function show similar patterns to another theoretical calculation based on density functional theory and a non-equilibrium Green's function [12]. In the above equation, however, the shift of Coulomb and exchange interactions between electrons induced by the perturbation are not considered. In other words, perturbed MOs are not consistent with themselves. As shown by Langhoff et al. in 1960's [16], the interactions between electrons are important to obtain physical properties which are qualitatively and quantitatively comparable to experimental results or real values. Hence, these interactions should be considered in calculations of perturbed MO energies.

In this work, we study the calculation method of electronic structures perturbed by non-Hermitian Hamiltonians. Our method is based on coupled perturbed HF (CPHF) equations [17, 18], and we apply them to the case with a non-Hermitian perturbation. MOs obtained by the CPHF method are consistent with Coulomb and exchange interactions between electrons at the HF level.

In order to investigate conductive properties of a molecule based on its electronic structure, the perturbation Hamiltonian inducing the electronic structure with an electric current must be considered. The self-energies of electrodes do not directly correspond to appropriate perturbation Hamiltonians, since they do not include effects of the passing through a channel molecule. In this study, we consider the density of states (DOS) for the molecule and Fermi-Dirac (FD) distributions of electrons in the electrodes in order to include the passing effects. By combining the DOS and the FD distributions with the self-energies, the perturbation Hamiltonian is constructed. It is noted that the DOS for the molecule is calculated by using perturbed orbital energies, and hence the perturbation Hamiltonian must be determined by a self-consistent field method.

In this work, we focus on BDT and dihydro-benzenedithiol (DHBDT). Perturbed MO energies of these molecules are computed.

2. Theory

In this section, CPHF equations for a non-Hermitian perturbation Hamiltonian are introduced. By using our method, perturbation effects of the Coulomb and exchange interactions between electrons can be included at the HF level. It is emphasized that our method shown in this section can be applied not only to the perturbation Hamiltonian introduced in this paper but also to other general non-Hermitian perturbation Hamiltonians. Although equations only for the first order perturbation are shown, equations for higher order perturbation are probably derived by using conventional CPHF schemes with appropriate conditions. Furthermore, it is considered that the CPHF equations for a non-Hermitian perturbation Hamiltonian can be applied to post-HF computations as well as the conventional ones can be done.

The indices i and j represent those of occupied spin-orbitals. The indices a and b are those of virtual (unoccupied) spin-orbitals. The indices $p, q, r, \cdots$ are used for both occupied and unoccupied ones. Unperturbed canonical MOs, $|p^{(0)}\rangle$, are used as basis. They satisfy the following HF equations without perturbation,

$$\boldsymbol{F}^{(0)}|p^{(0)}\rangle = \varepsilon_p^{(0)}|p^{(0)}\rangle, \tag{2}$$

$$\langle p^{(0)}|\boldsymbol{F}^{(0)} = \langle p^{(0)}|\varepsilon_p^{(0)}, \tag{3}$$

where $\boldsymbol{F}^{(0)}$ is the Fock matrix without perturbation and $\varepsilon_p^{(0)}$ is the energy of $|p^{(0)}\rangle$. Provided that $\boldsymbol{F}^{(0)}$ is Hermitian, (2) and (3) are equal to each other. Next, the change by the addition of perturbation Hamiltonian, $\boldsymbol{H}^{(1)}$, is discussed. In this study, $\boldsymbol{H}^{(1)}$ is not assumed to be Hermitian. For a non-Hermitian matrix, its right and left eigenvectors are different from each other. In this section, the superscripts L and R are used for the left and right eigenvectors, respectively. Perturbed MOs $|p^R\rangle$ and $\langle p^L|$ satisfy the following HF equations with perturbation, respectively,

$$\boldsymbol{F}|p^R\rangle = \varepsilon_p|p^R\rangle, \tag{4}$$

$$\langle p^L|\boldsymbol{F} = \langle p^L|\varepsilon_p, \tag{5}$$

where $\boldsymbol{F}$ is the Fock matrix with perturbation and ε_p is the energy of $|p^R\rangle$ and $\langle p^L|$. It is noted that $\boldsymbol{F}$ cannot be assumed to be Hermitian because of $\boldsymbol{H}^{(1)}$, and hence the right and left eigenvectors are generally different from each other. The zeroth and first order Fock matrices,

$\boldsymbol{F}^{(0)}$ and $\boldsymbol{F}^{(1)}$, respectively, can be written as,

$$\boldsymbol{F}^{(0)} = \boldsymbol{H}^{(0)} + \boldsymbol{G}(\boldsymbol{P}^{(0)}), \tag{6}$$

$$\boldsymbol{F}^{(1)} = \boldsymbol{H}^{(1)} + \boldsymbol{G}(\boldsymbol{P}^{(1)}), \tag{7}$$

where $\boldsymbol{H}^{(0)}$ is one-electron interactions without perturbation. The matrix $\boldsymbol{G}(\boldsymbol{P})$ represents two-electron interactions and is written as,

$$\boldsymbol{G}_{pq}(\boldsymbol{P}) = \sum_{r,s} P_{sr} \langle pr||qs\rangle, \tag{8}$$

where $\boldsymbol{P}$ is a density matrix and $\langle pq||rs\rangle$ is a two electron integral. The zeroth and the first order density matrices, $\boldsymbol{P}^{(0)}$ and $\boldsymbol{P}^{(1)}$, respectively, can be obtained as,

$$\boldsymbol{P}^{(0)} = \sum_i |i^{(0)}\rangle\langle i^{(0)}|, \tag{9}$$

$$\boldsymbol{P}^{(1)} = \sum_i \Big(|i^{(0)}\rangle\langle i^{L(1)}| + |i^{R(1)}\rangle\langle i^{(0)}|\Big). \tag{10}$$

It is emphasized that $\boldsymbol{P}^{(1)}$ consists of both the left and right eigenvectors. $|p^R\rangle$ is written by using a matrix $\boldsymbol{\kappa}$ as,

$$|p^R\rangle = \sum_q \exp(\boldsymbol{\kappa})_{qp}|q^{(0)}\rangle. \tag{11}$$

Specifically, the first order perturbed MOs, $|p^{R(1)}\rangle$, can be written as,

$$|p^{R(1)}\rangle = \sum_q \kappa_{qp}|q^{(0)}\rangle. \tag{12}$$

Due to properties of determinants, occupied-occupied and virtual-virtual elements of $\boldsymbol{\kappa}$ can be neglected in the rest of this section, since these elements do not contribute to the Slater determinant of the ground state. The left and right eigenvectors satisfy the following biorthonormal condition,

$$\langle p^L|q^R\rangle = \delta_{pq}. \tag{13}$$

In order to satisfy (13), $\langle p^L|$ should take a simple form with $\boldsymbol{\kappa}$,

$$\langle p^L| = \sum_q \langle q^{(0)}|\exp(-\boldsymbol{\kappa})_{pq}, \tag{14}$$

and the first order perturbed $\langle p^{L(1)}|$ can be written as,

$$\langle p^{L(1)}| = \sum_q \langle q^{(0)}|(-\kappa_{pq}). \tag{15}$$

The following equation is obtained by multiplying (4) $\langle p^L|$ from the left,

$$\langle p^L|\boldsymbol{F}|q^R\rangle = \varepsilon_p\delta_{pq}. \tag{16}$$

In order to satisfy (16) at the first order level, the virtual-occupied elements of $\boldsymbol{\kappa}$ are required to satisfy the following equation,

$$\langle a^{L(1)}|\boldsymbol{F}^{(0)}|i^{(0)}\rangle + \langle a^{(0)}|\boldsymbol{F}^{(1)}|i^{(0)}\rangle + \langle a^{(0)}|\boldsymbol{F}^{(0)}|i^{R(1)}\rangle = 0. \tag{17}$$

By substituting (7), (12), and (15) in (17), the following CPHF equation for a non-Hermitian Hamiltonian is derived,

$$(\varepsilon_a^{(0)} - \varepsilon_i^{(0)})\kappa_{ai} = -H_{ai}^{(1)} - \sum_{b,j}\Big(- \kappa_{jb}\langle ab||ij\rangle + \kappa_{bj}\langle aj||ib\rangle\Big). \tag{18}$$

By the same manner, we can obtain the following equation for occupied-virtual elements of $\boldsymbol{\kappa}$,

$$(\varepsilon_a^{(0)} - \varepsilon_i^{(0)})\kappa_{ia} = \quad H_{ia}^{(1)} + \sum_{b,j}\Big(- \kappa_{jb}\langle ib||aj\rangle + \kappa_{bj}\langle ij||ab\rangle\Big). \tag{19}$$

Actually, we can simplify the calculation of (18) and (19) by dividing $\boldsymbol{H}^{(1)}$ into the Hermitian and anti-Hermitian parts, $\boldsymbol{H}^{(1)\mathrm{H}}$ and $\boldsymbol{H}^{(1)\mathrm{A}}$, respectively. For $\boldsymbol{H}^{(1)\mathrm{H}}$, (18) and (19) give the same equation, and anti-Hermitian $\boldsymbol{\kappa}$ is obtained. Hence the number of simultaneous equations to be solved decreases by half. In a similar way, Hermitian $\boldsymbol{\kappa}$ can be obtained for $\boldsymbol{H}^{(1)\mathrm{A}}$. Due to the linearity of the first order perturbation, $\boldsymbol{\kappa}$ for the original $\boldsymbol{H}^{(1)}$ can be obtained as the sum of the ones for $\boldsymbol{H}^{(1)\mathrm{H}}$ and $\boldsymbol{H}^{(1)\mathrm{A}}$.

By using (16) and $\boldsymbol{\kappa}$ calculated from (18) or (19), $\varepsilon_p^{(1)}$ can be obtained as,

$$\begin{aligned}\varepsilon_p^{(1)} &= \langle p^{L(1)}|\boldsymbol{F}^{(0)}|p^{(0)}\rangle + \langle p^{(0)}|\boldsymbol{F}^{(1)}|p^{(0)}\rangle + \langle p^{(0)}|\boldsymbol{F}^{(0)}|p^{R(1)}\rangle \\ &= \langle p^{(0)}|\boldsymbol{F}^{(1)}|p^{(0)}\rangle \\ &= H_{pp}^{(1)} + \sum_{a,i}\Big(- \kappa_{ia}\langle pa||pi\rangle + \kappa_{ai}\langle pi||pa\rangle\Big).\end{aligned} \tag{20}$$

The improvement on (1) exists in the second term. The first order perturbation for the expectation value of an one body operator, $\hat{\boldsymbol{O}}$, can be calculated as,

$$\begin{aligned}\langle\hat{\boldsymbol{O}}\rangle^{(1)} &= \sum_i\Big(\langle i^{L(1)}|\hat{\boldsymbol{O}}|i^{(0)}\rangle + \langle i^{(0)}|\hat{\boldsymbol{O}}|i^{R(1)}\rangle\Big) \\ &= \sum_{a,i}\Big(- \kappa_{ia}\langle a^{(0)}|\hat{\boldsymbol{O}}|i^{(0)}\rangle + \kappa_{ai}\langle i^{(0)}|\hat{\boldsymbol{O}}|a^{(0)}\rangle\Big).\end{aligned} \tag{21}$$

Due to the Hermitian part of $\boldsymbol{\kappa}$ derived from $\boldsymbol{H}^{(1)\mathrm{A}}$, the value of $\langle\hat{\boldsymbol{O}}\rangle^{(1)}$ is generally a complex number. It should be noted that even if non-Hermitian perturbation is applied, the total number of electrons n_e does not change up to the first order level. This can easily be seen by the following equation,

$$\begin{aligned}n_e^{(1)} &= \sum_i\Big(\langle i^{L(1)}|i^{(0)}\rangle + \langle i^{(0)}|i^{R(1)}\rangle\Big) \\ &= \sum_{a,i}\Big(- \kappa_{ia}\langle a^{(0)}|i^{(0)}\rangle + \kappa_{ai}\langle i^{(0)}|a^{(0)}\rangle\Big) \\ &= 0.\end{aligned} \tag{22}$$

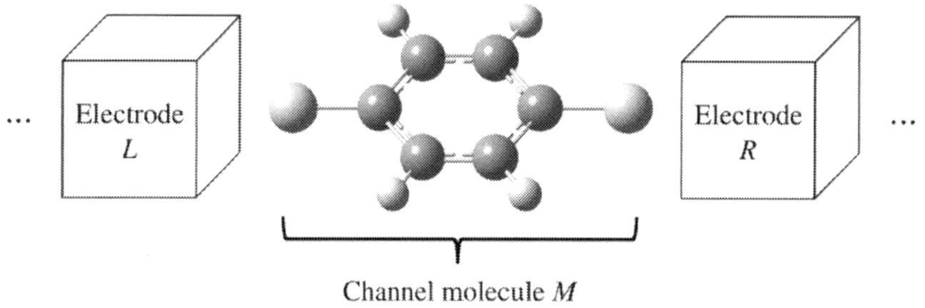

Figure 1. A whole system consisting of a channel molecule and two electrodes.

In terms of $\boldsymbol{\kappa}$ and $\varepsilon_p^{(1)}$, several approximate Green's functions are reported. One of them is proposed by Shimazaki et al. [13], which is given as,

$$\boldsymbol{G}_1(E) = \sum_p \frac{|p^{(0)}\rangle\langle p^{(0)}|}{E - (\varepsilon_p^{(0)} + \varepsilon_p^{(1)})}. \tag{23}$$

While (1) is used in [13] for obtaining $\varepsilon_p^{(1)}$, we can use (20) instead. $\varepsilon_p^{(1)}$ of (20) adds the effects of $\boldsymbol{\kappa}$ to the Green's function. Another improvement is given as,

$$\begin{aligned}\boldsymbol{G}_2(E) &= \boldsymbol{G}_1(E) + \sum_p \frac{|p^{(0)}\rangle\langle p^{L(1)}| + |p^{R(1)}\rangle\langle p^{(0)}|}{E - (\varepsilon_p^{(0)} + \varepsilon_p^{(1)})} \\ &= \boldsymbol{G}_1(E) + \sum_{p,q} \frac{-\kappa_{pq}|p^{(0)}\rangle\langle q^{(0)}| + \kappa_{qp}|q^{(0)}\rangle\langle p^{(0)}|}{E - (\varepsilon_p^{(0)} + \varepsilon_p^{(1)})}.\end{aligned} \tag{24}$$

In this improvement, perturbation effects on MOs through $\boldsymbol{\kappa}$ are added. If the self-energies of electrodes are given as perturbation Hamiltonian, transmission probabilities can be calculated by the conventional method based on these Green's functions.

3. Calculation Methods

3.1. Perturbation Hamiltonian inducing electronic structures with electric currents

In this subsection, we show how we construct a perturbation Hamiltonian giving electronic structures with electric currents. We focus on a channel molecule between two electrodes (see Fig. 1). The Hamiltonian of the whole system, $\boldsymbol{H}_{\text{whole}}$, can be written as,

$$\boldsymbol{H}_{\text{whole}} = \begin{pmatrix} \boldsymbol{H}_{LL} & \boldsymbol{H}_{LM} & \boldsymbol{H}_{LR} \\ \boldsymbol{H}_{ML} & \boldsymbol{H}_{MM} & \boldsymbol{H}_{MR} \\ \boldsymbol{H}_{RL} & \boldsymbol{H}_{RM} & \boldsymbol{H}_{RR} \end{pmatrix}, \tag{25}$$

where the indices L, R, and M represent the left electrode, the right electrode, and the channel molecule, respectively. Since interactions between the two electrodes are negligibly small, $\boldsymbol{H}_{LR} = \boldsymbol{H}_{RL} = 0$. $\boldsymbol{H}_{MM}$ is regarded as $\boldsymbol{F}^{(0)}$ in (6). The self-energy of the left electrode, $\boldsymbol{\Sigma}_L(E)$, can be calculated by using the following equation [15],

$$\boldsymbol{\Sigma}_L(E) = \boldsymbol{H}_{ML}\, \boldsymbol{g}_L(E)\, \boldsymbol{H}_{LM}, \tag{26}$$

where E is the energy of electrons passing through the channel molecule, and $\boldsymbol{g}_L(E)$ is the surface Green's function of the left electrode. $\boldsymbol{\Sigma}_R(E)$ for the right electrode can be obtained as the same form. The derivation of $\boldsymbol{g}_{L,R}(E)$ is discussed in the later part of this subsection. $\boldsymbol{\Sigma}_{L,R}(E)$ generates the outflow of electrons from the channel molecule, while $\boldsymbol{\Sigma}^{\dagger}_{L,R}(E)$ represents the inflow. We define the matrix $\boldsymbol{\Sigma}(E)$ as,

$$\begin{aligned}\boldsymbol{\Sigma}(E) =& \Big(\boldsymbol{\Sigma}_L(E) + \boldsymbol{\Sigma}^{\dagger}_R(E)\Big)\Big(1 - f_L(E)\Big) f_R(E) \\ &+ \Big(\boldsymbol{\Sigma}^{\dagger}_L(E) + \boldsymbol{\Sigma}_R(E)\Big) f_L(E)\Big(1 - f_R(E)\Big),\end{aligned} \tag{27}$$

where $f_L(E)$ and $f_R(E)$ are FD distribution functions for the left and right electrodes, respectively. The first term induces electrons which pass from the right electrode to the left one, while the second term induces the opposite flow. Both terms are weighted by $f_L(E)$ and $f_R(E)$. In this study, zero Kelvin is assumed in $f_L(E)$ and $f_R(E)$. The perturbation Hamiltonian inducing electronic currents is given by the following equation,

$$\boldsymbol{H}^{(1)} = \int_{-\infty}^{\infty} \boldsymbol{\Sigma}(E) w(E) dE, \tag{28}$$

where $w(E)$ is the DOS for the channel molecule. $\boldsymbol{H}^{(1)}$ includes transmission effects through $f_{L,R}(E)$ and $w(E)$. $w(E)$ is given by using the orbital energy, ε_p, and its width, γ_p, as,

$$w(E) = \frac{1}{\pi} \sum_p \frac{\gamma_p}{(E - \varepsilon_p)^2 + \gamma_p^2}. \tag{29}$$

ε_p and γ_p are determined as,

$$\varepsilon_p = \varepsilon_p^{(0)} + \mathrm{Re}(\varepsilon_p^{(1)}), \tag{30}$$

$$\gamma_p = \mathrm{Im}(\varepsilon_p^{(1)}), \tag{31}$$

and hence $\boldsymbol{H}^{(1)}$ depends on itself. Therefore, perturbation calculations are repeated iteratively until $\boldsymbol{H}^{(1)}$ becomes self-consistent.

There are various calculation methods for $\boldsymbol{g}_{L,R}$. In this study, we adopt the method proposed by Shimazaki et al. [13] due to its simplicity. In this method, a one-dimensional electrode is assumed. Then $\boldsymbol{g}_L$ is given as,

$$\boldsymbol{g}_L = \frac{\exp(ikc)}{\beta_L}, \tag{32}$$

$$E = \alpha_L + 2\beta_L \cos(kc), \tag{33}$$

where α_L is the Fermi energy of the left electrode, and β_L is the nearest-neighbor electronic coupling element [15, 19]. The sign of k is determined so that it represents the outflow of electrons. $\boldsymbol{g}_R$ is given in the same form. In this study, $\beta_{L,R}$ is fixed as $\beta_{L,R} = -3.00$ eV [13]. α_L is given by the following equation,

$$\alpha_L = E_F - \frac{eV_D}{2}, \tag{34}$$

where E_F is the Fermi energy of the channel molecule, e is the elementary charge, and V_D is the bias voltage. In this study, E_F is fixed as $E_F = -4.77$ eV [13]. Similarly, α_R is given as,

$$\alpha_R = E_F + \frac{eV_D}{2}. \tag{35}$$

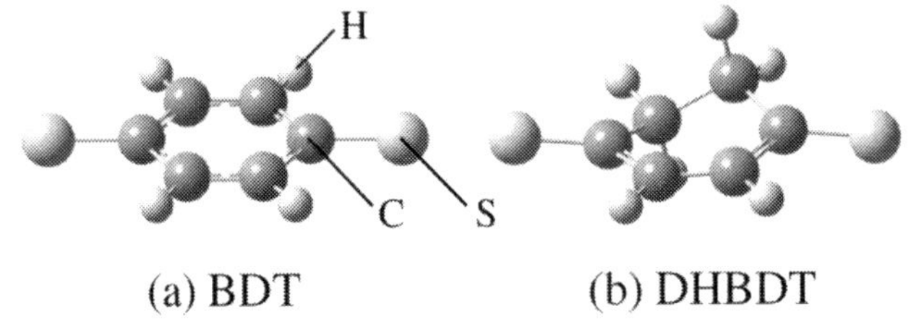

Figure 2. Optimized structures of (a) BDT and (b) DHBDT.

Table 1. Energies of MOs in eV.

	BDT			DHBDT		
	$\varepsilon^{(0)}$	$\mathrm{Re}(\varepsilon^{(1)})$	$\mathrm{Im}(\varepsilon^{(1)})$	$\varepsilon^{(0)}$	$\mathrm{Re}(\varepsilon^{(1)})$	$\mathrm{Im}(\varepsilon^{(1)})$
LUMO	2.7330	−0.0647	−0.0140	3.9434	−0.0451	−0.0098
HOMO	−5.8459	−0.0580	−0.0125	−6.4909	−0.0456	−0.0099
HOMO-1	−6.6850	−0.0616	−0.0133	−6.4959	−0.0470	−0.0102
HOMO-2	−6.7335	−0.0617	−0.0133	−6.6100	−0.0409	−0.0088

3.2. Computational models

In this work, the method explained above is applied to BDT and DHBDT. Optimized structures of them are shown in Fig. 2. DHBDT is composed of BDT and two additional hydrogen atoms. This molecule is used for the comparison between the effects of π and σ electrons. The sulfur atoms at the ends of the molecules are implicitly assumed to be connected to one-dimensional gold nanowire electrodes with coupling constants, −2.0 eV, through $3p_x$ and $3p_y$ atomic orbitals. The STO-6G basis set [20, 21] is used for all the atoms. Optimized structures are obtained by using Gaussian 09 [22] at the HF level, and CPHF computations are performed by using our original code.

4. Results and Discussion

Computed energies of the lowest unoccupied MO (LUMO), the highest occupied MO (HOMO), HOMO-1, and HOMO-2 for BDT and DHBDT are summarized in Table 1. As for occupied MOs, HOMO-1 and HOMO-2 of BDT have larger absolute values of $\mathrm{Im}(\varepsilon^{(1)})$ than that of HOMO. For DHBDT, on the other hand, HOMO-1 has a larger value, followed by HOMO and HOMO-2. This is investigated in relation to electronic structures of MOs shown in Fig. 3. HOMO of BDT and HOMO-2 of DHBDT, which have smaller values of $\mathrm{Im}(\varepsilon^{(1)})$, have similar structures. Hence, it can be speculated that the MOs consisting of π bonds have small $\mathrm{Im}(\varepsilon^{(1)})$. It is noted that the smaller values of $\mathrm{Im}(\varepsilon^{(1)})$ of DHBDT than those of BDT are mainly due to wider HOMO-LUMO gap. The wide HOMO-LUMO gap induces small $w(E)$ in (29) around E_F, and the perturbation effects are smaller.

In order to improve the results in this study, a larger basis set should be used. Actually, the STO-6G basis set provides poor results for orbital energies of BDT. Results from our method strongly depend on orbital energies of the channel molecule through (29), and hence a larger basis set probably improves the results. In addition, several theoretical reports show the importance of electron correlation [14, 23]. Therefore, our method should be extended to

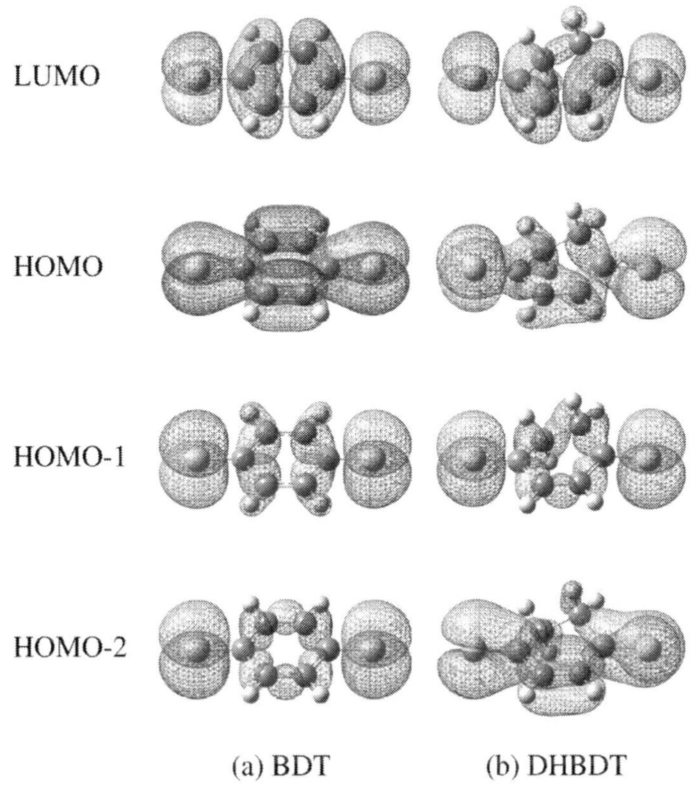

Figure 3. Electronic structures of the LUMO, HOMO, HOMO-1, and HOMO-2.

post-HF computations. This is our future task.

Previously, we have reported the local currents and conductivities of nanomaterials [24, 25, 26, 27]. These local quantities are defined in rigged quantum electrodynamics [28, 29] and useful for understanding effects of defects, impurities, or interfaces on conductive properties. In our future study, our method will be applied in order to obtain these local quantities.

5. Conclusion

In this study, we report on a method for investigating conductive properties of a system composed of a channel molecule and electrodes. Our method is based on CPHF equations for non-Hermitian perturbation Hamiltonians. In our method, perturbed MOs can be determined so as to be consistent with perturbed Coulomb and exchange interactions between electrons at the HF level. A non-Hermitian perturbation Hamiltonian which describes interactions with electrodes is also introduced. For this Hamiltonian, the effects of electrodes are included through the statistical distributions of electrons. The Hamiltonian also includes the effect of the DOS of a channel molecule. The BDT and DHBDT are analyzed by our method, and the first order perturbations of MO energies of them are computed.

Acknowledgments

This work was partially supported by a Grant-in-Aid for Scientific Research from the Japan Society for the Promotion of Science (JSPS) (22550011). M. S. is supported by a Grant-in-Aid for Young Scientists (B) from JSPS (24760028) and Mizuho Foundation for the Promotion of Sciences. Y. I. is supported by the Sasakawa Scientific Research Grant from the Japan Science Society.

References

[1] Reed M A, Zhou C, Muller C J, Burgin T P and Tour J M 1997 *Science* **278** 252–254
[2] Chen J, Reed M A, Rawlett A M and Tour J M 1999 *Science* **286** 1550–1552
[3] Donhauser Z J, Mantooth B A, Kelly K F, Bumm L A, Monnell J D, Stapleton J J, Price Jr D W, Rawlett A M, Allara D L, Tour J M and Weiss P S 2001 *Science* **292** 2303–2307
[4] Reed M A, Chen J, Rawlett A M, Price D W and Tour J M 2001 *Appl. Phys. Lett.* **78** 3735–3737
[5] Kushmerick J G, Holt D B, Yang J C, Naciri J, Moore M H and Shashidhar R 2002 *Phys. Rev. Lett.* **89** 086802
[6] Nitzan A and Ratner M A 2003 *Science* **300** 1384–1389
[7] Chao Li *et al* 2003 *Appl. Phys. Lett.* **82** 645–647
[8] Lörtscher E, Weber H B and Riel H 2007 *Phys. Rev. Lett.* **98** 176807
[9] Taylor J, Guo H and Wang J 2001 *Phys. Rev.* B **63** 121104
[10] Taylor J, Guo H and Wang J 2001 *Phys. Rev.* B **63** 245407
[11] Damle P S, Ghosh A W and Datta S 2001 *Phys. Rev.* B **64** 201403
[12] Xue Y and Ratner M A 2003 *Phys. Rev.* B **68** 115406
[13] Shimazaki T, Maruyama H, Asai Y and Yamashita K 2005 *J. Chem. Phys.* **123** 164111
[14] Shimazaki T and Yamashita K 2009 *Int. J. Quantum Chem.* **109** 1834–1840
[15] Datta S 1995 *Electron Transport in Mesoscopic Systems* (Cambridge: Cambridge University Press)
[16] Langhoff P W, Karplus M and Hurst R P 1966 *J. Chem. Phys.* **44** 505–514
[17] Stevens R M, Pitzer R M and Lipscomb W N 1963 *J. Chem. Phys.* **38** 550–560
[18] Gerratt J and Mills I M 1968 *J. Chem. Phys.* **49** 1719–1729
[19] Ferry D K and Goodnick S M 1997 *Transport in Nanostructures* (Cambridge: Cambridge University Press)
[20] Hehre W J, Stewart R F and Pople J A 1969 *J. Chem. Phys.* **51** 2657–2664
[21] Hehre W J, Ditchfield R, Stewart R F and Pople J A 1970 *J. Chem. Phys.* **52** 2769–2773
[22] M J Frisch *et al* Gaussian 09 Revision C.1 Gaussian Inc. Wallingford CT 2010
[23] Wang C K, Fu Y and Luo Y 2001 *Phys. Chem. Chem. Phys.* **3** 5017–5023
[24] Senami M, Ikeda Y, Fukushima A and Tachibana A 2010 *Jpn. J. Appl. Phys.* **49** 115002
[25] Senami M, Ikeda Y and Tachibana A 2011 *Jpn. J. Appl. Phys.* **50** 010103
[26] Senami M, Ikeda Y, Hara T and Tachibana A 2011 *Key Eng. Mater.* **470** 66–71
[27] Ikeda Y, Senami M and Tachibana A 2012 *AIP Adv.* **2** 042168
[28] Tachibana A 2005 *J. Mol. Model.* **11** 301–311
[29] Tachibana A 2010 *J. Mol. Struct.: THEOCHEM* **943** 138–151

24th IUPAP Conference on Computational Physics (IUPAP-CCP 2012) IOP Publishing
Journal of Physics: Conference Series **454** (2013) 012054 doi:10.1088/1742-6596/454/1/012054

Simulation of heavy ion collision using a time-dependent density functional theory including nuclear superfluidity

Shuichiro Ebata

Center for Nuclear Study, University of Tokyo, Hongo, Bunkyo, Tokyo 113-0033, Japan

E-mail: ebata@cns.s.u-tokyo.ac.jp

Abstract. We carried out a simulation of heavy ion collision using a time-dependent density functional theory. We call it the canonical-basis time-dependent Hartree-Fock-Bogoliubov theory (Cb-TDHFB) which can describe nuclear dynamics in three-dimensional coordinate space, treating nuclear pairing correlation. We simulate $^{20}O+^{20}O$ collision using the Cb-TDHFB with a contact-type pairing functional, and show the behavior of gap energy which is decreasing and vibrating while colliding.

1. Introduction

The atomic nucleus is a self-binding finite quantum many-body system whose size is several femtometers (= fm = 10^{-15}m). It is composed of two kinds of fermions (protons and neutrons) with spin 1/2 that are called "nucleons". The number of nucleons in a nucleus is a few hundreds at most. There are about 3,000 nuclei observed in experiments, and yet more to be discovered. In order to study the structure of nuclei, we investigate many excited modes of many nuclei. To study nuclear collision microscopically is also very important. Nuclei have "magic numbers", which are analogous to those in atoms. This suggests that the nucleons in the nucleus are freely moving in an attractive average potential. The nucleons create the mean field by themselves.

In the time-dependent mean-field theory, a heavy-ion collision is expressed as the collision of two mean fields. We should investigate the behavior of single-particle states while colliding two fields, to predict the structure of nuclei after the heavy-ion collision. Nuclear pairing correlation has a significant role to define single-particle states. It allows a fractional occupation probability of single-particle states due to the pair scattering, which is important to reconstruct a new mean field. The time-dependent Hartree-Fock theory (TDHF) is well-known as a useful tool to study nuclear dynamics [1], however it can not include effects of pairing correlation.

There is a theory to treat pairing correlation in nuclear dynamics self-consistently, which is called the time-dependent Hartree-Fock-Bogoliubov theory (TDHFB). But, there is no study of heavy ion collision using TDHFB with a modern effective interaction in three-dimensional space, because it is hard to prepare converged a initial state, and huge computational resources are needed [2]. We proposed a new time-dependent mean-field theory to study pairing treated as nuclear dynamics. This is named the canonical-basis TDHFB (Cb-TDHFB) [3]. The Cb-TDHFB is derived from full TDHFB equations represented in canonical basis which diagonalizes the density matrix, with Bardeen-Cooper-Schrieffer (BCS)-like approximation for pairing functional (*pp*,*hh*-channel). We confirmed that linear response calculations of the Cb-TDHFB are a good approximation of the quasi-particle random phase approximation (QRPA) which is a small amplitude limit of TDHFB [3].

In the present work, we apply the Cb-TDHFB to heavy ion collisions in the three-dimensional Cartesian coordinate space using a modern effective interaction (Skyrme: SkM*) with a contact pairing. In the following, we introduce the Cb-TDHFB equations and the contact pairing functional. Then, we show results of $^{20}O+^{20}O$ which indicate a collision of the gap-energy.

2. Formulation and Procedure

2.1. Basic equations and contact pairing

The Cb-TDHFB equations can be derived from the TDHFB equations represented in canonical basis with a simple approximation for the pairing functional [3]. The TDHFB equations can be written in terms of the generalized density matrix $\mathcal{R}(t)$ and the generalized Hamiltonian $\mathcal{H}(t)$ [4]. $\mathcal{R}(t)$ composes the one-body density matrix $\rho_{\mu\nu}(t) \equiv \langle\Psi(t)|\hat{c}_\nu^\dagger\hat{c}_\mu|\Psi(t)\rangle$ and the pairing-tensor $\kappa_{\mu\nu}(t) \equiv \langle\Psi(t)|\hat{c}_\nu\hat{c}_\mu|\Psi(t)\rangle$; μ,ν mean arbitrary basis. $\mathcal{H}(t)$ has $h(t)$ and $\Delta(t)$ corresponding to single-particle Hamiltonian and pair potential, respectively. We can express the TDHFB state at any time in the canonical (BCS) form as

$$|\Phi(t)\rangle \equiv \prod_{l>0}(u_l(t) + v_l(t)\hat{c}_l^\dagger(t)\hat{c}_{\bar{l}}^\dagger(t))|0\rangle, \tag{1}$$

where $u_l(t), v_l(t)$ are time-dependent BCS factors and $\hat{c}_l^\dagger$ is a creation operator of canonical basis which diagonalizes density matrix $\rho(t)$. Then, we choose the BCS form of pair potential as

$$\Delta_l(t) = -\sum_{k>0}\kappa_k(t)\bar{\mathcal{V}}_{l\bar{l},k\bar{k}}\ , \tag{2}$$

where $\kappa_k(t) \equiv u_k(t)v_k(t)$ corresponds to the pair tensor $\kappa(t)$ in the canonical-basis and $\bar{\mathcal{V}}_{l\bar{l},k\bar{k}}$ is the anti-symmetric two-body matrix element. The subscripts $\bar{l}$ and $\bar{k}$ mean the pair of canonical basis l and k, respectively. We can obtain the Cb-TDHFB equations from the canonical form (1) of TDHFB and with the pair potential (2), as follows:

$$\begin{aligned}
i\hbar\frac{\partial\phi_l(t)}{\partial t} &= (\hat{h}(t) - \eta_l(t))\phi_l(t), \quad i\hbar\frac{\partial\phi_{\bar{l}}(t)}{\partial t} = (\hat{h}(t) - \eta_{\bar{l}}(t))\phi_{\bar{l}}(t), \\
i\hbar\frac{\partial\rho_l(t)}{\partial t} &= \kappa_l(t)\Delta_l^*(t) - \kappa_l^*(t)\Delta_l(t), \\
i\hbar\frac{\partial\kappa_l(t)}{\partial t} &= (\eta_l(t) + \eta_{\bar{l}}(t))\kappa_l(t) + \Delta_l(t)(2\rho_l(t) - 1),
\end{aligned} \tag{3}$$

where $\eta_l(t) \equiv \langle\phi_l(t)|\,\hat{h}(t)\,|\phi_l(t)\rangle + i\hbar\langle\frac{\partial\phi_l}{\partial t}|\phi_l(t)\rangle$. The Cb-TDHFB equations compose the time-evolutions of the canonical basis $\phi_l(t), \phi_{\bar{l}}(t)$, the occupation probability $\rho_l(t) \equiv |v_l(t)|^2$ and the pair probability $\kappa_l(t)$. They conserve the orthonormal property of the canonical basis and average particle number. When we choose a special gauge condition $\eta_l(t) = \varepsilon_l(t) = \langle\phi_l(t)|\,\hat{h}(t)\,|\phi_l\rangle(t)$, they conserve the average total energy.

We introduce neutron-neutron and proton-proton BCS pairing. The BCS pairing matrix elements $\mathcal{V}_{l\bar{l},k\bar{k}}^\tau$ are written as

$$\begin{aligned}
\mathcal{V}_{l\bar{l},k\bar{k}}^\tau = \int d\boldsymbol{r}_1 d\boldsymbol{r}_2 \sum_{\sigma_1,\sigma_2} &\phi_l^*(\boldsymbol{r}_1,\sigma_1)\phi_{\bar{l}}^*(\boldsymbol{r}_2,\sigma_2)\hat{\mathcal{V}}^\tau(\boldsymbol{r}_1,\sigma_1;\boldsymbol{r}_2,\sigma_2) \\
&\times\left[\phi_k(\boldsymbol{r}_1,\sigma_1)\phi_{\bar{k}}(\boldsymbol{r}_2,\sigma_2) - \phi_{\bar{k}}(\boldsymbol{r}_1,\sigma_1)\phi_k(\boldsymbol{r}_2,\sigma_2)\right].
\end{aligned} \tag{4}$$

We introduce the spin-singlet contact interaction to (4):

$$\hat{\mathcal{V}}^{\tau}(\boldsymbol{r}_1, \sigma_1; \boldsymbol{r}_2, \sigma_2) \equiv V_0^{\tau} \frac{1 - \hat{\boldsymbol{\sigma}}_1 \cdot \hat{\boldsymbol{\sigma}}_2}{4} \delta(\boldsymbol{r}_1 - \boldsymbol{r}_2), \quad (5)$$

where τ indicates neutron or proton channel and V_0^{τ} is a strength of pairing [5]. In this present work, we do not use the time-reversal relation between l- and $\bar{l}$-canonical basis.

2.2. *Procedure and calculation space*

At first, we prepare the initial states (HF or HF+BCS ground states) using the wave functions for projectile and target with some impact parameters b corresponding to a distances where they feel only Coulomb interaction. Then, we boost the wave functions and translationally calculate the time-evolution of nuclear densities to obey (3). In this present work, we use the three-dimensional Cartesian coordinate-space representation for the canonical states, $\phi_l(\boldsymbol{r}, \sigma; t) = \langle \boldsymbol{r}, \sigma | \phi_l(t) \rangle$ with $\sigma = \pm 1/2$. The coordinate space is a box whose x- and y-sides are 30 fm and whose z-side is 50 fm, discretized in the square mesh of $\Delta x = \Delta y = \Delta z = 1.0$ fm.

3. Results

We simulate the ^{20}O+^{20}O collision with incident energy $E_{\rm int} = 40$ MeV which is higher than the Coulomb barrier in this system. Initial distance between projectile and target is 12 fm with $b = 0$ fm. ^{20}O (Z=8, N=12) has superfluidity for neutrons. The number of canonical basis is 48 (1.2 times that of TDHF) in total. The average gap-energy $\bar{\Delta} \equiv \sum_{l>0} \Delta_l / \sum_{l>0}$ is 1.901 MeV.

Figure 1 shows the time evolution of neutron density distributions. After the touch ($t = 98.66$ fm/c: panel (b) in figure 1), the density-distribution vibrates keeping the prolate shape. This shape of a new mean field corresponds to the shape of ^{40}S (Z=16, N=24) ground state in intrinsic frame. Figure 2 shows the time-evolution of neutron gap-energy for the projectile defined in (2). The behavior of the gap energy for the target is the same as that for the projectile, because the kind of nucleus is same in the projectile and target. From the touching configuration, the gap energies are decreasing significantly. The decrease is similar to the results of an old work [6]. We can see the vibration of gap energies related with nuclear density behavior.

We simulated the ^{20}O+^{20}O collision using time-dependent mean field theory including the effects of nuclear superfluidity and showed the vibration of gap energies in the collision. Currently, we are investigating the details of gap energy vibration and its relation with the excited states of ^{40}S. To study the pairing effects for collision phenomena, we should make our program code parallel. We parallelize the code using MPI for the canonical basis, because the canonical basis conserve orthogonality in the Cb-TDHFB, which means they do not need Gram-Schmidt orthonormalization in the real time-evolution. We can expect the parallelization efficiency of the method to be good, due to the relatively small communication-loss.

Acknowledgments

This work is supported by HPCI Strategic Program Field 5 "The origin of matter and the universe" and, for computational resources, by the RIKEN Integrated Cluster of Clusters(RICC) and by the SR16000 at YITP in Kyoto University.

References

[1] Koonin S E, Davies K T R, Maruhn-Rezwani V, Feldmeier H, Krieger S J and Negele J W 1977 *Phys. Rev.* C**15** 1359
[2] Bulgac A, Luo Y L, Magierski P, Roche K J, Stetcu I and Yu Y 2010 *UNEDF SciDAC Collaboration Research Highlight* http://unedf.org/content/highlights/tdslda-highlight-ascr_v3.pdf
[3] Ebata S, Nakatsukasa T, Inakura T, Yoshida K, Hashimoto Y and Yabana K 2010 *Phys. Rev.* C**82** 034306

[4] Blaizot J P and Ripka G 1986 *Quantum Theory of Finite Systems* (Cambridge, MA: MIT Press)
[5] Krieger S J, Bonche P, Flocard H, Quentin P and Weiss M S 1990 *Nucl. Phys.* A**517** 275
[6] Cusson R Y, Maruhn J A and Stöcker H 1980 *Z. Physik.* A**294** 257

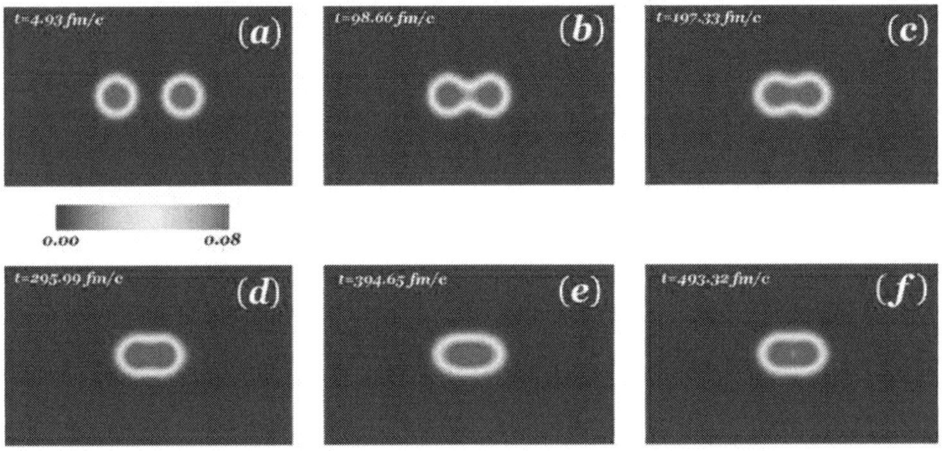

Figure 1. Neutron density distributions of XZ-plane in $^{20}O+^{20}O$ collision at t=(a)4.93, (b)98.66, (c)197.33, (d)295.99, (e)394.65 and (f)493.32 fm/c. The horizontal direction corresponds to Z direction.

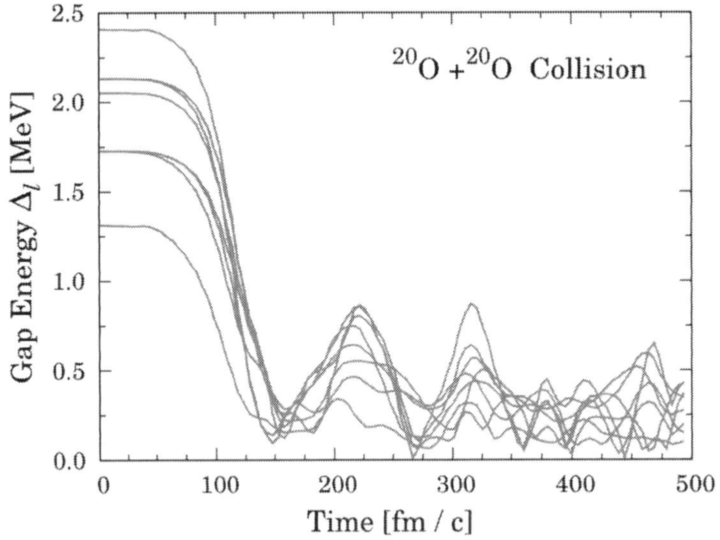

Figure 2. Time-evolution of neutron gap-energy for each canonical state of projectile, which defined in (2). Note that after the collision, we cannot distinguish the states of projectile from those of target, basically.

Journal of Physics: Conference Series **454** (2013) 012055 doi:10.1088/1742-6596/454/1/012055

How to determine boundaries for QM/MM calculations: A guideline based on linear response function for glutathione

Shusuke Yamanaka[1], Koki Ueda[1], Kazuto Nakata[2], Mitsutaka Okumura[1], Kizashi Yamaguchi[3] and Haruki Nakamura[4]

1. Graduate School of Science, Osaka University, Machikaneyama 1-1, Toyonaka, Osaka, Japan.
2. NEC Corporation, 1-10, Nisshincho, Fuchu, 183-8501, Japan.
3. TOYOTA Physical & Chemical Research Institute Nagakute, Aichi, Japan.
4. Protein Institute, Osaka University, 3-2, Yamadaoka Suita, Osaka, Japan.

syama@chem.sci.osaka-u.ac.jp

Abstract. Quantum mechanics/molecular mechanics (QM/MM) methods have grown to be a standard tool for chemical reactions in biological systems. Still, the remaining problem is that the MM point charges induce artificial polarizations in QM regions, spoiling the quality of the QM calculations. Thus, how to determine boundaries between QM and MM regions is an essential issue for QM/MM calculations. Recently, we proposed the use of a linear response function as an indicator to examine the validity of the replacement of QM peripheral ligands with MM point charges. In this study, we examine the glutathione molecule, for which protonation models have been proposed so far. The calculated results are discussed in relation to the QM/MM modeling of this system.

1. Introduction

The ab initio quantum chemical approach has been established for calculations of chemical reactions of various molecules. For most molecules consisting of several dozen of atoms, the chemical accuracy in estimating the activation barrier heights and formation enthalpies has been attained [1]. In contrast, the enzyme reactions are still tough targets because not only quantum fluctuations in the reaction site, but also thermal fluctuations around the site play important roles to determine quantitative values of such chemical properties. In order to tackle this problem, many researchers have developed quantum mechanics/molecular mechanics (QM/MM) methods, and applied them to biomolecular systems [2-12]. In QM/MM methods, the core parts of biomolecular systems, in which quantum events such as chemical reactions and electronic excitations occur, are treated with the QM method, while the peripheral regions such as protein environments, lipids, and waters are treated with the MM method. The QM method, in particular the ab intio QM method, is now a reliable tool to estimate chemical reactions by which we reproduce experimental results within the chemical accuracy (0.1-1 kcal/mol), and the MM method is also well trained to reproduce the experimental results of proteins [13]. However, the problems of QM/MM methods remain in the seam between QM and MM regions, where the MM point charges induce artificial overpolarizations in QM regions near the QM/MM boundaries. More explicitly, the difference between the QM potential and the MM potential, $\delta v = v_{QM} - v_{MM}$, which is just the error due to the replacement of the QM electronic structure by the MM point charges, causes the error of calculated density

$$\delta\rho(\mathbf{r}) = \int d\mathbf{r}' \frac{\delta\rho(\mathbf{r})}{\delta v(\mathbf{r}')} \delta v(\mathbf{r}'). \tag{1}$$

If $\delta\rho(\mathbf{r})$ at the QM event site is not negligible, it leads to significant failures of computational results in describing the event. Equation (1) implies that there are two ways to improve the QM/MM methods. The first one is the standard approach that many researchers have taken: a device on the QM/MM boundaries such as link-atom [2-4], generalized hybrid orbital [6] local self-consistent field [7], and effective potential approaches [9]. Although many efforts have been devoted to minimizing $\delta v = v_{QM} - v_{MM}$ in the right-hand side of (1), the complete solution has not been obtained. The second one consists in referring to $\delta\rho(\mathbf{r})/\delta v(\mathbf{r}')$, which indicates how far the effects of the MM point charges propagate. In striking contrast to many studies for the first one, there are few studies for the second one. In particular, the systematic study to determine QM/MM boundaries has never been implemented so far. Previously, we proposed to use $\delta\rho(\mathbf{r})/\delta v(\mathbf{r}')$ to examine how the effects of QM/MM errors depend on the intermediated molecular units and chemical bonds [15]. In that study, we examined typical covalent bonds, π -conjugated systems, and dialanine dipeptide systems.

In this study, we implemented a similar analysis for glutathione, which is known to function as an antioxidant working with glutathione (GSH) peroxidase/reductase [16]. Even without these enzymes, GSH can also react with radical species such as OH·, RO·, NO_2·, and carbon-centered radicals, which is why the GSH itself is considered as an antioxidant [16]. The glutathione radical (GS·) is produced by the scavenging reaction,

$$\mathrm{GSH} + \mathrm{X}\cdot \rightarrow \mathrm{GS}\cdot + \mathrm{XH}, \tag{2}$$

which is considered to be detoxified in the processes in vivo as follows:

$$\mathrm{GS}\cdot + \mathrm{GS}^- + \mathrm{O}_2 \rightarrow \mathrm{GSSG} + \mathrm{O}_2^-, \tag{3a}$$

$$2\mathrm{O}_2^- + 2\mathrm{H}^+ \xrightarrow{SOD} \mathrm{H}_2\mathrm{O}_2 + \mathrm{O}_2, \tag{3b}$$

$$2\mathrm{H}_2\mathrm{O}_2 \xrightarrow{catalase} 2\mathrm{H}_2\mathrm{O} + \mathrm{O}_2. \tag{3c}$$

Although the S-H bond has been usually assumed to be dissociated homolytically as in (2), Fiser et al. presented using density functional theory calculations with a polarizable continuum model that the N-H bond in the ammonia group is easier to dissociate than the S-H bond [17]. In their calculations, they partitioned the GSH into several parts by using "hydrogen caps". The dehydrogenation of GSH in water is a critically complicated problem, because there are various possible modes concerning protonation/deprotonations of carboxyl and amino groups, as well as dehydrogenation modes [18]. Further, the stabilities of the modes must depend on environmental effects, in particular on pH [18]. Thus, instead of studying the various modes, we focus our attention on examination of the validity of the partition methods of glutathione molecule on the basis of the computational results of the linear response function. The results are also discussed in relation to fundamental features of the LRF of molecules, as well as the implications of our results in the context of the usual QM/MM modeling tactics.

2. Methodology

So far, ab initio calculations of linear response functions (LRF) have been implemented by several groups [19, 20]. Morita and Kato calculated $\delta\rho(\mathbf{r})/\delta v(\mathbf{r}')$ on the basis of the coupled perturbed

Hartree-Fock equation [19]. Recently, Geerlings and his coworkers have investigated this property in the context of conceptual density functional theory [20]. They used a perturbation theory to estimate $\delta\rho(\mathbf{r})/\delta v(\mathbf{r}')$ and compute the linear response (LR) to the specific site with using Becke's fuzzy cell scheme [21]. Independently we also calculated LR to the specific site, X, i.e., $\delta\rho(\mathbf{r})/\delta v_X$ based on the perturbation theory, but using the linear combination of atomic orbital (LCAO) expansion [14,15]. The LCAO based method is suitable for our purpose: the QM/MM modeling is applied for specific sites, which would be described as quantum electronic structures using atomic orbitals (AOs) if all systems are treated with the QM method. Thus the perturbations and the responses expressed in terms of AOs are efficiently used in the discussion of the validity of the QM/MM boundary setting.

In this study, we computed the linear response for each site of the glutathione, $\delta\rho(\mathbf{r})/\delta v_X$. As discussed in the introduction, we focus our attention on the LRF of glutathione molecule. Thus we fixed a standard hydration mode with the geometry that was obtained by Fiser et al. [17], as shown in figure 1.

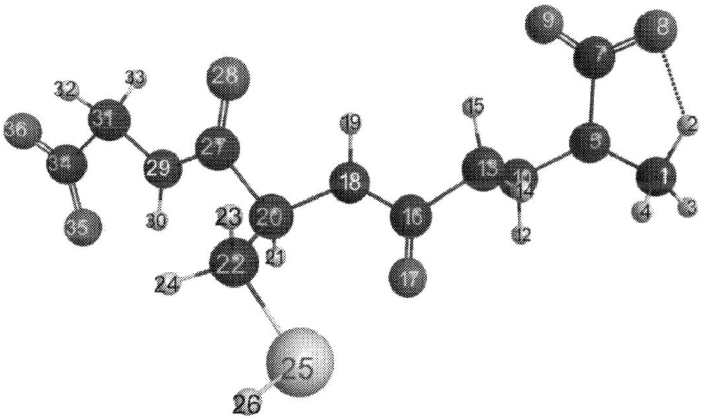

Figure 1. The structure of glutathione.

The B3LYP functional and the 6-31G** basis set were used for all calculations. Using the B3LYP solutions, we calculated the LRF based on first-order perturbation theory [14]. The code we used is a locally extended version of GAMESS [22], in which our LRF module is implemented. The results will be presented in the next section.

3. Results and discussion

Each of the 36 parts of figure 2 shows the isosurface of the linear response to the perturbation that is applied to the specific site X, $\delta\rho(\mathbf{r})/\delta v_X$ of the glutathione molecule. The number described below each figure corresponds to the atom number showed in figure 1. The isosurfaces for $\delta\rho(\mathbf{r})/\delta v_X = 0.001$ and $= -0.001$ are described as blue and red surfaces, respectively. The plus value (blue region) of $\delta\rho(\mathbf{r})/\delta v_X$ implies that, if the attractive perturbation, $\delta v_X\ (<0)$, is applied for the X site, the density decreases. Contrastingly, the minus value (red region) of $\delta\rho(\mathbf{r})/\delta v_X$ indicates that the density increases for the attractive perturbation.

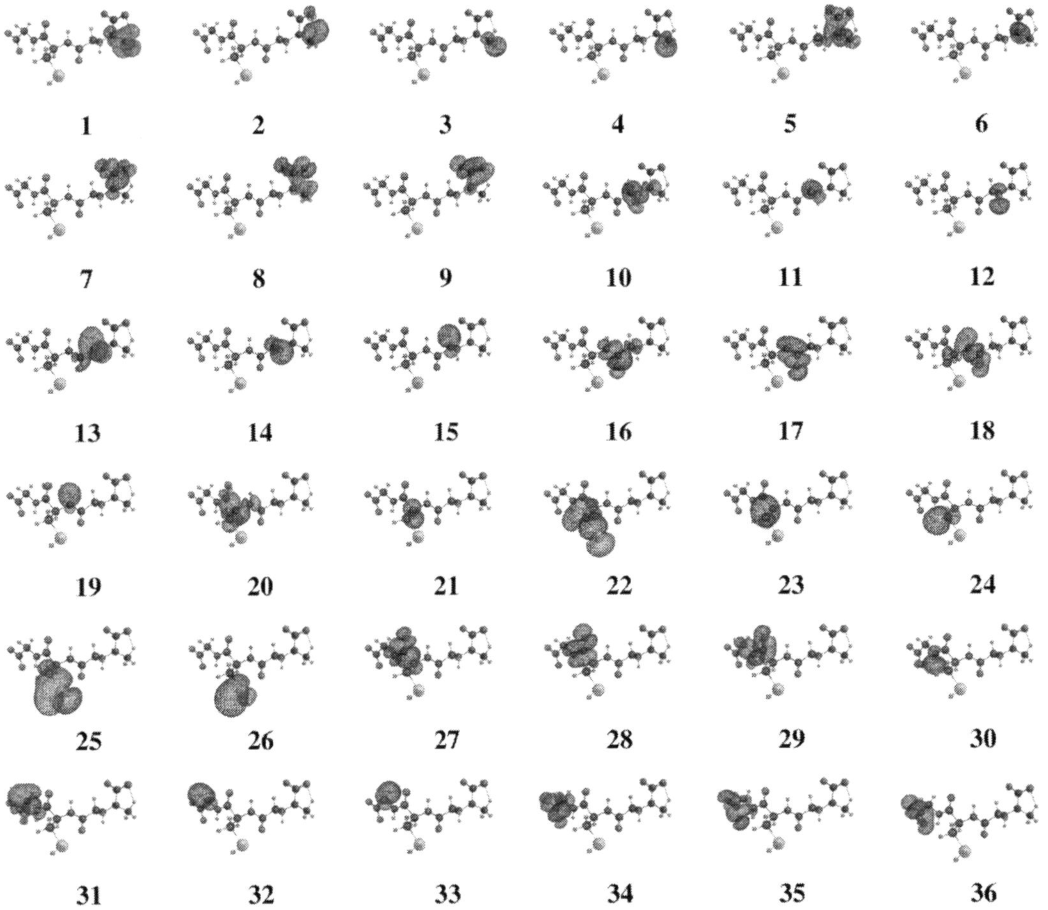

Figure 2. The linear response to the perturbation that is applied to the specific site X, $\delta\rho(\mathbf{r})/\delta v_X$ of the glutathione molecule. The numbers below the figures are the sites to which the perturbations are applied.

Because we assume that the number of electrons conserves, the integral of $\delta\rho(\mathbf{r})/\delta v_X$ is always zero:

$$\int d\mathbf{r}\,\delta\rho(\mathbf{r})/\delta v_X = 0\,. \tag{4}$$

Thus blue and red regions appear for all perturbations, δv_X. A noteworthy point is that, not a monotonically increase or decrease, but oscillations of $\delta\rho(\mathbf{r})$ are observed: for instance see $\delta\rho(\mathbf{r})/\delta v_{22}$ (22-th) in figure 2. If this phenomenon were dominated by classical mechanics, the behavior of $\delta\rho(\mathbf{r})$ should be monotonic. Then, it is obvious that the quantum characteristics, for instance phases of orbitals, of this system lead to oscillations of $\delta\rho(\mathbf{r})/\delta v_X$. A reader might notice that this is not a size effect: a similar oscillation phenomena is known as "Friedel oscillation" in solid state physics since more than 60 years ago [23, 24], and is recently observed as STM images for condensed matter systems [25]. Previous researchers, who compute the LRF of molecular systems, also presented that there are oscillation behaviors of the LRF [19, 20]. This type of oscillations must be universal for molecules, and macromolecules, and condensed phases.

For our primary purpose, it is important that the linear responses do not propagate over the C_α carbons (atom numbers are 5, 10, 13, 20,22, 31): if the perturbation is applied to the right side of a specific C_α, the region defined by $|\delta\rho(\mathbf{r})/\delta v_X| \le 0.001$ is mostly localized within the right side of the C_α carbon and is not delocalized the left side of the C_α carbon, and vice versa. However, if the perturbation is applied to an atom next to the C_α carbon, the response might pierce the C_α carbon, because the sp3 bonding orbital that connects the atom and the C_α carbon is delocalized over another side of the C_α carbon (for instance, see **18** in figure 2). In addition, the perturbation on the C_α carbon itself affects to the four directions of sp3 orbitals (see **5, 10, 13, 20, 22, 31** in figure 2), possibly spreading over two amide planes. Thus when a C_α carbon directly contributes to the QM event, the atoms in these two amide planes should not be replaced with the MM charges.

We should note the reason why we choose the threshold $|\delta\rho(\mathbf{r})/\delta v_X| = 0.001$ to illustrate the linear response surfaces. This surface provides boundaries for errors of order 0.001 on the density if the QM/MM potential yields the errors, δv_X, of the order 1.0 Hartree. Then the regions illustrated in figure 2 are actually useful to judge where we should set the QM/MM boundary. Of course, this estimation is not accurate enough to predict the accuracy of the specific QM/MM calculation plans. The LRF is calculated for the whole system of the glutathione molecule using a perturbation scheme [14,15]. In actual QM/MM calculations, the electronic structures of the model systems are relaxed with the self-consistent manner, in which the QM/MM errors might be enlarged (or rarely reduced). Higher order effects are completely neglected in the present calculations. How strongly the higher-order effects affect the results depends on how large the errors due to the QM/MM model, δv_X, are. Thus we need to specify the concrete QM/MM scheme in order to refine this type of error analyses for the QM/MM theories.

Nevertheless, the rough estimate described above could certainly provide valuable information for QM/MM applications. In fact, modeling of C_α carbons as hydrogen atoms is widely used. We expect that the discussion above becomes a guideline to determine the QM region in QM/MM calculations. We are now applying our method to small peptides having secondary structures, such as α – helix or extended β – conformations. A larger set of comprehensive guidelines will be presented in the future.

Acknowledgments

S. Yamanaka acknowledges the financial support by the Ministry of Education, Culture, Sports, Science and Technology (MEXT) (Grant-in-Aid for Scientific Research (C) No. 23550016 and (B) No. 23350064) and Research and Development of the Next-Generation Integrated Simulation of Living Matter, as a part of the Development and Use of the Next-Generation Supercomputer Project.

This work was partially supported by MEXT program "Elements Strategy Initiative to Form Core Research Center", MEXT, Ministry of Education Culture, Sports, Science and Technology, Japan.

References

[1] Frank J 2007 Introduction to Computational Chemistry (Chichester England, John Wiley and

Sons)
[2] Warshel A and Levitt M J 1976 *J. Mol. Biol.* **103** 227-249
[3] Senn H M and Thiel W 2009 *Angew. Chem. Int. Ed.* **48** 1198-1229
[4] Lin H and Truhlar D G 2007 *Theor. Chem. Acc.* **117** 185-199
[5] Hu L, Soderhjelm P and Ryde U 2011 *J. Chem. Theor. Comput.* **7** 761-777
[6] Dilabio G A, Hurley M H and Christiansen P A 2002 *J. Chem. Phys.* **116** 9578 –9584
[7] Murphy R. B. D. M. Philipp and R. A. Friesner 2000 *Chem. Phys. Lett.* **321** 113-120
[8] Gao J, Amara P, Alhambra C and Field M J 1998 *J. Phys. Chem.* A **102** 4714-4721
[9] Reuter N, Dejaegere A, Maigret B and Karplus M 2000 J. Phys. Chem. A **104** 1720-1735
[10] Dittrich M, Hayashi S and K. Schulten K 2004 *Biophys. J.* **87** 2954-2967
[11] Yasuda K and Yamaki D 2004 *J. Chem. Phys.* **121** 3964.3972
[12] Takahashi H, Matubayasi N, Nakahara M and Nitta T 2004 *J. Chem. Phys.* **121** 3989-3999
[13] Mackerell AD 2004 Empirical force fields for biological macromolecules: overview and issues *J. Comput. Chem.* **25** 1584–604
[14] Ueda K, Yamanaka S, Nakata K, Ehara M, Okumura M, Yamaguchi K and Nakamura H 2013 *Int. J. Quantum Chem.* **113** 336-341
[15] Yamanaka S, Yonezawa Y, Nakata K, Nishihara S, Okumura M, Takada T, Yamaguchi K and Nakamura H 2013 AIP Conf Proc **1504** 916-919
[16] Halliwell B and Gutteridge J C 2007 Free radicals in biology and medicine fourth Edition Oxford University Press
[17] Fiser B, Szori M, Jojart B, Csizmadia I. G and Viskolcz B 2011 *J. Phys. Chem.* **B 115** 11269-11277
[18] Lampela O, Juffer A H and Rauk A 2003 *J. Phys. Chem.* A **107** 9208-9220
[19] Morita A and Kato S 1997 *J. Am. Chem. Soc.* **119** 4021-4032
[20] For instance, Sablon N, Proft F D and Gerrlings P 2010 *J. Phys. Chem. Lett.* **1** 1228-1234
[21] Becke A 1988 *J. Chem. Phys.* **88** 2547-2553
[22] Schmidt MW, Baldridge K K, Boatz J A, Elbert S T, Gordon M S, Jensen J H, Koseki S, Matsunaga N, Nguyen K A, Su S, T L, Windus, Dupuis M and Montgomery Jr J A 1993 *J. Comp. Chem.* **14** 1347-1363
[23] Friedel J 1952 *Philos. Mag.* **43** 153-189
[24] Kohn W and Vosko S H 1960 *Phys. Rev.* **119** 912-918
[25] Hasegawa Y, Ono M, Nishigata Y, Nishino T and Eguch T 2006 *Phys. Rev. Lett.* **96** 016801-4

Checking the validity of the correlated Thomas-Fermi functional in the pair density functional theory

Katsuhiko Higuchi[1] and Masahiko Higuchi[2]

[1]Graduate School of Advanced Sciences of Matter, Hiroshima University, Higashi-Hiroshima 739-8527, Japan
[2]Department of Physics, Faculty of Science, Shinshu University, Matsumoto 390-8621, Japan

E-mail: khiguchi@hiroshima-u.ac.jp

Abstract. To check the validity of the recently proposed kinetic energy (KE) functional of the pair density (PD) functional theory, we perform the numerical calculation for the neutral Ne atom. This is the first computational trial for the PD functional theory which includes the KE functional that is approximately derived on the basis of the rigorous expression with the coupling-constant integration. In this work we adopt the correlated Thomas-Fermi functional as the approximate KE functional. Although it is shown that non-negligible errors exist in the resulting PD, the present results will be useful as reference data in cases of developing further the approximate KE functional.

1. Introduction

The pair density (PD) functional theory [1-33] is expected to be a promising theory to go beyond the conventional density functional theory (DFT) [34, 35]. In developing the PD functional theory, there exist two kinds of problems that need to be overcome [3-33]. One is how we devise to extend the search region of PDs within the set of N-representable PDs [3-22, 27-33]. The other is how we develop the approximate form of the kinetic energy (KE) functional [23-33].

Concerning the first problem, we have recently proposed a correction method for obtaining the variationally best ground-state PD [31]. It has been shown that the search region of PDs is substantially extended by adding the scaled PDs as elements of the search region [31].

To overcome the second problem, we can take two strategies in a similar way to the development of the exchange-correlation (xc) energy functional of the conventional DFT and its extensions [36-41]:

(i) Rigorous relations and bounds that should be fulfilled by the KE functional are utilized as restrictive conditions on the approximate form.

(ii) On the basis of the coupling-constant expression for the KE functional, the approximate form is developed.

Several types of approximate KE functionals have already been developed along the above strategy (i)[30]. Specifically, approximate KE functionals are developed by requiring them to satisfy the scaling property of the KE functional [5] and the Hohenberg-Kohn theorem of the PD functional theory [30]. We have furthermore begun numerical calculations to check their validities [30].

In this paper, we attempt the second strategy. We have recently derived the coupling-constant expression for the KE functional [32]. Since it is rigorous, this expression would be a good starting point for developing the approximate functional. Actually, using the perturbation theory, we have also proposed the approximate form on the basis of this expression [32]. The resultant form is written as the sum of the noninteracting KE functional and correlation energy functional [32]. In this paper, we adopt as such functionals the Thomas-Fermi (TF) functional and correlation energy functional of the local density approximation (LDA), respectively. Hereafter, this approximate form is called the *correlated TF functional*. In this paper, we check the validity of the correlated TF functional by means of numerical calculations of the Ne atomic structure.

Organization of this paper is as follows. In section 2, we review the coupling-constant expression for the KE functional of the PD functional theory, and give an explicit form of the approximate KE functional [32]. In order to perform numerical calculations for the present scheme that includes the correlated TF functional, the self-consistent procedure is indispensable. We shall explain in section 3 the details of numerical calculations. The results and discussions are given in section 4. Finally, we present the concluding remarks in section 5.

2. Approximate form of the kinetic energy functional

In this section, we present the correlated TF functional that has been developed on the basis of the coupling-constant expression [32]. The coupling-constant expression for the KE functional $T\left[\gamma^{(2)}\right]$ is given by [32]

$$T\left[\gamma^{(2)}\right]=T_s\left[\gamma_{SD}^{(2)}\left[\gamma^{(2)}\right]\right]+\int_0^1\langle\Psi_\lambda|\hat{W}-\hat{V}_{eff}^{\text{int}}|\Psi_\lambda\rangle_\Omega \mathrm{d}\lambda-W\left[\gamma^{(2)}\right]-V_{eff}^{\text{int}}\left[\gamma_{SD}^{(2)}\left[\gamma^{(2)}\right]\right], \quad (1)$$

where $\gamma^{(2)}$ and $\gamma_{SD}^{(2)}\left[\gamma^{(2)}\right]$ denote the ground-state PDs of the real and noninteracting reference systems [27,29], respectively, and where $\hat{W}$ is the operator of the electron-electron interaction energy, and where $\hat{V}_{eff}^{\text{int}}$ is the effective potential of the electron-electron interaction in the noninteracting reference system [27,29]. In the right-hand side of (1), $T_s\left[\gamma_{SD}^{(2)}\left[\gamma^{(2)}\right]\right]$, $V_{eff}^{\text{int}}\left[\gamma_{SD}^{(2)}\left[\gamma^{(2)}\right]\right]$ and $W\left[\gamma^{(2)}\right]$ denote the ground-state KE of the noninteracting reference system, the expectation value of $\hat{V}_{eff}^{\text{int}}$ with respect to the ground-state of the noninteracting reference system and the ground-state electron-electron interaction energy of the real system, respectively [32]. The second term of the right-hand side of (1) represents the coupling-constant integration. In this integration, $|\Psi_\lambda\rangle$ is the ground-state wave function for the Hamiltonian

$$\hat{H}_\lambda=\hat{T}+\hat{V}_{eff}^{\text{int}}+\hat{V}+\lambda\left(\hat{W}-\hat{V}_{eff}^{\text{int}}\right), \quad (2)$$

where $\hat{T}$ and $\hat{V}$ are operators of the KE and external potential energy, respectively, and where λ is the coupling constant. Equation (2) coincides with the Hamiltonian of the noninteracting reference system [27, 29] in the case of $\lambda=0$ and coincides with that of the real system in the case of $\lambda=1$.

In order to develop the approximate form of $T\left[\gamma^{(2)}\right]$, $|\Psi_\lambda\rangle$ is approximated by means of the perturbation theory. Namely, we take $\hat{H}_0$ and $\lambda\left(\hat{W}-\hat{V}_{eff}^{\text{int}}\right)$ as the nonperturbative Hamiltonian and perturbation, respectively [32]. As the lowest-order approximation, by neglecting the second- and higher-order terms of $\hat{W}-\hat{V}_{eff}^{\text{int}}$, the coupling-constant expression (1) is approximated to [32]

$$T\left[\gamma^{(2)}\right]\approx T_s\left[\gamma_{SD}^{(2)}\left[\gamma^{(2)}\right]\right]-\left\{W\left[\gamma^{(2)}\right]-W\left[\gamma_{SD}^{(2)}\left[\gamma^{(2)}\right]\right]\right\}, \quad (3)$$

where $W\left[\gamma_{SD}^{(2)}\left[\gamma^{(2)}\right]\right]$ is the expectation value of $\hat{W}$ with respect to the ground state of the noninteracting reference system. Since $T_s\left[\gamma_{SD}^{(2)}\left[\gamma^{(2)}\right]\right]$ is the KE of the noninteracting system, it may be reasonably approximated by the TF functional, $T_{TF}\left[\rho\right]$[42, 43], as an example. The magnitude of $W\left[\gamma^{(2)}\right]-W\left[\gamma_{SD}^{(2)}\left[\gamma^{(2)}\right]\right]$ of (3) is expected to be comparable to that of the correlation energy. Therefore, the second term may be approximated by the correlation energy functional of the LDA,

$E_c^{LDA}[\rho]$[44], as an example. Thus, we get the approximate form:

$$T_{cTF}\left[\gamma^{(2)}\right] = T_{TF}[\rho] - E_c^{LDA}[\rho] \tag{4}$$

with

$$\rho(\boldsymbol{r}) = \frac{2}{N-1}\int \gamma^{(2)}(\boldsymbol{rr}';\boldsymbol{rr}')\mathrm{d}^3r'. \tag{5}$$

Equation (4) is referred to as the correlated TF functional. The correlated TF functional is the first one that is developed on the basis of the coupling-constant expression for the KE functional [32].

3. Calculation method

In order to test the correlated TF functional, the atomic structure of the neutral Ne is calculated on the basis of the correction method that has recently been developed for the purpose of extending the search region of PDs [31]. The first step of this method is to prepare the *seed PD* that corresponds to the variationally best PD before extending the search region of PDs [31]. In order to prepare the seed PD, we employ as the search region the set of PDs that are constructed from the linear combination of the prescribed Slater determinants (SDs)[30]. If the coefficient of the linear combination is denoted by C_i, then the PDs that belong to the search region are expressed by

$$\gamma^{(2)}\left[\{C_i\}\right](\boldsymbol{rr}';\boldsymbol{rr}') = \sum_i |C_i|^2 \gamma^{(2)}_{SSD,i}(\boldsymbol{rr}';\boldsymbol{rr}') + \frac{1}{2}\sum_i \sum_{j\neq^1 i} C_i^* C_j \Gamma^{ij}_{i_1 j_1}(\boldsymbol{rr}') + \frac{1}{2}\sum_i \Gamma \sum_{j\neq^2 i} C_i^* C_j \,{}^{ij}_{i_1 i_2 j_1 j_2}(\boldsymbol{rr}') \tag{6}$$

with

$$\sum_i |C_i|^2 = 1, \tag{7}$$

where $\gamma^{(2)}_{SSD,i}(\boldsymbol{rr}';\boldsymbol{rr}')$ denotes the PD calculated from $i-$th SD, and where $\Gamma^{ij}_{i_1 j_1}(\boldsymbol{rr}')$ and $\Gamma^{ij}_{i_1 i_2 j_1 j_2}(\boldsymbol{rr}')$ correspond to the components which come from single and doubly-excitations, respectively. In (6), the symbol $i \neq^1 j$ denotes that $i-$th SD is different from $j-$th SD by one constituent orbital. The symbol i_1 (j_1) stands for the quantum number of such a different orbital of $i-$th ($j-$th) SD. In the same way, the symbol $i \neq^2 j$ in (6) denotes that $i-$th SD is different from $j-$th SD by two constituent orbitals. The symbols i_1 and i_2 (j_1 and j_2) stand for the quantum numbers of such two different orbitals of $i-$th ($j-$th) SD. Note that $\gamma^{(2)}_{SSD,i}(\boldsymbol{rr}';\boldsymbol{rr}')$, $\Gamma^{ij}_{i_1 j_1}(\boldsymbol{rr}')$ and $\Gamma^{ij}_{i_1 i_2 j_1 j_2}(\boldsymbol{rr}')$ are determined by the prescribed SDs [30]. By changing $C_i's$ with keeping the condition (7), the PDs given by (6) vary within a set of N-representable PDs. With the aid of the Hohenberg-Kohn theorem with respect to the PD, we obtain the following equations for $C_i's$ in a similar way to the previous work [30]:

$$\sum_j A_{ij}\left[\{C_i\}\right] C_i = \mu C_i \tag{8}$$

with

$$A_{ij}\left[\{C_i\}\right]=\begin{cases}\iint\left\{\frac{2}{N-1}f\left[\{C_i\}\right](\boldsymbol{r})+\frac{e^2}{|\boldsymbol{r}-\boldsymbol{r}'|}+\frac{v_{ext}(\boldsymbol{r})+v_{ext}(\boldsymbol{r}')}{N-1}\right\}\gamma^{(2)}_{SSD,i}(\boldsymbol{rr}';\boldsymbol{rr}')d^3rd^3r' & (i =^{1} j)\\ \iint\left\{\frac{1}{N-1}f\left[\{C_i\}\right](\boldsymbol{r})+\frac{e^2}{|\boldsymbol{r}-\boldsymbol{r}'|}+\frac{v_{ext}(\boldsymbol{r})+v_{ext}(\boldsymbol{r}')}{N-1}\right\}\Gamma^{ij}_{i_1j_1}(\boldsymbol{rr}')d^3rd^3r' & (i \neq^{1} j)\\ \iint\left\{\frac{e^2}{|\boldsymbol{r}-\boldsymbol{r}'|}+\frac{v_{ext}(\boldsymbol{r})+v_{ext}(\boldsymbol{r}')}{N-1}\right\}\Gamma^{ij}_{i_1i_2j_1j_2}(\boldsymbol{rr}')d^3rd^3r' & (i \neq^{2} j)\\ 0 & \text{otherwise}\end{cases} \tag{9}$$

Here, the function $f\left[\{C_i\}\right](\boldsymbol{r})$ is given by

$$f\left[\{C_i\}\right](\boldsymbol{r})=(3\pi^2)^{2/3}\rho\left[\{C_i\}\right](\boldsymbol{r})^{2/3}-v_c^{LDA}\left[\rho\left[\{C_i\}\right]\right](\boldsymbol{r}) \tag{10}$$

with

$$\rho\left[\{C_i\}\right](\boldsymbol{r})=\frac{2}{N-1}\int\gamma^{(2)}\left[\{C_i\}\right](\boldsymbol{rr}';\boldsymbol{rr}')d^3r', \tag{11}$$

where v_c^{LDA} denotes the correlation potential of the LDA. The solution $C_i's$ of the above equations provides the variationally best PD, i.e., the seed PD.

Equations (8)–(11) should be solved in a self-consistent way. The flowchart of numerical calculations is shown in figure 1. First, we give initial values of $C_i's$. Then, matrix elements $A_{ij}\left[\{C_i\}\right]$ are calculated through (9). By diagonalizing $A_{ij}\left[\{C_i\}\right]$, we get a new set of $C_i's$. This

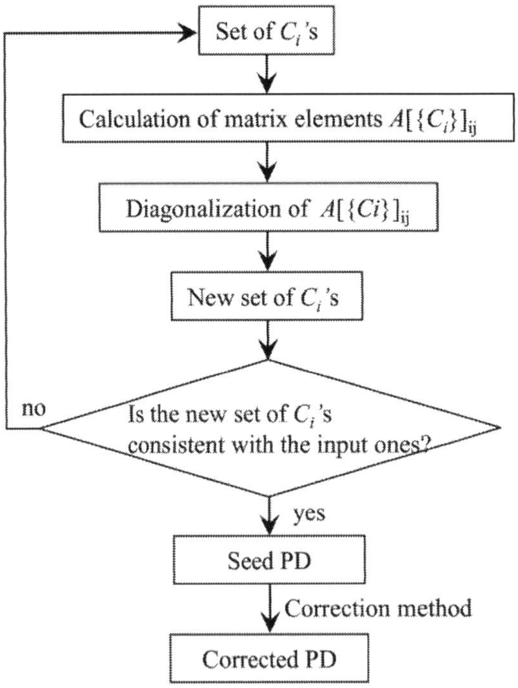

Figure 1. Flowchart of numerical calculations.

procedure is repeated until the input $C_i's$ are consistent with output ones. Thus, the coefficients that provide the seed PD are obtained.

Next, according to the correction method [31], the corrected PD, which corresponds to the variationally best PD after extending the search region of PDs, is calculated from the seed PD. By using the corrected PD, we evaluate errors of the KE, electron-electron interaction and potential energies, which are denoted by ΔT, ΔW and ΔV, respectively. These errors are calculated as the differences between the present results and the standard data. As the standard data, we use the results of the $1/Z$ expansion method [45] that is based on data of configuration interaction (CI) calculations and experiments [46].

4. Results of test calculations

Atomic structure calculations are performed for the neutral Ne atom. The virial ratio, ΔT, ΔW and ΔV are summarized in the first row of Tab. 1. For the purpose of reference, results that are calculated from the seed PD are also listed in the second row of Tab. 1 (without the correction).

It is found from Tab. 1 that the KE without the correction is lower than that of the CI method. This tendency is consistent with the fact that the usual TF functional is apt to underestimate the KE [47]. The KE, electron-electron interaction and potential energies change in a favorable way by means of the correction method. However, a certain amount of error remains in the KE, electron-electron interaction and potential energies. This means that there is a room for improvement in the approximate form of the KE functional.

Table 1. Errors of the KE, electron-electron interaction energy and potential energy that are denoted by ΔT, ΔW and ΔV, respectively, are shown together with the virial ratio. Results after (before) the correction method are shown in the first (second) row. The standard data of the KE, electron-electron interaction and potential energies are 257.874 (ryd.), 106.438 (ryd.) and -622.186 (ryd.), respectively [45].

	$\Delta T(\%)$	$\Delta W(\%)$	$\Delta V(\%)$	Virial ratio
Present results	7.922	2.634	7.208	-2.004
without the correction	-10.039	-6.344	-2.171	-2.161

It is also found that the virial ratio is much improved by the correction method. This is because the main component of the correlated TF functional, which is the usual TF functional $T_{TF}[\rho]$, satisfies the scaling property of the KE functional [48]. Therefore, we obtain the virial ratio that is close to the correct value -2.0.

It is well known that the atomic shell structure is absent in the profile of the electron density if the usual TF and related models are employed [47]. Figure 2 shows the profile of the electron density of the present scheme. The atomic shell structure clearly appears even though there remains the difference between the CI result [49] and the present one. This also seems to be an advantage of the present PD functional theory.

These test calculations are the first evaluations of the approximate KE functional that is developed on the basis of the coupling-constant expression. Therefore, calculation results obtained here are expected to be used as the reference data for comparison in developing more accurate functionals.

5. Concluding Remarks

In this paper, the validity of the correlated TF functional, which has recently been proposed for the KE functional of the PD functional theory on the basis of the coupling-constant expression, is checked by atomic structure calculations of the neutral Ne atom. Although it has been well known that the usual TF functional tends to underestimate the KE [47], the correlated TF functional inversely overestimates the KE by about 8 percent in our PD functional theory. The correction method [31] actually works toward reducing the errors, but there yet still remain some of them. This means that it is necessary to adopt or develop more appropriate functional as the noninteracting KE functional.

However, since this work is the first computational trial of the second strategy (ii) mentioned in section 1, the present results may become the reference data in judging whether the approximate form is valid or not. It can be said that we make the first but significant step toward developing the approximate KE functional which is based on the coupling-constant expression.

Using these reference data, in conjunction with the correction method [31], we will be able to improve on the approximate KE functional in future.

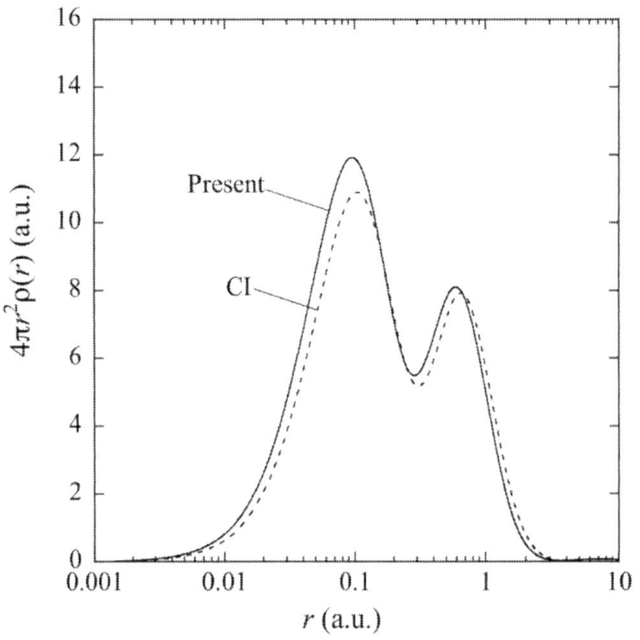

Figure 2. Profiles of the electron density of the neutral Ne atom. Solid and dashed lines denote the present and CI results[49], respectively.

Acknowledgements
This work was partially supported by Grants-in-Aid for Scientific Research (No. 22540390 and No. 23540446) of the Ministry of Education, Culture, Sports, Science, and Technology, Japan.

References
[1] Ziesche P 1994 *Phys. Lett.* A **195** 213
[2] Ziesche P 1996 *Int. J. Quantum Chem.* **60** 1361
[3] Gonis A, Schulthess T C, van Ek J and Turchi P E A 1996 *Phys. Rev. Lett.* **77** 2981
[4] Gonis A, Schulthess T C, Turchi P E A and van Ek J 1997 *Phys. Rev.* B **56** 9335
[5] Levy M and Ziesche P 2001 *J. Chem. Phys.* **115** 9110
[6] Furche F 2004 *Phys. Rev.* A **70** 022514
[7] Nagy Á 2002 *Phys. Rev.* A **66** 022505
[8] Nagy Á and Amovilli C 2004 *J. Chem. Phys.* **121** 6640
[9] Hetényi B, Brualla L and Fantoni S 2004 *Phys. Rev. Lett.* **93** 170202
[10] Ayers P W and Levy M 2005 *J. Chem. Sci.* **117** 507
[11] Nagy Á 2006 *Int. J. Quantum Chem.* **106** 1043
[12] Ayers P W, Golden S and Levy M 2006 *J. Chem. Phys.* **124** 054101
[13] Davidson E R 1995 *Chem. Phys. Lett.* **246** 209

[14] Samvelyan S Kh 1997 *Int. J. Quantum Chem.* **65** 127
[15] Pistol M –E 2004 *Chem. Phys. Lett.* **400** 548
[16] Gori-Giorgi P and Savin A 2006 *Philos. Mag.* **86** 2643
[17] Pistol M -E 2006 *Chem. Phys. Lett.* **422** 363
[18] Pistol M -E 2006 *Chem. Phys. Lett.* **431** 216
[19] Ayers P W and Liu S 2007 *Phys. Rev.* A **75** 022514
[20] Ayers P W and Davidson E R 2007 *Adv. Chem. Phys.* **134** 443
[21] Pistol M –E 2007 *Chem. Phys. Lett.* **449** 208
[22] Hetényi B and Hauser A W 2008 *Phys. Rev.* B **77** 155110
[23] March N H and Santamaria R 1991 *Int. J. Quantum Chem.* **39** 585
[24] Ayers P W 2005 *J. Math. Phys.* **46** 062107
[25] Ayers P W and Levy M 2005 *Chem. Phys. Lett.* **415** 211
[26] Cuevas-Saavedra R and Ayers P W 2009 *Int. J. Quantum Chem.* **109** 1699
[27] Higuchi M and Higuchi K 2007 *Physica B* **387** 117
[28] Higuchi M and Higuchi K 2007 *Phys. Rev.* A **75** 042510
[29] Higuchi M and Higuchi K 2008 *Phys. Rev.* B **78** 125101
[30] Higuchi K and Higuchi M 2010 *Phys. Rev.* B **82** 155135
[31] Higuchi M and Higuchi K 2011 *Phys. Rev.* A **84** 044502
[32] Higuchi K and Higuchi M 2012 *Phys. Rev.* A **85** 062508
[33] Higuchi M and Higuchi K 2012 *Comput. Theor. Chem.*, published online (DOI: 10.1016/j.comptc. 2012.09.015)
[34] Hohenberg P and Kohn W 1964 *Phys. Rev.* **136** B864
[35] Kohn W and Sham L J 1965 *Phys. Rev.* **140** A1133
[36] For instance, see, Dreitzler R M and Gross E K U 1990 *Density Functional Theory* (Berlin: Springer-Verlag).
[37] Higuchi M and Higuchi K 2002 *Phys. Rev.* B **65** 195122
[38] Higuchi K and Higuchi M 2005 *Phys. Rev.* B **71** 035116
[39] Higuchi K and Higuchi M 2006 *Phys. Rev.* B **74** 195122; Erratum 2007 *Phys. Rev.* B **75** 159902(E)
[40] Higuchi M and Higuchi K 2007 *Phys. Rev.* B **75** 195114
[41] Higuchi M and Higuchi K 2010 *Phys. Rev.* A **81** 042505
[42] Thomas L H 1927 *Proc. Cambrige Phil. Soc.* **23** 542
[43] Fermi E 1927 *Rend. Accad. Naz. Linzei* **6** 602
[44] Gunnarsson O and Lundqvist B I 1976 *Phys. Rev.* B **13** 4274
[45] Wang J and Smith Jr V H 1995 *Phys. Rev.* A **52** 1060
[46] Chakravorty S J, Gwaltney S R, Davidson E R, Parpia F A and Fischer C F 1993 *Phys. Rev.* A **47** 3649
[47] Parr R G and Yang W 1989 *Density-Functional Theory of Atoms and Molecules* (New York: Oxford University Press)
[48] As shown in [31], if the approximate form satisfies the scaling property of the KE energy functional, then the virial ratio is changed to the correct value -2.0 by the correction method.
[49] Bunge A V and Esquivel R O 1986 *Phys. Rev.* A **34** 853

Journal of Physics: Conference Series **454** (2013) 012057 doi:10.1088/1742-6596/454/1/012057

Structural relaxation and binding energy calculations of FK506 binding protein complexes using the large-scale DFT code CONQUEST

T Otsuka[1], N Okimoto[1], M Taiji[1], DR Bowler[2] and T Miyazaki[3]
[1]Quantitative Biology Center (QBiC), RIKEN, 1-6-5 Minatojima-Minamimachi, Chuo-ku, Kobe, Hyoho, 650-0047, Japan
[2]Department of Physics & Astronomy, University College London, Gower St, London, WC1E 6BT, UK
[3]Computational Materials Science Center, National Institute for Materials Science (NIMS), 1-2-1 Sengen, Tsukuba, Ibaraki, 305-0047, Japan

E-mail: totsuka@riken.jp

Abstract. We have performed structural relaxation and calculated binding energy of protein-ligand complex systems, FK506 binding protein (FKBP) and some ligand molecules, using a large-scale density functional theory (DFT) code CONQUEST. Detailed comparison of the calculated binding energies of FKBP with various ligand molecules is reported including the effects of the full geometry relaxation.

1. Introduction

In the field of drug discovery, computational methods are now commonly used to evaluate the binding affinities of drug molecules to proteins. Such computational methods are expected to improve the efficiency of drug design in the near future [1, 2]. Recently, *first-principles* (or *ab initio*) electronic structure approaches are also used in this field in order to calculate protein-ligand interactions accurately [3]. In the structure-based drug design (SBDD), however, (accurate) experimental atomic positions are not available in most cases. In general, less reliable experimental structures and/or partially relaxed structures by QM/MM or MM method are usually used in such studies. It is a great challenge for theorists to clarify the effects on the binding energy by relaxing all atomic positions solely by QM.

In the present study, we report fully relaxed structure and binding energy of the complex systems consisting of FK506 binding protein (FKBP) and a ligand molecule, using our density functional theory (DFT) code CONQUEST. The code has a high parallel efficiency and enables us to perform DFT studies on very large systems. The systems we have calculated in this work contain about 1700 atoms. Experimental structures of FKBP and various ligands can be obtained from the PDB codes: 1FKF, 1FKG, and 1FKI. To our knowledge, there have been no theoretical reports of the fully relaxed structures of these systems by first-principles DFT calculations. We believe it is important to collect such basic information, for example, to analyze the experimental or theoretical structural data for a protein-ligand system. The CONQUEST code can be used as a linear scaling code [4], and so it will be possible to model these systems in aqueous solution; the present work is a preparation also for such future studies.

2. Computational Methods

In this study, all calculations are performed with our own DFT code CONQUEST [5]. Details of the methods used in the code are explained in our previous papers [6]. Here we describe briefly some key points. In CONQUEST, we use the Kohn-Sham density matrix defined as

$$\rho(\boldsymbol{r},\boldsymbol{r}') = \sum_n f_n \psi_n(\boldsymbol{r}) \psi_n^*(\boldsymbol{r}'),$$

where $\psi_n(\boldsymbol{r})$ is the Kohn-Sham orbital for band index $\boldsymbol{n}$, and $\boldsymbol{f}_n$ is its occupation number between 0 and 1. The DFT total energy of system can be calculated from density matrix, using the pseudopotential technique and the standard exchange-correlation functionals such as the local density approximation (LDA) or the generalized gradient approximation (GGA). The density matrix is represented by localized orbitals called "support functions",

$$\rho(\boldsymbol{r},\boldsymbol{r}') = \sum_{i\alpha,j\beta} \phi_{i\alpha}(\boldsymbol{r}) K_{i\alpha,j\beta} \phi_{j\beta}(\boldsymbol{r}'),$$

where $\phi_{i\alpha}(\boldsymbol{r})$ are the support function that are non-zero only inside "support region" centered on the atoms $\boldsymbol{i}$, and $\boldsymbol{\alpha}$ runs over the support functions on a given atom. The matrix $K_{i\alpha,j\beta}$ is the density matrix expressed by non-orthogonal basis of support functions. In the CONQUEST code, two types of basis function are prepared for the support functions, one is B-splines on regular grids [7] and the other is numerical pseudo atomic orbitals (PAOs)[8]. In order to achieve the density matrix using the support functions, CONQUEST can employ two methods: conventional diagonalization or order-N methods. Although we only use the former method in this work, one of the advantage of using the code is we can treat extremely large systems, up to million atom systems [9], if we use the latter method.

In this work, we consider the FK506 binding protein (FKBP) complexes, which have been well studied both experimentally [10] and theoretically [11-13]. There is also our previous related work [14]. The systems consist of 107 residues for FKBP and small ligands. Experimental structures of FKBP and various ligands can be obtained from the PDB codes, 1FKF, 1FKG, and 1FKI, summarized in table 1. The initial (that is, experimental) structures as the reference structures are prepared by the following procedure: some missing hydrogen atoms are added to the PDB structures, and only hydrogen atoms are relaxed by using the classical force field of parm99. The fully relaxed structures are obtained by using our large-scale DFT code CONQUEST with double-zeta with polarization (DZP) basis set, Perdew–Burke–Ernzerhof (PBE) functional [15] for GGA, the diagonalization with non-self-consistent calculations, and the numerical integration grid cutoff of 100 Ha. The binding energy between a FKBP and the ligand molecule is simply evaluated by

$$\Delta E_{binding} = E_{complex} - \left(E_{protein} + E_{ligand}\right).$$

Here we calculate the total energy of protein, $E_{protein}$, using the structure of the protein in the complex system. In this study, we use only PBE functional in our DFT calculations for the evaluation of binding energy. The PBE functional reproduces the structure and the interaction energy of hydrogen bonds of biological systems (for example, see ref. [16,17]), but in some cases its lack of the description of long-range dispersion interactions [18,19] may cause a serious problem. We are preparing for DFT-D2 [18] and vdW-DF [19] scheme, and the results of dispersion effects using those methods will be presented in a future publication. Entropic contributions are also neglected in our DFT calculations.

Journal of Physics: Conference Series **454** (2013) 012057 doi:10.1088/1742-6596/454/1/012057

Table 1. PDB code of FKBP complex systems, the name of ligand, and the total number of atoms.

PDB code	Ligand	Number of atoms[a]
1FKF	L20	1792
1FKG	LG8	1734
1FKI	L13	1736

[a] The total number of atoms of FKBP is 1666.

3. Results and Discussion

In the protein and ligand complex system, the binding structure of the ligand molecule is significantly affected by the side chains of the amino acids surrounding the ligand. In order to collect the structural information of the ligand in such an environment, we first investigate the fully relaxed structures of FKBP-ligand systems. Figures 1, 2, and 3 show the structure of FKBP-L20, FKBP-LG8, and FKBP-L13 molecules before (blue line) and after (red line) the structure relaxation.

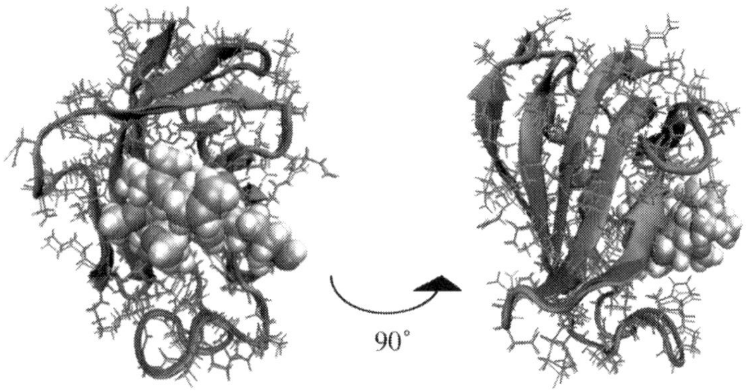

Figure 1. Structure of FKBP-L20 complex; before (blue) and after (red) fully structure relaxation.

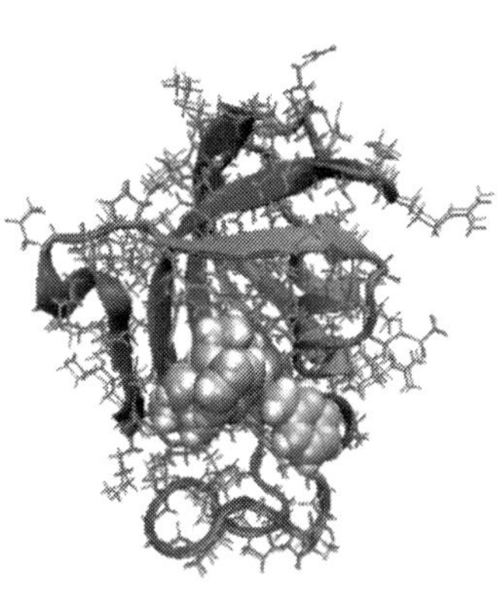

Figure 2. Structure of FKBP-LG8 complex; before (blue) and after (red) fully structure relaxation.

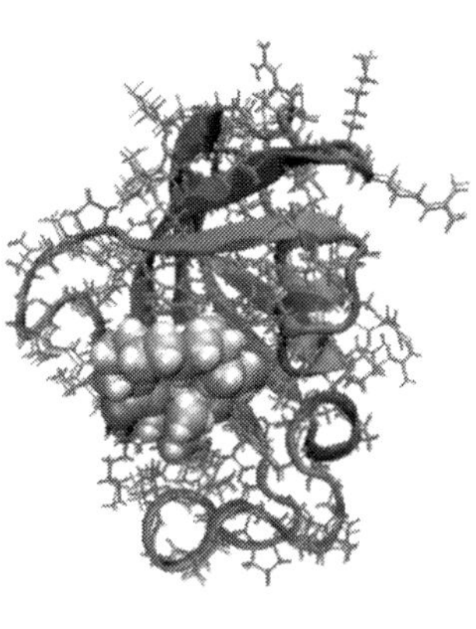

Figure 3. Structure of FKBP-L13 complex; before (blue) and after (red) fully structure relaxation.

Table 2 shows the root mean square deviation (RMSD) of the relaxed structure from the initial structure. In table 2, the RMSD of the backbone is much smaller than that of the whole protein including the residues. From table 2 and these figures (1, 2 and 3), we can see that the change of the backbone structure of the protein is small, while the structure of the side chains are largely changed by the full relaxation.

Table 2. Root mean square deviation (RMSD) of fully relaxed structure from the experimental one.

System	$RMSD_{backbone}$	$RMSD_{residue}$
FKBP-L20	0.27125	0.45675
FKBP-LG8	0.24112	0.38802
FKBP-L13	0.30690	0.45652

Next we compare the binding energies between FKBP and the ligand molecules. Table 3 shows the binding energies with or without the structure relaxation of FKBP-ligand complex systems. Note that the experimental values shown here are the binding affinities obtained by Holt et al [10]. In the present study, we regard the values as the approximate binding free energies. Table 3 shows that the order of the calculated binding energies is the same as the experimental one, for both cases using the initial or relaxed structures. The calculated binding energies are slightly overestimated in the present PBE calculations. It also shows that the binding energies calculated by using the relaxed structures are larger than those obtained by the initial structures. We can conclude that the ligand molecule can bind to the residues in FKBP more tightly in the fully relaxed structure. For the overestimation in the calculated binding energy, we should consider the hydration effects, entropic contribution in the future.

Table 3. Calculated biding energy of FKBP-ligand complex systems.

System	Experimental[a]	Initial structure	Relaxed structure
FKBP-L20	-12.8	-18.95	-26.62
FKBP-LG8	-10.9	-17.25	-24.37
FKBP-L13	-9.5	-14.02	-21.88

[a] Experimental values indicate the experimental binding affinities as referred in [10]

4. Conclusion

In this study, we have performed the structural relaxation of FKBP-ligand complexes using the large-scale DFT code CONQUEST. We have calculated the binding energies of FKBP and various ligands and found that the energy of the relaxed structure is stronger than that obtained by using the experimental structure. We have also analyzed the fully relaxed structure of the FKBP complexes in detail and have found that the structure of the side chains of the amino acids surrounding the ligand is largely changed. From these results, it seems that the structure of the entire protein, not just the area around the ligand, should be modelled when performing a study on SBDD. The results in this work assist in promoting the study of the "relaxed" SBDD. In order to clarify these aspects theoretically, very large-scale electronic structure calculations on FKBP-ligand complex systems in aqueous solution are necessary. Linear scaling DFT codes such as CONQUEST form the tools to perform these studies and the results will be presented in a future publication.

Acknowledgments

This work was partly supported by KAKENHI: 20540401, 22790122, 22104005, as well as Strategic Programs for Innovative Research (SPIRE), MEXT, and The Computational Materials Science Initiative (CMSI), Japan. The calculations were performed in part on the NIMS material numerical simulator and the RIKEN Integrated Cluster of Clusters (RICC) system. The authors also acknowledge Prof. M. J. Gillan (University College London), Dr. F. Pietrucci (EPFL Lausanne) and Prof. S. Tanaka (Kobe University) for useful discussions.

References

[1] Gilson M K, Given J A, Bush B and McCammon J A 1997 *Biophys. J.* **72**.1047-1069
[2] Gilson M K, and Zhou H 2007 *Annu. Rev. Biophys. Biomol. Struct.* **36** 21-42
[3] Kitaura K, Ikeo E, Asada T, Nakano T and Uebayasi M 1999 *Chem. Phys. Lett.* **313** 701–706
[4] Bowler D R and Miyazaki T 2012 *Rep. Prog. Phys.* **75** 036503
[5] See our web page; http://www.order-n.org/
[6] Bowler D R, Miyazaki T and Gillan M J 2002 *J. Phys.: Condens. Matter* **14** 2781; Bowler D R, Choudhury R, Gillan M J and Miyazaki, T 2006 *Phys. Stat. Sol.* B **243** 989
[7] Hernandez E, Gillan M J, and Goringe C M 1997 *Phys. Rev. B* **55** 13485
[8] Torralba A S, Todorovic M, Brazdova V, Choudhury R, Miyazaki T, Gillan M J and Bowler D R 2008 *J. Phys.: Condens. Matter* **20** 294206
[9] Bowler D R and Miyazaki T 2010 *J. Phys.: Condens. Matter* **22** 074207
[10] Holt D A, Luengo J I, Yamashita D S, Oh H J, Konialian A L, Yen H K, Rozamus L W, Brandt M, Bossard M J, Levy M A, Eggleston D S, Liang J, Schultz L W, Stout T J and Clardy J 1993 *J. Am. Chem. Soc.* **115** 9925-9938
[11] Fujitani H, Tanida Y and Matsuura A 2009 *Phys. Rev. E* **79** 021914
[12] Nakanishi I, Fedorov D G and Kitaura K 2007 *Proteins* **68** 145-158
[13] Xu Y and Wang R 2006 *Proteins* **64** 1058-1068
[14] Watanabe H, Tanaka S, Okimoto N, Hasegawa A, Taiji M, Tanida Y, Mitsui T, Katsuyama M and Fujitani H 2010 *CBI Journal* **10** 32-45
[15] Perdew J P, Burke K and Ernzerhof M 1996 *Phys. Rev. Lett.* **77** 3865, Errata: 1997 Phys Rev. Lett. **78** 1396
[16] Otsuka T, Miyazaki T, Ohno T, Bowler D R and Gillan M J 2008 *J. Phys. Condens. Matter* **20** 294201
[17] Sugiki S, Matsuoka M, Usuki R, Sengoku Y, Kurita N, Sekino H and Tanaka S 2005 *J. Theor. Comput. Chem.* **4** 183
[18] Grimme S 2006 *J. Comput. Chem.* **27** 1787-1799
[19] Dion M, Rydberg H, Schröder E, Langreth D C and Lundqvist B I 2004 *Phys. Rev. Lett.* **92** 246401

Journal of Physics: Conference Series **454** (2013) 012058 doi:10.1088/1742-6596/454/1/012058

Quantum chemical study of conformational preferences of intermediates and transition states in the alkaline hydrolysis of dimethyl phosphate

Y Takano, M Kita and H Nakamura

Institute for Protein Research, Osaka University, 3-2 Yamadaoka, Suita, Osaka 565-0871, Japan

E-mail: ytakano@protein.osaka-u.ac.jp

Abstract. The phosphate diester is a basic structure in DNA and RNA. The mechanism of phosphate diester hydrolysis is important for understanding the decomposition reactions of nucleic acids. In this study, we have explored the reaction pathway of alkaline hydrolysis of dimethyl phosphate, which is the simplest phosphate diester, with a hydroxide ion. Since the conformations of the intermediates and transition states reportedly influence the reaction mechanism of transacylation of methyl acetate with methoxide, we considered the conformational preferences on the alkaline hydrolysis of dimethyl phosphate, by using the most stable conformer as a reactant. Upon the reaction with hydroxide, a concerted reaction pathway was obtained in the gas phase, whereas a stepwise reaction pathway was obtained in water. As compared to the earlier study, our computation shows more stable conformations in the hydrolysis reactions than the previous study.

1. Introduction

DNA and RNA play essential roles in life, such as the storage and decoding of genomic information [1]. In this series of processes, called the central dogma, the decomposition reactions of DNA and RNA; namely, the hydrolysis of the phosphate ester bonds in DNA and RNA strands, is important. For example, the production and decomposition of RNA primers in DNA replication and the modification of mRNA in transcription are accomplished by enzymes that catalyze the decomposition of nucleic acids [1, 2]. Besides the central dogma, some restriction enzymes in bacteria distinguish and decompose foreign DNA, and prevent their host from being infected with a virus. In addition, the hydrolysis of a phosphate ester bond is involved in the removal of damaged DNA strands. Decomposition reactions of DNA and RNA are essential to maintain life. Therefore, it is important to fully understand the mechanism of this reaction in biology.

Recently, the hydrolysis reaction of phosphate diester has been investigated, as a model of the decomposition reactions of DNA and RNA. In addition, the hydrolysis of the phosphate diester bond is also involved in the energy utilization from ATP and the mechanism of intracellular signal transfer by GTP [2]. However, the mechanism is still being debated. Iche-Tarrat [3] and Ribeiro [4] calculated the hydrolysis reaction pathway of dimethyl phosphate with a hydroxide ion as a nucleophile, by using B3LYP [5–8] with a double-ζ plus polarization valence basis set [9] and B3LYP/6-311++G(2d,2p), respectively, and showed that the reaction is concerted. However, Imhof [10] explored it by using the B3LYP/6-31+G(d,p) method, and showed that the reaction is stepwise. These results suggest that the computed hydrolysis reaction pathway depends on the basis set. Dejaegere [11] calculated the hydrolysis reaction pathway under the same conditions as Iche-Tarrat, Ribeiro, and Imhof, by using three types of basis sets, HF/STO-3G, HF/3-21G(d), and HF/3-21+G(d). In all three calculations, one intermediate was observed in the pathway. However, the greater the accuracy of the basis set, the higher the energy of that intermediate.

In our previous study [12], we performed a conformational analysis for the transition states and the tetrahedral intermediates of the transacylation of methyl acetate, and compared the conformations involved in the transacylation to those of the phosphate inhibitors found in lipases and esterases. We concluded that the three-dimensional shapes of the active sites of lipases and esterases accommodate the conformational preferences of the tetrahedral intermediates and inhibitors.

In the present study, we have explored the alkaline hydrolysis pathway of phosphate diester at the CCSD/6-311++G(2d,2p)//B3LYP/6-311++G(2d,2p) level of theory, considering the conformational preferences of the stationary points. We used dimethyl phosphate, the simplest phosphate diester, as the reactant and examined the hydrolysis pathway with a hydroxide ion in vacuum and water.

2. Computational procedure

2.1. QM calculation

All calculations were done using the Gaussian09 program packages [13]. We optimized all possible conformations of dimethyl phosphate, and investigated the most stable conformation. Then, we explored the hydrolysis pathway of dimethyl phosphate in the most stable conformation with a hydroxide ion in the gas phase and the aqueous environment at the CCSD/6-311++G(2d,2p)//B3LYP/6-311++G(2d,2p) level of theory, because it is time-consuming to perform geometry optimizations, frequency calculations, and IRC calculations at the coupled cluster (CC) level [14–16]. In addition, Gaussian09 does not implement the geometry optimization in the implicit water model such as the polarizable continuum model (PCM) [17, 18] at the CC level of theory, and it is known that geometrical parameters converged very quickly with improving theoretical methods. However, since it is shown that B3LYP is not adequate functional to study the energetics of phosphodiester systems [4], single point energy calculations were performed by using the CCSD/6-311++G(2d,2p)//B3LYP/6-311++G(2d,2p) method [5–8].

All stationary points were characterized by frequency calculations. The reaction pathway was determined by following the IRC procedure [19–21]. Zero-point energies and thermal corrections at 298K were also included, to obtain the Gibbs free energies. Aqueous solvation energies were computed with the IEF-PCM [14, 15], and we also included the non-electrostatic term to estimate the aqueous solvation free energies. To remove the excessive estimation of entropies attendant upon the formation of complexes, we did not include the translational and rotational entropies in the thermal corrections, in the hydrolysis pathway in water [4], because the complex water–dimethylphosphate should always exist in solution and the nucleophile that attacks dimethylphosphate will substitute that water molecule.

2.2. Definition of conformations

Dimethyl phosphate has two methoxy groups, and they adopt various conformations depending on the orientation of these methoxy groups. The definitions of these conformations are shown in figure 1 [12]. The conformations of dimethyl phosphate are decided by the two dihedral angles of the methoxy groups, O1-C1 and O2-C2, with respect to the P-C2 and P-C1 bonds, respectively. We defined three notations, gauche+ (G+), gauche– (G–), and anti (A), to represent dihedral angles of 0° to 120°, –120° to 0°, and 120° to 180° or –180° to –120°, respectively.

For an unsymmetrical case, there are nine conformers at the P atom of a phosphate, while the nine possible conformers can be reduced for a symmetrical case. Dimethyl phosphate has four distinct conformers (AA, AG–, G+G–, and G–G–), because the AG– conformer is the same as the G+A, AG+, and G–A ones, the G+G– conformation is the same as the G–G+ one, and the G–G– and G+G+ conformers have the same structures, respectively.

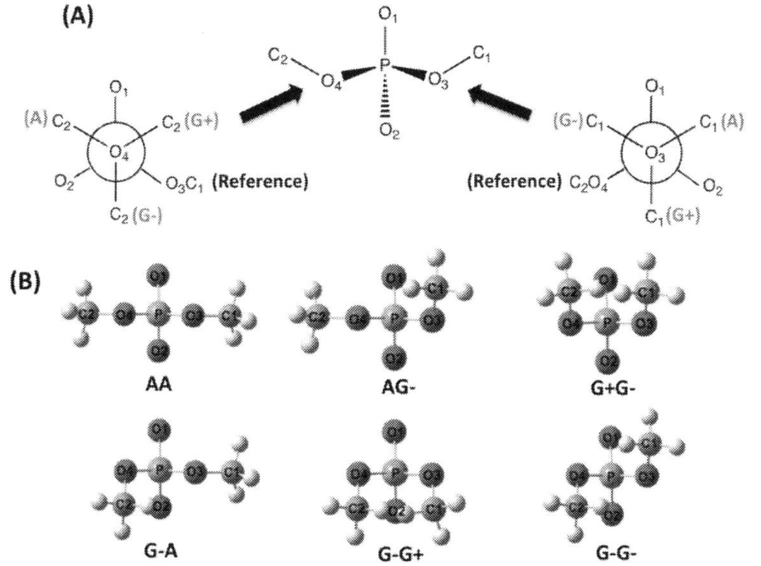

Figure 1. Definitions of the conformations of dimethyl phosphate (A), and the geometry of each conformation (B).

3. Results and discussion

3.1. Comparison of possible conformations of dimethyl phosphate

We optimized all four conformers of dimethyl phosphate, and calculated the Gibbs free energies. The bond lengths and dihedral angles of the optimized structures are shown in table 1. The relative Gibbs free energies of the optimized structures are shown in table 2.

3.1.1. Comparison of theoretical methods

We first assessed the validity of the theoretical methods with regard to the optimized geometrical parameters and the relative Gibbs free energies of the conformers of dimethyl phosphate. Table 1 indicates that the bond lengths computed by the B3LYP method are agreement with those by the MP2 method within 0.01Å, and that the dihedral angles are close to each other within 15°, implying that geometry optimization of dimethyl phosphate does not depend on the theoretical methods. As shown in table 2, the B3LYP method provides smaller relative Gibbs free energies than the CCSD and CCSD(T) methods by 0.3–0.7 kcal mol^{-1} for the AA conformers and by about 0.3 kcal mol^{-1} for the AG– conformers, while the MP2 method shows the equivalent relative free energies to the CCSD and CCSD(T) methods with the error within 0.1 kcal mol^{-1}. The difference in the relative Gibbs free energies between the CCSD/6-311++G(2d,2p)//B3LYP/6-311++G(2d,2p) and CCSD/6-311++G(2d,2p)//MP2/6-311++G(2d,2p) were evaluated to be within 0.1 kcal mol^{-1} in the gas phase and at most 0.25 kcal mol^{-1} in water, implying that the theoretical methods for the geometry optimization are insensitive to the relative Gibbs free energies. These findings indicate that our procedure (CCSD/6-311++G(2d,2p)//B3LYP/6-311++G(2d,2p)) is appropriate for the exploration of the alkaline hydrolysis pathway of phosphate diester. Hereafter, we use the CCSD/6-311++G(2d,2p)//B3LYP/6-311++G(2d,2p) method for the investigation.

3.1.2. Conformational preference of dimethyl phosphate

As shown in table 1, after optimization, conformational changes were observed in both vacuum (G+G– to AG–) and water (G+G– to AG–). The AA, AG–, and G–G– conformers were found as the energy minima in both the gas phase and water. The two close methoxy groups of the G+G–

conformer show a large steric hindrance, resulting in the conformational change to AG–. Table 2 shows that the G–G– conformer is the most stable conformation, in both vacuum and water. Since the two methoxy groups of the G–G– conformer point in opposite directions, the steric hindrance in the G–G– conformation is smaller than that in the others. The AG– and AA conformers are less stable than the G–G– conformer by about 1.2 and 2.2 kcal mol^{-1}.

Table 1. Bond lengths (P-O1, P-O3, and P-O4)[a] and dihedral angles (C2-O4-P-O3 and C1-O3-P-O4)[b] of the optimized structures of dimethyl phosphate in the gas phase and water at the B3LYP/6-311++G(2d,2p) level of theory[c]

	initial	optimized	P–O1	P–O3	P–O4	C2-O4-P-O3	C1-O3-P-O4
gas phase	AA	AA	1.495 (1.499)	1.658 (1.654)	1.658 (1.654)	+153.7 (+166.5)	+153.7 (+166.5)
	AG–	AG–	1.497 (1.501)	1.648 (1.645)	1.673 (1.671)	+171.0 (+170.3)	–73.23 (–72.05)
	G+G–	AG–	1.497 (1.501)	1.648 (1.645)	1.673 (1.671)	+171.0 (+170.3)	–73.23 (–72.05)
	G–G–	G–G–	1.489 (1.492)	1.665 (1.664)	1.665 (1.664)	–73.26 (–71.01)	–73.26 (–71.01)
water	AA	AA	1.502 (1.507)	1.636 (1.629)	1.636 (1.629)	–180.0 (–180.0)	–180.0 (–180.0)
	AG–	AG–	1.500 (1.505)	1.629 (1.623)	1.644 (1.636)	–172.9 (172.5)	–71.36 (–71.78)
	G+G–	AG–	1.500 (1.505)	1.629 (1.623)	1.644 (1.636)	–172.9 (172.5)	–71.36 (–71.78)
	G–G–	G–G–	1.497 (1.502)	1.639 (1.632)	1.639 (1.632)	–67.04 (–62.92)	–67.04 (–62.92)

[a] Bond lengths are shown in angstroms.
[b] Dihedral angles are shown in degrees.
[c] The geometrical parameters computed at the MP2/6-311++G(2d,2p) level of theory are shown in parentheses.

Table 2. Relative Gibbs free energies of the optimized structures in the gas phase and water[c]

	optimized	ΔG					
		B3LYP/6-311++G(2d,2p)-optimized			MP2/6-311++G(2d,2p)-optimized		
		CCSD(T)	CCSD	B3LYP	CCSD(T)	CCSD	MP2
gas phase	AA	2.276	2.207	1.557	2.395	2.288	2.364
	AG–	1.271	1.224	0.932	1.203	1.203	1.277
	G–G–	0.000	0.000	0.000	0.000	0.000	0.000
water	AA	—[b]	2.505	2.187	—[b]	2.251	2.252
	AG–	—[b]	1.419	1.105	—[b]	1.278	1.195
	G–G–	—[b]	0.000	0.000	—[b]	0.000	0.000

[a] Free energies are with respect to the G-G- conformer, and shown in kcal mol^{-1}.
[b] Relative Gibbs free energies are not shown, because Gaussian09 does not implement the CCSD(T) in the PCM.

3.2. Hydrolysis of phosphate diester

Since the G–G– conformer is the most stable conformation of dimethyl phosphate in both vacuum and water, we explored the hydrolysis reaction pathway of the G–G– conformer of dimethyl phosphate with hydroxide ion.

3.2.1. Gas phase

Figure 2 shows the lowest energy pathway for the gas-phase reaction at the B3LYP/6-311++G(2d,2p) level of theory, and the bond lengths (cleaved bond: P-O3 and formed bond: P-O5) and the dihedral angles (C2-O4-P-O3 and C1-O3-P-O4) are listed in table 3. As illustrated in figure 2, the reaction proceeds in a concerted manner. The hydrogen atom of the hydroxide ion approaches the phosphorus atom of the G–G– dimethyl phosphate, forming a transition state (TS) with an activation barrier of 99.1 kcal mol^{-1}. In this process, the O4C2 methoxy group slightly rotates to the leaving group, resulting in the G–G– transition state. The transition state decomposes to the isolated products, methoxide and methyl hydrogen phosphate. The overall reaction is exothermic, $\Delta G = -7.55$ kcal mol^{-1}.

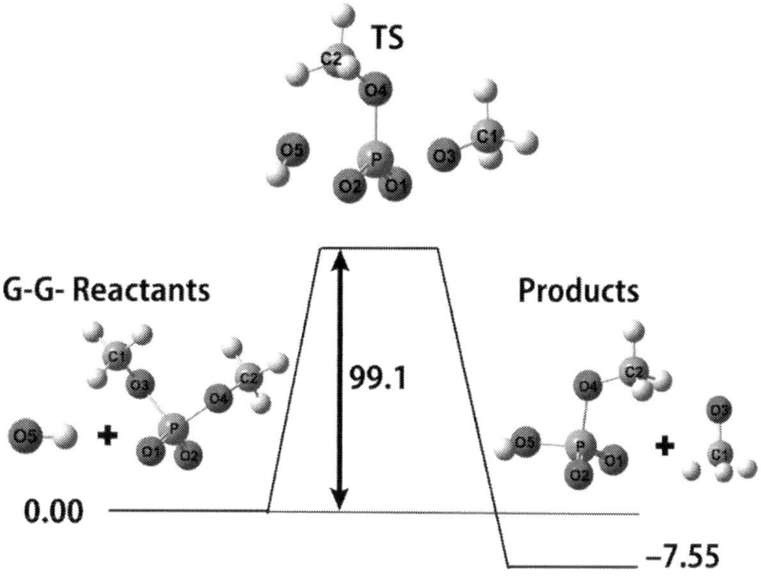

Figure 2. Hydrolysis reaction pathway of dimethyl phosphate with a hydroxide ion in the gas phase and the optimized structures of stationary points (reactants, transition states, and products) at the CCSD/6-311++G(2d,2p)//B3LYP/6-311++G(2d,2p) level of theory. Gibbs free energies of the stationary points are plotted, relative to the reactants. The relative Gibbs free energies are given in units of kcal mol^{-1}.

Table 3. The lengths of the cleaved (P-O3) and formed bonds (P-O5)[a], and the dihedral angles for the definition of the conformation (C2-O4-P-O3 and C1-O3-P-O4)[b].

Structure	P-O3	P-O5	C2-O4-P-O3	C1-O3-P-O4
Dimethyl phosphate (Reactant)	1.63	–	–73.26	–73.26
TS	1.87	2.18	–109.2	–67.95
Methyl hydrogen phosphate (Product)	–	1.65	–	–

[a] Bond lengths are shown in Å.
[b] Dihedral angles are shown in degrees.

However, when the CCSD/6-31+G(d)//B3LYP/6-31+G(d) method was used for the investigation of the gas-phase hydrolysis pathway, we found a stepwise reaction pathway with one intermediate, as shown in figure 3. Since the energy differences between TS1, TS2 and INT in these pathways are 1.9 kcal mol^{-1} at most, we concluded that the greater the accuracy of the basis set, the higher the energy of the INT. This is consistent with the previous studies [10, 11].

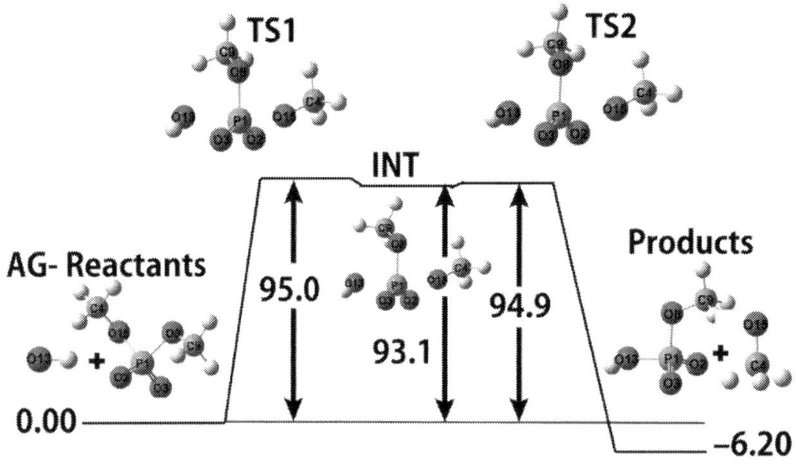

Figure 3. Hydrolysis reaction pathway of dimethyl phosphate with a hydroxide ion in the gas phase and the optimized structures of stationary points (reactants, transition states, and products) at the CCSD/6-311++G(2d,2p)//B3LYP/6-31+G(d) level of theory. Gibbs free energies of the stationary points are plotted, relative to the reactants. The relative Gibbs free energies are given in units of kcal mol^{-1}.

3.2.2. Aqueous phase

The reaction pathway and the geometrical parameters (the bond lengths of P-O3 and P-O5 and the dihedral angles of C2-O4-P-O3 and C1-O3-P-O4) in water are shown in figure 4 and table 4, respectively. As illustrated in figure 4, stepwise hydrolysis proceeds in water at the CCSD/6-311++G(2d,2p)//B3LYP/6-311++G(2d,2p) level of theory. The B3LYP/6-31+G(d) computation also provides the stepwise reaction pathway of the hydrolysis of dimethyl phosphate in water. In this pathway, the hydroxide ion attacks the phosphorus atom of dimethyl phosphate, resulting in the formation of an intermediate (INT) via a transition state (TS1). Then, the dissociation of the methoxide occurs to form the isolated products, methoxide and methyl hydrogen phosphate, through another transition state (TS2). The aqueous environment favors the stepwise pathway and remarkably reduces the activation barriers, by about 74 kcal mol^{-1}, due to the reduction in the electrostatic repulsion between the hydroxide and dimethyl phosphate.

The difference of the dipole moments between the transition states (TS1 and TS2) and the intermediate (INT) is considered to be responsible for the stepwise mechanism in water. In general, a molecule with a larger dipole moment becomes more stable in water. In this reaction, the large dipole moment of INT contributes to the stepwise reaction pathway, because INT has the largest dipole moment (TS1: 8.23 D, INT: 10.8 D, and TS2: 8.55 D).

3.3. Comparison with the previous study

Ribeiro and colleagues [4] tested many DFT exchange–correlation functionals by considering the hydrolysis reaction of dimethyl phosphate. They explored the hydrolysis reaction pathway of dimethyl phosphate by the B3LYP/6-311++G(2d,2p) calculations, and then calculated the energy of each structure in the pathway by using various DFT exchange–correlation functionals. We compared the reaction pathway in this study to that in the Ribeiro's study.

In the Ribeiro's study, the reactant has the G–G– geometry. In vacuum, the computed pathway in the Ribeiro's study is also a concerted one, with an activation energy of 98.4 kcal mol^{-1}, equivalent to our results. However, the transition state was not G–G–, but AA. The differences in the conformations along the reaction pathway hardly influence the number of the intermediates and transition states.

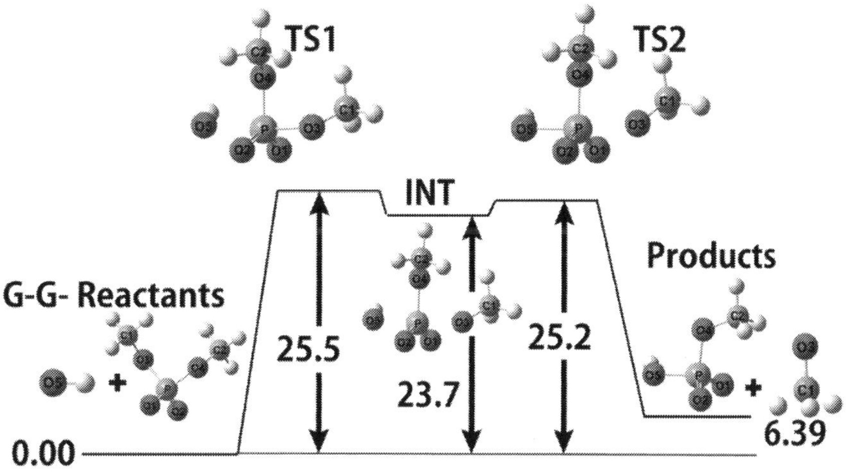

Figure 4. Hydrolysis reaction pathway of dimethyl phosphate with a hydroxide ion in water and the optimized structures of stationary points (reactants, transition states, and products) at the CCSD/6-311++G(2d,2p)//B3LYP/6-311++G(2d,2p) level of theory. Gibbs free energies of the stationary points are plotted, relative to the reactants. The relative Gibbs free energies are given in units of kcal mol^{-1}.

Table 4. The lengths of the cleaved (P-O3) and formed bonds (P-O5)[a], and the dihedral angles for the definition of the conformation (C2-O4-P-O3 and C1-O3-P-O4)[b].

Structure	P-O3	P-O5	C2-O4-P-O3	C1-O3-P-O4
Dimethyl phosphate (Reactant)	1.65	–	–67.40	–67.40
TS1	1.81	2.07	–67.48	–58.97
INT	1.91	1.85	–63.21	–57.82
TS2	2.02	1.79	–65.28	–54.51
Methyl hydrogen phosphate (Product)	–	1.64	–	–

[a] Bond lengths are shown in Å.
[b] Dihedral angles are shown in degrees.

In water, the components of the reaction pathway, such as the Gibbs free energies and the conformation of each structure, are remarkably different from those in the Ribeiro's study. Although the stepwise pathway was obtained in the present study, the concerted pathway was shown in the Ribeiro's study. A clear conformational change was not observed in this study (figure 4), but in the Ribeiro's study, the conformation of the reactant (G–G–) is quite different from that of the transition state (AA). Since both the reaction pathway (stepwise or concerted) and the conformation of each structure are different between this study and the Ribeiro's study, it is possible that the differences in the conformations along the reaction pathway influence the number of the intermediates and transition states. Moreover, the first activation energy of the hydrolysis reaction in this study is lower than that in the Ribeiro's study, by 1.8 kcal mol^{-1}. Therefore, if the conformations do not largely affect the reaction pathway, the current stepwise reaction pathway is considered to be more favorable than the concerted reaction pathway in water.

Since we determined the hydrolysis pathways by following the IRC procedure, the pathways obtained in this study are more reliable than the Ribeiro's study.

4. Concluding remarks

In this study, we have investigated the hydrolysis reaction mechanism of the most stable conformation (G–G–) of dimethyl phosphate. We examined the hydrolysis pathway with a hydroxide ion in vacuum and in water.

In the reaction with hydroxide, a concerted reaction pathway was obtained in vacuum, whereas a stepwise reaction pathway was obtained in water conditions. Regardless of the conformational difference of the transition state, the reaction pathway shown in the previous study is concerted as well as our study, in the gas phase. On the other hand, the reaction mechanism is quite different from those of the previous study in water, suggesting that the conformations in the reaction pathway affect the number of the intermediates and transition states.

Acknowledgements

This work was supported by grants from Research and Development of the Next-Generation Integrated Simulation of Living Matter, which is part of the Development and Use of the Next-Generation Supercomputer Project of MEXT. We are also grateful to MEXT for a Grant-in Aid for Scientific Research on Innovative Areas "Materials Design through Computics" (23104506) to JSPS for a Grant-in-Aid for Exploratory Research (23657103). The computations were performed using the Research Center for Computational Science, Okazaki, Japan and the Cybermedia Center at Osaka University, Japan.

References

[1] Alberts B, Johnson A, Lewis J, Raff M, Roberts K, Walter P 2007 *Molecular Biology of the Cell*, 5th edition (Abingdon: Garland Science)
[2] Voet D, Voet J G 2004 *Biochemistry*, 3rd edition (Hoboken NJ: John Wiley & Sons Inc.)
[3] Iche-Tarrat N, Barthelat J, Rinaldi D, Vigrous A 2005 *J. Phys. Chem. B* **109** 22570
[4] Ribeiro A J M, Ramos M J, Fernandes P A 2010 *J. Chem. Theory Comput.* **6** 2281
[5] Perdew J P, Chevary J A, Vosko S H, Jackson K A, Pederson M R, Singh D J 1992 *Phys. Rev. B* **46** 6671
[6] Becke A D 1988 *Phys. Rev. A* **38** 3098
[7] Becke A D 1993 *J. Chem. Phys.* **98** 5648
[8] Lee C, Yang W, Par R G *Phys. Rev. B* **37** 785
[9] Bouteiller Y, Mijoule C, Nizam M, Barthelat J-C, Daudey J-P, Pélissier M, Silvi B 1988 *Mol. Phys.* **65** 295
[10] Imhof P, Fischer S, Kramer R, Smith C J 2005 *J. Mol. Struct. (Theochem)* **713** 1
[11] Degaegere A, Lim C, Karplus M 1991 *J. Am. Chem. Soc.* **113** 4353
[12] Takano Y, Houk K N 2004 *J. Phys. Chem. A* **108** 11740
[13] Frisch M J, Trucks G W, Schlegel H B, Scuseria G E, Robb M A, Cheeseman J R, Scalmani G, Barone V, Mennucci B, Petersson G A, Nakatsuji H, Caricato M, Li X, Hratchian H P, Izmaylov A F, Bloino J, Zheng G, Sonnenberg J L, Hada M, Ehara M, Toyota K, Fukuda R, Hasegawa J, Ishida M, Nakajima T, Honda Y, Kitao O, Nakai H, Vreven T, Montgomery Jr J A, Peralta J E, Ogliaro F, Bearpark M, Heyd J J, Brothers E, Kudin K N, Staroverov V N, Kobayashi R, Normand J, Raghavachari K, Rendell A, Burant J C, Iyengar S S, Tomasi J, Cossi M, Rega N, Millam J M, Klene M, Knox J E, Cross J B, Bakken V, Adamo C, Jaramillo J, Gomperts R, Stratmann R E, Yazyev O, Austin A J, Cammi R, Pomelli C, Ochterski J W, Martin R L, Morokuma K, Zakrzewski V G, Voth G A, Salvador P, Dannenberg J J, Dapprich S, Daniels A D, Farkas Ö, Foresman J B, Ortiz J V, Cioslowski J, Fox D J 2009 *Gaussian09* (Wallingford CT: Gaussian Inc.)
[14] Urban M, Noga J, Cole S J, Bartlett R J 1985 *J. Chem. Phys.* **83** 4041
[15] Pople J A, Head-Gordon M, Raghavachari K 1987 *J. Chem. Phys.* **87** 5968
[16] Pople J A, Head-Gordon M, Raghavachari K, Trucks G W 1989 *Chem. Phys. Lett.* **164** 185
[17] Cancès E, Mennucci B, Tomasi J 1997 *J. Chem. Phys.* **107** 3032

[18] Tomasi J, Mennucci B, Cancès E 1999 *J. Mol. Struct. (Theochem)* **464** 211
[19] Fukui K 1981 *Acc. Chem. Res.* **14** 363
[20] Hratchian H P, Schlegel H B 2004 *J. Chem. Phys.* **120** 9918
[21] Hratchian H P, Schlegel H B 2006 *J. Chem. Theory Comput.* **1** 61

Effect of atomic adsorption of catalytic metals on mechanical properties of graphene

Ryosuke Iwata and Masato Aoki

Graduate School of Engineering, Gifu University,
1-1 Yanagido, Gifu 501-1193, Japan

E-mail : gifuu.iwtrsk@gmail.com

Abstract. The effect of the adsorption of catalytic metal atoms on tensile strength and structural deformation of graphene is examined using the density functional theory. In the case when a line of Ni atoms is adsorbed at hollow sites aligned along zigzag direction of the honeycomb structure, the tensile strength against the uniaxial strain in armchair direction is calculated to be 82.64 GPa, which is 21% smaller than that of the pristine graphene. The reduction of the tensile strength monotonically depends on the line density of Ni. It is predicted that if the stress reaches the tensile strength, the system will undergo a stress-induced structural transformation that involves the breaking of C-C bond adjacent to Ni atoms, the formation of Ni-C bonds and a resultant jump in the strain.

1. Introduction

Graphene, the monolayer graphite sheet, is one of the most promising materials for post-silicon nano-electronics devices and nano-sensors for its characteristic properties [1-3]. The honeycomb structure of graphene is built up of very strong sp^2 covalent bonding and the π bonding between carbon atoms, thus this thinnest membrane is the most stable allotrope of carbon with extremely high tensile strength along the sheet. In fabrications of nano-sized graphene such as the graphene ribbons, the development of new techniques to perform a cutting as designed in atomic scale are desired and are a challenging problem.

Researchers have experimentally and theoretically investigated the mechanical property of pristine graphene [4-6], and also the graphene with vacancies [7-9] or Stone-Wales defect [8,9]. It has been found that the introduction of defects promotes bond breaking and degrades the tensile strength of the graphene. Similar effects may be expected for catalytic 3d-transision metal atoms, which are known to exhibit strong adsorption to graphene and carbon nanotubes [10,11]. Ci *et al.* has reported a pioneering study that a Ni nanoparticle can be used to cut graphene in pieces with discrete armchair or zigzag edges [12].

In the present study, we examine the effect of the adsorption of a line of Ni atoms on atomic structures and mechanical properties of graphene using density functional calculations. The total energy, stress and bond lengths are calculated as functions of strain. The calculated tensile strengths and stress-induced structural transformation will be discussed.

2. Methods

The present calculations have been performed on the basis of the density functional theory using the VASP code [13]. The Kohn-Sham wave function is represented in the planewave basis with the projector augmented wave (PAW) method [14,15]. An optimized cut-off kinetic energy of 400 eV is used for planewave basis throughout the present study. The generalized gradient approximation with Perdew-Burke-Ernzerhof functional [16,17] is employed for the exchange-correlation energy.

Three kinds of orthorhombic unit cells with different periodicities in x (zigzag) direction are considered in order to examine systematically the effect of adsorption of Ni atoms. Figure 1 shows the initial structures containing one Ni adatom in (a) the eight C atoms unit cell (referred to as Period 1), (b) the sixteen C atoms unit cell (Period 2), and (c) the twenty four C atoms unit cell (Period 3). Along the c-axis, a vacuum separation, Z (> 10 Å), that is large enough to eliminate interactions between graphene layers was used in each unit cell. The k-points meshes of $18\times5\times1$, $9\times5\times1$, and $6\times5\times1$ are employed in the cases of period 1, 2 and 3, respectively.

Structures are relaxed using the conjugate gradient (CG) algorithm until the interatomic forces are less than 0.05 eV/Å. We first optimized the equilibrium structure with no strain for each unit cell. Then, structural relaxation was performed for given strains starting from the initial structures depicted in figure 1, uniformly expanded only in the y (armchair) direction. The cell size along x direction was kept at the equilibrium value (e_{xx}=0). Following the commonly used prescription [4], the stress of monolayer graphene is defined by rescaling the supercell stress by Z /3.34 Å (interlayer distance of graphite). Tensile strength was determined from the maximum value of the calculated stress.

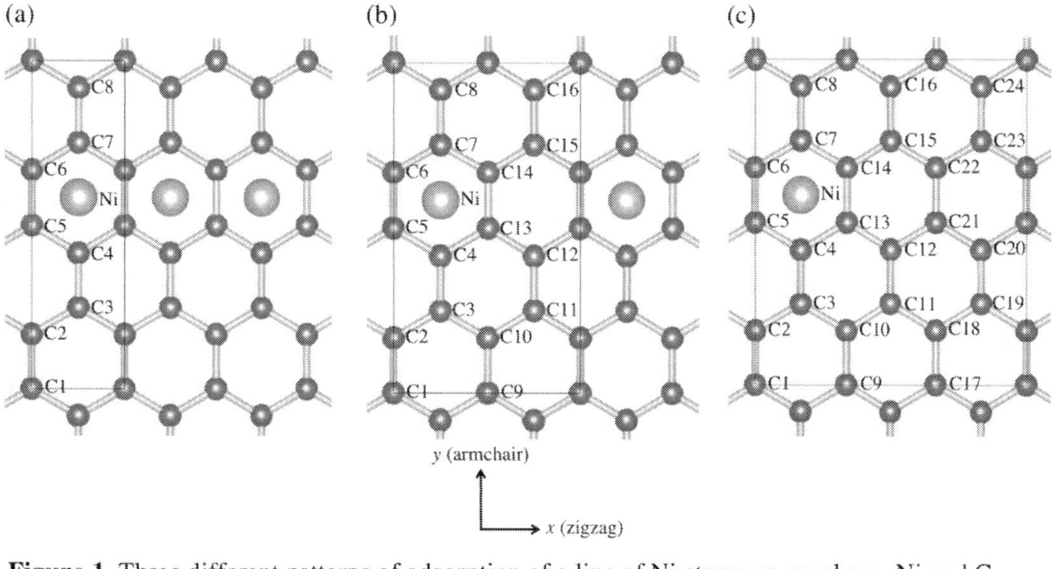

Figure 1. Three different patterns of adsorption of a line of Ni atoms on graphene. Ni and C atoms are indicated by larger and smaller spheres, respectively. These are referred to as (a) Period 1, (b) Period 2, and (c) Period 3.

3. Results and discussions

3.1. Tensile test of graphene with a line of adsorbed Ni: Period 1

Figure 2 shows (a) the calculated stress σ_{yy} and total energy, (b) the lengths of C-C bonds adjacent to the Ni atom (denoted as C5-C6) and the distance between the Ni and the nearest C atoms (denoted as C-Ni), (c) the lengths of other C-C bonds in y direction (denoted as C1-C2, C3-C4, C7-C8) versus strain e_{yy} in y direction.

As the strain increases from zero, the stress, total energy and C-C bond lengths increase and the stress takes the maximum value at e_{yy}=0.123, giving the tensile strength of 82.64 GPa. A drastic change takes place on going from e_{yy}=0.127 to 0.128. We see discontinuous change in figures 2(a) and (b) that the stress and total energy decrease and the C5-C6 bond length increases significantly,

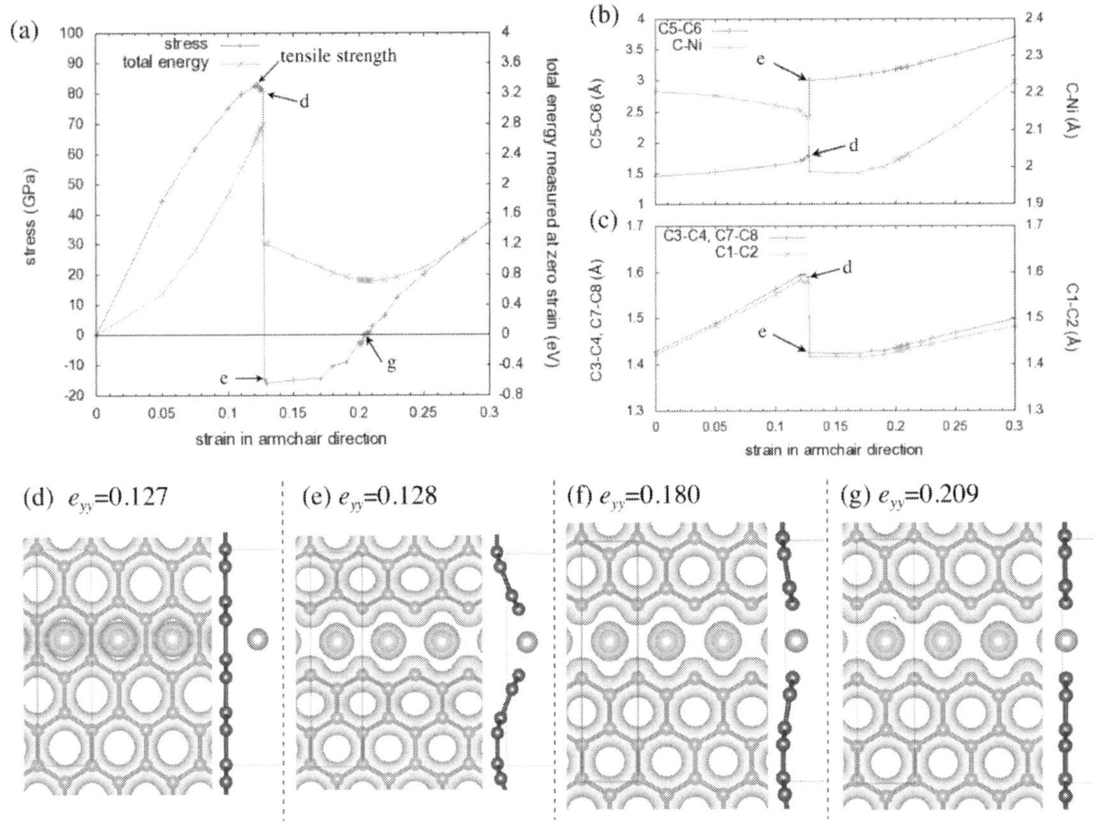

Figure 2. Calculated results for Period 1. (a) The stress σ_{yy} and total energy measured from that at zero strain. (b) The C-C bond length adjacent to Ni atom and the distance between a Ni and the nearest C atoms. (c) Lengths of the other C-C bonds in y direction versus strain in the y direction. (d)-(g) illustrate the structure and isosurface (at 0.1 electrons/Å^3) of the charge density projected on x-y (left) and z-y (right) plane, each corresponding to the data points indicated by arrows in panels (a)-(c).

whereas the C-Ni distance shrinks to create C-Ni bonds. Figure 2(c) shows that the other C-C bonds recover the bond length at the equilibrium. Figures 2(d) and (e) visually illustrate the change occurred between e_{yy}=0.127 and 0.128. We see a distinct reduction in bond charges when C5-C6 bonds break to cut the graphene sheet apart. At the same time, the sheet warps so that C atoms move in closer alignment with adjacent Ni atom to create C-Ni bonds and the other stretched C-C bonds in the sheet are virtually relaxed.

It should be noted that this warped structure exhibits a negative stress and therefore it is unstable. The instability however is to be dissolved spontaneously by further increase in the strain, which brings the total energy to the local minimum at e_{yy}=0.209, forming a zero stress, metastable planar structure as shown in the figure 2(g). Therefore, a stress-induced structural transformation, in which the line of Ni atoms adsorbed on graphene in zigzag direction cuts into and is embedded in the sheet to form an overstretched structure, will take place when an applied stress reaches the tensile strength. When a further increase in strain is given, C-Ni bonds are to be stretched predominantly as we see in figure 2(b), and the total energy starts to increase again, giving a positive stress. The stiffness in this high strain phase is much weaker than that in the original phase, since the average gradient of the stress around e_{yy}=0.209, 0.45 TPa, is nearly a half of the Young's modulus 1.03 TPa around zero strain.

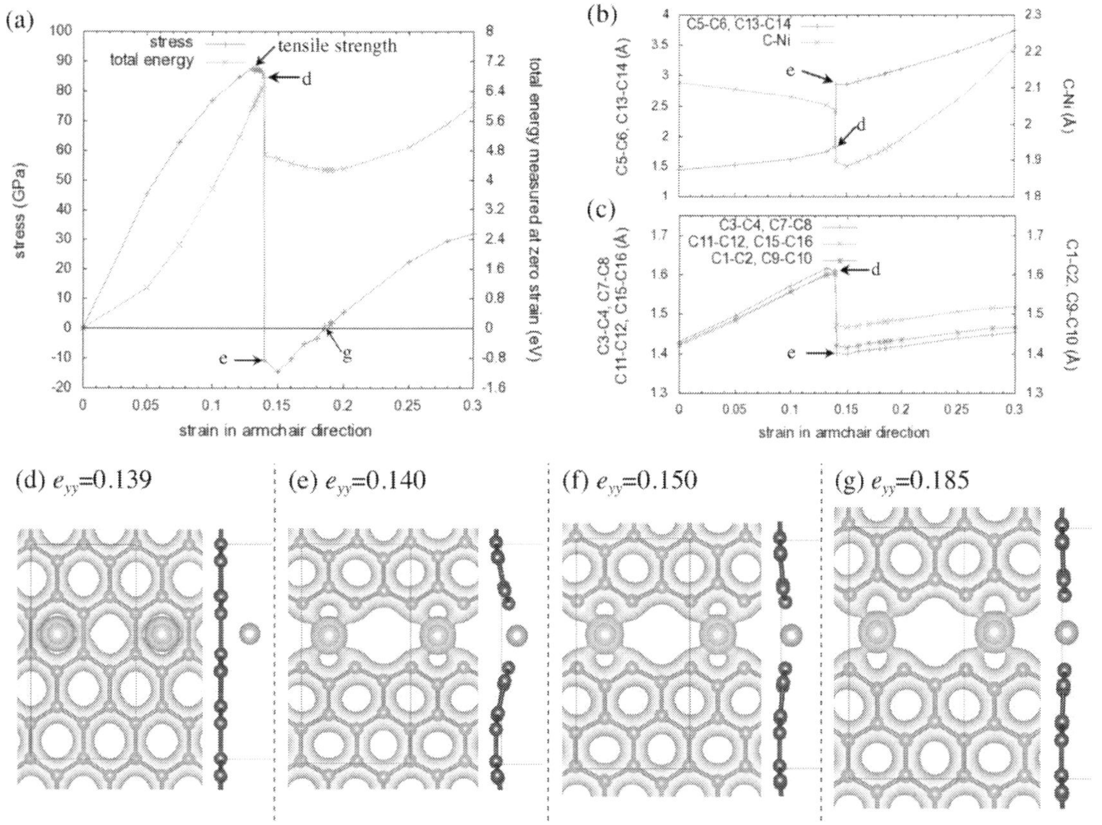

Figure 3. Calculated results for Period 2. (a) The stress σ_{yy} and total energy measured from that at zero strain. (b) The C-C bond length adjacent to Ni atom and the distance between a Ni and the nearest C atoms. (c) Lengths of the other C-C bonds in y direction versus strain in the y direction. (d)-(g) illustrate the structure and isosurface (at 0.1 electrons/Å^3) of the charge density projected on x-y (left) and z-y (right) plane, each corresponding to the data points indicated by arrows in panels (a)-(c).

3.2. Tensile test of graphene with a line of adsorbed Ni: Period 2

Next we show the results of calculated tensile test for the case of a halved density of Ni atoms in the line. Figure 3 shows (a) the calculated stress σ_{yy} and total energy, (b) the lengths of C5-C6, C13-C14 and C-Ni bonds, (c) the lengths of C1-C2, C3-C4, C7-C8, C9-C10, C11-C12 and C15-C16 bonds versus strain in y direction. The tensile strength in this case is calculated to be 87.62 GPa at e_{yy}=0.131, which is slightly higher than in the case of Period 1 reflecting the reduced density of the Ni adatoms.

The optimum structure as a function of the strain exhibits a discontinuous change between e_{yy}=0.139 and 0.140, where the stress drops to be negative and C-C bonds adjacent to Ni atom, *i.e.* C5-C6 and C13-C14, break and the bonding between C and Ni is created as we see in figures 3(a)-(e). The negative stress is to be dissolved spontaneously by flattening the sheet as depicted in figures 3(f) and (g). Thus, in common with the case of the Period 1 model, a stress-induced structural transformation is predicted to occur when an applied stress in the armchair direction reaches the tensile strength. Further increase in the strain causes predominantly a stretching of C-Ni bonds. The stiffness in this high strain phase is 83% of that in the previous case of Period 1, reflecting the less number of C-Ni bonds.

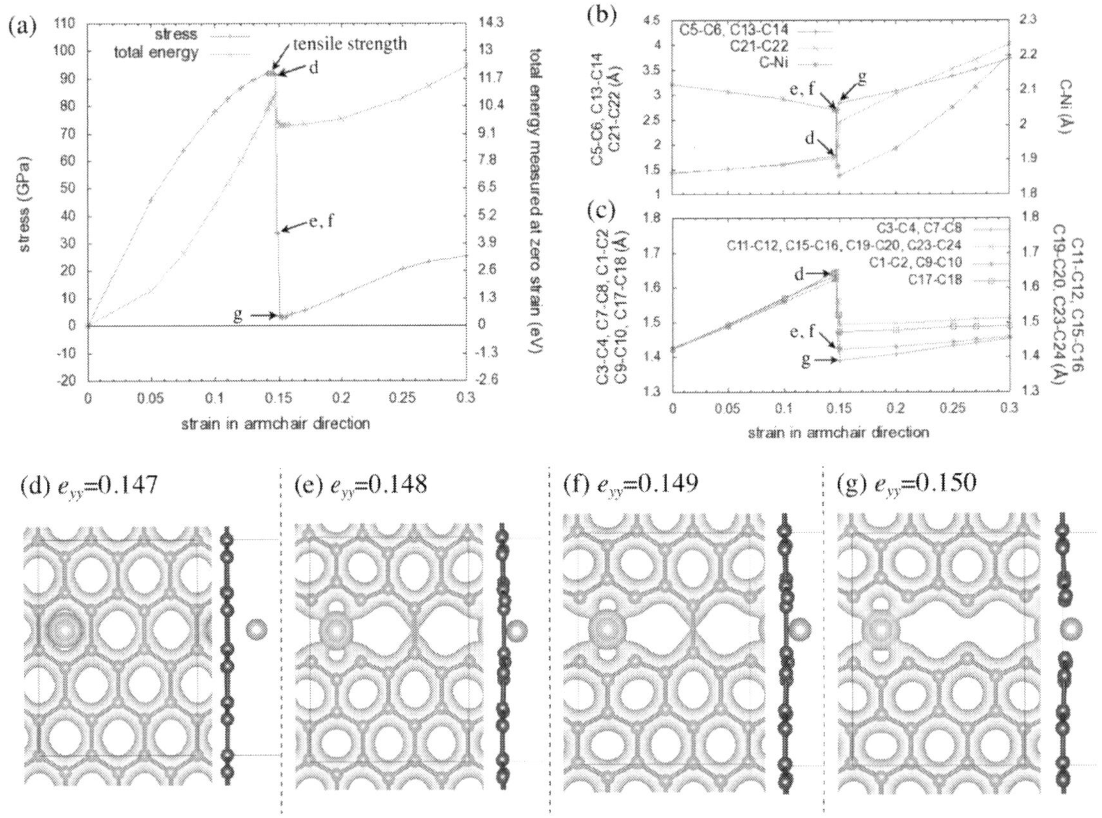

Figure 4. Calculated results for Period 3. (a) The stress σ_{yy} and total energy measured from that at zero strain. (b) The C-C bond length adjacent to Ni atom and the distance between a Ni and the nearest C atoms. (c) Lengths of the other C-C bonds in y direction versus strain in the y direction. (d)-(g) illustrate the structure and isosurface (at 0.1 electrons/Å^3) of the charge density projected on x-y (left) and z-y (right) plane, each corresponding to the data points indicated by arrows in panels (a)-(c).

3.3. Tensile test of graphene with a line of adsorbed Ni: Period 3

Finally the results of calculated tensile test are shown for the case that Ni atoms are adsorbed on only one in every three hollow sites in the line. Figure 4 shows (a) the calculated stress σ_{yy} and total energy, (b) the lengths of C5-C6, C13-C14, C21-C22 and C-Ni bonds, (c) the lengths of C1-C2, C3-C4, C7-C8, C9-C10, C11-C12, C15-C16, C17-C18, C19-C20 and C23-C24 bonds versus strain in y direction. Tensile strength is evaluated to be 92.05 GPa at e_{yy}=0.145. As a function of the strain, the stress shows a steep drop between e_{yy}=0.147 and 0.150, but it does not overshoot to negative value, unlike in the previous cases of more Ni atoms. The breaking of C-C bonds along y direction and formation of C-Ni bonds are illustrated in figures 4(d)-(g). When going from e_{yy}=0.147 up to 0.148, the Ni adatom comes closer to the graphene plane to create bonds with nearest C atoms, breaking the C-C bonds *i.e.* C5-C6 and C13-C14, that are adjacent to the Ni adatom. It follows that the remaining C-C bond, *i.e.*, C21-C22, breaks on going from e_{yy}=0.149 to 0.150. These calculated results indicate the stress-induced structural transformation that involves the formation of C-Ni bond and successive breaking of the stretched C-C bonds. Further increase in the strain causes predominantly a stretching of C-Ni bonds.

The stiffness in this phase at high strains is only 41% of that in the case of Period 1, reflecting a very low line density of Ni atoms that join the discrete zigzag edges.

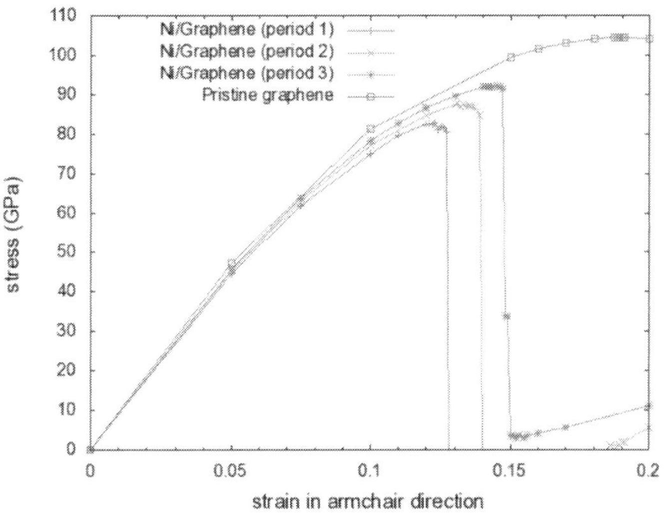

Figure 5. Comparison of the strain-stress curves for three patterns of Ni-adsorbed and pristine graphene.

3.4. Comparison of tensile strength

Table 1 with figure 5 summarizes tensile strengths for three cases of adsorption patterns we have shown and that for a pristine graphene. It is clearly shown that presence of a line of adsorbed Ni weakens the tensile strength monotonically depending on the line density of Ni. The tensile strength shows decrease by 21.0 % (Period 1), 16.2 % (Period 2), and 12.0 % (Period 3) compared with that for the pristine graphene. The effect is more prominent for higher density of adsorbed Ni atoms in the line. The breaking of C-C bonds adjacent to Ni atom occurs between e_{yy}=0.127 and 0.128 (Period 1), e_{yy}=0.139 and 0.140 (Period 2), and e_{yy}=0.147 and 0.148 (Period 3). Table 1 also shows the predicted strain, which is the result of the jump in strain caused by the predicted stress-induced structural transformation.

Table 1. Comparison of the tensile strengths in pristine graphene and each adsorption patterns. In the second column, the strain corresponding to the tensile strength is given. The third column shows the strain at the lowest non-negative stress after the stress-induced structural transformation.

	Tensile strength (GPa)	**Strain**	
		tensile strength	after structural transformation
Ni/Graphene (Period 1)	82.64	0.123	0.209
Ni/Graphene (Period 2)	87.62	0.131	0.185
Ni/Graphene (Period 3)	92.05	0.145	0.150
Pristine graphene	104.55	0.189	–

Journal of Physics: Conference Series **454** (2013) 012059 doi:10.1088/1742-6596/454/1/012059

4. Conclusions

Using the density functional calculations, we have performed uniaxial tensile tests of the graphene that is modified by adsorption of a line of Ni along zigzag direction. For three different cases of Ni line densities, the structure and total energy, stress and bond lengths are calculated as functions of strain in armchair direction. The calculated tensile strengths exhibited reduction of 21%, compared with that of the pristine graphene, for the case of highest Ni line density.

It is predicted that when the stress reaches the tensile strength, the system will undergo a stress-induced structural transformation, in which the graphene is disjoined to create zigzag cut edges along the line of Ni by the breaking of stretched C-C bonds adjacent to Ni atoms, and at the same time the zigzag cut edges are re-joined by Ni-C bonds. These calculated results should be useful in further understanding and exploring of new techniques for catalyst-assisted cutting of the graphene and graphene-based materials. Simulated tensile test for larger strains is under progress. VESTA [18] was used for visualization of structures and charge density.

References

[1] Novoselov K S, Geim A K, Morozov S V, Jiang D, Zhang Y, Dubonos S V, Grigorieva I V and Firsov A A 2004 *Science* **306** 666
[2] Geim A K and Novoselov K S 2007 *Nature Mat.* **6** 183
[3] Castro Neto A H, Guinea F, Peres N M R, Novoselov K S and Geim A K 2009 *Rev. Mod. Phys.* **81** 109
[4] Liu F, Ming P and Li J 2007 *Phys. Rev. B* **76** 064120
[5] Sakhaee-Pour A 2009 *Solid State Commun.* **149** 91
[6] Lee C, Wei X, Kysar J W and Hone J 2008 *Science* **321** 385
[7] Tapia A, Peón-Escalante R, Villanueva C and Avilés F 2012 *Comp. Mater. Sci.* **55** 255
[8] Wang M C, Yan C, Ma L, Hu N and Chen M W 2012 *Comp. Mater. Sci.* **54** 236
[9] Ansari R, Ajori S and Motevalli B 2012 *Superlattices and Microstructures* **51** 274
[10] Hu L, Hu X, Wu X, Du C, Dai Y and Deng J 2010 *Physica B* **405** 3337
[11] Meng F Y, Zhou L G, Shi S Q and Yang R 2002 *Carbon* **41** 2009
[12] Ci L, Xu Z, Wang L, Gao W, Ding F, Kelly K F, Yakobson B I, and Ajayan P M 2008 *Nano Res.* **1** 116
[13] Kresse G and Furthmüller J 1996 *Comput. Mater. Sci.* **6** 15
[14] Blöchl P E 1994 *Phys. Rev. B* **50** 17953
[15] Kresse G and Joubert D 1999 *Phys. Rev. B* **59** 1758
[16] Perdew J P, Burke K and Ernzerhof M 1996 *Phys. Rev. Lett.* **77** 3685
[17] Perdew J P, Burke K and Ernzerhof M 1997 *Phys. Rev. Lett.* **78** 1396
[18] Momma K and Izumi F 2011 *J. Appl. Crystallogr.* **44** 1272

Journal of Physics: Conference Series **454** (2013) 012060 doi:10.1088/1742-6596/454/1/012060

Ab-initio calculations of the photoelastic constants of the cubic SiC polytype

P Djemia[1] **and Kh Bouamama**[2]
[1]LSPM-CNRS, Sorbonne Paris Cité, Université Paris 13, 99 Avenue J.B. Clément, 93430 Villetaneuse, France.
[2]LOC, Département de Physique, Université Ferhat Abbas, 19000 Sétif, Algeria.

E-mail: khaled_bouamama@univ-setif.dz

Abstract: Residual defects after growth of semiconductors crystals is a hot issue to be solved for manufacturing new efficient electronic or optic devices. These defects can be conveniently observed using birefringence optical microscopy for extended defects that will create a local strain field which in turn can cause a nominally isotropic optical material to become anisotropic and induce birefringence. In order to perform a quantitative analysis, the knowledge of the photoelastic constants (P_{ij}) of the material that measure the strength of the change of the refractive index under application of strains or stresses is necessary. As an experimental determination of the whole set of constants is not always possible, a theoretical evaluation can be of valuable interest. In this work, we propose *ab-initio* calculations by the WIEN2k program of the optical properties of the zinc blende silicon carbide polytype with a self-consistent scheme by solving the Kohn-Sham equations using a full potential linearized augmented plane waves (FPLAPW) method in the framework of the density functional theory (DFT) along with the generalized gradient approximation (GGA) pseudo-potentials. A combination of specific compressive and tensile strains is applied to the two atoms unit cell and the tensor containing each specific combination of the P_{ij} constants is extracted.

1. Introduction

People pay more attention to SiC because of its great technological importance for high temperature, high power, high critical breakdown field and high frequency applications in microelectronic and photoelectronic devices [1]. Elaboration is principally done via the seeded physical vapor transport (PVT) technique [2, 3] on silicon substrate or 3C-SiC platelets. Many extended defects can appear during the growth of the crystals such as dislocations [4-8] that constitute a major issue that hampers the development of electronic-grade semiconductors also for diamond [9], which should bring an undisputable advantage over their GaN or SiC counterparts. In case of $LaBr_3$ that is a viable scintillator for gamma ray, the knowledge of the dislocations [10] that can appear in this material during the elaboration process is essential in order to minimize the fracture during the growth and increase further its size. These defects can be conveniently observed using birefringence optical microscopy [4-8] for extended defects such as dislocations that will create a local strain field which in turn can cause a nominally isotropic optical material to become anisotropic and induce birefringence

[2] To whom any correspondence should be addressed.

in light passing between cross-polarizers. In order to perform a quantitative analysis, the knowledge of the photoelastic constants (P_{ij}) of the material that measure the strength of the change of the refractive index under application of strains is necessary. Experimental determination of the (P_{ij}) constants is possible for large enough transparent anisotropic crystals, as for example, silicon [11], diamond [12], ZnTe [13] and ZnO [14], by the Brillouin light scattering technique [15], and other acousto-optical method as demonstrated for silicon, diamond, germanium and gallium-arsenide crystals [16-18].

It is also very important to know the optical properties of SiC due to their contributions to the photoelectronic device design and as fundamental inputs for radiative transfer models of astrophysical dust environments [19]. It is also useful for electron energy loss spectroscopy (EELS) analysis that can be used to image the dislocations as it was done for diamond [20]. However, the theoretical studies on the optical properties of zinc blende SiC are scarce up to now [21-24] in comparison to experimental works [25-28], and photoelastic data are still not available in the literature.

In this paper, we report a first-principle predictive calculation of the optical and photoelastic properties of β–SiC by using the FPLAPW method based on density functional theory. Firstly, calculations are performed for silicon (Si) and diamond (C), those are compared to the available experimental and theoretical results.

2. Calculation method

The calculation of the optical properties of silicon, carbon and silicon carbide in zinc blende structure, were carried out with a self-consistent scheme by solving the Kohn-Sham equation using a full potential linearized augmented plane waves (FPLAPW) method in the framework of the DFT along with the GGA method [29] as implemented in WIEN2k package [30, 31]. In the calculations, $R_{MT}K_{max}$=7 is used, which determines the matrix size, where K_{max} is the plane wave cut-off and R_{MT} is the smallest of all atomic sphere radii. Since calculations of the optical properties require denser *k*-space matrix, we used 28000 k-points in whole Brillouin zone. We chose the muffin–tin radii of 1.4, 2.1 Bohr for C and Si, respectively. We used the experimental values of the lattice parameters a_0 = 0.356685 nm [32], 0.5429 nm [33], 0.43596 nm [34] for C, Si and β-SiC, respectively.

The dielectric function of a solid is usually described in terms of complex $\varepsilon(\omega) = \varepsilon_1(\omega) + i\varepsilon_2(\omega)$, the imaginary part of the dielectric function $\varepsilon_2(\omega)$ being obtained directly from the band structure calculation. The real part of the dielectric function $\varepsilon_1(\omega)$ can be derived from the imaginary part $\varepsilon_2(\omega)$ by using *Kramers–Kronig* dispersion relationship. When a strain is applied to a crystal there is a corresponding change in the refractive index, arising from a change in the dielectric function. The photoelastic tensor (P_{ij}) relates the change of the inverse dielectric function with the applied strain. The presence of a dislocation or of an extended defect in a crystal causes a local strain field which in turn can lead a nominally isotropic material to become anisotropic and induce birefringence in light passing through cross-polarizers [4]. The accidental birefringence due to dislocations in silicon has been observed [35] and modelled theoretically [36]. However, these calculations require the photoelastic tensor which is not well known in diamond [12, 37-39], compared to silicon [18] or even unknown in the case of the β-SiC polytype.

The change in the inverse dielectric function due to strain is given by equation (1):

$$\delta(\varepsilon^{-1})_{ij} = P_{ijkl}\mu_{kl} \approx -\frac{1}{\varepsilon_b^2}\delta\varepsilon_{ij} \quad (1)$$

where P_{ijkl} is the fourth rank Pockels photoelastic tensor, μ_{kl} is the strain tensor and ε_b is the value of the dielectric function in the unstrained material. Adopting a contracted notation where 11 → 1, 22 → 2, 33 → 3 and 23, 32 → 4, P_{ijkl} reduces to a symmetric matrix with three non-zero elements for cubic structures:

$$P = \begin{pmatrix} P_{11} & P_{12} & P_{12} & 0 & 0 & 0 \\ P_{12} & P_{11} & P_{12} & 0 & 0 & 0 \\ P_{12} & P_{12} & P_{11} & 0 & 0 & 0 \\ 0 & 0 & 0 & P_{44} & 0 & 0 \\ 0 & 0 & 0 & 0 & P_{44} & 0 \\ 0 & 0 & 0 & 0 & 0 & P_{44} \end{pmatrix} \qquad (2)$$

To determine the P_{11}, P_{12} and P_{44}, we calculate the change in the dielectric function for three different strains [μ]:

1) Hydrostatic strain $[\mu] = \left(\frac{\delta a}{a_0}, \frac{\delta a}{a_0}, \frac{\delta a}{a_0}, 0,0,0\right)^T$

$$\frac{-\delta\varepsilon_{ij}}{(\varepsilon_b^2)} = \left(\frac{\delta a}{a_0}\right)\begin{pmatrix} P_{11}+2P_{12} & 0 & 0 \\ 0 & P_{11}+2P_{12} & 0 \\ 0 & 0 & P_{11}+2P_{12} \end{pmatrix} \qquad (3)$$

where a_0 is the lattice constant and $\delta a/a_0$ is the relative strain. The $P_{11} + 2P_{12}$ can be obtained from the difference between strained and unstrained dielectric constant.

2) Strain along the [001] direction $[\mu] = \left(0,0,\frac{\delta a}{a_0}, 0,0,0\right)^T$

$$\frac{-\delta\varepsilon_{ij}}{(\varepsilon_b^2)} = \left(\frac{\delta a}{a_0}\right)\begin{pmatrix} P_{12} & 0 & 0 \\ 0 & P_{12} & 0 \\ 0 & 0 & P_{11} \end{pmatrix} \qquad (4)$$

3) The P_{44} can be determined by changing the length $\frac{\delta l}{l}$ in the [111] direction by applying the strain $[\mu] = \frac{1}{3}\frac{\delta l}{l}(1,1,1,2,2,2)^T$

$$\frac{-\delta\varepsilon_{ij}}{(\varepsilon_b^2)} = \frac{1}{3}\left(\frac{\delta l}{l}\right)\begin{pmatrix} P_{11}+2P_{12} & 2P_{44} & 2P_{44} \\ 2P_{44} & P_{11}+2P_{12} & 2P_{44} \\ 2P_{44} & 2P_{44} & P_{11}+2P_{12} \end{pmatrix} \qquad (5)$$

The unit cells with two atoms are taken and the lattice vectors changed to apply strain in the [001] and [111] directions, and hydrostatically. The values of strain applied do not exceed ±0.4 % and the considered energy is below the first gap, typically below 3 eV. The values of P_{11}, P_{12} and P_{44} are found for an average of these compressive and tensile strains. In the case of strain in the [111] direction, we consider both results from relaxed and un-relaxed unit cell. The relaxed positions of the two atoms along this direction are obtained by minimization of the forces that act on them, for no symmetry arguments can fix them.

3. Results and discussion

3.1. Permittivity function

Figure 1a shows the real and imaginary parts of the calculated dielectric function of silicon (Si). These results compare well to the experimental data from Biegelsen *et al.* [18] and the previous *ab-initio* studies [40, 41] on the dielectric response of silicon, those are partially reported in the table 1. The direct band gap is found to be ~ 2.76 that underestimates the experimental value ~3.45 eV [42]. For theoretical DFT calculations, the use of a rigid shift to compensate for the LDA underestimation of the band gap affects all levels above the valence band top, to the same extent. It was used with some success by Levine *et al.* [40] and Hounsome *et al.* [41], and is often referred to as a 'scissors shift'. The scissors shift becomes important for defective crystals where states are introduced into the band gap and lead to additional optical transitions. In the work of Hounsome *et al.* [41] a shift of 0.52 eV was used to correct the silicon band gap calculated with the AIMPRO program. Nevertheless, it has been shown that the scissors shift technique of correcting the LDA underestimation of the band gap for one material does not in the same time, uniformly improve the calculated photoelastic constants that will be discussed in the next section. Hence, we will not try to use this technique in this work despite improvement of the DFT+LDA that is called generalized density functional theory (GDFT) [22] and provides appropriate corrections to the energy of conduction bands, to investigate the band structure of semiconductors or insulators, and thus obtains band gaps that are in agreement with experimental results.

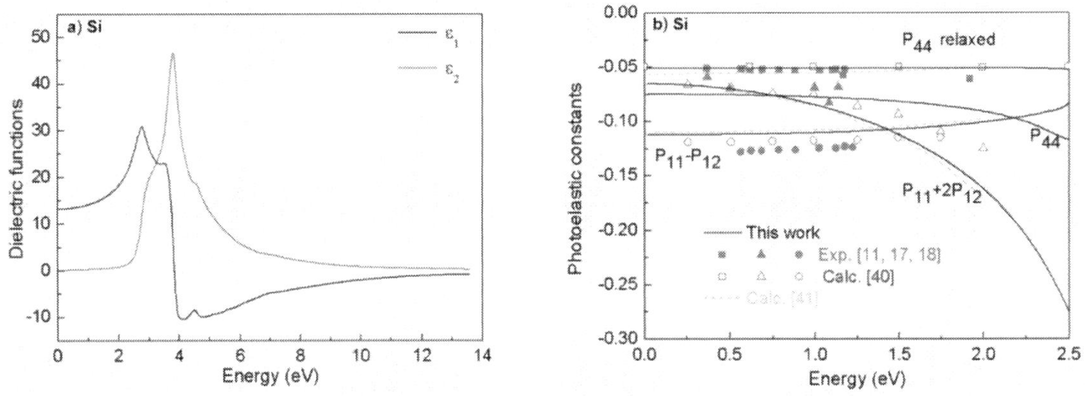

Figure 1. (a) Calculated dielectric function for silicon: real part $\varepsilon_1(\omega)$ and imaginary part $\varepsilon_2(\omega)$. (b) Calculated photoelastic constants for silicon compared to experimental data from [11, 17, 18] and calculated data from [40, 41].

Figure 2a shows the real and imaginary parts of the dielectric function of diamond (C) that compare well to the experimental [43, 44] and theoretical [41, 45] data. Some results are partially reported in the table 1 and compared to available data. In ref. [43], the experimental spectrum for ε_2 shows a threshold around the direct band gap at ~7 eV and a broad peak at ~12 eV due to band-band transitions. It then falls slowly with increasing energy. The theoretical curve displays a threshold at ~5.5 eV, and a peak around ~11.5 eV, which are downshifted from the experimental values due to the underestimation of the gap. In contrast with other semiconductors, diamond does not exhibit additional sharp peaks in this region, due to its band structure. Note that in the long wavelength limit, the calculated value of the real part of the dielectric function ε_1 = 5.77 agree well with the experimental value of the static dielectric function 5.82 [43] and other theoretical ones 5.89 [41] and 5.86 [45]. The experimental peak at 7 eV [43] appears much sharper than the calculated one due to the neglect of excitonic effects in our calculations. However the energy where $\varepsilon_1 = 0$ is in good agreement and corresponds to maximum absorption of light.

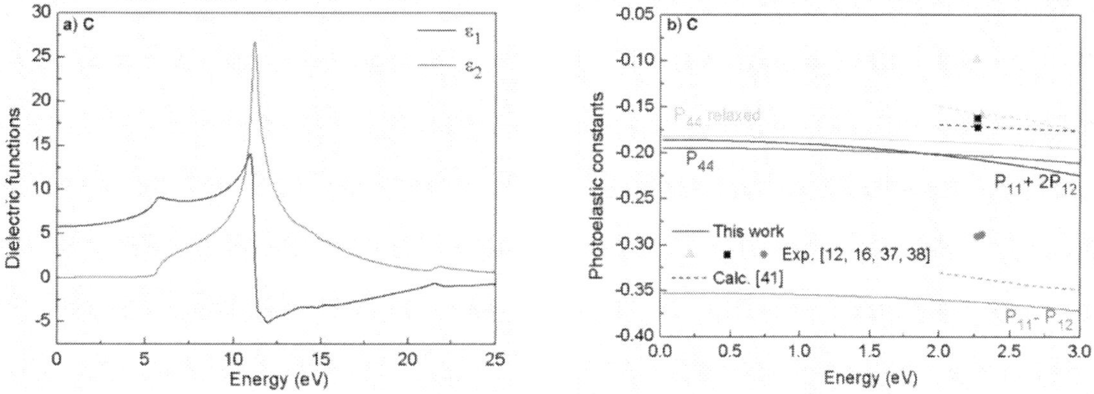

Figure 2. (a) Calculated dielectric function for diamond: real part $\varepsilon_1(\omega)$ and imaginary part $\varepsilon_2(\omega)$. (b) Calculated photoelastic constants for diamond compared to experimental data from [12, 16, 37, 38] and calculated data from [41].

Figure 3a shows the real and imaginary parts of the dielectric function of zinc blende silicon carbide (SiC) that compare well to the experimental [25-28] and theoretical [21-24] data. Some results are partially reported in the table 1 and compared to available data. In refs. [26, 28], the experimental spectrum for ε_2 shows a threshold around the direct band gap at ~5 eV and one broad peak at ~7.2 eV due to band-band transitions. It then falls slowly with increasing energy. The theoretical ε_2 curve displays a threshold at ~5 eV, one peak at ~7 eV which are not downshifted from the experimental values as surprisingly noticed by Logothetidis *et al.* [26] and a shoulder at 8 eV,. Note that in the long wavelength limit, the calculated value of the real part of the dielectric function $\varepsilon_1 = 7.05$ agree well with the experimental value of the static dielectric function of 7.29 [25, 26] and other theoretical ones 7.1 [22], 5.03 [24] and 6.95 [45]. The energy ~7.5 eV where $\varepsilon_1 = 0$ is also in good agreement.

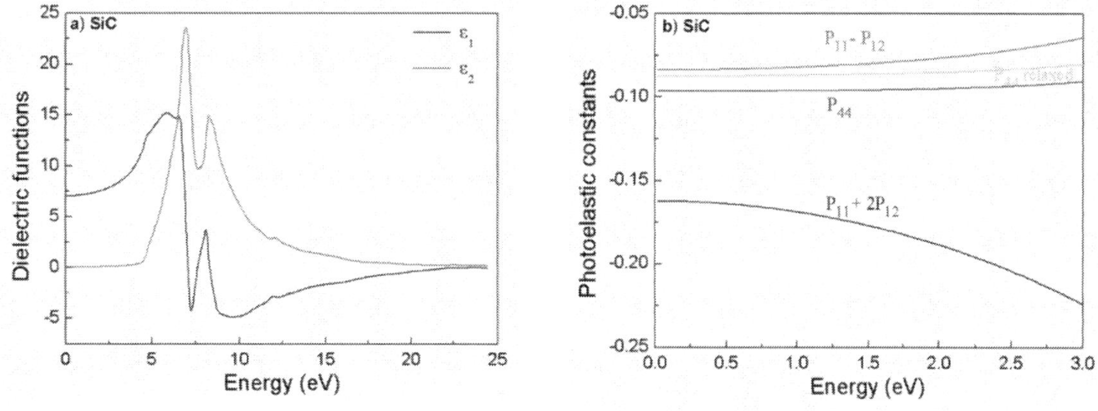

Figure 3. (a) Calculated dielectric function for zinc blende silicon carbide: real part $\varepsilon_1(\omega)$ and imaginary part $\varepsilon_2(\omega)$. (b) Calculated photoelastic constants for zinc blende silicon carbide.

Table 1. Permittivity (ε_b) and photoelastic constant (P_{ij}) for silicon, diamond and zinc blende silicon carbide. Experimental and previous *ab-initio* work is shown for comparison. All values for silicon refer to a wavelength of 3542 nm (0.35 eV), for diamond (2.27 eV) and for β-SiC (0.35, 2.27 and 3 eV).

Material	ε_b	P_{44}	$P_{11}+2P_{12}$	$P_{11}-P_{12}$	Ref.
Silicon (Si)					
0.35 eV	13.33	-0.051	-0.067	-0.112	Wien2k (this work)
0.35 eV	13.67	-0.051	-0.058	-0.109	AIMPRO [41]
0.35 eV	13.2	-0.046	-0.085	-0.105	DFT+LDA [40]
0.35 eV	11.83	-0.051	-0.054	-0.112	Exp. [18]
Diamond (C)					
2.27 eV	6.01	-0.189	-0.207	-0.362	Wien2k (this work)
2.27 eV	6.17	-0.171	-0.160	-0.335	AIMPRO [41]
2.27 eV	5.819	-0.172	-0.164	-0.292	Exp. [12, 39, 43, 44]
2.3 eV	-	-0.172	-0.160	-0.292	Exp. [38]
Silicon carbide (β-SiC)					
0.35 eV	7.06	-0.087	-0.163	-0.083	Wien2k (this work)
2.27 eV	7.71	-0.084	-0.196	-0.074	Wien2k (this work)
3 eV	8.34	-0.080	-0.224	-0.063	Wien2k (this work)
3 eV	7.29				Exp. [25, 26]

3.2. Photoelastic constants

Initially, values of P_{11}, P_{12} and P_{44} are calculated for silicon and diamond for these properties are unknown for the zinc blende polytype of SiC (β-SiC). These should demonstrate the reliability of the methodology, as the photoelastic constants for silicon are more studied [11, 17, 18, 40, 41] and in a less extent those of diamond [12, 38, 39, 41, 43, 44]. Previous *ab-initio* studies on the dielectric response of silicon and diamond [40, 41] have reviewed the available experimental data. The photoelastic constants are found using the calculated values of ε_b for various strains and considering an averaged combination of tensile and compressive strain for accuracy.

Figure 1b shows the variation of the calculated photoelastic constants with frequency for silicon. No scissors shift is applied, as described above. Comparisons with experimental values [11, 17, 18] and previous *ab-initio* calculations [40, 41] are shown. Some results are partially reported in the table 1 and compared to available data. The values of P_{11} - P_{12} are slightly larger than experiment by about 3% but the slow dispersion with frequency is in good agreement. Excellent agreement is achieved for P_{44}. The calculated values for P_{11} + $2P_{12}$ are in good agreement with experiment until ~ 1 eV but disperse quite strongly with increasing frequency. The calculated constants for silicon display on overall agreement with the results of Hounsome *et al.* [41] and relatively little dispersion except for the calculated values of P_{11} + $2P_{12}$. As stated above, this is due to the lowering of band gap and can be reduced if a shift is used [41] or if excitonic effects are taken into account [22]. One should also notice that calculations of this constant are very sensitive to the value of the lattice parameter used as input [40].

The values of the photoelastic constants for diamond in the literature differ widely and have been reviewed recently [39], where there is very little information on the variation of the constants with frequency. Grimsditch *et al.* [12] evaluate one set of results as being "most reliable", these results are from [38]. In figure 2b, it is evident that P_{11} - P_{12} and P_{11} + $2P_{12}$ are underestimated by ~20-25 % whereas P_{44} is underestimated by ~10 %. Some results are partially reported in the table 1 and compared to available data. The calculated constants for diamond underestimate the results of Hounsome *et al.* [41] by ~10% and show relatively little dispersion. Further experiments are suitable to confirm this behavior.

Figure 3b shows the variation of the calculated predictive photoelastic constants with frequency for β-SiC. P_{11} - P_{12} and P_{44} have closed values (~-0.08) and increase slowly with frequency. P_{11} + $2P_{12}$ is decreasing rapidly as in the case of silicon with values in the same range as for diamond [-0.15, -0.25] for energy below 3 eV.

4. Summary and conclusion

The dielectric function and photoelastic constants for silicon, diamond and β-SiC have been calculated for a large energy range using density functional theory applied to strained unit cells. Experimental [18] and other theoretical [40, 41] data for silicon are accurately reproduced with absolute values and variation with frequency matched for all the constants. Experimental data for diamond is still limited to a very few number of frequencies and recently re-discussed by Lang *et al.* [39]. One particular experimental set [16] has previously been regarded as "most reliable" [12, 38]. However, our calculated values underestimate this particular result from ~10 % for P_{44} and ~25 % for P_{11} - P_{12} and P_{11} + $2P_{12}$.

The agreement for silicon [18, 40, 41] and diamond [12, 16] gives confidence that the β-SiC theoretical photoelastic values should reasonably predict experimental results, unfortunately not yet available. Undoubtedly, synthesis of larger crystal size will motivate this work. Satisfactory agreement is found between our theoretical value of the dielectric function with the experimental ones [26, 28].

Acknowledgments

Kh Bouamama wishes to acknowledge the university Paris 13, Sorbonne Paris Cité for its fellowship as invited Professor and the access to the cluster of PC's of the university Paris 13.

References

[1] Carter Jr CH, Tsvetkov VF, Glass RC, Henshall D, Bardy M, Muller SG, Kordina O, Irvine K, Edmond JA, Kong HS, Singh R, Allen ST and Palmour JW 1999 Mater. Sci. Eng. B **61–62** 1
[2] Semmelroth K, Krieger M, Pensl G, Nagasawa H, Püsche R, Hundhausen M, Ley L, Nerding M and Strunk HP 2007 Journal of Crystal Growth **308** 241
[3] Chaussende D, Blanquet E, Baillet F, Ucar M and Chichignoud G 2006 Chemical vapor deposition **12** 541
[4] Mendelson S 1961 J. Appl. Phys. **32** 1999
[5] Tanner BK and Fathers BJ 1974 Phil. Mag. **29** 1081
[6] Ming N and GE C 1990 Journal of Crystal Growth **99** 1309
[7] Kato T, Ohsato H, Okamoto A, Sugiyama N and Okuda T 1999 Materials Science and Engineering B **57** 147
[8] Ouisse T, Chaussende D and Auvray L 2005 J. Appl. Cryst. **43** 122
[9] Tallaire A, Barjon J, Brinza O, Achard J, Silva F, Mille V, Issaoui R, Tardieu A and Gicquel A 2011 Diamond & Related Materials **20** 875
[10] Zhou XW, Doty FP and Yang P 2010 J. Appl. Phys. **107** 123509
[11] Chandrasekhar M, Grimsditch M and Cardona M 1978 Phys. Rev. B. **18** 4301
[12] Grimsditch MH and Ramdas AK 1975 Phys. Rev. B. **11** 3139
[13] Yamada M, Yamamoto K and Abe K 1977 J. Phys. D: Appl. Phys. **10** 1309

Journal of Physics: Conference Series **454** (2013) 012060 doi:10.1088/1742-6596/454/1/012060

[14] Azuhatu T, Takesada M, Yagi T, Shikanai A, Chichibu SF, Torii K, Nakamura A, Sota T, Cantwell G, Eason DB and Litton CW 2003 J. Appl. Phys. **94** 968
[15] Vacher R and Boyer L 1972 Phys. Rev. B. **6** 639
[16] Denning RM, Giardini AA, Poindexter E and Slawson CB 1957 The American Mineralogist **42** 556
[17] Higginbotham CW, Cardona M and Pollak FH 1969 Phys. Rev. Lett. **184** 821
[18] Biegelsen DK 1974 Phys. Rev. Lett. **32** 1196
[19] Pitman KM, Hofmeister AM, Corman AB, and Speck AK 2008 Astronomy &Astrophysics **483** 661
[20] Fall CJ, Blumenau AT, Jones R, Briddon PR, Frauenheim T, Gutiérrez-Sosa A, Bangert U, Mora AE, Steeds JW, and Butler JE 2002 Phys. Rev. B. **65** 205206
[21] Lee K, Park CH, Cheong B and Chang KJ 1994 Solid State Communication **92** 869
[22] Theodorou G, Tsegas G and Kaxiras E 1999 J. Appl. Phys. **85** 2179
[23] Xu P, Xie C, Pan H and Xu F 2005 Journal of Electron Spectroscopy and Related Phenomena **144–147** 593
[24] Ching WY, Xu Y, Rulis P, Ouyang L 2006 Materials Science and Engineering A **422** 147
[25] Choyke WJ and Palik ED 1985 Handbook of Optical Constants of Solids, ed. E.D. Palik (Academic Press, Orlando, FL) 587
[26] Logothetidis S, Polatoglou HM, Petalas J, Fuchs D and Johnson RL 1993 Physica B **185** 389
[27] Petalas J, Logothetidis S, Gioti M and Janowitz C 1998 Phys. Stat. Sol. B **209** 499
[28] Cobet C, Wilmers K, Wethkamp T, Edwards NV, Esser N and Richter W 2000 Thin Solid Films **364** 111
[29] Perdew JP, Burke S and Ernzerhof M 1996 Phys. Rev. Lett. **77** 3865
[30] Blaha P, Schwarz K and J. Luitz 1990 Comput. Phys. Commun. **59** 399
[31] Ambrosch-Draxl C and Sofo J 2006 Comp. Phys. Comm. **175** 1
[32] Skinner BJ 1957 Am. Mineral **42** 39
[33] Becker P, Seyfried P and Siegert H 1982 Z. Physik B **48** 17
[34] Taylor A and Jones RM 1960 Silicon Carbide – A High Temperature Semiconductor (eds. O'Connor J.R and Smiltens J), Pergamon Press, New York, 24
[35] Bond W and Andrus J 1956 Phys. Rev. **101** 1211
[36] Bullough R1958 Phys. Rev. **110** 620
[37] Ramachandran G 1950 Proc. Indian Acad. Sci. **32** 171
[38] Schneider W 1970 Ph.D. thesis, Materials Research Institute, Pennsylvania State University,United States of America
[39] Lang AR 2009 Diamond & Related Materials **18** 1
[40] Levine ZH, Zhong H, Wei S, Allan DC and Wilkins JW 1992 Phys. Rev. B. **45** 4131
[41] Hounsome LS, Jones R, Shaw MJ and Briddon PR 2006 phys. stat. sol. (a) **203** 3088
[42] Welkowsky M and Braunstein R 1972 Phys. Rev. B. **5** 497
[43] Philipp H and Taft E 1962 Phys. Rev. **127** 159
[44] Roberts R and Walker W 1967 Phys. Rev. **161** 730
[45] Chen J, Levine ZH and Wilkins JW 1994 J. Appl. Phys. **50** 11514

Journal of Physics: Conference Series **454** (2013) 012061 doi:10.1088/1742-6596/454/1/012061

Hybrid functional study on diffusion of silicate cathode material Li_2NiSiO_4

Kieu My Bui[1,2], **Van An Dinh**[2] **and Takahisa Ohno**[1,2,3]
[1] Graduate School of Pure and Applied Sciences, University of Tsukuba, 1-1-1 Tennodai, Tsukuba, Ibaraki 305-8577, Japan
[2] Global Research Center of Nanomaterials Science for Energy and Environment (GREEN), Sengen 1-2-1, Tsukuba, Ibaraki 305-0047, Japan
[3] Computational Materials Science Unit (CMSU), National Institute for Materials Science (NIMS), Sengen 1-2-1, Tsukuba, Ibaraki 305-0047, Japan

Email: BUI.Thikieumy@nims.go.jp

Abstract. The orthosilicate systems Li_2MSiO_4 (M = Fe, Mn, Co and Ni) are recently believed to be a promising alternative to the olivine phosphates. In this paper, we present an interpretation of the diffusion mechanism for polaron-Li vacancy complexes in Li_2NiSiO_4 based on the hybrid functional method HSE06. A comparison between the results obtained by GGA+U and HSE06 methods is carried out. The results confirm that the HSE06 method succeeds in describing the polaron localization in Li_2NiSiO_4. A polaron-Li vacancy complex model (V. A. Dinh *et al., Appl. Phys. Express*, **5**, 045801 (2012)) is used to handle the explanation of the diffusion mechanism in this material. Four elementary diffusion processes for the polaron-Li vacancy complex are investigated and the preferable diffusion pathways are constructed by combining the possible elementary processes. It is found that the Li diffusion may proceed along the two preferable pathways in the [100] and [001] directions with the activation barriers of 1.17 and 0.96 eV, respectively. Furthermore, the accompanied migration of polaron can enhance the diffusion rate of Li ion in Li_2NiSiO_4.

1. Introduction

Li ion batteries are considered to be one of the most powerful, convenient and safe power sources [1]. However, the cathode material is still facing many disadvantages that should be overcome. One of the highly appreciated cathode materials is $LiFePO_4$ due to its good thermal and chemical stability [2,3]. More recently, many researches have clarified that the orthosilicate system Li_2MSiO_4 (M = Fe, Mn, Co and Ni) can be an alternative to the olivine phosphate by reason of its strong Si-O bond which promotes a similar structure stability as in olivine phosphate [4]. Li_2MSiO_4 possesses many characteristics of an appealing cathode material such as high capacities, safe and low cost [5-9]. Especially the possibility of extraction of two lithium ions per formula unit makes this material the most promising cathode material. Hence, studying well the orthosilicate system is becoming an urgent issue. The ability of getting high capacity of Li_2FeSiO_4 and Li_2MnSiO_4 systems have been reported [10,11]. Having the same structure, Ni system is believed to have such advantages. However, to our knowledge, there is very few research on Ni system, hence, having a thorough study on Ni system is an urgent issue to have a complete picture on the hot material Li_2MSiO_4.

Ionic conductivity is one of the important features related to the performance of batteries and hence upgrading the ionic conductivity becomes a necessary mission. Having a thorough interpretation of the diffusion in the material is required for an understanding of conductivity. In olivine phosphate system, introducing a vacancy into the cell should lead to the formation of a bound polaron. Recently, a polaron-Li vacancy complex model has been proposed by Dinh *et al* [12] for the olivine system and this model has been successfully applied for Li_2FeSiO_4 [13]. In Li_2NiSiO_4 system, the Li diffusion could also be regarded as a polaron-Li vacancy complex diffusion. In this paper, we give out an understanding of the nature of the diffusion mechanism by revealing elementary diffusion processes in Li_2NiSiO_4. Aiming to the task, we first figure out the formation of the bound polaron as a Li ion is extracted from the lattice. Then, we explore the diffusion of Li ion in the manner accompanying with the migration of its bound polaron. During each step, those two accompanied motions would form an elementary diffusion process. Finally, the preferable diffusion pathways will be revealed by combining the possible elementary diffusion processes.

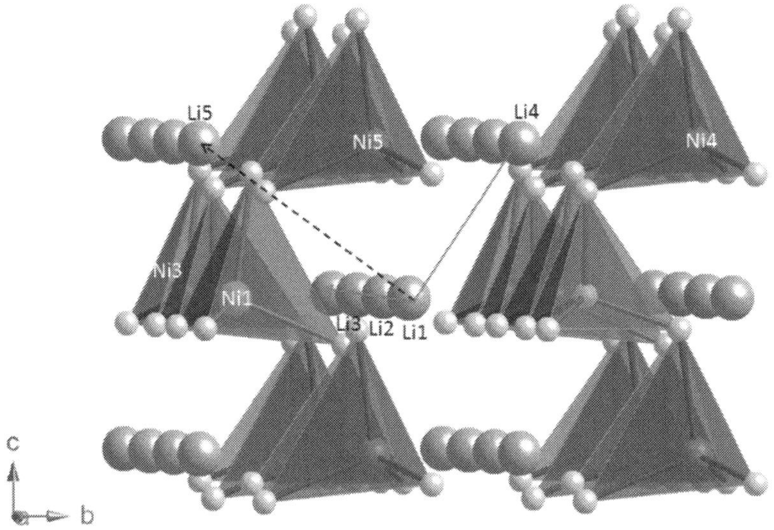

Figure 1. a) Crystal structure of Li_2NiSiO_4 in the *Pmn2₁* symmetry, and schematic of diffusion paths viewed in the [100] direction. Ni tetrahedrals and Si tetrahedrals are illustrated in pink and green, respectively. Li atoms are shown in cyan. The red and black dash lines represent the moving paths of the Li ions in the elementary diffusion processes.

2. Methodology

During our calculation, the Vienna ab initio simulation package (VASP) [14-17] is used within density functional theory (DFT). Two methods are applied to deal with the calculation: Generalized gradient approximation (GGA) [18] and Hybrid density functional (HSE06) [19,20]. Within GGA method, the on-site coulomb term U has been taken into account to deal with electron-electron interaction on the Ni site. The on-site Coulomb term U and the exchange term J can be embedded into equation $U_{eff} = U - J$. U_{eff} is set at 6eV [9]. To our knowledge, GGA+U can solve effectively the diffusion problem in Li_2FeSiO_4 as well as $LiFePO_4$, however, it fails in describing the polaron localization in $LiMnPO_4$ [12,21]. Recently, the HSE06 method is proved to be a reliable tool that can predict such properties as crystal structure, bulk moduli, and band gaps effectively [19,20]. Especially, it can solve the problem of polaron localization in $LiMnPO_4$ in a proper way [12]. Through our calculation, a 2x2x2 k-points grid [22] in GGA+U calculation and the center Γ-point in HSE06 calculation for a 2x2x2-supercell were used, and the full relaxation was performed for all of the perfect and defect configurations. The cut-off energy was set at 500eV. The activation energy of each diffusion pathways was calculated by using the climbing image nudged elastic band (CINEB) [23] method within the HSE06 framework.

Similar to the other materials in the family of Li_2MSiO_4, Li_2NiSiO_4 is most frequently observed in the *Pmn2₁* space group with the orthorhombic structure [7,8]. The previous GGA+U calculation [9] showed that the lattice parameters of Li_2NiSiO_4 are: a = 6.274 (Å), b = 5.365 (Å), c = 4.921 (Å). The cell is arranged in corrugated layers where tetrahedral SiO_4 and NiO_4 lying alternatively on *ab*-plane linked by LiO_4 tetrahedral along the c-axis as shown in figure 1. The LiO_4, SiO_4 or NiO_4 tetrahedral compose of one Li, Si or Ni atom locates in the central site surrounded by the oxygen tetrahedral. As mention above, there is only corner sharing between tetrahedral, each SiO_4 or NiO_4 tetrahedral shares its four corners with each other. LiO_4 tetrahedral lies between the two [SiO_4 and NiO_4] layers. In this space group, there are two formula units per unit cell which includes four Li, two Ni, two Si and eight O atoms.

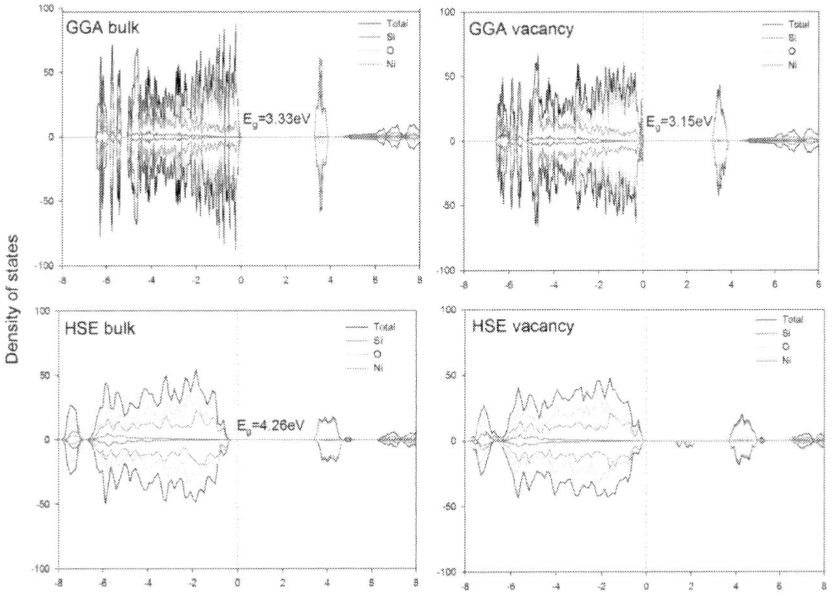

Figure 2. DOS of $Li_{32}Ni_{16}Si_{16}O_{64}$ and $Li_{31}Ni_{16}Si_{16}O_{64}$ obtained by GGA+U and HSE methods. Positive (negative) value of DOS indicates the majority (minority) spin.

3. Results and discussion

3.1 Structural, electronic properties and polaron formation

Table 1 illustrates the structure parameters obtained by GGA+U and HSE06 methods for the perfect structure of Li_2NiSiO_4. The result of HSE06 is compared with GGA+U in a percentage. For the perfect structure, the GGA+U and HSE06 methods give the similar structure parameter in x-direction. The lattice parameters obtained by HSE06 are: a = 6.290 (Å), b = 5.299 (Å), c = 4.856 (Å) and the volume is slightly smaller than that by GGA+U (2.67%) as indicated in table 1.

We present here the GGA+U and HSE06 study of density of states (DOS) in figure 2 for further understanding of the electronic structure of the material. The total DOS, partial DOS (PDOS) of Ni, O, and Si atoms are illustrated. The two approaches yield quite similar band gap in perfect structure. The material exhibits the characteristic of an insulator with a band gap of 3.33eV obtained by GGA+U, or 4.26eV obtained by HSE06. Apparently in vacancy structure, the two calculation method produces different DOS pictures. When a Li vacancy is introduced to the lattice, the Ni^{2+} ion which is a neighbor to the vacancy will be oxidized to Ni^{3+} resulting in the 3d localized states of Ni^{3+} appearing in the band gap. The appearance of these bound states in the band gap is confirmed by the electronic structure picture of HSE06 whereas those bound states could not be found in GGA+U result. This expresses the possibility of localizing the polaron formation of HSE method. Therefore, from now on, only the result within HSE06 method is represented.

Table 1. Calculated lattice parameters and band gap energy.

Li_2NiSiO_4	a/Å	b/ Å	c/ Å	V/ Å^3	Band gap/eV
HSE	6.290	5.299	4.856	161.85	4.26
GGA+U (U=6)	6.291 (0.02%)	5.368 (1.30%)	4.921 (1.34%)	166.17 (2.67%)	3.33

Generally, the removal of a Li atom would lead to the change of the oxidization of one of its neighboring Ni atom from 2+ to 3+; hence, the average Ni-O bond becomes shorter. Particularly, the shrinking of the average Ni-O bonds from 1.98 Å in Ni^{2+}-O bond to 1.87 Å in Ni^{3+}-O would lead to a local lattice distortion in which the charge carrier can be trapped. Consequently, a small polaron appears at a Ni^{3+} site. Around each Li site, there are four nearest Ni neighbors and our energy calculation predicted that the polaron most favorably forms at the third nearest Ni neighbor.

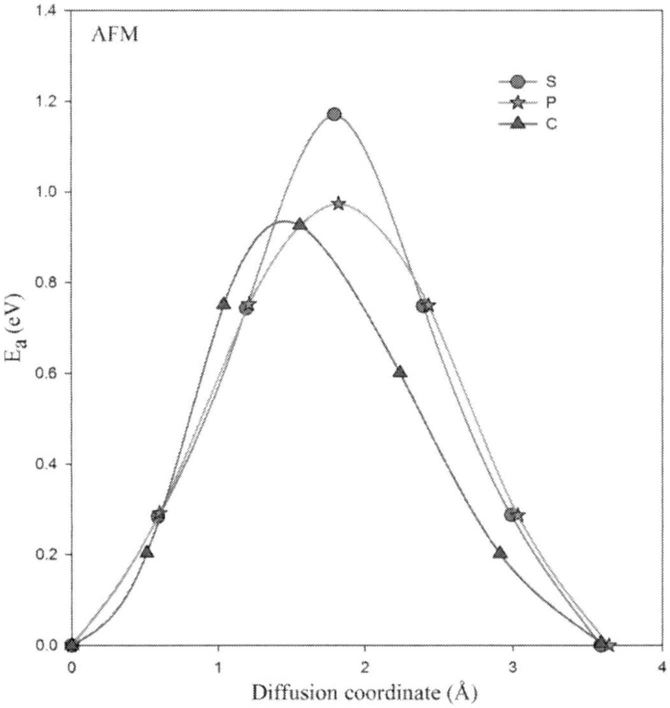

Figure 3. Calculated activation energy of polaron-Li vacancy complex diffusion in Li_2NiSiO_4. The green, pink and violet curves depict the activation energy profile of the preferable *single Li diffusion, parallel* and *crossing* processes, respectively.

3.2 Diffusion

As a vacancy is introduced to the lattice, some of neighboring Li atom would tend to hop to the vacancy site and cause the Li migration which takes places along an opposite path to the vacancy. During diffusion, the movement of the Li vacancy would be accompanied by the small polaron migration and would cause the polaron – Li vacancy complex diffusion [12]. Aiming at figuring out the diffusion properties of the material, we first investigate the elementary diffusion processes of the complex and then combine those processes to form the most preferable diffusion pathways. To make it easier to visualize, we mark the Li diffusion path as red and dashed black arrows as shown in figure 1. Four elementary processes are investigated: two intra-layer processes along the [100] direction (S and P) and two inter-layer processes in the (100) plane (C and T). As shown in figure 1, The Li atom are denoted as Li_i (i = 1→5) and its accompanied polaron sites are indexed as Ni_j (j = 1, 3→5), correspondingly. Figure 4 illustrates possible diffusion pathways by combining elementary processes while figure 3 presents activation barriers of elementary processes. The diffusion mechanism of four elementary processes and possible pathways are described as follow:

i) Process S and P: letters S and P are denoted "*Single*" and "*Parallel*". In process S, the Li vacancy diffuses from the site Li_1 to Li_2, and then in process P, the vacancy moves from Li_2 to Li_3. Since the vacancy at Li_1 and Li_2 possess the same accompanied polaron; Ni_1, then the polaron may stand still during the Li diffusion in process S. There is only the Li vacancy diffusion in this process and that is the reason why it was called *Single Li diffusion* (S) [13]. The activation barrier of this process is 1.17 eV (figure 3). Within P process, the polaron migration is included. When the Li vacancy jumps from Li_2 to Li_3, the polaron at Ni_1 should consequently hop to Ni_3 along parallel route to the one taken by the Li vacancy. Then, this process is named as *Parallel diffusion* (P) [13]. According to the activation barrier profile in figure 3, it costs 0.98 eV for the polaron - Li vacancy complex to diffuse in this process. The complex parallel diffusion has a lower barrier than the single Li ion diffusion. Thus, in contrast to the Li_2FeSiO_4 [13], the accompanied polaron migration in Li_2NiSiO_4 can enhance the diffusion of Li ions. In these processes, the Li vacancy diffuses through a parabolic path with an equal distance of 3.60 Å as illustrated by the green and pink dash curves in figure 4(a). By joining these processes, we can obtain a possible diffusion pathway along which Li ions diffuse through a straight line in the [100] direction.

ii) Process C: the letter C stands for "*Crossing*". When the vacancy at the Li_1 site jumps diagonally upward to occupy the Li_4 site, the accompanied polaron at Ni_1 would consequently jump to Ni_4. The polaron migrates across the Li diffusion route through neighboring Fe layers, hence, the name *Crossing diffusion* is given for this path [13]. The Li vacancy has to travel a distance of 3.60 Å. The activation barrier of the polaron – Li vacancy complex diffusion through this process is 0.93 eV. A zigzag pathway as illustrated in figure 4(b) can be formed by replicating this process within the (100) plane along the [001] direction.

iii) Process T: the letter T stands for "*Through*". Along with the diffusion of the Li vacancy from Li_1 to Li_5, the polaron will simultaneously jump from Ni_1 to Ni_4. *Through layered* (T) was named to this process because the Li vacancy diffuses through Ni-Si layers, as seen in figure 1. The diffusion distance for this process is 4.64 Å which is quite far for Li to travel. Because the complex has to travel a long distance through the screening Ni-Si layers, it costs as high as 2.12 eV to diffuse. It would be hardly for the complex to diffuse with such high activation energy. Hence, this process should not contribute to the possible pathway and we can ignore it.

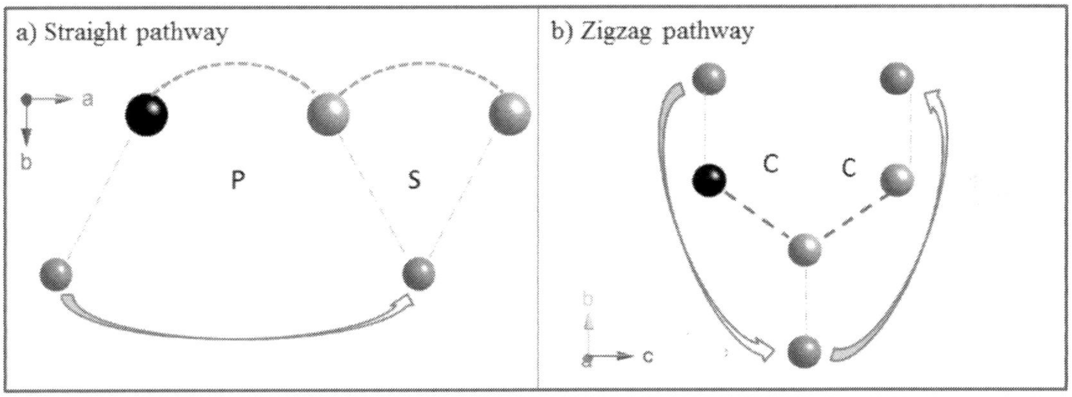

Figure 4. Schematic of possible diffusion pathways. Li vacancy and its accompanying polaron are "connected" by blue dash-dotted-dash lines. The green, pink and violet dash curves correspond to the single Li diffusion, parallel and crossing processes, respectively. The polaron migration is indicated by red arrows.

In summary, we have carefully studied the formation of polaron and the diffusion mechanism of the complex polaron-Li vacancy in Li_2NiSiO_4. One more time, the HSE06 method is proved to succeed in describing the polaron localization in the cathode materials. An investigation of the four elementary diffusion processes was carried out for an insight view of diffusion mechanism, and three processes were found to be preferable. Combining three favorable processes, we figured out the two most preferable diffusion pathways: straight and zigzag pathways. The complex diffusion through the [100] and [001] directions have the activation barriers of 1.17 eV and 0.94 eV, respectively. Similar to Li_2FeSiO_4, the diffusion in the Ni system is also two dimensional. However, in contrast to Li_2FeSiO_4, the Li ion diffusion in Li_2FeSiO_4 can be enhanced by its accompanied polaron migration. Finally, by means of HSE06, we have successfully investigated the structural and electronic structures, polaron formation, and the diffusion of the polaron-Li vacancy complex in Li_2NiSiO_4. With the possibility of the two-redox-potential and the two-dimensional diffusion of Li ions, Li_2NiSiO_4 can be considered as a promising cathode material for the rechargeable Li ion batteries.

Acknowledgments
This work was partially supported by the MEXT Program for Development of Environment Technology using Nanotechnology. One of the authors (K. M. Bui) wishes to thank the National Institute for Materials Science, Japan for financial support

References
[1] Tarascon J M and Armand M J 2001 *Nature* **414** 359
[2] Padhi A K, Nanjundaswamy K S and Goodenough J B 1997 *J. Electrochem. Soc.* **144** 1188
[3] Chung S Y, Bloking J T and Chiang Y M 2002 *Nat. Mater.* **1** 123
[4] Islam M S, Dominko R and Masquilier C 2011 *J. Mater. Chem.* **21** 9811
[5] Nytén A, Abouimrane A, Armand M, Gustafsson T and Thomas J O 2005 *Electrochem. Commun.* **7** 156
[6] Zaghib K, Salah A A, Ravet N, Mauger A, Gendron F and Julien C J 2006 *Power Sources.* **160** 1381
[7] Saracibar A, Van der Ven A and Arroyo-de Dompablo M E 2012 *Chem. Mater.* **24** 495
[8] Nishimura S, Hayase S, Kanno R, Yashima M, Nakayama N and Yamada A 2008 *J. Am. Chem. Soc.* **130** 13212
[9] Wu S, Zhu Z, Yang Y and Hou Z 2009 *Comp. Mater. Sci.* **44** 1243
[10] Nytén A, Kamali S, Häggström L, Gustafsson T and Thomas J O 2006 *J. Mater. Chem.* **16** 2266
[11] Li Y X, Gong Z L and Yang Y 2007 *J. Power Sources* **174** 528
[12] Dinh V A, Nara J and Ohno T 2012 *Appl. Phys. Express* **5** 045801
[13] Bui K M, Dinh V A and Ohno T 2012 *Appl. Phys. Express* **5** 125802
[14] Kresse G and Hafner J 1993 *Phys. Rev. B.* **47** 558
[15] Kresse G and Hafner J 1994 *Phys. Rev. B* **49** 14251
[16] Kresse G and Furthmuller J 1996 *Phys. Rev. B* **54** 11169
[17] Kresse G and Furthmuller J 1996 *Comput. Mat. Sci.* **6** 15
[18] Perdew J, Burke K and Ernzerhof M 1996 *Physc. Rev. Lett.* **77** 3865
[19] Heyd J, Scuseria G E and Ernzerhof M 2003 *J. Chem. Phys.* **118** 8207
[20] Paier J, Marsman M, Hummer K, Kresse G, Gerber I C and Ángyán J G 2006 *J. Chem. Phys.* **124** 154709
[21] Ong S P, Chevrier V L and Ceder G 2011 *Phys. Rev. B* **83** 075112
[22] Monkhorst H J and Pack J D 1976 *Phys. Rev. B* **13** 5188
[23] Henkelman G and Jonsson H 2000 *J. Chem. Phys.* **113** 9978

Journal of Physics: Conference Series **454** (2013) 012062 doi:10.1088/1742-6596/454/1/012062

Density functional study for optical properties of blue silicate phosphor: $BaCa_2MgSi_2O_8$

M Ishida[1], Y Imanari[2], T Isobe[2,3], S Kuze[2], T Ezuhara[2,4], T Umeda[2,3], K Ohno[2,5] and S Miyazaki[2,6]

[1]Advanced Materials Research Laboratory, Sumitomo Chemical Co., Ltd. 6 Kitahara, Tsukuba, 300-3294 Japan
[2] Tsukuba Material Development Laboratory, Sumitomo Chemical Co., Ltd. 6, Kitahara, Tsukuba, 300-3294 Japan
[3] Basic Chemicals Research Laboratory, Sumitomo Chemical Co., Ltd. 5-1, Sobiraki-cho, Niihama, Ehime, 792-8521 Japan
[4] Battery Materials Division, IT-Related Chemicals Sector, Sumitomo Chemical Co., Ltd. 1-1, Ohe-cho, Niihama, Ehime, 792-0015 Japan
[5] Intellectual Property Department, Sumitomo Chemical Co., Ltd. 5-33, 4 Kitahama, Chuo-ku, Osaka, 541-8550 Japan
[6] Japan Association for Chemical Innovation (JACI), 2, Sanban-cho, Chiyoda-ku, Tokyo, 102-0075 Japan

E-mail: ishida@tuc.sumitomo-chem.co.jp

Abstract. A density functional calculation is performed to investigate optical properties of blue silicate phosphor, $BaCa_2MgSi_2O_8$:Eu (BCMS:Eu). The optical absorption and emission processes are discussed based on electronic band structures and density of states. Our calculation indicates that hybridization of the wave function plays an important role for nonradiative migration of electrons and holes. Our results also explain adequately the luminescence feature of BCMS which exhibits two components of blue emission from Eu ion substituted for Ba and Ca site. In addition, it is shown that the structural relaxation around Eu ion does work effectively to describe appropriately the difference of crystal field strength between Ba and Ca sites in BCMS host.

1. Introduction

Inorganic phosphors are widely used in daily life such as fluorescent lamps for lighting and components of plasma display panels (PDPs) and are usually activated by doping with rare-earth elements (the activators). It is generally accepted that, in the optical absorption and emission processes, the absorption of vacuum ultraviolet (VUV: λ=147, 172nm) excitation energy by the host crystal is followed by energy transfer to the activators and then the activators can emit visible light [1]. However, the general interpretation is often based on conventional band scheme and, thus, a detailed description would necessitate further theoretical consideration.

From the viewpoint of industrial application, blue phosphors particularly play an important role for the development of PDPs to realize a full and clear color display. At present, Barium magnesium aluminate $BaMgAl_{10}O_{17}$(BAM), which is activated by Eu^{2+} doping, is widely used as the blue-light-emitting phosphor for the PDPs because of its high quantum efficiency, good chromaticity and strong absorption for VUV excitation [2, 3]. On the other hand, a new silicate phosphor, Calcium magnesium silicate $CaMgSi_2O_6$ (CMS), has been recently reported [4] with blue luminescence by Eu^{2+} doping under VUV excitation. The degradation of luminous efficiency of CMS from the VUV irradiation and the baking process compared with BAM was decreased [5,6].

In our previous report [7], we discussed the optical and emission properties of CMS in detail based on electronic band calculation and proposed an excitation mechanism in which hybridization of wave function plays an important role for nonradiative migration of electrons and holes. In addition, by means of computational materials screening to investigate the optical band gap of various silicate hosts, we also revealed that $BaCa_2MgSi_2O_8$ (BCMS) has an optical band gap similar to that of BAM and, therefore, BCMS has potential of light response to 172nm excitation. Our discussion on BCMS in the previous study, however, was restricted to the optical band gap only and did not make specific mention of other optical properties and its excitation mechanism.

In this work, density functional band calculation has been carried out to investigate the optical properties of BCMS and BCMS:Eu. The optical excitation process is discussed based on the electronic band structures and the density of states (DOS). In our previous study, the calculations on BCMS were carried out with reference to crystal structure data determined by XRD measurements and Rietveld refinements made by Yonesaki *et al.* [8]. They, however, recently reinvestigated the crystal structure using combined powder X-ray and neutron Rietveld techniques and, as a result, determined that BCMS crystallizes in the trigonal space group $P\bar{3}$ instead of the previously believed space group $P\bar{3}m1$ [9]. Thus, we have also recalculated the electronic band structures and DOS of BCMS and BCMS:Eu based on their reinvestigated crystallographic data and discuss the optical properties and its excitation mechanisms.

2. Details of computation

Crystallographic data including atomic positions of BCMS is taken from the report made by Park *et al.* [9]. The host has a trigonal space group with the $P\bar{3}$ (S.G. No.147, a_0=5.427, c_0=6.7946Å) as shown in Figure 1. The primitive cell used in band calculation for BCMS contains 14 atoms, composed by one Ba, two Ca, one Mg, two Si and eight O. On the other hand, to construct a model of BCMS including Eu ion (BCMS:Eu), a 1×1×2 super-cell containing 28 atoms extended to c-axis is used; this super-cell has two Ba and four Ca sites, and then two kinds of models, $BCMS:Eu_{(Ba)}$ and $BCMS:Eu_{(Ca)}$, are constructed by which one of two Ba and two of four Ca are substituted by Eu ion, respectively. The same space group of the primitive cell, $P\bar{3}$ (S.G. No.147), are imposed on these substituted models.

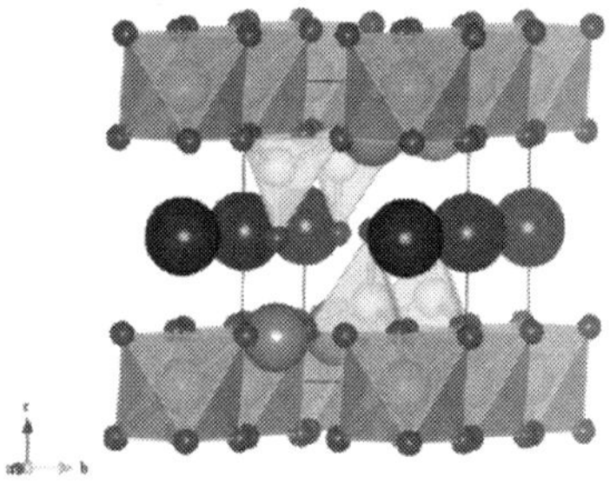

Figure 1. Crystal structure of non-doped BCMS. The white tetrahedron and the gray octahedron stand for SiO_4 and MgO_8 unit, respectively; the black and gray spheres stand for Ba and Ca, respectively; the small black spheres stand for O. The crystal structure is visualized by using VESTA [10].

In both models, the substitution ratio corresponds to 50% concentration of Eu and there could be a concern if such high concentration of Eu ion affects the electronic structures of BCMS:Eu as discussed in our previous study on CMS:Eu [7]. We, hence, have also checked the Eu concentration dependency by taking into account a model with 25% substitution of Eu ion for $BCMS:Eu_{(Ca)}$. The cell with 25% substitution model of $BCMS:Eu_{(Ca)}$ has lower symmetry of P3 (S.G. No. 143) including 28 atoms and one fourth Ca sites is substituted by a Eu ion. Table 1 shows the results for the band energies and the bandwidths using the different Eu concentration models. As clearly shown, those band energies and the bandwidths are independent of Eu concentration. Thus we will further use in this paper the results of the 1×1×2 super-cell (50% concentration model) to discuss the electronic structures and the optical properties of BCMS:Eu.

Table 1. Band energies ε of BCMS:$Eu_{(Ca)}$ for valence band top (VBT), 4f-band and conduction band bottom (CBB); band widths for O2p, Eu 4f, Ba 5d, Ca 3d and Eu 5d bands for the different Eu concentrations using the full-potential linearized augmented plane-wave (FLAPW) method.

Eu concentration (%)	Band energy (eV)			Band Width (eV)				
	ε(VBT)	ε(4f)	ε(CBB)	O 2p (occ.)	Eu 4f (occ.)	Ba 5d (unocc.)	Ca 3d (unocc.)	Eu 5d (unocc.)
25	-3.9	-0.88	0.9	6.80	0.24	> 7.1	> 7.1	> 7.1
50	-3.9	-0.90	0.9	6.78	0.26	> 7.1	> 7.1	> 7.1

The band calculation of BCMS and BCMS:Eu has been performed by means of the full-potential linearized augmented plane-wave (FLAPW) method [11-14] within the local density approximation (LDA) and the LDA with the on-site Coulomb repulsion (LDA+U) [15,16]. The linearization for the APW method is done by the scheme of Takeda and Kübler. The muffin-tin (MT) radii are 2.5 a.u. for Eu, 2.7 for Ba, 2.6 a.u. for Ca, 2.2 a.u for Mg, 1.55 a.u. for Si and 1.4 a.u. for O. The energy cutoff parameters are 12 Ry for the wave function and 48 Ry for charge density and the potential. The angular momentum expansion is truncated at l_{max}=7 for the wave function, and at l_{max}=6 for the charge density and the potential. We take 30***k*** points in the irreducible Brillouin zone, which corresponds to 144***k*** points in the whole Brillouin zone.

In the LDA+U calculation of BCMS:Eu, ferromagnetic state of Eu ($4f^7$; S=7/2) is assumed and U_{eff}=5.99 eV for Eu 4f state is used in line with our previous study on CMS:Eu [7]. Here, we have mentioned only about U parameter dependency in the range of 4.35 to 7.62eV for band energy difference between Eu 4f band and VBT of BCMS:Eu; we have obtained a result that the dependence is almost linear as is the case with the report by Hölsä *et al.* [17]. In addition, we can see that BCMS:$Eu_{(Ba)}$ and BCMS:$Eu_{(Ca)}$ exhibit the different values of the band energy difference which is decreased with an increase in U parameters such as *ca.* 2.8 to 1.5 eV and *ca.* 3.7 to 2.6 eV, respectively. The band energy differences of BCMS:$Eu_{(Ba)}$ and BCMS:$Eu_{(Ca)}$ for U_{eff}=5.99 eV are 2.0 and 3.0 eV respectively and those values are slightly small compared with that of CMS obtained in our previous study [7].

Structural optimization for BCMS:Eu has been performed by using the VASP code [18] within the GGA-PBEsol scheme [19]. The energy cutoff of 400eV and 5x5x5 ***k***-point grid are used. We set the convergence criteria of $1 \cdot 10^{-5}$ eV for SCF iterations and $1 \cdot 10^{-2}$ eV for the net forces acting on the atoms. During the optimization calculations, the positions of all atoms are relaxed within the fixed lattice parameters of the super cell.

3. Results and discussions

3.1. *Band structure and density of states (DOS)*

Figure 2 shows the LDA band structure and partial DOS for non-doped BCMS in the nonmagnetic state. The origin of the band energy is taken at the Fermi level. The valence band (VB) is mostly composed of the occupied O 2p bands between -9.1 and -2.4eV, whereas the unoccupied bands, which are located at +2.4eV or above, mostly consist of Ca 3d and Ba 5d bands mixing with O 2p bands. The lowest dispersive unoccupied band possesses mostly s-band character composed of Ba 6s, Ca 4s, O 2s, Mg and Si 3s bands, in which Ba 5d and Ca 3d components are slightly mixed.

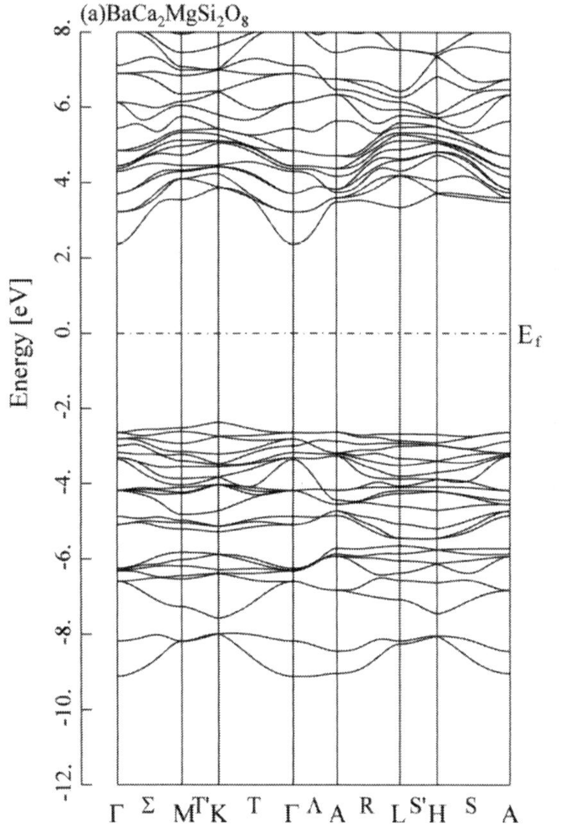

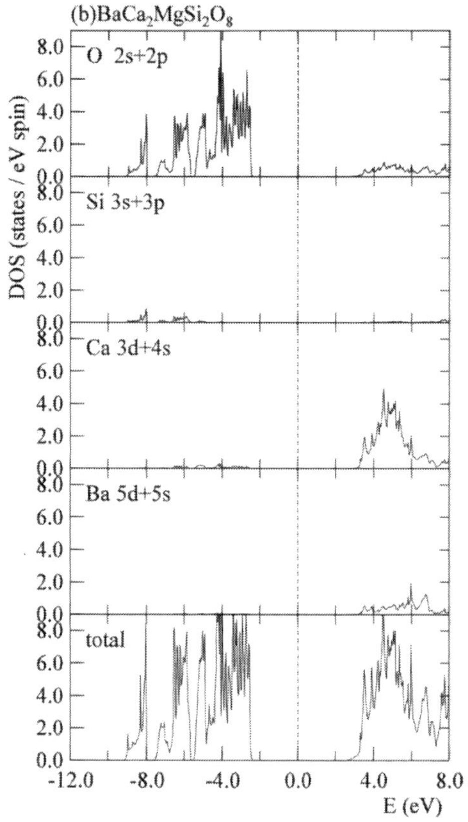

(c) BZ of $BaCa_2MgSi_2O_8$

Figure 2. The band structure (a) and DOS (b) for non-doped BCMS in the nonmagnetic state obtained from FLAPW method within the LDA scheme. The Brillouin zone (BZ) is depicted in (c) with the special k-points.

On the other hand, Figure 3 and 4 show the LDA+U band structures and partial DOS in the ferromagnetic up spin state for BCMS:Eu, respectively; the left and right illustration in each figure represent BCMS:$Eu_{(Ba)}$ and BCMS:$Eu_{(Ca)}$, respectively. Those of the down spin state are not presented here because those are almost the same as that of non-doped BCMS. In Figure 3, the localized flat bands are Eu 4f bands, which are located at -1.3eV in BCMS:$Eu_{(Ba)}$ and -0.9eV in BCMS:$Eu_{(Ca)}$ below the Fermi level, respectively. The magnetization in both Eu substitution models is $7\mu_B$, which corresponds to the ground state of Eu ($4f^7$; S=7/2). For both models, the DOS shown in Figure 4 are very similar; the VB between -10.0 and *ca.*-4.0 eV are mostly composed of the O 2p bands and the unoccupied bands located at *ca.* +2.0eV or above are mostly composed of the Ca 3d, Ba 5d and Eu 5d bands mixing with O 2p bands. Furthermore, it is noteworthy that the component from the Eu 5d states is mixing slightly with the lowest dispersive unoccupied band possessing s-band character.

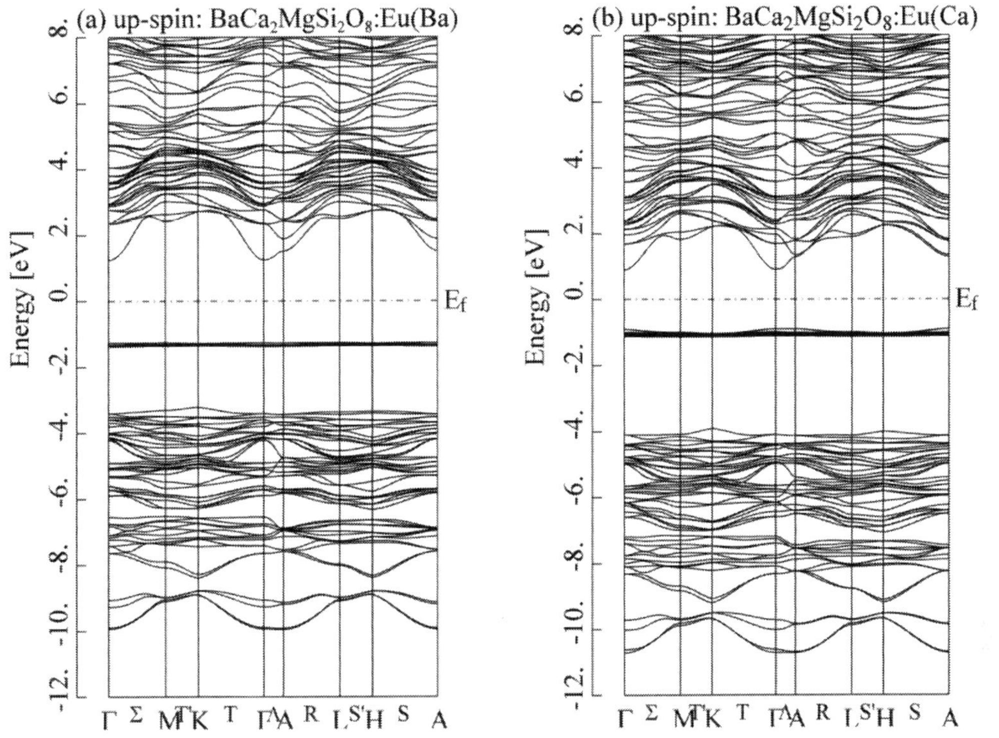

Figure 3. Band structures for ferromagnetic up-spin state of BCMS:Eu obtained from FLAPW method within LDA+U scheme; (a) BCMS:$Eu_{(Ba)}$ (left) and (b) BCMS:$Eu_{(Ca)}$ (right). The Coulomb U parameter of 5.99eV is used for Eu 4f state.

These results shown in Figure 3 and 4 exhibit that there are two possible optical absorptions of BCMS:Eu; one is the direct transition from Eu 4f bands and the other is the host interband transition which occurs from O 2p bands. In the direct transition, electrons could be excited into the unoccupied bands mainly consisting of the Ba 5d, Ca 3d and Eu 5d bands. However, the occupied Eu 4f wave function is strongly localized on Eu atom and the overlap of the wave functions between Eu 4f and Ba 5d and/or Ca 3d orbital might be small. Therefore, by symmetry, the optical transition of Eu direct excitation should be almost restricted to the Eu 4f → 5d transition. On the other hand, the host interband transition is attributed to charge transfer from O 2p bands to cation d bands and the optical transition induces electron-hole pair on the host lattice; electrons are excited into the unoccupied bands mainly consisting of the Ba 5d, Ca 3d and Eu 5d bands and holes are generated into the O 2p bands.

In order to produce the effective emission followed by the VUV excitation, it is necessary that excited electrons and holes on the host lattice transfer efficiently to Eu ions. In the partial DOS of Figure 4, it should be noted that the unoccupied Ba 5d and Ca 3d bands are mutually well hybridized with the Eu 5d bands; the hybridization could assist the efficient electron transfer from the host lattice to the Eu ion. Furthermore, interestingly, hybridization between the occupied Eu 4f and O 2p bands can also be seen in the partial DOS; it could also help for transferring the hole generated on the host lattice to the Eu ion. Consequently, it can be concluded that, even in BCMS:Eu, the hybridization of the wave functions plays an important role for the nonradiative migration of excited electrons and holes from the host lattice to the luminescence centre.

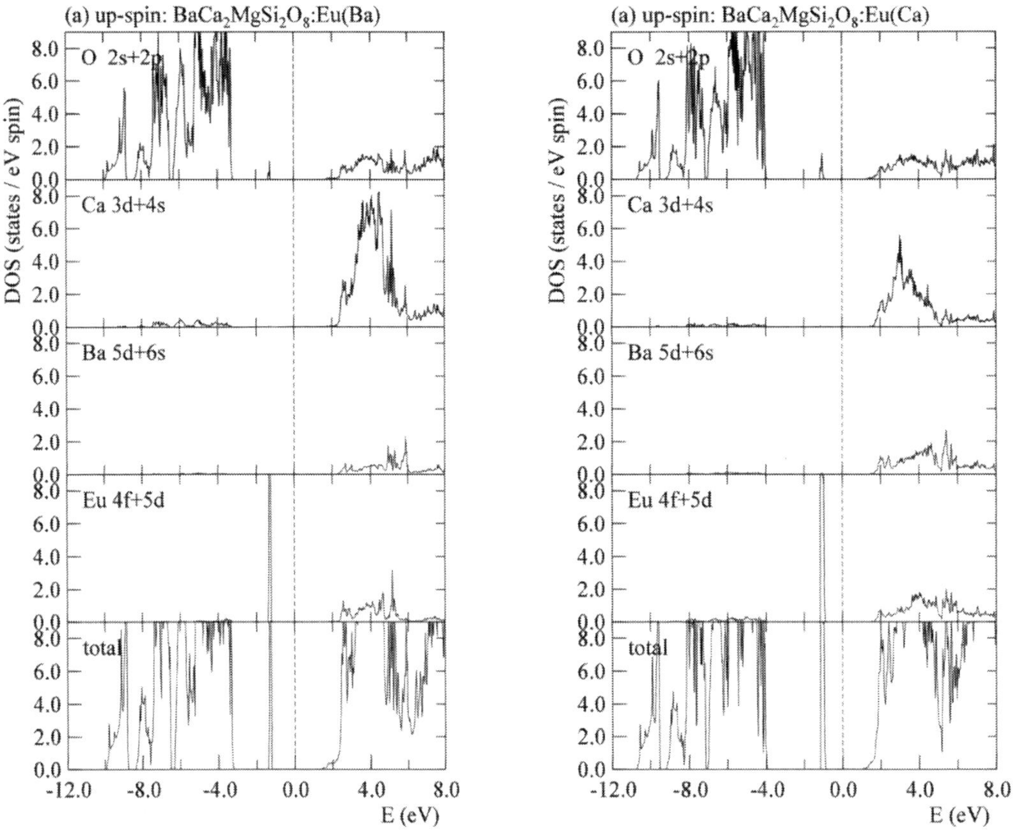

Figure 4. Total and partial density of states for ferromagnetic up spin state of BCMS:Eu obtained from FLAPW method within the LDA+U scheme; (a) BCMS:$Eu_{(Ba)}$ (left) and (b) BCMS:$Eu_{(Ca)}$ (right). The Coulomb U parameter 5.99eV is used for Eu 4f state.

The emission process of Eu^{2+} is usually explained in terms of $4f^6 5d \rightarrow 4f^7$ transition [4,9]. The present calculation supports that the assignment is reasonable from the viewpoint of the band structure, that is, the intra-atomic transition corresponds to the transition between band edges of Eu 4f band and the lowest dispersive unoccupied band. As was pointed out previously, the lowest unoccupied band possesses some Eu 5d component mixing with the main s-band characters. Thus, from the viewpoint of the symmetry, the dipole transition between Eu 4f and 5d states should be allowed in the emission process. Indeed, our FLAPW method takes account of symmetry of crystal and we have confirmed that the optical dipole transition at Γ point is allowed by selection rule based on the symmetry. This is a qualitative discussion and, in fact, it might be necessary to take into account many-body effect such as excitonic effect for further accurate description of the emission process. Here, however, it is important for us to grasp that the lowest unoccupied band of BCMS:Eu is available to participate in the emission process due to possessing non-zero Eu 5d component. This interpretation is essentially in common with that of CMS discussed in our previous study [7] and we can, therefore, conclude that the optical absorption and emission process in both CMS and BCMS are attributed to the same mechanism.

3.2. *Luminescence feature of BCMS:Eu*

We, here, discuss the luminescence feature of BCMS:Eu in more detail because basic understanding of such a luminescence feature is important to materials design and development in industry. BCMS has two different substitution sites, Ba and Ca sites, for Eu ion and it is therefore suggested that BCMS:Eu exhibits two components of luminescence; as shown in the band structures and the DOS in Figures 3 and 4, the relative position of Eu 4f bands and CBB measured from Fermi level are slightly different between BCMS:$Eu_{(Ba)}$ and BCMS:$Eu_{(Ca)}$. Hence this subsection is devoted to a discussion of such a luminescence feature of BCMS:Eu from the viewpoint of band energy.

It is well known that the LDA calculation tends to underestimate the optical band gap. For a qualitative discussion about optical emission energy, a uniform shift of unoccupied band energy, so-called "scissors operator", is often applied; that is, all of the energy levels of the unoccupied bands are uniformly shifted to adjust the energy difference between band edges associated with the luminescence under consideration to the observed emission energy. Validity of the approach using the scissors operator for unoccupied band of silicate phosphors was already discussed in our previous report in detail [7]. We, here, intend to investigate the relationship between the energies of the two emission components and the local structure environment around Eu substitution sites.

Table 2 shows the band edge energies and its energy differences of BCMS:Eu derived from LDA+U calculation. The uncorrected band energy differences of BCMS:$Eu_{(Ba)}$ and BCMS:$Eu_{(Ca)}$ are 2.51 and 1.81 eV, respectively, and those energies are smaller than those of experimental values due to the underestimation problem of band gap in LDA. Then, by means of the scissors operator, those energy differences are corrected; a value of 0.9 eV is applied to adjust the energy difference of BCMS:$Eu_{(Ca)}$ to the experimental emission energy, 2.69 eV (460 nm). As a result, the other energy difference of BCMS:$Eu_{(Ba)}$ is estimated at *ca*. 3.4 eV (365 nm) and the estimated energy is, however, too large (too short in wavelength) compared with the experimental emission energy, 2.88 eV (430 nm).

Table 2. Band edge energies and its energy differences of BCMS:Eu derived from FLAPW LDA+U calculation. A correction value of the scissors operator, 0.9eV, is used to adjust the band energy difference of BCMS:$Eu_{(Ca)}$ to the experimental value of the emission energy. Values in parenthesis stand for wavelength of the band energy difference.

System	Band edge energy (eV)			Band energy difference (eV)		Exp. [9]
	uncorrected ε(CBB)	corrected ε(CBB)	ε(4f)	uncorrected ε(CBB)-ε(4f)	corrected ε(CBB)-ε(4f)	
BCMS:$Eu_{(Ba)}$	+1.25	+2.15	-1.25	2.51 (494)	3.41 (364)	2.88 (430)
BCMS:$Eu_{(Ca)}$	+0.90	+1.80	-0.90	1.81 (685)	2.71 (458)	2.69 (460)

The discrepancy may arise from the difference of crystal field strength between Ba and Ca cations for Eu substitution; indeed, we can see that the Eu 4f bandwidth of BCMS:$Eu_{(Ca)}$ is slightly wider than that of BCMS:$Eu_{(Ba)}$ due to stronger crystal field. In general, Ba cation has a larger ionic radius compared with that of Ca cation. Coordination number of Ba and Ca site in BCMS host are twelve and seven, respectively [9] and, thus, the corresponding ionic radii by Shannon [20] are 1.61 and 1.06Å, respectively. On the other hand, the ionic radii of Eu ion are 1.35 and 1.2 Å for ten and seven coordination number, respectively. Thus the coordination environment of Ba (Ca) site is looser (tighter) than that of Ca (Ba) site for Eu substitution.

Hence we should take into account structure relaxation effect on Eu substitution and have carried out the structure optimization using VASP code based on GGA+U calculation [18,19]. The optimized bond distances of Ba-O, Eu-O and Ca-O are summarized in Table 3. As would be expected, Eu substitution for Ba site shorten Eu-O bond compared with the original Ba-O bond distance and the substitution for Ca site makes Eu-O bond longer compared with the original Ca-O bond. Then, by means of adjusting the uncorrected band energy difference of BCMS:$Eu_{(Ca)}$ to the experimental emission energy using the correction value of 0.7 eV, the other band energy difference of BCMS:$Eu_{(Ba)}$ are estimated at *ca*. 2.94 eV (422 nm) as shown in Table 3. This result describes adequately the relative energy difference in the experimental luminescence of Eu ion for Ba and Ca site. In addition, we can see that spectrum width of the experimental luminescence profile of Eu for Ca site is slightly wider than that of Eu for Ba site. The fact might reflect that BCMS:$Eu_{(Ca)}$ possesses a slightly wider bandwidth of Eu 4f bands compared with that of BCMS:$Eu_{(Ba)}$ as shown in Figures 3 and 4.

Journal of Physics: Conference Series **454** (2013) 012062 doi:10.1088/1742-6596/454/1/012062

Consequently, we conclude that the structural relaxation around Eu ion does work effectively to describe appropriately the difference of crystal field strength between Ba and Ca sites in BCMS host.

Table 3. Optimized bond distances of Ba-O, Eu-O and Ca-O bond, and band energy differences derived from VASP/GGA+U calculation. U=5.99 eV is used. Values in parenthesis stand for wavelength of band energy difference. The original bond distances in BCMS are 2.772 and 3.134 Å for Ba-O, and 2.246, 2.421 and 2.733 Å for Ca-O, respectively [9]. A correction value of the scissors operator, 0.7 eV, is used here.

System	Optimized bond distance (Å)			Band energy difference (eV)	
	Ba-O(12)	Eu-O	Ca-O(7)	uncorrected	corrected
BCMS:$Eu_{(Ba)}$	2.799 3.134	2.719 3.135	2.180-2.261 2.368-2.402 2.730-2.783	2.24 (554)	2.94 (422)
BCMS:$Eu_{(Ca)}$	2.738-2.782 3.133-3.136	2.335 2.554 2.731	2.195 2.402 2.699	2.00 (620)	2.71 (458)

3.3. *Revised result of computational materials screening for silicate phosphors*

Finally, we intend to add a short comment on our previous result of computational materials screening for silicate phosphors; we presented a result of host excitation energy of various silicate hosts using computational materials screening in our previous study [7].

However as the calculation of BCMS was carried out by using the crystallographic data reported by Yonesaki *et al.* in ref. 8, the host excitation energy of BCMS should be checked with the revised crystallographic data. We, thus, have recalculated the host excitation energy of BCMS based on the revised crystallographic data using the same calculation procedure described in ref. 7. As a result, we have obtained *ca.* 6.05 eV as the host excitation energy of BCMS and it is almost same as the previous calculated value, 6.1 eV. We, therefore, can see that the relative position of the host excitation energy of BCMS among that of various silicate phosphors does not change.

Although our approach presented in this study might not be totally based on first-principles, we have demonstrated that such a practical manner using empirical manipulation combined with first-principles calculation is effective in basic understanding of luminescence feature of phosphors and in developing industrial phosphors. Probably, it would be one of great theoretical challenge to be able to predict the optical absorption and emission features of materials beyond current density functional theory.

4. Conclusion

The LDA and LDA+U band calculation have been performed for blue silicate phosphor, BCMS:Eu. From the electronic band structure and the partial DOS, we conclude that the optical absorption arises from the host interband transition and the Eu 4f → 5d transition and, on the other hand, the optical emission attributes to the Eu 5d → 4f transition which mainly contributes to band edge luminescence by symmetry. In particular, the results indicate that the hybridization of wave functions plays an important role for nonradiative migration of electrons and holes from the host to the luminescence centre. Thus, our results have revealed that the optical absorption and emission process in both CMS and BCMS are attributed to the same mechanism.

BCMS, however, exhibits two components of blue luminescence attributed to Eu 5d → 4f transition for Ba and Ca substitution sites. Our results explained adequately such a luminescence feature of BCMS based on the band structures and the DOS analysis. In addition, it is shown that the structural optimization does work effectively to describe appropriately the difference of crystal field strength between Ba and Ca sites in BCMS host.

Acknowledgments
One of the authors (M.I.) is grateful to Prof. N. Hamada at Tokyo University of Science, who permitted the use of his own FLAPW code, ABCAP. Also, he continued to encourage and support.

References

[1] Bizarri G and Moine B 2005 *J. Lumin.* **115** 53
[2] Koike J, Kojima T, Toyonaga R, Kagami A, Hase T and Inaho S 1979 *J. Electrochem. Soc.* **126** 1008
[3] Kim K B, Kim Y I, Chun H G, Cho T Y, Jung J S and Kan J G 2002 *Chem. Mater.* **14** 5045
[4] Kunimoto T, Yoshimatsu R, Ohmi K, Tanaka S and Kobayashi H 2002 *IEICE Trans. Electron.* **E85-C** 1888
[5] Im W B, Kan J H, Lee D C, Lee S, Jeon D Y, Kang Y C and Jung K Y 2005 *Solid State Comm.* **133** 197
[6] Im W B, Kim Y I, and Jeon D Y 2005 *J. Mater. Res.* **20** 2061
[7] Ishida M, Imanari Y, Isobe T, Kuze S, Ezuhara T, Umeda T, Ohno K and Miyazaki S 2010 *J.Phys.:Cond.Mat.* **22** 384202
[8] Yonesaki Y, Takei T, Kumada N and Kinomura N 2008 *J. Lumin.* **128** 1507
[9] Park C-H, Kim T-H, Yonesaki Y and Kumada N 2011 *J. Lumin.* **184** 1566
[10] Momma K. and Izumi F. 2011 *J. Appl. Crystallogr.* **44** 1272 A following URL is available: http://jp-minerals.org/vesta/en/
[11] Hamada N, Sawada H, Solovyev I and Terakura K 1997 *Physica B* **11** 237
[12] Hohenberg P and Kohn W 1964 *Phys. Rev.* **136** B864
[13] Kohn W and Sham L J 1965 *Phys. Rev.* **140** A1133
[14] Vosko S H, Wilk L and Nusair M 1980 *Can. J. Phys.* **58** 1200
[15] Anisimov V I, Zaanen J and Andersen O K 1991 *Phys. Rev.B* **44** 943
[16] Anisimov V I, Solovyev I V, Korotin M A, Czyżyk M T and Sawatzky G A 1993 *Phys. Rev. B* **48** 16929
[17] Hölsä J, Kirm M, Laamanen T, Lastusaari M, Niittykoski J and Novák P 2009 *J. Lumin.* **129** 1560
[18] Kresse G and Hafner J 1999 *Phys. Rev. B* **47** 558
[19] VASP is the abbreviation for Vienna Ab-initio Simulation Package. The version 4.4.5 and 5.2.21 of VASP code were employed.
[20] Perdew J P, Ruzsinszky A, Csonka G I, Vydrov O A, Scuseria G E, Constantin L A, Zhou X, and Burke K 2008 *Phys. Rev. Lett.* **100** 136406
[21] Shannon R D 1976 *Acta Cryst.* **A32** 751

Journal of Physics: Conference Series **454** (2013) 012063 doi:10.1088/1742-6596/454/1/012063

No Core CI calculations for light nuclei with chiral 2- and 3-body forces

Pieter Maris[1], H Metin Aktulga[2], Sven Binder[3], Angelo Calci[3], Ümit V Çatalyürek[4,5], Joachim Langhammer[3], Esmond Ng[2], Erik Saule[4], Robert Roth[3], James P Vary[1] and Chao Yang[2]

[1] Department of Physics and Astronomy, Iowa State University, Ames, IA 50011, USA
[2] Computational Research Division, Lawrence Berkeley National Lab, Berkeley, CA 94720, USA
[3] Institut für Kernphysik, Technische Universität Darmstadt, 64289 Darmstadt, Germany
[4] Dept. of Biomedical Informatics, The Ohio State University, Columbus, OH 43210, USA
[5] Dept. of Electrical and Computer Engineering, The Ohio State University, Columbus, OH 43210, USA

E-mail: pmaris@iastate.edu

Abstract. The atomic nucleus is a self-bound system of strongly interacting nucleons. In No-Core Configuration Interaction calculations, the nuclear wavefunction is expanded in Slater determinants of single-nucleon wavefunctions (Configurations), and the many-body Schrödinger equation becomes a large sparse matrix problem. The challenge is to reach numerical convergence to within quantified numerical uncertainties for physical observables using finite truncations of the infinite-dimensional basis space. We discuss strategies for constructing and solving the resulting large sparse matrices for a set of low-lying eigenvalues and eigenvectors on current multicore computer architectures. Several of these strategies have been implemented in the code MFDn, a hybrid MPI/OpenMP Fortran code for ab initio nuclear structure calculations that scales well to over 200,000 cores. We discuss how the similarity renormalization group can be used to improve the numerical convergence. We present results for excitation energies and other selected observables for ^{8}Be and ^{12}C using realistic 2- and 3-body forces obtained from chiral perturbation theory. Finally, we demonstrate that collective phenomena such as rotational band structures can emerge from these microscopic calculations.

1. No-Core Configuration Interaction approach

Solving for nuclear properties with realistic potentials, using a quantum many-particle framework that respects all the known symmetries of these potentials, is recognized to be computationally hard. A nucleus with Z protons and N neutrons is a self-bound quantum many-body system with $A = N + Z$ strongly interacting nucleons. The interactions are strong (otherwise nuclei would not be bound) and feature both attractive and repulsive contributions along with significant spin and angular momentum dependence. Furthermore there are both short-range and long-range terms in the interaction, and in addition to two-body forces (2NF) [1, 2, 3], one also needs suitable three-nucleon forces (3NF) [4, 5, 6, 7, 8] (and possibly higher A-body forces).

A commonly used approach to address this problem in nuclear physics is the Configuration Interaction (CI) method for solving the many-body nuclear Hamiltonian in a (sufficiently large) single-particle basis space. In this approach, the many-body Schrödinger equation

$$H \, \Psi_i(\vec{r}_1, \ldots, \vec{r}_A) \;=\; E_i \, \Psi_i(\vec{r}_1, \ldots, \vec{r}_A) \qquad (1)$$

becomes a large sparse matrix problem with eigenvalues E_i. In No-Core CI (NCCI) calculations [9], the wavefunction Ψ of a nucleus consisting of A nucleons is expanded in an A-body basis of Slater determinants Φ_k of single-particle wavefunctions ϕ_i

$$\Psi(\vec{r}_1, \ldots, \vec{r}_A) = \sum c_k \Phi_k(\vec{r}_1, \ldots, \vec{r}_A) , \qquad (2)$$

with $\Phi_k(\vec{r}_1, \ldots, \vec{r}_A) = \mathcal{A}[\phi_{n_1 l_1 j_1 m_1 tz_1}(\vec{r}_1)\, \phi_{n_2 l_2 j_2 m_2 tz_2}(\vec{r}_2) \ldots \phi_{n_A l_A j_A m_A tz_A}(\vec{r}_A)]$. Here $\mathcal{A}$ is the antisymmetrization operation that insures compliance with the Pauli exclusion principle for identical fermions. Conventionally, one uses a harmonic oscillator (HO) basis for the single-particle wavefunctions, but it is straightforward to extend this approach to a more general single-particle basis [10]. The single-particle wavefunctions are labelled by the quantum numbers n, l, j, m, and tz, where n and l are the radial and orbital HO quantum numbers, with $N = 2n+l$ the number of HO quanta; j is the total single-particle spin, with m its projection along the z-axis; and tz is the isospin projection ($+\frac{1}{2}$ for protons and $-\frac{1}{2}$ for neutrons). The many-body basis states have well-defined total spin projection M, which is simply the sum of m_k of the single-particle states, $M = \sum m_k$, hence the name M-scheme. However, the many-body basis states do not have a well-defined total spin J. Some of the benefits of this scheme are that it is very simple to implement, and that in two runs (one for positive and one for negative parity states), we get the complete low-lying spectrum, including the ground state, even if the spin of the ground state is unknown.

In a complete basis, this method would give exact results for a given input interaction V. However, practical calculations can only be done in a finite-dimensional truncation of a complete basis. Different truncation schemes have different convergence rates: Full Configuration Interaction employs a widely-know truncation in which all many-body basis states that can be constructed from a finite set of single-particle states are retained. However, this truncation method has a rather slow convergence rate in typical ab initio nuclear structure calculations [11, 12]. The so-called $N_{\max}$ truncation, in which the total number of HO quanta in a basis state is limited: $\sum N_k \leq N_0 + N_{\max}$, is generally much more efficient for NCCI calculations. Here, N_k is the number of quanta of each single-particle state in the many-body basis state; N_0 is the minimal number of quanta for that nucleus; and $N_{\max}$ is the truncation parameter. For HO single-particle states, or more general, for any basis in which the single-particle states have radial and orbital quantum numbers n and l, we have $N_k = 2n_k + l_k$. Furthermore, in combination with a HO basis as single-particle basis, this $N_{\max}$ truncation leads to an exact factorization of the center-of-mass wavefunction and the relative wavefunction.

2. Computational Challenges and Solutions

For a very light nucleus like ^{4}He or ^{6}Li one can achieve convergence of the ground state energy and other observables with suitable 2NFs by simply going to a sufficiently large basis, but this is not practical for larger nuclei and with 3NFs. The computational effort depends not so much on the dimension of the matrix, but rather on the actual number of nonzero matrix elements in the sparse matrix. The number of nonzero matrix elements in a calculation including 3NFs is between one and two orders of magnitude larger than the number of nonzero matrix elements in a calculation for the same nucleus in the same basis space, but using 2NFs only [13].

For p-shell nuclei ($N, Z \leq 8$), the dimension of the M-scheme basis space at $N_{\max} = 8$ reaches about one billion basis states. With 2NFs plus 3NFs, the resulting symmetric matrix contains up to 40 trillion nonzero real matrix elements in the upper (or equivalently, in the lower) triangle; see figure 1. Storing these nonzero matrix elements in compressed-row format (or

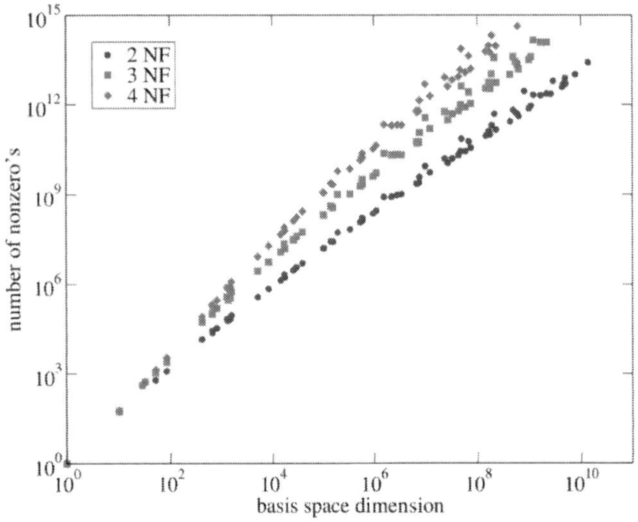

Figure 1. Number of nonzero matrix elements as function of the basis dimension, with 2-body forces (2NF, blue circles,), 2- and 3-body forces (3NF, red squares) and up to 4-body forces (4NF, green diamonds) in an M-scheme basis for $N = Z$ nuclei with $M = 0$.

equivalently, compressed-column format) using single-precision requires an aggregate memory of about 320 TB. That is slightly over half the total memory of Jaguar, a Cray XK6 with 18,688 16-core AMD processors at Oak Ridge National Laboratory (currently being upgraded to Titan, a Cray XK7); slightly less than half the total memory of Mira, an IBM Blue Gene/Q with 49,152 16-core processors at Argonne National Laboratory; and about a quarter of the total memory of the K computer, a Fujitsu supercomputer with 88,128 SPARC64 processors at the RIKEN Advanced Institute for Computational Science in Kobe, Japan. In order to make use of such resources, which are available for open science, we clearly need a suitable code that distributes the matrix evenly over the available processors, and that performs well at such large scales.

For the results presented here we used the code MFDn, which is a hybrid MPI/OpenMP parallel Fortran 90 code for nuclear structure calculations, that has been in development for over two decades. Significant improvements in its performance have been made over the last five years [14, 15, 16, 17] under the DOE SciDAC-2 UNEDF project. MFDn constructs the many-body basis states in M-scheme and the corresponding many-body matrix, and solves for the lowest eigenstates using an iterative Lanczos algorithm. Each Lanczos iteration consists of a sparse matrix-vector multiplication (SpMV) followed by orthogonalization. At the end of a run MFDn writes the eigenvectors to file, and evaluates selected physical observables which can be compared to experimental data (the wavefunctions themselves are not experimentally observable). It also writes out the one-body density matrix elements (in the underlying single-particle basis), which can be used for further analysis. By far the most compute-intensive parts of the calculation are constructing the large sparse matrix and obtaining the lowest eigenvalues and eigenvectors of this matrix using a Lanczos algorithm.

2.1. Two-dimensional distribution of sparse matrix

Each vector is distributed over n processors, and the sparse symmetric matrix itself is distributed in a two-dimensional fashion over $n(n + 1)/2$ processors. We keep the matrix in core, using compressed-row or compressed-column format, and we store only half the matrix, because the matrix is symmetric; we use the same data structure for both a SpMV and a transpose SpMV. The local work load of the SpMV is proportional to the number of nonzero matrix elements on each processor. Thus we want to distribute the matrix in such a way that each processor has approximately the same number of nonzero matrix elements. Note that the overall performance also depends on the communication overhead, and that the mapping of the matrix blocks onto

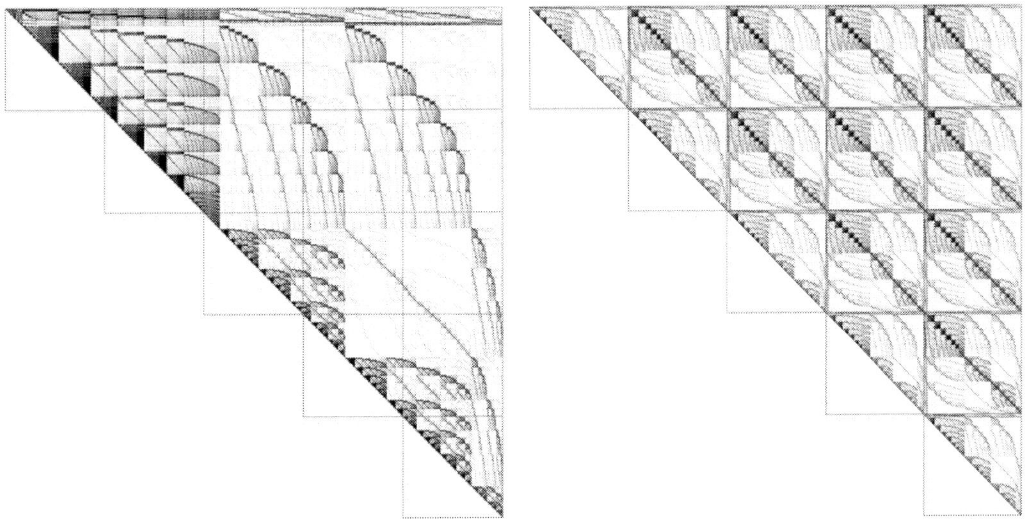

Figure 2. Sparsity structure for ^{8}Be at $N_{\max} = 4$ with 3NF, with a dimension of 143,792 M-scheme basis states and 402,513,272 nonzero matrix elements in the upper triangle: sparsity structure on a single processor (left); and evenly distributed over 15 processors after a round-robin distribution of the basis states over 5 processor (right).

the MPI processors has a large influence on this communication overhead [16, 17].

On a single processor, a lexicographical enumeration of the M-scheme many-body basis states leads to a very structured sparse matrix, as can be seen in the left panel of figure 2, although it is a rather nontrivial structure. A simple distribution of this structured matrix over 15 processors (see the red grid in figure 2) leads to rather poor load balancing and therefore very poor scaling.

On the other hand, with a round-robin distribution of individual M-scheme basis states over n processors, we achieve almost perfect load balancing over $n(n+1)/2$ processors for the matrix, both in memory usage and the number of local floating-point operations during the Lanczos iterations, as is indicated in the right panel of figure 2. This scheme has been implemented and used in various versions of MFDn, up through the current production Version 13, and indeed leads to very good scaling of the SpMV – in fact, sometimes better than ideal scaling, as we will show below. However, a disadvantage of a round-robin distribution of individual basis states is that on any given processor there is little correlation between neighboring states. Although this distribution leads to an overall sparsity pattern for each submatrix on one of the $n(n+1)/2$ processors that looks similar to that of the entire matrix on a single processor, see figure 2, at the level of individual matrix elements we are effectively dealing with a randomly sparse matrix; see the top panel of figure 3. For large runs with 3NFs, we have sparse matrices with less than 1 in 10^3 nonzero matrix elements. Determining exactly which matrix elements are nonzero becomes an extremely time-consuming task for such large sparse matrices [14]. Furthermore, any local SpMV with a randomly sparse matrix will suffer from poor cache performance, in particular if the input and output vectors are too large to remain in cache.

In the latest version of MFDn (Version 14, currently under development), we distribute groups of M-scheme basis states in a round-robin fashion over the n processors, instead of individual basis states. One of the advantages of this scheme is that on any given processor at least some of the neighboring basis states are correlated (namely those within a group), leading to a more regular and predictable sparsity structure down to the level of individual matrix elements; see the bottom panel of figure 3. This greatly improves the efficiency of

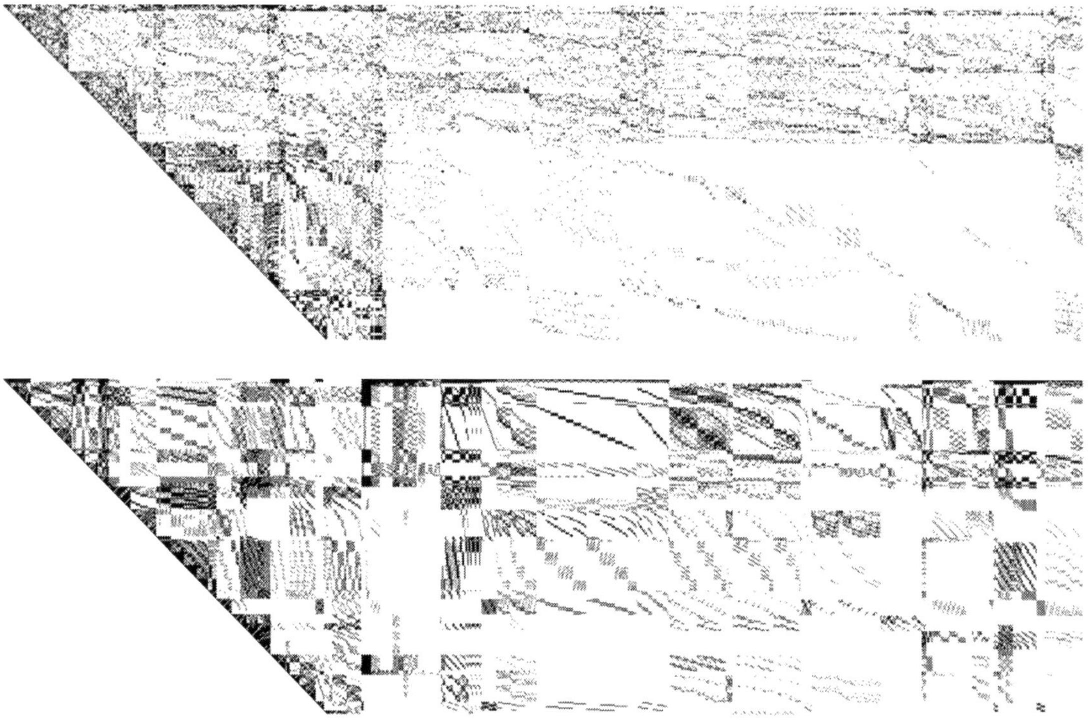

Figure 3. Sparsity structure for ^{8}Be at $N_{\max} = 4$ with 3NF on 120 processors: detailed structure of the upper-left (1000×300) submatrix on the first (diagonal) processor (each dot represents a single matrix element). Top: using round-robin distribution of individual M-scheme basis states (Version 13 of MFDn); Bottom: using round-robin distribution of groups of states (Version 14, under development).

determining which matrix elements are nonzero, while at the same time improving the cache performance. The main disadvantage is that the load-balancing is not as good as with the round-robin distribution of individual states. By changing the definition of a 'group of states' one can trade-off between more structure in the matrix but less load-balanced, and vice versa: smaller groups lead to better load-balancing but less structure, larger groups give less load-balancing but more structure. The 'ideal' definition of a group of states depends on a combination of the nucleus under consideration, the truncation and basis space, and the specific hardware for the calculation (available aggregate memory, memory per core, cache sizes, and memory bandwidth).

2.2. Input 3NF matrix elements

Another recent improvement concerns the input 3NF matrix elements. For the construction of the many-body matrix we need 2NF and 3NF matrix elements in M-scheme in a proton-neutron basis. These can be read in from file either in M-scheme or in a more compact representation known as the coupled-J (or coupled-JT) format [18]. The angular momenta j of the 2 (or 3) interacting nucleons are coupled to form 2-nucleon (or 3-nucleon) states of well-defined combined angular momentum J, using Clebsch–Gordan coefficients $(j_\alpha m_\alpha, j_\beta m_\beta | JM)$

$$|n_\alpha l_\alpha j_\alpha m_\alpha; n_\beta l_\beta j_\beta m_\beta\rangle = \sum_J (j_\alpha m_\alpha, j_\beta m_\beta | JM) \; |(n_\alpha l_\alpha j_\alpha, n_\beta l_\beta j_\beta)^J\rangle \,, \qquad (3)$$

with $M = m_\alpha + m_\beta$, and the sum runs over the allowed values $|j_\alpha - j_\beta| \leq J \leq j_\alpha + j_\beta$. Since the interactions are rotationally invariant, and thus preserve total angular momentum J, we only have nonzero matrix elements between 2-nucleon (or 3-nucleon) states of the same combined angular momentum J. (In addition to combining the angular momenta j of the interacting nucleons to a total angular momentum J, we can also combine the isospin $t = \frac{1}{2}$ of each nucleon to a total isospin of $T = 0$ or 1 in the 2NF case, or to $T = \frac{1}{2}$ or $\frac{3}{2}$ in the 3NF case.)

Thus for 2NFs we can either read in matrix elements between two nucleons in M-scheme

$$\langle n_\alpha l_\alpha j_\alpha m_\alpha ; n_\beta l_\beta j_\beta m_\beta | \, V \, | n_\gamma l_\gamma j_\gamma m_\gamma ; n_\delta l_\delta j_\delta m_\delta \rangle \ ,$$

with $m_\alpha + m_\beta = m_\gamma + m_\delta$, or we can read in matrix elements using the in coupled-J scheme

$$\left\langle (n_\alpha l_\alpha j_\alpha, n_\beta l_\beta j_\beta)^J \right| V \left| (n_\gamma l_\gamma j_\gamma, n_\delta l_\delta j_\delta)^J \right\rangle \ .$$

The latter is a more compact form and requires less memory, but when we construct the many-body matrix we need to decouple (or 'decompress') the coupled-J matrix elements. That is, for every nonzero matrix element needed for the construction of the many-body matrix, we need to perform a sum over the allowed values of J

$$\begin{aligned} \langle n_\alpha l_\alpha j_\alpha m_\alpha ; n_\beta l_\beta j_\beta m_\beta | V | n_\gamma l_\gamma j_\gamma m_\gamma ; n_\delta l_\delta j_\delta m_\delta \rangle \ = \\ \sum_J (j_\alpha m_\alpha, j_\beta m_\beta | JM) \ (j_\gamma m_\gamma, j_\delta m_\delta | JM) \ \left\langle (n_\alpha l_\alpha j_\alpha, n_\beta l_\beta j_\beta)^J \right| V \left| (n_\gamma l_\gamma j_\gamma, n_\delta l_\delta j_\delta)^J \right\rangle \ , \quad (4) \end{aligned}$$

with $|j_\alpha - j_\beta| \leq J \leq j_\alpha + j_\beta$, which involves either (repeatedly) calculating the corresponding Clebsch–Gordan coefficients, or (more efficient in terms of number of floating-point operations) use a look-up table for these Clebsch–Gordan coefficients.

For 3NFs we end up with triple sums over allowed J values. Schematically

$$\langle j_\alpha j_\beta j_\gamma | V | j_{\alpha'} j_{\beta'} j_{\gamma'} \rangle \ = \ \sum_{J_\delta, J_{\delta'}, J} \cdots \left\langle \left((j_\alpha, j_\beta)^{J_\delta}, j_\gamma \right)^J \right| V \left| \left((j_{\alpha'}, j_{\beta'})^{J_{\delta'}}, j_{\gamma'} \right)^J \right\rangle \quad (5)$$

with the dots representing a product of four Clebsch–Gordan coefficients. In principle, one could extend this further: For 4NFs one would get a quintuple sum involving a product of six Clebsch–Gordan coefficients.

Not surprisingly, the number of input matrix elements grows with the basis space. For p-shell nuclei ($N, Z \leq 8$), the number of 3NF input matrix elements in M-scheme in a proton-neutron basis is 741,823,056 at $N_{\rm max} = 6$, corresponding to an input file of 3 GB (using single-precision real numbers), and at $N_{\rm max} = 8$ the input file is 33 GB, containing over 8 billion input M-scheme matrix elements. Thus, up to $N_{\rm max} = 6$ we can work with the input data in M-scheme on current architectures, but at $N_{\rm max} = 8$ and above we quickly run out of memory. We can deal with such large input files by processing, for example, the input data for an $N_{\rm max} = 8$ run in eight to twelve 'chunks' of a few GB each, but this is not very efficient, as we demonstrate in the next section, and becomes prohibitively expensive above $N_{\rm max} = 8$.

In contrast, the number of 3NF input matrix elements in coupled-JT format is only 111,804,075 at $N_{\rm max} = 8$ for p-shell nuclei, that is, less than 0.5 GB. Even at $N_{\rm max} = 10$, the total memory required for the 3NF input in coupled-JT format is only about 2.4 GB [18], which is manageable on current multi-core (and future many-core) architectures. The price we pay for using the coupled-JT format is that we need to decompress the input data whenever we need to evaluate a matrix element, but this price may be tolerable since the trend in hardware is that threads for floating-point operations are becoming more abundant, whereas memory is expected to remain at a premium.

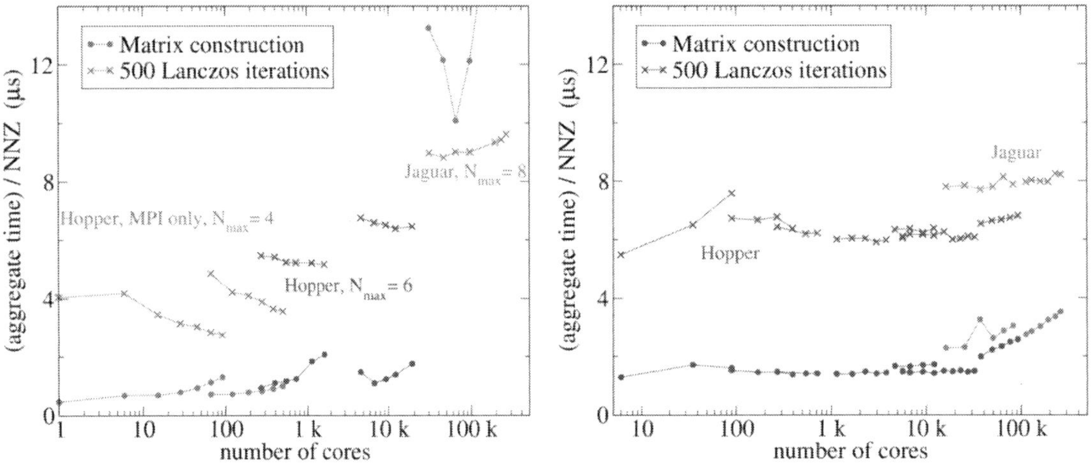

Figure 4. Scaling of MFDn Version 13 (left) and Version 14 (right) on Hopper at NERSC with MPI only (green symbols) and hybrid MPI/OpenMP (blue symbols), and on Jaguar at ORNL (red symbols) with two- plus three-body forces. (For technical reasons, we use the PGI compiler for Version 13 and the gnu compiler with Version 14, both with optimization.)

2.3. Strong and weak scaling

In figure 4 we show the strong and weak behavior of MFDn on Hopper at the National Energy Research Scientific Computing Center (NERSC) and on Jaguar at Oak Ridge National Laboratory (ORNL). Since the cost for the SpMV is largely proportional to the number of nonzero matrix elements, we show the aggregate time to construct the matrix (defined as wall clock time multiplied by the number of cores) divided by the number of nonzero matrix elements. Also the cost for the construction of the matrix is proportional to the number of nonzero matrix elements, once the sparsity structure is known (which is not included in the timing data of figure 4). Ideal scaling corresponds to a straight horizontal line in these plots; and the lower this line is, the better the performance. Symbols connected by a line are for the same nucleus and basis space, and thus indicate the strong scaling behavior.

Hopper is a Cray XE6 at NERSC consisting of 6,384 nodes with two 12-core AMD 'MagnyCours' processors per node and 32 GB memory per node. Each of the 24 cores have access to all of the 32 GB memory on the node, but the memory access time is not uniform: the compute nodes are so-called NUMA nodes (Non Uniform Memory Access). Each node has four memory banks of 8 GB each; 6 cores share a memory bank (as well as the L3 cache). Accessing memory from one of the three other memory banks is significantly slower than accessing local memory. For MFDn we therefore typically use four MPI processors per node, with six threads per MPI processor, on Hopper, though for relatively small runs our MPI-only implementation performs better. Jaguar was a Cray XK6 at ORNL, and consisted of 18,688 nodes with a 16-core AMD 'Interlagos' processor and 32 GB of memory per node. Again, each node is a NUMA node, with, on Jaguar, two NUMA domains and 8 cores per NUMA domain. On Jaguar we therefore typically ran with two MPI processors per node and 8 threads per MPI processor. Both Hopper and Jaguar use the Gemini interconnect between nodes. (Jaguar is currently being upgraded to Titan by adding a GPU to each node.)

In the left panel of figure 4 we see that for relatively small $N_{\max} = 4$ runs on Hopper using MPI only, the Lanczos iteration time actually decreases with increasing number of processors (see the green connected crosses): we have better than ideal strong scaling for these cases! It is not entirely clear what causes this behavior, but most likely it is due to a better cache performance as the local matrix blocks, as well as the local vector segments, on each processor get smaller. The minimal number of processors needed for a specific calculation is to a large extent dictated by the memory needed to store the matrix (distributed evenly over all processors). And since we perform both a SpMV and a transpose SpMV with the same compressed-row data structure for the matrix, our cache performance is far from ideal on the minimal number of processors. As we increase the number of processors for the same problem (strong scaling), the size of the matrix blocks on each processor decreases proportionally, and more of the in- and out-vectors of the SpMV can be kept in cache. Hence the cache performance during the Lanczos iterations is likely to improve as the number of processors increases for a fixed problem size.

For larger problem sizes using a hybrid MPI/OpenMP implementation (red and blue symbols), we do not see this effect. Most likely, the in- and out-vectors of the SpMV remain too large to fit in cache, even as we increase the number of processors: if we distribute the matrix such that we have four matrix blocks (one per NUMA node) on each node on Hopper, instead of 24 matrix blocks (one per core), then the dimensions of those blocks, and thus the length of the local vector segments, will be significantly larger. In addition, the memory needed for the local vector segments increases relative to the memory needed to store the local matrix block as both the problem size and the number of processors increase.

The scaling of the construction of the matrix (solid circles) is not as good as the scaling during the Lanczos iterations. This is likely to be caused by the fact that in Version 13 (left panel of figure 4) the time to distribute the input M-scheme matrix elements from the n processors on which they are read in to all $n(n+1)/2$ processors is included in the time to construct the matrix (the actual IO time is not included in figure 4). For $N_{\max} = 6$ calculations, with an M-scheme input file size of 3 GB, this is not too much of an issue, but for $N_{\max} = 8$ calculations, with an M-scheme input file size of 33 GB, we see that the matrix construction suddenly requires an order of magnitude more time than at $N_{\max} = 4$ and $N_{\max} = 6$ (red solid circles). This is partly due to the amount of input data to be redistributed (33 GB), and partly because we have to go through the construction of the matrix typically 8 to 12 times in order to process all input data.

The 3NF input matrix elements in coupled-JT format, as described in the previous subsection, resolves this issue of the size of the input file. Indeed, with Version 14 (right panel of figure 4) the construction of the matrix scales almost perfectly from 6 to 30,000 cores. The aggregate time is about twice as large as the most efficient $N_{\max} = 4$ and $N_{\max} = 6$ construction times using the M-scheme input matrix elements. Given the fact that we have to decouple the input matrix elements for every nonzero matrix element in the many-body matrix, see (5), this is actually a rather modest increase in aggregate time.

The scaling of the construction of the matrix breaks down above 40,000 cores, both on Hopper and Jaguar. Since there is no communication in this phase (at least not in Version 14 of MFDn), this has to be due to a load-imbalance. There are two sources of load imbalance: one is simply the number of nonzero matrix elements, and the other is the fact that for a subset of the nonzero matrix elements, we have to perform a single, double, or even triple sum over 'spectator' nucleons. In particular, all diagonal matrix elements involve such a triple sum. We are currently investigating this load imbalance problem.

Note that the Lanczos iterations with Version 14 show excellent strong and weak scaling over the entire range from 6 to 300,000 cores. Below 1,000 cores, the MPI-only implementation of Version 13 is more efficient, but for large runs, above 10,000 cores, Version 14 performs better, both in the construction of the matrix and in the Lanczos iterations, than Version 13. In addition, the time needed to determine the sparsity structure of the sparse matrix is typically, for large runs, an order of magnitude smaller with Version 14 than with Version 13, due to fact that Version 14 works with groups of M-scheme basis states, rather than individual states.

3. Convergence and renormalization of nuclear interactions

Even with all the recent improvements in code performance and efficiency, combined with the rapid increase in available computing power, convergence in a finite HO basis remains a challenge. Realistic 2NF potentials, such as Argonne V18 [1] and chiral N^3LO [2, 3], generate a combination of short-range and long-range correlations that cannot be represented in many-nucleon HO bases accessible at present computers. In order to account for the strong short-range (high-momenta) correlations and to improve convergence with increasing basis spaces, one can utilize a renormalization procedure specified by a unitary transformation of the Hamiltonian to a renormalized Hamiltonian

$$H_{\text{renormalized}} = U^\dagger \, H \, U \,, \tag{6}$$

that softens the interactions and generates effective operators for observables, while preserving all experimental quantities in the low-energy domain. The derived 'effective' interactions still act among all A nucleons and preserve all the symmetries of the initial 2NF and 3NF interactions.

One of these renormalization procedures is based on the Similarity Renormalization Group (SRG) [19, 20]. In the SRG approach the unitary transformation of the Hamiltonian is formulated in terms of a flow equation

$$\frac{dH_\alpha}{d\alpha} = [\eta_\alpha, H_\alpha] \,, \tag{7}$$

with a continuous 'SRG flow parameter' α, starting from $\alpha = 0$ fm^4 for the initial Hamiltonian $H_{\alpha=0}$. The physics of the SRG evolution is governed by the choice of the generator η_α. With a suitable choice of η_α, one can consistently evolve 2NFs and 3NFs (and even higher-body forces) to soften the short-range repulsion and tensor components of available initial interactions, so convergence of nuclear structure calculations is greatly accelerated [21, 22, 23, 24, 25, 26, 27]. The SRG evolution produces an 'effective' Hamiltonian that is independent of the nucleus and basis space, and generally enables smooth extrapolation of results [22, 27, 28].

A convenient and commonly used form in nuclear physics is

$$\eta_\alpha = m_N^2 \, [T_{\text{rel}}, H_\alpha] \,, \tag{8}$$

where T_{rel} is the kinetic energy operator in relative coordinates, and m_N is the nucleon mass. Thus the SRG evolution equation becomes

$$\frac{dH_\alpha}{d\alpha} = m_N^2 \left[[T_{\text{rel}}, H_\alpha], H_\alpha \right] . \tag{9}$$

This choice for the generator drives the renormalized Hamiltonian H_α towards the diagonal in a basis of eigenstates of the relative kinetic energy.

The SRG evolution generates many-body forces beyond those present in the initial Hamiltonian. The transformation is unitary only if all the induced terms up to the A-body level are included for an A-body calculation. Only in that case are the eigenvalues of the transformed Hamiltonian H_α identical to those of the initial Hamiltonian $H_{\alpha=0}$, independent of the SRG parameter α. These induced many-nucleon forces typically come in a hierarchy of decreasing strength with increasing number of interacting nucleons (at least for very small values of α). In practice one truncates the induced many-body terms either at the 2NF or at the 3NF level. A dependence of the obtained spectrum on α can therefore be attributed to the induced many-body forces that have been ignored in the calculation.

Calculations with the SRG evolved 2NF interactions, omitting induced 3NFs (as well as higher many-body forces) indeed show a significant dependence of the binding energies on the SRG flow parameter [22]. Most of this dependence is removed once the induced 3NF is properly taken into account [24, 25, 26, 27], at least for light nuclei and as long as the transformation is not pushed too far. With explicit 3NFs in the initial Hamiltonian however, there may be a need for induced 4-body forces [26].

4. Emergence of rotational band structures from ab initio calculations

Rotational bands arise quite naturally in α-cluster models for nuclei like ^{8}Be and ^{12}C, but it is not at all obvious that such structure also emerges from ab initio calculations, which describe these nuclei as bound states of eight and twelve nucleons respectively. The rotational energy for states in a rotational band with an axial symmetry is given by [29]

$$E_{\text{rotational}}(J) = \frac{\hbar^2}{2\mathcal{I}} \left(J(J+1) - K^2\right), \tag{10}$$

where $\mathcal{I}$ is the moment of inertia, K is the projection of J along the intrinsic symmetry axis, and J is the total angular momentum. Furthermore, the rotational model predicts for the quadrupole moments of members of a rotational band

$$Q(J) = \frac{3K^2 - J(J+1)}{(J+1)(2J+3)} Q_0, \tag{11}$$

and for the strength of the E2 transitions within a rotational band

$$B(E2; i \to f) = \frac{5}{16\pi} (e\,Q_0)^2 \,(J_i, K; 2, 0|J_f, K)^2, \tag{12}$$

where Q_0 is the 'intrinsic' quadrupole moment of the rotational band (in a body-fixed, non-inertial reference frame).

For even-even nuclei with a $J^\pi = 0^+$ ground state, we anticipate that we might have a $K = 0$ ground state rotational band. Rotational bands starting from $J = 0$ are restricted to even values of J only. Thus we expect a $J^\pi = 2^+$ and a 4^+ excited state, with a ratio of excitation energies $E_x(4)/E_x(2) = 10/3 = 3.33$.

Recent calculations using a realistic but nonlocal 2NF potential indicate that one does indeed obtain rotational band structures from ab initio nuclear structure calculations [30]. For the calculations presented here, we use the chiral N^{3}LO 2NF potential of [2] in combination with the chiral N^{2}LO 3NF potential of [5, 31]. For the two low-energy constants in the N^{2}LO 3NF potential we use the values that have been determined from the triton binding energy and β-decay [32]. In order to improve numerical convergence in a finite HO basis, we use the SRG renormalization procedure described in the previous section, and monitor the dependence of our results on the SRG parameter α.

4.1. Beryllium-8

Experimentally the two lowest excited states in ^{8}Be are a 2^+ and a 4^+ state, at 3.03 MeV and 11.35 MeV respectively, that is, with a ratio of excitation energies $E_x(4)/E_x(2) = 3.75$, somewhat larger than one would expect based on a rotational model.

We indeed find that the lowest two excited states are a 2^+ and a 4^+ state, as expected. With the chiral N^{3}LO 2NF potential, using SRG evolution including the induced 3NF, the excitation energies for these states are 3.2 MeV and 11.1 MeV respectively; see figure 5. We estimate these results to be converged to within about 5%, and they are virtually independent of the SRG parameter α. The ratio of the excitation energies is better converged, at $E_x(4)/E_x(2) = 3.45$, about halfway between the experimental value and that of a rotational model.

The effect of the chiral N^{2}LO 3NF on these excitation energies is rather small: including the initial 3NF we find 3.3 MeV and 11.5 MeV for the excitation energies of the 2^+ and 4^+ states respectively. The convergence of the excitation energies with the initial 3NF is similar to that with induced 3NFs only, and the ratio $E_x(4)/E_x(2)$ is converged to within a fraction of a percent at $N_{\text{max}} = 8$. For this ratio of the excitation energies we now find 3.48, only one percent larger than without initial 3NFs.

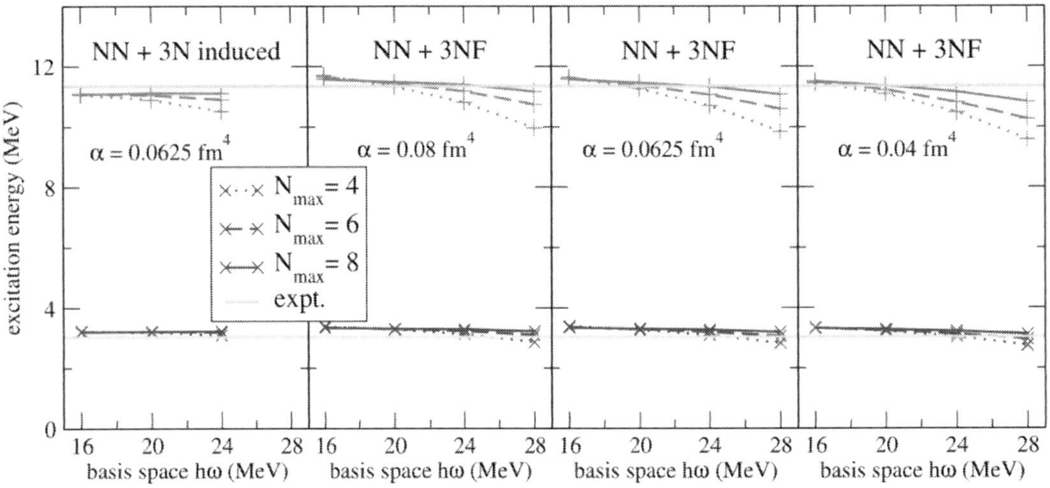

Figure 5. Excitation energies of the 2^+ (blue crosses) and 4^+ states (red plusses) for ^{8}Be with SRG evolved chiral N^{3}LO 2NF plus induced 3NF at $\alpha = 0.0625$ fm^4 (left-most panel) and with SRG evolved chiral N^{3}LO 2NF plus chiral N^{2}LO 3NF. Experimental values are indicated by the horizontal green lines.

The quadrupole moments and B(E2) transition strengths are not as well converged as the (excitation) energies. This is not surprising, since the E2 operator is a long-range operator, and therefore sensitive to the long-range behavior of the nuclear wavefunction. However, the long-range tail of the wavefunction is not well-represented in a HO basis: HO wavefunctions fall off like $\exp(-c\,r^2)$, whereas the nuclear wavefunction falls off like $\exp(-c\,r)$. Hence the convergence of observables like the quadrupole moments and B(E2)'s is rather slow in a HO basis.

Nevertheless, the ratios of the quadrupole moments seem to be in semi-quantitative agreement with the notion of these states being members of a rotational band. Although our results show a significant dependence on the basis space parameter $\hbar\omega$, see the middle panels of figure 6, the ratio $Q(4)/Q(2)$ is within about 10% of the rotational prediction of 1.27 at $N_{\max} = 8$. Note that the $\hbar\omega$ dependence is stronger with the initial chiral N^{2}LO 3NF than with the induced 3NFs only; and that the $\hbar\omega$ dependence is almost the same for the different values of α.

The B(E2) transition strength from the 2^+ to the ground state is in remarkable agreement with the rotational model, in terms of the quadrupole moment of the 2^+ state. Both without and with the initial chiral 3NFs the ratio $\mathrm{B(E2)}/Q(2)^2$ is very well converged and almost independent of both $N_{\max}$ and $\hbar\omega$, and very close to the rotational value of 0.243; see the lower panels of figure 6. On the other hand, for the transition from the 4^+ to the 2^+ state the ratio $\mathrm{B(E2)}/Q(4)^2$ is about 10% below what one would expect. Also note that this ratio appears to be somewhat better converged with initial chiral 3NFs than with the induced 3NFs only.

These results support the notion that the lowest 2^+ and the 4^+ excited states are, to a large extent, rotational excitations of the 0^+ ground state. Furthermore, the initial chiral 3NFs have little to no influence on the rotational nature of these excited states. Note that the total binding energy does depend on the 3NFs; also the question whether or not the ground state is bound with respect to breakup into two α particles with these interactions remains to be seen. At $N_{\max} = 8$ the binding energy of ^{8}Be is still below twice the binding energy of ^{4}He, calculated with the same interaction. Nevertheless, the excitation energies of lowest the 2^+ and the 4^+ states are very well converged, and in excellent agreement with the data.

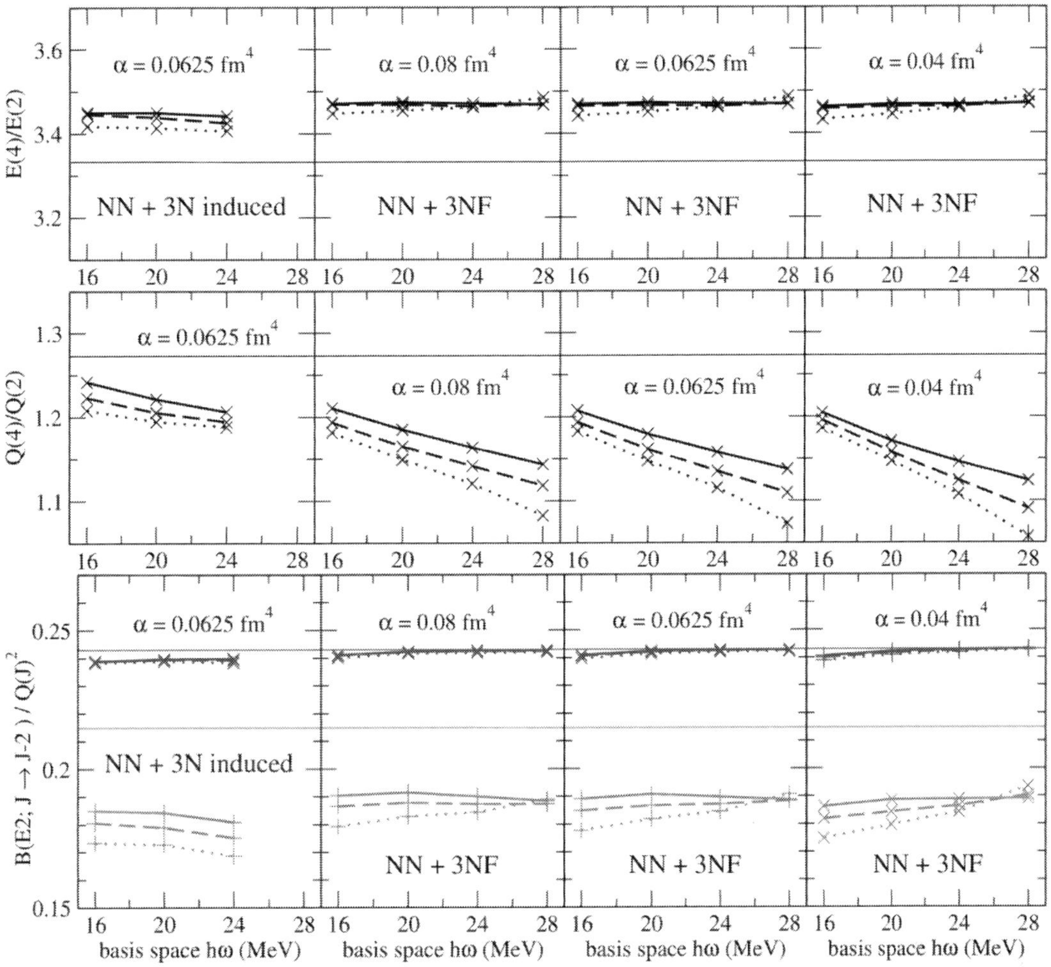

Figure 6. Ratios of results for ^{8}Be with SRG evolved chiral N^{3}LO 2NF plus induced 3NF at $\alpha = 0.0625$ fm^4 (left-most panel) and with SRG evolved chiral N^{3}LO 2NF plus chiral N^{2}LO 3NF. Horizontal lines represent the predictions from a rotational model. See figure 5 for the legend of the line styles. Top: excitation energies $E_x(4)/E_x(2)$; Middle: quadrupole moments $Q(4)/Q(2)$; Bottom: E2 transition strength B(E2)$/Q(J)^2$ for the $2^+ \to 0^+$ transition (blue crosses) and for the $4^+ \to 2^+$ transition (red plusses).

4.2. Carbon-12

Experimentally, the first excited state in ^{12}C is a 2^+ state, just as in ^{8}Be above, at an excitation energy of 4.44 MeV, and the lowest 4^+ state is at 14.08 MeV, so the ratio of excitation energies, $E_x(4)/E_x(2) = 3.17$, is quite close to what one would expect based on a rotational model for these states. A difference with ^{8}Be is that in ^{12}C there are several other excited states between the lowest 2^+ state and the lowest 4^+ state. Here, however, we only concentrate on these two excited states.

With the SRG evolved chiral N^{3}LO 2NF potential the lowest 2^+ and 4^+ states are very well converged [33], with excitation energies of 3.0 MeV and 10.7 MeV respectively; see the left panel of figure 7. Again, we estimate these results to be converged to within a few percent, and they are basically independent of the SRG parameter α. The ratio of the excitation energies is $E_x(4)/E_x(2) = 3.6$, about 8% above the rotational value, and more than 10% above the

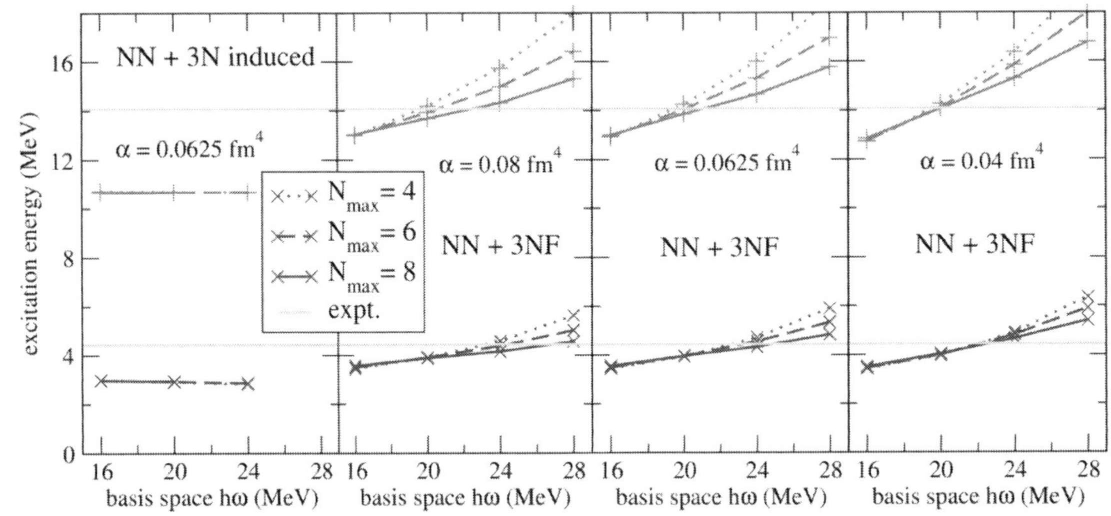

Figure 7. Excitation energies of the 2^+ (blue crosses) and 4^+ states (red plusses) for ^{12}C with SRG evolved chiral N^3LO 2NF plus induced 3NF at $\alpha = 0.0625$ fm^4 (left-most panel) and with SRG evolved chiral N^3LO 2NF plus chiral N^2LO 3NF. Experimental values are indicated by the horizontal green lines.

experimental value. Furthermore, the actual excitation energies are 25% to 30% below the data; also, there are no additional excited states below the lowest 4^+ state with the chiral N^3LO 2NF potential (at least no positive parity excited states), in contrast to the experimental situation.

However, in this case adding the chiral N^2LO 3NF to the N^3LO 2NF potential changes the results drastically; see the right three panels of figure 7. The initial chiral 3NF not only changes the excitation energies (and, for that matter, the spectrum in general), but it also spoils the excellent convergence of the excitation energies that we found with only the induced 3NFs. Though the weak dependence on $N_{\rm max}$ suggests that the excitation energies are almost converged at $\hbar\omega = 16$ MeV or $\hbar\omega = 20$ MeV, both the 2^+ and 4^+ states show a strong dependence on the basis space $\hbar\omega$ parameter: as $\hbar\omega$ increases from 16 MeV to 28 MeV, the excitation energies increase by 20% to 50%, depending on the state and on the SRG parameter α. It does mean that our results are now in better agreement with the data, but that is partly because of the rather large numerical uncertainty due to the lack of convergence. As expected, the convergence is somewhat better for the larger SRG value $\alpha = 0.08$ fm^4, but even at this SRG value we do not have converged results for the excitation energies. It appears that we need to go to larger bases ($N_{\rm max} = 10$ or even more) and/or larger values of α in order to obtain better converged results. With additional truncations on the basis space, such as the Importance-Truncated No-Core Shell Model [34], we can already perform calculations with 3NFs up to $N_{\rm max} = 12$ [26]. Alternatively, the Symmetry-Adapted No-Core Shell Model [35] may be used which aims to expand the range of accessible $N_{\rm max}$ values by implementing a truncation retaining those SU(3) configurations that are important for low-lying collective states.

The question remains: are these two states rotational excitations of the ground state? Without the initial chiral 3NF, the ratio of the excitation energies is about 3.6, that is, about 10% above the rotational value; see the top-left panel of figure 8. The ratio of the quadrupole moments, $Q(4)/Q(2)$ is very well converged at 1.3, only a few percent above the rotational value. Furthermore, the B(E2) transition strength for the transition from the 2^+ state to the ground state agrees to within 10% of a rotational model.

When we include the initial chiral N^2LO 3NF, the convergence patterns change, not only for

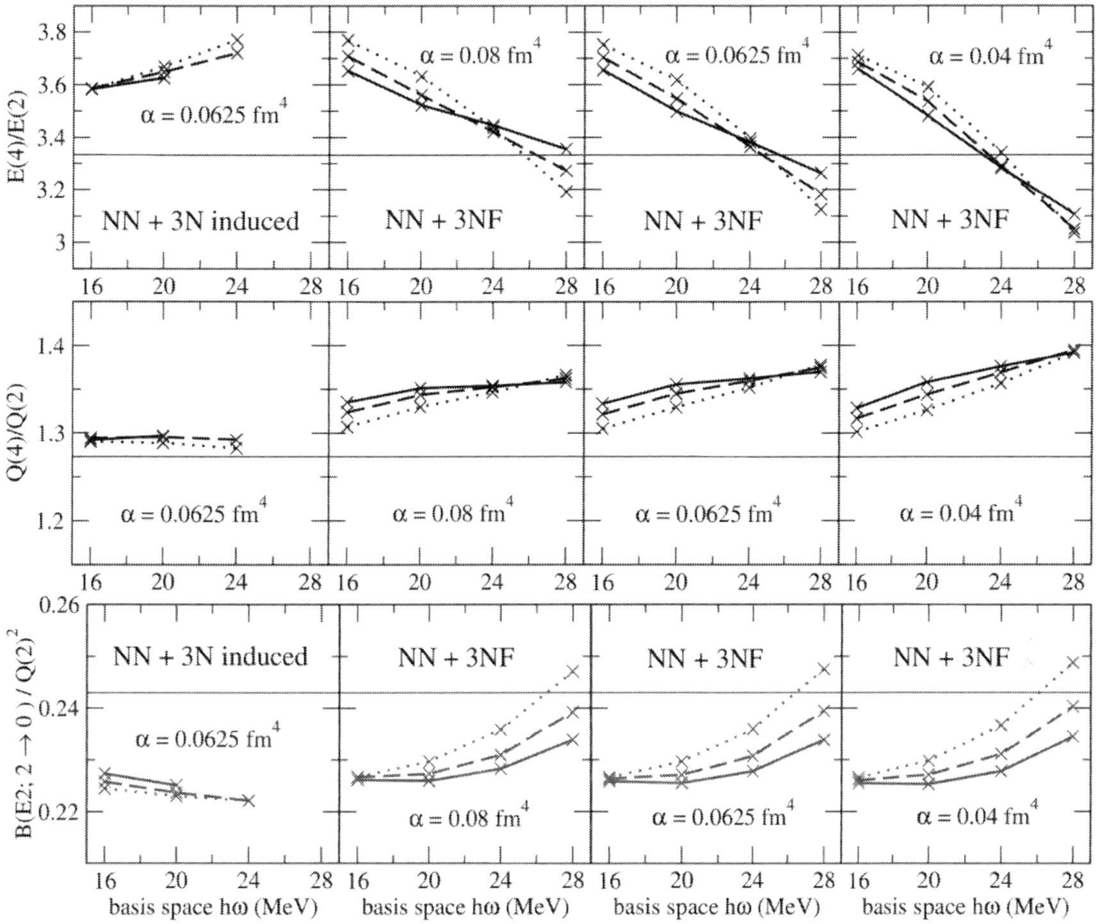

Figure 8. Ratios of results for ^{12}C with SRG evolved chiral N^{3}LO 2NF plus induced 3NF at $\alpha = 0.0625$ fm^4 (left-most panel) and with SRG evolved chiral N^{3}LO 2NF plus chiral N^{2}LO 3NF. Horizontal lines represent the predictions from a rotational model. See figure 7 for the legend of the line styles. Top: excitation energies $E_x(4)/E_x(2)$; Middle: quadrupole moments $Q(4)/Q(2)$; Bottom: E2 transition strength $\mathrm{B(E2)}/Q(2)^2$ for the $2^+ \to 0^+$ transition.

the excitation energies, but also for the other observables. Although the excitation energies of these two states increases by as much as 50% as $\hbar\omega$ increases from 16 MeV to 28 MeV, the ratio $E_x(4)/E_x(2)$ changes by at most 20% over this range in $\hbar\omega$, and at $\alpha = 0.08$ fm^4 this ratio seems to converge to a value of about 3.5 ± 0.1, which is only 5% above the rotational value. Note that this is quite similar to the result we found without initial 3NFs, but the $\hbar\omega$ dependence is very different with and without initial 3NFs. The ratios $Q(4)/Q(2)$ and $\mathrm{B(E2)}/Q(2)^2$ seem to converge to within about 10% of their rotational values; see figure 8. The convergence rate is slightly better at $\alpha = 0.08$ fm^4 than at $\alpha = 0.04$ fm^4, as one would expect.

Thus it seems that the rotational nature of the lowest 2^+ and the 4^+ states for ^{12}C emerges from ab initio calculations using chiral interactions. The chiral 3NFs do have a significant effect on the convergence rate of the excitation energies and on the spectrum of ^{12}C in general, as well as on other observables. However, the ratios of excitation energies, quadrupole moments, and $\mathrm{B(E2)}/Q(J)^2$ transition strength seem to converge to within 10% of their rotational values, both with the chiral N^{3}LO 2NF only and with the chiral N^{3}LO 2NF plus N^{2}LO 3NF.

Acknowledgments

This work was supported in part by U.S. Department of Energy Grants DE-FC02-09ER41582 (SciDAC-2/UNEDF), DESC0008485 (SciDAC-3/NUCLEI), DE-FG02-87ER40371, DE-FC02-06ER2775, by the U.S. NSF Grants 0904782, CNS-0643969, OCI-0904809, OCI-0904802, by U.S. Department of Energy Contract DE-AC02-05CH11231, by GAUSTEQ (Germany and U.S. Nuclear Theory Exchange Program for QCD Studies of Hadrons and Nuclei) under contract number DE-SC0006758, and by the DFG through SFB 634 and by HIC for FAIR. Computational resources were provided by the National Energy Research Scientific Computing Center (NERSC), which is supported by the DOE Office of Science, and by an INCITE award, "Nuclear Structure and Nuclear Reactions", from the DOE Office of Advanced Scientific Computing. This research also used resources of the Oak Ridge Leadership Computing Facility at ORNL, which is supported by the DOE Office of Science under Contract DE-AC05-00OR22725.

References

[1] Wiringa R B, Stoks V G J and Schiavilla R 1995 *Phys. Rev.* **C51** 38
[2] Entem D R and Machleidt R 2003 *Phys. Rev.* **C68** 041001
[3] Machleidt R and Entem D R 2011 *Phys. Rept.* **503** 1
[4] Pieper S C, Pandharipande V R, Wiringa R B and Carlson J 2001 *Phys. Rev.* **C64** 014001
[5] Epelbaum E, Nogga A, Gloeckle W, Kamada H, Meissner U G and Witala H 2002 *Phys. Rev.* **C66** 064001
[6] Epelbaum E 2006 *Prog. Part. Nucl. Phys.* **57** 654
[7] Nogga A, Navrátil P, Barrett B and Vary J P 2006 *Phys. Rev.* **C73** 064002
[8] Navrátil P, Gueorguiev V G, Vary J P, Ormand W E and Nogga A 2007 *Phys. Rev. Lett.* **99** 042501
[9] Navrátil P, Vary J P and Barrett B R 2000 *Phys. Rev. Lett.* **84** 5728
[10] Caprio M A, Maris P and Vary J P 2012 *Phys. Rev.* **C86** 034312
[11] Abe T, Maris P, Otsuka T, Shimizu N, Utsuno Y and Vary J P 2012 *Phys. Rev.* **C86** 054301
[12] Maris P 2012 *J. Phys. Conf. Ser.* **402** 012031
[13] Vary J P, Maris P, Ng E, Yang C and Sosonkina M 2009 *J. Phys. Conf. Ser.* **180** 012083
[14] Sternberg P, Ng E G, Yang C, Maris P, Vary J P, Sosonkina M and Le H V 2008 in *Proc. of the 2008 ACM/IEEE conf. on Supercomputing* SC '08 (Piscataway, NJ: IEEE Press) 15:1
[15] Maris P, Sosonkina M, Vary J P, Ng E G and Yang C 2010 *Procedia Computer Science* **1** 97
[16] Aktulga H M, Yang C, Ng E G, Maris P and Vary J P 2012 in *Lecture Notes in Computer Science* 7484 (Springer) 830
[17] Aktulga H M, Yang C, Ng E G, Maris P and Vary J P 2012 Improving the Scalability of a Symmetric Iterative Eigensolver for Multi-core Platforms, submitted to *Concurrency and Computation: Practice and Experience*
[18] Roth R, Calci A, Langhammer J and Binder S 2013 (in preparation)
[19] Glazek S D and Wilson K G 1993 *Phys. Rev.* **D48** 5863
[20] Wegner F J 2001 *Physics Reports* **348** 77
[21] Bogner S, Furnstahl R and Perry R 2007 *Phys. Rev.* **C75** 061001
[22] Bogner S K *et al.* 2008 *Nucl. Phys.* **A801** 21
[23] Roth R, Neff T and Feldmeier H 2010 *Prog. Part. Nucl. Phys.* **65** 50
[24] Jurgenson E D, Navrátil P and Furnstahl R J 2009 *Phys. Rev. Lett.* **103** 082501
[25] Jurgenson E D, Navrátil P and Furnstahl R J 2011 *Phys. Rev.* **C83** 034301
[26] Roth R, Langhammer J, Calci A, Binder S and Navrátil P 2011 *Phys. Rev. Lett.* **107** 072501
[27] Jurgenson E D, Maris P, Furnstahl R J, Navrátil P, Ormand W E and Vary J P 2013 *Phys. Rev.* **C87** 054312
[28] Furnstahl R, Hagen G and Papenbrock T 2012 *Phys. Rev.* **C86** 031301
[29] Rowe D J 1970 *Nuclear Collective Motion* (London: Methuen, London)
[30] Caprio M A, Maris P and Vary J P 2013 *Phys. Lett. B* **719** 179
[31] Navrátil P 2007 *Few Body Syst.* **41** 117
[32] Gazit D, Quaglioni S and Navrátil P 2009 *Phys. Rev. Lett.* **103** 102502
[33] Maris P *et al* 2012 *J. Phys. Conf. Ser.* **403** 012019
[34] Roth R 2009 *Phys. Rev.* **C79** 064324
[35] Dytrych T *et al* 2013 (in preparation)

Journal of Physics: Conference Series **454** (2013) 012064 doi:10.1088/1742-6596/454/1/012064

A multiprecision C++ library for matrix-product-state simulation of quantum computing: Evaluation of numerical errors

Akira SaiToh

Quantum Information Science Theory Group, National Institute of Informatics, 2-1-2 Hitotsubashi, Chiyoda, Tokyo 101-8430, Japan

E-mail: `akirasaitoh@nii.ac.jp`

Abstract. The time-dependent matrix-product-state (TDMPS) simulation method has been used for numerically simulating quantum computing for a decade. We introduce our C++ library ZKCM_QC developed for multiprecision TDMPS simulations of quantum circuits. Besides its practical usability, the library is useful for evaluation of the method itself. With the library, we can capture two types of numerical errors in the TDMPS simulations: one due to rounding errors caused by the shortage in mantissa portions of floating-point numbers; the other due to truncations of nonnegligible Schmidt coefficients and their corresponding Schmidt vectors. We numerically analyze these errors in TDMPS simulations of quantum computing.

1. Introduction

Simulations of time-evolving quantum states using a matrix-product-state (MPS) [1] representation have been widely used in a variety of physical systems [2, 3]. We have been developing a C++ library, named ZKCM_QC [4], for multiprecision time-dependent MPS (TDMPS) simulations of quantum computing using Vidal's representation [5] for MPS. Here, we define the precision by the length of a mantissa portion for each floating point number.

Let us begin with a brief description of conventions. We employ the computational basis $\{|0\rangle, |1\rangle\}^n$ for n-qubit quantum states, where $|0\rangle = \begin{pmatrix} 1 & 0 \end{pmatrix}^{\mathrm{T}}$ and $|1\rangle = \begin{pmatrix} 0 & 1 \end{pmatrix}^{\mathrm{T}}$. An n-qubit quantum state is represented as $|\Psi\rangle = \sum_{i_0\cdots i_{n-1}=0\cdots 0}^{1\cdots 1} c_{i_0\cdots i_{n-1}} |i_0\cdots i_{n-1}\rangle$ with complex amplitudes $c_{i_0\cdots i_{n-1}}$. In this paper, we employ the following MPS form [5, 6] of the state, for our TDMPS simulations.

$$|\Psi\rangle = \sum_{i_0\cdots i_{n-1}=0\cdots 0}^{1\cdots 1}\Bigg[\sum_{v_0=0}^{m_0-1}\sum_{v_1=0}^{m_1-1}\cdots\sum_{v_{n-2}=0}^{m_{n-2}-1} Q_0(i_0,v_0)V_0(v_0)Q_1(i_1,v_0,v_1)V_1(v_1)\cdots$$
$$Q_s(i_s,v_{s-1},v_s)V_s(v_s)\cdots V_{n-2}(v_{n-2})Q_{n-1}(i_{n-1},v_{n-2})\Bigg]|i_0\cdots i_{n-1}\rangle, \tag{1}$$

where we use tensors $\{Q_s\}_{s=0}^{n-1}$ with parameters i_s, v_{s-1}, v_s (v_{-1} and v_{n-1} are excluded) and $\{V_s\}_{s=0}^{n-2}$ with parameter v_s; m_s is a suitable number of values for v_s, with which the state is represented precisely or well approximated. [Tensor $V_s(v_s)$ stores the Schmidt coefficients for the splitting between the sth site and the $(s+1)$th site.]

With the MPS form, the cost to update data in accordance with a time evolution is reduced considerably in comparison to the brute-force method. We have only to update tensors corresponding to the sites under the influence of a unitary operator for each step. The cost to simulate an evolution by a single unitary operation $\in \mathrm{U}(4)$ acting on some consecutive sites is $O(m_{\max}^3)$ floating-point operations where $m_{\max}$ is the largest value of m_s among the sites s [5]. The total cost of a TDMPS simulation thus grows polynomially in the largest Schmidt rank among those for the splittings, which highly depends on the instance of the problem of one's concern. (Here, each splitting we consider is, of course, a bipartite splitting between consecutive sites.)

The TDMPS simulation method in double precision [*i.e.*, (52+1)-bits-long mantissa for a floating point number] has already been used commonly in the community of computational physics [7]. There are, however, known cases for general computational methods where more accurate computation is needed to obtain a reliable results describing physical phenomena [8]. Accumulations of rounding errors of basic arithmetic operations are the main cause in such cases. This should be true also in TDMPS simulations of quantum computing where many matrix diagonalizations are involved unless the depth of quantum circuits is very small. Besides the precision of basic operations, another factor of losing accuracy is the truncation of Schmidt coefficients. When many of Schmidt coefficients are nonnegligible at a certain step of a TDMPS simulation, imposing a threshold to the number of Schmidt coefficients for each splitting may truncate out important data affecting simulation results [9]. So far truncations of nonvanishing Schmidt coefficients have been uncommon[1] and the largest Schmidt rank and its upper bound in the absence of truncations have been of main concern when MPS and related data structures are used for handling quantum and/or classical computational problems [5, 11, 12, 6, 13, 14, 15, 16, 17]. This is in contrast to MPS and TDMPS simulations in condensed matter physics where truncations are very commonly employed [18].

In this report, we evaluate numerical errors in actual TDMPS simulations of quantum computing. We first begin with a brief introduction to the ZKCM_QC library in section 2. Then we conduct numerical evaluations in section 3: an error due to precision shortage is investigated in section 3.1 and that due to truncations of Schmidt coefficients is investigated in section 3.2. We discuss and summarize obtained results in sections 4 and 5, respectively.

2. ZKCM_QC library

The ZKCM_QC library has been developed as a C++ library for TDMPS simulations of quantum computing with an emphasis on multiprecision computation. It is an extension package of the ZKCM library [19] which uses the GMP [20] and MPFR [21] libraries for basic arithmetic operations. We briefly explain the usage of the ZKCM_QC library.

The library can be installed by the standard process: "`./configure`", "`make`", and "`make install`" in any Unix-like system with GNU tools. Once it is installed, a user will write a C++ program using the header file "`zkcm_qc.hpp`" and compile the program with the library flags "`-lzkcm_qc -lzkcm -lmpfr -lgmp -lgmpxx -lm`" using a C++ compiler.

In the main part of a program, a user will firstly specify the default precision. For example "`zkcm_set_default_prec(256);`" will set the default precision to 256 bits. Then, a user will call a constructor of the class "`mps`" to create the object of an MPS kept in the form of (1). For example, "`mps M(5);`" will generate an object keeping the data of $|0_0 0_1 0_2 0_3 0_4\rangle$ in the MPS form. Then a user will apply certain unitary operations by using some member functions of the class, such as "`applyU`" and "`applyU8`" with predefined and/or user-defined unitary matrices. For example, "`M.applyU(tensor2tools::Hadamard, 2);`" applies an Hadamard gate $H =$

[1] We are discussing on the standard circuit model of quantum computing. It was reported [10] that truncations of nonzero Schmidt coefficients were employed in a TDMPS simulation in a Hamiltonian model for adiabatic quantum computing.

$\begin{pmatrix} 1 & 1 \\ 1 & -1 \end{pmatrix}/\sqrt{2}$ to qubit 2. For another example, "`M.applyU(tensor2tools::CNOT, 2, 4);`" applies the CNOT gate to qubits 2 and 4 that are the control bit and the target bit, respectively. Here, CNOT is a conditional bit flip; the target bit is flipped when the control bit is in $|1\rangle$. There are other member functions for unitary operations. For instance, "`CCNOT`" is used for a CCNOT operation; "`M.CCNOT(0, 2, 4);`" applies a bit flip on qubit 4 under the condition that qubits 0 and 2 are both in $|1\rangle$. In addition, it is possible to simulate a projective measurement by calling the member function "`pmeasure`". For example, "`int event = M.pmeasure(zkcm_matrix("[1,0;0,-1]"), 2, -1);`" will simulate a projective measurement on qubit 2 with the observable Pauli Z, as a random process. The returned value "`event`" will be 0 when the event corresponding to the larger eigenvalue of Z occurs and will be 1 otherwise. There are other useful functions documented in the reference manual of the library. Among them, functions to achieve a reduced density matrix of specified qubits will be frequently used. One example to use such a function is "`std::string s = M.RDO_block(2, 4).str_dirac_b();`" which obtains the reduced density matrix of the block of qubits 2, 3, and 4 as a "`std::string`"-type string.

For each call of unitary operations and/or projective measurements, we update involved tensors of the MPS. This is a tedious process, but is concealed by the library. User programs usually do not pay attention to the background simulation process. One may still set a threshold $m_{\rm trunc}$ on the number of Schmidt coefficients of a bipartite splitting at one's risk. A user will write, *e.g.*, "`M.m_trunc(12);`" to specify the threshold (it is set to 12 for this example, *i.e.*, only largest 12 Schmidt coefficients will be kept at each time tensors are updated during the TDMPS simulation).

More detailed instructions on the installation and the usage of the ZKCM_QC library are found in the documents placed in the "`doc`" directory of the package.

3. Numerical errors in TDMPS simulations of quantum computing

In this section, we firstly investigate the influence of precision shortage in TDMPS simulations of a simple quantum search [22] and a quantum Fourier transform [23]; we secondly investigate the influence of the truncation of Schmidt coefficients in a TDMPS simulation of the Deutsch-Jozsa algorithm [24] for a simple boolean function.

3.1. Errors due to the precision shortage

A TDMPS simulation is in fact sensitive to the precision of floating-point computation. Here, we show typical examples.

3.1.1. Example 1 For the first example, we consider Grover's quantum search [22] (or quantum amplitude amplification) for a simple oracle[2] with a three-qubit input. The initial state is set to $|s\rangle = \frac{1}{\sqrt{8}}\sum_{x_0x_1x_2=000}^{111}|x_0x_1x_2\rangle$ and the target state is set to $|101\rangle$. We successively apply the so-called Grover routine[3] $R = (1-2|s\rangle\langle s|)(1-2|101\rangle\langle 101|)$ to the state starting from $|s\rangle$. This results in an oscillation of the population of $|101\rangle$, $\mathrm{Prob}_{101}(t) = |\langle 101|R^t|s\rangle|^2$, as shown in figure 1 ($t = 0, 1, 2, \ldots$). For the TDMPS simulation, we used five qubits including a single oracle qubit and a single ancilla qubit. The oracle circuit was simply constructed by three CCNOT gates and two NOT gates. This is just for simplified demonstration; a meaningful quantum search

[2] A TDMPS simulation of the Grover search for a simple oracle was first performed by Kawaguchi *et al.* [11] in 2004.

[3] It is in general $(1-2|s\rangle\langle s|)(1-2\sum_{x\in X}|x\rangle\langle x|)$ with $|s\rangle$ the equally-weighted superposition of states in a parent set and X the set of target states. It is highly dependent on a problem instance how this operation is constructed as a quantum circuit.

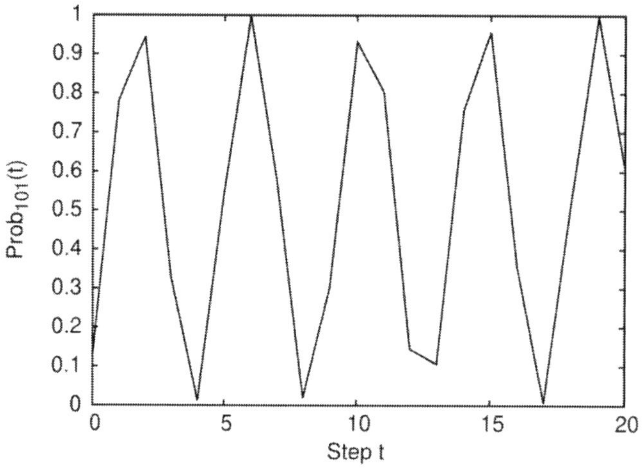

Figure 1. Theoretical graph of the oscillation of $\mathrm{Prob}_{101}(t)$ in the three-qubit quantum search described in the text.

should be performed for a realistic computational problem like a satisfiability problem [25] with a large input size. This will be hopefully realized in future with a real quantum computer. The quantum circuit of the total process of our present concern is quite simple as illustrated in figure 2. In addition, here, truncations of nonzero Schmidt coefficients were not employed.

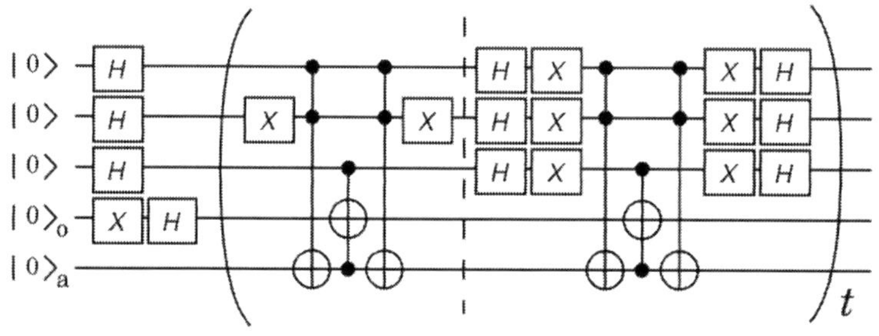

Figure 2. Quantum circuit of the three-qubit quantum search described in the text. The circuit in the parentheses illustrates R, in which the left half corresponds to $(1 - 2|101\rangle\langle 101|)$ and the right half corresponds to $(1 - 2|s\rangle\langle s|)$. Symbol "o" stands for the oracle qubit and "a" stands for the ancilla qubit.

We plot the computational error $|\widetilde{\mathrm{Prob}}_{101}(20) - \mathrm{Prob}_{101}(20)|$ after 20 iterations of the Grover routine against the precision in figure 3, where $\widetilde{\mathrm{Prob}}_{101}(20)$ is a computed value and $\mathrm{Prob}_{101}(20) = |\langle 101|R^{20}|s\rangle|^2 = \frac{5327874951961}{8796093022208}$ is the exact value calculated symbolically by the Maxima system [26] with the Qcomp.mac package [27]. It is manifest that the double precision (53 bits) is not enough and more than 70 bits precision is preferable for an accurate simulation. In addition, it is interesting that the computational error does not look like a simple elementary function of the precision; there is a sudden drop at a certain value of the precision.

We next evaluate the computational cost by increasing the precision of the TDMPS simulation. The real CPU time used for simulating the above Grover search is plotted against the precision in figure 4. In the figure, we employed a polynomial for coarse fitting, since the

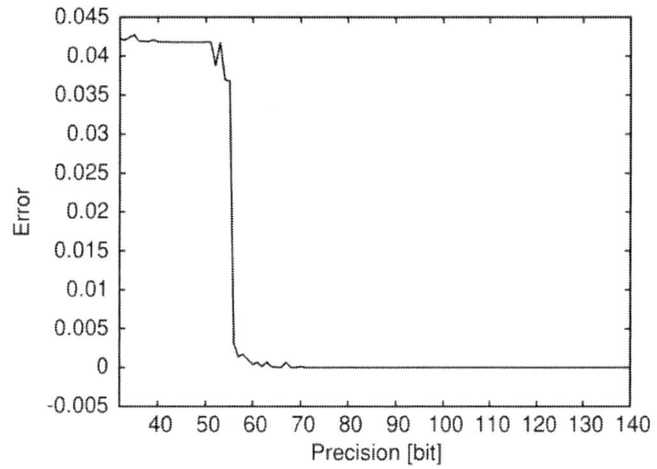

Figure 3. Computational error $|\widetilde{\mathrm{Prob}}_{101}(20) - \mathrm{Prob}_{101}(20)|$ plotted against the precision.

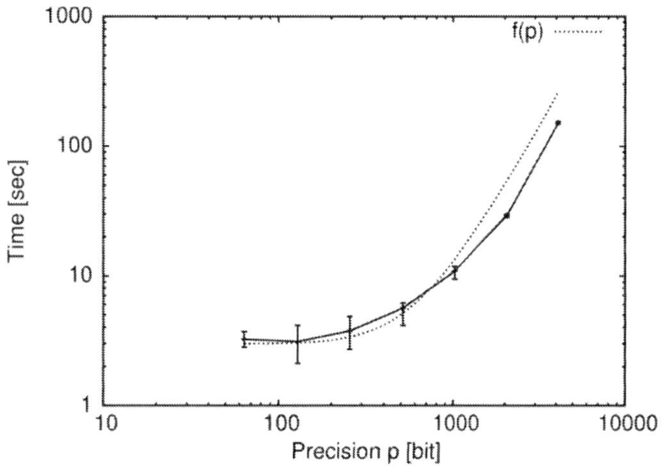

Figure 4. Computation time to simulate 20 iterations of the Grover routine, plotted against the precision. The average was taken over 10 trials for each data point. Error bars span from the minimum to the maximum. Here, $f(p) = 1.5 \times 10^{-9}(p - 64)^3 + 1.0 \times 10^{-5}(p - 64)^2 + 3.0$. Environment: Red Hat Enterprise Linux 6 on Intel Xeon X7542 CPU 2.67GHz, 132GB memory.

computational cost of a TDMPS simulation of a quantum algorithm is guaranteed to be some polynomial in the floating-point precision p (in bits) because each basic arithmetic operation is performed within polynomial time. It can be exponential in the input size of a problem instance, but this is not of our present interest. One might expect that the cost grows quasi-linearly in p considering the cost of each floating-point multiplication. However, the value of p does effect on the number of inverse iterations to guarantee enough accuracy in each matrix diagonalization. In this sense, the fitting seems plausible. The maximum Schmidt rank of the MPS at the time steps between basic quantum gates was 4 during the above simulation. One may find it smaller than expected. This is because operations $\in$ U(8) are simulated without decomposition in terms of U(4) and U(2) operations, in ZKCM_QC.

3.1.2. Example 2 The sudden drop of the numerical error found in figure 3 is not specific to the example. We will now find a similar behavior in another example. Here, we consider an n-qubit quantum circuit shown in figure 5. A Bell state $|\phi\rangle_{0,n-1} = (|0\rangle_0|0\rangle_{n-1} + |1\rangle_0|1\rangle_{n-1})/\sqrt{2}$ of qubits 0 and $n-1$ is initially created and then the state $|\phi\rangle_{0,n-1}|0\cdots0\rangle_{1,\dots,n-2}$ goes through the quantum Fourier transform (QFT) and its inverse. Thus, the resultant state of the qubits 0 and $n-1$ should be the initial Bell state. Here, we employ Fowler *et al.*'s linear-nearest-neighbour construction for QFT [28] exactly as it is.

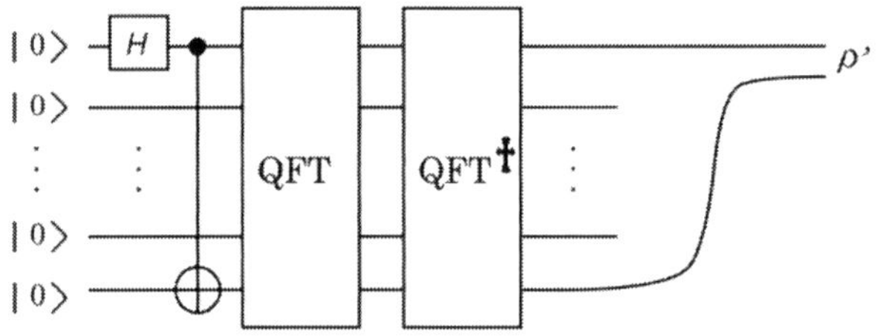

Figure 5. Quantum circuit with QFT and its inverse (QFT†). This circuit has no practical meaning but to examine a numerical error caused by precision shortage in a TDMPS simulation.

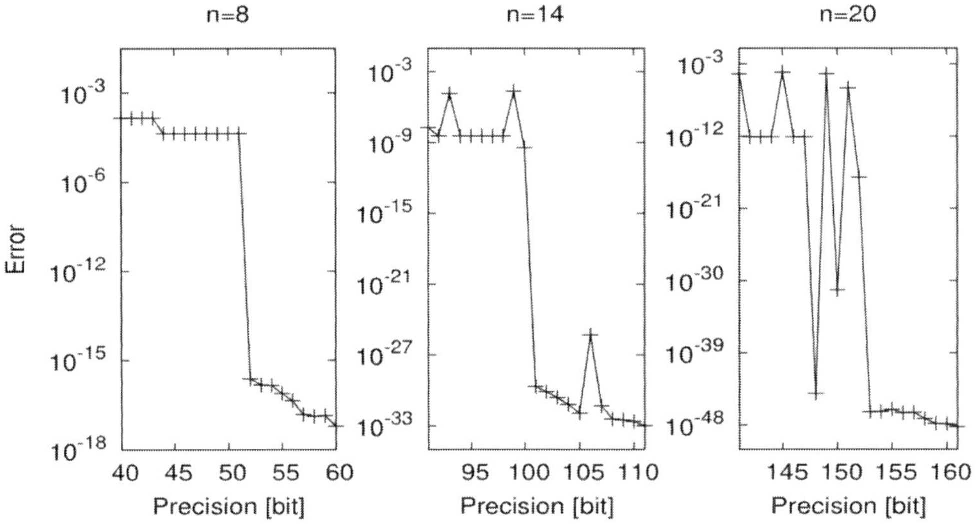

Figure 6. Precision dependence of the numerical error $|\langle 00|\widetilde{\rho'}|11\rangle - 1/2|$ for the three values of n.

We performed TDMPS simulations of the circuit for $n = 8$, 14, and 20 and examined the precision dependence of the numerical error in the $|00\rangle\langle 11|$ element of the reduced density matrix ρ' of the qubits 0 and $n-1$ in the resultant state.[4] Truncations of nonzero Schmidt coefficients were not employed. Ideally, we should have $\langle 00|\rho'|11\rangle = 1/2$. Let $\widetilde{\rho'}$ be the computed matrix;

[4] It took approximately 1.88 seconds for each run of the simulation to obtain $\widetilde{\rho'}$ when $n = 20$ and precision was set to 150 bits, in the environment: Red Hat Enterprise Linux 6 on Intel Xeon X7542 CPU 2.67GHz, 132GB memory. This fast simulation is not surprising since QFT maps a computational basis state to a product state (see, *e.g.*, reference [29]).

the numerical error is quantified as $|\langle 00|\widetilde{\rho'}|11\rangle - 1/2|$. We found that the sudden drop of the error appeared for all of the three values of n as shown in figure 6 (Note that the vertical axes are in the logarithmic scale). The figure also indicates that the precision required for avoiding an obvious numerical error tends to grow as the system size grows.[5]

So far we have seen two examples, which have clarified the importance of high-precision computation for accurate TDMPS simulations of quantum computing. We will next examine the truncation error of TDMPS.

3.2. Errors due to truncations

It is uncommon to make use of a truncation of nonzero Schmidt coefficients in TDMPS simulations of quantum computing unlike those used for condensed matter physics as we mentioned in section 1. We however investigate the case we dare to impose truncations. Let us denote the limit on the number of Schmidt coefficients that can be kept for each splitting as $m_{\rm trunc}$. Thus, among Schmidt coefficients, larger $m_{\rm trunc}$ coefficients are kept and the remaining small coefficients are truncated out at each step in the simulation of a quantum circuit.

Here we study the effect of a truncation in a TDMPS simulation of the Deutsch-Jozsa algorithm [24] for a function $f : \{0,1\}^9 \to \{0,1\}$ with a nine-bit input $\mathbf{x}$. Recall that f is either balanced (*i.e.*, $\sharp\{\mathbf{x}|f(\mathbf{x}) = 0\} = \sharp\{\mathbf{x}|f(\mathbf{x}) = 1\}$) or constant (*i.e.*, $f(\mathbf{x})$ is same for all $\mathbf{x}$) by the promise of the problem definition for the algorithm (see, *e.g.*, section 3.1.2 of [30]). The Deutsch-Jozsa algorithm for nine qubits is interpreted as the following process. (i) Apply $H^{\otimes 9} V_f H^{\otimes 9}$ to $|0_0 \ldots 0_8\rangle$ (here, V_f is an operation to put the factor $(-1)^{f(\mathbf{x})}$ to the states $|\mathbf{x}\rangle$); (ii) Measure the nine qubits of the resultant state. When f is balanced, the probability of having the nine qubits in $|0\rangle$'s simultaneously in the resultant state vanishes; when f is constant, the probability is exactly unity.

We employ the following function for our TDMPS simulation. $f(\mathbf{x}) = g(x_0x_1x_2) \oplus g(x_3x_4x_5) \oplus g(x_6x_7x_8)$ with $g(\mathbf{y}) = [(\neg y_0) \wedge (\neg y_1)] \vee (y_1 \wedge y_2)$. This function is balanced. In addition, more specifically, we have $\mathrm{Prob}(0_0 0_1 0_2) = 0$ in the resultant state of (i) for this particular function. The quantum circuit of the Deutsch-Jozsa algorithm for this function is illustrated in figure 7. We set the floating-point precision to 256 bits and tried several different values of $m_{\rm trunc}$ in the TDMPS simulation of the quantum circuit.[6] As shown in figure 8, the error in the computed value of $\mathrm{Prob}(0_0 0_1 0_2)$ (namely, a discrepancy from zero in the present context) vanishes for $m_{\rm trunc} \geq 12$ while a considerable error exists for $m_{\rm trunc} \leq 11$. The largest value among m_s's during the simulation is shown as $m_{\rm max}$ in the figure. It indicates that $m_{\rm trunc}$ should be set to at least the maximum possible value of $m_{\rm max}$ which is 12 in the present case, so as to avoid a nonnegligible error.

The above result shows that we cannot truncate out any nonzero Schmidt coefficient during the simulation. This phenomenon should be related to the distribution of Schmidt coefficients if discussions of reference [9] apply to the present case. We show the distribution at the points we had the maximum Schmidt rank 12 in the simulation, in figure 9 (this is same for both of the points). These points were the second CNOT gate from the left and one from the right among the seven CNOT gates in the middle part of circuit (a) of figure 7. It is now manifest that none of the twelve Schmidt coefficients is negligible. This is how even a single truncation caused a significant error.

[5] This is a rare example which exhibits such a clear size dependence of the required precision, among many quantum circuits so far as we tried. It is important to notice the existence of such an example; it indicates that one may perhaps encounter a *hard instance* for which a significantly high precision is mandatory for a relatively large system size.

[6] It took less than 10.2 seconds for each run of the simulation in the aforementioned environment.

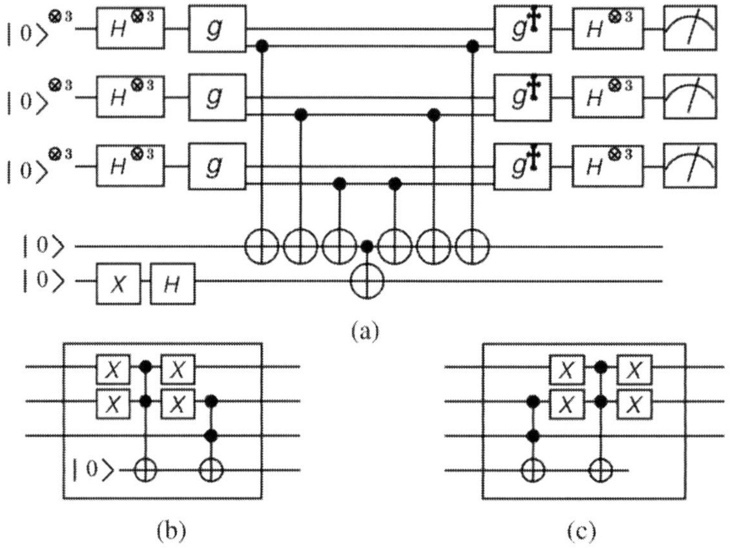

Figure 7. Quantum circuit of the Deutsch-Jozsa algorithm for the function f. Circuits (b) and (c) describe the gates g and $g^\dagger$ of circuit (a), respectively.

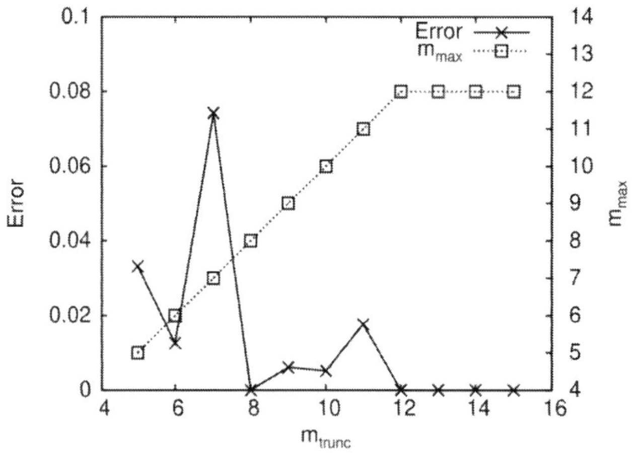

Figure 8. Plot of the error in the computed value of $\mathrm{Prob}(0_0 0_1 0_2)$ (crosses) and the plot of m_{max} (boxes) as functions of m_{trunc}. See the text for the details of the simulation.

4. Discussion

The TDMPS method is regarded as an approximation method in condensed matter physics, but is usually an exact method in simulation studies of quantum computing. Truncations of nonzero small Schmidt coefficients are thus unemployed, usually, for the purpose of simulating quantum computation. In a known case in condensed matter physics [9] where we should avoid truncations, a dominant number of Schmidt coefficients are nonnegligible. This should be true also in TDMPS simulations of quantum circuits. In fact, we found that this is the case and the threshold m_{trunc} for the largest number of surviving Schmidt coefficients should be at least

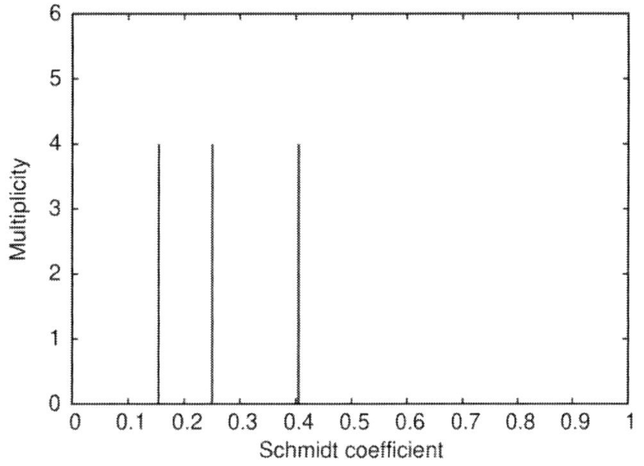

Figure 9. Distribution of Schmidt coefficients for the splitting at the points we had the maximum Schmidt rank in the simulation. The distribution was same for both of the points.

the maximum possible number of nonzero Schmidt coefficients, in the simulation results for the Deutsch-Jozsa algorithm presented in section 3.2.

Besides, we have also evaluated the effect of precision shortage. Rounding errors of individual basic arithmetic operations may accumulate to a large error in the resultant state of a TDMPS simulation of a quantum circuit. In our simulation of a simple quantum search presented in section 3.1.1, a nonnegligible error was observed in the resultant population of a target state unless we increased the precision beyond the double precision. Furthermore, we found that the error does not behave like a smooth curve as a function of the precision in bits, but behaves like a step function with a sudden drop, as shown in figure 3. This phenomenon was also observed in the TDMPS simulation of a quantum circuit containing the quantum Fourier transform, for three different system sizes, as presented in section 3.1.2. In addition, this is not a particular phenomenon for a TDMPS simulation but a rather commonly observed one for matrix computation (see the document of the ZKCM library [19]).

With our simulation results for investigating numerical errors, it is suggested that the precision should be preferably at least 70 bits and the truncation of nonzero Schmidt coefficients is not encouraged for a TDMPS simulation of quantum computation. As for the cost of multiprecision computation, figure 4 indicates that more than 1000 bits precision is quite expensive. As a matter of fact, the current computer architecture does not have a good hardware support for high precision computation. Thus the practical precision we may employ should be several hundreds bits for the time being.

5. Conclusion

We have utilized our multiprecision TDMPS library ZKCM_QC to investigate numerical errors in TDMPS simulations of quantum computing. To avoid a nonnegligible rounding error within a practical cost, it is suggested that the length of the mantissa portion of each floating-point number should be beyond the double-precision length but not more than 1000 bits in the present technology. It is also suggested that a truncation of nonzero Schmidt coefficients is discouraged.

6. Software information

The simulations were performed with ZKCM_QC ver. 0.0.9beta put on the repository "`https://sourceforge.net/p/zkcm/sublibqc`".

References

[1] White S R 1993 *Phys. Rev. B* **48** 10345
[2] Schollwöck U 2005 *Rev. Mod. Phys.* **77** 259–315
[3] Schollwöck U 2011 *Ann. Phys.* **326** 96–192
[4] SaiToh A ZKCM_QC https://sourceforge.net/p/zkcm/home/Extension Packages/; see also *Preprint* arXiv:1111.3124
[5] Vidal G 2003 *Phys. Rev. Lett.* **91** 147902
[6] SaiToh A and Kitagawa M 2006 *Phys. Rev. A* **73** 062332
[7] Bauer B *et al.* 2011 *J. Stat. Mech.* **2011(05)** P05001; http://alps.comp-phys.org/
[8] Bailey D H, Barrio R and Borwein J M 2012 *Appl. Math. Comput.* **218** 10106–10121
[9] Venzl H, Daley A J, Mintert F and Buchleitner A 2009 *Phys. Rev. E* **79** 056223
[10] Bañuls M C, Orús R, Latorre J I, Pérez A and Ruiz-Femenía P 2006 *Phys. Rev. A* **73** 022344
[11] Kawaguchi A, Shimizu K, Tokura Y and Imoto N Classical simulation of quantum algorithms using the tensor product representation *Preprint* arXiv:quant-ph/0411205
[12] Markov I L and Shi Y 2008 *SIAM J. Comput.* **38** 963–981
[13] Yoran N and Short A J 2006 *Phys. Rev. Lett.* **96** 170503
[14] Jozsa R On the simulation of quantum circuits *Preprint* arXiv:quant-ph/0603163
[15] Chamon C and Mucciolo E R 2012 *Phys. Rev. Lett.* **109** 030503
[16] Temme K and Wocjan P Efficient computation of the permanent of block factorizable matrices *Preprint* arXiv:1208.6589
[17] Johnson T H, Biamonte J D, Clark S R and Jaksch D 2013 *Sci. Rep.* **3** 1235
[18] Hallberg K A 2006 *Adv. Phys.* **55** 477–526
[19] Documents linked from http://zkcm.sf.net/
[20] The GNU multiple precision arithmetic library http://gmplib.org/
[21] Fousse L, Hanrot G, Lefèvre V, Pélissier P and Zimmermann P 2007 *ACM Trans. Math. Software* **33** 13; http://www.mpfr.org/
[22] Grover L K 1996 *Proceedings of the 28th Annual ACM Symposium on Theory of Computing (STOC 1996)* (Philadelphia, PA: ACM Press, New York, 1996) pp 212–219
[23] Shor P W 1997 *SIAM J. Comput.* **26** 1484–1509
[24] Deutsch D and Jozsa R 1992 *Proc. Royal Soc. London A* **439** 553–558
[25] Garey M R and Johnson D S 1979 *Computers and Intractability: A Guide to the Theory of NP-Completeness* (New York: W.H. Freeman and Co.)
[26] Maxima, a computer algebra system http://maxima.sourceforge.net/
[27] SaiToh A Qcomp.mac http://silqcs.org/~saitoh/qcomp.mac
[28] Fowler A G, Devitt S J and Hollenberg L C L 2004 *Quant. Inf. Comput.* **4** 237–251
[29] Draper T G Addition on a quantum computer *Preprint* arXiv:quant-ph/0008033
[30] Gruska J 1999 *Quantum Computing* (Berkshire, UK: McGraw-Hill)

Journal of Physics: Conference Series **454** (2013) 012065 doi:10.1088/1742-6596/454/1/012065

Computational issues of configuration interaction frameworks describing open quantum systems

Nicolas Michel
University of Tennessee / Michigan State University (USA)
E-mail: nicolasm@msu.edu

Abstract. Open quantum systems (OQS), extended in space (halo nuclei) or even unbound, differ from closed quantum systems (CQS), for which the methods of standard shell model (SM) [1] can be utilized in order to expand their wave function in a configuration interaction framework. Configuration interaction methods based on the use of Berggren bases [2], comprising bound, resonant and scattering states, which have the ability to generate the very long range asymptotic behavior of OQSs, are used instead for that matter. This demands the introduction of new computational techniques, including the optimization and discretization of the Berggren basis [3], the development of an algorithm to efficiently calculate their two-body matrix elements [4], and an overall optimization of memory storage absent from SM, where, for instance, all data related to proton and neutron spaces only can be precalculated and stored [1]. In order to diagonalize the very large induced matrices, the Density Matrix Renormalization Group (DMRG) method [5] extended to OQSs has been developed [6, 7]. A renormalization procedure which generates more and more correlated many-body basis states iteratively is used therein, so that the Hamiltonian matrix to diagonalize is very small compared to that occurring with a many-body basis of independent particles [6, 7]. Parallelization of presented methods will also be discussed.

1. Introduction

The precise description of drip-line nuclei is a challenge for both theoretical and experimental nuclear physics. Along with the advent of radioactive beams provided by accelerators of new generation, it is now possible to synthesize very short-lived nuclei, bearing important neutron-to-proton ratio, whose study will provide new experimental data which will have to be explained at theoretical level. The most exotic nuclei, contrary to those situated in the valley of stability, exhibit special features originating from the proximity of proton or neutron emission threshold. One of the most interesting phenomenon is the appearance of halos, i.e. nuclei which are extended beyond classically allowed regions. Due to the strong coupling with the continuum of scattering states, nuclear correlations and continuum effects mix together and give rise to complex nuclear structures. One can also mention the existence of Borromean nuclei, for which all subsystems are unbound, but so that the whole system is bound because of the correlations between its subparts. While these properties exist in the vicinity of proton and neutron drip-lines, the proton drip-line itself has unique properties, such as two-proton radioactivity, discovered a few years ago [8]. Another peculiarity is the redistribution of shell closures close to drip-lines [9]. Considering the very different natures of the nuclei of the valley of stability and drip-lines, new theoretical models must be formulated in order to provide a uniform vision of the nuclear chart. For this, the standard configuration interaction model of nuclear physics, called SM [1], must be extended. Indeed, SM is based on the use of the one-body harmonic oscillator basis, whose one-body states generate the many-body Slater determinant basis, upon which many-body nuclear states are expanded. Despite its numerous analytical advantages (analytical matrix elements, exact treatment of the center of mass motion, ...), the use of the harmonic oscillator basis is confined to the description of well-bound many-body states, as the necessary truncation of the many-body Slater determinant basis and the Gaussian fall-off of harmonic oscillator states implies that the asymptotic of many-body nuclear states is always falling very rapidly. Hence, far-extended states such as halo and resonant nuclear states cannot be described with this formalism. A solution to this problem is the use of the Gamow Shell Model (GSM) [10, 11, 12, 13, 14, 15, 16, 17]. GSM is an extension of the SM, where a basis of complex energy states, the Berggren basis [2], is used instead of a basis generated by a harmonic oscillator potential. The Berggren basis consists of bound, resonant (or Gamow states [18]), and of complex-energy scattering states. This model allows to expand A-body halo and resonant correlated nuclear states. Consequently, extended many-body asymptotic can be handled with the Berggren basis. However, due to the fact that it contains a continuous part, which must be discretized, the number of one-body states to consider in a numerical calculation becomes very large, so that the standard numerical techniques of many-body diagonalization, based on the Lanczos method, can no longer be used. Indeed, matrices to diagonalize become very much larger than in SM, whose current limit is around one billion and demands advanced numerical methods to be handled. In order to counter this state of affairs, the DMRG method [15] has been introduced, where the direct diagonalization of large matrices is replaced by an iterative construction of correlated basis states, whose number is sufficiently small so as to allow standard diagonalization techniques to be used. It is thus the object of this article to describe configuration interaction models using the Berggren basis on the hand, and to emphasize the numerical problems that it generates and how they are solved using parallel computation techniques on the other hand.

2. Berggren basis of one-body states

The standard completeness relation of Newton [19] is the decomposition of unity formed by the bound and real-energy scattering states generated by a finite-range potential:

$$\sum_{n \in b} u_n(r) u_n(r') + \int_0^{+\infty} u(k,r) u(k,r') \; dk = 1, \tag{1}$$

where $u_n(r)$ is a normalized bound state and $u(k,r)$ a scattering state of linear momentum k, normalized so that in the asymptotic region, where $u(k,r) = C_+ H^+_{\ell,\eta}(kr) + C_- H^-_{\ell,\eta}(kr)$ ($H^\pm_{\ell,\eta}$ is a Hankel or Coulomb wave function of orbital angular momentum ℓ and Sommerfeld parameter η), $2\pi C_+ C_- = 1$. This is equivalent to the normalization of scattering states to a Dirac delta. The completeness relation of (1) allows to expand integrable states. By deforming the real k-axis in the complex plane, thus enclosing narrow resonant states (see figure 1), which are poles of the S-matrix and are defined with outgoing wave function condition only ($C_- = 0$), Cauchy theorem implies that:

$$\sum_{n \in (b,d)} u_n(r) u_n(r') + \int_{L^+} u(k,r) u(k,r') \; dk = 1, \tag{2}$$

where $u_n(r)$ is at present a bound or decaying state lying between the real axis and the contour of complex linear momenta L^+. This is the Berggren completeness relation, introduced for the first time in [2]. One can expand bound or resonant states with (2). Note that resonant states are unbound, so that they cannot be normalized on the real axis as bound states, as their norm would diverge. To solve this problem, exterior complex scaling is utilized to define the norm of resonant states, where integration is effected on a contour of the complex plane where the resonant state decreases exponentially [20].

The completeness relations of Newton and Berggren have been demonstrated for potentials decreasing quickly on the real axis, i.e. for the neutron case in [2, 19], and they have been extended to the proton case, where the generating potential possesses an infinite-range Coulomb part, in [21]. In order to use the Berggren basis numerically, one has to discretize the continuum of complex-energy states. For this, the Gauss-Legendre quadrature is employed. Denoting (k_i, w_i) the N_c linear momenta discretized and their respective weight, (2) writes after discretization:

$$\sum_{n \in (b,d)} u_n(r) u_n(r') + \sum_{i=1}^{N_c} u(k_i, r) u(k_i, r') \; w_i \sim 1. \tag{3}$$

In order to render (3) formally identical to that of a discrete basis, as that generated by a harmonic oscillator potential, one redefines the complex-energy scattering states by $u_{i+N_d}(r) = u(k_i, r)\sqrt{w_i}$, where N_d is the number of discrete bound and resonant states. Hence, writing the bound and resonant states as $u_i(r) = u_n(r)$, where $i \leq N_d$, one obtains the requested formulation of (3):

$$\sum_{i=1}^{N} u_i(r) u_i(r') \sim 1, \tag{4}$$

where $N = N_d + N_c$. Equation (4) is the approximate completeness relation which is used in all GSM calculations.

Parallelization in the context of basis generation is meaningful only if one uses Hartree-Fock (HF) potentials, where one has to recalculate many times the one-body states of the basis, which can become lengthy as one can have tens of scattering states per partial wave contour. The use of HF potentials (or more generally of potentials tailored to the nucleus in consideration) has been seen to be necessary with the particular cases of ^{8}He [17] and the Li chain [13]. Indeed, the use of the basis generated by the core has been noticed to be insufficient, as the size of the nucleus changes significantly with the number of nucleons, so that convergence with the basis generated by the core potential becomes very slow compared to that of the HF basis. For the calculation of the basis, all nodes calculate all HF potentials independently, but calculate only part of the basis. Thus, MPI communication never occurs during the HF process, as it is only at the end of the calculation that HF potentials and wave functions are distributed to all nodes. However, OpenMP parallelization can be applied for the calculation of HF potentials, as it does not lead to communication overhead, the different points of HF potentials being calculated independently from each other.

3. Realistic interactions in GSM

The inclusion of nuclear realistic interactions (see [22, 23, 24, 25] for the ones most widely used) in GSM cannot be done directly, as in SM (see [26] where SM calculations using realistic interactions are effected for light nuclei), because the calculation of their matrix elements relies on the mathematical properties of the harmonic oscillator basis used in SM, absent from GSM. However, it is possible to expand the nuclear interaction only with harmonic oscillator states, even though the GSM matrix is represented with the Berggren basis. This is due to the finite-range character of the nuclear interaction, which is important only in the vicinity of the nucleus. The used nuclear interaction $\hat{V}$ writes using this method [14]:

$$\langle ab|\hat{V}|cd\rangle = \sum_{\alpha\beta\gamma\delta}^{N_{\max}} \langle \alpha\beta|\hat{V}|\gamma\delta\rangle \langle a|\alpha\rangle \langle b|\beta\rangle \langle c|\gamma\rangle \langle d|\delta\rangle, \tag{5}$$

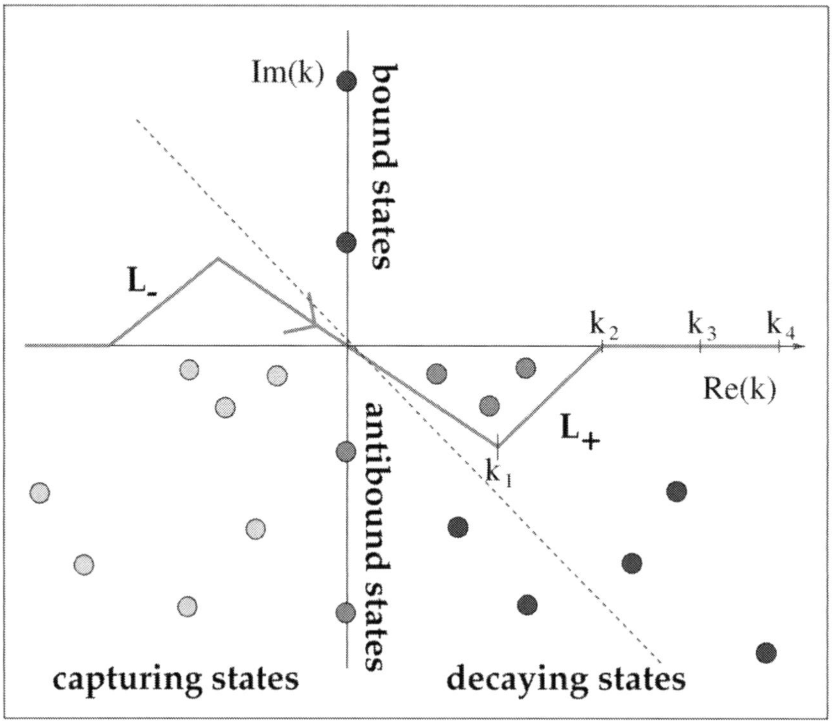

Figure 1. Complex plane of linear momenta k, where the positions of bound, resonant, antiresonant and antibound states are illustrated. L_+ is the contour of non-resonant states. The Berggren completeness relation includes bound states and the resonant states lying between the real axis and the L_+ contour. The L_+ contour must be chosen so that the poles of the discrete sum of (2) are enclosed between L^+ and the real k-axis.

where $N_{\max}$ is the number of harmonic oscillator states and Greek and Latin letters refer respectively to harmonic oscillator states and Berggren basis states. As $\hat{V}$ appears only in the right-hand side of (5) through matrix elements of harmonic oscillator states, the Moshinsky-Brody transformation can be used to calculate them. Berggren basis states can be found only in overlaps of the form $\langle a|\alpha\rangle$, necessitating no complex scaling as harmonic oscillator states always decrease like Gaussians for $r \to +\infty$ whereas Berggren basis states increase at most exponentially in modulus. Note that the one-body kinetic part of the Hamiltonian is directly expressed with the Berggren basis, hence without harmonic oscillator basis expansion, so that this scheme is not equivalent to a SM calculation.

This method has been applied to the ^{6}He and ^{18}O nuclei, modelled by two neutrons above a core, ^{4}He for ^{6}He and ^{16}O for ^{18}O, using the N^{3}LO realistic interaction renormalized with the low-momentum interaction method [14]. The convergence of the energies of ground and excited states of the ^{6}He and ^{18}O nuclei, as well as that of the density of the halo state 0_1^+ of the ^{6}He nucleus, are illustrated respectively in Tabs.1,2 and figure 2. One notices therein that convergence is attained for $N_{\max} \sim 10$ in Tabs.1,2, and that even for resonant states. Moreover, the radial density of valence neutrons of the ground state of the ^{6}He nucleus corresponds to that of a halo state, as shown by linear and logarithmic scales (see figure 2, issued from [16]), which confirms the precision of the method. Hence, realistic interactions can be included in practice in the GSM formalism, which would be impossible without the decomposition described in (5).

Parallelization of the implementation of two-body matrix elements is trivial as they are all calculated independently from each other. MPI is used for distribution of two-body matrix elements among nodes while OpenMP is used for the loop calculation of (5). Parallelization at this level is useful because, on the one hand, calculation of two-body matrix elements using firstly the Moshinsky-Brody transformation and secondly (5) is lengthy, and, on the other hand, the Berggren basis can contain tens of one-body basis states, so that the number of two-body matrix elements to calculate can be very large.

Table 1. Convergence of the energies of the $0_1{}^+$ and $2_1{}^+$ states of the ^{6}He nucleus as a function of the number of nodes of the harmonic oscillator states used in the expansion of the N^{3}LO interaction, renormalized with the low-momentum method. The momentum cut is effected at $\Lambda = 1.9$ fm^{-1} in the space of relative coordinates. The harmonic oscillator parameter is equal to $b = 2$ fm. Energies are given in MeV.

	$J^\pi = 0_1{}^+$		$J^\pi = 2_1{}^+$	
$n_{\max}$	Re[E]	Im[E]	Re[E]	Im[E]
4	-0.4760	0.0000	0.9504	-0.0467
6	-0.4714	0.0000	0.9546	-0.0461
8	-0.4719	0.0000	0.9597	-0.0453
10	-0.4721	0.0000	0.9602	-0.0452
12	-0.4721	0.0000	0.9600	-0.0452
14	-0.4721	0.0000	0.9601	-0.0452
16	-0.4721	0.0000	0.9601	-0.0453
18	-0.4721	0.0000	0.9601	-0.0453
20	-0.4721	0.0000	0.9601	-0.0453

Table 2. Same as in Tab.1, but for the $0_1{}^+, 0_2{}^+, 4_1{}^+$ and $4_2{}^+$ states of the ^{18}O nucleus.

	$J^\pi = 0_1{}^+$		$J^\pi = 0_2{}^+$		$J^\pi = 4_1{}^+$		$J^\pi = 4_2{}^+$	
$n_{\max}$	Re[E]	Im[E]	Re[E]	Im[E]	Re[E]	Im[E]	Re[E]	Im[E]
4	-12.225	0.000	-8.438	0.000	-11.0641	0.0000	-1.4373	-0.8275
6	-12.226	0.000	-8.498	0.000	-11.0907	0.0000	-1.4292	-0.7600
8	-12.228	0.000	-8.499	0.000	-11.0922	0.0000	-1.4380	-0.7405
10	-12.229	0.000	-8.499	0.000	-11.0921	0.0000	-1.4400	-0.7390
12	-12.228	0.000	-8.499	0.000	-11.0923	0.0000	-1.4393	-0.7401
14	-12.228	0.000	-8.499	0.000	-11.0923	0.0000	-1.4394	-0.7401
16	-12.228	0.000	-8.499	0.000	-11.0923	0.0000	-1.4394	-0.7401
18	-12.228	0.000	-8.499	0.000	-11.0923	0.0000	-1.4394	-0.7401
20	-12.228	0.000	-8.499	0.000	-11.0923	0.0000	-1.4394	-0.7401

4. Diagonalization of giant GSM matrices with the density matrix renormalization group method

The presence of numerous scattering states in the Berggren basis within GSM generates very large valence space dimensions, which cannot be handled with Krylov methods, unless drastic truncations are imposed, as was the case for the first applications of GSM. In order to be able to determine eigenstates almost exactly in GSM, the DMRG method is being developed in the non-Hermitian case of GSM. The DMRG method originates from solid states physics [27] and allows to calculate eigenstates of matrices whose dimensions are beyond standard Krylov methods capabilities. It has been tested in nuclear physics in the frame of SM [28], and generalized to

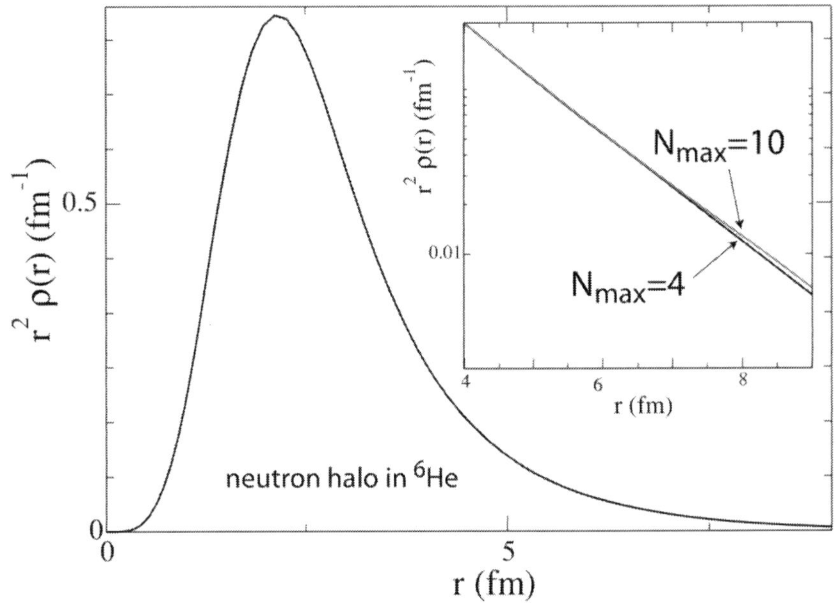

Figure 2. Radial density of valence particles of the ground state 0^+ of the ^{6}He nucleus for three different harmonic oscillator expansions of the nuclear interaction: $n_{\max} = 4$, 10 and 16. The asymptotic of the radial density is illustrated in the insert in logarithmic scale. Results for $n_{\max} = 10$ and 16 are indistinguishable.

GSM firstly for a space of valence neutrons and then for GSM space containing both protons and neutrons afterwards [15].

The aim of this method is to build iteratively bases becoming more and more correlated, so that the GSM eigenstates can be expanded in a DMRG basis containing $N_d \sim 1000$ basis states [15]. For this, the GSM valence space is separated in two spaces A and B. The A space is constructed from the one-body resonant states of the Berggren basis, and from a few one-body non-resonant states of quantum numbers (ℓ, j) different from those of resonant states, necessary to generate all the possible couplings present in the Hamiltonian [15]. The B space is generated during the DMRG process, by adding one by one the remaining scattering states of the Berggren basis. Its elements are numbered by $i_B = 0, 1, \cdots$. One then starts the "warm-up" phase [15]. For this, the configurations for all number of particles n_A and total angular momentum j_A, denoted as $|k_A\rangle$, are built in the A space, as well as all the associated suboperators of the Hamiltonian, whose list reads $a^\dagger$, $(a^\dagger \tilde{a})^K$, $((a^\dagger a^\dagger)^K \tilde{a})^L$, $(a^\dagger a^\dagger)^K (\tilde{a}\tilde{a})^K$, where $a^+/\tilde{a}$ is a creation/annihilation operator and L, K angular momenta. Then, a new scattering shell (n, ℓ, j) is added, the configurations $|i_B\rangle$ for all possible number of particles n_B and total angular momentum j_B are constructed, which augments the size of the B space and generates a set of basis states $(|k_A(n_A, j_A)\rangle \otimes |i_B(n_B, j_B)\rangle)^J$, where $n_{A(B)}$ and $j_{A(B)}$ are respectively the numbers of particles and angular momenta of the $k_A(i_B)$ states and J is the total angular momentum of the calculated eigenstate. When the number of configurations i_B of the B space reaches N_{opt}, of the order of 10 to 100 typically, the Hamiltonian is diagonalized to provide an approximation of the calculated nuclear state:

$$|\Psi_J\rangle = \sum_{k_A, i_B} c_{i_B}^{k_A} \{|k_A\rangle \otimes |i_B\rangle\}^J, \tag{6}$$

where the Lanczos or Davidson method is used for diagonalization. $|\Psi_J\rangle$ of (6) has been determined by the overlap method [11], which consists in diagonalizing firstly the GSM matrix without scattering basis states, so that a zeroth-order eigenstate can be determined with the Lanczos method, and secondly diagonalizing the full GSM matrix, whose sought bound or resonant eigenstate is the one which maximizes the overlap with the previously calculated zeroth-order eigenstate at pole approximation. The reduced density matrix $\rho^{B}_{i_B i'_B}$ is thus calculated from (6):

$$\rho^{j_B}_{i_B i'_B} = \sum_{k_A} c^{k_A}_{i_B} c^{k_A}_{i'_B}, \tag{7}$$

where the angular momentum j_B, which is the same for $|i_B\rangle$ and $|i'_B\rangle$, is fixed. The reduced density matrix is then diagonalized, and the resulting eigenvectors which possess the largest eigenvalues in modulus (at most N_{opt} of them) are kept in the model space, the other ones being rejected. This method is indeed motivated by the variational principle, which states that the wave function bearing the largest overlap with the exact nuclear state is the one built from the aforementioned eigenvectors [27]. The eigenvectors then generate the new B space, and the suboperators associated to the Hamiltonian are recalculated in this new space. When all the shells of the Berggren basis have been taken into account, the "warm-up" phase terminates and the "sweep" phases start [15]. They are similar to the "warm-up" phase, except that the states of the A space are now replaced by the $(|k_A(n_A, j_A)\rangle \otimes |i_B(n_B, j_B)\rangle)^J$ states, issued from the previous iteration. The scattering states (n, ℓ, j) of the Berggren basis are then reintroduced one by one, with the index associated to (n, ℓ, j) firstly decreasing ("sweep down") and afterwards increasing ("sweep up") [15]. Indeed, the truncation applied after each diagonalization of the reduced density matrix suppresses correlations, which must be recovered during sweeps. Convergence is obtained after a few "sweeps", of the order of five [15]. Every time a new scattering shell is added during a sweep, which translates into diagonalizations of new reduced density matrices and recalculations of Hamiltonian suboperators, by definition, a "step" occurs. The speed of convergence is measured via their number N_{step}.

The DMRG method has been tested with the calculation of the $3/2^-_1$ and $1/2^-_1$ eigenstates of ^{7}He and to the ground state of the ^{7}Li nucleus [15]. The convergence of the DMRG method is illustrated in Figs.3,4 for the ^{7}He and ^{7}Li nuclei. One can see that convergence is very rapid. The precision with respect to the number of discretized scattering states of the Berggren basis has also been studied (see figure 4). One can see that the energy and width of the ground state $3/2^-_1$ of the ^{7}He nucleus converge to a finite value, while the width of the ground state 0^+_1 of the ^{6}He nucleus goes to zero as the number of discretized scattering states of the basis increases, which is expected as the ground states of ^{6}He is bound. Even though it is bound, the ground state $3/2^-_1$ of the ^{7}Li nucleus has a non-zero width. The discretization of contours has indeed been chosen to be coarse for the ^{7}Li nucleus, so that GSM matrix dimensions are small enough so as to be able to calculate its ground state with the Davidson method and compare the exact result with that produced by the DMRG method. The imaginary part of the ground state of the ^{7}Li nucleus, which is bound, eventually disappears when the number of discretized scattering states augments [15].

The parallelization of the DMRG method has to be effected at two different levels, the first one at the level of the calculation of sub-operators (see above) and density/Hamiltonian matrix elements, and the second one at the level of the application of matrices on vectors. The latter is trivial to diagonalize. However, that of the sub-operators poses problem. Indeed, due to the numerous states of the Berggren basis, the matrix elements involving creation and annihilation operators become very large, contrary to standard shell model, where the proton and neutron parts of the Hamiltonian can be fully stored and hence efficiently handled, the proton-neutron part of the Hamiltonian taking virtually all the time of calculation [1]. For neutron-rich nuclei, for which the Berggren basis consists mainly of discretized neutron contours, the neutron part is

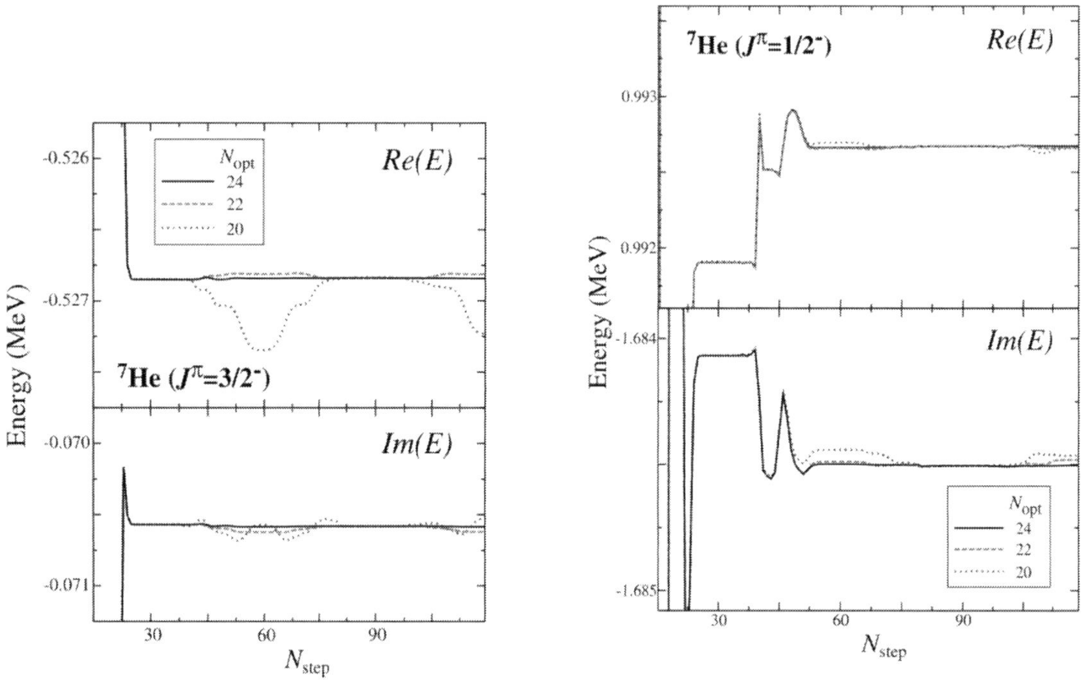

Figure 3. Convergence of the real (upper part) and imaginary (lower part) parts of the energy of the ground state $3/2^-_1$ (left) and excited state $1/2^-_1$ (right) of the ^{7}He nucleus as a function of N_{step} for several values for N_{opt} (see text for their definition).

the most expensive to treat, proton and proton-neutron parts being secondary, if not negligible. The parallelization of the calculation of sub-operators is by itself trivial to consider as they are independent from each other. However, due to their number, they must be stored on hard disk, and the parallelization of input/output has not yet been done, due to the complex distribution of used sub-operators matrix elements in the generation of the Hamiltonian matrix. Hence, the saving and loading of sub-operators matrix elements is for the moment the bottleneck of the DMRG method as it is coded now, as calculations demanding more than six valence particles are precluded. The parallelization of the input/output process is currently investigated.

5. Conclusion

Exotic nuclei, begin OQSs, are radically different from stable nuclei, of CQS nature. Hence, they must be studied with models incorporating their uncommon features, for example the presence of a halo or resonant character at ground state level. For this, the Gamow Shell Model has been developed, which makes use of the Berggren basis, with which the direct inclusion of bound, resonant and scattering states allows to describe the long-range character of the many-body wave functions of nuclei far from the valley of stability. However, the mathematical properties of the standard harmonic oscillator basis being absent in the Berggren basis, a new mathematical method had to be developed in order to calculate efficiently the matrix elements of the nuclear interaction in the Berggren basis. The most difficult part to handle in this model is the diagonalization of matrices much larger than in SM, as the Lanczos and Davidson are insufficient to treat such large dimensions. For this, the DMRG has been proved to be able to

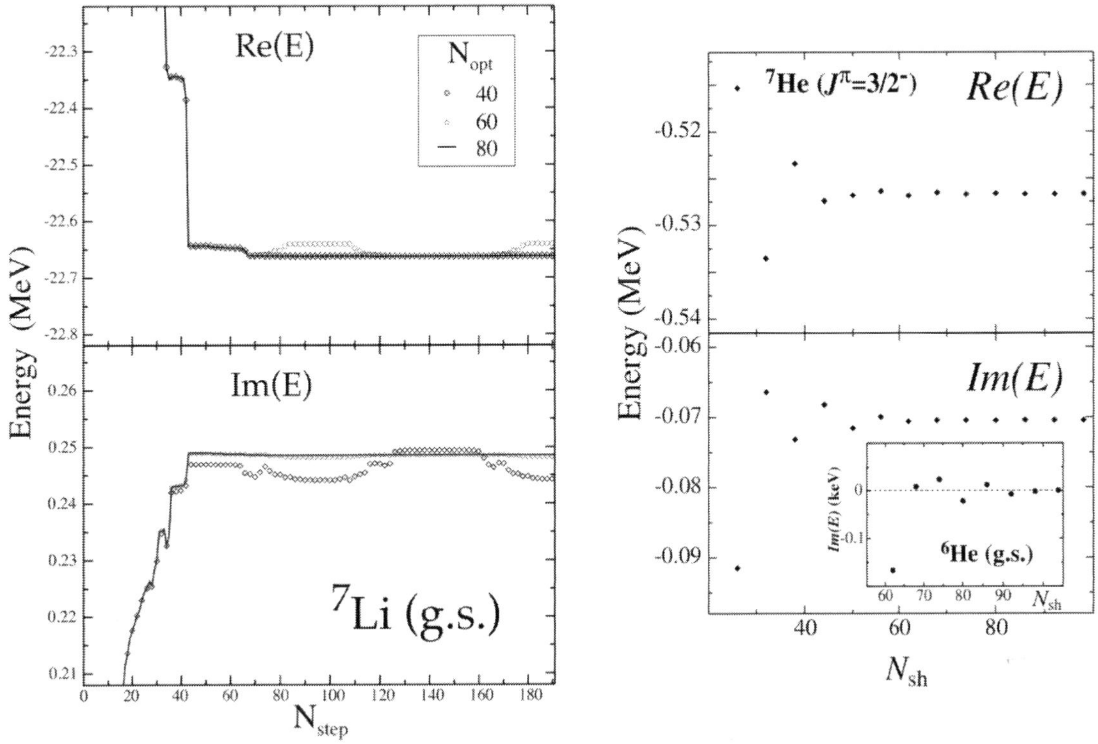

Figure 4. On the left, same as figure 3, but for the ground state $3/2_1^-$ of the ^{7}Li nucleus. On the right, the convergence of the energy (upper part) and width (lower part) of the ground state $3/2_1^-$ of the ^{7}He nucleus as a function of the number of discretized scattering states of the Berggren basis is depicted. The width of the ground state 0_1^+ of the ^{6}He nucleus as a function of the number of discretized scattering states of the Berggren basis is shown in the insert of the lower part.

solve this problem. While parallelization of introduced methods can usually be effected without difficulty, the necessary disk input/output of large tables in the DMRG is, on the contrary, problematic for that matter. This problem will have to be solved in order to be able to consider more valence particles with this powerful method.

References

[1] Caurier E, Martínez-Pinedo G, Nowacki F, Poves A and Zuker A P 2005 *Rev. Mod. Phys.* **77** 427
[2] Berggren T 1968 *Nucl. Phys.* A **109** 265
[3] Michel N 2009 *Eur. Phys. J.* A **42** 523
[4] Hagen G, Hjorth-Jensen M and Michel N 2006 *Phys. Rev.* C **73** 064307
[5] Schollwök U 2005 *Rev. Mod. Phys.* **77** 259
[6] Rotureau J, Michel N, Nazarewicz W, Płoszajczak M and Dukelsky J 2006 *Phys. Rev. Lett.* **97** 110603
[7] Rotureau J, Michel N, Nazarewicz W, Płoszajczak M and Dukelsky J 2009 *Phys. Rev.* C **79** 014304
[8] Giovinazzo J 2002 *et al.*, *Phys. Rev. Lett.* **89** 102501 ; Pfützner M 2002 *et al.*, *Eur. Phys. J.* A **14** 279
[9] Stanoiu M et al. 2008 Phys. Rev. C **78** 034315
[10] Michel N, Nazarewicz W, Płoszajczak M and Bennaceur K 2002 *Phys. Rev. Lett.* **89** 042502
[11] Michel N, Nazarewicz W, Płoszajczak M and Okołowicz J 2003 *Phys. Rev.* C **67** 054311
[12] Michel N, Nazarewicz W, Płoszajczak M and Rotureau J 2004 *Rev. Mex. Fis.* **50** Suppl. 2, 74
[13] Michel N, Nazarewicz W and Płoszajczak M 2004 *Phys. Rev.* C **70** 064313
[14] Hagen G, Hjorth-Jensen M and Michel N 2006 *Phys. Rev.* C **73** 064307

[15] Rotureau J, Michel N, Nazarewicz W, Płoszajczak M and Dukelsky J 2006 *Phys. Rev. Lett.* **97** 110603; ibid. 2009 *Phys. Rev.* C **79** 014304
[16] Michel N, Nazarewicz W, Płoszajczak M and Vertse T 2009 *J. Phys. G: Nucl. Part. Phys.*, Topical Review **36** 013101
[17] Papadimitriou G, Kruppa A T, Michel N, Nazarewicz W, Płoszajczak M and Rotureau J 2011 *Phys. Rev.* C **84** 051304
[18] Gamow G 1928 *Z. Phys.* **51** 204; Gurney R W, Condon E U 1929 *Phys. Rev.* **33** 127
[19] Newton R 1982 *Scattering Theory of Waves and Particles* Springer-Verlag, New York, Heidelberg
[20] Gyarmati B and Vertse T 1971 *Nucl. Phys.* **A 160** 523
[21] Michel N 2008 *J. Math. Phys.* **49** 022109
[22] Stoks V G , Klomp R A , Terheggen C P and de Swart J J 1994 *Phys. Rev.* C **49** 2950
[23] Wiringa R B , Stoks V G and Schiavilla R 1995 *Phys. Rev.* C **51** 38
[24] Machleidt R 2001 *Phys. Rev.* C **63** 024001
[25] Entem D R and Machleidt R 2003 *Phys. Rev.* C **68** 041001(R)
[26] Navràtil P and Ormand W E 2003 *Phys. Rev.* C **68** 034305
[27] White S R 1993 *Phys. Rev.* B **48** 10345
[28] Dukelsky J, Pittel S, Dimitrova S S and Stoitsov M V 2002 *Phys. Rev.* C **65** 054319

Recent development of Monte Carlo shell model and its application to no-core calculations

T Abe[1], P Maris[2], T Otsuka[1,3,4], N Shimizu[3], Y Tsunoda[1], Y Utsuno[5], J P Vary[2] and T Yoshida[3]

[1] Department of Physics, the University of Tokyo, Hongo, Tokyo 113-0033, Japan
[2] Department of Physics and Astronomy, Iowa State University, Ames, Iowa 50011, USA
[3] Center for Nuclear Study, the University of Tokyo, Hongo, Tokyo 113-0033, Japan
[4] National Superconducting Cyclotron Laboratory, Michigan State University, East Lansing, Michigan 48824, USA
[5] Advanced Science Research Center, Japan Atomic Energy Agency, Tokai, Ibaraki 319-1195, Japan

E-mail: `tabe@nt.phys.s.u-tokyo.ac.jp`

Abstract. One of the major challenges in nuclear theory is to reproduce and to predict nuclear structure from *ab initio* calculations with realistic nuclear forces. As the current limitation of direct diagonalization of Hamiltonian matrices by Lanczos iteration method is around the order of matrix dimensionality 10^{10} in shell-model calculations, it is difficult to access heavier nuclei beyond the p shell with sufficiently large basis spaces. It is possible to overcome this difficulty by utilizing efficient approximate methods to reproduce full *ab initio* solutions with good precision and quantified uncertainties. Following the major success of the Monte Carlo shell model (MCSM) with an assumed inert core in the *sd*- and pf-shell regions and also by recent developments in the MCSM algorithm, the no-core MCSM is expected to be one of the most powerful tools to meet these conditions. We have performed benchmark calculations in the p-shell region. Results of energies are compared with those in the full configuration interaction and no-core full configuration methods. These are found to be consistent with each other within quoted uncertainties when they could be quantified. We also compare and discuss the radial density of the helium-4 ground state extracted from the MCSM and FCI many-body wave functions.

1. Introduction

One of the major challenges in nuclear physics is to understand nuclear structure and reactions from *ab initio* calculations with realistic nuclear forces. Such calculations have recently become feasible for nuclear many-body systems beyond $A = 4$ due to the rapid evolution of computational technologies. Together with the Green's Function Monte Carlo [1] and Coupled Cluster theory [2], the No-Core Shell Model (NCSM) is one of the relevant *ab initio* methods and has been emerging for about a decade. It is now available for the study of nuclear structure and reactions in the p-shell nuclei [3].

As the NCSM treats all the nucleons democratically, computational demands for the calculations explode exponentially as the number of nucleons increases. Current computational resources limit the direct diagonalization of the Hamiltonian matrix using the Lanczos algorithm to M-scheme basis spaces (total angular momentum projection is fixed in the basis) with a

Journal of Physics: Conference Series **454** (2013) 012066 doi:10.1088/1742-6596/454/1/012066

dimension of around 10^{10} as indicated in figure 1 (left). In order to access heavier nuclei beyond the p-shell region with larger basis dimensions, many efforts have been devoted to the NCSM calculations. One of these approaches is the Importance-Truncated NCSM [4] where the model spaces are extended by using an importance measure evaluated with perturbation theory. Another approach is the Symmetry-Adapted NCSM [5] where the model spaces are truncated by the selected symmetry groups. Similar to these attempts, the no-core Monte Carlo Shell Model (MCSM) [6, 7, 8] is one of the promising candidates to go beyond the Full Configuration Interaction (FCI) method which is a different truncation of the basis states than the one commonly used in the NCSM. Shell-model calculations with an assumed inert core by the MCSM have succeeded in obtaining the approximated solutions where the direct diagonalization is difficult due to large dimensionalities as described in figure 1 (right).

In these proceedings, we focus on the latest application of the MCSM toward the *ab initio* no-core calculations, which has become viable recently with the aid of major developments in the MCSM algorithm [8, 9, 10] and also a remarkable growth in the computational power of state-of-the-art supercomputers. The overview of the benchmarks in the no-core MCSM is based on the results presented in [7, 8]. Moreover, we compare the radial density of the ^{4}He 0^+ ground state from the MCSM and FCI methods. An initial attempt to draw the intrinsic density from no-core MCSM wave functions can be found in [8].

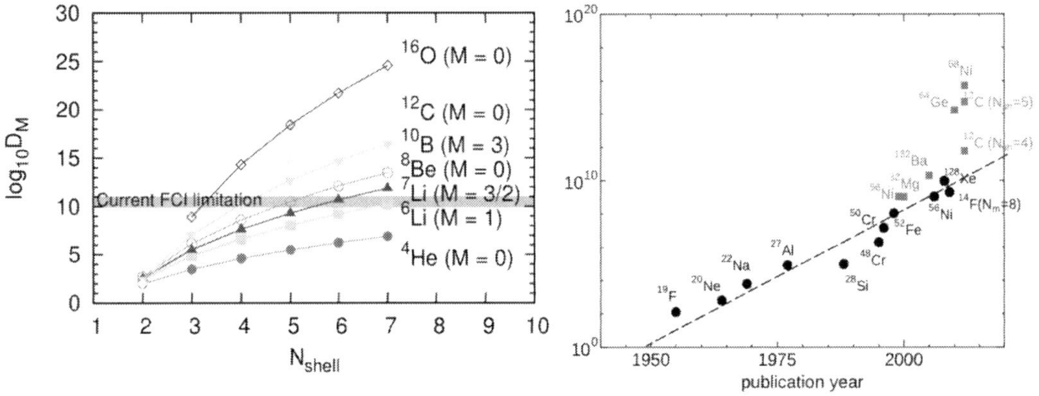

Figure 1. M-scheme dimension for light nuclei as a function of basis space cutoff, N_{shell} (left) and M-scheme dimension for conventional shell-model calculations with an assumed inert core as a function of publication year (right). Red squares are for the MCSM results, and black circles are for the conventional shell-model results by the direct diagonalization with the Lanczos technique.

2. MCSM

The MCSM has been developed mainly for conventional shell-model calculations with an assumed inert core [11]. Recently the algorithm and code itself have been heavily revised and rewritten so as to accommodate massively parallel computing environments [8, 9, 10]. In this section, we briefly overview the MCSM and introduce some of recent developments. An initial attempt for the application on K computer is also presented shortly.

2.1. Brief overview

The MCSM approach [11] proceeds through a sequence of diagonalization steps within the Hilbert subspace spanned by the deformed Slater determinants in the harmonic oscillator (HO) single-particle basis as the selected importance-truncated bases. Here in this subsection, we describe our many-body wave function and Hamiltonian adopted. Some technical details and equations needed for the discussion in the following subsections are also presented shortly. See regular and review articles [8, 9, 10, 11] for a detailed account.

2.1.1. Many-body wave function A many-body basis state $|\Psi^{J^\pi M}\rangle$ is approximated as a linear combination of non-orthogonal angular-momentum (J) and parity (π) projected deformed Slater determinants with good total angular momentum projection (M),

$$|\Psi^{J^\pi M}\rangle = \sum_{n=1}^{N_b} f_n \sum_{K=-J}^{J} g_{nK} P^J_{MK} P^\pi |\phi_n\rangle, \tag{1}$$

where N_b is the number of the Slater determinants selected stochastically. P^J_{MK} is the projection operator for the total angular momentum J with its z-projection in the laboratory (body-fixed) frame, M (K). P^π is the projection operator for the parity.

In the actual computations, the integrations for the angular-momentum and parity projections are performed in parallel by discretizing the integrals on discrete mesh points. The angular-momentum projection operator is discretized as

$$P^J_{MK} = \frac{2J+1}{8\pi^2} \int d\Omega D^{J}_{MK}{}^{*} R(\Omega) \to \sum_{\lambda}^{N_\lambda} W^{J(\lambda)}_{MK} R^{(\lambda)}, \tag{2}$$

with the Euler angles, $\Omega \equiv (\alpha, \beta, \gamma)$, and the Wigner D matrix, D^J_{MK}. $R(\Omega)$ is the rotation operator, $R(\alpha, \beta, \gamma) \equiv e^{i\alpha J_z} e^{i\beta J_y} e^{i\gamma J_z}$. N_λ is the number of mesh points for discretized Euler angles. In the discretized summation over λ,

$$W^{J(\lambda)}_{MK} = \frac{1}{N_\lambda} \frac{2J+1}{8\pi^2} D^{J}_{MK}{}^{*}(\alpha_\lambda, \beta_\lambda, \gamma_\lambda), \tag{3}$$

is the weight for the sum and $R^{(\lambda)} = R(\alpha_\lambda, \beta_\lambda, \gamma_\lambda)$ is the rotational operator at the Euler angle $\Omega_\lambda = (\alpha_\lambda, \beta_\lambda, \gamma_\lambda)$. The parity projection operator is written as

$$P^\pi = \frac{1+\pi\Pi}{2} = \sum_{\nu=1,2} W^{\pi(\nu)} \Pi^{(\nu)}, \tag{4}$$

with the parity transformation, Π, and the weight for positive and negative parity,

$$W^{\pi(\nu)} = \pi^{(\nu)}. \tag{5}$$

Note that $\pi^{(1)} = 1/2$, $\pi^{(2)} = \pi/2$, $\Pi^{(1)} = 1$, and $\Pi^{(2)} = \Pi$.

The deformed Slater determinant in (1) is described as

$$|\phi\rangle = \prod_{i=1}^{A} a_i^\dagger |-\rangle, \tag{6}$$

with the vacuum $|-\rangle$ and the creation operator,

$$a_i^\dagger = \sum_{\alpha=1}^{N_{\rm sp}} c_\alpha^\dagger D_{\alpha i}. \tag{7}$$

$N_{\rm sp}$ is specified by the cutoff of the single particle basis space, $N_{\rm shell}$. The transformation coefficients $D_{\alpha i}$ form the complex $N_{\rm sp} \times A$ matrix with the normalization condition, $D^\dagger D = 1$.

2.1.2. Hamiltonian In usual *ab initio* shell-model calculations, the intrinsic Hamiltonian is expressed as

$$H = T_{\rm rel} + V_{\rm NN} + V_{\rm 3N} + \cdots + V_{\rm Coulomb}, \tag{8}$$

where $T_{\rm rel}$ is the relative kinetic energy. The $V_{\rm NN}$ and $V_{\rm 3N}$ represent two- and three-nucleon interactions, respectively. The $V_{\rm Coulomb}$ denotes the Coulomb interaction. As explicit three-nucleon forces have not yet been treated in the MCSM algorithm, the Hamiltonian considered in this work consists of one- and two-body interactions only,

$$\begin{aligned} H &= H^{(1)} + H^{(2)} = T_{\rm rel} + V_{\rm NN} \\ &= \sum_{ij} t_{ij} c_i^\dagger c_j + \frac{1}{2}\sum_{ijkl} v_{ijkl} c_i^\dagger c_j^\dagger c_l c_k, \end{aligned} \tag{9}$$

where $H^{(1)}$ and $H^{(2)}$ are one- and two-body terms. $c_i^\dagger$ and c_i are the creation and annihilation operators of single-particle state i, respectively. Here, the Coulomb interaction is also omitted in this study.

In order to obtain eigenvalues and eigenfunctions, the coefficients, f_n and g_{nK}, in (1) are determined by solving the generalized eigenvalue problem,

$$\sum_{nK} \mathcal{H}_{mM,nK} f_n g_{nK} = \mathcal{E} \sum_{nK} \mathcal{N}_{mM,nK} f_n g_{nK}, \tag{10}$$

with the normalization condition, $\langle \Psi^{J^\pi M} | \Psi^{J^\pi M} \rangle = 1$, so that $\sum_n f_n^* f_n = 1$ and $\sum_K g_{nK}^* g_{nK} = 1$. $\mathcal{N}_{mM,nK}$ and $\mathcal{H}_{mM,nK}$ are the norm and Hamiltonian matrix spanned by the deformed Slater determinants,(6), respectively.

From (2) and (4), the norm matrix can be expressed as

$$\mathcal{N}_{mM,nK} = \langle \phi_m | P^J_{MK} P^\pi | \phi_n \rangle \to \sum_{\lambda\nu} W^{J(\lambda)}_{MK} W^{\pi(\nu)} \langle \phi_m | \phi_n^{(\lambda\nu)} \rangle, \tag{11}$$

with

$$\langle \phi_m | \phi_n^{(\lambda\nu)} \rangle = \det \left(D^{m\dagger} D^{n(\lambda\nu)} \right). \tag{12}$$

Note that $|\phi_n^{(\lambda\nu)}\rangle = R^{(\lambda)} \Pi^{(\nu)} |\phi_n\rangle$ and the matrix D^n characterizes the n-th single Slater determinant $|\phi_n\rangle$ as in (7).

The Hamiltonian matrix can be also written as

$$\begin{aligned} \mathcal{H}_{mM,nK} &= \langle \phi_m | H P^J_{MK} P^\pi | \phi_n \rangle \\ &\to \sum_{\lambda\nu} W^{J(\lambda)}_{MK} W^{\pi(\nu)} \langle \phi_m | H R^{(\lambda)} \Pi^{(\nu)} | \phi_n^{(\lambda\nu)} \rangle \\ &= \sum_{\lambda\nu} W^{J(\lambda)}_{MK} W^{\pi(\nu)} \langle \phi_m | \phi_n^{(\lambda\nu)} \rangle {\rm Tr} \left(\rho^{(\lambda\nu)} \left(t + \frac{1}{2} \Gamma^{(\lambda\nu)} \right) \right), \end{aligned} \tag{13}$$

where the generalized density matrix, $\rho^{(\lambda\nu)}$, and the self-consistent field, $\Gamma^{(\lambda\nu)}$, are defined, respectively, as

$$\rho_{ij}^{(\lambda\nu)} = \frac{\langle \phi_m | c_j^\dagger c_i | \phi_n^{(\lambda\nu)} \rangle}{\langle \phi_m | \phi_n^{(\lambda\nu)} \rangle} = \left(D^{n(\lambda\nu)} \left(D^{m\dagger} D^{n(\lambda\nu)} \right)^{-1} D^{m\dagger} \right)_{ij}, \tag{14}$$

$$\Gamma_{ik}^{(\lambda\nu)} = \sum_{jl} \bar{v}_{ijkl} \rho_{lj}^{(\lambda\nu)}, \tag{15}$$

with the antisymmetrized two-body matrix element (TBME), $\bar{v}_{ijkl} = v_{ijkl} - v_{ijlk}$. The most-time consuming part in the MCSM is the computation of $\Gamma_{ik}^{(\lambda\nu)}$. Recent improvement for the computation related to the TBMEs is discussed in section 2.2.1.

2.1.3. Basis search First, we stochastically sample the coefficients $D_{\alpha i}$ in (7) among all possible many-body basis states around the mean field solutions through the auxiliary fields $\vec{\sigma}$ with the aid of the imaginary-time evolution of the state vector,

$$|\phi(\vec{\sigma})\rangle \propto e^{-\Delta\beta h(\vec{\sigma})}|\phi^{(0)}\rangle, \tag{16}$$

where $\Delta\beta$ is the time slice of the evolution and h is the one-body Hamiltonian derived by using the Hubbard-Stratonovich transformation [11]. We then diagonalize the Hamiltonian matrix within the subspace spanned by these N_b bases, and obtain the coefficients f_n and g_{nK} in (1). Stochastically sampled bases are accepted so as to minimize the energy variationally. Therefore the MCSM can evade the so-called negative sign problem, which is the fundamental issue that cannot be avoided in quantum Monte Carlo methods. With increasing MCSM basis dimension, N_b, the ground state energy of a MCSM calculation converges from above to the exact value. The energy, therefore, always gives the variational upper bound in this framework. Recently, in the step of the basis search, we adopt the conjugate gradient (CG) method by using D as a gradient vector for minimizing the energy in addition to the initial basis search with a stochastic sampling of D through $\vec{\sigma}$ [8, 10]. The CG process gives the reduction of the number of basis states around 30% to obtain the same level of accuracy in the existing code which generates candidates of the Slater determinant only by the stochastic sampling.

An exploratory no-core MCSM investigation demonstrating a proof-of-principle has been performed for the low-lying states of the beryllium isotopes by applying the existing MCSM algorithm without introducing the recent developments [6]. Recent improvements on the MCSM algorithm have enabled significantly larger calculations [8, 9, 10]. We adopt these improvements in the present work [7, 8].

2.2. Recent developments

Among the recently achieved developments of our MCSM algorithm [8, 9, 10], in this subsection, we focus on two improvements; (1) the efficient computation of the TBMEs for the most time-consuming part in our calculations [8, 9] and (2) the energy-variance extrapolation for our MCSM (approximated) results to the FCI (exact) ones [8, 10]. There are other improvements, such as the CG method in the process of the basis search and the reordering technique in the energy-variance extrapolations. We refer for the details of these improvements to [8, 10].

2.2.1. Efficient computation of the TBMEs One of the main issues in the shell-model calculations is to evaluate TBMEs efficiently. As the matrix for the TBMEs is sparse in general, the indirect-index (list-vector) method is usually adopted by storing only the values of the non-zero matrix element and their indices. The indirect-index method is also known as the coordinate format (COO) in computer science, which is one of the most flexible and simplest formats for the sparse matrix representation. However, it tends to give slow performance due to the irregular memory access patterns.

Alternatively, in our recent MCSM code, we transform the sparse matrix to a block matrix with dense blocks by utilizing the symmetries of the two-body interaction [9]. The one-body density-matrix elements ρ_{ij} are grouped as $\tilde{\rho}(\Delta m)$ according to $\Delta m \equiv j_z(j) - j_z(i)$ where i and j are the labels for the state. The TBMEs, $\bar{v}$'s, are also similarly categorized into $\tilde{v}$. Then, the two-body part of the Hamiltonian overlap can be expressed as

$$\sum_{ik}^{N_s} \rho_{ki}\,\Gamma_{ik}^{(\lambda\nu)} = \sum_{ijkl}^{N_s} \rho_{ki}\,\bar{v}_{ijkl}\,\rho_{lj}^{(\lambda\nu)} = \sum_{\Delta m}\sum_{\alpha\beta} \tilde{\rho}(-\Delta m)_\alpha \tilde{v}(-\Delta m, \Delta m)_{\alpha\beta}\tilde{\rho}^{(\lambda\nu)}(\Delta m)_\beta, \tag{17}$$

where the last equality is derived from the necessary condition for $\bar{v}_{ijkl}$ being non-zero: $j_z(i) + j_z(j) = j_z(k) + j_z(l)$, i.e., $\Delta m \equiv j_z(k) - j_z(i) = -(j_z(l) - j_z(j))$. N_s represents

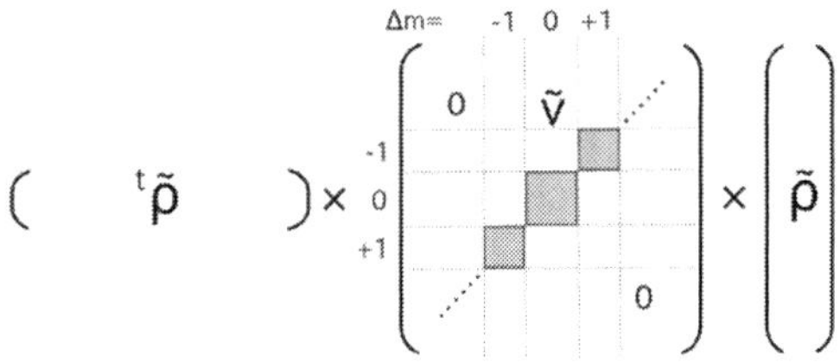

Figure 2. Schematic illustration of the $(\text{vector})^t \times (\text{matrix}) \times (\text{vector})$ operation.

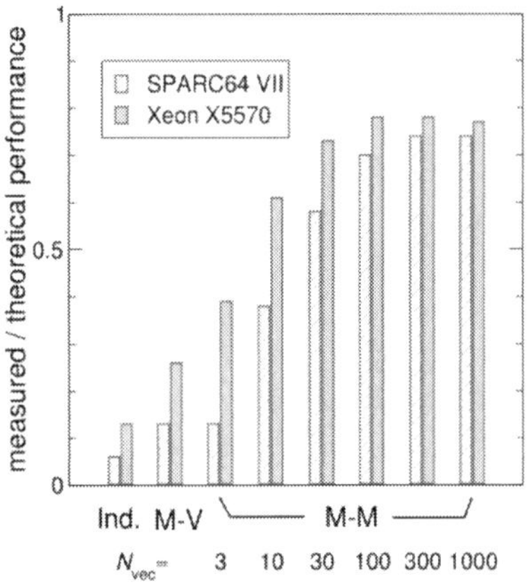

Figure 3. Comparison of the computational performance among the indirect-index method (Ind.), matrix-vector method (M-V) and matrix-matrix method (M-M) with different $N_{\rm vec}$ measured on the SPARC64 VII and Xeon X5570 systems. The values are normalized by their theoretical peak performance. See [9] for the details.

the single-particle state in a given model space. Note that $\bar{v}$ is the antisymmetrized TBME. Figure 2 is the schematic illustration of the $(\text{vector})^t \times (\text{matrix}) \times (\text{vector})$ operation expressed with (17), $(\text{matrix}) \times (\text{vector})$. As seen in figure 2, the sparse matrix $\bar{v}$ is transformed to a block-antidiagonal matrix $\tilde{v}$ whose blocks are dense submatrices.

Furthermore, most of the computational time is devoted to the $(\text{matrix}) \times (\text{vector})$ operation. It is usually repeated a number of times for different vectors, $\tilde{\rho}^{(\lambda\nu)}$ with the same matrix, $\tilde{v}$. By binding $N_{\rm vec}$ $\tilde{\rho}^{(\lambda\nu)}$-vectors into a matrix, $\theta^{(\lambda\nu)}$, repeated $(\text{matrix})\times(\text{vector})$ operations are replaced by a $(\text{matrix}) \times (\text{matrix})$ operation at once,

$$(\tilde{\Gamma}_1^{(\lambda\nu)}, \tilde{\Gamma}_2^{(\lambda\nu)}, \ldots, \tilde{\Gamma}_{N_{\rm vec}}^{(\lambda\nu)}) = (\tilde{v}\tilde{\rho}_1^{(\lambda\nu)}, \tilde{v}\tilde{\rho}_2^{(\lambda\nu)}, \ldots, \tilde{v}\tilde{\rho}_{N_{\rm vec}}^{(\lambda\nu)}) \equiv \tilde{v}\theta^{(\lambda\nu)}, \tag{18}$$

where the number of columns $N_{\rm vec}$ can be chosen arbitrarily. As shown in figure 3, this matrix product is computed with the BLAS level 3 library and we can achieve $70-80\%$ of the peak performance with $N_{\rm vec} \sim 30-100$ in the test case of the $(\text{matrix}) \times (\text{matrix})$ operation [9].

2.2.2. Energy-variance extrapolation to the FCI results With increasing Monte Carlo basis dimension N_b, the MCSM results converge to the FCI results from above. In order to estimate the exact FCI answer, we extrapolate the energy and other observables evaluated by MCSM wave functions using the energy variance [8, 10]. That is, the MCSM results are plotted as a function of the evaluated energy variance,

$$\Delta E_2 = \langle\Psi|H^2|\Psi\rangle - (\langle\Psi|H|\Psi\rangle)^2 \tag{19}$$

and then extrapolated to zero variance.

This kind of extrapolation has been enabled recently by the growth of the computational power and the factorization of the formula. In order to evaluate the expectation value of H^2, we have to compute

$$\langle\Psi^{J^\pi M}|H^2|\Psi^{J^\pi M}\rangle = \sum_{m,M,n,K,\lambda,\nu} f_m^* g_{mM}^* f_n g_{nK} W_{MK}^{J(\lambda)} W^{\pi(\nu)} \langle\phi_m|H^2|\phi_n^{(\lambda\nu)}\rangle. \tag{20}$$

The obstacle in the implementation was the large amount of computation to evaluate $\langle\phi_m|H^2|\phi_n^{(\lambda\nu)}\rangle$. Suppose H^2 is a general four-body operator, the evaluation of the matrix element involves the eightfold-loop summation of the 24 terms of products of four generalized one-body density matrices, ρ. However, due to the separability of H^2, the evaluation of the matrix element can be factorized as

$$\begin{aligned}\frac{\langle\phi_m|H^2|\phi_n^{(\lambda\nu)}\rangle}{\langle\phi_m|\phi_n^{(\lambda\nu)}\rangle} &= \sum_{i<j,\alpha<\beta}\left\{\sum_{k<l}\bar{v}_{ijkl}\left[(1-\rho^{(\lambda\nu)})_{k\alpha}(1-\rho^{(\lambda\nu)})_{l\beta}-(1-\rho^{(\lambda\nu)})_{l\alpha}(1-\rho^{(\lambda\nu)})_{k\beta}\right]\right\}\\ &\times\left\{\sum_{\gamma<\delta}\bar{v}_{\alpha\beta\gamma\delta}(\rho_{\gamma i}^{(\lambda\nu)}\rho_{\delta j}^{(\lambda\nu)}-\rho_{\delta i}^{(\lambda\nu)}\rho_{\gamma j}^{(\lambda\nu)})\right\}\\ &+\mathrm{Tr}\left[(t+\Gamma^{(\lambda\nu)})(1-\rho^{(\lambda\nu)})(t+\Gamma^{(\lambda\nu)})\rho^{(\lambda\nu)}\right]\\ &+\left\{\mathrm{Tr}\left[\rho^{(\lambda\nu)}\left(t+\frac{1}{2}\Gamma^{(\lambda\nu)}\right)\right]\right\}^2.\end{aligned} \tag{21}$$

Here, the trivial summations and their indices for the matrix products are omitted for readability. This factorization reduces the eightfold loop into a sixfold loop and decreases the computation time drastically. Note that the similar transformation to block matrices discussed in section 2.2.1 is also utilized in the evaluation of (21) so that the computation time is further reduced.

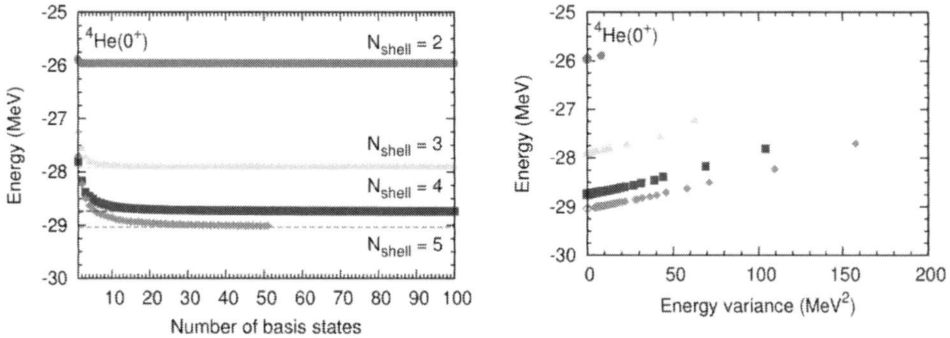

Figure 4. ^{4}He ground-state energies as functions of number of basis states (left) and energy variance (right). From the above to the bottom, the symbols (horizontal dashed lines in the left figure and open symbols at the zero energy variance in the right figure) are the MCSM (FCI) results in $N_{shell} = 2, 3, 4$ and 5, respectively. See [7] for the details.

As a typical example of the implementation, the behavior of the ground-state energies of ^{4}He (0^+) with respect to the number of basis states and to the energy variance are shown in figure 4.

From figure 4, one can see that the MCSM results can be extrapolated into the FCI ones by using the quadratic fit function of $E(\Delta E_2) = E(\Delta E_2 = 0) + c_1 \Delta E_2 + c_2(\Delta E_2)^2$ with the fit parameters, $E(\Delta E_2 = 0)$, c_1, and c_2.

2.3. Test runs on K computer

At the initial stage of the implementation of K computer, we have performed some test calculations to measure our code performance. In this subsection, we show some of the test calculations: the ratio to the peak performance and the parallel efficiency of our code.

2.3.1. Ratio to the peak performance In order to measure our code performance on K computer, we have chosen the optimization of 15th basis dimension of the wave function in $N_{\rm shell} = 5$ with 100 CG iterations without the preprocessing as a test case. The code has run on K computer by using MPI/OpenMP with 8 threads.

Figure 5 illustrate our recent MCSM code performance. The left panel of figure 5 shows the ratio to the peak performance in the calculation of the ^{4}He 0^+ ground state. Although the performance decreases as the number of CPU cores increases, it is around 30 - 40% up to 30720 cores (8 cores per node). The right panel of figure 5 shows the ratio to the peak performance as a function of the atomic numbers. The nuclear states listed in the figure are for the ground state of each nucleus. From the figure, the dependence of the performance on atomic number A is relatively weak for the number of nucleons at least up to $A = 12$.

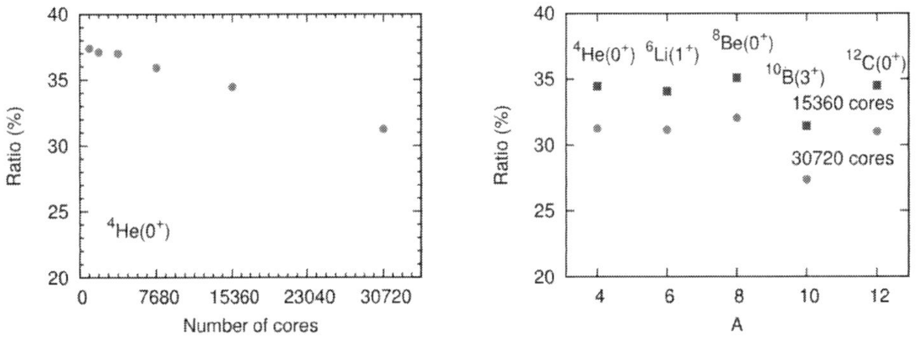

Figure 5. Ratio to the peak performance of the MCSM test calculations. Peak ratio of the calculation for the ^{4}He (0^+) ground state as a function of the number of cores (left). Peak ratio of the calculation for the ground states as a function of the number of nucleons (right). Red circles denotes the results with 30720 cores, and blue squares are with 15360 cores.

2.3.2. Parallel efficiency For testing the parallel efficiency, we have measured the dependence on the number of CPU cores. Figure 6 demonstrates the speedup (left) and the strong scaling (right) of our MCSM code on K computer as a function of the cores. The test case is the optimization of the 15th (48th) basis for ^{4}He 0^+ ground state in $N_{\rm shell} = 5$ (6) with 100 CG iterations without the preprocessing. Each setup has been chosen so that the number of MPI tasks is divisible by $N_{\rm procs}$, for simplicity. $32 \times 32 \times 30$ mesh points are used for the angular momentum projection, and 2 for the parity projection.

The left panel describes the speedup with arbitrary unit. In figure 6, the dotted line describes the perfect (ideal) scaling. The right panel of figure 6 is about the strong scaling. Here $\alpha_{\rm strong}$ is defined by the ratio of the time T with the number of CPU cores $N_{\rm procs}$ as $\alpha_{\rm strong} \equiv T(N_{\rm procs})/(T(N_{\rm procs}/2) \times 2)$. In this definition, $\alpha_{\rm strong} = 1$ describes the perfect strong scaling. As seen in figure 6, the strong scaling is nearly perfect up to 98304 cores both in $N_{\text{shell}} = 5$ and 6.

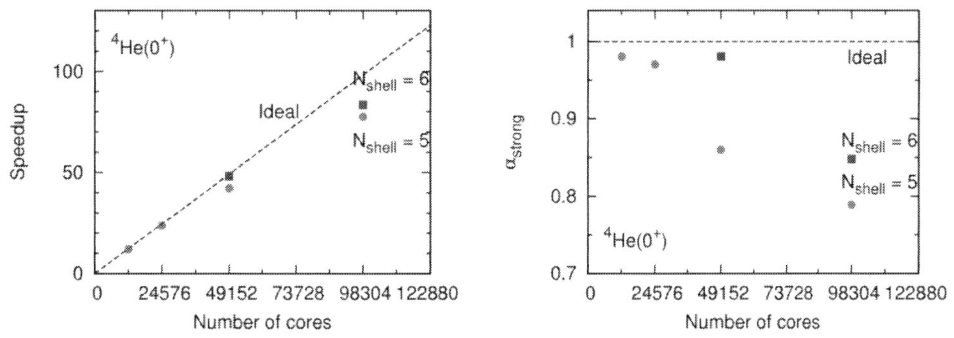

Figure 6. Speedup of the parallel computation with arbitrary unit (left), and the strong scaling (right).

3. Application

In this section, we show some of benchmark results for light nuclei. Also, the comparison of the radial density from the MCSM and FCI many-body wave functions are briefly presented in the case of ^{4}He ground state as an example.

3.1. Benchmarks

Owing to the recent development of the MCSM algorithm [8, 9, 10], we have performed the benchmark of no-core MCSM calculations [7, 8]. Figure 7 is the main outcome of the benchmark. In figure 7, we illustrate the comparisons of the energies for each state and model space between the MCSM and FCI methods. The FCI gives the exact energies in the finite model spaces, while the MCSM gives approximated ones. Thus the comparisons between them show how well the MCSM works in no-core calculations. Furthermore, we also plot the No-Core Full Configuration (NCFC) [12] results for the states of $4 \leq A \leq 10$ as the fully converged energies in the infinite model space, with assessed uncertainties visible in some cases.

For this benchmark comparison, the JISP16 two-nucleon interaction [13] is adopted and the Coulomb force is turned off. Isospin symmetry is assumed. The energies are evaluated for the optimal HO frequencies where the calculated energies are minimized for each state and model space. Here the contributions from the spurious center-of-mass (CM) motion are ignored for simplicity. The comparisons are made for the states; ^{4}He (0^+), ^{6}He (0^+), ^{6}Li (1^+), ^{7}Li $(1/2^-$, $3/2^-)$, ^{8}Be (0^+), ^{10}B $(1^+$, $3^+)$ and ^{12}C (0^+). The model space ranges from $N_{shell} = 2$ to 5 for $A \leq 6$ (4 for $A \geq 7$). Note that the energies of ^{10}B $(1^+$, $3^+)$ and ^{12}C (0^+) in $N_{shell} = 4$ are available only from the MCSM results. The M-scheme dimensions for these states are already close to or above the current computational limitation in the FCI approach. The numbers of MCSM basis states are taken up to 100 in $N_{shell} = 2-4$ and 50 in $N_{shell} = 5$.

As seen in figure 7, the energies are consistent with each other where the FCI results are available to within ~ 100 keV (~ 500 keV) at most of the MCSM results with(out) the energy-variance extrapolation of the MCSM results. The other observables; the point-particle root-mean-square matter radii and electromagnetic moments also give reasonable agreements between the MCSM and FCI results. The detailed comparisons among the MCSM, FCI, and NCFC methods are discussed in [7].

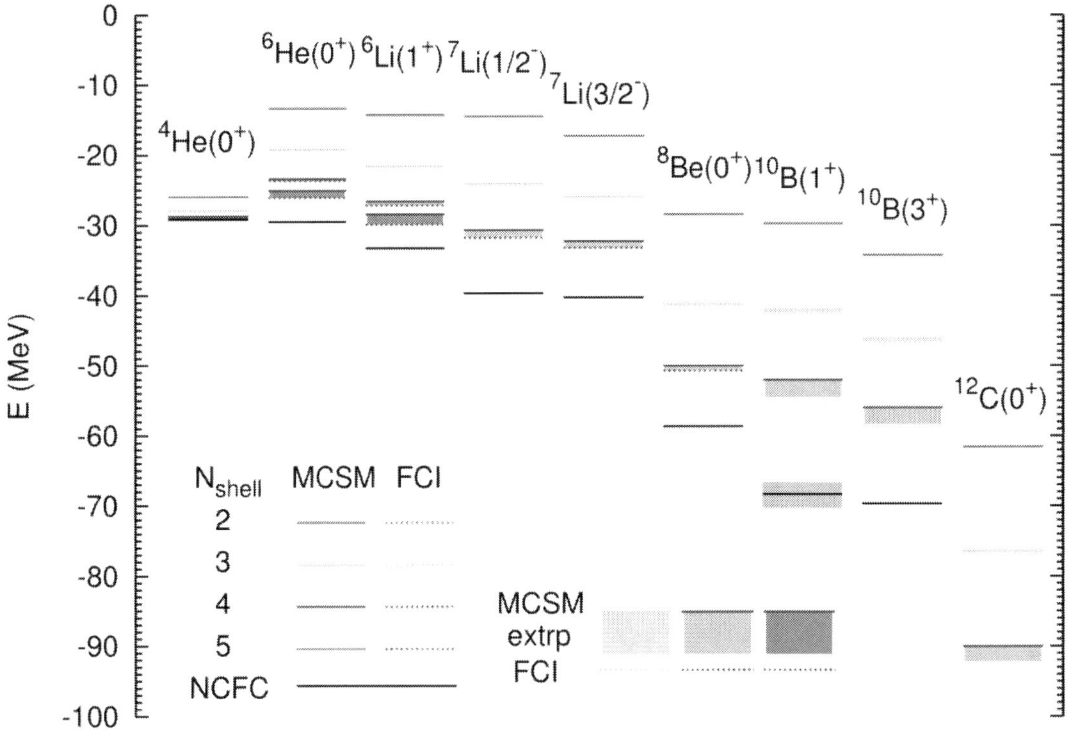

Figure 7. Comparisons of the energies between the MCSM and FCI along with the fully converged NCFC results where available. The NCFC result for the ^{10}B(1^+) state has a large uncertainty indicated by the grey band. The MCSM (FCI) results are shown as the solid (dashed) lines that nearly coincide where both are available. The extrapolated MCSM results are illustrated by bands. From top to bottom, the truncation of the model space is $N_{\rm shell} = 2$ (red), 3 (green), 4 (blue), and 5 (purple). Note that the MCSM results are extrapolated by the energy variance with the second-order polynomials. Also note that all of the results of ^{10}B and ^{12}C at $N_{\rm shell} = 4$ were obtained only with MCSM. Taken from [7].

3.2. Radial density of the ^{4}He ground state

The intrinsic local density is one of the basic quantities to understand the structure of nuclei intuitively. It is, however, challenging to obtain it from *ab initio* calculations. For example, the α-cluster structure of the ground and two excited states of ^{8}Be has been studied by the Variational Monte Carlo (VMC) and GFMC calculations with realistic nuclear forces [14]. They have attempted to obtain the intrinsic density in body-fixed coordinates by computing the moment of inertia matrix for each configuration and then diagonalizing them to rotate the configurations along the principal axis. Motivated by these analyses, we try to extract an intrinsic density from MCSM wave functions in a similar way adopted in the VMC and GFMC calculations [8, 15].

In shell-model calculations, we use single-particle coordinates, and the A-body wave function describes both the CM and relative motions. Using HO single-particle wave functions in combination with the $N_{\rm max}$ truncation (a truncation on the total number of HO quanta in the many-body basis), as is commonly done in the NCSM, the CM and relative wave functions factorize exactly. This allows for a deconvolution of the CM and translationally-invariant densities using three-dimensional Fourier transformations [16]. In [17], this method is used to show α-cluster structure for the ground state of ^{9}Be. Neither in the FCI calculations nor in the MCSM, however, do we have such an exact factorization.

In the MCSM, the one-body density is computed by using the MCSM wave function, equation (1),

$$\rho(\vec{r}) = \langle \Psi^{J^\pi M} | \sum_i \delta^3(\vec{r} - \vec{r}_i) | \Psi^{J^\pi M} \rangle \tag{22}$$

with the coordinate of the i-th nucleon, $\vec{r_i}$. Integrated over the angular directions, this gives the radial density, $\rho(r) = \int d\Omega \rho(\vec{r})$. Note that this density includes the CM motion. In principle, we could obtain the translationally-invariant one-body density in relative coordinates from the two-body density in single-particle coordinates, analogous to the calculation of the intrinsic rms radius by using a two-body operator, $\hat{r}^2 = (\hat{r}_i - \hat{r}_j)^2$, whereas the naive one-body operator, $\hat{r}_i^2$, gives us the rms radius that includes the effects of the CM motion.

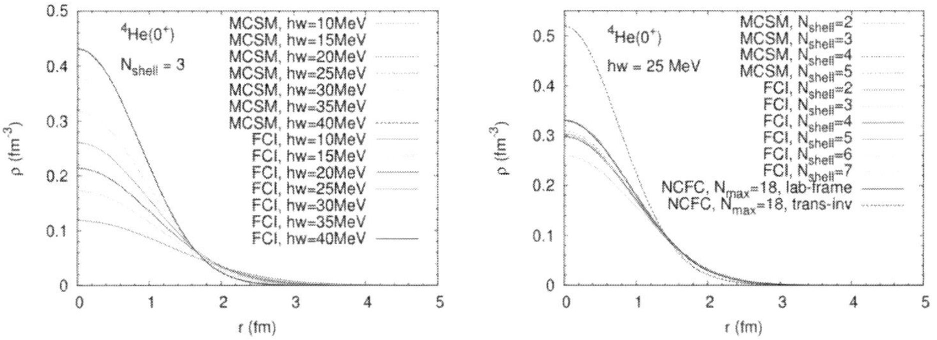

Figure 8. The radial density ρ of the ^{4}He ground state as a function of radius r for various HO energies $\hbar\omega$'s in the basis cutoff $N_{\rm shell} = 3$ (left), and for various $N_{\rm shell}$'s with $\hbar\omega = 25$ MeV (right). Both in the figures, the dashed (solid) curves denote the MCSM (FCI) results. In the left figures, the MCSM and FCI results agree well, and the differences are unrecognizable. For comparison, in the right figure, we include NCFC results before and after the deconvolution of the CM effect, corresponding to the lab-frame and translationally-invariant densities, respectively.

Figure 8 is the comparison of the radial densities of ^{4}He ground state using MCSM, and FCI, including the effects of the CM motion. The left figure shows the dependence of the radial density ρ on the HO parameter $\hbar\omega$, while the right figure is for the dependence on the basis cutoff $N_{\rm shell}$. For comparison, we also include the NCFC result in the right figure, both before and after the deconvolution of the CM motion. In the figure, these are labeled by "lab-frame" and "trans-inv", respectively. From figure 8, the MCSM and FCI results agree quite well. The small discrepancies (especially for the case of $N_{\rm shell} = 5$ in the right figure) come from the fact that MCSM results for ρ are not extrapolated to the FCI basis cutoff by using the energy variances. The left figure clearly shows that there is a strong dependence on HO parameter $\hbar\omega$. This is due to the fact not only that ρ is not converged in such a small basis space, $N_{\rm shell} = 3$, but also that the CM motion is strongly $\hbar\omega$ dependent. Even in the limit $N_{\rm shell}$ to infinity, these densities will depend strongly on $\hbar\omega$. Furthermore, the convergence with $N_{\rm shell}$ is not monotonic, as can be seen in the right figure, most likely because the CM motion is different for different $N_{\rm shell}$ values. Note that the density is comparable to the density in lab-frame coordinates obtained with the NCFC approach, suggesting that in the MCSM and FCI approaches the CM motion is approximately a $0s$ HO wave function. The translationally-invariant density, however, has a significantly larger central density, and falls off more rapidly: the effect of the CM motion is to smear out the translationally-invariant density [16].

4. Summary

By exploiting the recent development in the computation of the Hamiltonian matrix elements between non-orthogonal Slater determinants and the technique of energy-variance extrapolation, the no-core calculations with the MCSM algorithm can be performed efficiently on massively parallel supercomputers. From the benchmark calculations, the observables give good agreement between the MCSM and FCI results in the p-shell nuclei. The code performance of our MCSM on K computer was tested at the initial stage of implementation. We also present the comparison of the radial density of ^{4}He ground state among MCSM, FCI, and NCFC methods as an initial demonstration towards the intuitive description of nuclear shapes from shell-model calculations.

Acknowledgments

This work was supported in part by the SPIRE Field 5 from MEXT, Japan. We also acknowledge Grants-in-Aid for Young Scientists (Nos. 20740127 and 21740204), for Scientific Research (Nos. 20244022 and 23244049), and for Scientific Research on Innovative Areas (No. 20105003) from JSPS, and the CNS-RIKEN joint project for large-scale nuclear structure calculations. This work was also supported in part by the US DOE Grants No. DESC0008485 (SciDAC-3/NUCLEI), and DE-FG02-87ER40371, by US NSF grant 0904782, and through JUSTIPEN under grant No. DE-FG02-06ER41407. A part of the MCSM calculations was performed on the T2K Open Supercomputer at the University of Tokyo and University of Tsukuba, and the BX900 Supercomputer at JAEA. Computational resources for the FCI and NCFC calculations were provided by the National Energy Research Supercomputer Center (NERSC), which is supported by the Office of Science of the U.S. Department of Energy under Contract No. DE-AC02-05CH11231, and by the Oak Ridge Leadership Computing Facility at the Oak Ridge National Laboratory, which is supported by the Office of Science of the U.S. Department of Energy under Contract No. DE-AC05-00OR22725.

References

[1] Pieper S C, Wiringa R B and Carlson J 2004 *Phys. Rev.* C **70** 054325; Nollett K M, Pieper S C, Wiringa R B, Carlson J and Hale G M 2007 *Phys. Rev. Lett.* **99** 022502
[2] Hagen G, Hjorth-Jensen M, Jansen G R, Machleidt R and Papenbrock T 2012 *Phys. Rev. Lett.* **108** 242501; 2012 *Phys. Rev. Lett.* **109** 032502 and references therein
[3] Navrátil P, Vary J P and Barrett B R 2000 *Phys. Rev. Lett.* **84** 5728; 2000 *Phys. Rev.* C **62** 054311; Quaglioni S and Navrátil P 2008 *Phys. Rev. Lett.* **101** 092501; 2009 *Phys. Rev.* C **79** 044606; Barrett B R, Navrátil P and Vary J P 2013 *Prog. Part. Nucl. Phys.* **69** 131
[4] Roth R 2009 *Phys. Rev.* C **79** 064324; Roth R, Binder S, Vobig K, Calci A, Langhammer J and Navrátil P 2012 *Phys. Rev. Lett.* **109** 052501
[5] Dytrych T, Sviratcheva K D, Bahri C, Draayer J P and Vary J P 2007 *Phys. Rev. Lett.* **98** 162503; 2008 *J. Phys.* G. **35** 095101; Dytrych T, Sviratcheva K D, Draayer J P, Bahri C and Vary J P 2008 *J. Phys.* G. **35** 123101
[6] Liu L, Otsuka T, Shimizu N, Utsuno Y and Roth R 2012 *Phys. Rev.* C **86** 014304
[7] Abe T, Maris P, Otsuka T, Shimizu N, Utsuno Y and Vary J P 2011 *AIP Conf. Proc.* **1355** 173; 2012 *Phys. Rev.* C **86** 054301
[8] Shimizu N, Abe T, Tsunoda Y, Utsuno Y, Yoshida T, Mizusaki T, Honma M and Otsuka T 2012 *Prog. Theor. Exp. Phys.* 01A205
[9] Utsuno Y, Shimizu N, Otsuka T and Abe T 2013 *Commput. Phys. Commun.* **184** 102
[10] Shimizu N, Utsuno Y, Mizusaki T, Otsuka T, Abe T and Honma M 2010 *Phys. Rev.* C **82** 061305(R); Shimizu N, Utsuno Y, Mizusaki T, Honma M, Tsunoda Y and Otsuka T 2012 *Phys. Rev.* C **85** 054301
[11] Honma M, Mizusaki T and Otsuka T 1995 *Phys. Rev. Lett.* **75** 1284; 1996 *Phys. Rev. Lett.* **77** 3315; Otsuka T, Honma M and Mizusaki T 1998 *Phys. Rev. Lett.* **81** 1588; Otsuka T, Honma M, Mizusaki T, Shimizu N and Utsuno Y 2001 *Prog. Part. Nucl. Phys.* **47** 319
[12] Maris P, Vary J P and Shirokov A M 2009 *Phys. Rev.* C **79** 14308; Maris P, Shirokov A M and Vary J P 2010 *Phys. Rev.* C **81** 021301
[13] Shirokov A M, Vary J P, Mazur A I and Weber T A 2007 *Phys. Lett.* B **644** 33; Shirokov A M, Vary J P, Mazur A I, Zaytsev S A and Weber T A 2005 *Phys. Lett.* B **621** 96

[14] Wiringa R B, Pieper S C, Carlson J and Pandharipande V R 2000 *Phys. Rev.* C **62** 014001
[15] Yoshida T, Shimizu N, Abe T and Otsuka T contribution to the CCP2012 proceedings
[16] Cockrell C, Vary J P and Maris P 2012 *Phys. Rev.* C **86** 034325
[17] Maris P 2012 *J. Phys. Conf. Ser.* **402** 012031

Superfluidity of one-dimensional trapped fermionic optical lattices with spatially alternating interactions

Atsushi Yamamoto[1] **and Seiji Yunoki**[1,2,3,4]

[1]Computational Materials Science Research Team, RIKEN AICS, Kobe, Hyogo 650-0047, Japan
[2]Computational Condensed Matter Physics Laboratory, RIKEN ASI, Saitama 351-0198, Japan
[3]CREST, Japan Science and Technology, Kawaguchi, Saitama 332-0012, Japan
[4]Computational Quantum Matter Research Team, RIKEN CEMS, Saitama 351-0198, Japan

E-mail: a-yamamoto@riken.jp

Abstract. Recent experiments have made it possible to spatially control local two-body interactions between ultracold atoms in inhomogeneous optical lattices. Motivated by this experimental progress, here we study theoretically one-dimensional trapped fermionic optical lattices with spatially modulated on-site interactions. The density matrix renormalization group method is used to examine the ground state phase diagram of a harmonically trapped Hubbard model with spatially alternating on-site repulsive interactions, U_1 and U_2. With varying the strength of the harmonic trapping potential, we find a pair correlated metallic state in the ground state phase diagram, which can be stable only when $U_1 \neq U_2$. We discuss the properties of this possibly new state by examining various static physical quantities.

1. Introduction

Since Bose-Einstein condensation (BEC) was observed in a ^{87}Rb system [1], it has been recognized that ultracold atoms trapped in optical lattices can ideally emulate experimentally quantum models for interacting bosons or/and fermions in lattices. The most important advantage of ultracold atoms in optical lattices is high controllability of physical parameters such as interaction strength, particle number, and atomic hybridization. Therefore, these tunability enable us to simulate experimentally a wide range of quantum many body phenomena which might be difficult to observe in solid materials. In fact, many interesting phenomena including Mott insulator-metal transition [2, 3, 4] and BCS-BEC crossover [5, 6, 7] have been observed in fermionic optical lattice systems. Very recently, it has been reported that even the strength of the on-site interaction can be spatially modulated in ^{174}Yb gas [8]. This system provides an interesting possibility to explore the quantum effects of spatially modulated on-site interactions, which is certainly not possible in solid materials.

In this paper, we study theoretically one-dimensional (1D) trapped fermionic optical lattices with spatially alternating on-site interactions. We use the simplest model, i.e., the 1D Hubbard model with spatially alternating on-site repulsive interactions, U_1 and U_2, for odd and even sites, respectively. We also include in the model a harmonic trapping potential and the strength of this potential is varied as one of the model parameters. Using the density matrix renormalization group (DMRG) method [9], we study static properties to establish the ground state phase diagram. Among several different states identified in the phase diagram, we find as increasing

$|U_1 - U_2|$ a parameter region where a charge density wave like state and a metallic like state coexist. More interestingly, we find a pair correlated metallic state close to this coexisting region. This new state is characterized with zero charge gap but with finite binding energy. We also find that the pair correlation function increases in this region. Thus, we attribute this state to a superfluid state.

The rest of this paper is organized as follows. In section 2, we describe the model Hamiltonian and give a briefly summary of the numerical methods. In section 3, we show the results of ground state phase diagram for different values of the harmonic trapping potential. We also show numerical evidence of a possible superfluid state which appears by controlling the alternating repulsive on-site interactions and the trapping potential. The summary is given in section 4.

2. Model and Method

We consider two-component spin 1/2 fermions in a harmonically trapped 1D optical lattice with spatially alternating repulsive interactions. Our model is described by the following Hamiltonian,

$$\begin{aligned} \mathcal{H} &= -J \sum_{i=1}^{L} \sum_{\sigma} \left(c_{i,\sigma}^{\dagger} c_{i+1,\sigma} + H.c. \right) \\ &+ U_1 \sum_{i\ (\mathrm{odd})} n_{i,\uparrow} n_{i,\downarrow} + U_2 \sum_{i\ (\mathrm{even})} n_{i,\uparrow} n_{i,\downarrow} \\ &- V_c \left(\frac{2}{L-1} \right)^2 \sum_{i=1}^{L} \sum_{\sigma} \left(i - \frac{L+1}{2} \right)^2 n_{i,\sigma}, \end{aligned} \tag{1}$$

where $c_{i,\sigma}^{\dagger}(c_{i,\sigma})$ creates (annihilates) a fermion with spin $\sigma(=\uparrow,\downarrow)$ at site i on a L site lattice, and $n_{i,\sigma} = c_{i,\sigma}^{\dagger} c_{i,\sigma}$. The hopping parameter between the nearest neighbor sites is described by J, the repulsive on-site interaction for odd (even) sites is denoted by U_1 (U_2), and the strength of the harmonic trapping potential is controlled by V_c.

Because of finite V_c, the system becomes inhomogeneous. Indeed, when $U_1 = U_2$, it is known that the ground state show a coexisting state of a Mott insulating like state and a metallic like state with increasing the trap strength V_c [10]. Recently, the homogeneous system with $V_c = 0$ but with $U_1 \neq U_2$ has been also studied, revealing an unexpected metallic state [11]. Here, we focus on the effects of V_c on the ground state phase diagram. We employ the DMRG method [9] to study static properties such as local density distribution and binding energy. The number of states kept in DMRG calculations is 512 and we have checked that the results are well converged with this number of states. In the following, we set $J = 1$ as the energy unit.

3. Results

We first discuss the local density profile with varying V_c. The results are summarized in Fig. 1 for five different values of V_c. Here, we set $L = 90$, the numbers of up and down fermions $n_\uparrow = n_\downarrow = 45$ (i.e., half-filling), and the alternating repulsive interactions $U_1 = 2$ and $U_2 = 10$.

In the case of $U_1 = U_2$, it has been known that a coexisting phase appears, consisting of a Mott insulating like state and a metallic like state, with increasing V_c [10]. Even when $U_1 \neq U_2$, we expect the similar coexisting phase appears with increasing V_c. As we can see in Fig. 1, this is indeed what happens in the case of $U_1 = 2$ and $U_2 = 10$. The local density distribution $\langle n_i \rangle$ gradually changes with increasing V_c and becomes nonuniform. We can clearly see in Fig. 1 (d) and (e) that for large values of V_c the local density in the central region of the system has its value of either 1 or 2, i.e., $\langle n_i \rangle \approx 1$ for even sites with $U_2 = 10$ and $\langle n_i \rangle \approx 2$ for odd sites with $U_1 = 2$, strongly indicating a charge density wave (CDW) like state in the central region. Thus, when the trapping potential and the difference of the on-site interactions are both large, a CDW like state and a metallic like state coexist. Although this coexisting states are interesting, in this paper we focus on a parameter region of smaller values of V_c [Figs. 1 (b) and (c)], where an intriguing phase may also appear such as a possible superfluid state as discussed below.

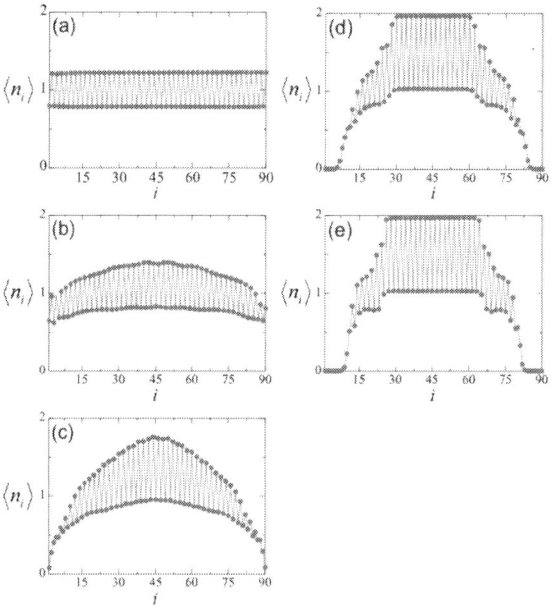

Figure 1. (Color online). Local density profile $\langle n_i \rangle$ for (a) $V_c = 0$, (b) $V_c = 2$, (c) $V_c = 5$, (d) $V_c = 10$, and (e) $V_c = 20$. The system size is $L = 90$ and the numbers of up and down fermions are $n_\uparrow = 45$ and $n_\downarrow = 45$, respectively. Here, we set $U_1 = 2$ and $U_2 = 10$.

Next, we evaluate binding energy E_B defined by

$$E_B = E(n_\uparrow + 1, n_\downarrow + 1) + E(n_\uparrow, n_\downarrow) - 2E(n_\uparrow, n_\downarrow + 1), \tag{2}$$

where $E(n_\uparrow, n_\downarrow)$ is the ground state energy for $n_\uparrow$ up and $n_\downarrow$ down fermions [12, 13]. As shown in Fig. 2 (a), we find that the binding energy becomes large for a region where U_1 and U_2 are very different when $V_c = 0$. The binding energy reaches its maximum value in a parameter region of $U_1 >> U_2 \approx 2$ (also equivalently $U_2 >> U_1 \approx 2$). Introducing the harmonic trapping potential, the parameter region where the binding energy becomes maximum moves to $U_1 >> U_2 \approx 2.5$ (also equivalently $U_2 >> U_1 \approx 2.5$) for $V_c = 2$ and $U_1 >> U_2 \approx 4$ (also equivalently $U_2 >> U_1 \approx 4$) for $V_c = 5$, as shown in Figs. 2 (b) and (c), respectively. The most interesting and important observation here is that with increasing V_c, the binding energy become negative in a region around $U_1 >> U_2 \approx 2 - 3$ (and $U_2 >> U_1 \approx 2 - 3$), as clearly seen in Fig. 2 (c). The negative binding energy is strong indication of a possible superfluid state. It should be noted here that as opposed to the cases for $V_c = 10$ and 20, these values of V_c where the negative binding energy is found are not large enough to induce the coexisting states, as already discussed in Fig. 1.

Now, it is highly interesting to calculate pair correlation functions. The pair correlation function $\Delta_i^\dagger \Delta_j$ is defined by

$$\Delta_i^\dagger \Delta_j = \langle \Psi_0 | c_{i,\uparrow}^\dagger c_{i,\downarrow}^\dagger c_{j,\uparrow} c_{j,\downarrow} | \Psi_0 \rangle, \tag{3}$$

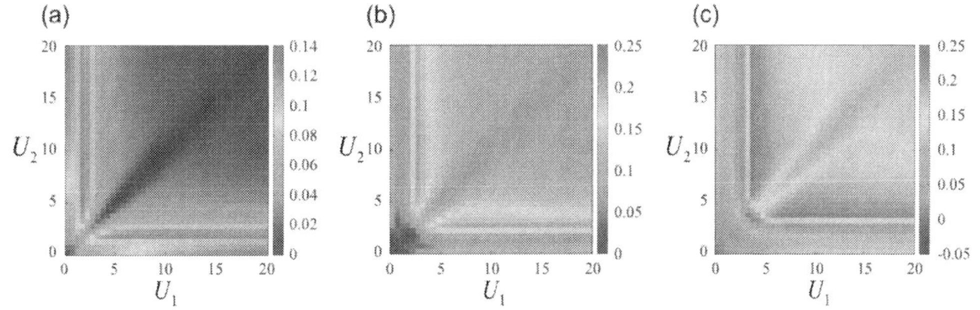

Figure 2. (Color online). Contour plots of the binding energy E_B as functions of U_1 and U_2 for (a) $V_c = 0$, (b) $V_c = 2$, and (c) $V_c = 5$. The system size is $L = 90$ and the numbers of up and down fermions are $n_\uparrow = 45$ and $n_\downarrow = 45$, respectively.

where $|\Psi_0\rangle$ is the ground state. The results are summarized in Fig. 3 for different values of V_c and $L = 90$. The pair correlation function $\Delta_i^\dagger \Delta_j$ is calculated for a given site j, i.e., for $j = 15$ [Fig. 3 (a)] and for $j = 29$ [Fig. 3 (b)]. Notice that, as shown in Fig. 1, the local state at site $j = 15$ is always within a metallic like state for all values of V_c studied here, while the local state at site $j = 29$ changes gradually from a metallic like state to a CDW like state with increasing V_c. Figure 3 clearly show that the pair correlation functions decay exponentially for large vales of $V_c \geq 10$. However, for small values of V_c, we notice that the long range tail of the pair correlation function increases with increasing V_c from 0 up to 5 as compared with the results for $V_c = 0$. The enhancement of the pair correlations strongly indicates that a state with the negative binding energy is indeed superfluid.

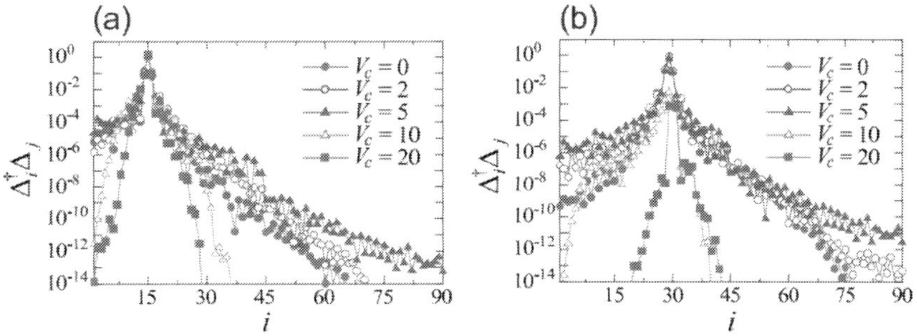

Figure 3. (Color online). Pair correlation function $\Delta_i^\dagger \Delta_j$ as a function of i for (a) $j = 15$ and for (b) $j = 29$ with different V_c indicated in the figures. The system size is $L = 90$ and the numbers of up and down fermions are $n_\uparrow = 45$ and $n_\downarrow = 45$, respectively. Here, we set $U_1 = 2$ and $U_2 = 10$.

4. Summary

We have studied the harmonically trapped 1D fermionic optical lattice with the spatially alternating on-site repulsive interactions. Using the DMRG method, we have obtained the ground state phase diagram and found that the ground state phase diagram is affected strongly by the harmonic trapping potential. We have found that a possible superfluid state can be stable in the phase diagram by tuning both the spatially modulated on-site interactions and the harmonic trapping potential. It is highly interesting to study whether a similar superfluid state appears in two spatial dimensions as well.

Acknowledgments
We wish to thank A. Koga, T. Saito, T. Shirakawa, H. Watanabe, M. Machida, and S. Yamada for valuable discussions. The calculations were performed mostly using the RIKEN Integrated Cluster of Clusters (RICC).

References

[1] Anderson M H, Ensher J R, Matthews M R, Wieman C E and Cornell E A 1995 *Science* **269**, 198
[2] Greiner M, Mandel O, Esslinger T, Hänsch T W and Bloch I 2002 *Nature* **415** 39
[3] Jördens R, Strohmaier N, Günter K, Moritz H and Esslinger T 2008 *Nature* **455** 204
[4] Schneider U, Hackermuller L, Will S, Best Th, Bloch I, Costi T A, Helmes R W, Rasch D and Rosch A 2008 *Science* **322** 1520
[5] Jochim S,Bartenstein M, Altmeyer A, Hendl G, Riedl S, Chin C,Hecker Denschlag J and Grimm R 2003 *Science* **302** 2101
[6] Zwierlein M W,Stan C A, Schunck C H, Raupach S M F, Gupta S, Hadzibabic Z and Ketterle W 2003 *Phys. Rev. Lett.* **91** 250401
[7] Bourdel T,Khaykovich L, Cubizolles J, Zhang J, Chevy F, Teichmann M, Tarruell L, Kokkelmans S J J M F and Salomon C 2004 *Phys. Rev. Lett.* **93** 050401
[8] Yamazaki R, Taie S, Sugawa S and Takahashi Y 2010 *Phys. Rev. Lett.* **105**, 050405
[9] White S R 1992 *Phys. Rev. Lett.* **69** 2863; 1993 *Phys. Rev. B* **48** 10345
[10] Rigol M, Muramatsu A, Batrouni G G and Scalettar R T 2003 *Phys. Rev. Lett.* **91** 130403; Rigol M and Muramatsu A 2004 *Phys. Rev. A* **69** 053612
[11] Yamamoto A, Yamada S and Machida M 2012 (unpublished).
[12] Noack R M, White S R and Scalapino D J 1994 *Phys. Rev. Lett.* **73** 882; 1995 *Europhys. Lett.* **30** 163; 1996 *Physica C* **270** 281
[13] Hayward C A, Poilblanc D, Noack R M, Scalapino D J and Hanke W 1995 *Phys. Rev. Lett.* **75** 926

Journal of Physics: Conference Series **454** (2013) 012068 doi:10.1088/1742-6596/454/1/012068

Theoretical studies of a three-band Hubbard model with a strong spin-orbit coupling for 5d transition metal oxide Sr_2IrO_4

Tomonori Shirakawa[1,2,3], **Hiroshi Watanabe**[1,2,3] **and Seiji Yunoki**[1,2,3]

[1]Computational Materials Science Research Team, RIKEN AICS, Kobe, Hyogo 650-0047, Japan

[2]Computational Condensed Matter Physics Laboratory, RIKEN ASI, Saitama 351-0198, Japan

[3]CREST, Japan Science and Technology (JST), Kawaguchi, Saitama 332-0012, Japan

E-mail: t-shirakawa@riken.jp

Abstract. Motivated by recent experiments on Sr_2IrO_4, we study the ground state properties of a two-dimensional three-band Hubbard model with a strong relativistic spin-orbit coupling. Using the exact diagonalization technique, the dynamical magnetic structure factor $M(\mathbf{q}, \omega)$ is calculated to examine the low-energy magnetic excitations. We find that the low-energy excitations in $M(\mathbf{q}, \omega)$ are well described by an effective Heisenberg model composed of a local Kramers doublet of an effective total angular momentum $J_{\text{eff}} = |\vec{s} - \vec{\ell}| = 1/2$. The antiferromagnetic exchange interaction estimated from $M(\mathbf{q}, \omega)$ is as large as ~ 80 meV, which is in good quantitative agreement with experiments. To study a possible long-range ordered state in the thermodynamic limit, we use the variational cluster approximation based on the self-energy functional theory, which is parallelized to accelerate the calculations. We find the ground state where the local Kramers doublet is in-plane antiferromagnetically ordered.

1. Introduction

5d transition metal oxides in a layered perovskite structure such as Sr_2IrO_4 [1] and Ba_2IrO_4 [2] have attracted much attention because of their unique properties caused by a strong relativistic spin-orbit coupling (SOC) of $5d$ transition element. It has been observed experimentally that these materials behave like an effective total angular momentum $J_{\text{eff}} = |\vec{s} - \vec{\ell}| = 1/2$ Mott insulator at low temperatures [2, 3, 4], where $\vec{s}$ ($\vec{\ell}$) is the spin (orbital) angular momentum operator. Moreover, very recent experiments using muon spin rotation (μSR), magnetic susceptibility, and resonant inelastic x-ray scattering (RIXS) have revealed that the low-energy magnetic excitations can be described by an "isospin"-1/2 Heisenberg model with an effective exchange interaction as large as $\sim$ 60-100 meV [2, 5, 6], which is almost comparable to the ones in the parent compounds of high-T_c cuprate superconductors [7].

Motivated by these experiments, we study theoretically the nature of magnetic excitations for Sr_2IrO_4 using a three-band Hubbard model with the SOC. For this purpose, here the numerically exact diagonalization technique is employed. We find that the low-energy magnetic excitations are described by an antiferromagnetic Heisenberg model with the nearest-neighbor exchange coupling as large as $\sim$ 80 meV. Furthermore, to study the stability of this antiferromagnetically ordered state, we use the variational cluster approximation (VCA) [8] based on the self-energy functional theory (SFT) [9]. The VCA is efficiently parallelized up to ~ 300 cores. We find that the ground state is in-plane antiferromagnetically ordered with the $J_{\text{eff}} = 1/2$ local Kramers doublet. This finding is in good agreement with experiments [3, 4, 5].

This paper is organized as follows. In Section 2, we introduce an effective three-band Hubbard model for Sr_2IrO_4. The magnetic excitations of this model are next discussed in Section 3. After giving brief explanation of VCA and description of its parallelization scheme, the stability of an antiferromagneticallty ordered state is discussed in Section 4. Finally, Section 5 concludes this paper.

2. Model

The local electronic configuration of Ir^{4+} ion in Sr_2IrO_4 is a low-spin state with $(t_{2g})^5$ because of the large crystalline field effect, which is estimated much larger than the SOC and the local Coulomb interactions [10]. Thus, one of the simplest models which capture the low-energy physics of Sr_2IrO_4 is the following effective Hubbard model, consisting of three t_{2g} orbitals (d_{xy}, d_{yz}, and d_{zx}), defined on a square lattice [11],

$$H = H_{\text{kin}} + H_{\text{so}} + H_U, \tag{1}$$

where

$$H_{\text{kin}} = \sum_{\mathbf{k}} \sum_{\alpha,\sigma} \varepsilon_{\mathbf{k}}^{\alpha} c_{\mathbf{k},\alpha,\sigma}^{\dagger} c_{\mathbf{k},\alpha,\sigma}, \tag{2}$$

$$H_{\text{so}} = \lambda \sum_{\mathbf{r}} \sum_{\alpha,\beta} \sum_{\sigma,\sigma'} \langle \phi_{\mathbf{r}\alpha\sigma} | \vec{\ell} \cdot \vec{s} | \phi_{\mathbf{r}\beta\sigma'} \rangle \, c_{\mathbf{r},\alpha,\sigma}^{\dagger} c_{\mathbf{r},\beta,\sigma'}, \tag{3}$$

$$\begin{aligned} H_U = {} & U \sum_{\mathbf{r}\alpha} n_{\mathbf{r},\alpha,\uparrow} n_{\mathbf{r},\alpha,\downarrow} + \frac{1}{2}(U' - J) \sum_{\mathbf{r}\sigma} \sum_{\alpha \neq \beta} n_{\mathbf{r},\alpha,\sigma} n_{\mathbf{r},\beta,\sigma} + \frac{U'}{2} \sum_{\mathbf{r}\sigma} \sum_{\alpha \neq \beta} n_{\mathbf{r},\alpha,\sigma} n_{\mathbf{r},\beta,\bar{\sigma}} \\ & - J \sum_{\mathbf{r}} \sum_{\alpha \neq \beta} c_{\mathbf{r},\alpha,\uparrow}^{\dagger} c_{\mathbf{r},\alpha,\downarrow} c_{\mathbf{r},\beta,\downarrow}^{\dagger} c_{\mathbf{r},\beta,\uparrow} + J' \sum_{\mathbf{r}} \sum_{\alpha \neq \beta} c_{\mathbf{r},\alpha,\uparrow}^{\dagger} c_{\mathbf{r},\alpha,\downarrow}^{\dagger} c_{\mathbf{r},\beta,\downarrow} c_{\mathbf{r},\beta,\uparrow}. \end{aligned} \tag{4}$$

Here, $c_{\mathbf{r},\alpha,\sigma}^{\dagger}$ ($c_{\mathbf{r},\alpha,\sigma}$) represents the creation (annihilation) operator of an electron with spin $\sigma = (\uparrow, \downarrow)$ and orbital $\alpha = (xy, yz, zx)$ on site $\mathbf{r}$, $n_{\mathbf{r},\alpha,\sigma} = c_{\mathbf{r},\alpha,\sigma}^{\dagger} c_{\mathbf{r},\alpha,\sigma}$, $c_{\mathbf{k},\alpha,\sigma}^{\dagger}$ is the Fourier transform of $c_{\mathbf{r},\alpha,\sigma}^{\dagger}$, and $\bar{\sigma}$ indicates the opposite spin of σ. The number N_e (N_h) of electrons (holes) is set to be $5N$ (N), where N is the number of sites. $\varepsilon_{\mathbf{k}}^{\alpha}$ is the energy dispersion of orbital α defined by

$$\varepsilon_{\mathbf{k}}^{xy} = -2t_1(\cos k_x + \cos k_y) - 4t_2 \cos k_x \cos k_y - 2t_3(\cos 2k_x + \cos 2k_y) + \Delta, \tag{5}$$

$$\varepsilon_{\mathbf{k}}^{yz} = -2t_4 \cos k_y - 2t_5 \cos k_x, \tag{6}$$

$$\varepsilon_{\mathbf{k}}^{zx} = -2t_4 \cos k_x - 2t_5 \cos k_y, \tag{7}$$

where Δ is the energy level difference between d_{xy} and the others [10, 11]. The relativistic SOC term is described by H_{so} in (3), where $\phi_{\mathbf{r}\alpha\sigma}$ is the Wannier orbital with spin σ and orbital α at site $\mathbf{r}$. We assume the matrix elements $\langle \phi_{\mathbf{r}\alpha\sigma} | \vec{\ell} \cdot \vec{s} | \phi_{\mathbf{r}'\beta\sigma'} \rangle$ are finite only for $\mathbf{r} = \mathbf{r}'$. H_{so} is then written in the matrix form

$$H_{\text{so}} = \frac{\lambda}{2} \sum_{\mathbf{r}\sigma} \left(c_{\mathbf{r},xy,\bar{\sigma}}^{\dagger}, c_{\mathbf{r},yz,\sigma}^{\dagger}, c_{\mathbf{r},zx,\sigma}^{\dagger} \right) \begin{pmatrix} 0 & -s_\sigma & -i \\ -s_\sigma & 0 & is_\sigma \\ i & -is_\sigma & 0 \end{pmatrix} \begin{pmatrix} c_{\mathbf{r},xy,\bar{\sigma}} \\ c_{\mathbf{r},yz,\sigma} \\ c_{\mathbf{r},zx,\sigma} \end{pmatrix}, \tag{8}$$

where $s_\uparrow = +1$ ($s_\downarrow = -1$). Equation (4) describes the Coulomb interaction term which includes intra-orbital Coulomb repulsion U, inter-orbital Coulomb repulsion U', Hund's coupling J, and pair hopping term J' with $U = U' + 2J$ and $J = J'$ [12].

A set of parameters $\mathbf{t} = (t_1, t_2, t_3, t_4, t_5, \Delta, \lambda) = (0.36, 0.18, 0.09, 0.37, 0.06, -0.36, 0.37)$ eV in $H_\mathbf{t} = H_{\text{kin}} + H_{\text{so}}$ for Sr_2IrO_4 [11] is determined to reproduce the energy band dispersion calculated by the first-principles electronic structure calculations based on the density functional theory [10]. In this paper, we set $U/t_1 = 6.111$ and $J/U = 0.1$. These values are similar to the ones estimated by the constrained random phase approximation [13, 14].

3. Exact diagonalization study on magnetic excitations

In this section, the magnetic excitations are studied using the numerically exact diagonalization technique. Because of the exponential increase of the Hilbert space with N, the system sizes that can be treated are limited. However, this method provides the unbiased results, which are valuable and complementary to other approximate calculations such as VCA.

3.1. Method

We employ the numerically exact diagonalization technique for a cluster of $\sqrt{8}\times\sqrt{8}$ with periodic boundary conditions. To reduce the Hilbert space dimensions, we fully use the symmetry of Hamiltonian, i.e., translational symmetry and spin inversion symmetry. We first use the Lanczos algorithm [15] to calculate the ground state $|\psi_0\rangle$ and the ground state energy E_0. Then, the continued fraction expansion [15] is used to calculate the dynamical correlation function for an operator $\mathcal{O}$, which is defined by

$$C_\mathcal{O}(\omega) = -\frac{1}{\pi}\text{Im}\,\langle\psi_0|\,\mathcal{O}^\dagger(\omega + i\eta - H + E_0)^{-1}\mathcal{O}\,|\psi_0\rangle\,, \tag{9}$$

where $\eta(>0)$ is a broadening factor. Note that the static correlation function $S_\mathcal{O} = \langle\psi_0|\,\mathcal{O}^\dagger\mathcal{O}\,|\psi_0\rangle$ is related to the dynamical correlation function by $S_\mathcal{O} = \int d\omega C_\mathcal{O}(\omega)$. Following the notation in (9), the dynamical magnetic structure factor is given by $M_\xi(\mathbf{q},\omega) = C_\mathcal{O}(\omega)$ with $\mathcal{O} = M_\xi(\mathbf{q}) = 2S_\xi(\mathbf{q}) + L_\xi(\mathbf{q})$, where $S_\xi(\mathbf{q})$ $[L_\xi(\mathbf{q})]$ is the Fourier transform of the ξ $(= x, y, z)$ component of the local spin (orbital) angular momentum operator $S^\xi_\mathbf{r}$ $(L^\xi_\mathbf{r})$,

$$S^\xi_\mathbf{r} = \frac{1}{2}\sum_\alpha\sum_{\sigma_1\sigma_2} c^\dagger_{\mathbf{r},\alpha,\sigma_1}(\hat{\sigma}_\xi)_{\sigma_1,\sigma_2}c_{\mathbf{r},\alpha,\sigma_2}, \tag{10}$$

$$L^\xi_\mathbf{r} = \begin{cases} i\sum_\sigma(c^\dagger_{\mathbf{r},zx,\sigma}c_{\mathbf{r},xy,\sigma} - c^\dagger_{\mathbf{r},xy,\sigma}c_{\mathbf{r},zx,\sigma}) & (\xi = x) \\ i\sum_\sigma(c^\dagger_{\mathbf{r},xy,\sigma}c_{\mathbf{r},yz,\sigma} - c^\dagger_{\mathbf{r},yz,\sigma}c_{\mathbf{r},xy,\sigma}) & (\xi = y) \\ i\sum_\sigma(c^\dagger_{\mathbf{r},yz,\sigma}c_{\mathbf{r},zx,\sigma} - c^\dagger_{\mathbf{r},zx,\sigma}c_{\mathbf{r},yz,\sigma}) & (\xi = z) \end{cases}, \tag{11}$$

and $\hat{\sigma}_\xi$ is the ξ component of the Pauli matrix.

3.2. Results

The results for the dynamical magnetic structure factor $M_\xi(\mathbf{q},\omega)$ for Sr_2IrO_4 are shown in the left panels of figure 1. Although there are several peaks in each momentum $\mathbf{q}$, we will show below that the low-energy spectral peaks below $\sim 0.5t_1$ are identified as the magnetic excitations of the local Kramers doublet.

Let us first define a quasi-particle operator by

$$a^\dagger_{\mathbf{r},\theta,\sigma} = s_\sigma\cos\theta c^\dagger_{\mathbf{r},xy,\sigma} + \frac{\sin\theta}{\sqrt{2}}(c^\dagger_{\mathbf{r},yz,\bar{\sigma}} + is_\sigma c^\dagger_{\mathbf{r},zx,\bar{\sigma}}), \tag{12}$$

characterizing a slightly generalized $J_{\text{eff}} = 1/2$-like local Kramers doublet [16, 17]. The value of θ is determined by maximizing the hole density $n_{\theta,\sigma} = \langle\psi_0|\,a_{\mathbf{r},\theta,\sigma}a^\dagger_{\mathbf{r},\theta,\sigma}\,|\psi_0\rangle$. Next, using this

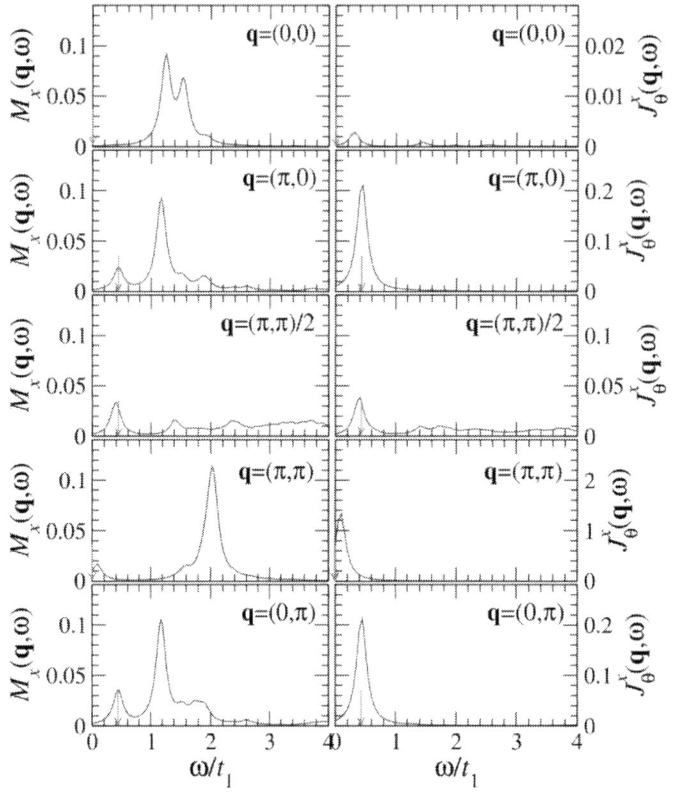

Figure 1. Dynamical magnetic structure factor $M_x(\mathbf{q},\omega)$ (left panels) and dynamical correlation function $J^x_\theta(\mathbf{q},\omega)$ (right panels) (see the text for definition) for several momenta $\mathbf{q}$ indicated in the figures. A set of parameter used is for Sr_2IrO_4 mentioned in text. The broadening factor η used here is $\eta = 0.1t_1$. Red arrows indicate the position of the magnon dispersion of an isotropic "isospin"-1/2 Heisenberg model with the nearest neighbor exchange interaction $0.22t_1$ (= 79 meV).

quasi-particle operator, we construct a pseudo-spin operator $J^\xi_\theta(\mathbf{r})$:

$$J^\xi_\theta(\mathbf{r}) = \frac{1}{2}\sum_{\sigma_1,\sigma_2} a^\dagger_{\mathbf{r},\theta,\sigma_1}(\hat{\sigma}_\xi)_{\sigma_1,\sigma_2} a_{\mathbf{r},\theta,\sigma_2}. \tag{13}$$

Then, we calculate the dynamical correlation function for $J^\xi_\theta(\mathbf{q})$, i.e., $J^\xi_\theta(\mathbf{q},\omega) = C_\mathcal{O}(\mathbf{q},\omega)$ with $\mathcal{O} = J^\xi_\theta(\mathbf{q})$, where $J^\xi_\theta(\mathbf{q})$ is the Fourier transform of $J^\xi_\theta(\mathbf{r})$. The results for $J^\xi_\theta(\mathbf{q},\omega)$ are shown in the right panels of figure 1. Comparing these two spectra $M_x(\mathbf{q},\omega)$ and $J^x_\theta(\mathbf{q},\omega)$ in figure 1, we find that the position of the low-lying peak at each $\mathbf{q}$ in $M_x(\mathbf{q},\omega)$ are exactly the same as the one in $J^x_\theta(\mathbf{q},\omega)$ [18]. This is the direct numerical evidence that the low-energy magnetic excitations correspond nicely to the excitations of the local Kramers doublet.

Assuming that the low-energy excitation dispersions are expressed by the magnon dispersion $\omega_\mathbf{k}$ of an isotropic "isospin"-1/2 Heisenberg model, i.e., $\omega_\mathbf{k} = 2J\sqrt{1-(\cos k_x + \cos k_y)^2/4}$, we can estimate an effective exchange interaction J. As shown in figure 1, we find that J is as large as $\sim 0.22t_1$ (= 79 meV). This value is in good agreement with experimental observations [2, 6, 5].

Journal of Physics: Conference Series **454** (2013) 012068 doi:10.1088/1742-6596/454/1/012068

4. Variational Cluster Approximation (VCA) study on the long-range ordered ground state

In the previous section, we have found that the low-energy magnetic excitation is well described by the effective antiferromagnetic Heisenberg model composed of the local Kramers doublet. However, the method used in Section 3 is not suitable to study properties in the thermodynamic limit. To access the thermodynamic limit and study a possible long-rage ordered state, here we employ an approximation technique called VCA.

4.1. Method

The VCA [8] based on the SFT [9] is one of cluster approaches to study interacting fermion systems on lattices. In this method, we first divide the original system into small clusters, such that each of them can be treated by the exact diagonalization technique. This set of small clusters is called a "reference system" below. In order to apply the VCA, the interacting part in the reference system must be the same as that in the original system, which is certainly satisfied in (1). However, the non-interacting part such as the kinetic energy term and the SOC term can be different. Here, we shall denote by $\mathbf{t}'$ the parameter set of the non-interacting part of the reference system. Within the VCA, the grand potential Ω of the original system per site is written by the following equation:

$$\Omega = \Omega_{\mathbf{t}'} - \frac{1}{N\pi}\int_0^\infty \mathrm{d}x \sum_{\mathbf{K}} \ln \left|\det(1-\hat{V}(\mathbf{K}))\hat{G}_{\mathbf{t}'}(ix)\right| + \frac{1}{2N}\sum_{\mathbf{K}} \mathrm{tr}\hat{V}(\mathbf{K}), \tag{14}$$

where $\Omega_{\mathbf{t}'}$ is the exact ground potential of the reference system and $\hat{G}_{\mathbf{t}'}(z)$ is the Green's function of a decoupled cluster of the reference system. The matrix elements of $\hat{G}_{\mathbf{t}'}(z)$ are simply given by

$$(\hat{G}_{\mathbf{t}'}(z))_{a,a'} = \langle\psi_{\mathbf{t}'}|\, c_a(z+\mu'-H_{\mathbf{t}'}+E_{\mathbf{t}'})^{-1}c_{a'}^\dagger\,|\psi_{\mathbf{t}'}\rangle + \langle\psi_{\mathbf{t}'}|\, c_{a'}^\dagger(z+\mu'+H_{\mathbf{t}'}-E_{\mathbf{t}'})^{-1}c_a\,|\psi_{\mathbf{t}'}\rangle\,, \tag{15}$$

where a is the shorthand notation of $(\mathbf{r}_s, \alpha, \sigma)$ and $\mathbf{r}_s$ is the site position in a decoupled cluster of the reference system. $H_{\mathbf{t}'}$ is the Hamiltonian in a decoupled cluster of the reference system with the ground state $|\psi_{\mathbf{t}'}\rangle$ and the ground state energy $E_{\mathbf{t}'}$. The chemical potential for the reference system is denoted by μ'. $\hat{G}_{\mathbf{t}'}(z)$ can be calculated using the continued fraction method. The matrix elements of $\hat{V}(\mathbf{K})$ is given by

$$(\hat{V}(\mathbf{K}))_{a,a'} = \frac{1}{N_c}\sum_{\mathbf{r}_c,\mathbf{r}'_c}(t_{A,A'}-t'_{A,A'})e^{\mathbf{K}\cdot(\mathbf{r}_c-\mathbf{r}'_c)}, \tag{16}$$

where N_c is the number of clusters, A is the shorthand notation of $(\mathbf{r}_c+\mathbf{r}_s,\alpha,\sigma)$, $\mathbf{r}_c$ is the position of the cluster, and $t_{A,A'}$ ($t'_{A,A'}$) indicates the coefficient of the term $c_{\mathbf{r},\alpha,\sigma}^\dagger c_{\mathbf{r}',\alpha',\sigma'}$ in the non-interacting part of the Hamiltonian for the original (reference) system, which includes the hopping parameter t_i ($i=1$–5), the SOC λ, and the potential difference Δ. $\mathbf{K}$ is the momentum for the superlattice of clusters defined by $\mathbf{r}_c$.

The choice of the parameter set $\mathbf{t}'$ is arbitrary and, in principle, the best result should be obtained by searching for the optimized $\mathbf{t}'$ which satisfies the stationary condition for Ω. However, computing directly Ω is impractical. For this reason, it is more convenient to start with an appropriate Weiss field. In previous section, we found that the low-energy magnetic excitations is similar to the ones of the effective Heisenberg model composed of the effective local Kramers doublet. Therefore, it is natural to introduce the following Weiss field for the reference system:

$$H' = h'\cos\phi' \sum_{\mathbf{r}} e^{i\mathbf{Q}\cdot\mathbf{r}} J_{\theta'}^x(\mathbf{r}) + h'\sin\phi' \sum_{\mathbf{r}} e^{i\mathbf{Q}\cdot\mathbf{r}} J_{\theta'}^z(\mathbf{r}), \tag{17}$$

with $\mathbf{Q} = (\pi, \pi)$. The variational parameters in $\mathbf{t}'$ treated in this study are thus h', θ', and ϕ', in addition to chemical potential μ'. For the optimization of multi-variables (h',θ', ϕ',μ'), we use the direction set method by optimizing one parameter at a time (with the remaining parameters fixed) iteratively until the full optimization is achieved for all parameters. The Brent's method is used for the optimization [19].

The most time consuming part in the VCA is the calculation for all the matrix elements of $\hat{G}(z)$. Because this calculation is independent, it is readily parallelized by using MPI without any trouble. We also parallelize the integral on x and summation over $\mathbf{K}$ in (14). Figure 2 (a) shows the total elapsed time (t_e) versus the total number $N_{\rm core}$ of cores used to optimize two parameters, μ' and θ', with a fixed system size. It is clearly seen in figure 2 (a) that t_e monotonically decrease with increasing $N_{\rm core}$ up to $N_{\rm core} \sim 300$ without saturation. We define the following quantity r_s to evaluate the efficiency on the strong scaling of the parallelization:

$$r_s = \frac{t_e^*}{t_e}\frac{N_{\rm core}}{N_{\rm core}^*}, \tag{18}$$

where $N_{\rm core}^*$ is a reference number of cores and t_e^* is the total elapsed time when $N_{\rm core}^*$ number of cores are used. Thus, $r_s = 1$ corresponds to 100% efficiency and usually r_s is less than one. As shown in fig. 2 (b), we find that our numerical scheme keeps $r_s > 0.8$ up to $N_{\rm core} \sim 300$.

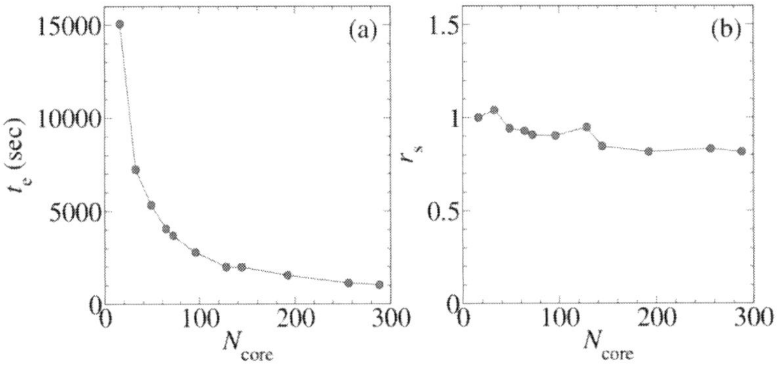

Figure 2. (a) Number of cores ($N_{\rm core}$) dependence of the elapsed time t_e to optimize two parameters (μ', θ') for fixed h', ϕ'. The model parameters used here are given in the text, and a 4-site cluster is used. (b) $N_{\rm core}$ dependence of r_s [defined by (18)] for the parallel performance. The reference number of cores is $N_{\rm core}^* = 8$.

4.2. Results

Figure 3 shows the variational parameter h' dependence of the ground state energy $E(h')$ obtained by the VCA using a 4-site cluster. The optimal h' is found to be $h' = 0.1892t_1$. A finite value of the optimal h' means that the system prefers to be symmetry broken. For all calculations, we find $\phi' = 0$, which indicates the effective Kramers doublet prefers to order antiferromagnetically with easy xy-plane anisotropy. This anisotropy is caused by the presence of the SOC and the Hund's coupling J. This finding is consistent with other theoretical results obtained by a strong-coupling theory [16], a 2-site exact diagonalization study [20], and a variational Monte Carlo study [11]. We also find in the VCA calculation that θ determined by minimizing the hole density $n_{\mathbf{r}\theta\sigma}$ is $\theta \sim 0.35\pi$, which is close the the value for the exact $J_{\rm eff} = 1/2$ limit, i.e., $\theta = \arctan\sqrt{2} \sim 0.304\pi$.

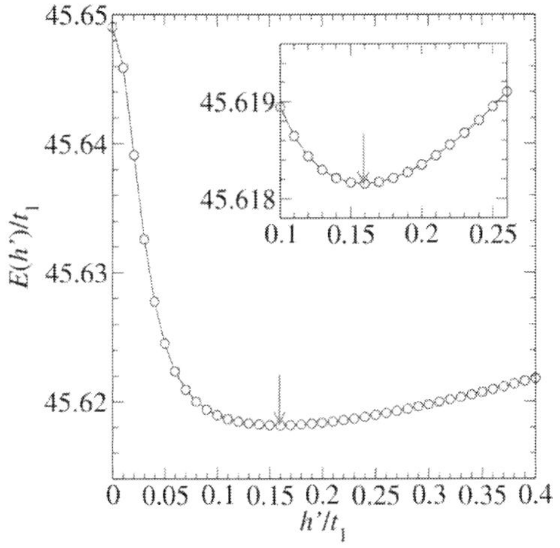

Figure 3. Variational parameter h' dependence of the energy $E(h')$ for the three-band Hubbard model [(1)] with a 4-site cluster. The other variational parameters (θ', ψ', μ') are optimized at each h'. The model parameters used are given in the text. The inset is an enlarged figure around the energy minima. Arrows indicate h' which gives the global minima of $E(h')$.

5. Conclusion

Using the numerically exact diagonalization technique, we have studied the magnetic excitations of the three-band Hubbard model with the strong SOC for Sr_2IrO_4. We have found that the low-energy magnetic excitations are well described by those of the $J_{\rm eff} = 1/2$ Kramers doublet, and that the lowest excitation dispersions can be reproduced by the antiferromagnetic Heisenberg model. We have estimated the effective exchange interaction as large as ~ 80 meV, which is in good agreement with recent experimental observations. We have also discussed the stability of the antiferromagnetically ordered ground state by using the VCA. We have found the ground state where the local Kramers doublet is in-plane antiferromagnetically ordered. These results strongly support the SOC induced $J_{\rm eff} = 1/2$ Mott insulator for Sr_2IrO_4.

Acknowledgments

The authors would like to thank H. Onishi and S. Fujiyama for fruitful discussions. The computation has been done using the RIKEN Cluster of Clusters (RICC) facility. This work has been supported by Grant-in-Aid for Scientific Research from MEXT Japan under the grant numbers 24740251 and 24740269.

References

[1] Randall Jr J J, Katz L and Ward R 1957 *J. Am. Chem. Soc.* **79** 266
[2] Okabe H, Isobe M, Takayama-Muromachi E, Koda A, Takeshita S, Hiraishi M, Miyazaki M, Kadono R, Miyake Y, and Akimitsu J 2011 *Phys. Rev. B* **83** 115118
[3] Kim B J, Ohsumi H, Komesu T, Sakai S, Morita T, Takagi H and Arima T 2009 *Science* **323** 1329
[4] Kim B J, Jin H, Moon S J, Kim J-Y, Park B-G, Leem C S, Yu J, Noh T W, Kim C, Oh S-J, Park J-H, Durairaj V, Cao G and Rotenberg E 2008 *Phys. Rev. Lett.* **101** 076402
[5] Fujiyama S, Ohsumi H, Komesu T, Matsuno J, Kim B J, Takata M, Arima T and Takagi H 2012 *Phys. Rev. Lett.* **108** 247212
[6] Kim J, Casa D, Upton M H, Gog T, Kim Y J, Mitchell J F, van Veenendaal M, Daghofer M, van den Brink J, Khaliullin G and Kim B J 2012 *Phys. Rev. Lett.* **108** 177003
[7] For instance, Lyons K B, Fleury P A, Schneemeyer L F and Waszczak J V 1988 *Phys. Rev. Lett.* **60** 732
[8] Potthoff M, Aichhorn M and Dahnken C 2003 *Phys. Rev. Lett.* **91** 206402
[9] Potthoff M 2003 *Eur. Phys. J.* B **32** 429
[10] Jin H, Jeong H, Ozaki T and Yu J 2009 *Phys. Rev.* B **80** 075112
[11] Watanabe H, Shirakawa T and Yunoki S 2010 *Phys. Rev. Lett.* **105** 216410
[12] Kanamori J 1963 *Prog. Theor. Phys.* **30** 275
[13] Martins C, Aichhorn M, Vaugier L and Biermann S 2011 *Phys. Rev. Lett.* **107** 266404
[14] Arita R, Kuneš J, Kozhevnikov A V, Eguiluz A G and Imada M 2012 *Phys. Rev. Lett.* **108** 086403

[15] Dagotto E 1994 *Rev. Mod. Phys.* **66** 763
[16] Jackeli G and Khaliullin G 2009 *Phys. Rev. Lett.* **102** 017205
[17] Shirakawa T, Watanabe H, and Yunoki S 2011 *J. Phys. Soc. Jpn. Suppl.* **80** SB010
[18] The low-lying peak in $M_\xi(\mathbf{q},\omega)$ at $\mathbf{q}=0$ corresponding to $J_\theta^\xi(\mathbf{q},\omega)$ is too small to see in this scale, but there exists the corresponding peak in $M_\xi(\mathbf{q},\omega)$ even at $\mathbf{q}=0$.
[19] Press W H, Teukolsky S A, Vetterling W T and Flannery B P 2007 *Numerical Recipes: The Art of Scientific Computing, Third Edition* (New York: Cambridge University Press) chapter 10 pp 497-515
[20] Kim B H, Khaliullin G and Min B I 2012 *Phys. Rev. Lett.* **109** 167205

Journal of Physics: Conference Series **454** (2013) 012069 doi:10.1088/1742-6596/454/1/012069

Correlated theory of linear optical absorption of octacene and nonacene

Himanshu Chakraborty and Alok Shukla

Physics Department, Indian Institute of Technology, Bombay, Powai, Mumbai 400076, India

E-mail: chakraborty.himanshu@gmail.com

Abstract. The technological importance of higher acenes has led to resurgence of interest in synthesizing higher acenes such as octacene, nonacene etc. Recently, Tönshoff and Bettinger [2010 *Angew. Chem. Int. Ed.* **49** 4125] have synthesized octacene and nonacene. Motivated by their work, we have performed large-scale calculations of linear optical absorption of octacene and nonacene. Methodology adopted in our work is based upon Pariser-Parr-Pople model (PPP) Hamiltonian, along with large-scale multi-reference singles-doubles configuration interaction (MRSDCI) approach.

1. Introduction

Polyacenes are linear fused benzene rings known for their well-defined structures and crystalline forms [1-2]. Their applications in novel opto-electronic devices such as light-emitting field effect transistors make them experimentally and theoretically very important [3-5]. As compared to smaller acenes, the electronic structure and excited state properties of heptacene, octacene and larger acenes have not been fully explored because of increasing reactivity along the conjugation length [5-6]. Payne et al. [7] have synthesized heptacene single crystals. Octacene and nonacene have been synthesized by Tönshoff and Bettinger by using cryogenic matrix-isolation technique and a protection group strategy [8]. Although pentacene has often been reported to be the best available organic p-type semiconductor, larger acenes could be even more useful [9] for material applications [10-11].

In this paper we present the singlet linear optical absorption calculations of octacene and nonacene and compare the results with the experimental spectra [8].

2. Theory

The schematic structures of octacene and nonacene are shown in Figure 1. The molecule is assumed to lie in the xy -plane with the conjugation direction taken to be along the x-axis. The carbon-carbon bond length has been fixed at 1.4 Å, and all bond angles have been taken to be 120°. The reason of choosing this symmetric geometry, against various other possibilities has already been discussed in our earlier paper [21]. It can be noted that these structures can also be seen as two polyene chains of suitable lengths, coupled together along the y-direction.

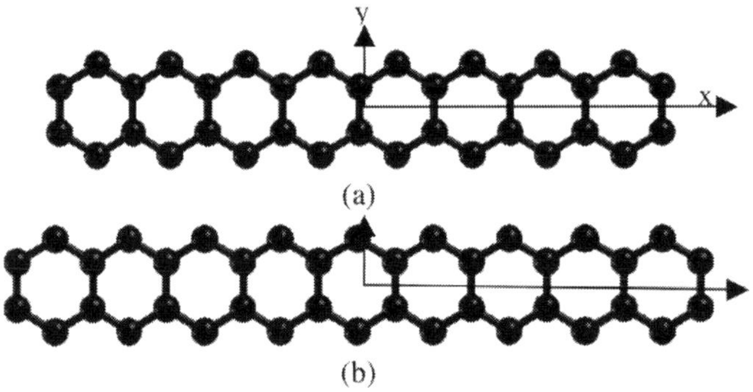

Figure 1. (a) Octacene and (b) Nonacene.

The correlated calculations are performed using the PPP model Hamiltonian [14], which can be written as

$$H = \sum_{i,j,\sigma} -t\,(c^{+}_{i\sigma}c_{j\sigma} + c^{+}_{j\sigma}c_{i\sigma}) + U\sum_{i} n_{i_\uparrow} n_{i_\downarrow} + \sum_{i<j} V_{i,j}(n_i - 1)(n_j - 1) \qquad (1)$$

$t = 2.4$ eV is nearest neighbour hopping; U and $V_{i,j}$ are on site and long-range Coulomb interactions respectively.

The Coulomb interactions are parameterized according to the Ohno relationship [16]

$$V_{i,j} = \frac{1}{\kappa_{i,j}\sqrt{(1 + 0.6117R^2_{i,j})}} \qquad (2)$$

$\kappa_{i,j}$ depicts the dielectric constant of the system which can simulate the effects of screening and R_{ij} is the distance in Å between i^{th} and the jth carbon atoms. We have performed calculations using the "standard parameters (std. par.)" with $U = 11.13$ eV and $\kappa_{i,j} = 1$ as well as the "screened parameters (scr. par.)" with $U = 8$ eV and $\kappa_{i,j} = 2$ $(i \neq j)$ and $\kappa_{i,i} = 1$ [15].

The starting point of the correlated calculations for the molecules is the Restricted Hartree-Fock (RHF) calculations, using the PPP Hamiltonian. All the resultant HF molecular orbitals are treated as active orbitals. The many-body effects beyond RHF are computed using the Multi-Reference Singles Doubles Configuration Interaction (MRSDCI) method [17-20] in the following manner. After RHF calculations of the ground state 1^1A_g are performed, the MRSDCI calculation of the ground state, 1^1A_g and excited states, $^1B_{2u}$ and $^1B_{3u}$ by taking the lowest energy configuration of the D_{2h} symmetry (1^1B_{2u} is H→L and 1^1B_{3u} is H→L+1 and H-1→L where H and L corresponds to Highest Occupied Molecular Orbital (HOMO) and Lowest Unoccupied Molecular Orbital (LUMO)). From the CI calculations, we obtain the eigenfunctions and eigenvalues corresponding to the correlated ground and excited states of the examined molecules. Using these eigenfunctions the dipole matrix elements between the ground state and various excited states are computed. These dipole matrix elements, along with the energies of the excited states are, in turn, utilized to calculate linear optical absorption spectra. The above process is repeated with the MRSDCI calculations of ground and excited states with different references added to the previous ones based on a coefficient value, say (0.1 or more) and also the references corresponding to the important states in the previous optical spectra. This procedure is repeated until satisfactory convergence is achieved. Using the PPP model, the MRSDCI calculations of linear optical absorption of smaller acenes from naphthalene up to heptacene have already been performed in our group and it was shown that the screened parameter results are in better agreement with experimental results [21-22].

3. Results and Discussion

In this section we present the calculations on the optical absorption of octacene and nonacene from their lowest-singlet state (1^1A_g), and compare the results with the experimental ones.

3.1. Singlet Linear Optical Absorption Calculations

Here, we present the linear optical absorption spectra of octacene and nonacene computed using the standard (std.) and the screened (scr.) parameters in Figure 2 and Figure 3.

Tables 2-5 in the Appendix displays the detailed results of our theoretical singlet linear optical absorption calculations of octacene and nonacene, for the most important low-lying odd parity states of B_{2u} (y-polarized) and B_{3u} (x-polarized) states with respect to the ground state ($1^1A_g^-$) using both the parameters.

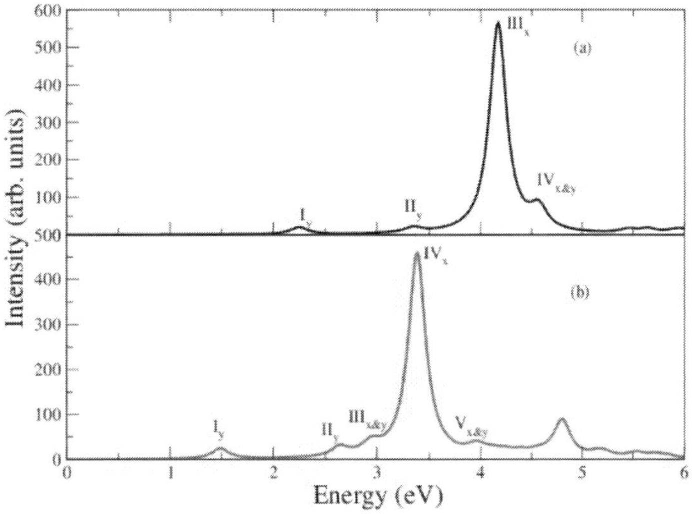

Figure 2. (color online) Linear optical absorption spectra of octacene using the (a) standard (black) and (b) screened (red) parameters. A uniform line width of 0.1 eV was assumed while plotting the spectra. The subscripts attached to the peak labels indicate the polarization directions x and y.

The singlet linear optical absorption of octacene and nonacene contains following features:

- Most of the intensity is concentrated in the x-polarized (long-axis polarized) spectra to the absorption into the $^1B_{3u}^+$ type of states, while the y-polarized (short-axis polarized) absorption into the $^1B_{2u}^+$ type states is very faint.
- The first peak is a weak peak, corresponds to the y-polarized, $1^1B_{2u}^+$ excited state of the system. The most important configuration contributing to the many-particle wave function of the state corresponds to $|H\rightarrow L\rangle$ excitation irrespective of the choice of the parameters.
- The second peak is a weak peak, corresponds to the y-polarized, $2^1B_{2u}^+$ excited state of the system. The most important configuration contributing to the many-particle wave function of the state corresponds to $|H\text{-}1\rightarrow L\text{+}1\rangle$ excitation irrespective of the choice of the parameters.
- The third peak corresponds to the x-polarized, $1^1B_{3u}^+$ excited state of the system and is the most intense state for the standard parameter case, whereas for the screened parameter results, it is a faint peak containing mixture of x and y polarized states, $1^1B_{3u}^+$ and $3^1B_{2u}^+$. The most important configuration contributing to the many-particle wave function for the standard parameter case, of the $1^1B_{3u}^+$ state, corresponds to $|H\rightarrow L\text{+}3\rangle$ and its charge conjugate, $|H\text{-}3\rightarrow L\rangle$ of octacene and $|H\text{-}4\rightarrow L\rangle$ and its charge conjugate, $|H\rightarrow L\text{+}4\rangle$ of nonacene, while that for the screened parameter case, of the $1^1B_{3u}^+$ state, is $|H\rightarrow L;\ H\text{-}1\rightarrow L\rangle$ and its charge

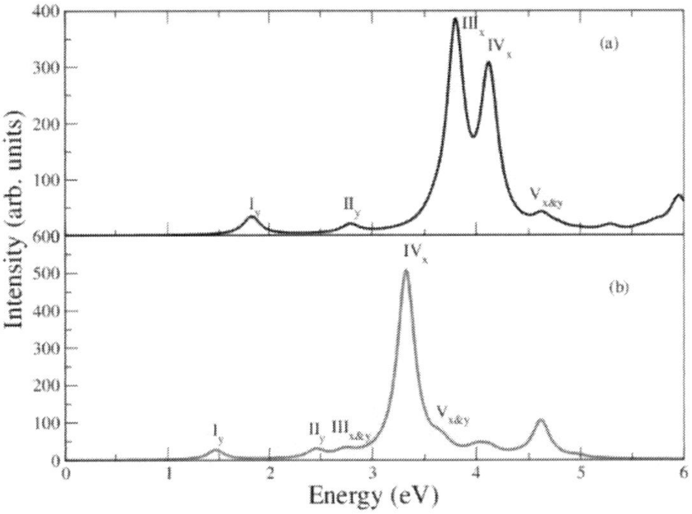

Figure 3. (color online) Linear optical absorption spectra of nonacene using (a) standard (black) and (b) screened (red) parameters. The rest of the information is same as in the caption of Figure 2.

conjugate |H→L; H→L+1⟩, and of the $3^1B_{2u}^+$ state, is |H-2→L⟩ and its charge conjugate |H→L+2⟩ excitations for both octacene and nonacene.

- The fourth peak corresponds to the x-polarized, $2^1B_{3u}^+$ excited state of the system which is the most intense state of the system for the screened parameter case whereas for the standard parameter results, for octacene, it is a faint peak containing mixture of x and y polarized states, $2^1B_{3u}^+$ and $4^1B_{2u}^+$ while for nonacene, it corresponds to the x-polarized state, $2^1B_{3u}^+$ which is a higher intense state than that of octacene because of the splitting of the most intense state $1^1B_{3u}^+$ for the standard parameter case in nonacene, it has the same features as the most intense state. The most important configuration contributing to the many-particle wave function for the screened parameter case, of the $2^1B_{3u}^+$ state, is |H-4→L⟩ and its charge conjugate, |H→L+4⟩ for both octacene and nonacene, while that for the standard parameter case of octacene, for the $2^1B_{3u}^+$ state, is |H→L; H-1→L⟩ and its charge conjugate |H→L; H→L+1⟩ and of the $4^1B_{2u}^+$ state, is |H-2→L+2⟩ whereas that for nonacene for the $2^1B_{3u}^+$ state, is same as the most intense state, |H→L+4⟩ and its charge conjugate, |H-4→L⟩.
  In general, the most important configuration contributing to the many-particle wave function of the most intense state for the standard parameter case, for n = even oligomers e.g. octacene (n=8), are excitations |H→L+(n/2-1)⟩ and its charge conjugate, |H-(n/2-1)→L⟩, and for n= odd (n=9) e.g. nonacene, the excitations |H→L+(n-1)/2⟩ and its charge conjugate, |H-(n-1)/2)→L⟩ while for the screened parameter case, for n = even oligomers e.g. octacene, are excitations |H→L+n/2⟩ and its charge conjugate, |H-n/2→L⟩, and for n= odd e.g. nonacene, the excitations |H→L+(n-1)/2⟩ and its charge conjugate, |H-(n-1)/2)→L⟩.
- Another important state, namely $1^1B_{3u}^-$ state exists for all oligomers. Because it has the same particle-hole symmetry (-) as the ground state, in PPP calculations it does not contribute to the absorption spectrum. But many experiments report this state as a very weak feature in the absorption spectrum. It is at higher excitation energy than the $1^1B_{2u}^+$ state. The important configurations contributing to the wave function of this state are the doubly excited configurations, |H→L; H-1→L⟩ which contribute significantly to this state. Thus, it is the electron-correlation effects which are responsible for its distinct location in the spectrum as compared to the $1^1B_{3u}^+$ state.

3.2. Comparison of Singlet Linear Optical Absorption with Experimental Absorption

We present comparison of the energies (eV) of most intense state of our singlet linear optical absorption with that of the experimental absorption spectra [8] of octacene and nonacene in Table 1. Hence, on the basis of the energies of the most intense state in the optical absorption spectra of octacene and nonacene, it is clear that our results match qualitatively with the experimental spectra irrespective of the parameters used. Quantitatively, our standard parameters results overshoot the excitation energies as compared to the experiments whereas the results based on the screened parameters show an opposite trend. This trend is reasonable as the standard parameters used to overestimate the gaps in general, so the screened parameters [15] were designed to include solid state or solvation effects (inter-chain screening). Therefore, our standard parameter results predict higher excitation energies as compared to the experiment and the same has been shown earlier in our group [21-22] for optical absorption of the lower acenes i.e. naphthalene up to heptacene. It has also been shown that the screened parameters agree well with experiments and other theoretical works as compared to the standard parameters. So, in our results, the screened parameters show lower energies with respect to experiments.

Further work on detailed comparison of our results with the experimental spectra and triplet excited state absorption calculations of octacene, nonacene and decacene is in progress [23].

The errors in terms of electron correlation (MRSDCI) and Hamiltonian (PPP model) are discussed below:

- MRSDCI approximation method is used for determination of the excitation energies of low-lying excited states which relies on an assumed cancellation of errors and thereby accurate prediction of the spectra [16-19]. As it is a variational method and any truncated CI has lesser no. of configurations than the full CI, so, energies predicted by them will be higher than those by full CI and hence, underestimates the correlations in ground and excited states. But the left out configurations in the MRSDCI method of both the ground and excited states will largely cancel out in the optical absorption calculations. Therefore, it gives a good description of the excitation state energies with respect to the ground state.
- In PPP model, only π electrons are considered explicitly, while the effect of σ-electrons are included in an implicit manner in terms of various parameters. Moreover, long range electron-electron interactions of density–density type are taken into account by means of suitable Coulomb parameters [12] while the rest are ignored. However, in spite of so many approximations involved, PPP model based calculations are extremely successful in describing the electronic structure of planar hydrocarbons in general, and their optical properties, in particular [13].

Table 1. Comparison of the energies of the most intense state of the absorption spectra of the theoretical and the experimental [8] work.

	Octacene		Nonacene	
Expt.	3.78		3.66	
	std. par.	scr. par.	std. par.	scr. par.
Theory	4.17	3.38	3.80	3.32

Acknowledgments

Ms. Himanshu Chakraborty would like to thank the Council of Scientific and Industrial Research (CSIR), India, for the Senior Research Fellowship (SRF).

Journal of Physics: Conference Series **454** (2013) 012069 doi:10.1088/1742-6596/454/1/012069

Appendix

Table 2. Excited states contributing to the linear absorption spectrum of octacene computed using MRSDCI method coupled with the standard parameters in the PPP model Hamiltonian. The table includes many particle wave functions, excitation energies, and dipole matrix elements of various states with respect to the ground state. DF corresponds to the dipole forbidden state. Below, '+c.c.' indicates that the coefficient of charge conjugate of a given configuration has the same sign, while '-c.c.' implies that the two coefficients have opposite signs.

Peak	State	E(eV)	Transition Dipole (°A)	Wave Functions
DF	$1^1B_{3u}^-$	2.31	0	$\|H\rightarrow L;H-1\rightarrow L\rangle$-c.c.(0.4948)
I	$1^1B_{2u}^+$	2.24	0.905	$\|H\rightarrow L\rangle$(-0.8471)
II	$2^1B_{2u}^+$	3.34	0.641	$\|H-1\rightarrow L+1\rangle$(-0.5826)
III	$1^1B_{3u}^+$	4.17	3.622	$\|H\rightarrow L+3\rangle$+c.c.(0.4622)
IV	$2^1B_{3u}^+$	4.57	1.079	$\|H\rightarrow L;H\rightarrow L+1\rangle$+c.c.(0.3214)
	$4^1B_{2u}^+$	4.51	0.367	$\|H-2\rightarrow L+2\rangle$(-0.4148)

Table 3. Excited states contributing to the linear absorption spectrum of octacene computed using MRSDCI method coupled with the screened parameters in the PPP model Hamiltonian. The table includes many particle wave functions, excitation energies, and dipole matrix elements of various states with respect to the ground state. DF corresponds to the dipole forbidden state. Below, '+c.c.' indicates that the coefficient of charge conjugate of a given configuration has the same sign, while '-c.c.' implies that the two coefficients have opposite signs.

Peak	State	E(eV)	Transition Dipole (°A)	Wave Functions
DF	$1^1B_{3u}^-$	1.59	0	$\|H\rightarrow L;H-1\rightarrow L\rangle$–c.c.(0.5078)
I	$1^1B_{2u}^+$	1.49	1.241	$\|H\rightarrow L\rangle$(0.8503)
II	$2^1B_{2u}^+$	2.65	0.897	$\|H-1\rightarrow L+1\rangle$(-0.7244)
III	$1^1B_{3u}^+$	2.97	0.845	$\|H\rightarrow L;H-1\rightarrow L\rangle$-c.c.(0.4942)
	$3^1B_{2u}^+$	2.87	0.440	$\|H\rightarrow L+2\rangle$+c.c.(0.2638)
IV	$2^1B_{3u}^+$	3.38	3.675	$\|H-4\rightarrow L\rangle$+c.c.(0.5831)
V	$4^1B_{2u}^+$	3.97	0.641	$\|H-2\rightarrow L+2\rangle$(-0.5280)
	$3^1B_{3u}^+$	3.91	0.410	$\|H-2\rightarrow L; H-1\rightarrow L\rangle$-c.c.(0.4028)

Table 4. Excited states contributing to the linear absorption spectrum of nonacene computed using MRSDCI method coupled with the standard parameters in the PPP model Hamiltonian. The table includes many particle wave functions, excitation energies, and dipole matrix elements of various states with respect to the ground state. DF corresponds to the dipole forbidden state. Below, '+c.c.' indicates that the coefficient of charge conjugate of a given configuration has the same sign, while '-c.c.' implies that the two coefficients have opposite signs.

Peak	State	E(eV)	Transition Dipole (°A)	Wave Functions
DF	$1^1B_{3u}^-$	1.86	0	$\|H\rightarrow L;H-1\rightarrow L\rangle$+c.c.(0.4872)
I	$1^1B_{2u}^+$	1.82	1.328	$\|H\rightarrow L\rangle$(+0.8290)
II	$2^1B_{2u}^+$	2.79	0.733	$\|H-1\rightarrow L+1\rangle$(+0.5506)
III	$1^1B_{3u}^+$	3.80	3.037	$\|H-4\rightarrow L\rangle$-c.c.(0.3757)
IV	$2^1B_{3u}^+$	4.12	2.583	$\|H\rightarrow L+4\rangle$-c.c.(0.3943)
V	$3^1B_{3u}^+$	4.63	0.549	$\|H\rightarrow L+1;H-1\rightarrow L+1\rangle$-c.c.(0.3339)
	$5^1B_{2u}^+$	4.62	0.439	$\|H\rightarrow L;H\rightarrow L; H-1\rightarrow L+1\rangle$(+0.6215)

Table 5. Excited states contributing to the linear absorption spectrum of nonacene computed using MRSDCI method coupled with the screened parameters in the PPP model Hamiltonian. The table includes many particle wave functions, excitation energies, and dipole matrix elements of various states with respect to the ground state. DF corresponds to the dipole forbidden state. Below, '+c.c.' indicates that the coefficient of charge conjugate of a given configuration has the same sign, while '-c.c.' implies that the two coefficients have opposite signs.

Peak	State	E(eV)	Transition Dipole (°A)	Wave Functions
DF	$1^1B_{3u}^-$	1.51	0	$\|H\rightarrow L;H\rightarrow L+1\rangle$+c.c.(0.5143)
I	$1^1B_{2u}^+$	1.46	1.316	$\|H\rightarrow L\rangle$(0.8551)
II	$2^1B_{2u}^+$	2.45	0.935	$\|H-1\rightarrow L+1\rangle$(0.7260)
III	$3^1B_{2u}^+$	2.77	0.507	$\|H\rightarrow L+2\rangle$-c.c.(0.5434)
	$1^1B_{3u}^+$	2.75	0.611	$\|H\rightarrow L; H\rightarrow L+1\rangle$-c.c.(0.4901)
IV	$2^1B_{3u}^+$	3.32	3.887	$\|H-4\rightarrow L\rangle$-c.c.(0.5689)
V	$4^1B_{2u}^+$	3.70	0.559	$\|H-2\rightarrow L+2\rangle$(-0.4715)
	$3^1B_{3u}^+$	3.64	0.782	$\|H-2\rightarrow L; H-1\rightarrow L\rangle$+c.c.(0.3441)

References

[1] Clar E 1964 *Polycyclic Hydrocarbons* vol 1 (London: Academic Press)
[2] Mullen K and Wegner G 1998 *Electronic Materials: The Oligomer Approach* (Weinheim Germany: Wiley-VCH)
[3] Dimitrakopoulos C D and Malenfant P R L 2002 *Adv. Mater.* **14** 99
[4] Hegmann F A, Tykwinski R R, Lui K P H, Bullock J E and Anthony J E 2002 *Phys. Rev. Lett.* **89** 227403
[5] Hepp A, Heil H, Weise W, Ahles M, Schmechel R and Seggern H 2003 *Phys. Rev. Lett.* **91** 157406
[6] Biermann D and Schmidt W 1980 *J. Am. Chem. Soc.* **102** 3163
[7] Payne M M, Parkin S R and Anthony J E 2005 *J. Am. Chem. Soc.* **127** 8028
[8] Tönshoff C and Bettinger H F 2010 *Angew. Chem. Int. Ed.* **49** 4125
[9] Antony J E 2008 *Angew. Chem. Int. Ed.* **47** 452
[10] Chun D, Cheng Y and Wudl F 2008 *Angew. Chem. Int. Ed.* **47** 8380
[11] Zade S S and Bendikov M 2010 *Angew. Chem. Int. Ed.* **49**, 4012
[12] Pariser R and Parr R G 1953 *J. Chem. Phys.* **21** 466 and 767; Pople J A 1953 *Trans Faraday Soc.* **49** 1375
[13] Salem L 1996 *Molecular orbital theory of conjugated systems* (New York: W. A. Benjamin Inc.)
[14] Sony P and Shukla A 2010 *Comp. Phys. Comm.* **181** 821
[15] Chandross M and Mazumdar S 1997 *Phys. Rev. B* **55** 1497
[16] Ohno K 1964 *Theor. Chim. Acta* **2** 219
[17] Ghosh H, Shukla A and Mazumdar S 2000 *Phys. Rev. B* **62** 12763
[18] Shukla A 2002 *Phys. Rev. B* **65** 12504
[19] Shukla A 2004 *Chem. Phys.* **300** 177
[20] Shukla A 2004 *Phys. Rev. B* **69** 165218
[21] Sony P and Shukla A 2007 *Phys. Rev.* B **75** 155208
[22] Sony P and Shukla A 2009 *J. Chem.Phys.* **131** 014302
[23] Chakraborty H and Shukla A in progress

Percolation phenomena in disordered topological networks

V Rodriguez, Y Diao and J Arsuaga

Department of Physics and Astronomy, San Francisco State University, San Francisco, CA 94116, USA

Department of Mathematics and Statistics, University of North Carolina at Charlotte, Charlotte, NC 28223, USA

Department of Mathematics, San Francisco State University, San Francisco, CA 94116, USA

E-mail: jarsuaga@sfsu.edu

Abstract. Topological polymer networks consist of circular polymers that are topologically linked. Topological networks made of small circular DNA or protein molecules are of great interest in biology and nanotechnology because they are found in living organisms and can be constructed in-vitro. The physical factors that determine the topology of a network as well as the pathways that are followed for its formation remain poorly understood. In our previous work we proposed a novel biophysical/computational approach to model the formation of planar DNA minicircle networks in trypanosomatid parasites. This model suggests that minicircle networks in trypanosomatid parasites emerged from topologically free minicircles upon space confinement through a percolation pathway. Our model however is somewhat idealized because it assumes that the centers of the minicircles in the network are positioned following a regular planar lattice. Here we propose an extension of the model by allowing the centers of the minicircles to be randomly displaced from the vertices of the lattice. We numerically show that networks form following a percolation pathway upon increasing minicircle density. Our model suggests that the critical percolation density increases as $D^{perc} = 0.8357 - 1.4297\exp(0.6439x)$ with x is the maximum displacement of the centers of the minicircles. Our results therefore show that the plane distribution of minicircles does not dramatically affect the percolation of minicircles and therefore supports they hypothesis that DNA minicircle networks in trypanosomes evolved through a percolation pathway.

1. Introduction

Topological polymer networks are networks that are held together through topological interactions of their subunits. Traditionally, topological polymer networks have been studied in the context of olympic gels where short linear polymers are cyclized in the presence of minicircles [14, 15, 17]. However topological networks are also found in diverse biological systems such as the protein network that makes up the capsid of bacteriophage HK97 [11, 20] or the DNA network found in the mitochondria of trypanosomatid parasites (reviewed in [18]). In vitro assays also can produce complex topological networks as in the case of type II topoisomerases assays [13] and of DNA nanotechnology assays (e.g. [14]).

Theoretical studies of topological networks were pioneered by de Gennes [15]. More recently, and motivated by DNA networks found in the kinetoplast of trypanosomatids [9], parasitic organisms that cause fatal diseases in human and livestock, we have proposed a new theoretical approach to study the formation of topological networks. Our model incorporates several features that are specific to the networks found in Trypanosomatid parasites. First, networks are made of small DNA molecules ($\leq$ 2.5kb), second the centers of the minicircles are confined to a plane (hence the name planar network) [12]; and third, any two minicircles are topologically linked by a single interlock [16].

In [9] we proposed that space confinement (or density) of DNA minicircles is a key factor in the formation of minicircle networks in Trypanosomatids. To test this hypothesis we introduced a model in which randomly oriented, geometric minicircles are placed on the vertices of a square lattice. We found that increasing density of minicircles triggers the formation of percolating clusters, that is clusters of topologically linked minicircles that span the length of the lattice. This model however may be an oversimplification of the network found in trypanosomatid parasites since for instance, the density at which percolation occurs depends on the specific lattice employed [7]. We here consider a model in which the placement of the minicircles is random. We find that under this condition networks still form by percolation and conclude that a percolation pathway is a rather robust phenomena that is observed upon minicircle confinement. We suggest that is a likely mechanism for the evolution of minicircle networks in trypanosomes.

2. Methods

2.1. Lattice generation and minicircle placement

Two dimensional square lattice models were generated as described in [9]. In brief, the center of each minicircle was placed at a unique lattice point and no two minicircles shared the same center; the orientation of each minicircle was determined by its normal vector which was sampled uniformly over the unit sphere. Disordered (irregular) lattices were generated by displacing each lattice point uniformly over a disc centered at the lattice vertex with radius given by a predetermined maximum displacement (M_{disp}), measured in terms of the distance between two adjacent lattice points. Similarly to the model of [9], the density of minicircles in this model is defined by the number of minicircles per unit area where the radius of the minicircles is used as the unit length. Consequently, if the distance between two adjacent lattice points in the square lattice is r, then the density of minicircles of the model is given by $1/r^2$.

2.2. Topological linking between minicircles

Let C_1 and C_2 be two disjoint minicircles in $\mathbb{R}^3$ (namely two circles that do not intersect each other). We say that C_1 and C_2 form an unsplittable link/non-trivial topological link (of two components) if no topological 2-sphere separates them. The same definition can be extended to the general case where there are n disjoint simple closed curves in $\mathbb{R}^3$. To determine whether two minicircles were linked we used a geometrical criterion that we introduced in [9]. Given two minicircles C_1 and C_2 of fixed radius r centered at P and Q respectively, the planes that contain the minicircles, together with the plane that bisects P and Q (so that it goes through the midpoint O of P and Q and is perpendicular to the line segment PQ) intersect at a single point x with probability one. As illustrated by figure 1, C_1 and C_2 form an unsplittable link if and only if $R > |Px| = |Qx|$.

2.3. An approach to the estimation of the percolation density in disordered topological networks

In order to determine the growth properties of minicircle networks we represent each minicircle as a lattice point and the linking between two minicircles as a lattice bond between two points. In our previous studies we determined percolation based on the existence of clusters that spanned the length the lattice. Disordered lattices do not have well defined lattice edges therefore these methods are not suitable. To calculate the critical percolation probabilities on two-dimensional random point distributions we extended the methods developed by Becker and Ziff [2]. In

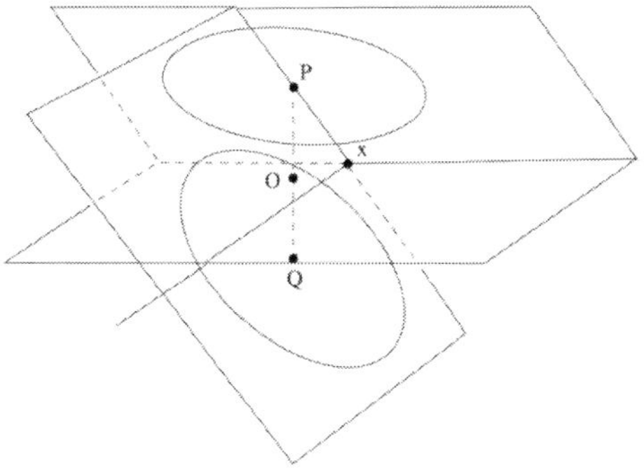

Figure 1. The relative positions of two minicircles and the bisecting plane of their centers. In the shown case the radius of the minicircles is less than $|Px| = |Qx|$ and they are clearly unlinked.

their work, randomized lattices with periodic boundary conditions were first generated using partitions of the plane. Clusters were grown from randomly selected points (called seed points) by connecting nearby points to the cluster according to a preselected probability p. This process was iterated until no more bonds could be formed or a pre-determined cluster size limit was reached. Percolation was determined by the distribution of cluster sizes as discussed next.

If P_s denotes the probability that a cluster grows to be at least size s and p_c the critical percolation probability then we may assume that P_s grows as $As^{2-\tau}f[B(p-p_c)s^{\sigma}]$ where $\tau = 187/91$ is the Fisher exponent for the two-dimensional percolation cluster universality class, $\sigma = 63/91$ and f is a the universal scaling function which measures the probability of occurrence of large clusters. By computing the Taylor expansion of f near the p_c we obtain $P_s \sim s^{2-\tau}[A + D(p-p_c)s^{\sigma} + ...]$ and if we define $C_s = P_s s^{\tau-2}$ then $C_s \sim [A + D(p-p_c)s^{\sigma}]$ which means that for large values of s, C_s grows linearly with s^{σ} and its slope vanishes when p is equal to the critical percolation probability. In our calculations we use D_c, the critical minicircle density, instead p_c to estimate the critical percolation density.

3. Results

3.1. *Estimation of Fisher exponent τ_{mini} in minicircle networks*

The Becker-Ziff method was developed to estimate the critical percolation probability for site and bond percolation on random lattices. To test that our model is consistent with these models and therefore the methods proposed by Becker-Ziff can be used in our minicircle model we first aimed at determining whether minicircle clusters grow in accordance with the critical exponents assumed in the Becker-Ziff method. For this reason we estimated the value of the Fisher exponent τ by studying the size distribution of minicircle clusters grown on the regular square lattice.We denote the estimated value τ_{mini}.

In independent bond and site percolation problems, n_s, the number of clusters of size s per lattice site, follows the general scaling relation $n_s(p) = s^{-\tau}f[(p-p_c)s^{\sigma}]$, where the behavior of the scaling function f can be approximated numerically [19]. As $p \to p_c$, f approaches a constant value. Thus, for $p \sim p_c$, $n_s(p) \sim k \cdot s^{-\tau}$ for some positive constant k. Using this relation, τ_{mini} can then be approximated by the slope of the linear fit of a log-log plot of the observed cluster size distribution N_s ($N_s = n_s \times K^2$ where the lattice is of dimension $K \times K$) using the critical percolation density $D_c = 0.637$ [9]. Figure 2 shows our numerical results for minicircle clusters generated on 1000×1000 regular square lattices with a sample of 1.2×10^7 minicircle clusters. The graph shows the linear regression fit to the log of the cluster size s and

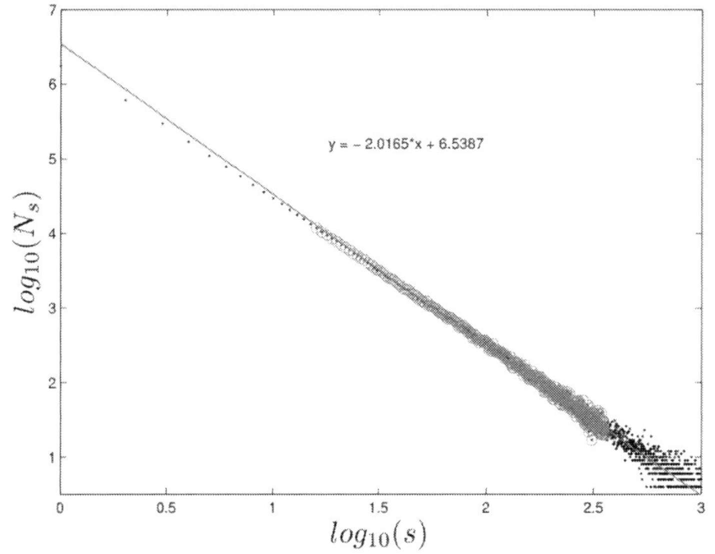

Figure 2. Log-log plot of N_s vs. s. The linear regression fit to $\log(N_s)$ vs. $\log(s)$ yields a τ_{mini} value which approximates the accepted traditional site and bond percolation $\tau = 187/91 \approx 2.055$ within one standard deviation.

the number of clusters of that size N_s. Our estimated value for τ_{mini} is 2.02 ± 0.06, which is in agreement with the value of τ for the 2-dimensional percolation cluster universality class. From this result we conclude that the methods developed in [2] are amenable for our study.

3.2. Estimation of the critical percolation density in the square lattice

To validate our approach, we first estimated the square lattice critical percolation density and compared it with the value of $D_c \sim 0.637$ obtained in [9]. As described in Subsection 2.3 we expect that for large clusters the slope of C_s is a constant and that at $D = D_c$ the slope of $C_s = 0$. We then first estimated C_s as a function of s^σ.

Figure 3 shows our numerical results. Notice that each curve in the figure shows the value of C_s for a different density values near 0.637. As expected the slope of each curve is constant for large clusters and by mere inspection one can appreciate that the value closest to the known percolation density seems to have zero slope.

To confirm this observation we proceeded as in [2]. We computed the slope of the best linear fit to each curve C_s for $s^\sigma \geq 10$. Since the value at which percolation occurs (i.e. slope of $C_s = 0$) may be in between any two of the tested density values, the best linear fit relating the growth of the slope of C_s with respect to the density and the x-intercept of the best fit was computed. Figure 4 (bottom) displays the best linear fit and the x-intercept highlighted. Our estimated value for $D_{perc} = 0.6370$ which is in exact agreement with the critical percolation density obtained in [9].

In order to test the dependence of our results on the s^σ value used on our linear fit we repeated the study with the linear fits for C_s for large s^σ for the cases $s^\sigma \geq 9.5$, $s^\sigma \geq 10$, $s^\sigma \geq 10.5$ and $s^\sigma \geq 11$. Our results are shown in Table 1. The data in the table shows that when we restrict the fitting range to larger s^σ values, the changes in the slopes of C_s vs. s^σ regression lines are

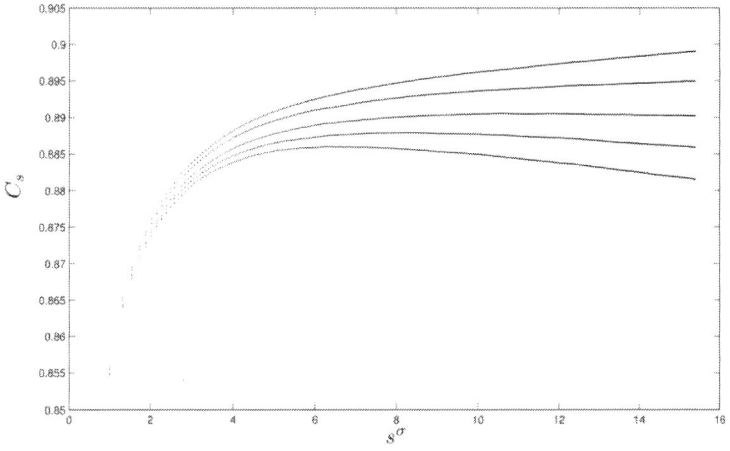

Figure 3. C_s curves for minicircle clusters grown for $D = 0.6363, 0.6366, 0.6370, 0.6373, 0.6376$ from bottom to top. The sample size is 10^7 for each density value.

small.

Table 1. C_s linear slope dependence on s^σ range.

minicircle density	min. s^σ	slope of lin. fit $\cdot 10^{-4}$
0.6363	9.5	−6.4851
	10.0	−6.5843
	10.5	−6.6722
	11.0	−6.7593
0.6366	9.5	−3.4024
	10.0	−3.5171
	10.5	−3.6150
	11.0	−3.7172
0.6370	9.5	−0.5547
	10.0	−0.6666
	10.5	−0.7397
	11.0	−0.7824
0.6373	9.5	2.4597
	10.0	2.3775
	10.5	2.3046
	11.0	2.2004
0.6376	9.5	5.3423
	10.0	5.2980
	10.5	5.2518
	11.0	5.1642

Journal of Physics: Conference Series **454** (2013) 012070 doi:10.1088/1742-6596/454/1/012070

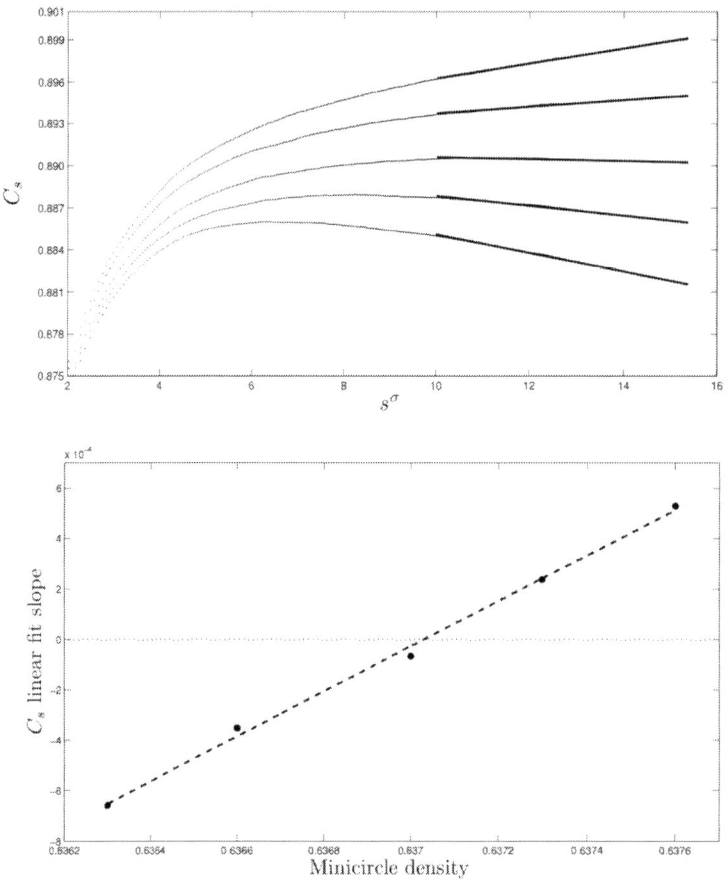

Figure 4. *Top:* Linear relationship between C_s and s^σ for large s in the regular square lattice. Curves were generated using 10^7 independent clusters grown at minicircle densities of $D = 0.6363, 0.6366, 0.6370, 0.6373, 0.6376$. Best fit lines for $s^\sigma \geq 10$ are shown to highlight the near linear behavior of C_s for large values of s^σ. *Bottom:* The linear regression line of the corresponding slopes of C_s best-fit lines from the top figure ($y = 0.89598x - 0.57077$, $R^2 = 0.9992$). The best fit line intersects the horizontal line at a minicircle density of 0.63703.

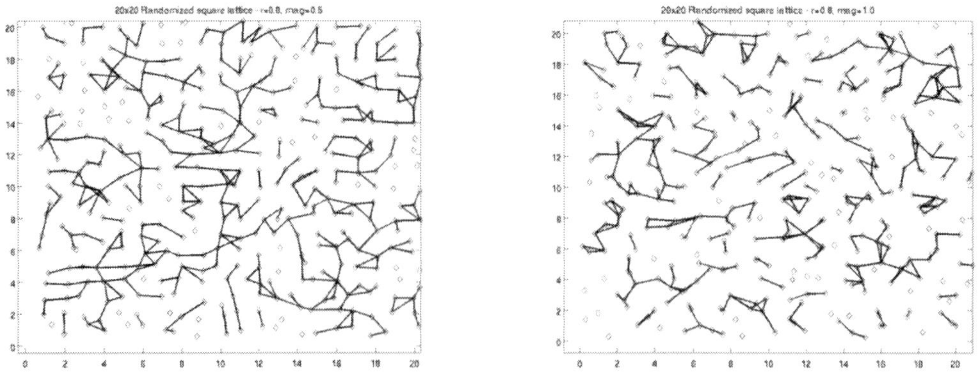

Figure 5. Minicircle clusters in disordered lattices with maximum displacement of 0.5 (left) and 1.0 (right) and density of 0.64.

Journal of Physics: Conference Series **454** (2013) 012070 doi:10.1088/1742-6596/454/1/012070

3.3. Estimation of the percolation density in disordered lattices

Our previous results validate our approach. Next we aimed to employ these methods to determine the critical percolation densities in disordered lattices. We generated disordered lattices as described in the methods section with maximum displacement magnitudes of 0.5, 1.0, 1.5, 2.0, 2.5 and 3.0. Figure 5 shows examples of two disordered lattices with different maximum displacements and same minicircle density. The figure highlights the different clusters.

As in the previous section the growth of C_s was calculated and linearly approximated for $s^\sigma \geq 10$ (similar results would be obtained if a different range of s^σ were used, so long as it is large enough, as indicated by table 1). Our results are shown in figures 6 to 11 for the cases of maximum displacements 0.5, 1.0, 1.5, 2.0, 2.5 and 3.0. Results for each displacement are consistent with those obtained in the previous section and also with those obtained for other displacements. A visual inspection of these results already reveals that C_s tends to be noisier for larger displacements and that the percolation density grows with the displacement. The growth of the percolation density is clearly appreciated in the overall range of the slopes of the linear approximations of C_s.

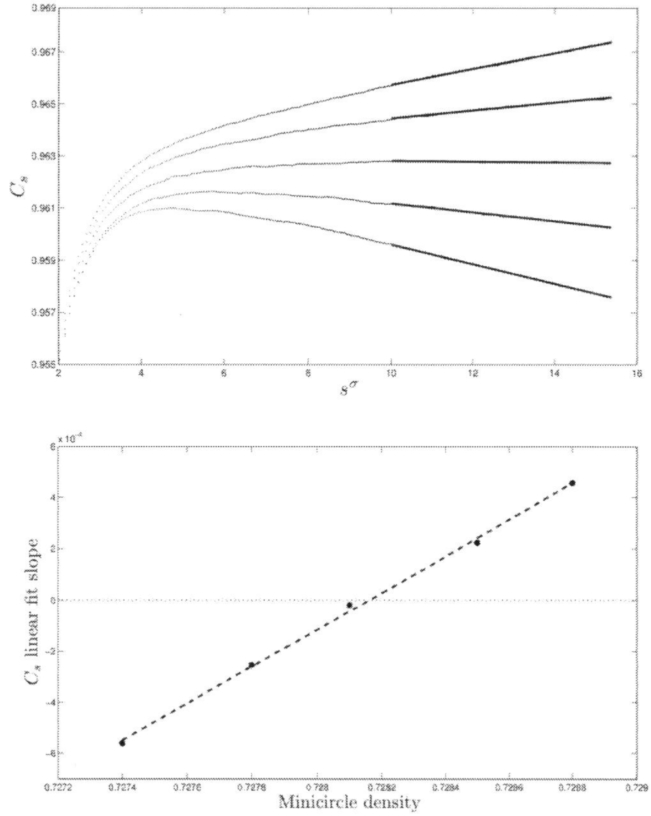

Figure 6. *Top:* C_s curves for lattices of maximum displacement 0.5 evaluated at minicircle densities 0.7274, 0.7278, 0.7281, 0.7285 and 0.7288 (from bottom to top). 5×10^6 clusters were sampled for each curve. *Bottom:* The linear regression line of the corresponding slopes of C_s best-fit lines ($y = 0.71742x - 0.52240$, $R^2 = 0.9995$). The best fit line intersects the horizontal line at a minicircle density of 0.72816.

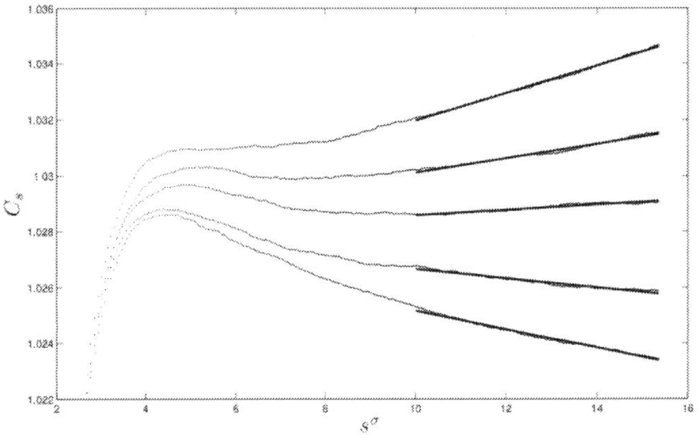

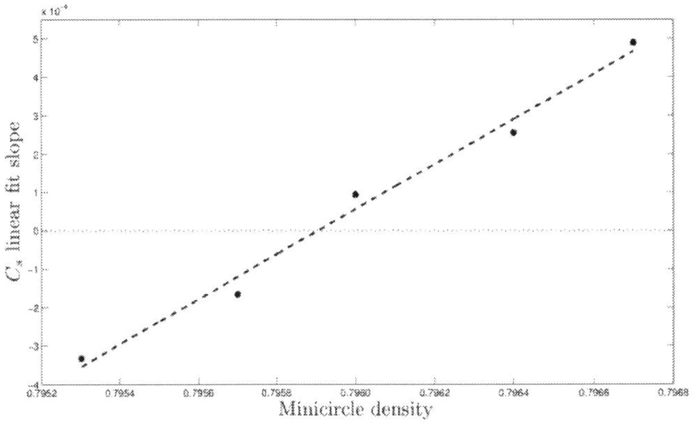

Figure 7. *Top:* C_s curves for lattices of maximum displacement 1.0 evaluated at minicircle densities 0.7953, 0.7957, 0.7960, 0.7964 and 0.7967 (from bottom to top). 5×10^6 clusters were sampled for each curve. *Bottom:* The linear regression line of the corresponding slopes of C_s best-fit lines ($y = 0.58672x - 0.46697$, $R^2 = 0.9948$). The best fit line intersects the horizontal line at a minicircle density of 0.79590.

Notice that while the minicircle densities sampled for the square lattice without displacement range from 0.6363 to 0.6376, those corresponding for a maximum displacement of 0.5 range from 0.7274 to 0.7288 and those corresponding to a maximum displacement of 3.0 range from 0.8299 to 0.8314. The critical percolation density for each disordered lattice is estimated using a linear approximation of the growth of the slopes as a function of minicircle density as shown on the bottom panels of figures 6 to 11. The estimated critical percolation densities are 0.7282 for $M_{disp} = 0.5$, 0.7959 for $M_{disp} = 1.0$, 0.8154 for $M_{disp} = 1.5$, 0.8239 for $M_{disp} = 2.0$, 0.8282 for $M_{disp} = 2.5$ and 0.8309 for $M_{disp} = 3.0$. Next we characterized the growth of the critical percolation density as a function of the maximum displacement. Results are shown in figure 12 where the critical percolation densities are plotted against the maximum displacements. These results seem to suggest that the critical percolation density will level off as the maximum displacement increases. The fitted curve in figure 12 has the form $y = a - b\exp(-cx)$ where a, b and c are real-valued constants. As x grows large, y approaches $a = 0.83579$, which is our estimate of the limiting value for the critical percolation density as a function of maximum displacement.

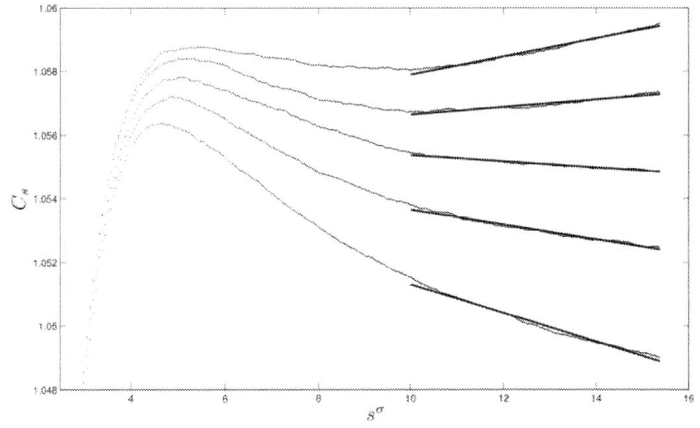

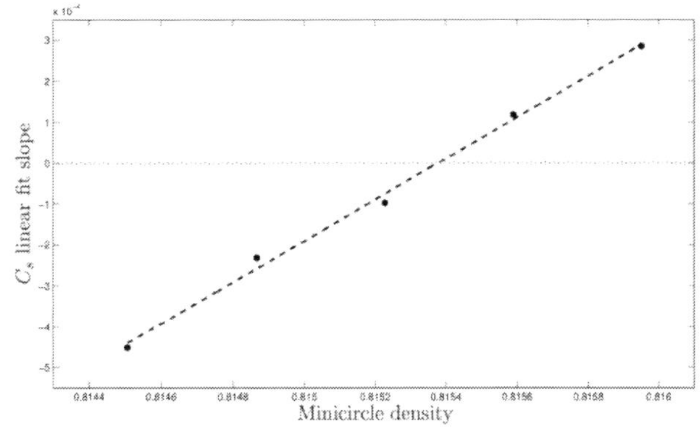

Figure 8. *Top:* C_s curves for lattices of maximum displacement 1.5 evaluated at minicircle densities 0.8145, 0.8149, 0.8152, 0.8156 and 0.8160 (from bottom to top) using 5×10^6 clusters for each curve. *Bottom:* The linear regression line of the corresponding slopes of C_s best-fit lines ($y = 0.50491x - 0.41170$, $R^2 = 0.9990$) intersects the x-axis at a minicircle density of 0.81537.

4. Conclusions

Topological networks made of small minicircles are important in the engineering of olympic gels and of nanomaterials [14, 15, 17] and in the formation of biologically relevant structures [11, 13, 18, 20]. In [9] we introduced a new theoretical model to study the formation of DNA minicircle networks in trypanosomatid parasites. Our model allowed us to characterize the effect of confinement [9], of the relative minicircle position[7], of the relative orientation of minicircles [1] and of DNA volume exclusion [8] in the formation of networks. All these models confirm the existence of a percolation pathway which suggest that the same pathway may be present in the formation of large topological networks in biological systems. The density at which percolation occurs depends on the parameters of each model. Interestingly our results show that the relative orientation of the minicircles has the most dramatic effect on percolation. These studies assume that minicircles are regularly positioned across a planar lattice. This assumption limits the applications of the model to structures that are crystalline like and even in those cases they still represent a somewhat idealized situation. In this study we have investigated the growth of networks on randomized lattices extending the methods developed by Becker and Ziff [2]. Our results show that disordered topological networks also show a percolation pathway when

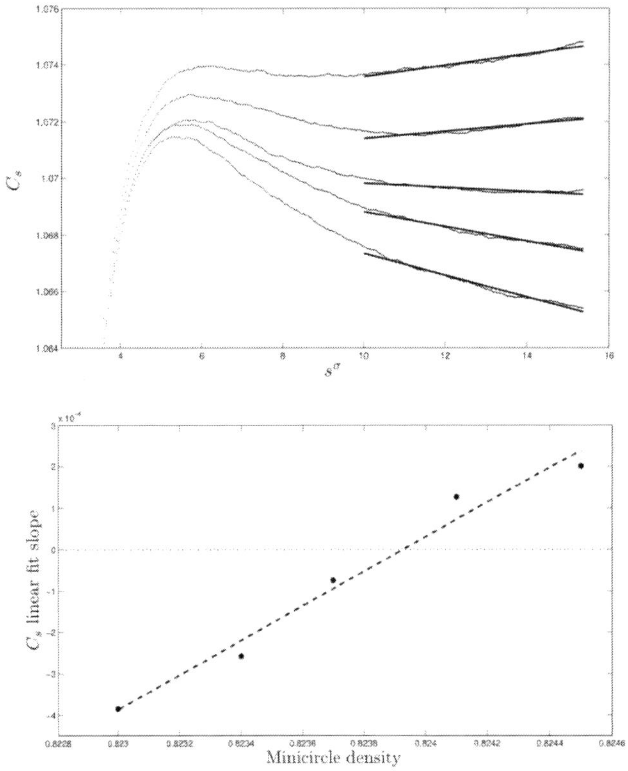

Figure 9. *Top:* C_s curves for lattices of maximum displacement 2.0 evaluated at minicircle densities 0.8230, 0.8234, 0.8237, 0.8241 and 0.8245 (from bottom to top), using 5×10^6 clusters for each curve. *Bottom:* The linear regression line of the corresponding slopes of C_s best-fit lines ($y = 0.41766x - 0.34412$, $R^2 = 0.9839$) intersects the x-axis at a minicircle density of 0.82393.

the density of minicircles is increased and that the critical percolation density increases as $D^{perc} = 0.8357 - 1.4297\exp(-0.6439x)$ with the maximum displacement of minicircles. These results therefore support our hypothesis that networks in trypanosomatid parasites grew through a percolation pathway and serve as a first step in the modeling of networks grown in more diluted conditions as those grown in-vitro in the presence of type II topoisomerases [13].

Acknowledgments

V. Rodriguez is supported by the MBRS-RISE Program 5R25-GM048972 of the National Institutes of Health (to Dr. F. Bayliss) and by NSF grant DMS 0920887, Y Diao is currently supported by grants DMS-0920880 and DMS-1016460 from National Science Foundation of USA and J Arsuaga is partially supported by NSF grants DMS 0920887 and DMS 1217324.

References

[1] Arsuaga J, Diao Y and Hinson K 2012 The effect of angle restriction on the topological characteristics of minicircle networks *J. of Stat. Phys.* **146**(2) 434–445

[2] Becker AM and Ziff RM (2009) Percolation thresholds on two-dimensional Voronoi networks and Delaunay triangulations *Phys. Rev.* E **80** 041101

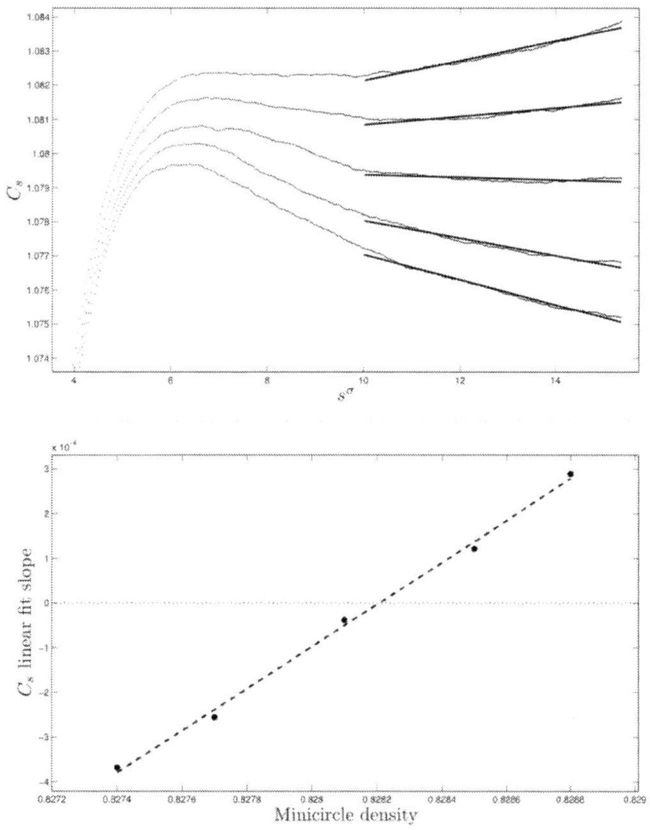

Figure 10. *Top:* C_s curves for lattices of maximum displacement 2.5 evaluated at minicircle densities 0.8274, 0.8277, 0.8281, 0.8285 and 0.8288 (from bottom to top). 5×10^6 clusters were sampled for each curve. *Bottom:* The linear regression line of the corresponding slopes of C_s best-fit lines ($y = 0.47026x - 0.38947$, $R^2 = 0.9983$). The best fit line intersects the horizontal line at a minicircle density of 0.82821.

[3] Chen J, Rauch CA, White JH, Englund PT and Cozzarelli NR 1995 The topology of the kinetoplast DNA network *Cell* **80**(1) 61–69
[4] Chen J, Englund PT and Cozzarelli NR Changes in network topology during the replication of kinetoplast DNA 1995 *EMBO J.* **14**(24) 6339–6347
[5] De Gennes PG 1979 *Scaling Concepts in Polymer Physics* Cornel University Press
[6] Diao Y 1994 Unsplittability of Random Links *J. Knot Theor. Ramif.* **3**(3) 379–389
[7] Diao Y, Hinson K and Arsuaga J 2012 The growth of minicircle networks on regular lattices *J. Phys. A: Math. Theor.* **45** 035004
[8] Diao Y, Hinson K and Arsuaga J 2013 The effect of volume exclusion in the formation of minicircle networks *Submitted*
[9] Diao Y, Hinson K, Kaplan R, Vazquez M and Arsuaga J 2012 The effects of minicircle density on the topological structure of the mitochondrial DNA from trypanosomes *J. Math. Biol.* **64**(6) 1087–1108
[10] Diao Y and van Rensburg J 1998 The Percolation of Linked Circles; *Topology and Geometry in Polymer Science* editors S G Whittington et al, the IMA Volumes in Mathematics and its Applications **103** 79–88
[11] Duda RL 1998 Protein Chainmail: Catenated Protein in Viral Capsids *Cell* **94**(1) 55–60
[12] Ferguson M, Torri A F, Perez-Morga D, Ward D C and Englund PT 1992 In situ hybridization to the Crithidia fasciculata kinetoplast reveals two antipodal sites involved in kinetoplast DNA replication *Cell* **70** 612–629

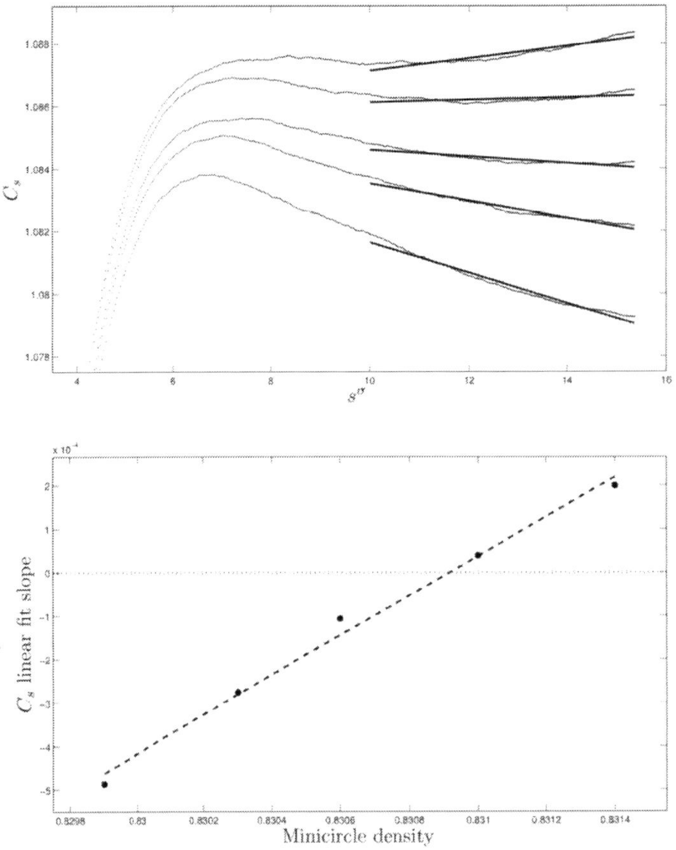

Figure 11. *Top:* C_s curves for lattices of maximum displacement 3.0 evaluated at minicircle densities 0.8299, 0.8303, 0.8306, 0.8310 and 0.8314 (from bottom to top). 5×10^6 clusters were sampled for each curve. *Bottom:* The linear regression line of the corresponding slopes of C_s best-fit lines ($y = 0.45449x - 0.37764$, $R^2 = 0.9958$). The best fit line intersects the horizontal line at a minicircle density of 0.83092.

[13] Kreuzer K N and Cozzarelli N R 1980 Formation and resolution of DNA catenanes by DNA gyrase *Cell* **20**(1) 245–54
[14] Pickett G T 2006 DNA-origami technique for olympic gels *Europhys. Lett.* **76** 616–22
[15] Raphaël E, Gay C and de Gennes P G 1997 Progressive Construction of an Olympic gel *J. Stat. Phys.* **89** 111–8
[16] Rauch C A, Perez-Moga D, Cozzarelli N R and Englund P T (1993) The absence of of supercoiling in kinetoplast DNA minicircles *EMBO J.* **12** 403–411
[17] Shafi KVPM et al 1999 Olympic Ring formation from newly prepared barium hexaferrite nanoparticle suspension *J. Phys. Chem.* B **103** 3358–60
[18] Shapiro T and Englund P 1995 The Structure and replication of kinetoplast DNA *Annu Rev Microbiol* **49** 117–43
[19] Stauffer D and Aharony A 1994 *Introduction to Percolation Theory, Revised Second Edition* CRC Press, New York, 41-49
[20] Wikoff W R et al 2000 Topologically linked protein rings in the bacteriophage HK97 capsid *Science* **289**(5487) 2129–33

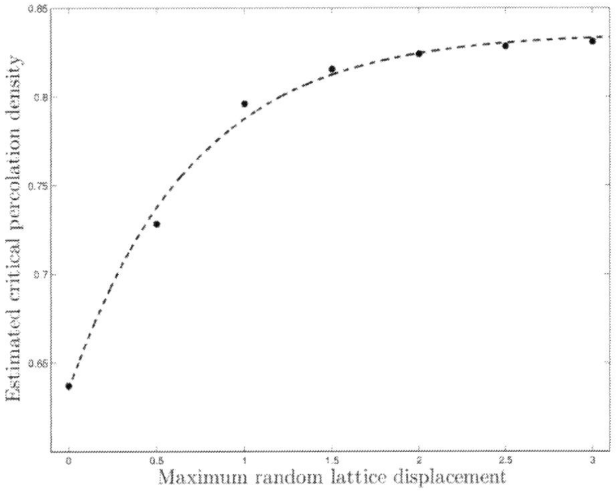

Figure 12. Percolation densities for maximum randomization magnitudes between 0.0 and 3.0. The critical percolation density increases monotonically as the maximum displacement magnitude is increased. The equation of the fitted curve is $y = 0.83579 - 1.42980e^{-0.64392x}$ ($R^2 = 0.9944$).

Journal of Physics: Conference Series **454** (2013) 012071 doi:10.1088/1742-6596/454/1/012071

Thermodynamics and structural properties of a confined HP protein determined by Wang-Landau simulation

Busara Pattanasiri[1,2], **Ying Wai Li**[3,7], **David P. Landau**[3], **Thomas Wüst**[4] **and Wannapong Triampo**[1,5,6]

[1] R&D Group of Biological and Environmental Physics, Department of Physics, Faculty of Science, Mahidol University, Bangkok 10400, Thailand
[2] Department of Chemistry, Faculty of Science, Mahidol University, Bangkok 10400, Thailand
[3]Center for Simulational Physics, Department of Physics and Astronomy, University of Georgia, Athens, GA 30602, USA
[4] Swiss Federal Research Institute WSL, Zürcherstrasse 111, CH-8903 Birmensdorf, Switzerland
[5] ThEP Center, CHE, 328 Si Ayutthaya Road, Bangkok 10400, Thailand
[6] Institute for Innovative Learning, Mahidol University, 999, Phuttamonthon 4 Road, Salaya, Nakorn Pathom 73170, Thailand
[7] National Center for Computational Sciences, Oak Ridge National Laboratory, Oak Ridge, TN 37831, USA

E-mail: bairng@physast.uga.edu, ywli@physast.uga.edu, dlandau@physast.uga.edu, thomas.wuest@wsl.ch, wtriampo@gmail.com

Abstract. Understanding protein folding confined by surfaces is important for both biological sciences and the development of nanomaterials. In this work, we study the properties of a confined HP model protein by three different types of surfaces, namely, surfaces that attract: (a) all monomers; (b) only P monomers; and (c) only H monomers. The thermodynamic and structural quantities, such as the specific heat, number of surface contacts, and number of hydrophobic pairs, are obtained by using Wang-Landau sampling. The conformational "transitions", specifically, the debridging process and hydrophobic core formation, can be identified based on an analysis of these quantities. We found that these transitions take place at different temperatures, and the ground state configurations show variations in structural properties when different surface type is used. These scenarios are confirmed by snapshots of typical states of the systems. From our study, we conclude that the thermodynamics of these transitions and the structural changes depend on the combined actions of both the composition of the H monomers and the P monomers in the HP chain and the surface types.

1. Introduction

Proteins are important biological molecules that perform most of the vital functions in living cells. Some proteins provide structural support, while others are involved in defending the body from antigens, or in muscle contraction [1]. In order to perform their specific functions, proteins must fold to a correct three dimensional shape or conformation, known as the native state. Incorrect protein folding are shown to be associated with a number of diseases, including Alzheimers, Huntington's, Parkinsons, prion and other degenerative diseases [2–4]. The native conformation of a protein is determined by its amino acid sequence; however, understanding how proteins find their own native structure remains one of the most fundamental problems in the biological sciences.

It is known that within the living cell, the environment is extraordinarily complex, many other molecules exist and folding spaces are restricted [5,6]. Therefore, the effects of the surrounding environment on protein folding has drawn much attention from the research community; one of which being the confinement effect. This includes not only protein folding in a cell but also its practical applications, such as protein crystallization [7], biotechnology [8–10] and the food industry [11,12].

Experimental studies and simulations have suggested that protein stability and folding dynamics are strongly correlated with the geometry and degree of confinement [13–16]. The effect of confinement can be absent if the confinement space is too large, whereas the native state is very restricted if the confinement space is too small [17]. Furthermore, hydrophobicity of the confining surfaces also plays an important role in protein folding. For example, weak hydrophobic confining surfaces can accelerate the folding of proteins [18,19].

In this work, we investigate the effect of surface properties on protein folding under confinement. We focus on the thermodynamics and structural behavior using the hydrophobic-polar (HP) lattice protein model [20]. This simple, coarse-grained protein model is chosen because it is difficult to obtain the "big picture" of this problem using, for instance, atomistic simulations, force-field or molecular dynamics of specific proteins where too many degrees of freedom have to be accounted for. Monte Carlo methods are in general amenable tools to study equilibrium properties. However, the seemingly simple HP model is unexpectedly challenging to study due to the rough energy landscapes. On the contrary, Wang-Landau sampling [21–24], coupled with inventive Monte Carlo moves, is shown to be robust to overcome this problem. It is, therefore, used to estimate the density of states from which the thermodynamic and structural properties can be calculated. The conformational "transitions" are also discussed in this paper.

2. Methods

2.1. Hydrophobic-polar (HP) protein folding model

The HP model is a simplified lattice model which is used to simulate protein folding to its native state. It was first proposed by K. A. Dill in 1985 [20]. The protein chain is defined as a sequence of two types of amino acid residues, hydrophobic (H) or polar (P), and is modeled as a self-avoiding walk in a 3-dimensional cubic lattice. The driving force for protein folding is imitated by assigning a favorable interaction ε_{HH} between two non-bonded, nearest-neighbor H residues.

To study protein folding under confinement, we construct two parallel surfaces in the xy plane at $z = 0$ and $z = h_w$. The boundary conditions are periodic only in the x and y directions. The energy can then be calculated by:

$$E = -\varepsilon_{HH}n_{HH} - \varepsilon_{SH}n_{SH} - \varepsilon_{SP}n_{SP}, \tag{1}$$

where n_{HH} is the number of hydrophobic pairs; n_{SH} and n_{SP} are the number of H monomers and P monomers adjacent to the surfaces; ε_{SH} and ε_{SP} are the surface attractive strengths for the H and P monomers, respectively. We have used a 36mer ($P_3H_2P_2H_2P_5H_7P_2H_2P_4H_2P_2HP_2$) that consists of 16 H monomers and 20 P monomers. The 36mer is one of the HP benchmark sequences for testing a genetic algorithm [25]. This short sequence is already able to exhibit features of protein folding and it has been used to investigate protein-surface interaction [26,27]. The distance between the two confining surfaces was set to be smaller than the chain length ($h_w = 10$). A schematic diagram of this model is shown in figure 1.

Journal of Physics: Conference Series **454** (2013) 012071 doi:10.1088/1742-6596/454/1/012071

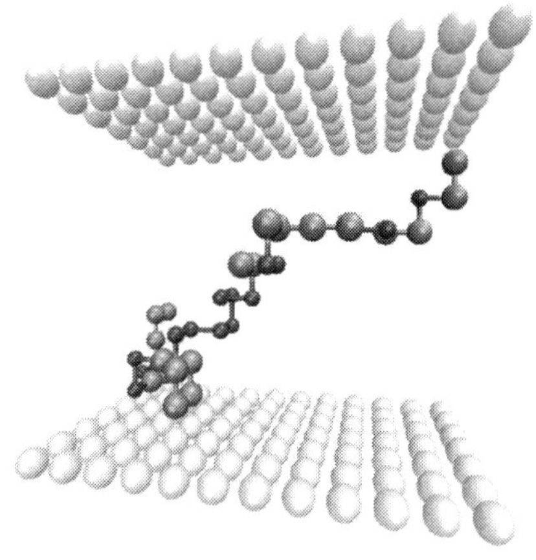

Figure 1. (Color online) A schematic diagram of the HP model under confinement. The smaller black spheres and the larger orange spheres represent the H monomers and the P monomers in the HP chain, respectively. The molecules of two parallel, confining surfaces are represented by light spheres.

2.2. Wang-Landau sampling

Wang-Landau (WL) sampling [21–24] is a Monte Carlo algorithm to calculate the relative density of states for a physical system. In the first step of the simulation, we applied WL sampling to determine the density of states as a function of the energy of the system, $g(E)$. The partition function at a particular temperature T can be obtained by summing the density of states over all energies weighted by the Boltzmann factor, i.e.,

$$Z(T) = \sum_E g(E) e^{-E/kT}, \tag{2}$$

where E is the total energy of the system described by equation (1), and k is the Boltzmann constant. Since $g(E)$ is independent of the temperature of the system, we can then compute any desired thermodynamic quantities at all temperatures after obtaining $g(E)$. For example, the average energy $\langle E\rangle(T)$ and the heat capacity $C_V(T)$ are calculated as:

$$\langle E\rangle(T) = \frac{1}{Z(T)} \sum_E E g(E) e^{-E/kT}, \tag{3}$$

$$C_V(T) = \frac{1}{kT^2}(\langle E^2\rangle - \langle E\rangle^2). \tag{4}$$

The specific heat is then defined by C_V/N, where N is the number of monomers in the chain. Traditionally speaking, the specific heat is an indicator of phase transitions as they are usually associated with large energy fluctuations. However, as we will show from our results, it is no longer adequate to signal all the phase transitions for our system as some structural transitions do not necessarily affect the total energy of the system.

In the WL method, $g(E)$ is iteratively refined during the simulation. The algorithm starts with an initial guess of $g(E) = 1$ for all possible energies E. The histogram in energy, $H(E)$, which keeps track of the number of visits to an energy within an iteration, is set to $H(E) = 0$ at the beginning. During a Monte Carlo step, a trial configuration is generated and accepted with probability

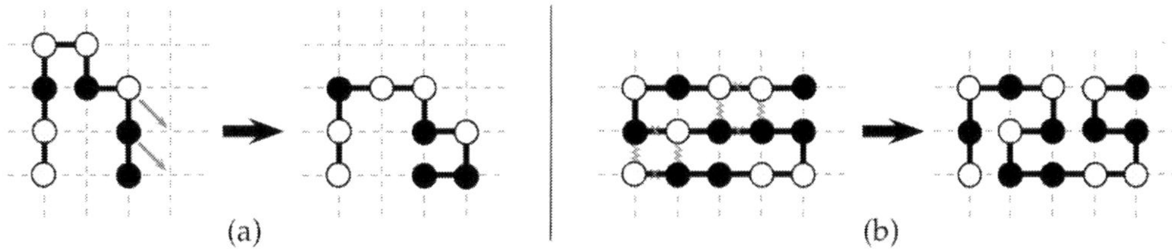

Figure 2. (Color online) Examples of Monte Carlo trial moves used in this study (illustrated on a two-dimensional square lattice): (a) a pull move, and (b) a bond-rebridging move.

$$p(E_{old} \to E_{new}) = \min\left(\frac{g(E_{old})}{g(E_{new})}, 1\right). \quad (5)$$

If the trial move is rejected, the old configuration is retained and it also enters into the statistics. Once the configuration is determined, the density of states is modified by $g(E) \to f \times g(E)$, where f is a modification factor which is slowly reduced to unity during the simulation. In this work, the initial and the final modification factor are $f_0 = e^1 \approx 2.71828$, and $f_{\text{final}} = \exp(10^{-8})$, respectively. Every time $g(E)$ is updated, $H(E)$ is also accumulated by $H(E) \to H(E) + 1$. This procedure repeats until a flat histogram is obtained, i.e., the simulation time spent at each energy is approximately the same. The histogram is then reset $H(E) = 0$, the density of states $g(E)$ is kept, and a new iteration begins with a reduced modification factor, $f_{i+1} = \sqrt{f_i}$. A "flat" histogram is attained when all the entries of $H(E)$ are greater than 80% of the average number of total visits: $\frac{1}{N_b}\sum_E H(E)$, where N_b is the total number of energy bins.

We have adopted two non-traditional Monte Carlo trial moves, pull moves [28] and bond-rebridging moves [29], to generate new configurations. Examples of these moves are shown in figure 2. The combination of WL sampling with these moves has been proven to be an efficient iterative method to estimate an accurate $g(E)$ [30, 31]. The acceptance rate of the bond-rebridging moves is rather low because it is not always possible to propose this move from a configuration (the configuration needs to be dense enough to have at least a pair of parallel bonds; sometimes two pairs of parallel bonds are necessary for this move to be feasible). As a consequence, we assigned a calling probability for pull move and bond-rebridging move to be 0.2 and 0.8, respectively.

2.3. Production run

The second step of the simulation is making a production run from the $g(E)$ that we obtained from the first step. Here, we perform a multicanonical simulation [32] to estimate a two-dimensional density of states $g(E, Q)$ in energy E and structural quantity Q. In this process, the acceptance probability is the same as equation (5), except that $g(E)$ is held fixed during this stage. Every time a configuration is selected, the structural quantities Qs are calculated while the corresponding two-dimensional histograms $H(E, Q)$ are updated. This has an advantage that a number of structural quantities can be calculated within a single run. Examples of structural quantities are n_{HH}, n_{SP}, and n_{SH}. At the end of the simulation, the $g(E, Q)$ is obtained by re-weighting $g(E)$ with $H(E, Q)$:

$$g(E, Q) = g(E)H(E, Q). \quad (6)$$

The partition function and the average of the structural quantity Q are therefore calculated as

$$Z(T) = \sum_{E,Q} g(E,Q)e^{-E/kT}, \tag{7}$$

$$\langle Q\rangle(T) = \frac{1}{Z_Q(T)} \sum_{E,Q} Qg(E,Q)e^{-E/kT}. \tag{8}$$

3. Results and Discussion

To investigate the effects of surface characteristics on the physical behavior of a confined HP protein, we considered three types of confining surfaces:

(i) non-specific surfaces that attract all monomers ($\varepsilon_{SH} = \varepsilon_{SP} = \varepsilon_{HH} = 1$);

(ii) polar surfaces ($\varepsilon_{SH} = 0, \varepsilon_{SP} = \varepsilon_{HH} = 1$);

(iii) hydrophobic surfaces ($\varepsilon_{SP} = 0, \varepsilon_{SH} = \varepsilon_{HH} = 1$).

Here, 15 independent runs were performed in each case to obtain statistical errors. On an AMD Opteron 2.2 GHz CPU, a single simulation took around 3-21 hours, 5-20 hours, and 1-12 hours for the 36mer confined between non-specific surfaces, polar surfaces, and hydrophobic surfaces, respectively.

Generally, the conformational "phase transitions" can be identified by locating the peak of the specific heat C_V/N. However, it is very important to investigate the structural quantities together with C_V/N so as to identify individual transition processes, since structural transitions do not necessarily cause large energy fluctuations. In these cases, the transition does not incur a pronounced peak in C_V/N but merges it into another transition peak caused by other processes, or it occurs as a subtle shoulder.

Figure 3 shows the thermodynamics for the 36mer confined between non-specific surfaces which attract all monomers ($\varepsilon_{SH} = \varepsilon_{SP} = \varepsilon_{HH} = 1$). At high temperature, the protein tends to stretch out. Since the distance between two confining surfaces was set to be smaller than the chain length, parts of the chain form contacts with both surfaces and the chain acts as a "bridge" between the two surfaces. To maximize the surface contacts as the temperature lowers, the chain detaches from one surface to adsorb on the other until most of the adsorbents sit on the same surface. We call the destruction of the bridge a "debridging" process. This process causes a small amount of energy fluctuations, leading to a shoulder observed in the specific heat at $kT/\varepsilon_{HH} \approx 1.6$. Further decrease in temperature induces the hydrophobic core (H-core) formation, where the adsorbed, extended chain collapses to form a compact structure, in which the hydrophobic monomers cluster together inside and the polar monomers reside on the outside. Unlike the debridging process, the H-core formation causes a large decrease in energy. A peak in specific heat is then observed at $kT/\varepsilon_{HH} \approx 0.3$. More examples of these behaviors can be found in [33]. The debridging process and the H-core formation can be identified more clearly by the thermal derivatives of $\langle n_{SH}\rangle$, $\langle n_{SP}\rangle$, and $\langle n_{HH}\rangle$. The negative peaks in $d\langle n_{SH}\rangle/dT$ and $d\langle n_{SP}\rangle/dT$ are associated with the formation of surface contacts, i.e., debridging and adsorption; while the negative peak in $d\langle n_{HH}\rangle/dT$ signifies the construction of the hydrophobic interactions. See [27, 34, 35] for more details of the identification of phase transitions.

Figure 4 and figure 5 show the thermodynamics for the 36mer confined between polar surfaces ($\varepsilon_{SH} = 0, \varepsilon_{SP} = \varepsilon_{HH} = 1$) and hydrophobic surfaces ($\varepsilon_{SP} = 0, \varepsilon_{SH} = \varepsilon_{HH} = 1$), respectively. The same basic transitions, debridging at a higher temperature and the H-core formation at a lower temperature, are also observed as in the previous case. However, they occur at different temperatures. With the same surface attractive strength, having a larger number of monomers attracted by the surface makes adsorption to occur easier, resulting in a higher debridging temperature. For this reason, the debridging process of the 36mer confined between surfaces that attract all monomers occurs at a higher temperature ($kT/\varepsilon_{HH} \approx 1.6$). It takes place at a

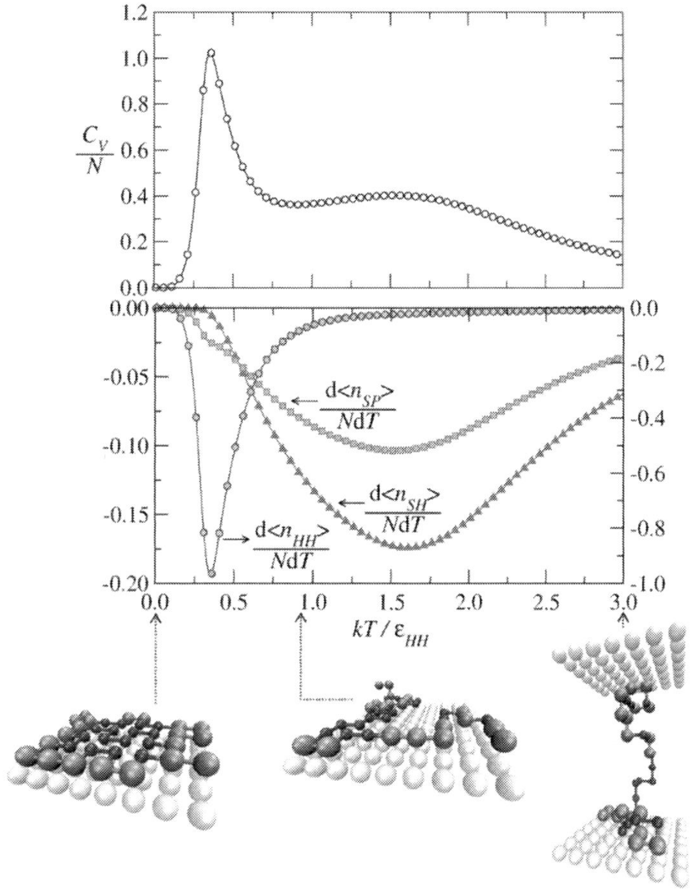

Figure 3. (Color online) Thermodynamic and structural quantities of the 36mer confined between two attractive surfaces that attract all monomers ($\varepsilon_{SH} = \varepsilon_{SP} = \varepsilon_{HH} = 1$, $h_w = 10$). Typical structures are displayed with smaller black spheres and larger orange spheres for the H monomers and the P monomers, respectively. The horizontal arrows beside the labels indicate the axes to which the quantities refer. Errors are smaller than the size of the data points.

lower temperature with the polar surfaces ($kT/\varepsilon_{HH} \approx 1.0$), and at an even lower temperature with the hydrophobic surfaces ($kT/\varepsilon_{HH} \approx 0.75$).

The details of forming an H-core are different from case to case, resulting in different ground state structures. Typical configurations of the ground states are shown at $kT/\varepsilon_{HH} = 0$. For the case that the confining surfaces attract all types of monomers, the ground state is a two-dimensional structure with a planar H-core, and it is formed after the chain adsorbed to the surface. For the other two cases, the ground states are three-dimensional structures, but the hydrophobic cores are three-(or two-)dimensional structures that adsorb on a polar (or hydrophobic) surface.

The lower panel of figure 4 reveals a negative peak in $d\langle n_{SH}\rangle/dT$ and a positive peak in $d\langle n_{SP}\rangle/dT$ when the H-core is formed at a very low temperature ($kT/\varepsilon_{HH} \approx 0.25$). This suggests that during the H-core formation process, some surface contacts with the P monomers need to be broken to facilitate the construction of the H-H contacts. Similar behavior is observed in figure 5 for the case of hydrophobic confining surfaces, except that there is no negative peak in $d\langle n_{SH}\rangle/dT$ at low temperature. It is because all the H monomers have already adsorbed

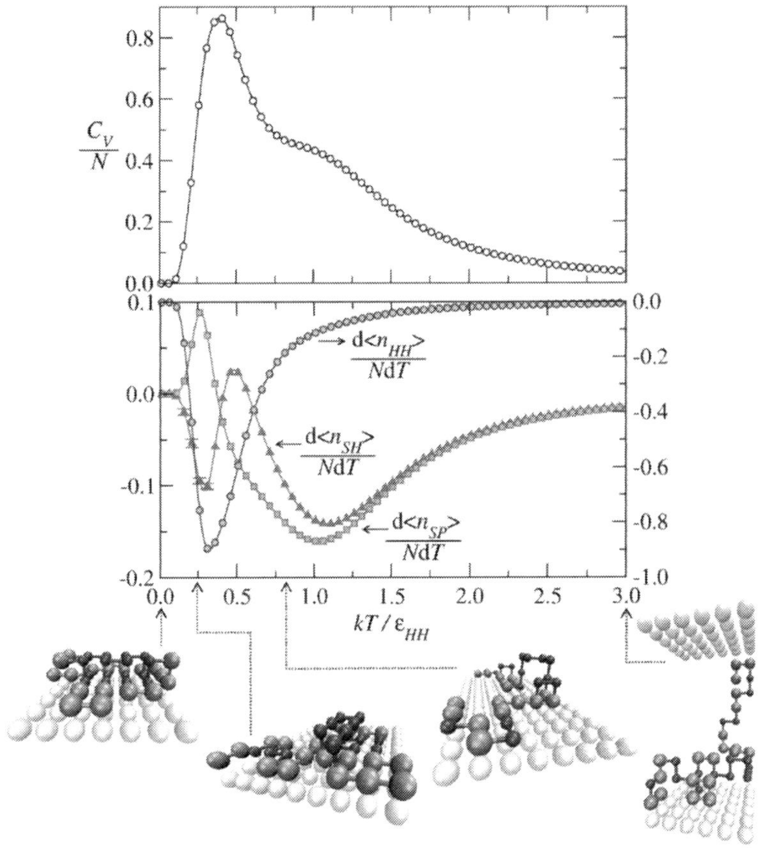

Figure 4. (Color online) Thermodynamic and structural quantities of the 36mer confined between two attractive surfaces that attract only P monomers ($\varepsilon_{SH} = 0, \varepsilon_{SP} = 1, \varepsilon_{HH} = 1$, h_w = 10). Typical structures are displayed with smaller black spheres and larger orange spheres for the H monomers and the P monomers, respectively. The horizontal arrows beside the labels indicate the axes to which the quantities refer. Error bars smaller than the data points are not shown.

on the surface before H-core formation in the latter case. A subtle shoulder in both C_V and $d\langle n_{HH}\rangle/dT$ in the very low temperature regime ($kT/\varepsilon_{HH} \approx 0.25$) signals a two-stage acquisition of the ground state, where the chain maximizes the number of H-H contacts by uncoiling the dense, but still slightly disordered, globule before collapsing again to form a highly ordered, rectangular H-core.

4. Conclusion

We studied the physical behavior of a protein confined between two attractive surfaces using a coarse-grained lattice model, the HP model. Wang-Landau sampling with inventive Monte Carlo moves, pull moves and bond-rebridging moves, were employed. In this work we considered three different types of surfaces with different affinities to the H or P monomers. The thermodynamic and structural quantities were used to identify conformational "transitions". It was found that at high temperature the chain can touch both surfaces. As temperature decreases, two basic transitions, debridging and H-core formation, were observed. We further discovered the number of adsorbates, i.e., monomers that are attracted by the surfaces, played a major role in these

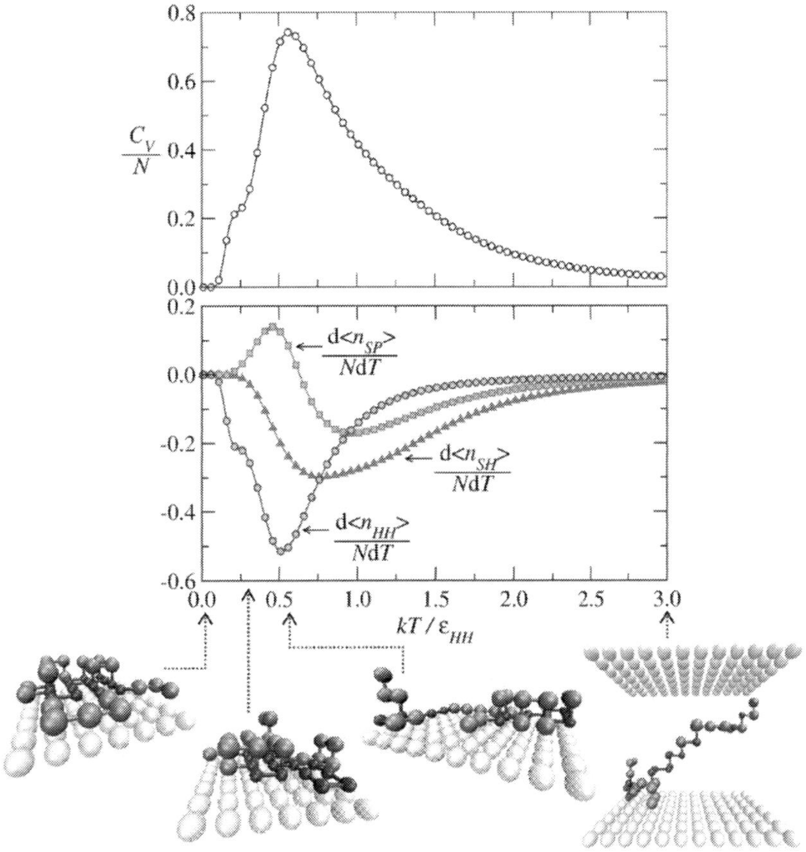

Figure 5. (Color online) Thermodynamic and structural quantities of the 36mer confined between two attractive surfaces that attract only H monomers ($\varepsilon_{SH} = 1, \varepsilon_{SP} = 0, \varepsilon_{HH} = 1$, $h_w = 10$). Typical structures are displayed with black spheres and orange spheres for the H monomers and the P monomers, respectively. The horizontal arrows beside the labels indicate the axes to which the quantities refer. Errors are smaller than the size of the data points.

transition processes. This, in turn, depends on both the surface types and the proportion of the H and P monomers in the chain. As a consequence, the debridging and H-core formation temperatures varied as these systems parameters change, and different structural properties for the ground states were resulted. All these imply that the materials of the confining surfaces, as well as the protein chain composition, are the dominant factors in determining the physical properties of this kind of composites.

Acknowledgments

This work was supported by the National Science Foundation under Grants DMR-0810223. B. Pattanasiri is supported by The Royal Thai Government Scholarship under the Development and Promotion of Science and Technology Talent Project (DPST), and the grant from The Institute for the Promotion of Teaching Science and Technology (IPST). Y.W. Li was partly sponsored by the Office of Advanced Scientific Computing Research; U.S. Department of Energy. Part of the work was performed at the Oak Ridge National Laboratory, which is managed by UT-Battelle, LLC under Contract No. De-AC05-00OR22725.

References

[1] Creighton T E 1993 *Proteins: Structures and Molecular Properties* (New York: W. H. Freeman)
[2] Taubes G 1996 *Science* **271** 1493–95
[3] Selkoe D J 2004 *Nat. Cell. Biol.* **6** 1054–61
[4] Prusiner S B 1998 *Proc. Natl. Acad. Sci. USA* **95** 13363–83
[5] Zhou H -X 2008 *Arch. Biochem. Biophys.* **469** 76–82
[6] Ebbinghaus S, Dhar A, McDonald D J and Gruebele M 2010 *Nature Methods* **7** 319–23
[7] Ng J D, Stevens R C and Kuhn P 2008 *Methods Mol. Biol.* **426** 363–76
[8] Birnbaum S 1993 *Immobilised Macromolecules: Application Potentials (Springer Series in Applied Biology)* ed U B Sleytr *et al* (London: Springer)
[9] Martin C R and Kohli P 2003 *Nat. Rev. Drug Discov.* **2** 29–37
[10] Oh K J, Cash K J and Plaxco K W 2009 *Chemistry* **15** 2244–51
[11] Foegeding E A, Luck P J and Davis J P 2006 *Food Hydrocolloids* **20** 284–92
[12] Schmidt I, Novales B, Boué F and Axelos M 2010 *J. Colloid Interface Sci.* **345** 316–24
[13] Campanini B, Bologna S, Cannone F, Chirico G, Mozzarelli A and Bettati S 2005 *Protein Sci.* **14** 1125–33
[14] Peterson R W, Anbalagan K, Tommos C and Wand A J 2004 *J. Am. Chem. Soc.* **126** 9498–99
[15] Ping G, Yuan J M, Vallieres M, Dong H, Sun Z, Wei Y, Li F Y and Lin S H 2003 *J. Chem. Phys.* **118** 8042
[16] Takagi F, Koga N and Takada S 2003 *Proc. Natl. Acad. Sci. USA.* **100** 11367
[17] Rathore N, Knotts T A and Pablo J J D 2006 *Biophys. J.* **90** 1767–73
[18] Jewett A I, Baumketner A and Shea J -E 2004 *Proc. Natl. Acad. Sci. U.S.A.* **101** 13192–97
[19] Jewett A I, Baumketner A and Shea J -E 2006 *J. Mol. Biol.* **363** 945–57
[20] Dill K A 1985 *Biochemistry* **24** 1501–9
[21] Wang F and Landau D P 2001 *Phys. Rev. Lett.* **86** 2050–53
[22] Wang F and Landau D P 2001 *Phys. Rev.* E **64** 056101
[23] Wang F and Landau D P 2002 *Comput. Phys. Commun.* **147** 674–77
[24] Landau D P, Tsai S -H and Exler M 2004 *Am. J. Phys.* **72** 1294
[25] Unger R and Moult J 1993 *J. Mol. Biol.* **231** 75–81
[26] Swetnam A D and Allen M P 2009 *Phys. Chem. Chem. Phys.* **11** 2046–55
[27] Li Y W, Wüst T and Landau D P 2011 *Comput. Phys. Commun.* **182** 1896–99
[28] Lesh N, Mitzenmacher M and Whitesides S 2003 *Proc. Seventh Annual Int. Conf. on Computational Molecular Biology*, ed T Lengauer *et al.* (Berlin: RECOMB) p 188
[29] Deutsch J M 1997 *J. Chem. Phys.* **106** 8849
[30] Wüst T and Landau D P 2009 *Phys. Rev. Lett.* **102** 178101
[31] Wüst T and Landau D P 2012 *J. Chem. Phys.* **137** 064903
[32] Berg B A and Neuhaus T 1992 *Phys. Rev. Lett.* **68** 9–12
[33] Pattanasiri B, Li Y W, Landau D P and Wüst T 2012 *Int. J. Modern Phys.* C **23** 1240008
[34] Wüst T, Li Y W and Landau D P 2011 *J. Stat. Phys.* **144** 638–51
[35] Li Y W, Wüst T and Landau D P 2013 *Phys. Rev.* E **87** 012706

Challenge toward the prediction of typhoon behaviour and down pour

K Takahashi[1], R Onishi[1], Y Baba[1], S Kida[1], K Matsuda[1], K Goto[2] and H Fuchigami[3]

[1] The Earth Simulator Center, JAMSTC, 3173-25 Showa-machi, Kanazawa-ku, Yokohama, 236-0001 Japan
[2] NEC Corporation, 1-10 Nisshin-cho, Fuchu-shi, Tokyo, 183-5801, Japan
[3] NEC Informatec Systems LTD, 3-2-1 Sakato, Takatsu-ku, Kawasaki-shi, Kanagawa, Japan

E-mail: takahasi@jamstec.go.jp

Abstract. Mechanisms of interactions among different scale phenomena play important roles for forecasting of weather and climate. Multi-scale Simulator for the Geoenvironment (MSSG), which deals with multi-scale multi-physics phenomena, is a coupled non-hydrostatic atmosphere-ocean model designed to be run efficiently on the Earth Simulator. We present simulation results with the world-highest 1.9km horizontal resolution for the entire globe and regional heavy rain with 1km horizontal resolution and 5m horizontal/vertical resolution for urban area simulation. To gain high performance by exploiting the system capabilities, we propose novel performance evaluation metrics introduced in previous studies that incorporate the effects of the data caching mechanism between CPU and memory. With a useful code optimization guideline based on such metrics, we demonstrate that MSSG can achieve an excellent peak performance ratio of 32.2% on the Earth Simulator with the single-core performance found to be a key to a reduced time-to-solution.

1. Introduction

Intense research effort is focused on understanding the climate/weather system using coupled atmosphere-ocean models. It is widely accepted that the most powerful tools available for assessing future weather/climate are fully coupled general circulation models. Not only interactions between atmosphere or ocean components, but also various other components have been coupled with various interactive ways and influence on the Earth system. In order to get further information on perspectives of future weather/climate and the Earth system, the whole Earth system should be simulated using coupled models.

The Earth Simulator Center has been developing coupled non-hydrostatic atmosphere-ocean-land general circulation models to be run on the Earth Simulator with ultra high resolution and really high performance computing architectures. When development of the coupled model has been completed and various simulations are possible on the Earth Simulator, the ambitious task of simulating and understanding the Earth system should bring us further detail information on the Earth System. Simulations with target scale shown in figure 1 will be planned in near future.

In this paper, outline of the Multi-scale Simulator for the Geoenvironment (MSSG) is described in section 2. Computational performance is reported in section 3. Validation results for several time/space scale phenomena are presented in section 4 and conclusions and near future plans are shown in section 5.

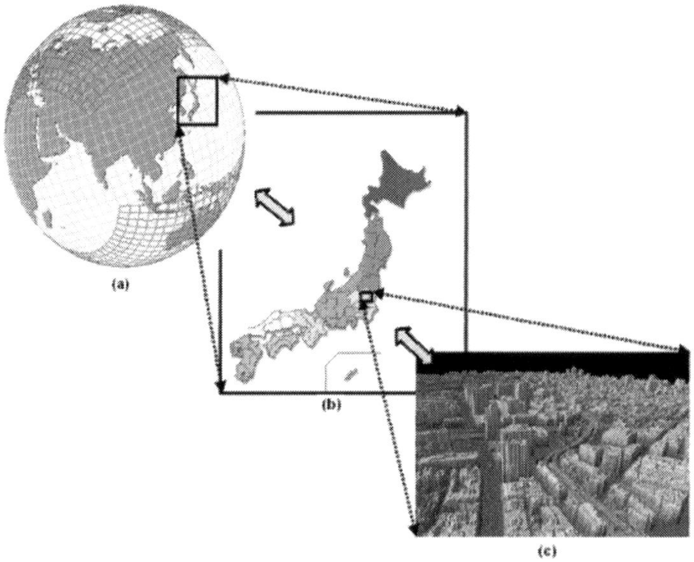

Figure 1. (a): Yin-Yang girds system for global simulations. The coloured grids are the Yin and Yang grids. (b): Japan region is nested with two-way interaction to the global grid. (c): Urban scale weather/climate simulations will be allowed with two-way interactions to the global/regional scale simulations. Urban topography was given by the Geographical Survey Institute.

2. Model description

2.1. The atmosphere component: MSSG-A

The atmosphere component consists of non-hydrostatic, fully compressive flux form of dynamics [1] and Smagorinsky-Lilly type parameterizations [2][3] for subgrid scale mixing, surface fluxes [4][5], cloud microphysics with mixed phases [6], cumulus convective processes [7][8] and simple radiation scheme. The set of the prognostic equations is presented as follows:

$$\frac{\partial \rho'}{\partial t} + \frac{1}{G^{\frac{1}{2}} a \cos\varphi} \frac{\partial (G^{\frac{1}{2}} G^{13} \rho u)}{\partial \lambda} + \frac{1}{G^{\frac{1}{2}} a \cos\varphi} \frac{\partial (G^{\frac{1}{2}} G^{23} \cos\varphi \rho v)}{\partial \varphi} + \frac{1}{G^{\frac{1}{2}}} \frac{\partial (\rho w^*)}{\partial z^*} = 0, \tag{1}$$

$$\frac{\partial \rho u}{\partial t} + \frac{1}{G^{\frac{1}{2}} a \cos\varphi} \frac{\partial (G^{\frac{1}{2}} G^{13} P')}{\partial \lambda} = -\nabla \bullet (\rho u \mathbf{v}) + 2 f_r \rho v - 2 f_\varphi \rho w + \frac{\rho v u \tan\varphi}{a} - \frac{\rho w u}{a} + F_\lambda, \tag{2}$$

$$\frac{\partial \rho v}{\partial t} + \frac{1}{G^{\frac{1}{2}} a} \frac{\partial (G^{\frac{1}{2}} G^{23} P')}{\partial \varphi} = -\nabla \bullet (\rho v \mathbf{v}) + 2 f_\lambda \rho w - 2 f_r \rho u - \frac{\rho u u \tan\varphi}{a} - \frac{\rho w v}{a} + F_\varphi, \tag{3}$$

$$\frac{\partial \rho w}{\partial t} + \frac{1}{G^{\frac{1}{2}}} \frac{\partial P'}{\partial z^*} + \rho' \mathbf{g} = -\nabla \bullet (\rho w \mathbf{v}) + 2 f_\varphi \rho u - 2 f_\lambda \rho v + \frac{\rho u u}{a} + \frac{\rho v v}{a} + F_r, \tag{4}$$

$$\frac{\partial P'}{\partial t} + \nabla \bullet (P \mathbf{v}) + (\gamma - 1) P \nabla \bullet \mathbf{v} = (\gamma - 1) \nabla \bullet (\kappa \nabla T) + (\gamma - 1)\phi, \tag{5}$$

$$P = \rho R T, \tag{6}$$

$$\rho w^* = \frac{1}{G^{\frac{1}{2}}} \left(G^{\frac{1}{2}} G^{13} \rho u + G^{\frac{1}{2}} G^{23} \rho v + \rho w \right). \tag{7}$$

In (1)-(7), prognostic valuables are momentum $\rho\mathbf{v} = (\rho u, \rho v, \rho w)$, ρ' , which is calculated as $\rho' = \rho - \bar{\rho}$, and P' , defined by $P' = P - \overline{P}$. ρ is the density; P is the pressure; $\bar{P}$ is a constant reference pressure; T is the temperature; R is the gas constant; $\mathbf{f}, \kappa$ and γ are the Coriolis force, the diffusion coefficient and the ratio of specific heat, respectively. ϕ is the heat source term and F is the viscosity term, respectively. G is the metric term for vertical coordinate; ϕ is latitude; λ is longitude. In addition, a is the radius of the Earth and g is the gravitational acceleration.

The treatment of cloud and precipitation is controlled by selecting a parameterization scheme due to horizontal resolution. For grid spacing greater than 10 km, the Kain and Fritsh scheme [7][8] is used and cloud microphysics based on mixed phase micro cloud physics [6] is used for below 5km spacing. Over land, the ground temperature and ground moisture are computed by using a bucket model as a simplified land model. As upper boundary condition, Rayleigh friction layer is set. For the lateral boundary condition of regional version, a sponge type boundary condition [9] is used.

Regional version of the atmosphere component is utilized with one way nesting scheme by choosing the target region on the sphere, although two-way nesting is available as an option. Any large regions can be selected from the global, because both Coriolis and metric terms are introduced in the regional formulation. As another option, multiple regions are allowed to be selected at the same time and computed in parallel.

2.2. The ocean component: MSSG-O

In the ocean component: MSSG-O, the in-compressive and hydrostatic equations with the Boussinesq approximation are used based on the description in [10][11]. The set of equations in the ocean component becomes as follows:

$$\frac{\partial c}{\partial t} = -\mathbf{v}grad c + F_c \tag{8}$$

$$\frac{\partial T}{\partial t} = -\mathbf{v}grad T + F_T \tag{9}$$

$$0 = \nabla \bullet \mathbf{v} = \left(\frac{1}{r\cos\varphi}\frac{\partial u}{\partial \lambda} + \frac{1}{r\cos\varphi}\frac{\partial(\cos\varphi v)}{\partial \varphi} + \frac{1}{r^2}\frac{\partial(r^2 w)}{\partial r} \right) \tag{10}$$

$$\frac{\partial u}{\partial t} = -\mathbf{v}grad u + 2f_r v - 2f_\varphi w + \frac{vu\tan\varphi}{r} - \frac{wu}{r} - \frac{1}{\rho_0 r\cos\varphi}\frac{\partial P'}{\partial \lambda} + F_\lambda \tag{11}$$

$$\frac{\partial v}{\partial t} = -\mathbf{v}grad v + 2f_\lambda w - 2f_r u - \frac{uu\tan\varphi}{r} - \frac{wv}{r} - \frac{1}{\rho_0 r}\frac{\partial P'}{\partial \varphi} = +F_\varphi \tag{12}$$

$$\frac{\partial w}{\partial t} = -\mathbf{v}grad w + 2f_\varphi u - 2f_\lambda v + \frac{uu}{r} + \frac{vv}{r} - \frac{1}{\rho_0}\frac{\partial P'}{\partial r} - \frac{\rho'}{\rho_0}\mathbf{g} + F_r \tag{13}$$

$$\rho = \rho(T, c, P_0) \tag{14}$$

$$\frac{d}{dr}P_0 = -\rho_0 g(r) \tag{15}$$

where the Boussinesq approximation is adopted in (9) and all variables are defined as above for the atmospheric component. In (14), UNESCO scheme [12] is used.

A Smagorinsky type scheme [2][3] is used as the subgrid-scale mixing in identical experiments with the ocean component. The level-2 turbulence closure of Mellor Yamada [13] has been also introduced to the ocean component as one of optional schemes.

In MSSG-O, sponge layers are used for lateral boundary in the open ocean. The lateral boundary condition between ocean and land is defined as $\partial T/\partial t = \partial S/\partial t = 0$ and $\mathbf{v} = 0$. Bottom condition is defined by the Neumann condition without vertical velocity. The upper boundary conditions are given as momentum fluxes by wind and heat fluxes from observational data of atmosphere.

2.3. Grid configuration

The Yin-Yang grid system [14] is used both for the atmosphere-land and ocean components. The Yin-Yang gird system as shown in figure 1(a) is characterized by two overlapping panels that cover the sphere. Basically, one component grid is defined as a part of low-latitude region covered between 45N and 45S and 270 in longitude of the usual latitude-longitude grid system and the other component of the grid system is defined in the same way but in different spherical coordinates. The region covered by a panel can change by rotating the axes of the panels.

By using the Yin-Yang grid system, we can find a solution to the issue of how to avoid singular points such as the south and north poles on a latitude/longitude grid system. In addition, the advantage to enlarge the time step is compared to conventionally utilized latitude/longitude grid systems.

2.4. Differencing schemes

In both the MSSG-A and MSSG-O, the Arakawa C grid is adopted. The MSSG-A utilizes the terrain following vertical coordinate with Lorenz type variables distribution [15]. The MSSG-O uses the z-coordinate system for the vertical direction. In discretization of time, the 2nd, 3rd and 4th Runge-Kutta schemes and leap-flog schemes with Robert-Asselin time filter are available; the 3rd Runge-Kutta schemes is adopted for the atmosphere component. In this study, leap-flog schemes with Robert-Asselin time filter is used for the ocean component.

For momentum and tracer advection computations, several discretization schemes are available [16][21]. In this study, the 5th order upwind scheme is used for the atmosphere and central difference is utilized in the ocean component.

The vertical speed of sound in the atmosphere is dominant comparing the horizontal speed, because vertical discretization tends to be finer than horizontal discretization. For those reasons, a horizontally explicit vertically implicit (HEVI) scheme [21] is adopted in the atmosphere component. The speed of sound in the ocean is three times faster than it is in the atmosphere, an implicit method is introduced, and the Poisson equation (16) is solved with that method. The Poisson equation is described as

$$\nabla \bullet gradP' = B, \tag{16}$$

$$B = \rho_0 \nabla \bullet \mathbf{G}_v = \frac{\rho_0}{r\cos\varphi}\frac{\partial}{\partial\lambda}G_u + \frac{\rho_0}{r\cos\varphi}\frac{\partial}{\partial\varphi}(\cos\varphi G_v) + \frac{\rho_0}{r^2}\frac{\partial}{\partial r}(r^2 G_w), \tag{17}$$

which are solved under Neumann boundary condition of $n \bullet gradP = n \bullet \mathbf{G}_v$.

Computational noise over the interface of an overlapped grid system such as Yin-Yang grid was avoided by using a new interpolation scheme validated by various benchmark experiments [16][17][18][19][20].

2.5. Algebraic multigrid method in a Poison solver

Algebraic Multi-Grid (AMG) method [22] is used in order to solve the Poisson equation mentioned in section 2.4. AMG is well known as an optimal solution method. We used the AMG library which has been developed by Fuji Research Institute Corporation. The AMG library has the following characteristics.

- AGM in the library has been developed based on aggregation-type AMG [22].
- In the library, AMG is used as a pre-conditioner in Krylov subspace algorithms.
- Incomplete LU decomposition (ILU) is adopted as a smoother in the library, which shows good computational performance even for ill-structured matrixes.
- Local ILU is used for parallelization, in addition, fast convergence speed has been kept.
- Aggregation without smoothing is adopted with recalling procedure, because remarkably fast convergence has been obtained by using the aggregation.

2.6. Coupling between the atmosphere and ocean components

The coupling interface between the atmosphere and ocean should be taken into account to maintain a self-consistent representation in the coupled model. Generally, the time step in the ocean component is set longer than that of the atmosphere component. The heat fluxes, moisture and momentum fluxes are computed and averaged over larger time step. The averaged fluxes are used as the upper boundary condition of the ocean component. Precipitation computed in the atmosphere component is transferred to the ocean as a source term as fresh water. Sea surface temperature is defined in the uppermost layer in the ocean component and is transferred to the atmospheric component as one heat source The SST is fixed in the atmosphere during all time substeps within any large step.

2.7. Distribution architecture and communications

The architecture and data structures are based on domain decomposition methods. In the Yin-Yang grid system, communication cost imbalance might occur by adopting one-dimensional decomposition. The case of decomposition with 16 processes is considered in figure 2. Each color corresponds to a process. The number of arrows linking different colored areas corresponds to the amount of communication between processes. For example, in figure 2(a) for one dimensional domain decomposition, black colored process called A communicates to 8 different colored processes. In figure 2(b) for two-dimensional decomposition, a black colored process called A communicates two processes. In figure 2(a), communication data size is small and the number of communications is increased. If the same number for decompositions is used in both (a) and (b), it is clear that the amount of communications is smaller in figure 2(b). Two dimensional decomposition was adopted for both atmosphere-land and ocean components due to these reasons.

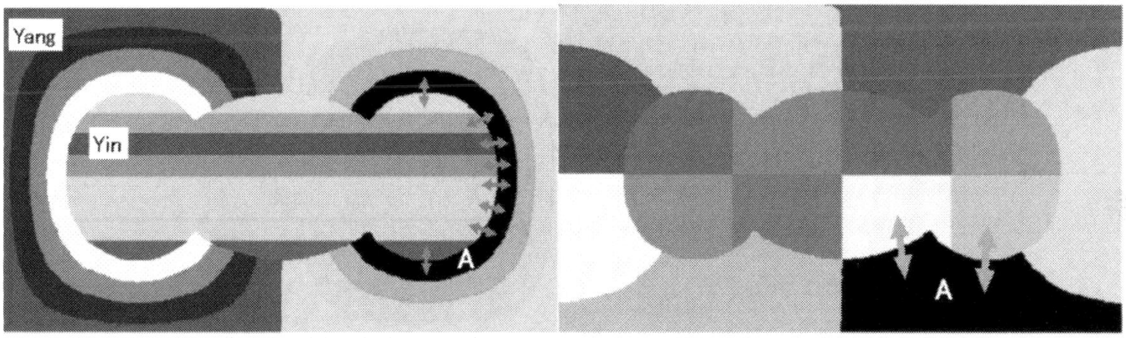

(a) One dimensional domain decomposition (b) Two dimensional domain decomposition

Figure 2. Schematic features of domain decomposition on Yin-Yang grid system.

3. Formatting the text performance of MSSG on the Earth Simulator

3.1. System configuration of Earth Simulator

The Earth Simulator (hereinafter called ES) upgraded and put into operational use in 2009 is a distributed-memory type parallel vector supercomputer that inherits the NEC SX-9 architecture [23]. The ES consists of 160 processor nodes interconnected by the (two-stage) fat-tree network [24]. One processor node is made up of 8 CPUs, each having a vector-type arithmetic processing unit and sharing 128GB of memory within a node. The CPU consists of the world fastest single core of 102.4GFLOPS. The whole ES is comprised of 1280CPUs with 20TB of main memory, thus giving a peak performance of 131TFLOPS (figure 3). The ES features Assignable Data Buffer (ADB) within a CPU for the selective buffering of data, enabling the effective use of memory bandwidth by suppressing the transfer of reusable data assigned to the ADB. The ADB is software-controllable, which makes possible the effective use of the ADB.

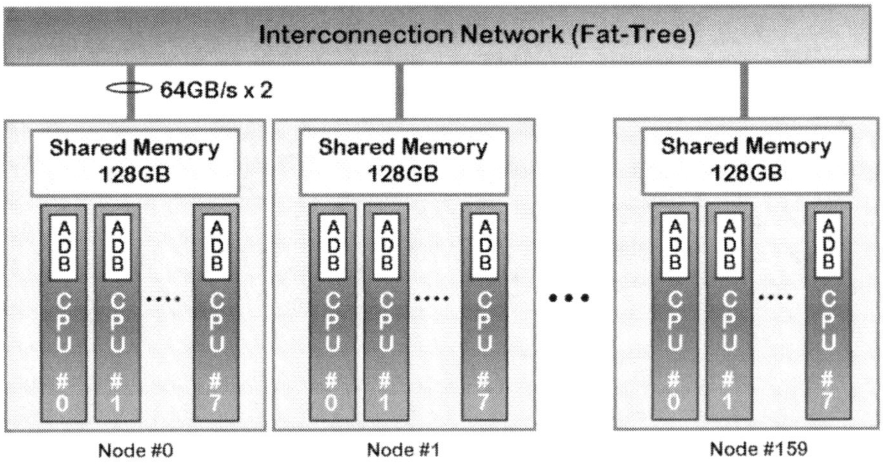

Figure 3. System configuration of the Earth Simulator.

3.2. Performance of MSSG-A on Earth Simulator

By tuning the performance using the novel performance evaluation metrics introduced in the previous studies [25], the wall-clock time for the entire MSSG program on the 160 ES nodes (1280 cores) was successfully reduced by 37% from 172.0 sec to 108.2 sec with the achieved sustained performance of 42.2 TFLOPS (peak performance ratio of 32.2%). The 3km global simulation run was also conducted with 80 ES nodes (640 cores). The measured wall-clock time was 108.2 sec and 205.7 sec for 160 and 80 nodes, respectively with the parallelization ratio of 99.9915%, which can be derived from the Amdahl's law. Figure 4 shows the sustained performance measured with 1280 and 640 cores, the resulting performance curve using the parallelization ratio based on the Amdahl's law and the line representing the ideal parallelization ratio of 100%.

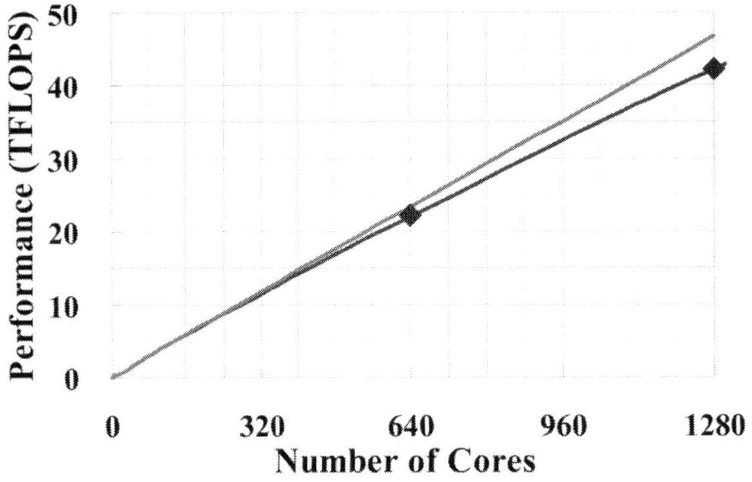

Figure 4. Strong scaling of MSSG-A on the ES. Blue marker: sustained performance measured with 1280 and 640 cores; blue line: the resulting performance curve using the parallelization ratio based on Amdahl's law; red line: the line representing the ideal parallelization ratio.

4. Simulation results

4.1. Stand alone simulations with atmospheric component of MSSG-A

A global simulation has been performed to validate physical performance under the condition of 5.5 km horizontal resolution and 32 vertical layers. A 72-hours integration was executed with the atmospheric component. Initialized data was interpolated at 00UTC08Aug2003 from Grid Point Value (GPV) data provided by Japan Meteorological Business Support Center. Sea surface data was also made by GPV data at 00UTC08Aug2003 and fixed during the simulation. Figure 5 shows averaged precipitation distribution one hour before 00UTC10Aug2003. The unit is mm per hour. Precipitation distribution has been computed through cloud microphysics and is comparable to observation data. In this simulation, the diurnal cycle of precipitation in Indonesian region and fine structure of fronts are captured.

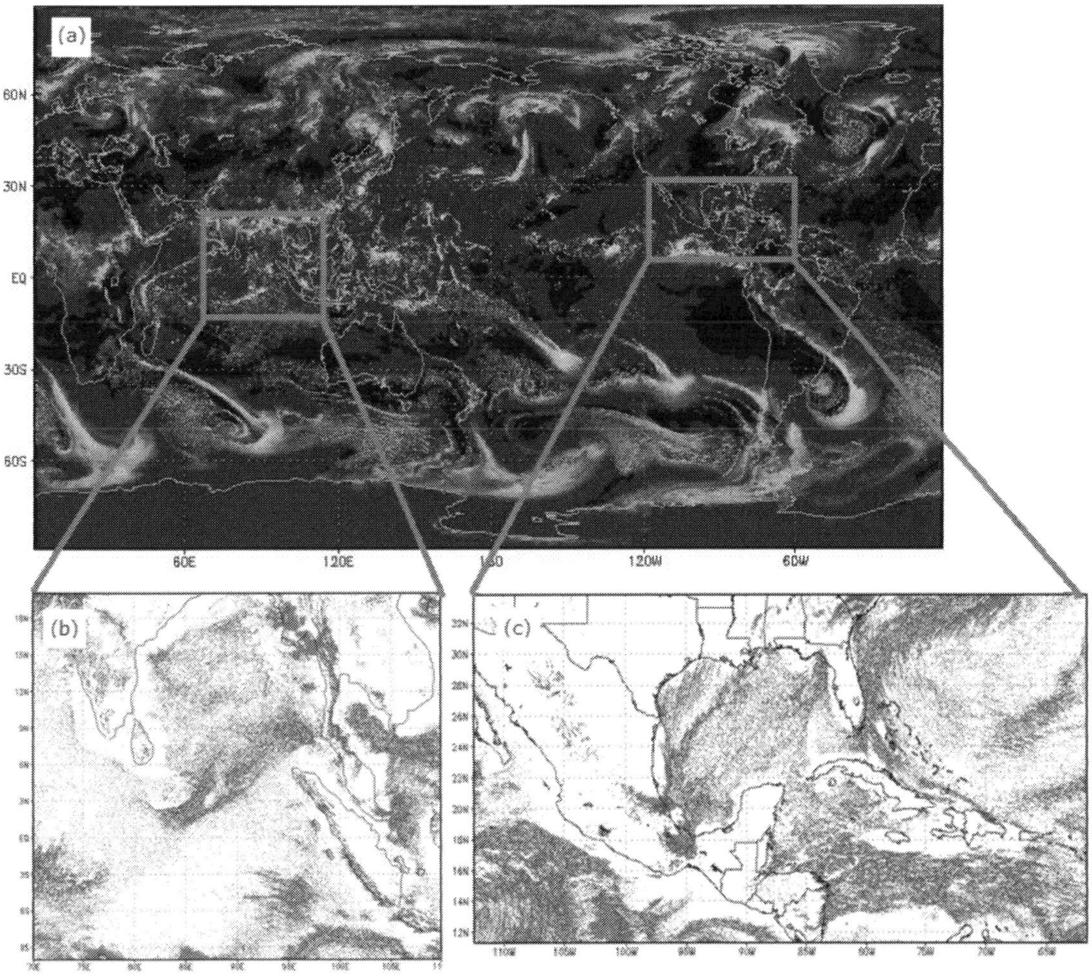

Figure 5. Global precipitation distribution simulated with MSSG-A for a 1.9km horizontal resolution.

4.2. Synoptic simulation results

Regional coupled simulations with MSSG for physical validation has been performed with one way nesting from 11 km global to Japanese region with 2.7km horizontal resolution. Horizontal resolution was set to the same condition in the oceanic component of MSSG. Initialized data was interpolated by using GPV data at 00UTC08Aug2003 provided by Japan Meteorological Business Support Center the boundary condition was made by interpolation the above simulation with 5.5km horizontal resolution. Sea surface temperature was also fixed to data at 00UTC08Aug2003 during the simulation. 72-hours integration has been performed. Figure 6 shows the result of the 72-hours integration. White gradation distribution in the typhoon shows cloud water distribution corresponding to cloud distribution. The fine rain band structure has been captured and is shown in figure 6. In the ocean part, distribution of sea surface temperature (SST) is presented as well in figure 8. SST responses to a strong wind due to typhoon and disturbance of SST are simulated. Not only SST but also vertical velocity in the ocean has been dramatically changed in the Kuroshio region (data not shown).

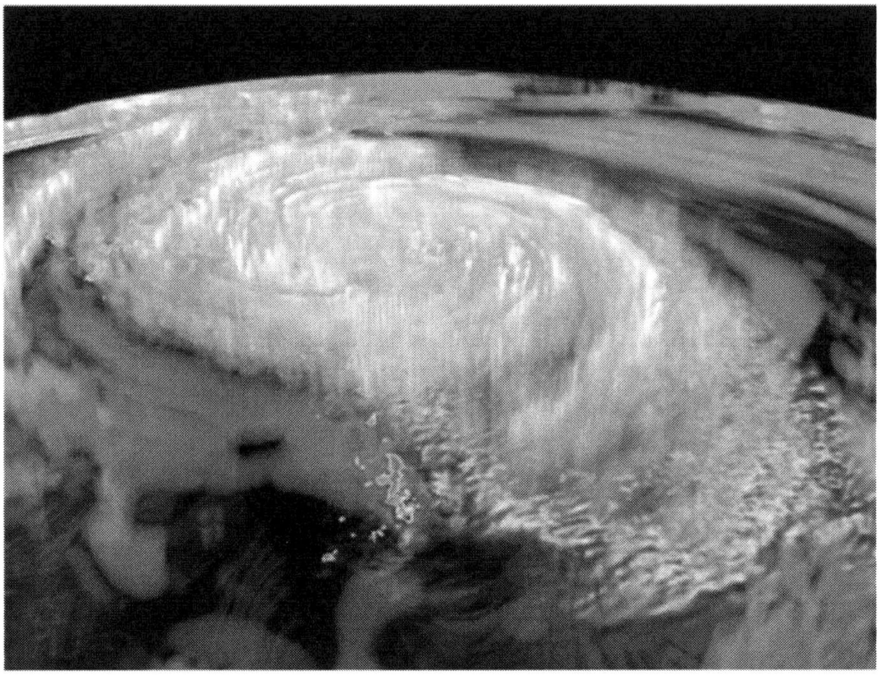

Figure 6. Regional validation simulation results of typhoon.

4.3. Regional heavy rain

Validation simulation with 1km horizontal resolution and 32 vertical layers for regional heavy rain was performed for the case during 21:00-24:00 JST on 4th September in 2005. Initial data were interpolated from the Grid Point Value (GPV) data at 06UTC04Sep2005 provided by Japan Meteorological Business Support Center and results after 6 model hours integration was used for initial data for prediction validation. Figure 7 and figure 8 show the observation data and simulation results of precipitation distribution averaged over one hour. The simulated precipitation distribution results were weak and the line shaped distribution of the observational data had not been captured. Initial data and representation of instability in the boundary layers must be considered.

4.4. Toward weather and climate variability prediction in urban area

We especially focus on developing MSSG for prediction of urban scale weather/climate phenomena which will be one of the key parts of *seamless* simulation from global to urban scale. As the first step of validation, the Marunouchi area in the center of Tokyo was selected for the simulation with MSSG-A due to the impacts of Large Eddy Simulation scheme for boundary layer. The initial state was settled at 15:00 on 5th August in 2005 and initial thermal condition was set taking account of shade in a day. In simulations with 5m horizontal resolution (figure 9), buildings can be resolved with anthropogenic heating source. The result shows a snap shot of horizontal temperature distribution during non-stationary computation from the initial time 15:00. Dynamics of thermal plume have been well represented.

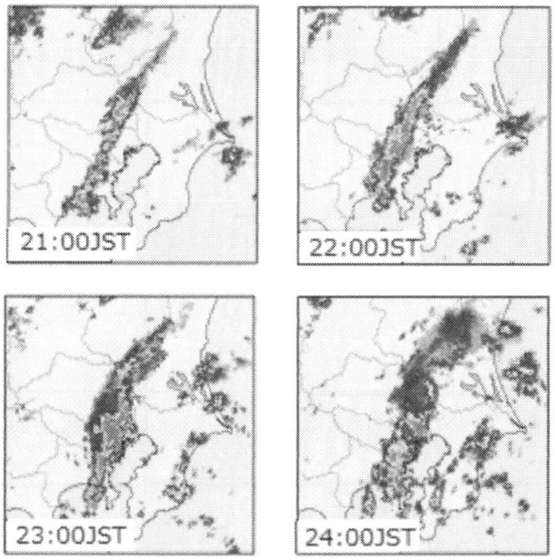

Figure7. Observation data from radar echo (mm/h).

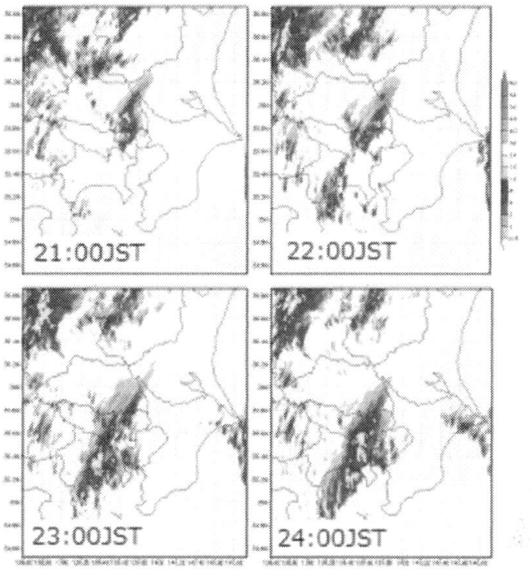

Figure 8. Simulation results (mm/h).

Figure 9. Temperature distribution of urban scale simulations for Marunouchi area in Tokyo.

5. Conclusions and perspectives

Development of MSSG was almost completed with high computational performance, although further optimization remains to be achieved in communications. Simulation results with each time/space scale encourage us toward seamless simulations with MSSG. As challenging issues, more various forecasting experiments are going to be performed and longer integration will be executed in order to estimate the accuracy of forecasting.

References

[1] Satomura T and Akiba S 2003. *Annuals of Disas. Prev. Res. Inst.,* Kyoto Univ. **46B** 331-336
[2] Lilly D K., 1962. On the numerical simulation of buoyant convection. Tellus **14** 148-172
[3] SmagorinskyJ, ManabeS. and HollowayJLJr. 1965 Monthly Weather Review ***93*** 727-768
[4] Zhang D and Anthes RA 1982 Journal of Applied Meteorology **21** 1594-1609
[5] Blackadar A K 1979 Advances in Environmental Science and Engineering, 1, Pfafflin and Ziegler, Eds., Gordon and Breach Publ. Group, Newark, p 50-85
[6] Reisner J, Ramussen RJ and Bruintjes RT 1998 Explicit forecasting of supercoolez liquid water in winter storms using the MM5 mesoscale model *Quart. J. Roy. Meteor. Soc.* **124B** 1071–1107
[7] Kain JS and Fritsch JM 1993. Convective parameterization for mesoscale models: The Kain-Fritsch Scheme. The Representation of Cumulus Convection in Numerical Models of the Atmosphere *Meteor. Monogr. Amer. Meteor. Soc.* **46**165-170
[8] Fritsch JM and Chappel CF 1980 *J. Atmos. Sci.* **37** 1722-1733
[9] Davies HC 1976 *Quart. J. R. Met. Soc.* **102**, 405-418
[10] Marshall J Hill C Perelman L and Adcroft A 1997 Journal of Geophysical Research **102** 5733-5752
[11] Marshall J Adcroft A Hill C Perelman L and Heisey C 1997 *Journal of Geophysical Research* **102** 5753-5766
[12] Gill A 1982 Atmosphere-Ocean dynamics, Academic Press Inc
[13] Mellor GL and Yamada T 1974 *Journal of Atmospheric Sciences* **31** 1791-1806
[14] Kageyama A and Sato T 2004 *Geochem.Geophys.Geosyst.* **5** Q09005, doi:10.1029/2004GC000734
[15] Gal-Chen T and Somerville RCJ 1975 *Journal of Computational Physics* **17** 209-228
[16] Peng X Xiao F Takahashi K and Yabe T 2004 *JSME international Journal (Series B)* **47**(4) 725-734
[17] Komine K 2005 Validation Results from Non-hydrostatic Atmospheric Simulations. Proc. *2005* SIAM Conference on Computational Science and Engineering, Florida, USA
[18] Ohdaira M Takahashi K and Watanabe K 2004 Validation for the Solution of Shallow Water Equations in Spherical Geometry with Overset Grid System" in Spherical Geometry. Proc. The 2004 Workshop on the Solution of Partial Differential Equations on the Sphere, Yokohama, July 2004 p 71
[19] Takahashi K et al Proc. 2004 7th International Conference on High Performance Computing and Grid in Asia Pacific Region p487
[20] Takahashi K et al 2004 Non-Hydrostatic Atmospheric GCM Development and its computationalperformance http://www.ecmwf.int/newsevents/meetings/workshops/2004/high_performance_computing-11th/presentations.html
[21] Wicker LJ Skamarock WC 2002 *Monthly Weather Review* **130** 2088-2097
[22] Stuben K 1999 A Review of Algebraic Multigrid, GMD Report 96
[23] Soga T Musa A Shimomura Y Egawa R Itakura K Takizawa H, Okabe K, and Kobayashi H. 2009 Performance evaluation of NEC SX-9 using real science and engineering applications. In Proceedings of the Conference on High Performance Computing Networking, Storage and Analysis (SC '09). ACM, New York, NY, USA, Article 28
DOI=10.1145/1654059.1654088 http://doi.acm.org/10.1145/1654059.1654088
[24] System Overview (ES2) http://www.jamstec.go.jp/es/en/system/index.html
[25] Takahashi K et al World-highest Resolution Global Atmospheric Model and Its Performance on the Earth Simulator, Proceeding of SC '11 State of the Practice Reports, Doi: 10.1145/2063348.2063376, 2011

Super high-resolution mesoscale weather prediction

K Saito[1,2], T Tsuyuki[1], H Seko[1,2], F Kimura[2], T Tokioka[2], T Kuroda[2,1], L Duc[2,1], K Ito[2,1], T Oizumi[2], G Chen[3], J Ito[4] and the SPIRE Field 3 Mesoscale NWP group

[1]Meteorological Research Institute, Tsukuba, Ibaraki 305-0052, Japan
[2]Japan Agency for Marine-Earth Science and Technology, Yokohama, Japan
[3]Tohoku University, Sendai, Japan
[4]Atmosphere and Ocean Research Institute /University of Tokyo, Japan

Email: ksaito@mri-jma.go.jp

Abstract. A five-year research project of high performance regional numerical weather prediction is underway as one of the five research fields of the Strategic Programs for Innovative Research (SPIRE). The ultimate goal of the project is to demonstrate feasibility of precise prediction of severe weather phenomena using the K-computer. Three sub-themes of the project are shown with achievements at the present and developments in the near future.

1. Introduction

Accuracy of numerical weather prediction (NWP) in world advanced weather forecast centers has been considerably improved recently. For example, the quantitative precipitation forecast (QPF) performance of the operational mesoscale model (MSM) of the Japan Meteorological Agency (JMA) has been increased almost twofold in threat scores for moderate rains in the recent ten years [1]. Continuous advance in numerical modelling and data assimilation techniques has contributed to the achievement. However, many difficulties remain in numerically predicting small scale severe weathers (e.g., torrential rains, local strong winds, tornadoes) with specifications of intensity, location and timing. Spatial and temporal scales of these severe phenomena are generally small, and evolution of these phenomena is sometimes very sensitive to trivial differences in the initial condition, because they often occur in convectively unstable atmospheric conditions. To overcome above problems, data assimilation and the ensemble forecast with the cloud-resolving resolution are required, as well as development of high accuracy high-resolution atmospheric models.

As one of the sub-subjects of the Field 3 of the Strategic Programs for Innovative Research (SPIRE), a research on high performance regional NWP has been carried out since April 2010. The sub-subject has the following three themes:

1) Development of cloud resolving 4-dimensional data assimilation systems,
2) Development and validation of a cloud resolving ensemble analysis forecast system, and
3) Basic research using very high-resolution atmospheric models.

2. Goals of the research subject

2.1. *Development of cloud resolving 4-dimensional data assimilation systems (by MRI, JAMSTEC, NPD, NIED, DPRI and ISM)*

The goal of the first theme is to dynamically predict deep convection and associated local severe weathers by modifying the initial condition of the cloud-resolving model. So far, high resolution data assimilation methods based on the JMA nonhydrostatic model (NHM; [1-3]) have been developed, and applied to case studies of cloud resolving forecast experiments of local precipitation (e.g., NHM-4DVAR [4]) and NHM-LETKF [5]). Dense observation data such as radar reflectivity, Doppler radar radial winds, GPS-derived total precipitable water vapor data are assimilated in storm scale to obtain more accurate initial conditions. A two-way nested LETKF (figure 1), which employs a low resolution LETKF as a no-cost smoother to supply lateral boundary conditions to inner high resolution LETKF, has been developed [6] and is tested for the K-computer. A displaced ensemble variational assimilation method has also been developed and tested for a data assimilation experiment on satellite microwave imager data [7]. As a more ambitious trial for the K-computer, development of a particle filter is underway at ISM [8].

Another storm scale LETKF system and a hybrid ensemble-variational data assimilation system for deep convection based on a cloud resolving model CReSS [9] are also under development at DPRI [10] and NIED [11], respectively.

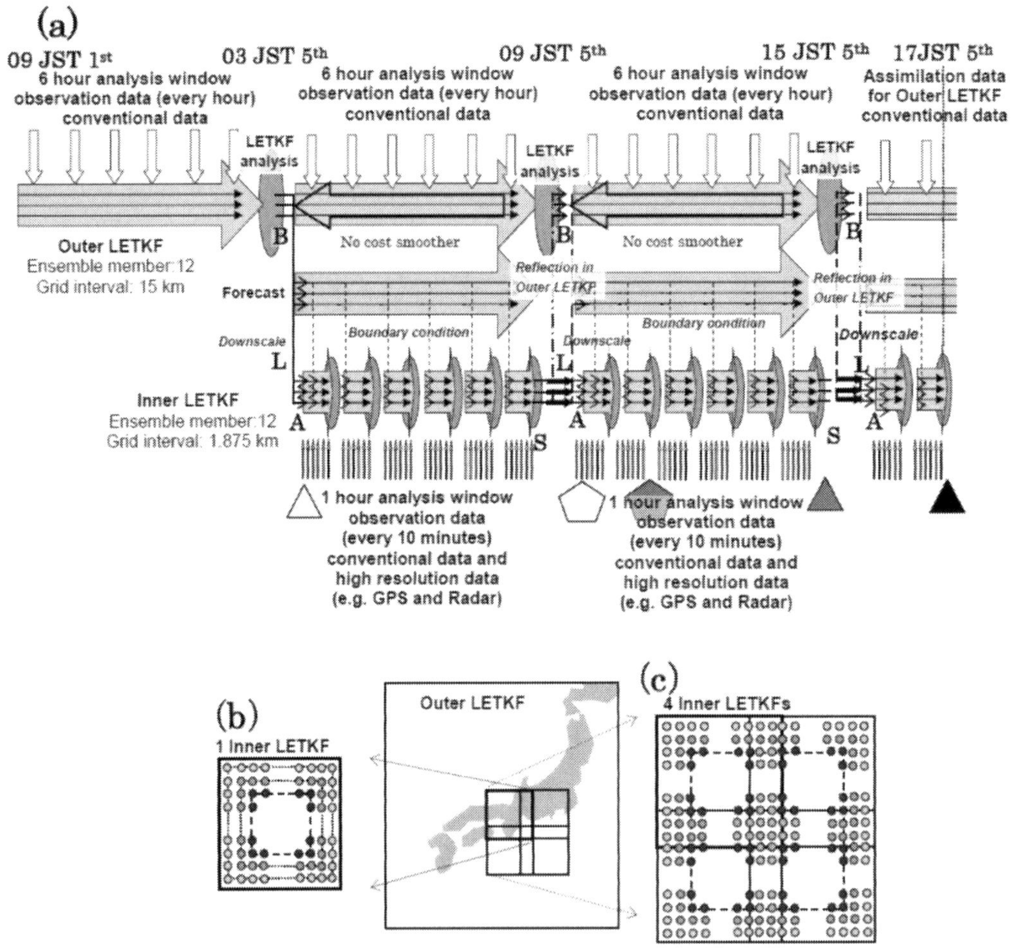

Figure 1. Concept of the two-way nested LETKF. After Seko et al. [6].

2.2. *Development and validation of a cloud resolving ensemble analysis and forecast system (by MRI, JAMSTEC, NPD, DPRI, Kobe University and Tohoku University)*

Ensemble prediction is another approach to extend the predictability of meteorological phenomena. MRI and its collaborators once participated in the international intercomparison of mesoscale ensemble prediction which was held in the WWRP Beijing Olympics 2008 Research and Development Project (B08RDP [12-14]). Initial perturbation methods and effect of lateral boundary perturbations were investigated for the mesoscale ensemble forecast with a horizontal resolution of 15 km [15, 16].

In SPIRE, the goal of the second theme is to demonstrate the plausibility of the probabilistic quantitative forecast of severe weathers for disaster prevention by cloud resolving ensemble NWP. An NHM-LETKF system using the incremental approach method developed at NPD has been tested [17] and applied to the K-computer. This ensemble data assimilation system shares observation operators with the JMA's operational nonhydrostatic 4DVAR system (JNoVA), while its target is the quantitative probabilistic forecast for heavy rainfalls using cloud resolving ensemble prediction. Duc et al. [18] applied the LETKF system to the ensemble numerical simulation of Myanmar cyclone Nargis with the K-computer.

Cloud resolving (2 km) ensemble forecast experiments were performed for the summers of 2010 and 2011 as the test trials using the MRI's super computer and were validated with observed rainfall [19]. The JNoVA 4DVAR analyses were used as the initial conditions, while the JMA one-week ensemble prediction was used as the lateral boundary perturbations. Saito et al. [20] applied this system to the 2011 Fukushima-Niigata heavy rainfall and discussed the predictability of the event. Origuchi et al. [21] tested the similar system for the heavy rainfall case of typhoon T1112 (Talas), and compared with the LETKF-derived analyses and perturbations. Results of the probabilistic forecast are validated and used as the input data for application systems for disaster prevention such as the ensemble flood forecasting at Kobe University [22].

As the ultra high resolution forecast system, the LETKF nested system was linked with a computational fluid dynamics (CFD) model with a horizontal resolution of 10 m at Tohoku University [23]. Figure 2 shows the model result that captures well an invasion of the sea breeze front and the 3-dimensional structure of horizontal convective rolls over Sendai downtown with resolved buildings.

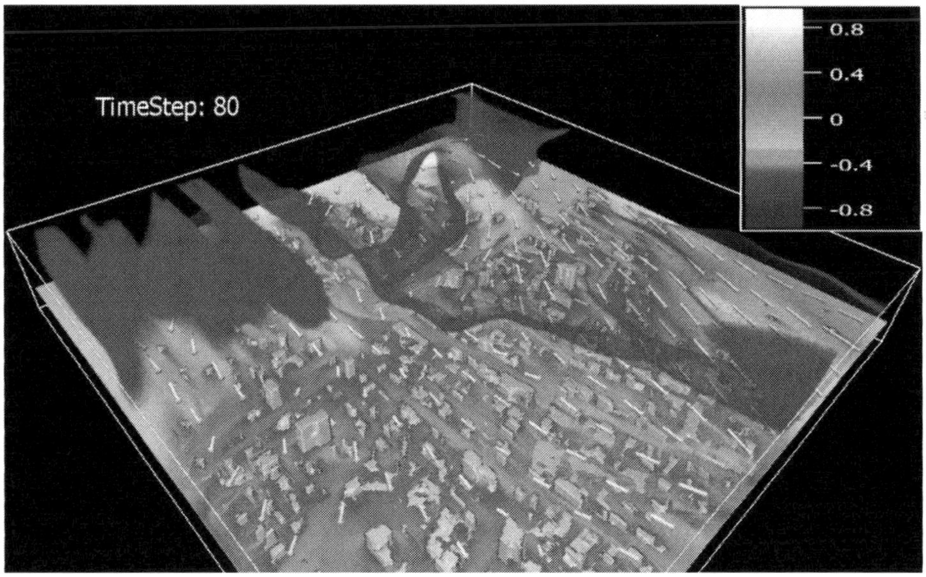

Figure 2. The invasion of sea breeze over Sendai downtown. The warm/cold air is plotted in red/blue, respectively. Near surface winds are shown by yellow vectors. After Chen et al. [23].

2.3 *Basic research using very high resolution atmospheric models (by JAMSTEC, MRI, NPD, AORI, Tohoku University, DPRI, HyARC and NDA)*

In this theme, uncertainties of the current atmospheric models due to subgrid scale parameterization in physical processes are evaluated using very high-resolution numerical models. Uncertainties of the bulk cloud microphysics scheme for cloud resolving models are evaluated using the spectral BIN method models (e.g., [24, 25]).

To challenge the 'grey zone' problem in atmospheric models with horizontal resolutions around 1 km, so-called 'Terra Incognita', the planetary boundary layer parameterization scheme is evaluated by the large eddy simulation (LES) models [26]. Figure 3 shows an example of LES simulation by AORI with a grid distance of 50 m. To reduce all-to-all communications in the MPI parallelization, Fast Fourier Transform to solve the Poisson equation is conducted within a single computational node of the K-computer. Development of regional versions of NICAM (Nonhydrostatic ICosahedral Atmospheric Model) is also underway at JAMSTEC to utilize the model for very high-resolution experiments [28].

High-resolution simulations of typhoons and tornadoes are conducted to study mechanisms of their development (e.g., [29, 30]).

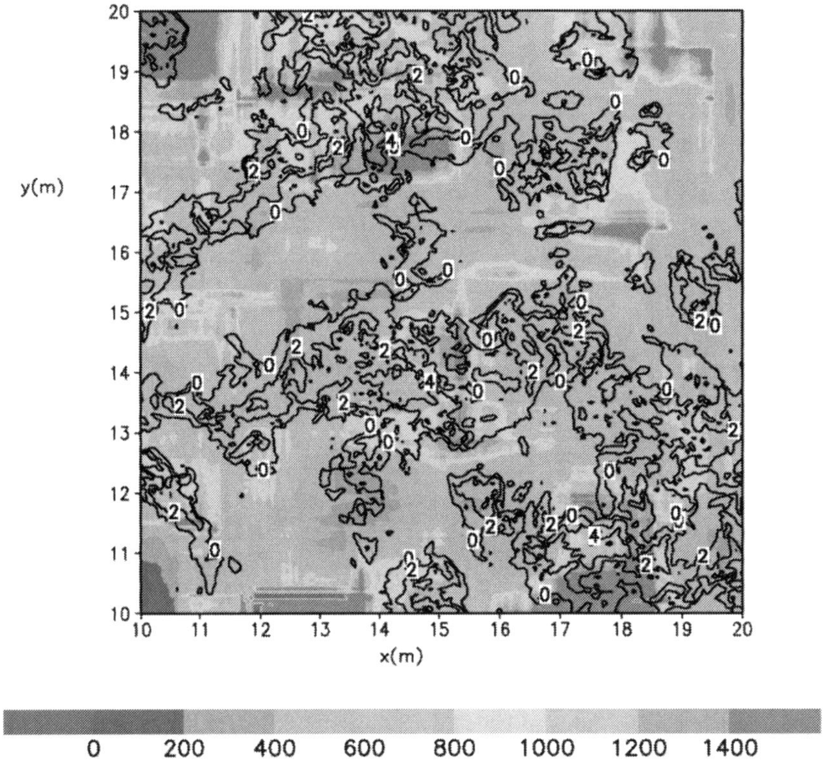

Figure 3. Horizontal distribution of turbulent mixing length (relative values) with a horizontal subfilter size of 2 km estimated by LES. Isolines denote upward vertical velocity (m/s). After Ito et al. [27].

Acknowledgments

A part of this research is supported by the Strategic Programs for Innovative Research funded by the Ministry of Education, Culture, Sports, Science and Technology. The authors are indebted to members of mesoscale NWP group, especially T. Kawabata, S. Origuchi, M. Kunii, A. Aonashi, A. Hashimoto, W. Mashiko of the Meteorological Research Institute, T. Fujita and H. Eito of the Numerical Prediction Division of the Japan Meteorological Agency, M. Hara, K. Nakamura, A. Noda and M. Sugi of the Japan Agency for Marine-Earth Science and Technology, S. Shimizu of the National Research Institute for Earth Science and Disaster Prevention, K. Yamaguchi, E. Nakakita, T. Takemi and Y. Yamashiki of Kyoto University, K. Kobayashi of Kobe University, G. Ueno and S. Nakano of the Institute of Statistical Mathematics. W. Sha, and T. Iwasaki of Tohoku University, H. Niino and N. Kuba of the Atmosphere and Ocean Research Institute, K. Tsuboki of the Hydrospheric Atmospheric Research Center, M. Nakanishi of National Defense Academy, for their contributions to the research project. Thanks are extended to S. Imawaki of Japan Agency for Marine-Earth Science and Technology, a principal investigator of Field 3 of the SPIRE project, for his leadership.

Appendix

List of acronyms

AORI	Atmosphere and Ocean Research Institute/University of Tokyo
CReSS	Cloud Resolving Storm Simulator
DPRI	Disaster Prevention Research Institute/Kyoto University
HyARC	Hydrospheric Atmospheric Research Center/Nagoya University
ISM	Institute of Statistical Mathematics
JAMSTEC	Japan Agency for Marine-Earth Science and Technology
LETKF	Local Ensemble Transform Kalman Filter
MRI	Meteorological Research Institute/Japan Meteorological Agency
NDA	National Defense Academy
NIED	National Research Institute for Earth Science and Disaster Prevention
NPD	Numerical Prediction Division/Japan Meteorological Agency
WWRP	World Weather Research Programme
4DVAR	Four dimensional variational (data assimilation)

References

[1] Saito K (2012) JMA nonhydrostatic model. –Its application to operation and research *InTech.* Atmospheric Model Applications 85-110 doi:10.5772/35368.

[2] Saito K, Fujita T, Yamada Y, Ishida J, Kumagai Y, Aranami K, Ohmori S, Nagasawa R, Kumagai S, Muroi C, Kato T, Eito H and Yamazaki Y (2006) The operational JMA Nonhydrostatic Mesoscale Model. *Mon. Wea. Rev.* **134** 1266-1298

[3] Saito K, Ishida J, Aranami K, Hara T, Segawa T, Narita M and Honda Y (2007) Nonhydrostatic atmospheric models and operational development at JMA. *J. Meteor. Soc. Japan* **85B** 271-304

[4] Kawabata T, Kuroda T, Seko H and Saito K (2011) A cloud-resolving 4D-Var assimilation experiment for a local heavy rainfall event in the Tokyo metropolitan area *Mon. Wea. Rev.* **139** 1911-1931

[5] Seko H, Miyoshi T, Shoji Y and Saito K (2011) A data assimilation experiment of PWV using the LETKF system -Intense rainfall event on 28 July (2008- *Tellus* **63A** 402-414

[6] Seko H, Tsuyuki T, Saito K and Miyoshi T (2013) Development of a two-way nested LETKF system for cloud resolving model *Data Assimilation for Atmospheric, Oceanic and Hydrological Applications* **2** 489-505

[7] Aonashi K and Eito H (2011) Displaced Ensemble Variational Assimilation Method to Incorporate Microwave Imager Brightness Temperatures into a Cloud-resolving Model *J. Meteoro.Soc. Japan* **89** 175-194

[8] Nakano S and Higuchi T (2012) Weight adjustment of the particle filter on distributed computing systems *Proceedings of 15th International Conference on Information Fusion*

[9] Tsuboki K and Sakakibara A (2007) Numerical prediction of high-impact weather systems *Text book for Seventeenth IHP training course in 2007 HyARC, Nagoya University, Japan, and UNESCO*, p 273

[10] Yamaguchi K and Nakakita E (2009) A suggestion for a data assimilation method of number concentration of graupel using polarimetric radar *Ann. J. Hydr. Engineer.* **53** 355-360 (in Japanese with English abstract)
[11] Shimizu S (2012) Development of a CReSS-Hybrid data assimilation system *Achievement Report, SPIRE Field 3*, p 46 (in Japanese)
[12] Saito K, Kunii M, Hara M, Seko H, Hara T, Yamaguchi M, Miyoshi T and Wong W (2010) WWRP Beijing (2008 Olympics Forecast Demonstration / Research and Development Project (B08FDP/RDP) *Tech. Rep. MRI* **62** 210
[13] Kunii M, Saito K, Seko H, Hara M, Hara T, Yamaguchi M, Gong J, Charron M, Du J, Wang Y and Chen D (2011) Verifications and intercomparisons of mesoscale ensemble prediction systems in B08RDP *Tellus* **63A** 531-549
[14] Duan Y, Gong J, Charron M, Chen J, Deng G, DiMego G, Du J, Hara M, Kunii M, Li X, Li Y, Saito K, Seko H, Wang Y and Wittmann C (2012) An overview of Beijing 2008 Olympics Research and Development Project (B08RDP) *Bull. Amer. Meteor. Soc.* **93** 381-403
[15] Saito K, Hara M, Kunii M, Seko H and Yamaguchi M (2011) Comparison of initial perturbation methods for the mesoscale ensemble prediction system of the Meteorological Research Institute for the WWRP Beijing 2008 Olympics Research and Development Project (B08RDP) *Tellus* **63A** 445-467
[16] Saito K, Seko H, Kunii M and Miyoshi T (2012) Effect of lateral boundary perturbations on the breeding method and the local ensemble transform Kalman filter for mesoscale ensemble prediction *Tellus* **64** 11594
[17] Kuroda T, Fujita T, Seko H and Saito K (2012) Construction of Mesoscale LETKF Data Assimilation Experiment System *CAS/JSC WGNE Res. Act. Atmos. Ocea. Model.* **42** 1.15-1.16
[18] Duc L, Koruda T, Saito K and Fujita T (2012) Data assimilation experiments of Myanmar cyclone Nargis based on NHM-LETKF *CAS/JSC WGNE Res. Act. Atmos. Ocea. Model.* **42** 1.03-1.04
[19] Duc L, Saito K and Seko H (2013) Spatial-temporal fractions verification for high resolution ensemble forecasts *Tellus* **65** 18171
[20] Saito K, Origuchi S, Duc L, and Kobayashi K (2013) Mesoscale ensemble predction of the 2010 Niigata-Fukushima heavy rainfall *Tech. Rep. JMA* **134** 170-184 (in Japanese)
[21] Origuchi S, Saito K and Kobayashi K (2013) Mesoscale ensemble predction of the typhoon T1122 (Tallas) *Tech. Rep. JMA* **134** 184-194 (in Japanese)
[22] Kobayashi K, Otsuka S, Takara K, Origuchi S and Saito K (2013) Ensemble forecasting of rainfall and flood in a small river basin *Ann. J. Hydr. Engineer.* **57** I_1597-I_1602 (in Japanese with English abstract)
[23] Chen G, Sawada M, Sha W, Iwasaki T, Iwai H, Seko H and Saito K (2012) A challenge to realize the ultra-high-resolution weather forecast for big city *Proceeding, K-computer symposium 2012*, p 71-72
[24] Kuba N and Fujiyoshi Y (2006) Development of a cloud microphysical model and parameterizations to describe the effect of CCN on warm cloud *Atmos. Chem. Phys.* **6** 2793–2810
[25] Misumi R, Hashimoto A, Murakami M, Kuba N, Orikasa N, Saito A, Tajiri T, Yamashita K and Chen J-P (2010) Microphysical structure of a developing convective snow cloud simulated by an improved version of the multi-dimensional bin model *Atmos. Sci. Lett.* **11** 186-191
[26] Ito J, Tanaka R, Niino H and Nakanishi M (2010) Large Eddy Simulation of Dust Devils in a Diurnally-Evolving Convective Mixed Layer *J. Meteoro.Soc. Japan* **88** 63-77
[27] Ito J, Niino H and Nakanishi M (2012) Unknown region in the atmospheric model *Proceeding, K-computer symposium 2012*, p 69-70
[28] Noda A, Hara M, Iga S, Ohno T, Tomita H, Satoh M, Nakamura K and Kimura F (2012) Development of regional atmospheric models based on NICAM *Achievement Report, SPIRE Field 3*, 83-86 (in Japanese)
[29] Akter N and Tsuboki K (2012) Numerical simulation of cyclone Sidr using a cloud-resolving model) Characteristics and formation process of an outer rainband *Mon. Wea. Rev.* **140** 789-810
[30] Mashiko W, Niino H and Kato T (2009) Numerical simulation of tornadogenesis in an outer-rainband minisupercell of typhoon Shanshan on 17 September 2006 *Mon. Wea. Rev.* **137** 4238–4260

Probing the extensive nature of entropy

T Salagaram[1] and N Chetty[1,2]

[1] Department of Physics, University of Pretoria, Pretoria, 0001, South Africa
[2] National Institute for Theoretical Physics, Gauteng, 2000, South Africa

E-mail: trisha.salagaram@up.ac.za

Abstract. We have devised a general numerical scheme applied to a system of independent, distinguishable, non-interacting particles, to demonstrate in a direct manner the extensive nature of statistical entropy. Working within the microcanonical ensemble, our methods enable one to directly monitor the approach to the thermodynamic limit ($N \to \infty$) in a manner that has not been known before. We show that $(s_N - s_\infty) \to N^{-\alpha}$ where s_N is the entropy per particle for N particles and s_∞ is the entropy per particle in the thermodynamic limit. We demonstrate universal behaviour by considering a number of different systems each defined by its unique single-particle spectrum. Various thermodynamic quantities as a function of N may be computed using our methods; in this paper, we focus on the entropy, the chemical potential and the temperature. Our results are applicable to systems of finite size, e.g. nano-particle systems. Furthermore, we demonstrate a new phenomenon, referred to as entropic interference, which manifests as a cancellation of terms in the thermodynamic limit and which results in the additive nature of entropy.

1. Introduction

In 1871, Ludwig Boltzmann wrote down his celebrated equation (which is inscribed on his tombstone) relating statistical entropy S of an isolated atomistic system to the number of accessible system microstates W subject to macroscopic constraints,

$$S = k_B \log W, \tag{1}$$

where k_B is the Boltzmann constant. This is one of the most fundamental equations in physics. Given the importance of the second law of thermodynamics [1] in governing the flow of entropy in physical and chemical processes [2], including in biological systems and ultimately in processes that shape life and the evolution of the universe, there is little question about the significance of Boltzmann's work in endeavours that go well beyond physics. Equation (1) connects the microscopic details of a system with its macroscopic properties in a simple and elegant way.

This equation, which forms the basis of the statistical interpretation of thermodynamics [3], has been the subject of rigourous verification over the past more than 100 years, but was initially met with strong resistance by so-called energeticists including some of the most famous physicists of that time who felt that energy conservation alone was necessary to describe spontaneous processes and the arrow of time. This fierce opposition ultimately led to the depressive Boltzmann taking his own life in 1906 [4].

In our paper we make observations concerning (1) that have important implications for physics.

We demonstrate in a direct manner the extensive nature of statistical entropy. A property of any system is extensive if, in the thermodynamic limit (infinite system size limit), the property depends on the size of the system. We are able to do this now in a manner that was not possible 100 years ago because of our access to computation. Boltzmann only *concluded* that statistical entropy is extensive by first equating statistical temperature with thermodynamic temperature, and by then inferring the equivalence of statistical entropy with thermodynamic entropy. The extensive nature of thermodynamic entropy was already well established in the early part of the 19th century.

An important step for us was the formulation of an efficient computational algorithm that enables us to calculate statistical entropy in the microcanonical ensemble as a function of the number of particles N in a recursive manner. Our algorithm is quick and, despite the exponential increase of W as a function of N, we are able to derive converged results in real time on a modern desktop computer with no overflow errors that have dogged previous attempts.

We also demonstrate the manner in which entropy approaches the thermodynamic limit. To the best of our knowledge, this has not been done before. We demonstrate universality by considering a number of different systems each defined by its unique single-particle spectrum. Despite the vast differences in the microscopic details of the single particle energy states for differing systems, we show that $(s_N - s_\infty) \sim N^{-\alpha}$, where s_N is the entropy per particle for N particles, s_∞ is the entropy per particle in the thermodynamic limit (i.e. in the infinite system size limit where $N \to \infty$), and α is the exponent. This slow $N^{-\alpha}$ powerlaw behaviour of statistical entropy, as it converges in the thermodynamic limit, is an important subject of this paper.

Our results have important implications for finite system sizes. Using our results, we are able to derive in a consistent manner, various thermodynamic properties in the microcanonical ensemble for finite systems of size N, such as temperature, pressure, chemical potential, magnetism, the heat capacities, the various thermodynamic potentials, etc. The corollary is that our work enables finite size corrections for systems that are large but not so large that the thermodynamic limit has been reached. Finite systems are more the rule than the exception for typical systems that are measured in the laboratory, and so finite size corrections would usually be very necessary. Our work is of particular relevance to nano-particle physical, chemical and biological systems, and is thus applicable in the quantitative study of such systems at finite temperature.

In addition, we demonstrate entropic interference, which manifests as a cancellation of terms in the thermodynamic limit and which results in the additive nature of entropy.

2. Extensive behaviour

We consider an isolated system comprising N independent, distinguishable, non-interacting particles with total energy E. The number of microstates for such a system within the microcanonical ensemble can be calculated recursively by

$$W_N(E) = \sum_{E'=0}^{E} W_{N-1}(E - E') \times w(E'), \tag{2}$$

where $w(E')$ is the degeneracy of the single-particle state with energy E'. The entropy for such a system may then be determined recursively by

$$S_N(E) = S_{N-1}(E) + k_B \sum_{E'=0}^{E} \exp\left[\frac{S_{N-1}(E - E') - S_{N-1}(E')}{k_B}\right] \times w(E'). \tag{3}$$

A derivation of (3) from (2) is given in [5]. These equations may be used to calculate statistical entropy recursively at energy E systematically for $N = 2, 3, 4, \ldots, N_{max}$, where N_{max} can be

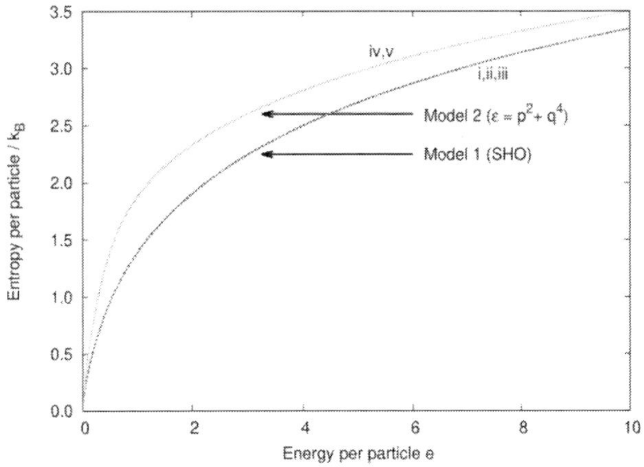

Figure 1. Entropy per particle s_N versus energy per particle e for Models 1 and 2.

made sufficiently large. In so-doing we are able to analyse the *approach* to the thermodynamic limit in a manner that has not been possible before.

Equation (3) is an extremely efficient means to calculate entropy on a modern desktop computer. We have applied and successfully tested and verified (3) for a range of analytically solvable models including a system of harmonic oscillators; free particles in one, two and three dimensions; spin 1/2 and spin 1 particles, etc., and for a range of analytically intractable problems including a host of hypothetical models, each defined by its unique single-particle spectrum (for example: $\varepsilon = p^2$; $\varepsilon = p^3$; $\varepsilon = p^4$; $\varepsilon = p + q$; $\varepsilon = p + q^2$, etc., where $p, q = 0, 1, 2, 3, \ldots$). For the purposes of this paper and to demonstrate our main ideas, we shall primarily focus only on two models, namely the well-known analytically solvable system of harmonic oscillators with single-particle spectrum (in appropriate units) given by $\varepsilon(p) = p$ where $p = 0, 1, 2, 3, \ldots$ (Model 1); and the analytically insolvable model system with single-particle spectrum (in appropriate units) given by $\varepsilon(p, q) = p^2 + q^4$ where $p, q = 0, 1, 2, 3, \ldots$ (Model 2). The second model is sufficiently complex (two-parameter quantum model with unequally-spaced energy levels and with degenerate states) that it may be considered to be representative of the general class of discrete-level problems, and therefore a resolution of this problem is a measure of the power of our methods.

In figure 1 we plot for Model 1 the entropy per particle s_∞ (in units of k_B) versus the energy per particle e in the thermodynamic limit. We actually present three plots which, to within numerical error, coincide with each other, namely (i) $s_\infty(e) = (1 + e)\log(1 + e) - e \log e$, (ii) $s_\infty(e)$ calculated in the canonical ensemble, and (iii) $s_{3000}(e)$ calculated using (3). From (iii) we conclude that $N = 3000$ is a sufficiently large number of particles to estimate, for our purposes, the thermodynamic limit. The plot involving (i) is possible only because this model is solvable in the thermodynamic limit using the Stirling approximation. This is not the case for the second model so the equivalence of plot (i) for Model 2 does not exist. For Model 2 we plot in figure 1 (iv) $s_\infty(e)$ calculated in the canonical ensemble, and (v) $s_{3000}(e)$ calculated using (3) where once again, to within numerical error, these two plots coincide.

This demonstrates the accuracy and the efficiency with which (3) may be used to calculate the entropy of any system within the microcanonical ensemble. We have shown here that independent of the single particle spectrum of Models 1 and 2, $s_N(e) \to$ constant real number, designated as $s_\infty(e)$, as $N \to \infty$. Similarly (3) may be used to show that this is true for *any* single particle spectrum. This establishes in a general manner the extensive nature of statistical

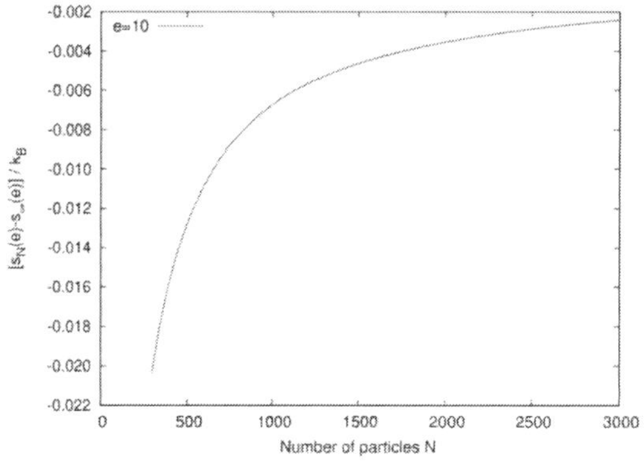

Figure 2. The deviation of the entropy per particle from its value in the thermodynamic limit $(s_N - s_\infty)$ versus the number of particles N for Model 1 for energy per particle $e = 10$.

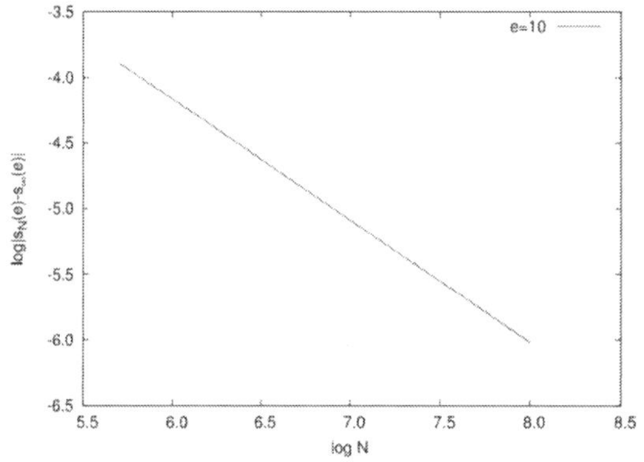

Figure 3. $\log|s_N - s_\infty|$ versus $\log N$ for Model 1 for $e = 10$.

entropy.

3. Powerlaw behaviour

In the course of our numerical work we made the fortuitous discovery that the approach of (3) to the thermodynamic limit is slow. For example, a ten-fold increase in the number of particles from $N = 100$ to 1000 for Model 1 at $e = 10$ results in a reduction in the deviation of the entropy per particle from its value in the thermodynamic limit by less than 1.5%. What proved to be frustrating for us initially, on closer inspection turned out to be revealing of the powerlaw behaviour of entropy as we approached the thermodynamic limit.

Using (3), we are able to monitor the approach of the entropy per particle to the thermodynamic limit. In figure 2, we plot $(s_N - s_\infty)$ versus N for Model 1 for $e = 10$. The curve is negative which indicates that the thermodynamic limit is approached from below. The asymptotic behaviour is observed to be slow (powerlaw rather than exponential) and may be analysed more carefully by plotting $\log|s_N - s_\infty|$ versus $\log N$ which we do in figure 3. The behaviour is linear which confirms powerlaw behaviour, with a negative slope of $\alpha = 0.92623 \pm 0.00003$.

We observe similar behaviour for $\log|s_N - s_\infty|$ versus $\log N$ for Model 1 separately for each value of e. We are therefore able to trace the curve $\alpha(e)$, which we present in figure 4. In figure 5

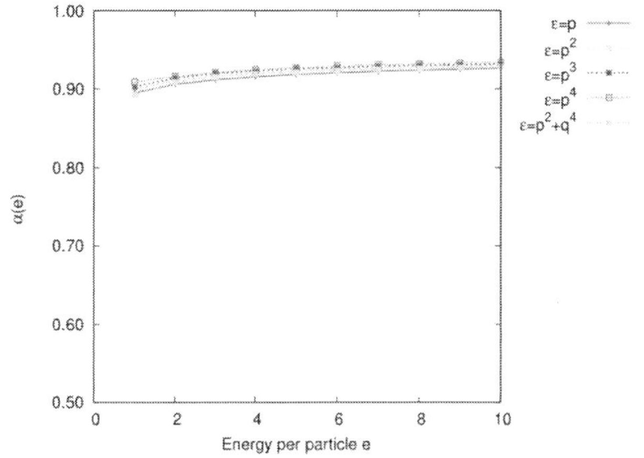

Figure 4. The exponent $\alpha(e)$ versus the energy per particle e for a range of different models. $\alpha(e)$ exhibits only a very weak dependence on e and a very weak dependence on the microscopic details of the system single particle levels.

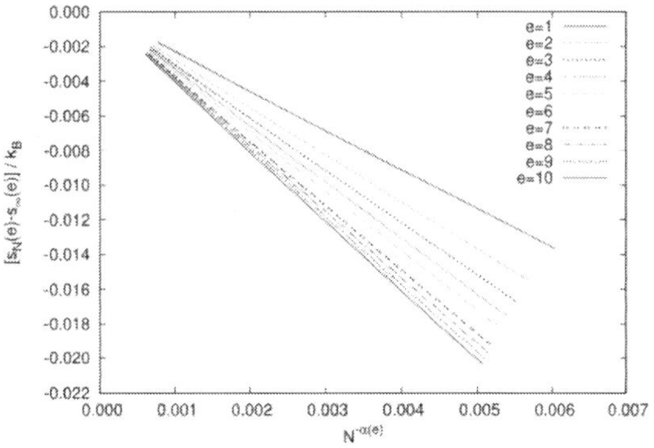

Figure 5. $(s_N(e) - s_\infty(e))$ versus $N^{-\alpha(e)}$ for Model 1 for $e = 1, 2, 3, \ldots, 10$.

we plot $(s_N - s_\infty)$ versus $N^{-\alpha(e)}$ for Model 1. The linear behaviour of this series of curves is revealing.

We discovered the same behaviour for Model 2 and for other models studied in the general class of similar problems each defined by its unique single-particle spectrum. It is significant also that α is only weakly dependent on e, and only weakly dependent on the details of the single particle spectrum as shown in figure 4.

4. Thermodynamic properties to order $N^{-\alpha}$

We conclude that the first order correction to $s_N(e)$ may be written as

$$s_N(e) = s_\infty(e) + \eta(e)N^{-\alpha(e)} \qquad \text{as } N \to \infty, \tag{4}$$

where $\eta(e)$ is the slope of each curve indicated in figure 5, which we have calculated and plotted in figure 6.

We find smooth behaviour for $\eta(e)$ and we expect $\eta(e) \to 0$ as $e \to 0$ by the third law of thermodynamics. The function $\eta(e)$ is unique for each system under consideration, and

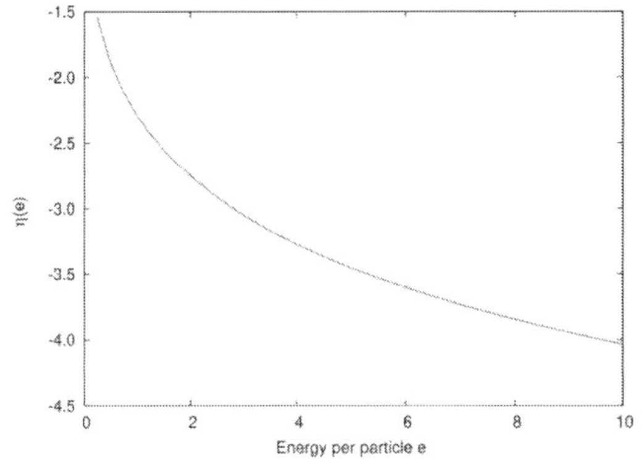

Figure 6. $\eta(e)$ versus e for Model 1.

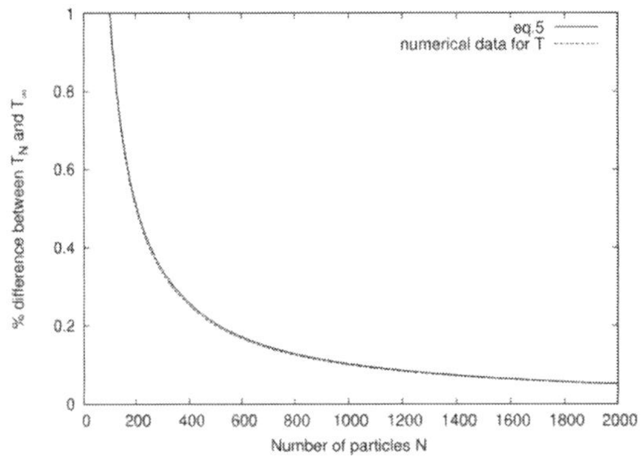

Figure 7. Percentage difference between T_N and T_∞ as a function of N (up to 2000) for Model 1 at energy per particle $e = 10$. The results in this graph are calculated using (5) and the derivative of (3) with respect to E. The range on the vertical axis is from 0 to 1%.

must be evaluated separately for each system to enable one to proceed to calculate the various thermodynamic properties of the system to order $N^{-\alpha}$. For example, the temperature of a finite system of size N is given by

$$\frac{1}{T_N} = \frac{1}{T_\infty} + \left[\eta'(e) - \eta(e)\alpha'(e)\log N\right] N^{-\alpha(e)}, \tag{5}$$

where the prime indicates the first derivative with respect to e, and T_∞ represents the temperature in the thermodynamic limit. This establishes the expression for temperature for a finite system of size N.

For Model 1 at $e = 10$ we plot in figure 7 the percentage difference between T_N and T_∞ as a function of N. We note only about $\sim 1\%$ *elevation* in the temperature for $N = 100$ compared with the temperature in the infinite system size limit. This is a surprising result: even through the entropy per particle converges rather slowly to its value in the thermodynamic limit, the derivative of entropy with respect to the energy converges much more rapidly. This is a general feature of our conclusions - the physically measurable quantities that are expressible as a gradient of the entropy, such as temperature, converge much more rapidly to the thermodynamic limit than does the entropy itself.

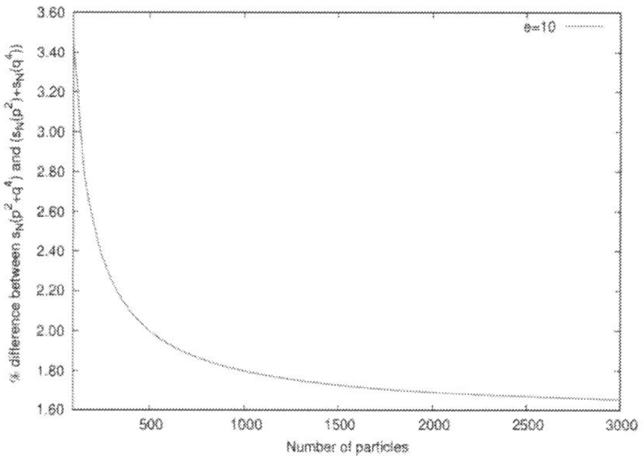

Figure 8. Interference term $J_N^{p^2+q^4}(e) \to 0$ in the thermodynamic limit

Similarly, the ratio of chemical potential to temperature $\frac{\mu}{T}$ for a finite system of size N is given by

$$\left(\frac{\mu}{T}\right)_N = \left(\frac{\mu}{T}\right)_\infty + \left\{-e\eta'(e) + \eta(e)\left[1 - \alpha(e) + e\alpha'(e)\log N\right]\right\} N^{-\alpha(e)}, \tag{6}$$

where $\left(\frac{\mu}{T}\right)_\infty$ defines the ratio of the chemical potential to temperature in the thermodynamic limit.

All thermodynamic quantities such as pressure, magnetism, the heat capacities, the various thermodynamic potentials, etc. may now be calculated for any finite system of size N.

5. Entropic interference

To demonstrate our ideas of entropic interference, we focus on Model 2 for which $\varepsilon = p^2 + q^4$. To start with, we note that the partition function is separable

$$z = \sum_{p,q} e^{-\beta(p^2+q^4)} = \sum_p e^{-\beta p^2} \times \sum_q e^{-\beta q^4}, \tag{7}$$

and therefore entropy is additive in the thermodynamic limit

$$s_\infty^{p^2+q^4}(e) = s_\infty^{p^2}(e) + s_\infty^{q^4}(e). \tag{8}$$

However we don't expect $\Omega^{p^2+q^4}(N,E)$ to be related to $\Omega^{p^2}(N,E)$ and $\Omega^{q^4}(N,E)$ in a simple way since Ω corresponds to the counting of states! For finite N, we may therefore write

$$s_N^{p^2+q^4}(e) = s_N^{p^2}(e) + s_N^{q^4}(e) + J_N^{p^2+q^4}(e), \tag{9}$$

where $J_N^{p^2+q^4}(e)$ may be referred to as an interference term. This can only happen if

$$J_N^{p^2+q^4}(e) \to 0 \,\text{as}\, N \to \infty. \tag{10}$$

In figure 8 we plot $J_N^{p^2+q^4}(e)$ as a function of N which verifies this result.

We therefore conclude that entropy is only additive in the thermodynamic limit.

6. Further theoretical notions

Boltzmann is sometimes credited with being a founding father of quantum mechanics because he was first to assume a discrete set of states (which, unbeknownst to him at that time, was the levels of the quantum harmonic oscillator; this is the same Model 1 that we presented above) for an atomistic system, which he used to derive (1). However, it should be kept in mind that Boltzmann did this for mathematical convenience, and in the end he tried to extract the classical limit and he argued that his results still applied in this continuum limit. There is no evidence to suggest that Boltzmann genuinely believed that the discrete set of levels was anything more than an intermediate step in his arguments for essentially classical systems.

Using (3), we can now see why Boltzmann's argument was valid. Equation (3) is valid independent of the overall scaling factor for energy (which we stated earlier was measured in appropriate units). This energy scale factor may be finite, which is appropriate for quantum systems and is a measure of the quantum unit of energy for the system, or the scale factor may approach zero which is appropriate for classical systems, whereupon (3) will then need to be re-written in terms of an integral rather than a summation over E. The overall conclusions that we have drawn in this paper stand independent of the actual value of this energy scale factor.

A further point worth noting is that Boltzmann used physical arguments to conclude that statistical entropy is an extensive quantity. We have derived and used what may be referred to as a direct numerical method embodied in (3), to show that statistical entropy is an extensive quantity for any arbitrary system defined by its own unique single particle spectrum. In this process, we discovered universal scaling behaviour for statistical entropy.

7. Conclusions

We conclude that (3) is a computationally efficient means to calculate the entropy of any system comprising N independent, indistinguishable, non-interacting particles in the microcanonical ensemble. We have shown that the entropy per particle converges to a constant real number in the thermodynamic limit independent of the details of the single-particle spectrum, thus verifying for the first time in a direct manner, the extensive nature of statistical entropy. We investigated the manner in which the entropy per particle approaches the thermodynamic limit and discovered universal scaling behaviour for $(s_N - s_\infty) \sim N^{-\alpha}$, where α is the exponent. For the range of systems that we have considered, $\alpha(e)$ exhibits only a very weak dependence on e and a very weak dependence on the microscopic details of the system single particle levels. Our work is applicable in the quantitative study of nano-particle physical, chemical and biological systems at finite temperature. We showed that entropy is only additive in the thermodynamic limit.

Acknowledgments

NC thanks the University of Pretoria and the National Institute for Theoretical Physics for support. TS is supported by a National Research Foundation postdoctoral fellowship. We thank Wolfgang Christian (Davidson College, North Carolina, USA) for useful discussions.

References

[1] Carnot S 1824 *Reflexions sur la puissance motrice de feu et sur les machines propres à developper cette puissance* (Bachelier, Paris)
[2] Maxwell J C 1867 On the dynamical theory of gases *Philosophical Transactions of the Royal Society of London* **157** 49–88
[3] Boltzmann L 1872 Weitere Studien über das Wärmegleichgewicht unter Gasmolekülen *Sitzungsberichte Akad. Wiss., Vienna, part II* **66**, 275–370
[4] Brush S G 1976 *The kind of motion we call heat I* (North Holland Publishing Co.) 244–47
[5] Salagaram T and Chetty N 2011 Enhancing the understanding of entropy through computation *Am. J. Phys.* **79** 1127–32

Journal of Physics: Conference Series **454** (2013) 012075 doi:10.1088/1742-6596/454/1/012075

The challenges of developing computational physics: the case of South Africa

T Salagaram[1] and N Chetty[1,2]

[1] Department of Physics, University of Pretoria, Pretoria, 0001, South Africa
[2] National Institute for Theoretical Physics, Gauteng, 2000, South Africa

E-mail: trisha.salagaram@up.ac.za

Abstract. Most modern scientific research problems are complex and interdisciplinary in nature. It is impossible to study such problems in detail without the use of computation in addition to theory and experiment. Although it is widely agreed that students should be introduced to computational methods at the undergraduate level, it remains a challenge to do this in a full traditional undergraduate curriculum. In this paper, we report on a survey that we conducted of undergraduate physics curricula in South Africa to determine the content and the approach taken in the teaching of computational physics. We also considered the pedagogy of computational physics at the postgraduate and research levels at various South African universities, research facilities and institutions. We conclude that the state of computational physics training in South Africa, especially at the undergraduate teaching level, is generally weak and needs to be given more attention at all universities. Failure to do so will impact negatively on the countrys capacity to grow its endeavours generally in the field of computational sciences, with negative impacts on research, and in commerce and industry.

1. Introduction

Computational physics has been an important research tool in physics for over 40 years. It has enabled physicists to understand complex problems more completely compared to using theoretical and experimental methods alone. A generation ago, many postgraduate students working in computational physics research groups around the world would have been required to develop computational codes as part of their thesis work. This was during the infancy of computational physics, and this endeavour was confined only to a small number of specialised research groups around the world. Computational physics then was largely a research endeavour with very little formal training at the undergraduate level. Research students were simply expected to acquire the necessary computational skills as they proceeded with their research work. This resulted in highly skilled computational physics graduates, but these individuals were very few and far between.

This trend has changed significantly over recent times where more and more now do computational physics research groups rely on more sophisticated commercial software, or freeware computer programmes produced by more specialised groups for their research work. The advantage is that more researchers now have access to more efficient codes that incorporate more complex modelling of physical systems of interest. An ever-increasing body of researchers is now able to use such codes in their research endeavours, including experimental scientists. The downfall is that we are now producing students who are mostly schooled in utilising large computational codes by applying these to physical systems of interest with very little fundamental knowledge and understanding of the intricate details of the underlying theoretical, algorithmic and programming aspects. The research focus now has shifted away from computational methods toward applications. It is reasonable to ask whether these graduates should be referred to as computational physicists or computational technicians.

It is our view that computational physics students should be involved in all aspects of the computational project: theoretical modelling, algorithmic design, numerical methods, programming development, applications to physical problems of interest, analysis of results, and the graphical presentation of the results which could involve animation and visualisation (see Table 1). In so doing, computational physics students learn useful marketable skills that are transferable to other disciplines involving the modelling of problems in engineering, chemistry, biology, ecology, finance, economics, and so on. These skills are absolutely vital in a developing country such as South Africa. This is important to enable commerce and industry to move significantly ahead in terms of developing in-house computational applications, for example in the South African mining industry, rather than to rely on commercial codes developed abroad which sometimes are of questionable relevance for local conditions.

Aside from preparing students for the workplace and being an important research tool, computational physics is also of great pedagogical value. Evidence suggests that allowing students to study complex problems using computation improves their analytical skills and enhances their understanding of the underlying physics[1]-[4]. Getting students to design physical models to explain and predict phenomena can also correct weaknesses in traditional approaches to teaching physics, which often involves using analytical techniques to solve simple problems.

The 2012 IUPAP Conference on Computational Physics held at Kobe, Japan, held a panel discussion on the importance of computational physics, and the question was asked whether computational physics should be viewed separately from theoretical and experimental physics. We believe that the methods of computational physics are best learnt by considering applications to real physical systems and therefore the endeavour of computational physics should not be viewed in isolation of theoretical and experimental physics.

Despite the benefits, it remains a challenge to include computational physics in the undergraduate physics curriculum. Although some physics departments at South African universities teach computational physics at the undergraduate level this trend is yet to fully develop.

There is enormous potential for computational physics to develop rapidly on the African continent. With the advent of cheaper computers, faster networking, access to foreign supercomputing centres and the availability of freeware software, computational physics will be able to thrive in Africa if a greater effort is made at developing this subject more fundamentally at the pedagogical level. This can be a strong basis for developing physics more generally in Africa where investing in expensive experimental infrastructure is always going to be a challenge for many years into the future.

We conducted an in-depth survey of the teaching programmes in all physics departments to draw our conclusions about the development of computational physics in South Africa. We were especially interested to learn whether there was any computational physics being taught at each of the South African universities and, if so, what is the mode, approach and content of the teaching of this subject at these institutions. We also used the survey to determine whether the undergraduate courses are equipping students with the widest range of computational skills, and to determine the local challenges of including computational physics in the undergraduate physics curriculum.

We also considered the pedagogy of computational physics at the postgraduate and research levels at various South African universities, and research facilities and institutions. Here,

Table 1. Different approaches to the teaching and practice of computational physics

Fundamental computational physics	**Algorithmic physics**	**Applied computational physics or Modelling**
The practitioner develops large production codes or short computational exercises *ab initio*	The practitioner uses high-level languages, e.g. MathLab, Mathematica, Easy Java Simulations, etc, to develop short computational exercises	The practitioner operates large production codes in black box-mode
The practitioner engages in all aspects of the computational project: theoretical and mathematical modelling, algorithmic development, programming, graphical visualisation of the data, analysis of the results	The practitioner develops good algorithmic understanding of the problem, but the detailed numerical methods are glossed over	The practitioner only has superficial understanding of the algorithmic and programming details since the focus is on the application and on the analysis of the results
⇒ Computational physicist	⇒ Algorithmic physicist	⇒ Computational technician or Modeller

we were especially keen to scrutinise those research programmes that were considered to be computationally intensive. How do these programmes ensure that the postgraduate training goes well beyond simply teaching students to utilise computational codes in producing the ubiquitous computational technician, to imparting the full range of skills that are absolutely essential for the complete development of the computational physicist?

The data for this survey was obtained from course information on physics department websites, telephone interviews of staff members at various physics departments, personal visits to physics departments and feedback from a questionnaire which was sent to university physics departments. The questionnaire was based on the details given in reference[5]. We also surveyed most of the major research institutions in South Africa to determine their approach to the training in computational physics at the postgraduate and research levels.

2. Undergraduate training in computational physics

There are 17 traditional and comprehensive universities in total in South Africa. Information on computational physics could not be obtained for one traditional and four comprehensive universities. We excluded universities of technology from our investigation since these universities primarily offer physics only as a service course to other diploma or degree courses. Our results show that computational physics is taught in some way or form at 8 universities at the undergraduate level.

The most comprehensive degree programme in computational physics is offered at the University of KwaZulu-Natal (UKZN). Students who enrol for a degree in computational physics take physics as a full major, computational physics as a half major and one other half major subject drawn from mathematics or computer science. At second year level students are given an introductory computational physics course with a focus on basic numerical methods using the Fortran programming language where they display their results using Gnuplot in an open source environment. The applications are mostly in classical mechanics where the problems quickly become fairly complex, for example chaotic systems. At the third year level there is a close mirroring of modules taught in the mainstream physics programme and the computational physics programme. In this way students cover the relevant theory in the mainstream physics modules, and they are able to focus on algorithmic development and programming of problems in a wide range of topics including problems in quantum mechanics, statistical physics and solid state physics. By the end of the third year, the students acquire very advanced computational skills that enable them to embark on very complex projects at the postgraduate level, for example in astronomy. These skills have made these graduates highly employable in a wide range of industries, such as at the CSIR, Element Six, ISSI, etc., and many graduates have gone on to successfully pursue postgraduate studies at other universities and at other research institutions.

The University of Stellenbosh offers separate computational physics modules within its mainstream physics programme. Two computational physics modules are taught at second year level and one at third year level. The second year modules are designed to teach students the basics of computational methods in the context of solving classical mechanics and electrodynamics problems. Students use either Matlab or the Python programming language, and students may choose to work either in a Windows or a Linux environment. The third year module places a strong emphasis on the statistical skills necessary to do experimental analysis in the laboratory. Students write code either in C or Java within the Ubuntu Linux environment. A major computational project involves the students studying the Ising model.

The University of Pretoria teaches computational physics within the mainstream physics laboratory programme at second and third year levels. The course instructors have chosen to use symbolic programming in the teaching of this subject. In second year, students are introduced to symbolic programming using the Maple software package. They can work either in the Windows or Linux environments. In second and third year, students use the default solvers within the Maple environment to study various computational problems from classical and quantum mechanics, statistical physics and non-linear physics.

The University of Witwatersrand includes elements of computational physics in the laboratory courses from first to third year levels. Students are introduced to the use of computers in physics from data capture, processing and the presentation of results to the solution of mathematical models to fit data. At third year level, students have the option of choosing a computationally intensive project using Mathematica, Sigma Plot or Excel. Students are free to program in Python, C++ or Fortran, and may use either the Windows or Linux operating systems.

The University of the Western Cape includes computational physics within the second year laboratory course. The basics of computational physics are taught in the context of solving problems in classical mechanics. The computational problems are similar to the experiments which students conduct so that they can compare experimental data with numerical data. Students program in Python within the Windows environment. At third year level students are given a substantial computational physics project in each semester. They are taught to write scripts to solve problems using Octave, which is freeware, and they work within the Windows environment. The problems for these projects are taken from the mainstream theory courses taught in each semester thereby allowing students to focus on learning computational methods in the laboratory course.

The University of Cape Town includes elements of computational physics in the first year mainstream course, the second year laboratory course, and offers computational physics projects over the vacation period. In first year, students are shown how mathematical equations can be translated into computer code. This is based on the Matter and Interactions curriculum by Chabay and Sherwood[6, 7]. Using vPython simulations, students are able to visualize their dynamical results in real time. In second year, students learn to formulate algorithms and numerical methods by solving classical mechanics problems on the computer. Students have the option to use Matlab or they may program in Python or Java. There is no formal computational physics module taught at third-year level, however students are required to use computers to analyse their experimental results using techniques such as Fourier transforms where the students are required to write a short piece of code which reads in their experimental data and performs a

Fourier analysis. Undergraduate students may take an optional vacation computational project which is usually based on some aspect of research conducted within the department.

The North West University offers computational physics at both its Potchefstroom and Mafikeng campuses. The physics department at the Potchefstroom campus includes computational physics in the third year laboratory course. Two laboratory sessions have been allocated to teaching computational physics. In the first laboratory session students are taught the syntax of Fortran and the Interactive Data Language (or IDL, which is used for data analysis especially in astronomy). They can work either in the Windows or Linux environments. In the second laboratory session they write a program to solve a partial differential equation using the explicit and implicit numerical methods for numerical solutions of partial differential equations, and study the effects of various boundary conditions.

At the Mafikeng campus, computational physics is taught in the laboratory course at second and third year levels. In the second year laboratory course, there are several laboratory sessions allocated to the numerical and experimental study of a ball moving through a viscous medium. Students develop algorithms for the numerical solution of this problem and program in Fortran. The numerical results are compared to actual experimental results to test the accuracy and stability of the numerical methods used. At third year level, students are given mini computational projects on non-linear physics, stellar structure and the one dimensional Schrdinger equation. Students write their own codes in Fortran to solve these problems numerically.

At the University of Johannesburg the approach to teaching computational physics is to give students homework problems based on material covered in the theory courses which have to be solved numerically. These homework problems are designed to teach algorithmic development and programming in a systematic way, and students may use Matlab or Fortran for programming. They are required to work in the Linux environment.

Some of the challenges in not being able to properly teach computational physics at the undergraduate level include: it is difficult to include computational physics in the already full physics curriculum; finding the right balance between taking a fundamental approach to the teaching of computational physics which is time-consuming, and using ready-to-go packages in black box-mode is difficult; there is a lack of qualified lecturers and support staff with expertise in computational physics; there is not sufficient support from physics colleagues, i.e. the teaching of computational physics is still not recognised as being important by some senior colleagues and heads of departments; faculties do not appreciate the distinction between computer science and computational physics; there is not a proper understanding of what the teaching of computational physics entails; there are not a sufficient number of good computational physics text books appropriate at the undergraduate level; there are no dedicated computational facilities, or there is limited laboratory space for a computational physics local area network (LAN); there is a lack of staff to assist with LAN administration; South African physics students lack basic mathematical and programming skills that makes teaching undergraduate computational physics very challenging; many colleagues believe that Fortran is out of date and not worth using in the classroom.

2.1. Discussion

South African universities need to take a greater interest in developing computational physics at the undergraduate level. Computational physics combines the mathematical, theoretical, numerical and physical modelling of real systems in a unique manner that makes computational physics distinctly different from computer science. The wide range of problem solving skills that are developed with a training in computational physics make these graduates highly marketable and productive in endeavours that go well beyond physics, including in mainstream commerce and industry.

While Fortran has been traditionally used in the teaching of computational physics, we observe that Python is gaining in popularity in South Africa, with some departments using C, C++ and Java as their preferred programming languages. It should be noted, however, that many of the large production codes in physics totalling hundreds of thousands of lines of code are written in Fortran.

Time is needed to develop computational physics more completely at the undergraduate level. However, the undergraduate curriculum at many universities is generally very crowded with traditional topics in physics. It is very difficult to properly integrate computational physics into the current degree structure. There is much resistance to doing away with standard physics topics in lieu of introducing computational physics.

The most effective option is to introduce additional modules at the second and third year levels to teach computational physics fundamentally. This requires careful negotiation within the faculty, as this will impact on the number of credits that a student will be able to take in the co-major, and may impact also on the number of prerequisites and co-requisites. Not many South African university physics departments have been able to adopt this approach to date.

A lesser option is to integrate computational physics into physics laboratory programmes[8]. This requires that some experimental laboratory exercises be done away with, which requires careful negotiation amongst colleagues. Some physics departments have a dearth of good quality experimental equipment, and so introducing computational exercises as a substitute could be seen as a valuable improvement to the training of undergraduate students. With less time available for the development of computational physics, it is pertinent to ask what is realistically achievable?

The greatest flaw in the undergraduate teaching of computational physics at South African universities is to simply use ready-to-go packages in black box-mode (see Table 1). Here, students adopt a dropdown-drag-click approach and the entire exercise is often reduced to changing parameters and producing pictures. This may be instructive insofar as the application is concerned, for example in the understanding of a physical problem such as the changes of the magnetic field lines when a charge is accelerated in vacuum, but there is little gained in terms of developing real computational skills. This is the primary mode in which postgraduate training is taking place in South Africa in producing what we refer to as the computational technician, and this should be avoided as much as possible.

It is realistic within the South African university undergraduate system to introduce computational physics using higher level languages such as Easy Java Simulations or symbolic manipulation programmes such as Maple and Mathematica or interpreted languages such as Octave or Perl. This way, the students can solve complex computational problems applied to real physical systems without necessarily having to pay a lot of attention to detailed numerical matters such as determining the most stable algorithm for the solution of a set of coupled partial differential equations. We refer to this as algorithmic physics rather than computational physics (see Table 1). Maple, for example, has features that enable one to solve such numerical problems with ease. Still, the students have useful experiences in computation, especially in algorithmic development even if they are not called upon to develop large and complex programs. This mode of training is intermediate to taking a fundamental approach to computing which we have seen is not always possible given severe time constraints and teaching computational physics in black box-mode, which we believe should be avoided as much as possible.

3. Postgraduate training and research in computational physics

Eleven out of twelve universities train postgraduate students through computational physics research projects in various branches of physics. Very often students use large commercial or opensource production codes to conduct their research. At nine universities some postgraduate students write short pieces of code or some sections of their own codes or make some modifications of production codes to carry out their research, by incrementally improving on mathematical models, algorithms, computational speed and data storage. However, there is little evidence that postgraduate students are routinely required to develop substantial sections of codes or to make substantial changes to existing production codes in their research. This is especially evident in the field of parallel computing. Many large production codes have been parallelised to run on massively parallel computers, and this aspect appears to be least understood or dissected in the training of postgraduate students in South Africa. Several physics departments have access to local computational clusters for high performance computing and many others make use of the supercomputers at the Centre for High Performance Computing (CHPC) to run large, complex, parallelisable calculations, for example in computational solid state physics, astrophysics, cosmology and particle physics. Postgraduate students working at two universities are taught the theory of parallel computing and are required to write message passing interface (MPI) codes to improve and manage their calculations on the cluster. CHPC runs specialised workshops annually for postgraduate students on the theory of parallel computing and how to parallelise and optimise codes using the CHPC platforms. But this endeavour is very far from developed in South Africa.

The African Institute for Mathematical Sciences (AIMS) has a strong focus on computational physics training. The students, who are drawn from various African countries, may be considered to be intermediate to the Honours and Masters levels. They have access to a state-of-the-art computational LAN working in an opensource environment and use the Python programming language. Depending on the lecturers who are selected each year, the students work on a variety of problems in applied mathematics, theoretical physics, computational biology and computational finance. The students are expected to develop their own computer codes during structured computational classes.

The Centre for Research in Computational and Applied Mathematics (CERECAM), at the University of Cape Town, provides a multidisciplinary research environment for postgraduates comprising theoretical mechanics, computational mechanics and high performance computing. Research programmes include projects of a fundamental nature such as numerical analysis and partial differential equations, and projects which are relevant to industry and other applications such as computational solid-, structural- and particulate-mechanics, computational electromagnetics, computational fluid dynamics and biomechanics. Computational work within research projects consists of the development and implementation of finite element (FE) approximations. Researchers use commercial software packages such as Abaqus FEA and ANSYS FLUENT CFD to implement FE solutions, and develop their own FE codes ab initio using Matlab, C++ or a C++ program library called deal.II.

The National Astronomy and Space Science Programme (NASSP) is hosted at the University of Cape Town and offers degree programmes in astrophysics and space science at the honours, masters and PhD levels. The program places strong emphasis on training in computational astronomy and astrophysics at the honours and masters levels. Students are taught to program in Fortran and Python. They also use commercial software such as Maple and Mathematica as well as open source astronomy and astrophysics software packages. Students typically write their own codes or extend existing in-house codes to perform parameter fitting, standard statistical and Bayesian statistical analysis of data and N-body simulations.

Astronomers are substantial users of computer resources, for reducing observational data, for robotically operating telescopes, for interfacing with hardware, and also for theoretical modelling and simulating astrophysical processes, especially in cosmology. Scientists at SALT have developed PySALT[9], which is written in Python, for reducing the SALT data. The software is opensource and therefore easily available for the entire international SALT community. The pathfinder KAT7 helped win South Africas bid to host the major share of the Square Kilometre Array (SKA) primarily because all aspects of the technical design, development and manufacture of KAT7 were conducted very successfully and on schedule. In here, computation played a significant role in the design, and in the reduction and analysis of the astrophysical data. KAT7 science commissioners are routinely writing their own codes in Python, C and C++ to turn the raw radio data into meaningful quantitative astrophysical images of the sky. The SKA will require thousands of times more processing power than is currently available in the fastest supercomputers. The boundaries of data compression and on-site processing will be pushed by the SKA to limits that require a completely new approach to computing. This has spurred the interest by major hardware and software companies such as IBM; this has the potential to significantly improve South Africas capability in the field of computing.

The Nuclear Corporation of South Africa (NECSA) has developed its own software package to study the interaction between neutrons and matter within their research nuclear reactors. Their software package solves a partial differential equation, known as the neutron transport equation, using deterministic and stochastic techniques. This package is also being used by a research facility in the Netherlands with a similar nuclear reactor. Commercial codes are used to study other neutron transport problems involving criticality and shielding.

The medical physics research department of iThemba Labs studies the use of neutrons in the treatment of various diseases. Scientists there develop their own codes in Fortran to solve the neutron transport equation and they also use commercial software packages to calculate neutron dosages for treatment.

Mintek and the Advanced Mathematical Modelling and Simulation Group at the CSIR use commercial and opensource software packages for computational solid state research to predict the properties of new materials. Mintek also writes code using Java, C++ or C to study diffusion in alloys and to determine stable geometries of nanoparticles.

3.1. Discussion

Some of the large production codes, for example computational fluid dynamical codes or quantum mechanical electronic structure codes or astrophysical magneto-hydrodynamical codes total in excess of several hundred thousand lines of code. It is unrealistic and unproductive to expect any single research student to develop such production codes in its entirety starting from a clean slate. Today, it is more realistic to expect a postgraduate student to start with a working piece of code and to make some modifications to integrate new theoretical, mathematical or physical models and to test these models with new applications. Being able to accomplish this with ease is very central to the modern computational physics research endeavour. This requires that the student develops a level of competence in reading computer codes and making algorithmic and programmatic changes to these codes. These interventions are very much needed if we want our research students to move the boundaries of their disciplines significantly in new directions and to other disciplines. Using standard off-the-shelf codes with no recourse to making changes to the code ties the researcher to a restricted class of problems that would already have been conceived of by the original programme developers. There is little or no room for innovating in ones research discipline by only being able to execute code. For example, if a student develops a new model in the field of ecology, how does the student test this model in a comprehensive way if the student is unable to do a realistic simulation using the model? Writing the entire code ab initio is not necessarily a productive way to proceed. On the other hand, being able to modify an existing piece of similar code is a preferred way to progress if this is at all possible. Some commercial codes come with exorbitant licensing fees and, as a part of the licensing agreement, there is no access to the source code. This may be acceptable in an industrial setting, but is not helpful in an academic environment and this should be avoided whenever possible.

Developing paradigms for more complex scientific codes should be pursued more vigourously in our teaching of computational physics. This is relevant for (i) the pedagogy of computational physics at the postgraduate and research levels, and is (ii) an endeavour that is ripe for research productivity in the field of computational physics education research, and finally, (iii) this is also a basis for developing new material for the teaching of computational physics at the undergraduate level. The idea is very simple: if research students today are not in a position to develop their production codes in its entirety from a clean slate and if they are primarily operating these production codes in black box-mode then, for pedagogical reasons, these students should be tasked with solving a series of self-contained computational exercises, each mimicking particular critical elements of the production code. There is a fair amount of creativity that is required to develop such computational exercises, and this in itself has the potential to become the subject of publishable research in computational physics educational journals. These exercises may also be utilised in teaching at the undergraduate level. This has the potential to stimulate a new and exciting research endeavour in computational physics.

4. Conclusions

The training in computational physics at the undergraduate level, and at the postgraduate and research levels at South African universities, and research facilities and institutions is generally weak and needs to be taken more seriously. South Africa is more in the mode of producing computational technicians rather than computational physicists. Failure to improve the culture of computational physics in South Africa will impact negatively on South Africas capacity to grow its endeavours generally in the field of computational sciences, with negative impacts on research, and in commerce and industry.

Acknowledgments

TS thanks the NRF and the department of Physics at the UP for funding. We thank the lecturers at all the physics departments and scientists at the research institutions who participated in the survey. TS thanks Norman Chonacky for assistance with the survey questions. NC acknowledges support from the National Institute for Theoretical Physics.

References

[1] Christian W O and Esquembre F (2007) Modeling physics with easy java simulations *The Physics Teacher* **45** 475
[2] Wieman C E, Perkins K K and Adams W K (2008) Oersted Medal Lecture 2007: Interactive simulations for teaching physics: what works, what doesn't and why Âm. J. Phys. **76** (4 and 5) 393
[3] Singh C (2008) Interactive learning tutorials on quantum mechanics *Am. J. Phys.* **76** (4 and 5) 400
[4] McKagan S B, Perkins K K, Dubson M, Mally C, Reid S, LeMaster R and Weimann C E (2008) Developing and researching PhET simulations for teaching quantum mechanics *Am. J. Phys.* **76** (4 and 5) 406
[5] Chonacky N and Winch D (2008) Integrating computation into the undergraduate curriculum *Am. J. Phys.* **76** (4 and 5) 327
[6] R Chabay and B Sherwood (2007) *Matter and Interactions* 2nd ed, Wiley, Hoboken, New Jersery. See also http://www.matterandinteractions.org
[7] Beichner R, Chabay R and Sherwood B (2010) Labs for the Matter and Interactions curriculum *Am. J. Phys.* **78** (5) 456
[8] Spencer R L (2005) Teaching computational physics as a laboratory sequence *Am. J. Phys.* **73** (2) 151
[9] Crawford SM, Still M, Schellart P, Balona L, Buckley D A H, Gulbis A A S, Kniazev A, Kotze M, Loaring N, Nordsieck KH, Pickering TE, Potter S, Romero Colmenero E, Vaisanen P, Wiliams T and Zietsman E (2010) PySALT: the SALT Science Pipeline, *SPIE Astronomical Instrumentation* 7737

3d visualization of atomistic simulations on every desktop

Dan Peled, Amihai Silverman and Joan Adler
Technion - IIT, Haifa, Israel, 32000
E-mail: phr76ja@tx.technion.ac.il

Abstract.
Once upon a time, after making simulations, one had to go to a visualization center with fancy SGI machines to run a GL visualization and make a movie. More recently, OpenGL and its mesa clone have let us create 3D on simple desktops (or laptops), whether or not a Z-buffer card is present. Today, 3D a la Avatar is a commodity technique, presented in cinemas and sold for home TV. However, only a few special research centers have systems large enough for entire classes to view 3D, or special immersive facilities like visualization CAVEs or walls, and not everyone finds 3D immersion easy to view. For maximum physics with minimum effort a 3D system must come to each researcher and student. So how do we create 3D visualization cheaply on every desktop for atomistic simulations? After several months of attempts to select commodity equipment for a whole room system, we selected an approach that goes back a long time, even predating GL. The old concept of anaglyphic stereo relies on two images, slightly displaced, and viewed through colored glasses, or two squares of cellophane from a regular screen/projector or poster. We have added this capability to our AViz atomistic visualization code in its new, 6.1 version, which is RedHat, CentOS and Ubuntu compatible. Examples using data from our own research and that of other groups will be given.

1. Introduction

Our group carries out atomistic simulations in carbon systems, with an emphasis on geometric structure and vibrations, especially near defects and interfaces. We need to observe the system development, and are often concerned with the structure deep inside the sample, which may not be accessible to non-destructive experiment. Analog models are helpful for gaining insight into 3D structures, but it is hardly practical to construct precise analog models for each timestep frame of a simulation. The next best is a 3D stereo visualization.

We will introduce anaglyphic stereo in the next section, our AViz code in the following one, and conclude with a description of the new AViz 6.1 and some sample results.

2. Anaglyphic stereo

Let us review two early stereoscopic computer visualization approaches: one is the 1965 ORTEP (Oak Ridge Thermal-Ellipsoid Plot Program) software [1, 2] that drew two stereoscopic images using a plotter (predating laser printers and inkjet printers) that were placed side by side in a system with two lenses. One relaxed the focus of ones eyes till they came into a view with depth perception. A single frame of two ORTEP plots from our group's research (M. Sc. thesis of A. Silverman [3]) is shown in Figures 1 and 2. The sample shown on the left is of crystalline silicon and in the center there is an image of an amorphous silicon sample, created from the crystal via a melt and quench using molecular dynamics. The opportunity to see the amorphous structure that would have been hard, if not impossible, to achieve with an analog model was exhilarating way back then, with or without the stereo implementation. This approach was popular for molecules, but the limited number of atoms meant that for simulations of solids, it was not practical. (We were able to draw only 78 atoms of our 216 atom sample). The size restriction of many chemical visualization programs continued to be limiting for applications to solids for a long time.

The second stereo example is the anaglyphic process which uses simple colored glasses or two squares of cellophane. An example of this method is the image of a random walk on a sphere, shown in Figure 3. This was drawn with a FORTRAN/PGPLOT code written in 1993 as part of a graduate Computational Physics class project by Dimitry Gangardt, based on an article by J.C. Sprott[4]. Sprott's original code, was presented in BASIC and won the first annual Computers in Physics Software award. Viewing this and similar analglyphic images without the two-colored glass shows some red and cyan lines. Unfortunately while showing depth clearly this specific implementation removes the possibility to use color in the images and thus an important route to information is missed.

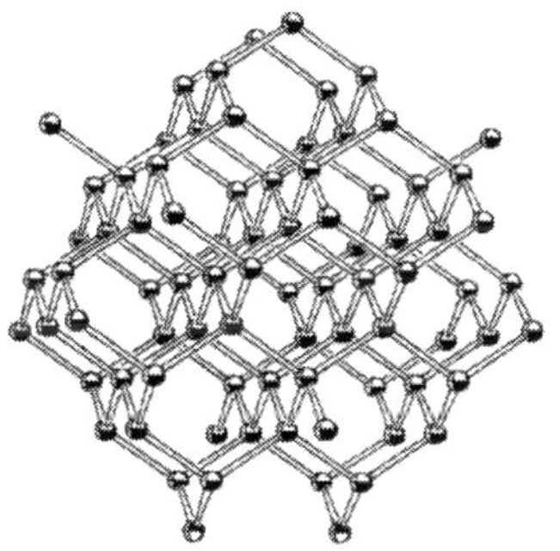

Figure 1. Crystalline silicon with ORTEP.

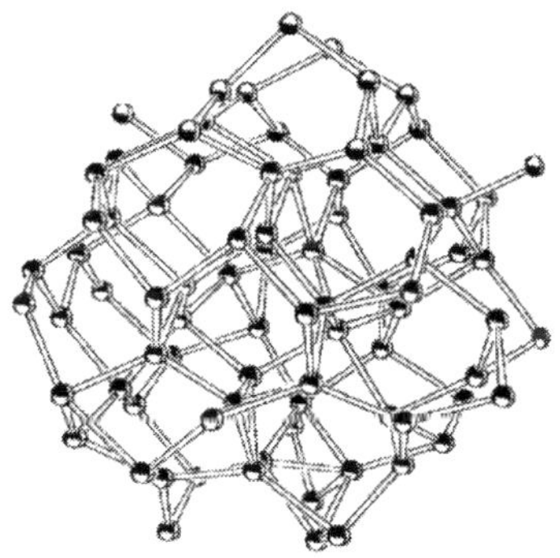

Figure 2. Amorphous silicon with PGPLOT.

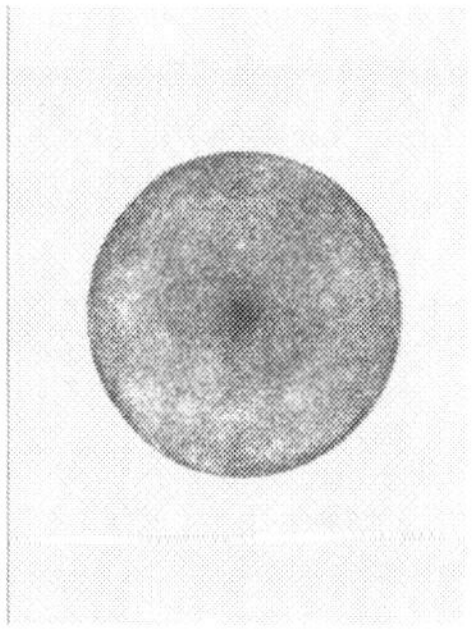

Figure 3. Analglyphic random walk with PGPLOT.

3. AViz basics and history

With no apology, AViz [5, 6, 7] was created in 2001 to meet the visualization needs of the Technion Computational Physics Group, within an opensource, LINUX/HPC oriented framework. There are other systems that could probably supply some of these, and we recommend that everyone uses what suits them. Our visualization needs are many, but the most important requirement from the start has been for cheap, legal, 3D visualization on every group member's desktop, with a system that is robust to the graduation of an individual student. Interactive or postprocessing options, and a way to prepare movies for presentation are also desirable.

There have been three main releases. Original AViz, (4.6, for Red Hat 7.0) prepared by G. Wagner and A. Hashibon, for which A. Sorkin wrote the manual, was replaced by AViz 5.6 in 2002 with an improved interface. Various additional features (sequential polymers etc) were added to specific implementations, and in 2011, Y. Koenka and A. Silverman prepared AViz 6.0 with several major updates and features from the specific implementations[8]. The newest version, described below, is called AViz 6.1 and has been shown to be compatible with Ubuntu, CentOS, and Red Hat Enterprise.

The underlying methodology of all versions is as follows: prepare a data file of atomic site types and locations (or a set of datafiles called a filelist), tune the visualization with the buttons on the interface, and record a set of images. Tuning includes manipulation thru space, changing the size of objects or connecting them with "bonds"or adjusting color or lighting. There is a wide selection of websites giving instructions and examples for specific features. Much of the effort has come from undergraduate course or final projects, where preparation of a website presenting the project has been one of the requirements. A project of consolidation of the explanatory material is now underway.

4. AViz 6.1

The first stereo AViz was prepared by H. Zilken in Juelich. This version was very exciting, but the hardware rendering, projection and manpower demands were not feasible to reproduce at the Technion in even a single realization. This may be possible in the future, but since we are making visualizations for High Performance Computing (HPC) atomistic simulations, in recent months we have focused on creating a hardware environment for parallel HPC, and could not find resources for a powerful visualization system as well. Such a powerful system is not possible for every desktop, and the most effective way to ensure researchers visualize is to enable them to do so from their desktops. Our need for improved 3D insight has become stronger than ever and so we returned to the older style analyglyphic stereo.

AViz 6.1 [9, 10] has incorporated the possibility of Stereo Vision (SV), and although more than two colors are possible there remains some color washout, depending on color selection. SV works by showing a different image to each eye, thus creating the illusion of a 3D image. The SV images generated by AViz, such as those in this paper, are best viewed using red-cyan anaglyphic glasses, Figure 4, or two squares of red (left eye) and cyan (right eye) cellophane. For quick viewing, a cardboard frame of approximately 14 by 5 cm with two 3.5 by 2.5 cm holes whose centers are about 6cm apart, over which cellophane is stuck, can be easily constructed.

In order to operate the SV abilities, 3 buttons in the shape of stereoscopic glasses were added in the bottom of the right hand toolbar of the main AViz viewing window. In order to enable/disable the SV mode, press the "Enable/Disable" button, as shown in Figures 5 (enabled) and 6 (disabled). It is easily seen (even if this paper is printed in black and white) that in the enabled case of Figure 5 there is a double frame and (cyan/red) bleeding from the spheres that is not present in Figure 6. Readers who are viewing this in black and white are recommended to refer to the color versions of the images that can be found together with animations at the website[10]. These images can be enlarged for easier viewing.

Figure 4. Two-color glasses.

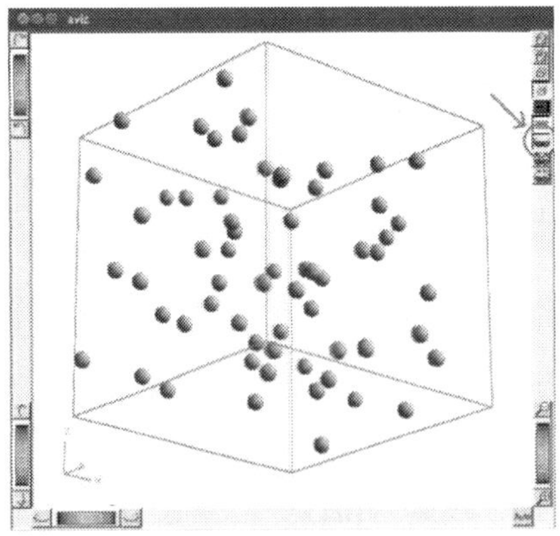

Figure 5. AViz 6.1 interface with stereo enabled.

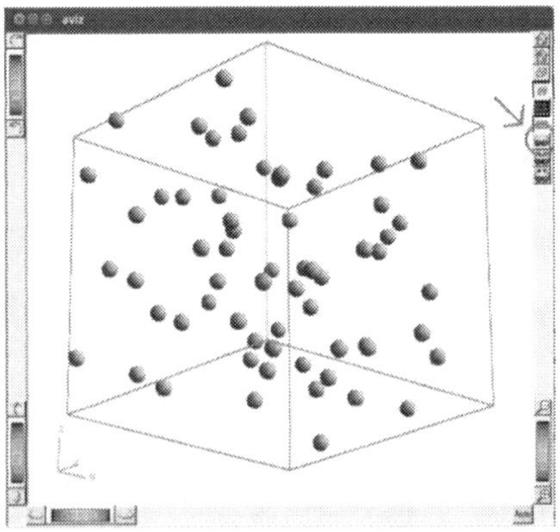

Figure 6. AViz 6.1 interface with stereo disabled.

In order to achieve a proper prediction of how the model would look for each eye, the parameter of the distance between the two eyes, known as the pupillary distance (PD), needs to be set. In reality the PD is almost fixed for everyone, but since the models shown in AViz are re-scaled to fit the screen, the pre-configured PD may not be right for all models. Therefore, the 2 buttons may be pressed continuously to increase/decrease the PD as a function of time, and to increase/decrease the 3D effect.

Links to the source code as well as installation instructions for several Ubuntu, RedHat, and CentOS versions of LINUX can be found on the AViz 6.1 website[9]. On this site some animated 3D movies can also be viewed.

5. Results

A selection of still stereo realizations are given in Figures 7-9. Figures 7 and 8 show two different views of a nanotube. The third figure presents data from an off-label implementation of AViz 6.1, [11] extending our visualizations of the electron density of a hydrogen atom to the stereo case.

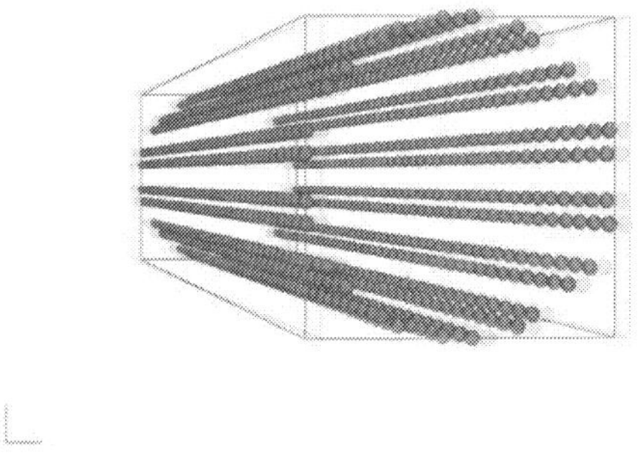

Figure 7. Nanotube side-view.

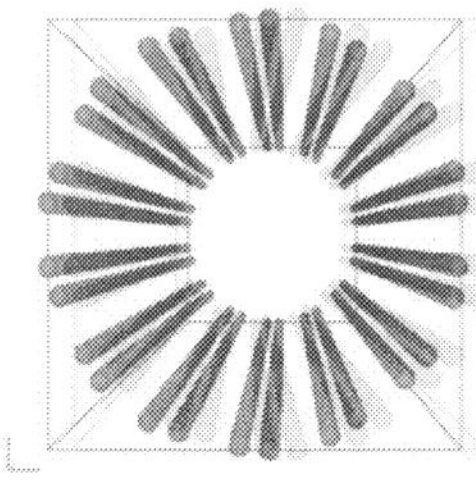

Figure 8. View along the axis of a nanotube.

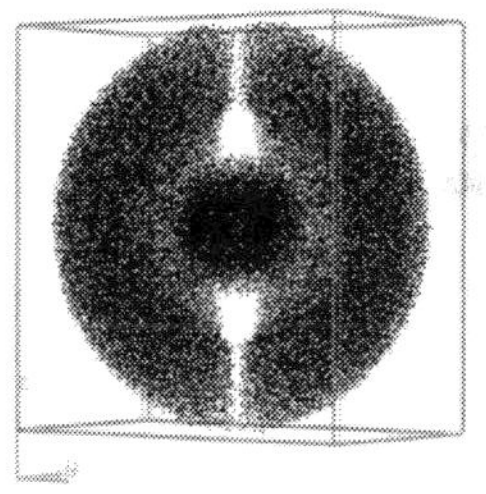

Figure 9. Orbital electronic density of the (3,1,0) state of the hydrogen atom.

The next set of images, in Figures 10 -12, use the dataset for the split interstitial published in [7, 8], but the center one shows a larger number of atoms in order to provide relative bond thicknesses. In [7] and [8], instructions are given for visualizations of this dataset with earlier AViz implementations using different colors (AViz 5.6) and different bond widths (AViz 6.0) respectively. The new image was prepared as described above. It shows that a substantial amount of color is retained. The final image, Figure 13, shows some 5120 atoms of an implanted diamond sample [12].

Acknowledgments

We thank H. Zilken of Juelich for developing the first stereo AViz, and the Juelich-Technion-Aachen Umbrella project for enabling our collaboration. The new code was written as an undergraduate project by D.P. based on earlier visualizations by previous members of the Computational Physics Group at the Technion. Some of the data used in the testing of AViz 6.1 was generated by researchers and students at CiNaM, CNRS, Marseilles and J. A. thanks A. Saul for hospitality there. We also thank R. Kreiselman and Y. Katz for assistance in clarifying and enhancing the installation instructions.

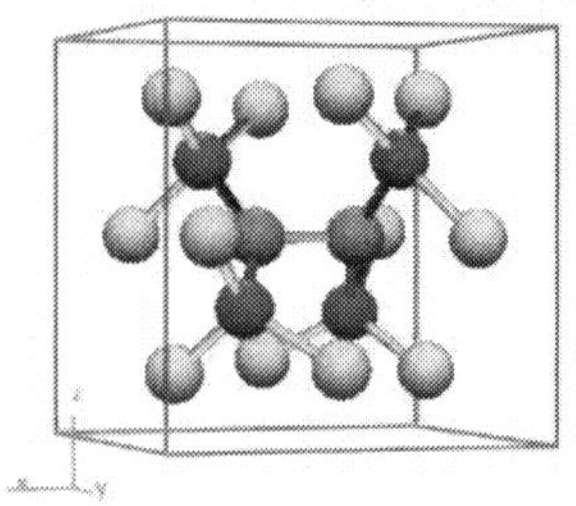

Figure 10. Split interstitial from [7] using color for differently coordinated atoms and the bond between them.

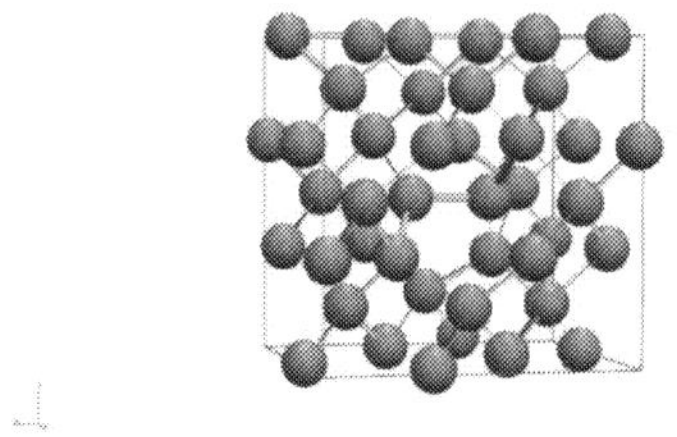

Figure 11. Split interstitial from [8] with variable bond thickness.

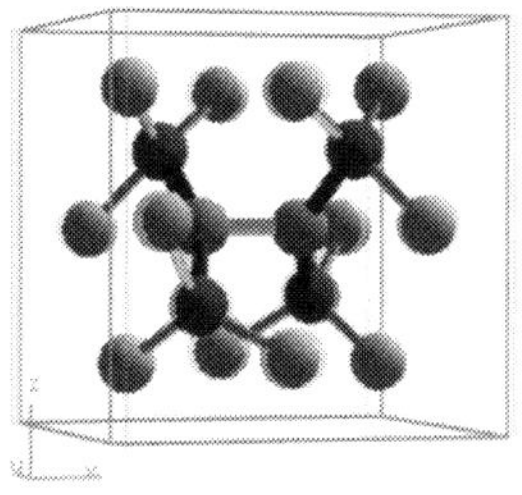

Figure 12. Split interstitial from AViz 6.1 showing that this type of analyglyphic stereo can retain much of the color.

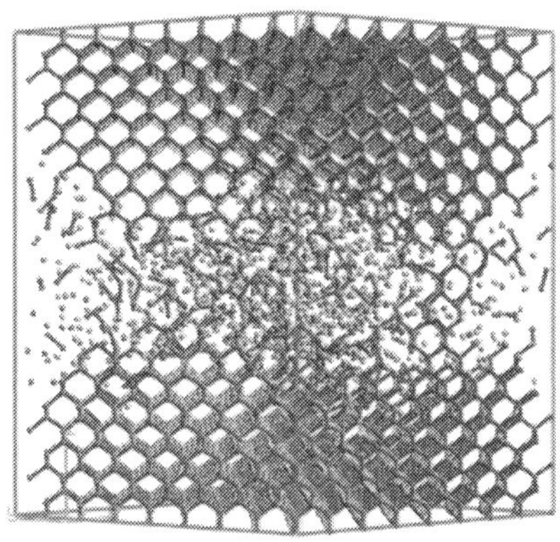

Figure 13. Stereo rendering of data from [12] of an implanted diamond sample modelling creation of a diamond membrane, showing some 5120 atoms.

References

[1] Carroll K.Johnson, 1965. "ORTEP: A FORTRAN Thermal-Ellipsoid Plot Program for Crystal Structure Illustrations". ONRL Report no. 3794. Oak Ridge, Ten, Oak Ridge National Laboratory)
[2] Michael N. Burnett and Carroll K. Johnson, ORTEP-III: Oak Ridge Thermal Ellipsoid Plot Program for Crystal Structure Illustrations, Oak Ridge National Laboratory Report ORNL-6895, 1996.
[3] Silverman A M. Sc. Thesis, Technion, 1990.
[4] Sprott J C "Simple programs create 3D Images" 1992 *Computers in Physics* **6** pp 132-8
[5] http://phycomp.technion.ac.il/~aviz
[6] Adler J, Hashibon A, . Schreiber N, Sorkin A, Sorkin S and Wagner G 2002 "Visualization of MD and MC Simulations for Atomistic Modeling" *Computer Physics Communications*, **147** pp 665-9
[7] Adler J 2003 "Visualization in Atomistic and Spin Simulations" *Computers in Science and Engineering*, **5** pp 61-65
[8] Adler J, Koenka Y and Silverman A 2011 "Adventures in carbon visualization with AViz" *Physics Procedia* **15** pp 7-16
[9] http://phelafel.technion.ac.il/~peledan
[10] http://phycomp.technion.ac.il/~newaviz
[11] Adler J, Fox J, Kalish R, Mutat T, Sorkin A and Warszawski E 2007 "The essential role of visualization for modeling nanotubes and nanodiamond" *Computer Physics Communications* **177** pp 19-20
[12] Silverman A, Adler J and Kalish R 2011 "Diamond membrane surface after ion implantation induced graphitization for graphite removal:molecular dynamics simulation" *Phys. Rev. B* **83** 155410

Applications and a three-dimensional desktop environment for an immersive virtual reality system

Akira Kageyama and Youhei Masada

Graduate School of System Informatics, Kobe University, Kobe 657-8501, Japan

E-mail: kage@cs.kobe-u.ac.jp, ymasada@harbor.kobe-u.ac.jp

Abstract. We developed an application launcher called Multiverse for scientific visualizations in a CAVE-type virtual reality (VR) system. Multiverse can be regarded as a type of three-dimensional (3D) desktop environment. In Multiverse, a user in a CAVE room can browse multiple visualization applications with 3D icons and explore movies that float in the air. Touching one of the movies causes "teleportation" into the application's VR space. After analyzing the simulation data using the application, the user can jump back into Multiverse's VR desktop environment in the CAVE.

1. Introduction

CAVE is a room-sized virtual reality (VR) system, which was developed in the early 1990s at the University of Illinois, Chicago [1]. In a CAVE room, the viewer is surrounded by wall screens and a floor screen. Stereo-images are projected onto the surfaces. Tracking systems are used to capture the viewer's head position and direction. The wide viewing angle provided by the surrounding screens on the walls and floor generates a high-quality immersive VR experience. The viewer can interact with three-dimensional (3D) virtual objects using a portable controller known as wand, in which the tracking system is installed.

CAVE systems have been used for scientific visualizations from the first system [1] until the latest generation (StarCAVE) [2]. For example, visualization applications in CAVE systems have been developed to analyze general computational fluid dynamics (CFD) [3], turbulence simulations [4], CFD of molten iron [5], CFD of wind turbines [6], seismic simulation [7], meteorological simulation [8], biomedical fluid simulation [9], magnetic resonance imaging [10], geomagnetic fields [11], archaeological studies [12], and geophysical surveys [13].

Recently, a new CAVE system was installed at the Integrated Research Center (IRC) at Kobe University. This CAVE system was named "π-CAVE" after the IRC's location on Port Island (PI). Figure 1 shows a front view of the π-CAVE while figure 2 shows the configuration of its projectors and mirrors.

The original CAVE system had a cubic geometry with a side length of 3 m. A straightforward extension to enlarge the VR space of a CAVE is to use a rectangular parallelepiped shape. More sophisticated configurations have been proposed for advanced CAVE systems, such as StarCAVE [2], but we used the rectangular parallelepiped approach for π-CAVE to maximize the VR volume in the space allowed in the IRC building. The side lengths of π-CAVE were 3 m $\times$ 3 m $\times$ 7.8 m. As far as we know, this is the largest CAVE system in Japan.

Figure 1. Overview of the π-CAVE system installed at Kobe University.

Figure 2. Projectors and mirrors used by the π-CAVE system.

We have developed several VR applications for the scientific visualization of large-scale simulation data. Of these, Virtual LHD [14] was our first VR visualization application. This application was developed for the CompleXcope CAVE system installed at the National Institute for Fusion Science, Japan. Currently, Virtual LHD is used to visualize the magnetohydrodynamic (MHD) equilibrium state of a nuclear fusion experiment. We also developed a general-purpose visualization application, VFIVE [15, 16, 17, 17], for 3D scalar/vector field data. Recently, we added a new visualization method to VFIVE at π-CAVE for visualizing magnetic field lines frozen into a fluid [18]. The original VFIVE only accepted a structured grid data format as the input, but an extension of VFIVE for unstructured grid data was developed at Chuo University [19]. The development and its applications of VFIVE are summarized in our recent papers [20, 21].

In addition to improvements of VFIVE, we also developed the following four types of novel CAVE visualization applications for π-CAVE. (1) IonJetEngine: for VR visualization of plasma particle in cell (PIC) simulations of an ion jet engine in space probes (2) RetinaProtein: for molecular dynamics (MD) simulations of proteins (3) SeismicWave: for the simulation of seismic wave propagation (4) CellDivision: to simulate three-dimensional time sequence microscope images of mouse embryos. All of these new CAVE visualization programs were written using OpenGL and CAVElib. We started developing these visualization applications when the construction of π-CAVE was underway.

Several problems occur if multiple CAVE visualization applications are executed one after another, as follows. First, the command has to be typed in to launch the first application using the keyboard beside the CAVE room. The user then enters the CAVE room wearing stereo glasses. After analyzing the data from the first application in the CAVE, the user leaves the CAVE room and takes off the glasses. Next, the user types in the command to launch the second application and enters the CAVE room wearing the stereo glasses. These steps have to be repeated if there are many applications. This inconvenience occurs because the CAVE must be used for single tasks.

To resolve this inconvenience, we developed an application launcher for CAVE. This program, Multiverse, is a CAVE application written in CAVElib and OpenGL. Multiverse can control other VR applications. These sub-applications are depicted in CAVE's VR space using 3D icons or panels. If the user in the CAVE room touches one of the panels using the wand, they are "teleported" to the corresponding VR application.

In this paper, we report the hardware used by the π-CAVE system in section 2, and we describe the design and implementation of Multiverse in section 3. The visualization applications loaded into Multiverse are described in section 4.

2. π-CAVE system

π-CAVE has a rectangular parallelepiped configuration with side lengths of 3 m $\times$ 3 m $\times$ 7.8 m (figure 3). The large width (7.8 m) is one of the characteristic features of the CAVE system. The large volume of π-CAVE allows several people to stand on the floor at the same time, without any mutual occlusion of the screen views in the room.

Figure 3. Alternative view of the π-CAVE system.

Like many other CAVE systems, π-CAVE has four screens: three wall screens (front, right, and left) and a floor screen. Soft, semi-transparent screens are used on the walls. The images are rear-projected onto these screens. The floor is a hard screen where the stereo image is projected from the ceiling. Two projectors are used to generate the front wall image (figure 4). An edge blending technique is applied to the interface between the two images. Another pair of projectors is used for the floor screen. Each side wall screen (right and left) is projected onto using a projector. In total, six projectors are used.

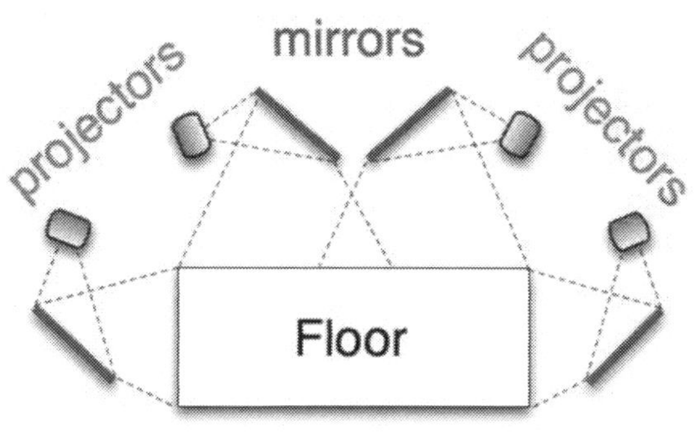

Figure 4. Projector settings for π-CAVE, viewed from above.

The resolution of the projector (Christie WU12K-M) shown in figure 5 with the counterpart mirror, is 1920 $\times$ 1200 pixels. The brightness is 10,500 lumens. An optical motion tracking system (Vicon) is used for head and wand tracking. Ten cameras with 640 $\times$ 480 resolution are installed on top of the wall screens. A commonly used API (Trackd) is used for the interface to CAVElib.

Figure 5. Pair of projectors and mirrors.

Two computer systems are used for computations and for rendering π-CAVE. One is a Linux PC (HP Z800) with 192 GB of shared memory. Three sets of GPUs (NVIDIA QuadroPLEX) are used for real-time stereoscopic image generation by the six projectors. The other computer system is a Windows PC cluster system.

We used OpenGL for the graphics API and CAVElib for the VR API. We are also aiming to use VR Juggler [22] for the VR API. Some of our first trials using VR Juggler can be found in our report [23].

3. Multiverse

We developed an applications launcher, Multiverse, for the π-CAVE system. At the start of this Multiverse environment, the viewer in the π-CAVE stands in the virtual building in IRC where π-CAVE is installed. The 3D CAD model data of the IRC building (figure 6) is loaded into Multiverse and rendered in 3D in real time. This is the Multiverse's start-up environment known as *World*. In the *World* mode of Multiverse, the viewer can walk through the building. Figure 7(a) shows a snapshot where the user is approaching the IRC building. In figure 7(b), the viewer is (literally) walking into the (virtual) IRC building. Some fine structures of the building, including the virtual π-CAVE is shown in figure 7(c) and (d), are also loaded from CAD data files.

Figure 6. Three-dimensional CAD data for the IRC building loaded in Multiverse.

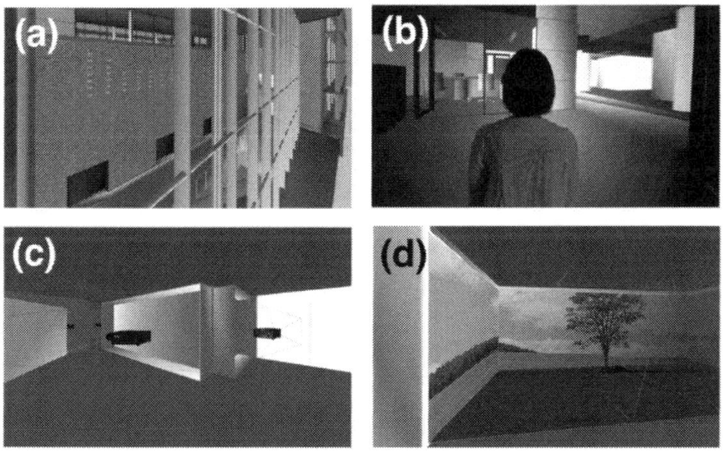

Figure 7. A snapshot sequence of Multiverse in the World mode. (a) The viewer is entering the virtual IRC building. (b) The viewer walks into the building. (c) Rear of the (virtual) π-CAVE. The CAD models of the projectors behind the CAVE screens can be seen. (d) Virtual π-CAVE in the real π-CAVE.

In Multiverse, there are two methods of showing the application list loaded in Multiverse. The first is to use "ribbons" that connect the wand and application icons. In the "ribbons" mode, the user in the *World* finds one or more curves or wires that start from the wand tip. Each wire is a type of guide that leads the user to a Gate.

A Gate is an entrance to the VR world of the corresponding application. If multiple visualization applications are loaded into Multiverse, this automatically generates the corresponding number of Gates. All of these are connected to the user (or the wand) via guide wires (figure 8). If the user walks or "flies" into a place in front of a Gate, they will find an exploratory movie near the Gate (see the rectangular panel in the center of the blue, torus-shaped Gate in figure 8). This explains the type of application that will be executed when the user selects the Gate. To select the application, the user (literally) walks through the Gate when the corresponding VR application program loads and the user feels as if they have been "teleported" to the visualization space. Each VR world is known as a *Universe* in Multiverse.

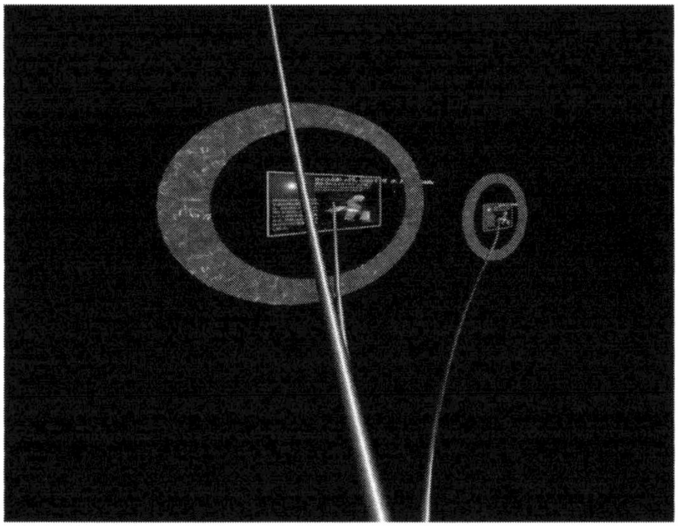

Figure 8. Gates to the individual visualization applications floating in the *World*.

Another method of showing the application lists loaded in the Multiverse is to use a virtual elevator. When the user enters the elevator in the (virtual) IRC building, they are automatically taken upward by the elevator into the sky above the IRC building. The spatial scale of the view changes rapidly from the building, to the city, country, and finally the globe. The user finds that they are "floating" in space surrounded by stars. Several panels then appear in front of the viewer. Each panel represents a visualization application (figure 9).

Figure 9. Virtual touch screens in the *World* mode. Each panel represents a VR visualization program or *Universe*. A *Universe* is executed when the user touches the panel using the wand.

When the user touches one of the panels, the corresponding VR application is launched and the user is "teleported" to the selected visualization *Universe*.

In short, Multiverse is composed of the *World* and several *Universes*. *World* is a type of 3D desktop environment and a *Universe* is a visualization application loaded onto Multiverse.

In the program code, each *Universe* is simply a standard CAVE application with a unified interface to the Multiverse class. A *Universe* is an instance of a class that is derived from a virtual class known as Vacuum. Vacuum represents an empty space, which only has an interface to the Multiverse class through the member functions `initialize()`, `draw()`, `update_per_frame()`, and `compute()`. These function names convey their roles to readers who are familiar with CAVElib programming.

4. Applications

In this section, we describe five applications, or *Universes*, which we developed as the first applications for the Multiverse environment.

4.1. Universe::GeomagField

We converted VFIVE, which is described in section 1, into a class of *Universe*. VFIVE is a general-purpose visualization tool, so we can visualize any vector/scalar field provided that the data are legitimate for VFIVE's input data format in the Multiverse framework.

Figure 10 shows a snapshot of an example of a *Universe* based on VFIVE, known as GeomagField. The input data used by GeomagField was a geodynamo simulation performed by one of the authors and his colleagues [24, 25, 26]. The purpose of this simulation was to understand the mechanism that generates the Earth's magnetic field (or geomagnetic field).

Figure 11 shows another snapshot of GeomagField in which two VFIVE visualization methods were applied. The temperature distribution was visualized by volume rendering (colored in orange to yellow). The 3D arrow glyphs show the flow velocity vectors around the wand position. The arrows followed the motion when the viewer moved the wand, which changed the directions and lengths (vector amplitudes) in real time. The white balls are tracer particles that also visualized the flow velocity. These balls were highlighted in a spotlight or cone-shaped

Figure 10. GeomagField, a *Universe* in Multiverse. This VR application was used to visualize an MHD simulation of geomagnetic field generation in Earth's liquid core.

Figure 11. Another snapshot of GeomagField. The scalar field distribution of the temperature is visualized by volume rendering. The convection flow velocity is visualized using 3D arrow glyphs while tracer particles are under the spotlight.

region, the apex of which was the wand. This visualization method is known as Snowflakes in VFIVE. The viewer can change the focus of the flow visualization by changing the direction of the spotlight via wand direction movements.

4.2. Universe::IonJetEngine

The second example of a *Universe* is known as IonJetEngine and a snapshot is shown in figure 12.

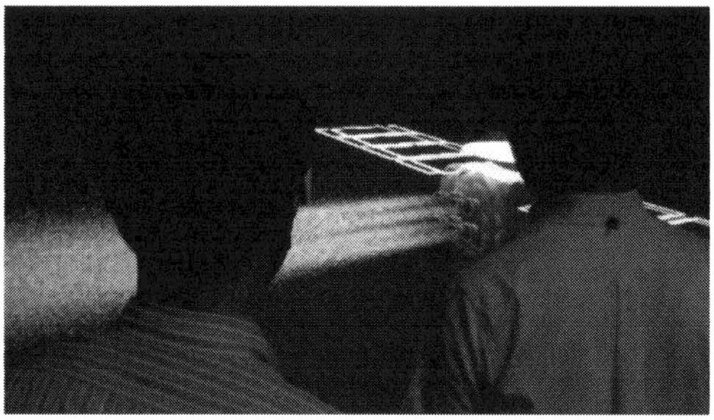

Figure 12. A snapshot of the Universe known as IonJetEngine. Plasma jets from a space probe were visualized by moving ions (yellow) and electrons (blue) as particles. The simulation data were provided by Prof. Usui.

This *Universe* visualized a plasma PIC simulation of the ion jet engine of a space probe. The positions of the particles (ions and electrons) were represented by balls (yellow for ions and blue for electrons). The velocity distribution of the jet was visualized as the set of the individual motions of the particles. A 3D model of the virtual space probe from which the plasma jet beams were ejected is also shown in figure 12.

4.3. Universe::RetinaProtein
Figure 13 shows a *Universe* known as RetinaProtein, which was a molecular dynamics simulation of rhodopsin [27], a protein in the human retina. At the start of this *Universe*, the viewer observed a 3D model of a human (see the top panel of figure 13). As the viewer approached the model's face, the fine structures of the eyes became visible until MD simulation visualization appeared.

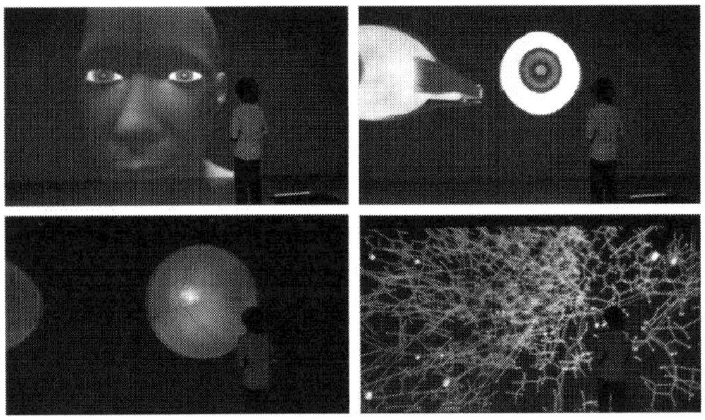

Figure 13. Snapshots of the RetinaProtein Universe. The molecular structure of rhodopsin was visualized in the human retina. The MD simulation data were provided by Prof. Ten-no of Kobe University and his colleagues.

4.4. Universe::SeismicWave
In this *Universe*, a simulation of seismic wave propagation [28]was visualized, which was performed by Prof. Furumura of the University of Tokyo by animated volume rendering (see figure 14). In this *Universe*, we implemented rapid volume rendering based on the 3D texture mapping technique in CAVEs. The full details of this implementation will be reported elsewhere.

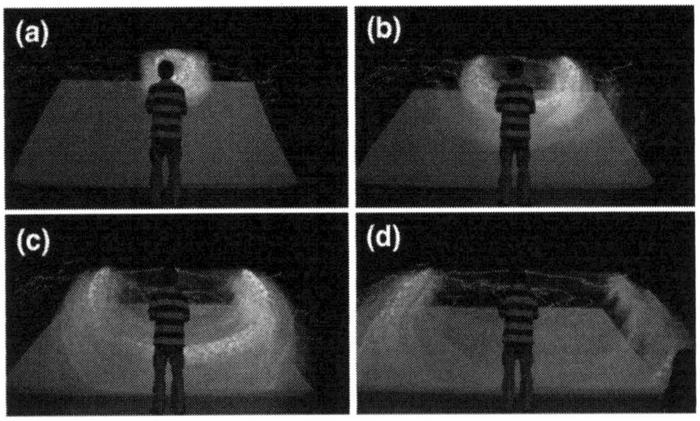

Figure 14. Time sequence of snapshots of the SeismicWave *Universe*.

4.5. Universe::CellDivision
The final *Universe* described here is CellDivision and a snapshot is shown in figure 15. The target data used for this visualization were not simulation data. Instead, they were microscope images of live mouse embryos. The data were provided by Dr. Yamagata of Osaka University. The time sequence of microscope images was visualized as an animated volume rendering using the same tool used for SeismicWave in the previous subsection.

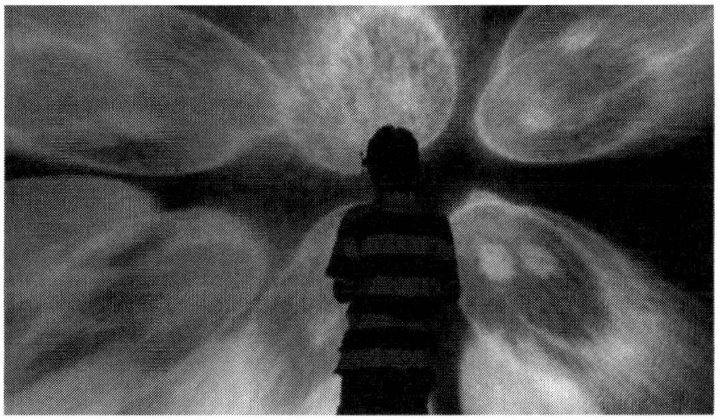

Figure 15. A snapshot of CellDivision. Animated volume rendering of live cell images of a mouse embryo. The data were provided by Dr. K. Yamagata, Osaka University.

5. Summary

In many CAVE systems, VR applications are executed as single tasks. Thus, the user has to type in each command one after another outside the CAVE room. To convert a CAVE into a more convenient tool for scientific visualization, we developed an application launcher known as Multiverse. Multiverse comprises a *World* and *Universes*. *World*, which correspond to the desktop of a PC operating system, where the user can select visualization applications by touching icons floating in the *World*. Using the virtual touch screen interface, the specified application program is launched and the user is "teleported" to another VR space containing the corresponding visualization application, which is known as a *Universe*. We developed five *Universes*, which can be launched from the Multiverse environment. Multiverse was designed as a general application framework, so it can read and control other *Universes*. A user can jump back to a *World* and switch to another *Universe* at any time from any *Universe*.

During the implementation of Multiverse, we developed several new fundamental tools and methods for the CAVE environment, such as a fast speed volume renderer, a 3D model (CAD) data loader/renderer, and a 2D movie file loader/renderer. Details of these fundamental tools and methods will be reported elsewhere.

Acknowledgments

We thank the undergraduate students at our laboratory at Kobe University (Toshiaki Morimoto, Yasuhiro Nishida, Yuta Ohno, Tomoki Yamada, and Mana Yuki) for contributing to the development of Multiverse. The plasma particle simulation data were provided by Prof. H. Usui, Dr. Y. Miyake, and Mr. A. Hashimoto (Kobe University). The MD simulation data were provided by Prof. S. Ten-no and Dr. Y. Akinaga. The simulation data for seismic wave propagation were provided by Prof. T. Furumura (University of Tokyo). The microscope images were provided by Dr. K. Yamagata (Osaka University).

This work was supported by JSPS KAKENHI Grant Numbers 23340128 and 30590608, and also by the Takahashi Industrial and Economic Research Foundation.

References

[1] Cruz-neira C, Sandin D J and Defanti T A 1993 *Proc. SIGGRAPH '93* pp 135–142
[2] Defanti T, Dawe G, Sandin D, Schulze J, Otto P, Girado J, Kuester F, Smarr L and Rao R 2009 *Future Generation Computer Systems* **25** pp 169–178
[3] Jaswal V 1997 *Proc. Visualization '97* pp 301–308
[4] Tufo H M, Fischer P F, Papka M E and Blom K 1999 *Proc. ACM/IEEE Conf. Supercomputing 1999* pp 62–76
[5] Fu D, Wu B, Chen G, Moreland J, Tian F, Hu Y and Zhou C Q 2010 *Proc. 14th Int. Heat Transfer Conf.*, pp 1–8

[6] Yan N, Okosun T, Basak S K, Fu D, Moreland J and Zhou C Q 2011 *Proc. ASME 2011 Int. Design Engineering Technical Conf.& Computers and Information in Engineering Conf.*, pp 1–8
[7] Chopra P, Meyer J and Fernandez A 2002 *IEEE Visualization, 2002*, pp 497–500
[8] Ziegeler S, Moorhead R J, Croft P J and Lu D 2001 *Proc. Conf. Visualization '01*, pp 489–493
[9] Forsberg A, Laidlaw D, Van Dam A, Kirby R, Kafniadakis G and Elion J 2000 *Proc. Conf. Visualization '00*, pp 457–460
[10] Zhang S, Demiralp C, Keefe D, DaSilva M, Laidlaw D, Greenberg B, Basser P, Pierpaoli C, Chiocca E and Deisboeck T 2001 *Proc. Visualization, 2001*, pp 437–584
[11] Bidasaria H B 2005 *Proc. 43rd Annual Southeast Regional Conf. on ACM-SE 43*, p 355
[12] Acevedo D, Vote E, Laidlaw D and Joukowsky M 2001 *Proc. Visualization*, pp 493–597
[13] Lin A Y m, Novo A, Weber P P, Morelli G and Goodman D 2011 *Advances in Visual Computing* (Springer Berlin Heidelberg) pp 229–238
[14] Kageyama A, Hayashi T, Horiuchi R, Watanabe K and Sato T 1998 *Proc. 16th Int. Conf. Numerical Simulation Plasmas* (Santa Barbara, CA, USA) pp 138–142
[15] Kageyama A, Tamura Y and Sato T 2000 *Prog. Theor. Phys., Suppl.* **138** pp 665–673
[16] Ohno N and Kageyama A 2007 *Phys. Earth Planet. Inter.* **163** pp 305–311
[17] Ohno N and Kageyama A 2010 *Comput. Phys. Comm.* **181** pp 720–725
[18] Murata K and Kageyama A 2011 *Plasma Fusion Res.* **6** 2406023–1–5
[19] Kashiyama K, Takada T, Yamazaki T, Kageyama A, Ohno N and Miyachi H 2009 *Proc. 9th Int. Conf. Construction Applications of Virtual Reality* (Sydney) pp 1–6
[20] Kageyama A, and Ohno N *submitted to Int. J. Modeling Simulation & Scientific Comput.*
[21] Kageyama A, Ohno N, Kawahara S, Kashiyama K and Ohtani H *submitted to Int. J. Modeling Simulation & Scientific Comput.*
[22] Bierbaum A, Just C, Hartling P, Meinert K, Baker A and Cruz-Neira C 2001 *Proc. IEEE Virtual Reality 2001*, pp 89–96
[23] Meno D, Kageyama A and Masada Y 2012 *Proc. Int. Conf. Simulation Technology* pp 387–389
[24] Kageyama A, Miyagoshi T and Sato T 2008 *Nature* **454** pp 1106–1109
[25] Miyagoshi T, Kageyama A and Sato T 2010 *Nature* **463** pp 793–796
[26] Miyagoshi T, Kageyama A and Sato T 2011 *Phys. Plasmas* **18** p 072901
[27] Akinaga Y, Jung J and Ten-no S 2011 *Phys. Chem. Chem. Phy.* **13** pp 14490-14499
[28] Furumura T, Kennett B L N and Koketsu K 2003 *Bul. Seismological Soc. America* **93** pp 870–881

Journal of Physics: Conference Series **454** (2013) 012078 doi:10.1088/1742-6596/454/1/012078

Irreversible data compression concepts with polynomial fitting in time-order of particle trajectory for visualization of huge particle system

H Ohtani[1,2], K Hagita[3], A M Ito[1], T Kato[4], T Saitoh[5] and T Takeda[6]

[1] Department of Helical Plasma Research, National Institute for Fusion Science, 322-6 Oroshi-cho, Toki 509-5292, Japan

[2] Department of Fusion Science, The Graduate University for Advanced Studies (SOKENDAI), 322-6 Oroshi-cho, Toki 509-5292, Japan

[3] Department of Applied Physics, National Defense Academy of Japan, 1-10-20 Hashirimuzu, Yokosuka 239-8686, Japan

[4] Department of Physical Science, Graduate School of Science, Hiroshima University, 1-3-1 Kagamiyama, Higashi Hiroshima 739-8526, Japan

[5] Interactive Research Center of Science, Graduate School of Science and Engineering, Tokyo Institute of Technology, 2-12-1 Ookayama, Meguro-ku, Tokyo 152-8550, Japan

[6] Center for Computational Astrophysics, National Astronomical Observatory of Japan, 2-21-1 Osawa, Mitaka 181-8588, Japan

E-mail: `ohtani.hiroaki @ nifs.ac.jp`

Abstract. We propose in this paper a data compression scheme for large-scale particle simulations, which has favorable prospects for scientific visualization of particle systems. Our data compression concepts deal with the data of particle orbits obtained by simulation directly and have the following features: (i) Through control over the compression scheme, the difference between the simulation variables and the reconstructed values for the visualization from the compressed data becomes smaller than a given constant. (ii) The particles in the simulation are regarded as independent particles and the time-series data for each particle is compressed with an independent time-step for the particle. (iii) A particle trajectory is approximated by a polynomial function based on the characteristic motion of the particle. It is reconstructed as a continuous curve through interpolation from the values of the function for intermediate values of the sample data. We name this concept "TOKI (Time-Order Kinetic Irreversible compression)". In this paper, we present an example of an implementation of a data-compression scheme with the above features. Several application results are shown for plasma and galaxy formation simulation data.

1. Introduction

The peta-scale computers, e.g., Titan, Sequoia, and K computer, have started operation, and ultra-huge-scale simulations can be performed. Such simulations will discover many new features in the various fields of science. Though many scientific discoveries are expected, there are big urgent issues, that are, data input and output (data I/O), transmission, saving, and analysis including visualization for the enormous volumes of data. The calculation speed of computers has become very large, but the speed of data I/O onto a storage system is not as high as the CPU calculation speed. Because the data size of huge-scale simulations is enormous, in some cases not

Journal of Physics: Conference Series **454** (2013) 012078 doi:10.1088/1742-6596/454/1/012078

all data can be saved onto the storage system. Supercomputer systems are, in general, remotely connected, and the generated data is transmitted to a local machine. Data transmission needs very long time. Since the local machine does not have as large memory as the supercomputer, it is impossible to analyze and/or visualize all the data at once. These problems are a major bottleneck in huge-scale simulation in many fields.

In order to resolve the problems of data size, several data compression schemes were proposed for the time-sequential data of process computers [1], electrocardiogram [2, 3], moving objects [4], and GPS trajectories [5], compression of images [6, 7], video and sound [8, 9], and particle trajectory data of molecular dynamics simulation [10, 11, 12]. H.264/MPEG-4 AVC is one of the video compression standards [8, 9]. It adopts the integer transform operation with 16 bit integer accuracy, which enables it to operate with only adding, subtraction and a bit shift. The advantage of this operation is that it does not generate the difference in operation results by the different implementations of a real number operation. It also adopts the motion compensation which describes an image in terms of the transformation of a reference image at an earlier time or even an image from the future, to the current image. Davisson presented a method of theoretically analyzing the data compression techniques which relied on the approximation of the source output by polynomial segments, and considered straight-line interpolation, specifically [7]. Omeltchenko et al. used octree indexing and sorted atoms accordingly on the resulting space-filling curve for scalable I/O of molecular dynamics data [12]. By storing differences of successive atomic coordinates and using an adaptive, variable-length encoding to handle exceptional values, the I/O size was reduced by an order-of-magnitude with a user-controlled error bound. The GROMACS XTC format is a portable format for particle trajectories in the molecular dynamics simulation [10]. The trajectories are written using a reduced precision algorithm which works in the following way: the coordinates are multiplied by a scale factor (typically 1000). These are rounded to integer values. The algorithms by Spangberg et al. for XTC algorithm are based on the observation that differences in atomic coordinates and velocities, in either time or space, are generally smaller than the absolute values of the coordinates and velocities [11]. They also developed an algorithm which used a block sorting. By using an interactive polynomial interpolation, Daehlen and Floater described the decomposition algorithm and presented some numerical examples, such as Franke's function, which was a well-known standard function used in the investigation of methods for interpolating scattered data, and a discrete height data above sea-level on a uniform grid of coordinates in an area of land [13].

A characteristic particle motion plays an important role in the several phenomena under the simulations of plasma physics, astrophysics, and so on. In the collisionless magnetic reconnection, un-magnetized particles execute a complex thermal motion, and this complex motion is one of the main causes to trigger the reconnection [14]. Several particles are selectively accelerated in the collisionless shock wave [15]. In order to capture these behaviors, it is important to trace the particle trajectory or local averages of physical quantities in detail instead of statistical averages, such as temperature and average velocity. On the other hand, galaxy formation simulation deals with huge-scale spatiotemporal phenomena, such as 10^{10} years for age of the universe, 10^8 years for the galaxy dynamics, 10^6 years for the giant molecular clouds dynamics, and 10^4 years for the supernova remnants. Small-scale temporal phenomena take place in the small compact region. In order to see what phenomena occur in the galaxy formation, it is needed to record the simulation data in the small-time scale since the interstellar medium evolves on the smaller scale. If we know how large the scale of the phenomena is and/or, where it takes place in advance, before starting the simulation, it is possible to perform on-the-fly visualization. However, in the case that we know those only after the simulation has proceeded, it is needed to store and visualize all the data as a post process. When we log all data at each uniform time step, the size of stored data becomes enormous, and it needs a long time to input and output the large data. Data compression is indispensable in performing the huge-scale simulation.

We propose in this paper that a compression of simulation data from a large-scale particle simulation holds favorable prospects for scientific visualization of particle systems. Our concepts deal with the data of particle orbits obtained by simulation directly. Switching from a real to an integer representation, as proposed in the field of molecular dynamics simulation, is a kind of approximation, and the data compression of binary digit strings is one of the powerful tools. In this paper, we approximate a particle trajectory with a polynomial function based on the characteristic motion of the particle. The parameters of each polynomial are stored in place of the original data. Moreover, we can use arbitrary mathematical functions based on known behavior of the particle dynamics. The function is changed according to the characteristic of the particle dynamics, or the simulation data of the particle trajectory itself is recorded without the approximation by the polynomial. This is expected to reduce significantly the data size saved onto the storage and improve I/O performance for viewing software. Accordingly, it becomes possible to visualize a larger-scale particle system. We name this concept "TOKI (Time-Order Kinetic Irreversible compression)".

In section 2, we propose the concepts of how to compress the data. One implementation of our concept is explained as a flow chart. We apply the compression scheme to plasma and galaxy formation simulations. In section 3, we show the data compression rate, reproducibility, and the error between the simulation results and the reconstructed data from the compressed data. We discuss the results and the future plan in section 4. Finally, we conclude in section 5.

2. Concepts of a compression scheme

Our proposal "TOKI" adopts a data-compression method for particle trajectory data which has following features:

- Through control over the compression scheme, the difference between the simulation variables and the reconstructed values for the visualization from the compressed data becomes smaller than given constant.
- The particles in the simulation are regarded as independent particles and the time-series data for each particle is compressed with independent time-step for the particle.
- A particle trajectory is approximated by a polynomial function based on the characteristic motion of the particle. In other words, the parameters of each polynomial are stored in place of the original data. It is reconstructed as a continuous curve by the interpolation from the values of the function for intermediate values of the sample data.

In short, the continuous curve which is established by a small number of parameters of the interpolation function approximates the original particle trajectory, and the amount of necessary data can be reduced. We consider the following procedures in order to realize these concepts.

(i) In the case that a particle moves rapidly with a high-frequency oscillating motion, the simulation data is stored as raw data without a change, since such motion cannot be approximated by the polynomial. The data size becomes a product of the number of the fast oscillating particles and the number of time steps.

(ii) In the case that a particle moves with a slow-frequency oscillation motion, the trajectory can be traced by the interpolation function of the polynomial. The coefficients of the polynomial are stored in this case. The data size becomes a product of the number of particles, the number of the coefficients, and the number of the approximation calculations. The polynomial adopted for the interpolation function in this compression scheme depends on the physical system we investigate. For example, if a particle executes circular motion, a circular function is suitable for expressing such a motion. After obtaining the coefficients of the interpolation function, we compare the simulation data and the interpolated values. If the difference between those is smaller than a given constant, which is an input parameter for the error control (we call this an allowed error in the following), the coefficients are adopted and stored. If the difference is larger than the allowed error, we reduce the number of the simulation data which is interpolated in this procedure and calculate the interpolation again. Until the difference becomes smaller than the allowed error, reduction of the data size and calculation of the interpolation are repeated. When the size of the simulation data is equal or smaller than the number of coefficients of the polynomial, the calculation is stopped, and the raw simulation data is stored.

By these procedures, the following data compression is performed. For example, by using a cubic function as an interpolation polynomial, the number of coefficients is four. If the number of the raw data is larger than four, and it is interpolated by a cubic function, four coefficients are stored in place of the data. It means a decrease in the number of the stored data. We call this a data compression. In this paper, we adopt a cubic function for the test of the compression code. If the number of rapid particles which cannot be represented by the interpolation function is much smaller compared to the number in the whole system, an efficient data compression is expected by this scheme. We call the above procedure "encoding", and reproduction of the particle trajectory from compressed data "decoding", respectively, in the following.

Figure 1 shows one implementation of our scheme as a flow chart in which the encoder performs the procedures in order to realize the "TOKI" procedure.

Initialization allocates working arrays for storing the time-sequential data of particles, the coefficients and the raw data. The particle position is stored in the working array sequentially (In figure 1, the position data x is stored in the array $work$, and the data number is set as i). If we want to record the compressed data (we have a forcing-output flag set), or the working array is full, the encoder calculates the coefficients of the interpolation function by the least-squares method. When those are obtained, we compare the simulation data with the interpolation values. If the difference between them is larger than the allowed error, the data length of the time-sequential particle data for the interpolation in the working array is reduced, and the least-squares method calculates the interpolation again. When the difference is smaller than the allowed error, the data length is increased, and the calculation of the least-squares method is performed again, too. In this procedure, we find the longest data length n for interpolation. Or, when the number of the data for interpolation is equal or smaller than the number of the coefficients of the polynomial, this procedure is stopped.

At this time, if the number of data which is interpolated is larger than the number of the coefficients, the coefficients are stored. Otherwise, the raw simulation data is stored. The particle position data in the work array is shifted by the number n of the raw data, which is interpolated or stored. Because the size of the working array is limited, the data which is not interpolated is shifted to the beginning of the array. After the data shift, new data is added after the last data in order to find the coefficients for the data as large as needed. If we have a forcing-output flag set, then the stored compression data is recorded, and we determine whether the data compression procedure continues or not. If the procedure continues, the position data is stored in the working array again. Finally the working arrays are de-allocated.

In the initialization of the encoder, several initial parameters are input; basename, digit number and extension of the output file, allowed error, order of the polynomial, number of particles and size of working array. The digit number specifies the number of digits when we assign a number to the output file. In the body of the encoder, the simulation time, the enforcing-output flag and the position of particles are inputted. In the output part of the encoder, several data are recorded: particle number, dimension (x, y and z), number of interpolation calculations, time when the interpolation starts, interpolated data length, and coefficients of the interpolation functions. When the interpolation function cannot represent the particle trajectory, the raw data of the simulation is recorded, and the number of the recorded raw data is recorded instead of the interpolated data length.

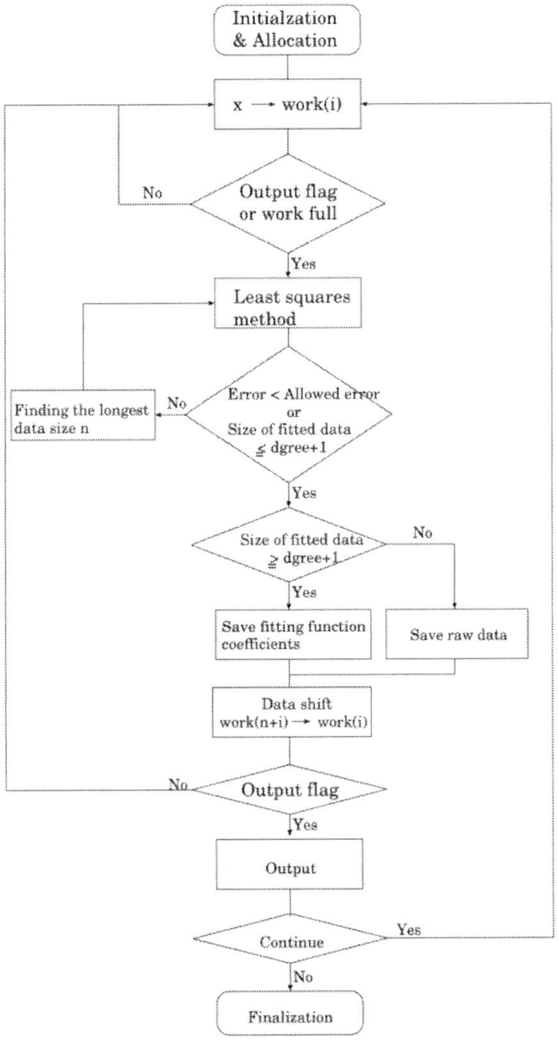

Figure 1. Flow chart of the encoder. In the procedures of storing the particle position in the working array and of shifting data, i is a data number, and n is the number of the raw data which is interpolated or the raw data itself.

3. Applications

We aim to apply this compression method "TOKI" to the simulations in plasma physics and galaxy formation. Since the characteristic time scales in the particle dynamics are different in these systems, it is practically impossible to store visualization data which is logged simply at every smallest time step. We consider that those are suitable systems for application of the scheme.

In this paper, we encode the simulation data as a post process to examine whether our concepts of data compression and the encoder work well or not. Implementing the encoder in the simulation code itself is left as future work. Moreover, we use a cubic function as interpolation polynomial for the encoder test.

As mentioned above, the simulation data is represented by an interpolation function in the encoder. The decoder reconstructs the data from the interpolation function at each time given in the time in the simulation. In the following, we compare the reconstructed or decoded data with the raw simulation data and examine the differences. Moreover, we compare the data size of encoded data with that of raw data. The raw data contains the simulation time and the three components of particle positions. Compression rate is defined as the ratio of the encoded data size to the raw data size. We study the dependence of the rate on the size of the working array and the allowed error values. As a first preliminary test, we choose a Fortran binary with double precision as data recording format.

3.1. Plasma particle simulation

We then apply our compression scheme to the plasma particle simulation. The simulation code is a three-dimensional electromagnetic particle-in-cell code [16, 17]. The boundary conditions are periodic. The number of particles per cell is not uniform in space and changes from 10 to 100. The data for compression is a part of the full simulation data, and the number of particles in the encoder test is 800. The data was calculated for 10,000 time steps, and the time step of the simulation satisfies the Courant-Friedrich-Lewy (CFL) condition [18]. Under the CFL condition, in the plasma particle simulation the time step dt needs to be much smaller than the plasma frequency $1/\omega_p$ or the cyclotron frequency $1/\omega_c$. As a result, for calculation stability, the time step dt and space interval dx are defined by the electron plasma frequency ω_{pe} and the electron Debye length λ_e, respectively, in the full plasma particle simulation. In fact, in order to express the dynamics of both electrons and ions, dt and dx are smaller than $1/\omega_{pe}$ and λ_e, respectively. Since dt is small, the size of the output data becomes large.

Figure 2 shows the comparison between the simulation data and the decoded data reconstructed from the encoded data by the decoder for one electron in three-dimension, and the error between the simulation and the decoded data. The size of the working array is 64 and the allowed error is 0.1. The time evolution of x and z coordinates has a discontinuity because of the periodic boundary condition used for analyzing and visualizing the particle trajectory in this research area. In this plasma particle simulation, the particle goes out from the simulation box, and the position of the particle is modified; $x \to x \pm L_x$, where L_x is the simulation box length. It is found that the decoded data reproduces the high-frequency motion of the particle, and that the error between the simulation data and the decoded data is less than the allowed error value (=0.1). Figure 3 displays the dependence of the error between the simulation and the decoded data on the allowed error. In all cases, the error is smaller than the allowed error. From this result, it is shown that the error introduced by the encoder is under control. Table 1 shows the dependence of CPU time, the data size and the data compression rate on the working array size and on the allowed error. The larger the size of working array becomes, the longer the CPU time that the calculation needs and the smaller the compressed data size becomes. It is because the encoder tries to obtain an interpolation function for time-sequential data as long as possible.

3.2. N-body/gas simulation

We next apply our compression scheme to N-body/gas simulation of galaxy formation [19, 20]. Gravitational interaction is calculated through a tree method with GRAPE, and the gas is simulated by smoothed particle hydrodynamics under the open boundary condition. The number of sequential simulation data is 31. The number of particles in the original simulation data is 1,005,600, and we sample 10,560 particles from it for the first test of the encoder.

Table 2 shows the size of compressed data and the data compression rate. When the allowed error value is too small, the compressed data size becomes almost the same as the raw data size. In such a simulation case, it is convenient that the allowed error value is 0.005 to 0.0001. Figure 4 shows the comparison between the simulation data and the decoded data reconstructed from the encoded data by the decoder for two particles with high- and low-frequency motions,

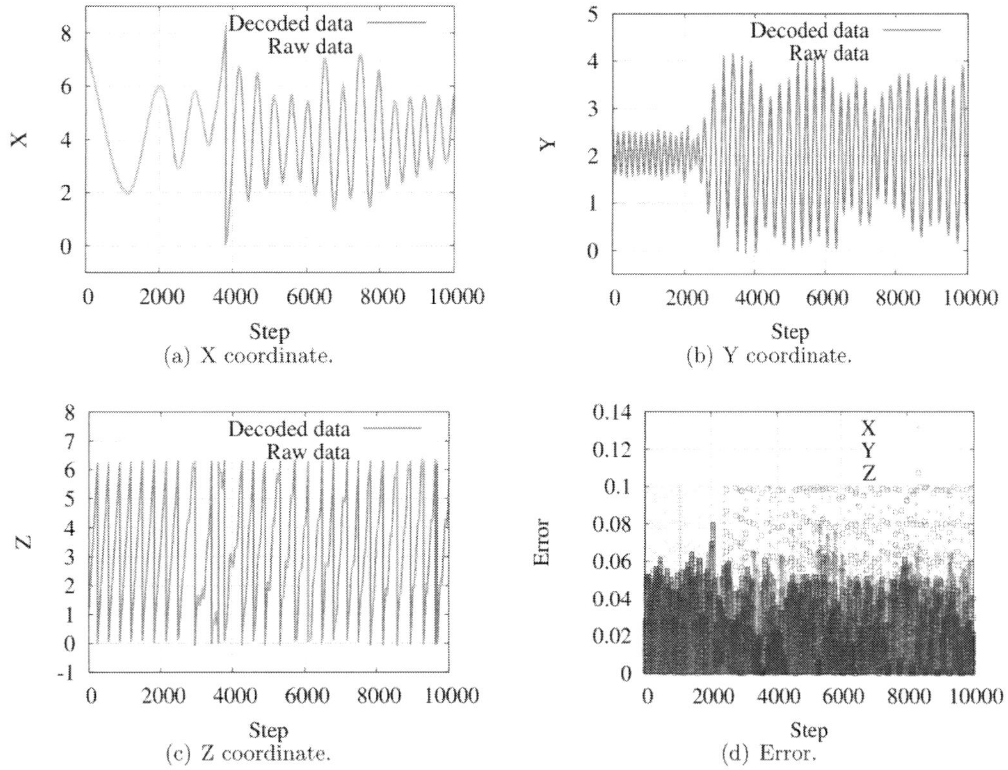

Figure 2. Reproducibility and error between the simulation and the decoded data in the case of plasma particle simulation. The time evolution of the particle coordinates ((a) x, (b) y, (c) z) is shown. The red line indicates the decoded data, and the green plus symbols show the simulation data. (d) Errors (difference) between the simulation and decoded data are shown. Plus, cross, and square symbols are the errors for x, y, and z coordinates, respectively.

Table 1. Dependence of CPU time, data size and data compression rate on work array size and on allowed error in the case of plasma particle simulation.

Work array size	Allowed error	CPU time (sec)	Data size (MB)	Compression rate (%)
Raw data	-	-	183.45	-
32	0.1	2.67	50.13	27.3
64	0.1	2.28	25.81	14.1
128	0.1	3.44	14.01	7.6
256	0.5	4.11	7.59	4.1
256	0.1	7.75	10.54	5.7
256	0.05	8.88	12.39	6.8

respectively. The original data was sampled at coarse time intervals when the simulation ran. That is, the simulation time step is small enough, but the time step of the output is coarse. For the simulation data obtained with the coarse output intervals, the encoder cannot obtain a

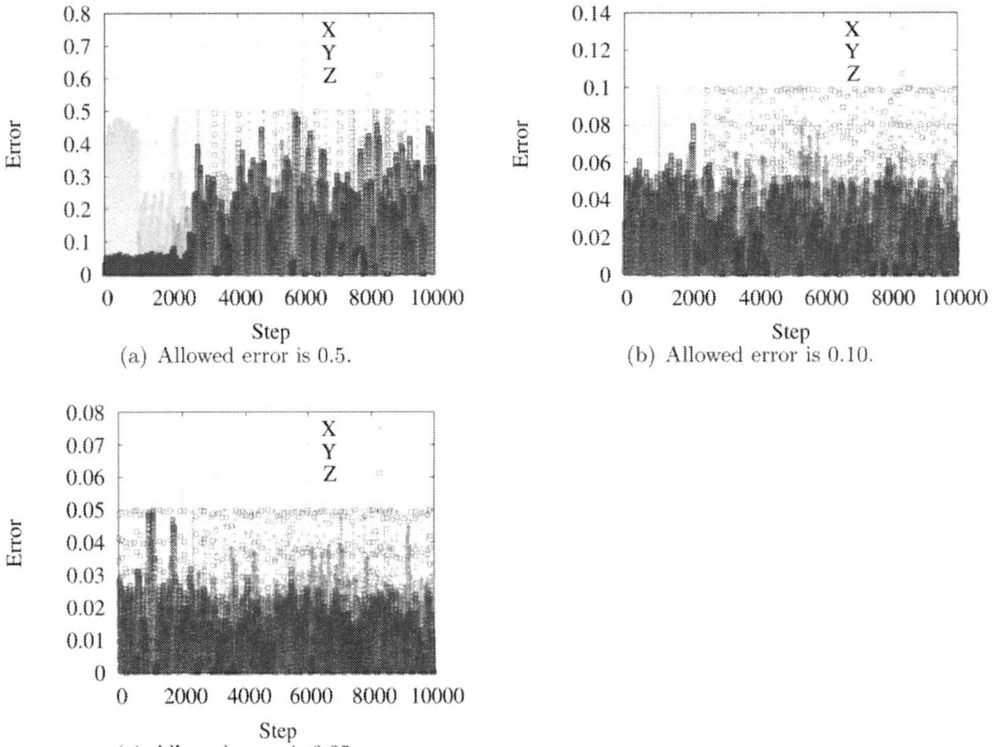

(a) Allowed error is 0.5.

(b) Allowed error is 0.10.

(c) Allowed error is 0.05.

Figure 3. Dependence of the error between the simulation and the decoded data on the allowed error. Plus, cross. and square symbols show the error for x, y, and z coordinates, respectively.

suitable interpolation function of the rapid particle, as shown in figure 4(a). On the other hand, for the low-frequency motion shown in figure 4(b), the encoder can produce an interpolation function to represent the motion. If the encoder used the sufficiently large data set obtained with small time-step output, it is expected that the fitting function can represent the high-frequency wave motion, and that the data compression ratio becomes better. In order to deal with such large data set directly, we will implement the encoder into the simulation code and calculate simultaneously the interpolation and the running simulation. Then, it would not be necessary to sample the simulation data only at coarse time intervals, instead the full data set will be stored.

4. Discussion

In this paper, we study how to encode simulation data as a post process and examine whether our concepts of data compression and our encoder work well or not. The compression ratios are about 5% in the plasma simulation and about 30% in the galaxy formation simulation, respectively. These values are good for practical applications. As next stage, we want to implement this encoder into the simulation code and encode simultaneously the data while performing the simulation. In this case, there are some problems: The encoder needs memory for the working array and calculation time for obtaining the interpolation function. The issue is whether these costs are comparable to the I/O time for the raw data and whether the memory size of the working array can be reduced. Those are important points in trade-off. We need to

Table 2. Dependence of data size and data compression rate on allowed error in the case of N-body/gas simulation.

Number of particles	Allowed error	Data size (MB)	Compression rate (%)
10,560	Raw data	11.24	-
10,560	0.00005	11.21	99.4
10,560	0.0001	3.64	32.4
10,560	0.005	3.26	29.0
1,005,600	Raw data	1070.26	-
1,005,600	0.001	344.49	32.2

optimize the encoder for high-performance computing.

We define dt as the time step of data output in the simulation here. In this case, the output data size is proportional to $\sim 1/dt$. On the other hand, in this paper, we trace the particle trajectory through an interpolation function. We define Δt as a time interval in which a single interpolation function can represent the trajectory. Δt is dependent on the character of particle dynamics, since the determination of the interpolation function depends on the aspect of the dynamics. Accordingly, Δt takes different values from one particle to to the other, that is, Δt time step is an independent time interval. The data size which is recorded every Δt is proportional to $\sim 1/\Delta t$. Since Δt is larger than the uniform dt defined in the simulation, the data size for Δt becomes smaller than that for dt.

For the test of the encoder, we used a cubic function for fitting the sequential simulation data. As mentioned in section 2, we consider that the polynomial adopted for the interpolation function in this compression scheme should depend on the actual physical system we investigate. Selecting the preferred function represents a problem. If we change the interpolation fitting function from a cubic function to higher-order functions, complex data can be expressed better, but the function may over-fit an occasional noise, and the compression rate may become worse. If Δt becomes N times as large as dt because of the interpolation of an n-order polynomial function, the data compression ratio is N/n. When n increases, the ratio becomes better and the objective of the data compression is achieved. Moreover, even if it is concluded by the examination that the data should be interpolated by a high-order polynomial function, we also need to investigate the conclusion from the engineering view point, such as calculation cost and memory size.

As shown in section 3, the size of the working array has an influence on the compression rate, the calculation time, and the memory size of the encoder. Optimization of the size of the working array by auto-tuning is also important. As mentioned above, the reductions of the memory size of the working array and of the calculation cost of interpolation remain technical problems. And, it is important to compare the schemes as mentioned in section 1 with same simulation data in order to discuss advantages and disadvantages of our method. In the future we plan to realize this task by applying our concepts to a variety of systems.

The concept "TOKI" can be applied to other vector and scalar simulation data, such as electromagnetic field, temperature, etc, which does not change with a high-frequency oscillation. Needless to say, ultra-huge-scale simulation data are a common issue also for fields other than plasma physics and astrophysics. Further we are going to apply our concepts to other simulation variables and to a variety of systems.

In future, we will implement a decoder for our compression method on viewer applications, for example, Zindaiji [21] and realize visualization of a large-scale particle system.

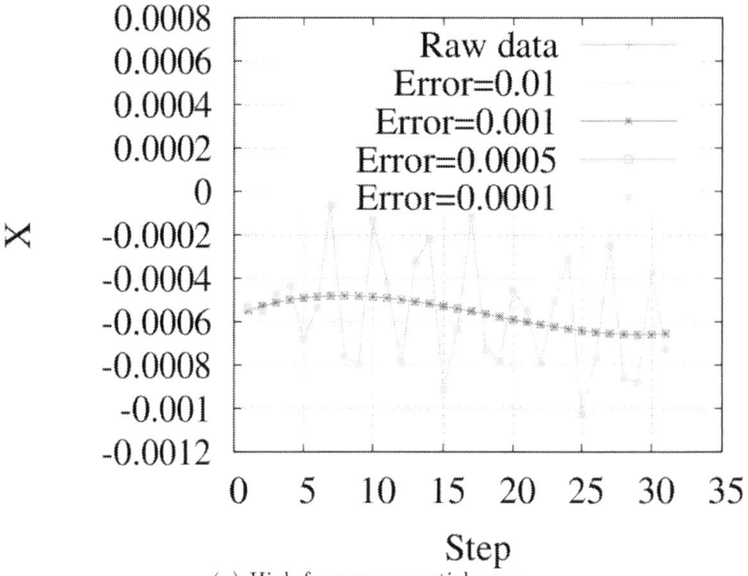

(a) High-frequency particle case.

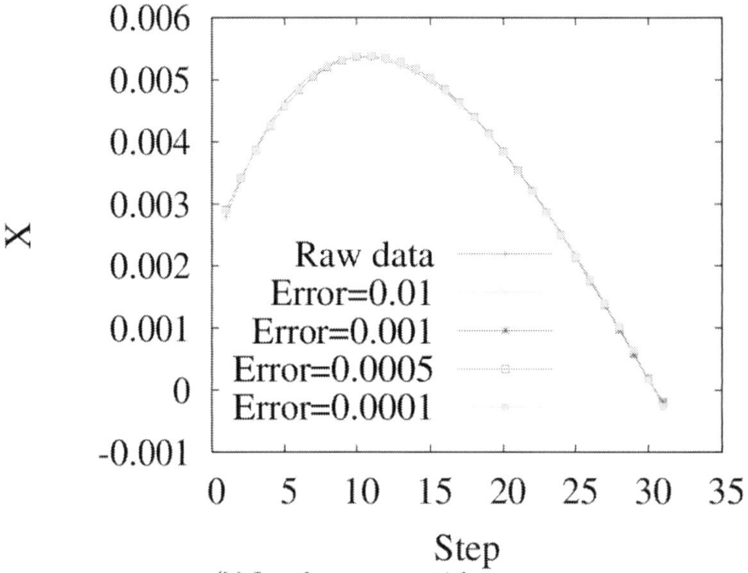

(b) Low-frequency particle case.

Figure 4. Reproducibility in the case of N-body/gas simulation data. We show only the x coordinate of a particle which moves in the three-dimensional space. Red, green, blue, magenta, and sky blue lines show the simulation raw data, decoded data under the conditions of the allowed errors 0.01, 0.001, 0.005, and 0.0001, respectively. In the high-frequency particle case (a), raw data and data for error=0.0001 overlap, and data for error=0.01, 0.001 and 0.005 overlap.

5. Conclusions

We proposed the data compression method "TOKI" for particle trajectory data produced by simulation. It was possible to realize a good reproducibility from encoded data and to control the error between the decoded values and simulation data.

By using this compression method, the data size saved onto the storage is expected to be reduced and I/O performance for visualization software significantly improved. As a result, it will be possible to visualize more large-scale simulation data and to realize reactive animation visualization. It will be possible to visualize time-sequential simulation data of a large-scale systems and to analyze them by exploring in space. Plasma phenomena, such as, reconnection and shock waves, time evolution of plasma particles and large-scale structure, will be simultaneously studied in such real-time visualization. In research on galaxy formation, it will be possible to describe a fully dynamical evolution from a region of high-density star formation to a whole galaxy. The data compression scheme could open up a new path to understand the physics of these systems through capturing phenomena occurring in local regions with the visualization of the detailed behavior of particles.

Acknowledgments

This work was supported by National Institutes of Natural Sciences (NINS) Program for Cross-Disciplinary Study.

References

[1] Hale J C and Sellars H L 1981 *Chem. Eng. Prog.* **77** 38
[2] Jalaleddine S M S, Hutchens C G, Strattan R D and Coberly W A 1990 *IEEE Tras. Biomedical Eng.* **37** 329
[3] Nygaard R, Melnikov G and Katsaggelos A K 2001 *IEEE Tras. Biomedical Eng.* **48** 28
[4] Meratnia N and de By R A 2004 *Lecture Notes in Computer Science* **2992** 765
[5] Chen M, Xu M and Franti P, 2012 *Proc. Data Compression Conference* 62
[6] Staadt O G, Gross M H and Weber R 1997 *Proc. IEEE Vis. '97* 337
[7] Davisson L D 1968 *IEEE Trans. Information Theory* **IT-14** 390
[8] http://www.itu.int/
[9] http://www.iso.org/iso/jtc1_home.html
[10] Green D G, Meacham K E, Surridge M, van Hoesel F and Berendsen H J C, 1955 *Metecc-95 Proceedings*
[11] Spangberg D, Larsson D S D and van der Spoel D 2011 *J.Mol.Model* **17** 2669
[12] Omeltchenko A, Campbell T J , Kalia R K, Liu X and Nakano A 2000 *Comp. Phys. Comm.* **131** 78
[13] Daehlen M and Floater M 1993 *Numerical Algorithms* **5** 165
[14] For example, Horton W and Tajima T 1991 *J. Geophys. Res.* **96** 15811, Ishizawa A and Horiuchi R 2005 *Phys. Rev. Lett.* **95** 045003, and Ohtani H, Horton W, Petrosky T and Horiuchi R 2009 *J. Plasma Fusion Res. SERIES* **8** 203
[15] For example, Sugiyama T 2011 *Phys. Plasmas* **18** 022302, Kato T and Takabe H 2010 *Astrophys. J.* **721** 828, Riquelme M A and Spitkovsky A 2011 *Astrophys. J.* **733** 63, and Matsumoto Y, Amano T and Hoshino M 2012 *Astrophys. J.* **755** 109
[16] Birdsall C K and Langdon A B 1985 *Plasma Physics Via Computer Simulation* (McGraw-Hill, New York)
[17] Ohtani H and Horiuchi R 2009 *Plasma Fusion Res.* **4** 024
[18] Courant R, Friedrichs K O, and Lewy H 1928 *Math. Ann.* **100** 32
[19] Saitoh T R, Koda J, Okamoto T, Wada K, and Habe A 2006 *Astrophysical J.* **640** 22
[20] Takeda T 2007 *Proc. SIGGRAPH '07 ACM SIGGRAPH 2007 computer animation festival* 62
[21] http://www.cfca.nao.ac.jp/~takedatk/COMPUTER/ZINDAIJI3/Zindaiji3Top_E.html

Photoinduced spin-order destructions in one-dimensional extended Hubbard model

Hantao Lu[1], **Shigetoshi Sota**[2], **Hiroaki Matsueda**[3], **Janez Bonča**[4,5] and **Takami Tohyama**[1]

[1] Yukawa Institute for Theoretical Physics, Kyoto University, Kyoto, 606-8502, Japan
[2] Computational Materials Science Research Team, RIKEN AICS, Kobe, Hyogo 650-0047, Japan
[3] Sendai National College of Technology, Sendai, 989-3128, Japan
[4] Faculty of Mathematics and Physics, University of Ljubljana, SI-1000 Ljubljana, Slovenia
[5] J. Stefan Institute, SI-1000 Ljubljana, Slovenia

E-mail: luht@yukawa.kyoto-u.ac.jp

Abstract. By employing the time-dependent Lanczos method, the nonequilibrium process of the half-filled one-dimensional extended Hubbard model under the irradiation of a transient laser pulse is investigated. We show that in the spin-density-wave (SDW) phase, the antiferromagnetic spin correlations are impaired by the photoinduced charge carriers. Near the phase boundary between the SDW and charge-density-wave (CDW) phases, a local enhancement of charge (spin) order that is absent in the original SDW (CDW) phase can be realized with proper laser frequency and strength. The possibility of restoration of spin orders from the CDW phase by optical means is discussed.

1. Introduction

New insights into the dynamics of strongly correlated electron systems can be achieved by investigating their nonequilibrium processes, which demands the knowledge and understanding of the properties of excited states. The process have to be treated nonperturbatively if the system is driven far from the equilibrium.

In experiments, one way to achieve nonequilibrium states is by applying various external electric fields, including pulses of electromagnetic field, which couple with charge degrees of freedom of the system. Well-known experimental facts on (quasi) one-dimensional (1D) materials include dielectric breakdown [1, 2], insulator to metal transitions and so on [3, 4, 5].

In Mott insulators, we have to deal with the states above the Mott gap in order to induce charge carriers optically. It is well known that the ground states of Mott insulators have quasi-long-range antiferromagnetic spin orders. The charge carriers created by photons are, thus, expected to affect the spin correlations. It is desirable to obtain the knowledge of spin dynamics, as well as that of the charge, during and after the doping. In many 1D materials, it has been shown that besides the on-site Coulomb interaction U, nearest-neighbor (NN) interaction V has to be taken into account [6, 7]. With increase of V, a phase transition from the Mott insulator to a charge-order phase can occur. The nonequilibrium dynamics near the phase boundary is also important.

In this proceeding, we investigate the nonequilibrium process of a 1D strongly correlated system under the irradiation of a laser pulse. The model we are working on is the half-filled extended Hubbard model, where the on-site, and NN interactions (U and V), are included. Its phase diagram in equilibrium is well-known and understood [8, 9, 10, 11]. Here we would like to reiterate some known properties of the model, concerning our work. In strong-coupling regime (large $U(>0)$), with increase of $V(>0)$, a first-order phase transition from the spin-density-wave (SDW) to charge-density-wave (CDW) phase is realized when V reaches around $U/2$. The transition can be understood as driven by the competition between the energetic costs for doublon generation and the benefits due to the attraction between doublons and holons. The SDW and CDW phases are recognized by algebraic decay of spin correlations, and a long-range (staggered) charge order, respectively. In both phases, the charge sector is always gapped except on the boundary. When spin degrees of freedom are concerned, gapless spin modes can be found in the SDW phase, while absent in the CDW phase.

In the following sections, by studying time-dependent spin and charge correlation functions numerically, we show that spin correlations are suppressed by the photoinduced charge carriers in the SDW phase. Near the phase boundary from the SDW side, with proper laser frequency and strength, a sustainable charge order enhancement can be realized [12] but local spin correlations remains. Analogously, from the CDW side, the suppression of long-range charge order is accompanied with a local spin correlation enhancement.

2. Model and Numerical Method

The time-dependent Hamiltonian for the 1D extended Hubbard model, where the external field is incorporated by means of the Peierls substitution for the hopping coefficients, can be written as

$$\begin{aligned} H(t) &= -t_h \sum_{i,\sigma} \left(e^{iA(t)} c^\dagger_{i,\sigma} c_{i+1,\sigma} + \text{H.c.} \right) \\ &\quad + U \sum_i \left(n_{i,\uparrow} - \frac{1}{2} \right) \left(n_{i,\downarrow} - \frac{1}{2} \right) + V \sum_i (n_i - 1)(n_{i+1} - 1), \end{aligned} \tag{1}$$

where $c^\dagger_{i,\sigma}$ ($c_{i,\sigma}$) creates (annihilates) electrons with spin σ at site i, $n_{i,\sigma} = c^\dagger_{i,\sigma} c_{i,\sigma}$, $n_i = n_{i,\uparrow} + n_{i,\downarrow}$, t_h is the hopping constant. Particularly, with temporal gauge, we use a time-dependent vector potential $A(t)$ to describe the laser pulse [13]

$$A(t) = A_0 e^{-(t-t_0)^2/2t_d^2} \cos\left[\omega_{\text{pump}} (t - t_0)\right], \tag{2}$$

where A_0 controls the laser amplitude, which reaches its full strength at $t = t_0$; t_d characterizes the duration time of light action. Note that, due to finite t_d, the incoming photon frequency is broadened into a Gaussian-like distribution, with the variance of $1/t_d^2$ around the central value ω_{pump}. Throughout this paper, we set t_h and t_h^{-1} as energy and time units.

Starting from the Schrödinger equation $i\partial\psi(t)/\partial t = H(t)\psi(t)$, in order to obtain $\psi(t)$, we employ the time-dependent Lanczos method, which is originally described in [14], and followed by its applications in nonequilibrium dynamics of strongly correlated systems recently in [15]. The basic idea is that we approximate the time evolution of $|\psi(t)\rangle$ by a step-vise change of time t in small increments δt. At each step, the Lanczos basis with dimension M is generated resulting in the time evolution

$$|\psi(t + \delta t)\rangle \simeq e^{-iH(t)\delta t} |\psi(t)\rangle \simeq \sum_{l=1}^{M} e^{-i\epsilon_l \delta t} |\phi_l\rangle \langle \phi_l | \psi(t) \rangle, \tag{3}$$

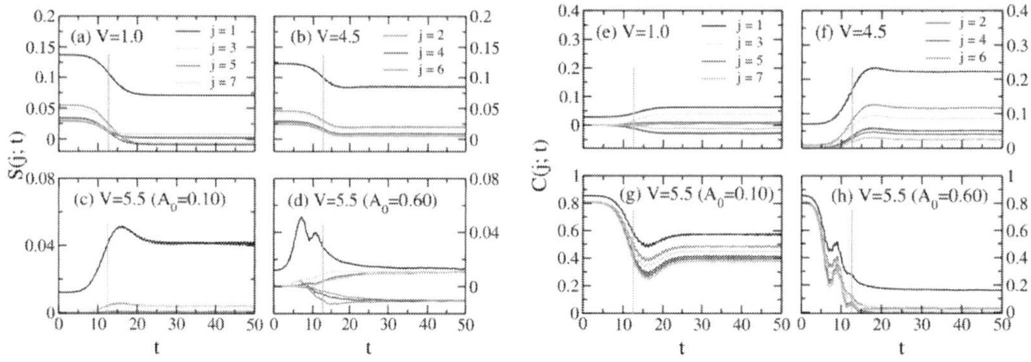

Figure 1. Time dependence of the spin-spin (left, from (a) to (d)) and charge-charge (right, from (e) to (h)) correlations as functions of distance (labeled by j) for 14-site lattice. The laser pulse with Gaussian magnitude modulation reaches its full strength at $t = 12.5$, as indicated by solid lines. Without exception, the pumping frequencies are set to match the resonance peaks of the optical absorption spectra. Parameters: $\delta t = 0.02$, $M = 100$. In each subfigure, (a),(e): $V = 1.0$, $\omega_{\rm pump} = 7.1$, $A_0 = 0.10$; (b),(f): $V = 4.5$, $\omega_{\rm pump} = 4.0$, $A_0 = 0.07$; (c),(g): $V = 5.5$, $\omega_{\rm pump} = 4.1$, $A_0 = 0.10$; (d),(h): the same as (c) (or (g)), except for $A_0 = 0.60$.

where ϵ_l and $|\phi_l\rangle$, respectively, are eigenvalues and eigenvectors of the tridiagonal matrix generated in Lanczos iteration. As one of the advantages, the Lanczos method preserves the unitarity of the time-evolution operator at each time step, with relative low cost for a desired accuracy. The convergence check shows that for time length t around hundred, taking $M = 30$ with $\delta t \lesssim 0.1$ can provide adequate accuracy for the parameter regime we are working in. For even smaller δt, M can be smaller.

In the succeeding numerical calculations, periodic boundary conditions are employed to avoid possible boundary effects. We fix the parameters $t_d = 5$, $t_0 = 12.5$, and chose $U = 10$, where the first-order transition occurs around $V \approx 5.1$ [11].

3. Numerical Results

In order to investigate the time evolutions of the spin and charge orders under the pulse, we define the spin-spin and charge-charge correlations, respectively:

$$S(j;t) = \frac{(-1)^j}{L} \sum_{i=0}^{L-1} \langle \psi(t) | \mathbf{S}_{i+j} \cdot \mathbf{S}_i | \psi(t) \rangle, \tag{4a}$$

$$C(j;t) = \frac{(-1)^j}{L} \sum_{i=0}^{L-1} \langle \psi(t) | (n_{i+j} - 1)(n_i - 1) | \psi(t) \rangle, \tag{4b}$$

where L is the lattice size, and $\mathbf{S}_i$ is the spin operator on site i. Intuitively, the antiferromagnetic spin and staggered charge orders are competitive.

Figure 1 shows the results of correlations $S(j;t)$, $C(j;t)$, up to $t = 50$, for the system with 14 lattice sites, where the largest distance between two sites is 7. Total crystal momentum P, as a good quantum number, is implemented. The size of the Hilbert space for $P = 0$ is 841332. For 2600 time step, it took around 30 hours for 12 cores. In succeeding discussions, we put more emphasis on spin. Detailed discussions on the charge correlations can be found in [12].

In figure 1, four cases are shown: (a) and (e), deep into the SDW phase ($V = 1$); (b) and (f), the ground state is near the phase boundary, but set on the SDW side ($V = 4.5$); (c) and (g),

(d) and (h), at the CDW side ($V = 5.5$) with different laser strength. Several observations can be made. First, notice that for the ground states (corresponding to $t = 0$) in the SDW phase, we can observe the gradually decay of spin correlations (Figs. 1(a), (b)) while no long-range charge order exists (Figs. 1(e), (f)); in the CDW phase, the presence of a long-range charge order can be clearly recognized (Figs. 1(g) or (h)), and the spin correlations largely vanish except for the NN one (See Figs. 1(c) or (d), where $S(j;0) \lesssim 10^{-4}$ for $j > 1$, and $S(1;0) \sim 0.01$). Second, for the case of $V = 1.0$, where the system is deep inside the SDW phase, introducing charge carriers into the system by the pulse cannot produce charge order (figure 1(e)), while the spin correlations are severely suppressed except for the NN one (See figure 1(a), where $S(1;t)$ decreases from 0.14 at $t = 0$ to 0.07 at $t = 50$, while some correlations with other distances even become negative). Notice that in this case, after the pulse, the total number of doublons (defined as $N_{\mathrm{db}} = \sum_i n_{i,\uparrow} n_{i,\downarrow}$) is around 2.2, corresponding to 30% doping [16]. Third, we would like to pick up figure 1(f), which shows that proper incoming laser pulse can enhance the charge order considerably in the SDW near the phase boundary, where we can observe that the NN correlation increases from 0.07 at $t = 0$ to 0.22 at $t = 50$, together with a clear enhancement of other correlations in larger distances, which start from almost zero magnitudes at $t = 0$; meanwhile, the spin order does not suffer heavy suppression (See figure 1(b), where the NN correlation drops from 0.12 to 0.08). Analogously, from the CDW side close to the boundary, the pulse can suppress the charge order (figure 1(g)) with local spin enhancement taking place at the same time (figure 1(c), where the NN correlation increases from 0.01 to 0.04). Finally, from figure 1(d) and (h), we show that strong enough laser pulse can destroy the charge order substantially without spin correlation enhancement.

The results of the time-dependent correlations suggest intimate connections between spin and charge degrees of freedom in the photoexcited states. Interesting phenomena occur around the phase boundary, where local spin and charge orders can coexist. For a more quantitative analysis, we compare the spectra of systems with different values of V in the time-independent Hamiltonian, i.e., equation (1) but $A(t) = 0$. In order to describe the spin and charge orders which have finite spatial extensions, we define

$$\mathcal{O}_{\mathrm{SDW}} = \frac{1}{LL_c} \sum_{i=0}^{L-1} \sum_{j=1}^{L_c} (-1)^j \mathbf{S}_{i+j} \cdot \mathbf{S}_i \tag{5a}$$

$$\mathcal{O}_{\mathrm{CDW}} = \frac{1}{LL_c} \sum_{i=0}^{L-1} \sum_{j=1}^{L_c} (-1)^j \left(n_{i+j} - 1\right) \left(n_i - 1\right). \tag{5b}$$

Here L_c is introduced as a cut off parameter for the correlation length, and we set $L_c = 5$, estimated from the results of the density-matrix renormalization group (DMRG) on larger system size up to 30. The expectation values of the order parameter $\langle \mathcal{O}_{\mathrm{CDW}} \rangle$ and $\langle \mathcal{O}_{\mathrm{SDW}} \rangle$ for eigenstates of a smaller 10-site system are shown in figure 2.

In figure 2 from (a) to (d), we can notice that for $V = 1.0$ to 5.5, states with relative large $\langle \mathcal{O}_{\mathrm{SDW}} \rangle$ are always situated in the low-energy regime (say, $E < 10$). On the contrary, states with prominent $\langle \mathcal{O}_{\mathrm{CDW}} \rangle$ only appear there when the system is approaching the phase boundary. In figure 2(g) for $V = 4.5$, around the resonance frequency of optical absorption spectrum ($\omega_R \approx 4.3$), states with noticeable values of charge order can be found. This explains why a considerable charge order enhancement can be observed when the frequency of laser with proper strength is tuned to ω_R (see figure 1(f), where $\omega_R \approx 4.0$ for 14-site lattice). However, in the complementary case, only the enhancement of the NN spin correlation is apparently visible, with rapid decrease with increase of distance. The same holds for $L = 10$, in spite of the fact that a state with excitation energy $E \approx 2.7$ has a distinguished $\langle \mathcal{O}_{\mathrm{SDW}} \rangle$ (shown by star in figure 2(d)).

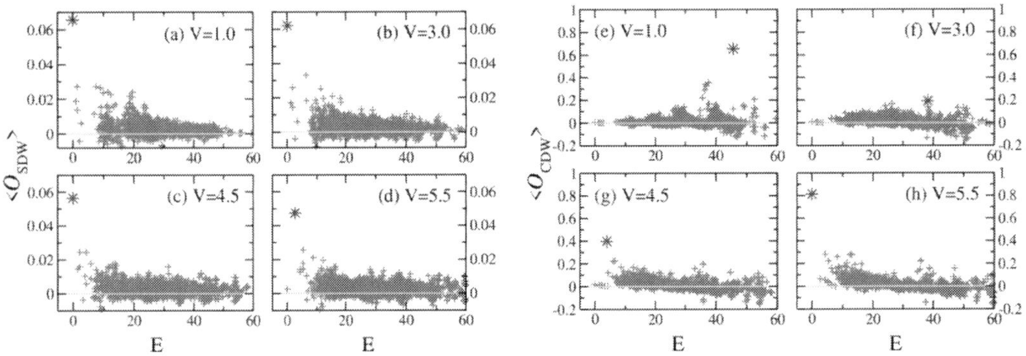

Figure 2. The expectations of SDW (from (a) to (d)) and CDW (from (e) to (h)) order parameters of eigenstates for 10-site lattice with $V = 1.0$, 3.0, 4.5, and 5.5. The energy E is measured from the ground state. Only the data of states with $P_{\rm tot} = S_{\rm tot} = 0$ are shown, up to $E = 60$. Note that the largest values of the order parameters in (a) through (h) are marked by stars.

In order to understand the different behavior of spin and charge correlations, and to further elaborate on the condition for the emergence of order enhancements induced by the laser pulse, we perform parameter-sweeping calculations in terms of $\omega_{\rm pump}$ and A_0 for 10-site lattice. For a given pair of $\omega_{\rm pump}$ and A_0, we carry out the time evolutions up to $t = 52.5$, then calculate the expectations of the SDW order , CDW order, and doublon number, noted as $\langle\mathcal{O}_{\rm SDW}\rangle_{\rm av}$, $\langle\mathcal{O}_{\rm CDW}\rangle_{\rm av}$, and $\langle n_{\rm db}\rangle_{\rm av}$, respectively, after averaging on the last 50 time steps (corresponding to time length $\Delta t = 5$). The results, presented in contour plots, are shown in figure 3.

In figure 3, we notice that for each V, there exist some frequencies with which the optical absorptions are most efficient, e.g., $\omega_{\rm pump} \approx 7.5$ in figure 3(i) for $V = 1.0$, $\omega_{\rm pump} \approx 6.3$ in figure 3(j) for $V = 3.0$. These are resonance frequencies of optical absorption spectra ω_R, which can be obtained from the imaginary parts of the dynamical current-current correlation functions.

We can see that in order to excite the system effectively with weak fields, we need to tune the pulse frequency $\omega_{\rm pump}$ close to ω_R. In general, in the SDW phase, accompanied with photodoping, spin correlations are suppressed (see the correspondence between the bright regions in Figs. 3(i), (j), (k) for doublon generation, and the dark regions in Figs. 3(a), (b), (c) for $\langle\mathcal{O}_{\rm SDW}\rangle_{\rm av}$). On the other hand, in the CDW phase, the destruction of charge order alone does not sufficiently bring back the spin order (see figure 3(l), compared with (d)).

Things can be more subtle near the phase boundary. With A_0 less than 0.1, a local enhancement of CDW order in the SDW phase can be induced accompanied with a suppression of spin orders, and vice versa in the CDW phase near the boundary. For example, see around $\omega_{\rm pump} \approx 4.3$ in Figs. 3(g) and (c) for the case of $V = 4.5$, and $\omega_{\rm pump} \approx 4.4$ in Figs. 3(d) and (h) for the case of $V = 5.5$. Among them, the weakness of the SDW enhancement in figure 3(d) can be attributed to the fact that the states with distinguished $\langle\mathcal{O}_{\rm SDW}\rangle$ (around $E \approx 2.7$, see figure 2(d)) are in the off-resonance region, which makes it hard for them to be excited by the laser.

Going to the high-field region, we observe that in figure 3(d), at $\omega_{\rm pump} \approx 1.2$, there is a well resolved stripe when $A_0 > 0.4$, indicating an enhancement of the spin order. From the results of $S(j;t)$ (not shown here), we notice that the spin correlations there extend themselves beyond NN. This may come from nonlinear multiphoton optical process [17]. Similar phenomenon for the case of charge order, can be found in figure 3(g), at $\omega_{\rm pump} \approx 2$. Further investigation on the nonlinear effect is in progress.

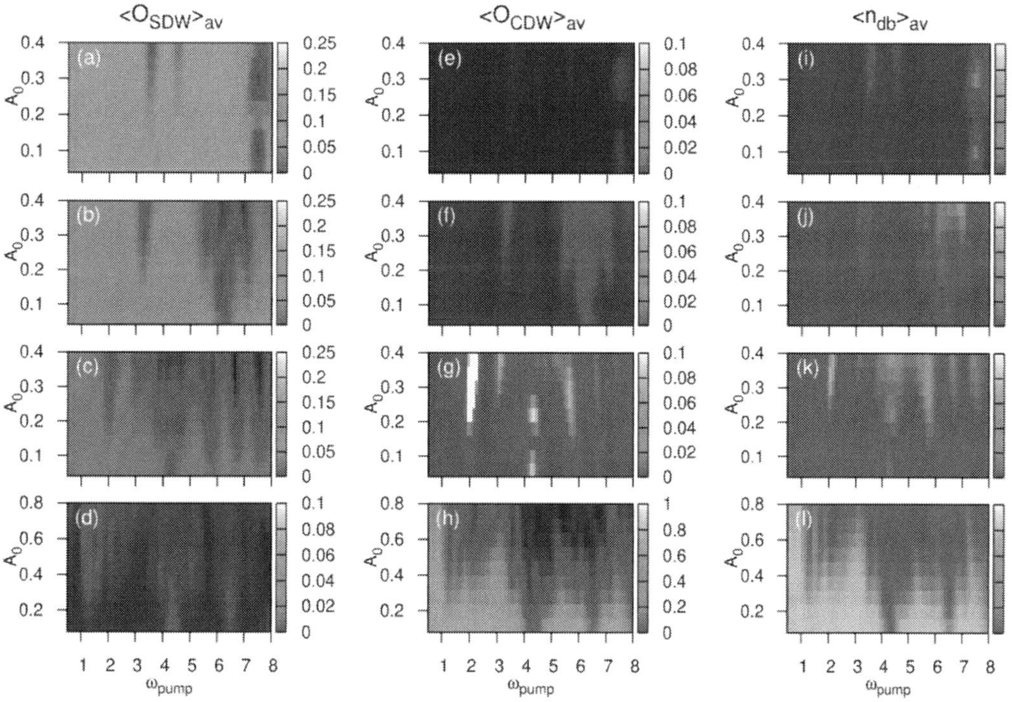

Figure 3. Contour plots of final time-evolution results of SDW order $\langle \mathcal{O}_{\rm SDW}\rangle_{\rm av}$ (first column), CDW order $\langle \mathcal{O}_{\rm CDW}\rangle_{\rm av}$ (second column) and doublon number $N_{\rm db}$ (last column) as functions of $\omega_{\rm pump}$ and A_0, obtained by averaging on the last 50 time steps (time length $\Delta t = 5$), for 10-site lattice. $\omega_{\rm pump} \in [0.5, 8]$; $A_0 \in [0.04, 0.4]$ for the first three rows, and $[0.08, 0.8]$ for the last row. From top to bottom, $V = 1.0$, 3.0, 4.5, and 5.5, respectively. Here, we take $\delta t = 0.1$, $M = 30$.

4. Conclusions

In summary, by studying time-dependent spin and charge correlation functions of the 1D extended Hubbard model with applied laser pulse, we have shown that in the SDW phase, the spin order is suppressed due to the optical doping. Starting from the SDW side not far from the phase boundary, when the incoming laser frequency matches the optical absorption peak, a sustainable charge order enhancement can be realized. Similarly, starting from the CDW side, an enhancement of spin correlations, though more localized, has been observed. In the off-resonance region, we find more extended recovery of spin correlations which may come from nonlinear effects.

References

[1] Tokura Y, Okamoto H, Koda T, Mitani T and Saito G 1988 *Phys. Rev. B* **38**(3) 2215–2218
[2] Taguchi Y, Matsumoto T and Tokura Y 2000 *Phys. Rev. B* **62**(11) 7015–7018
[3] Iwai S, Ono M, Maeda A, Matsuzaki H, Kishida H, Okamoto H and Tokura Y 2003 *Phys. Rev. Lett.* **91**(5) 057401
[4] Okamoto H, Matsuzaki H, Wakabayashi T, Takahashi Y and Hasegawa T 2007 *Phys. Rev. Lett.* **98** 037401
[5] Kimura K, Matsuzaki H, Takaishi S, Yamashita M and Okamoto H 2009 *Phys. Rev. B* **79** 075116
[6] Yamashita M, Manabe T, Kawashima T, Okamoto H and Kitagawa H 1999 *Coord. Chem. Rev.* **190-192** 309–330
[7] Kumar M, Topham B J, Yu R, Ha Q B D and Soos Z G 2011 *J. Chem. Phys.* **134** 234304 (pages 8)
[8] van Dongen P 1994 *Phys. Rev. B* **49** 7904–7915
[9] Nakamura M 2000 *Phys. Rev. B* **61** 16377–16392

[10] Tsuchiizu M and Furusaki A 2002 *Phys. Rev. Lett.* **88** 056402
[11] Ejima S and Nishimoto S 2007 *Phys. Rev. Lett.* **99** 216403
[12] Lu H, Sota S, Matsueda H, Bonča J and Tohyama T 2012 *Phys. Rev. Lett.* **109**(19) 197401
[13] Matsueda H, Sota S, Tohyama T and Maekawa S 2012 *J. Phys. Soc. Jpn.* **81** 013701
[14] Park T J and Light J C 1986 *J. Chem. Phys.* **85** 5870–5876
[15] Prelovšek P and Bonča J 2011 Ground State and Finite Temperature Lanczos Methods
[16] Takahashi A, Itoh H and Aihara M 2008 *Phys. Rev. B* **77** 205105
[17] Oka T 2012 *Phys. Rev. B* **86** 075148

Journal of Physics: Conference Series **454** (2013) 012080 doi:10.1088/1742-6596/454/1/012080

A fitting formula for radiative cooling based on non-local thermodynamic equilibrium population from weakly-ionized air plasma

Yousuke Ogino, Atsushi Nagano, Tomoaki Ishihara and Naofumi Ohnishi

Department of Aerospace Engineering, Tohoku University, 6-6-01 Aramaki-Aza-Aoba, Aoba-ku, Sendai 980-8579, Japan

E-mail: yogi@cfd.mech.tohoku.ac.jp

Abstract. A fitting formula for radiative cooling with collisional-radiative population for air plasma flowfield has been developed. Population number densities are calculated from rate equations in order to evaluate the effects of nonequilibrium atomic and molecular processes. Many elementary processes are integrated to be applied to optically-thin plasmas in the number density range of $10^{12}/\mathrm{cm}^3 \leq N \leq 10^{19}/\mathrm{cm}^3$ and the temperature range of 300 K $\leq T \leq$ 40,000 K. Our results of the total radiative emissivity calculated from the collisional-radiative population are fitted in terms of temperature and total number density. To validate the analytic fitting formula, numerical simulation of a laser-induced blast wave propagation with the nonequilibrium radiative cooling is conducted and successfully reproduces the shock and plasma wave front time history observed by experiments. In addition, from the comparison between numerical simulations with the radiation cooling effect based on the fitting formula and those with a gray gas radiation model that assumes local thermodynamic equilibrium, we find that the displacement of the plasma front is slightly different due to the deviation of population probabilities. By using the fitting formula, we can easily and more accurately evaluate the radiative cooling effect without solving detailed collisional-radiative rate equations.

1. Introduction

In most aerospace applications of air plasma flow (hypersonic flow around a reentry body [1–4], laser-driven blast wave [5–8], and plasma processing techniques using electric discharges), plasma internal states do not achieve local thermodynamic equilibrium (LTE). It is necessary to consider nonequilibrium properties of population distribution into each plasma internal state. The nonequilibrium properties directly affect radiative emissivities and opacities, partition functions of internal energy modes, and thermal relaxations. These effects of nonequilibrium processes are still a problem that remains to be solved in those applications.

For instance, we take a gas-driven laser-propulsion system whose thrust power is obtained through interaction with a propellant gas heated by external laser beam irradiation. After a focused laser breaks down the propellant gas, a blast wave is formed by the laser-produced plasma and propagates to the projectile through the surrounding gas. The performance of a gas-driven laser propulsion system depends heavily on the blast-wave dynamics and the plasma states sustaining it. Our previous work [8] numerically simulated a laser-driven blast wave coupled with rate equations based on a collisional-radiative (CR) model for argon plasma, in order to study the time-dependent effect of excitation and ionization processes during pulse heating. Computational results revealed that a laser-induced plasma driving a blast wave is still in ionization nonequilibrium over the pulse duration. The blast wave shell is accelerated more by the driving plasma, which is in the ionization relaxation state, because there is less radiative energy loss than indicated by plasma flow computations assuming LTE. Note that the radiation emitted from the driving plasma in the unsteady ionizing phase is weaker than that from the LTE population, since the number densities of highly-excited species are relatively lower.

As another example, we take a hypersonic flow around a space vehicle entering into the earth atmosphere. In front of the space vehicle, a strong shock wave with a high temperature layer is formed and then aerodynamic and radiative heating occurs. To protect the space vehicle from such a severe heating environment, an appropriate thermal protection system (TPS) must be equipped. For accurate prediction of the heating rates to design a proper TPS during the entry flight, the behavior of excited species has to be well known. This can be achieved through detailed description of nonequilibrium air plasma states through a CR model [1–4].

The objective of this study is to develop a curve-fitted radiative cooling formula for weakly-ionized air plasma. Population number densities are calculated from collisional-radiative rate equations in order to estimate the effects of nonequilibrium atomic and molecular processes. Our CR model consists of fifteen air species: e^-, N, N^+, N^{2+}, O, O^+, O^{2+}, O^-, N_2, N_2^+, NO, NO^+, O_2, O_2^+, and O_2^- with their major electronic excited states. Many elementary processes are included and applied to doubly-ionized air plasma in a number density range of $10^{12}/\text{cm}^3 \leq N \leq 10^{19}/\text{cm}^3$ and a temperature range of 300 K $\leq T \leq$ 40,000 K. For the validation of the developed radiative cooling formula, we compute a laser-induced blast wave propagation with radiative energy loss, and compare with experimentally observed time evolution of shock wave and plasma front displacement.

2. Collisional-radiative model and radiative cooling formula

The previously developed CR code for air plasma in the optically thin case is employed to calculate nonequilibrium population densities. More detailed descriptions of our code are in the literature [9].

2.1. Atomic and molecular model

We consider that air plasma is composed of fifteen species: e^-, N, N^+, N^{2+}, O, O^+, O^{2+}, O^-, N_2, N_2^+, NO, NO^+, O_2, O_2^+, and O_2^-. For atomic species, each fine structure is degenerated into terms, in addition, terms are coalesced into configurations for highly excited states. For molecular species, electronic states are included to encompass a group of major radiation bands from the infrared to the vacuum ultraviolet wavelength regions. Each rotational, vibrational, and nuclear spin states is coalesced into one electronic state after the manner of Bacri and Medani [10]. All of atomic energies and degeneracies of each electronic state and molecular spectroscopic constants are referred to Saruhashi's radiative database [11]. Our CR model comprises a total of 310 different levels [9].

2.2. Rate equations and rate constants

Number densities $N_{s,i}$ of the coalesced states i of the different chemical species s are calculated using the conservation equation:

$$\frac{dN_{s,i}}{dt} = \sum \text{populating terms} - \sum \text{depopulating terms}. \quad (1)$$

The source terms are populating and depopulating number densities per unit of time, summed over involved transition reactions. It is assumed two temperature model consist of the translation and rotation temperature T and the vibration, electronic excitation, and electron temperature T_e to take into account of thermal nonequilibrium. Transition rate constants included in the right hand side of (1) then are expressed as a function of (T, T_e). The number density of free electron n_e is calculated by the charge neutrality assumption for the partially ionized plasma. For the numerical time integration, we employ the backward differentiation formula (BDF) method included in the LSODE package [12]. This method is a one of the representative multi-step implicit method, and can stably solve the stiff system of ordinary differential equations.

The electron impact excitation and ionization cross sections are given by the Drawin's model [13, 14]. Rate coefficients of inverse processes, such as deexcitation and three body recombination, are derived using the detailed balance principle. For electron impact processes of diatomic molecules are calculated from Bacri's weighted-total cross section method [15], and tangible data of their rate constants are obtained from Teulet et al. [16] Backward transition rate constants obey the detailed balance in common with atomic rates. Additionally, we employ Park's rate constants for electron collisional dissociation and dissociative recombination [17]. For heavy particle impact processes, we consider fourteen chemical reactions as shown in Park's book [1]. Arrhenius parameters and the equilibrium constants for each chemical reaction are also addressed under Park's book.

Both atomic and molecular spontaneous emission coefficients $A(i,j)$ are used from Saruhashi's database [11]. We also take into account the radiative recombination rate $R_r\,(i,j;T_e)$ and the dielectronic recombination rate $R_d\,(i,j;T_e)$ of monatomic ions, using Nahar's recombination rate database [18, 19]. Furthermore, the radiative attachment rate constant for an oxygen atom (O + e→O^- + $h\nu$) is derived from Soon and Kunc [20].

2.3. Total radiative emissivity

The total radiative emissivity $Q_{\rm rad}$ can be expressed by the summation of following three different electronic transition processes.

$$Q_{\rm rad} = \sum_s J_{s,bb} + \sum_s J_{s,fb} + J_{ff}. \tag{2}$$

(i) atomic line and molecular band emission $J_{s,bb}$ is

$$J_{s,bb} = \sum_{i>j} N_{s,i} A(i,j) E_{ij}, \tag{3}$$

where E_{ij} represents energy level.

(ii) recombination radiation $J_{s,fb}$ can be expressed as [21]

$$J_{s,fb} = n_e N_{s,i} R_r(i,j;T_e) \left(I_{\infty,s} + \frac{3}{2} k T_e \right), \tag{4}$$

where $I_{\infty,s}$ represents ionization potential.

(iii) Bremsstrahlung radiation J_{ff} can be expressed as [22]

$$J_{ff} = 1.426 \times 10^{-28}\ G\ n_e^2 \bar{z} \sqrt{T_e}\,(1+D), \tag{5}$$

where $\bar{z}$ represents the averaged ionization degree. The averaged Gaunt factor G and the average nonhydrogenic correction factor D are set to 1.2 and -0.15, respectively.

2.4. *Appropriate conditions for local thermodynamic and Corona equilibrium*

In this subsection, we refer to quasi-steady state in the CR model as the CR steady state (CRSS). CRSS is an intermediate state between LTE and Corona equilibrium (CE) [21]. In LTE or CE, the population distribution among the various excitation states $\mathcal{P}_{s,i} = N_{s,i}/N_s$ can be determined by basic principles. For LTE, $\mathcal{P}_{s,i}$ obeys the Boltzmann distribution function,

$$\mathcal{P}_{s,i}^{\mathrm{LTE}} = \frac{N_{s,i}}{N_s} = \frac{G_i}{Q_s} \exp\left(-\frac{E_i}{kT_e}\right). \tag{6}$$

In contrast, in low-density CE plasmas, the upward excitation rate due to collisions is so low relative to the spontaneous decay rate that one can safely assume that an electron excitation to an upper level will most likely decay to the ground state before experiencing further excitation. Therefore, most chemical species are in their ground state; $\mathcal{P}_{s,i}^{\mathrm{CE}} = N_{s,1}/N_s = 1$. Unlike LTE or CE, in CRSS the population distribution $\mathcal{P}_{s,i}^{\mathrm{CRSS}}$ cannot be determined from these principles, and the density of a given excited species can be solved only from the whole set of rate equations. Thus, we must take into account all the processes that increase or decrease the excited species' abundance. Using CRSS is not always easy and convenient; whenever possible, it is preferable to use LTE or CE to model a plasma. It is therefore important and helpful to know temperature and density conditions for which these three models are valid.

Figure 1 indicates appropriate conditions to employ the above three models. Computations are carried out over a temperature range of 5,000 K to 40,000 K, and a total number density range of $10^{12}/\mathrm{cm}^3$ to $10^{19}/\mathrm{cm}^3$. Such conditions cover a range associated with many applications utilizing air plasmas (e.g., hypersonic flow around a reentry space vehicle, air-driven laser-propulsion system, electric discharges, plasma processing, and arcs in high-power electric thrusters). Figure 1 confirms that the discrepancy from LTE is large in a lower-density, higher-temperature region. We can find that the population probability $\mathcal{P}_{s,i}^{\mathrm{CRSS}}$ does not achieve Boltzmann equilibrium when the total number density is less than or equal to $10^{16}/\mathrm{cm}^3$. Therefore, the CRSS population must be considered when we analyze low-density plasmas.

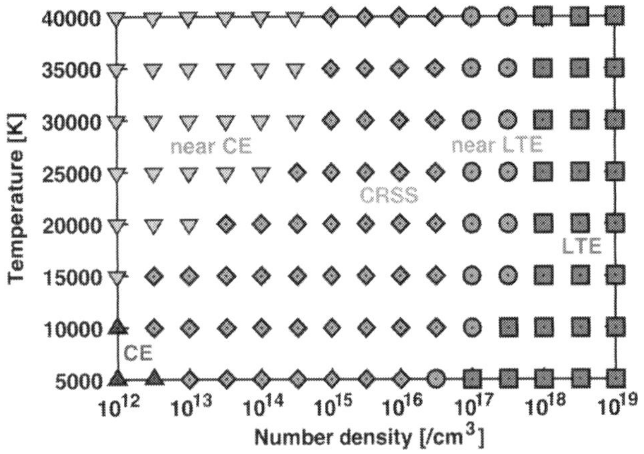

Figure 1. Summary of appropriate conditions for the air plasmas in the total number density range from $10^{12}/\mathrm{cm}^3$ to $10^{19}/\mathrm{cm}^3$ and the temperature range from 5,000 K to 40,000 K.

Note that, in this work, the quasi-steady state is defined as the numerical solution of the time-dependent rate equations at 1 μs with residuals of all time derivatives less than 10^{-8}. If time derivatives of number densities of the some excited states are not converged at 1 μs, the numerical time integration is continued until the residuals become lower than the threshold. Because long-lifetime excited species having a small number density are still emitting radiation in a CE or a near CE state. On the other hand, while in a near LTE state whose internal state distributions approach Boltzmann equilibrium function in the quasi-steady state, it would be required to solve the whole set of time-dependent rate equations when evaluating unsteady properties of air plasmas more accurately down to submicrosecond scale.

2.5. Fitting formula for radiative cooling with CR population

It is very helpful to develop a radiative cooling formula for nonequilibrium flow computation, since we can easily include the radiative energy loss with low computational cost. In this section, we present a curve fitted formula for radiative cooling and results of the total radiative emissivity $Q_{\rm rad}$ (see (2)–(5)) calculated from the collisional-radiative population that was shown in figure 1.

Obtained results are indicated in figure 2 in the total number density range from $10^{12}/\text{cm}^3$ to $10^{19}/\text{cm}^3$ and the temperatures from 5,000 K to 40,000 K. Vertical axis is the radiative emissivity in common logarithm scale, and each curve represents number densities from $10^{12}/\text{cm}^3$ to $10^{19}/\text{cm}^3$. For all the cases, the radiative emissivities are monotonically increasing function of temperature and total number density. We have fitted radiative emissivity $Q_{\rm rad}$ in terms of temperature T_e and total number density N as follows,

$$\log_{10} Q_{\rm rad}(T_e, N) = a + \frac{b}{Z} + \frac{c}{Y} + d\,Y + e\,Y^2, \tag{7}$$

where $Y = \log_{10} N$ and $Z = T_e/10000$. Fitting parameters are given by

$$a = -127.4,\ b = -5.931,\ c = 425.9,\ d = 10.16,\ e = -0.2148. \tag{8}$$

This curve fitted function means a modification to the gray gas radiation in accordance with the nonequilibrium population. By utilizing this relation, we can evaluate radiative energy loss without solving CR rate equations.

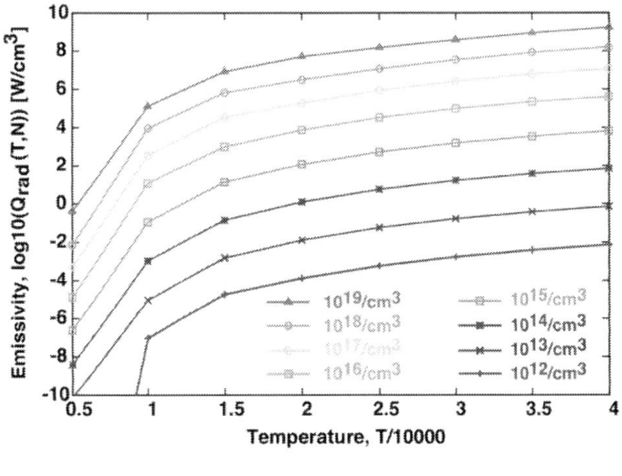

Figure 2. Total radiative emissivities computed from CR population in the total number density range from $10^{12}/\text{cm}^3$ to $10^{19}/\text{cm}^3$ and the temperature range from 5,000 K to 40,000 K.

3. Validation of the radiative cooling formula by laser-induced blast wave

Hereafter, to validate our fitting formula (7) for radiative cooling with CR population. We compute a laser-induced blast wave coupled with radiative energy loss and compare with experimentally observed shock wave and plasma front displacements [7]. The blast wave code is based on axisymmetric Navier-Stokes equations with eleven chemical species and chemical reactions including some ionization reactions. An amount of laser energy absorbed in flowfield is computed with the ray-tracing method [23]. Description of the code is summarized as follows.

3.1. Flow equations

The governing equations are the two-dimensional axisymmetric Navier-Stokes equations, which can be written as

$$\frac{\partial y\boldsymbol{Q}}{\partial t}+\frac{\partial y\left(\boldsymbol{F}-\boldsymbol{F_v}\right)}{\partial x}+\frac{\partial y\left(\boldsymbol{G}-\boldsymbol{G_v}\right)}{\partial y}+\boldsymbol{S}=y\boldsymbol{W}, \tag{9}$$

where $\boldsymbol{Q}$ is the conservative variables, $\boldsymbol{F}$ and $\boldsymbol{G}$ are the convective flux vector, $\boldsymbol{F_v}$ and $\boldsymbol{G_v}$ are the viscous flux vector, $\boldsymbol{S}$ is the vector from axisymmetric formulation, and $\boldsymbol{W}$ is the source term. The equation set (9) consists of ten species mass, momentum, total energy and vibrational, electronic excitation, and free electron energy conservation equations. The conservative variables $\boldsymbol{Q}$ thus becomes

$$\boldsymbol{Q}=\left(\tilde{\rho}_s,\rho_s,\rho u,\rho v,\epsilon,\epsilon_v+\epsilon_{ex}+\epsilon_{el}\right)^t, \tag{10}$$

where a charge neutrality is assumed to omit ρ_{e^-}. Here, $\tilde{\rho}_s$, ρ_s, u, v, ϵ, ϵ_v, ϵ_{ex} and ϵ_{el} are the density of elemental species s, the density of species s, the velocity in the x-direction, the velocity in the y-direction, the total energy per unit volume, the vibrational energy per unit volume and the electronic energy per unit volume, the free electron energy per unit volume. The total density ρ is given by the sum of the densities of all species. The viscosity of each species is given by a model for reacting flow developed by Blottner et al. [24]. The conductivities of translational-rotational temperature and vibrational temperature for each species are given by Eucken's relation [25]. Moreover, the viscosity and conductivity of the mixture gas are calculated using Wilke's semi-empirical mixing rule [26]. The mass diffusion coefficients are assumed to be constant for all species with a constant Schmitt number S_c of 0.5. Therefore the mass diffusion coefficients D_s for each species s can be given by $D_s=\mu/S_c\rho$, where μ is the viscosity of the mixture. We employ the reaction rate coefficients proposed in the two-temperature model of Park [27] as mentioned above. The vibrational-electronic-electron energy conservation equation accounts for the following eight physical processes.

(i) energy transfer between the translational and vibrational energy modes
(ii) vibrational energy source due to dissociation or recombination reactions
(iii) energy transfer between the molecular translational and electron translational energy modes
(iv) electron energy loss when a free electron strikes a neutral particle and frees another electron
(v) electronic excitation energy production
(vi) free electron energy source due to ionization reactions
(vii) free electron energy generated by laser absorption
(viii) energy loss by radiative cooling $Q_{\rm rad}$

The governing equations are discretized with a cell-centered finite volume method. We employ the AUSM-DV upwind scheme for obtaining the numerical flux [28]. A second-order spatial accuracy is obtained by the MUSCL approach [29]. An Euler explicit method is used for the time integration, while the diagonal point implicit method [30] is employed in the time integration of source terms. Note that a more detailed description about these numerical modeling can be found in previous a paper [31].

3.2. Numerical condition

At the beginning of the simulation, we make a focal spot which has a higher electron temperature because of the lack of the breakdown processes in our code. The electron temperature of the spot is distributed in a Gaussian form with FWHM of 10 μm and the peak value of $\simeq$ 7,000 K. A two-dimensional ray-tracing method is employed to find out the absorption rate by inverse bremsstrahlung of free electron. A laser pulse profile is assumed to be composed of a pulse and a tail part as indicated in figure 3. the properties of our laser are that the FWHM of a pulse part is 40 ns, the maximum power is 24 MW, the total laser energy is 10 J, and the wavelength of laser is 10.6 μm. These the properties are the same of the CO_2 laser that is employed in the experiment of laser-induced blast wave [7]. The simulation domain covers a 32 mm $\times$ 12 mm section with 161 $\times$ 61 grid points as shown in figure 4. The incident laser is divided into 160 rays and is traced from the right boundary with a Gaussian distribution whose FWHM is 40 mm. The divided rays are inclined to be focused at the distance of 25 mm from the boundary. Initial temperature and density are 288.15 K and 1.225 kg/m^3, the same as in the experimental conditions.

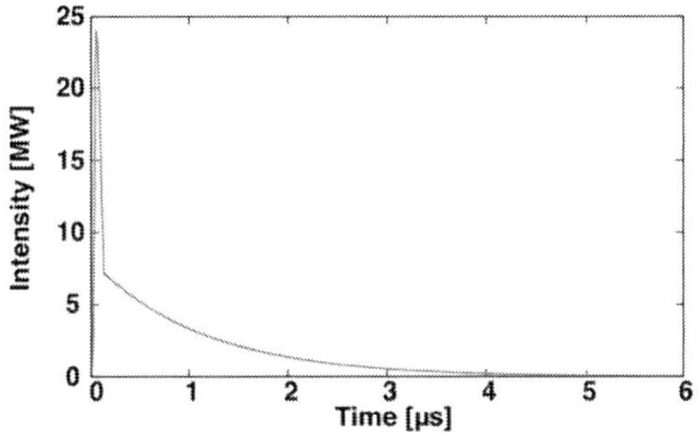

Figure 3. Incident laser profile.

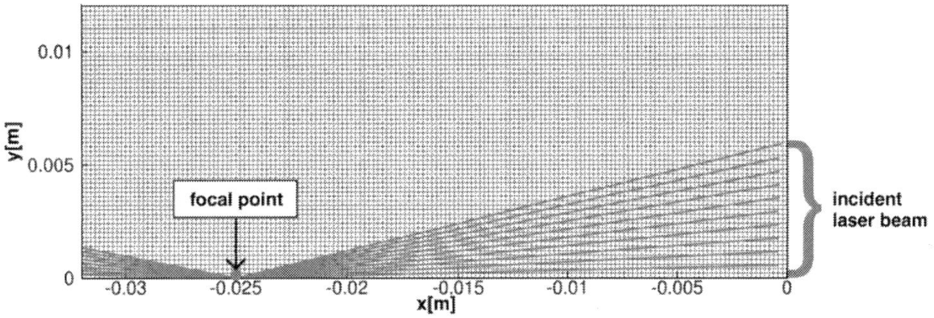

Figure 4. Computational domain and initial laser alignment.

3.3. Propagation speed of shock and ionization wave

Figures 5 (a)-(d) show the pressure (upper) and the electron number density (lower) contours at t = 1, 2, 3, and 6 μs after a laser beam irradiation. Seeds of free electrons generated by the initial ionization absorb laser energy at the focal point ($x = -0.025$ m). At first, the ionization wave propagates toward the incident direction of the laser, and the tear-drop-shaped blast wave is generated as shown in figure 5 (a). After that, the plasma region expands with a blast wave propagation, and the electron number density becomes lower. Therefore, more laser energy reaches the plasma region inside of blast wave, and the ionization wave propagates toward the direction of laser irradiation as indicated in figure 5 (b). From $t = 3$ μs to 6 μs, as indicated by figure 5 (c) and (d), the flowfield absorbs a little energy with low incident laser intensity as shown in figure 3. The total energy absorbed in the flowfield is 3.1 J, or 31 % of the total laser energy.

Figure 6 shows numerically and experimentally obtained results of the propagation speeds of the shock and ionization wave toward the direction of the incident laser pulse. We can find that

Journal of Physics: Conference Series **454** (2013) 012080 doi:10.1088/1742-6596/454/1/012080

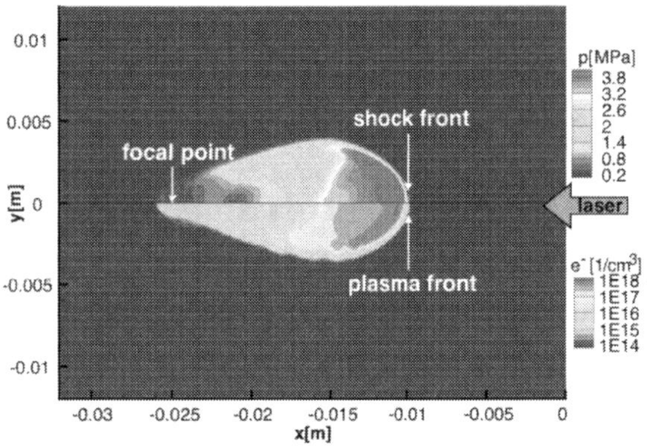

(a) $t = 1\ \mu$s

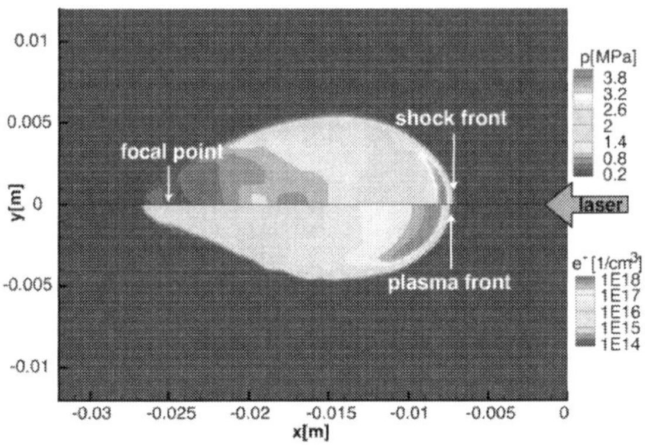

(b) $t = 2\ \mu$s

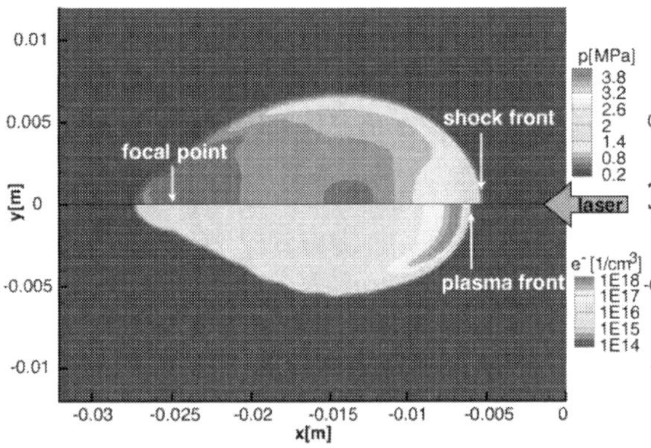

(c) $t = 3\ \mu$s

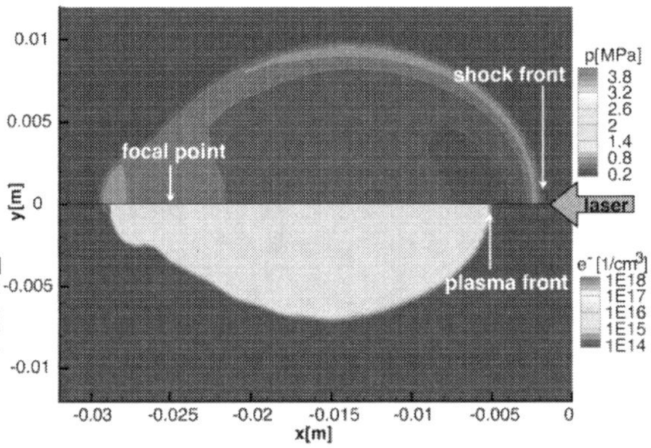

(d) $t = 6\ \mu$s

Figure 5. Time evolution of the pressure (upper) and the electron number density (lower) contours after the laser beam irradiation.

the speeds of the shock wave and the plasma front propagation agree well with experimental results except at an early stage of rapid expansion lasting until $t \simeq 0.4\ \mu$s. Moreover, each transition time from the configuration of laser-supported detonation (LSD) to laser-supported combustion (LSC) is almost the same. Additionally, a flowfield simulation with a gray gas radiation is also conducted to compare with the derived radiative cooling formula. Note that the computed optical depth along the symmetric axis at 3 μs after the irradiation of the laser pulse is about 0.1. Then some reabsorption of radiation emitted from the other site in the plasma region occurs but is neglected in this work. For the gray gas radiation $Q_{\rm rad} \equiv \kappa_p \sigma T_e^4$, the Planck mean absorption coefficient κ_p for equilibrium air plasma estimated by Taylor and Ali [32] is employed with the necessary modification of setting the values of κ_p to values ten times smaller than those of the original model, because their model is developed by assuming plasma of higher density (total number density $N \simeq 10^{19}/\text{cm}^3$) with respect to the density of the laser-induced plasma ($N \simeq 10^{18}/\text{cm}^3$). From the comparison of the non-LTE radiative cooling formula with the modified gray gas radiation, we see that the displacement of the plasma front is slightly different due to the deviation of population probabilities in each electronic excitation state, since the air density is high enough to achieve LTE as indicated in figure 1. The gray gas radiation model derived from LTE population could overestimate the radiative energy loss; therefore, we should use the curve fitted formula for radiative cooling with CR population to make reliable computational predictions.

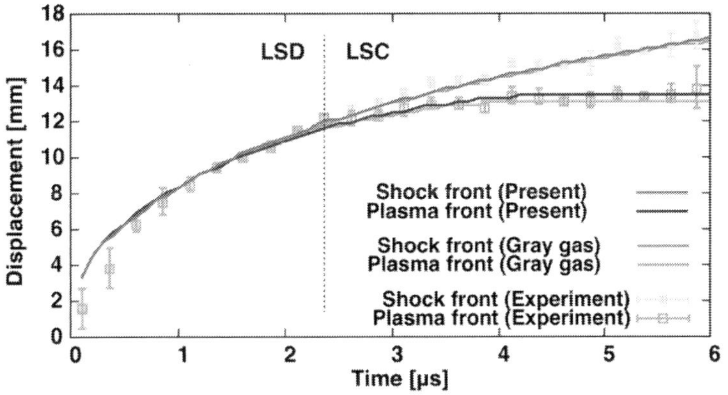

Figure 6. Comparison of numerically and experimentally obtained x–t diagrams of the shock and plasma front toward the direction of the incident laser pulse.

4. Conclusion

A fitting formula for radiative cooling with CR population for air plasma flowfield has been developed to estimate the radiative cooling rate with higher accuracy and lower computational cost. Population number densities were calculated from CR rate equations in order to evaluate the effects of nonequilibrium atomic and molecular processes. The CR model consists of fifteen species: e^-, N, N^+, N^{2+}, O, O^+, O^{2+}, O^-, N_2, N_2^+, NO, NO^+, O_2, O_2^+, and O_2^- with their major electronic excited states. Many elementary processes are integrated and applied to optically-thin plasmas in the number density range of $10^{12}/\text{cm}^3 \leq N \leq 10^{19}/\text{cm}^3$ and the temperature range of 300 K $\leq T \leq$ 40,000 K.

We have investigated the appropriate conditions for CRSS, LTE, and CE plasmas in the total number density range from $10^{12}/\text{cm}^3$ to $10^{19}/\text{cm}^3$ and at temperatures in the range from 5,000 K to 40,000 K. The results demonstrate that the LTE assumption is valid when the total number density exceeds $10^{17}/\text{cm}^3$. We should employ the CRSS model to analyze low-density plasmas in engineering applications.

Our results of the total radiative emissivity calculated from the CR population are fitted in terms of temperature and total number density. To validate this fitted radiation cooling formula, we have also computed a laser-induced blast wave propagation with radiative energy loss, and compared with experimentally observed shock and plasma front. We could fairly reproduce the thermochemical nonequilibrium flowfield for the blast wave induced by a pulse laser heating. From the comparison of the fitted radiative cooling formula with the corrected gray gas radiation assuming LTE, the displacement of the plasma front was slightly different due to the deviation of population probabilities in each electronic excitation state. The radiative cooling formula with CR model is an improvement on the gray gas radiation with nonequilibrium properties of population densities. Therefore, we can easily and appropriately evaluate the radiative energy loss by utilizing this fitting formula without solving detailed CR rate equations.

Acknowledgments

This work was supported by the Grant-in-Aid for Young Scientists (B), No. 24760657, from the Japan Society for the Promotion of Science (JSPS).

Journal of Physics: Conference Series **454** (2013) 012080 doi:10.1088/1742-6596/454/1/012080

References

[1] Park C 1990 *Nonequilibrium Hypersonic Aerothermodynamics* (New York: Wiley)
[2] Bultel A, Chéron B G, Bourdon A, Motapon O and Schneider I F 2006 *Physics of Plasmas* **13** 1–11
[3] Panesi M, Magin T, Bourdon A, Bultel A and Chazo O 2009 *Journal of Thermophysics and Heat Transfer* **23** 236–248
[4] Annaloro J, Morel V, Bultel A and Omaly P 2012 *Physics of Plasmas* **19** 1–15
[5] Sasoh A, Yu X, Ohtani T, Ogawa T, kawahara T, Kim S and Jeung I S 2004 AIAA Paper 2004-650
[6] Myrabo L N and Raizer Y P 1994 AIAA Paper 94-2451
[7] Mori K and Komurasaki K 2004 *Journal of Applied Physics* **95** 5979–5983
[8] Ogino Y, Ohnishi N and Sawada K 2008 *Journal of Physics: Conference Series* **112** 1–4
[9] Ogino Y and Ohnishi N 2011 *Shock Waves* **21** 289–299
[10] Bacri J, Lagreca M and Medani A 1982 *Physica* **113C** 403–418
[11] Saruhashi Y 2008 Master's thesis Department of Aerospace Engineering, Nagoya University
[12] Radhakrishnan K and Hindmarsh A C 1993 Tech. Rep. UCRL-ID-113855 LLNL
[13] Drawin H W 1967 Tech. Rep. EUR-CEA-FC-383 revised Fontenay-aux-Roses
[14] Drawin H W 1968 *Plasma Diagnostics* (Amsterdam: North Holland)
[15] Bacri J and Medani A 1980 *Physica* **101C** 399–409
[16] Teulet P, Sarrette J P and Gomes A M 1999 *Journal of Quantitative Spectroscopy & Radiative Transfer* **62** 549–569
[17] Park C 1993 *Journal of Thermophysics and Heat Transfer* **7** 385–398
[18] Nahar S N and Pradhan A K 1997 *Astrophysical Journal Supplement Series* **111** 339–355
[19] Nahar S N 1999 *Astrophysical Journal Supplement Series* **120** 131–145
[20] Soon W H and Kunc J A 1990 *Physical Review A* **41** 825–843
[21] Salzmann D 1998 *Atomic Physics in Hot Plasmas* (Oxford: Oxford University Press)
[22] Kunc J A and Soon W H 1989 *Physical Review A* **40** 5822–5843
[23] Edwards A L and Fleck Jr J A 1979 *Journal of Applied Physics* **50** 4307–4313
[24] Blottner F G, Johnson M and Ellis M 1971 Tech. Rep. SC-RR-70-754 Sandiana Laboratories
[25] Vincenti W G and Kruger C H 1965 *Introduction to Physical Gas Dynamics* (New York: Wiley)
[26] Wilke C R 1950 *Journal of Chemical Physics* **18** 517–519
[27] Park C 1986 AIAA paper 86-1347
[28] Wada Y and Liu M S 1994 AIAA Paper 94-0083
[29] van Leer B 1979 *Journal of Computational Physics* **32** 101–136
[30] Eberhardt S and Imlay S 1991 *Journal of Thermophysics and Heat Transfer* **5** 257–265
[31] Ohnishi N, Tate M and Ogino Y 2012 *Shock Waves* **22** 521–531
[32] Taylor R D and Ali A W 1987 *Journal of Quantitative Spectroscopy & Radiative Transfer* **38** 29–36

24th IUPAP Conference on Computational Physics (IUPAP-CCP 2012) IOP Publishing
Journal of Physics: Conference Series **454** (2013) 012081 doi:10.1088/1742-6596/454/1/012081

Acceleration of Feynman loop integrals in high-energy physics on many core GPUs

F Yuasa[1], T Ishikawa[1], N Hamaguchi[1], T Koike[2] and N Nakasato[3]
[1] High Energy Accelerator Research Organization (KEK), 1-1 Oho, Tsukuba, Ibaraki 305-0801, Japan
[2] Seikei University, Musashino, Tokyo 180-8633, Japan
[3] University of Aizu, Aizu-Wakamatsu, Fukushima 965-8580, Japan

E-mail: `fukuko.yuasa@kek.jp`

Abstract. The current and future colliders in high-energy physics require theorists to carry out a large scale computation for a precise comparison between experimental results and theoretical ones. In a perturbative approach several methods to evaluate Feynman loop integrals which appear in the theoretical calculation of cross-sections are well established in the one-loop level, however, more studies are necessary for higher-order levels. *Direct Computation Method* (DCM) is developed to evaluate multi-loop integrals. DCM is based on a combination of multi-dimensional numerical integration and extrapolation on a sequence of integrals. It is a fully numerical method and is applicable to a wide class of integrals with various physics parameters. The computation time depends on physics parameters and the topology of loop diagrams and it becomes longer for the two-loop integrals. In this paper we present our approach to the acceleration of the two-loop integrals by DCM on multiple GPU boards.

1. Introduction

Much of the success of the Standard Model in high-energy physics rests on the result of the various precision measurements[1, 2]. These precision measurements require the knowledge of higher order quantum corrections. In Feynman diagrammatic approach followed by most of the automated systems, the cross sections can be obtained by computing directly the squared of matrix elements or amplitudes. In order to include higher order corrections in the Standard model or beyond it is indispensable to handle evaluations of Feynman loop integrals.

The most general form of the scalar Feynman loop integral with L loops and n internal lines is given by

$$\mathcal{I} = \lim_{\epsilon \to 0+} \int \prod_{j=1}^{L} \frac{d^4 l_j}{i(2\pi)^4} \prod_{r=1}^{n} \frac{1}{D_r} \tag{1}$$

where l_j is the j-th loop momentum and we put the space-time dimension 4. Here

$$D_r = q_r^2 - m_r^2 + i\epsilon, \tag{2}$$

is the inverse of Feynman propagator with the momentum q_r and m_r the mass of the r-th particle. Using the generalized Feynman's identity,

$$\prod_{r=1}^{n} \frac{1}{D_r} = \Gamma(n) \int_0^1 \prod dx_r \frac{\delta(1 - \sum x_r)}{(\sum x_r D_r)^n}, \tag{3}$$

we find

$$\mathcal{I} = \left(\frac{1}{4\pi}\right)^{4L/2} \Gamma\left(n - \frac{4L}{2}\right) \times I, \tag{4}$$

where

$$I = (-1)^n \int_0^1 \prod_{i=1}^{n} dx_i \delta(1 - \sum x_i) \frac{C^{n-4(L+1)/2}}{(D - i\epsilon C)^{n-4L/2}}. \tag{5}$$

Here D function is a polynomial in Feynman parameters $\{x_i\}$ and it contains physics parameters such as external momenta and particle masses and C function is a polynomial in $\{x_i\}$. Both functions are determined by the topology of Feynman diagram. The details on how to get them are summarized in [3].

In the case of modern theoretical calculations in high-energy physics, loop integrals are performed with new mathematical methods and are based on an intense use of computer power to get theoretical predictions of cross-sections for particle reactions. Depending on the number of loops, legs, and physics parameters, it may take long computation time. In fact it is reported that the two-loop box (four legs) with arbitrary masses requires long computation time[3]. To reduce this practical problem, current hardware trends such as many-core/multi-core computation can be available. Recently GPU boards have been used in scientific applications which require a huge computation time. Also in high-energy physics successful computations on GPU have been reported in [4, 5]. In this paper, we show our approach to reduce the computation time of loop integrals on multiple GPUs and show its performance.

The plan of this paper is as follows; we explain numerical techniques of our program for loop integrals in Section 2. The environment for the acceleration of the computation and the results of the performance test are shown in Section 3. Section 4 gives summary of this paper.

2. Numerical Techniques

One of the difficult points for the numerical computation of loop integrals is vanishing denominators in (5). Our program code for loop integrals is a combination of numerical multi-dimensional integration and extrapolation in order to reduce the difficulty. It consists of the following three steps:

(i) Let ϵ in (5) be finite as $\epsilon_l = \epsilon_0/(const.)^l$, with $l = 0, 1, 2, \cdots$, where a starting ϵ, ϵ_0, and *constant* are positive numbers.

(ii) Evaluate the multi-dimensional integral I in (5) numerically. Thanks to ϵ_l we get finite value corresponding to each l. Repetition of integration gives a sequence of $I(\epsilon_l)$, $l = 0, 1, 2, \cdots$.

(iii) Extrapolate the sequence $I(\epsilon_l)$ to the limit and determine I.

We call above prescriptions *Direct Computation Method* (DCM). Please note that in the unphysical region extrapolation is not necessary and we can put $\epsilon = 0$ in (5) because singularities do not appear and the integration is not difficult in this region. On the other hand, singularities appear in the physical region and we should take the parameters for extrapolation, ϵ_0 and *constant*, appropriately corresponding to physics parameters such as masses. In current DCM, the way to define them automatically is not included and we take them empirically.

In DCM, the integration routine plays a central role. The numerical result of the integral should be as precise as possible for each ϵ_l, so that the extrapolation of $I(\epsilon_l)$ converges well. For multi-dimensional integration we can make use of any method when once it gives a good approximation. We have been using two different integration methods. One is an adaptive algorithm routine, `DQAGE`, in `QUADPACK`[6] and the other is *Double Exponential Formulas*[7], shortly `DE`. We call the combination of `DQAGE` and and the extrapolation technique DQ-DCM and call the combination of `DE` and the extrapolation technique DE-DCM.

We use Wynn's ε algorithm[8, 9, 10] for the extrapolation and we usually repeat the integration 15 times. Since computation time for the extrapolation of the sequence is negligible small compared to that for the integration, we do not go further about the extrapolation.

Throughout this paper, we concentrate exclusively on the acceleration of DE-DCM. Let us explain about DE formulas briefly. An integral with the interval (a, b) is given as

$$I = \int_a^b f(x)dx \tag{6}$$

and with $x = \phi(t)$, $a = \phi(-\infty)$ and $b = \phi(\infty)$.

$$I = \int_{-\infty}^{\infty} g(t)dt, \tag{7}$$

where

$$g(t) = f(\phi(t))\phi'(t). \tag{8}$$

Application of the trapezoidal rule with mesh size h gives

$$I_h = h \sum_{k=-\infty}^{\infty} f(\phi(kh))\phi'(kh). \tag{9}$$

Here note the infinite sum should be truncated appropriately as

$$I_h^N = h \sum_{k=-N_-}^{N_+} f(\phi(kh))\phi'(kh), \tag{10}$$

$$N = N_- + N_+ + 1, \tag{11}$$

where N is the number of function evaluation. In DE-DCM we take the following transformation

$$I = \int_0^1 f(x)dx, \tag{12}$$

$$\phi(t) = \frac{1}{2}(1 + \tanh(\frac{\pi}{2}\sinh(t))). \tag{13}$$

Through our experiences in DE-DCM[11], we point out following things:

- DE is effective for singularities appearing in the boundary of the integration domain,
- When singularities appear in the integration domain, we may split the integration domain,
- DE has been developed for the one-dimensional integration, we apply it in an iterative way,
- The parallelization of DE is straightforward.

2.1. *Two-loop self-energy integral*

In this paper, we take two-loop self-energy integrals (two legs) as an example because there is a rich literature on the analytic methods or semi-numerical methods such as [12, 13, 14] and we can compare numerical results by DE-DCM to analytic ones.

D and C functions corresponding to the diagram shown in Fig.1 are given as:

$$D = -s(x_5(x_1+x_3)(x_2+x_4) + (x_1+x_2)x_3x_4 + (x_3+x_4)x_1x_2) + C\tilde{M}^2, \quad (14)$$

$$C = (x_1+x_2+x_3+x_4)x_5 + (x_1+x_2)(x_3+x_4), \quad (15)$$

$$\tilde{M}^2 = \sum_{i=1}^{5} x_i m_i^2. \quad (16)$$

Here, m_i is a mass of an internal particle indicated Feynman parameter x_i and s is a physics parameter.

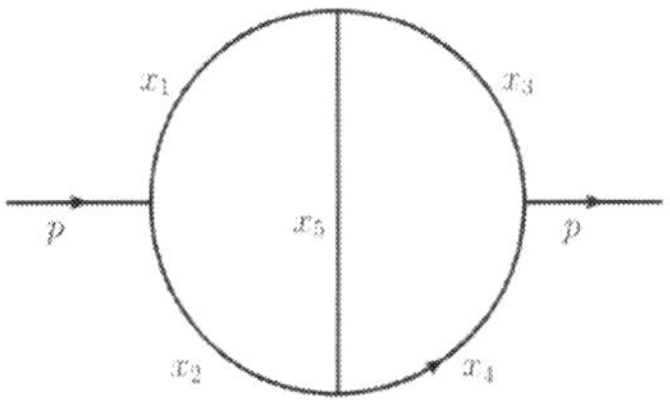

Figure 1. Two-loop self-energy diagram with five internal lines. x_i is Feynman parameter related to each line and p a physics parameter which determines s in (14) as p^2 is s.

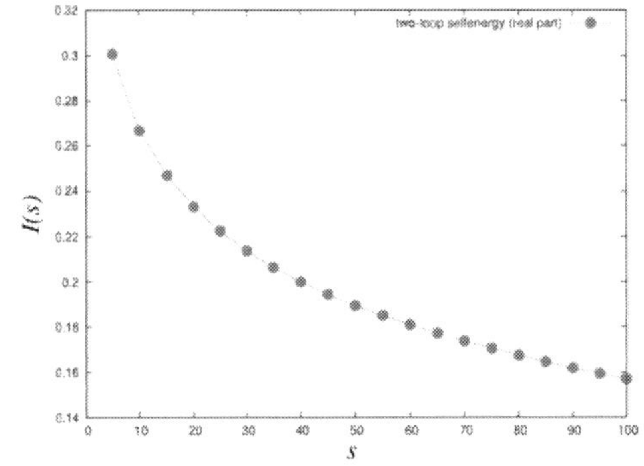

Figure 2. Numerical results of real part of two-loop self-energy integrals with $n = 1024$ and $h = 8.36181640625 \times 10^{-3}$. All filled circles are results after the extrapolation. The starting ϵ for the extrapolation, ϵ_0, is 1.2^{-15} and *constant* is 1.2. In this computation we did not split the integration domain.

The numerical results of the real part of the integrals for $4.0 \leq s \leq 10.0$ GeV^2 in the physical region with mass parameters as $m_1 = m_2 = m_5 = 10.0$ GeV and $m_3 = m_4 = 0.0$ GeV are plotted in Fig.2. For positive s, we repeated the integration to obtain $I(\epsilon_l)$for the extrapolation. For above physics parameter setting we don't have to split the integration domain because the values of masses given are not so small compared to s. All the computation shown here has been done in the double precision arithmetic because there is no infrared divergence in this case. For example, for the case $s = 10.0$ GeV^2, a numerical result is $0.26673229\times10^{-1} \pm 0.191\times10^{-8}$. The result using analytic formulas in [12] is 0.266732×10^{-1} and the agreement is excellent.

3. Acceleration of Computation

In this section firstly we explain the software environment and the modification of DE-DCM program code for the acceleration. Then we show the hardware environment for the computation time measurement and results.

3.1. Software Environment

We use `LSUMP` compiler[15] which has been developed by Nakasato *et al.* [15]. In addition to the `LSUMP`, we use `Goose`[16] software package which enables us parallel programming with `Goose directives` on the SIMD hardware accelerator. Both are originally developed for astrophysical N-body simulations and several very effective N-body simulation codes are available on GPU or special-purpose computers. With the combination of `LSUMP` and `Goose`, we only have to define the inner-most computation loop in the program. In order to use `LSUMP` and `Goose`, we transform the variables so as to make integration domain as $[0, 1]$ and then divide multiple computation loops for the multi-dimensional integration to double loops. Though the double precision and quadruple precision arithmetic are available with `LSUMP`[17] and `Goose`[18], all computation shown in this paper has been done in double precision arithmetic for the physics parameters given in Section 2.

3.2. Hardware Environment

We setup two kinds of hardware environment. One is a host computer (AMD Phenon II x6 1090T 3.2GHz) with two AMD/ATI HD5870 GPU boards and the other is a host computer (AMD FX-8150 3.6GHz) with four AMD/ATI HD6970 GPU boards. Parameters of each GPU board are shown in Table 1.

Table 1. Specification of each GPU board. "SP FP" stands for Single Precision floating-point arithmetic and "DP FP" Double Precision floating point, respectively.

	HD5870	HD6970
Clock [MHz]	850	880
number of cores	1600	1536
SP FP [GFLOP]	2740	2732
DP FP [GFLOP]	544	683

Table 2. Computation time for $s = 5.0$ GeV^2 with $N = 1024$ and $s = -1.0$ GeV^2 with $N = 2048$.

s	GPU	# of boards	time [sec]
5.0	HD5870	1	242.313
	HD5870	2	122.251
5.0	HD6970	1	221.059
	HD6970	2	110.715
	HD6970	3	74.519
	HD6970	4	55.638
-1.0	HD6970	4	882.288

3.3. Computation Time Measurement

We have measured the computation time using one to four GPU boards taking two-loop self-energy integral for $s = 5.0$ GeV^2 with $N = 1024$ and $h = 8.362 \times 10^{-3}$ as examples. The measured results are shown in Table 2. For positive s, in the physical region, we should repeat the integration to obtain the sequence $I(\epsilon_l)$ and extrapolate it to accomplish the loop integral. However, the values of computation time shown in Table 2 are time required for only one repetition for simplicity. Therefore the amount of time in total becomes 15 times longer in the practical computation. We also run the same program code compiled by `Goose` on a host computer using one core with same integration parameters for $s = 5.0$ GeV^2 and roughly speaking it takes about one day. For example on AMD Phenon II x6 1090T 3.2GHz, it took 79903.860 seconds.

In DE-DCM, the expected computation time for one repetition in both the unphysical region and the physical region is more or less the same and it depends on the value of N. For comparison we also show the computation time for $s = -1.0$ GeV^2 with $N = 2048$ and $h = 4.181 \times 10^{-3}$in Table 2 even though it is in the unphysical region. Please note that the computation time with $N = 2048$ is about 16 times longer compared to one with $N = 1024$, because the integration is 4-dimensional one.

4. Summary

We have been developing a fully numerical method for Feynman loop integrals, *Direct Computation Method* (DCM). DCM is based on the numerical integration and extrapolation method and it can handle the singularities due to the vanishing denominator in the expression of loop integrals. The numerical integration plays a central role in DCM and in this paper we used *Double Exponential Formulas* in an iterated way for the multi-dimensional integration. Since it is fully numerical, it can be applicable to multi-loop integrals with multi-leg and arbitrary physics parameters in principle. However, the computation time becomes longer for higher-order loops and reducing it is a crucial problem. In this paper we demonstrated the acceleration of the loop integrals taking two-loop self-energy integral as an example. In the performance test, it is clearly shown that the reduction of the computation time is excellent on multiple GPU boards. Concerning the further development we need to test various two-loop integrals with more legs and various physics parameters different from those in this paper.

Acknowledgments

We acknowledge the support from Grant-in-Aid for Scientific Research (No.24540292 and No. 23540328) of JSPS and the Large Scale Simulation Program No.12-07 of KEK. This work is supported in part by Strategic Programs for Innovative Research (SPIRE) Field 5 "The origin of matter and the universe". One of authors (F.Y.) wishes to thank Prof. E de Doncker for her continuous encouragement.

References

[1] D. Perret-Gallix, *Computational Particle Physics for Event Simulation and Analysis*, a talk in CCP2012.
[2] J. Blümlein, *Why precision*, arXiv:1205.4991v2 [hep-ph].
[3] F. Yuasa, E. de Doncker, N. Hamaguchi, T. Ishikawa, K. Kato, Y. Kurihara, J. Fujimoto, Y. Shimizu, *Numerical computation of two-loop box diagrams with masses*, Comput. Phys. Commun. 183 (2012) 2136.
[4] J. Kanzaki, *Monte Carlo integration on GPU*, DOI:10.1140/epjc/s10052-011-1559-8.
[5] M. Hayakawa, K.-I. Ishikawa, Y. Osaki, S. Takeda, S. Uno, N. Yamada, *Improving many flavor QCD simulations using multiple GPUs*, PoS Lattice2010:325,2010.
[6] R. Piessens, E. de Doncker, C. W. Ubelhuber and D. K. Kahaner, *QUADPACK, A Subroutine Package for Automatic Integration*, Springer Series in Computational Mathematics. Springer-Verlag, 1983.
[7] H. Takahasi and M. Mori, *Double Exponential Formulas for Numerical Integration*, RIMS Kyoto Univ. 9 (1974) 721.
[8] D. Shanks, *Non-linear transformations of divergent and slowly convergent sequences*, J. Math. Phys. 34 (1955) 1.
[9] P. Wynn, *On a device for calculating the $e_m(S_m)$ transformations*, Mathematical Tables Aids to Computing 10 (1956) 91.
[10] P. Wynn, *On the convergence and stability of the epsilon algorithm*, SIAM J. Numer. Anal. 3 (1966) 91.
[11] F. Yuasa and N. Hamaguchi, *Double Exponential Integration Method with algorithm for the Feynman Loop Integration Calculation* [in Japanese], 2008-HPC-114 No.6 pp. 31-36.
[12] D.J. Broadhurst, J. Fleischer and O.V. Tarasov, *Two-loop two-point functions with masses: asymptotic expansions and Taylor series, in any dimension*, Z.Phys. C60 (1993) 287.
[13] J. Fujimoto, Y. Shimizu, K. Kato and Y. Oyanagi, *Numerical Approach to Loop Integrals*, in Proc. of the Second International Workshop on Software Engineering, Artificial Intelligence and Expert Systems in High Energy and Nuclear Physics, January 13-18, 1992, l'Agelonde.
[14] G. Passarino and S. Uccirati, *Algebraic-numerical evaluation of Feynman diagrams: two-loop self-energies*, Nucl. Phys. B629 (2002) 97.
[15] N. Nakasato and J. Makino, *A Compiler for High Performance Computing With Many-Core Accelerators*, in: IEEE International Conference on Cluster Computing and Workshops, 2009, pp. 1-9.

[16] A. Kawai, T. Fukushige, N. Nakasato and T. Narumi, *Goose and DS-CUDA* [in Japanese], "ENSEMBLE", The Bulletin of The Molecular Simulation Society of Japan, Vol.14, No. 2, (2012.04.30) 81-84.
[17] N. Nakasato, T. Ishikasa, J. Makino and F. Yuasa, *Fast Quad-Precision Operations On Many-Core Accelerators* [in Japanese], Vol2009-HPC-121 No.39.
[18] T. Koike, *Speed-up of numerical Feynman loop integration by graphics processing unit* [in Japanese], unpublished Masters thesis, Seikei University, Musashino, Tokyo, Japan, 2012.

Multi-threaded adaptive extrapolation procedure for Feynman loop integrals in the physical region

E de Doncker[1], **F Yuasa**[2] **and R Assaf**[1]

[1] Department of Computer Science, Western Michigan University, Kalamazoo MI 49008, U. S.
[2] High Energy Accelerator Research Organization (KEK), Oho 1-1, Tsukuba, Ibaraki, 305-0801, Japan

E-mail: elise.dedoncker@wmich.edu, fukuko.yuasa@kek.jp, rida.assaf@wmich.edu

Abstract. Feynman loop integrals appear in higher order corrections of interaction cross section calculations in perturbative quantum field theory. The integrals are computationally intensive especially in view of singularities which may occur within the integration domain. For the treatment of threshold and infrared singularities we developed techniques using iterated (repeated) adaptive integration and extrapolation. In this paper we describe a shared memory parallelization and its application to one- and two-loop problems, by multi-threading in the outer integrations of the iterated integral. The implementation is layered over OpenMP and retains the adaptive procedure of the sequential method exactly. We give performance results for loop integrals associated with various types of diagrams including one-loop box, pentagon, two-loop self-energy and two-loop vertex diagrams.

1. Introduction

Feynman loop integrals are generally affected by non-integrable singularities, through vanishing denominators in the interior and/or at the boundaries of the integration domain. In order to handle singularities inside the domain, the value for the integral is calculated by introducing a parameter $i\delta$ in the integrand denominator, effectively moving the singularity into the complex plane, and by taking the limit of the integral as $\delta \to 0$. We make use of numerical iterated integration and a procedure for convergence acceleration or extrapolation of a sequence of integral values as δ decreases.

A loop integral for a diagram with L loops and N propagators is given in Feynman parameter space as

$$\mathcal{I} = \frac{\Gamma\left(N - \frac{nL}{2}\right)}{(4\pi)^{nL/2}} (-1)^N \int_{\mathcal{D}} \prod_{j=1}^{N} dx_j \, \delta(1 - \sum x_j) \frac{C^{N-n(L+1)/2}}{(D - i\,\delta\, C)^{N-nL/2}}, \qquad (1)$$

where C and D are polynomials in $x_1, x_2, \ldots, x_N$, determined by the topology of the corresponding diagram and the physical parameters. Here $\mathcal{D}$ represents the N-dimensional unit hypercube integration domain, and the delta function $\delta(1 - \sum x_j)$ reduces the integration to the $(N-1)$-dimensional unit simplex.

Loop integrals are generally divergent when denominators vanish in the integration region. We assume $\mathcal{I}$ does not suffer from other divergences, such as infrared (IR) divergence, and exists in the limit as the parameter $\delta \to 0$. To calculate the limit numerically [1], we generate a sequence of $\mathcal{I} = \mathcal{I}(\delta)$ for (geometrically) decreasing values of δ, and apply convergence acceleration or extrapolation to the limit with the ε-algorithm [2, 3].

We have implemented the basic method in DCM (*Direct Computational Method*) for loop integrals. The technique can achieve high accuracies compared to, e.g., Monte Carlo integration, and it has been shown to successfully handle some singular problems which are problematic for other adaptive multivariate integration programs. The obtained accuracy helps controlling the accumulation of numerical error in large-scale calculations, where the precision of the end-results is fundamental for comparisons between theoretical and experimental results.

In previous work [4] we introduced a technique utilizing multi-threading for a parallel computation of the integrals. In this paper we add a parallelization on possibly multiple levels of the iterated integral, and focus further on testing the application to loop integrals. The implementation is layered over OpenMP [5] and retains the adaptive procedure of the sequential method exactly. Thus, each parallel execution performs the same subdivisions and function evaluations as the corresponding sequential run and can be regenerated; which is an important issue in the verification and debugging of parallel programs, and for avoiding useless parallel work. The parallelization is with respect to the function evaluations in the iterated method, where each function evaluation is itself an integral (except at the lowest level). This is important to guarantee a large enough granularity for the parallel procedure.

Concepts underlying automatic adaptive integration are reviewed in section 2 below. Their incorporation within an extrapolation strategy as implemented in DCM is outlined in section 3. section 4 describes our parallelization of iterated integration for a multi-threaded environment. Applications to the computation of Feynman loop integrals are given in section 5, with timing results using the OpenMP Application Program Interface (API).

2. Automatic adaptive integration

An *automatic* integration procedure can be considered as a *black-box* approach to produce (as outputs) an approximation $Q(f)$ to an integral

$$If = \int_{\mathcal{D}} f(\vec{x})\, d\vec{x} \tag{2}$$

and an error estimate $\mathcal{E}f$ of the actual error $Ef = |Qf - If|$, in order to satisfy an accuracy requirement of the form

$$|\, Qf - If \,| \;\leq\; \mathcal{E}f \;\leq\; \max\{\, t_a \,,\, t_r\, |If|\,\}, \tag{3}$$

where the integrand function f, region $\mathcal{D}$ and (absolute/relative) error tolerances t_a and t_r, respectively, are specified as part of the input.

A versatile type of algorithm to implement the black-box approach performs an adaptive partitioning of the integration region as shown in figure 1. At each step, a region is subdivided, integral and error estimates are computed over the subregions, and the overall result and error estimate are updated. Various strategies are possible for the selection of the region to be subdivided at each step. A *global adaptive* strategy maintains a priority queue on the subregion collection (e.g., a linked list or a heap), keyed with the local error estimates of the subregions, and selects the region with the highest (absolute) error estimate for subdivision at each step. As a result of the region selection in adaptive procedures, sample points tend to be concentrated in the vicinity of irregular behavior such as singularities, discontinuities, peaks or ridges and troughs of the integrand function.

Examples of adaptive integrators include the 1D adaptive programs of the QUADPACK [6] package, the program DCUHRE [7] for multivariate integration over a cube, and programs in CUBPACK [8] for adaptive integration over cubes and simplices.

```
Evaluate initial region and update results
Initialize priority queue with initial region
while (evaluation limit not reached and
      estimated error too large)
    Retrieve region from priority queue
    Split region into subregions
    Evaluate new subregions and update results
    Insert new subregions into priority queue
```

Figure 1. Adaptive Integration Meta-Algorithm

The local integral (over a subregion/interval) is approximated by a (cubature/quadrature) rule which is a linear combination of function values, of the form $\sum_{k=1}^{K} w_k f(x_1^{(k)}, x_2^{(k)}, \ldots, x_N^{(k)})$. By evaluating more than one rule over the subregion, a local error estimate can be obtained as a function of the difference between local integral approximations. For example, the (1D) program DQAGE of the QUADPACK package uses a pair of approximations, given by an r-point Gauss rule and the interlacing $(2r+1)$-point Kronrod rule. The QUADPACK user has the option of choosing one of the pairs with $r = 7, 10, 15, 20, 25$ or 30. The Kronrod rule supplies the local integral approximation, and the Gauss rule (together with the Kronrod rule) serves to obtain the local error estimate. The r-point Gauss rule is of polynomial degree of accuracy $2r-1$ (i.e., it integrates all polynomials of degrees $0, \cdots, 2r-1$ exactly, and there are polynomials of degree $2r$ which are not integrated exactly). The Kronrod rule is of polynomial degree $3r+1$ if r is even, and of degree $3r+2$ when r is odd (because of symmetry). The multivariate program DCUHRE applies an embedded sequence of cubature rules [9] for the local integral and error approximations over a subregion.

3. Integration and extrapolation strategy

The infinitesimal parameter $i\delta$ in the denominator of (1) prevents the integral from diverging and in that sense plays the role of a regulator. We consider the integral as a function of δ and construct a sequence of approximations to $\mathcal{I} = \mathcal{I}(\delta)$ for a decreasing sequence of δ, in order to perform an extrapolation to the limit as $\delta \to 0$. Methods for an extrapolation of the sequence rely on the existence of an asymptotic expansion $I(\delta) \sim I + a_1\varphi_1(\delta) + a_2\varphi_2(\delta) + \cdots$, as $\delta \to 0$. Linear or nonlinear extrapolation methods can be explored under certain conditions on the functions $\varphi_\ell(\delta)$ (see, e.g., [1]). DCM implements a nonlinear extrapolation or convergence acceleration with the ε-algorithm of Wynn [2, 3], which can be applied under more general conditions than linear extrapolation.

A schematic view of the program flow of DCM is given in figure 2 [10]. Note that $\delta = 0$ is valid in cases with unphysical kinematics, that is, where no divergence appears and no extrapolation is required, so the multi-dimensional integration only can be carried out.

For the numerical integration, we use iterated integration with the QUADPACK programs DQAGE or DQAGSE [6]. The extrapolation is applied to a sequence of $\mathcal{I}(\delta_j)$ computed for a geometric progression of δ_j, such as $c(1.2^{-j}), j \geq 0$ where c is a constant. We have used DCM for automatic integration without explicit information about the location or nature of the singularity, for various cases of one- and two-loop diagrams including 3-point (*vertex*), 4-point (*box*) [1, 11, 12, 13], 5-point (*pentagon*) and, after reductions, 6-point (*hexagon*) diagrams [14]; as well as two-loop *self-energy* [15], *double box*, *ladder* and *crossed vertex* [16, 17, 18, 19, 20, 21] diagrams with masses. The techniques were further incorporated for semi-automatic calculations

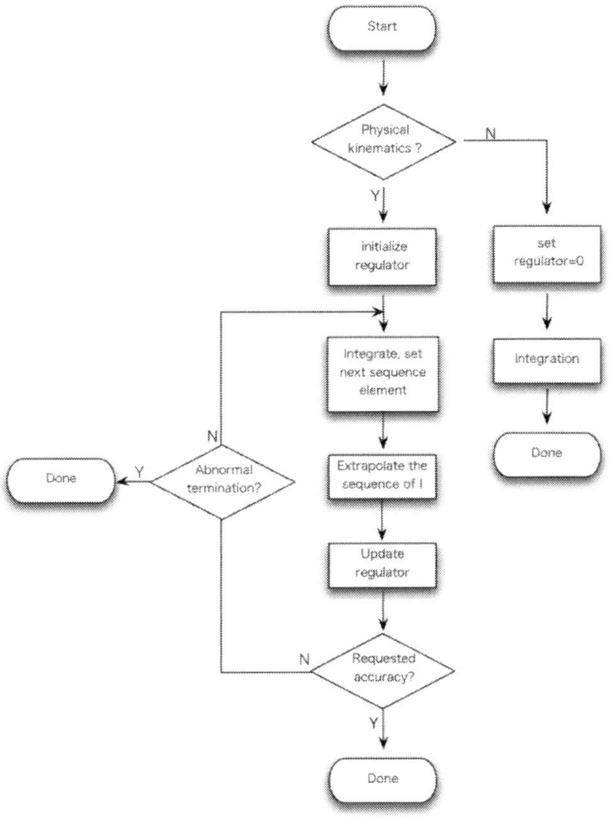

Figure 2. Program flow of DCM

of loop integrals with infrared divergences, in [18, 22].

4. Parallel numerical iterated integration

4.1. Iterated integration

In this paper we are dealing with iterated 1D integration over a finite d-dimensional product region, i.e., the integral If in (2) can be written as

$$If = \int_{\alpha_1}^{\beta_1} dx_1 \int_{\alpha_2}^{\beta_2} dx_2 \ldots \int_{\alpha_d}^{\beta_d} dx_d \; f(x_1, x_2, \ldots, x_d), \tag{4}$$

and the limits of integration may in general be functions, $\alpha_j = \alpha_j(x_1, x_2, \ldots, x_{j-1})$ and $\beta_j = \beta_j(x_1, x_2, \ldots, x_{j-1})$. The integration over the interval $[\alpha_j, \beta_j]$ will be performed with a 1D adaptive integration code, and different 1D integration programs may be applied in different directions $1 \le j \le d$.

If an interval $[a, b]$ arises in the subdivision of $[\alpha_j, \beta_j]$ for $1 \le j < d$, then the local integral approximation over $[a, b]$ is of the form

$$\int_a^b dx_j \; F(c_1, \ldots, c_{j-1}, x_j) \;\approx\; \sum_{k=1}^{K} w_k F(c_1, \ldots, c_{j-1}, x^{(k)}), \tag{5}$$

where the w_k and $x^{(k)}, 1 \leq k < K$, are the weights and abscissae of the local rule scaled to the interval $[a, b]$ and applied in the x_j-direction. For $j = 1$ this is the outer integration direction. The function evaluation

$$F(c_1, \ldots, c_{j-1}, x^{(k)}) = \int_{\alpha_{j+1}}^{\beta_{j+1}} dx_{j+1} \ldots \int_{\alpha_d}^{\beta_d} dx_d \, f(c_1, \ldots, c_{j-1}, x^{(k)}, x_{j+1}, \ldots, x_d), \qquad 1 \leq k \leq K, \tag{6}$$

is itself an integral in the $x_{j+1}, \ldots, x_d$-directions, and is computed by the method(s) for the inner integrations. For $j = d$, (6) is the evaluation of the integrand function

$$F(c_1, \ldots, c_{d-1}, x^{(k)}) = f(c_1, \ldots, c_{d-1}, x^{(k)}).$$

Note that the error incurred in the inner integration is subject to an error control condition of the form (3) and will contribute to the overall integration error. Work on the integration error interface is reported in [23, 24, 25].

Subsequently, in section 5, we give results obtained with 1D repeated integration by the program DQAGE from QUADPACK, where the local integration is performed with the (7, 15)- or the (10, 21)-points Gauss-Kronrod pairs. We will refer to the latter as GK15 and GK21, respectively.

The inner integrals given by (6), which appear in the local rule sum of (5), are independent and can thus be evaluated in parallel, by multiple threads. In previous work [4, 26] we reported results of this parallelization applied to the outer direction of integration using OpenMP. Important properties of this method include that: 1) the granularity of the parallel integration is large, especially when the inner integrals given by (6) are of dimension greater than two; 2) apart from possibly the order of the summation in the local rule evaluation, the parallel calculation is the same as the sequential evaluation.

Nesting of parallel constructs is also feasible, for a parallel evaluation of the rule in multiple coordinate directions of the iterated integral. The current implementation allows for a nested parallelization in the outer and the next to outer level (corresponding to the x_1 and x_2 directions in the iterated integral (4)); this can be extended to more or different levels. This feature is beneficial when the inner integral evaluations at the x_1-level are very time-consuming, so the work can be further distributed to new threads, on systems where a large number of threads are available.

5. Results for Feynman loop integrals

We implemented the methods of section 4 as a multi-threaded application layered over OPENMP [5], and report timing results on: a multi-core Intel workstation with Xeon X5680 @3.33GHz, 6-core (12 logical)-core CPU (*"minamivt005"*); the Hitachi SR16000/M1 with POWER7 @3.83GHz, 32-core processors at KEK; and the Intel cluster (*"thor"*) at WMU with Xeon E5-2670 processors, 16 cores per node.

The compilers used were: on *minamivt005*, the Intel *ifort* Fortran XE compiler with flag *-openmp*; the f90 compiler with flag *-omp* on SR16000; and the gfortran compiler with flag *-fopenmp* on *thor*. Furthermore we are working on a C version of the package, which is being tested with gcc *-fopenmp*.

In all cases the main loop in the rule routines DQK15 and DQK21 of QUADPACK was parallelized (for the outer two integrations), by means of an OpenMP *omp parallel do* for the Gauss and Kronrod rule evaluations. On comparing loop schedules, we found that the *dynamic* loop schedule outperforms the default *static* schedule in selected test problems, on *minamivt005* and on *thor*. In the *static* schedule, the participating threads are assigned their loop iterations at the beginning of the loop, in a round-robin fashion; thus the assigned iterations are determined

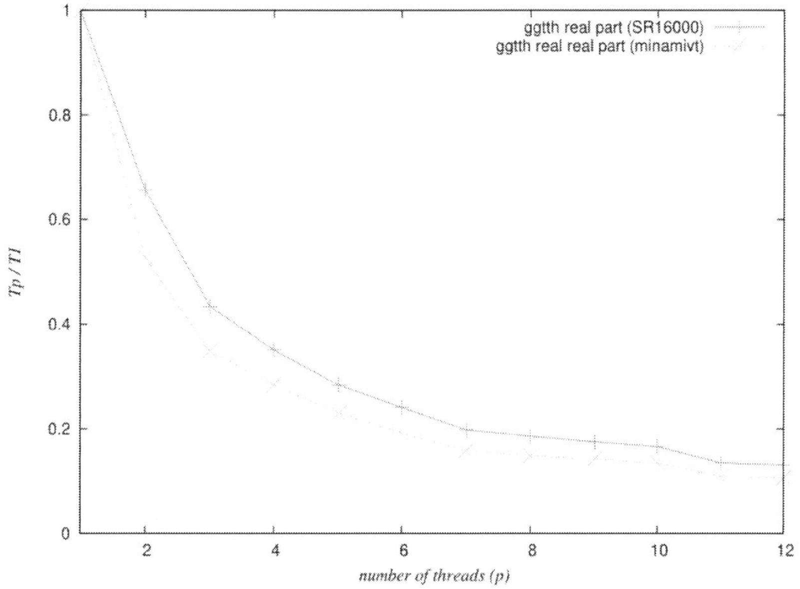

Figure 3. ρ_p vs. # threads p on SR16000 and *minamivt005* for $\gamma\gamma \to t\,\bar{t}\,H$ (1-loop pentagon)

only by the thread ID and the total number of iterations. The *dynamic* schedule adjusts the assignments at runtime. The threads can request more work at the completion of their previously assigned iterations, which is beneficial for load balancing in cases where the integrand behavior over the subregion is non-uniform. On SR16000 the *dynamic* schedule is not available, and neither is nested loop scheduling.

In [4] we reported timings for a one-loop non-scalar box problem, $M_4(f, g; \delta)$ from [27]. Presently we give timings for loop integral computations corresponding to the one-loop pentagon, two-loop self-energy, and two-loop ladder vertex diagrams and parameters specified below. The running times are total elapsed times spent in the integration code. The time for the extrapolation is negligible compared to the integration time.

For the one-loop pentagon diagram corresponding to the interaction $\gamma\gamma \to t\bar{t}H$, the integral (1) with $L = 1$, $N = 5$, $n = 4$, reduces to

$$\mathcal{I} = -2\int_0^1 dx \int_0^{1-x} dy \int_0^{1-x-y} dz \int_0^{1-x-y-u} du \,\frac{C}{(D - i\delta)^3}$$

via the simple transformation $x_1 = 1$, $x_2 = 1 - x - y - z - u$, $x_3 = y$, $x_4 = z$, $x_5 = u$ (and omitting the factor $\frac{1}{(4\pi)^{nL/2}}$). Here

$$\begin{aligned} D &= x_1 m_1^2 + x_2 m_2^2 + x_3 m_3^2 + x_4 m_4^2 + x_5 m_5^2 - x_1 x_2 s_1 - x_2 x_3 s_2 - x_3 x_4 s_3 \\ &\quad - x_4 x_5 s_4 - x_5 x_1 s_5 - x_1 x_3 s_{12} - x_2 x_4 s_{23} - x_3 x_5 s_{34} - x_4 x_1 s_{45} - x_5 x_2 s_{51}, \\ C &= 1, \end{aligned} \tag{7}$$

and the m_j are masses, the p_j are external momenta, $s_j = p_j^2$ and $s_{ij\cdots} = (p_i + p_j + \ldots)^2$, $0 \leq i, j \leq 5$. Figure 3 displays the parallel running time T_p scaled by the sequential time T_1, i.e., $\rho_p = T_p/T_1$ as a function of the number of threads p, on *minamivt* and SR16000. The parallel time for the integration decreases significantly particularly on the Intel system (from $T_1 = 2726$ sec to $T_{12} = 287$ sec, so $\rho_{12} \approx 0.105$). The extrapolation converges to about 5 digits.

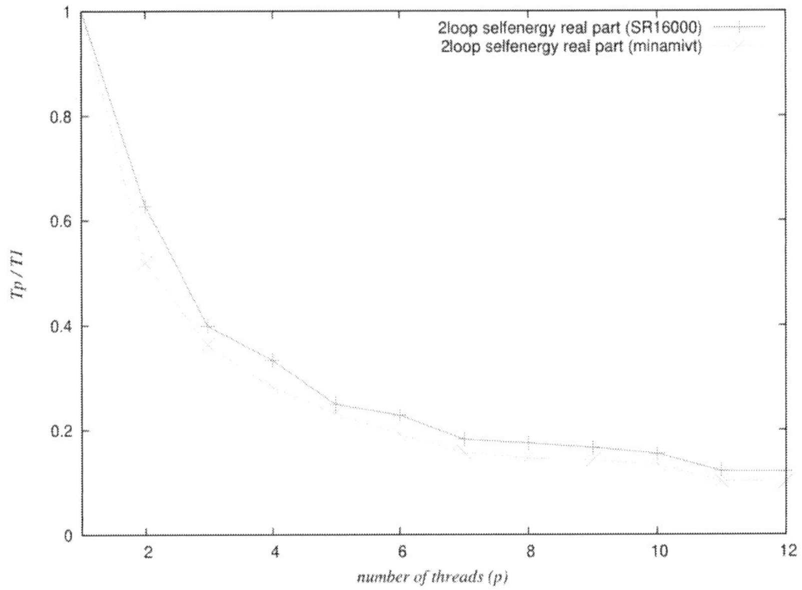

Figure 4. ρ_p vs. # threads p on SR16000 and *minamivt005* for 2-loop self-energy diagram

Two-loop self-energy integral results are reported for the integral

$$\mathcal{I} = \int_0^1 dx_1\, dx_2\, dx_3\, dx_4\, dx_5\, \delta(1 - \sum_{i=1}^{5} x_i)\, \frac{1}{CD},$$

with

$$\begin{aligned} D &= -k^2(x_5(x_1+x_3)(x_2+x_4) + (x_1+x_2)x_3x_4 + (x_3+x_4)x_1x_2) + C\tilde{M}^2, \\ C &= (x_1+x_2+x_3+x_4)x_5 + (x_1+x_2)(x_3+x_4), \\ \tilde{M}^2 &= \sum_{1=1}^{5} x_i m_i^2. \end{aligned}$$

The masses are $m_1 = m_2 = m_3 = m_4 = m_t$ (= 150 GeV, top quark mass), and $m_5 = m_Z$ (= 91.17 GeV, Z-boson mass). The integral is considered for an energy parameter $k^2 = 45000$ with $k^2/m_t^2 = 2$ and can be evaluated without extrapolation, in sequential time $T_1 \approx 27$ seconds on *minamivt005*. Figure 4 displays ρ_p as a function of the number of threads p on *minamivt* and SR16000.

The integral for the two-loop ladder vertex is (1) with $N = 6$, $n = 4$, and C, D given by

$$\begin{aligned} D &= -C(x_1(p_1^2 - m_1^2) + x_2(p_2^2 - m_2^2) - x_3m_3^2 + x_4(p_1^2 - m_4^2) + x_5(p_2^2 - m_5^2) - x_6m_6^2) \\ &+ C_1(x_5^2p_2^2 + x_4^2p_1^2 - x_4x_5(p_3^2 - p_1^2 - p_2^2)) + C_2(x_2^2p_2^2 + x_1^2p_1^2 - x_1x_2(p_3^2 - p_1^2 - p_2^2)) \\ &+ 2x_2x_3x_5p_2^2 + 2x_1x_3x_4p_1^2 - x_3(x_2x_4 + x_1x_5)(p_3^2 - p_1^2 - p_2^2), \\ x_6 &= 1 - x_1 - x_2 - x_3 - x_4 - x_5, \\ C_1 &= x_1 + x_2 + x_3, \\ C_2 &= 1 - x_1 - x_2, \\ C &= x_3(1 - x_1 - x_2 - x_3) + (x_1 + x_2)(1 - x_1 - x_2). \end{aligned} \quad (8)$$

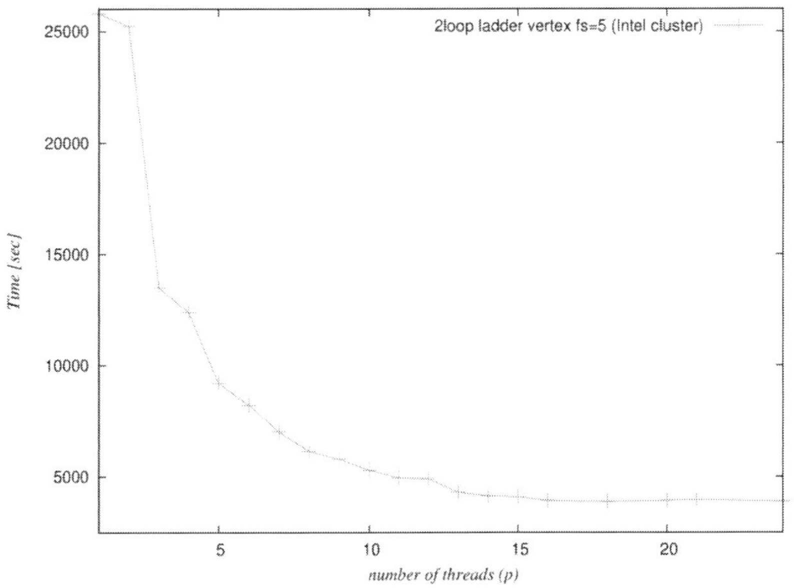

Figure 5. Time T_p vs. # threads p on WMU Intel cluster for 2-loop ladder vertex, $k^2/m_t^2 = 5$

The parameters are $m_1 = m_2 = m_4 = m_5 = m_t$ (= 150 GeV, top quark mass), and $m_3 = m_6 = m_Z$ (= 91.17 GeV, Z-boson mass). Figure 5 gives the running time T_p (on *thor*) for the real part integral (in seconds) as a function of the number of threads p, for parameter $k^2 = 112500$ with $k^2/m_t^2 = 5$ (denoted by $fs = 5$ in the key of figure 5). It is shown that the parallelization yields a spectacular savings in time for this compute-intensive problem (from $T_1 = 25800$ sec to $T_{24} = 3897$ sec).

Figure 6 gives corresponding times for the case $k^2/m_t^2 = 1$ which is part of the unphysical region for this problem; it is thus obtained without extrapolation by setting $\delta = 0$. The two curves in figure 6 represent the computations carried out with the GK15 and GK21 Gauss-Kronrod pairs in the x_1 dimension; the pair GK15 is used in the other dimensions. The (5D) integral is computed using the transformation derived in [1].

6. Concluding remarks

In addition to the lowest order or tree level, higher order corrections are required for an accurate theoretical prediction of the cross section of particle interactions, and for checking its agreement with the data observed at colliders. Feynman loop diagrams need to be taken into account, necessitating the calculation of *loop integrals*. While packages based on symbolic integration are available for one-loop integrals, symbolic reductions may lead to large sets of integrals, and analytic integration is not generally possible. The intensive nature of the direct computation performed by DCM motivates a parallelization of the individual problems, in particular for 3D and higher-dimensional integrals.

We have developed a multi-core parallelization on one or more function evaluation levels in the iterated integration procedure. In this paper we give timing results of the method implemented in OpenMP, which show good parallel performance for problems executed on a Xeon X5680, 6-(12 logical)-core CPU workstation, and furthermore on the (Hitachi) SR16000 system at KEK (Japan), and the Intel cluster (based on Xeon E5-2670 processors with 16 cores per node) at WMU. The Intel cluster at WMU and the SR16000 are suited for multi-threaded computations on each node and message passing between nodes. Future plans include hybrid (distributed

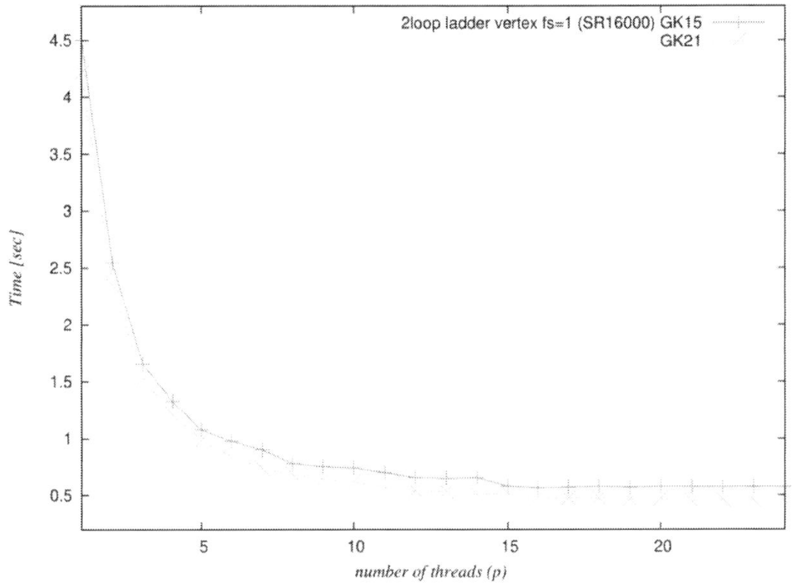

Figure 6. Time T_p vs. # threads p on SR16000 for 2-loop ladder vertex, $k^2/m_t^2 = 1$

shared memory) parallelizations using OpenMP and MPI.

Our future work will naturally involve implementations and testing for other and more computational intensive loop integrals.

Acknowledgments

We acknowledge the support from the National Science Foundation under Award Number 1126438. This work is further supported by Grant-in-Aid for Scientific Research (24540292) of JSPS and the Large Scale Simulation Program No.12-07 of KEK. We also want to thank PhD student Vasilije Perovic at WMU, who generated a first version of the C code.

References

[1] de Doncker E, Shimizu Y, Fujimoto J and Yuasa F 2004 *Computer Physics Communications* **159** 145–156
[2] Shanks D 1955 *J. Math. and Phys.* **34** 1–42
[3] Wynn P 1956 *Mathematical Tables and Aids to Computing* **10** 91–96
[4] de Doncker E and Yuasa F 2012 *IUPAP C20 Conference on Computational Physics (CCP 2011)* vol 402 (The Journal of Physics: Conference Series Dec. 2012)
[5] OpenMP web site, http://www.openmp.org
[6] Piessens R, de Doncker E, Überhuber C W and Kahaner D K 1983 *QUADPACK, A Subroutine Package for Automatic Integration* Springer Series in Computational Mathematics (Springer-Verlag)
[7] Berntsen J, Espelid T O and Genz A 1991 *ACM Trans. Math. Softw.* **17** 452–456
[8] Cools R and Haegemans A 2003 *ACM Transactions on Mathematical Software* **29** 287–296
[9] Berntsen J, Espelid T O and Genz A 1991 *ACM Trans. Math. Softw.* **17** 437–451
[10] de Doncker E and Yuasa F 2012 *Measurements in Quantum Mechanics* ISBN 978–953–51–0058–4
[11] de Doncker E, Shimizu Y, Fujimoto J, Yuasa F, Cucos L and Van Voorst J 2004 *Nuclear Instruments and Methods in Physics Research A* **539** 269–273 hep-ph/0405098
[12] de Doncker E, Shimizu Y, Fujimoto J and Yuasa F 2007 *PAMM - Wiley InterScience Journal* **7**
[13] Yuasa F, de Doncker E, Fujimoto J, Hamaguchi N, Ishikawa T and Shimizu Y 2007 *XI Adv. Comp. and Anal. Tech. in Phys. Res.* PoS (ACAT07) 087, arXiv:0709.0777v2 [hep-ph]
[14] de Doncker E, Fujimoto J, Kurihara Y, Hamaguchi N, Ishikawa T, Shimizu Y and Yuasa F 2010 *XIV Adv. Comp. and Anal. Tech. in Phys. Res.* PoS (ACAT10) 073

[15] Yuasa F, Ishikawa T, Fujimoto J, Hamaguchi N, de Doncker E and Shimizu Y 2008 *XII Adv. Comp. and Anal. Tech. in Phys. Res.* PoS (ACAT08) 122; arXiv:0904.2823
[16] de Doncker E, Shimizu Y, Fujimoto J and Yuasa F 2006 *LoopFest V, Stanford Linear Accelerator Center* http://www-conf.slac.stanford.edu/loopfestv/proc/present/DEDONCKER.pdf
[17] de Doncker E, Fujimoto J, Hamaguchi N, Ishikawa T, Kurihara Y, Shimizu Y and Yuasa F 2010 *Springer Lecture Notes in Computer Science (LNCS)* **6017** 139–154
[18] de Doncker E, Fujimoto J, Hamaguchi N, Ishikawa T, Kurihara Y, Ljucovic M, Shimizu Y and Yuasa F 2010 Extrapolation algorithms for infrared divergent integrals arXiv:hep-ph/1110.3587; PoS (CPP2010)011
[19] Yuasa F, Ishikawa T, Kurihara Y, Fujimoto J, Shimizu Y, Hamaguchi N, de Doncker E and Kato K 2010 Numerical approach to calculation of Feynman loop integrals arXiv:1109.4213v1 [hep-ph]; PoS (CPP2010)011
[20] de Doncker E, Fujimoto J, Hamaguchi N, Ishikawa T, Kurihara Y, Shimizu Y and Yuasa F 2011 *Journal of Computational Science (JoCS)* doi:10.1016/j.jocs.2011.06.003
[21] Yuasa F, de Doncker E, Hamaguchi N, Ishikawa T, Kato K, Kurihara Y and Shimizu Y 2012 *Journal Computer Physics Communications* **183** 2136–2144
[22] de Doncker E, Yuasa F and Kurihara Y 2012 *XV Adv. Comp. and Anal. Tech. in Phys. Res.* vol 368 (The Journal of Physics: Conference Series)
[23] Fritsch F N, Kahaner D K and Lyness J N 1981 *ACM TOMS* **7** 46–75
[24] Kahaner D, Moler C and Nash S 1988 *Numerical Methods and Software* (Prentice Hall)
[25] de Doncker E and Kaugars K 2010 *Procedia Computer Science* **1** 117–124
[26] de Doncker E and Yuasa F 2012 *Int. Conf. on Math. Modeling in Phys. Sciences*
[27] de Doncker E, Kaugars K, Cucos L and Zanny R 2001 *Proc. of Computational Particle Physics Symposium (CPP 2001)* pp 110–119

Journal of Physics: Conference Series **454** (2013) 012083 doi:10.1088/1742-6596/454/1/012083

Exact partition functions of a polymer on a square lattice up to chain length 38

Jae Hwan Lee[1], **Seung-Yeon Kim**[2] **and Julian Lee**[1]

[1]School of Systems Biomedical Science and Department of Bioinformatics and Life Science, Soongsil University, Seoul 156-743, Korea

[2]School of Liberal Arts and Sciences, Korea National University of Transportation, Chungju 380-702, Korea

E-mail: jul@ssu.ac.kr

Abstract. Collapse transition of the lattice polymer on a square lattice is studied by calculating the exact partition functions up to chain length 38; they are obtained by enumerating the number of possible conformations for each energy value. We observe that the locus of partition function zeros approaches the positive real axis as the chain length increases, providing evidence of the collapse transition. The crossover exponent and the transition temperature are estimated from the scaling behavior of the first partition function zeros with increasing chain length.

1. Introduction

A flexible polymer chain in a dilute solution is influenced by both hydrophobic interactions between the monomers and the excluded volume effect. The attractive interactions are neglected at high temperatures or in a good solvent, but become significant as the temperature T is lowered. As T reaches a special temperature θ, the linear polymer undergoes an abrupt change from an expanded conformation for $T > \theta$ to a fully compact conformation for $T < \theta$ [1, 2, 3]. A long polymer in a good solvent is a critical system, and the collapse transition at $T = \theta$ has been identified as a tricritical transition [3, 4]. The θ point behavior is well-described by self-avoiding walks with attractive interaction energy assigned for each pair of nonbonded nearest-neighbor (NN) monomers. The tricritical exponents take the mean-field values for $d > 3$, and there are logarithmic corrections at $d = 3$ [3, 4, 5, 6, 7, 8, 9]. A great deal of studies have been performed to understand the nature of the collapse transition in two dimensions, which is expected to exhibit much more non-trivial behavior than its higher dimensional counterparts.

In this work, we calculate the exact partition function of a polymer on a square lattice with chain length 38 for the first time, and obtain the crossover exponent and the tricritical temperature by examining their scaling behavior of the partition function zeros with increasing chain length.

2. Number of conformations

By using self-avoiding walks on a square lattice, conformations of a polymer chain with N monomers are modeled. The position of a monomer i is expressed as $\mathbf{r}_i = (a, b)$ with integer

Journal of Physics: Conference Series **454** (2013) 012083 doi:10.1088/1742-6596/454/1/012083

values of the coordinates a and b. Chain connectivity requires bond length is unity, $|\mathbf{r}_i - \mathbf{r}_{i+1}| = 1$, and the excluded volume effect does allow only one monomer on each lattice, $\mathbf{r}_i \neq \mathbf{r}_j$ for $i \neq j$.

We consider the Hamiltonian with the nearest-neighbor interaction:

$$\mathcal{H} = -\epsilon \sum_{i<j} \Delta(\mathbf{r}_i, \mathbf{r}_j), \tag{1}$$

where

$$\Delta(\mathbf{r}_i, \mathbf{r}_j) = \begin{cases} 1 & \text{if } |i-j| > 1 \text{ and } |\mathbf{r}_i - \mathbf{r}_j| = 1, \\ 0 & \text{otherwise,} \end{cases} \tag{2}$$

and ϵ is set to a positive value to incorporate the attractive interaction between the monomers.

Assuming that the polymer chain has an intrinsic direction, the conformations with reverse labels $i \leftrightarrow N - i + 1$ for all $(i = 1, 2, \cdots, N)$ are considered distinct. The total number of conformations generated by rotations and reflections from a given two-dimensional conformation is 8, since the rigid rotations and reflections in two dimensions form an 8-fold symmetry. The straight chain is an exception, however, a one-dimensional conformation invariant with respect to reflection perpendicular to the chain, where the total number of conformations generated by rotations and reflections is 4. Here, we define the reduced number of conformations $\omega_N(K)$, where conformations related by rigid rotations, reflections, and translations are regarded as equivalent, and counted only once. Consequently, the number of conformations with rigid rotations and reflections considered distinct, denoted by $\Omega_N(K)$, can be easily obtained by

$$\Omega_N(K) = \begin{cases} 8\omega_N(K) - 4 & \text{if } K = 0, \\ 8\omega_N(K) & \text{otherwise.} \end{cases} \tag{3}$$

Thus, one can achieve about 8-fold reduction in the computing time by enumerating the reduced number of conformations $\omega_N(K)$ instead of $\Omega_N(K)$ [10, 11]. We obtained $\omega_N(K)$ up to $N = 38$ by the help of a parallel algorithm classifying conformations by sizes of rectangles they span [12]. The number of conformations $\Omega_N(K)$ for $N = 38$ is presented in table 1.

3. Partition function zeros in the complex temperature plane

Partition function zeros have been the subject of interest as a sensitive indicator of a phase transition [10, 11, 13, 14, 15, 16]. Partition function zeros were introduced by Yang and Lee in the complex fugacity plane of a fluid system and the complex magnetic-field plane of the nearest-neighbor Ising ferromagnet (Yang-Lee zeros), to study the phase transition driven by the fugacity or the magnetic field [17]. Later, Fisher [18] used the partition function zeros in the complex temperature plane (Fisher zeros) of the square-lattice Ising model, to study the temperature driven transition. In the thermodynamic limit, the locus of zeros forms a continuous curve which crosses the real axis if a transition exists. Thus, the theory of partition function zeros provides the explanation on how the partition function, which is an analytic function of thermodynamic parameters at a finite size, acquires the singularities necessary for a phase transition in the thermodynamic limit. In the case of Fisher zeros, the transition temperature in the thermodynamic limit is the intersection point of the locus of zeros with the real temperature axis. Therefore, the conjugate pair of zeros closest to the positive real axis, called the *first zeros*, determine the leading singular behavior of the partition function. Since the behavior of the first zeros can be analyzed separately from the other zeros, the phase transition can be analyzed more accurately by computing the partition function zeros than studying real-valued quantities such as the specific heat which includes the effect from all the zeros [11].

The partition function of our model is

$$Z = \sum e^{-\beta\mathcal{H}} = \sum_{K=0}^{K_{\max}(N)} \Omega_N(K) y^K, \tag{4}$$

Table 1. The number of conformations $\Omega(K)$ on a square lattice as a function of the number of contacts K for chain length 38.

K	$\Omega(K)$
0	128296079455572
1	515427549335840
2	1128115046251464
3	1756218784960648
4	2184512179070336
5	2306182278125992
6	2144957763099744
7	1799694003510208
8	1385401931446192
9	990209382129048
10	663137244947784
11	418840297796560
12	250761159690416
13	142813585256872
14	77560661732336
15	40218582883856
16	19921507091832
17	9413632521400
18	4236279016096
19	1807558226256
20	725595746792
21	271663920168
22	93029010016
23	28148212408
24	7210320464
25	1082656960
26	45293464
Total	15968852281708724

where $y \equiv \exp(\beta\epsilon)$, $\beta \equiv 1/k_B T$, and $K_{\rm max}(N)$ is the maximum number of possible contacts for polymer length N [19],

$$K_{\rm max}(N) = \begin{cases} N - 2m & \text{for } m^2 < N \le m(m+1), \\ N - 2m - 1 & \text{for } m(m+1) < N \le (m+1)^2, \end{cases} \tag{5}$$

where m is a positive integer. Then, the partition function becomes a $K_{\rm max}$-th order polynomial of y and can be expressed in the form,

$$Z(y) = A(y) \prod_i (y - y_i), \tag{6}$$

where $A(y)$ is a function which is analytic in the whole complex plane and the partition function zeros y_i $(i = 1, 2, \cdots, K_{\rm max})$ are obtained by solving the polynomial equation $Z(y) = 0$ with MATHEMATICA. Figure 1 shows the partition function zeros for $N = 38$ in the complex temperature plane. One can realize that the first zeros approach the positive real axis in the complex temperature plane as polymer length increases (figure 2).

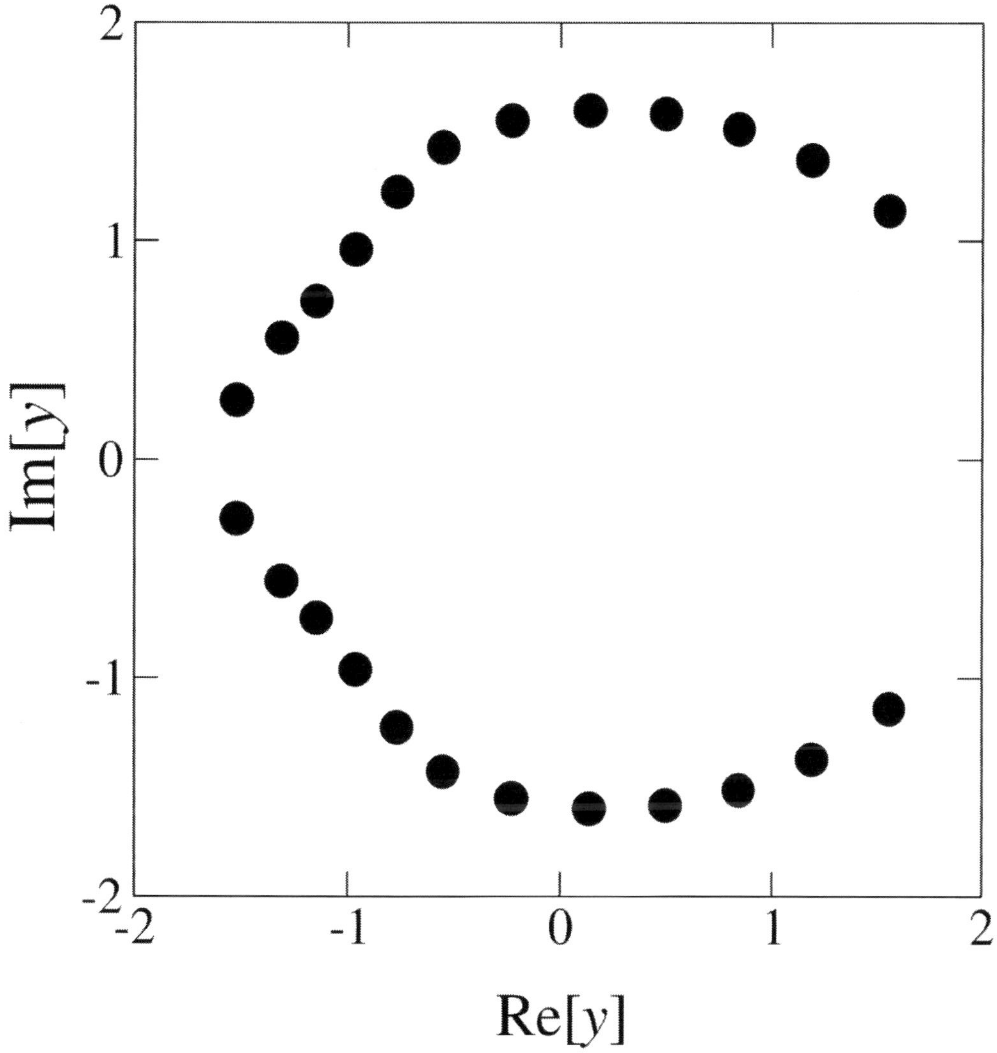

Figure 1. Positions of the partition function zeros in the complex temperature ($y = e^{\beta\epsilon}$) plane for $N = 38$. The first zeros are the ones closest to the positive real axis.

4. Crossover exponent and tricritical temperature

The crossover exponent ϕ, introduced in the Introduction, also describes how rapidly the first zeros approach the positive real axis as N increases [11, 13],

$$\mathrm{Im}[y_1(N)] \sim N^{-\phi}, \tag{7}$$

where $y_1(N)$ is a first zero for a polymer chain with N monomers. In finite-size systems (N = even), the crossover exponent is approximated as

$$\phi(N) = -\frac{\ln\{\mathrm{Im}[y_1(N+2)]/\mathrm{Im}[y_1(N)]\}}{\ln\{(N+2)/N\}}, \tag{8}$$

which reduces to the exact value of ϕ in $N \to \infty$ limit, estimated by using the Bulirsch-Stoer (BST) extrapolation [20]. We obtain $\phi = 0.430(29)$ from the data for even N with $N \geq 22$, where

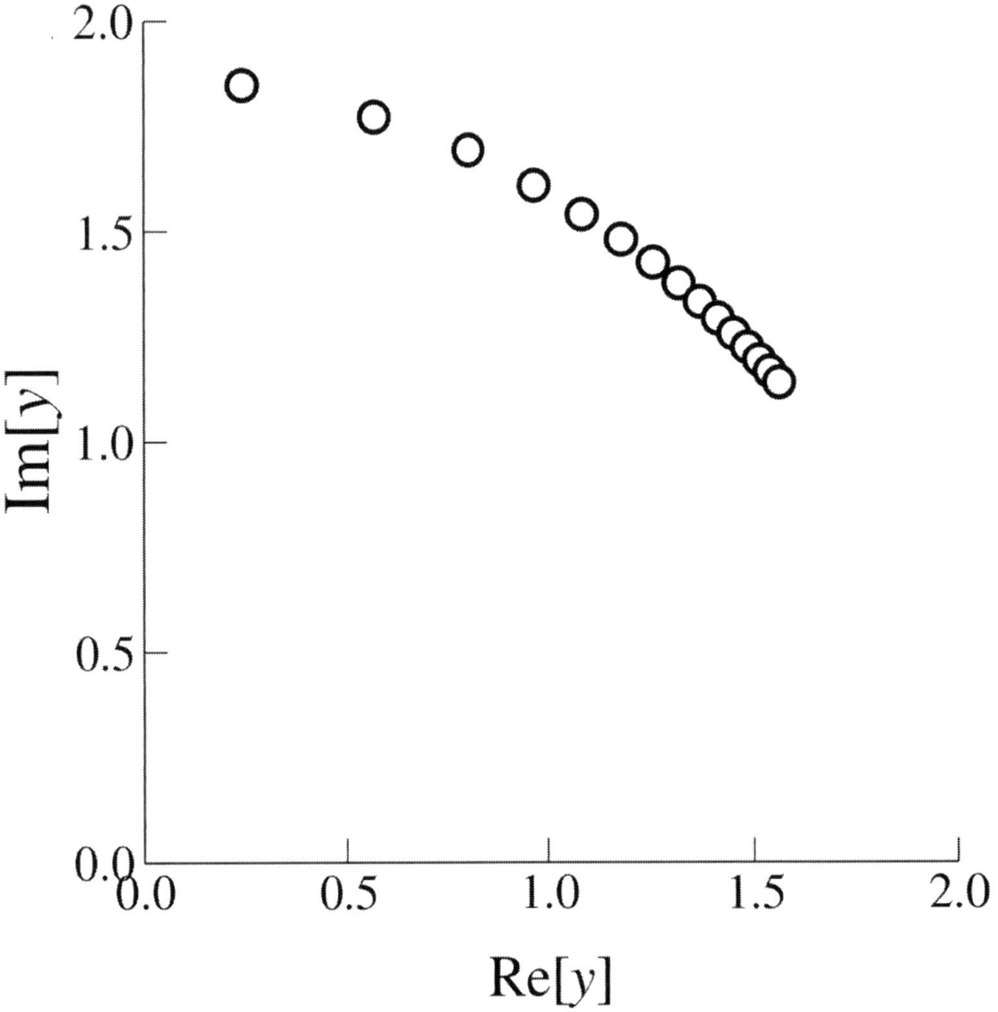

Figure 2. Positions of the first zeros in the first quadrant of the complex temperature ($y = e^{\beta\epsilon}$) plane for even lengths $N = 10, 12, 14, \cdots, 38$ from left to right. The first zeros approach the positive real axis as N increases.

the estimated error can be further reduced by removing unreliable data obtained from $N < 22$. The error is estimated by examining the robustness of the extrapolated value with respect to perturbations of the data points, but it is not a statistically rigorous confidence level [11, 20]. The value of ϕ we obtained agrees quite well with the exact value 3/7 obtained by analytic calculation on the polymers on the hexagonal lattice [8], which is believed to be in the same universality class as those on the square lattice.

With the value of ϕ, the tricritical temperature can be obtained by estimating the point on the positive real axis where the first zeros approach in the limit of $N \to \infty$,

$$\mathrm{Re}[y_1(N)] - y_c \sim N^{-\phi}. \tag{9}$$

The value of $y_c = 2.15(30)$, which corresponds to $\theta = 1.31(32)$, is obtained by extrapolating the data for even N with $N \geq 22$. In table 2, our results given in the first line are compared with those from the earlier works.

Table 2. The tricritical temperature θ and the crossover exponent ϕ obtained in the current work, displayed in the first line, are compared with those in the literature. θ is displayed only for the model of the current work.

Method	lattice	$N_{\max}$	θ	ϕ
Exact partition function zeros	square	38	1.31(32)	0.430(29)
Field theory [5]	N/A	N/A	-	$\frac{7}{11}$ (≈ 0.64)
Monte Carlo [6]	square	160	1.31(6)	-
Renormalization group [21]	N/A	N/A	-	$\frac{19}{22}$ (≈ 0.86)
Monte Carlo [22]	square	200	1.55(15)	0.6(1)
Transfer matrix [7, 23]	square	N/A	1.42(4)	0.48(7)
Series expansion [24]	triangular	16	-	0.64(5)
Coulomb gas method [8]	hexagonal	N/A	-	$\frac{3}{7}$ (≈ 0.43)
Monte Carlo and RG [25]	square	40	1.54(7)	0.52(7)
Monte Carlo [26]	hexagonal	300	-	0.5(1)
Scanning simulation [27]	square	240	1.52(1)	0.530(4)
Recursive enrichment method [28]	square	2048	1.504(5)	0.435(6)
The pruned-enriched Rosenbluth method [29]	square	256	1.4993(23)	-
Interacting growth walk [30]	square	2000	-	0.419(3)
Monte Carlo [31]	square	1600	1.50	0.545(4)
Monte Carlo [32]	square	300	1.505(18)	-
Monte Carlo [33]	square	3200	1.4986(11)	0.46(3)
Exact enumeration [34]	square	26	1.515(47)	-
Monte Carlo [34]	square	71	1.529	-

5. Discussion

With the number of all possible conformations obtained by exhaustive enumeration up to chain length $N = 38$, we studied the collapse transition of a polymer on a square lattice by calculating the partition function zeros. We observed that the first zeros tend to approach the positive real axis as the chain length increases, and we measured the crossover exponent ϕ and the tricritical temperature θ by the BST extrapolation.

The maximum chain length in our study is so much shorter in contrast to Monte Carlo approaches with polymer lengths up to several hundreds or thousands. According to the exactness of our data, however, we could use a powerful extrapolation method and then estimated reasonably accurate values of the crossover exponent and the tricritical temperature. Furthermore, the partition function zeros allow us to carry out much more accurate analysis of the collapse transition than examining the radius of gyration or specific heat.

References

[1] Flory P J 1967 *Principles of Polymer Chemistry* (Ithaca: Cornell University Press)
[2] Chan H S and Dill K A 1991 *Annu. Rev. Biophys. Biophys. Chem.* **20** 447
[3] de Gennes P-G 1975 *J. Physique Lett.* **36** 55
[4] Griffiths R B 1973 *Phys. Rev.* B **7** 545
[5] Stephen M J 1975 *Phys. Lett.* A **53** 363
[6] Baumgärtner A 1982 *J. Physique* **43** 1407
[7] H. Saleur H 1986 *J. Stat. Phys.* **45** 419
[8] Duplantier B and Saleur H 1987 *Phys. Rev. Lett.* **59** 539
[9] Lee J H, Kim S-Y and Lee J 2012 *Phys. Rev.* E **86** 011802
[10] Lee J 2004 *J. Korean Phys. Soc.* **44** 617

[11] Lee J H, Kim S-Y and Lee J 2010 *J. Chem. Phys.* **133** 114106
[12] Lee J H, Kim S-Y and Lee J 2011 *Comput. Phys. Commun.* **182** 1027
[13] Lee J H, Kim S-Y and Lee J 2011 *J. Chem. Phys.* **135** 204102
[14] Itzykson C, Pearson R B and Zuber J B 1983 *Nucl. Phys.* B **220** 415
[15] Alves N A and Hansmann U H E 2000 *Phys. Rev. Lett.* **84** 1836; 2001 *Physica* A **292** 509
[16] Bena I, Droz M and Lipowski A 2005 *Int. J. Mod. Phys.* B **19** 4269 and references therein.
[17] Yang C N and Lee T D 1952 *Phys. Rev.* **87** 404; Lee T D and Yang C N 1952 *Phys. Rev.* **87** 410
[18] Fisher M E 1965, in *Lectures in Theoretical Physics* edited by Brittin W E (Boulder: University of Colorado Press), Vol. 7c, p. 1.
[19] Chan H S and Dill K A 1989 *Macromolecules* **22** 4559
[20] Bulirsch R and Stoer J 1964 *Numer. Math.* **6** 413
[21] Kholodenko A L and Freed K F 1984 *J. Phys. A: Math. Gen.* **17** L191; 1984 *J. Chem. Phys.* **80** 900
[22] Birshtein T M, Buldyrev S V and Elyashevitch A M 1985 *Polymer* **26** 1814
[23] Derrida B and Saleur H 1985 *J. Phys. A: Math. Gen.* **18** L1075
[24] Privman V 1986 *J. Phys. A: Math. Gen.* **19** 3287
[25] Seno F and Stella A L 1988 *J. Physique* France **49** 739
[26] Poole P H, Coniglio A, Jan N and Stanley H E 1989 *Phys. Rev.* B **39** 495
[27] Chang I and Meirovitch H 1993 *Phys. Rev.* E **48** 3656
[28] Grassberger P and Hegger R 1985 *J. Phys. I France* **5** 597
[29] Barkema G T, Bastolla U and Grassberger P 1998 J. Stat. Phys. **90** 1311
[30] Narasimhan S L, Krishna P S R, Murthy K P N and Ramanadham M 2001 *Phys. Rev.* E **65** 010801(R)
[31] Zhou J, Ou-Yang Z-C and Zhou H 2008 *J. Chem. Phys.* **128** 124905
[32] Cunha-Netto A G, Dickman R and Caparica A A 2009 *Comput. Phys. Commun.* **180** 583
[33] Caracciolo S, Gherardi M, Papinutto M and Pelissetto A 2011 *J. Phys. A: Math. Gen.* **44** 115004
[34] Ponmurugan M and Satyanarayana S V M 2012 *J. Stat. Mech.* P06010

Double-layer evolutionary algorithm for distributed optimization of particle detection on the Grid

Adam Padée[1,2], Krzysztof Kurek[3] and Krzysztof Zaremba[1]

[1] Warsaw University of Technology, Institute of Radioelectronics,
ul. Nowowiejska 15/19, 00-665 Warsaw, Poland
[2] National Centre for Nuclear Research, Świerk Computing Centre,
ul. Andrzeja Sołtana 7, 05-400 Otwock-Świerk, Poland
and
University of Warsaw, Interdisciplinary Centre for Mathematical and Computational Modeling (ICM),
ul. Pawińskiego 5a, 02-106 Warsaw, Poland
[3] National Centre for Nuclear Research, Theoretical Physics Department,
ul. Andrzeja Sołtana 7, 05-400 Otwock-Świerk, Poland
and
Cardinal Stefan Wyszyński University in Warsaw,
Faculty of Mathematics and Natural Sciences, The School of Exact Sciences,
ul. Wóycickiego 1/3, 01-938 Warsaw, Poland

E-mail: adam.padee@ncbj.gov.pl

Abstract. Reconstruction of particle tracks from information collected by position-sensitive detectors is an important procedure in HEP experiments. It is usually controlled by a set of numerical parameters which have to be manually optimized. This paper proposes an automatic approach to this task by utilizing evolutionary algorithm (EA) operating on both real-valued and binary representations. Because of computational complexity of the task a special distributed architecture of the algorithm is proposed, designed to be run in grid environment. It is two-level hierarchical hybrid utilizing asynchronous master-slave EA on the level of clusters and island model EA on the level of the grid. The technical aspects of usage of production grid infrastructure are covered, including communication protocols on both levels. The paper deals also with the problem of heterogeneity of the resources, presenting efficiency tests on a benchmark function. These tests confirm that even relatively small islands (clusters) can be beneficial to the optimization process when connected to the larger ones. Finally a real-life usage example is presented, which is an optimization of track reconstruction in Large Angle Spectrometer of NA-58 COMPASS experiment held at CERN, using a sample of Monte Carlo simulated data. The overall reconstruction efficiency gain, achieved by the proposed method, is more than 4%, compared to the manually optimized parameters.

Journal of Physics: Conference Series **454** (2013) 012084 doi:10.1088/1742-6596/454/1/012084

1. Introduction

Detection of momenta of charged particles plays an important role in High Energy Physics (HEP) experiments. It is also one of the most complex steps of the data processing chain, because it requires reconstruction of the tracks of particles in 3 dimensions, on the basis of signals from a set of position sensitive detectors. Physical limitations of the detectors (precision, efficiency) make it necessary to control the process with parameterized tolerances. Setting of these parameters can be treated as an optimization task. Although there is no precise formula to evaluate the quality of each parameter set, it can be estimated using Monte Carlo simulation of the experiment. The drawback of using this kind of criterion is the lack of precision. The quality of the reconstruction is expressed in the ratio of properly reconstructed tracks to all tracks existing in the generated sample, so it may not reflect the effects of small changes in the parameters. The criterion is also dependent on the exact events that were generated, so it is only statistically correct when using different samples. These features of the objective function make the problem hard to solve using standard numerical optimization methods.

Evolutionary algorithms have been extensively used in many scientific and engineering applications over last decades. Their unique features make it possible to solve problems that are difficult for classical optimization methods. One of the main advantages of evolutionary algorithms is that they do not rely directly on any derivative-based information (e.g. gradient) [1]. Therefore they can be applied to problems where objective function is either imprecise or distorted by a random noise. The application described in this article belongs to this class of problems, because of the random nature of the evaluation process. This is the reason why evolutionary algorithms have been chosen as the optimization method.

One of the major difficulties of using evolutionary optimization is the relatively high demand for the computing power, at least in comparison to the analytical methods. This is the price for robustness coming from ignoring the derivative-based information [1]. The problem is even more severe when the objective function itself is computationally expensive, as it is with the reconstruction of Monte Carlo data, which is the application described in this paper. On the other hand, evolutionary algorithms are relatively easy to parallelize, because at each step they operate on large populations instead of individual points. This feature makes them well suited for usage in a broad spectrum of parallel and distributed environments. There were many industrial and scientific applications of evolutionary algorithms that make use of supercomputers, clusters or even computational grids. One of such applications is the previous version of the program presented in this paper, which solved the problem of reconstruction optimization using master-slave parallel genetic algorithm designed for cluster of PCs. Detailed description of that program can be found in [2]. The results achieved by that version were not entirely satisfactory, mainly because of lack of computing power. This was the inspiration for the creation of the algorithm described in this paper, designed to use computational grid instead of single cluster.

2. Objective function

The application described in this article was created to support the muon program of NA-58 COMPASS (Common Muon-Proton Apparatus for Spectroscopy and Structure) experiment that is currently held at CERN. Further details on the experiment can be found in [3] and [4]. As the work described in this article is a continuation and extension of the application described in [2], only a short summary of the information on momentum measurement, data reconstruction process and the objective function will be presented.

The measurement of the momentum of high-energy charged particles is done by checking the curvature of their tracks in a strong magnetic field. There are two sets of position-sensitive detector planes and a magnet between them. The data from the detectors is registered in raw format and, due to computational complexity, further processing is done offline. The reconstruction of the tracks is a major part of this processing and it can be split into the following steps:

- Conversion of the raw signals to „hits" (2D positions on the detector planes plus timestamps).

- Reconstruction of straight segments of the tracks in the areas with the detectors (the procedure assumes that the magnetic field in these regions is zero).
- Linking of the segments into tracks, extrapolating them in the magnet area.
- Finding the interaction vertex (common point from which the tracks originate)

There are several limitations that decrease the efficiency and precision of the procedure. The most notable examples are: the efficiency of the detectors, limited spatial and temporal resolution of the planes, high number of tracks in one event, background noise, pile-up tracks and simplified mathematical models (e.g. assumption that the segments in the detector areas are straight, which is not true because of fringe magnetic field in that area). This is why the reconstruction process is done iteratively, and each of the iterations is controlled by separate set of parameters, e.g. tolerances for the straight parts of the track (in both detector sets), the precision of the crossing point in the magnet area, tolerance for the interaction vertex and minimum number of hits needed to qualify a track as good. Although their values can be narrowed down to certain ranges by analyzing the physical properties of the detector, the choice of the exact values have to be done experimentally using Monte Carlo simulation of the detector. In COMPASS the detector is simulated using Geant with input from the event generators (PYTHIA, AROMA, LEPTO). The offline processing program is called CORAL (COmpass Reconstruction and anALysis).

The quality of particular set of parameters is expressed with the following criteria: the number of properly reconstructed tracks (denoted by Φ_1) and the number of false tracks introduced by the reconstruction routine (denoted by Φ_2). The overall criteria used in the algorithm is:

$$\Phi = \Phi_1 - c \cdot \Phi_2 \tag{1}$$

where c is an arbitrarily chosen constant (in the experiments it was set to 1.5). Additionally the reconstruction time is checked, if it exceeds a certain threshold, CORAL is aborted and the objective function is zeroed. More details on the reconstruction can be found in [5], [4]. The objective function is described in more detail in [2].

3. Computational grid in HEP

The growing complexity of HEP experiments results in higher demands for computing power. They often exceed the capabilities of the computing farms dedicated to the experiments. This is the reason why HEP is shifting towards distributed data analysis using computational grids. The best examples of this approach are the experiments built around LHC (Large Hadron Collider), which established their own grid environment named WLCG (Worldwide LHC Computing Grid [6]). Although the project itself does not provide any resources, it collaborates with many national and continental grid initiatives to make their infrastructures available for physicists and ensure the availability of a standard set of middleware components and protocols. Standards set by WLCG are widely accepted also by smaller, non-LHC experiments, COMPASS being one of them. The access to the resources in the grid uses the concept of Virtual Organizations (VOs). Each VO corresponds to the community of researchers. In HEP a separate VO is usually set up for each experiment. It is up to the resource providers to decide which VOs to support and how big share of their resources they are granted. The authorization of the individual users and the verification of their VO membership are done using x509 certificates.

There are two main types of the resources accessible via the grid: CPUs and storage. Computing resources in WLCG are served by machines called Computing Elements (CEs). Each CE is a frontend service to a cluster of Worker Nodes (WNs) managed by a queuing system (Torque, LSF, SGE, ...). The user does not need to know what type of a queuing system lies underneath a specific CE, because jobs are submitted using standard API or CLI. Storage resources are served by machines called Storage Elements (SEs). There are couple of types of SEs, but despite the differences in hardware and storage capacity, all of them offer similar set of protocols. The management of the storage space is done via SRM (Storage Resource Manager) protocol. File uploads and downloads are performed using

GridFTP standard. Each of the SEs also offers its own implementation of a low-level random I/O protocol (e.g. RFIO, dcap). Although they are different, there is a common API called GFAL (Grid File Access Library) that offers unified access to data with POSIX-like interface, hiding the actual access method from the user. Real life tests with WLCG infrastructure show that the delay for a single transaction with small file varies from 1.5 s to couple of seconds for GridFTP (depending on the distance between the communicating parties), and from 0.5 s to couple of seconds when using GFAL.

The information about the resources connected to the grid is gathered in a common information system called BDII (Berkeley Database Information Index). It contains both static (like a number of CPUs in a CE, total storage space of a SE, supported VOs) and dynamic data (e.g. current usage). Using this system the resources available to a VO can be easily discovered and centrally managed, which is done by the collection layer services. The optimal distribution of the load between CEs is managed by WMS (Workload Management Service). Unified access to all the storage resources available to a VO is possible using LFC (Logical File Catalogue), which constitutes a global virtual namespace, resembling a single filesystem. The names of files in LFC, called LFNs (Logical File Names) can be used directly in GFAL API calls.

Typical usage pattern of the grid by HEP communities is the submission of series of sequential jobs which share data sets using SEs. In these models there is no need to provide any direct communication protocol between the jobs, especially when they are running on different clusters. For this reason WLCG does not provide any tools for running inter-cluster parallel jobs. There are some software components compatible with WLCG middleware, developed by external projects (e.g. Int.EU.Grid [7]), that enable parallel communication between processes running at different clusters, but they did not become a widely accepted standard yet. Even the support for MPI jobs limited to a single cluster is not compulsory for the WLCG member sites and remains an optional feature.

The facts outlined in the paragraph above are a huge barrier for design and usage of the parallel algorithms that could use all the resources available in the grid in a coordinated way. Still this is potentially a very beneficial approach, because the large fraction of the computational resources in the grid is provided by many small clusters. If the numbers of CPU cores provided by the smallest clusters are summed up, they often prove to be larger than the contribution of the largest clusters. The example of such a situation can be seen in table 1, which contains distribution of the cluster sizes in EGI (European Grid Infrastructure), which is one of the major contributors of resources to WLCG.

Table 1. Number of CPU cores in clusters grouped by size (EGI, Oct. 2012, data source: [6]).

Size range	0-1024	1025-2048	2049-4096	4097-8192	8193-16384	16385+	Sum
Total cores	33346	32870	70615	57661	32476	16484	243452
No of clusters	167	22	24	9	3	1	226

None of the categories presented in the table is predominant. Although differences between the categories are significant, their contributions have the same order of magnitude. Designing an algorithm capable of using only large clusters would waste a considerable part of the resources. It was therefore essential to find an architecture capable of using the resources distributed in the smallest clusters. An additional requirement was to use file transfer protocols, as they are the only standard option for inter-cluster communication.

4. Evolutionary algorithms in numerical optimization

Evolutionary Algorithm (EA) is a general name for a class of metaheuristic optimization methods. Their common trait is that they use mechanisms inspired by biological evolution: selection, reproduction, mutation and recombination (crossover). Detailed description of EAs can be found in [8], so here only a short summary of the most important facts will be given. EAs are iterative algorithms, but unlike the classical optimization methods, they operate on N-element population in each step, instead of a single point. This is the main reason why they are relatively easy to parallelize, because many time-consuming operations can be done independently for each of the population members. An iteration is often called a generation. EAs use specific representations of the points in the search space, which are often called genotype or chromosome. The choice of representation is the most distinctive feature of different EA types, because it also determines the way in which the genotype is modified by mutation and crossover operators. In EAs used for numerical optimization, there are two main types of representations, historically originating from different types of algorithms:

- Genetic algorithm. Optimized parameters are coded as binary strings. Each parameter is sampled evenly within allowed range with desired precision and represented as a fixed-point number (sometimes using Gray code). Mutation and crossover operate on individual bits.
- Evolution strategy, evolutionary programming. Optimized parameters are coded as real values, usually with floating-point representation. Mutation adds a small perturbation with Gaussian distribution to the processed parameter. Crossover can exchange only whole parameters; in some implementations (called mixing crossover) it can also calculate weighted average of the respective parameters.

In the reconstruction optimization both representations are useful, because binary coding is better for enumerative parameters, and real better suits continuous variables. As stated in [1] there is a rationale that binary coding of variables for which the objective function has periodic component may be also beneficial. Therefore both representations were implemented in the program. More details on how the reconstruction parameters are coded in the genotype can be found in the description of the previous version in [2], because this part of the program did not change significantly. More general discussion about the representations and operators can be found in [8].

The most general rule of EAs is that better points should have more copies in subsequent generations than worse ones. This is ensured by the selection and reproduction operators. They usually do not work directly on the objective function values, but use its modified form, called fitness function. The conversion from the original objective function to the fitness function involves simple operations like negation (necessary for minimization problems) and/or linear scaling to ensure that the values are in the proper range (e.g. to avoid negative values which are not acceptable by proportional selection). The optimization of the reconstruction is a maximization problem, so it does not require modifications except simple linear scaling. The genotype together with a corresponding fitness value is called an individual in EA terminology.

5. Architectures of parallel evolutionary algorithms

According to [9] and [10] there are three main types of parallel evolutionary algorithms: master-slave, cellular and multiple-population. The simplest and the most popular one is master-slave. In this variant all the evolutionary operators are executed centrally within the single master process. Only the fitness calculation is distributed and delegated to multiple slave processes. The popularity of this variant is based on the fact that calculation of the fitness values is usually the most time-consuming part of the EA iteration. If the algorithm is executed in k-processor environment, T_g denotes time needed to execute the evolutionary operators for a single individual, T_c time of communication with one slave process and T_f time of fitness calculation for single individual, the overall time needed to perform a single iteration of the algorithm T_p is given by:

$$T_p = NT_g + kT_c + \frac{NT_f}{k} \tag{2}$$

In optimization of the reconstruction the assumption that $T_f >> T_g$ is fulfilled, but using grid protocols for communication results in very high T_c. Theoretical scaling boundary of the master-slave algorithm can be calculated as:

$$k^* = \left(\frac{NT_f}{T_c}\right)^{\frac{1}{2}} \tag{3}$$

Increasing the number of processors above k^* will increase the total execution time of an iteration (T_p) instead of decreasing it. Moreover, it has to be noted that the algorithm operating with $k \approx k^*$ is very inefficient, and almost half of the CPU time is wasted on waiting. E. Cantú-Paz in [10] calculated (with some simplifying assumptions) the efficient scaling boundary k_l, where the wasted CPU time equals roughly $1/k$ (the equivalent of 1 CPU core is wasted).

$$k_l = \left(\frac{NT_f}{T_c}\right)^{\frac{1}{3}} \tag{4}$$

Analyzing a real case of optimization of the reconstruction in the grid environment, T_f is about 3-10 minutes (depending on the sample size and the particular values of the parameters) and T_c 1-3 seconds. With 200-element population, $k_l \approx 30$ and $k^* \approx 200$ which makes the practical usability of the algorithm very limited.

There is also another problem concerning the classical master-slave design, which is variable T_f. In the analyzed case it depends not only on the particular parameter settings, but also on the differences in CPU processing speeds in different grid clusters. The master process have to wait for the slowest slave, so the expected duration of the fitness calculation phase for the whole algorithm will be much higher than just the mean value of T_f multiplied by N/k (for simplification it is assumed that N is divisible by k without the remainder and the time spent by a single slave on the fitness calculation is denoted by $T_{fs} \equiv T_f \cdot N/k$). If $f(T_{fs})$ denotes the probability density function of T_{fs} and $F(T_{fs})$ cumulative density function, then the probability density of the duration of fitness calculation phase on k processors will be given by the first member of the order statistics:

$$f_{MAX}(T_{fs},k) = k \cdot f(T_{fs}) \cdot [F(T_{fs})]^{k-1} \tag{5}$$

The exponential relation between the overall generation time PDF and slave calculation time CDF confirms the intuitive conviction that $f_{MAX}(T_{fs},k)$ will rise very quickly when increasing the number of processors.

To overcome both problems outlined in the paragraphs above a special asynchronous version of master-slave algorithm can be used. It doesn't have clearly marked generations when whole parent population is replaced by offspring population, but instead new individuals are constantly generated, sent to slaves and immediately inserted to the current population whenever they are returned. This requires the introduction of a new operator – succession. When new individual is received from a slave, this operator decides which place in the population it will be inserted to. This choice can be random or controlled by multiple criteria e.g. fitness measures, similarity of the genotypes, lifetime, etc.. As a matter of fact this type of the algorithm, called steady-state EA (abbreviated to ssEA), does not have to be connected with parallelism. It was invented and used first in a sequential version [14].

Asynchronous master-slave EA is not as efficient as previously described synchronous variant when it is used with low k. The reason is that in this model the master process does not do any fitness calculations itself, because it has to be constantly ready for the new individuals coming from the slaves. Therefore the resource usage efficiency has always an upper bound of $(k-1)/k$, regardless of the T_f , T_g and T_c. Nevertheless, with higher k one can expect higher efficiency, because the time T_{pa} needed by the algorithm to process N individuals (which is equivalent to processing one generation in the synchronous variant) is expressed by:

$$T_{pa} = T_c + \frac{NT_f}{k-1} \tag{6}$$

Although this equation sets the practical scaling boundary much higher than the value k_l for the synchronous variant, the absolute boundary of k^* stays the same, because the equation (6) holds up only within the range given by the following inequality:

$$T_g + \frac{(k-1)T_c}{N} < \frac{T_f}{k-1}$$

Journal of Physics: Conference Series **454** (2013) 012084 doi:10.1088/1742-6596/454/1/012084

(7)

Still the asynchronous variant is much better for higher number of CPUs, because it is practically usable up to k^* and its efficiency of resource usage at that threshold is close to $(k\text{-}1)\cdot N/k\cdot(N+1)$, which is much higher than ½. It is also not sensitive to random variations of T_f. For these reasons it has been used in the previous version of the optimization program, designed to run on a single cluster, which is described in [2]. Nevertheless, the scaling boundary of k^* is too low to use it efficiently in the grid. Higher number of the processing units can be used only in other variants of parallel EAs.

Cellular algorithm is not suitable for the grid, because it is communication-intensive and requires homogenous execution environment. Better results could be achieved with multiple-population algorithm, which is also called island model EA. This is a fully distributed algorithm, in which each processing unit holds its own subpopulation. The number of subpopulations (also called islands or demes) is denoted by r. The algorithm works in two phases. In the first one all the subpopulations evolve independently, so it is equal to r independent sequential algorithms. In the second phase the demes exchange selected individuals. This is done by an additional operator named migration. The connections between the islands form a topology graph. Topologies are mostly regular graphs, described by δ parameter, which denotes the degree of each vertex. It is useful when comparing different topologies, as it is closely related with the time spent on communication. Other parameters that control the island model EA are: migration coefficient ϱ, which defines percentage of individuals in each population that are exchanged in the migration phase, interval between the migration phases (also called epoch – measured in number of generations and denoted by g), number of individuals in each subpopulation (N_d) and number of epochs (τ). As it has been shown in [10], ϱ and δ do not have to be high to achieve the convergence speed comparable to single population with $r\cdot N_d$ individuals.

There is no absolute scaling boundary for island model EA, because communication to local calculations ratio is set by parameters g, ϱ and δ, which are defined arbitrarily by the user. Still there are some ranges in which these parameters should be contained in order to achieve good convergence speed. For example if the subpopulations are small, it makes little sense to set the epoch length too high, because at some point each subpopulation will be taken over by one dominant solution and from that moment further local calculations are pointless. In distributed EAs there is a general rule that the communications frequency should be inversely proportional to the size of subpopulations. This relation is often depicted in the form of 3-dimensional plot, binding the communication frequency, the amount of local calculations and the number of subpopulations [10], [9]. The plot is presented on figure 1. Optimally configured algorithms are placed on space diagonal of the cube. On one end of the diagonal there is sequential algorithm, on the other – cellular. All the configurations in between are island model EAs with different settings of the parameters. Using the island model EA with grid resources would be difficult, because a big number of small islands require frequent communication. Possible solution to this problem would be to use hybrid algorithm which is composed of two layers of communication and uses different architecture on each of them.

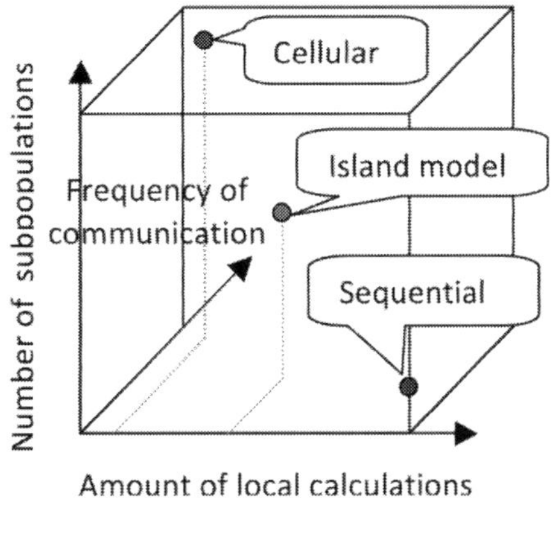

Figure 1. Relation between communication frequency, local calculations and number of subpopulations in the distributed EA.

6. Hybrid EA for the grid

The idea behind the design of hybrid, double-layer parallel EA is to use multiple parallel EAs and connect them in a bigger structure using the migration operator. EAs on the lower level could be of any type (master-slave, cellular or multiple-population) and possibly they could use different communication protocol than used by the migration on the higher level. The last feature is important when running the algorithm on non-uniform resources and allows adapting it to their physical layout (fast interconnect is used within the islands, slow one between them). The main advantage of the hybrid algorithm over the flat island model is that thanks to parallelism on the level of the islands it is possible to make them much bigger than in sequential case. This allows to set a lower migration intensity so the usage of slow interconnect can be lowered by orders of magnitude. Hybrid EAs have been successfully used in many applications over last two decades, there are also good performance analyses of this type of algorithm, e.g. in [12]. The facts outlined above make this algorithm perfectly suited for usage in the grid environment. Islands can be formed on whole clusters instead of single CPU cores and use fast MPI interconnect for internal communication wherever possible. Still there are several major problems when adapting hybrid model EA for the grid. First, the clusters have different sizes, varying from tens to thousands of cores. Although large clusters usually support more VOs, so the partition available for each of them is relatively smaller, but still the algorithm should be able to work with differences in the processing speeds of the islands reaching one or two orders of magnitude. Second problem is that the availability of the communication protocols is different over the grid. Not all of the clusters enable MPI usage and not all of the SEs allow the external clients (outside their own subnet) to use random I/O protocols (RFIO, dcap). The algorithm should allow using different protocols on both levels and be resistant to unpredictable communication delays.

The problems described above implicate the usage of asynchronous communication model both locally and globally. It is relatively easy to implement when using master-slave ssEA at the islands. The master process in ssEA is always ready to accept a newly rated individual from any of the slaves. Migrating individuals from other subpopulations can be treated in the same way. Using this design there is no need to synchronize migrations and local calculations phases between the islands. Migrants can be incorporated to any of the subpopulations at any time, and the local population benefits from their presence immediately (which is better than using the migration buffers).

The problem of huge differences in islands' processing speeds is a more serious one, and it can have potentially negative influence on the convergence of the whole algorithm. There are no theoretical approaches to this problem so far in the literature concerning parallel evolutionary algorithms. However, there is a thorough experimental study by J. Branke et al. in [13], where the authors analyze islands with processing speed differences more than one order of magnitude (the fastest one is 32 times faster than the slowest one). The tests are done with several topologies and migration strategies. The results show that such big differences in processing power of the islands indeed have negative impact on the convergence speed when compared to the uniform environment of the same summary size. The conclusion from the work is that the negative influence of non-uniform distribution of the resources can be minimized by using a topology where big and small islands are mixed (size differences between neighbours are maximized). Better results were also achieved when the migration was initiated by the receiving island (import strategy).

Another problem, specific for the grid environment, is the instability of the resources. There is no central resource manager (WMSes are not synchronized), so transient changes in the load of the clusters (including total congestion) are impossible to predict. Whole clusters can also appear and disappear from the information system (e.g. due to maintenance or failures) so the algorithm should be able to adapt to the actual state of the grid. It is therefore not advisable to use a static topology, especially with low δ (which is desired taking into account the slow interconnect), because the graph could even disintegrate. This was addressed in the algorithm by selecting random edges from fully connected topology at each migration. Dynamic changes of the computing power of the clusters have also another side effect. They should be reflected in appropriate changes of the migration intensity to maintain the optimal proportion between the communication and local calculations, like it is depicted on figure 1. Moreover, even without these changes, it would be a big burden on the user of the algorithm to tune the migration rates for each island manually, depending on the size of the island, number of islands and the optimized function. To automate this process it is possible to use adaptive migration. The idea is to run the migration operator when some criterion of convergence is met. The most popular indicator used in this kind of algorithms is based on changes of standard deviation of the fitness in the population. It has been proposed by Munetomo et al. in [11]. The standard deviation in the first generation is calculated and saved as σ_0. In every subsequent generation the current value of σ is calculated and the algorithm checks if σ/σ_0 did not fall below certain threshold (chosen arbitrarily between 0.1 and 0.3, depending on the optimized problem). If it is lower, the migration is performed. The algorithm can be also used in ssEA, despite that there are no designated generations in it – σ_0 is calculated immediately after first N_d individuals are collected, and subsequent checks of σ/σ_0 are done in regular intervals after each N_d processed individuals.

7. Technical details of the implementation

The algorithm outlined above has been implemented as an extension of the previous version, presented in [2]. The interface with CORAL works in the same way: a temporary file with the parameters is created and CORAL is run as an external program. The algorithm keeps track of all the running CORAL instances in the internal structure called execution queue. Fitness is calculated by parsing text output from CORAL and assigned to the proper genotype basing on the index in the queue. The master-slave part though uses different scheme of communication than its predecessor. The previous mechanism of dynamically launched slaves via cluster queuing system for each individual proven to be inefficient on heavily loaded clusters without special high-priority queues for short jobs. This was the case for the majority of the resources, because most of the clusters have just one queue per VO. Therefore the mechanism was changed to persistent slaves. Their communication with the master no longer uses queuing system mechanisms, but was changed to one of the following options:

- MPI. This is the preferred interconnect. The master process uses synchronous sends and asynchronous receives. The program can use any flavour of MPI (tested with OpenMPI 1.x and MVAPICH v2). To avoid version mismatches, the program can be compiled with local MPI library after submission to the target cluster in pre-run hook. No scaling problems were observed up to 1500 slaves, which was the maximum cluster share available (conditions of the test: DDR Infiniband network as a local interconnect, test fitness function with T_f=1s).
- GFAL. Backup interconnect option for clusters not having MPI. Local SE has to be given in the program options (automatic discovery was not implemented yet). Practical scaling boundary measured for DPM RFIO is about 150-170 slaves for the production fitness function, close to k^* estimated in chapter 5.

The upper layer of the algorithm works in fully connected logical topology with random drawing of the edge, which will be used in a particular migration call. In case of failure the procedure is repeated and another edge is drawn. Mutual discovery of the islands is addressed by using a central directory either on specified SE (faster, vulnerable to SE failure) or in LFC filesystem (slower, but with the possibility of creating replicas on different SEs). Total number of the islands has to be defined statically in the options before running the algorithm, but not all of them have to start at the same time. Migrations can use either of the two interconnects:

- GFAL. Requires the SE (or SEs) selected for migration to allow global incoming connections to random I/O ports (RFIO, gsidcap), so in certain configurations it may not work.
- GridFTP. Uses GridFTP protocol, slightly slower than random I/O but accessible for grid-wide connections on all the SEs. Actual transfers are performed by external client (uberftp), which is part of standard WN installation.

8. Test on a benchmark function

Initial tests were performed on slightly modified 10-dimensional Rastrigin function. It is popular benchmark to test optimization methods. The function is composed of low frequency paraboloid component modulated by higher frequency periodic function, so it has many local optima. The exact formula that was used in the test is following:

$$\Phi^{test}(x_1,...,x_{10}) = 100 - \sum_{i=1}^{10}[x_i^2 - 10\cdot\cos(2\pi\cdot x_i)] \quad (7)$$

All the variables were limited to <-10,10>. The main goal of the test (apart from verifying the performance of the algorithm on a well-known multi-modal problem) was to check how the slow interconnect and the non-uniform distribution of the resources affect the optimization process. It was not meant to examine all the possible configurations and topologies, because this kind of tests were already performed in [13], but rather to check if the smallest clusters separated from the rest of the algorithm by a barrier of slow interconnect can be of any benefit to the search process. The test conditions were set to amplify possible problems. Only two islands were used in the test, bigger one with 200 individuals and smaller one with 20. The testing program (emulating reconstruction program with Rastrigin formula) introduced only 1 second delay, so the proportion between T_f and T_c was unfavourable for the algorithm, compared to the real life case. The results are presented on figure 2.

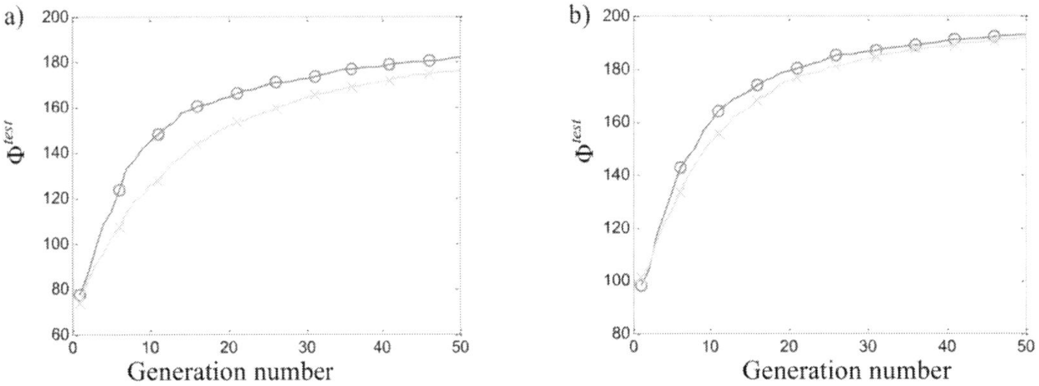

Figure 2. Values of Φ^{test} of the best individual averaged over 400 runs (o refers to migration on, x to migration off). a) small island with 20 slaves; b) big island with 200 slaves. Statistical significance level < 5% is maintained from generation 7 upwards for both graphs.

Migration was performed using GFAL and RFIO on DPM SE. Both islands used MPI for local communication. Figure 2a confirms that the smaller of the islands benefits considerably from turning on the migration. Noticeably better results with the migration turned on, can be also seen on the bigger island, on Figure 2b. These values were achieved with relatively low migration frequency of about 20 migrations per 10000 individuals processed locally. The test confirmed that using even the smallest clusters in the algorithm is beneficial for the quality of the results. The difference in the results diminishes in later generations, which is probably caused by being close to the global optimum at that stage.

9. Test with COMPASS reconstruction program

The test on the real case was performed using CORAL. The optimization concerned 28 parameters concurrently; all of them were geometrical tolerances and were represented by floating point numbers. Each individual was rated on the same set of 200 events. For the sake of the test, the configuration of the algorithm was the same as in the benchmark case. There were two islands, one with 200 slaves and the other with 20, both using MPI for local communication and GFAL for migration with the same migration intensity. The optimization in each configuration was executed 30 times. This is less than in the test case (because of time constraints), so the graphs are slightly more noisy. The results are presented on figure 3:

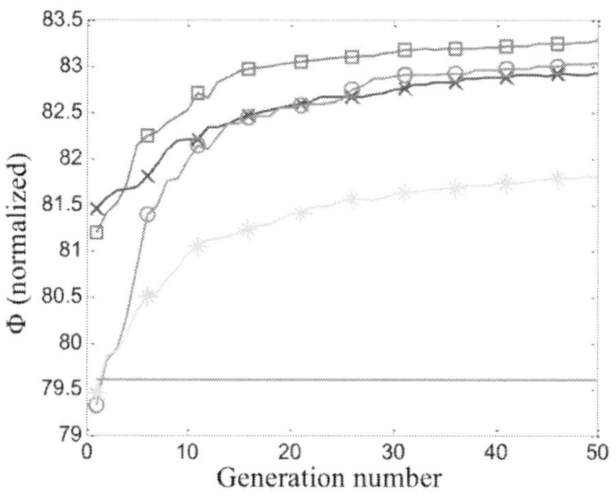

Figure 3. Values of Φ of the best individual, normalized to the percent of tracks possible to reconstruct and averaged over 30 runs. Line with red □ shows the results from big island with the migration, blue x from big island without the migration, magenta o from small island with the migration and green * from small island without the migration. Flat orange line at the bottom marks the level set by the default CORAL options (manually optimized). Statistical significance level < 5% is maintained from generation 13 upwards for both islands.

The results show that the optimization of the reconstruction is a problem more difficult, than the multimodal benchmark function used in the test run. The benefits from using the migration are higher and they stay high to the end of the run. The results achieved by the small island without the migration are low and the graph flattens at later generations, which is symptom of being stuck in local optima.

Big differences between the runs with the migration on and off lead to the conclusion that there is still room for improvement with adding more resources. Several tests were performed with 3-island setup (about 500 slaves in total, the same migration intensity) with noticeably better result. The detailed comparison between the default options and the best set found by the algorithm is following:

- reference options: efficiency 79,600% + 20 errors.
- best optimized results: efficiency 84,457% + 19 errors.

10. Conclusions and future work

The values achieved by the algorithm, when compared to the default CORAL options, confirm that the reconstruction can be successfully optimized using distributed EA. The overall performance gain over manually optimized parameters is more than 4,8% with 1 error less on the reference sample. It was also confirmed that the optimization of the reconstruction is a multimodal problem and should be treated by heuristic methods. The initial tests with derivative-based method (hill climbing which was added to the program as an operator) show no performance gain at all. Both benchmark and production runs show that the chosen communication model is effective for a broad spectrum of objective function execution times, despite the slow interconnect that is used between the islands. The main strength of the algorithm lies in automatic adaptation of migration frequency and random selection of migration partners. Thanks to these features the migration to local calculations ratio stays the same despite the number of islands used. Therefore the algorithm has no theoretical scaling boundary. However it can be expected that at some threshold the best solution will not have time to spread all over the algorithm until the end of the run. Theoretically this threshold can be estimated using simplified approach presented in [10], where it was shown that in order to reach the takeover condition, the number of epochs should be proportional to r divided by square root of δ. This calculation, although it was done with many simplifying assumptions, shows that the exact barrier should be checked experimentally in the production environment. This makes the analysis of the behaviour of the algorithm with big number of islands one of the most interesting topics for future work. Unfortunately it was impossible to make this kind of tests in the current setup, because COMPASS VO is a small one and it is supported only by couple of clusters in WLCG. It could be done though if the algorithm is used for a bigger experiment, with more complicated detector and larger computational resources. The drawback of using the method depicted in this paper is that the program optimizes the reconstruction parameters for a particular sample and detector alignment, used for the evaluation. This may result in lack of generality of the optimized parameters and lower performance for different samples and setups. The problem can be corrected, though, with randomization of the sample. The program has partial support for this kind of procedure – every evaluation can use random set of events selected from a larger file. High automation of the program and good results that are achieved even with relatively small resources make it also a good tool for continuous optimization, so the parameters could be optimized for every change in the detector setup. From the point of view of the algorithm itself, it may be useful to replace the σ criterion in migration adaptation with some other check that would be more sensitive to the changes on the later stages of population development. There is also some space for performance improvements lying in the implementation of better evolutionary operators and more automation in resources discovery. The work presented here was concentrated on the communication aspects in hybrid EA working in grid environment, so the rest of the EA framework used the most popular, simple algorithms. As there are very comprehensive studies on these topics available, including the classical [8], this would be more of an engineering work than research, but the real application that lies behind this work would surely benefit from it.

Acknowledgements
This work was supported by the EU and MSHE grant nr POIG.02.03.00-00-013/09.

References

[1] Goldberg D 1989 *Genetic Algorithms in Search, Optimization & Machine Learning* (Addison Wesley, Reading, MA)
[2] Padée A, Kurek K, Zaremba K 2007 *Meas. Sci. Technol.* **18** 2472
[3] The COMPASS Collaboration 1996 *Proposal: Common Muon and Proton Apparatus for Structure and Spectroscopy*. CERN/SPSLC 96-14, SPSC/P 297
[4] Abbon P et.al. 2007 *Nucl. Instr. Meth.* A **577** 455-518
[5] Kurek K et.al. 2002 *Nucl. Instr. Meth. Phys. Res.* A **485** 720-738
[6] http://wlcg.web.cern.ch/
[7] http://grid.ifca.es/int.eu.grid/
[8] Bäck T, Fogel D, Michalewicz Z 1996 *Handbook of Evolutionary Computation* (Oxford University Press)
[9] Tomassini M, Alba E 2002 *IEEE Trans. Evol. Comp.* **6(5)** 443-462
[10] Cantú-Paz E 2000 *Efficient and Accurate Parallel Genetic Algorithms* (Kluwer Academic Publishers, Norwell, MA)
[11] Munetomo M, Takai Y, Sato Y 1993 An Efficient Migration Scheme for Subpopulation-Based Asynchronously Parallel Genetic Algorithms *Proc. 5th Int. Conf. on Genetic Algorithms (Urbana-Champaign, IL, USA, June 1993)* ed S Forrest (Morgan Kaufmann) p 649
[12] Alba E, Troya J 1999 *Lecture Notes in Computer Science* **1586** 248-256
[13] Branke J, Kamper A, Schmeck H 2004 *Lecture Notes in Computer Science* **3102** 923-934
[14] Whitley D 1989 The GENITOR Algorithm and Selective Pressure: Why Rank-Based Allocation of Reproductive Trials is Best *Proc. 3rd Int. Conf. on Genetic Algorithms (Fairfax, USA, June 1989)* ed J Schaffer (Morgan Kaufmann) p 116-121

A numerical calculation method for frequency-shift of a plasma wave using wave digital filters

Toshio Utsunomiya and Masashi Shimokawabe

Department of Communications Engineering, National Defense Academy,
1-10-20 Hashirimizu, Yokosuka 239-0811, Japan

E-mail: utnomiya@nda.ac.jp

Abstract. It is difficult to measure any quantity in plasmas, since the physical quantity is usually measured only under very severe experimental conditions. For instance, it is often observed that the frequency of the wave changes during propagation and also different waves from the incident one are newly stimulated. Therefore, proposed is the most suitable method for computer experiment using technique of signal processing. In plasmas, various phenomena are described by a fluid theory. The fluid theory consists of a system of Maxwell's equations and that of Euler's equations. Applying the technique of wave digital filters (WDF) to these equation systems, the partial differential equations are integrated directly. Considering all physical quantities as currents, a system of partial differential equation systems is transformed into an equivalent Kirchhoff circuit. Wave-flow diagrams corresponding to their Kirchhoff circuits are obtained by the ordinary WDF procedures, which are algorithms themselves for numerical calculations. Therefore, we may safely say that these algorithms construct a computer experimental apparatus, which simulates many plasma experiments concerning propagation properties of plasma waves. Here, we simulate frequency-shift phenomena of some plasma wave by means of the function with respect to spectrum analysis of this apparatus.

1. Introduction

Some normal form of the ordinary differential equation of the first order can be solved by only integration as inverse manipulation of differentiation. Even if the problem cannot be solved analytically, the problem is solved by integrating numerically with the aid of trapezoidal rule and so on. This technique cannot be directly applied to partial differential equations from the mathematical point of view.

However, from the viewpoint of signal processing, it is realized to get the numerical solution for the partial differential equations by considering multidimensional operator of integration as an inverse operator, because mathematical manipulations such as differentiation and integration and so on are considered to be operators.

Some system of partial differential equations describing physical system can be transformed into an electric circuit. According to the Maxwell's analogy proposed in 1873, there has a merit that it is easy for us to treat the physical systems in computer if the physical systems are transformed into the circuits, since the mathematical models for the circuits corresponding to the various physical systems become the same ones when we treat them in computer.

From this point of view, a new method for numerically solving partial differential equations that describe physical systems has been proposed so far [1-3]. The fundamental principle of this method is as follows: In plasmas, various phenomena are described by a fluid theory. In the fluid theory, a system of plasma equations consists of a system of Maxwell's equations and that of Euler's equations. Euler's equation system is nonlinear, while Maxwell's one is linear. However, this method can treat their equations consistently whether they are linear or not. An outline of this method is as follows: At first, a system of partial differential equations describing some physical system is transformed into equivalent Kirchhoff circuits as electric circuits. And these circuits are transformed into wave digital filter (WDF) structure by digitizing based on wave digital filter theory. Obtained WDF structure leads to a numerical algorithm which is reduced to a solution in time-domain. Since the sampling of all variables is done along the time axis in the field of digital signal processing by nature, this method is considered to be a kind of FDTD method. Therefore, it is considered that this method is justified and also suitable for the analysis of time-dependent or transient phenomena [4,5]. In addition, the stability of numerical calculation is guaranteed, because the numerical calculation is carried out after constructing the digital filter structure.

Furthermore, it is found that this numerical method is very suitable for some computer experiments. Here, using this virtual plasma experimental apparatus, many plasma experiments concerning propagation properties of plasma waves are carried out and this method is furthermore applied to plasma scattering phenomena.

2. Basic plasma equations

First, Maxwell's equations are shown as

$$\nabla \times \boldsymbol{H} = \boldsymbol{j} + \frac{\partial(\varepsilon_0 \boldsymbol{E})}{\partial t}, \tag{1}$$

$$\nabla \times \boldsymbol{E} = -\frac{\partial(\mu_0 \boldsymbol{H})}{\partial t}, \tag{2}$$

where $\boldsymbol{j}$ is given by

$$\boldsymbol{j} = -n_e q_e \boldsymbol{v}_e \tag{3}$$

and n_e, $\boldsymbol{v}_e$ and $-q_e$ stand for density, velocity and charge of electron, respectively.

Second, as for a system of Euler's equations, ion is considered to be static as background for high frequency phenomena like around 0.1GHz, because the mass of ion is much heavier than that of electron. From this, we have only to consider the electron fluid alone for Euler's equation, continuity equation and equation of state, which are shown as

$$m_e n_e \left[\frac{\partial \boldsymbol{v}_e}{\partial t} + (\boldsymbol{v}_e \cdot \nabla)\boldsymbol{v}_e\right] + \nabla p_e = \boldsymbol{f}, \tag{4}$$

$$\frac{\partial n_e}{\partial t} + \nabla \cdot (n_e \boldsymbol{v}_e) = 0, \tag{5}$$

$$p_e = C\,(m_e n_e)^{\chi_e}, \tag{6}$$

where m_e, p_e and $\varkappa_e$ stand for mass, pressure and the ratio of specific heats of electron fluid, respectively. $\boldsymbol{f}$ is an external force against electron particle, which expresses the electric force and the Lorentz force shown as

$$\boldsymbol{f} = - n_e q_e [\boldsymbol{E} + \boldsymbol{v}_e \times \mu_0 (\boldsymbol{H}_0 + \boldsymbol{H})], \tag{7}$$

where $\boldsymbol{H}_0$ denotes a static magnetic field.

3. Equivalent Kirchhoff circuits

At first, usual coordinates (x, y, z) are replaced with (t_1, t_2, t_3) and furthermore t_4 is added in place of time, t, where the relation between t_4 and t is given by

$$v_4 = \frac{dt_4}{dt}. \tag{8}$$

Treating the problem in four-dimensional manifold [6, 7], physical variables are functions of four-dimensional vector, $\boldsymbol{t} = (t_1, t_2, t_3, t_4)^{\mathrm{T}}$. When partial derivative with respect to t_i is denoted as D_i, the operator vector $\mathbf{D}$ is defined as

$$\mathbf{D} = (D_1, D_2, D_3, D_4)^{\mathrm{T}}. \tag{9}$$

And furthermore introducing offset sampling [8], $\boldsymbol{t}$ and $\mathbf{D}$ are transformed into

$$t' = (t'_1, t'_2, t'_3, t'_4)^{\mathrm{T}} \tag{10}$$

$$\mathbf{D}' = (D'_1, D'_2, D'_3, D'_4)^{\mathrm{T}}, \tag{11}$$

respectively.

Here, problems are analyzed from a viewpoint of multidimensional signal processing [9,10]. Therefore, when we transform plasma equations into equivalent electric circuits, the supplementary operator vector as

$$\mathbf{D}'' = (D''_1, D''_2, D''_3, D''_4, D''_5, D''_6, D''_7), \tag{12}$$

is also introduced, where 4-dimensional space is embedded into 7-dimensional one [3].

From inverse point of view, embedding is considered to be an extension into higher dimensional space. It is said that lower dimensional problems can be often solved by transforming them into higher dimensional problems, which is safely said to be called as "Oka's lifting principle" [11]. We have transformed a system of Maxwell's equations as follows:

$$D''_7(\varepsilon''\hat{E}_1) + (D''_3 - D''_6)\hat{H}_2 + (D''_5 - D''_2)\hat{H}_3 + \hat{j}_1 = 0, \tag{13}$$

$$D''_7(\varepsilon''\hat{E}_2) + (D''_1 - D''_4)\hat{H}_3 + (D''_6 - D''_3)\hat{H}_1 + \hat{j}_2 = 0, \tag{14}$$

$$D''_7(\varepsilon''\hat{E}_3) + (D''_2 - D''_5)\hat{H}_1 + (D''_4 - D''_1)\hat{H}_2 + \hat{j}_3 = 0, \tag{15}$$

$$D''_7(\mu''\hat{H}_1) + (D''_6 - D''_3)\hat{E}_2 + (D''_2 - D''_5)\hat{E}_3 = 0, \tag{16}$$

$$D''_7(\mu''\hat{H}_2) + (D''_4 - D''_1)\hat{E}_3 + (D''_3 - D''_6)\hat{E}_1 = 0, \tag{17}$$

$$D''_7(\mu''\hat{H}_3)+(D''_5-D''_2)\hat{E}_1+(D''_1-D''_4)\hat{E}_2=0, \tag{18}$$

where hatted physical quantities stand for normalized ones.

On the other hand, a system of Euler's equations are also transformed as follows [12]:

$$\sum_{k=1}^{6}\eta_{0k}D''_k(\eta_{0k}\hat{p}_e)+\sum_{k=1}^{3}[D''_k(\hat{p}_e+\hat{v}_{ek})+D''_{k+3}(\hat{p}_e-\hat{v}_{ek})]=0, \tag{19}$$

$$\sum_{k=1}^{6}\eta_k D''_k(\eta_k\hat{v}_{el})+D''_l(\hat{v}_{el}+\hat{p}_e)+D''_{l+3}(\hat{v}_{el}-\hat{p}_e)=\hat{f}_{el} \qquad (l=1,2,3). \tag{20}$$

Considering all physical quantities as currents and also replacing the differential operators with inductances, a system of partial differential equations is transformed into what is called an equivalent Kirchhoff circuit, which is named by considering that each equation of the system mentioned above represents the Kirchhoff's second law. Kirchhoff circuit of Maxwell's equations is shown in figure 1. On the other hand, Kirchhoff circuit of Euler's equations is shown in figure 2. Resistances are not used in these circuits because of the losslessness of the physical system. A pair of zigzag lines facing each other stands for an ideal (lossless) transformer traditionally [13].

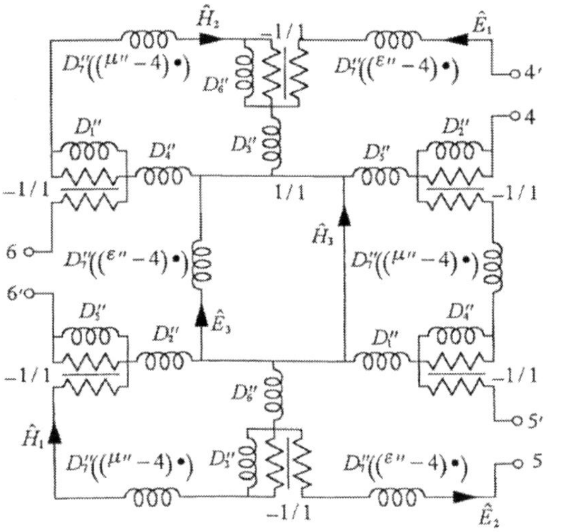

Figure 1. Kirchhoff circuit for a system of Maxwell's equations.

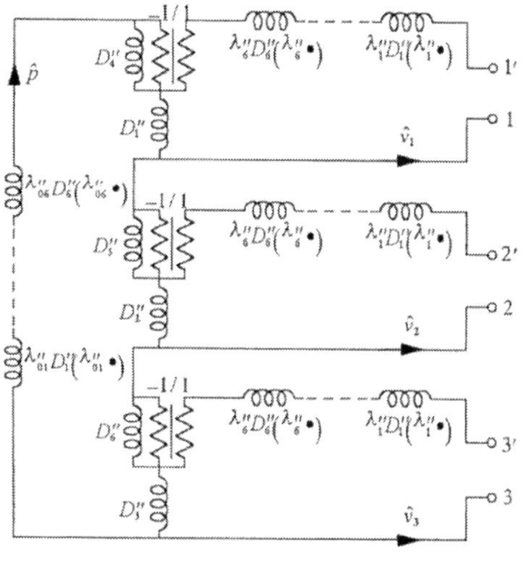

Figure 2. Kirchhoff circuit for a system of Euler's equations.

4. Derivation of WDF algorithm

In the wave digital filters, input and output variables correspond to wave quantities such as forward wave, a, and backward wave, b, respectively. Therefore, the wave-flow diagram plays the role of an algorithm of difference equation with respect to wave quantity in the time step cycle.

On the other hands, in figures 1 ~ 5, endpoint is drawn as to be connected to another circuit or flow-graph, because Kirchhoff circuit or wave-flow diagram of the plasma equations is drawn in a single figure by nature.

5. Computer experiments and results

As an example, analyzed are propagation properties of some plasma wave in addition to scattering phenomena from a cylindrical conductor as an obstacle immersed in a plasma which is magnetized perpendicularly to the paper plane in figure 6 [14,15]. When a plane electromagnetic wave with stimulated longitudinal waves is incident from left-hand side on a cylindrical conductor whose radius a is given by $ka = 1$, calculations are done in the area $3\lambda \times 3\lambda$, where λ is wavelength of the incident electromagnetic wave at initial time.

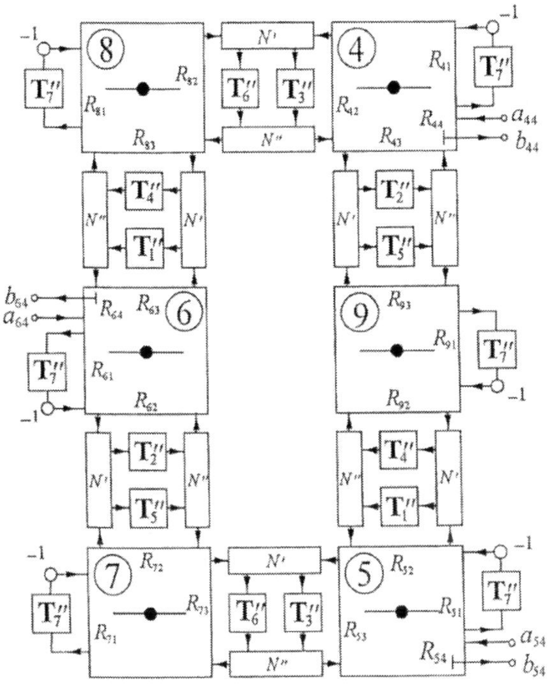

Figure 3. Wave-flow diagram for Maxwell's Euler's equations.

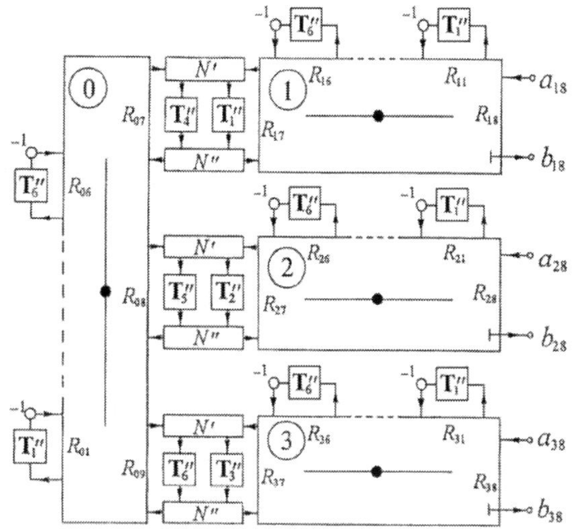

Figure 4. Wave-flow diagram for equations.

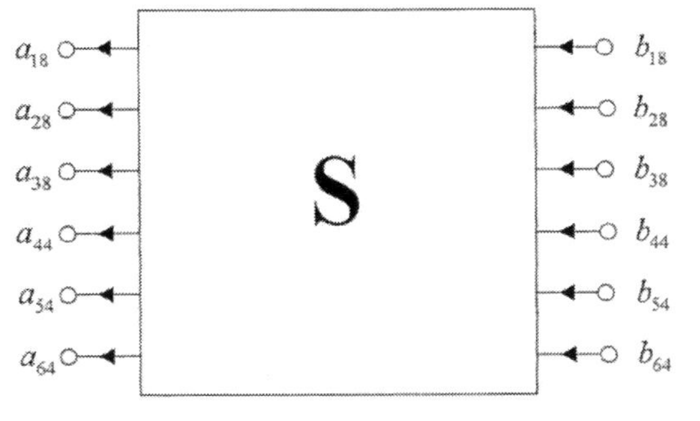

Figure 5. Wave-flow diagram for connecting network.

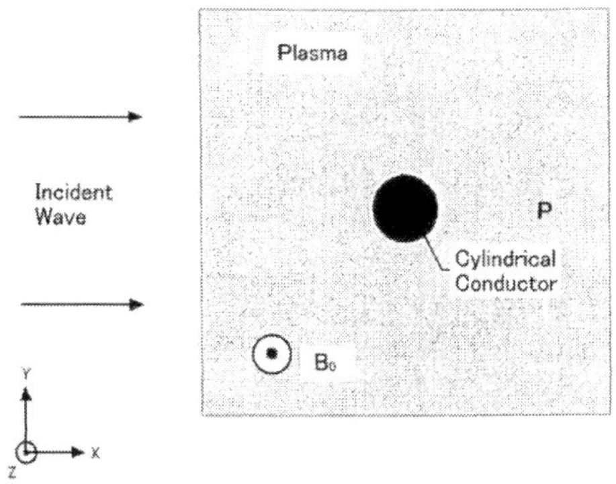

Figure 6. Setup for numerical calculations.

Boundary conditions in this case are as follows: For Maxwell's equations, tangential component of the electric field on the boundary surface is null because of continuity condition of tangential components of electric fields between both media. And for Euler's equations, normal components of both velocity and external force are null, because the cylindrical conductor is considered merely as a rigid body. Furthermore, by similar reason, null is the first derivative of both the tangential component of velocity and the pressure with respect to normal direction to the boundary surface.

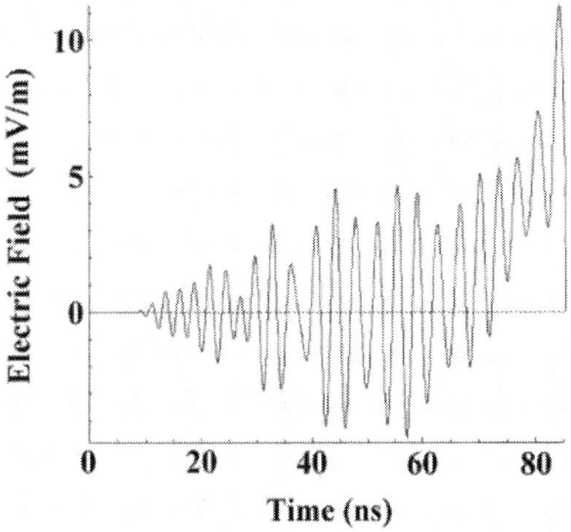

Figure 7. Observed waveform at the point P.

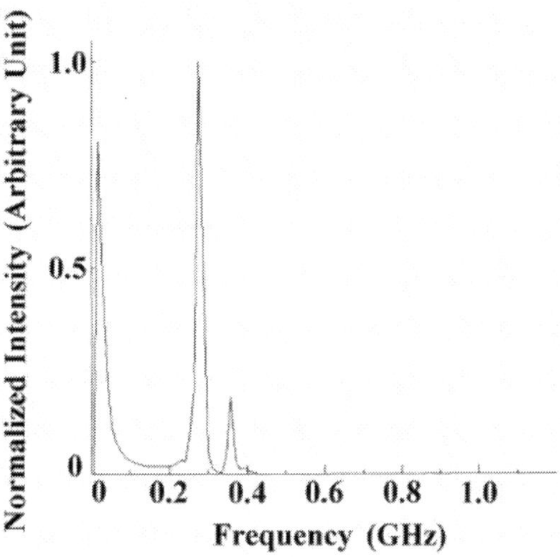

Figure 8. Frequency spectrum of the observed waveform

It is often observed that the frequency of the wave changes during propagation and also different waves from the incident one are newly stimulated. To inspect these effects, the electric field is observed at a point P behind the conductor cylinder in the plasma area of the plasma experimental apparatus as shown in figure 6. Electric field observed at the point P is recorded automatically and is drawn as in figure 7. After all, frequency spectrum as shown in figure 8, which corresponds to the waveform in figure 7, is also obtained. Such a function like this is contained in this virtual plasma experimental apparatus.

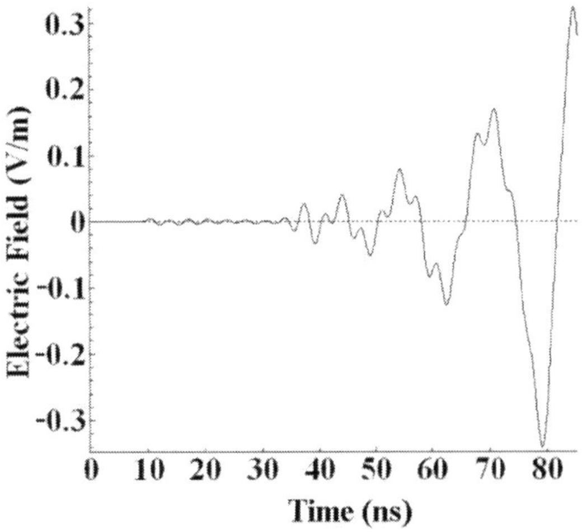

Figure 9. Observed waveform for $\omega_c / \omega = 2.5$.

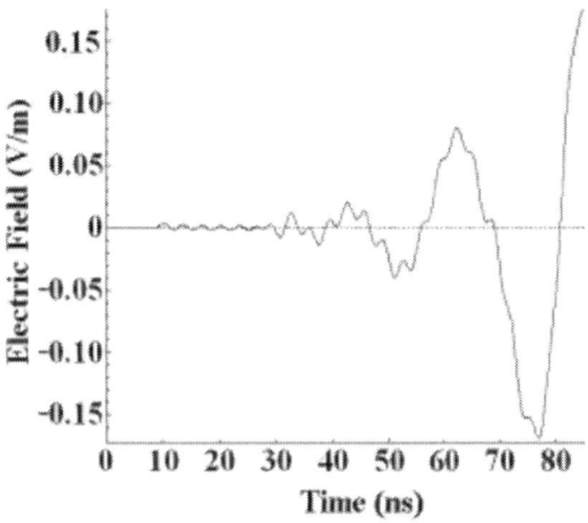

Figure 10. Observed waveform for $\omega_c / \omega = 4.0$.

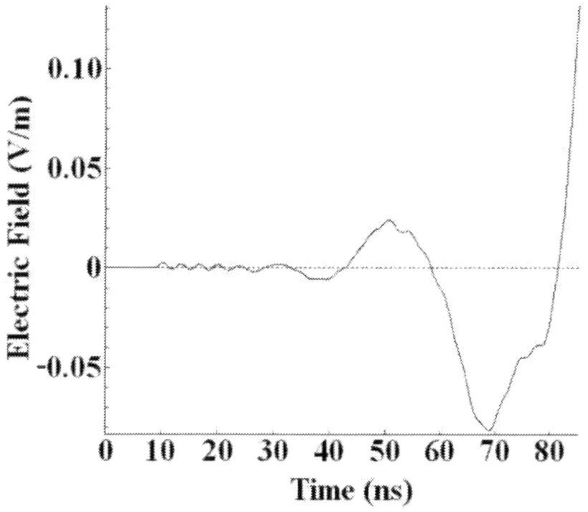

Figure 11. Observed waveform for $\omega_c / \omega = 7.0$.

Since figures 7 and 8 show waveform and spectrum, respectively when the frequency of the incident wave corresponds to the hybrid resonant one, the calculated results are very interesting, where electron density, static magnetic field and wavelength of incident wave are 1.0×10^{15} [m^{-3}], ω_c / ω = 0.321 and 1.0 [m], respectively.

As a result, in the frequency spectrum of figure 8, it is clearly found that the third peak is appeared whose frequency is higher than that of the second peak which corresponds to that of the incident wave. In addition, the frequency of the first peak is lower than that of any of other two peaks. That is why the frequency of the first peak corresponds to that of the envelope of the observed waveform.

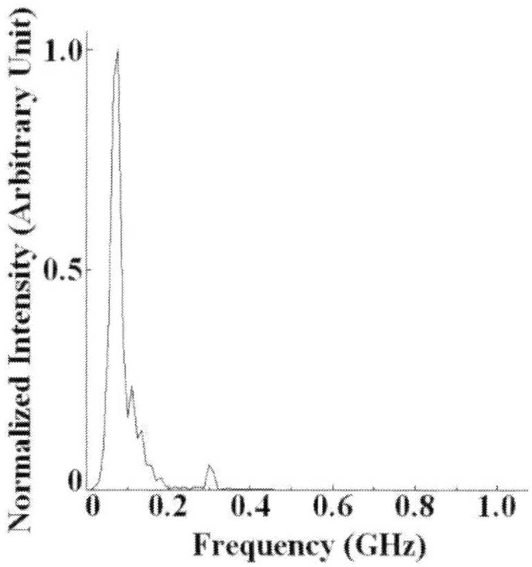

Figure 12. Corresponding frequency spectrum to the waveform for $\omega_c / \omega = 2.5$.

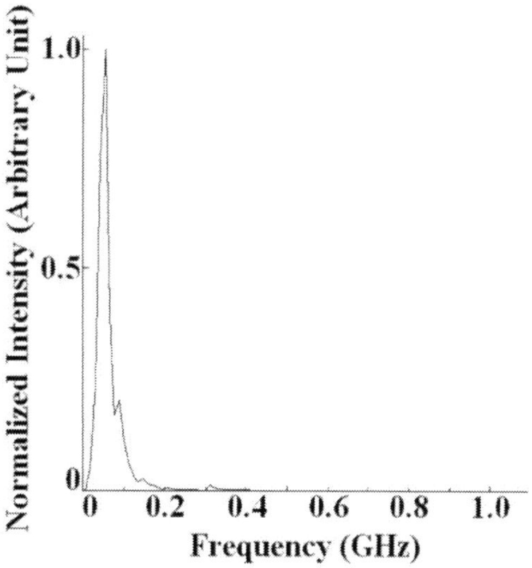

Figure 13. Corresponding frequency spectrum to waveform for $\omega_c / \omega = 4.0$.

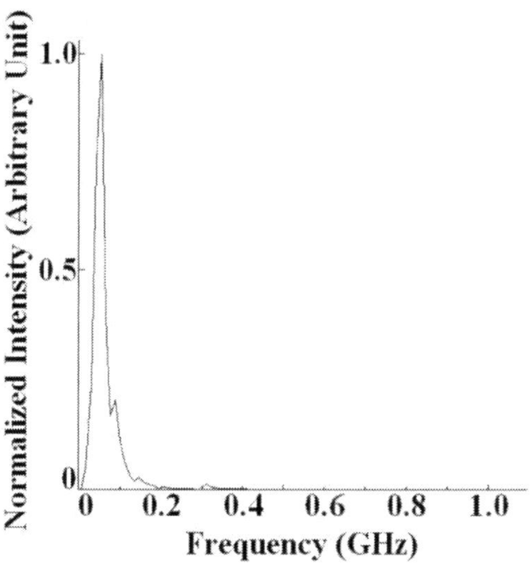

Figure 14. Corresponding frequency spectrum to the waveform for $\omega_c / \omega = 7.0$.

And furthermore, figures 9 ~ 11 show the observed waveforms corresponding to the static magnetic fields $\omega_c / \omega = 2.5$, 4.0 and 7.0, respectively, where the electron density is 5.574×10^{14} [m^{-3}]. Here, in figures 9 ~ 11, only longitudinal components of the electric field are drawn, because separated effects from the incident transverse electromagnetic wave are considered to be emphasized.

On the other hand, figures 12 ~ 14 show the frequency spectra corresponding to the waveforms for figures 9 ~ 11, respectively.

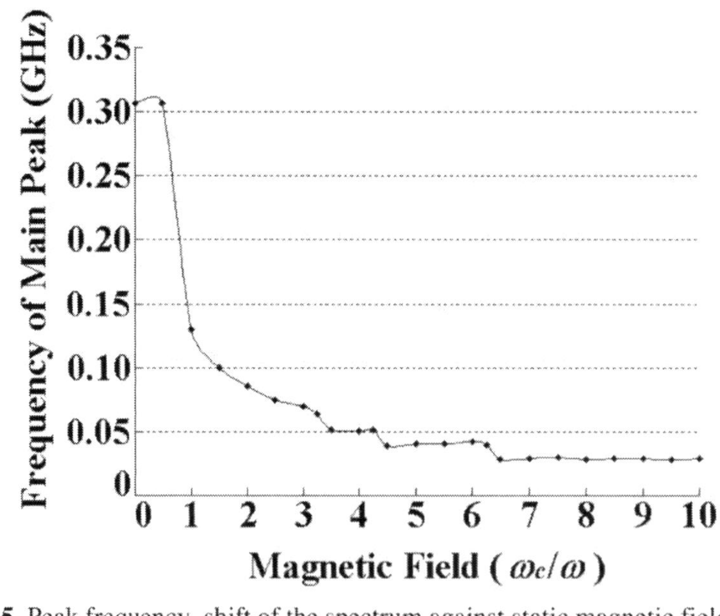

Figure 15. Peak frequency–shift of the spectrum against static magnetic field ω_c / ω.

Watching these figures, it is easily found that ripples on the swell become smaller according as applied magnetic field becomes stronger. This corresponds to the fact that the peak of the frequency spectrum around 0.3GHz as the incident wave becomes smaller similarly.

Journal of Physics: Conference Series **454** (2013) 012085 doi:10.1088/1742-6596/454/1/012085

From these figures, it is clearly found that the stronger the static magnetic field becomes, the lower the frequency of the main peak becomes. Figure 15 shows that the frequency of the main peak in the spectrum of the waveform of the propagating plasma wave shifts lower against increasing magnetic field from $\omega_c / \omega = 0$ to $\omega_c / \omega = 10$.

Since this apparatus is a virtual experimental one, it is natural that this method is suitable for the time-dependent phenomena. However, even in steady state, this method is applicable if the time is sufficiently elapsed. The scattering pattern is also actually obtained by this manner.

Figure 16. A scattering pattern of the longitudinal plasma wave.

Figure 16 shows an example of scattering pattern as a steady state of the longitudinal plasma wave against the conductor cylinder when electron density, static magnetic field and wavelength of incident wave are 1.0×10^{15} [m^{-3}], $\omega_c / \omega = 2.0$ and 1.0 [m], respectively.

6. Conclusion

In calculating the frequency-shift phenomena of a plasma wave during propagation, used is the so-called wave-digital method which utilizes the principle of the multidimensional wave digital filters.

Considering all physical quantities as currents, transforming partial differential equations into equivalent electric circuits by using Kirchhoff's second law and moreover digitizing them, various parameters and derived phenomena in plasmas are calculated with time. In other words, this method is a kind of FDTD method, because all variables are sampled along the time axis as usual procedure in digital signal processing, and spatial and physical variables are calculated every sampling time. Therefore, the validity of this method is guaranteed.

Furthermore, other many advantages are added by transforming a system of differential equations into a circuit network system, because differential equations themselves in the field of mathematics can be grasped as a physical system based on the energy principle. Therefore, singular types of partial differential equations, as is often shown in the field itself of mathematics, are naturally excluded. There are some instances: one is that, by introducing the power wave, this method can treat the partial differential equations consistently whether they are linear or not, another is that inconveniences inherent in mathematical or numerical manipulations are avoided by introducing circuit concepts like stability of the circuit system as an example, and so forth.

From this, applying the wave-digital method to the plasma equation system consisting of a system of Maxwell's equations and that of Euler's equations, a plasma simulator is actualized. As plasma is the dispersive medium, it is often observed that the frequency of the wave changes during propagation and also different waves from the incident one are newly stimulated.

Observing the electric field at an arbitrary point in the plasma area of the virtual experimental apparatus, we can inspect them.

Accordingly, we may safely say that this virtual plasma experimental apparatus is called wave-digital plasma simulator. It is considered that this wave-digital plasma simulator is promising for the virtual experiments such as nonlinear interaction among various plasma waves.

References

[1] Fischer H D 1984 Wave digital filters for numerical integration *ntz Archiv* **6** 37-40

[2] Fettweis A and Nitsche G 1991 Numerical integration of partial differential equation using principles of multidimensional wave digital filters *J. VLSI Signal Processing* **3** 7-24

[3] Fettweis A and Nitsche G 1991 Transformation approach to numerically integrating PDEs by means of WDF principles *Multidimensional System and Signal Processing* **2** 127-159

[4] Okayama M and Utsunomiya T 1996 Transient analysis of electromagnetic waves using principles of wave digital filters *Proc. 1996 International Symposium on Antenna and Propagation* **3** 613-616

[5] Naka Y, Ikuno H , Nishimoto M and Yata A 1998 FDTD method with PMLs ABC based on the principles of multidimensional wave digital filters for discrete-time modelling of Maxwell's equations *IEICE Trans.Electron.* **E81-C** 305-314

[6] Matsushima Y 1971 *Introduction to manifold* (Tokyo: Shokabo)

[7] Komatsu A and Nagata M 1966 *Algebra and geometry* (Tokyo: Kyoritsushuppan)

[8] Liu X and Fettweis A 1990 Multidimensional digital filtering by using parallel algorithms based on diagonal processing *Multidimensional Systems and Signal Processing* **1** 51-66

[9] Fettweis A 1986 Wave digital filters: theory and practice *proc. IEEE* **74** 270-327

[10] Fettweis A 1992 Multidimensional wave digital filters for discrete-time modelling of Maxwell's equations *Int. J. Numerical Modelling* **5** 183-201

[11] Nishino T 2001 *Function Theory in Several Complex Variables*(Province, Rhode Island: American Mathematical Society)

[12] Fettweis A 1992 Discrete modelling of lossless fluid dynamic systems *Archiv fuer Elektronik und Uebertragungtechnik* **46** 209-218

[13] Sakaki Y, Oono Y and Ozaki H 1968 *Electric circuit (1)* (Tokyo: Ohmsha)

[14] Utsunomiya T 2000 A plasma simulation method using technique of wave digital filters *Progress of Theoretical Physics* Suppl. **138** 741-743

[15] Utsunomiya T 2006 Simulations of plasmas and waves by using technique of wave digital filters *Journal of The Japan Society for Simulation Technology* **25** 43-52

Wavelet transform and Huffman coding based electrocardiogram compression algorithm: Application to telecardiology

S A Chouakri[1], O Djaafri[1] and A Taleb-Ahmed[2]

[1]Laboratoire de Télécommunications et Traitement Numérique du Signal, University of Sidi Bel Abbes, Algeria

[2]Laboratoire LAMIH, Université de Valenciennes et du Hainaut Cambrésis, France

E-mail: sa_chouakri@hotmail.com

Abstract. We present in this work an algorithm for electrocardiogram (ECG) signal compression aimed to its transmission via telecommunication channel. Basically, the proposed ECG compression algorithm is articulated on the use of wavelet transform, leading to low/high frequency components separation, high order statistics based thresholding, using level adjusted kurtosis value, to denoise the ECG signal, and next a linear predictive coding filter is applied to the wavelet coefficients producing a lower variance signal. This latter one will be coded using the Huffman encoding yielding an optimal coding length in terms of average value of bits per sample. At the receiver end point, with the assumption of an ideal communication channel, the inverse processes are carried out namely the Huffman decoding, inverse linear predictive coding filter and inverse discrete wavelet transform leading to the estimated version of the ECG signal. The proposed ECG compression algorithm is tested upon a set of ECG records extracted from the MIT-BIH Arrhythmia Data Base including different cardiac anomalies as well as the normal ECG signal. The obtained results are evaluated in terms of compression ratio and mean square error which are, respectively, around 1:8 and 7%. Besides the numerical evaluation, the visual perception demonstrates the high quality of ECG signal restitution where the different ECG waves are recovered correctly.

1. Introduction

Telemedicine is defined as a part of medicine that uses the transmission, by telecommunication means, of medical information (images, recording, etc.), to obtain a diagnosis, from a specialized, remote monitoring of a patient and a therapeutic decision [1]. For the World Health Organization (WHO) [2], telemedicine is a branch of medicine. It enables to bring the health to where the distance and isolation are factors limiting its access. They use NICT (New Information and Communication Technologies) for diagnosis, treatment and prevention, research and ongoing prevention.

The history of telemedicine began in the 1960s in the United States, with particular networking tele-consultation and tele-education around the Nebraska Psychiatric Institute programs [3-6]. Technological developments have helped remote transmission, in real time, of medical data that involved, firstly, voice and image transmission.

Technically, telemedicine is based on the exchange of data between several terminals. Several technical solutions have been tested with varying degrees of success to enable the transmission of information. The main physical supports for data transmission are the twisted pair (Ethernet, for example), coaxial cables, telecommunications, satellite links network and optical fiber.

Telecardiology, a widespread discipline of telemedicine, is a method of remote sensing a cardiovascular pathology. Compression is a method to manipulate an initial set of data by coding to save storage space. Electrocardiogram (ECG) signal compression has become more important with the development of telemedicine; it can significantly reduce the costs of transmission of medical information through telecommunication channels. There are two broad categories of the ECG signal compression algorithms: lossy and lossless.

Numerous works have been developed in the context of analysis and compression of the ECG signal, using various techniques of processing of the signal such as AZTEC, SAPA, KLT, DCT and the Fourier transform; we quote principally [7-11].

The time-frequency joint analysis (and we focus mainly on the wavelet transform) seemed one of the alternatives to the theory of Fourier. Wavelets are applied in the analysis of the signal (for example, the location of discontinuities in a signal), functional analysis (e.g. numerical analysis) and the processing of the signal (the coding into sub-bands) [12]. Wavelets are considered to be one of the most interesting tools for various ECG signal processing applications; we cite a few techniques: compression, denoising, analysis of the ECG signal. Mentioning a few works in the wavelet domain [13-15], who used the wavelet transform for compression of the signal ECG, Chen et al. (1993) [16] got a compression ratio of the ECG signal up to 22.9:1, Bradie (1996) [17] completed the compression by wavelet packets for the ECG signal compression. Miaou and Lin (2002) [18] have worked on bio-orthogonal wavelets [19]. Ku CT et al. (2010) [20] conducted research on the ECG signal compression by the wavelet transform based on the calculation of linear control quality.

Advanced telemedicine applications require a sophisticated and expensive telecommunications infrastructure. The main technologies used in wireless systems, namely GSM, 3G (W-CDMA, CDMA2000 and TD-CDMA), wireless GPRS, satellite and local area network wireless [21-24].

2. Telemedicine system

A telemedicine system can essentially be composed of a material with a computer that transmits signal/image to another computer remotely. Each of the computers must be connected to a modem that ensures the interface of the PC an analog channel. The images can be viewed, stored and transmitted to the remote computer. Software includes tools to scan, transmit, process and compress images. A communication network connects the two computers between which data is to be exchanged. A classic telemedicine channel, used in biomedical devices, is represented in figure 1 [25].

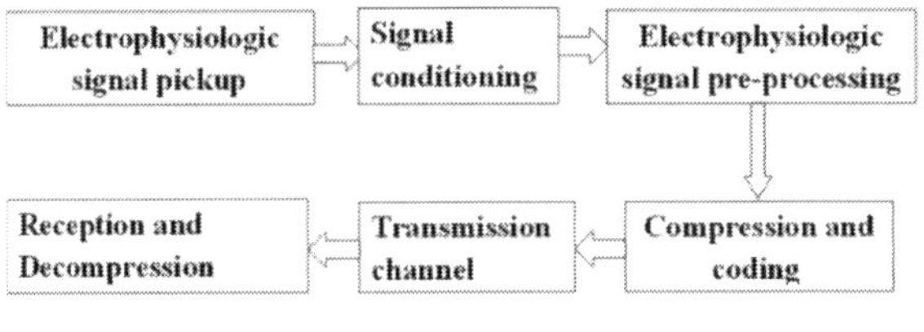

Figure 1. Synoptic scheme of telemedecine channel.

The components of the telemedicine channel, from the patient until the distant doctor, are first the collection of physiological signals by electrodes then the processing of data (decorrelation). The physiological signal decorrelation requires transformation methods, which operate a transformation of the original data. The physiological signal is decomposed on a basis of functions, and then the coefficients of the decomposition are quantified. The quantized coefficients are stored in a vector in a pre-determined order. The coding is the last step in the chain of compression.

3. Compression

Typical signal processing for automated medical systems acquired a large amount of data difficult to store and transmit [26-27]. The compression is to reduce the amount of data required for the description of the information. It is a technique to manipulate a set of initial data by a given encoding to save storage space. Note that compression can be divided into two types: lossy compression and lossless compression. In order to achieve these effects, it is desirable to find a method to reduce the amount of data without loss of important information [6, 28].

In the case of lossless compression, information (signal, sound, image, etc.) rebuilt after decompression is identical to the original data. No distortion is introduced on the bits of information. Thus, the process of compression will not alter the original signal. In contrast, lossy compression introduces irreversible disruption but allows a much greater compression rate than that obtained by lossless methods.

There is a wide variety of information compression techniques. In order to compare these methods, different criteria are necessary.

3.1. The measurement of the compression

The measurement of the compression and the reconstruction error are usually mutually dependent and are generally used for the calculation of the rate of deformation of the compressed signal induced by the algorithm.

3.1.1. Error criteria

Most of ECG compression algorithms use the Percentage of Root Mean Square Difference (PRMSD) between the original signal and the compressed one [26]

$$PRMSD = \sqrt{\frac{\sum_{n=1}^{N}(x(n)-\tilde{x}(n))^2}{\sum_{n=1}^{N}x^2(n)}}\times 100, \tag{1}$$

where $x(n)$ is the original signal, $\tilde{x}(n)$ is the reconstructed signal, and N represents the number of samples of the signal. There are other measures of error to compare original signal versus the reconstructed one, such as the root mean square error (RMS)

$$RMS = \sqrt{\frac{\sum_{n=1}^{N}(x(n)-\tilde{x}(n))^2}{N}}. \tag{2}$$

Another measure of deformation is the SNR signal-to-noise ratio, which is been expressed as

$$SNR = 10\log\left(\frac{\sum_{n=1}^{N}(x(n)-\bar{x}(n))^2}{\sum_{n=1}^{N}(x(n)-\tilde{x}(n))^2}\right), \tag{3}$$

where $\bar{x}$ is the average value of the original signal.

3.1.2. Compression evaluation

The compression rate (CR) is defined as the ratio to the bitrate of the original signal to the compressed signal bitrate.

4. Materials

4.1. Wavelet analysis

Signal analysis dates back to 1807, when J. Fourier developed his famous theory [29-30]. New technologies pushed researchers to introduce new techniques; among those one called the time-frequency representation, mainly presented by Gabor in 1946, and Wiegner-Ville in 1948. However these techniques had some limitations. A. Grossman and J. Morlet, in 1983, introduced the "wavelet"

Journal of Physics: Conference Series **454** (2013) 012086 doi:10.1088/1742-6596/454/1/012086

the main feature of which was the variable scale. After that date many works enriched the wavelet theory; we limit ourselves to quote the work of Y. Meyer, in 1985, who introduced the orthogonal basis for DWT, that of S. Mallat, in 1988, who introduced the multiresolution analysis (MRA), and that of I. Daubechies, in 1988, who developed regular wavelets with compact support [12, 31-33]. One of the various fields of application of wavelet theory is biomedical signal. There are, generally, two principal applications: analysis and filtering. The first one mainly includes the detection and recognition of forms of different waves of ECG signals, while the second one includes the ECG signal denoising and compression techniques.

4.1.1. The continuous wavelet transformation (CWT)

The CWT of a signal f(t) $\in L^2(R)$ is the inner product of the function f(t) and the set of the functions $\psi_{s,\tau}(t)$, called wavelet and given by [34]

$$Wf(s,\tau) = < f, \psi_{s,\tau} > = \frac{1}{\sqrt{s}} \int f(t) \psi^* (\frac{t-\tau}{s}) dt \tag{4}$$

The wavelet functions $\psi_{s,\tau}(t)$ are generated from a prototype function called 'mother wavelet' $\psi(t)$ by scaling and translation:

$$\psi_{s,\tau}(t) = \frac{1}{\sqrt{s}} \psi(\frac{t-\tau}{s}) \tag{5}$$

where s is the scaling factor, with $s > 0$, and 'τ' is the shifting factor.

4.1.2. The discrete wavelet transform DWT

The CWT allows the time-frequency analysis of a signal f (t) with its fine details as the factor of scale and translation are iterated continuously. However, the CWT is very redundant. Y. Meyer has shown that they exist of wavelet functions $\psi_{s,\tau}(t)$ such as $s = 2^j$ and $\tau = k.2^j$ with j and k $\in Z$. The functions $\psi_{j,k}(t)$ constitute an orthonormal basis in L^2 (R). These wavelets $\psi_{j,k}(t)$ are given by:

$$\psi_{j,k}(t) = 2^{-j/2} \psi(2^{-j}t - k), \tag{6}$$

which means that any function f (t) $\in L^2$ (R) can be decomposed upon the bases $\psi_{j,k}(t)$ as follows:

$$f(t) = \Sigma_j \Sigma_k d_{j,k}(t) \psi_{j,k}(t) \tag{7}$$

$$\text{with } d_{j,k}(t) = <f, \psi_{j,k}> = \int f(t). \psi_{j,k}(t)\, dt. \tag{8}$$

More generally, the function f (t) $\in L^2$ (R) can be expressed as follows:

$$f(t) = \sum_{k \in Z} a_L(k) \phi_{L,k}(t) + \sum_{j=1}^{L} \sum_{k \in Z} d_j(k) \psi_{j,k}(t) \tag{9}$$

$$\text{with } a_L(k) = <f(t), \phi_{L,k}(t)> \tag{10}$$

The DWT consists of applying the discrete signal to a bank of octave band filters. Figure 2 illustrates the DWT process.

4.2. Thresholding

Wavelet thresholding is a non-linear method to suppress noise corrupted to the signal of interest. A threshold is estimated and referring to it the noise is removed. The choice of the threshold is a very delicate and important statistical problem. On the one hand, a big threshold leads to a large bias of the estimator. But on the other hand, a small threshold increases the variance of the smoother.

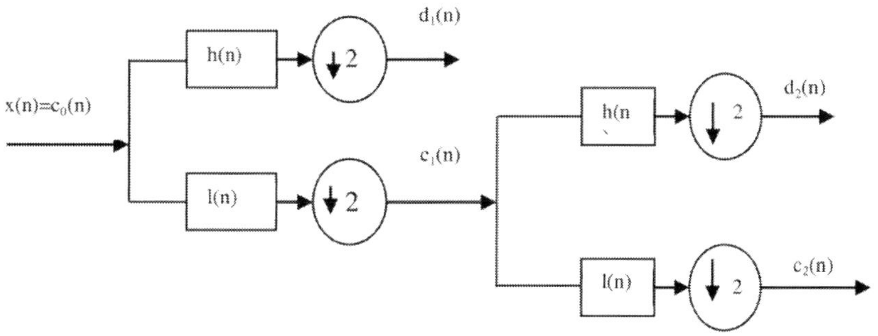

Figure 2. Schematic wavelet series decomposition.

The threshold value is computed, using a set of strategies, according to the model of the signal to denoised - s(i) - and the corrupted noise - e(i) -. The underlying model for the noisy signal is the superposition of the signal s(i) and a Gaussian white noise with a noise power of σ, that is [35-36]:

$$f(i) = s(i) + \sigma e(i) \tag{11}$$

Two strategies of thresholding are used: 'hard' and 'soft' (T_{soft} and T_{hard} respectively). In hard thresholding, known as crude thresholding, the coefficients whose absolute values are lower than the threshold *Thr* are eliminated that is set to zero. The soft thresholding, known as shrinkage algorithm, is a nonlinear operation that sets to zero the coefficients whose absolute values are lower than the threshold *Thr*, and then shrinks the nonzero coefficients towards zero.
Mathematically saying, this can be expressed as follows:

$$\begin{aligned} T_{soft} &= sign(x) .(|x|-Thr) \\ T_{hard} &= x.1(|x|>Thr) \end{aligned} \tag{12}$$

4.3. High order statistics (HOS)
High Order Statistics measures are extensions of second-order measures (such as the autocorrelation function and power spectrum) to higher orders. The second-order measures work fine if the signal has a Gaussian (Normal) probability density function, but many real-life signals are non-Gaussian.

In probability theory and statistics, kurtosis is any measure of the "peakedness" of the probability distribution of a real-valued random variable. *Kurtosis* is a descriptor of the shape of a probability distribution [37-39].

For univariate data $Y_1, Y_2, ..., Y_N$, the formula for kurtosis is:

$$kurtosis = \frac{\sum_{i=1}^{N}(Y_i-\bar{Y})^4}{(N-1)s^4}, \tag{13}$$

where $\bar{Y}$ is the mean, s is the standard deviation, and N is the number of data points.

The kurtosis for a standard normal distribution is three. For this reason, some sources use the following definition of kurtosis (often referred to as "excess kurtosis"):

$$kurtosis = \frac{\sum_{i=1}^{N}(Y_i-\bar{Y})^4}{(N-1)s^4} - 3. \tag{14}$$

This definition is used so that the standard normal distribution has a kurtosis of zero. In addition, with the second definition positive kurtosis indicates a "peaked" distribution and negative kurtosis indicates a "flat" distribution.

4.4. Huffman coding

In Huffman coding, the data are represented as a set of variable length binary words. The lengths depend on the occurrence frequency of the symbols used for representing each signal value. Characters that are used often are represented with fewer bits, and those that are seldom used are represented with more bits. This is a statistical data compression method that allows reducing the average length of the coding of an alphabet [40].

The Huffman coding algorithm allows building a variable length coding with an average length that is close to the entropy of the source. The principle illustrated by figure 3 is as follows:

1) The symbols are listed in order of decreasing probability. Is then assigned two low probability symbols 0 and 1 bits.

2) The two symbols are combined into a fictional symbol which the probability is the sum of the probabilities of elementary symbols. A new bit 0 and 1 are arbitrarily attributed to each branch.

3) The procedures 1) and 2) is repeated until the list has more than two elements to which we affect bits 0 and 1.

4) The code for each initial symbol is then obtained from the root of the tree to the leaves.

Figure 3 illustrates an example of Huffman coding of an alphabet with five symbols $\{a_1, a_2, a_3, a_4, a_5\}$.

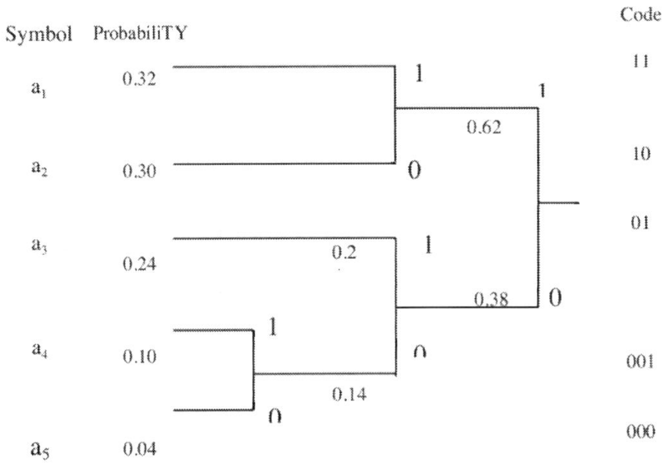

Figure 3. An example of Huffman coding.

Linear prediction

The applications of the linear prediction coding are multiple from signal compression to digital control. In General, the calculation of the coefficients of the linear prediction [41] is summarized as follows.

The problem is set as follows: If one knows N samples of a signal x[n], how to determine the p "optimal" coefficients a_j such that the prediction error is the smallest possible on the block duration N.

$$e[n] = x[n] - \sum_{j=1}^{p} a_j x[n-j] \quad \text{where} \quad p[n] = \sum_{j=1}^{p} a_j x[n-j] \tag{15}$$

is the prediction at time n of the sample $x[n]$.

We can see the Predictor as a filter that takes $x[n]$ input, and which produces the prediction error output $e[n]$. By applying the Z-transform to the difference equation, we obtain

$$E(z) = \left(1 - \sum_{j=1}^{p} a_j z^{-j}\right). \tag{16}$$

So $A(z)$ of this predictor filter transfer function is given by

$$A(z) = 1 - \sum_{j=1}^{p} a_j z^{-j} \tag{17}$$

and the inverse predictor transfer function is given by:

$$\frac{1}{A(z)} = \frac{1}{1-\sum_{j=1}^{p} a_j z^{-j}} \tag{18}$$

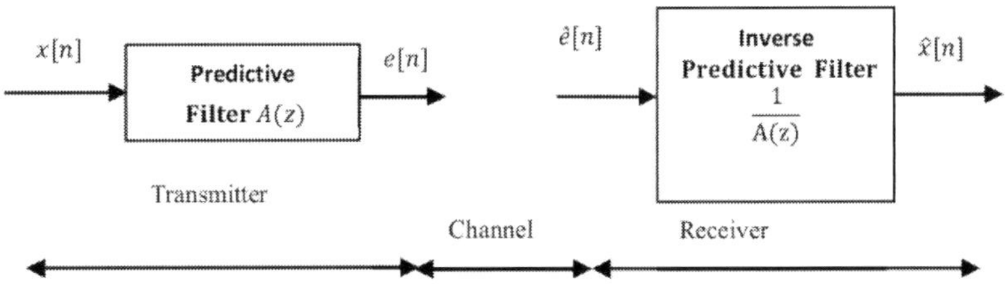

Figure 4. Typical scheme of the linear prediction encoder.

In the diagram of figure 4, $x(n)$ is the original signal, $e(n)$ is the output of the Predictor filter "analysis", $\hat{e}(n)$ is the inverse Predictor filter entry "synthesis" and $\hat{x}(n)$ is the reconstructed signal with $A(z) = 1 + a_1 z^{-1} + \cdots + a_p z^{-p}$.

5. Algorithm description

We propose in this paper, an ECG signal compression algorithm based on the wavelet transform and higher-order statistics (HOS), using the linear prediction coding followed by Huffman encoding to transmit the compressed signal via an ideal channel, and evaluate the performance of the compression system. The particularity of the ECG signal is to be compressed in the irreversible cases (non-conservative compression) under severe quality constraints that can be measured in terms of the PRD (Percentage of the RMS Difference). Our work is registered in the thesis of the telecardiology. The WHOSC method was proposed in [42-43], where the encoding of the ECG data is performed by linear prediction coding (LPC) of wavelet coefficients. Our contribution is the introduction of Huffman coding whose the objective is an improvement in quantifying quality gains by reducing the number of bits representing the samples. Figure 5 shows the proposed algorithm.

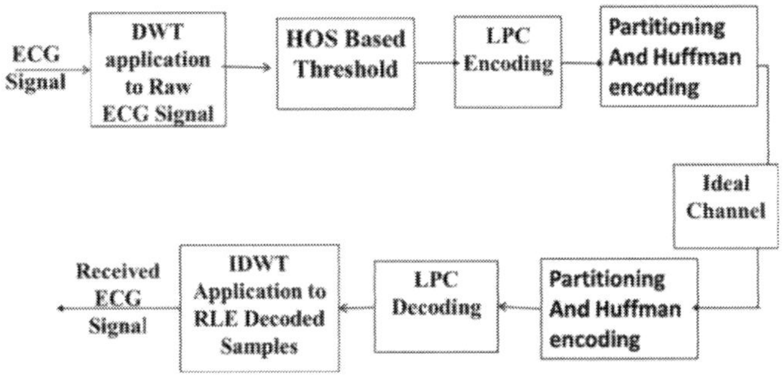

Figure 5. ECG signal compression algorithm scheme.

A more detailed flowchart of the proposed algorithm is shown in figure 6. The algorithm is applied to the ECG signals extracted from the MIT-BIH ARRYTHMIA database. Two criteria are used as metrics of the quality of the obtained signal: CR compression ratio and the mean square error RMSE (MSE).

5.1. *Transmission*

The transmission phase of our algorithm consists of the following steps:

A) Decoding of MIT-BIH ARRYTHMIA records

The decoding of the MIT-BIH ARRHYTHMIA records phase allows extracting and decoding. Each record contains 560000 samples and which would be partitioned into segments of 8192 = (2^{13}) samples.

B) ECG signal pre-processing

The preprocessing phase consists of ECG signal high pass filtering by the 4th order Butterworth filter model with a normalized very low cutoff frequency signal frequency of 0.001 to remove the DC component presented in the original ECG signal. The Butterworth filter transfer function is given by [44]:

$$H(f) = \frac{1}{\sqrt{1+(f/f_c)^{2n}}} \tag{25}$$

$H(f)$: Transfert function of the filter

f_c : normalized cutoff frequency and n: Butterworth filter order

C) DWT based ECG signal decomposition

The wavelet transform can effectively represent the non-stationarities and the localized time components of the ECG signal. The ECG signal could well be coded with a scheme of wavelet-based coding; the wavelet function "db6" was chosen because it offers a better result, likewise we chose the level three of decomposition of the ECG signal based on DWT because it gives a good quality of the reconstructed ECG signal.

D) HOS based thresholding

The significant DWT coefficients selection procedure involves the application of HOS based thresholding. In particular, each DWT scale j (j∈z) is estimated by the flattening (kurtosis) k_j, and the used kurtosis is multiplied by an adjustment factor F_{aj} as a criterion of thresholding at each scale of DWT. The method used to calculate the threshold is the kurtosis, given by:

$$w_j^s > THR_j = k_j F_{aj}, \tag{26}$$

where w_j^s represent the significant details DWT coefficients at scale j.

The choice of the value adjustment factor was set at $F_{aj} = 1/10$.

E) LPC coding

In our algorithm, we use the linear prediction coding (LPC) applied to the thresholded coefficients. The basic idea behind this method in the analysis of the ECG signal is that the thresholded ECG data can be approximated as a linear combination of samples from ECG signal. At the output of the predictor filter output it is seen a decrease on the ripples as it is presented in figure 7.

We remark, well, there is a decrease in the mean amplitude of the samples of the prediction error, and a greater correlation between the prediction filter input signal and the prediction error.

F) Huffman coding

Huffman coding is to associate with each symbol an optimized code. The probability of occurrence of the symbol in the message is taken into account, combining the shortest possible code for the 'common' symbol i.e. with the highest probability.

Journal of Physics: Conference Series **454** (2013) 012086 doi:10.1088/1742-6596/454/1/012086

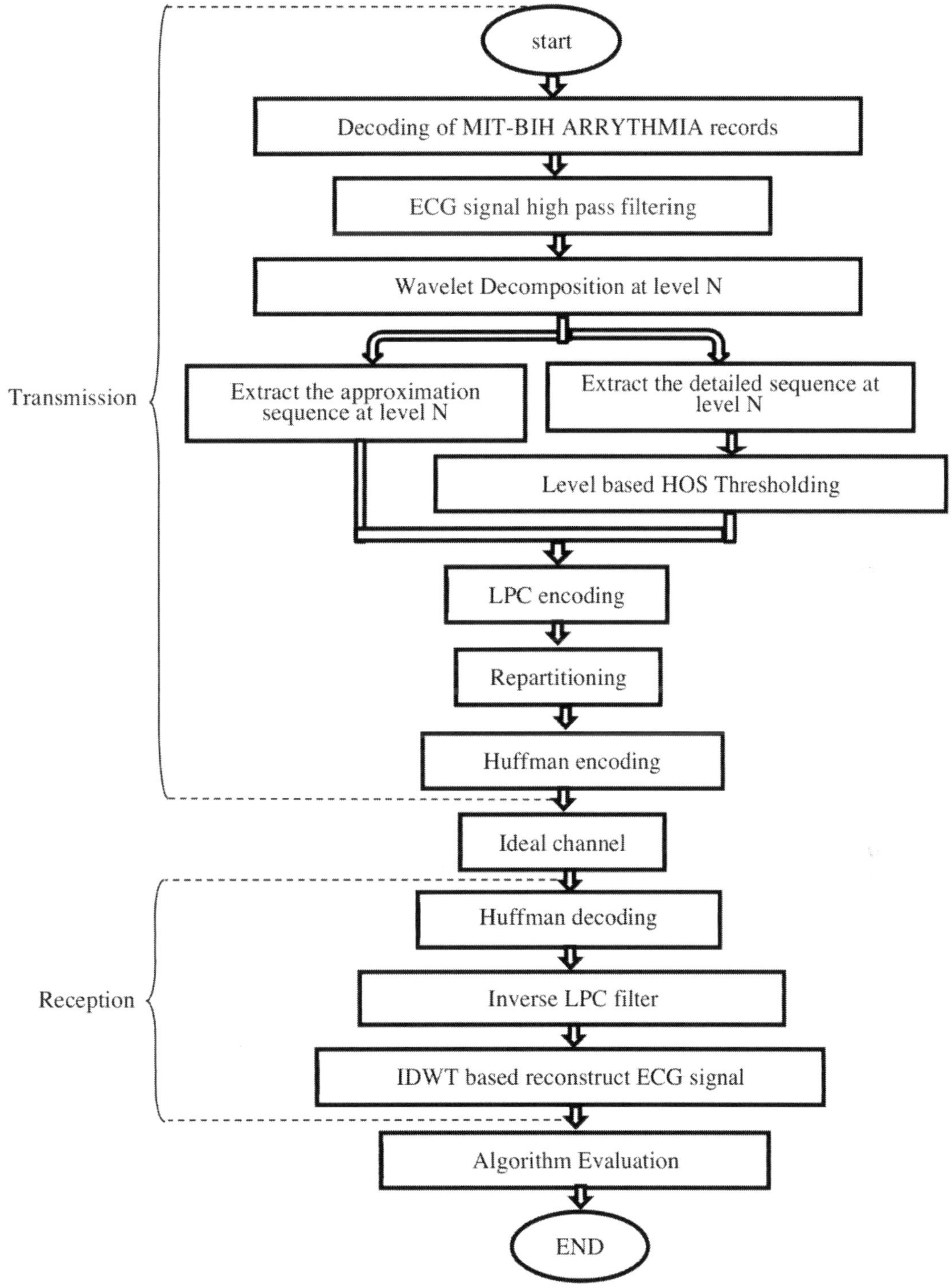

Figure 6. ECG signal compression algorithm flowchart.

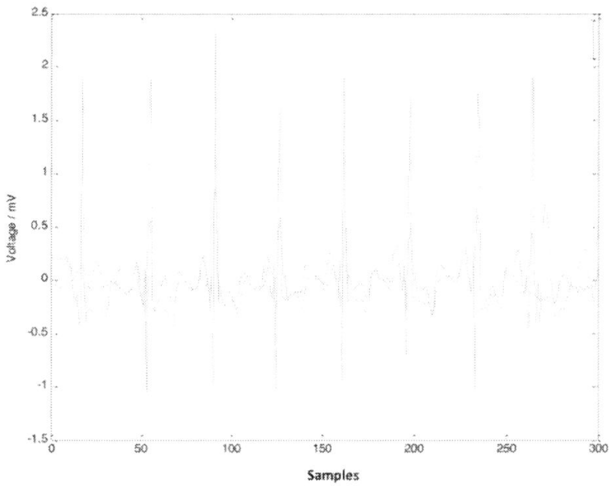

Figure 7. The thresholded ECG signal (blue line) in superposition with the ECG signal coded by the LPC filter (red line).

The quantification is the primary step in the encoding of the source in which the volume of information is reduced in a significant way at the cost of a loss of information. Quantization rule is imposed by the processing system at each voltage levels is associated with a binary value bits; N voltage levels ranging from peak-to-peak value. There is thus a uniform step which is equal to V_{pp}/N.

In the course of the first scan is posed the following problem: how to define the closest samples at each level and the evolution of the quantization noise? The possible answer to result an accurate and best resolution is to establish a second scan based on centered quantification in order to reduce the quantization error. Dividing the, yet established, intervals in two and choose the Ceiling function to find the samples nearest the first scan levels resulting from (N + 1) ranging from peak-to-peak voltage value levels. We represent the final portioning with its two sweeps in figure 8.

The phase of portioning arises the quantified values and each vector, containing the number of samples with the same value of quantification, corresponds a value of probability of occurrence.

The probability of occurrence relative to a ‘given’ level is the ratio of the number of samples close to this level and the total number of samples of the ECG signal. The next step is to apply the Huffman coding resulting binary 'optimal' train and will be passed through an ideal channel.

5.2. Reception

At the reception end-point, the inverse operations of those at the transmission phase are applied to reconstruct the original ECG signal. They are the Huffman decoding to render the bit stream of the quantified ECG signal, the inverse linear prediction filter and finally the inverse wavelet transform (IDWT). The obtained result of the reconstructed original ECG signal is represented in figure 9.

The ECG signal reconstruction error, illustrated in figure 10, is the difference between the original signal and the reconstructed version. This error is produced, mainly, because of the effect of thresholding and quantization error and the effect of filtering etc.

The numerical study of our algorithm is based on the calculation of the compression ratio (CR) and the mean square error (MSE), i.e. the number of bits that we were earning when transmitting the compressed ECG signal. As the MIT-BIH ARRYTHMIA database records have a resolution of 11 bits and the ECG signal was transmitted with a resolution of almost 1.5 bits (Avglen: average length of words: Avglen = 1.5083 bits), so we have a significant gain on the number of transmitted bits.

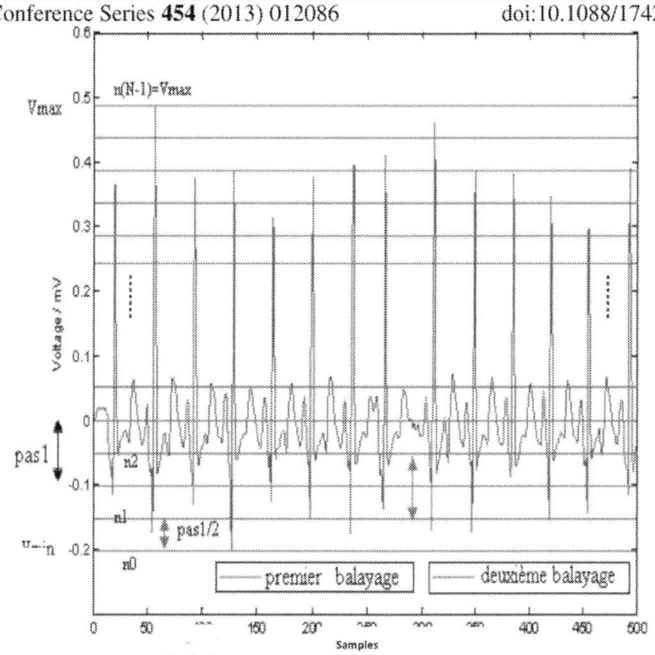

Figure 8. Double scanning of the ECG signal; 1st scanning in red and 2nd one in green.

The number of bits to be transmitted is equal to 12403 bits, so the compression ratio is 1:7.2653. The mean square error MSE = 0.0741 and the percentage of the difference of the mean square error (Percentage of Root mean square Difference, PRD) is 19.35%. The PRD is most important, when receiving the ECG signal, the set of waves {P, T, Q and S} are well restored while the R waves are attenuated. However, this distortion of the ECG signal does not destroy the useful information carried by that signal as it does not reveal non-existent pathological cases.

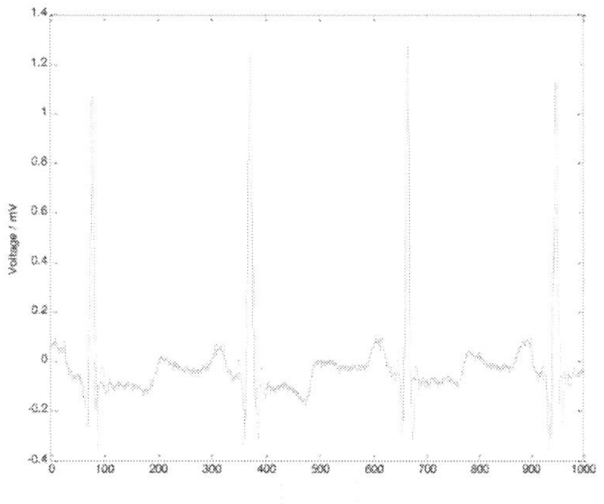

Figure 9. The reconstructed (red line) in superposition with the original ECG signal (blue line).

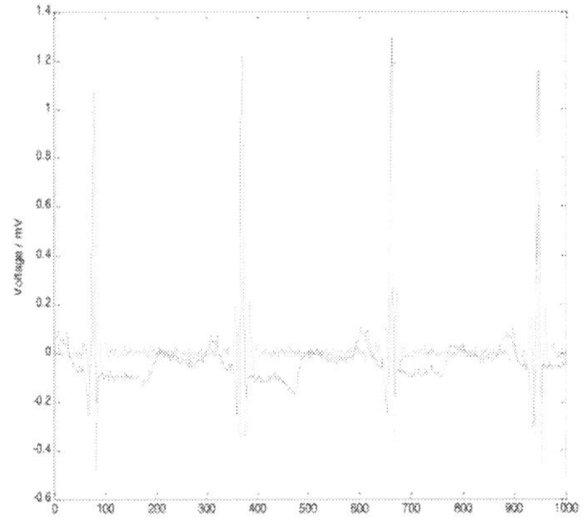

Figure 10. Reconstruction error (red line) in superposition with the original ECG signal (black line).

6. Evaluation of algorithm

6.1. Numerical evaluation

Table 1 lists the rate of compression CR values, root mean square error RMS, average length of words (Avglen: Averagelength) corresponding to a set, chosen arbitrarily, of pathological case extracted from the MIT-BIH ARRHYTHMIA database records.

Table 1. Assessment of the ECG compression algorithm applied to the MIT-BIH ARRYTHMIA records exhibited pathological cases.

File name	Number of segment	CR	MSE	Avglen	Pathological Case
100.dat	67	1:8.0342	0.0719	1.3640	Premature Ventricular Contraction
101.dat	5	1:7.1780	0.0795	1.5267	Huge base line wandering
228.dat	2	1:7.4117	0.0412	1.4785	Artefacts
233.dat	1	1:7.1055	0.0740	1.5423	Premature Ventricular/auricular Contraction

6.2. Visual evaluation

Despite the interest of the numerical evaluation of the ECG signal compression algorithm, the quantitative one based upon the visual perception is essential for clinical cases. Figure 11 shows the good performance of the application of the algorithm to the ECG signal with premature ventricular contractions.

7. Comparative study

The ECG signal compression algorithms are several. In order to promote our algorithm it is indispensable to compare it to works, already realized, and that the results are satisfactory. Among these algorithms for which the sampling rate is 360 Hz, 11-bit resolution and the MIT-BIH ARRHYTHMIA database records are used, we quote:

Djohan algorithm used the discrete wavelet transform [27]. It carried a value of PRD 3.9% with a compression ratio of 12.5% for the MIT-BIH ARRHYTHMIA database 117.dat record.

The algorithm developed by Robert S. H. Istepanian et al. is based on the wavelet transform and using two methods of encoding; the first named Optimal Zonal Wavelet Coding (OZWC) and the other one is called wavelet transform Higher Order Statistics based Coding (WHOSC) [43]. They carried out a compression ratio of 1:8.16 and 1:17.51 for OZWC and WHOSC, respectively, applied to the 100.dat of the MIT-BIH ARRHYTHMIA database record.

The algorithm based on the combination of wavelet transform and Set Partitioning in Hiearchical Tree (SPIHT) [45]; it provided a PRD value of 1.31% and a compression ratio of 45% applied to the 117.dat record of the MIT-BIH ARRHYTHMIA database. The set of the results are shown in table 2.

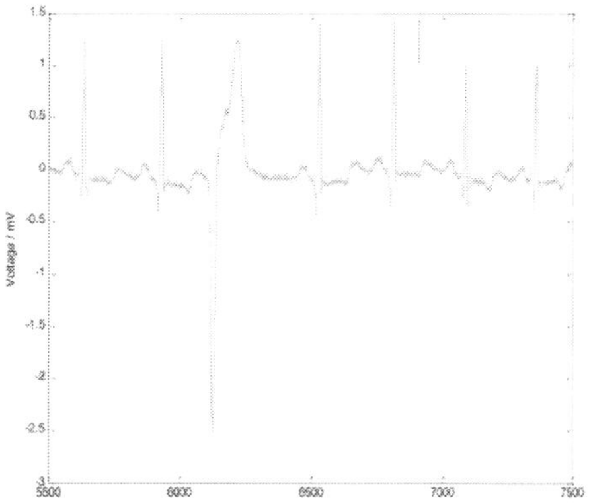

Figure 11. Application of the algorithm to the ECG signal with premature ventricular contractions.

Table 2. Comparative study of the proposed algorithm.

Algorithm		PRD (%)	CR (%)	F_s (Hz)
Djohan 117.dat		3.9	12.5%	360
Istepanian	OZWC 100.dat	NRMSE=0.5778%	8.16	360
	WHOSC 100.dat	NRMSE 1.7399%	17.51	360
SPIHT117.dat		1.31	45%	360
Our algorithm	100.dat	19.35	13.76	360
	117.dat	4.75	14.01	360

The comparative study, illustrated by the table above, shows, that even though our algorithm does not outperform the most of the compared algorithms, the obtained results are acceptable and encouraging. Examining the table of the comparative study shows, in other hands, as examples, that the proposed algorithm outperforms Djohan's and Istepanian based on OZWC algorithms in sense of compression ratio.

8. Conclusion

The ECG signal compression finds more importance with the development of telemedicine. Indeed, compression can significantly reduce the cost of the transmission of medical information through telecommunications channels. Higher-order statistics are tools which have played a very important role in the field of signal processing. We can envisage utilizing the HOS for denoising and compression. With the use of wavelet and probabilistic and statistical mathematical tools, the methodology is appropriate for the future integration of mobile applications of telemedicine in providing one of the mechanisms of compression improved for the coding of ECG.

The proposed ECG compression algorithm combining wavelet and higher-order statistics was discussed. This offered a good compression ratio and quality of the rendered ECG signal. Our proposed ECG compression algorithm has strengths and weaknesses. We believe that the strength of our proposed algorithm is the double scanning of the portioning used for Huffman encoding; this double scanning has considerably lessened the quantification error. However, the weakness of the algorithm is due to the overall LPC filtering. Adaptive LPC filtering based on optimal segmentation will, surely, improve the performance of the algorithm. But, on the other hand, this overall LPC filtering process lowers, considerably, the complexity of the algorithm and, consequently, gains time of processing.

References

[1] Guedat C and Gindis D 2007 Services de télésanté' département télécommunications services et usages insa de lyon services de télésanté p 1-43
[2] Lavaill L 2011 THESIS, Télémédecine Définition, applications et enquête auprès des médecins généralistes de Franche-Comté UNIVERSITE DE FRANCHE-COMTE FRANCE
[3] Sörnmo L and Laguna P 2006 Electrocardiogram (ECG) signal processing. **2** 1298–1313
[4] Pandian P S 2007 Store and Forward Applications in Telemedicine for Wireless IP Based Networks **2**(6) 58-65
[5] Pattichis C S, Kyriacou E, Pattichis M S, Panayides A and Pitsillides A 2006 A review of m-Health e-Emergency Systems 1-8
[6] Alesanco A and Jose Garcıa 2007 A SimpleMethod for Guaranteeing ECG Quality in Real timeWavelet Lossy Coding Article ID 93195, 9 pages
[7] Cox J R, Nolle F M, Fozzard H A, and Oliver G C 1968, AZTEC, a preprocessing program for real-time ECG rhythm analysis **15** 128-129
[8] Ishijima M, Shin S B, Hostetter G H and Sklansky J 1983 Scan along polygon approximation for data compression of Electrocardiograms **30** 723-729
[9] Ahmed N, Milne P J and Harris S G, Nov. *1975 Electrocardiographic data compression via orthogonal transforms* **22** 484-487
[10] Philips W and Jonghe G D 1992 Data Compression of ECG's by High-Degree polynomial approximation **39** 330-337
[11] Madhukar B and Murthy I S N 1993 ECG Data Compression by Modeling *AMIA, Inc.* 586-591.
[12] Graps A 1995 An introduction to wavelets. *IEEE Computational Sciences and Engineering.* **2**(2) 50-61.
[13] Crowe J A, Gibson N M, Woolfson M S and Somekh M G 1992 Wavelet transform as a potential tool for ECG analysis and compression *J. Biomed. Eng.* **14** 268–72
[14] Jalaleddine S, Hutchens C, Strattan R and Coberly W 1990 ECG data compression techniques—a unified approach *IEEE Trans. Biomed. Eng.* **37** 329–343.

[15] Thakor N V, Sun Y C, Rix H and Caminal P 1993 Multiwave: a wavelet-based ECG data compression algorithm *IEICE Trans. Inf. Syst. E76D* 1462–9.
[16] Chen J, Itoh S and Hashimoto T 1993 ECG data compression by using wavelet transform *IEICE Trans. Inf. Syst. E76D* 1454–61
[17] Bradie B 1996 Wavelet packet-based compression of single lead ECG *IEEE Trans. Biomed. Eng.* **43** 493–501
[18] Miaou S G and Lin C L 2002 A quality-on-demand algorithm for wavelet-based compression of electrocardiogram signals *IEEE Trans. Biomed. Eng.* **49** 233–9
[19] Addison P S 2005 Wavelet transforms and the ECG: a review *INSTITUTE OF PHYSICS PUBLISHING physiological measurement* **26** 155–199
[20] Ku C T, Hung K C, Wu T C and Wang HS 2010 Wavelet-Based ECG Data Compression System With Linear Quality Control Scheme *IEEE engineering in medicine and biology society*
[21] Pattichis C S, Kyriacou E, Voskarides S, Pattichis M S, Istepanian R and Schizas C N 2002 Wireless Telemedicine Systems *IEEE Antennas & Propagation Magazine*, **44**(2) 143-153
[22] Laxminarayan S and Istepanian R 2000 Unwired e-Med: The next generation of wireless and Internet telemedicine systems [Editorial*] IEEE Transactions on Information Technology in Biomedicine* **4** 189-193
[23] Perez R 1998 Wireless Communications Design Handbook, Volume I: Space (Interference: Aspects of Noise, Interference, and Environmental Concerns), Spacecraft Design Jet Propulsion Laboratory, California Institute of Technology, Academic Press. San Diego, CA
[24] Varshney U 2006 Patient monitoring using infrastructure oriented wireless LAN. *Int. J. Electronic Healthcare* **2** 149-163
[25] Brims W 2002 WIRELESS ECG Bachelor of Engineering In the Division of Electrical Engineering University of Queensland 1-84
[26] Tompkins W J 1993 *Biomedical Digital Signal Processing*. Englewood Cliffs, New Jersey, Prentice-Hall.
[27] Djohan A, Nguyen T Q and Tompkins W J 1997 ECG compression using discrete symmetric wavelets transform *IEEE –EMBC and CMBEC*, 167-168
[28] Bottou L and Pigeon S 1998 Lossy Compression of Partially Masked Still Images 1-10
[29] Bourdreaux-Bartels G 1996 *Chapter 12 Classified Time-Frequency Representations* The transforms and applications handbook, IEEE Press.
[30] Qian S and Chen D 1999 Understanding the nature of signals whose power spectra change with time: Joint Analysis. *IEEE Signal Processing Magazine*. 52-67.
[31] Cohen A and Kovačevič A 1996 Wavelets: The mathematical background. *Proceeding of the IEEE*. **4** 514-522.0
[32] Mallat S 1999 *A Wavelet Tour of Signal Processing*. Academic Press, 2nd edition.
[33] Truchetet F 1998 *Ondelettes pour le signal numérique*, Edition HERMES, Paris.
[34] Sheng Y 1996 Chapter 10. Wavelet Transform Handbook CRC Press Inc. 748-827.
[35] Donoho D 1995 De-noising by soft-thresholding *IEEE Trans. Inform.Theory* **41**(3): 612-627
[36] Sardy S, Tseng P and Brune P 2001 Robust Wavelet Denoising IEEE Transactions on Signal Processing **49**(6) 1146-1152
[37] McLaughlin S, Stogioglou A and Fackrell J 1995 Statistiques Présentation ordre supérieur (HOS) pour la détection des non-linéarités.
[38] Lacoume J L, Amblard O and Comon P 1999 Statistiques d'ordres supérieurs pour le Traitement du Signal.
[39] Ravier P 1992 Détection de transitoires par ondelettes adaptées -Critères d'adaptation fondés sur les statistiques d'ordre supérieur-, Thèse, INSTITUT NATIONAL POLYTHECHNIQUE DE GRENOBLE FRANCE
[40] Wickramasinghe N and Geisler E 2008 'Encyclopedia of Healthcare Information Systems' Volume I A-D Medical Information science reference 2008, p 157-166 Pedro de A. Berger, Francisco A de O Nascimento, Leonardo R A X de Menezes, Adson F da Rocha, Joao L A

Journal of Physics: Conference Series **454** (2013) 012086 doi:10.1088/1742-6596/454/1/012086

Carvalho 'Biomedical Signal Compression B'
[41] Moreau N 1995 Techniques de compression des signaux. ENST-Masson
[42] Ramakrishnam A G and Saha S 1997 ECG coding by wavelet-based linear prediction. *IEEE Trans. Biomed. Eng.* **44**(12) 1253–1261
[43] Istepanian R S, Leontios J Hadjileontiadis and Stavros M Panas 2001 ECG Data Compression Using Wavelets and Higher Order Statistics Methods, *IEEE TRANSACTIONS ON INFORMATION TECHNOLOGY IN BIOMEDICINE*, **5**(2) 108-115
[44] Bellanger M 1990 Traitement numérique du signal. MASSON 4th edition
[45] Ktata S, Ouni K, and Ellouze N 2009 A Novel Compression Algorithm for Electrocardiogram Signals based on Wavelet Transform and *SPIHT World Academy of Science, Engineering and Technology* **35** 855-860

AUTHOR INDEX

AUTHOR INDEX

AUTHOR INDEX